Prealgebra

Second Edition

Alan S. Tussy
Citrus College

R. David Gustafson
Rock Valley College

BROOKS/COLE
THOMSON LEARNING™

Australia • Canada • Mexico • Singapore • Spain • United Kingdom • United States

This book is dedicated

To my wife, Liz, for your love and encouragement, and to my three sons, Kevin, Glenn, and Brandon, who always bring me joy and make me proud — *AST*

And to four grandboys, Daniel, Tyler, Spencer, and Skyler — *RDG*

BROOKS/COLE

THOMSON LEARNING™

Publisher: *Robert W. Pirtle*
Assistant Editor: *Rachael Sturgeon*
Editorial Assistant: *Lisa Jones*
Marketing Representative: *Leah Thomson*
Marketing Communications: *Samantha Cabaluna*
Marketing Assistant: *Maria Salinas*
Production Editor: *Ellen Brownstein*
Production Service: *Hoyt Publishing Services*
Manuscript Editor: *David Hoyt*
Permissions Editor: *Sue Ewing*
Interior Design: *Vernon T. Boes* and *John Edeen*
Cover Design: *Roy R. Neuhaus*
Cover Illustration: *George Abe*
Interior Illustration: *Lori Heckelman*
Print Buyer: *Vena Dyer*
Typesetting: *The Clarinda Company*
Cover Printing: *Phoenix Color Corp.*
Printing and Binding: *Quebecor World Book Services*

For more information about this or any other Brooks/Cole product, contact:
BROOKS/COLE
511 Forest Lodge Road
Pacific Grove, CA 93950 USA
www.brookscole.com
1-800-423-0563 (Thomson Learning Academic Resource Center)

Printed in the United States of America

10 9 8 7 6 5

Images provided by PhotoDisc © 2000

Library of Congress Cataloging-in-Publication Data
Tussy, Alan S., [date]
Prealgebra / Alan S. Tussy, R. David Gustafson—2nd ed.
p. cm.
Includes index.
ISBN 0-534-37642-8 (pbk. : acid-free paper)
1. Mathematics. I. Gustafson, R. David (Roy David), [date]. II. Title.
QA39.2.T88 2002
513′.14—dc21 00-052903

CONTENTS

3 *The Language of Algebra* 143

4 *Fractions and Mixed Numbers* 205

5 *Decimals* 291

PREFACE

For the Instructor

The purpose of this book is to prepare students for an introductory algebra course. Its content, pedagogy, and features result from our efforts to answer this question: What prerequisite skills are necessary for success in algebra? Our goal was to write a book that is interesting and enjoyable to read—one that will attract and keep the attention of college students of all ages.

Prealgebra aims to develop students' basic mathematical skills in the context of solving meaningful application problems. A variety of instructional approaches are used, reflecting the recommendations of NCTM and AMATYC. In combination with the student and instructor supplements that are available, *Prealgebra* can be used in lecture, laboratory, or self-study formats.

The second edition retains the basic philosophy and organization of the highly successful first edition. However, we have made several improvements as a direct result of the comments and suggestions we received from instructors and students who have used the first edition. To make the book more enjoyable to read, easier to understand, and more relevant, we have

- used a new and more spacious design.
- included additional examples and problems that involve real-life data.
- printed the Study Set problems in vertical columns.
- added more art and diagrams to help the visual learner.
- moved the section on Multiplication Rules for Exponents to Chapter 6.
- inserted a one-page group-work feature called Accent on Teamwork at the end of each chapter.
- included Cumulative Review Exercises at the end of each chapter (except the first).

Features of the text

A Blend of the Traditional and Reform Approaches

We have used a combination of instructional methods from the traditional and reform approaches, endeavoring to write a book that contains the best of both. You will find the vocabulary, practice, and well-defined pedagogy of a traditional prealgebra book. The text also features problem solving, reasoning, communicating, and technology, as emphasized by the reform movement.

Variables, Equations, and Problem Solving Appear Early

To prepare students for introductory algebra, this book provides a review of arithmetic while introducing basic algebraic concepts. For example, Chapter 1 covers whole-number arithmetic, but it also introduces the concept of a variable, develops the geometric formulas for perimeter and area, and shows how to simplify numerical expressions. In Chapter 1, we also lay the groundwork for rectangular coordinate graphing and solve some simple equations. Additionally, we establish a five-step problem-solving strategy and use it to solve real-world problems.

Arithmetic and Algebra Are Integrated Throughout

The integration of algebra and arithmetic continues throughout the book as algebraic concepts are introduced and reinforced in an arithmetic setting. In Chapter 2, after learning how to add, subtract, multiply, and divide integers, students solve equations and application problems involving integers. In Chapter 3, The Language of Algebra, we introduce the concept of algebraic expressions, develop a formal equation-solving strategy, and expand the problem-solving model. These topics are then revisited in the context of fractions, decimals, and percents in Chapters 4, 5, and 7, respectively.

Competency with Signed Numbers

The rules for adding, subtracting, multiplying, and dividing integers are introduced in Chapter 2. Students apply these rules again in Chapter 3 when working with algebraic expressions, in Chapter 4 with signed fractions, in Chapter 5 with signed decimals, and in Chapter 6 when they graph equations. We feel this spiral approach is instructionally superior to that of introducing signed numbers in a single chapter at, or near, the end of the textbook. Revisiting these rules in several different contexts builds a thorough understanding of signed numbers that pays great dividends when the prealgebra student takes the next mathematics course, Introductory Algebra.

Interactivity

Author's notes are used to explain the steps in the solutions of examples. The notes are extensive so as to increase the students' ability to read and write mathematics. Most worked examples in the text are accompanied by Self Checks. This feature allows students to practice skills discussed in the example by working a similar problem. Because the Self Check problems are adjacent to the worked examples, students can easily refer to the solution and author's notes of the example as they solve the Self Check.

Example titles highlight the concept being discussed. ▶

Author's notes explain the steps in the solution process. ▶

The Self Check answers are provided. ▶

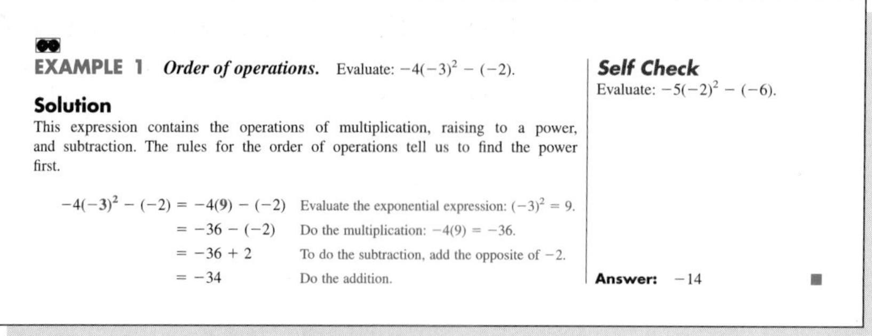

EXAMPLE 1 Order of operations. Evaluate: $-4(-3)^2 - (-2)$.
Solution
This expression contains the operations of multiplication, raising to a power, and subtraction. The rules for the order of operations tell us to find the power first.
$-4(-3)^2 - (-2) = -4(9) - (-2)$ Evaluate the exponential expression: $(-3)^2 = 9$.
$= -36 - (-2)$ Do the multiplication: $-4(9) = -36$.
$= -36 + 2$ To do the subtraction, add the opposite of -2.
$= -34$ Do the addition.
Self Check
Evaluate: $-5(-2)^2 - (-6)$.
Answer: -14

Study Sets—More Than Just Exercises

The problems at the end of each section are called Study Sets. Each Study Set includes Vocabulary, Notation, and Writing problems designed to help students improve their ability to read, write, and communicate mathematical ideas. The problems in the Concepts section of the Study Sets encourage students to engage in independent thinking and reinforce major ideas through exploration. In the Practice section of the Study Sets, students get the drill necessary to master the material. In the Applications section, students deal with real-life situations that involve the topics being studied. Each Study Set concludes with a Review section consisting of problems selected from previous sections.

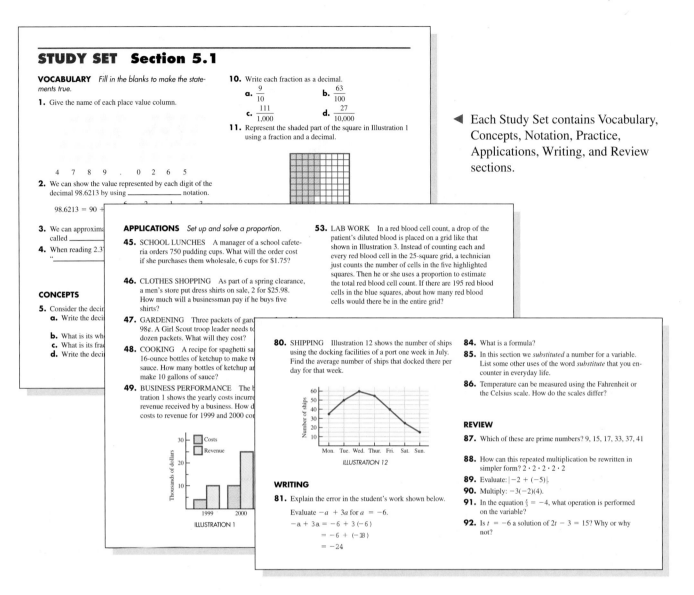

STUDY SET Section 5.1

VOCABULARY *Fill in the blanks to make the statements true.*

1. Give the name of each place value column.

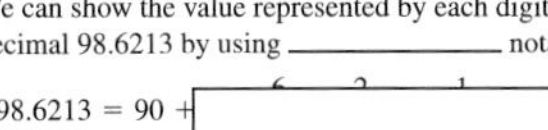

2. We can show the value represented by each digit of the decimal 98.6213 by using ______ notation.

98.6213 = 90 +

3. We can approxima... called ______
4. When reading 2.3... "______

CONCEPTS

5. Consider the deci...
 a. Write the deci...
 b. What is its wh...
 c. What is its fra...
 d. Write the deci...

10. Write each fraction as a decimal.
 a. $\frac{9}{10}$ b. $\frac{63}{100}$
 c. $\frac{111}{1,000}$ d. $\frac{27}{10,000}$

11. Represent the shaded part of the square in Illustration 1 using a fraction and a decimal.

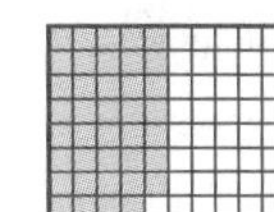

APPLICATIONS *Set up and solve a proportion.*

45. SCHOOL LUNCHES A manager of a school cafeteria orders 750 pudding cups. What will the order cost if she purchases them wholesale, 6 cups for $1.75?

46. CLOTHES SHOPPING As part of a spring clearance, a men's store put dress shirts on sale, 2 for $25.98. How much will a businessman pay if he buys five shirts?

47. GARDENING Three packets of gar... 98¢. A Girl Scout troop leader needs t... dozen packets. What will they cost?

48. COOKING A recipe for spaghetti sa... 16-ounce bottles of ketchup to make t... sauce. How many bottles of ketchup a... make 10 gallons of sauce?

49. BUSINESS PERFORMANCE The ... tration 1 shows the yearly costs incurr... revenue received by a business. How d... costs to revenue for 1999 and 2000 co...

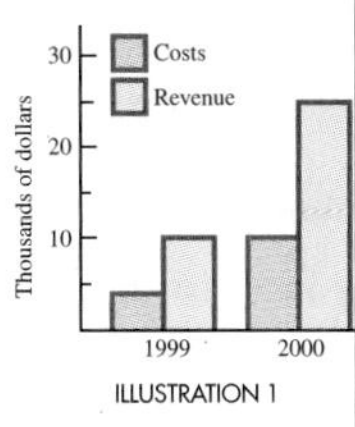

ILLUSTRATION 1

53. LAB WORK In a red blood cell count, a drop of the patient's diluted blood is placed on a grid like that shown in Illustration 3. Instead of counting each and every red blood cell in the 25-square grid, a technician just counts the number of cells in the five highlighted squares. Then he or she uses a proportion to estimate the total red blood cell count. If there are 195 red blood cells in the blue squares, about how many red blood cells would there be in the entire grid?

80. SHIPPING Illustration 12 shows the number of ships using the docking facilities of a port one week in July. Find the average number of ships that docked there per day for that week.

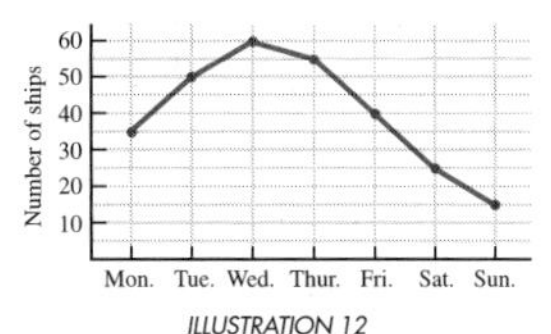

ILLUSTRATION 12

WRITING

81. Explain the error in the student's work shown below.

Evaluate $-a + 3a$ for $a = -6$.

$-a + 3a = -6 + 3(-6)$
$= -6 + (-18)$
$= -24$

84. What is a formula?

85. In this section we *substituted* a number for a variable. List some other uses of the word *substitute* that you encounter in everyday life.

86. Temperature can be measured using the Fahrenheit or the Celsius scale. How do the scales differ?

REVIEW

87. Which of these are prime numbers? 9, 15, 17, 33, 37, 41

88. How can this repeated multiplication be rewritten in simpler form? $2 \cdot 2 \cdot 2 \cdot 2 \cdot 2$

89. Evaluate: $|-2 + (-5)|$.

90. Multiply: $-3(-2)(4)$.

91. In the equation $\frac{x}{3} = -4$, what operation is performed on the variable?

92. Is $t = -6$ a solution of $2t - 3 = 15$? Why or why not?

◀ Each Study Set contains Vocabulary, Concepts, Notation, Practice, Applications, Writing, and Review sections.

Applications and Connections to Other Disciplines

A distinguishing feature of this book is its wealth of application problems. We have included numerous applications from disciplines such as science, economics, business, manufacturing, history, and entertainment, as well as mathematics.

Every application problem has a title. ▶

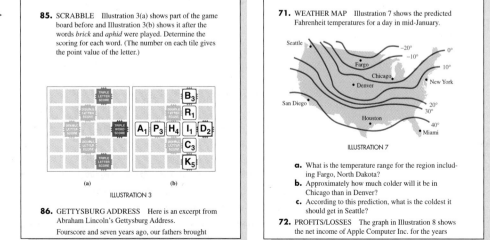

85. SCRABBLE Illustration 3(a) shows part of the game board before and Illustration 3(b) shows it after the words *brick* and *aphid* were played. Determine the scoring for each word. (The number on each tile gives the point value of the letter.)

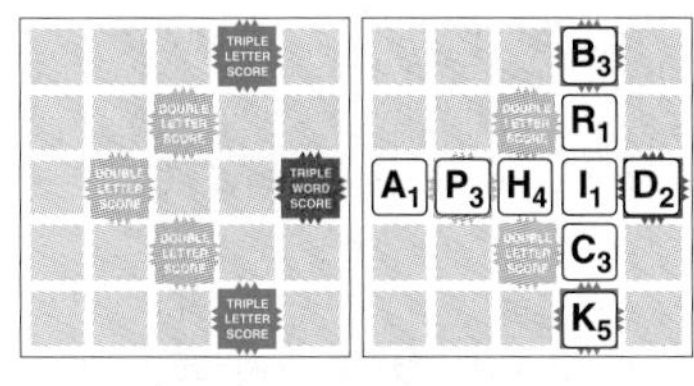

ILLUSTRATION 3

86. GETTYSBURG ADDRESS Here is an excerpt from Abraham Lincoln's Gettysburg Address.

Fourscore and seven years ago, our fathers brought

71. WEATHER MAP Illustration 7 shows the predicted Fahrenheit temperatures for a day in mid-January.

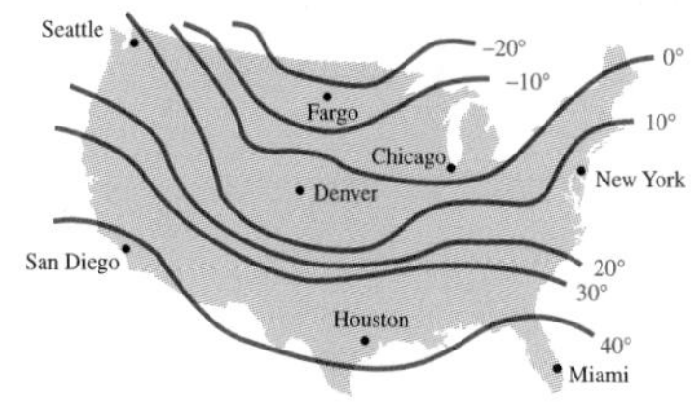

ILLUSTRATION 7

a. What is the temperature range for the region including Fargo, North Dakota?
b. Approximately how much colder will it be in Chicago than in Denver?
c. According to this prediction, what is the coldest it should get in Seattle?

72. PROFITS/LOSSES The graph in Illustration 8 shows the net income of Apple Computer Inc. for the years

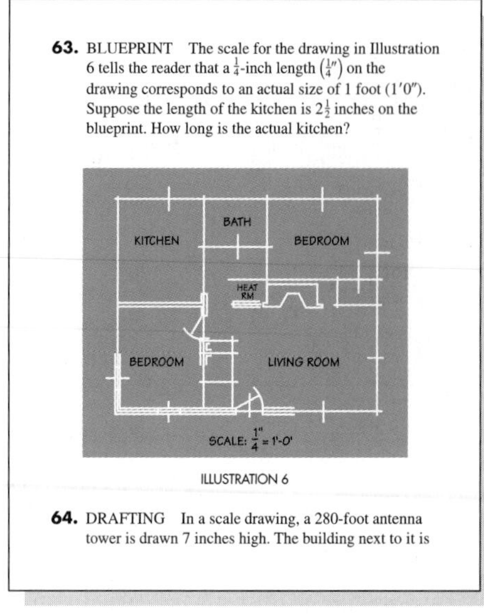

63. BLUEPRINT The scale for the drawing in Illustration 6 tells the reader that a $\frac{1}{4}$-inch length $\left(\frac{1}{4}''\right)$ on the drawing corresponds to an actual size of 1 foot (1′0″). Suppose the length of the kitchen is $2\frac{1}{2}$ inches on the blueprint. How long is the actual kitchen?

ILLUSTRATION 6

64. DRAFTING In a scale drawing, a 280-foot antenna tower is drawn 7 inches high. The building next to it is

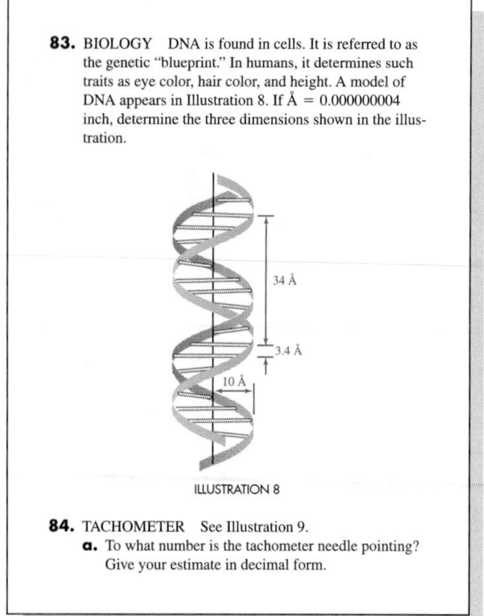

83. BIOLOGY DNA is found in cells. It is referred to as the genetic "blueprint." In humans, it determines such traits as eye color, hair color, and height. A model of DNA appears in Illustration 8. If Å = 0.000000004 inch, determine the three dimensions shown in the illustration.

ILLUSTRATION 8

84. TACHOMETER See Illustration 9.

a. To what number is the tachometer needle pointing? Give your estimate in decimal form.

Building a Foundation for Rectangular Coordinate Graphing

From the beginning, we build a foundation to prepare students for the study of rectangular coordinate graphing in Chapter 6. For example, Chapter 1 introduces the number line and bar and line graphs. In subsequent chapters, this preparation continues with graphing that involves integers, fractions, and decimals.

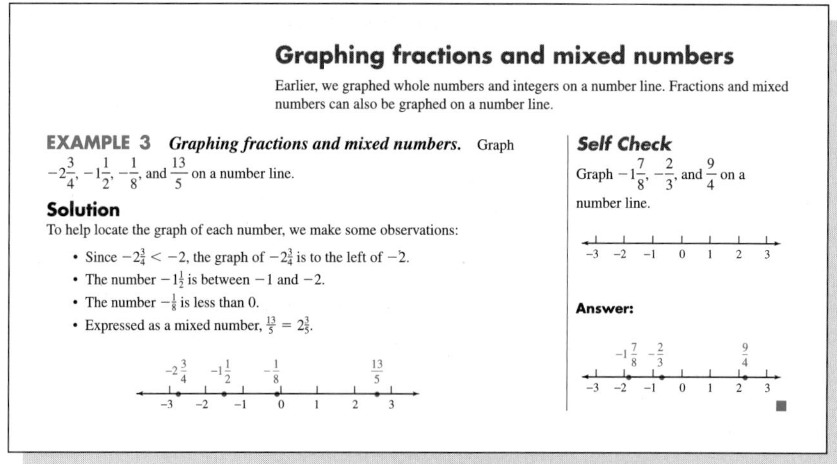

Graphing fractions and mixed numbers

Earlier, we graphed whole numbers and integers on a number line. Fractions and mixed numbers can also be graphed on a number line.

EXAMPLE 3 ***Graphing fractions and mixed numbers.*** Graph $-2\frac{3}{4}$, $-1\frac{1}{2}$, $-\frac{1}{8}$, and $\frac{13}{5}$ on a number line.

Solution

To help locate the graph of each number, we make some observations:

- Since $-2\frac{3}{4} < -2$, the graph of $-2\frac{3}{4}$ is to the left of -2.
- The number $-1\frac{1}{2}$ is between -1 and -2.
- The number $-\frac{1}{8}$ is less than 0.
- Expressed as a mixed number, $\frac{13}{5} = 2\frac{3}{5}$.

Self Check

Graph $-1\frac{7}{8}$, $-\frac{2}{3}$, and $\frac{9}{4}$ on a number line.

Answer:

Constructing Charts, Tables, and Graphs; Statistics

Many problems require students to present their solutions in the form of a table, graph, or chart. Often, students must examine such data displays to obtain the information necessary to solve a problem. Spreadsheets are informally discussed. The concepts of arithmetic mean, median, and mode are discussed in Chapter 5.

Measurement and Unit Analysis

In Chapter 8, we discuss measurement and unit conversion. In preparation for this chapter, some problems in earlier Study Sets require the student to read a scale or gauge.

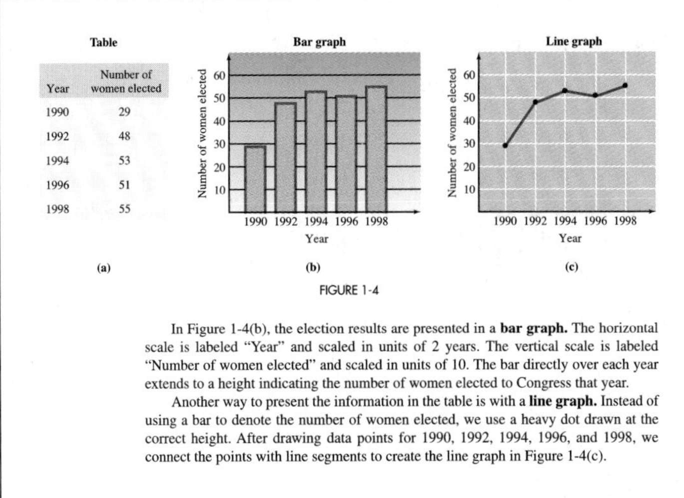

Year	Number of women elected
1990	29
1992	48
1994	53
1996	51
1998	55

FIGURE 1-4

In Figure 1-4(b), the election results are presented in a **bar graph.** The horizontal scale is labeled "Year" and scaled in units of 2 years. The vertical scale is labeled "Number of women elected" and scaled in units of 10. The bar directly over each year extends to a height indicating the number of women elected to Congress that year.

Another way to present the information in the table is with a **line graph.** Instead of using a bar to denote the number of women elected, we use a heavy dot drawn at the correct height. After drawing data points for 1990, 1992, 1994, 1996, and 1998, we connect the points with line segments to create the line graph in Figure 1-4(c).

◀ Real data are integrated throughout the text.

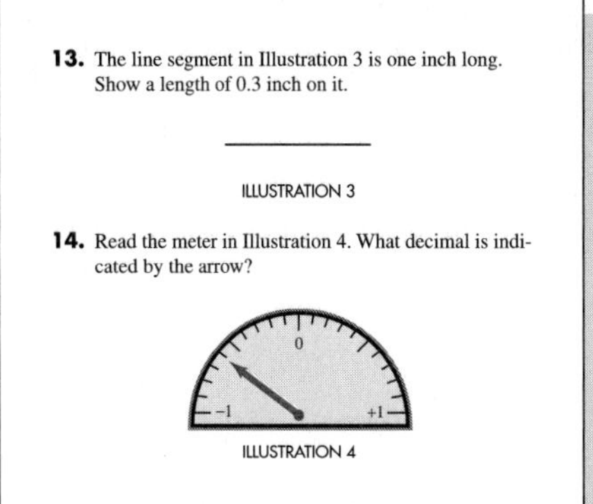

13. The line segment in Illustration 3 is one inch long. Show a length of 0.3 inch on it.

ILLUSTRATION 3

14. Read the meter in Illustration 4. What decimal is indicated by the arrow?

ILLUSTRATION 4

Calculators Are Optional

For those instructors who wish to use calculators as part of the instruction in this course, the text includes an Accent on Technology feature that introduces keystrokes and shows how scientific calculators can be used to solve application problems. Some Study Sets include problems that are to be solved using a calculator; these problems are indicated by the calculator logo. Instructors who do not wish to introduce calculators can skip that material without interrupting the flow of ideas.

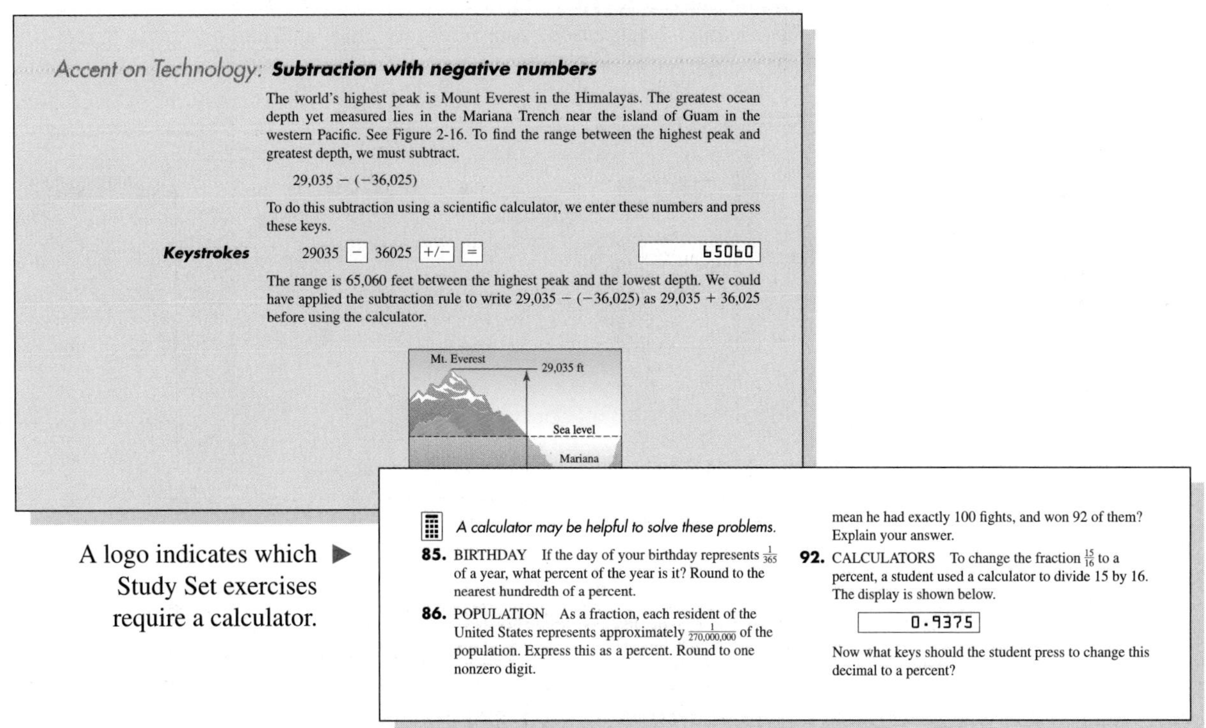

Accent on Technology: ***Subtraction with negative numbers***

The world's highest peak is Mount Everest in the Himalayas. The greatest ocean depth yet measured lies in the Mariana Trench near the island of Guam in the western Pacific. See Figure 2-16. To find the range between the highest peak and greatest depth, we must subtract.

$29{,}035 - (-36{,}025)$

To do this subtraction using a scientific calculator, we enter these numbers and press these keys.

Keystrokes 29035 [−] 36025 [+/−] [=] 65060

The range is 65,060 feet between the highest peak and the lowest depth. We could have applied the subtraction rule to write $29{,}035 - (-36{,}025)$ as $29{,}035 + 36{,}025$ before using the calculator.

A calculator may be helpful to solve these problems.

85. BIRTHDAY If the day of your birthday represents $\frac{1}{365}$ of a year, what percent of the year is it? Round to the nearest hundredth of a percent.

86. POPULATION As a fraction, each resident of the United States represents approximately $\frac{1}{270{,}000{,}000}$ of the population. Express this as a percent. Round to one nonzero digit.

mean he had exactly 100 fights, and won 92 of them? Explain your answer.

92. CALCULATORS To change the fraction $\frac{15}{16}$ to a percent, a student used a calculator to divide 15 by 16. The display is shown below.

0.9375

Now what keys should the student press to change this decimal to a percent?

A logo indicates which Study Set exercises require a calculator. ▶

Problem-Solving Strategy

One of the major objectives of this textbook is to make students better problem solvers. To this end, we use a five-step problem-solving strategy throughout the book. The five steps are: *Analyze the problem, Form an equation, Solve the equation, State the conclusion,* and *Check the result.*

A five-step problem-solving ▶ strategy is used.

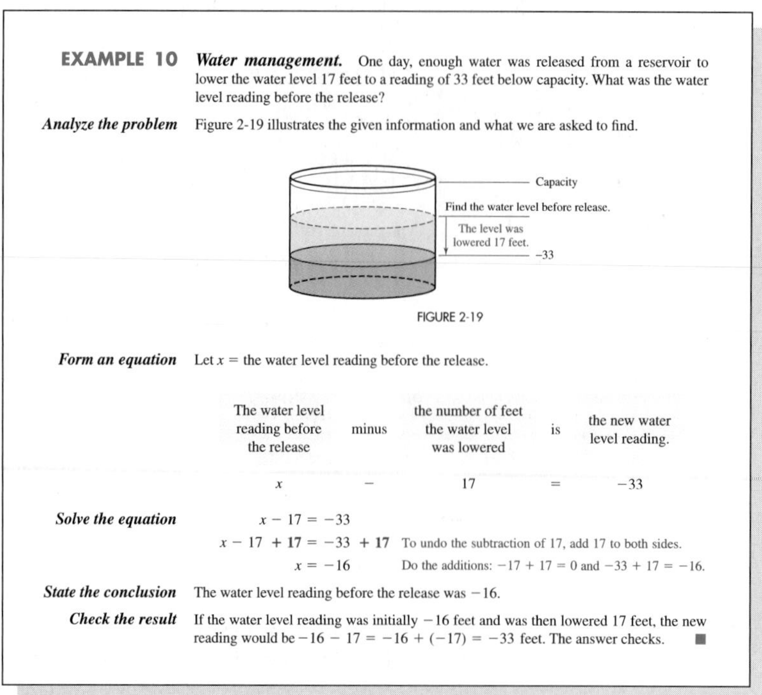

EXAMPLE 10 Water management. One day, enough water was released from a reservoir to lower the water level 17 feet to a reading of 33 feet below capacity. What was the water level reading before the release?
Analyze the problem Figure 2-19 illustrates the given information and what we are asked to find.
Capacity
Find the water level before release.
The level was lowered 17 feet.
−33
FIGURE 2-19
Form an equation Let x = the water level reading before the release.
The water level reading before the release minus the number of feet the water level was lowered is the new water level reading.
x − 17 = −33
Solve the equation
x − 17 = −33
x − 17 + 17 = −33 + 17 To undo the subtraction of 17, add 17 to both sides.
x = −16 Do the additions: −17 + 17 = 0 and −33 + 17 = −16.
State the conclusion The water level reading before the release was −16.
Check the result If the water level reading was initially −16 feet and was then lowered 17 feet, the new reading would be −16 − 17 = −16 + (−17) = −33 feet. The answer checks.

Estimation

Estimation is often used to check the reasonableness of answers. Special two-page Estimation features appear in the chapters on Whole Numbers, Decimals, and Percent. In these features, estimation procedures are introduced and put to use in real-life situations that require only approximate answers.

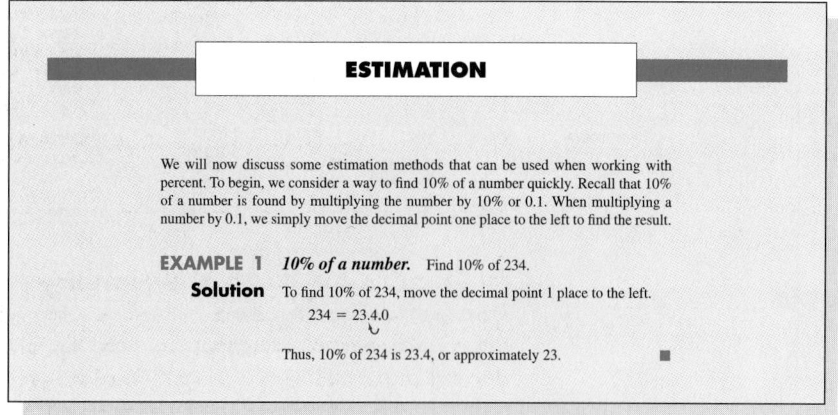

ESTIMATION
We will now discuss some estimation methods that can be used when working with percent. To begin, we consider a way to find 10% of a number quickly. Recall that 10% of a number is found by multiplying the number by 10% or 0.1. When multiplying a number by 0.1, we simply move the decimal point one place to the left to find the result.
EXAMPLE 1 10% of a number. Find 10% of 234.
Solution To find 10% of 234, move the decimal point 1 place to the left.
234 = 23.4.0
Thus, 10% of 234 is 23.4, or approximately 23.

Key Concepts

Nine key algebraic concepts are highlighted in one-page Key Concept features, appearing near the end of each chapter. Each Key Concept page summarizes a concept and gives students an opportunity to review the role it plays in the overall picture.

Group Work

A one-page feature called Accent on Teamwork appears near the end of each chapter. It gives the instructor a set of problems that can be assigned as group work or to individual students as outside-of-class projects.

KEY CONCEPT

Order of Operations

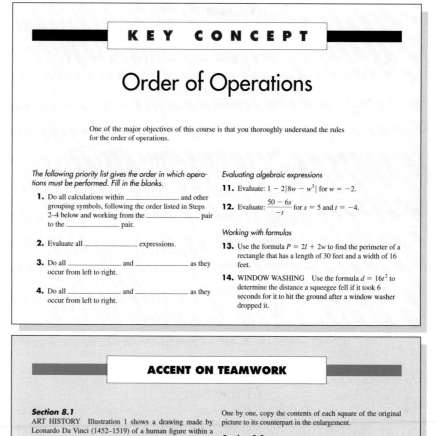

One of the major objectives of this course is that you thoroughly understand the rules for the order of operations.

The following priority list gives the order in which operations must be performed. Fill in the blanks.

1. Do all calculations within ______ and other grouping symbols, following the order listed in Steps 2–4 below and working from the ______ pair to the ______ pair.
2. Evaluate all ______ expressions.
3. Do all ______ and ______ as they occur from left to right.
4. Do all ______ and ______ as they occur from left to right.

Evaluating algebraic expressions

11. Evaluate: $1 - 2|8w - w^3|$ for $w = -2$.
12. Evaluate: $\frac{50 - 6s}{-t}$ for $s = 5$ and $t = -4$.

Working with formulas

13. Use the formula $P = 2l + 2w$ to find the perimeter of a rectangle that has a length of 30 feet and a width of 16 feet.
14. WINDOW WASHING Use the formula $d = 16t^2$ to determine the distance a squeegee fell if it took 6 seconds for it to hit the ground after a window washer dropped it.

ACCENT ON TEAMWORK

Section 8.1
ART HISTORY Illustration 1 shows a drawing made by Leonardo Da Vinci (1452–1519) of a human figure within a square. We see that the man's height and the span of his outstretched arms are the same. That is, their ratio is 1 to 1. Use a tape measure to determine this ratio for each member of your group. Work in terms of inches. Are any ratios exactly 1 to 1?

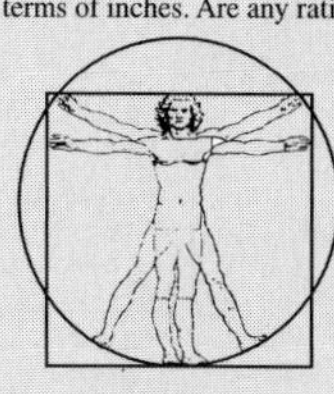

ILLUSTRATION 1

One by one, copy the contents of each square of the original picture to its counterpart in the enlargement.

Section 8.3
DISNEY CLASSICS The movie *20,000 Leagues Under the Sea* is a science fiction thriller about Captain Nemo and the crew of the submarine Nautilus. Use the fact that 1 mile = $\frac{1}{3}$ league to express a depth of 20,000 leagues in feet.

Section 8.4
METRIC MISHAP In 1999, NASA lost the $125 million Mars Climate Orbiter because a Lockheed Martin engineering team used American units of measurement, while NASA's team used the metric system. Use the Internet to research this incident and make a report to your class.

Section 8.4
TRUTH IN LABELING Have each member of your group

In-Depth Coverage of Geometry

Perimeter, area, and volume, as well as many other geometric concepts, are used in a variety of contexts throughout the book. We have included many drawings to help students improve their ability to spot visual patterns in their everyday lives.

Geometry topics are presented ▶ in a practical setting.

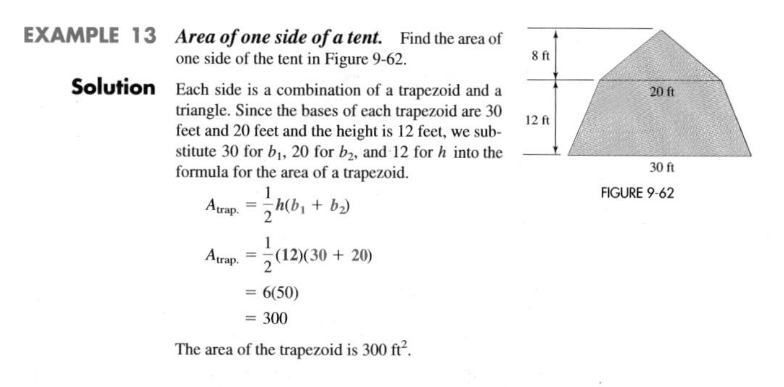

EXAMPLE 13 ***Area of one side of a tent.*** Find the area of one side of the tent in Figure 9-62.

Solution Each side is a combination of a trapezoid and a triangle. Since the bases of each trapezoid are 30 feet and 20 feet and the height is 12 feet, we substitute 30 for b_1, 20 for b_2, and 12 for h into the formula for the area of a trapezoid.

$$A_{\text{trap.}} = \frac{1}{2}h(b_1 + b_2)$$

$$A_{\text{trap.}} = \frac{1}{2}(12)(30 + 20)$$
$$= 6(50)$$
$$= 300$$

The area of the trapezoid is 300 ft^2.

FIGURE 9-62

Systematic Review

Each Study Set ends with a Review section that contains problems similar to those in previous sections. Each chapter ends with a Chapter Review and a Chapter Test. The chapter reviews have been designed to be user friendly. In a unique format, the reviews lists the important concepts of each section of the chapter in one column, with appropriate review problems running parallel in a second column. In addition, Cumulative Review Exercises appear at the end of each chapter (except Chapter 1).

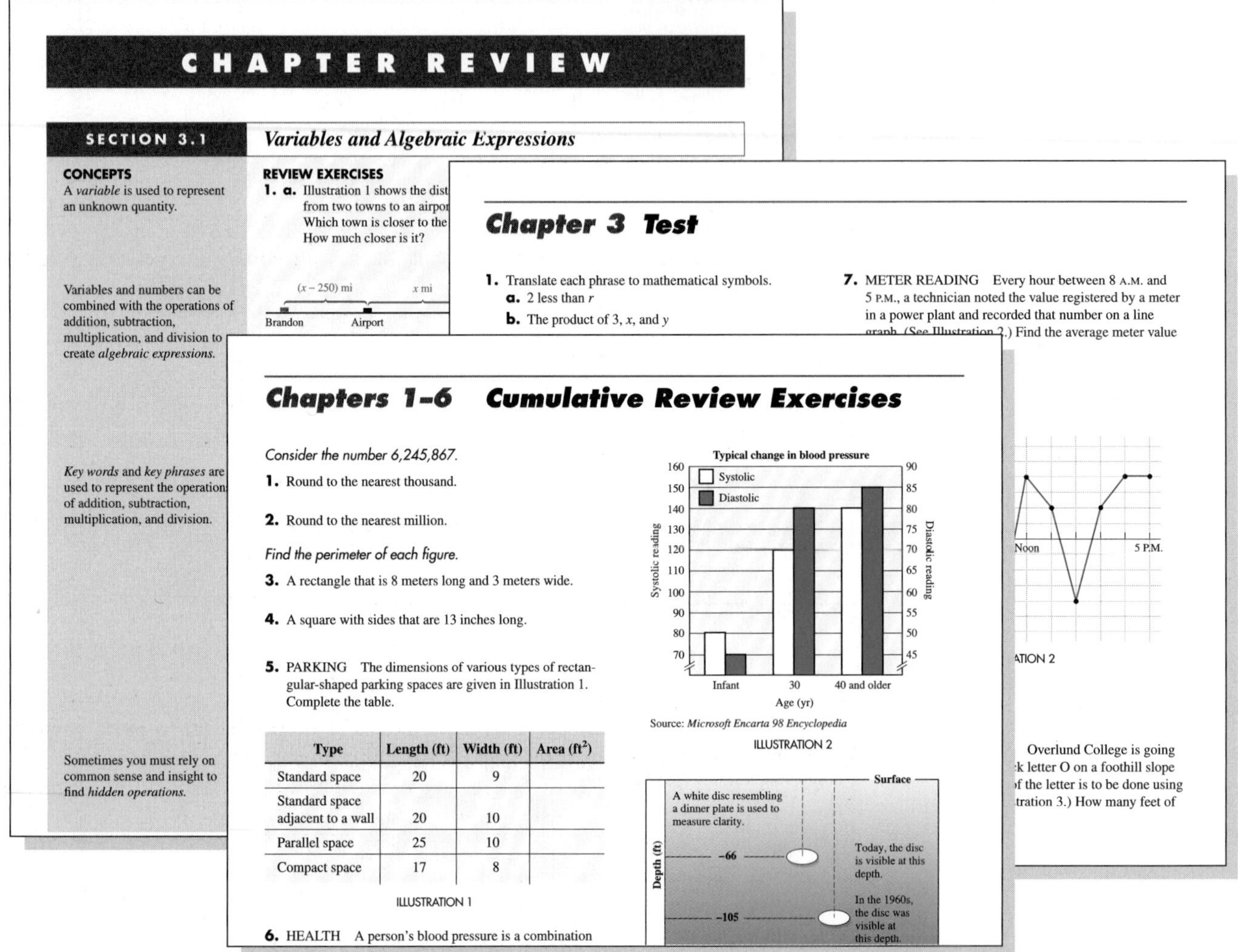

CHAPTER REVIEW

SECTION 3.1 | *Variables and Algebraic Expressions*

CONCEPTS

A *variable* is used to represent an unknown quantity.

Variables and numbers can be combined with the operations of addition, subtraction, multiplication, and division to create *algebraic expressions.*

Key words and *key phrases* are used to represent the operation of addition, subtraction, multiplication, and division.

Sometimes you must rely on common sense and insight to find *hidden operations.*

REVIEW EXERCISES

1. a. Illustration 1 shows the dist from two towns to an airpo Which town is closer to the How much closer is it?

Chapter 3 Test

1. Translate each phrase to mathematical symbols.

a. 2 less than r

b. The product of 3, x, and y

7. METER READING Every hour between 8 A.M. and 5 P.M., a technician noted the value registered by a meter in a power plant and recorded that number on a line graph. (See Illustration 2.) Find the average meter value

Overlund College is going k letter O on a foothill slope f the letter is to be done using tration 3.) How many feet of

Chapters 1–6 Cumulative Review Exercises

Consider the number 6,245,867.

1. Round to the nearest thousand.

2. Round to the nearest million.

Find the perimeter of each figure.

3. A rectangle that is 8 meters long and 3 meters wide.

4. A square with sides that are 13 inches long.

5. PARKING The dimensions of various types of rectangular-shaped parking spaces are given in Illustration 1. Complete the table.

Type	Length (ft)	Width (ft)	Area (ft^2)
Standard space	20	9	
Standard space adjacent to a wall	20	10	
Parallel space	25	10	
Compact space	17	8	

ILLUSTRATION 1

6. HEALTH A person's blood pressure is a combination

Source: *Microsoft Encarta 98 Encyclopedia*

ILLUSTRATION 2

Student support

We have included many features that make *Prealgebra* very accessible to students. (See the examples starting on page viii.)

Worked Examples

The text contains more than 425 worked examples, many with several parts. Explanatory notes make the examples easy to follow.

Author's Notes

Author's notes, printed in red, are used to explain the steps in the solutions of examples. The notes are extensive; complete sentences are used so as to increase the students' ability to read and write mathematics.

A special logo shows which examples are included in the videotape series. ▶

Each step is explained using detailed author's notes. ▶

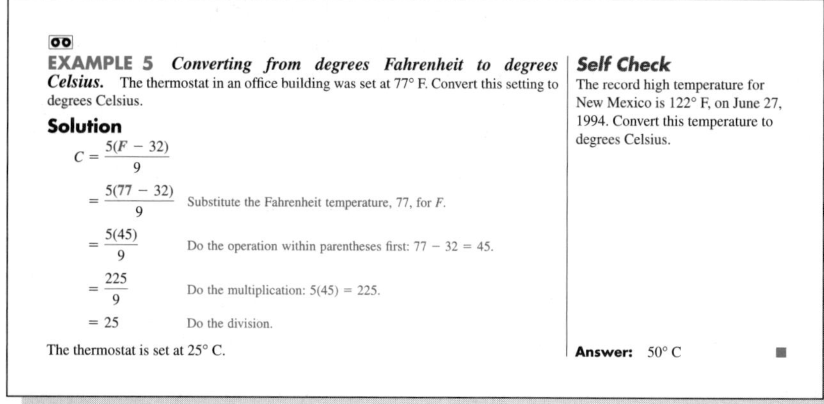

EXAMPLE 5 ***Converting from degrees Fahrenheit to degrees Celsius.*** The thermostat in an office building was set at 77° F. Convert this setting to degrees Celsius.

Solution

$C = \frac{5(F - 32)}{9}$

$= \frac{5(77 - 32)}{9}$ Substitute the Fahrenheit temperature, 77, for F.

$= \frac{5(45)}{9}$ Do the operation within parentheses first: $77 - 32 = 45$.

$= \frac{225}{9}$ Do the multiplication: $5(45) = 225$.

$= 25$ Do the division.

The thermostat is set at 25° C.

Self Check

The record high temperature for New Mexico is 122° F, on June 27, 1994. Convert this temperature to degrees Celsius.

Answer: 50° C ■

Self Checks

There are more than 365 Self Check problems that allow students to practice the skills demonstrated in the worked examples.

Comments

Throughout the text, Comments call attention to common mistakes and how to avoid them.

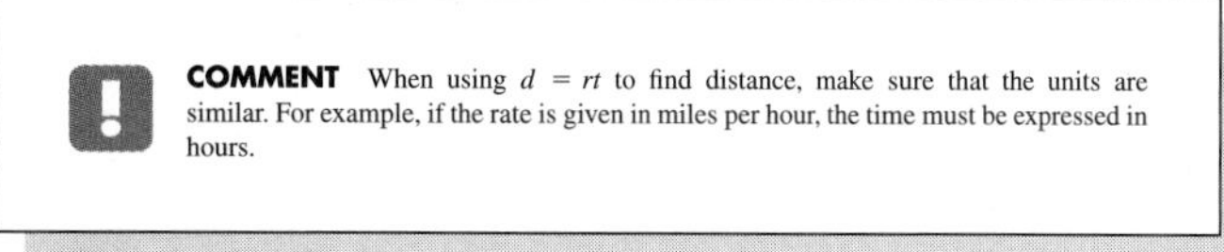

COMMENT When using $d = rt$ to find distance, make sure that the units are similar. For example, if the rate is given in miles per hour, the time must be expressed in hours.

Functional Use of Color

For easy reference, definitions, strategies, rules, and properties are printed in blue boxes. In addition, the book uses color to highlight terms and expressions that you would point to in a classroom discussion.

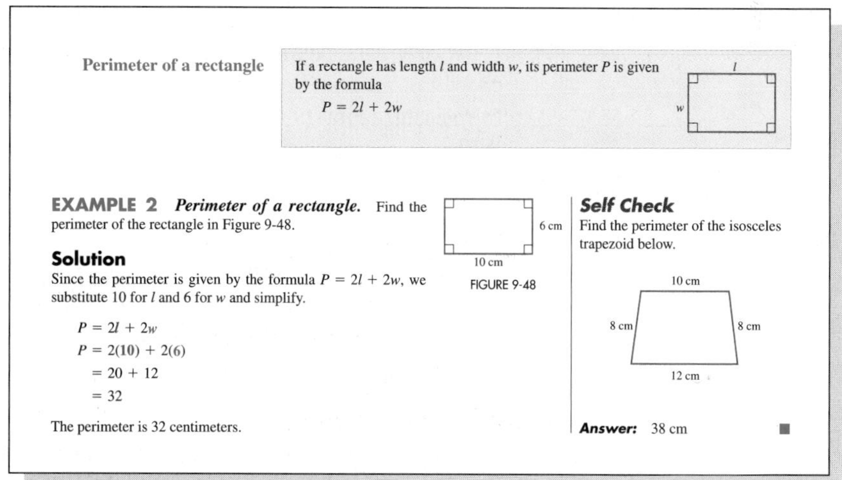

Perimeter of a rectangle If a rectangle has length l and width w, its perimeter P is given by the formula

$P = 2l + 2w$

EXAMPLE 2 ***Perimeter of a rectangle.*** Find the perimeter of the rectangle in Figure 9-48.

FIGURE 9-48

Solution

Since the perimeter is given by the formula $P = 2l + 2w$, we substitute 10 for l and 6 for w and simplify.

$P = 2l + 2w$

$P = 2(10) + 2(6)$

$= 20 + 12$

$= 32$

The perimeter is 32 centimeters.

Self Check

Find the perimeter of the isosceles trapezoid below.

Answer: 38 cm ■

Problems and Answers

The book includes thousands of carefully graded exercises. Appendix III provides the answers to the odd-numbered exercises in the Study Sets as well as all the answers to the Chapter Review, Chapter Test, and Cumulative Review problems.

Reading and Writing Mathematics

Also included (on pages xix–xx) are two features to help students improve their ability to read and write mathematics. "Reading Mathematics" helps students get the most out of the examples in this book by showing them how to read the solutions properly. "Writing Mathematics" highlights the characteristics of a well-written solution.

Study Skills and Math Anxiety

These two topics are discussed in detail in the section entitled "For the Student" at the end of this preface. In "Success in Prealgebra," students are asked to design a strategy for studying and learning the material. "Taking a Math Test," on page 74, helps students prepare for a test and then gives them suggestions for improving their performance.

Ancillaries for the instructor

Annotated Instructor's Edition

This is a special version of the complete student text, with all answers printed in blue next to the respective exercises.

Complete Solutions Manual

The *Complete Solutions Manual* provides worked-out solutions to all the exercises.

Test Bank

The test bank includes 8 tests per chapter as well as 3 final exams. The tests are made up of a combination of multiple-choice, free-response, true/false, and fill-in-the-blank questions.

BCA Testing

Brooks/Cole Assessment is a text-specific, Internet-ready testing suite that allows instructors to customize exams and track student progress in a browser-based format. BCA offers full algorithmic generation of problems and free-response questions. The testing and course-management components simplify routine tasks. Test results flow automatically into the gradebook, and the instructor can easily communicate with individuals, sections, or entire courses.

Text-Specific Videotapes

A set of videotapes is available free upon adoption of the text. Each tape covers one chapter of the text, broken into problem-solving sessions of 10 to 20 minutes. Examples from each section of the chapter are covered, as well as exercises from each Study Set. Where an example is taught on tape, a special logo is printed next to the example in the text.

Ancillaries for the student

Student Solutions Manual

The *Student Solutions Manual* provides worked-out solutions to the odd-numbered exercises in the text.

BCA Tutorial

This text-specific interactive software is delivered via the Web (at http://bca.brookscole.com). It is offered in both student's and instructor's versions. Because it is browser-based, it can serve as an intuitive guide even for students who have little technological proficiency. BCA Tutorial allows students to work with real math notation in real time, providing instant analysis and feedback. In the instructor's version, a built-in tracking program enables instructors to monitor student progress.

Interactive Video Skillbuilder CD

Packaged with each book, this is a single CD-ROM containing over eight hours of video instruction. There is one video lesson for each section of the book. The problems worked during each video lesson are listed next to the viewing screen, so that students can work them ahead of time, if they choose. In order to help students evaluate their progress, each section contains a 10-question web quiz and each chapter contains a chapter test, with answers to each problem on each test.

Acknowledgments

We are grateful to the instructors who have reviewed the text at various stages of its development. Their comments and suggestions have proven invaluable in making this a better book. We sincerely thank them for lending their time and talent to this project.

April Allen
Hartnell College

Laurette Blakey-Foster
Prairie View A & M

Julia Brown
Atlantic Community College

Mark Greenhalgh
Fullerton College

Maria Gushanas
Seton Hall University

Judith M. Jones
Valencia Community College

Mickey Levendusky
Pima Community College, Downtown

Donna E. Nordstrom
Pasadena City College

Martha Scarbrough
Motlow State Community College

Angelo Segalla
Orange Coast College

Allison Sloan
Valencia West Campus

Rita Sturgeon
San Bernardino Valley College

Terrie Teegarden
San Diego Mesa College

Gwen Terwilliger
University of Toledo

Jane Weber
University of Alaska, Fairbanks

Mary Jane Wolfe
University of Rio Grande

We want to express our gratitude to Elizabeth Morrison, Judy Jones, Linda Rottman, and Mike Murphy for their input and suggestions. We offer special thanks to Barbara Rugeley, Eagle Zhuang, Tom Gerfen, Arnold Kondo, Rob Everest, Sheila White, Doug Keebaugh, Ron Livingston, Eric Rabitoy, Robin Carter, Karl Hunsicker, Dennis Korn, and Laureen Breegle for their assistance with some of the application problems.

Without the talents and dedication of the editorial, marketing, and production staff of Brooks/Cole, this revision of *Prealgebra* could not have been so well accomplished. We express our sincere appreciation for the hard work of Bob Pirtle, Rachael Sturgeon, Leah Thomson, Samantha Cabaluna, Ellen Brownstein, Vernon Boes, Micky Lawler, and Vena Dyer, as well as the freelance talents of David Hoyt, Lori Heckelman, John Edeen, and Roy Neuhaus and the superb typesetting of the Clarinda Company.

Alan S. Tussy
R. David Gustafson

For the Student

Success in prealgebra

To be successful in mathematics, you need to know how to study it. The following checklist will help you develop your own personal strategy to study and learn the material. The suggestions listed below require some time and self-discipline on your part, but it will be worth the effort. This will help you get the most out of this course.

As you read each of the following statements, place a check mark in the box if you can truthfully answer Yes. If you can't answer Yes, think of what you might do to make the suggestion part of your personal study plan. You should go over this checklist several times during the semester to be sure you are following it.

Preparing for the Class

- ☐ I have made a commitment to myself to give this course my best effort.
- ☐ I have the proper materials: a pencil with an eraser, paper, a notebook, a ruler, a calculator, and a calendar or day planner.
- ☐ I am willing to spend a minimum of two hours doing homework for every hour of class.
- ☐ I will try to work on this subject every day.
- ☐ I have a copy of the class syllabus. I understand the requirements of the course and how I will be graded.
- ☐ I have scheduled a free hour after the class to give me time to review my notes and begin the homework assignment.

Class Participation

- ☐ I know my instructor's name.
- ☐ I will regularly attend the class sessions and be on time.
- ☐ When I am absent, I will find out what the class studied, get a copy of any notes or handouts, and make up the work that was assigned when I was gone.
- ☐ I will sit where I can hear the instructor and see the chalkboard.
- ☐ I will pay attention in class and take careful notes.
- ☐ I will ask the instructor questions when I don't understand the material.
- ☐ When tests, quizzes, or homework papers are passed back and discussed in class, I will write down the correct solutions for the problems I missed so that I can learn from my mistakes.

Study Sessions

- ☐ I will find a comfortable and quiet place to study.
- ☐ I realize that reading a math book is different from reading a newspaper or a novel. Quite often, it will take more than one reading to understand the material.
- ☐ After studying an example in the textbook, I will work the accompanying Self Check.
- ☐ I will begin the homework assignment only after reading the assigned section.
- ☐ I will try to use the mathematical vocabulary mentioned in the book and used by my instructor when I am writing or talking about the topics studied in this course.
- ☐ I will look for opportunities to explain the material to others.
- ☐ I will check all my answers to the problems with those provided in the back of the book (or with the *Student Solutions Manual*) and reconcile any differences.
- ☐ My homework will be organized and neat. My solutions show all the necessary steps.
- ☐ I will work some review problems every day.
- ☐ After completing the homework assignment, I will read the next section to prepare for the coming class session.
- ☐ I will keep a notebook containing my class notes, homework papers, quizzes, tests, and any handouts—all in order by date.

Special Help

- ☐ I know my instructor's office hours and am willing to go in to ask for help.
- ☐ I have formed a study group with classmates that meets regularly to discuss the material and work on problems.
- ☐ When I need additional explanation of a topic, I view the tutorial videos and check the website.
- ☐ I make use of extra tutorial assistance that my school offers for mathematics courses.
- ☐ I have purchased the *Student Solutions Manual* that accompanies this text, and I use it.

To follow each of these suggestions will take time. It takes a lot of practice to learn mathematics, just as with any other skill.

No doubt, you will sometimes become frustrated along the way. This is natural. When it occurs, take a break and come back to the material after you have had time to clear your thoughts. Keep in mind that the skills and discipline you learn in this course will help make for a brighter future. Good luck!

Reading mathematics

To get the most out of this book, you need to learn how to read it correctly. A mathematics textbook must be read differently than a novel or a newspaper. For one thing, you need to read it slowly and carefully. At times, you will have to reread a section to understand its content. You should also have pencil and paper, so that you can work along with the text to understand the concepts presented.

Perhaps the most informative parts of a mathematics book are its examples. Each example in this textbook consists of a problem and its corresponding solution. One form of solution that is used many times in this book is shown in the diagram below. It is important that you follow the "flow" of its steps if you are to understand the mathematics involved. For this solution form, the basic idea is this:

- A property, rule, or procedure is applied to the original expression to obtain an equivalent expression. We show that the two expressions are equivalent by writing an equals sign between them. The property, rule, or procedure that was used is then listed next to the equivalent expression in the form of an author's note, printed in red.
- The process of writing equivalent expressions and explaining the reasons behind them continues, step by step, until the final result is obtained.

The solution in the following diagram consists of three steps, but solutions have varying lengths.

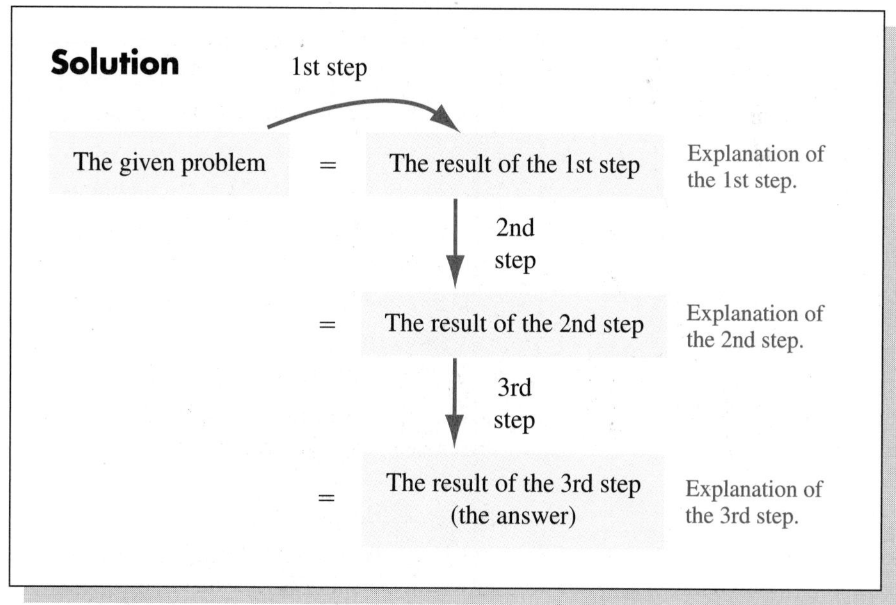

Writing mathematics

One of the major objectives of this course is for you to learn how to write solutions to problems properly. A written solution to a problem should explain your thinking in a series of neat and organized mathematical steps. Think of a solution as a mathematical essay—one that your instructor and other students should be able to read and understand. Some solutions will be longer than others, but they must all be in the proper format and use the correct notation. To learn how to do this will take time and practice.

To give you an idea of what will be expected, let's look at two samples of student work. In the first, we have highlighted some important characteristics of a well-written solution. The second sample is poorly done and would not be acceptable.

Evaluate $35 - 2^2 \cdot 3$.

A well-written solution:

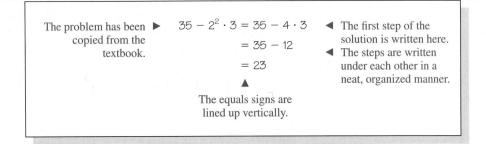

The problem has been copied from the textbook. ▶

$$\begin{aligned} 35 - 2^2 \cdot 3 &= 35 - 4 \cdot 3 \\ &= 35 - 12 \\ &= 23 \end{aligned}$$

▲ The equals signs are lined up vertically.

◀ The first step of the solution is written here.

◀ The steps are written under each other in a neat, organized manner.

A poorly written solution:

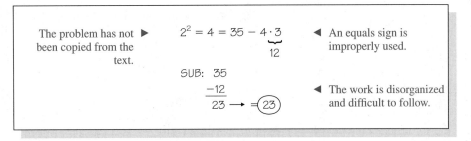

The problem has not been copied from the text. ▶

$2^2 = 4 = 35 - \underbrace{4 \cdot 3}_{12}$

SUB: $\begin{array}{r} 35 \\ -12 \\ \hline 23 \end{array} \longrightarrow = \text{(23)}$

◀ An equals sign is improperly used.

◀ The work is disorganized and difficult to follow.

Whole Numbers

1

IN THIS CHAPTER, WE BEGIN OUR MATHEMATICS STUDY BY EXAMINING THE PROCEDURES USED TO SOLVE PROBLEMS THAT INVOLVE WHOLE NUMBERS.

1.1 An Introduction to the Whole Numbers

In this section, you will learn about

- Sets of numbers • Place value • Expanded notation
- Graphing on the number line • Ordering of the whole numbers
- Rounding whole numbers • Tables and graphs

INTRODUCTION. In this section, we will discuss the natural numbers and the whole numbers. These numbers are used to answer questions such as "How many?" "How fast?" "How heavy?" "How far?"

- The movie *Saving Private Ryan* won 5 Academy Awards.
- The speed limit on interstate highways in Wisconsin is 65 mph.
- The Statue of Liberty weighs 225 tons.
- The driving distance between Chicago and Houston is 1,067 miles.

Sets of numbers

A **set** is a collection of objects. Two important sets in mathematics are the natural numbers (the numbers that we count with) and the whole numbers. When writing a set, we use **braces** { } to enclose the **members** (or **elements**) of the set.

The set of natural numbers	{1, 2, 3, 4, 5, 6, 7, 8, 9, 10, 11, 12, . . .}
The set of whole numbers	{0, 1, 2, 3, 4, 5, 6, 7, 8, 9, 10, 11, 12, . . .}

The three dots in the previous lists indicate that these sets continue on forever. There is no largest natural number or whole number.

Since every natural number is also a whole number, we say that the set of natural numbers is a **subset** of the set of whole numbers. However, not all whole numbers are natural numbers, because 0 is a whole number but not a natural number.

Place value

When we express a whole number with a *numeral* containing the *digits* 0, 1, 2, 3, 4, 5, 6, 7, 8, 9, we say that we have written the number in **standard notation.** The position of a digit in a numeral determines its value. In the numeral 325, the 5 is in the *ones column,* the 2 is in the *tens column,* and the 3 is in the *hundreds column.*

```
                    3 2 5
Hundreds column ————┘ | └———— Ones column
                      |
                 Tens column
```

To make a numeral easy to read, we use commas to separate its digits into groups of three, called **periods.** Each period has a name, such as *ones, thousands, millions,* and so on. The following table shows the place value of each digit in the numeral 345,576,402,897,415, which is read as

three hundred forty-five trillion, five hundred seventy-six billion, four hundred two million, eight hundred ninety-seven thousand, four hundred fifteen

345 trillion			576 billion			402 million			897 thousand			4 hundred fifteen		
3	4	5	5	7	6	4	0	2	8	9	7	4	1	5
Trillions			Billions			Millions			Thousands			Ones		
Hundreds	Tens	Ones	Hundreds	Tens	Ones	Hundreds	Tens	Ones	Hundreds	Tens	Ones	Hundreds	Tens	Ones

As we move to the left in this table, the place value of each column is 10 times greater than the column to its right. This is why we call our number system a *base-10 number system.*

EXAMPLE 1 ***TV news.*** The cable network CNN is carried by 11,528 cable systems. Which digit tells the number of hundreds?

Solution

In 11,528, the hundreds column is the third column from the right. The digit 5 tells the number of hundreds.

Self Check

The Fox Family Channel is carried by 13,820 cable systems. Which digit tells the number of ten thousands?

Answer: 1 ■

Expanded notation

In the numeral 6,352, the digit 6 is in the thousands column, 3 is in the hundreds column, 5 is in the tens column, and 2 is in the ones (or units) column. The meaning of 6,352 becomes clear when we write it in **expanded notation.**

6 thousands + 3 hundreds + 5 tens + 2 ones

We read the numeral 6,352 as "six thousand, three hundred fifty-two."

EXAMPLE 2 ***Expanded notation.*** Write each number in expanded notation: **a.** 63,427 and **b.** 1,251,609.

Solution

a. 6 ten thousands + 3 thousands + 4 hundreds + 2 tens + 7 ones

We read this number as "sixty-three thousand, four hundred twenty-seven."

b. 1 million + 2 hundred thousands + 5 ten thousands + 1 thousand + 6 hundreds + 0 tens + 9 ones

Since 0 tens is zero, the expanded notation can also be written as

1 million + 2 hundred thousands + 5 ten thousands + 1 thousand + 6 hundreds + 9 ones

We read this number as "one million, two hundred fifty-one thousand, six hundred nine."

Self Check

Write 808,413 in expanded notation.

Answer:
8 hundred thousands + 8 thousands + 4 hundreds + 1 ten + 3 ones. Read as "eight hundred eight thousand, four hundred thirteen." ■

EXAMPLE 3 ***Translating to standard notation.*** Write twenty-three thousand forty in standard notation.

Solution

In expanded notation, the number is written as

2 ten thousands + 3 thousands + 4 tens There are 0 hundreds and 0 ones.

In standard notation, this is written as 23,040.

Self Check

Write seventy-six thousand three in standard notation.

Answer: 76,003 ■

Graphing on the number line

Whole numbers can be illustrated by drawing points on a **number line.** A number line is a horizontal or vertical line that is used to represent numbers graphically. Like a ruler, a number line is straight and has uniform markings. (See Figure 1-1.) To construct a number line, we begin on the left with a point on the line representing the number 0. This point is called the **origin.** We then proceed to the right, drawing equally spaced marks and labeling them with whole numbers that increase progressively in size. The arrowhead at the right indicates that the number line continues forever.

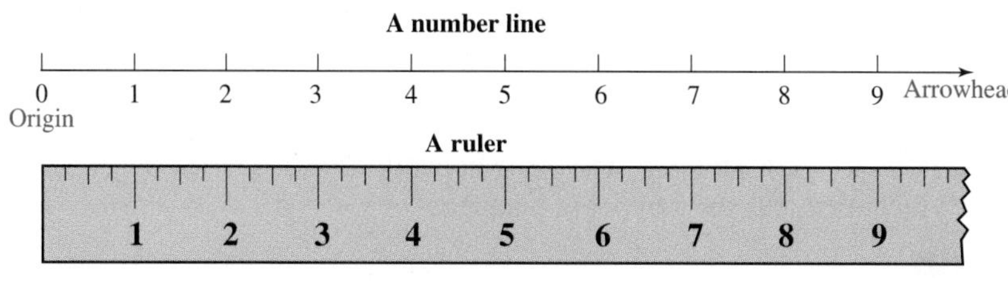

FIGURE 1-1

Using a process known as **graphing,** a single number or a set of numbers can be represented on a number line. *The graph of a number* is the point on the number line that corresponds to that number. *To graph a number* means to locate its position on the number line and then to highlight it using a heavy dot. Figure 1-2 shows the graphs of the whole numbers 5 and 8.

FIGURE 1-2

Ordering of the whole numbers

As we move to the right on the number line, the numbers get larger. Because 8 lies to the right of 5, we say that 8 is greater than 5. The **inequality symbol** $>$ ("is greater than") can be used to write this fact.

$8 > 5$ Read as "8 is greater than 5."

Since $8 > 5$, it is also true that $5 < 8$. (Read as "5 is less than 8.")

COMMENT To distinguish between these two inequality symbols, remember that they always point to the smaller of the two numbers involved.

$8 > 5$ $5 < 8$

Points to the smaller number.

EXAMPLE 4 ***Inequality symbols.*** Place an $<$ or an $>$ symbol in the box to make a true statement: **a.** 3 ☐ 7 and **b.** 18 ☐ 16.

Solution

a. Since 3 is to the left of 7 on the number line, $3 < 7$.

b. Since 18 is to the right of 16 on the number line, $18 > 16$.

Self Check

Place an $<$ or an $>$ symbol in the box to make a true statement.

a. 12 ☐ 4 **b.** 7 ☐ 10

Answers: **a.** $>$, **b.** $<$ ■

Rounding whole numbers

When we don't need exact results, we often round numbers. For example, when a teacher with 36 students in his class orders 40 textbooks, he has rounded the actual number to the *nearest ten,* because 36 is closer to 40 than it is to 30.

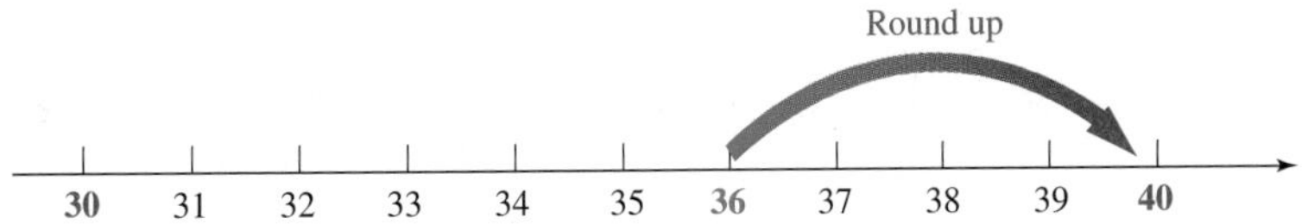

When a geologist says that the height of Alaska's Mount McKinley is "about 20,300 feet," she has rounded to the *nearest hundred,* because its actual height of 20,320 feet is closer to 20,300 than it is to 20,400.

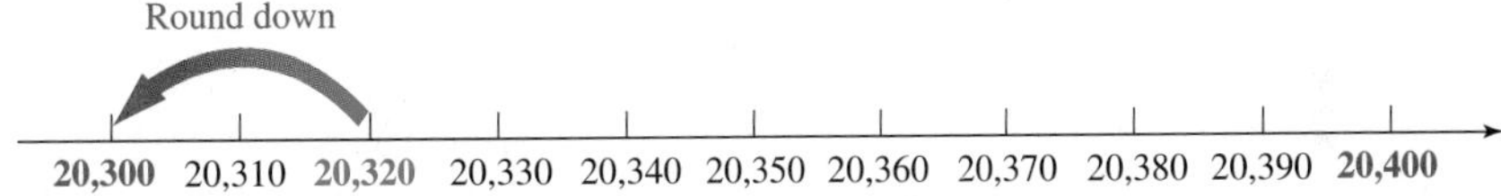

To round a whole number, we follow an established set of rules. To round a number to the nearest ten, for example, we begin by locating the **rounding digit** in the tens column. If the **test digit** to the right of that column (the digit in the ones column) is 5 or greater, we *round up* by increasing the tens digit by 1 and placing a 0 in the ones column. If the test digit is less than 5, we *round down* by leaving the tens digit unchanged and placing a 0 in the ones column.

EXAMPLE 5 ***Rounding to the nearest ten.*** Round each number to the nearest ten: **a.** 3,764 and **b.** 12,087.

Solution

a. We find the rounding digit in the tens column, which is 6.

Rounding digit
3,764
Test digit

We then look at the test digit to the right of 6, the 4 in the ones column. Since $4 < 5$, we round down by leaving the 6 unchanged and replacing the test digit with 0. The rounded answer is 3,760.

b. We find the rounding digit in the tens column, which is 8.

Rounding digit
12,087
Test digit

We then look at the test digit to the right of 8, the 7 in the ones column. Because $7 > 5$, we round up by adding 1 to 8 and replacing the test digit with 0. The rounded answer is 12,090.

Self Check

Round each number to the nearest ten:

a. 35,642

b. 3,756

Answers: **a.** 35,640, **b.** 3,760 ■

A similar procedure is used to round numbers to the nearest hundred, the nearest thousand, the nearest ten thousand, and so on.

Rounding a whole number

1. To round a number to a certain place, locate the rounding digit in that place.
2. Look at the test digit to the right of the rounding digit.
3. If the test digit is 5 or greater, round up by adding 1 to the rounding digit and changing all of the digits to the right of the rounding digit to 0.

   If the test digit is less than 5, round down by keeping the rounding digit and changing all of the digits to the right of the rounding digit to 0.

EXAMPLE 6 ***Rounding to the nearest hundred.*** Round 7,960 to the nearest hundred.

Solution

First, we find the rounding digit in the hundreds column. It is 9.

```
  ┌— Rounding digit
7,960
   └— Test digit
```

We then look at the 6 to the right of 9. Because $6 > 5$, we round up and increase 9 in the hundreds column by 1. Since the 9 in the hundreds column represents 900, increasing 9 by 1 represents increasing 900 to 1,000. Thus, we replace the 9 with a 0 and add 1 to the 7 in the thousands column. Finally, we replace the two rightmost digits with 0's. The rounded answer is 8,000.

Self Check

Round 365,283 to the nearest hundred.

Answer: 365,300 ■

EXAMPLE 7 ***U.S. cities.*** In 1999, Denver was the nation's 27th largest city. Round the population of Denver given in Figure 1-3 **a.** to the nearest thousand and **b.** to the nearest ten thousand.

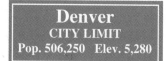

Denver
CITY LIMIT
Pop. 506,250 Elev. 5,280

FIGURE 1-3

Solution

a. The rounding digit in the thousands column is 6. The test digit, 2, is less than 5, so we round down. To the nearest thousand, Denver's population was 506,000 in 1999.

b. The rounding digit in the ten thousands column is 0. The test digit, 6, is more than 5, so we round up. To the nearest ten thousand, Denver's population was 510,000 in 1999. ■

Tables and graphs

The table in Figure 1-4(a) is an example of the use of whole numbers. It shows the number of women elected to the United States House of Representatives in the Congressional elections held every two years from 1990 to 1998.

Table

Year	Number of women elected
1990	29
1992	48
1994	53
1996	51
1998	55

(a)

Bar graph

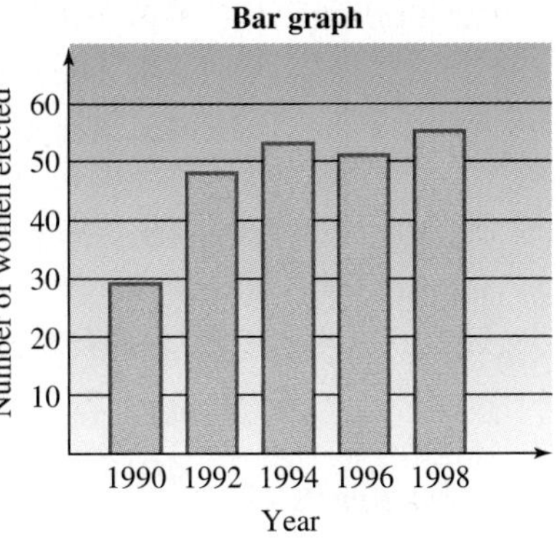

(b)

Line graph

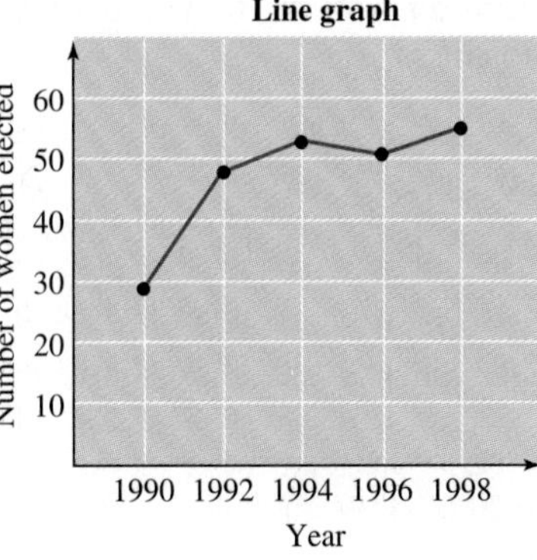

(c)

FIGURE 1-4

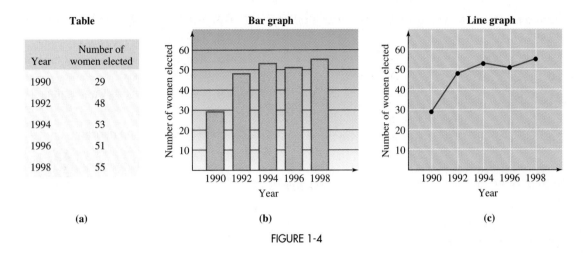

In Figure 1-4(b), the election results are presented in a **bar graph.** The horizontal scale is labeled "Year" and scaled in units of 2 years. The vertical scale is labeled "Number of women elected" and scaled in units of 10. The bar directly over each year extends to a height indicating the number of women elected to Congress that year.

Another way to present the information in the table is with a **line graph.** Instead of using a bar to denote the number of women elected, we use a heavy dot drawn at the correct height. After drawing data points for 1990, 1992, 1994, 1996, and 1998, we connect the points with line segments to create the line graph in Figure 1-4(c).

STUDY SET Section 1.1

VOCABULARY *Fill in the blanks to make the statements true.*

1. A ____________ is a collection of objects.

2. The set of ____________ numbers is {1, 2, 3, 4, 5, . . .}, and the set of ____________ numbers is {0, 1, 2, 3, 4, 5, . . .}.

3. When 297 is written as 2 hundreds + 9 tens + 7 ones, it is written in ____________ notation.

4. If we ____________ 627 to the nearest ten, we get 630.

5. Using a process known as graphing, whole numbers can be represented as points on a ____________ line.

6. The symbols > and < are ____________ symbols.

CONCEPTS *In Exercises 7–10, consider the numeral 57,634.*

7. What digit is in the tens column?

8. What digit is in the thousands column?

9. What digit is in the hundreds column?

10. What digit is in the ten thousands column?

11. What set of numbers is obtained when 0 is combined with the natural numbers?

12. Place the numbers 25, 17, 37, 15, 45 in order, from smallest to largest.

13. Graph 1, 3, 5, and 7.

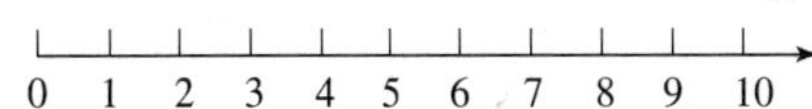

14. Graph 0, 2, 4, 6, and 8.

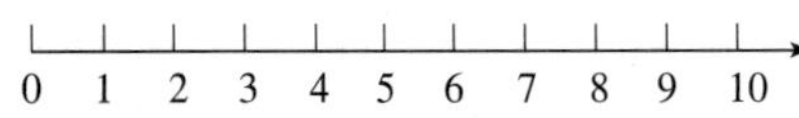

15. Graph the whole numbers less than 6.

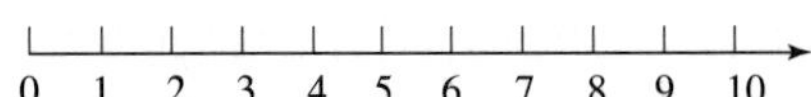

16. Graph the whole numbers between 2 and 8.

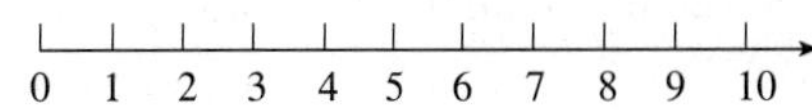

Place an > or an < symbol in the box to make a true statement.

17. 47 ☐ 41
18. 53 ☐ 67
19. 309 ☐ 300
20. 841 ☐ 814
21. 2,052 ☐ 2,502
22. 999 ☐ 998
23. Since 4 < 7, it is also true that 7 ☐ 4.
24. Since 9 > 0, it is also true that 0 ☐ 9.

NOTATION *Fill in the blanks to make the statements true.*

25. The symbols { }, called ______________, are used when writing a set.
26. The symbol > means ______________, and the symbol < means ______________.

PRACTICE *Write each number in expanded notation and then write it in words.*

27. 245
28. 508
29. 3,609
30. 3,960
31. 32,500
32. 73,009
33. 104,401
34. 570,003

Write each number in standard notation.

35. 4 hundreds + 2 tens + 5 ones
36. 7 hundreds + 7 tens + 7 ones
37. 2 thousands + 7 hundreds + 3 tens + 6 ones
38. 7 billions + 3 hundreds + 5 tens
39. Four hundred fifty-six
40. Three thousand seven hundred thirty-seven
41. Twenty-seven thousand five hundred ninety-eight
42. Seven million, four hundred fifty-two thousand, eight hundred sixty
43. Nine thousand one hundred thirteen
44. Nine hundred thirty
45. Ten million, seven hundred thousand, five hundred six
46. Eighty-six thousand four hundred twelve

Round 79,593 to the nearest . . .

47. ten
48. hundred
49. thousand
50. ten thousand

Round 5,925,830 to the nearest . . .

51. thousand
52. ten thousand
53. hundred thousand
54. million

Round $419,161 to the nearest . . .

55. $10
56. $100
57. $1,000
58. $10,000

APPLICATIONS

59. EATING HABITS The following list shows the ten countries with the largest per-person annual consumption of meat. Construct a two-column table that presents the data in order, beginning with the largest per-person consumption. (The abbreviation "lb" means "pounds.")

Australia: 239 lb	New Zealand: 259 lb
Austria: 229 lb	Saint Lucia: 222 lb
Canada: 211 lb	Spain: 211 lb
Cyprus: 236 lb	Uruguay: 230 lb
Denmark: 219 lb	United States: 261 lb

60. PRESIDENTS The following list shows the ten youngest U.S. presidents and their ages (in years/days) when they took office. Construct a two-column table that presents the data in order, beginning with the youngest president.

C. Arthur	50 yr/350 days	U. Grant	46 yr/236 days
G. Cleveland	47 yr/351 days	J. Kennedy	43 yr/236 days
W. Clinton	46 yr/154 days	F. Pierce	48 yr/101 days
M. Filmore	50 yr/184 days	J. Polk	49 yr/122 days
J. Garfield	49 yr/105 days	T. Roosevelt	42 yr/322 days

61. MISSIONS TO MARS The United States, Russia, and Japan have launched Mars space probes. The graph in Illustration 1 shows the success rate of the missions, by decade.

a. What decade had the greatest number of successful missions?

b. What decade had the greatest number of unsuccessful missions?

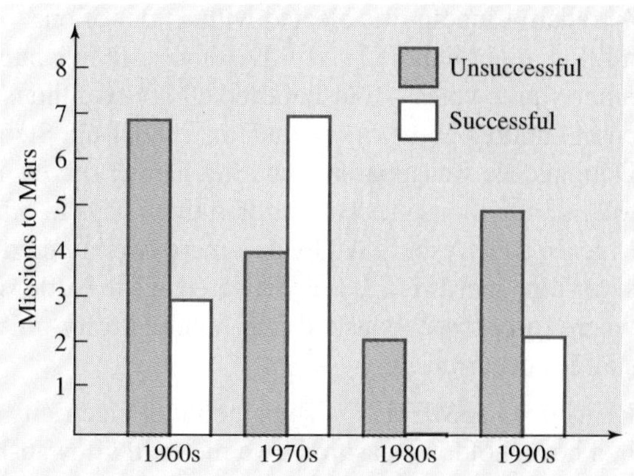

Source: *USA TODAY* (December 8, 1999)

ILLUSTRATION 1

62. BANKING CRISES Illustration 2 shows the number of banks that were closed or taken over by federal agencies during the years 1934–1995.

a. During what two time spans was there an upsurge in bank failures?

b. In what year were there the most bank failures? Estimate the number of banks that failed that year.

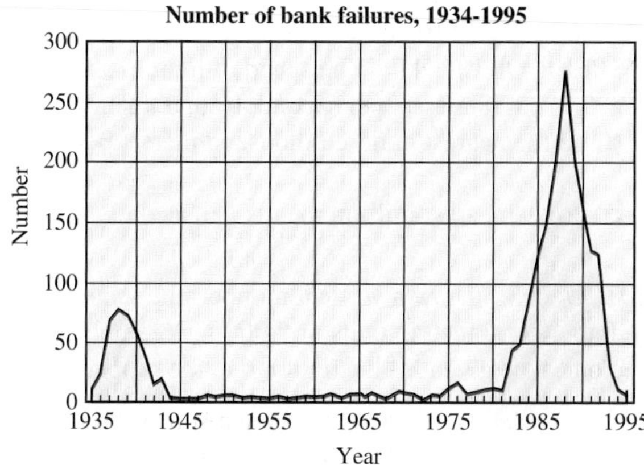

Source: FDIC Division of Research and Statistics

ILLUSTRATION 2

63. ENERGY RESERVES Construct a bar graph using the data shown in Illustration 3.

Natural gas reserves, 1998 (in trillion cubic feet)	
United States	167
Venezuela	143
Canada	65
Mexico	64
Argentina	24

Source: *Oil and Gas Journal*

ILLUSTRATION 3

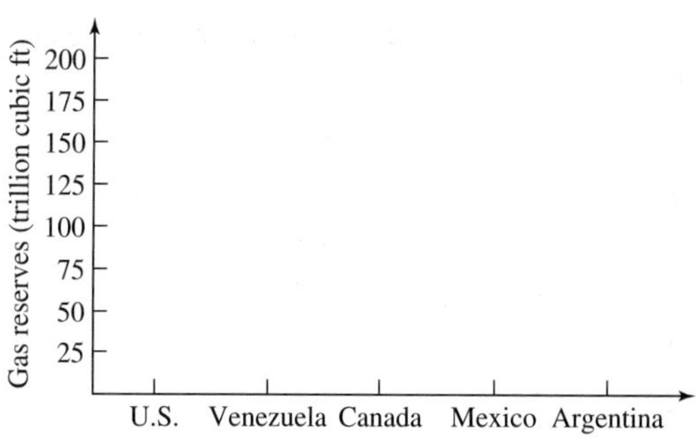

64. ENERGY RESERVES Construct a line graph using the data in Illustration 3.

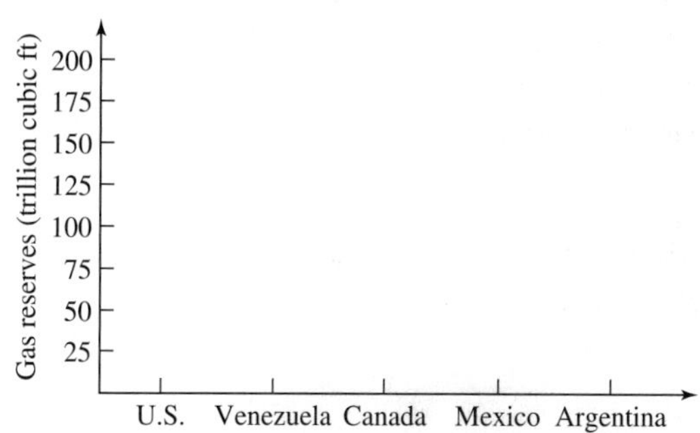

65. COFFEE Construct a line graph using the data shown in Illustration 4.

Starbucks locations	
Year	**Number**
1992	165
1993	272
1994	425
1995	676
1996	1,015
1997	1,412
1998	1,886
1999	2,200

Source: Starbucks Company

ILLUSTRATION 4

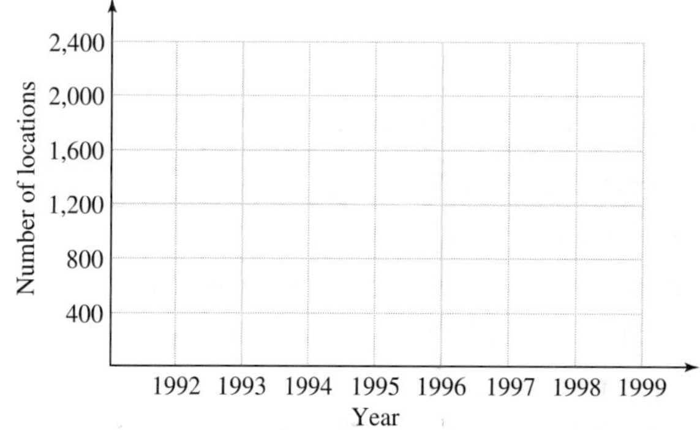

66. COFFEE Construct a bar graph using the data shown in Illustration 4 on the preceding page.

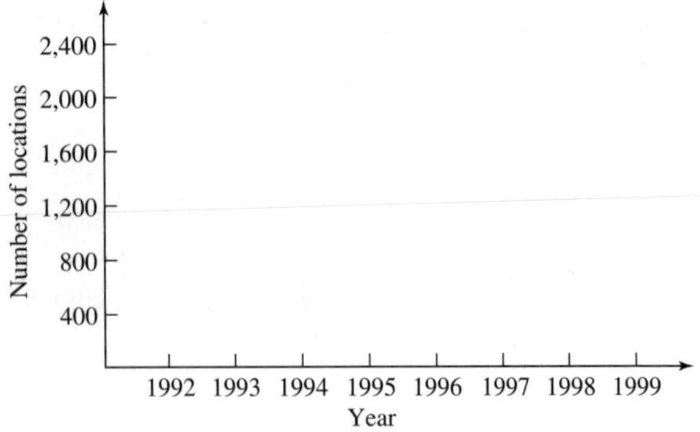

67. Complete each check by writing the amount in words on the proper line.

a.

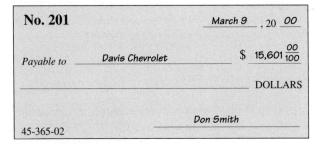

No. 201 — *March 9*, 20 *00*

Payable to *Davis Chevrolet* — $ *15,601* $\frac{00}{100}$

________________ DOLLARS

Don Smith

45-365-02

b.

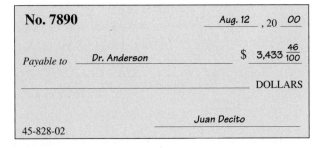

No. 7890 — *Aug. 12*, 20 *00*

Payable to *Dr. Anderson* — $ *3,433* $\frac{46}{100}$

________________ DOLLARS

Juan Decito

45-828-02

68. FORMAL ANNOUNCEMENTS One style used when printing formal invitations and announcements is to write all numbers in words. Use this style to write each of the following phrases.

a. This diploma awarded this 27th day of June, 1996.

b. The suggested contribution for the fundraiser is \$850 a plate, or an entire table may be purchased for \$5,250.

69. EDITING Edit this excerpt from a history text by circling all numbers written in words and rewriting them using digits.

Abraham Lincoln was elected with a total of one million, eight hundred sixty-five thousand, five hundred ninety-three votes—four hundred eighty-two thousand, eight hundred eighty more than the runner-up, Stephen Douglas. He was assassinated after having served a total of one thousand five hundred three days in office. Lincoln's Gettysburg Address, a mere two hundred sixty-nine words long, was delivered at the battle site where forty-three thousand four hundred forty-nine casualties occurred.

70. READING A METER The amount of electricity used in a household is measured by a meter in kilowatt-hours (kwh). Determine the reading on the meter shown in Illustration 5. (When the pointer is between two numbers, read the lower number.)

ILLUSTRATION 5

71. SPEED OF LIGHT The speed of light in a vacuum is 299,792,458 meters per second. Round this number

a. to the nearest hundred thousand meters per second.

b. to the nearest million meters per second.

72. CLOUDS Draw a vertical number line scaled from 0 to 40,000 feet, in units of 5,000 feet. Graph each cloud type given in Illustration 6 at the proper altitude.

Cloud type	Altitude (ft)
Altocumulus	21,000
Cirrocumulus	37,000
Cirrus	38,000
Cumulonimbus	15,000
Cumulus	8,000
Stratocumulus	9,000
Stratus	4,000

ILLUSTRATION 6

WRITING

73. Explain why the natural numbers are called the counting numbers.

74. Explain how you would round 687 to the nearest ten.

75. The houses in a new subdivision are priced "in the low 130s." What does this mean?

76. A million is a thousand thousands. Explain why this is so.

1.2 *Adding and Subtracting Whole Numbers*

In this section, you will learn about

- Properties of addition • Adding whole numbers
- Perimeter of a rectangle and a square • Subtracting whole numbers
- Combinations of operations

INTRODUCTION. Mastering the operations of addition and subtraction with whole numbers enables us to solve problems from geometry, business, and science. For example, to find the distance around a rectangle, we need to add the lengths of its four sides. To prepare an annual budget, we need to add separate line items. To find the difference between two temperatures, we need to subtract one from the other.

Properties of addition

Addition is the process of finding the total of two (or more) numbers. It can be illustrated using a number line (see Figure 1-5). For example, to compute 4 + 5, we begin at zero and draw an arrow 4 units long, extending to the right. This represents 4. From the tip of that arrow, we draw another arrow 5 units long, also extending to the right. The second arrow points to 9. This result corresponds to the addition fact 4 + 5 = 9, where 4 and 5 are called **addends** and 9 is called the **sum.**

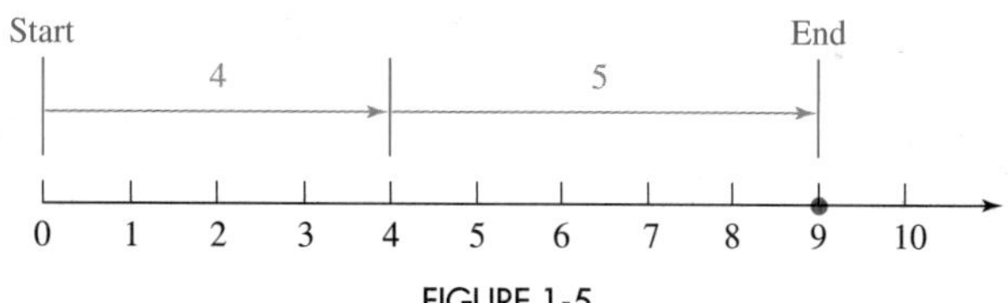

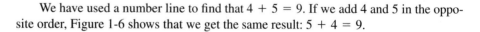

FIGURE 1-5

We have used a number line to find that 4 + 5 = 9. If we add 4 and 5 in the opposite order, Figure 1-6 shows that we get the same result: 5 + 4 = 9.

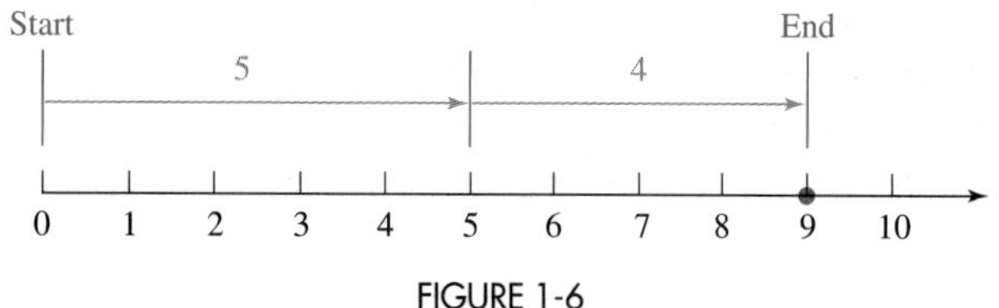

FIGURE 1-6

These examples illustrate that two whole numbers can be added in either order to get the same sum. The order in which we add two numbers does not affect the result. This property is called the **commutative property of addition.** To state the commutative property of addition concisely, we can use variables.

Variables

A **variable** is a letter that is used to stand for a number.

We now use the variables a and b to state the commutative property of addition.

Commutative property of addition

If a and b represent numbers, then

$$a + b = b + a$$

To find the sum of three whole numbers, we add two of them and then add the third to that result. In the following examples, we add $3 + 4 + 2$ in two ways. We will use the **grouping symbols** ( ), called **parentheses,** to show this. We must do the operation within parentheses first.

Method 1: Group 3 + 4

$(3 + 4) + 2 = 7 + 2$ Because of the parentheses, add 3 and 4 first to get 7.

$= 9$ Then add 7 and 2 to get 9.

Method 2: Group 4 + 2

$3 + (4 + 2) = 3 + 6$ Because of the parentheses, add 4 and 2 to get 6.

$= 9$ Then add 3 and 6 to get 9.

Either way, the sum is 9. It does not matter how we group (or associate) numbers in addition. This property is called the **associative property of addition.**

Associative property of addition

If a, b, and c represent numbers, then

$$(a + b) + c = a + (b + c)$$

Whenever we add 0 to a number, the number remains the same. For example,

$3 + 0 = 3$, $5 + 0 = 5$, and $9 + 0 = 9$

These examples suggest the **addition property of 0.**

Addition property of 0

If a represents any number, then

$a + 0 = a$ and $0 + a = a$

EXAMPLE 1 ***Properties of addition.*** Find each sum: **a.** $8 + 9$ and $9 + 8$, **b.** $5 + (1 + 8)$ and $(5 + 1) + 8$, and **c.** $(3 + 0) + 4$.

Solution

a. $8 + 9 = 17$ and $9 + 8 = 17$. The results are the same.

b. In each case, we do the addition within parentheses first.

$5 + (1 + 8) = 5 + 9$ $(5 + 1) + 8 = 6 + 8$

$= 14$ $= 14$

The results are the same.

c. $(3 + 0) + 4 = 3 + 4$ Do the addition within parentheses first: $3 + 0 = 3$.

$= 7$

Self Check

Find each sum.

a. $6 + 7$ and $7 + 6$

b. $2 + (6 + 3)$ and $(2 + 6) + 3$

c. $3 + (0 + 4)$

Answers: **a.** 13, 13, **b.** 11, 11, **c.** 7

Adding whole numbers

We can add whole numbers greater than 10 by using a vertical format that adds digits with the same place value. Because the additions within each column often exceed 9, it is sometimes necessary to *carry* the excess to the next column to the left. For example, to add 27 and 15, we write the numerals with the digits of the same place value aligned vertically.

$$\begin{array}{r} 27 \\ +\underline{15} \end{array}$$

We begin by adding the digits in the ones column: 7 + 5 = 12. Because 12 = 1 ten and 2 ones, we place a 2 in the ones column of the answer and carry 1 to the tens column.

$$\begin{array}{r} 1 \\ 27 \\ +\underline{15} \\ 2 \end{array}$$

Add the digits in the ones column: 7 + 5 = 12. Carry 1 (shown in blue) to the tens column.

Then we add the digits in the tens column.

$$\begin{array}{r} 1 \\ 27 \\ +\underline{15} \\ 42 \end{array}$$

Add 1, 2, and 1. Place the result, 4, in the tens column of the answer.

Thus, 27 + 15 = 42.

EXAMPLE 2 *Carrying.* Add: 9,834 + 692.

Solution

We write the numerals with their corresponding digits aligned vertically. Then we add the numbers, one column at a time, working from right to left.

$$\begin{array}{r} 9{,}834 \\ +\underline{692} \\ 6 \end{array}$$

Add the digits in the ones column and place the result in the ones column of the answer.

$$\begin{array}{r} 1 \\ 9{,}834 \\ +\underline{692} \\ 26 \end{array}$$

Add the digits in the tens column. The result, 12, exceeds 9. Place the 2 in the tens column of the answer and carry 1 (shown in blue) to the hundreds column.

$$\begin{array}{r} 11 \\ 9{,}834 \\ +\underline{692} \\ 526 \end{array}$$

Add the digits in the hundreds column. Since the result, 15, exceeds 9, place the 5 in the hundreds column of the answer and carry 1 (shown in green) to the thousands column.

$$\begin{array}{r} 11 \\ 9{,}834 \\ +\underline{692} \\ 10{,}526 \end{array}$$

Since the sum of the digits in the thousands column is 10, write 0 in the thousands column and 1 in the ten thousands column of the answer.

Thus, 9,834 + 692 = 10,526.

Self Check

Add: 675 + 1,497.

Answer: 2,172

To see if the result in Example 2 is reasonable, we can **estimate** the answer. 9,834 is a little less than 10,000, and 692 is a little less than 700. We estimate that the answer will be a little less than 10,000 + 700, or 10,700. An answer of 10,526 is reasonable. Estimation is discussed in more detail later in this chapter.

Words such as *increase, gain, credit, up, forward, rise, in the future,* and *to the right* are used to indicate addition.

EXAMPLE 3 ***Calculating temperatures.*** At noon, the temperature in Helena, Montana, was 31°. By 1:00 P.M., the temperature had increased 5°, and by 2:00 P.M., it had risen another 7°. Find the temperature at 2:00 P.M.

Solution To the temperature at noon, we add the two increases.

$$31 + 5 + 7$$

The two additions are done working from left to right.

$$\begin{aligned} \mathbf{31 + 5} + 7 &= \mathbf{36} + 7 \\ &= 43 \end{aligned}$$

The temperature at 2:00 P.M. was 43°. ■

EXAMPLE 4 ***U.S. history.*** The populations of four American colonies in 1630 are shown in Figure 1-7. Find the total population.

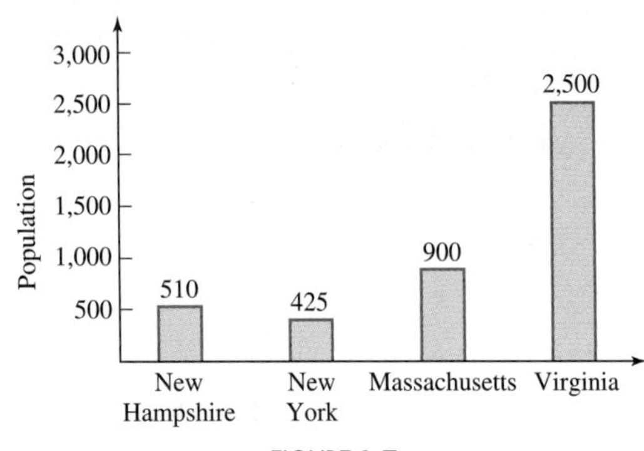

FIGURE 1-7

Solution

The word *total* indicates that we must add the populations of the individual colonies.

$$\begin{array}{r} {}^{2} \\ 510 \\ 425 \\ 900 \\ +2{,}500 \\ \hline 4{,}335 \end{array}$$

Align the numerals vertically. Add the digits, one column at a time, working from right to left.

The total population was 4,335.

Self Check

By 1700, the populations of the four colonies were New Hampshire 5,000, New York 19,100, Massachusetts 55,900, and Virginia 58,600. Find the total population.

Answer: 138,600 ■

Perimeter of a rectangle and a square

A **rectangle** is a four-sided figure (like a dollar bill) whose opposite sides are of equal length. Either of the longer sides is called its **length,** and either of the shorter sides is called its **width.** A rectangle with all four sides of equal length is called a **square.** (See Figure 1-8.)

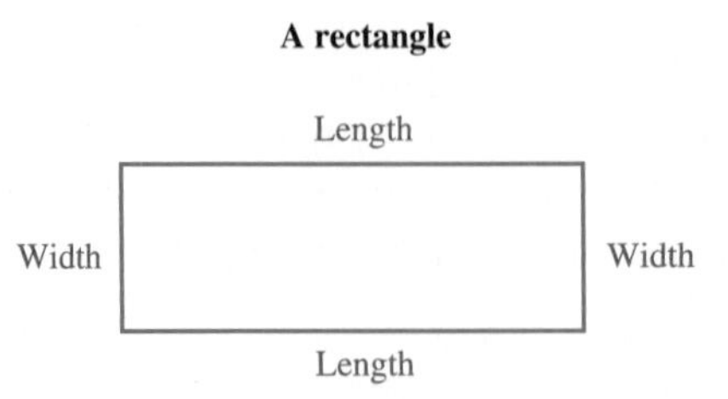

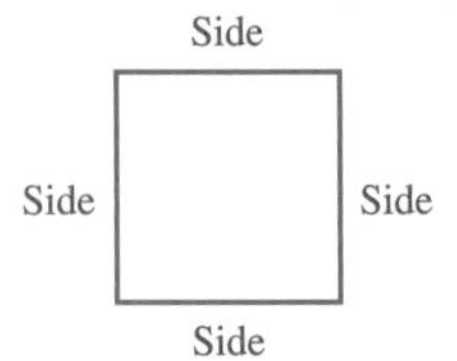

FIGURE 1-8

The distance around a rectangle or a square is called its **perimeter.** To find the perimeter of a rectangle, we add the lengths of its four sides.

The perimeter of a rectangle	=	length	+	length	+	width	+	width

To find the perimeter of a square, we add the lengths of its four sides.

The perimeter of a square	=	side	+	side	+	side	+	side

EXAMPLE 5 ***Perimeter.*** Find the perimeter of the dollar bill shown in Figure 1-9 (mm stands for millimeters).

Solution

To find the perimeter of the rectangular bill, we add the lengths of its four sides.

FIGURE 1-9

$$\begin{array}{r} \scriptstyle 22 \\ 156 \\ 156 \\ 65 \\ +\ 65 \\ \hline 442 \end{array}$$

The perimeter is 442 mm.

To see whether this result is reasonable, we estimate the answer. Because the rectangle is about 150 mm by 70 mm, its perimeter is approximately 150 + 150 + 70 + 70, or 440 mm. An answer of 442 mm is reasonable.

Self Check

A Monopoly game board is square-shaped, with sides 19 inches long. Find the perimeter of the board.

Answer: 76 in.

Subtracting whole numbers

Subtraction is the process of finding the difference between two numbers. It can be illustrated using a number line (see Figure 1-10). For example, to compute 9 − 4, we begin at zero and draw an arrow 9 units long, extending to the right. From the tip of that arrow, we draw another arrow 4 units long, but extending to the left. (This represents taking away 4.) The second arrow points to 5, indicating that 9 − 4 = 5. In this subtraction fact, 9 is called the **minuend,** 4 is called the **subtrahend,** and 5 is called the **difference.**

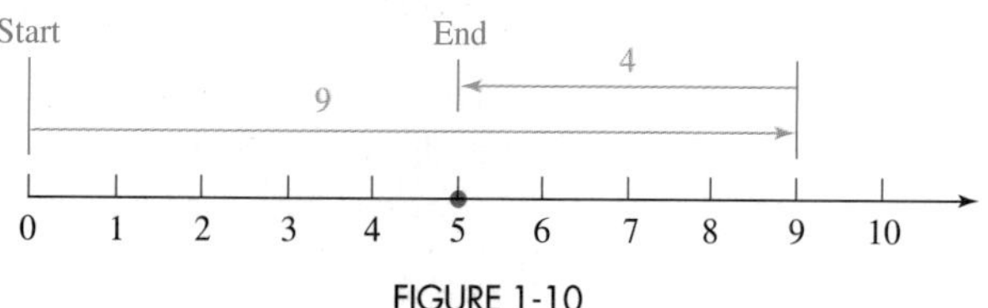

FIGURE 1-10

With whole numbers, we cannot subtract in the opposite order and find the difference 4 − 9, because we cannot take away 9 objects from 4 objects. Since subtraction of whole numbers cannot be done in either order, subtraction is not commutative.

Subtraction is not associative either, because if we group in different ways, we get different answers.

$$(9 - 5) - 1 = 4 - 1 \quad \text{but} \quad 9 - (5 - 1) = 9 - 4$$
$$= 3 \qquad\qquad\qquad\qquad = 5$$

EXAMPLE 6 ***Subtracting whole numbers.*** Do the subtractions: $9 - (6 - 3)$ and $(9 - 6) - 3$.

Solution

In each case, we do the subtraction in parentheses first.

$$9 - (\mathbf{6 - 3}) = 9 - \mathbf{3} \qquad (\mathbf{9 - 6}) - 3 = \mathbf{3} - 3$$
$$= 6 \qquad\qquad\qquad = 0$$

We note that the results are different.

Self Check

Do the subtractions: $8 - (5 - 2)$ and $(8 - 5) - 2$.

Answers: 5, 1

Whole numbers can be subtracted using a vertical format. Because subtractions often require subtracting a larger digit from a smaller digit, we may need to *borrow.* For example, to subtract 15 from 32, we write the minuend, 32, and the subtrahend, 15, in a vertical format, aligning the digits with the same place value.

$$\begin{array}{r} 3\ 2 \\ -1\ 5 \\ \hline \end{array}$$

Write the numerals in a column, with corresponding digits aligned vertically.

Since 5 can't be subtracted from 2, we borrow from the tens column of 32.

$$\begin{array}{r} \mathbf{2}\ \mathbf{12} \\ \cancel{3}\ \cancel{2} \\ -1\ \ 5 \\ \hline \mathbf{7} \end{array}$$

To subtract in the ones column, borrow 1 ten from the tens column. We show this by drawing a slash through the 3 and writing a 2 above it. Add 10 to the 2 in the ones column, which gives 12. Then subtract: $12 - 5 = 7$.

$$\begin{array}{r} 2\,12 \\ \cancel{3}\ \cancel{2} \\ -1\ 5 \\ \hline \mathbf{1}\ 7 \end{array}$$

Subtract in the tens column: $2 - 1 = 1$.

Thus, $32 - 15 = 17$. To check the result, we add the difference, 17, and the subtrahend, 15. We should obtain the minuend, 32.

$$\text{Check:}\quad \begin{array}{r} 1 \\ 17 \\ +15 \\ \hline 32 \end{array}$$

EXAMPLE 7 ***Borrowing.*** Subtract 576 from 2,021.

Solution

$$\begin{array}{r} 2{,}0\ 2\ 1 \\ -\quad 5\ 6\ 7 \\ \hline \end{array}$$

Write the numerals in a column, with the digits of the same place value aligned vertically.

$$\begin{array}{r} \mathbf{1}\ \mathbf{11} \\ 2{,}0\ \cancel{2}\ \cancel{1} \\ -\quad 5\ 7\ 6 \\ \hline \mathbf{5} \end{array}$$

To subtract in the ones column, borrow 1 ten from the tens column and add it to the ones column. Then subtract: $11 - 6 = 5$.

Since we can't subtract 7 from 1 in the tens column, we borrow. Because there is a 0 in the hundreds column of 2,021, we must borrow from the thousands column. We can take 1 thousand from the thousands column (leaving 1 thousand behind) and write it as 10 hundreds, placing a 10 in the hundreds column. From these 10 hundreds, we take 1 hundred (leaving 9 hundreds behind) and think of it as 10 tens. We add these 10 tens to the 1 ten that is already in the tens column to get 11 tens. From these 11 tens, we subtract 7 tens: $11 - 7 = 4$.

$$\begin{array}{r} 9\quad\ \ \\ 1\ \cancel{10}\,11\,11 \\ \cancel{2}{,}\ \cancel{0}\ \cancel{2}\ \cancel{1} \\ -\quad 5\ 7\ 6 \\ \hline \mathbf{4}\ 5 \end{array}$$

To subtract in the tens column, borrow 10 hundreds from the thousands digit and add it to the hundreds digit. Borrow 10 tens from the hundreds digit and add it to the tens digit. Then subtract: $11 - 7 = 4$.

Self Check

Subtract 1,445 from 2,021. Then check the result using addition.

$$\begin{array}{r} 9\\ 1\;\cancel{10}\;11\;11\\ \cancel{2},\cancel{0}\;\cancel{2}\;\cancel{1}\\ -\quad 5\;7\;6\\ \hline 4\;4\;5 \end{array}$$ Subtract in the hundreds column: 9 − 5 = 4.

$$\begin{array}{r} 9\\ 1\;\cancel{10}\;11\;11\\ \cancel{2},\cancel{0}\;\cancel{2}\;\cancel{1}\\ -\quad 5\;7\;6\\ \hline 1,4\;4\;5 \end{array}$$ Subtract in the thousands column: 1 − 0 = 1.

Thus, 2,021 − 576 = 1,445. Check the result using addition.

Answer: 576; 576 + 1,445 = 2,021 ■

Words such as *minus, decrease, loss, debit, down, backward, fall, reduce, in the past,* and *to the left* indicate subtraction.

EXAMPLE 8 ***Drinking and driving.*** In 1996, there were 17,126 alcohol-related traffic deaths in the United States. That number declined in 1997, dropping by 937. In 1998, it fell by an additional 253. How many alcohol-related traffic fatalities were there in 1998?

Solution

The words *dropping* and *fell* indicate subtraction. We can show the calculations necessary to solve this example in a single expression:

17,126 − 937 − 253

The two subtractions are done working from left to right.

$$\begin{aligned} \mathbf{17{,}126 - 937} - 253 &= \mathbf{16{,}189} - 253\\ &= 15{,}936 \end{aligned}$$

There were 15,936 alcohol-related traffic deaths in 1998. ■

Combinations of operations

Additions and subtractions often appear in the same problem. It is important to read the problem carefully, extract the useful information, and organize it correctly.

EXAMPLE 9 ***Bus passengers.*** Twenty-seven people were riding a bus on Route 47. At the Seventh Street stop, 16 riders got off the bus and 5 got on. How many riders were left on the bus?

Solution

The route and street number are not important. The phrase *got off the bus* indicates subtraction, and the phrase *got on* indicates addition. The number of riders on the bus can be found by calculating 27 − 16 + 5. Working from left to right, we have

$$\begin{aligned} \mathbf{27 - 16} + 5 &= \mathbf{11} + 5\\ &= 16 \end{aligned}$$

There were 16 riders left on the bus.

Self Check

One share of ABC Corporation stock cost $75. The price fell $7 per share. However, it recovered and rose $13 per share. What is its current price?

Answer: $81 ■

COMMENT When doing the calculation in Example 9, we must perform the subtraction first. If the addition is done first, we obtain an incorrect answer of 6. For expressions containing addition and subtraction, perform them as they occur from left to right.

$$\begin{aligned} 27 - 16 + 5 &= 27 - 21\\ &= 6 \end{aligned}$$

Accent on Technology: **Calculators**

A calculator can be helpful when checking an answer or when performing a tedious computation. Before making regular use of one, make sure that you have mastered the fundamentals of arithmetic.

Several brands of calculators are available. For specific details about the operation of your calculator, please consult the owner's manual.

To check the addition done in Example 4 (U.S. history) using a scientific calculator, we enter these numbers and press these keys.

Keystrokes 510 [+] 425 [+] 900 [+] 2500 [=] **4335**

The display shows that in 1630, the total population of the four colonies was 4,335.

We can use a scientific calculator to check the subtraction performed in Example 8 (drinking and driving) by entering these numbers and pressing these keys.

Keystrokes 17126 [−] 937 [−] 253 [=] **15936**

The display shows that there were 15,936 alcohol-related traffic deaths in 1998.

STUDY SET Section 1.2

VOCABULARY *Fill in the blanks to make the statements true.*

1. When two numbers are added, the result is called a ______________. The numbers that are to be added are called ______________.

2. A ______________ is a letter that stands for a number.

3. A ______________ is a four-sided figure (like a dollar bill) whose opposite sides are of equal length.

4. A ______________ is a rectangle with all sides of equal length.

5. When two numbers are subtracted, the result is called a ______________. In a subtraction problem, the ______________ is subtracted from the ______________.

6. The property that guarantees that we can add two numbers in either order and get the same sum is called the ______________ property of addition.

7. The property that allows us to group numbers in an addition in any way we want is called the ______________ property of addition.

8. The distance around a rectangle (or a square) is called its ______________.

CONCEPTS *In Exercises 9–12, tell which property of addition guarantees that the quantities are equal.*

9. $3 + 4 = 4 + 3$

10. $(3 + 4) + 5 = 3 + (4 + 5)$

11. $7 + (8 + 2) = (7 + 8) + 2$

12. $(8 + 5) + 1 = 1 + (8 + 5)$

13. **a.** Use the variables x and y to write the commutative property of addition.

b. Use the variables x, y, and z to write the associative property of addition.

14. Show how to check the result:

$$\begin{array}{r} 74 \\ -\ 29 \\ \hline 45 \end{array}$$

15. Any number added to ______ stays the same.

16. **a.** In calculating $12 + (8 + 5)$, which numbers should be added first?

b. In calculating $60 - 15 + 4$, which operation should be performed first?

17. What addition fact is illustrated below?

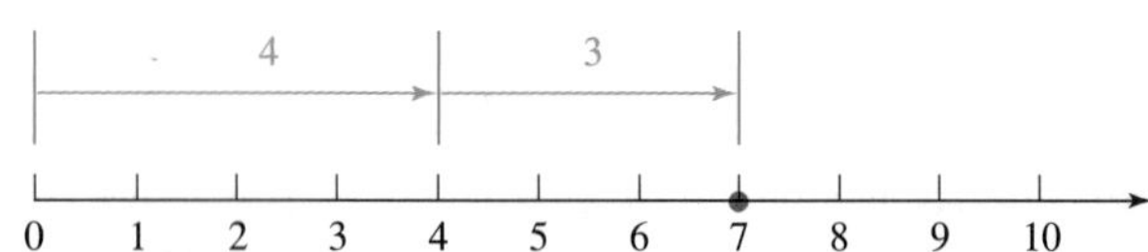

18. What subtraction fact is illustrated below?

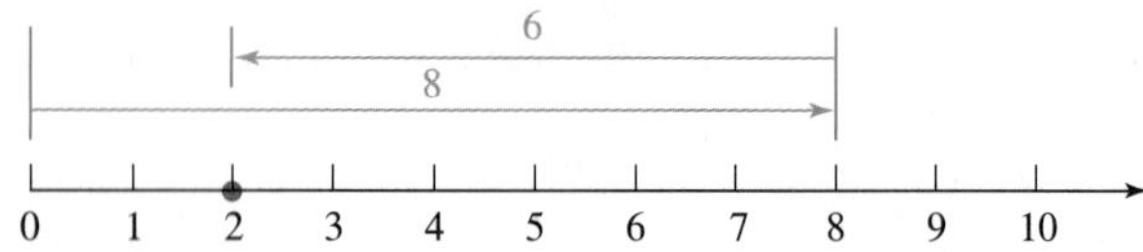

NOTATION *Fill in the blanks to make a true statement.*

19. The grouping symbols ( ) are called ________.

20. The minus sign − means ________.

Complete each solution.

21. $(36 + 11) + 5 = \_\_\_ + 5$
$= 52$

22. $12 + (15 + 2) = 12 + \_\_\_$
$= 29$

PRACTICE *Do each addition.*

23. 25 + 13

24. 47 + 12

25. 156 + 305

26. 647 + 38

27. 19 + 39 + 53

28. 27 + 16 + 48

29. (95 + 16) + 39

30. 832 + (97 + 27)

31. 25 + (321 + 17)

32. (4,231 + 213) + 5,234

33. $\begin{array}{r} 632 \\ +347 \\ \hline \end{array}$

34. $\begin{array}{r} 423 \\ +570 \\ \hline \end{array}$

35. $\begin{array}{r} 1{,}372 \\ +\ \ 613 \\ \hline \end{array}$

36. $\begin{array}{r} 2{,}477 \\ +\ \ 693 \\ \hline \end{array}$

37. $\begin{array}{r} 6{,}427 \\ +3{,}573 \\ \hline \end{array}$

38. $\begin{array}{r} 3{,}567 \\ +8{,}778 \\ \hline \end{array}$

39. $\begin{array}{r} 8{,}539 \\ +7{,}368 \\ \hline \end{array}$

40. $\begin{array}{r} 5{,}799 \\ +6{,}879 \\ \hline \end{array}$

41. $\begin{array}{r} 1{,}246 \\ 578 \\ +\ \ \ 37 \\ \hline \end{array}$

42. $\begin{array}{r} 4{,}689 \\ 3{,}422 \\ +\ \ \ 26 \\ \hline \end{array}$

43. $\begin{array}{r} 3{,}156 \\ 1{,}578 \\ +\ \ 578 \\ \hline \end{array}$

44. $\begin{array}{r} 2{,}379 \\ 4{,}779 \\ +2{,}339 \\ \hline \end{array}$

Find the perimeter of each rectangle or square.

45. 32 feet (ft)

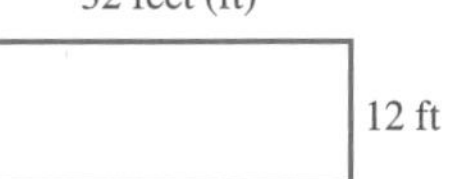

12 ft

46. 127 meters (m)

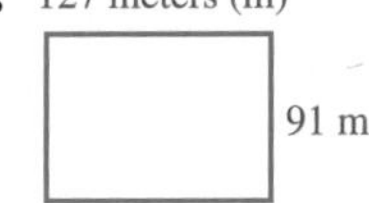

91 m

47. 17 inches (in.)

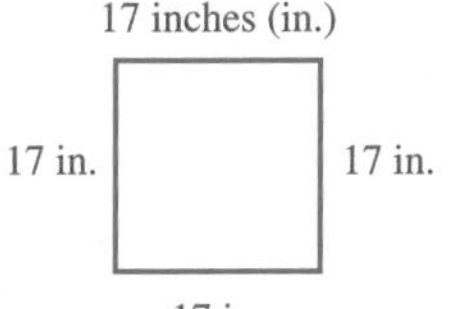

17 in. 17 in. 17 in.

48. 5 yards (yd)

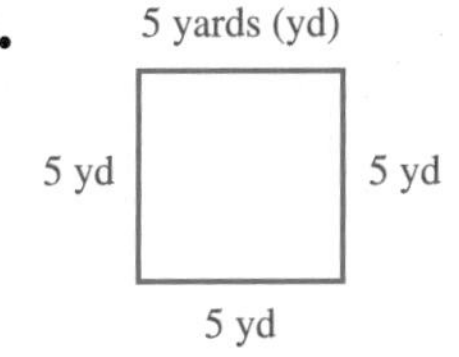

5 yd 5 yd 5 yd

Do each subtraction.

49. 17 − 14

50. 42 − 31

51. 39 − 14

52. 45 − 32

53. 174 − 71

54. 257 − 155

55. 633 − (598 − 30)

56. 600 − (497 − 60)

57. 160 − 15 − 4

58. 498 − 17 − 162

59. 29 − 17 − 12

60. 53 − 26 − 27

61. $\begin{array}{r} 367 \\ -343 \\ \hline \end{array}$

62. $\begin{array}{r} 224 \\ -122 \\ \hline \end{array}$

63. $\begin{array}{r} 423 \\ -305 \\ \hline \end{array}$

64. $\begin{array}{r} 330 \\ -270 \\ \hline \end{array}$

65. $\begin{array}{r} 1{,}537 \\ -\ \ 579 \\ \hline \end{array}$

66. $\begin{array}{r} 2{,}470 \\ -\ \ 863 \\ \hline \end{array}$

67. $\begin{array}{r} 4{,}267 \\ -2{,}578 \\ \hline \end{array}$

68. $\begin{array}{r} 7{,}356 \\ -3{,}578 \\ \hline \end{array}$

69. $\begin{array}{r} 17{,}246 \\ -\ \ 6{,}789 \\ \hline \end{array}$

70. $\begin{array}{r} 34{,}510 \\ -27{,}593 \\ \hline \end{array}$

71. $\begin{array}{r} 15{,}700 \\ -15{,}397 \\ \hline \end{array}$

72. $\begin{array}{r} 35{,}021 \\ -23{,}999 \\ \hline \end{array}$

Do the computations.

73. 43 − 12 + 9

74. 59 − 16 + 2

75. 120 + 30 − 40

76. 600 + 99 − 54

APPLICATIONS

77. TAXIS For a 17-mile trip, Wanda paid the taxi driver $23. If $5 was a tip, how much was the fare?

78. SPACE FLIGHT Astronaut Walter Schirra's first space flight orbited the Earth 6 times and lasted 9 hours. His second flight orbited the Earth 16 times and lasted 26 hours. How long was Schirra in space?

79. MAGAZINE CIRCULATION In 1997, the monthly circulation of *Ebony* magazine grew by 15,865. In 1998, the monthly circulation decreased by 69,404. If the monthly circulation in 1996 was 1,803,566, what was it in 1998?

80. JEWELRY MAKING Gold melts at about 1,947°F. The melting point of silver is 183°F lower. What is the melting point of silver?

81. BANKING A savings account contained $370. After a deposit of $40 and a withdrawal of $197, how much is in the account?

82. TRAVEL A student wants to make the 2,221-mile trip from Detroit to Seattle in three days. If she drives 751 miles on the first day and 875 miles on the second day, how far must she travel on the third day?

83. TAX DEDUCTION For tax purposes, a woman kept the mileage records shown in Illustration 1. Find the total number of miles that she drove in the first 6 months of the year.

84. COMPANY BUDGET A department head prepared an annual budget with the line items shown in Illustration 2. Find the projected number of dollars to be spent.

Month	Miles driven
January	2,345
February	1,712
March	1,778
April	445
May	1,003
June	2,774

ILLUSTRATION 1

Line item	Amount
Equipment	17,242
Contractual	5,443
Travel	2,775
Supplies	10,553
Development	3,225
Maintenance	1,075

ILLUSTRATION 2

In Exercises 85–86, refer to Illustration 3. To use this salary schedule, note that the annual salary of a third-year teacher with 15 units of course work beyond a Bachelor's degree is $30,887 (Step 3/Column 2).

Teachers' Salary Schedule

Years teaching	Column 1: B.D.	Column 2: B.D. + 15	Column 3: B.D. + 30
Step 1	$26,785	$28,243	$29,701
Step 2	$28,107	$29,565	$31,023
Step 3	$29,429	$30,887	$32,345
Step 4	$30,751	$32,209	$33,667
Step 5	$32,073	$33,531	$34,989

ILLUSTRATION 3

85. INCOME How much money will a new teacher make in his first five years of teaching if he begins at
a. Step 1/Column 1?
b. Step 1/Column 3?

86. PAY INCREASE If a teacher is now on Step 2/Column 2, how much more money will she make next year when she
a. gains one year of experience?
b. completes 15 units of course work?

87. BLUEPRINT Find the length of the house shown in Illustration 4.

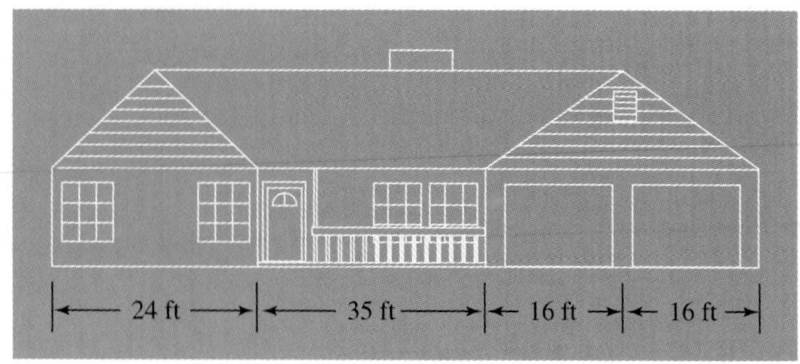

ILLUSTRATION 4

88. MACHINERY Find the length of the motor on the machine shown in Illustration 5. (cm stands for centimeters.)

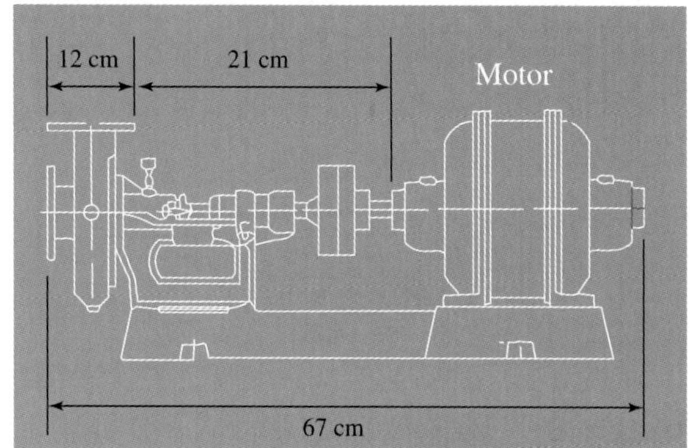

ILLUSTRATION 5

89. CAR EMISSIONS Illustration 6 shows the number of tons of hydrocarbons and nitrogen oxides that have been removed from the air because of antismog legislation in California. How many tons have been removed daily because of this legislation?

Step taken	Tons removed daily
State gasoline reformulation	215
Federal gasoline reformulation	85
Auto nitrogen oxide standard	117
Auto hydrocarbon standard	35
Diesel fuel reformulation	70
Smog check	150
Gas pump nozzles	120

ILLUSTRATION 6

90. DALMATIANS See Illustration 7. How many fewer Dalmatians were registered in 1998 than in the year when they were at their height of popularity?

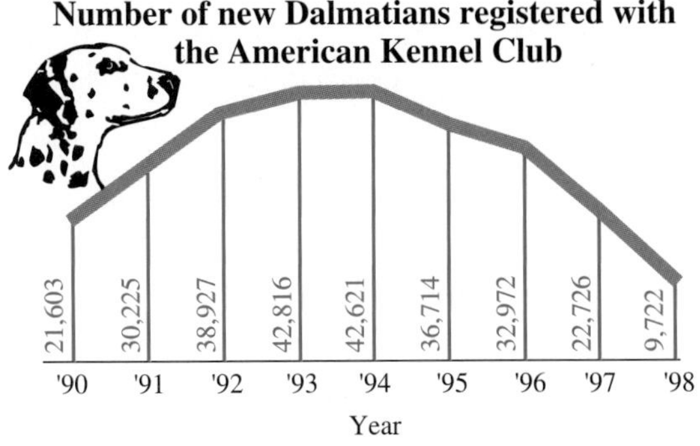

ILLUSTRATION 7

91. CITY FLAG To decorate a city flag, yellow fringe is to be sewn around its outside edges, as shown in Illustration 8. The fringe comes on long spools and is sold by the inch. How many inches of fringe must be purchased to complete the project?

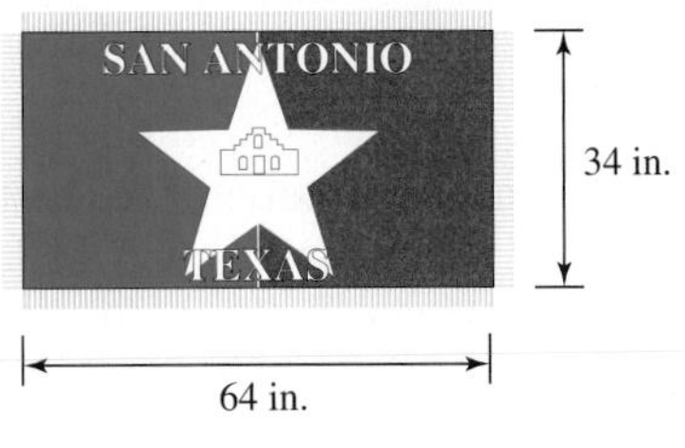

ILLUSTRATION 8

92. BOXING How much padded rope is needed to create the square boxing ring in Illustration 9 if each side is 24 feet long?

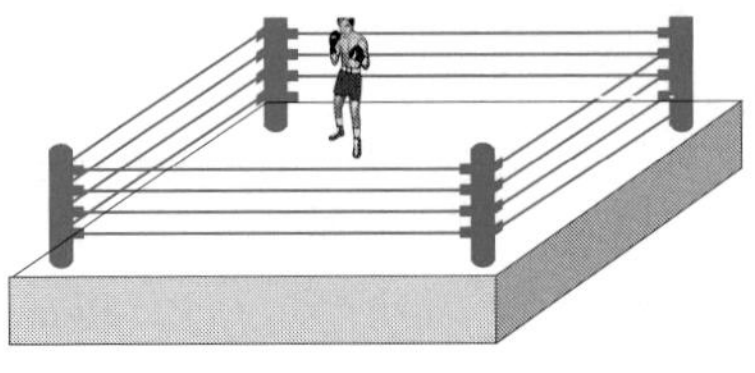

ILLUSTRATION 9

WRITING

93. Explain why the operation of addition is commutative.

94. Explain how addition can be used to check subtraction.

REVIEW *Write each numeral in expanded notation.*

95. 3,125

96. 60,037

Round 6,354,784 to the specified place.

97. nearest ten

98. nearest hundred

99. nearest ten thousand

100. nearest hundred thousand

1.3 *Multiplying and Dividing Whole Numbers*

In this section, you will learn about

- Properties of multiplication • Multiplying whole numbers
- Finding the area of a rectangle • Division • Properties of division
- Dividing whole numbers

Introduction. Mastering the operations of multiplication and division with whole numbers enables us to find areas of geometric figures and to solve business and transportation problems. For example, to find the area of a rectangle, we need to multiply its length by its width. To figure a paycheck, we need to multiply the number of hours worked by the hourly rate of pay. To calculate the fuel economy of a bus, we need to divide the miles it travels by the number of gallons of gas that are used.

Properties of multiplication

There are several symbols used to indicate multiplication.

Symbols that are used for multiplication

Symbol		Examples
$\times$	times sign	4×5 or $\begin{array}{r} 4 \\ \times\, \underline{5} \end{array}$
$\cdot$	raised dot	$4 \cdot 5$
( )	parentheses	(4)(5) or 4(5) or (4)5

Recall that a variable is a letter that stands for a number. We will often multiply a variable by another number or multiply a variable by another variable. When we do this, we don't need to use a symbol for multiplication.

$5a$ means $5 \cdot a$, ab means $a \cdot b$, and xyz means $x \cdot y \cdot z$

COMMENT In this book, we will seldom use the $\times$ sign, because it can be confused with the letter x.

Multiplication is repeated addition. For example, $4 \cdot 5$ means the sum of four 5's:

$$\begin{aligned} 4 \cdot 5 &= \overbrace{5 + 5 + 5 + 5}^{\text{The sum of four 5's.}} \\ &= 20 \end{aligned}$$

In the above multiplication, the result of 20 is called a **product.** The numbers that were multiplied (4 and 5) are called **factors.**

$$\begin{array}{ccccc} \text{Factor} & & \text{Factor} & & \text{Product} \\ \downarrow & & \downarrow & & \downarrow \\ 4 & \cdot & 5 & = & 20 \end{array}$$

The multiplication $5 \cdot 4$ means the sum of five 4's:

$$\begin{aligned} 5 \cdot 4 &= \overbrace{4 + 4 + 4 + 4 + 4}^{\text{The sum of five 4's.}} \\ &= 20 \end{aligned}$$

We see that $4 \cdot 5 = 20$ and $5 \cdot 4 = 20$. The results are the same. These examples illustrate that the order in which we multiply two numbers does not affect the result. This property is called the **commutative property of multiplication.**

Commutative property of multiplication

If a and b represent numbers, then

$a \cdot b = b \cdot a$ or, more simply, $ab = ba$

Table 1-1 summarizes the basic multiplication facts.

- To find the product of 6 and 8 using the table, we find the intersection of the 6th row and the 8th column. The product is 48.
- To find the product of 8 and 6, we find the intersection of the 8th row and the 6th column. Once again, the product is 48.

In the table, we see that the set of answers above and the set of answers below the diagonal line in bold print are identical. This further illustrates that multiplication is commutative.

·	0	1	2	3	4	5	6	7	8	9
0	**0**	0	0	0	0	0	0	0	0	0
1	0	**1**	2	3	4	5	6	7	8	9
2	0	2	**4**	6	8	10	12	14	16	18
3	0	3	6	**9**	12	15	18	21	24	27
4	0	4	8	12	**16**	20	24	28	32	36
5	0	5	10	15	20	**25**	30	35	40	45
6	0	6	12	18	24	30	**36**	42	48	54
7	0	7	14	21	28	35	42	**49**	56	63
8	0	8	16	24	32	40	48	56	**64**	72
9	0	9	18	27	36	45	54	63	72	**81**

TABLE 1-1

From the table, we see that whenever we multiply a number by 0, the product is 0. For example,

$0 \cdot 5 = 0,\qquad 0 \cdot 8 = 0,\qquad \text{and}\qquad 9 \cdot 0 = 0$

We also see that whenever we multiply a number by 1, the number remains the same. For example,

$3 \cdot 1 = 3,\qquad 7 \cdot 1 = 7,\qquad \text{and}\qquad 1 \cdot 9 = 9$

These examples suggest the multiplication properties of 0 and 1.

Multiplication properties of 0 and 1

If a represents any number, then

$a \cdot 0 = 0\quad \text{and}\quad 0 \cdot a = 0$

$a \cdot 1 = a\quad \text{and}\quad 1 \cdot a = a$

EXAMPLE 1 ***Computing daily wages.*** Raul worked an 8-hour day at an hourly rate of \$9. How much money did he earn?

Solution

For each of the 8 hours, Raul earned \$9. His total pay for the day is the sum of eight 9's: $9 + 9 + 9 + 9 + 9 + 9 + 9 + 9$. This can be calculated by multiplication.

Total wages $= 8 \cdot 9$

$= 72$ See the multiplication table.

Raul earned \$72.

Self Check

At a rate of \$8 per hour, how much will a school bus driver earn if she works from 8:00 A.M. until noon?

Answer: \$32 ■

To multiply three numbers, we first multiply two of them and then multiply that result by the third number. In the following examples, we multiply $3 \cdot 2 \cdot 4$ in two ways. The parentheses show us which multiplication to do first.

Method 1: Group 3 · 2

$(\mathbf{3 \cdot 2}) \cdot 4 = \mathbf{6} \cdot 4$ Multiply 3 and 2 to get 6.

$= 24$ Then multiply 6 and 4 to get 24.

Method 2: Group 2 · 4

$3 \cdot (\mathbf{2} \cdot \mathbf{4}) = 3 \cdot \mathbf{8}$ Multiply 2 and 4 to get 8.

$= 24$ Then multiply 3 and 8 to get 24.

The answers are the same. This illustrates that changing the grouping when multiplying numbers does not affect the result. This property is called the **associative property of multiplication.**

Associative property of multiplication

If a, b, and c represent numbers, then

$(a \cdot b) \cdot c = a \cdot (b \cdot c)$ or, more simply, $(ab)c = a(bc)$

Multiplying whole numbers

To find the product $8 \cdot 47$, it is inconvenient to add up eight 47's. Instead, we find the product by a multiplication process.

$$\begin{array}{r} 47 \\ \times\ \ 8 \\ \hline \end{array}$$ Write the factors in a column, with the corresponding digits aligned vertically.

$$\begin{array}{r} 5 \\ 47 \\ \times\ \ 8 \\ \hline 6 \end{array}$$ Multiply 7 by 8. The product is 56. Place 6 in the ones column of the answer and carry 5 (in blue) to the tens column.

$$\begin{array}{r} 5 \\ 47 \\ \times\ \ 8 \\ \hline 376 \end{array}$$ Multiply 4 by 8. The product is 32. To the 32, add the carried 5 to get 37. Place the 7 in the tens column and the 3 in the hundreds column of the answer.

The product is 376.

To find the product $23 \cdot 435$, we use the multiplication process. Because $23 = 20 + 3$, we multiply 435 by 20 and by 3 and then add the products. To do this, we write the factors in a column, with the corresponding digits aligned vertically. We then begin the process by multiplying 435 by 3:

$$\begin{array}{r} 1 \\ 435 \\ \times\ 23 \\ \hline 5 \end{array}$$ Multiply 5 by 3. The product is 15. Place 5 in the ones column and carry 1 (in blue) to the tens column.

$$\begin{array}{r} 11 \\ 435 \\ \times\ 23 \\ \hline 05 \end{array}$$ Multiply 3 by 3. The product is 9. To the 9, add the carried 1 to get 10. Place the 0 in the tens column and carry the 1 (in green) to the hundreds column.

$$\begin{array}{r} 11 \\ 435 \\ \times\ 23 \\ \hline 1305 \end{array}$$ Multiply 4 by 3. The product is 12. Add the 12 to the carried 1 to get 13. Write 13.

We continue by multiplying 435 by 2 tens, or 20:

$$\begin{array}{r} 1 \\ 435 \\ \times\ 23 \\ \hline 1305 \\ 0 \end{array}$$ Multiply 5 by 2. The product is 10. Write 0 in the tens column and carry 1 (in purple).

$$\begin{array}{r} 1 \\ 435 \\ \times\ 23 \\ \hline 1305 \\ 70 \end{array}$$ Multiply 3 by 2. The product is 6. Add 6 to the carried 1 to get 7. Write the 7. There is no carry.

```
   1
  435    Multiply 4 by 2. The product is 8. There is no carry to add. Write the 8.
 × 23
 1305
 870
```

```
  435    Draw another line beneath the two completed rows. Add the two rows. This
 × 23    sum gives the product of 435 and 23.
 1305
 870
10005
```

Thus, $23 \cdot 435 = 10{,}005$.

EXAMPLE 2 ***Mileage.*** Specifications for a 1999 Ford Explorer 4 × 4 are shown in the table below. For city driving, how far can it travel on a tank of gas? (The abbreviation "mpg" means "miles per gallon.")

Engine	4.0 L V6
Fuel capacity	21 gal
Fuel economy (mpg)	15 city/19 hwy

Solution

For city driving, each of the 21 gallons of gas that the tank holds enables the Explorer to go 15 miles. The total distance it can travel is the sum of twenty-one 15's. This can be calculated by multiplication: $21 \cdot 15$.

```
  1
  15
 ×21
  15
 30
 315
```

For city driving, the Explorer can go 315 miles on a tank of gas.

Self Check

For highway driving, how far can the Explorer travel on a tank of gas?

Answer: 399 mi ■

EXAMPLE 3 ***Calculating production.*** The labor force of an electronics firm works two 8-hour shifts each day and manufactures 53 television sets each hour. Find how many sets will be manufactured in 5 days.

Solution

The number of TV sets manufactured in 5 days is given by the product:

2 shifts per day		8 hr per shift		53 sets per hr		5 days	
↓		↓		↓		↓	
2	·	8	·	53	·	5	This could also be written 2(8)(53)(5).

We perform the multiplications working from left to right.

$\mathbf{2 \cdot 8} \cdot 53 \cdot 5 = \mathbf{16} \cdot 53 \cdot 5$ Multiply 2 and 8.

$= 848 \cdot 5$ Multiply 16 and 53.

$= 4{,}240$

4,240 television sets will be manufactured in 5 days. ■

Accent on Technology: Checking an answer

We can use a scientific calculator to check the multiplication performed in Example 3. To find the product $2 \cdot 8 \cdot 53 \cdot 5$, we enter these numbers and press these keys.

Keystrokes 2 [×] 8 [×] 53 [×] 5 [=] 4240

The display verifies that the multiplication was done correctly in Example 3.

We can use multiplication to count objects arranged in rectangular patterns. For example, the display on the left below shows a rectangular array consisting of 5 rows of 7 stars. The product $5 \cdot 7$, or 35, indicates the total number of stars.

Because multiplication is commutative, the array on the right below, consisting of 7 rows of 5 stars, contains the same number of stars.

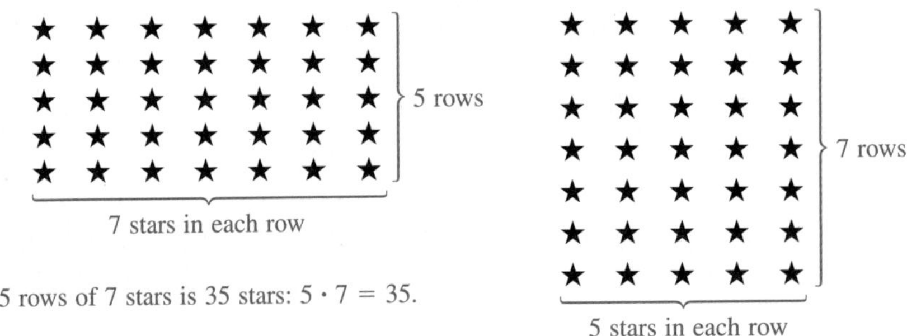

5 rows of 7 stars is 35 stars: $5 \cdot 7 = 35$.

7 rows of 5 stars is 35 stars: $7 \cdot 5 = 35$.

EXAMPLE 4 ***Computer science.*** To draw graphics on a computer screen, a computer controls each *pixel* (one dot on the screen). See Figure 1-11. A standard computer graphics image is 800 pixels wide and 600 pixels high. How many pixels does the computer control?

Solution

The graphics image is a rectangular array of pixels. Each of its 600 rows consists of 800 pixels. The total number of pixels is the product of 600 and 800:

$$600 \cdot 800 = 480{,}000$$

The computer controls 480,000 pixels.

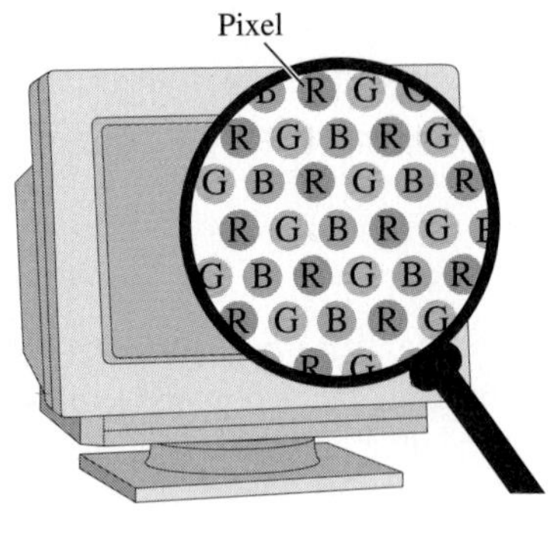

FIGURE 1-11

Self Check

On a color monitor, each of the pixels can be red, green, or blue. How many colored pixels does the computer control?

Answer: 1,440,000

Finding the area of a rectangle

One important application of multiplication is finding the area of a rectangle. The **area of a rectangle** is the measure of the amount of surface it encloses. Area is measured in square units, such as square inches (denoted as in.^{2}) or square centimeters (denoted as cm^2). (See Figure 1-12.)

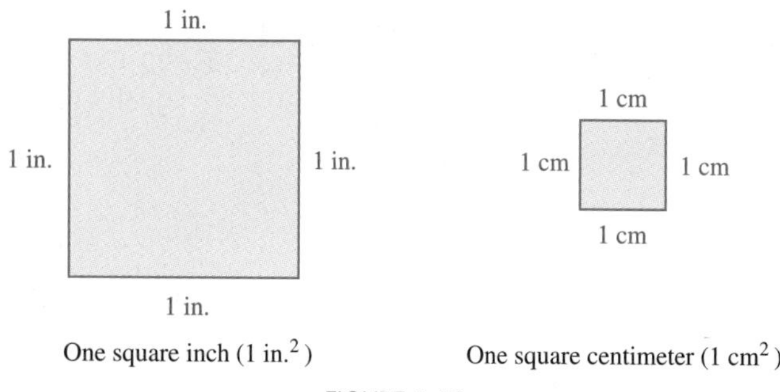

FIGURE 1-12

The rectangle in Figure 1-13 has a length of 5 centimeters and a width of 3 centimeters. Each small square covers an area of one square centimeter (1 cm^2). The small squares form a rectangular pattern, with 3 rows of 5 squares.

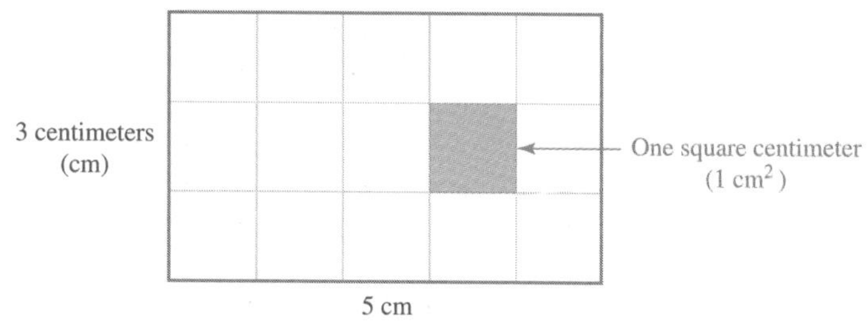

FIGURE 1-13

Because there are $5 \cdot 3$, or 15, small squares, the area of the rectangle is 15 cm^2. This suggests that the area of any rectangle is the product of its length and its width.

Area of a rectangle = length · width

Using the variables l and w to represent the length and width, we can write this formula in simpler form.

Area of a rectangle

The area A of a rectangle is the product of the rectangle's length l and its width w:

$A = l \cdot w$ or, more simply, $A = lw$

EXAMPLE 5 ***Wrapping paper.*** When completely unrolled, a long sheet of gift wrapping paper has the dimensions shown in Figure 1-14. How many square feet of gift wrap are on the roll?

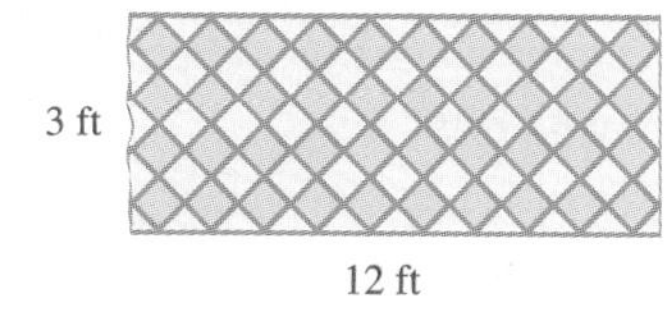

FIGURE 1-14

Solution

To find the number of square feet of paper, we need to find the area of the rectangle shown in the figure.

$A = lw$ The formula for the area of a rectangle.

$A = 12 \cdot 3$ Replace l with 12 and w with 3.

$= 36$ Do the multiplication.

There are 36 square feet (ft^2) of wrapping paper on the roll.

Self Check

A package of 9-inch-by-12-inch paper contains 500 sheets. How many square inches of paper does one package contain?

Answer: 54,000 in.^{2}

COMMENT Remember that the perimeter of a rectangle is the distance around it. The area of a rectangle is a measure of the surface it encloses.

Division

If \$12 is distributed equally among 4 people, we must divide to see that each person would receive \$3.

$$4\overline{)12}\quad \text{quotient } 3$$

Several symbols can be used to indicate division.

Symbols that are used for division

Symbol		Example
$\div$	division sign	$12 \div 4$
$\overline{)\ \ }$	long division	$4\overline{)12}$
—	fraction bar	$\frac{12}{4}$

In a division, the number that is being divided is called the **dividend.** The number that we are dividing by is called the **divisor,** and the answer is called the **quotient.**

$$\text{Dividend} \div \text{divisor} = \text{quotient} \qquad \text{Divisor}\overline{)\text{dividend}}\ \ (\text{quotient}) \qquad \frac{\text{Dividend}}{\text{Divisor}} = \text{quotient}$$

Division can be thought of as repeated subtraction. To divide 12 by 4 is to ask, "How many 4's can be subtracted from 12?" Exactly three 4's can be subtracted from 12 to get 0:

$$12 - \overbrace{4 - 4 - 4}^{\text{Three 4's}} = 0$$

Thus, $12 \div 4 = 3$.

Division is also related to multiplication.

$$\frac{12}{4} = 3 \quad \text{because} \quad 4 \cdot 3 = 12 \qquad \text{and} \qquad \frac{20}{5} = 4 \quad \text{because} \quad 5 \cdot 4 = 20$$

Properties of division

We will now consider three types of division that involve zero. In the first case, we will examine a division of zero; in the second, a division by zero; in the third, a division of zero by zero.

Division statement	Corresponding multiplication statement	Result
$\frac{0}{2} = ?$	$2(?) = 0$ ↑ This must be 0 if the product is to be 0.	$\frac{0}{2} = \mathbf{0}$
$\frac{2}{0} = ?$	$0(?) = 2$ ↑ There is no number that gives 2 when multiplied by 0.	There is no quotient.
$\frac{0}{0} = ?$	$0(?) = 0$ ↑ Any number times 0 is 0.	Any number can be the quotient.

We see that $\frac{0}{2} = 0$. This suggests that the quotient of 0 divided by any nonzero number is 0. Since $\frac{2}{0}$ does not have a quotient, we say that division of 2 by 0 is *undefined.* In general, division of any nonzero number by 0 is undefined. Since $\frac{0}{0}$ can be any number, we say that $\frac{0}{0}$ is undetermined.

Division with 0

1. If a represents any nonzero number, $\dfrac{0}{a} = 0$.
2. If a represents any nonzero number, $\dfrac{a}{0}$ is undefined.
3. $\dfrac{0}{0}$ is undetermined.

The example $\frac{12}{1} = 12$ illustrates that *any number divided by 1 is the number itself.* The example $\frac{12}{12} = 1$ illustrates that *any number (except 0) divided by itself is 1.*

Division properties

If a represents any number,

$\dfrac{a}{1} = a$ and $\dfrac{a}{a} = 1$ (provided that $a \neq 0$) Read $\neq$ as "is not equal to."

Dividing whole numbers

We can use a process called **long division** to divide whole numbers. To divide 832 by 23, for example, we proceed as follows:

Quotient ⟶
Divisor ⟶ $23\overline{)832}$
↑
Dividend

Place the divisor and the dividend as indicated. The quotient will appear above the long division symbol.

We will find the quotient using the following division process.

$$\begin{array}{r} 4 \\ 23\overline{)832} \end{array}$$

Ask: "How many times will 23 divide 83?" Because an estimate is 4, place 4 in the tens column of the quotient.

$$\begin{array}{r} 4 \\ 23\overline{)832} \\ 92 \end{array}$$

Multiply $23 \cdot 4$ and place the answer, 92, under the 83. Because 92 is larger than 83, our estimate of 4 for the tens column of the quotient was too large.

$$\begin{array}{r} 3 \\ 23\overline{)832} \\ 69\downarrow \\ \hline 142 \end{array}$$

Revise the estimate of the quotient to be 3. Multiply $23 \cdot 3$ to get 69, place 69 under the 83, draw a line, and subtract.

Bring down the 2 in the ones column.

$$\begin{array}{r} 37 \\ 23\overline{)832} \\ 69 \\ \hline 142 \\ 161 \\ \hline \end{array}$$

Ask, "How many times will 23 divide 142?" The answer is approximately 7. Place 7 in the ones column of the quotient. Multiply $23 \cdot 7$ to get 161. Place 161 under 142. Because 161 is larger than 142, the estimate of 7 is too large.

$$\begin{array}{r} 36 \\ 23\overline{)832} \\ 69 \\ \hline 142 \\ 138 \\ \hline 4 \end{array}$$

Revise the estimate of the quotient to be 6. Multiply: $23 \cdot 6 = 138$.

Place 138 under 142 and subtract.

The quotient is 36, and the leftover 4 is the **remainder.** We can write this result as 36 R 4.

To check the result of a division, we multiply the divisor by the quotient and then add the remainder. The result should be the dividend.

Check: Quotient · divisor + remainder = dividend

$$36 \cdot 23 + 4 = 832$$
$$828 + 4 = 832$$
$$832 = 832$$

EXAMPLE 6 ***Managing a soup kitchen.*** A soup kitchen plans to feed 1,990 people. Because of space limitations, only 165 people can be served at one time. How many seatings will be necessary to feed everyone? How many will be served at the last seating?

Solution

The 1,990 people can be fed 165 at a time. To find the number of seatings, we divide.

$$\begin{array}{r} 12 \\ 165\overline{)1{,}990} \\ 1\,65\downarrow \\ \hline 340 \\ 330 \\ \hline 10 \end{array}$$

The quotient is 12, and the remainder is 10. Thirteen seatings will be needed: 12 full-capacity seatings and one partial seating to serve the remaining 10 people.

Self Check

Each gram of fat in a meal provides 9 calories. A fast-food meal contains 243 calories from fat. How many grams of fat does the meal contain?

Answer: 27

Accent on Technology: Yard sale

At a yard sale, a family sold old shirts, pants, blouses, and dresses for 75¢ each. They made $80.25 from the sale of the clothes. To find the number of items they sold, we divide the total sales (8,025¢) by the cost of each item (75¢). We can use a scientific calculator to find 8,025 ÷ 75 by entering these numbers and pressing these keys.

Keystrokes 8025 [÷] 75 [=] 107

They sold 107 items of clothing at the yard sale.

STUDY SET Section 1.3

VOCABULARY *Fill in the blanks to make the statements true.*

1. ______________ is repeated addition.
2. Numbers that are to be multiplied are called ______________. The result of a multiplication is called a ______________.
3. The statement $ab = ba$ expresses the ______________ property of multiplication.
4. The statement $(ab)c = a(bc)$ expresses the ______________ property of multiplication.
5. If a square measures one inch on each side, its area is 1 ______________.
6. In a division, the dividend is divided by the ______________. The result of a division is called a ______________.

CONCEPTS

7. Write 8 + 8 + 8 + 8 as a multiplication.

8. **a.** Use the variables x and y to write the commutative property of multiplication.
b. Use the variables x, y, and z to write the associative property of multiplication.

9. How do we find the amount of surface enclosed by a rectangle?

10. Tell whether *perimeter* or *area* is the concept that should be applied to find each of the following.
a. The amount of floor space to be carpeted
b. The amount of clear glass to be tinted
c. The amount of lace needed to trim the sides of a handkerchief

11. Do each multiplication.
a. $1 \cdot 25$ **b.** $62(1)$
c. $10 \cdot 0$ **d.** $0(4)$

12. Do each division.
a. $25 \div 1$ **b.** $\frac{7}{1}$
c. $\frac{0}{1}$ **d.** $\frac{5}{0}$
e. $\frac{0}{0}$ **f.** $\frac{0}{2,757}$

13. Write a multiplication statement that finds the number of red squares.

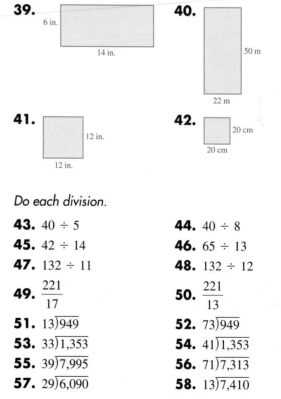

14. Consider

$$\begin{array}{r} 12 \\ 15\overline{)182} \\ \underline{15} \\ 32 \\ \underline{30} \\ 2 \end{array}$$

Fill in the blanks:

12 · ___ + ___ = ___

NOTATION

15. **a.** Write three symbols that are used for multiplication.
b. Write three symbols that are used for division.

16. Write each multiplication in simpler form.
a. $8 \cdot x$ **b.** $l \cdot w$

17. What does ft^2 mean?

18. Draw a figure having an area of 1 square inch.

PRACTICE *Do each multiplication.*

19. $12 \cdot 7$
20. $15 \cdot 8$
21. $27(12)$
22. $35(17)$
23. $9 \cdot (4 \cdot 5)$
24. $(3 \cdot 5) \cdot 12$
25. $5 \cdot 7 \cdot 3$
26. $7 \cdot 6 \cdot 8$
27. $\begin{array}{r} 99 \\ \underline{\times 77} \end{array}$
28. $\begin{array}{r} 73 \\ \underline{\times 59} \end{array}$
29. $\begin{array}{r} 20 \\ \underline{\times 53} \end{array}$
30. $\begin{array}{r} 78 \\ \underline{\times 20} \end{array}$
31. $\begin{array}{r} 112 \\ \underline{\times \ 23} \end{array}$
32. $\begin{array}{r} 232 \\ \underline{\times \ 53} \end{array}$
33. $\begin{array}{r} 207 \\ \underline{\times \ 97} \end{array}$
34. $\begin{array}{r} 768 \\ \underline{\times \ 70} \end{array}$
35. $13,456 \cdot 217$
36. $17,456 \cdot 257$
37. $3,302 \cdot 358$
38. $123,112 \cdot 46$

Find the area of each rectangle or square. (m stands for meters and cm stands for centimeters.)

39.

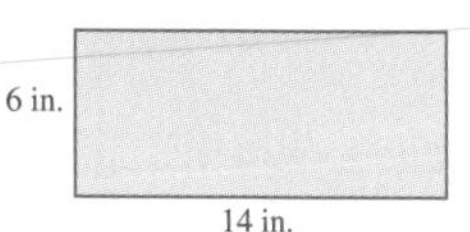

40.

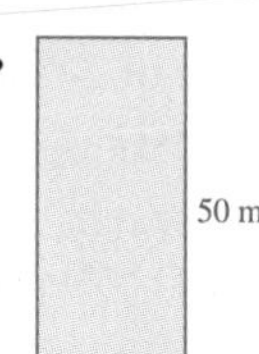

41.

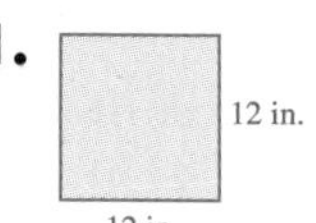

42.

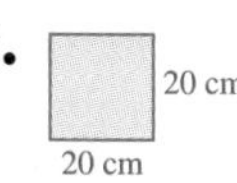

Do each division.

43. $40 \div 5$
44. $40 \div 8$
45. $42 \div 14$
46. $65 \div 13$
47. $132 \div 11$
48. $132 \div 12$
49. $\frac{221}{17}$
50. $\frac{221}{13}$
51. $13\overline{)949}$
52. $73\overline{)949}$
53. $33\overline{)1,353}$
54. $41\overline{)1,353}$
55. $39\overline{)7,995}$
56. $71\overline{)7,313}$
57. $29\overline{)6,090}$
58. $13\overline{)7,410}$

Do each division and give the quotient and the remainder.

59. $31\overline{)273}$
60. $25\overline{)290}$
61. $37\overline{)743}$
62. $79\overline{)931}$

63. $42\overline{)1,273}$ **64.** $83\overline{)3,280}$

65. $57\overline{)1,795}$ **66.** $99\overline{)9,876}$

APPLICATIONS

67. FIGURING WAGES A cook worked 12 hours at $11 per hour. How much did she earn?

68. SQUARES ON A CHESSBOARD A chessboard consists of 8 rows, with 8 squares in each row. How many squares are there on a chessboard?

69. FINDING DISTANCE A car with a tank that holds 14 gallons of gasoline goes 29 miles on one gallon. How far can the car go on a full tank?

70. RENTING APARTMENTS Mia owns an apartment building with 18 units. Each unit generates a monthly income of $450. Find her total monthly income.

71. CONCERT ATTENDANCE A jazz quartet gave two concerts in each of 37 cities. Approximately 1,700 fans attended each concert. How many persons heard the group?

72. CEREAL A cereal maker advertises "Two cups of raisins in every box." Find the number of cups of raisins in a case of 36 boxes of cereal.

73. ORANGE JUICE It takes 13 oranges to make one can of orange juice. Find the number of oranges used to make a case of 24 cans.

74. ROOM CAPACITY A college lecture hall has 17 rows of 33 seats. A sign on the wall reads, "Occupancy by more than 570 persons is prohibited." If the seats are filled and there is one instructor, is the college breaking the rule?

75. CAPACITY OF AN ELEVATOR There are 14 people in an elevator with a capacity of 2,000 pounds. If the average weight of a person on the elevator is 150 pounds, is the elevator overloaded?

76. CHANGING UNITS There are 12 inches in 1 foot. How many inches are in 80 feet?

77. WORD PROCESSING A student used the option shown in Illustration 1 when typing a report. How many entries will the table hold?

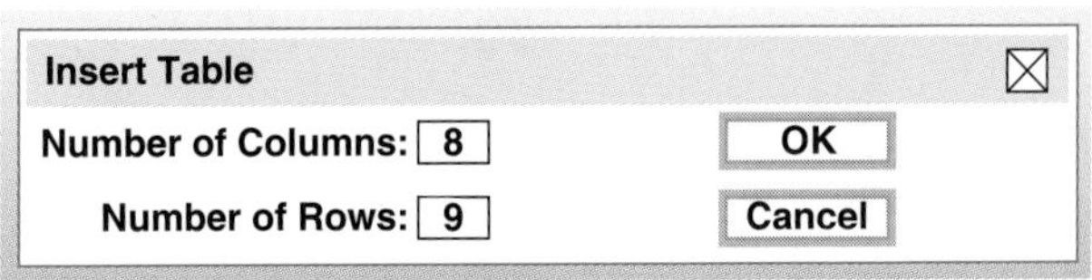

ILLUSTRATION 1

78. FILLING PRESCRIPTIONS How many tablets should a pharmacist put in the container shown in Illustration 2?

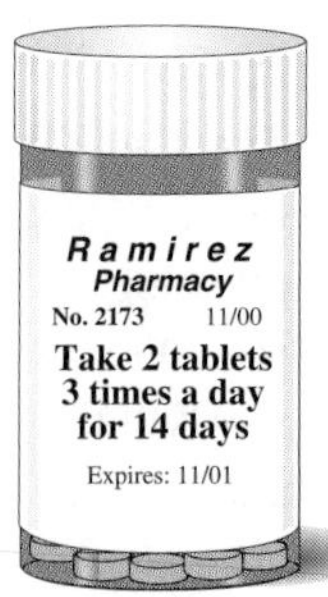

ILLUSTRATION 2

79. DISTRIBUTING MILK A first grade class received 73 half-pint cartons of milk to distribute evenly to the 23 students. How many cartons were left over?

80. LIFT SYSTEM If the bus shown in Illustration 3 weighs 58,000 pounds, how much weight is on each jack?

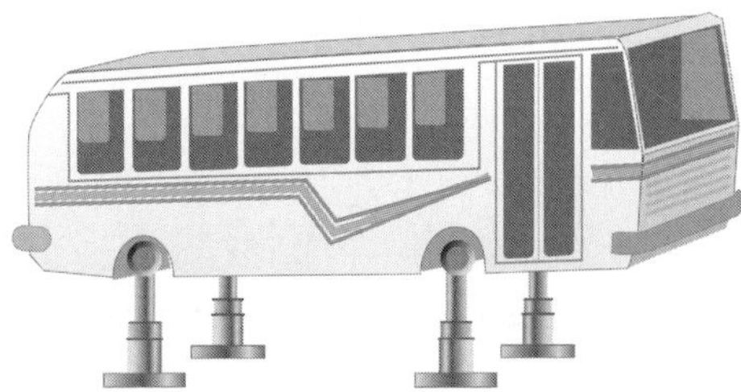

ILLUSTRATION 3

81. MILEAGE A touring rock group travels in a bus that has a range of 700 miles on one tank (140 gallons) of gasoline. How far can the bus travel on one gallon of gas?

82. RUNNING Brian runs 7 miles each day. In how many days will Brian run 371 miles?

83. How many feet more than two miles is 11,000 feet? (*Hint:* 5,280 feet = 1 mile.)

84. ORDERING DOUGHNUTS How many dozen doughnuts must be ordered for a meeting if 156 people are expected to attend, and each person will be served one doughnut?

85. PRICE OF A TEXTBOOK An author knows that her publisher received $954,193 on the sale of 23,273 textbooks. What is the price of each book?

86. WATER DISCHARGE The Susquehanna River discharges 38,200 cubic feet of water per second into the Chesapeake Bay. How long will it take for the river to discharge 1,719,000 cubic feet?

87. VOLLEYBALL LEAGUE A total of 216 girls tried out for a city volleyball program. How many girls should be put on each team roster if the following requirements must be met?

- All the teams are to have the same number of players.
- A reasonable number of players on a team is 7 to 10.
- For scheduling purposes, there must be an even number of teams.

88. AREA OF WYOMING The state of Wyoming is a rectangle 360 miles long and 270 miles wide. Find its perimeter and its area.

89. COMPARING ROOMS Which has the greater area, a rectangular room that is 14 feet by 17 feet or a square room that is 16 feet on each side? Which has the greater perimeter?

90. MATTRESSES A queen-size mattress measures 60 inches by 80 inches, and a full-size mattress measures 54 inches by 75 inches. How much more sleeping surface is there on a queen-size mattress?

91. GARDENING A rectangular garden is 27 feet long and 19 feet wide. A path in the garden uses 125 square feet of space. How many square feet are left for planting?

92. TENNIS See Illustration 4.

a. Find the number of square feet of court area a singles tennis player must defend.

b. Do the same for a doubles player.

c. What is the difference between the two results?

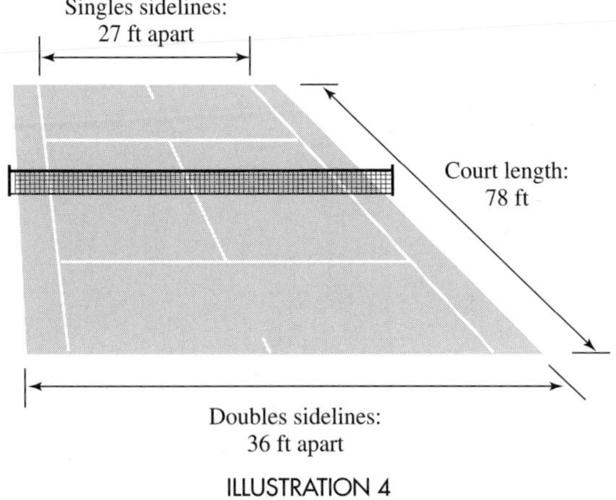

ILLUSTRATION 4

WRITING

93. Explain why the division of two numbers is not commutative.

94. Explain the difference between what perimeter measures and what area measures.

95. Explain the difference between 1 foot and 1 square foot.

96. When two numbers are multiplied, the result is 0. What conclusion can be drawn about the numbers?

REVIEW

97. Consider 372,856. What digit is in the hundreds column?

98. Round 45,995 to the nearest thousand.

99. Add 357, 39, and 476.

100. DISCOUNT A car, originally priced at $17,550, is being sold for $13,970. By how many dollars has the price been decreased?

ESTIMATION

In the previous two sections, we have used **estimation** as a means of checking the reasonableness of an answer. We now take a more in-depth look at the process of estimating.

Estimation is used to find an *approximate* answer to a problem. Estimates can be helpful in two ways. First, they serve as an accuracy check that can detect major computational errors. If an answer does not seem reasonable when compared to the estimate, the original problem should be reworked. Second, some situations call for only an approximate answer rather than the exact answer.

There are several ways to estimate, but there is one overriding theme of all the methods: The numbers in the problem are simplified so that the computation can be made easily and quickly. The first method we will study uses what is called **front-end rounding.** Each number is rounded to its largest place value, so that all but the first digit of each number is zero.

EXAMPLE 1 ***Estimating sums, differences, and products.***

a. Estimate the sum: 3,714 + 2,489 + 781 + 5,500 + 303.

Solution: Use front-end rounding.

3,714 →	4,000	Each number is rounded to its largest place value. All but the first digit is zero.
2,489 →	2,000	
781 →	800	
5,500 →	6,000	
+ 303 →	+ 300	
	13,100	

The estimate is 13,100.

If we compute 3,714 + 2,489 + 781 + 5,500 + 303, the sum is 12,787. We can see that our estimate is close; it's just 313 more than 12,787. This example illustrates the tradeoff when using estimation: The calculations are easier to perform and they take less time, but the answers are not exact.

b. Estimate the difference: 46,721 − 13,208.

Solution: Use front-end rounding.

46,721 →	50,000	Only the first digit is nonzero.
−13,208 →	−10,000	
	40,000	

The estimate is 40,000.

c. Estimate the product: 334 · 59.

Solution: Use front-end rounding.

334 →	300	334 rounds to 300, and 59 rounds to 60.
× 59 →	× 60	
	18,000	

The estimate is 18,000.

Self Check

a. Estimate the sum.

6,780
3,278
566
4,230
+1,923

b. Estimate the difference.

89,070
−15,331

c. Estimate the product.

707
×251

Answers: **a.** 16,600, **b.** 70,000, **c.** 210,000

To estimate quotients, we will use a method that approximates both the dividend and the divisor so that they will divide easily. With this method, some insight and intuition are needed. There is one rule of thumb for this method: If possible, round both numbers up or both numbers down.

EXAMPLE 2 ***Estimating quotients.*** Estimate the quotient: 170,715 ÷ 57.

Solution

Both numbers are rounded up. The division can then be done in your head.

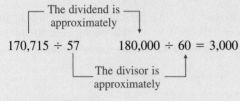

The dividend is approximately

170,715 ÷ 57 → 180,000 ÷ 60 = 3,000

The divisor is approximately

The estimate is 3,000.

Self Check

Estimate: 33,642 ÷ 42.

Answer: 800 ■

STUDY SET *Use front-end rounding to find an estimate to check the reasonableness of each answer. Write yes if it appears reasonable and no if it does not.*

1.
```
  25,405
  11,222
   8,909
   1,076
  14,595
 +33,999
  73,206
```

2.
```
  568,334
 − 31,225
  497,109
```

3.
```
    451
  × 73
 39,923
```

4.
```
    616
  × 98
 60,368
```

Use estimation to check the reasonableness of each answer.

5. 57,238 ÷ 28 = 200

6. $322\overline{)13{,}202}$ = 41

Use an estimation procedure to answer each problem.

7. CAMPAIGNING The number of miles flown each day by a politician on a campaign swing are shown here. Estimate the number of miles she flew during this time.

Day 1	3,546 miles
Day 2	567
Day 3	1,203
Day 4	342
Day 5	2,699

8. SHOPPING MALL The total sales income for a downtown mall in its first three years in operation are shown here.

1998	$5,234,301
1999	$2,898,655
2000	$6,343,433

Estimate the difference in income for 1999 and 2000 as compared to the first year, 1998.

9. GOLF COURSE Estimate the number of bags of grass seed needed to plant a fairway whose area is 86,625 square feet if the seed in each bag covers 2,850 square feet.

10. CENSUS Estimate the total population of the ten largest counties in the United States as of 1998.

Largest counties, by population	
1. Los Angeles, CA	9,213,533
2. Cook, IL	5,189,689
3. Harris, TX	3,206,063
4. Maricopa, AZ	2,784,075
5. San Diego, CA	2,780,592
6. Orange, CA	2,721,701
7. Kings, NY	2,267,942
8. Miami-Dade, FL	2,152,437
9. Wayne, MI	2,118,129
10. Dallas, TX	2,050,865

11. CURRENCY Estimate the number of $5 bills in circulation as of March 1, 1999, if the total value of the currency was $7,733,317,335.

12. REVENUES In 1998, the revenues of IBM were $81,667,000,000. Approximately how many times larger was this than the revenues of Sun Microsystems, which took in $9,791,000,000?

1.4 Prime Factors and Exponents

In this section, you will learn about

- Factoring whole numbers • Even and odd whole numbers • Prime numbers
- Composite numbers • Finding prime factorizations with the tree method
- Exponents • Finding prime factorizations with the division method

Introduction. In this section, we will learn how to represent whole numbers in alternative forms. The procedures used to find these forms involve multiplication and division. We will then discuss exponents, a shortcut way to represent repeated multiplication.

Factoring whole numbers

The statement $3 \cdot 2 = 6$ has two parts: the numbers that are being multiplied, and the answer. The numbers that are being multiplied are *factors,* and the answer is the *product.* We say that 3 and 2 are factors of 6.

Factors

Numbers that are multiplied together are called **factors.**

EXAMPLE 1 ***Finding factors of a whole number.*** Find the factors of 12.

Solution

We need to find the possible ways that we can multiply two whole numbers to get a product of 12.

$$1 \cdot 12 = 12, \quad 2 \cdot 6 = 12, \quad \text{and} \quad 3 \cdot 4 = 12$$

In order, from least to greatest, the factors of 12 are 1, 2, 3, 4, 6, and 12.

Self Check
Find the factors of 20.

Answer: 1, 2, 4, 5, 10, and 20 ■

Example 1 shows that 1, 2, 3, 4, 6, and 12 are the factors of 12. This observation was established by using multiplication facts. Each of these factors is related to 12 by division as well. Each of them divides 12, leaving a remainder of 0. Because of this fact, we say that 12 is **divisible** by each of its factors. When a division ends with a remainder of 0, we say that the division comes out even or that one of the numbers divides the other *exactly.*

Divisibility

One number is **divisible** by another if, when dividing them, the remainder is 0.

When we say that 3 is a factor of 6, we are using the word *factor* as a noun. The word *factor* is also used as a verb.

Factoring a whole number

To **factor** a whole number means to express it as the product of other whole numbers.

EXAMPLE 2 ***Factoring a whole number.*** Factor 40 using **a.** two factors and **b.** three factors.

Solution

a. There are several possibilities:

$$40 = 1 \cdot 40, \quad 40 = 2 \cdot 20, \quad 40 = 4 \cdot 10, \quad \text{or} \quad 40 = 5 \cdot 8$$

b. Again, there are several possibilities. Two of them are

$$40 = 5 \cdot 4 \cdot 2 \quad \text{and} \quad 40 = 2 \cdot 2 \cdot 10$$

Self Check

Factor 18 using **a.** two factors and **b.** three factors.

Answers: **a.** $1 \cdot 18, 2 \cdot 9, 3 \cdot 6$, **b.** $2 \cdot 3 \cdot 3$ ■

Even and odd whole numbers

Even and odd whole numbers

> If a whole number is divisible by 2, it is called an **even** number.
> If a whole number is not divisible by 2, it is called an **odd** number.

The even whole numbers are the numbers

0, 2, 4, 6, 8, 10, 12, 14, 16, 18, . . .

The odd whole numbers are the numbers

1, 3, 5, 7, 9, 11, 13, 15, 17, 19, . . .

There are infinitely many even and infinitely many odd whole numbers.

Prime numbers

EXAMPLE 3 ***Finding the factors of a whole number.*** Find the factors of 17.

Solution

$1 \cdot 17 = 17$

The only factors of 17 are 1 and 17.

Self Check

Find the factors of 23.

Answer: 1 and 23 ■

In Example 3 and its Self Check, we saw that the only factors of 17 are 1 and 17, and the only factors of 23 are 1 and 23. Numbers that have only two factors, 1 and the number itself, are called **prime numbers.**

Prime numbers

> A **prime number** is a whole number, greater than 1, that has only 1 and itself as factors.

The prime numbers are the numbers

2, 3, 5, 7, 11, 13, 17, 19, 23, 29, 31, . . .

The dots at the end of the list indicate that there are infinitely many prime numbers.

Note that the only even prime number is 2. Any other even whole number is divisible by 2, and thus has 2 as a factor, in addition to 1 and itself. Also note that not all odd whole numbers are prime numbers. For example, since 15 has factors of 1, 3, 5, and 15, it is not a prime number.

Composite numbers

The set of whole numbers contains many prime numbers. It also contains many numbers that are not prime.

Composite numbers

> The **composite numbers** are whole numbers, greater than 1, that are not prime.

The composite numbers are the numbers

4, 6, 8, 9, 10, 12, 14, 15, 16, 18, . . .

The three dots at the end of the list indicate that there are infinitely many composite numbers.

EXAMPLE 4 ***Prime and composite numbers.***

a. Is 37 a prime number? **b.** Is 45 a prime number?

Solution

a. Since 37 is a whole number greater than 1 and its only factors are 1 and 37, it is prime.

b. The factors of 45 are 1, 3, 5, 9, 15, and 45. Since there are factors other than 1 and 45, 45 is not prime. It is a composite number.

Self Check

a. Is 57 a prime number?

b. Is 39 a prime number?

Answers: **a.** no, **b.** no ■

COMMENT The numbers 0 and 1 are neither prime nor composite, because neither is a whole number greater than 1.

Finding prime factorizations with the tree method

Every composite number can be formed by multiplying a specific combination of prime numbers. The process of finding that combination is called **prime factorization.**

Prime factorization

> To find the **prime factorization** of a whole number means to write it as the product of only prime numbers.

Two methods can be used to find the prime factorization of a number. The first is called the **tree method.** We will use the tree method to find the prime factorization of 90 in two ways.

1. Factor 90 as $9 \cdot 10$.
2. Factor 9 and 10.
3. The process is complete when only prime numbers appear.

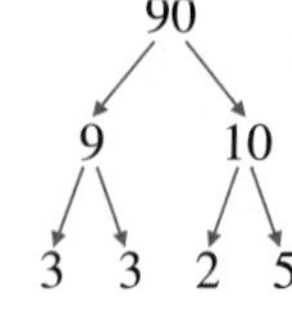

1. Factor 90 as $6 \cdot 15$.
2. Factor 6 and 15.
3. The process is complete when only prime numbers appear.

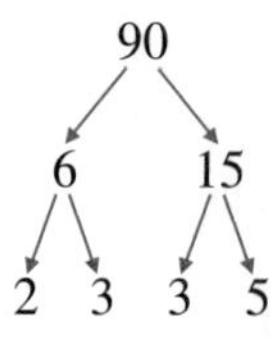

In either case, the prime factors are $2 \cdot 3 \cdot 3 \cdot 5$. Thus, the prime-factored form of 90 is $2 \cdot 3 \cdot 3 \cdot 5$. As we have seen, it does not matter how we factor 90. We will always get the same set of prime factors. No other combination of prime factors will multiply together and produce 90. This example illustrates an important fact about composite numbers.

Fundamental theorem of arithmetic

> Any composite number has exactly one set of prime factors.

EXAMPLE 5 ***Factoring whole numbers with factor trees.*** Use a factor tree to find the prime factorization of 210.

Solution

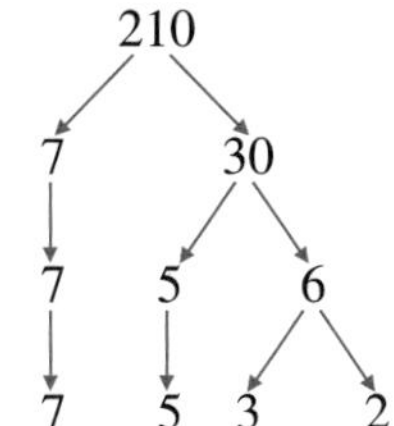

Factor 210 as $7 \cdot 30$.

Bring down the 7. Factor 30 as $5 \cdot 6$.

Bring down the 7 and the 5. Factor 6 as $3 \cdot 2$.

The prime factorization of 210 is $7 \cdot 5 \cdot 3 \cdot 2$. Writing the prime factors in order, from least to greatest, we have $210 = 2 \cdot 3 \cdot 5 \cdot 7$.

Self Check

Use a factor tree to find the prime factorization of 120.

Answer: $2 \cdot 2 \cdot 2 \cdot 3 \cdot 5$ ■

Exponents

In the Self Check of Example 5, we saw that the prime factorization of 120 is $2 \cdot 2 \cdot 2 \cdot 3 \cdot 5$. Because this factorization has three factors of 2, we call 2 a *repeated factor.* To express a repeated factor, we can use an **exponent.**

Exponent and base

> An **exponent** is used to indicate repeated multiplication. It tells how many times the **base** is used as a factor.

The exponent is 3.

$$\underbrace{2 \cdot 2 \cdot 2}_{\text{Repeated factors.}} = 2^3$$

The base is 2.

Read 2^3 as "2 to the third power" or "2 cubed."

The prime factorization of 120 can be written in a more compact form using exponents: $2 \cdot 2 \cdot 2 \cdot 3 \cdot 5 = 2^3 \cdot 3 \cdot 5$.

In the **exponential expression** a^n, a is the base, and n is the exponent. The expression is called a **power of** ***a*****.**

EXAMPLE 6 ***Exponents.*** Use exponents to write each prime factorization: **a.** $5 \cdot 5 \cdot 5$, **b.** $7 \cdot 7 \cdot 11$, and **c.** $2(2)(2)(2)(3)(3)(3)$

Solution

a. $5 \cdot 5 \cdot 5 = 5^3$ — 5 is used as a factor 3 times.

b. $7 \cdot 7 \cdot 11 = 7^2 \cdot 11$ — 7 is used as a factor 2 times.

c. $2(2)(2)(2)(3)(3)(3) = 2^4(3^3)$ — 2 is used as a factor 4 times, and 3 is used as a factor 3 times.

Self Check

Use exponents to write each prime factorization.

a. $3 \cdot 3 \cdot 7$

b. $5(5)(7)(7)$

c. $2 \cdot 2 \cdot 2 \cdot 3 \cdot 3 \cdot 5$

Answers: **a.** $3^2 \cdot 7$, **b.** $5^2(7^2)$, **c.** $2^3 \cdot 3^2 \cdot 5$ ■

EXAMPLE 7 ***Exponential expressions.*** Find the value of each expression.

a. $7^2 = 7 \cdot 7 = 49$ — Read 7^2 as "7 to the second power" or "7 squared."

b. $2^5 = 2 \cdot 2 \cdot 2 \cdot 2 \cdot 2 = 32$ — Read 2^5 as "2 to the fifth power."

c. $10^4 = 10 \cdot 10 \cdot 10 \cdot 10 = 10{,}000$ — Read 10^4 as "10 to the fourth power."

d. $6^1 = 6$ — Read 6^1 as "6 to the first power."

Self Check

Which of the numbers 3^5, 4^4, and 5^3 is the largest?

Answer: $4^4 = 256$ ■

EXAMPLE 8 ***Evaluating exponential expressions.*** The prime factorization of a number is $2^3 \cdot 3^4 \cdot 5$. What is the number?

Solution

To find the number, we find the value of each power and then do the multiplication.

$$\begin{aligned} 2^3 \cdot 3^4 \cdot 5 &= \mathbf{8 \cdot 81} \cdot 5 && 2^3 = 8 \text{ and } 3^4 = 81. \\ &= 648 \cdot 5 && \text{Do the multiplications, working from left to right.} \\ &= 3{,}240 \end{aligned}$$

The number is 3,240.

Self Check

The prime factorization of a number is $3^3 \cdot 4^2 \cdot 5^2$. What is the number?

Answer: 10,800 ■

COMMENT Note that 5^3 means $5 \cdot 5 \cdot 5$. It does not mean $5 \cdot 3$. That is, $5^3 = 125$ and $5 \cdot 3 = 15$.

Accent on Technology: Bacterial growth

At the end of one hour, a culture contains two bacteria. Suppose the number of bacteria doubles every hour thereafter. Use exponents to determine how many bacteria the culture will contain after 24 hours.

We can use Table 1-2 to help model the situation. From the table, we see a pattern developing: The number of bacteria in the culture after 24 hours will be 2^{24}. We can evaluate this exponential expression using the exponential key $\boxed{y^x}$ on a scientific calculator ($\boxed{x^y}$ on some models).

To find the value of 2^{24}, we enter these numbers and press these keys.

Time	Number of bacteria
1 hr	$2 = 2^1$
2 hr	$4 = 2^2$
3 hr	$8 = 2^3$
4 hr	$16 = 2^4$
24 hr	$? = 2^{24}$

TABLE 1-2

Keystrokes 2 $\boxed{y^x}$ 24 $\boxed{=}$ 16777216

Since $2^{24} = 16{,}777{,}216$, there will be 16,777,216 bacteria after 24 hours.

Finding prime factorizations with the division method

We can also find the prime factorization of a whole number by division. For example, to find the prime factorization of 363, we begin the division method by choosing the *smallest* prime number that will divide the given number exactly. We continue this "inverted division" process until the result of the division is a prime number.

Step 1: The prime number 2 doesn't divide 363 exactly, but 3 does. The result is 121, which is not prime. We continue the division process.

$$\begin{array}{r|l} 3 & 363 \\ \hline & 121 \end{array}$$

Step 2: Next, we choose the smallest prime number that will divide 121. The primes 2, 3, 5, and 7 don't divide 121 exactly, but 11 does. The result is 11, which is prime. We are done.

$$\begin{array}{r|l} \mathbf{3} & 363 \\ \mathbf{11} & 121 \\ & \mathbf{11} \end{array}$$

$$363 = \mathbf{3 \cdot 11 \cdot 11}$$

Using exponents, we can write the prime factorization of 363 as $3 \cdot 11^2$.

EXAMPLE 9 ***Factoring with the division method.*** Use the division method to find the prime factorization of 100. Use exponents to express the result.

Solution

2 divides 100 exactly. The result is 50, which is not prime. → 2⌊100
2 divides 50 exactly. The result is 25, which is not prime. → 2⌊50
5 divides 25 exactly. The result is 5, which is prime. We are done. → 5⌊25
5

The prime factorization of 100 is $2^2 \cdot 5^2$.

Self Check

Use the division method to find the prime factorization of 108. Use exponents to express the result.

Answer: $2^2 \cdot 3^3$

STUDY SET Section 1.4

VOCABULARY *Fill in the blanks to make the statements true.*

1. Numbers that are multiplied together are called ______________.
2. One number is ______________ by another if the remainder is 0 when they are divided. When a division ends with a remainder of 0, we say that one of the numbers divides the other ______________.
3. To ______________ a whole number means to express it as the product of other whole numbers.
4. A ______________ number is a whole number, greater than 1, that has only 1 and itself as factors.
5. Whole numbers, greater than 1, that are not prime numbers are called ______________ numbers.
6. An ______________ whole number is exactly divisible by 2. An ______________ whole number is not exactly divisible by 2.
7. To prime factor a number means to write it as a product of only ______________ numbers.
8. An ______________ is used to represent repeated multiplication.
9. In the exponential expression 6^4, 6 is called the ______________, and 4 is called the ______________.
10. Another way to say "5 to the second power" is 5 ______________. Another way to say "7 to the third power" is 7 ______________.

CONCEPTS

11. Write 27 as the product of two factors.
12. Write 30 as the product of three factors.
13. The complete list of the factors of a whole number is given. What is the number?
    a. 2, 4, 22, 44, 11, 1
    b. 20, 1, 25, 100, 2, 4, 5, 50, 10
14. **a.** Find the factors of 24.
    b. Find the prime factorization of 24.
15. Find the factors of each number.
    a. 11
    b. 23
    c. 37
    d. From the results obtained in parts a–c, what can be said about 11, 23, and 37?
16. Suppose a number is divisible by 10. Is 10 a factor of the number?
17. If 4 is a factor of a whole number, will 4 divide the number exactly?
18. Give examples of whole numbers that have 11 as a factor.

In Exercises 19–22, the prime factorization of a whole number is given. Find the number.

19. $2 \cdot 3 \cdot 3 \cdot 5$
20. $3^3 \cdot 2$
21. $11^2 \cdot 5$
22. $2 \cdot 2 \cdot 2 \cdot 7$
23. Can we change the order of the base and the exponent in an exponential expression and obtain the same result? In other words, does $3^2 = 2^3$?
24. Find the prime factors of 30 and 165. What prime factors do they have in common?
25. Find the prime factors of 30 and 242. What prime factor do they have in common?
26. Find the prime factors of 20 and 35. What prime factor do they have in common?
27. Find the prime factors of 20 and 50. What prime factors do they have in common?
28. Find 1^2, 1^3, and 1^4. From the results, what can be said about any power of 1?
29. Finish the process of prime factoring 150. Compare the results.

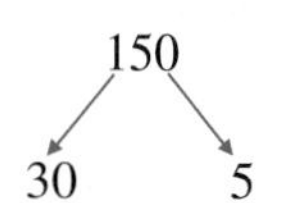

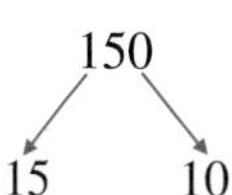

30. Find three whole numbers, less than 10, that would fit at the top of this tree diagram.

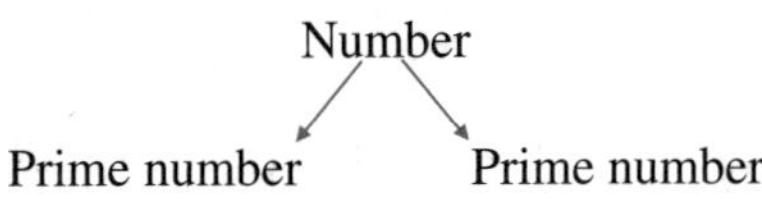

31. Complete the table

Product of factors of 12	Sum of factors of 12
$1 \cdot 12$	
$2 \cdot 6$	
$3 \cdot 4$	

32. Consider 1, 4, 9, 16, 25, 36, 49, 64, 81, 100. Of the numbers listed, which is the *largest* factor of
a. 18 **b.** 24 **c.** 50

33. When using the division method to find the prime factorization of an even number, what is an obvious choice with which to start the division process?

34. When using the division method to find the prime factorization of a number ending in 5, what is an obvious choice with which to start the division process?

NOTATION *Write each expression without using exponents.*

35. 7^3
36. 8^4
37. 3^5
38. 4^6
39. $5^2(11)$
40. $2^3 \cdot 3^2$
41. 10^1
42. 2^1

Use exponents to write each expression in simpler form.

43. $2 \cdot 2 \cdot 2 \cdot 2 \cdot 2$
44. $3 \cdot 3 \cdot 3 \cdot 3 \cdot 3 \cdot 3$
45. $5 \cdot 5 \cdot 5 \cdot 5$
46. $9 \cdot 9 \cdot 9$
47. $4(4)(5)(5)$
48. $12 \cdot 12 \cdot 12 \cdot 16$

PRACTICE *Find the factors of each whole number.*

49. 10
50. 6
51. 40
52. 75
53. 18
54. 32
55. 44
56. 65
57. 77
58. 81
59. 100
60. 441

Write each number in prime-factored form.

61. 39
62. 20
63. 99
64. 105
65. 162
66. 400
67. 220
68. 126
69. 64
70. 243
71. 147
72. 98

Evaluate each exponential expression.

73. 3^4
74. 5^3
75. 2^5
76. 10^5
77. 12^2
78. 7^3
79. 8^4
80. 9^5
81. $3^2(2^3)$
82. $3^3(4^2)$
83. $2^3 \cdot 3^3 \cdot 4^2$
84. $3^2 \cdot 4^3 \cdot 5^2$
85. 234^3
86. 51^4
87. $23^2 \cdot 13^3$
88. $12^3 \cdot 15^2$

APPLICATIONS

89. PERFECT NUMBERS A whole number is called a **perfect number** when the sum of its factors that are less than the number equals the number. For example, 6 is a perfect number, because $1 + 2 + 3 = 6$. Find the factors of 28. Then use addition to show that 28 is also a perfect number.

90. CRYPTOGRAPHY Information is often transmitted in code. Many codes involve writing products of large primes, because they are difficult to factor. To see how difficult, try finding two prime factors of 7,663. (*Hint:* Both primes are greater than 70.)

91. LIGHT Illustration 1 shows that the light energy that passes through the first unit of area, 1 yard away from the bulb, spreads out as it travels away from the source. How much area does that energy cover 2 yards, 3 yards, and 4 yards from the bulb? Express each answer using exponents.

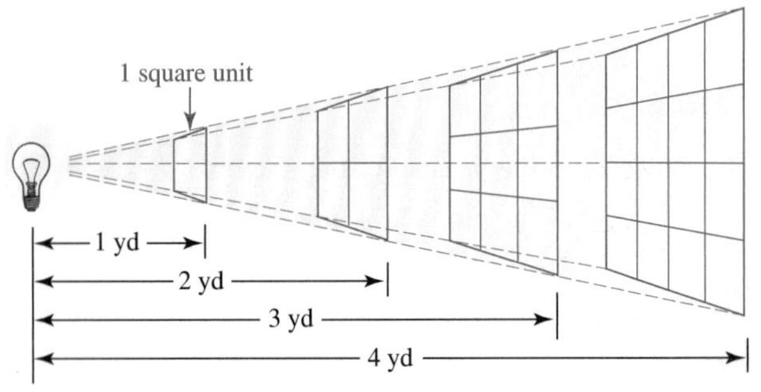

ILLUSTRATION 1

92. CELL DIVISION After one hour, a cell has divided to form another cell. In another hour, these two cells have divided so that four cells exist. In another hour, these four cells divide so that eight exist.

a. How many cells exist at the end of the fourth hour?

b. The number of cells that exist after each division can be found using an exponential expression. What is the base?

c. Use a calculator to find the number of cells after 12 hours.

WRITING

93. Explain how to test a number to see whether it is prime.

94. Explain how to test a number to see whether it is even.

95. Explain the difference between the *factors* of a number and the *prime factorization* of the number.

96. Explain why it would be incorrect to say that the area of the square shown in Illustration 2 is 25^2 ft. How should we express its area?

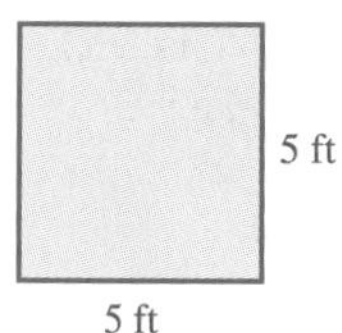

ILLUSTRATION 2

REVIEW

97. Round 230,999 to the nearest thousand.

98. Write the set of whole numbers.

99. What is $0 \div 15$?

100. Find $15 \cdot (6 \cdot 9)$.

101. What is the formula for the area of a rectangle?

102. MARCHING BANDS When a university band lines up in eight rows of 15 musicians, there are five musicians left over. How many band members are there?

1.5 Order of Operations

In this section, you will learn about

- Order of operations
- Evaluating expressions with no grouping symbols
- Evaluating expressions with grouping symbols
- The arithmetic mean (average)

INTRODUCTION. Punctuation marks, such as commas, quotations, and periods, serve an important purpose when writing compositions. They determine the way in which sentences are to be read and interpreted. To read and interpret mathematical expressions correctly, we must use an agreed-upon set of priority rules for the *order of operations.*

Order of operations

Suppose you are asked to contact a friend if you see a certain type of watch for sale while you are traveling in Europe. While in Switzerland, you spot the watch and send the following E-mail message.

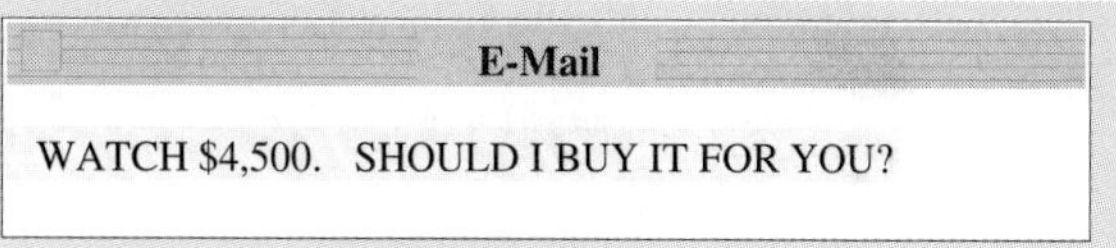

E-Mail

WATCH $4,500. SHOULD I BUY IT FOR YOU?

The next day, you get this response from your friend.

E-Mail

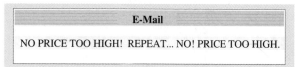

NO PRICE TOO HIGH! REPEAT... NO! PRICE TOO HIGH.

Something is wrong. One statement says to buy the watch at any price. The other says not to buy it, because it's too expensive. The placement of the exclamation point makes us read these statements differently, resulting in different interpretations.

When reading a mathematical statement, the same kind of confusion is possible. For example, we consider

$$3 + 2 \cdot 5$$

This expression contains two operations: addition and multiplication. We can evaluate it (find its value) in two different ways. We can do the addition first and then do the multiplication. Or we can do the multiplication first and then do the addition. However, we get different results.

Method 1

$3 + 2 \cdot 5 = \mathbf{5} \cdot 5$ Add first: $3 + 2 = 5$.

$= 25$ Multiply 5 and 5.

Method 2

$3 + \mathbf{2 \cdot 5} = 3 + \mathbf{10}$ Multiply first: $2 \cdot 5 = 10$.

$= 13$ Add 3 and 10.

Different results

If we don't establish an order of operations, the expression $3 + 2 \cdot 5$ will have two different answers. To avoid this possibility, we will evaluate expressions in the following order.

Order of operations

1. Do all calculations within parentheses and other grouping symbols following the order listed in Steps 2–4, working from the innermost pair to the outermost pair.
2. Evaluate all exponential expressions.
3. Do all multiplications and divisions as they occur from left to right.
4. Do all additions and subtractions as they occur from left to right.

When all grouping symbols have been removed, repeat Steps 2–4 to complete the calculation.

If a fraction bar is present, evaluate the expression above the bar (called the **numerator**) and the expression below the bar (called the **denominator**) separately. Then do the division indicated by the fraction bar, if possible.

To evaluate $3 + 2 \cdot 5$ correctly, we must apply the rules for the order of operations. Since there are no grouping symbols and there are no powers, we perform the multiplication first and then do the addition.

Ignore the addition for now — and do the multiplication first: $2 \cdot 5 = 10$.

$3 + \mathbf{2 \cdot 5} = 3 + \mathbf{10}$

$= 13$ Now do the addition.

Using the rules for the order of operations, we see that the correct answer is 13.

Evaluating expressions with no grouping symbols

EXAMPLE 1 *Order of operations.* Evaluate $2 \cdot 4^2 - 8$.

Solution

Since the expression does not contain any grouping symbols, we begin with Step 2 of the rules for the order of operations.

$2 \cdot \mathbf{4^2} - 8 = 2 \cdot \mathbf{16} - 8$ Evaluate the exponential expression: $4^2 = 16$.

$= 32 - 8$ Do the multiplication: $2 \cdot 16 = 32$.

$= 24$ Do the subtraction.

Self Check

Evaluate $4 \cdot 3^3 - 6$.

Answer: 102

EXAMPLE 2 ***Order of operations.*** Evaluate $8 - 3 \cdot 2 + 16$.

Solution

Since the expression does not contain grouping symbols and since there are no powers to find, we look for multiplications or divisions to perform.

$$\begin{aligned} 8 - \mathbf{3 \cdot 2} + 16 &= 8 - \mathbf{6} + 16 && \text{Do the multiplication: } 3 \cdot 2 = 6. \\ &= 2 + 16 && \text{Working from left to right, do the subtraction: } 8 - 6 = 2. \\ &= 18 && \text{Do the addition.} \end{aligned}$$

Self Check

Evaluate $10 - 2 \cdot 3 + 24$.

Answer: 28 ■

COMMENT Some students incorrectly think that additions are always done before subtractions. As Example 2 shows, this is not true. Working from left to right, we do the additions or subtractions *in the order in which they occur.* The same is true for multiplications and divisions.

EXAMPLE 3 ***Order of operations.*** Evaluate $192 \div 6 - 5(3)2$.

Solution

Although this expression contains parentheses, there are no calculations to perform within them. Since there are no powers, we do multiplications and divisions as they are encountered from left to right.

$$\begin{aligned} \mathbf{192 \div 6} - 5(3)2 &= \mathbf{32} - 5(3)2 && \text{Working from left to right, do the division: } 192 \div 6 = 32. \\ &= 32 - 15(2) && \text{Working from left to right, do the multiplication: } 5(3) = 15. \\ &= 32 - 30 && \text{Do the multiplication: } 15(2) = 30. \\ &= 2 && \text{Do the subtraction.} \end{aligned}$$

Self Check

Evaluate $36 \div 9 + 4(2)3$.

Answer: 28 ■

EXAMPLE 4 ***Phone bill.*** Figure 1–15 shows the rates for international telephone calls charged by a 10-10 long-distance company. A businesswoman calls Germany for 20 minutes, South Korea for 5 minutes, and Mexico City for 35 minutes. What is the total cost of the calls?

All rates are per minute.	
Canada	10¢
Germany	23¢
Jamaica	68¢
Mexico City	42¢
South Korea	29¢

FIGURE 1–15

Solution

We can find the cost of a call (in cents) by multiplying the rate charged per minute by the length of the call (in minutes). To find the total cost, we add the costs of the three calls.

The cost of the call to Germany. ↓ $23(20)$ + The cost of the call to South Korea. ↓ $29(5)$ + The cost of the call to Mexico City. ↓ $42(35)$

$$23(20) + 29(5) + 42(35)$$

To evaluate this expression, we apply the rules for the order of operations.

$$\begin{aligned} 23(20) + 29(5) + 42(35) &= 460 + 145 + 1{,}470 && \text{Do the multiplications.} \\ &= 2{,}075 && \text{Do the additions.} \end{aligned}$$

The total cost of the calls is 2,075 cents, or \$20.75. ■

Evaluating expressions with grouping symbols

Grouping symbols serve as mathematical punctuation marks. They help determine the order in which an expression is to be evaluated. Examples of grouping symbols are parentheses ( ), brackets [ ], and the fraction bar ——.

In the next example, we have two similar-looking expressions. However, because of the parentheses, we evaluate them in a different order.

EXAMPLE 5 *Working with grouping symbols.* Evaluate each expression: **a.** $12 - 3 + 5$ and **b.** $12 - (3 + 5)$.

Solution

a. We perform the additions and subtractions as they occur, from left to right.

$$12 - 3 + 5 = 9 + 5 \quad \text{Do the subtraction: } 12 - 3 = 9.$$
$$= 14 \quad \text{Do the addition.}$$

b. This expression contains parentheses. We must do the calculation within the parentheses first.

$$12 - (3 + 5) = 12 - 8 \quad \text{Do the addition: } 3 + 5 = 8.$$
$$= 4 \quad \text{Do the subtraction.}$$

Self Check

Evaluate each expression.

a. $20 - 7 + 6$

b. $20 - (7 + 6)$

Answers: **a.** 19, **b.** 7 ■

EXAMPLE 6 *Working with grouping symbols.* Evaluate $(2 + 6)^3$.

Solution

We begin by doing the calculation within the parentheses.

$$(2 + 6)^3 = 8^3 \quad \text{Do the addition.}$$
$$= 512 \quad \text{Evaluate the exponential expression: } 8^3 = 8 \cdot 8 \cdot 8 = 512.$$

Self Check

Evaluate $(1 + 2)^4$.

Answer: 81 ■

EXAMPLE 7 *Order of operations within grouping symbols.* Evaluate $5 + 2(13 - 5 \cdot 2)$.

Solution

This expression contains grouping symbols. We will apply the rules for the order of operations within the parentheses first, to evaluate $13 - 5 \cdot 2$.

$$5 + 2(13 - 5 \cdot 2) = 5 + 2(13 - 10) \quad \text{Do the multiplication within the parentheses.}$$
$$= 5 + 2(3) \quad \text{Do the subtraction within the parentheses.}$$
$$= 5 + 6 \quad \text{Do the multiplication: } 2(3) = 6.$$
$$= 11 \quad \text{Do the addition.}$$

Self Check

Evaluate $25 - 2(12 - 5 \cdot 2)$.

Answer: 21 ■

Sometimes an expression contains two or more sets of grouping symbols. Since it can be confusing to read an expression such as $16 + 2(14 - 3(5 - 2))$, we often use brackets in place of the second pair of parentheses.

$$16 + 2[14 - 3(5 - 2)]$$

If an expression contains more than one pair of grouping symbols, we always begin by working within the innermost pair and then work to the outermost pair.

Innermost parentheses
↓ ↓
$$16 + 2[14 - 3(5 - 2)]$$
↑ ↑
Outermost brackets

EXAMPLE 8 ***Grouping symbols within grouping symbols.*** Evaluate $16 + 2[14 - 3(5 - 2)]$.

Solution

$16 + 2[14 - 3(5 - 2)] = 16 + 2[14 - 3(3)]$ Do the subtraction within the parentheses.

$= 16 + 2(14 - 9)$ Do the multiplication within the brackets. Since only 1 set of grouping symbols is needed, write $14 - 9$ within parentheses.

$= 16 + 2(5)$ Do the subtraction within the parentheses.

$= 16 + 10$ Do the multiplication: $2(5) = 10$.

$= 26$ Do the addition.

Self Check

Evaluate $46 - 2[4 + 3(6 - 2)]$.

Answer: 14

EXAMPLE 9 ***Working with a fraction bar.*** Evaluate $\frac{2(13) - 2}{3(2^3)}$.

Solution

A fraction bar is a grouping symbol. We evaluate the numerator and denominator separately and then do the indicated division.

$\frac{2(13) - 2}{3(2^3)} = \frac{26 - 2}{3(8)}$ In the numerator, do the multiplication. In the denominator, do the calculation within the parentheses.

$= \frac{24}{24}$ In the numerator, do the subtraction. In the denominator, do the multiplication.

$= 1$ Do the division.

Self Check

Evaluate $\frac{3(14) - 6}{2(3^2)}$.

Answer: 2

The arithmetic mean (average)

The **arithmetic mean,** or **average,** of several numbers is a value around which the numbers are grouped. It gives you an indication of the "center" of the set of numbers. When finding the mean of a set of numbers, we usually need to apply the rules for the order of operations.

Finding an arithmetic mean

To find the mean of a set of scores, divide the sum of the scores by the number of scores.

EXAMPLE 10 ***NCAA basketball.*** In 1998, the Lady Vols of the University of Tennessee won the women's basketball championship, capping a perfect 39-0 season. Find their average margin of victory in their last four tournament games shown below.

Regional	Regional final	Semifinal	Championship
beat Rutgers by 32 points	beat North Carolina by 6 points	beat Arkansas by 28 points	beat Louisiana Tech by 18 points

Solution

To find the average margin of victory, add the margins of victory and divide by 4.

$$\text{Average} = \frac{32 + 6 + 28 + 18}{4}$$

$$= \frac{84}{4}$$

$$= 21$$

Their average margin of victory was 21 points.

Self Check

The University of Kentucky won the 1998 NCAA men's basketball championship. Find their average margin of victory in their last five tournament games, which they won by 27, 26, 2, 1, and 9 points.

Answer: 13 points

Accent on Technology: **Order of operations and parentheses**

Scientific calculators have the rules for order of operations built in. Even so, some evaluations require the use of a left parenthesis key [( ] and a right parenthesis key [ )]. For example, to evaluate $\frac{240}{20-15}$, we enter these numbers and press these keys.

Keystrokes 240 [÷] [( ] 20 [−] 15 [ )] [=] — 48

STUDY SET Section 1.5

VOCABULARY *Fill in the blanks to make the statements true.*

1. The grouping symbols ( ) are called ______________, and the symbols [ ] are called ______________.
2. The expression above a fraction bar is called the ______________. The expression below a fraction bar is called the ______________.
3. To ______________ $2 + 5 \cdot 4$ means to find its value.
4. To find the ______________ of several values, we add the values and divide by the number of values.

CONCEPTS

5. Consider $5(2)^2 - 1$. How many operations need to be performed to evaluate the expression? List them in the order in which they should be performed.
6. Consider $15 - 3 + (5 \cdot 2)^3$. How many operations need to be performed to evaluate this expression? List them in the order in which they should be performed.
7. Consider $\frac{5 + 5(7)}{2 + (8 - 4)}$. In the numerator, what operation should be done first? In the denominator, what operation should be done first?
8. In the expression $\frac{3 - 5(2)}{5(2) + 4}$, the bar is a grouping symbol. What does it separate?
9. Explain the difference between $2 \cdot 3^2$ and $(2 \cdot 3)^2$.
10. Use brackets to write $2(12 - (5 + 4))$ in better form.

NOTATION *Complete each solution.*

11. $28 - 5(2)^2 = 28 - 5(\quad)$
    $= 28 - \quad$
    $= 8$
12. $2 + (5 + 6 \cdot 2) = 2 + (5 + \quad)$
    $= 2 + \quad$
    $= 19$
13. $[4(2 + 7)] - 6 = [4(\quad)] - 6$
    $= \quad - 6$
    $= 30$
14. $\frac{5(3) + 12}{9 - 6} = \frac{\quad + 12}{\quad}$
    $= \frac{\quad}{\quad}$
    $= 9$

PRACTICE *Evaluate each expression.*

15. $7 + 4 \cdot 5$
16. $10 - 2 \cdot 2$
17. $2 + 3(0)$
18. $5(0) + 8$
19. $20 - 10 + 5$
20. $80 - 5 + 4$
21. $25 \div 5 \cdot 5$
22. $6 \div 2 \cdot 3$
23. $7(5) - 5(6)$
24. $4 \cdot 2 + 2 \cdot 4$
25. $4^2 + 3^2$
26. $12^2 - 5^2$
27. $2 \cdot 3^2$
28. $3^3 \cdot 5$
29. $3 + 2 \cdot 3^4 \cdot 5$
30. $3 \cdot 2^3 \cdot 4 - 12$
31. $5 \cdot 10^3 + 2 \cdot 10^2 + 3 \cdot 10^1 + 9$
32. $8 \cdot 10^3 + 0 \cdot 10^2 + 7 \cdot 10^1 + 4$
33. $3(2)^2 - 4(2) + 12$
34. $5(1)^3 + (1)^2 + 2(1) - 6$
35. $(8 - 6)^2 + (4 - 3)^2$
36. $(2 + 1)^2 + (3 + 2)^2$
37. $60 - (6 + \frac{40}{8})$
38. $7 + (5^3 - \frac{200}{2})$
39. $6 + 2(5 + 4)$
40. $3(5 + 1) + 7$
41. $3 + 5(6 - 4)$
42. $7(9 - 2) - 1$
43. $(7 - 4)^2 + 1$
44. $(9 - 5)^3 + 8$
45. $6^3 - (10 + 8)$
46. $5^2 - (9 + 3)$
47. $50 - 2(4)^2$
48. $30 + 2(3)^3$
49. $16^2 - 4(2)(5)$
50. $8^2 - 4(3)(1)$

51. $39 - 5(6) + 9 - 1$

52. $15 - 3(2) - 4 + 3$

53. $(18 - 12)^3 - 5^2$

54. $(9 - 2)^2 - 3^3$

55. $2(10 - 3^2) + 1$

56. $1 + 3(18 - 4^2)$

57. $6 + \frac{25}{5} + 6(3)$

58. $15 - \frac{24}{6} + 8 \cdot 2$

59. $3\left(\frac{18}{3}\right) - 2(2)$

60. $2\left(\frac{12}{3}\right) + 3(5)$

61. $(2 \cdot 6 - 4)^2$

62. $2(6 - 4)^2$

63. $4[50 - (3^3 - 5^2)]$

64. $6[15 + (5 \cdot 2^2)]$

65. $80 - 2[12 - (5 + 4)]$

66. $15 + 5[12 - (2^2 + 4)]$

67. $2[100 - (5 + 4)] - 45$

68. $8[6(6) - 6^2] + 4(5)$

69. $\frac{10 + 5}{6 - 1}$

70. $\frac{18 + 12}{2(3)}$

71. $\frac{5^2 + 17}{6 - 2^2}$

72. $\frac{3^2 - 2^2}{(3 - 2)^2}$

73. $\frac{(3 + 5)^2 + 2}{2(8 - 5)}$

74. $\frac{25 - (2 \cdot 3 - 1)}{2 \cdot 9 - 8}$

75. $\frac{(5 - 3)^2 + 2}{4^2 - (8 + 2)}$

76. $\frac{(4^3 - 2) + 7}{5(2 + 4) - 7}$

77. $12{,}985 - (1{,}800 + 689)$

78. $\frac{897 - 655}{88 - 77}$

79. $3{,}245 - 25(16 - 12)^2$

80. $\frac{24^2 - 4^2}{22 + 58}$

APPLICATIONS *In Exercises 81–86, write an expression to solve each problem. Then evaluate the expression.*

81. BUYING GROCERIES At the supermarket, Carlos has 2 cases of soda, 4 bags of potato chips, and 2 cans of dip in his cart. Each case of soda costs \$6, each bag of chips costs \$2, and each can of dip costs \$1. Find the total cost of the groceries.

82. JUDGING The scores received by a junior diver are as follows:

5	2	4	6	3	4

The formula for computing the overall score for the dive is as follows:

1. Throw out the lowest score.
2. Throw out the highest score.
3. Divide the sum of the remaining scores by 4.

Find the diver's score.

83. BANKING When a customer deposits cash, a teller must complete a "currency count" on the back of the deposit slip. In Illustration 1, what is the total amount of cash being deposited?

Currency count, for financial use only			
24	x 1's		
—	x 2's		
6	x 5's		
10	x 10's		
12	x 20's		
2	x 50's		
1	x 100's		
	TOTAL \$		

ILLUSTRATION 1

84. WRAPPING GIFTS How much ribbon is needed to wrap the package shown in Illustration 2 if 15 inches of ribbon are needed to make the bow?

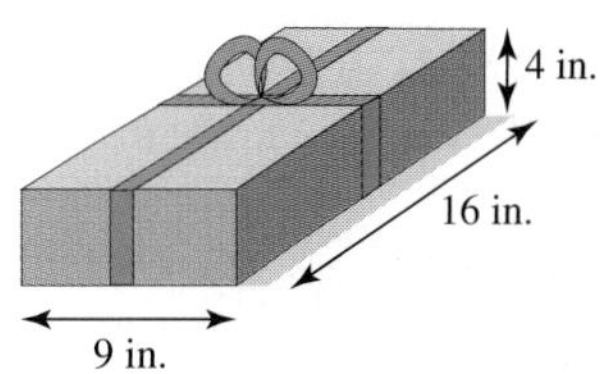

ILLUSTRATION 2

85. SCRABBLE Illustration 3(a) shows part of the game board before and Illustration 3(b) shows it after the words *brick* and *aphid* were played. Determine the scoring for each word. (The number on each tile gives the point value of the letter.)

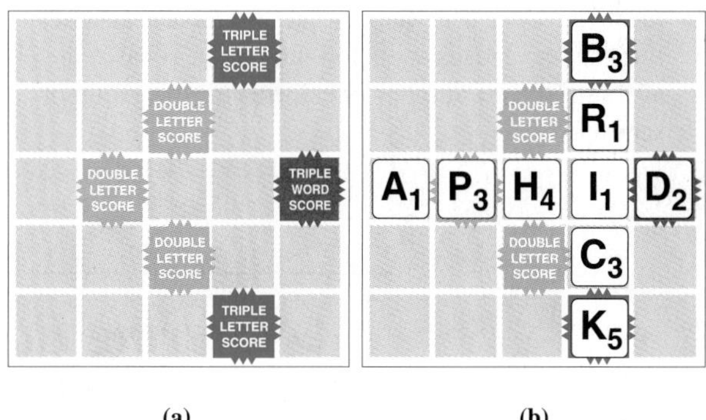

ILLUSTRATION 3

86. GETTYSBURG ADDRESS Here is an excerpt from Abraham Lincoln's Gettysburg Address.

Fourscore and seven years ago, our fathers brought forth on this continent a new nation, conceived in liberty, and dedicated to the proposition that all men are created equal.

Lincoln's comments refer to the year 1776, when the United States declared its independence. If a score is 20 years, in what year did Lincoln deliver the Gettysburg Address?

87. CLIMATE One December week, the temperatures in Honolulu, Hawaii were 75°, 80°, 83°, 80°, 77°, 72°, and 86°. Find the week's average (mean) temperature.

88. GRADES In a psychology class, a student had test scores of 94, 85, 81, 77, and 89. He also overslept, missed the final exam, and received a 0 on it. What was his test average in the class?

89. NATURAL NUMBERS What is the average (mean) of the first nine natural numbers?

90. ENERGY USAGE See Illustration 4. Find the average number of therms of natural gas used per month. Then draw a dashed line across the graph showing the average.

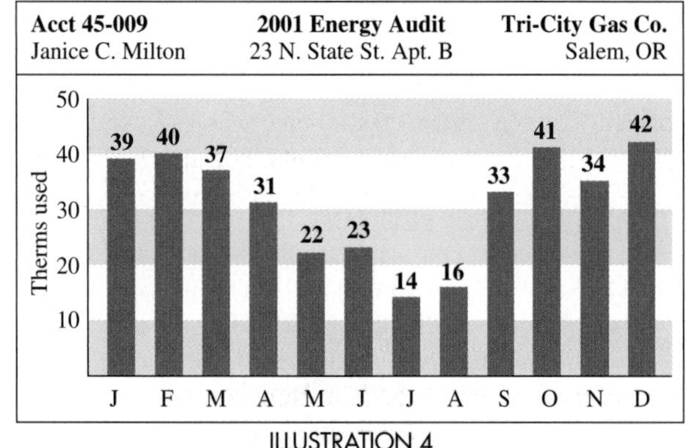

ILLUSTRATION 4

91. FAST FOOD Illustration 5 shows the sandwiches Subway advertises as its low-fat menu. What is the average (mean) number of calories for the group of sandwiches?

6-inch subs	Calories	Fat (g)
Veggie Delite	237	3
Turkey Breast	289	4
Turkey Breast & Ham	295	5
Ham	302	5
Roast Beef	303	5
Subway Club	312	5
Roasted Chicken Breast	348	6

ILLUSTRATION 5

92. TV RATINGS The list below shows the number of people watching *Who Wants to Be a Millionaire?* on five weeknights in November of 1999. How large was the average audience?

Monday	26,800,000
Tuesday	24,900,000
Wednesday	22,900,000
Thursday	25,900,000
Friday	21,900,000

WRITING

93. Explain why rules for the order of operations are necessary.

94. Explain the difference between the steps used to evaluate $5 \cdot 2^3$ and $(5 \cdot 2)^3$.

95. Explain the process of finding the mean of a large group of numbers. What does an average tell you?

96. What does it mean when we say to do all additions and subtractions *as they occur from left to right*?

REVIEW *Do the operations.*

97. $\begin{array}{r} 4{,}029 \\ +3{,}271 \\ \hline \end{array}$

98. $\begin{array}{r} 4{,}263 \\ -3{,}764 \\ \hline \end{array}$

99. $\begin{array}{r} 417 \\ \times\ 23 \\ \hline \end{array}$

100. $82\overline{)50{,}430}$

1.6 *Solving Equations by Addition and Subtraction*

In this section, you will learn about

- Equations • Checking solutions • Solving equations
- Problem solving with equations

Introduction. The language of mathematics is *algebra.* The word *algebra* comes from the title of a book written by the Arabian mathematician Al-Khowarazmi around

A.D. 800. Its title, *Ihm aljabr wa'l muqabalah,* means restoration and reduction, a process then used to solve equations. In this section, we will begin discussing equations, one of the most powerful ideas in algebra.

Equations

An **equation** is a statement indicating that two expressions are equal. Some examples of equations are

$$x + 5 = 21, \qquad 16 + 5 = 21, \qquad \text{and} \qquad 10 + 5 = 21$$

Equations

Equations are mathematical sentences that contain an $=$ sign.

In the equation $x + 5 = 21$, the expression $x + 5$ is called the **left-hand side,** and 21 is called the **right-hand side.** The letter x is the **variable** (or the **unknown**).

An equation can be true or false. For example, $16 + 5 = 21$ is a true equation, whereas $10 + 5 = 21$ is a false equation. An equation containing a variable can be true or false, depending upon the value of the variable. If $x = 16$, the equation $x + 5 = 21$ is true, because

$\mathbf{16} + 5 = 21$ Substitute 16 for x.

However, this equation is false for all other values of x.

Any number that makes an equation true when substituted for its variable is said to *satisfy* the equation. Such numbers are called **solutions** or **roots.** Because 16 is the only number that satisfies $x + 5 = 21$, it is the only solution of the equation.

Checking solutions

EXAMPLE 1 ***Checking a solution.*** Verify that 18 is a solution of the equation $x - 3 = 15$.

Solution

We substitute 18 for x in the equation and verify that both sides of the equation are equal.

$x - 3 = 15$ The given equation.

$\mathbf{18} - 3 \stackrel{?}{=} 15$ Substitute 18 for x. Read $\stackrel{?}{=}$ as "is possibly equal to."

$15 = 15$ Do the subtraction.

Since $15 = 15$ is a true equation, 18 is a solution of $x - 3 = 15$.

Self Check

Is 8 a solution of $x + 17 = 25$?

Answer: yes ■

EXAMPLE 2 ***Checking a solution.*** Is 23 a solution of $32 = y + 10$?

Solution

We substitute 23 for y and simplify.

$32 = y + 10$ The given equation.

$32 \stackrel{?}{=} \mathbf{23} + 10$ Substitute 23 for y.

$32 \neq 33$ Do the addition.

Since the left-hand and right-hand sides are not equal, 23 is not a solution.

Self Check

Is 5 a solution of $20 = y + 17$?

Answer: no ■

Solving equations

Since the solution of an equation is usually not given, we must develop a process to find it. This process is called *solving the equation.* To develop an understanding of the properties and procedures used to solve an equation, we will examine $x + 2 = 5$ and make some observations as we solve it in a practical way.

We can think of the scales shown in Figure 1-16(a) as representing the equation $x + 2 = 5$. The weight (in grams) on the left-hand side of the scales is $x + 2$, and the weight (in grams) on the right-hand side is 5. Because these weights are equal, the scales are in balance. To find x, we need to isolate it. That can be accomplished by removing 2 grams from the left-hand side of the scales. Common sense tells us that we must also remove 2 grams from the right-hand side if the scales are to remain in balance. In Figure 1-16(b), we can see that x grams will be balanced by 3 grams. We say that we have *solved* the equation and that the *solution* is 3.

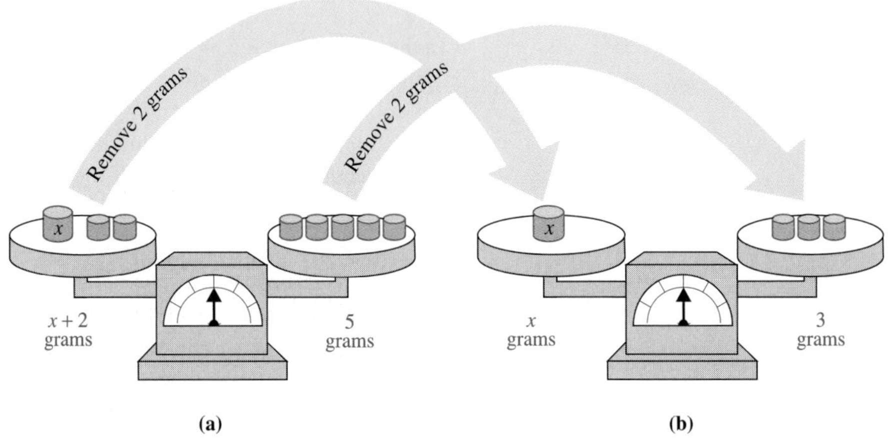

FIGURE 1-16

From this example, we can make some observations about solving an equation.

- To find the value of x, we needed to isolate it on the left-hand side of the scales.
- To isolate x, we had to undo the addition of 2 grams. This was accomplished by subtracting 2 grams from the left-hand side.
- We wanted the scales to remain in balance. When we subtracted 2 grams from the left-hand side, we subtracted the same amount from the right-hand side.

The observations suggest a property of equality: *If the same quantity is subtracted from equal quantities, the results will be equal quantities.* We can express this property in symbols.

Subtraction property of equality

Let a, b, and c represent numbers.

If $a = b$, then $a - c = b - c$.

When we use this property, the resulting equation will be equivalent to the original equation.

Equivalent equations

Two equations are **equivalent equations** when they have the same solutions.

In the previous example, we found that $x + 2 = 5$ is equivalent to $x = 3$. This is true because these equations have the same solution, $x = 3$.

We now show how to solve $x + 2 = 5$ using an algebraic approach.

EXAMPLE 3 ***Solving an equation.*** Solve $x + 2 = 5$.

Self Check

Solve $x + 7 = 14$ and check the result.

Solution

To isolate x on the left-hand side of the equation, we undo the addition of 2 by subtracting 2 from both sides of the equation.

$x + 2 = 5$ The equation to solve.

$x + 2 \mathbf{- 2} = 5 \mathbf{- 2}$ Subtract 2 from both sides.

$x = 3$ On the left-hand side, subtracting 2 undoes the addition of 2 and leaves x. On the right-hand side, $5 - 2 = 3$.

We check by substituting 3 for x in the original equation and simplifying. If 3 is the solution, we will obtain a true statement.

$x + 2 = 5$

$\mathbf{3} + 2 \stackrel{?}{=} 5$ Substitute 3 for x.

$5 = 5$ Do the addition.

Since the resulting equation is true, 3 is a solution.

Answer: 7

A second property that we will use to solve equations involves addition. It is based on the following idea: *If the same quantity is added to equal quantities, the results will be equal quantities.* In symbols, we have the following property.

Addition property of equality

Let a, b, and c represent numbers.

If $a = b$, then $a + c = b + c$.

We can think of the scales shown in Figure 1-17(a) as representing the equation $x - 2 = 3$. To find x, we need to use the addition property of equality and add 2 grams of weight to each side. The scales will remain in balance. From the scales in Figure 1-17(b), we can see that x grams will be balanced by 5 grams. The solution of $x - 2 = 3$ is therefore $x = 5$.

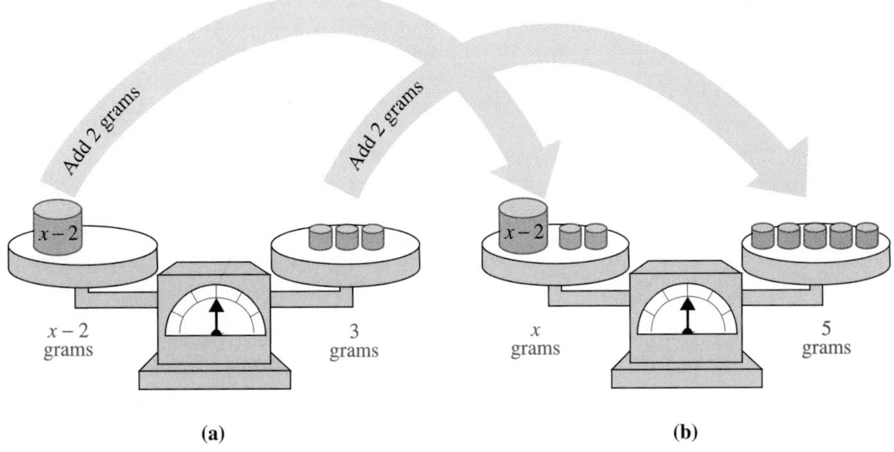

FIGURE 1-17

To solve $x - 2 = 3$ algebraically, we apply the addition property of equality. We can isolate x on the left-hand side of the equation by adding 2 to both sides.

$x - 2 = 3$

$x - 2 \mathbf{+ 2} = 3 \mathbf{+ 2}$ To undo the subtraction of 2, add 2 to both sides.

$x = 5$ On the left-hand side, adding 2 undoes the subtraction of 2 and leaves x. On the right-hand side, $3 + 2 = 5$.

To check this result, we substitute 5 for x in the original equation and simplify.

$x - 2 = 3$

$5 - 2 \stackrel{?}{=} 3$ Substitute 5 for x.

$3 = 3$ Do the subtraction.

Since this is a true statement, $x = 5$ is a solution.

EXAMPLE 4 ***Isolating the variable on the right-hand side.*** Solve $19 = y - 7$ and check the result.

Solution

To isolate the variable y on the right-hand side, we use the addition property of equality. We can undo the subtraction of 7 by adding 7 to both sides.

$19 = y - 7$

$19 + 7 = y - 7 + 7$ Add 7 to both sides.

$26 = y$ On the left-hand side, $19 + 7 = 26$. On the right-hand side, adding 7 undoes the subtraction of 7 and leaves y.

$y = 26$ When stating a solution, it is common practice to write the variable first. If $26 = y$, then $y = 26$.

We check by substituting 26 for y in the original equation and simplifying.

$19 = y - 7$ The original equation.

$19 \stackrel{?}{=} 26 - 7$ Substitute 26 for y.

$19 = 19$ Do the subtraction.

Since this is a true statement, 26 is a solution.

Self Check

Solve $75 = b - 38$ and check the result.

Answer: 113

Problem solving with equations

The key to problem solving is to understand the problem and then to devise a plan for solving it. The following list of steps provides a good strategy to follow.

Strategy for problem solving

1. **Analyze the problem** by reading it carefully to understand the given facts. What information is given? What vocabulary is given? What are you asked to find? Often a diagram will help you visualize the facts of a problem.
2. **Form an equation** by picking a variable to represent the quantity to be found. Then express all other unknown quantities as expressions involving that variable. Finally, write an equation expressing a quantity in two different ways.
3. **Solve the equation.**
4. **State the conclusion.**
5. **Check the result** in the words of the problem.

We will now use this five-step strategy to solve problems. The purpose of the following examples is to help you learn the strategy, even though you can probably solve these examples without it.

EXAMPLE 5 ***Financial data.*** Figure 1-18 shows the 1999 quarterly net income for Nike, the athletic shoe company. What was the company's total net income for 1999?

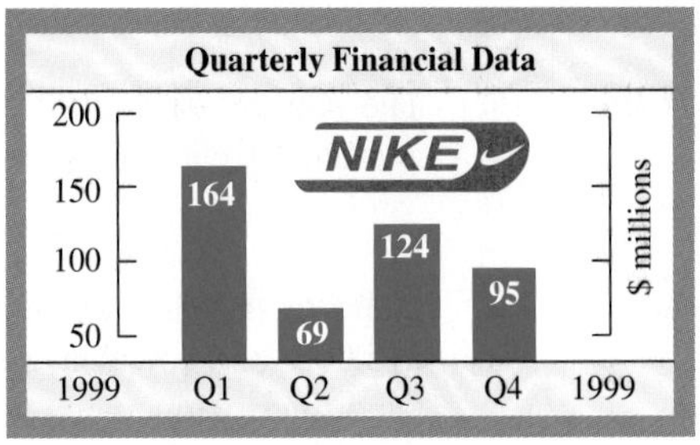

FIGURE 1-18

Analyze the problem

- We are given the net income for each quarter.
- We are asked to find the total net income.

Form an equation Let n = the total net income for 1999. To form an equation involving n, we look for a key word or phrase in the problem.

Key word: *total* **Translation:** *addition*

We can express the total income in two ways.

Total net income	is	1st qtr. net income	plus	2nd qtr. net income	plus	3rd qtr. net income	plus	4th qtr. net income.
n	$=$	164	$+$	69	$+$	124	$+$	95

Solve the equation

$n = 164 + 69 + 124 + 95$ We are working in millions of dollars.

$n = 452$ Do the additions.

State the conclusion Nike's total net income was $452 million.

Check the result We can check the result by estimation. To estimate, we round the net income from each quarter and add.

$$160 + 70 + 120 + 100 = 450$$

The answer, 452, is reasonable. ■

EXAMPLE 6 ***Small business.*** Last year a hairdresser lost 17 customers who moved away. If he now has 73 customers, how many did he have originally?

Analyze the problem

- We know that he started with some unknown number of customers, and after 17 moved away, 73 were left.
- We are asked to find the number of customers he had before any moved away.

Form an equation We can let c = the original number of customers.
To form an equation involving c, we look for a key word or phrase in the problem.

Key phrase: *moved away* **Translation:** *subtract*

We can express the remaining number of customers in two ways.

The original number of customers	minus	17	is	the remaining number of customers.
c	$-$	17	$=$	73

Solve the equation

$c - 17 = 73$

$c - 17 + \mathbf{17} = 73 + \mathbf{17}$ To undo the subtraction of 17, add 17 to both sides.

$c = 90$ Simplify each side of the equation.

State the conclusion He originally had 90 customers.

Check the result The hairdresser had 90 customers. After losing 17, he has $90 - 17$, or 73 left. The answer, 90, checks. ■

EXAMPLE 7 ***Buying a house.*** Sue wants to buy a house that costs $87,000. Since she has only $15,000 for a down payment, she will have to borrow some additional money by taking a mortgage. How much will she have to borrow?

Analyze the problem

- The house costs $87,000.
- Sue has $15,000 for a down payment.
- We must find how much money she needs to borrow.

Form an equation We can let x = the money that she needs to borrow.
To form an equation involving x, we look for a key word or phrase in the problem.

Key phrase: *borrow some additional money* **Translation:** *addition*

We can express the total cost of the house in two ways.

The amount Sue has	plus	the amount she borrows	is	the total cost of the house.
15,000	$+$	x	$=$	87,000

Solve the equation

$$15{,}000 + x = 87{,}000$$

$$15{,}000 + x \mathbf{- 15{,}000} = 87{,}000 \mathbf{- 15{,}000}$$ To undo the addition of 15,000, subtract 15,000 from both sides.

$$x = 72{,}000$$ Do the subtractions.

State the conclusion She must borrow $72,000.

Check the result With a $72,000 mortgage, she will have $15,000 + $72,000, which is the $87,000 that is necessary to buy the house. The answer, 72,000, checks. ■

STUDY SET Section 1.6

VOCABULARY *Fill in the blanks to make the statements true.*

1. An equation is a statement that two expressions are ______________. An equation contains an ______________ sign.
2. A ______________ of an equation is a number that satisfies the equation.
3. The answer to an equation is called a ______________ or a ______________.
4. A letter that is used to represent a number is called a ______________.
5. ______________ equations have exactly the same solutions.
6. To solve an equation, we ______________ the variable on one side of the equals sign.

CONCEPTS *In Exercises 7–8, complete the rules.*

7. If $x = y$ and c is any number, then $x + c =$ ______________.
8. If $x = y$ and c is any number, then $x - c =$ ______________.
9. In $x + 6 = 10$, what operation is performed on the variable? How do we undo that operation to isolate the variable?
10. In $9 = y - 5$, what operation is performed on the variable? How do we undo that operation to isolate the variable?

NOTATION *Complete each solution to solve the given equation.*

11. $x + 8 = 24$

$x + 8 - \_\_ = 24 - \_\_$

$x = 16$

Check: $x + 8 = 24$

$\_\_ + 8 \stackrel{?}{=} 24$

$\_\_ = 24$

So __ is a solution.

12. $x - 8 = 24$

$x - 8 + \_\_ = 24 + \_\_$

$x = 32$

Check: $x - 8 = 24$

$\_\_ - 8 \stackrel{?}{=} 24$

$\_\_ = 24$

So __ is a solution.

PRACTICE *Tell whether each statement is an equation.*

13. $x = 2$
14. $y - 3$
15. $7x < 8$
16. $7 + x = 2$
17. $x + y = 0$
18. $3 - 3y > 2$
19. $1 + 1 = 3$
20. $5 = a + 2$

For each equation, is the given number a solution?

21. $x + 2 = 3; 1$
22. $x - 2 = 4; 6$
23. $a - 7 = 0; 7$
24. $x + 4 = 4; 0$
25. $8 - y = y; 5$
26. $10 - c = c; 5$
27. $x + 32 = 0; 16$
28. $x - 1 = 0; 4$
29. $z + 7 = z; 7$
30. $n - 9 = n; 9$
31. $x = x; 0$
32. $x = 2; 0$

Use the addition or subtraction property of equality to solve each equation. Check each answer.

33. $x - 7 = 3$
34. $y - 11 = 7$
35. $a - 2 = 5$
36. $z - 3 = 9$
37. $1 = b - 2$
38. $0 = t - 1$
39. $x - 4 = 0$
40. $c - 3 = 0$
41. $y - 7 = 6$
42. $a - 2 = 4$
43. $70 = x - 5$
44. $66 = b - 6$
45. $312 = x - 428$
46. $x - 307 = 113$
47. $x - 117 = 222$
48. $y - 27 = 317$
49. $x + 9 = 12$
50. $x + 3 = 9$
51. $y + 7 = 12$
52. $c + 11 = 22$
53. $t + 19 = 28$
54. $s + 45 = 84$
55. $23 + x = 33$
56. $34 + y = 34$
57. $5 = 4 + c$
58. $41 = 23 + x$
59. $99 = r + 43$
60. $92 = r + 37$
61. $512 = x + 428$
62. $x + 307 = 513$
63. $x + 117 = 222$
64. $y + 38 = 321$
65. $3 + x = 7$
66. $b - 4 = 8$
67. $y - 5 = 7$
68. $z + 9 = 23$
69. $4 + a = 12$
70. $5 + x = 13$
71. $x - 13 = 34$
72. $x - 23 = 19$

APPLICATIONS *Complete each solution.*

73. ARCHAEOLOGY A 1,700-year-old manuscript is 425 years older than the clay jar in which it was found. How old is the jar?

Analyze the problem

- The manuscript is ________ old.
- The manuscript is ________ older than the jar.
- We are asked to find ________.

Form an equation Since we want to find the age of the jar, we can let $x =$ ________. Now we look for a key word or phrase in the problem.

Key phrase: ________
Translation: ________

We can express the age of the manuscript in two ways.

	is	425	plus	the age of the jar.
	=	425	+	

Solve the equation

$= 425 + x$

$1{,}700 - \quad = 425 + x -$

$= x$

State the conclusion ________.

Check the result If the jar is years old, then the manuscript is $1{,}275 + 425 =$ years old. The answer checks.

74. BANKING After a student wrote a \$1,500 check to pay for a car, he had a balance of \$750 in his account. How much did he have in the account before he wrote the check?

Analyze the problem

- A ________ check was written.
- The balance became ________.
- We are asked to find ________.

Form an equation Since we want to find his balance before he wrote the check, we let
$x =$ __________________________ Now we look for a key word or phrase in the problem.

Key phrase: ____________
Translation: ____________

We can express the balance now in the account in two ways.

The original balance in the account	minus	1,500	is	750.
x	$-$		$=$	750

Solve the equation

$$\quad - 1{,}500 = 750$$
$$x - 1{,}500 + \quad = 750 + $$
$$x =$$

State the conclusion ______________________

Check the result The original balance was ______. After writing a check, his balance was \$2,250 − \$1,500, or ______. The answer checks.

In Exercises 75–86, let a variable represent the unknown quantity. Then write and solve an equation to answer the question.

75. ELECTIONS Illustration 1 shows the votes received by the three major candidates running for President of the United States in 1996. Find the total number of votes cast for them.

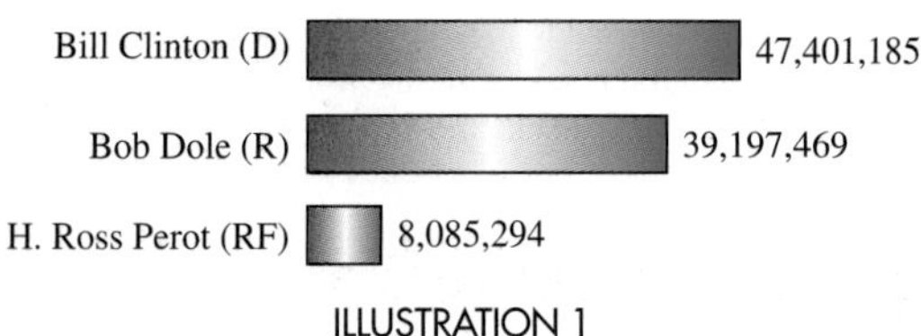

ILLUSTRATION 1

76. HIT RECORDS The oldest artist to have a number 1 single was Louis Armstrong at age 67, with *Hello Dolly.* The youngest artist to have the number 1 single was 12-year-old Jimmy Boyd, with *I Saw Mommy Kissing Santa Claus.* What is the difference in their ages?

77. PARTY INVITATIONS Three of Mia's party invitations were lost in the mail, but 59 were delivered. How many invitations did she send?

78. HEARING PROTECTION The sound intensity of a jet engine is 110 decibels. What noise level will an airplane mechanic experience if the ear plugs she is wearing reduce the sound intensity by 29 decibels?

79. FAST FOOD The franchise fee and startup costs for a Taco Bell restaurant are \$287,000. If an entrepreneur has \$68,500 to invest, how much money will she need to borrow to open her own Taco Bell restaurant?

80. BUYING GOLF CLUBS A man needs \$345 for a new set of golf clubs. How much more money does he need if he now has \$317?

81. CELEBRITY EARNINGS *Forbes* magazine estimates that in 1998, Celine Dion earned \$56 million. If this was \$69 million less than Oprah Winfrey's earnings, how much did Oprah earn in 1998?

82. HELP WANTED From the following advertisement from the classified section of a newspaper, determine the value of the benefit package. (\$45K means \$45,000.)

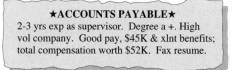

★ACCOUNTS PAYABLE★
2-3 yrs exp as supervisor. Degree a +. High vol company. Good pay, \$45K & xlnt benefits; total compensation worth \$52K. Fax resume.

83. POWER OUTAGE The electrical system in a building automatically shuts down when the meter shown in Illustration 2 reads 85. By how much must the current reading increase to cause the system to shut down?

ILLUSTRATION 2

84. VIDEO GAMES After a week of playing Sega's *Sonic Adventure,* a boy scored 11,053 points in one game—an improvement of 9,485 points over the very first time he played. What was his score for his first game?

85. AUTO REPAIR A woman paid \$29 less to have her car fixed at a muffler shop than she would have paid at a gas station. At the gas station, she would have paid \$219. How much did she pay to have her car fixed?

86. BUS RIDER A man had to wait 20 minutes for a bus today. Three days ago, he had to wait 15 minutes longer than he did today, because four buses passed by without stopping. How long did he wait three days ago?

WRITING

87. Explain what it means for a number to satisfy an equation.

88. Explain how to tell whether a number is a solution of an equation.

89. Explain what Figure 1-16 (page 52) is trying to show.

90. Explain what Figure 1-17 (page 53) is trying to show.

91. When solving equations, we *isolate* the variable. Write a sentence in which the word *isolate* is used in a different setting.

92. Think of a number. Add 8 to it. Now subtract 8 from that result. Explain why we will always obtain the original number.

REVIEW

93. Round 325,784 to the nearest ten

94. Find 1^5.

95. Evaluate: $2 \cdot 3^2 \cdot 5$.

96. Represent $4 + 4 + 4$ as a multiplication.

97. Evaluate: $8 - 2(3) + 1^3$.

98. Write 1,055 in words.

1.7 *Solving Equations by Division and Multiplication*

In this section, you will learn about

- The division property of equality • The multiplication property of equality
- Problem solving with equations

INTRODUCTION. In the previous section, we solved equations of the forms

$$x - 4 = 10 \qquad \text{and} \qquad x + 5 = 16$$

by using the addition and subtraction properties of equality. In this section, we will learn how to solve equations of the forms

$$2x = 8 \qquad \text{and} \qquad \frac{x}{3} = 25$$

by using the division and multiplication properties of equality.

The division property of equality

To solve many equations, we must divide both sides of the equation by the same nonzero number. The resulting equation will be equivalent to the original one. This idea is summed up in the division property of equality: *If equal quantities are divided by the same nonzero quantity, the results will be equal quantities.*

Division property of equality

Let a, b, and c represent numbers.

If $a = b$, then $\frac{a}{c} = \frac{b}{c}$. $\quad (c \neq 0)$

We will now consider how to solve the equation $2x = 8$. You will recall that $2x$ means $2 \cdot x$. Therefore, the given equation can be rewritten as $2 \cdot x = 8$. We can think of the scales in Figure 1-19(a) as representing the equation $2 \cdot x = 8$. The weight (in grams) on the left-hand side of the scales is $2 \cdot x$, and the weight (in grams) on the right-hand side is 8. Because these weights are equal, the scales are in balance. To find x, we need to isolate it. That can be accomplished by using the division property of equality to remove half of the weight from each side. The scales will remain in balance. From the scales shown in Figure 1-19(b), we see that x grams will be balanced by 4 grams.

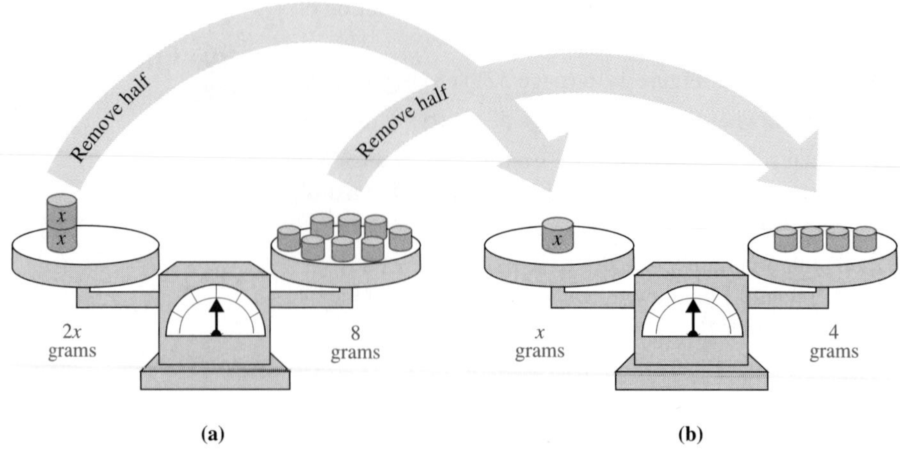

FIGURE 1-19

We now show how to solve $2x = 8$ using an algebraic approach.

EXAMPLE 1 ***Solving equations.*** Solve $2x = 8$ and check the result.

Solution

Recall that $2x = 8$ means $2 \cdot x = 8$. To isolate x on the left-hand side of the equation, we undo the multiplication by 2 by dividing both sides of the equation by 2.

$2x = 8$ The equation to solve.

$\frac{2x}{2} = \frac{8}{2}$ To undo the multiplication by 2, divide both sides by 2.

$x = 4$ When x is multiplied by 2 and that product is then divided by 2, the result is x. Do the division: $8 \div 2 = 4$.

To check this result, we substitute 4 for x in $2x = 8$.

$2x = 8$

$2 \cdot \mathbf{4} \stackrel{?}{=} 8$ Substitute 4 for x.

$8 = 8$ Do the multiplication.

Since $8 = 8$ is a true statement, 4 is a solution.

Self Check

Solve $17x = 153$ and check the result.

Answer: 9

The multiplication property of equality

We can also multiply both sides of an equation by the same nonzero number to get an equivalent equation. This idea is summed up in the multiplication property of equality: *If equal quantities are multiplied by the same nonzero quantity, the results will be equal quantities.*

Multiplication property of equality

Let a, b, and c represent numbers.

If $a = b$, then $c \cdot a = c \cdot b$ or, more simply, $ca = cb$ $(c \neq 0)$.

We can think of the scales shown in Figure 1-20(a) as representing the equation $\frac{x}{3} = 25$. The weight on the left-hand side of the scales is $\frac{x}{3}$ grams, and the weight on the right-hand side is 25 grams. Because these weights are equal, the scales are in balance. To find x, we can use the multiplication property of equality to triple (or multiply by 3) the weight on each side. The scales will remain in balance. From the scales shown in Figure 1-20(b), we can see that x grams will be balanced by 75 grams.

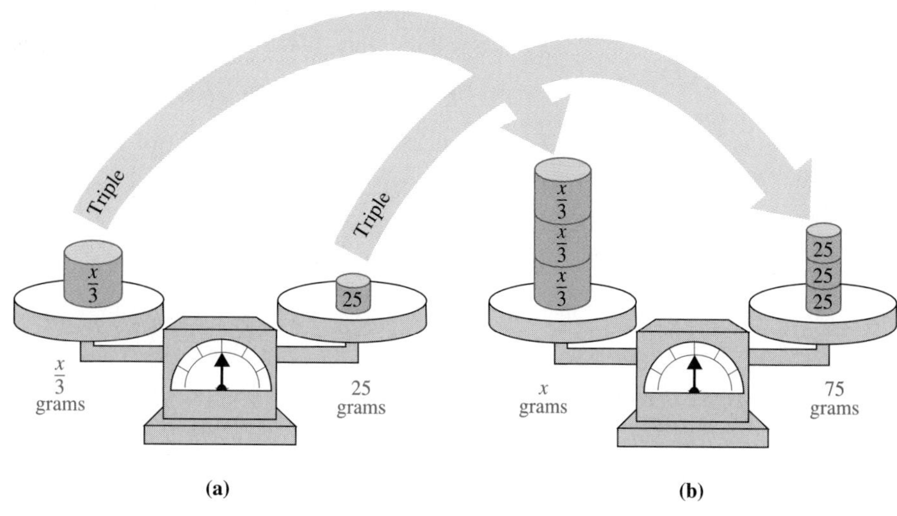

FIGURE 1-20

We now show how to solve $\frac{x}{3} = 25$ using an algebraic approach.

EXAMPLE 2 ***Solving equations.*** Solve $\frac{x}{3} = 25$ and check the result.

Solution

To isolate x on the left-hand side of the equation, we undo the division of the variable by 3 by multiplying both sides by 3.

$\frac{x}{3} = 25$ The equation to solve.

$3 \cdot \frac{x}{3} = 3 \cdot 25$ To undo the division by 3, multiply both sides by 3.

$x = 75$ When x is divided by 3 and that quotient is then multiplied by 3, the result is x. Do the multiplication: $3 \cdot 25 = 75$.

Check:

$\frac{x}{3} = 25$

$\frac{75}{3} \stackrel{?}{=} 25$ Substitute 75 for x.

$25 = 25$ Do the division: $75 \div 3 = 25$. The answer checks.

Self Check

Solve $\frac{x}{12} = 24$ and check the result.

Answer: 288

Problem solving with equations

As before, we can use equations to solve problems. Remember that the purpose of these early examples is to help you learn the strategy, even though you can probably solve the examples without it.

EXAMPLE 3 ***Buying electronics.*** The owner of an apartment complex bought six television sets that were on sale for \$499 each. What was the total cost?

Analyze the problem

- 6 television sets were bought.
- They cost $499 each.
- We are asked to find the total cost.

Form an equation We let c = the total cost of the TVs. To form an equation, we look for a key word or phrase in the problem. We can add 499 six times or multiply 499 by 6. Since it is easier, we will multiply.

Key phrase: *six TVs, each costing $499* **Translation:** *multiplication*

We can express the total cost in two ways.

The total number of TVs	multiplied by	the cost of each TV	is	the total cost.
6	$\cdot$	499	=	c

Solve the equation

$6 \cdot 499 = c$

$2{,}994 = c$ Do the multiplication.

State the conclusion The total cost will be $2,994.

Check the result We can check by estimation. Since each TV costs a little less than $500, we would expect the total cost to be a little less than 6 · $500, or $3,000. An answer of $2,994 is reasonable. ■

EXAMPLE 4 ***Splitting an inheritance.*** If seven brothers inherit $343,000 and split the money evenly, how much will each brother get?

Analyze the problem

- There are 7 brothers.
- They split $343,000 evenly.
- We are asked to find how much each brother will get.

Form an equation We can let g = the number of dollars each brother will get. To form an equation, we look for a key word or phrase in the problem.

Key phrase: *split the money evenly* **Translation:** *division*

We can express the share each brother will get in two ways.

The total amount of the inheritance	divided by	the number of brothers	is	the share each brother will get.
343,000	$\div$	7	=	g

Solve the equation

$\frac{343{,}000}{7} = g$ $343{,}000 \div 7$ can be written as $\frac{343{,}000}{7}$.

$49{,}000 = g$ Do the division.

State the conclusion Each brother will get $49,000.

Check the result If we multiply $49,000 by 7, we get $343,000. ■

EXAMPLE 5 ***Traffic violations.*** For a speeding ticket, a motorist had to pay a fine of $592. The violation occurred on a stretch of highway posted with special signs like that shown in Figure 1-21. What would the fine have been if such signs were not posted?

TRAFFIC FINES DOUBLED IN CONSTRUCTION ZONE

FIGURE 1-21

Analyze the problem

- The motorist was fined \$592.
- The fine was double what it would normally have been.
- We are asked to find what the fine would have been, had the area not been posted.

Form an equation We can let f = the amount that the fine would normally have been. To form an equation, we look for a key word or phrase in the problem or analysis.

Key word: *double* **Translation:** *multiply by 2*

We can express the amount of the new fine in two ways.

Two	times	the normal speeding fine	is	the new fine.
2	$\cdot$	f	=	592

Solve the equation

$2f = 592$ Write $2 \cdot f$ as $2f$.

$\frac{2f}{2} = \frac{592}{2}$ To undo the multiplication by 2, divide both sides by 2.

$f = 296$ Do the division: $592 \div 2 = 296$.

State the conclusion The fine would normally have been \$296.

Check the result If we double \$296, we get 2 (\$296) = \$592. The answer checks. ■

EXAMPLE 6 ***Entertainment costs.*** A five-piece band worked on New Year's Eve. If each player earned \$120, what fee did the band charge?

Analyze the problem

- There were 5 players in the band.
- Each player made \$120.
- We are asked to find the band's fee. We know that the fee divided by the number of players will give each person's share.

Form an equation We can let f = the band's fee. To form an equation, we look for a key word or phrase. In this case, we find it in the analysis of the problem.

Key phrase: *divided by* **Translation:** *division*

We can express each person's share in two ways.

The band's fee	divided by	the number in the band	is	each person's share.
f	$\div$	5	=	120

Solve the equation

$\frac{f}{5} = 120$ Write $f \div 5$ as $\frac{f}{5}$.

$5 \cdot \frac{f}{5} = 5 \cdot 120$ To undo the division by 5, multiply both sides by 5.

$f = 600$ Do the multiplication: $5 \cdot 120 = 600$.

State the conclusion The band's fee was \$600.

Check the result If we divide \$600 by 5, we get each person's share: \$120. ■

STUDY SET Section 1.7

VOCABULARY *Fill in the blanks to make the statements true.*

1. According to the ______________ property of equality, "If equal quantities are divided by the same nonzero quantity, the results will be equal quantities."
2. According to the ______________ property of equality, "If equal quantities are multiplied by the same nonzero quantity, the results will be equal quantities."

CONCEPTS *In Exercises 3–6, fill in the blanks to make the statements true.*

3. If we multiply x by 6 and then divide that product by 6, what is the result?
4. If we divide x by 8 and then multiply that quotient by 8, what is the result?
5. If $x = y$, then $\frac{x}{z} = \quad (z \neq 0)$.
6. If $x = y$, then $zx = \quad (z \neq 0)$.
7. In the equation $4t = 40$, what operation is being performed on the variable? How do we undo it?
8. In the equation $\frac{t}{15} = 1$, what operation is being performed on the variable? How do we undo it?
9. Name the first step in solving each of the following equations.
   a. $x + 5 = 10$ **b.** $x - 5 = 10$
   c. $5x = 10$ **d.** $\frac{x}{5} = 10$
10. For each of the following equations, check the given answer.
    a. $16 = t - 8$; $t = 33$
    b. $16 = t + 8$; $t = 8$
    c. $16 = 8t$; $t = 128$
    d. $16 = \frac{t}{8}$; $t = 2$

NOTATION *Complete each solution.*

11. $3x = 12$
    $\frac{3x}{3} = \frac{12}{3}$
    $x = 4$

    Check: $3x = 12$
    $3 \cdot 4 \stackrel{?}{=} 12$
    $12 = 12$
    So 4 is a solution.

12. $\frac{x}{5} = 9$
    $5 \cdot \frac{x}{5} = 45 \cdot 9$
    $x = 45$

    Check: $\frac{x}{5} = 9$
    $\frac{45}{5} \stackrel{?}{=} 9$
    $9 = 9$
    So 45 is a solution.

PRACTICE *Use the division or the multiplication property of equality to solve each equation. Check each answer.*

13. $3x = 3$
14. $5x = 5$
15. $2x = 192$
16. $4x = 120$
17. $17y = 51$
18. $19y = 76$
19. $34y = 204$
20. $18y = 90$
21. $100 = 100x$
22. $35 = 35y$
23. $16 = 8r$
24. $44 = 11m$
25. $\frac{x}{7} = 2$
26. $\frac{x}{12} = 4$
27. $\frac{y}{14} = 3$
28. $\frac{y}{13} = 5$
29. $\frac{a}{15} = 5$
30. $\frac{b}{25} = 5$
31. $\frac{c}{13} = 3$
32. $\frac{d}{100} = 11$
33. $1 = \frac{x}{50}$
34. $1 = \frac{x}{25}$
35. $7 = \frac{t}{7}$
36. $4 = \frac{m}{4}$
37. $9z = 90$
38. $3z = 6$
39. $7x = 21$
40. $13x = 52$
41. $86 = 43t$
42. $288 = 96t$
43. $21s = 21$
44. $31x = 155$
45. $\frac{d}{20} = 2$
46. $\frac{x}{16} = 4$
47. $400 = \frac{t}{3}$
48. $250 = \frac{y}{2}$

APPLICATIONS *Complete each solution.*

49. NOBEL PRIZE In 1998, three Americans, Louis Ignarro, Robert Furchgott, and Dr. Fred Murad, were awarded the Nobel Prize for Medicine. They shared the prize money. If each person received \$318,500, what was the Nobel Prize cash award?

Analyze the problem

- people shared the cash award.
- Each person received .
- We are asked to find the ______________________.

Form an equation

Since we want to find what the Nobel Prize cash award was, we let c = ______________________. To form an equation, we look for a key word or phrase in the problem.

Key phrase: ________________

Translation: ______________

We can now form the equation.

The Nobel Prize cash award	divided by	the number of recipients	was	$318,500.
	÷	3	=	$318,500

Solve the equation

$$\frac{x}{3} = 318{,}500$$

$$\cdot \frac{x}{3} = \quad \cdot 318{,}500$$

$$x =$$

State the conclusion

____

Check the result

If we divide the Nobel Prize cash award by 3, we have

$\frac{\qquad}{3} =$. This was the amount each person received. The answer checks.

50. INVESTING An investor has watched the value of his portfolio double in the last 12 months. If the current value of his portfolio is $274,552, what was its value one year ago?

Analyze the problem

- The value of the portfolio in 12 months.
- The current value is .
- We must find ______________.

Form an equation

We can let x = ______________________________. We now look for a key word or phrase in the problem.

Key phrase: ______________

Translation: ______________

We can now form the equation.

2	times	the value of the portfolio one year ago	is	the current value of the portfolio.
2	·		=	$274,552

Solve the equation

$$2x =$$

$$\frac{2x}{\quad} = \frac{274{,}552}{\quad}$$

$$x =$$

State the conclusion

____

Check the result

If the value of the portfolio one year ago was and it doubled, its current value would be . The answer checks.

In Exercises 51–60, let a variable represent the unknown quantity. Then write and solve an equation to answer the question.

51. SPEED READING An advertisement for a speed reading program claimed that successful completion of the course could triple a person's reading rate. If Alicia can currently read 130 words a minute, at what rate can she expect to read after taking the classes?

52. COST OVERRUN Lengthy delays and skyrocketing costs caused a rapid-transit construction project to go over budget by a factor of 10. The final audit showed the project costing $540 million. What was the initial cost estimate?

53. STAMPS Large sheets of commemorative stamps honoring Marilyn Monroe are to be printed. See Illustration 1. On each sheet, there are 112 stamps, with 8 stamps per row. How many rows of stamps are on a sheet?

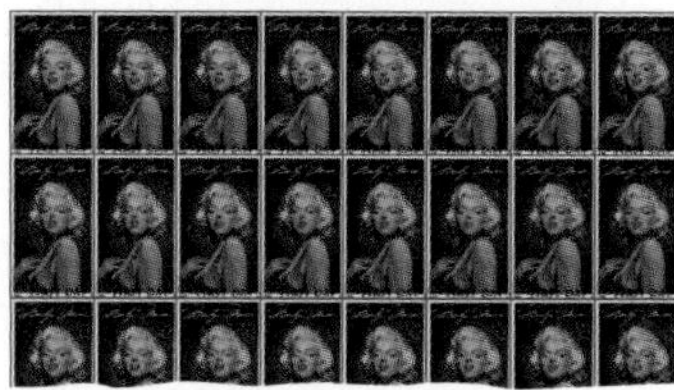

ILLUSTRATION 1

54. SPREADSHEET The grid shown in Illustration 2 is a Microsoft Excel spreadsheet. The rows are labeled with numbers, and the columns are labeled with letters. Each empty box of the grid is called a *cell.* Suppose a certain project calls for a spreadsheet with 294 cells, using columns A through F. How many rows will need to be used?

Microsoft Excel-Book 1

File Edit View Insert Format Tools

	A	B	C	D	E	F
1						
2						
3						
4						
5						
6						
7						
8						

Sheet 1 Sheet 2 Sheet 3 Sheet 4 Sheet 5

ILLUSTRATION 2

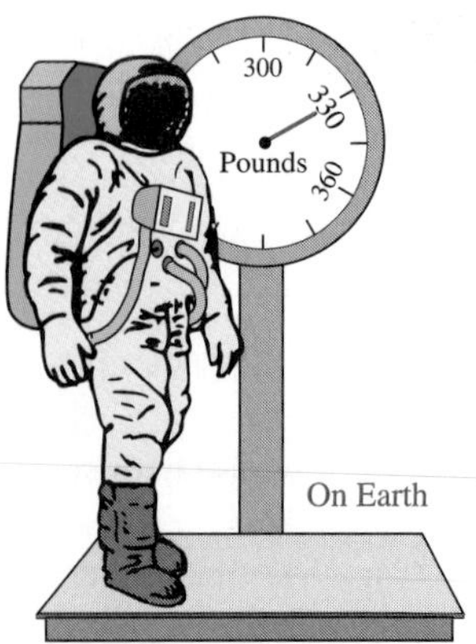

ILLUSTRATION 3

55. PHYSICAL EDUCATION A high school PE teacher had the students in her class form three-person teams for a basketball tournament. Thirty-two teams participated in the tournament. How many students were in the PE class?

56. LOTTO WINNERS The grocery store employees listed below pooled their money to buy \$120 worth of lottery tickets each week, with the understanding they would split the prize equally if they happened to win. One week they did have the winning ticket and won \$480,000. What was each employee's share of the winnings?

Sam M. Adler	Ronda Pellman	Manny Fernando
Lorrie Jenkins	Tom Sato	Sam Lin
Kiem Nguyen	H. R. Kinsella	Tejal Neeraj
Virginia Ortiz	Libby Sellez	Alicia Wen

57. ANIMAL SHELTER The number of phone calls to an animal shelter quadrupled after the evening news aired a segment explaining the services the shelter offered. Before the publicity, the shelter received 8 calls a day. How many calls did the shelter receive each day after being featured on the news?

58. OPEN HOUSE The attendance at an elementary school open house was only half of what the principal had expected. If 120 people visited the school that evening, how many had she expected to attend?

59. GRAVITY The weight of an object on Earth is 6 times greater than what it is on the moon. The situation shown in Illustration 3 took place on the Earth. If it took place on the moon, what weight would the scale register?

60. INFOMERCIAL The number of orders received each week by a company selling skin care products increased fivefold after a Hollywood celebrity was added to the company's infomercial. After adding the celebrity, the company received about 175 orders each week. How many orders were received each week before the celebrity took part?

WRITING

61. Explain what Figure 1-19 (page 60) is trying to show.

62. Explain what Figure 1-20 (page 61) is trying to show.

63. What does it mean to solve an equation?

64. Think of a number. Double it. Now divide it by 2. Explain why you always obtain the original number.

REVIEW

65. Find the perimeter of a rectangle with sides measuring 8 cm and 16 cm.

66. Find the area of a rectangle with sides measuring 23 inches and 37 inches.

67. Find the prime factorization of 120.

68. Find the prime factorization of 150.

69. Evaluate $3^2 \cdot 2^3$.

70. Evaluate $5 + 6 \cdot 3$.

71. FUEL ECONOMY Five basic models of automobiles made by Saturn have city mileage ratings of 24, 22, 28, 29, and 27 miles per gallon. What is the average (mean) city mileage for the five models?

72. Solve the equation $x - 4 = 20$.

KEY CONCEPT

Variables

One of the major objectives of this course is for you to become comfortable working with **variables.** You will recall that a variable is a letter that stands for a number.

The application problems of Sections 1.6 and 1.7 were solved with the help of a variable. In these problems, we let the variable represent an unknown quantity such as the number of customers a hairdresser used to have, the age of a jar, and the cash award given a Nobel Prize winner. We then wrote an equation to describe the situation mathematically and solved the equation to find the value represented by the variable.

In Exercises 1–6, suppose that you are going to solve the following problems. What quantity should be represented by a variable? State your response in the form "Let $x = \ldots .$"

1. The monthly cost to lease a van is $120 less than to buy it. To buy it, the monthly payments are $290. How much does it cost to lease the van each month?

2. One piece of pipe is 10 feet longer than another. Together, their lengths total 24 feet. How long is the shorter piece of pipe?

3. The length of a rectangular field is 50 feet. What is its width if it has a perimeter of 200 feet?

4. If one hose can fill a vat in 2 hours and another can fill it in 3 hours, how long will it take to fill the vat if both hoses are used?

5. Find the distance traveled by a motorist in three hours if her average speed was 55 miles per hour.

6. In what year was a couple married if their 50th anniversary was in 1988?

Variables can also be used to state properties of mathematics in a concise, "shorthand" notation. In Exercises 7–14, state each property using mathematical symbols and the given variable(s).

7. Use the variables a and b to state that two numbers can be added in either order to get the same sum.

8. Use the variable x to state that when 0 is subtracted from a number, the result is the same number.

9. Use the variable b to state that the result when dividing a number by 1 is the same number.

10. Use the variable x to show that the sum of a number and 1 is greater than the number.

11. Using the variable n, state the fact that when 1 is subtracted from any number, the difference is less than the number.

12. State the fact that the product of any number and 0 is 0, using the variable a.

13. Use the variables r, s, and t to state that the way we group three numbers when adding them does not affect the answer.

14. Using the variable n, state the fact that when a number is multiplied by 1, the result is the number.

ACCENT ON TEAMWORK

Section 1.1

PLACE VALUE Have each student in your group bring a calculator to class so that you can examine several different models. For each model, determine the largest number (if there is one) that can be entered on the display of the calculator. Then press the appropriate calculator keys to add 1 to that number. What does the display show?

LARGE NUMBERS Bill Gates, founder of Microsoft Corporation, is said to be a billionaire. How many millions make one billion?

Section 1.2

READING THE PROBLEM CAREFULLY In reading Example 9 of Section 1.2, you will notice that it contains several facts that are not used in the solution of the problem. Have each person in your group write a similar problem that requires careful reading to extract the useful information. Then have each person share his or her problem with the other students in the group.

Section 1.3

DIVISIBILITY TESTS Certain tests can help us decide whether one whole number is divisible by another.

- A number is divisible by 2 if the last digit of the number is 0, 2, 4, 6, or 8.
- A number is divisible by 3 if the sum of the digits is divisible by 3.
- A number is divisible by 4 if the number formed by the last two digits is divisible by 4.
- A number is divisible by 5 if the last digit of the number is 0 or 5.
- A number is divisible by 6 if the last digit of the number is 0, 2, 4, 6, or 8 and the sum of the digits is divisible by 3.
- A number is divisible by 8 if the number formed by the last three digits is divisible by 8.
- A number is divisible by 9 if the sum of the digits is divisible by 9.
- A number is divisible by 10 if the last digit of the number is 0.
- Determine whether each number is divisible by 2, 3, 4, 5, 6, 8, 9, and/or 10.

a. 660 **b.** 2,526
c. 11,523 **d.** 79,503
e. 135,405 **f.** 4,444,440

Section 1.4

COMMON FACTORS The prime factorizations of 36 and 126 are shown below. The prime factors that are common to 36 and 126 (highlighted in color) are 2, 3, and 3.

$$36 = 2 \cdot 2 \cdot 3 \cdot 3$$
$$126 = 2 \cdot 3 \cdot 3 \cdot 7$$

Find the common prime factors for each of the following pairs of numbers.

a. 25, 45 **b.** 24, 60
c. 18, 45 **d.** 40, 112
e. 180, 210 **f.** 242, 198

Section 1.5

ORDER OF OPERATIONS Consider the expression

$$5 + 8 \cdot 2^3 - 3 \cdot 2$$

Insert a set of parentheses somewhere in the expression so that, when it is evaluated, you obtain

a. 63 **b.** 132
c. 21 **d.** 127

Section 1.6

SOLVING EQUATIONS Borrow a scale and some weights from the chemistry department. Use them as part of a class presentation to explain how the subtraction property of equality is used to solve the equation $x + 2 = 5$. See the discussion and Figure 1-16 on page 52 for some suggestions on how to do this.

Section 1.7

FORMING AN EQUATION Reread Example 4 in Section 1.7. This problem could have been solved by forming an equation involving the operation of multiplication instead of the operation of division.

The number of brothers	times	the share each brother will get	is	the total amount of the inheritance.
7	$\cdot$	g	=	343,000

For Examples 5 and 6 in Section 1.7, write another equation that could be used to solve the problem. Then solve the equation and state the result.

CHAPTER REVIEW

SECTION 1.1 An Introduction to the Whole Numbers

CONCEPTS

A *set* is a collection of objects. The set of *natural numbers* is {1, 2, 3, 4, 5, . . .} The set of *whole numbers* is {0, 1, 2, 3, 4, 5, . . .}

REVIEW EXERCISES

1. Graph each set.

a. The natural numbers less than 5

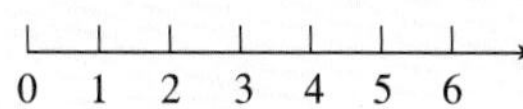

b. The whole numbers between 0 and 3

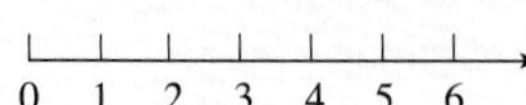

Whole numbers are often used in tables, bar graphs, and line graphs.

2. FARMING The table below shows the size of the average U.S. farm (in acres) for the period 1940–2000, in 20-year increments.

Year	1940	1960	1980	2000
Average size (acres)	174	297	426	432

a. Construct a bar graph of the data.

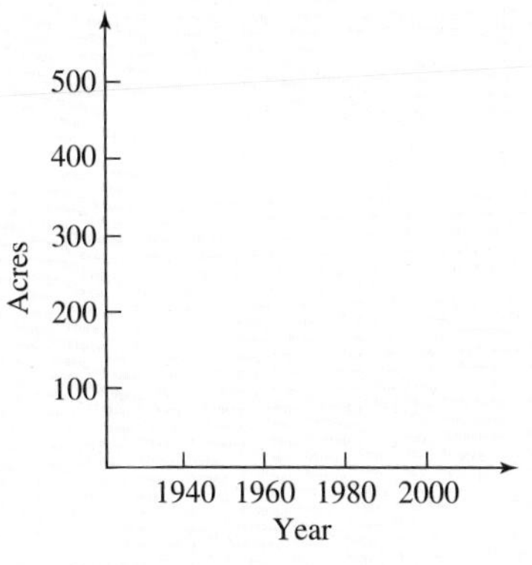

b. Construct a line graph of the data.

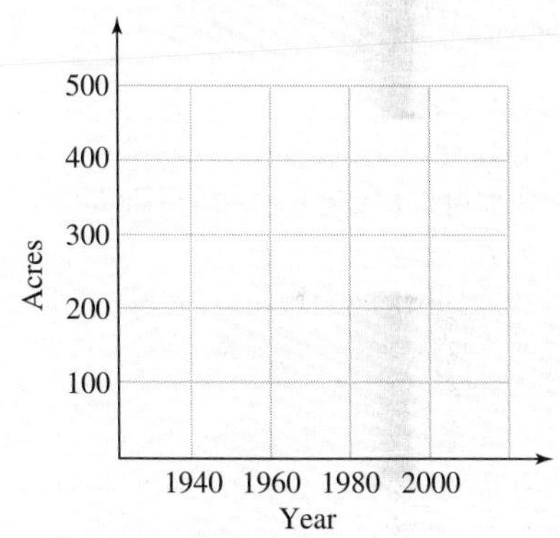

The digits in a whole number have *place value.*

3. Consider the number 2,365,720. Which digit is in the

a. ten thousands column?

b. hundreds column?

A whole number is written in *expanded notation* when its digits are written with their place values.

4. Write each number in expanded notation.

a. 570,302

b. 37,309,054

We use the digits 0, 1, 2, 3, 4, 5, 6, 7, 8, and 9 to write a number in *standard notation.*

5. Write each number in standard notation.

a. 3 thousands + 2 hundreds + 7 ones

b. twenty-three million, two hundred fifty-three thousand, four hundred twelve

c. sixteen billion

The symbol $<$ means "is less than." The symbol $>$ means "is greater than."

6. Place an $<$ or $>$ symbol in the box to make a true statement.

a. 9 ☐ 7

b. 3 ☐ 5

To give approximate answers, we often use *rounded numbers.*

7. Round 2,507,348 to the specified place.

a. nearest hundred
b. nearest ten thousand
c. nearest ten
d. nearest hundred thousand

SECTION 1.2 Adding and Subtracting Whole Numbers

Addition is the process of finding the total of two (or more) numbers. Do additions within parentheses first.

8. Find each sum.

a. $56 + 22$
b. $137 + 0$
c. $15 + (27 + 13)$
d. $82 + 17 + 50$
e. $(111 + 222) + 444$
f. $0 + 2,332$

The *commutative* and *associative properties of addition:*

$a + b = b + a$

$(a + b) + c = a + (b + c)$

9. Do each addition.

a. $\begin{array}{r} 236 \\ +282 \\ \hline \end{array}$
b. $\begin{array}{r} 5,345 \\ +\ 655 \\ \hline \end{array}$
c. $135 + 213 + 615 + 47$
d. $4,447 + 7,478 + 13,061$

10. What property of addition is shown?

a. $12 + 8 = 8 + 12$
b. $12 + (8 + 2) = (12 + 8) + 2$

The *perimeter* of a rectangle or a square is the distance around it.

11. Find the perimeter of a square with sides 24 inches long.

Subtraction is the process of finding the difference between two numbers.

12. Do each subtraction.

a. $18 - 5$
b. $9 - (7 - 2)$
c. $22 - 5 - 6$
d. $5,231 - 5,177$
e. $\begin{array}{r} 343 \\ -269 \\ \hline \end{array}$
f. $\begin{array}{r} 17,800 \\ -15,725 \\ \hline \end{array}$

13. Give the addition or subtraction fact that is illustrated by each figure.

a.

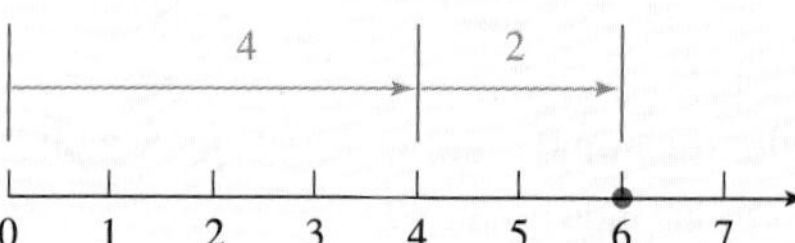

b.

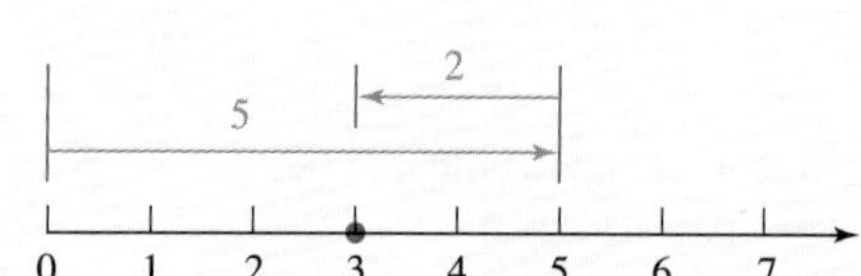

14. TRAVEL A direct flight from Omaha to San Francisco costs \$237. Another flight with one stop in Reno costs \$192. How much can be saved by taking the less expensive flight?

15. SAVINGS ACCOUNTS A savings account contains \$931. If the owner deposits \$271 and makes withdrawals of \$37 and \$380, find the final balance.

16. REBATE The price of a new Chevrolet Camaro was advertised in a newspaper as \$21,991*. A note at the bottom of the ad read, "*Reflects \$1,550 factory rebate." What was the car's original sticker price?

SECTION 1.3 Multiplying and Dividing Whole Numbers

Multiplication is repeated addition. For example,

$$\overbrace{4 \cdot 6 = 6 + 6 + 6 + 6}^{\text{The sum of four 6's}}$$
$$= 24$$

The result, 24, is called the *product,* and the 4 and 6 are called *factors.*

The *commutative* and *associative properties of multiplication.*

$a \cdot b = b \cdot a$

$(a \cdot b) \cdot c = a \cdot (b \cdot c)$

The *area A of a rectangle* is the product of its length l and its width w.

$A = l \cdot w$

Division is an operation that determines how many times a number (the *divisor*) is contained in another number (the *dividend*). *Remember that you can never divide by 0.*

17. Do each multiplication.

a. $8 \cdot 7$ **b.** $7(8)$

c. $8 \cdot 0$ **d.** $7 \cdot 1$

e. $10 \cdot 8 \cdot 7$ **f.** $5 \cdot (7 \cdot 6)$

18. Do each multiplication.

a. $157 \cdot 21$ **b.** $3{,}723 \cdot 48$

c. 356×89 **d.** $5{,}624 \times 81$

19. What property of multiplication is shown?

a. $12 \cdot (8 \cdot 2) = (12 \cdot 8) \cdot 2$

b. $12 \cdot 8 = 8 \cdot 12$

20. WAGES If a math tutor worked for 38 hours and was paid $9 per hour, how much did she earn?

21. HORSESHOES Find the perimeter and the area of the rectangular horseshoe court shown in Illustration 1.

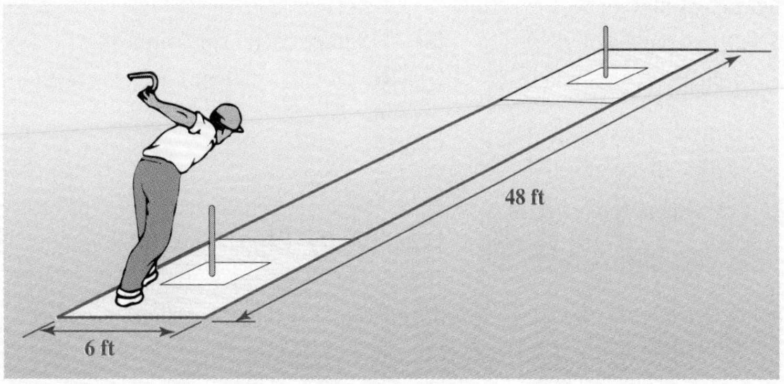

ILLUSTRATION 1

22. PACKAGING There are 12 eggs in one dozen, and 12 dozen in one gross. How many eggs are in a shipment of 5 gross?

23. Do each division, if possible.

a. $\frac{6}{3}$ **b.** $\frac{15}{1}$

c. $73 \div 0$ **d.** $\frac{0}{8}$

e. $357 \div 17$ **f.** $1{,}443 \div 39$

g. $21\overline{)405}$ **h.** $54\overline{)1{,}269}$

24. TREATS If 745 candies are divided equally among 45 children, how many will each child receive? How many candies will be left over?

25. COPIES An elementary school teacher had copies of a 3-page social studies test made at Quick Copy Center. She was charged for 84 sheets of paper. How many copies of the test were made?

SECTION 1.4 Prime Factors and Exponents

Numbers that are multiplied together are called *factors.*

A *prime number* is a whole number greater than 1 that has only 1 and itself as factors. Whole numbers greater than 1 that are not prime are called *composite numbers.*

Whole numbers divisible by 2 are *even* numbers. Whole numbers not divisible by 2 are *odd* numbers.

The *prime factorization* of a whole number is the product of its prime factors.

An *exponent* is used to indicate repeated multiplication. In the *exponential expression* a^n, a is the *base,* and n is the exponent.

26. Find all of the factors of each number.
a. 18 **b.** 25

27. Identify each number as a prime, composite, or neither.
a. 31 **b.** 100
c. 1 **d.** 0
e. 125 **f.** 47

28. Identify each number as an even or odd number.
a. 171 **b.** 214
c. 0 **d.** 1

29. Find the prime factorization of each number.
a. 42 **b.** 375

30. Write each expression using exponents.
a. $6 \cdot 6 \cdot 6 \cdot 6$ **b.** $5 \cdot 5 \cdot 5 \cdot 13 \cdot 13$

31. Evaluate each expression.
a. 5^3 **b.** 11^2
c. $2^3 \cdot 5^2$ **d.** $2^2 \cdot 3^3 \cdot 5^2$

SECTION 1.5 Order of Operations

Do mathematical operations in the following order:

1. Do all calculations within parentheses and other grouping symbols.
2. Evaluate all exponential expressions.
3. Do all multiplications and divisions in order from left to right.
4. Do all additions and subtractions in order from left to right.

To evaluate an expression containing grouping symbols, do all calculations within each pair of grouping symbols, working from the innermost pair to the outermost pair.

The *arithmetic mean* (average) is a value around which numbers are grouped.

32. Evaluate each expression.
a. $13 + 12 \cdot 3$ **b.** $35 - 15 \div 5$
c. $(13 + 12)3$ **d.** $(8 - 2)^2$
e. $8 \cdot 5 - 4 \div 2$ **f.** $8 \cdot (5 - 4 \div 2)$
g. $2 + 3(10 - 4 \cdot 2)$ **h.** $4(20 - 5 \cdot 3 + 2) - 4$
i. $3^3\left(\frac{12}{6}\right) - 1^4$ **j.** $\frac{12 + 3 \cdot 7}{5^2 - 14}$
k. $7 + 3[10 - 3(4 - 2)]$ **l.** $5 + 2[(15 - 3 \cdot 4) - 2]$

33. DICE GAME Write an expression which finds the total of all the dice shown in Illustration 2. Then evaluate the expression.

34. YAHTZEE See Illustration 3. Find the player's average (mean) score for the 6 games.

ILLUSTRATION 2

yahtzee SCORE CARD

Game #1	Game #2	Game #3	Game #4	Game #5	Game #6
159	244	184	240	166	213

ILLUSTRATION 3

SECTION 1.6 Solving Equations by Addition and Subtraction

An *equation* is a statement that two expressions are equal. A *variable* is a letter that stands for a number.

Two equations with exactly the same solutions are called *equivalent equations.*

To solve an equation, isolate the variable on one side of the equation by undoing the operation performed on it.

If the same number is added to (or subtracted from) both sides of an equation, an equivalent equation results:

If $a = b$, then $a + c = b + c$.
If $a = b$, then $a - c = b - c$.

Problem-solving strategy:

1. Analyze the problem.
2. Form an equation.
3. Solve the equation.
4. State the conclusion.
5. Check the result.

35. Tell whether the given number is a solution of the equation. Explain why or why not.

a. $x + 2 = 13; x = 5$
b. $x - 3 = 1; x = 4$

36. Identify the variable in each equation.

a. $y - 12 = 50$
b. $114 = 4 - t$

37. Solve the equation and check the result.

a. $x - 7 = 2$
b. $x - 11 = 20$
c. $225 = y - 115$
d. $101 = p - 32$
e. $x + 9 = 18$
f. $b + 12 = 26$
g. $175 = p + 55$
h. $212 = m + 207$
i. $x - 7 = 0$
j. $x + 15 = 1{,}000$

In Exercises 38–39, let a variable represent the unknown quantity. Then write and solve an equation to answer the question.

38. FINANCING A newly married couple made a $25,500 down payment on a $122,750 house. How much did they need to borrow?

39. DOCTOR'S PATIENTS After moving his office, a doctor lost 13 patients. If he had 172 patients left, how many did he have originally?

SECTION 1.7 Solving Equations by Division and Multiplication

If both sides of an equation are divided by (or multiplied by) the same nonzero number, an equivalent equation results:

If $a = b$, then $\frac{a}{c} = \frac{b}{c}$ $(c \neq 0)$.
If $a = b$, then $a \cdot c = b \cdot c$ $(c \neq 0)$.

40. Solve the equation and check the result.

a. $3x = 12$
b. $15y = 45$
c. $105 = 5r$
d. $224 = 16q$
e. $\frac{x}{7} = 3$
f. $\frac{a}{3} = 12$
g. $15 = \frac{s}{21}$
h. $25 = \frac{d}{17}$
i. $12x = 12$
j. $\frac{x}{12} = 12$

In Exercises 41–42, let a variable represent the unknown quantity. Then write and solve an equation to answer the question.

41. CARPENTRY If you cut a 72-inch board into three equal pieces, how long will each piece be? Disregard any loss due to cutting.

42. JEWELRY Four sisters split the cost of a gold chain evenly. How much did the chain cost if each sister's share was $32?

TAKING A MATH TEST

The best way to relieve anxiety about taking a mathematics test is to know that you are well-prepared for it and that you have a plan. Before any test, ask yourself three questions. When? What? How?

When will I study?

1. When is the test?
2. When will I begin to review for the test?
3. What are the dates and times that I will reserve for studying for the test?

What will I study?

1. What sections will the test cover?
2. Has the instructor indicated any types of problems that are guaranteed to be on the test?

How will I prepare for the test?

Put a check mark by each method you will use to prepare for the test.

- ☐ Review the class notes.
- ☐ Outline the chapter(s) to see how the topics relate to one another.
- ☐ Recite the important formulas, definitions, vocabulary, and rules into a tape recorder.
- ☐ Make flash cards for the important formulas, definitions, vocabulary, and rules.
- ☐ Rework problems from the homework assignments.
- ☐ Rework each of the Self Check problems in the text.
- ☐ Form a study group to discuss and practice the topics to be tested.
- ☐ Complete the appropriate Chapter Review(s) and the Chapter Test(s).
- ☐ Review the Comments given in the text.
- ☐ Work on improving my speed in answering questions.
- ☐ Review the methods that can be used to check my answers.
- ☐ Write a sample test, trying to think of the questions the instructor will ask.
- ☐ Complete the appropriate Cumulative Review Exercises.
- ☐ Get organized the night before the test. Have materials ready to go so that the trip to school will not be hurried.
- ☐ Take some time to relax immediately before the test. Don't study right up to the last minute.

Taking the test

Here are some tips that can help improve your performance on a mathematics test.

- Write down any formulas or rules as soon as you receive the test.
- When you receive the test, scan it, looking for the types of problems you had expected to see. Do them first.
- Read the instructions carefully.
- Don't spend too much time on any one problem until you have attempted all the problems.
- If your instructor gives partial credit, at least try to begin a solution.
- Don't be afraid to skip a problem and come back to it later.
- Save the most difficult problems for last.
- If you finish early, go back over your work and look for mistakes.

Chapter 1 Test

1. Graph the whole numbers less than 5.

0 1 2 3 4 5 6 7

2. Write "five thousand two hundred sixty-six" in expanded notation.

3. Write "7 thousands + 5 hundreds + 7 ones" in standard notation.

4. Round 34,752,341 to the nearest million.

In Problems 5–6, refer to the data in the table.

Lot number	1	2	3	4
Defective bolts	7	10	5	15

5. Use the data to make a bar graph.

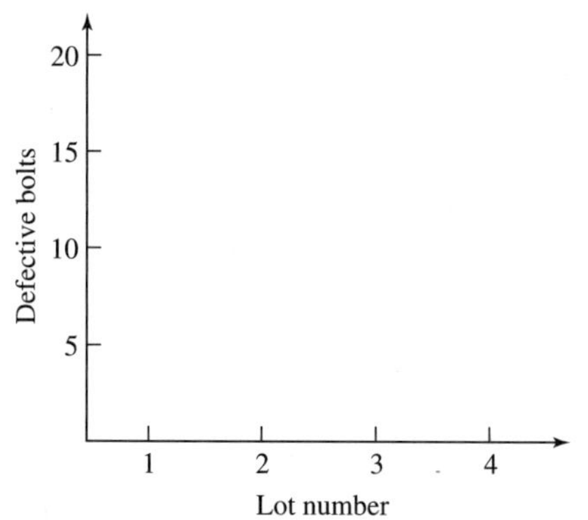

6. Use the data to make a line graph.

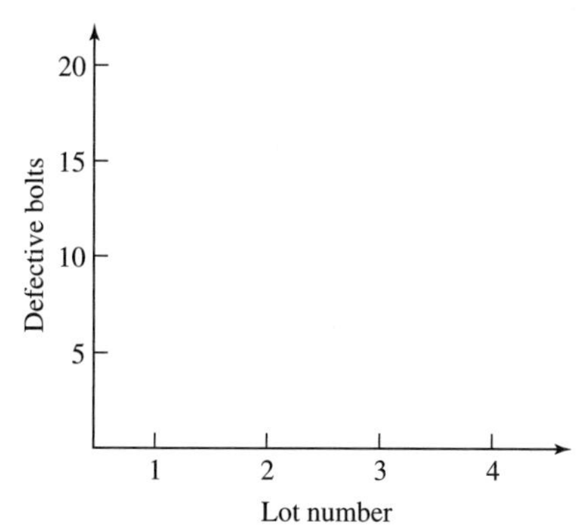

Place an $<$ or $>$ symbol in the box to make a true statement.

7. 15 10

8. 12 17

9. Add 327 + 435 + 123 + 606.

10. Subtract 287 from 535.

11. Add: $\begin{array}{r} 44{,}526 \\ +\underline{13{,}579} \end{array}$

12. Subtract: $\begin{array}{r} 4{,}521 \\ -\underline{3{,}579} \end{array}$

13. STOCKS On Tuesday, a share of KBJ Company was selling at \$73. The price rose \$12 on Wednesday and fell \$9 on Thursday. Find its price on Thursday.

14. List the factors of 20 in order, from least to greatest.

Do each operation.

15. Multiply: $\begin{array}{r} 53 \\ \times\underline{\ 8} \end{array}$

16. Multiply: 367(73).

17. Divide: $63\overline{)4{,}536}$

18. Divide: $73\overline{)8{,}379}$

19. FURNITURE SALE See the advertisement in Illustration 1. Find the perimeter of the rectangular space under the tent. Then fill in the blank in the advertisement.

The Greatest Parking Lot Tent
Sale in Our History!
? square feet
in our "outdoor showroom"
2 DAYS ONLY!
Thomastown Home
Furnishings
105 ft
75 ft

ILLUSTRATION 1

20. If 3,451 students are placed in groups of 74, how many will be left over?

21. COLLECTIBLES There are 12 baseball cards in every pack. There are 24 packs in every box. There are 12 boxes in every case. How many cards are in a case?

22. Find the prime factorization of 252.

23. Evaluate $9 + 4 \cdot 5$.

24. Evaluate $\frac{3 \cdot 4^2 - 2^2}{(2 - 1)^3}$.

25. Evaluate $10 + 2[12 - 2(6 - 4)]$.

26. GRADES A student scored 73, 52, and 70 on three exams and received 0 on two missed exams. Find his average (mean) score.

27. Is 3 a solution of the equation $x + 13 = 16$? Explain why or why not.

Solve each equation. Check the result.

28. $100 = x + 1$

29. $y - 12 = 18$

30. $5t = 55$

31. $\frac{q}{3} = 27$

In Problems 32–33, let a variable represent the unknown quantity. Then write and solve an equation to answer the question.

32. PARKING After many student complaints, a college decided to commit funds to double the number of parking spaces on campus. This increase would bring the total number of spaces up to 6,200. How many parking spaces does the college have at this time?

33. LIBRARY A library building is 6 years shy of its 200th birthday. How old is the building at this time?

34. Explain what it means to *solve* an equation.

The Integers

2

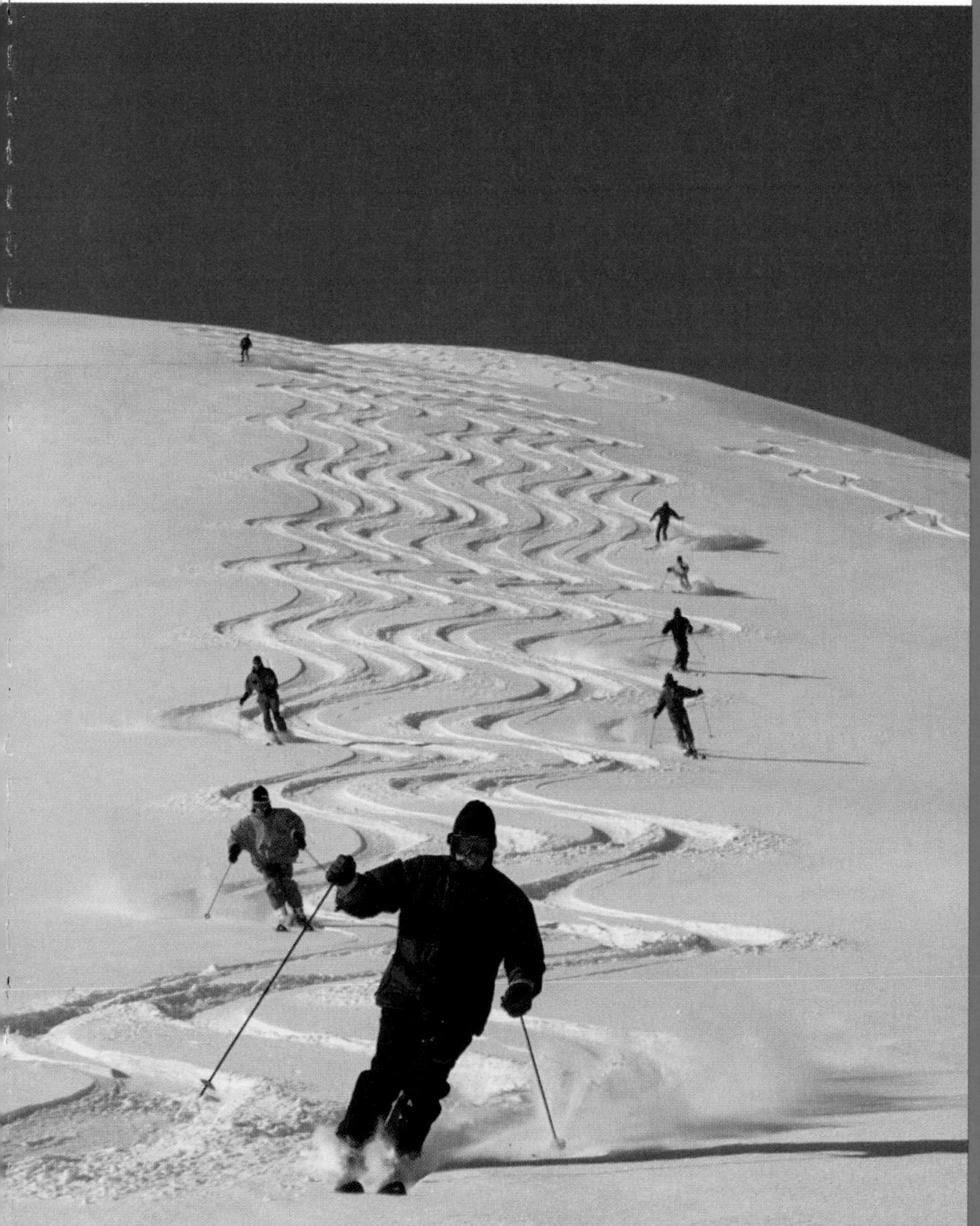

IN THIS CHAPTER, THE CONCEPT OF A NEGATIVE NUMBER IS INTRODUCED, AS WE EXPLORE AN EXTENSION OF THE SET OF WHOLE NUMBERS CALLED THE INTEGERS.

2.1 An Introduction to the Integers

In this section, you will learn about

- The integers • Extending the number line • Inequality symbols
- Absolute value • The opposite of a number • The − symbol

INTRODUCTION. Whole numbers are not adequate to describe many situations that arise in everyday life.

The record cold temperature in the state of Florida was 2 degrees *below* zero on February 13, 1899, in Tallahassee.

RECORD ALL CHARGES OR CREDITS THAT AFFECT YOUR ACCOUNT

NUMBER	DATE	DESCRIPTION OF TRANSACTION	PAYMENT/DEBIT (−)	√ T	FEE (IF ANY) (−)	DEPOSIT/CREDIT (+)	BALANCE
							$ 450 00
1207	5/2	Wood's Auto Repair Transmission	$ 500 00		$	$	

A check for $500 was written when there was only $450 in the account. The checking account is *overdrawn.*

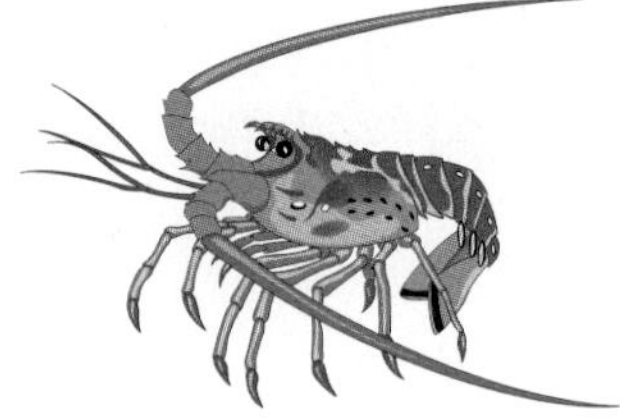

The American lobster is found off the East Coast of North America at depths as much as 600 feet *below* sea level.

In this section, we will see how negative numbers called *integers* (read as "int-i-jers") can be used to describe these three situations, as well as many others.

The integers

To describe a temperature of 2 degrees below zero, or $50 overdrawn, or 600 feet below sea level, we need to use negative numbers. **Negative numbers** are numbers less than 0, and they are written using a **negative sign, −.**

In words	In symbols	Read as
2 degrees below zero	-2	"negative two"
$50 overdrawn	-50	"negative fifty"
600 feet below sea level	-600	"negative six hundred"

Positive and negative numbers

Positive numbers are greater than 0. **Negative numbers** are less than 0.

COMMENT Zero is neither positive nor negative.

Positive and negative numbers are often referred to as **signed numbers.** Negative numbers must always be written with a negative sign. However, positive numbers are not always written with a **positive sign,** $+$. For example, the elevation of Mexico City, which is 7,110 feet above sea level, can be written as $+7{,}110$ feet or as 7,110 feet.

The collection of all positive whole numbers, negative whole numbers, and 0 is called the set of **integers.**

The set of integers $\{\ldots, -5, -4, -3, -2, -1, 0, 1, 2, 3, 4, 5, \ldots\}$

Since every natural number is also an integer, the natural numbers form a subset of the integers. Likewise, the whole numbers form a subset of the integers.

COMMENT Note that not all integers are natural numbers, and that not all integers are whole numbers. For example, -2 is an integer, but it is neither a natural number nor a whole number.

Extending the number line

An excellent way to learn about negative numbers is with the help of a number line. Negative numbers can be represented on a number line by extending the line to the left. Beginning with zero, we move left, marking equally spaced points and then labeling them with progressively smaller negative whole numbers. (See Figure 2-1.) As you move to the right on a number line, the values of the numbers increase. As you move to the left, the values decrease.

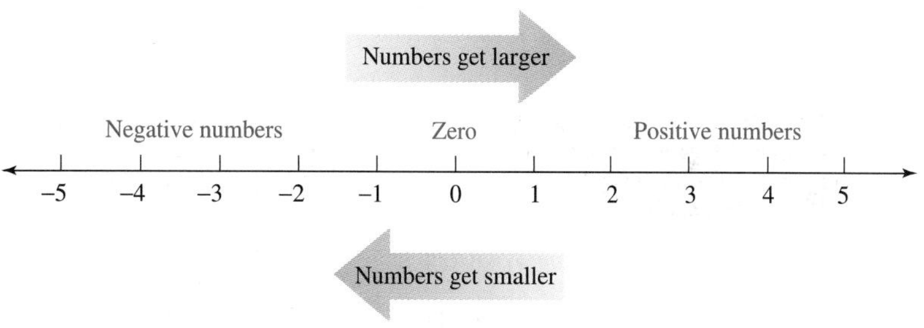

FIGURE 2-1

A thermometer is an example of a number line. The thermometer on the left is scaled in degrees, and it shows a temperature of $-10°$. In the study set, you will see examples of number lines illustrating historical and scientific situations that involve negative numbers.

EXAMPLE 1 ***Graphing.*** Graph the integers 2, -3, 4, and -1.

Solution

To graph each integer, we locate its position on the number line and draw a heavy dot.

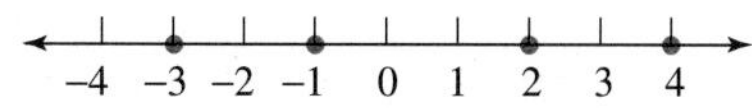

Self Check

Graph the integers 3, -4, 1, and -2.

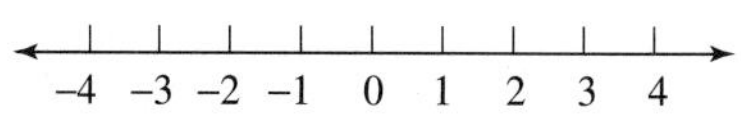

Answer:

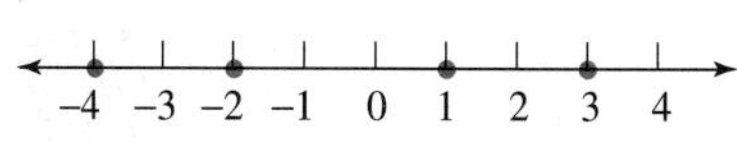

By extending the number line to include negative numbers, we can represent more situations using bar graphs and line graphs. For example, the bar graph shown in Figure 2-2 illustrates the annual profits *and losses* of Toys R Us over a five-year period. Note that the profit in 1998 was \$490 million, and that the loss in 1999 was \$132 million.

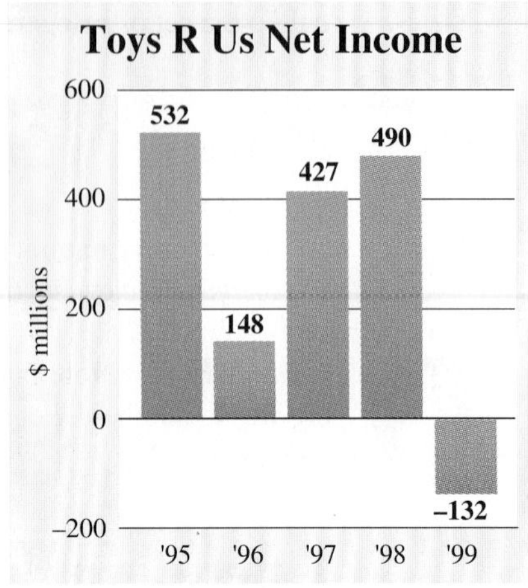

FIGURE 2-2

Inequality symbols

Figure 2-3 shows the graph of the integers −2 and 1. Because −2 is to the left of 1 on the number line, −2 is less than 1. We can use an inequality symbol to state this fact: $-2 < 1$. Since $-2 < 1$, it is also true that $1 > -2$.

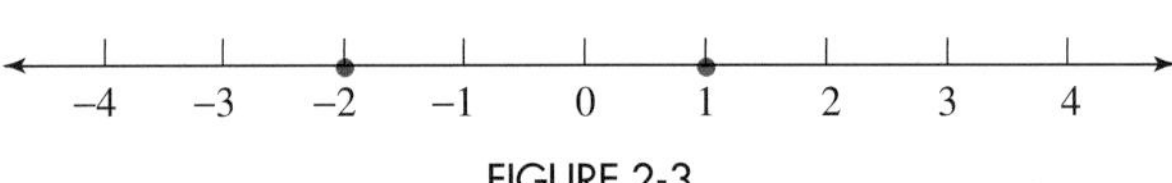

FIGURE 2-3

EXAMPLE 2 ***Inequality symbols.*** Use one of the symbols > or < to make each statement true: **a.** 4 ☐ −5 and **b.** −4 ☐ −2.

Solution

a. Since 4 is to the right of −5 on the number line, $4 > -5$.

b. Since −4 is to the left of −2 on the number line, $-4 < -2$.

Self Check

Use one of the symbols > or < to make each statement true.

a. 6 ☐ −6

b. −6 ☐ −5

Answers: **a.** >, **b.** <

Absolute value

Using a number line, we can see that the numbers 3 and −3 are both a distance of 3 units away from 0, as shown in Figure 2-4.

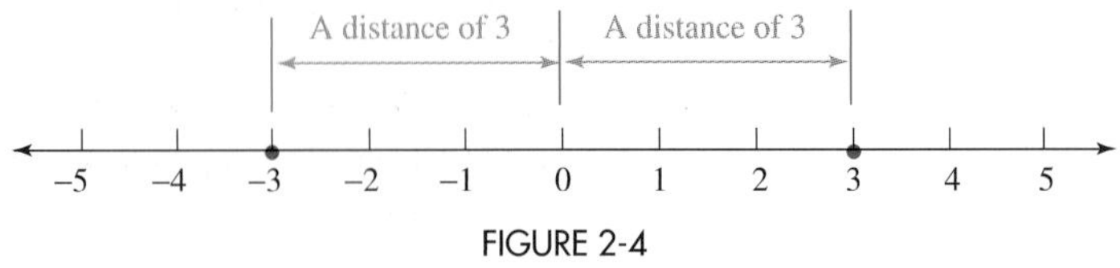

FIGURE 2-4

The **absolute value** of a number gives the distance between the number and 0 on a number line. To indicate absolute value, the number is inserted between two vertical bars. For the previous example, we would write $|-3| = 3$. This is read as "The absolute value of negative 3 is 3," and it tells us that the distance between -3 and 0 is 3 units. In the example, we also see that $|3| = 3$.

Absolute value

> The **absolute value** of a number is the distance on a number line between the number and 0.

COMMENT Absolute value expresses distance. The absolute value of a number is always positive or zero, but never negative!

EXAMPLE 3 ***Absolute value.*** Evaluate each absolute value: **a.** $|8|$, **b.** $|-5|$, and **c.** $|0|$.

Solution

a. On a number line, the distance between 8 and 0 is 8. Therefore,

$$|8| = 8$$

b. On a number line, the distance between -5 and 0 is 5. Therefore,

$$|-5| = 5$$

c. On a number line, the distance between 0 and 0 is 0. Therefore,

$$|0| = 0$$

Self Check

Evaluate each absolute value:

a. $|-9|$

b. $|4|$

Answers: **a.** 9, **b.** 4

The opposite of a number

Opposites or negatives

> Two numbers represented by points on a number line that are the same distance from 0, but on opposite sides of it, are called **opposites** or **negatives.**

The numbers 4 and -4 are opposites because they are the same distance from zero. (See Figure 2-5.)

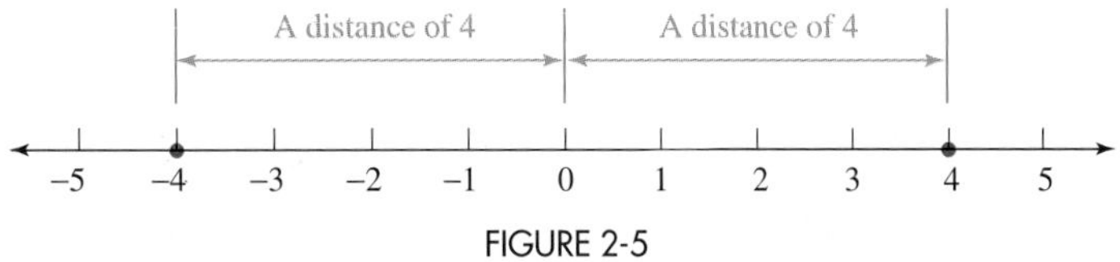

FIGURE 2-5

To write the opposite of a number, a $-$ symbol is used. For example, the opposite of 5 is -5 (read as "negative 5"). Parentheses are needed to express the opposite of a negative number. The opposite of -5 is written as $-(-5)$. Since 5 and -5 are the same distance from zero, the opposite of -5 is 5. Therefore, $-(-5) = 5$. This leads to the following conclusion.

Opposite of an opposite

> If a is any number, then $-(-a) = a$.

In words, this rule says that *the opposite of the opposite of a number is that number.*

Number	Opposite	
57	-57	Read as "negative fifty-seven."
-8	$-(-8) = 8$	Read as "the opposite of negative eight." Apply the double negative rule.
0	$-0 = 0$	The opposite of 0 is 0.

The concept of opposite can also be applied to an absolute value. For example, the opposite of the absolute value of -8 can be written as $-|-8|$. Think of this as a two-step process. Find the absolute value first, and then attach the $-$ to that result.

First, find the absolute value.

$$-|-8| = -8$$

Then attach a $-$ sign.

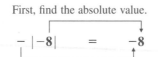

EXAMPLE 4 ***The opposite of a number.*** Simplify each expression: **a.** $-(-44)$ and **b.** $-|-225|$.

Solution

a. $-(-44)$ means the opposite of -44. Since the opposite of -44 is 44, we write

$$-(-44) = 44$$

b. $-|-225|$ means the opposite of $|-225|$. Since $|-225| = 225$, and the opposite of 225 is -225, we write

$$-|-225| = -225$$

Self Check

Simplify each expression:
a. $-(-1)$ and **b.** $-|-99|$.

Answers: **a.** 1, **b.** -99 ■

The − symbol

The $-$ symbol is used to indicate a negative number, the opposite of a number, and the operation of subtraction. The key to interpreting the $-$ symbol correctly is to examine the context in which it is used.

Interpreting the − symbol

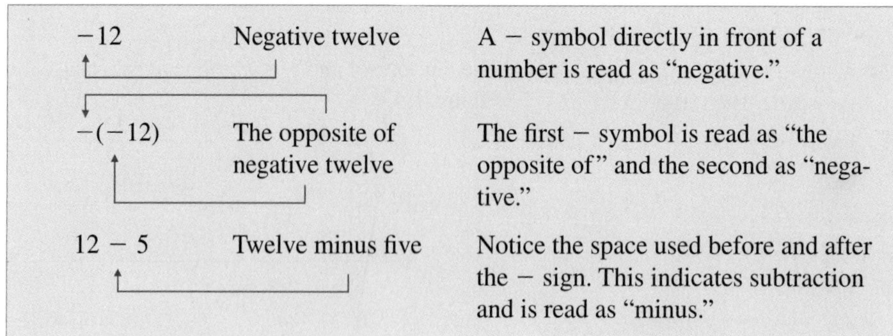

-12	Negative twelve	A $-$ symbol directly in front of a number is read as "negative."
$-(-12)$	The opposite of negative twelve	The first $-$ symbol is read as "the opposite of" and the second as "negative."
$12 - 5$	Twelve minus five	Notice the space used before and after the $-$ sign. This indicates subtraction and is read as "minus."

STUDY SET Section 2.1

VOCABULARY *Fill in the blanks to make the statements true.*

1. ______________ numbers are less than 0.

2. The collection of all positive whole numbers, negative whole numbers, and 0 is called the set of ______________.

3. Numbers can be represented by points equally spaced on a ______________ line.

4. To ______________ a number means to locate it on a number line and highlight it with a dot.

5. The symbols $>$ and $<$ are called ______________ symbols.

6. The ______________ value of a number is the distance between it and zero on a number line.

7. Two numbers on a number line that are the same distance from zero, but on opposite sides of the origin, are called ______________.

8. The opposite of the ______________ of a number is that number.

CONCEPTS

9. As we move to the left on a number line, how do the values of the numbers change?

10. Tell what is wrong with each number line.

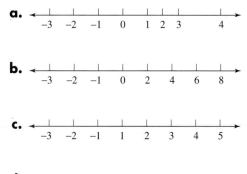

a.

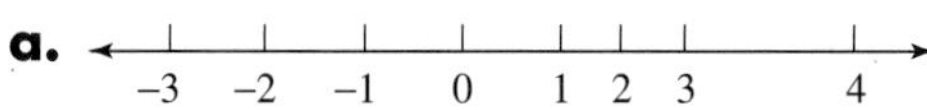

b. –3 –2 –1 0 2 4 6 8

c. –3 –2 –1 1 2 3 4 5

d. –3 –2 –1 0 1 2 3 4

11. Does every integer have an opposite?

12. Is the absolute value of a number ever negative?

13. Which of the following contains a minus sign: $15 - 8$, $-(-15)$, or -15?

14. Is there a number that is both greater than 10 and less than 10 at the same time?

15. Express the fact $12 < 15$ using the $>$ symbol.

16. Express the fact $5 > 4$ using the $<$ symbol.

17. Represent each of these situations using a signed number.

a. \$225 overdrawn
b. 10 seconds before liftoff
c. 3 degrees below normal
d. A trade deficit of \$12,000

18. Represent each of these situations using a signed number, and then describe its opposite in words.

a. A bacteria count 70 more than the standard

b. A profit of \$67
c. A business \$1 million in the "black"

d. 20 units over their quota

19. On a number line, what number is 3 units to the right of -7?

20. On a number line, what number is 4 units to the left of 2?

21. What two numbers on a number line are a distance of 5 away from -3?

22. What two numbers on a number line are a distance of 4 away from 3?

23. Which number is closer to -3 on the number line, 2 or -7?

24. Which number is farther from 1 on the number line, -5 or 8?

25. Give examples of the $-$ symbol used in three different ways.

26. What is the opposite of 0?

NOTATION

27. Translate each phrase to mathematical symbols.

a. The opposite of negative eight
b. The absolute value of negative eight
c. Eight minus eight
d. The opposite of the absolute value of negative eight

28. Write the set of integers.

PRACTICE *Simplify each expression.*

29. $\lvert 9 \rvert$	**30.** $\lvert 12 \rvert$
31. $\lvert -8 \rvert$	**32.** $\lvert -1 \rvert$
33. $\lvert -14 \rvert$	**34.** $\lvert -85 \rvert$
35. $-\lvert 20 \rvert$	**36.** $-\lvert 110 \rvert$
37. $-\lvert -6 \rvert$	**38.** $\lvert 0 \rvert$
39. $\lvert 203 \rvert$	**40.** $-\lvert -11 \rvert$
41. -0	**42.** $-\lvert 0 \rvert$
43. $-(-11)$	**44.** $-(-1)$
45. $-(-4)$	**46.** $-(-9)$
47. $-(-1{,}201)$	**48.** $-(-255)$

Graph each set of numbers on a number line labeled from -5 to 5.

49. $\{-3, 0, 3, 4, -1\}$

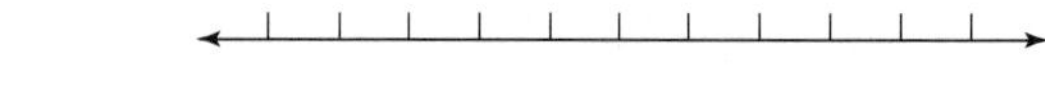

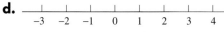

50. $\{-4, -1, 2, 5, 1\}$

51. The opposite of -3, the opposite of 5, and the absolute value of -2

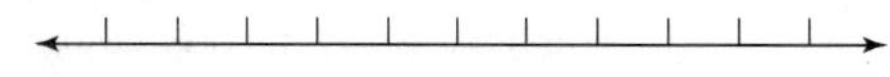

52. The absolute value of 3, the opposite of 3, and the number that is 1 less than -3

Insert one of the symbols $>$ or $<$ in the blank to make a true statement.

53. -5 ____ 5

54. 0 ____ -1

55. -12 ____ -6

56. -6 ____ -7

57. -10 ____ -11

58. -11 ____ -20

59. $|-2|$ ____ 0

60. $|-30|$ ____ -40

61. $-1{,}255$ ____ $-(-1{,}254)$

62. 0 ____ -3

63. $-|-3|$ ____ 4

64. $-|-163|$ ____ -150

APPLICATIONS

65. FLIGHT OF A BALL A boy throws a ball from the top of a building, as shown in Illustration 1. At the instant he does this, his friend starts a stopwatch and keeps track of the time as the ball rises to a peak and then falls to the ground. Use the vertical number line to complete the table at the top of the page by finding the position of the ball at the specified times.

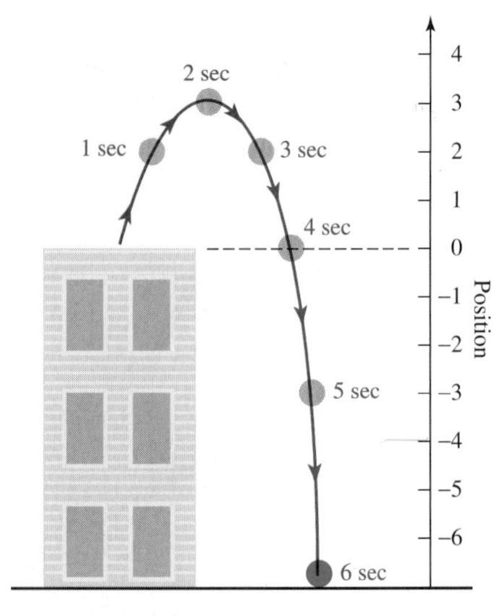

ILLUSTRATION 1

Time	Position of ball
1 sec	
2 sec	
3 sec	
4 sec	
5 sec	
6 sec	

66. SHOOTING GALLERY At an amusement park, a shooting gallery contains moving ducks. The path of one duck is shown in Illustration 2, along with the time it takes the duck to reach certain positions on the gallery wall. Complete the table using the horizontal number line in the illustration.

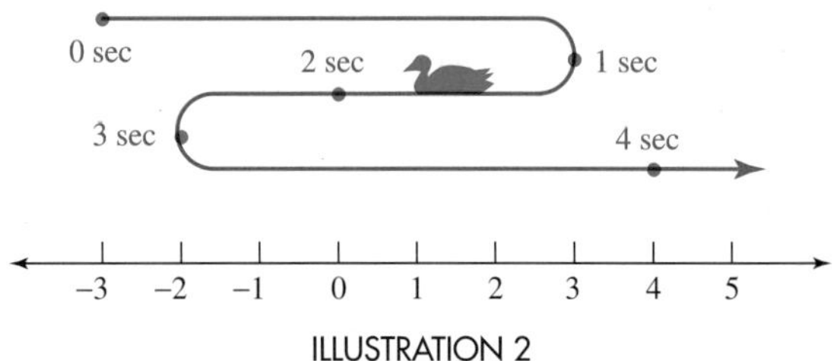

ILLUSTRATION 2

Time	Position of duck
0 sec	
1 sec	
2 sec	
3 sec	
4 sec	

67. TECHNOLOGY The readout from a testing device is shown in Illustration 3. It is important to know the height of each of the three "peaks" and the depth of each of the three "valleys." Use the vertical number line to find these numbers.

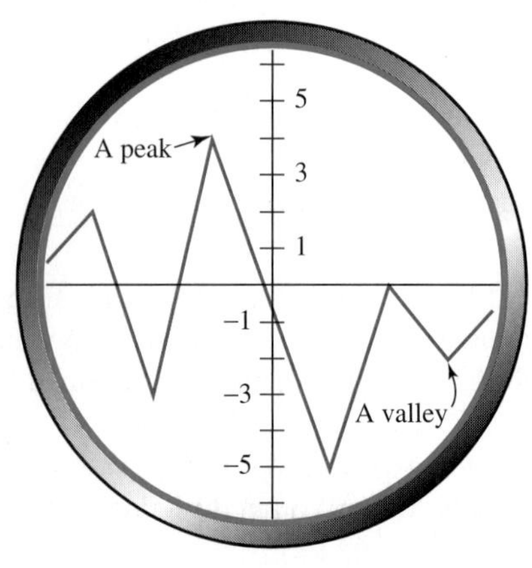

ILLUSTRATION 3

68. FLOODING A week of daily reports listing the height of a river in comparison to flood stage is given in the table. Complete the bar graph in Illustration 4.

Flood stage report	
Sun.	2 ft below
Mon.	3 ft over
Tue.	4 ft over
Wed.	2 ft over
Thu.	1 ft below
Fri.	3 ft below
Sat.	4 ft below

ILLUSTRATION 4

69. GOLF In golf, *par* is the standard number of strokes considered necessary on a given hole. A score of -2 indicates that a golfer used 2 strokes less than par. A score of $+2$ means 2 more strokes than par were used. In Illustration 5, each golf ball represents the score of a professional golfer on the 16th hole of a certain course.

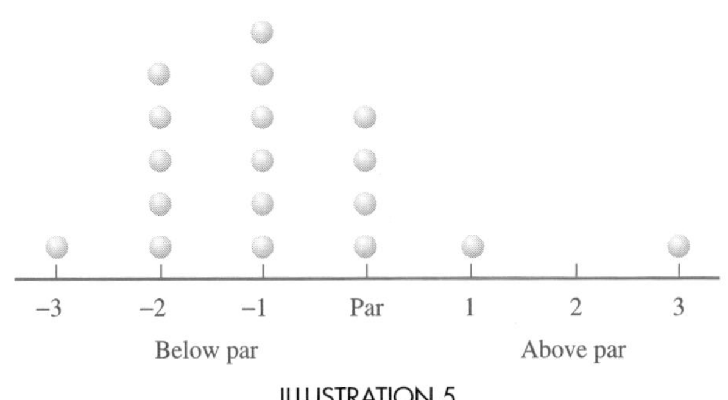

ILLUSTRATION 5

a. What score was shot most often on this hole?

b. What was the best score on this hole?

c. Explain why this hole appears to be too easy for a professional golfer.

70. PAYCHECK Examine the items listed on the paycheck stub in Illustration 6. Then write two columns on your paper—one headed "positive" and the other "negative." List each item under the appropriate heading.

Tom Dryden Dec. 01		Christmas bonus	$100
Gross pay	$2,000	**Reductions**	
Overtime	$300	Retirement	$200
Deductions		**Taxes**	
Union dues	$30	Federal withholding	$160
U.S. Bonds	$100	State withholding	$35

ILLUSTRATION 6

71. WEATHER MAP Illustration 7 shows the predicted Fahrenheit temperatures for a day in mid-January.

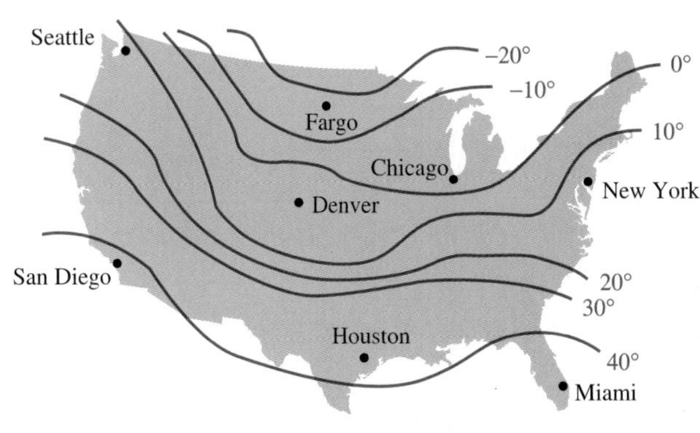

ILLUSTRATION 7

a. What is the temperature range for the region including Fargo, North Dakota?

b. Approximately how much colder will it be in Chicago than in Denver?

c. According to this prediction, what is the coldest it should get in Seattle?

72. PROFITS/LOSSES The graph in Illustration 8 shows the net income of Apple Computer Inc. for the years 1990–1999.

a. In what years did the company suffer a loss? Estimate each loss.

b. Explain why an article in the business section of a newspaper, when referring to this graph, would say, "Apple earnings turned the corner in 1998."

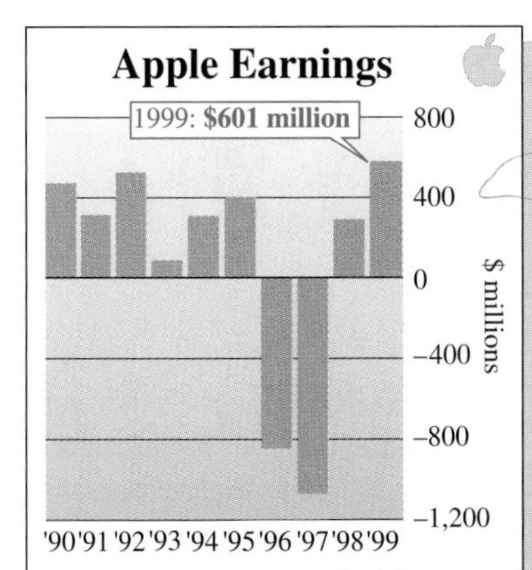

Based on data from *Los Angeles Times* (October 14, 1999)

ILLUSTRATION 8

73. HISTORICAL TIME LINE Number lines can be used to display historical data. Some important world events are shown on the time line in Illustration 9.

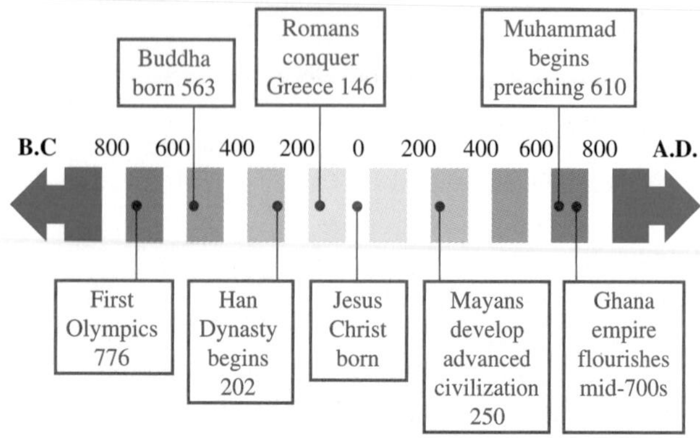

ILLUSTRATION 9

a. What basic unit was used to scale this time line?
b. What could be thought of as positive numbers?
c. What could be thought of as negative numbers?
d. What important event distinguishes the positive from the negative numbers?

74. ASTRONOMY Astronomers use a type of number line called the *apparent magnitude scale* to denote the brightness of objects in the sky. The brighter an object appears to an observer on Earth, the more negative is its apparent magnitude. Graph each of the following on the scale in Illustration 10.

- Full moon −12
- Pluto +15
- Sirius (brightest star) −2
- Sun −26
- Venus −4
- Visual limit of binoculars +10
- Visual limit of large telescope +20
- Visual limit of naked eye +6

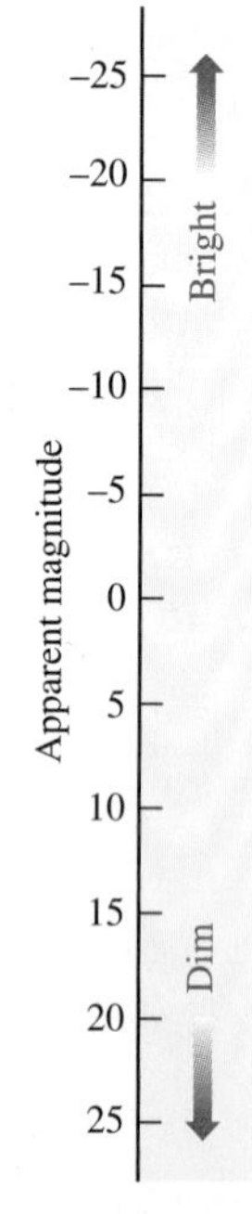

ILLUSTRATION 10

75. LINE GRAPH Each thermometer in Illustration 11 gives the daily high temperature in degrees Fahrenheit. Plot each daily high temperature on the grid and then construct a line graph.

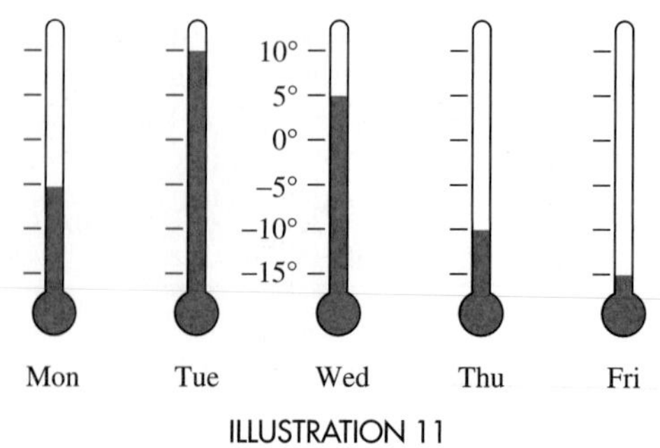

ILLUSTRATION 11

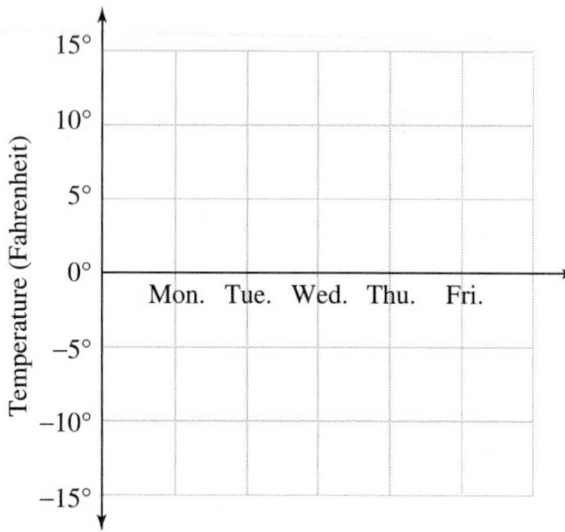

76. GARDENING Illustration 12 shows the depths at which the bottoms of various types of flower bulbs should be planted. (The symbol ″ represents inches.)

a. At what depth should a tulip bulb be planted?

b. How much deeper are hyacinth bulbs planted than gladiolus bulbs?

c. Which bulb must be planted the deepest? How deep?

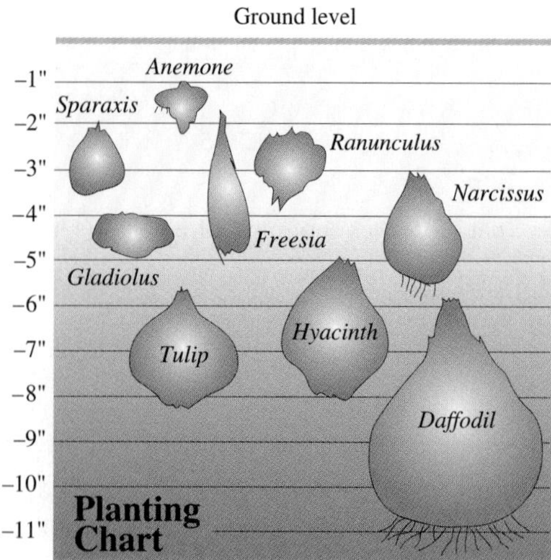

ILLUSTRATION 12

WRITING

77. Explain the concept of the opposite of a number.

78. What real-life situation do you think gave rise to the concept of a negative number?

79. Explain why the absolute value of a number is never negative.

80. Give an example of the use of a number line that you have seen in another course.

81. DIVING Divers use the terms *positive buoyancy, neutral buoyancy,* and *negative buoyancy* as shown in Illustration 13. What do you think each of these terms means?

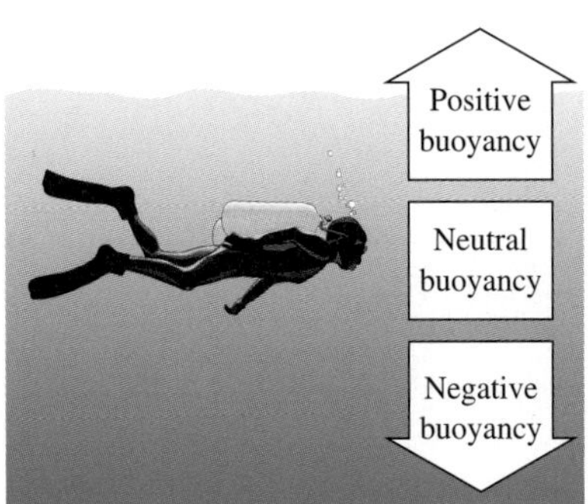

ILLUSTRATION 13

82. NEW ORLEANS The city of New Orleans, Louisiana, lies largely below sea level. Find out why the city is not under water.

REVIEW

83. Round 23,456 to the nearest hundred.

84. Evaluate: $19 - 2 \cdot 3$.

85. Solve $2x = 34$.

86. Is this statement true or false? $345 < 354$

87. Give the name of the property illustrated here: $(13 \cdot 2) \cdot 5 = 13 \cdot (2 \cdot 5)$

88. Write four times five using three different notations.

2.2 *Addition of Integers*

In this section, you will learn about

- Adding two integers with the same sign
- Adding two integers with different signs • The addition property of zero
- The additive inverse of a number

? – 7:32 A.M.
49° increase
−4° F – 7:30 A.M.

INTRODUCTION. A dramatic change in temperature occurred in 1943 in Spearfish, South Dakota. On January 22, at 7:30 A.M., the temperature was −4°F. In just two minutes, the temperature rose 49 degrees! To calculate the temperature at 7:32 A.M., we need to add 49 to −4.

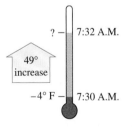

$$-4 + 49$$

To perform this addition, we must know how to add positive and negative integers. In this section, we will develop rules to help us make such calculations.

Adding two integers with the same sign

$4 + 3$
both positive

To explain addition of signed numbers, we can use a number line. (See Figure 2-6 on the next page.) To compute $4 + 3$, we begin at the **origin** (the point labeled 0) and draw an arrow 4 units long, pointing to the right. This represents positive 4. From that point, we draw an arrow 3 units long, pointing to the right, to represent positive 3. The second arrow points to the answer. Therefore, $4 + 3 = 7$.

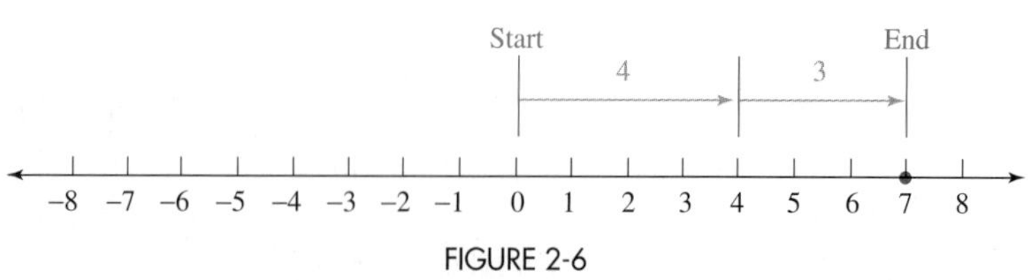

FIGURE 2-6

To check our work, let's think of the problem in terms of money. If you had \$4 and earned \$3 more, you would have a total of \$7.

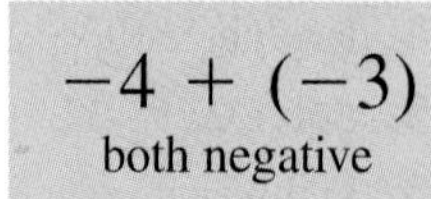

To compute $-4 + (-3)$, we begin at the origin and draw an arrow 4 units long, pointing to the left. (See Figure 2-7.) This represents -4. From there, we draw an arrow 3 units long, pointing to the left, to represent -3. The second arrow points to the answer: $-4 + (-3) = -7$.

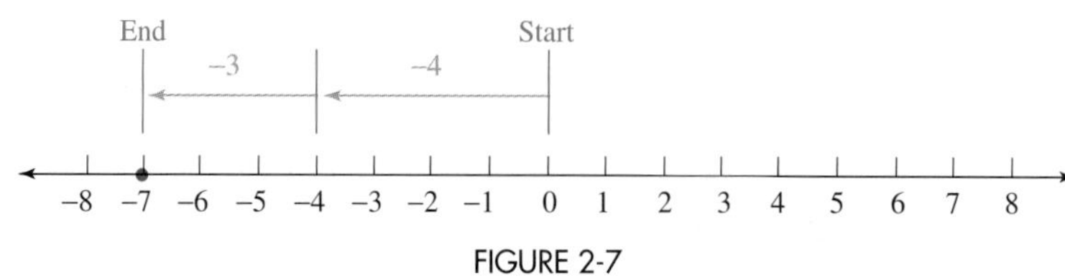

FIGURE 2-7

Let's think of this problem in terms of money. If you had a debt of \$4 (negative 4) and then incurred \$3 more debt (negative 3), you would be in debt \$7 (negative 7).

Here are some observations about the process of adding two numbers that have the same sign, using a number line.

- Both arrows point in the same direction and "build" on each other.
- The answer has the same sign as the two numbers being added.

4	+	3	=	7	−4	+	(−3)	=	−7
positive	+	positive	=	positive answer	negative	+	negative	=	negative answer

These observations suggest the following rule.

Adding two integers with the same sign

To add two integers with the **same sign,** add their absolute values and attach their common sign to the sum. If both integers are positive, their sum is positive. If both integers are negative, their sum is negative.

COMMENT When writing additions that involve signed numbers, write negative numbers within parentheses to separate the negative sign $-$ from the plus sign $+$.

$9 + (-4)$ ~~$9 + -4$~~ and $-9 + (-4)$ ~~$-9 + -4$~~

EXAMPLE 1 ***Adding two negative integers.*** Find the sum: $-9 + (-4)$.

Solution *Step 1:* To add two integers with the same sign, we first add the absolute values of each of the integers. Since $|-9| = 9$ and $|-4| = 4$, we begin by adding 9 and 4.

$$9 + 4 = 13$$

Step 2: We then attach the common sign (which is negative) to this result. Therefore,

$$-9 + (-4) = -13$$

Make the answer negative.

After some practice, you will be able to do this kind of problem in your head. It will not be necessary to show all the steps as we have done here. ■

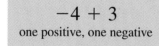

EXAMPLE 2 ***Adding two negative integers.*** Find the sum: $-80 + (-60)$.

Solution

Since both integers are negative, the answer will be negative.

$-80 + (-60) = -140$ Because the numbers have the same sign, add their absolute values, 80 and 60, to get 140. Then attach the common negative sign.

Self Check

Find the sum: $-300 + (-100)$.

Answer: -400

Adding two integers with different signs

$4 + (-3)$
one positive, one negative

To compute $4 + (-3)$, we start at the origin and draw an arrow 4 units long, pointing to the right. (See Figure 2-8.) This represents positive 4. From there, we draw an arrow 3 units long, pointing to the left, to represent -3. The second arrow points to the answer: $4 + (-3) = 1$.

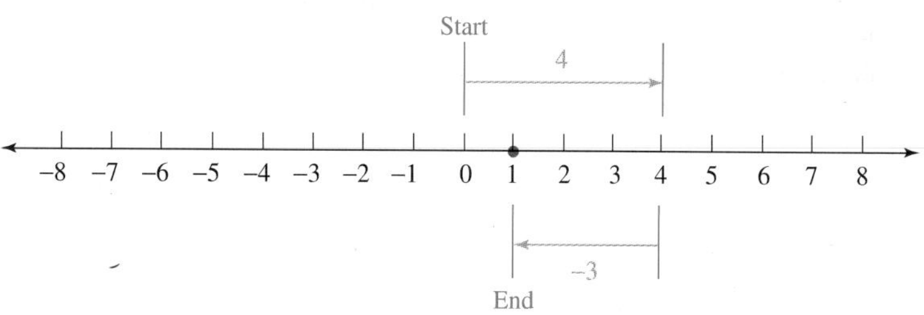

FIGURE 2-8

In terms of money, if you had $4 (positive 4) and then incurred a debt of $3 (negative 3), you would have $1 (positive 1) left.

$-4 + 3$
one positive, one negative

The problem $-4 + 3$ can be illustrated by drawing an arrow 4 units long from the origin, pointing to the left. (See Figure 2-9.) This represents -4. From there, we draw an arrow 3 units long, pointing to the right, to represent positive 3. The second arrow points to the answer: $-4 + 3 = -1$.

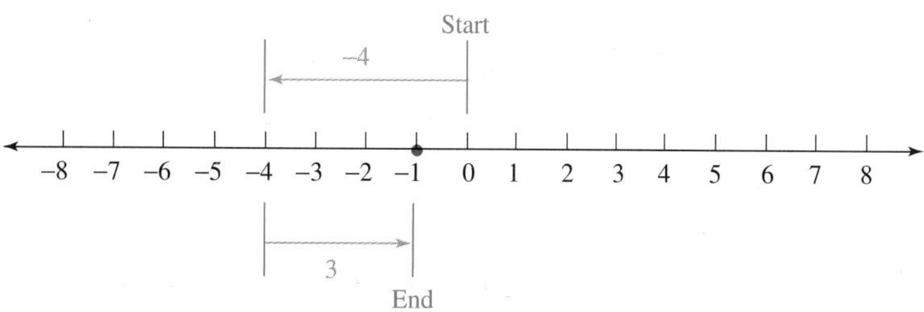

FIGURE 2-9

This problem can be thought of as owing $4 (negative 4) and then paying back $3 (positive 3). You will still owe $1 (negative 1).

The last two examples lead us to some observations about adding two integers with different signs, using a number line.

- The arrows representing the integers point in opposite directions.
- The longer of the two arrows determines the sign of the answer.

These observations suggest the following rule.

Adding two integers with different signs

To add two integers with **different signs,** subtract their absolute values, the smaller from the larger. Then attach to that result the sign of the integer with the larger absolute value.

EXAMPLE 3 *Adding a positive and a negative integer.* Find the sum: $5 + (-7)$.

Solution *Step 1:* To add two integers with different signs, we first subtract the smaller absolute value from the larger absolute value. Since $|5|$, which is 5, is smaller than $|-7|$, which is 7, we begin by subtracting 5 from 7.

$$7 - 5 = 2$$

Step 2: -7 has the larger absolute value, so we attach a negative sign to the result from Step 1. Therefore,

$$5 + (-7) = -2$$

Make the answer negative. ■

EXAMPLE 4 *Adding a positive and a negative integer.* Find the sum: **a.** $-8 + 5$ and **b.** $11 + (-5)$.

Solution

a. Since -8 has the larger absolute value, the answer will be negative.

$-8 + 5 = -3$ Because the signs of the numbers are different, subtract their absolute values, 5 from 8, to get 3. Attach a negative sign to that result.

b. Since 11 has the larger absolute value, the answer will be positive.

$11 + (-5) = 6$ Subtract the absolute values, 5 from 11, to get 6. The answer is positive.

Self Check
Find the sum:

a. $-2 + 7$

b. $6 + (-9)$

Answers: **a.** 5, **b.** -3 ■

EXAMPLE 5 *Temperature change.* In the introduction to this section, we learned that at 7:30 A.M. on January 22, 1943, in Spearfish, South Dakota, the temperature was $-4°$. The temperature then rose 49 degrees in just two minutes. What was the temperature at 7:32 A.M.?

Solution The phrase *temperature rose 49 degrees* indicates addition. We need to add 49 to -4.

$-4 + 49 = 45$ Subtract the smaller absolute value, 4, from the larger absolute value, 49. The sum is positive.

At 7:32 A.M., the temperature was 45°F. ■

EXAMPLE 6 *Adding several integers.* Add: $-3 + 5 + (-12) + 2$.

Solution

This expression contains four integers. We add them, working from left to right.

$$\begin{aligned} -3 + 5 + (-12) + 2 &= 2 + (-12) + 2 && \text{Add: } -3 + 5 = 2. \\ &= -10 + 2 && \text{Add: } 2 + (-12) = -10. \\ &= -8 \end{aligned}$$

Self Check
Add: $-12 + 8 + (-6) + 1$.

Answer: -9 ■

An alternate approach to problems like Example 6 is to add all the positive numbers, add all the negative numbers, and then add those results.

EXAMPLE 7 ***Adding several integers.*** Find the sum: $-3 + 5 + (-12) + 2$.

Solution

We can use the commutative property of addition to reorder the numbers and use the associative property of addition to group the positives together and the negatives together.

$-3 + 5 + (-12) + 2 = \mathbf{5 + 2 + (-3) + (-12)}$	Reorder the numbers.
$= \mathbf{(5 + 2) + [(-3) + (-12)]}$	Group the positives. Group the negatives.

We do the operations inside the grouping symbols first.

$\mathbf{(5 + 2) + [(-3) + (-12)]} = \mathbf{7 + (-15)}$	Add the positives. Add the negatives.
$= -8$	Add the numbers with different signs.

Self Check

Find the sum:

$$-12 + 8 + (-6) + 1$$

Answer: -9 ■

Accent on Technology: *Entering negative numbers*

The United States' largest trading partner in Africa is Nigeria. To calculate the 1998 U.S. trade balance with Nigeria, we add the \$816,800,000 worth of exports to Nigeria (considered positive) to the \$4,194,000,000 worth of imports from Nigeria (considered negative). We can use a scientific calculator to do the addition: $816{,}800{,}000 + (-4{,}194{,}000{,}000)$.

- We do not have to do anything special to enter a positive number. When we key in 816,800,000, a positive number is entered.
- To enter $-4{,}194{,}000{,}000$, we press the change-of-sign key [+/−] *after* entering 4,194,000,000. Note that the change-of-sign key is different from the subtraction key [−].

Keystrokes 816800000 [+] 4194000000 [+/−] [=] [-3377200000]

In 1998, the United States had a trade balance of $-\$3{,}377{,}200{,}000$ with Nigeria. Because the result is negative, it is called a *trade deficit.*

The addition property of zero

When 0 is added to a number, the number remains the same. For example, $5 + 0 = 5$, and $0 + (-4) = -4$. Because of this, we call 0 the **additive identity.**

Addition property of zero

For any number a,

$$a + 0 = a \quad \text{and} \quad 0 + a = a$$

The additive inverse of a number

A second fact concerning 0 and the operation of addition can be demonstrated by considering the sum of a number and its opposite. To illustrate this, we use the number line in Figure 2-10 on page 92 to add 6 and its opposite, -6. We see that $6 + (-6) = 0$.

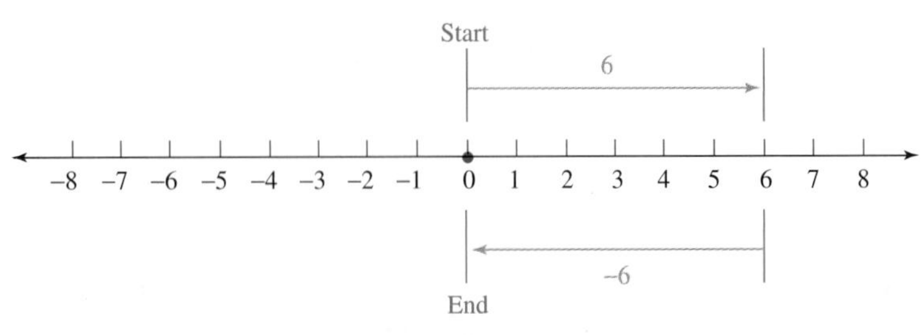

FIGURE 2-10

If the sum of two numbers is 0, the numbers are said to be **additive inverses** of each other. Since $6 + (-6) = 0$, we say that 6 and -6 are additive inverses.

We can now classify a pair of numbers such as 6 and -6 in three ways: as opposites, negatives, or additive inverses.

The additive inverse of a number

For any numbers a and b, if $a + b = 0$, then a and b are called **additive inverses.**

EXAMPLE 8 ***The additive inverse.*** What is the additive inverse of -3? Justify your result.

Solution

The additive inverse of -3 is its opposite, 3. To justify the result, we add and show that the sum is 0.

$$-3 + 3 = 0$$

Self Check

What is the additive inverse of 12? Justify your result.

Answer: -12; $12 + (-12) = 0$

STUDY SET Section 2.2

VOCABULARY *Fill in the blanks to make the statements true.*

1. When 0 is added to a number, the number remains the same. We call 0 the additive ______.

2. Since $-5 + 5 = 0$, we say that 5 is the additive ______ of -5. We can also say that 5 and -5 are ______.

CONCEPTS *In Exercises 3–6, find each answer using a number line.*

3. $-3 + 6$

4. $-3 + (-2)$

5. $-5 + 3$

6. $-1 + (-3)$

7. a. Is the sum of two positive integers always positive?

b. Is the sum of two negative integers always negative?

8. a. What is the sum of a number and its additive inverse?

b. What is the sum of a number and its opposite?

9. Find each absolute value.

a. $|-7|$

b. $|10|$

10. If the sum of two numbers is 0, what can be said about the numbers?

Fill in the blanks to make the statements true.

11. To add two integers with unlike signs, ______ their absolute values, the smaller from the larger. Then attach to that result the sign of the number with the ______ absolute value.

12. To add two integers with like signs, add their ______ values and attach their common ______ to the sum.

NOTATION *Complete each solution.*

13. Evaluate $-16 + (-2) + (-1)$.

$-16 + (-2) + (-1) = \_\_\_\_ + (-1)$

$= -19$

14. Evaluate $-8 + (-2) + 6$.

$-8 + (-2) + 6 = \_\_\_\_ + 6$

$= -4$

15. Evaluate $(-3 + 8) + (-3)$.

$(-3 + 8) + (-3) = \_\_\_ + (-3)$

$= 2$

16. Evaluate $-5 + [2 + (-9)]$.

$-5 + [2 + (-9)] = -5 + (\_\_\_)$

$= -12$

17. Explain why the expression $-6 + -5$ is not written correctly. How should it be written?

18. What mathematical symbol is implied when the word *sum* is used?

PRACTICE *Find the additive inverse of each number.*

19. -11 **20.** 9

21. -23 **22.** -43

23. 0 **24.** 1

25. 99 **26.** 250

Find each sum.

27. $-6 + (-3)$ **28.** $-2 + (-3)$

29. $-5 + (-5)$ **30.** $-8 + (-8)$

31. $-6 + 7$ **32.** $-2 + 4$

33. $-15 + 8$ **34.** $-18 + 10$

35. $20 + (-40)$ **36.** $25 + (-10)$

37. $30 + (-15)$ **38.** $8 + (-20)$

39. $-1 + 9$ **40.** $-2 + 7$

41. $-7 + 9$ **42.** $-3 + 6$

43. $5 + (-15)$ **44.** $16 + (-26)$

45. $24 + (-15)$ **46.** $-4 + 14$

47. $35 + (-27)$ **48.** $46 + (-73)$

49. $24 + (-45)$ **50.** $-65 + 31$

Evaluate each expression.

51. $-2 + 6 + (-1)$

52. $4 + (-3) + (-2)$

53. $-9 + 1 + (-2)$

54. $5 + 4 + (-6)$

55. $6 + (-4) + (-13) + 7$

56. $8 + (-5) + (-10) + 6$

57. $9 + (-3) + 5 + (-4)$

58. $-3 + 7 + 1 + (-4)$

59. Find the sum of -6, -7, and -8.

60. Find the sum of -11, -12, and -13.

Find each sum.

61. $-7 + 0$ **62.** $6 + 0$

63. $9 + 0$ **64.** $0 + (-15)$

65. $-4 + 4$ **66.** $18 + (-18)$

67. $2 + (-2)$ **68.** $-10 + 10$

69. What number must be added to -5 to obtain 0?

70. What number must be added to 8 to obtain 0?

Evaluate each expression.

71. $2 + (-10 + 8)$

72. $(-9 + 12) + (-4)$

73. $(-4 + 8) + (-11 + 4)$

74. $(-12 + 6) + (-6 + 8)$

75. $[-3 + (-4)] + (-5 + 2)$

76. $[9 + (-10)] + (-7 + 9)$

77. $[6 + (-4)] + [8 + (-11)]$

78. $[5 + (-8)] + [9 + (-15)]$

79. $-2 + [-8 + (-7)]$

80. $-8 + [-5 + (-2)]$

81. $789 + (-9{,}135)$

82. $2{,}701 + (-4{,}089)$

83. $-675 + (-456) + 99$

84. $-9{,}750 + (-780) + 2{,}345$

APPLICATIONS *Use signed numbers to help answer each question.*

85. G FORCES As a fighter pilot dives and loops, different forces are exerted on the body, just like the forces you experience when riding on a roller coaster. Some of the forces, called G's, are positive and some are negative. The force of gravity, 1G, is constant. Complete the diagram in Illustration 1.

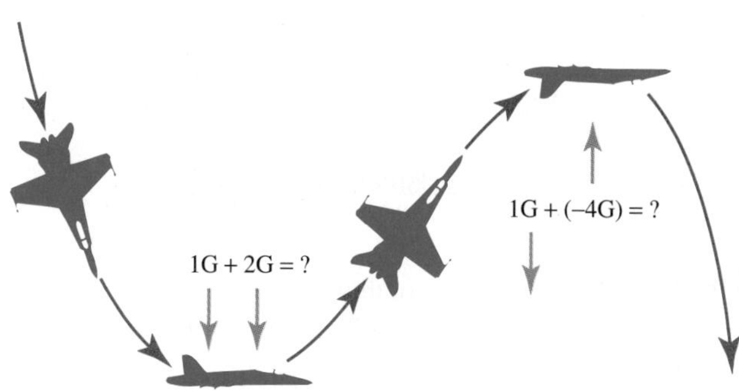

ILLUSTRATION 1

86. CHEMISTRY The first several steps of a chemistry lab experiment are listed here. The experiment begins with a compound that is stored at −40° F.

Step 1: Raise the temperature of the compound 200°.

Step 2: Add sulfur and then raise the temperature 10°.

Step 3: Add 10 milliliters of water, stir, and raise the temperature 25°.

What is the resulting temperature of the mixture after step 3?

87. CASH FLOW The maintenance costs, utilities, and taxes on a duplex are \$900 per month. The owner of the apartments receives monthly rental payments of \$450 and \$380. Does this investment produce a positive cash flow each month?

88. JOGGING A businessman's lunchtime workout includes jogging up 10 stories of stairs in his high-rise office building. If he starts on the fourth level below ground in the underground parking garage, on what story of the building will he finish his workout?

89. MEDICAL QUESTIONNAIRE Determine the risk of contracting heart disease for the man whose responses are shown in Illustration 2.

Age		Total Cholesterol	
Age	Points	Reading	Points
34	−1	150	−3

Cholesterol		Blood Pressure	
HDL	Points	Systolic/Diastolic	Points
62	−2	124/100	3

Diabetic		Smoker	
	Points		Points
Yes	2	Yes	2

10-Year Heart Disease Risk

Total Points	**Risk**	Total Points	**Risk**
−2 or less	1%	5	4%
−1 to 1	2%	6	6%
2 to 3	3%	7	6%
4	4%	8	7%

Source: National Heart, Lung, and Blood Institute

ILLUSTRATION 2

90. SPREADSHEET Monthly rain totals for four counties are listed in the spreadsheet shown in Illustration 3. The −1 entered in cell B1 means that the rain total for Suffolk County for a certain month was one inch below average. We can analyze this data by asking the computer to perform various operations.

a. To ask the computer to add the numbers in cells C1, C2, C3, and C4, we type SUM(C1:C4). Find this sum.

b. Find SUM(B4:F4).

	A	B	C	D	E	F
1	Suffolk	−1	−1	0	+1	+1
2	Marin	0	−2	+1	+1	−1
3	Logan	−1	+1	+2	+1	+1
4	Tipton	−2	−2	+1	−1	−3

ILLUSTRATION 3

91. ATOMS An atom is composed of protons, neutrons, and electrons. A proton has a positive charge (represented by +1), a neutron has no charge, and an electron has a negative charge (−1). Two simple models of atoms are shown in Illustration 4. What is the net charge of each atom?

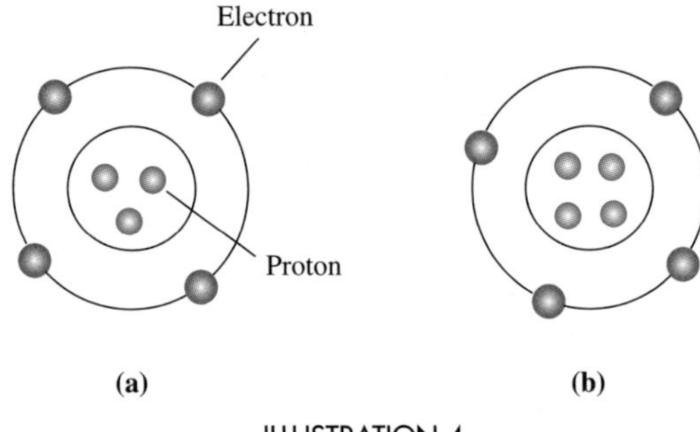

ILLUSTRATION 4

92. POLITICAL POLLS Six months before a general election, the incumbent senator found himself trailing the challenger by 18 points. To overtake his opponent, the campaign staff decided to use a four-part strategy. Each part of this plan is shown below, with the anticipated point gain.

1. Intense TV ad blitz	+10
2. Ask for union endorsement	+2
3. Voter mailing	+3
4. Get-out-the vote campaign	+1

With these gains, will the incumbent overtake the challenger on election day?

93. FLOODING After a heavy rainstorm, a river that had been 4 feet under flood stage rose 11 feet in a 48-hour period. Find the height of the river after the storm in comparison to flood stage.

94. MILITARY SCIENCE During a battle, an army retreated 1,500 meters, regrouped, and advanced 3,500 meters. The next day, it had to retreat 1,250 meters. Find the army's net gain.

95. FILM PROFITS A movie studio produced four films, two financial successes and two failures. The profits and losses of the films are shown in Illustration 5. Find the studio's profit, if any, for the year.

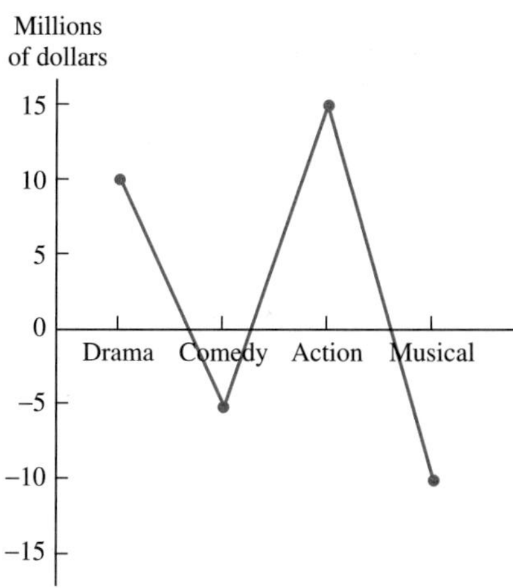

ILLUSTRATION 5

96. ACCOUNTING On a financial balance sheet, debts (negative numbers) are denoted within parentheses. Assets (positive numbers) are written without parentheses. What is the 1999 fund balance for the preschool whose financial records are shown in Illustration 6?

Community Care Preschool Balance Sheet, June 1999	
Fund balances	
Classroom supplies	$ 5,889
Emergency needs	927
Holiday program	(2,928)
Insurance	1,645
Janitorial	(894)
Licensing	715
Maintenance	(6,321)
BALANCE	?

ILLUSTRATION 6

WRITING

97. Is the sum of a positive and a negative number always positive? Explain why or why not.

98. How do you explain the fact that when asked to *add* -4 and 8, we must actually *subtract* to obtain the result?

99. Why is the sum of two negative numbers a negative number?

100. Write an application problem that will require adding -50 and -60.

REVIEW

101. Find the area of the rectangle in Illustration 7.

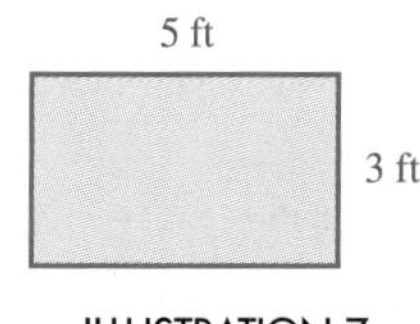

ILLUSTRATION 7

102. A car with a tank that holds 15 gallons of gasoline goes 25 miles on one gallon. How far can it go on a full tank?

103. Solve $x - 7 = 20$.

104. Is $t = 4$ a solution of $3t = 12$?

105. Prime factor 125. Use exponents to express the result.

106. Do the division: $\frac{144}{12}$.

2.3 *Subtraction of Integers*

In this section, you will learn about

- Adding the opposite • Order of operations • Applications of subtraction

Introduction. In this section, we will study another way to think about subtraction. This new procedure is helpful when subtraction problems involve negative numbers.

Adding the opposite

The subtraction problem $6 - 4$ can be thought of as taking away 4 from 6. We can use a number line to illustrate this. (See Figure 2-11 on the next page.) Beginning at the origin, we draw an arrow of length 6 units in the positive direction. From that point, we move back 4 units to the left. The answer, called the **difference,** is 2.

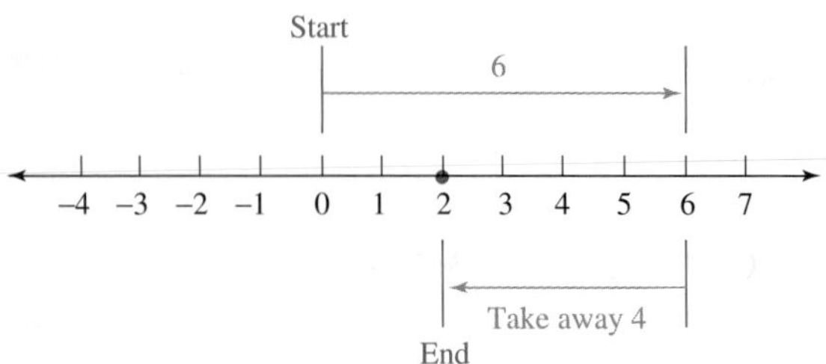

FIGURE 2-11

The work shown in Figure 2-11 looks like the illustration for the *addition* problem $6 + (-4) = 2$, shown in Figure 2-12.

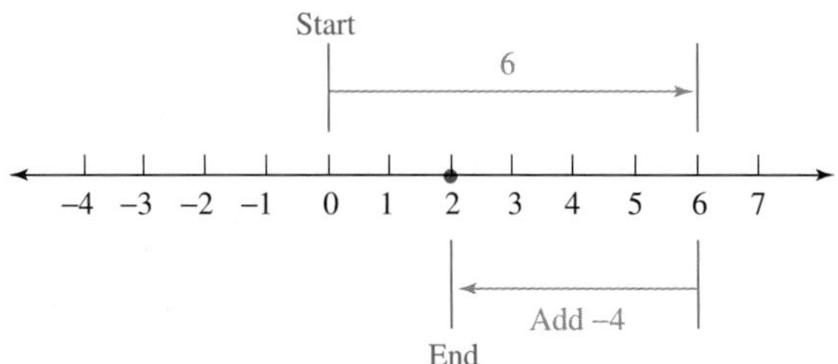

FIGURE 2-12

In the first problem, $6 - 4$, we subtracted 4 from 6. In the second, $6 + (-4)$, we added -4 (which is the opposite of 4) to 6. In each case, the result was 2.

Subtracting 4.	Adding the opposite of 4.
↓	↓
$6 - 4 = 2$	$6 + (-4) = 2$

The same result.

This observation helps to justify the following rule for subtraction.

Rule for subtraction

> If a and b are any numbers, then
>
> $$a - b = a + (-b)$$
>
> In words, this rule says that subtraction is the same as adding the opposite of the number to be subtracted.

You won't need to use this rule for every subtraction problem. For example, $6 - 4$ is obviously 2; it does not need to be rewritten as adding the opposite. But for more complicated problems such as $-6 - 4$ or $-6 - (-4)$, where the result is not obvious, the subtraction rule will be quite helpful.

EXAMPLE 1 ***Adding the opposite.*** Find $-6 - 4$.

Solution

The number to be subtracted is 4. Applying the subtraction rule, we write

$-6 - 4 = -6 + (-4)$	Write the subtraction as an addition of the opposite of 4, which is -4. Write -4 within parentheses.
$= -10$	To add -6 and -4, apply the rule for adding two negative numbers.

Self Check

Find $-2 - 3$ and check the result.

-2-3 =

-2+-3 = -5

To check the result, we add the difference, -10, and the subtrahend, 4. We should obtain the minuend, -6.

$$-10 + 4 = -6$$

The answer, -10, checks.

Answer: -5 ■

EXAMPLE 2 ***Adding the opposite.*** Find $3 - (-5)$.

Solution

The number being subtracted is -5.

$3 - (-5) = 3 + 5$	Write the subtraction as an addition of the opposite of -5, which is 5.
$= 8$	Do the addition.

Self Check

Find $3 - (-2)$.

3-(-2)
3+2=5

Answer: 5 ■

EXAMPLE 3 ***Adding the opposite.*** Subtract -3 from -8.

Solution

The number being subtracted is -3, so we write it after -8.

$-8 - (-3) = -8 + 3$	Add the opposite of -3, which is 3.
$= -5$	Do the addition.

Self Check

Find $-2 - (-6)$.

-2-(-6)
-2+6=4

Answer: 4 ■

Remember that any subtraction problem can be rewritten as an equivalent addition. We just add the opposite of the number that is to be subtracted.

Subtraction can be written as addition . . .

$$4 - 8 = 4 + (-8) = -4$$
$$4 - (-8) = 4 + 8 = 12$$
$$-4 - 8 = -4 + (-8) = -12$$
$$-4 - (-8) = -4 + 8 = 4$$

of the opposite of the number to be subtracted.

Order of operations

Expressions can contain repeated subtraction or subtraction in combination with grouping symbols. To work these problems, we apply the rules for the order of operations.

EXAMPLE 4 ***Repeated subtraction.*** Evaluate: $-1 - (-2) - 10$.

Solution

This subtraction problem involves three numbers. We work from left to right, rewriting each subtraction as an addition of the opposite.

$-1 - (-2) - 10 = -1 + 2 + (-10)$	Add the opposite of -2, which is 2. Add the opposite of 10, which is -10. Write -10 in parentheses.
$= 1 + (-10)$	Work from left to right. Add $-1 + 2$.
$= -9$	Do the addition.

Self Check

Evaluate: $-3 - 5 - (-1)$.

-3-5-(-1)=
-3+(-5)+1=-7

Answer: -7 ■

EXAMPLE 5 ***Order of operations.*** Evaluate: $-8 - (-2 - 2)$.

Solution

We must do the subtraction within the parentheses first.

$$-8 - (-2 - 2) = -8 - [-2 + (-2)]$$ Add the opposite of 2, which is -2. Since -2 must be written within parentheses, we write $-2+(-2)$ within brackets.

$$= -8 - (-4)$$ Add -2 and -2. Since only one set of grouping symbols is needed, we write -4 within parentheses.

$$= -8 + 4$$ Add the opposite of -4, which is 4.

$$= -4$$ Do the addition.

Self Check

Evaluate: $-2 - (-6 - 5)$.

Answer: 9

Applications of subtraction

Things are constantly changing in our daily lives. The temperature, the amount of money we have in the bank, and our ages are a few examples. In mathematics, the operation of subtraction is used to measure change. In general, to find the change in a quantity, subtract the earlier value from the later value.

EXAMPLE 6 ***Change of water level.*** On Monday, the water level in a city storage tank was 6 feet above normal. By Friday, the level had fallen to a mark 4 feet below normal. Find the change in the water level from Monday to Friday. (See Figure 2-13.)

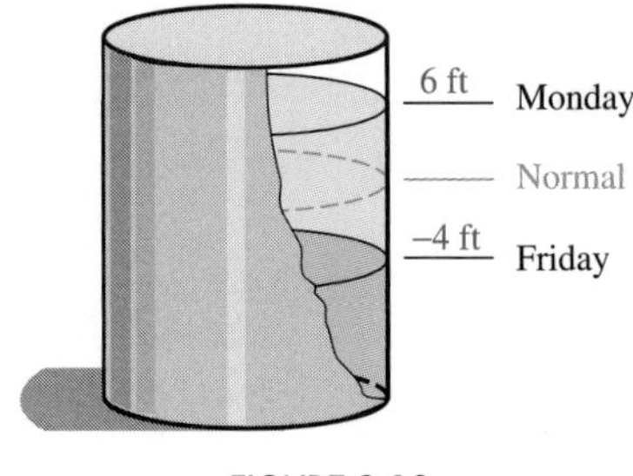

FIGURE 2-13

Solution

We will use subtraction to find the amount of change. The water levels of 4 feet below normal (the later value) and 6 feet above normal (the earlier value) can be represented by -4 and 6, respectively.

The water level Friday	minus	the water level Monday	is	the change in the water level.

$$-4 - 6 = -4 + (-6)$$ Add the opposite of 6, which is -6.

$$= -10$$ Do the addition. The negative result indicates that the water level fell.

The water level fell 10 feet from Monday to Friday.

In the next example, a number line will serve as a mathematical model of a real-life situation. You will see how the operation of subtraction can be used to find the distance between two points on a number line.

EXAMPLE 7 ***Artillery accuracy.*** In a practice session, an artillery group fired two rounds at a target. The first landed 65 yards short of the target, and the second landed 50 yards past it. (See Figure 2-14.) How far apart were the two impact points?

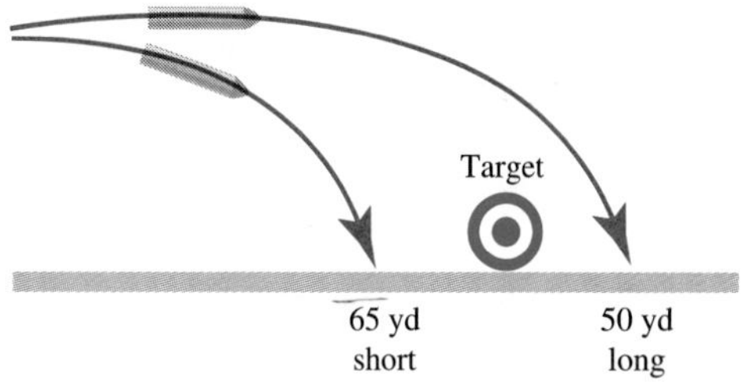

FIGURE 2-14

Solution We can use a number line to model this situation. The target is the origin. The words *short of the target* indicate a negative number, and the words *past it* indicate a positive number. Therefore, we graph the impact points at -65 and 50 in Figure 2-15.

FIGURE 2-15

The phrase *how far apart* tells us to subtract.

The position of the long	minus	the position of the short	is	the distance between impact points.

$50 - (-65) = 50 + 65$ Add the opposite of -65.

$= 115$ Do the addition.

The impact points are 115 yards apart. ■

Accent on Technology: *Subtraction with negative numbers*

The world's highest peak is Mount Everest in the Himalayas. The greatest ocean depth yet measured lies in the Mariana Trench near the island of Guam in the western Pacific. See Figure 2-16. To find the range between the highest peak and greatest depth, we must subtract.

$29{,}035 - (-36{,}025)$

To do this subtraction using a scientific calculator, we enter these numbers and press these keys.

Keystrokes 29035 [−] 36025 [+/−] [=] 65060

The range is 65,060 feet between the highest peak and the lowest depth. We could have applied the subtraction rule to write $29{,}035 - (-36{,}025)$ as $29{,}035 + 36{,}025$ before using the calculator.

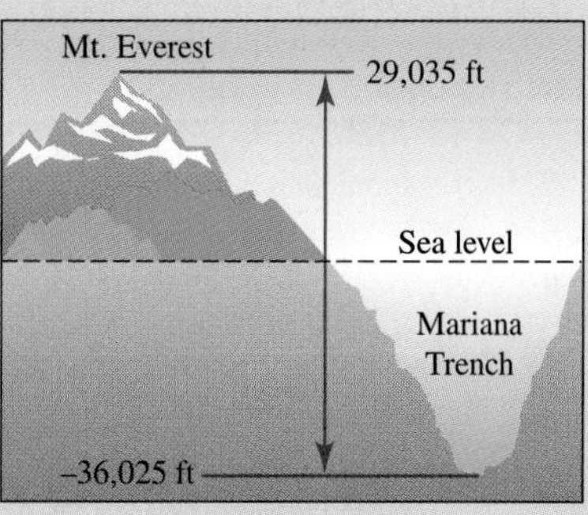

FIGURE 2-16

STUDY SET Section 2.3

VOCABULARY *Fill in the blanks to make the statements true.*

1. The answer to a subtraction problem is called the ______________.

2. Two numbers represented by points on a number line that are the same distance away from the origin, but on opposite sides of it, are called ______________.

CONCEPTS *In Exercises 3–10, fill in the blanks to make the statements true.*

3. ______________ is the same as adding the opposite of the number to be subtracted.

4. Subtracting 3 is the same as adding ____.

5. Subtracting -6 is the same as adding ____.

6. The opposite of -8 is ____.

7. For any numbers x and y, $x - y =$ ______________.

8. **a.** $2 - 7 = 2 +$ ____
b. $2 - (-7) = 2 +$ ____
c. $-2 - 7 = -2 +$ ____
d. $-2 - (-7) = -2 +$ ____

9. After using parentheses as grouping symbols, if another set of grouping symbols is needed, we use ______________.

10. We can find the ______________ in a quantity by subtracting the earlier value from the later value.

11. Write this problem using mathematical symbols: negative eight minus negative four.

12. Write this problem using mathematical symbols: negative eight subtracted from negative four.

13. Find the distance between -4 and 3 on a number line.

14. Find the distance between -10 and 1 on a number line.

15. Is subtracting 3 from 8 the same as subtracting 8 from 3? Explain.

16. Evaluate each expression.
a. $-2 - 0$ **b.** $0 - (-2)$

NOTATION *Complete each solution.*

17. Evaluate: $1 - 3 - (-2)$.

$$\begin{aligned} 1 - 3 - (-2) &= 1 + (\quad) + 2 \\ &= -2 + \quad \\ &= 0 \end{aligned}$$

18. Evaluate: $-6 + 5 - (-5)$.

$$\begin{aligned} -6 + 5 - (-5) &= -6 + 5 + \quad \\ &= \quad + 5 \\ &= 4 \end{aligned}$$

19. Evaluate: $(-8 - 2) - (-6)$.

$$\begin{aligned} (-8 - 2) - (-6) &= [-8 + (\quad)] - (-6) \\ &= \quad - (-6) \\ &= -10 + \quad \\ &= -4 \end{aligned}$$

20. Evaluate: $-5 - (-1 - 4)$.

$$\begin{aligned} -5 - (-1 - 4) &= -5 - [-1 + (\quad)] \\ &= -5 - (\quad) \\ &= -5 + \quad \\ &= 0 \end{aligned}$$

PRACTICE *Find each difference.*

21. $8 - (-1)$
22. $3 - (-8)$
23. $-4 - 9$
24. $-7 - 6$
25. $-5 - 5$
26. $-7 - 7$
27. $-5 - (-4)$
28. $-9 - (-1)$
29. $-1 - (-1)$
30. $-4 - (-3)$
31. $-2 - (-10)$
32. $-6 - (-12)$
33. $0 - (-5)$
34. $0 - 8$
35. $0 - 4$
36. $0 - (-6)$
37. $-2 - 2$
38. $-3 - 3$
39. $-10 - 10$
40. $4 - 4$
41. $9 - 9$
42. $4 - (-4)$
43. $-3 - (-3)$
44. $-5 - (-5)$

Evaluate each expression.

45. $-4 - (-4) - 15$
46. $-3 - (-3) - 10$
47. $-3 - 3 - 3$
48. $-1 - 1 - 1$
49. $5 - 9 - (-7)$
50. $6 - 8 - (-4)$
51. $10 - 9 - (-8)$
52. $16 - 14 - (-9)$
53. $-1 - (-3) - 4$
54. $-2 - 4 - (-1)$
55. $-5 - 8 - (-3)$
56. $-6 - 5 - (-1)$
57. $(-6 - 5) - 3$
58. $(-2 - 1) - 5$
59. $(6 - 4) - (1 - 2)$
60. $(5 - 3) - (4 - 6)$
61. $-9 - (6 - 7)$
62. $-3 - (6 - 12)$
63. $-8 - [4 - (-6)]$
64. $-1 - [5 - (-2)]$
65. $[-4 + (-8)] - (-6)$
66. $[-5 + (-4)] - (-2)$
67. Subtract -3 from 7.
68. Subtract 8 from -2.
69. Subtract -6 from -10.

70. Subtract -4 from -9.

71. $-1{,}557 - 890$

72. $-345 - (-789)$

73. $20{,}007 - (-496)$

74. $-979 - (-44{,}879)$

75. $-162 - (-789) - 2{,}303$

76. $-787 - 1{,}654 - (-232)$

APPLICATIONS *Use signed numbers to help answer each question.*

77. SCUBA DIVING After descending 50 feet, a scuba diver paused to check his equipment before descending an additional 70 feet. Use a signed number to represent the diver's final depth.

78. TEMPERATURE CHANGE Rashawn flew from his New York home to Hawaii for a week of vacation. He left blizzard conditions and a temperature of $-6°$, and stepped off the airplane into $85°$ weather. What temperature change did he experience?

79. READING PROGRAM In a state reading test administered at the start of a school year, an elementary school's performance was 23 points below the county average. The principal immediately began a special tutorial program. At the end of the school year, retesting showed the students to be only 7 points below the average. How many points did the school's reading score improve over the year?

80. SUBMARINE A submarine was traveling 2,000 feet below the ocean's surface when the radar system warned of an impending collision with another sub. The captain ordered the navigator to dive an additional 200 feet and then level off. Find the depth of the submarine after the dive.

81. AMPERAGE During normal operation, the ammeter on a car reads $+5$. (See Illustration 1.) If the headlights, which draw a current of 7 amps, and the radio, which draws a current of 6 amps, are both turned on, what number will the ammeter register?

ILLUSTRATION 1

82. GIN RUMMY After a losing round, a card player must subtract the value of each of the cards left in his hand (shown in Illustration 2) from his previous point total of 21. If face cards are counted as 10 points, what is his new score?

ILLUSTRATION 2

83. GEOGRAPHY Death Valley, California, is the lowest land point in the United States, at 283 feet below sea level. The lowest land point on the earth is the Dead Sea, which is 1,290 feet below sea level. How much lower is the Dead Sea than Death Valley?

84. LIE DETECTOR TEST On one lie detector test, a burglar scored -18, which indicates deception. However, on a second test, he scored -1, which is inconclusive. Find the difference in the scores.

85. FOOTBALL A college football team records the outcome of each of its plays during a game on a stat sheet. (See Illustration 3.) Find the net gain (or loss) after the 3rd play.

Down	Play	Result
1st	run	lost 1 yd
2nd	pass—sack!	lost 6 yd
penalty	delay of game	lost 5 yd
3rd	pass	gained 8 yd
4th	punt	—

ILLUSTRATION 3

86. ACCOUNTING Complete the balance sheet in Illustration 4. Then determine the overall financial condition of the company by subtracting the total liabilities from the total assets.

Walker Corporation
Balance Sheet 2001

Assets				
Cash	$11	1	0	9
Supplies	7	8	6	2
Land	67	5	4	3
Total assets	$			
Liabilities				
Accounts payable	$79	0	3	7
Income taxes	20	1	8	1
Total liabilities	$			

ILLUSTRATION 4

87. DIVING A diver jumps from a platform. After she hits the water, her momentum takes her to the bottom of the pool. (See Illustration 5.)

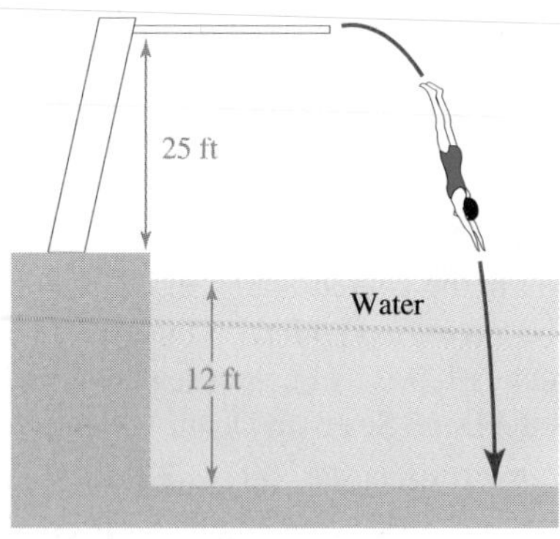

ILLUSTRATION 5

a. Use a number line and signed numbers to model this situation. Show the top of the platform, the water line, and the bottom of the pool.

b. Find the total length of the dive from the top of the platform to the bottom of the pool.

88. TEMPERATURE EXTREMES The highest and lowest temperatures ever recorded in several cities are shown in Illustration 6. List the cities in order, from the largest to smallest range in temperature extremes.

City	Extreme temperatures Highest	Lowest
Atlantic City, NJ	106	−11
Barrow, AK	79	−56
Kansas City, MO	109	−23
Norfolk, VA	104	−3
Portland, ME	103	−39

ILLUSTRATION 6

89. CHECKING ACCOUNT Michael has $1,303 in his checking account. Can he pay his car insurance premium of $676, his utility bills of $121, and his rent of $750 without having to make another deposit? Explain your answer.

90. HISTORY Two of the greatest Greek mathematicians were Archimedes (287–212 B.C.) and Pythagoras (569–500 B.C.). How many years apart were they born?

WRITING

91. Explain what is meant when we say that subtraction is the same as addition of the opposite.

92. Give an example showing that it is possible to subtract something from nothing.

93. Explain how to check the result: $-7 - 4 = -11$.

94. Explain why students don't need to change every subtraction they encounter into an addition of the opposite. Give some examples.

REVIEW

95. Solve: $5x = 15$.

96. Round 5,999 to the nearest hundred.

97. List the factors of 20.

98. When solving the equation $6x = 24$, what operation on the variable must be undone?

99. It takes 13 oranges to make one can of orange juice. Find the number of oranges used to make 12 cans.

100. Evaluate: $12^2 - (5 - 4)^2$.

101. Write 4,502 in expanded notation.

102. What property does the following illustrate? $a \cdot b = b \cdot a$

2.4 *Multiplication of Integers*

In this section, you will learn about

- Multiplying two positive integers
- Multiplying a positive and a negative integer
- Multiplying a negative and a positive integer • Multiplying by zero
- Multiplying two negative integers • Powers of integers

Introduction. We now turn our attention to multiplication of integers. When we multiply two nonzero integers, the first factor can be positive or negative. The same is true for the second factor. This means that there are four possible combinations to consider.

Positive · positive	Positive · negative
Negative · positive	Negative · negative

In this section, we will discuss these four combinations and use our observations to establish rules for multiplying two integers.

Multiplying two positive integers

4(3)
like signs
both positive

We begin by considering the product of two positive integers, 4(3). Since both factors are positive, we say that they have *like* signs. In Chapter 1, we learned that multiplication is repeated addition. Therefore, 4(3) represents the sum of four 3's.

$4(3) = 3 + 3 + 3 + 3$ Multiplication is repeated addition. Write 3 four times.

$4(3) = 12$ The result is 12, which is a positive number.

This result suggests that *the product of two positive integers is positive.*

Multiplying a positive and a negative integer

4(−3)
unlike signs
one positive, one negative

Next, we will consider $4(-3)$. This is the product of a positive and a negative integer. The signs of these factors are *unlike.* According to the definition of multiplication, $4(-3)$ means that we are to add -3 four times.

$4(-3) = (-3) + (-3) + (-3) + (-3)$ Use the definition of multiplication. Write -3 four times.

$4(-3) = (-6) + (-3) + (-3)$ Work from left to right. Apply the rule for adding two negative numbers.

$4(-3) = (-9) + (-3)$ Work from left to right. Apply the rule for adding two negative numbers.

$4(-3) = -12$ Do the addition.

This result is -12, which suggests that *the product of a positive integer and a negative integer is negative.*

Multiplying a negative and a positive integer

−3(4)
unlike signs
one negative, one positive

To develop a rule for multiplying a negative and a positive integer, we will consider $-3(4)$. Notice that the factors have *unlike* signs. Because of the commutative property of multiplication, the answer to $-3(4)$ will be the same as the answer to $4(-3)$. We know that $4(-3) = -12$ from the previous discussion, so $-3(4) = -12$. This suggests that *the product of a negative integer and a positive integer is negative.*

Putting the results of the last two cases together leads us to the rule for multiplying two integers with unlike signs.

Multiplying two integers with unlike signs

> To multiply a positive integer and a negative integer, or a negative integer and a positive integer, multiply their absolute values. Then make the answer negative.

EXAMPLE 1 ***Multiplying two integers with unlike signs.*** Find each product: **a.** $7(-5)$, **b.** $20(-8)$, and **c.** $-8 \cdot 5$.

Solution

To multiply integers with unlike signs, we multiply their absolute values and make the product negative.

a. $7(-5) = -35$ Multiply the absolute values, 7 and 5, to get 35. Then make the answer negative.

b. $20(-8) = -160$ Multiply the absolute values, 20 and 8, to get 160. Then make the answer negative.

c. $-8 \cdot 5 = -40$ Multiply the absolute values, 8 and 5, to get 40. Then make the answer negative.

Self Check

Find each product:

a. $2(-6)$

b. $30(-2)$

c. $-15 \cdot 2$

Answers: **a.** -12, **b.** -60, **c.** -30 ■

COMMENT When writing multiplication involving signed numbers, do not write a negative sign − next to a raised dot · (the multiplication symbol). Instead, use parentheses to show the multiplication.

$6(-2)$ ~~$6 \cdot -2$~~ and $-6(-2)$ ~~$-6 \cdot -2$~~

Multiplying by zero

Before we can develop a rule for multiplying two negative integers, we need to examine multiplication by zero. If $4(3)$ means that we are to find the sum of four 3's, then $0(-3)$ means that we are to find the sum of zero -3's. Obviously, the sum would be 0. Thus, $0(-3) = 0$.

The commutative property of multiplication guarantees that we can change the order of the factors in the multiplication problem without affecting the result.

$$(-3)(0) = 0(-3) = 0$$

↑ ↑ Change the order of the factors. ↑ The result is still 0.

We see that the order in which we write the factors 0 and -3 doesn't matter—their product is 0. This example suggests that the product of any number and zero is zero.

Multiplying by zero

> If a is any number, then
>
> $a \cdot 0 = 0$ and $0 \cdot a = 0$

EXAMPLE 2 ***Multiplication by zero.*** Find $-12 \cdot 0$.

Solution

Since the product of any number and zero is zero, we have

$$-12 \cdot 0 = 0$$

Self Check

Find $0(-56)$.

Answer: 0 ■

Multiplying two negative integers

$-3(-4)$
like signs
both negative

To develop a rule for multiplying two negative integers, we consider the pattern displayed below. There, we multiply -4 by a series of factors that decrease by 1. After determining each product, we graph each product on a number line (Figure 2-17). See if you can determine the answers to the last three multiplication problems by examining the pattern of answers leading up to them.

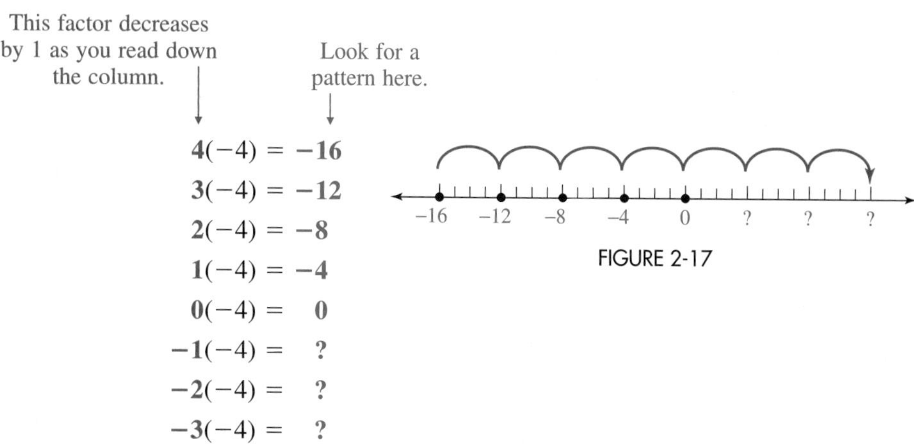

This factor decreases by 1 as you read down the column.
Look for a pattern here.
4(−4) = −16
3(−4) = −12
2(−4) = −8
1(−4) = −4
0(−4) = 0
−1(−4) = ?
−2(−4) = ?
−3(−4) = ?
−16 −12 −8 −4 0 ? ? ?

FIGURE 2-17

The products are graphed in Figure 2-17. From the pattern, we see that

$$-1(-4) = 4$$
$$-2(-4) = 8$$
$$-3(-4) = 12$$

For two negative factors, the product is a positive.

These results suggest that *the product of two negative integers is positive.* Earlier in this section, we saw that the product of two positive integers is also positive. This leads to the following conclusion.

Multiplying two integers with like signs

> To multiply two positive integers, or two negative integers, multiply their absolute values. The answer is positive.

EXAMPLE 3 ***Multiplying two negative integers.*** Find each product: **a.** $-5(-9)$ and **b.** $-8(-10)$.

Solution

To multiply two negative integers, we multiply their absolute values and make the result positive.

a. $-5(-9) = 45$ Multiply the absolute values, 5 and 9, to get 45. The answer is positive.

b. $-8(-10) = 80$ Multiply the absolute values, 8 and 10, to get 80. The answer is positive.

Self Check

Find each product:

a. $-9(-7)$

b. $-12(-2)$

Answers: **a.** 63, **b.** 24 ■

We now summarize the rules for multiplying two integers.

Multiplying two integers

> To multiply two integers, multiply their absolute values.
> 1. The product of two integers with *like* signs is positive.
> 2. The product of two integers with *unlike* signs is negative.

Accent on Technology: Multiplication with negative numbers

At Thanksgiving time, a supermarket offered customers a free turkey with every grocery purchase of \$100 or more. Each turkey cost the store \$6, and 375 people took advantage of the offer.

Since each of the 375 turkeys given away represented a loss of \$6 (which can be expressed as -6 dollars) the store lost $375(-6)$ dollars. To find this product, we enter these numbers and press these keys on a scientific calculator.

Keystrokes 375 [×] 6 [+/−] [=] -2250

The negative result indicates that with this promotion, the supermarket gave away \$2,250 in turkeys.

EXAMPLE 4 ***Multiplying three integers.*** Multiply $-4(5)(-3)$ in two ways.

Solution

First, we work from left to right.

$-4(5)(-3) = -20(-3)$ Multiply -4 and 5 first: $-4(5) = -20$.

$= 60$ Do the multiplication.

An alternate approach would be to multiply the two negative numbers first.

$-4(5)(-3) = -4(-3)(5)$ Apply the commutative property of multiplication. Change the order of the factors.

$= 12(5)$ Multiply the negative numbers: $-4(-3) = 12$.

$= 60$ Do the multiplication.

After some practice, you will be able to do this work in your head.

Self Check

Multiply $-5(5)(-2)$ in two ways.

Answer: 50 ■

Powers of integers

Recall that exponential expressions are used to represent repeated multiplication. For example, 2 to the third power, or 2^3, is a shorthand way of writing $2 \cdot 2 \cdot 2$. In this expression, 3 is the exponent, and the base is positive 2. In the next example, we evaluate exponential expressions with bases that are negative numbers.

EXAMPLE 5 ***Evaluating powers of integers.*** Find each power: **a.** $(-2)^4$ and **b.** $(-5)^3$.

Solution

a. $(-2)^4 = (-2)(-2)(-2)(-2)$ Write -2 as a factor 4 times.

$= 4(-2)(-2)$ Work from left to right. Multiply -2 and -2 to get 4.

$= -8(-2)$ Work from left to right. Multiply 4 and -2 to get -8.

$= 16$ Do the multiplication.

b. $(-5)^3 = (-5)(-5)(-5)$ Write -5 as a factor 3 times.

$= 25(-5)$ Work from left to right. Multiply -5 and -5 to get 25.

$= -125$ Do the multiplication.

Self Check

Find each power:

a. $(-3)^4$

b. $(-4)^3$

Answers: **a.** 81, **b.** -64 ■

In Example 5, part a, -2 was raised to an even power, and the answer was positive. In part b, another negative number, -5, was raised to an odd power, and the answer was negative. These results suggest a general rule.

Even and odd powers of a negative integer

When a negative integer is raised to an even power, the result is positive.

When a negative integer is raised to an odd power, the result is negative.

EXAMPLE 6 ***Evaluating a power of an integer.*** Find $(-1)^5$.

Solution

We have a negative integer raised to an odd power. The result will be negative.

$$(-1)^5 = (-1)(-1)(-1)(-1)(-1)$$
$$= -1$$

Self Check

Find the power: $(-1)^8$.

Answer: 1 ■

COMMENT Although the expressions -3^2 and $(-3)^2$ look somewhat alike, they are not. In -3^2, the base is 3 and the exponent 2. The $-$ sign in front of 3^2 means the opposite of 3^2. In $(-3)^2$, the base is -3 and the exponent is 2. When we evaluate them, it becomes clear that they are not equivalent.

-3^2 represents *the opposite of* 3^2.

$-3^2 = -(3 \cdot 3)$ Write 3 as a factor 2 times.

$= -9$ Multiply within the parentheses first.

$(-3)^2$ represents $(-3)(-3)$.

$(-3)^2 = (-3)(-3)$ Write -3 as a factor 2 times.

$= 9$ The product of two negative numbers is positive.

Notice that the results are different.

EXAMPLE 7 ***The opposite of a power.*** Evaluate: **a.** -2^2 and **b.** $(-2)^2$.

Solution

a. $-2^2 = -(2 \cdot 2)$ Since 2 is the base, write 2 as a factor two times.

$= -4$ Do the multiplication within the parentheses.

b. $(-2)^2 = (-2)(-2)$ The base is -2. Write it as a factor twice.

$= 4$ The signs are like, so the product is positive.

Self Check

Evaluate: **a.** -4^2 and **b.** $(-4)^2$.

Answers: **a.** -16, **b.** 16 ■

Accent on Technology: **Raising a negative number to a power**

Negative numbers can be raised to a power using a scientific calculator. We use the change-of-sign key [+/−] and the power key [y^x] (on some calculators, [x^y]). For example, to evaluate $(-5)^6$, we enter these numbers and press these keys.

Keystrokes 5 [+/−] [y^x] 6 [=] 15625

The result is 15,625.

STUDY SET Section 2.4

VOCABULARY *Fill in the blanks to make the statements true.*

1. In the multiplication $-5(-4)$, the integers -5 and -4, which are being multiplied, are called ____________. The answer, 20, is called the ____________.

2. The definition of multiplication tells us that $3(-4)$ represents repeated ____________: $-4 + (-4) + (-4)$.

3. In the expression -3^5, ____ is the base and 5 is the ____________.

4. In the expression $(-3)^5$, ____ is the base and ____ is the exponent.

CONCEPTS *In Exercises 5–8, fill in the blanks to make the statements true.*

5. The product of two integers with ____________ signs is negative.

6. The product of two integers with like signs is ____________.

7. The ____________ property of multiplication implies that $-2(-3) = -3(-2)$.

8. The product of zero and any number is ____.

9. Find $-1(9)$. In general, what is the result when we multiply a positive number by -1?

10. Find $-1(-9)$. In general, what is the result when we multiply a negative number by -1?

11. When we multiply two integers, there are four possible combinations of signs. List each of them.

12. When multiplying two integers, there are four possible combinations of signs. How can they be grouped into two categories?

13. If each of the following powers were evaluated, what would be the *sign* of the result?

a. $(-5)^{13}$ **b.** $(-3)^{20}$

14. A student claimed, "A positive and a negative is negative." What is wrong with this statement?

15. Find each absolute value.

a. $|-3|$ **b.** $|12|$

c. $|-5|$ **d.** $|9|$

e. $|10|$ **f.** $|-25|$

16. Find each product and then graph it on a number line. What is the distance between each product?

$2(-2), 1(-2), 0(-2), -1(-2), -2(-2)$

17. a. Complete the table in Illustration 1.

Problem	Number of negative factors	Answer
$-2(-2)$		
$-2(-2)(-2)(-2)$		
$-2(-2)(-2)(-2)(-2)(-2)$		

ILLUSTRATION 1

b. The answers entered in the table help to justify the following rule: The product of an ____________ number of negative integers is positive.

18. a. Complete the table in Illustration 2.

Problem	Number of negative factors	Answer
$-2(-2)(-2)$		-8
$-2(-2)(-2)(-2)(-2)$		-32
$-2(-2)(-2)(-2)(-2)(-2)(-2)$		-128

ILLUSTRATION 2

b. The answers entered in the table help to justify the following rule: The product of an ____________ number of negative integers is negative.

NOTATION *In Exercises 19–20, complete each solution.*

19. Find $-3(-2)(-4)$.

$$-3(-2)(-4) = \_\_\_(-4)$$
$$= -24$$

20. Find $(-3)^4$.

$$(-3)^4 = (-3)(-3)(-3)\_\_\_$$
$$= \_\_\_(-3)(-3)$$
$$= \_\_\_(-3)$$
$$= 81$$

21. Explain why the expression below is not written correctly. How should it be written?

$-6 \cdot -5$

22. Translate into mathematical symbols: the product of negative three and negative two.

PRACTICE *Find each product.*

23. $-9(-6)$ **24.** $-5(-5)$

25. $-3 \cdot 5$ **26.** $-6 \cdot 4$

27. $12(-3)$
28. $11(-4)$
29. $(-8)(-7)$
30. $(-9)(-3)$
31. $(-2)10$
32. $(-3)8$
33. $-40 \cdot 3$
34. $-50 \cdot 2$
35. $-8(0)$
36. $0(-27)$
37. $-1(-6)$
38. $-1(-8)$
39. $-7(-1)$
40. $-5(-1)$
41. $1(-23)$
42. $-35(1)$

Evaluate each expression.

43. $-6(-4)(-2)$
44. $-3(-2)(-3)$
45. $5(-2)(-4)$
46. $3(-3)(3)$
47. $2(3)(-5)$
48. $6(2)(-2)$
49. $6(-5)(2)$
50. $4(-2)(2)$
51. $(-1)(-1)(-1)$
52. $(-1)(-1)(-1)(-1)$
53. $-2(-3)(3)(-1)$
54. $5(-2)(3)(-1)$
55. $3(-4)(0)$
56. $-7(-9)(0)$
57. $-2(0)(-10)$
58. $-6(0)(-12)$
59. Find the product of -6 and the opposite of 10.
60. Find the product of the opposite of 9 and the opposite of 8.

Find each power.

61. $(-4)^2$
62. $(-6)^2$
63. $(-5)^3$
64. $(-6)^3$
65. $(-2)^3$
66. $(-4)^3$
67. $(-9)^2$
68. $(-10)^2$
69. $(-1)^5$
70. $(-1)^6$
71. $(-1)^8$
72. $(-1)^9$

Evaluate each expression.

73. $(-7)^2$ and -7^2
74. $(-5)^2$ and -5^2
75. -12^2 and $(-12)^2$
76. -11^2 and $(-11)^2$

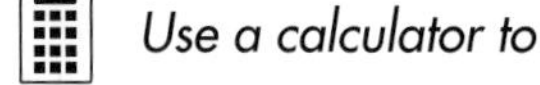
Use a calculator to evaluate each expression.

77. $-76(787)$
78. $407(-32)$
79. $(-81)^4$
80. $(-6)^5$
81. $(-32)(-12)(-67)$
82. $(-56)(-9)(-23)$
83. $(-25)^4$
84. $(-41)^5$

APPLICATIONS *Use signed numbers to help answer each problem.*

85. DIETING After giving a patient a physical exam, a physician felt that the patient should begin a diet. Two options were discussed. (See Illustration 3.)

	Plan #1	**Plan #2**
Length	10 weeks	14 weeks
Daily exercise	1 hour	30 min
Weight loss per week	3 lb	2 lb

ILLUSTRATION 3

a. Find the expected weight loss from each diet plan. Express each answer as a signed number.

b. With which plan should the patient expect to lose the most weight? Explain why the patient might not choose it.

86. INVENTORY A spreadsheet is used to record inventory losses at a warehouse. The items, their cost, and the number missing are listed in Illustration 4.

	A	B	C	D
1	Item	Cost	Number of units	$ losses
2	CD	$5	−11	
3	TV	$200	−2	
4	Radio	$20	−4	

ILLUSTRATION 4

a. What instruction should be given to find the total losses for each type of item? Find each of those losses and fill in column D.

b. What instruction should be given to find the *total* inventory losses for the warehouse? Find this number.

87. MAGNIFICATION Using an electronic testing device, a mechanic can check the emissions of a car. The results of the test are displayed on a screen. (See Illustration 5.)

a. Find the high and low values for this test as shown on the screen.

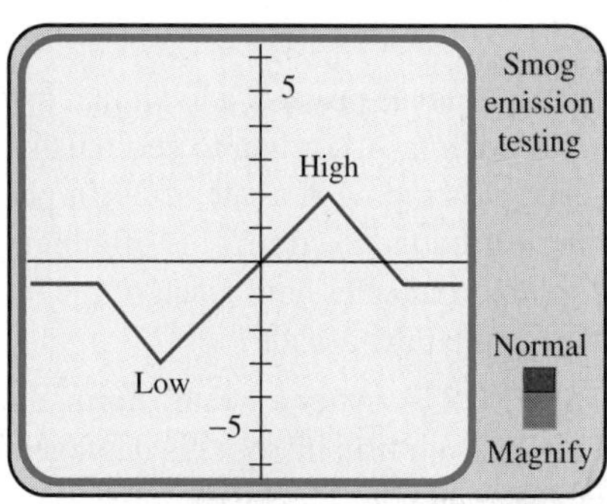

ILLUSTRATION 5

b. By switching a setting on the monitor, the picture on the screen can be magnified. What would be the new high and new low if every value were doubled?

88. LIGHT Sunlight is a mixture of all colors. When sunlight passes through water, the water absorbs different colors at different rates, as shown in Illustration 6.

a. Use a signed number to represent the depth to which red light penetrates water.

b. Green light penetrates 4 times deeper than red light. How deep is this?

c. Blue light penetrates 3 times deeper than orange light. How deep is this?

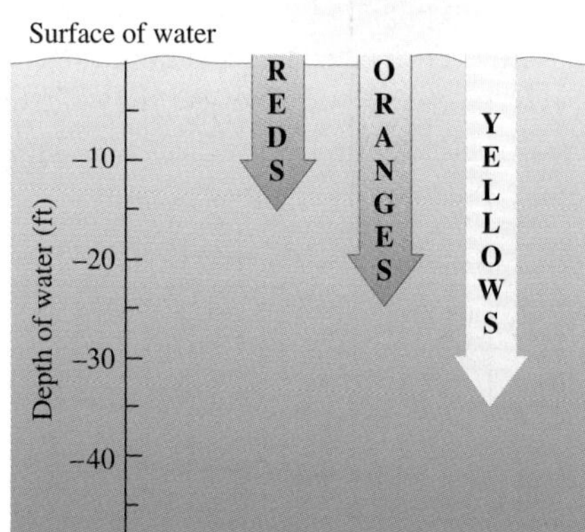

ILLUSTRATION 6

89. TEMPERATURE CHANGE A farmer, worried about his fruit trees suffering frost damage, calls the weather service for temperature information. He is told that temperatures will be decreasing approximately 4° every hour for the next 5 hours. What signed number represents the total change in temperature expected over the next five hours?

90. DEPRECIATION For each of the last four years, a businesswoman has filed a $200 depreciation allowance on her income tax return, for an office computer system. What signed number represents the total amount of depreciation written off over the four-year period?

91. EROSION A levee protects a town in a low-lying area from flooding. According to geologists, the banks of the levee are eroding at a rate of 2 feet per year. If something isn't done to correct the problem, what signed number indicates how much of the levee will erode during the next decade?

92. DECK SUPPORT After a winter storm, a homeowner has an engineering firm inspect his damaged deck. Their report concludes that the original pilings were not anchored deep enough, by a factor of 3. (See Illustration 7.) What signed number represents the depth to which the pilings should have been sunk?

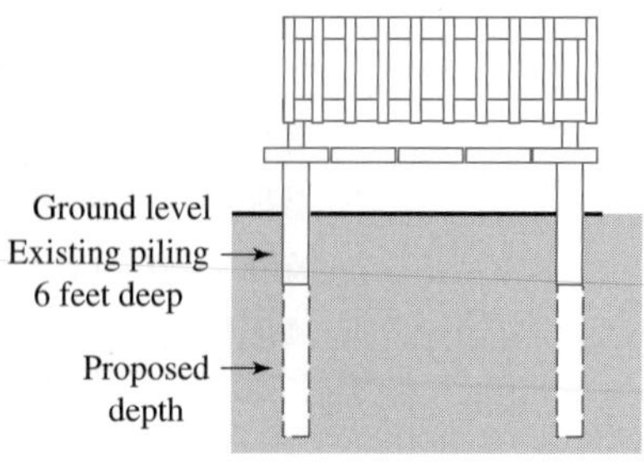

ILLUSTRATION 7

93. WOMEN'S NATIONAL BASKETBALL ASSOCIATION The average attendance for the WNBA Houston Comets is 11,906 a game. Suppose the team gives a sports bag, costing $3, to everyone attending a game. What signed number expresses the financial loss from this promotional giveaway?

94. HEALTH CARE A health care provider for a company estimates that 75 hours per week are lost by employees suffering from stress-related or preventable illness. In a 52-week year, how many hours are lost? Use a signed number to answer.

WRITING

95. If a product contains an even number of negative factors, how do we know that the result will be positive?

96. Explain why the product of a positive number and a negative number is negative, using $5(-3)$ as an example.

97. Explain why the result is the opposite of the original number when a number is multiplied by -1.

98. Can you think of any number that yields a nonzero result when it is multiplied by 0? Explain your response.

REVIEW

99. The prime factorization of a number is $3^2 \cdot 5$. What is the number?

100. Solve for y: $\frac{y}{8} = 10$.

101. The enrollment at a college went from 10,200 to 12,300 in one year. What was the increase in enrollment?

102. Find the perimeter of a square with sides 6 yards long.

103. What does the symbol $<$ mean?

104. List the first ten prime numbers.

2.5 *Division of Integers*

In this section, you will learn about

- The relationship between multiplication and division
- Rules for dividing integers • Division and zero

Introduction. In this section, we will develop rules for division of integers, just as we did for multiplication of integers. We will also consider two types of division involving zero.

The relationship between multiplication and division

When we solved equations in Chapter 1, multiplication was used to undo division, and division was used to undo multiplication. Because one operation undoes the other, multiplication and division are said to be *inverse operations.* Every division fact containing three numbers can be written as an equivalent multiplication fact involving the same three numbers. For example,

$$\frac{6}{3} = 2 \quad \text{because} \quad 3(2) = 6$$

Remember that in the division statement, 6 is the *dividend,* 3 is the *divisor,* and 2 is the *quotient.*

Rules for dividing integers

We will now use the relationship between multiplication and division to help develop rules for dividing integers. There are four cases to consider.

Case 1: In the first case, a positive integer is divided by a positive integer. From years of experience, we already know that the result is positive. Therefore, *the quotient of two positive integers is positive.*

Case 2: Next, we consider the quotient of two negative integers. As an example, consider the division $\frac{-12}{-2} = ?$ We can do this division by examining its related multiplication statement, $-2(?) = -12$. Our objective is to find the number that should replace the question mark. To do this, we use the rules for multiplying integers, introduced in the previous section.

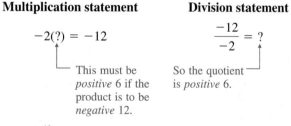

Multiplication statement	**Division statement**
$-2(?) = -12$	$\frac{-12}{-2} = ?$
This must be *positive* 6 if the product is to be *negative* 12.	So the quotient is *positive* 6.

Therefore, $\frac{-12}{-2} = 6$. From this example, we can see that *the quotient of two negative integers is positive.*

Case 3: The third case we will examine is the quotient of a positive integer and a negative integer. Let's consider $\frac{12}{-2} = ?$ Its equivalent multiplication statement is $-2(?) = 12$.

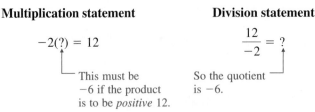

Multiplication statement	**Division statement**
$-2(?) = 12$	$\frac{12}{-2} = ?$
This must be -6 if the product is to be *positive* 12.	So the quotient is -6.

Therefore, $\frac{12}{-2} = -6$. This result shows that *the quotient of a positive integer and a negative integer is negative.*

Case 4: Finally, to find the quotient of a negative integer and a positive integer, let's consider $\frac{-12}{2} = ?$ Its equivalent multiplication statement is $2(?) = -12$.

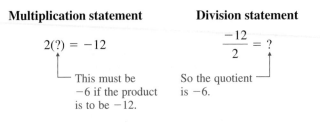

Multiplication statement	**Division statement**
$2(?) = -12$	$\frac{-12}{2} = ?$
This must be -6 if the product is to be -12.	So the quotient is -6.

Therefore, $\frac{-12}{2} = -6$. From this example, we can see that *the quotient of a negative integer and a positive integer is negative.*

We now summarize the results from the previous discussion.

Dividing two integers

To divide two integers, divide their absolute values.

1. The quotient of two integers with *like* signs is positive.
2. The quotient of two integers with *unlike* signs is negative.

The rules for dividing integers are similar to those for multiplying integers.

EXAMPLE 1 ***Dividing integers.*** Find each quotient:

a. $\frac{-35}{7}$ and **b.** $\frac{20}{-5}$.

Solution

To divide integers with unlike signs, we find the quotient of their absolute values and make the quotient negative.

a. $\frac{-35}{7} = -5$ Divide the absolute values of the numbers, 35 by 7, to get 5. The quotient is negative.

To check the result, we multiply the divisor, 7, and the quotient, -5. We should obtain the dividend, -35.

$$7(-5) = -35$$

The answer, -5, checks.

b. $\frac{20}{-5} = -4$ Divide the absolute values, 20 by 5, to get 4. The quotient is negative.

Self Check

Find each quotient. Then check the result.

a. $\frac{-45}{5}$

b. $\frac{60}{-20}$

Answers: **a.** -9, **b.** -3

EXAMPLE 2 ***Dividing integers.*** Divide: $\frac{-12}{-3}$.

Solution

The integers have like signs. The quotient will be positive.

$\frac{-12}{-3} = 4$ Divide the absolute values, 12 by 3, to get 4. The quotient is positive.

Self Check

Divide: $\frac{-21}{-3}$.

Answer: 7

EXAMPLE 3 ***Price reduction.*** Over the course of a year, a retailer reduced the price of a television set by an equal amount each month, because it was not selling. By the end of the year, the cost was \$132 less than at the beginning of the year. How much did the price fall each month?

Solution We will label the drop in price of \$132 for the year as -132. It occurred in 12 equal reductions. This indicates division.

$\frac{-132}{12} = -11$ The quotient of a negative number and a positive number is negative.

The drop in price each month was \$11.

Division and zero

In Chapter 1, we discussed division involving zero.

To review the concept of division of zero, we will look at $\frac{0}{2} = ?$ The equivalent multiplication statement is $2(?) = 0$.

Multiplication statement	Division statement
$2(?) = 0$	$\frac{0}{2} = ?$
This must be 0 if the product is to be 0.	So the quotient is 0.

Therefore, $\frac{0}{2} = 0$. This example suggests that *the quotient of zero divided by any nonzero number is zero.*

To review division by zero, let's look at $\frac{2}{0} = ?$ The equivalent multiplication statement is $0(?) = 2$.

Multiplication statement	Division statement
$0(?) = 2$	$\frac{2}{0} = ?$
There is no number that gives 2 when multiplied by 0.	There is no quotient.

Therefore, $\frac{2}{0}$ does not have an answer. We say that division by zero is **undefined.** This example suggests that *the quotient of any number divided by zero is undefined.*

Division with 0

1. If a represents any nonzero number, $\frac{0}{a} = 0$.
2. If a represents any nonzero number, $\frac{a}{0}$ is undefined.
3. $\frac{0}{0}$ is undetermined.

EXAMPLE 4 ***Division with zero.*** Find $\frac{-4}{0}$, if possible.

Solution

Since $\frac{-4}{0}$ is division by 0, the division is undefined.

Self Check

Find $\frac{0}{-4}$.

Answer: 0

Accent on Technology: Division with negative numbers

The Bureau of Labor statistics estimated that the United States lost 270,000 jobs in the manufacturing sector of the economy in 1999. Because the jobs were lost, we write this as $-270{,}000$. To find the average number of manufacturing jobs lost each month, we divide: $\frac{-270{,}000}{12}$. To do this division using a scientific calculator, we enter these numbers and press these keys.

Keystrokes 270000 [+/−] [÷] 12 [=] → -22500

The average number of manufacturing jobs lost each month in 1999 was 22,500.

STUDY SET Section 2.5

VOCABULARY *Fill in the blanks to make the statements true.*

1. In $\frac{-27}{3} = -9$, the number -9 is called the ____________, and the number 3 is the ____________.
2. Division by zero is ____________. Division ____________ zero by a nonzero number is 0.
3. The ____________ of a number is the distance between it and 0 on the number line.
4. $\{\ldots, -4, -3, -2, -1, 0, 1, 2, 3, 4, \ldots\}$ is the set of ____________.
5. The quotient of two negative integers is ____________.
6. The quotient of a negative integer and a positive integer is ____________.

CONCEPTS

7. Write the related multiplication statement for $\frac{-25}{5} = -5$.
8. Write the related multiplication statement for $\frac{0}{-15} = 0$.
9. Show that there is no answer for $\frac{-6}{0}$ by writing the related multiplication statement.
10. If x is any number except zero, what is $\frac{0}{x}$?
11. Write a related division statement for $5(-4) = -20$.
12. How do the rules for multiplying integers compare with the rules for dividing integers?
13. Tell whether each statement is always true, sometimes true, or never true.
    - **a.** The product of a positive integer and a negative integer is negative.
    - **b.** The sum of a positive integer and a negative integer is negative.
    - **c.** The quotient of a positive integer and a negative integer is negative.
14. Tell whether each statement is always true, sometimes true, or never true.
    - **a.** The product of two negative integers is positive.
    - **b.** The sum of two negative integers is negative.
    - **c.** The quotient of two negative integers is negative.

PRACTICE *Find each quotient, if possible.*

15. $\frac{-14}{2}$
16. $\frac{-10}{5}$
17. $\frac{-8}{-4}$
18. $\frac{-12}{-3}$
19. $\frac{-25}{-5}$
20. $\frac{-36}{-12}$
21. $\frac{-45}{-15}$
22. $\frac{-81}{-9}$

23. $\frac{40}{-2}$ **24.** $\frac{35}{-7}$

25. $\frac{50}{-25}$ **26.** $\frac{80}{-40}$

27. $\frac{0}{-16}$ **28.** $\frac{0}{-6}$

29. $\frac{-6}{0}$ **30.** $\frac{-8}{0}$

31. $\frac{-5}{1}$ **32.** $\frac{-9}{1}$

33. $-5 \div (-5)$ **34.** $-11 \div (-11)$

35. $\frac{-9}{9}$ **36.** $\frac{-15}{15}$

37. $\frac{-10}{-1}$ **38.** $\frac{-12}{-1}$

39. $\frac{-100}{25}$ **40.** $\frac{-100}{50}$

41. $\frac{75}{-25}$ **42.** $\frac{300}{-100}$

43. $\frac{-500}{-100}$ **44.** $\frac{-60}{-30}$

45. $\frac{-200}{50}$ **46.** $\frac{-500}{100}$

47. Find the quotient of -45 and 9.

48. Find the quotient of -36 and -4.

49. Divide 8 by -2.

50. Divide -16 by -8.

51. $\frac{-13{,}550}{25}$ **52.** $\frac{-3{,}876}{-19}$

53. $\frac{272}{-17}$ **54.** $\frac{-6{,}776}{-77}$

APPLICATIONS *Use signed numbers to help answer each problem.*

55. TEMPERATURE DROP During a five-hour period, the temperature steadily dropped. (See Illustration 1.) What was the average change in the temperature per hour over this five-hour time span?

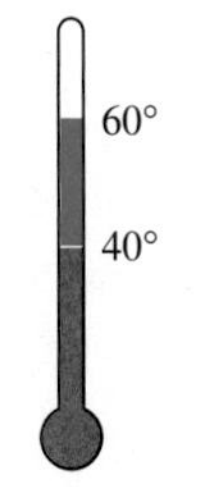

ILLUSTRATION 1

56. PRICE DROP Over a three-month period, the price of a VCR steadily fell. (See Illustration 2.) What was the average monthly change in the price of the VCR over this period?

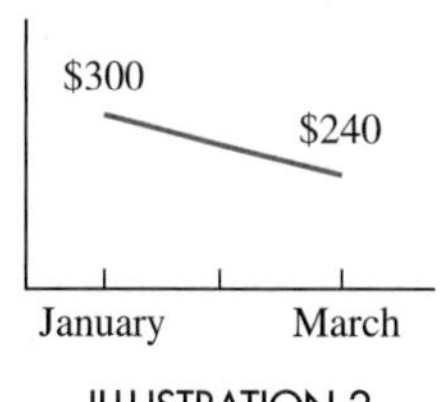

ILLUSTRATION 2

57. SUBMARINE DIVE In a series of three equal dives, a submarine is programmed to reach a depth of 3,000 feet below the ocean surface. (See Illustration 3.) What signed number describes how deep each of the three dives will be?

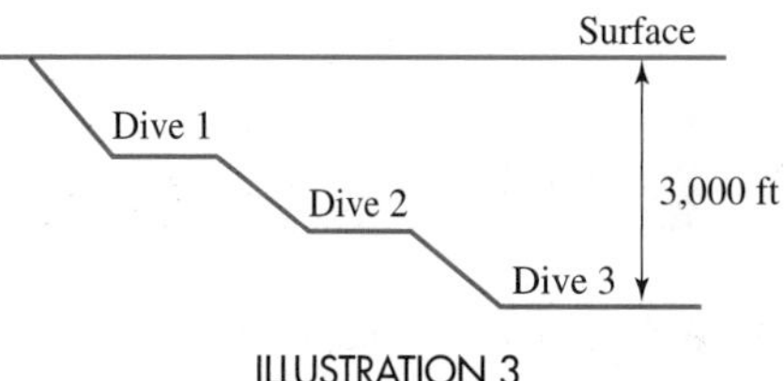

ILLUSTRATION 3

58. GRAND CANYON TRIP A mule train is to travel from a stable on the rim of the Grand Canyon to a camp on the canyon floor, approximately 5,000 feet below the rim. If the guide wants the mules to be rested after every 1,000 feet of descent, how many stops will be made on the trip?

59. BASEBALL TRADE At the midway point of the season, a baseball team finds itself 12 games behind the league leader. Team management decides to trade for a talented hitter, in hopes of making up at least half of the deficit in the standings by the end of the year. Where in the league standings does management expect to finish at season's end?

60. BUDGET DEFICIT A politician proposed a two-year plan for cutting a county's $20 million budget deficit, as shown in Illustration 4. If this plan is put into effect, what will be the change in the county's financial status in two years?

	Plan	Prediction
1st year	Raise taxes, drop subsidy programs	Will cut deficit in half
2nd year	Search out waste and fraud	Will cut remaining deficit in half

ILLUSTRATION 4

61. PRICE MARKDOWN The owner of a clothing store decides to reduce the price on a line of jeans that are not selling. She feels she can afford to lose \$300 of projected income on these pants. By how much can she mark down each of the 20 pairs of jeans?

62. WATER RESERVOIR Over a week's time, engineers at a city water reservoir released enough water to lower the water level 35 feet. On average, how much did the water level change each day during this period?

63. PAY CUT In a cost-cutting effort, a business decides to lower expenditures on salaries by \$9,135,000. To do this, all of the 5,250 employees will have their salaries reduced by an equal dollar amount. How big a pay cut will each employee experience?

64. STOCK MARKET On Monday, the value of Maria's 255 shares of stock was at an all-time high. By Friday, the value had fallen \$4,335. What was her per-share loss that week?

WRITING

65. Explain why the quotient of two negative numbers is positive.

66. Think of a real-life situation that could be represented by $\frac{0}{4}$. Explain why the answer would be zero.

67. Using a specific example, explain how multiplication can be used as a check for division.

68. Explain what it means when we say that division by zero is undefined.

REVIEW

69. Evaluate: $3\left(\frac{18}{3}\right)^2 - 2(2)$.

70. List the set of whole numbers.

71. Find the prime factorization of 210.

72. Complete the statement: The addition property of equality states that if $x = y$ and c is any number, then $x + c =$ ____________.

73. Solve $99 = r + 43$.

74. Does $8 - 2 = 2 - 8$?

75. Evaluate: 3^4.

76. Sharif has scores of 55, 70, 80, and 75 on four mathematics tests. What is his mean (average) score?

2.6 Order of Operations and Estimation

In this section, you will learn about

- Order of operations • Absolute value • Estimation

Introduction. In this section, we will evaluate expressions that involve more than one operation. To do this, we will apply the rules for the order of operations discussed in Chapter 1, as well as the rules for working with integers. We will also continue the discussion of estimating an answer. Estimation can be used when you need a quick indication of the size of the actual answer to a calculation.

Order of operations

In Section 1.5, we introduced the rules for the order of operations: an agreed-upon sequence of steps for completing the operations of arithmetic.

Order of operations

1. Do all calculations within parentheses and other grouping symbols, following the order listed in Steps 2–4 and working from the innermost pair to the outermost pair.
2. Evaluate all exponential expressions.
3. Do all multiplications and divisions as they occur from left to right.
4. Do all additions and subtractions as they occur from left to right.

When all grouping symbols have been removed, repeat Steps 2–4 to complete the calculation.

If a fraction bar is present, evaluate the expression above the bar (the *numerator*) and the expression below the bar (the *denominator*) separately. Then do the division indicated by the fraction bar, if possible.

EXAMPLE 1 ***Order of operations.*** Evaluate: $-4(-3)^2 - (-2)$.

Solution

This expression contains the operations of multiplication, raising to a power, and subtraction. The rules for the order of operations tell us to find the power first.

$-4(-3)^2 - (-2) = -4(9) - (-2)$	Evaluate the exponential expression: $(-3)^2 = 9$.
$= -36 - (-2)$	Do the multiplication: $-4(9) = -36$.
$= -36 + 2$	To do the subtraction, add the opposite of -2.
$= -34$	Do the addition.

Self Check

Evaluate: $-5(-2)^2 - (-6)$.

Answer: -14

EXAMPLE 2 ***Order of operations.*** Evaluate: $2(3) + (-5)(-3)(-2)$.

Solution

This expression contains the operations of multiplication and addition. By the rules for the order of operations, we do the multiplications first.

$2(3) + (-5)(-3)(-2) = 6 + (-30)$	Working from left to right, do the multiplications.
$= -24$	Do the addition.

Self Check

Evaluate: $4(2) + (-4)(-3)(-2)$.

Answer: -16

EXAMPLE 3 ***Order of operations.*** Evaluate: $40 \div (-4)5$.

Solution

This expression contains the operations of division and multiplication. We do the divisions and multiplications as they occur from left to right.

$40 \div (-4)5 = -10 \cdot 5$	Do the division first: $40 \div (-4) = -10$.
$= -50$	Do the multiplication.

Self Check

Evaluate: $45 \div (-5)3$.

Answer: -27

EXAMPLE 4 ***Order of operations.*** Evaluate: $-2^2 - (-2)^2$.

Solution

This expression contains the operations of raising to a power and subtraction. We are to find the powers first. (Recall that -2^2 means the *opposite* of 2^2.)

$-2^2 - (-2)^2 = -4 - 4$	Find the powers: $-2^2 = -4$ and $(-2)^2 = 4$.
$= -8$	Do the subtraction.

Self Check

Evaluate: $-3^2 - (-3)^2$.

Answer: -18

EXAMPLE 5 ***Working with grouping symbols.*** Evaluate: $-15 + 3(-4 + 7)$.

Solution

$-15 + 3(-4 + 7) = -15 + 3(3)$	Do the addition within the parentheses: $-4 + 7 = 3$.
$= -15 + 9$	Do the multiplication: $3(3) = 9$.
$= -6$	Do the addition.

Self Check

Evaluate: $-18 + 4(-7 + 9)$.

Answer: -10

EXAMPLE 6 ***Working with a fraction bar.*** Evaluate: $\dfrac{-20 + 3(-5)}{(-4)^2 - 21}$.

Solution

We first simplify the expressions in the numerator and the denominator, separately.

$$\frac{-20 + 3(-5)}{(-4)^2 - 21} = \frac{-20 + (-15)}{16 - 21}$$ In the numerator, do the multiplication: $3(-5) = -15$. In the denominator, evaluate the power: $(-4)^2 = 16$.

$$= \frac{-35}{-5}$$ In the numerator, add: $-20 + (-15) = -35$. In the denominator, subtract: $16 - 21 = -5$.

$$= 7$$ Do the division.

Self Check

Evaluate: $\dfrac{-9 + 6(-4)}{(-5)^2 - 28}$.

Answer: 11 ■

EXAMPLE 7 ***Grouping symbols within grouping symbols.*** Evaluate: $-5[-1 + (2 - 8)^2]$.

Solution

We begin by working within the innermost pair of grouping symbols and work to the outermost pair.

$-5[-1 + (2 - 8)^2] = -5[-1 + (-6)^2]$ Do the subtraction within the parentheses.

$= -5(-1 + 36)$ Evaluate the power within the brackets.

$= -5(35)$ Do the addition within the parentheses.

$= -175$ Do the multiplication.

Self Check

Evaluate: $-4[-2 + (5 - 9)^2]$.

Answer: -56 ■

Absolute value

You will recall that the absolute value of a number is the distance between the number and 0 on a number line. Earlier in this chapter, we evaluated simple absolute value expressions such as $|-3|$ and $|10|$. Absolute value symbols are also used in combination with more complicated expressions, such as $|-4(3)|$ and $|-6 + 1|$. When we apply the rules for the order of operations to evaluate these expressions, *the absolute value symbols are considered to be grouping symbols,* and any operations within them are to be completed first.

EXAMPLE 8 ***Absolute value.*** Find each absolute value: **a.** $|-4(3)|$ and **b.** $|-6 + 1|$.

Solution

We do the operations within the absolute value symbols first.

a. $|-4(3)| = |-12|$ Do the multiplication within the absolute value symbols: $-4(3) = -12$.

$= 12$ Find the absolute value of -12.

b. $|-6 + 1| = |-5|$ Do the addition within the absolute value symbols: $-6 + 1 = -5$.

$= 5$ Find the absolute value of -5.

Self Check

Find each absolute value:

a. $|(-6)(5)|$

b. $|-3 + (-4)|$

Answers: **a.** 30, **b.** 7 ■

COMMENT Just as $-5(8)$ means $-5 \cdot 8$, the expression $-5|8|$ (read as "negative 5 times the absolute value of 8") means $-5 \cdot |8|$. To evaluate such an expression, we find the absolute value first and then multiply.

$-5|8| = -5 \cdot 8$ Find the absolute value: $|8| = 8$.

$= -40$ Do the multiplication.

EXAMPLE 9 ***Grouping symbols.*** Evaluate: $8 - 4|-6 - 2|$.

Solution

We do the operation within the absolute value symbols first.

$8 - 4\|-6 - 2\| = 8 - 4\|-8\|$	Do the subtraction within the absolute value symbols: $-6 - 2 = -8$.
$= 8 - 4(8)$	Find the absolute value: $\|-8\| = 8$.
$= 8 - 32$	Do the multiplication: $4(8) = 32$.
$= -24$	Do the subtraction.

Self Check

Evaluate: $7 - 5|-1 - 6|$.

7-5|-1+6|: -7
7-5(7)
7-35
7+-35= -28

Answer: -28

Estimation

Recall that the idea behind estimation is to simplify calculations by using rounded numbers that are close to the actual values in the problem. When an exact answer is not necessary and a quick approximation will do, we can use estimation.

EXAMPLE 10 ***The stock market.*** The Dow Jones Industrial Average is announced at the end of each trading day to give investors an indication of how the New York Stock Exchange performed. A positive number indicates good performance, while a negative number indicates poor performance. Estimate the net gain or loss of points in the Dow for the first week of the year 2000, shown in Figure 2-18.

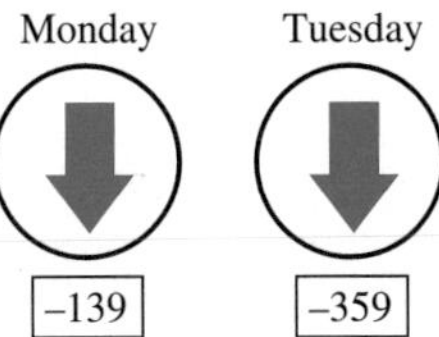

FIGURE 2-18

Solution

We will approximate each of these numbers. For example, -139 is close to -140, and $+124$ is close to $+120$. To estimate the net gain or loss, we add the approximations.

$-140 + (-360) + 120 + 130 + 270 = -500 + 520$	Add positive and negative numbers separately to get subtotals.
$= 20$	Do the addition.

This estimate tells us that there was a gain of approximately 20 points in the Dow.

STUDY SET Section 2.6

VOCABULARY *Fill in the blanks to make the statements true.*

1. When asked to evaluate expressions containing more than one operation, we should apply the rules for the ______________ of operations.
2. In situations where an exact answer is not needed, an approximation or ______________ is a quick way of obtaining a rough idea of the size of the actual answer.
3. Absolute value symbols, parentheses, and brackets are types of ______________ symbols.
4. If an expression involves two sets of grouping symbols, always begin working within the ______________ symbols and then work to the ______________.

CONCEPTS

5. Consider $5(-2)^2 - 1$. How many operations need to be performed to evaluate this expression? List them in the order in which they should be performed.

6. Consider $15 - 3 + (-5 \cdot 2)^3$. How many operations need to be performed to evaluate this expression? List them in the order in which they should be performed.

7. Consider $\dfrac{5 + 5(7)}{2 + (4 - 8)}$. In the numerator, what operation should be performed first? In the denominator, what operation should be performed first?

8. In the expression $4 + 2(-7 - 1)$, how many operations need to be performed? List them in the order in which they should be performed.

9. Explain the difference between -3^2 and $(-3)^2$.

10. In the expression $-2 \cdot 3^2$, what operation should be performed first?

NOTATION *Complete each solution.*

11. Evaluate: $-8 - 5(-2)^2$.

$$\begin{aligned} -8 - 5(-2)^2 &= -8 - 5(4) \\ &= -8 - 20 \\ &= -8 + (-20) \\ &= -28 \end{aligned}$$

12. Evaluate: $2 + (5 - 6 \cdot 2)$.

$$\begin{aligned} 2 + (5 - 6 \cdot 2) &= 2 + (5 - 12) \\ &= 2 + [5 + (-12)] \\ &= 2 + (-7) \\ &= -5 \end{aligned}$$

13. Evaluate: $[-4(2 + 7)] - 6$.

$$\begin{aligned} [-4(2 + 7)] - 6 &= [-4(9)] - 6 \\ &= -36 - 6 \\ &= -42 \end{aligned}$$

14. Evaluate: $\dfrac{|-9 + (-3)|}{9 - 6}$.

$$\begin{aligned} \frac{|-9 + (-3)|}{9 - 6} &= \frac{|12|}{3} \\ &= \frac{12}{3} \\ &= 4 \end{aligned}$$

PRACTICE *Evaluate each expression.*

15. $(-3)^2 - 4^2$

16. $-7 + 4 \cdot 5$

17. $3^2 - 4(-2)(-1)$

18. $2^3 - 3^3$

19. $(2 - 5)(5 + 2)$

20. $-3(2)^2 4$

21. $-10 - 2^2$

22. $-50 - 3^3$

23. $\dfrac{-6 - 8}{2}$

24. $\dfrac{-6 - 6}{-2 - 2}$

25. $\dfrac{-5 - 5}{2}$

26. $\dfrac{-7 - (-3)}{2 - 4}$

27. $-12 \div (-2)2$

28. $-60(-2) \div 3$

29. $-16 - 4 \div (-2)$

30. $-24 + 4 \div (-2)$

31. $|-5(-6)|$

32. $|-7 - 9|$

33. $|-4 - (-6)|$

34. $|-2 + 6 - 5|$

35. $5|3|$

36. $5|4|$

37. $-6|-7|$

38. $-6|-4|$

39. $(7 - 5)^2 - (1 - 4)^2$

40. $5^2 - (-9 - 3)$

41. $-1(2^2 - 2 + 1^2)$

42. $(-7 - 4)^2 - (-1)$

43. $-50 - 2(-3)^3$

44. $(-2)^3 - (-3)(-2)$

45. $-6^2 + 6^2$

46. $-9^2 + 9^2$

47. $3\left(\dfrac{-18}{3}\right) - 2(-2)$

48. $2\left(\dfrac{-12}{3}\right) + 3(-5)$

49. $6 + \dfrac{25}{-5} + 6 \cdot 3$

50. $-5 - \dfrac{24}{6} + 8(-2)$

51. $\dfrac{1 - 3^2}{-2}$

52. $\dfrac{-3 - (-7)}{2^2 - 3}$

53. $\dfrac{-4(-5) - 2}{-6}$

54. $\dfrac{(-6)^2 - 1}{-4 - 3}$

55. $-3\left(\dfrac{32}{-4}\right) - (-1)^5$

56. $-5\left(\dfrac{16}{-4}\right) - (-1)^4$

57. $6(2^3)(-1)$

58. $2(3^3)(-2)$

59. $2 + 3[5 - (1 - 10)]$

60. $12 - 2[1 - (-8 + 2)]$

61. $-7(2 - 3 \cdot 5)$

62. $-4(1 + 3 \cdot 5)$

63. $-[6 - (1 - 4)^2]$

64. $-[9 - (9 - 12)^2]$

65. $15 + (-3 \cdot 4 - 8)$

66. $11 + (-2 \cdot 2 + 3)$

67. $|-3 \cdot 4 + (-5)|$

68. $|-8 \cdot 5 - 2 \cdot 5|$

69. $|(-5)^2 - 2 \cdot 7|$

70. $|8 \div (-2) - 5|$

71. $-2 + |6 - 4^2|$

72. $-3 - 4|6 - 7|$

73. $2|1 - 8| \cdot |-8|$

74. $2(5) - 6(|-3|)^2$

75. $-2(-34)^2 - (-605)$

76. $11 - (-15)(24)^2$

77. $-60 - \dfrac{1{,}620}{-36}$

78. $\dfrac{2^5 - 4^6}{-42 + 58}$

Make a mental estimate.

79. $-379 + (-103) + 287$

80. $\dfrac{-67 - 9}{-18}$

81. $-39 \cdot 8$

82. $-568 - (-227)$

83. $-3{,}887 + (-5{,}106)$

84. $-333(-4)$

85. $\dfrac{6{,}267}{-5}$

86. $-36 + (-78) + 59 + (-4)$

APPLICATIONS

87. TESTING In an effort to discourage her students from guessing on multiple-choice tests, a professor uses the grading scale shown in Illustration 1. If unsure of an answer, a student does best to skip the question, because incorrect responses are penalized very heavily. Find the test score of a student who gets 12 correct and 3 wrong and leaves 5 questions blank.

Response	Value
Correct	+3
Incorrect	−4
Left blank	−1

ILLUSTRATION 1

88. THE FEDERAL BUDGET See Illustration 2. Suppose you were hired to write a speech for a politician who wanted to highlight the improvement in the federal government's finances during 1990s. Would it be better for the politician to refer to the average budget deficit/surplus for the last half, or for the last four years of that decade? Explain your reasoning.

U.S. Budget Deficit/Surplus
($ billions)

Deficit	Year	Surplus
−164	1995	
−107	1996	
−22	1997	
	1998	+70
	1999	+123

ILLUSTRATION 2

89. SCOUTING REPORT Illustration 3 shows a football coach how successful his opponent was running a "28 pitch" the last time the two teams met. What was the opponent's average gain with this play?

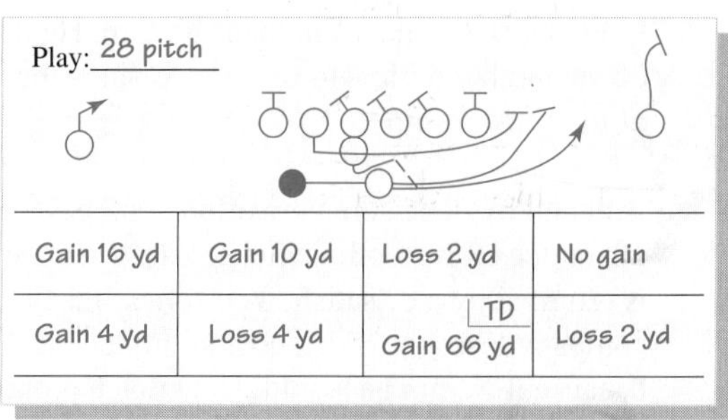

Play: 28 pitch
Gain 16 yd
Gain 10 yd
Loss 2 yd
No gain
Gain 4 yd
Loss 4 yd
TD
Gain 66 yd
Loss 2 yd

ILLUSTRATION 3

90. SPREADSHEET Illustration 4 shows the data from a chemistry experiment in spreadsheet form. To obtain a result, the chemist needs to add the values in row 1, double that sum, and then divide that number by the smallest value in column C. What is the final result of these calculations?

	A	B	C	D
1	12	−5	6	−2
2	15	4	5	−4
3	6	4	−2	8

ILLUSTRATION 4

Use estimation to answer each question.

91. OIL PRICES The price per barrel of crude oil fluctuates with supply and demand. It can rise and fall quickly. The line graph in Illustration 5 shows how many cents the price per barrel rose or fell each day for a week. Estimate the net gain or loss in the value of a barrel of crude oil for the week.

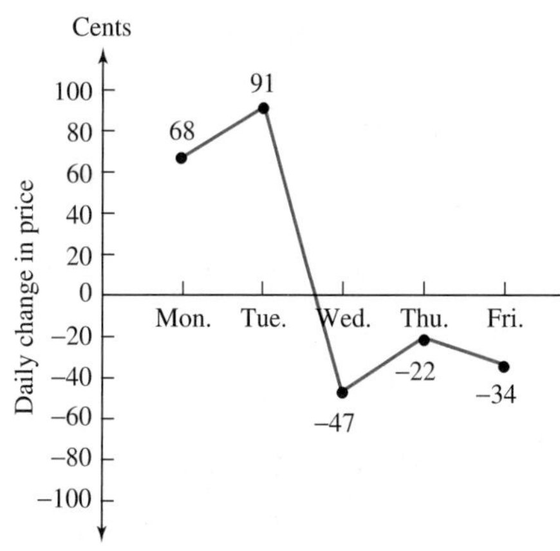

ILLUSTRATION 5

92. ESTIMATION Quickly determine a reasonable estimate of the exact answer in each of the following situations.

a. A diver, swimming at a depth of 34 feet below sea level, spots a sunken ship beneath him. He dives down another 57 feet to reach it. What is the depth of the sunken ship?

b. A dental hygiene company offers a money-back guarantee on its tooth whitener kit. When the kit is returned by a dissatisfied customer, the company loses the $11 it cost to produce it, because it cannot be resold. How much money has the company lost because of this return policy if 56 kits have been mailed back by customers?

c. A tram line makes a 7,561-foot descent from a mountaintop in 18 equal stages. How much does it descend in each stage?

WRITING

93. When evaluating expressions, why are rules for the order of operations necessary?

94. In the rules for the order of operations, what does the phrase *as they occur from left to right* mean?

95. Name a situation in daily life where you use estimation.

96. List some advantages and some disadvantages of the process of estimation.

REVIEW

97. Solve $8 = 2x$.

98. Evaluate: $6^2 - (10 - 8)$.

99. How do we find the perimeter of a rectangle?

100. Is this statement true or false? "When measuring perimeter, we use square units."

101. An elevator has a weight capacity of 1,000 pounds. Seven people, with an average weight of 140 pounds, are in it. Is it overloaded?

102. List the factors of 36.

2.7 *Solving Equations Involving Integers*

In this section, you will learn about

- The properties of equality • Solving equations involving $-x$
- Combinations of operations • Problem solving with equations

INTRODUCTION. In this section, we revisit the topic of solving equations. The equations that we will be solving involve negative numbers, and some of the solutions are negative numbers as well. The section concludes with some application problems. As before, we will use a five-step problem-solving strategy to solve them.

The properties of equality

Recall that the addition property of equality states: *If the same number is added to equal quantities, the results will be equal quantities.* When working with negative numbers, this property can be used in a new way.

To solve $x + (-8) = -10$, we need to isolate x on the left-hand side of the equation. We can do this by adding 8 to both sides.

$$x + (-8) = -10$$

$$x + (-8) + \mathbf{8} = -10 + \mathbf{8}$$ Apply the addition property of equality. To undo the addition of -8, add 8 to both sides.

$$x = -2$$ Do the additions: $(-8) + 8 = 0$ and $-10 + 8 = -2$.

From this example, we see that to undo addition, we can *add the opposite* of the number that is added to the variable.

EXAMPLE 1 ***The addition property of equality.*** Solve $-4 + x = -12$ and check the result.

Solution

We want to isolate x on one side of the equation. We can do this by adding the opposite of -4 to both sides.

$-4 + x = -12$	
$-4 + x \mathbf{+ 4} = -12 \mathbf{+ 4}$	To undo the addition of -4, add 4 to both sides.
$x = -8$	Do the additions: $-4 + 4 = 0$ and $-12 + 4 = -8$.

To check, we substitute -8 for x in the original equation and then simplify.

$-4 + x = -12$	
$-4 + (\mathbf{-8}) \stackrel{?}{=} -12$	Substitute -8 for x.
$-12 = -12$	Do the addition: $-4 + (-8) = -12$.

$x = -8$ is a solution.

Self Check

Solve $-6 + y = -8$ and check the result.

Answer: -2 ■

Recall that the subtraction property of equality states: *If the same quantity is subtracted from equal quantities, the results will be equal quantities.*

EXAMPLE 2 ***The subtraction property of equality.***

Solve $x + 16 = -8$.

Solution

$x + 16 = -8$	
$x + 16 \mathbf{- 16} = -8 \mathbf{- 16}$	To undo the addition of 16, subtract 16 from both sides.
$x = -8 + (-16)$	Simplify the left side of the equation. On the right side, write the subtraction as addition of the opposite.
$x = -24$	Do the addition.

Check the result.

Self Check

Solve $c + 4 = -3$.

Answer: -7 ■

EXAMPLE 3 ***Simplifying first.*** Solve $-3 + 7 = h + 11(-2)$ and check the result.

Solution

Equations should be simplified before we use the properties of equality. In this case, there is an addition on the left-hand side and a multiplication on the right-hand side of the equation. We compute those first.

$-3 + 7 = h + 11(-2)$	
$4 = h + (-22)$	Do the addition: $-3 + 7 = 4$. Do the multiplication: $11(-2) = -22$.
$4 \mathbf{+ 22} = h + (-22) \mathbf{+ 22}$	To undo the addition of -22, add 22 to both sides.
$26 = h$	Simplify: $4 + 22 = 26$ and $(-22) + 22 = 0$.
$h = 26$	Since $26 = h$, $h = 26$.

Check

$-3 + 7 = h + 11(-2)$	The original equation.
$-3 + 7 \stackrel{?}{=} \mathbf{26} + 11(-2)$	Substitute 26 for h.
$4 \stackrel{?}{=} 26 + (-22)$	Simplify both sides.
$4 = 4$	Do the addition.

Self Check

Solve $-2 + 8 = y + 3(-4)$ and check the result.

Answer: 18 ■

Recall that the division property of equality states: *If equal quantities are divided by the same nonzero quantity, the results will be equal quantities.*

EXAMPLE 4 ***The division property of equality.*** Solve each equation: **a.** $-3x = 15$ and **b.** $-16 = -4y$.

Solution

a. Recall that $-3x$ indicates multiplication: $-3 \cdot x$. We must undo the multiplication of x by -3. To do this, we divide by -3.

$-3x = 15$ The original equation.

$\dfrac{-3x}{-3} = \dfrac{15}{-3}$ Divide both sides by -3.

$x = -5$ -3 times x, divided by -3, is x. On the right-hand side, do the division: $15 \div (-3) = -5$.

Check:

$-3x = 15$ The original equation.

$-3(-5) \stackrel{?}{=} 15$ Substitute -5 for x.

$15 = 15$ Do the multiplication: $-3(-5) = 15$.

b. $-16 = -4y$

$\dfrac{-16}{-4} = \dfrac{-4y}{-4}$ To undo the multiplication by -4, divide both sides by -4.

$4 = y$ Do the divisions.

$y = 4$ If $4 = y$, $y = 4$.

Check the result.

Self Check

Solve each equation and check the result:

a. $-7k = 28$

b. $-40 = -8f$

Answers: **a.** -4, **b.** 5 ■

Recall that the multiplication property of equality states: *If equal quantities are multiplied by the same nonzero quantity, the results will be equal quantities.*

EXAMPLE 5 ***The multiplication property of equality.*** Solve $\dfrac{x}{-5} = -10$ and check the result.

Solution

In this equation, x is being divided by -5. To undo this division, we multiply both sides of the equation by -5.

$\dfrac{x}{-5} = -10$

$-5\left(\dfrac{x}{-5}\right) = -5(-10)$ Use the multiplication property of equality. Multiply both sides by -5.

$x = 50$ When x is divided by -5 and then multiplied by -5, the result is x. Do the multiplication: $-5(-10) = 50$.

Check:

$\dfrac{x}{-5} = -10$ The original equation.

$\dfrac{50}{-5} \stackrel{?}{=} -10$ Substitute 50 for x.

$-10 = -10$ Do the division.

Self Check

Solve $\dfrac{t}{-3} = 4$ and check the result.

Answer: -12 ■

Solving equations involving −x

Consider the equation $-x = 3$. The variable x is not isolated, because there is a $-$ in front of it. The symbol $-x$ means -1 times x. Therefore, the equation $-x = 3$ can be rewritten as $-1x = 3$. To isolate the variable, we can either multiply both sides by -1 or divide both sides by -1.

$$-x = 3$$
$$-1x = 3 \qquad -x = -1x.$$
$$(-1)(-1x) = (-1)3 \qquad \text{Multiply both sides by } -1.$$
$$1x = -3 \qquad -1(-1) = 1.$$
$$x = -3 \qquad 1x = x.$$

$$-x = 3$$
$$-1x = 3 \qquad -x = -1x.$$
$$\frac{-1x}{-1} = \frac{3}{-1} \qquad \text{Divide both sides by } -1.$$
$$x = -3 \qquad \text{Do the divisions.}$$

EXAMPLE 6 ***Multiplying both sides by −1.*** Solve $-x = -9$ and check the result.

Solution

$$-x = -9$$
$$-1x = -9 \qquad -x = -1x.$$
$$-1(-1x) = -1(-9) \qquad \text{Multiply both sides by } -1.$$
$$x = 9 \qquad \text{Do the multiplications: } -1(-1x) = x \text{ and } -1(-9) = 9.$$

Check:

$$-x = -9$$
$$-(9) \stackrel{?}{=} -9 \qquad \text{Substitute 9 for } x.$$
$$-9 = -9$$

Self Check

Solve $-h = -10$ and check the result.

$-1h = -10$

$\frac{-1h}{-1} = \frac{-10}{-1}$

$h = 10$

Answer: 10 ■

Combinations of operations

In the previous examples, each equation was solved by applying a single property of equality. Sometimes, if the equation is more complicated, it becomes necessary to use several properties of equality to solve it. For example, consider the equation $2x + 5 = 9$ and the operations performed on x.

$$2x + 5 = 9$$

The variable is multiplied by 2. (→ $2x$)

Then 5 is added. (→ $+ 5$)

To solve this equation, we use the rules for the order of operations in reverse.

- First, use the subtraction property of equality to undo the addition of 5.
- Second, apply the division property of equality to undo the multiplication by 2.

$$2x + 5 = 9$$
$$2x + 5 - 5 = 9 - 5 \qquad \text{To undo the addition of 5, subtract 5 from both sides.}$$
$$2x = 4 \qquad \text{Do the subtraction on each side: } 5 - 5 = 0 \text{ and } 9 - 5 = 4.$$
$$\frac{2x}{2} = \frac{4}{2} \qquad \text{To undo the multiplication by 2, divide both sides by 2.}$$
$$x = 2 \qquad \text{Do the divisions.}$$

EXAMPLE 7 ***Applying two properties of equality.*** Solve $-4x - 5 = 15$ and check the result.

Solution

The operations performed on x are multiplication by -4 and subtraction of 5. We undo these operations in the opposite order.

$$-4x - 5 = 15$$

$$-4x - 5 + 5 = 15 + 5 \quad \text{Add 5 to both sides.}$$

$$-4x = 20 \quad \text{Do the addition on each side: } -5 + 5 = 0 \text{ and } 15 + 5 = 20.$$

$$\frac{-4x}{-4} = \frac{20}{-4} \quad \text{Divide both sides by } -4.$$

$$x = -5 \quad \text{Do the divisions.}$$

Check:

$$-4x - 5 = 15$$

$$-4(-5) - 5 \stackrel{?}{=} 15 \quad \text{Substitute } -5 \text{ for } x.$$

$$20 - 5 \stackrel{?}{=} 15 \quad \text{Do the multiplication: } -4(-5) = 20.$$

$$15 = 15 \quad \text{Do the subtraction.}$$

■

Self Check

Solve $-6b - 1 = 11$ and check the result.

Answer: -2 ■

EXAMPLE 8 ***Applying two properties of equality.*** Solve $2 - 3p = -1$.

Solution

We begin by writing the subtraction on the left-hand side of the equation as addition of the opposite.

$$2 - 3p = -1$$

$$2 + (-3p) = -1 \quad \text{Add the opposite of } 3p \text{, which is } -3p.$$

$$2 + (-3p) - 2 = -1 - 2 \quad \text{To undo the addition of 2 on the left-hand side of the equation, subtract 2 from both sides.}$$

$$-3p = -3 \quad \text{Simplify.}$$

$$\frac{-3p}{-3} = \frac{-3}{-3} \quad \text{To undo the multiplication by } -3 \text{, divide both sides by } -3.$$

$$p = 1 \quad \text{Do the divisions.}$$

Self Check

Solve $6 - 8k = -34$.

Answer: 5 ■

EXAMPLE 9 ***Applying two properties of equality.*** Solve $\frac{y}{-2} - 6 = -18$.

Solution

The operations performed on y are division by -2 and subtraction of 6. We undo these operations in the opposite order.

$$\frac{y}{-2} - 6 = -18$$

$$\frac{y}{-2} - 6 + 6 = -18 + 6 \quad \text{To undo the subtraction of 6, add 6 to both sides.}$$

$$\frac{y}{-2} = -12 \quad \text{Simplify both sides of the equation.}$$

$$-2\left(\frac{y}{-2}\right) = -2(-12) \quad \text{To undo the division by } -2 \text{, multiply both sides by } -2.$$

$$y = 24 \quad \text{Do the multiplications.}$$

Self Check

Solve $\frac{m}{-8} - 10 = -14$.

Answer: 32 ■

Problem solving with equations

EXAMPLE 10 ***Water management.*** One day, enough water was released from a reservoir to lower the water level 17 feet to a reading of 33 feet below capacity. What was the water level reading before the release?

Analyze the problem Figure 2-19 illustrates the given information and what we are asked to find.

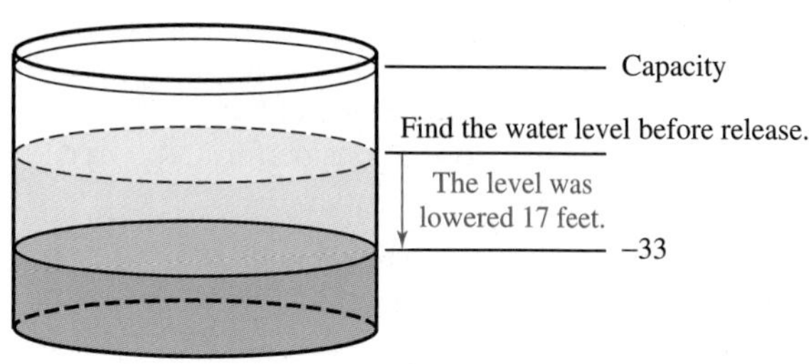
Capacity
Find the water level before release.
The level was lowered 17 feet.
−33

FIGURE 2-19

Form an equation Let $x =$ the water level reading before the release.

The water level reading before the release	minus	the number of feet the water level was lowered	is	the new water level reading.
x	$-$	17	$=$	-33

Solve the equation

$$x - 17 = -33$$

$$x - 17 + 17 = -33 + 17$$ To undo the subtraction of 17, add 17 to both sides.

$$x = -16$$ Do the additions: $-17 + 17 = 0$ and $-33 + 17 = -16$.

State the conclusion The water level reading before the release was -16.

Check the result If the water level reading was initially -16 feet and was then lowered 17 feet, the new reading would be $-16 - 17 = -16 + (-17) = -33$ feet. The answer checks. ■

STUDY SET Section 2.7

VOCABULARY *Fill in the blanks to make the statements true.*

1. To ____________ an equation, we isolate the variable on one side of the = sign.

2. The ____________ of -3 is 3.

CONCEPTS

3. If we multiply x by -3 and then divide that product by -3, what is the result?

4. If we divide x by -4 and then multiply that quotient by -4, what is the result?

5. In the equation $x + 3 = 10$, we can isolate x in two ways. Find the missing numbers.

a. $x + 3 - 3 = 10 -$

b. $x + 3 + (-3) = 10 +$

6. In the equation $12 + c = 10$, we can isolate c in two ways. Find the missing numbers.

a. $12 + c - \quad = 10 -$

b. $12 + c + (\quad) = 10 + (\quad)$

7. Tell what operations are performed on the variable x and the order in which they occur.

a. $-2x = -100$

b. $-6 + x = -9$

c. $-4x - 8 = 12$

d. $-1 = -6 + (-5x)$

8. Tell what operations are performed on the variable x and the order in which they occur.

a. $\dfrac{x}{-4} - 8 = 50$

b. $-16 = -5 + \dfrac{x}{-3}$

In Exercises 9–12, fill in the blanks to make the statements true.

9. When solving the equation $-4 + t = -8 - 2$, it is best to ____________ the right-hand side of the equation first before undoing any operations performed on the variable.

10. To solve the equation $-2x - 4 = 6$, we first undo the ____________ of 4. Then we undo the ____________ by -2.

11. When solving an equation, we isolate the variable by undoing the operations performed on it in the ____________ order.

12. To solve $-x = 6$, we can multiply or divide both sides of the equation by ______.

13. When solving each of these equations, which operation should you undo first?

a. $-2x - 3 = -19$

b. $-6 + \dfrac{h}{-3} = -14$

14. When solving each of these equations, which operation should you undo first?

a. $5 + (-9x) = -1$

b. $-16 = -9 + \dfrac{t}{7}$

NOTATION *Complete each solution to solve the equation.*

15.
$$y + (-7) = -16 + 3$$
$$y + (-7) = -13$$
$$y + (-7) + 7 = -13 + 7$$
$$y = -6$$

16.
$$x - (-4) = -1 - 5$$
$$x + 4 = -1 + 5$$
$$x + 4 = -6$$
$$x + 4 - 4 = -6 - 4$$
$$x = -6 + (-4)$$
$$x = -10$$

17.
$$-13 = -4y - 1$$
$$-13 + 1 = -4y - 1 + 1$$
$$-12 = -4y$$
$$\frac{-12}{-4} = \frac{-4y}{-4}$$
$$3 = y$$
$$y = 3$$

18. Solve $1 = \dfrac{m}{-5} + 6$.
$$1 - 6 = \frac{m}{-5} + 6 - 6$$
$$-5 = \frac{m}{-5}$$
$$-5(-5) = -5\left(\frac{m}{-5}\right)$$
$$25 = m$$
$$m = 25$$

19. What does $-10x$ mean?

20. What does $\dfrac{x}{-8}$ mean?

PRACTICE *Tell whether the number is a solution of the equation.*

21. $-3x - 4 = 2; -2$

22. $\dfrac{x}{-2} + 5 = -10; 20$

23. $-x + 8 = -4; 4$

24. $-3 + 2x = -3; 0$

Solve each equation. Check each result.

25. $x + 6 = -12$

26. $y + 1 = -4$

27. $-6 + m = -20$

28. $-12 + r = -19$

29. $-5 + 3 = -7 + f$

30. $-10 + 4 = -9 + t$

31. $h - 8 = -9$

32. $x - 1 = -7$

33. $0 = y + 9$

34. $0 = t + 5$

35. $r - (-7) = -1 - 6$

36. $x - (-1) = -4 - 3$

37. $t - 4 = -8 - (-2)$

38. $r - 1 = -3 - (-4)$

39. $x - 5 = -5$

40. $r - 4 = -4$

41. $-2s = 16$

42. $-3t = 9$

43. $-5t = -25$

44. $-6m = -60$

45. $-2 + (-4) = -3n$

46. $-10 + (-2) = -4x$

47. $-9h = -3(-3)$

48. $-6k = -2(-3)$

49. $\dfrac{t}{-3} = -2$

50. $\dfrac{w}{-4} = -5$

51. $0 = \dfrac{y}{8}$

52. $0 = \dfrac{h}{7}$

53. $\dfrac{x}{-2} = -6 + 3$

54. $\dfrac{a}{-5} = -7 + 6$

55. $\dfrac{x}{4} = -5 - 8$

56. $\dfrac{r}{2} = -5 - 1$

57. $2y + 8 = -6$

58. $5y + 1 = -9$

59. $-21 = 4h - 5$

60. $-22 = 7l - 8$

61. $-3v + 1 = 16$

62. $-4e + 4 = 24$

63. $8 = -3x + 2$

64. $15 = -2x + (-11)$

65. $-35 = 5 - 4x$

66. $12 = -9 - 3x$

67. $4 - 5x = 34$

68. $15 - 6x = 21$

69. $-5 - 6 - 5x = 4$

70. $-7 - 5 - 7x = 16$

71. $4 - 6x = -5 - 9$

72. $8 - 2d = -5 - 5$

73. $\frac{h}{-6} + 4 = 5$

74. $\frac{p}{-3} + 3 = 8$

75. $-2(4) = \frac{t}{-6} + 1$

76. $-2(5) = \frac{y}{-3} + 3$

77. $0 = 6 + \frac{c}{-5}$

78. $0 = -6 + \frac{s}{-3}$

79. $-1 = -8 + \frac{h}{-2}$

80. $-5 = 4 + \frac{g}{-4}$

81. $2x + 3(0) = -6$

82. $3x - 4(0) = -12$

83. $2(0) - 2y = 4$

84. $5(0) - 2y = 10$

85. $-x = 8$

86. $-y = 12$

87. $-15 = -k$

88. $-4 = -p$

APPLICATIONS *Complete each solution.*

89. SHARKS During a research project, a diver inside a shark cage made observations at a depth of 120 feet. For a second set of observations, the cage was raised to a depth of 75 feet. How many feet was the cage raised between observations?

Analyze the problem

- The first observations were at ft.
- The next observations were at ft.
- We must find __________________.

Form an equation

Let x = __________________

Key word: *raised* **Translation:** ________

We can express the second position of the cage in two ways.

The first position of the cage	plus	the amount the cage was raised	is	the second position of the cage.
	+		=	

Solve the equation

$$+ x =$$

$$-120 + x + \quad = -75 +$$

$$x = 45$$

State the conclusion

____

Check the result

If we add the number of feet the cage was raised to the first position, we get $-120 + \quad = -75$. The answer checks.

90. TRACK TIMES In one race, an athlete's time for the mile was 7 seconds under the school record. In a second race, her time continued to drop, to 16 seconds under the old school record. How much time did she drop between the first and second races?

Analyze the problem

- The 1st race was sec under the record.
- The 2nd race was sec under the record.
- We must find __________________

Form an equation

Let x = __________________

Key phrase: *dropped* **Translation:** ________

We can express her improvement in two ways.

First race performance	minus	amount of time dropped	is	second race performance.
	−		=	

Solve the equation

$$- x =$$

$$-7 - x + \quad = -16 +$$

$$-x =$$

$$x = 9$$

State the conclusion

____

Check the result

If we subtract the time she dropped in the second race from the time dropped in the first race, we get $-7 - \quad = -16$. The answer checks.

In Exercises 91–102, let a variable represent the unknown quantity. Then write and solve an equation to answer the question.

91. DREDGING A HARBOR In order to handle larger vessels, port officials are having a harbor deepened by dredging. The harbor bottom is already 47 feet below sea level. After the dredging, the bottom will be 65 feet below sea level. (See Illustration 1 on the next page.) How many feet must be dredged out?

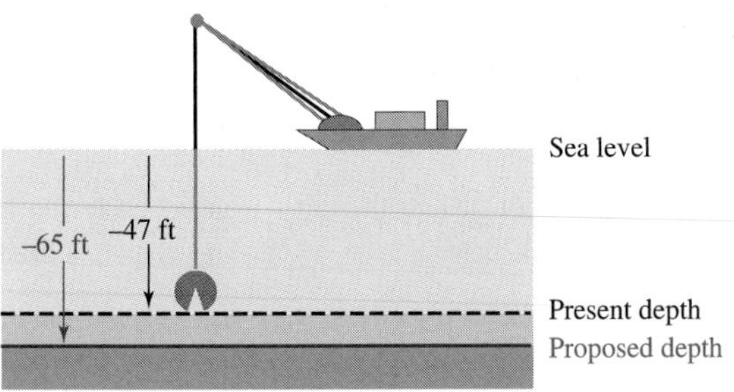

ILLUSTRATION 1

92. ROLLER COASTER DESIGN Engineers have decided that part of a roller coaster ride will consist of a steep plunge from a peak 145 feet high. The car will then come to a screeching halt in a cave. (See Illustration 2.) How far below ground should the cave be if engineers want the overall drop in height to be 170 feet?

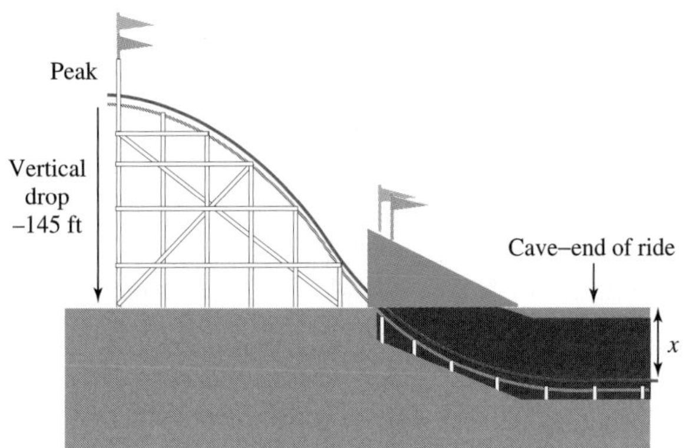

ILLUSTRATION 2

93. FOOTBALL During the first half of a football game, a team ran for a total of 43 yards. After a dismal second half, they ended the game with a total of -8 yards rushing. What was their rushing total in the second half?

94. WEATHER FORECAST The weather forecast for Fairbanks, Alaska warned listeners that the daytime high temperature of 2° below zero would drop to a nighttime low of 28° below. What was the overnight change in temperature?

95. MARKET SHARE After its first year of business, a manufacturer of smoke detectors found its market share 43 points behind the industry leader. Five years later, it trailed the leader by only 9 points. How many points of market share did the company pick up over this five-year span?

96. CHECKING ACCOUNT After he made a deposit of $220, a student's account was still $215 overdrawn. What was his balance before the deposit?

97. PRICE REDUCTION Over the past year, the price of a video game player has dropped each month. If the price fell $60 this year, how much did the price drop each month on average?

98. REBATES A store decided that it could afford to lose some money to promote a new line of sunglasses. A $9 rebate was offered to each customer purchasing these sunglasses. If the rebate program resulted in a loss of $225 for the store, how many customers took advantage of the offer?

99. ELECTION POLLS Six months before an election, a political candidate was 31 points behind in the polls. Two days before the election, polls showed that his support had skyrocketed; he found himself only 2 points behind. How much support had he gained over the six-month period?

100. HORSE RACING At the midway point of a 6-furlong horse race, the long shot was 3 lengths ahead of the pre-race favorite. In the last half of the race, the long shot lost ground and eventually finished 6 lengths behind the favorite. By how many lengths did the long shot lose to the favorite during the last half of the race?

101. INTERNATIONAL TIME ZONES The world is divided into 24 times zones. Each zone is one hour ahead of or behind its neighboring zones. In the portion of the world time zone map shown in Illustration 3, we see that Tokyo is in zone $+9$. What time zone is Seattle in if it is 17 hours behind Tokyo?

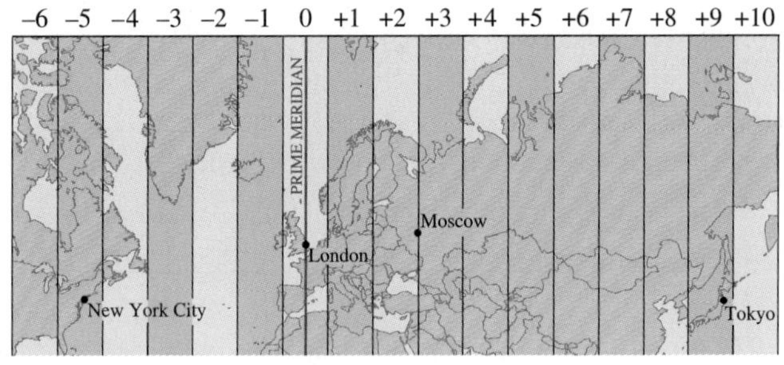

ILLUSTRATION 3

102. PROFITS AND LOSSES In its first year of business, a nursery suffered a loss due to frost damage, ending the year $11,560 in the red. In the second year, it made a sizable profit. If the total profit for the first two years in business was $32,090, how much profit was made the second year?

WRITING

103. Explain why the variable is not isolated in the equation $-x = 10$.

104. Explain how to check the result after solving an equation.

REVIEW

105. Write 5^6 without using exponents.

106. Give the definition of an even whole number.

107. Solve $7 + 3y = 43$.

108. How can the addition $2 + 2 + 2 + 2 + 2$ be represented using multiplication?

109. Write $16 \div 8$ using a fraction bar.

110. In the expression $5(6) - 3 + 2$, list each operation in the order in which it must be performed.

KEY CONCEPT

Signed Numbers

In algebra, we work with both positive and negative numbers. We study negative numbers because they are necessary to describe many situations in daily life.

Represent each of these situations using a signed number.

1. Stocks fell 5 points.

2. The river was 12 feet over flood stage.

3. 30 seconds before going on the air

4. A business \$6 million in the red

5. 10 degrees above normal

6. The year 2000 B.C.

7. \$205 overdrawn

8. 14 units under their quota

A number line can be used to illustrate positive and negative numbers.

9. On this number line, label the location of the positive integers and the negative integers.

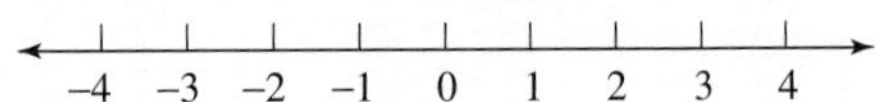

10. On this number line, graph 2 and its opposite.

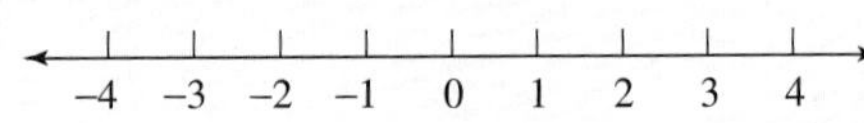

11. Two numbers, x and y, are graphed on this number line. What can you say about their relative sizes?

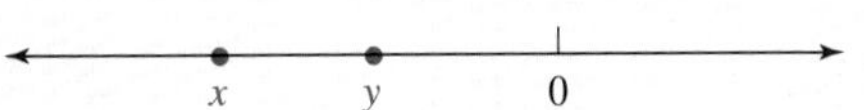

12. The absolute value of -3, written $|-3|$, is the distance between -3 and 0 on a number line. Show this distance on the number line.

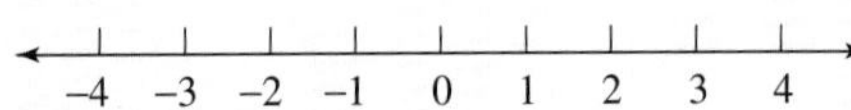

In the space provided, summarize how addition, multiplication, and division are performed with two integers having like signs and with two integers having unlike signs. Then explain the method that is used to subtract integers.

13. Addition
Like signs:

Unlike signs:

14. Multiplication
Like signs:
Unlike signs:

15. Division
Like signs:
Unlike signs:

16. Subtraction with integers

ACCENT ON TEAMWORK

Section 2.1

OPPOSITES Have everyone in the class get in pairs. The object of the game is for one member of a team, using one-word clues, to get the other member to say each of the words in Column I. This is done by giving your partner a word clue having the opposite meaning. For example, if you want your partner to say the word *up,* give the clue *down.* Go through all of the words in Column I. Keep track of how many words are guessed correctly.

Then switch assignments. The person who first gave the clues now receives the clues. Go through all of the words in Column II.

Column I		Column II	
below	surplus	win	positive
over	overdrawn	gain	increase
ahead	profit	forward	debt
before	deduct	retreat	withdrawal
less	liabilities	rise	accelerate

Section 2.2

ADDING INTEGERS To illustrate how to add $-5 + 3$, think of a hole 5 feet deep (-5) which then has 3 feet of dirt (3) added to it. Illustration 1 shows that the resulting hole would be 2 feet deep (-2). So $-5 + 3 = -2$.

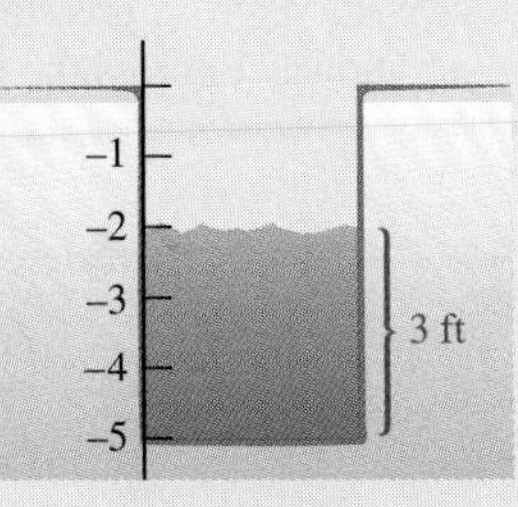

ILLUSTRATION 1

Draw a similar picture to help find each sum.

a. $-4 + 1$ **b.** $-5 + 4$

c. $-6 + 6$ **d.** $-3 + (-1)$

e. $-3 + (-3)$ **f.** $-3 + 5$

Section 2.3

SUBTRACTING INTEGERS Write a subtraction problem in which the difference of two negative numbers is

a. a positive number. **b.** a negative number.

Section 2.4

MULTIPLYING INTEGERS

a. Complete the multiplication table that follows.

b. Construct another table with a top row of -1 through -10 and a first column of 1 through 10.

c. Construct a table with a top row of -1 through -10 and a first column of -1 through -10.

Multiplication Table

·	1	2	3	4	5	6	7	8	9	10
−1										
−2										
−3										
−4										
−5										
−6										
−7										
−8										
−9										
−10										

Section 2.5

OPERATIONS WITH TWO INTEGERS For each operation listed in the table, tell whether the answer is always positive, always negative, or may be positive or negative.

Signs of the two integers	Add	Subtract	Multiply	Divide
Both positive				
Both negative				
One positive, one negative				

Section 2.6

ESTIMATION Estimate the answer to each problem.

a. $405 - 567$ **b.** $-2{,}564 - 2{,}456$

c. $989 - 898$ **d.** $-23{,}250 + 22{,}750$

e. $56(-87)$ **f.** $-40 - 30 - 45$

g. $608 \div (-2)$ **h.** $-94 + 90 - 45$

Section 2.7

SOLVING EQUATIONS What is wrong with each solution?

a. Solve $2x + 4 = 10$.

$$2x + 4 = 10$$
$$\frac{2x}{2} + 4 = \frac{10}{2}$$
$$x + 4 = 5$$
$$x + 4 - 4 = 5 - 4$$
$$x = 1$$

b. Solve $2x + 4 = 10$.

$$2x + 4 = 10$$
$$2x + 4 - 4 = 10$$
$$2x = 10$$
$$\frac{2x}{2} = \frac{10}{2}$$
$$x = 5$$

CHAPTER REVIEW

SECTION 2.1 An Introduction to the Integers

CONCEPTS

A *number line* is a horizontal or vertical line used to represent numbers graphically.

Integers: $\{. . . , -3, -2, -1, 0, 1, 2, 3, . . .\}$

Inequality symbols:
$>$ is greater than
$<$ is less than

A *negative* number is less than 0. A *positive* number is greater than 0.

The *absolute value* of a number is the distance between it and 0 on the number line.

The opposite of the opposite of a number is that number.

REVIEW EXERCISES

1. Graph each set of numbers.

a. $\{-3, -1, 0, 4\}$

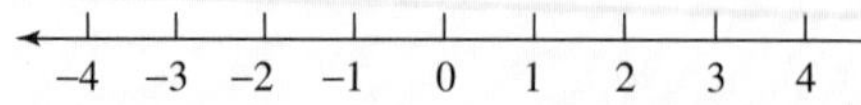

b. The integers greater than -3 but less than 4.

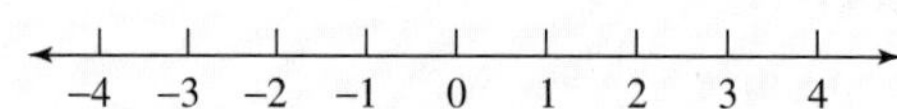

2. Insert one of the symbols $>$ or $<$ in the blank to make a true statement.

a. -7 ___ 0
b. -20 ___ -19
c. $|-16|$ ___ -16
d. 56 ___ 60

3. WATER PRESSURE Salt water exerts a pressure of 14.7 pounds per square inch at a depth of 33 feet. See Illustration 1. Express the depth using a signed number.

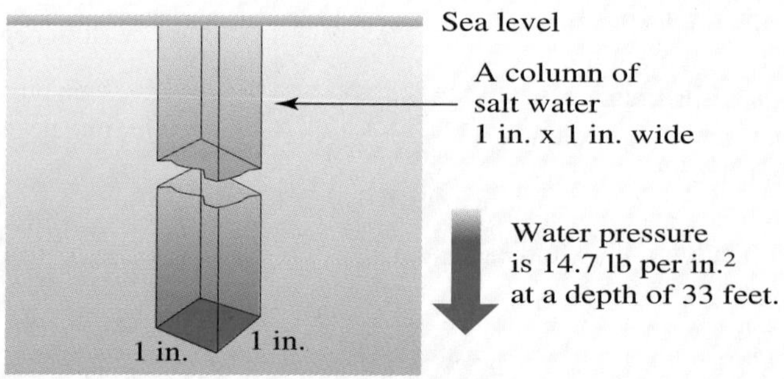

ILLUSTRATION 1

4. Represent each of these situations using a signed number.

a. A deficit of \$1,200
b. 10 seconds before going on the air

5. Evaluate each expression.

a. $|-4|$
b. $|0|$
c. $|-43|$
d. $-|12|$

6. Explain the meaning of each red $-$ sign.

a. -5
b. $-(-5)$
c. $-(-5)$
d. $5 - (-5)$

On a number line, two numbers the same distance away from 0, but on different sides of it, are called *opposites.*

7. Find each of the following.

a. $-(-12)$

b. The opposite of 8

c. The opposite of -8

d. -0

SECTION 2.2 Addition of Integers

To add two integers with *like signs,* add their absolute values and attach their common sign to that sum.

To add two integers with *unlike signs,* subtract their absolute values, the smaller from the larger. Attach the sign of the number with the larger absolute value to that result.

8. Use a number line to find each sum.

a. $4 + (-2)$

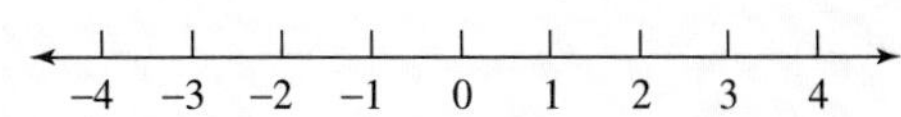

b. $-1 + (-3)$

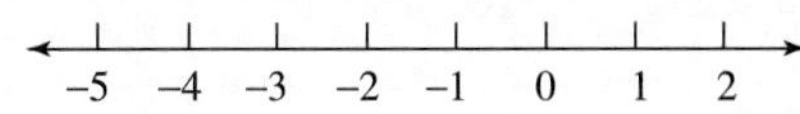

9. Add.

a. $-6 + (-4)$

b. $-23 + (-60)$

c. $-1 + (-4) + (-3)$

d. $-4 + 3$

e. $-28 + 140$

f. $9 + (-20)$

g. $3 + (-2) + (-4)$

h. $(-2 + 1) + [(-5) + 4]$

Addition property of zero: If a is any number, then

$a + 0 = a$ and $0 + a = a$

10. Add.

a. $-4 + 0$

b. $0 + (-20)$

c. $-8 + 8$

d. $73 + (-73)$

If $a + b = 0$, then a and b are called *additive inverses.*

11. Give the additive inverse of each number.

a. -11

b. 4

12. DROUGHT During a drought, the water level in a reservoir fell to a point 100 feet below normal. After two rainy months, it rose 35 feet. How far below normal was the water level after the rain?

SECTION 2.3 Subtraction of Integers

Rule for subtraction: If a and b are any numbers, then

$a - b = a + (-b)$

13. Subtract.

a. $5 - 8$

b. $-9 - 12$

c. $-4 - (-8)$

d. $-6 - 106$

e. $-8 - (-2)$

f. $7 - 1$

g. $0 - 37$

h. $0 - (-30)$

14. Fill in the blanks to make a true statement: Subtracting a number is the same as ______________ the ______________ of that number.

15. Evaluate each expression.

a. $-9 - 7 + 12$

b. $7 - [(-6) - 2]$

c. $1 - (2 - 7)$

d. $-12 - (6 - 10)$

16. Subtract 27 from -50.

17. GOLD MINING Some miners discovered a small vein of gold at a depth of 150 feet. This prompted them to continue their exploration. After descending another 75 feet, they came upon a much larger find. Use a signed number to represent the depth of the second discovery.

18. Evaluate: $2 - [-(-3)]$.

To find the *change* in a quantity, subtract the earlier value from the later value.

19. RECORD TEMPERATURES The lowest and highest recorded temperatures for Alaska and Virginia are shown here. For each state, find the difference in temperature between the record high and low.

Alaska: Low $-80°$ Jan. 23, 1971
High $100°$ June 27, 1915

Virginia: Low $-30°$ Jan. 22, 1985
High $110°$ July 15, 1954

SECTION 2.4 *Multiplication of Integers*

The product of two integers with *like signs* is positive. The product of two integers with *unlike signs* is negative.

20. Multiply.

a. $-9 \cdot 5$
b. $-3(-6)$
c. $7(-2)$
d. $(-8)(-47)$
e. $-20 \cdot 5$
f. $-1(-1)$
g. $-1(25)$
h. $(5)(-30)$

21. Multiply.

a. $(-6)(-2)(-3)$
b. $4(-3)3$
c. $0(-7)$
d. $(-1)(-1)(-1)(-1)$

22. TAX DEFICIT A state agency's prediction of a tax shortfall proved to be two times worse than the actual deficit of $3 million. The federal prediction of the same shortfall was even more inaccurate—three times the amount of the actual deficit. Complete Illustration 2, which summarizes these incorrect forecasts.

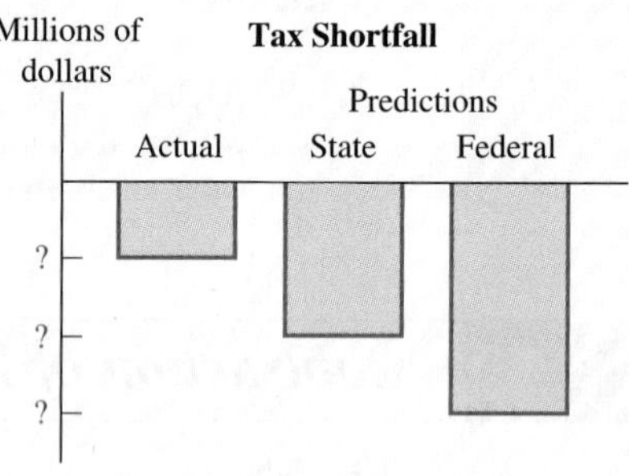

ILLUSTRATION 2

An *exponent* is used to represent repeated multiplication.

23. Find each power.

a. $(-5)^2$
b. $(-2)^5$
c. $(-8)^2$
d. $(-4)^3$

When a negative integer is raised to an *even* power, the result is positive. When it is raised to an *odd* power, the result is negative.

24. When $(-5)^9$ is evaluated, will the result be positive, or will it be negative?

25. Explain the difference between -2^2 and $(-2)^2$ and then evaluate each.

SECTION 2.5 Division of Integers

26. Fill in the box to make a true statement: We know that $\frac{-15}{5} = -3$ because $5(\quad) = -15$.

The quotient of two numbers with *like* signs is positive.

The quotient of two integers with *unlike* signs is negative.

27. Divide.

a. $\frac{-14}{7}$ **b.** $\frac{25}{-5}$

c. $-64 \div 8$ **d.** $\frac{-202}{-2}$

If a is any nonzero number,

$$\frac{0}{a} = 0$$

Division *by* 0 is undefined.

28. Find each quotient, if possible.

a. $\frac{0}{-5}$ **b.** $\frac{-4}{0}$

c. $\frac{-673}{-673}$ **d.** $\frac{-10}{-1}$

29. PRODUCTION TIME Because of improved production procedures, the time needed to produce an electronic component dropped by 12 minutes over the past six months. If the drop in production time was uniform, how much did it change each month over this period?

SECTION 2.6 Order of Operations and Estimation

The rules for the order of operations:

1. Do all calculations within parentheses and other grouping symbols.
2. Evaluate all exponential expressions.
3. Do all the multiplications and divisions, working from left to right.
4. Do all the additions and subtractions, working from left to right.

Always work from the *innermost* set of grouping symbols to the *outermost* set.

A fraction bar is a grouping symbol.

An *estimation* is an approximation that gives a quick idea of what the actual answer would be.

30. Evaluate each expression.

a. $2 + 4(-6)$ **b.** $7 - (-2)^2 + 1$

c. $2 - 5(4) + (-25)$ **d.** $-3(-2)^3 - 16$

e. $-2(5)(-4) + \frac{|-9|}{3^2}$ **f.** $-4^2 + (-4)^2$

g. $-12 - (8 - 9)^2$ **h.** $7|-8| - 2(3)(4)$

31. Evaluate each expression.

a. $-4\left(\frac{15}{-3}\right) - 2^3$ **b.** $-20 + 2(12 - 5 \cdot 2)$

c. $-20 + 2[12 - (-7 + 5)^2]$ **d.** $8 - |-3 \cdot 4 + 5|$

32. Evaluate each expression.

a. $\frac{10 + (-6)}{-3 - 1}$ **b.** $\frac{3(-6) - 11 + 1}{4^2 - 3^2}$

33. Estimate each answer.

a. $-89 + 57 + (-42)$ **b.** $\frac{-507}{-24}$

c. $(-681)(9)$ **d.** $317 - (-775)$

SECTION 2.7 Solving Equations Involving Integers

To *solve an equation* means to find all values of the variable that, when substituted into the original equation, make the equation a true statement.

34. Is $x = -4$ a solution of the equation? Explain why or why not.

a. $2x + 6 = -2$ **b.** $6 + \frac{x}{2} = -4$

To solve an equation, we undo the operations in the reverse order from that in which they were performed on the variable. The objective is to isolate the variable.

35. Solve each equation. Check the result.

a. $t + (-8) = -18$ **b.** $\frac{x}{-3} = -4$

c. $y + 8 = 0$ **d.** $-7m = -28$

36. Solve each equation. Check the result.

a. $-x = -15$ **b.** $4 = -y$

37. Solve each equation. Check the result.

a. $-5t + 1 = -14$ **b.** $3(2) = 2 - 2x$

c. $\frac{x}{-4} - 5 = -1 - 1$ **d.** $c - (-5) = 5$

The five-step *problem-solving strategy* can be used when solving application problems.

1. Analyze the problem.
2. Form an equation.
3. Solve the equation.
4. State the conclusion.
5. Check the result.

In Exercises 38–40, let a variable represent the unknown quantity. Then write and solve an equation to answer the question.

38. WIND-CHILL FACTOR If the wind is blowing at 25 miles per hour, an air temperature of 5° below zero will feel like 51° below zero. Find the perceived change in temperature that is caused by the wind.

39. CREDIT CARD PROMOTION During the holidays, a store offered an $8 gift certificate to any customer applying for its credit card. If this promotion cost the company $968, how many customers applied for credit?

40. BANK FAILURE When a group of 7 investors decided to acquire a failing bank, each had to assume an equal share of the bank's total indebtedness, which was $57,400. How much debt did each investor assume?

Chapter 2 Test

1. Insert one of the symbols > or < in the blank to make the statement true.

a. $-8 \quad -9$ **b.** $-8 \quad |-8|$

c. The opposite of 5 0

2. List the integers.

3. SCHOOL ENROLLMENT According to the projections in Illustration 1, which high school will face the greatest shortage in the year 2006?

High schools with shortage of classroom seats by 2006	
Sylmar	−669
San Fernando	−1,630
Monroe	−2,488
Cleveland	−350
Canoga Park	−586
Polytechnic	−2,379
Van Nuys	−1,690
Reseda	−462
North Hollywood	−1,004
Hollywood	−774

ILLUSTRATION 1

4. Use a number line to find the sum: $-3 + (-2)$.

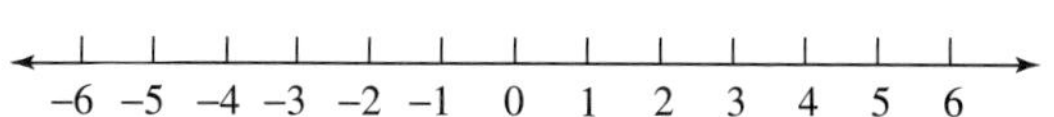

5. Add.

a. $-65 + 31$ **b.** $-17 + (-17)$

c. $[6 + (-4)] + [-6 + (-4)]$

6. Subtract.

a. $-7 - 6$ **b.** $-7 - (-6)$

c. $0 - 15$ **d.** $-60 - 50 - 40$

7. Find each product.

a. $-10 \cdot 7$ **b.** $-4(-2)(-6)$

c. $(-2)(-2)(-2)(-2)$ **d.** $-55(0)$

8. Write the related multiplication statement for $\frac{-20}{-4} = 5$.

9. Find each quotient, if possible.

a. $\frac{-32}{4}$ **b.** $\frac{8}{6-6}$

c. $\frac{-5}{1}$ **d.** $\frac{0}{-6}$

10. BUSINESS TAKEOVER Six businessmen are contemplating taking over a company that has potential, but they must retire the debt incurred by the company over the past three quarters. (See Illustration 2.) If they plan equal ownership, how much will each have to contribute to retire the debt?

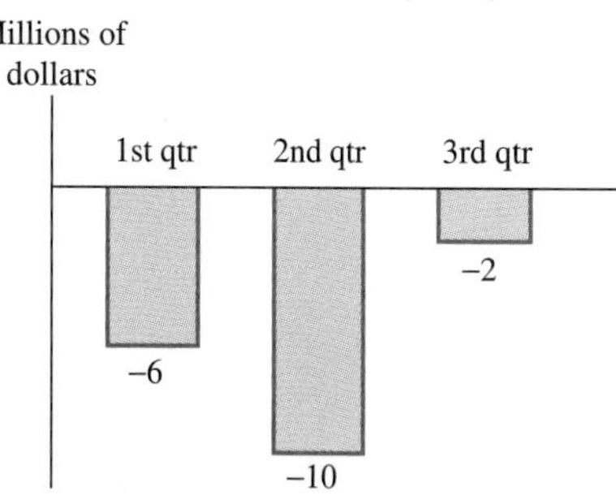

ILLUSTRATION 2

11. GEOGRAPHY The lowest point on the African continent is the Qattarah Depression in the Sahara Desert, 436 feet below sea level. The lowest point on the North American continent is Death Valley, California, 282 feet below sea level. What is the difference in these elevations?

12. Evaluate each expression.

a. $-(-6)$ **b.** $|-7|$

c. $|-9 + 3|$ **d.** $2|-66|$

13. Find each power.

a. $(-4)^2$ **b.** -4^2 **c.** $(-4 - 3)^2$

14. Evaluate: $-18 \div 2 \cdot 3$.

15. Evaluate: $4 - (-3)^2 + 6$.

16. Evaluate: $-3 + \left(\frac{-16}{4}\right) - 3^3$.

17. Evaluate: $-10 + 2[6 - (-2)^2(-5)]$.

18. Evaluate: $\frac{4(-6) - 2^2}{-3 - 4}$.

19. Solve $c - (-7) = -8$.

20. Solve $6 - x = -10$.

21. Solve $\frac{x}{-4} = 10$.

22. Solve $3x + (-7) = -11 + (-11)$.

23. Solve $-5 = -6a + 7$ and check the result.

24. Solve $\frac{x}{-2} + 3 = (-2)(-6)$.

In Problems 25–26, let a variable represent the unknown quantity. Then write and solve an equation to answer each question.

25. CHECKING ACCOUNT After making a deposit of \$225, a student's account was still \$19 overdrawn. What was her balance before the deposit?

26. HOSPITAL CAPACITY One morning, the number of beds occupied by patients in a hospital was 3 under capacity. By afternoon, the number of unoccupied beds was 21. If no new patients were admitted, how many patients were released to go home?

27. Multiplication means repeat addition. Use this fact to show that the product of a positive and a negative number, such as $5(-4)$, is negative.

28. Explain why the absolute value of a number can never be negative.

Chapters 1–2 Cumulative Review Exercises

Consider the numbers −2, −1, 0, 1, 2, $\frac{3}{2}$, 5, and 9.

1. List each natural number.

2. List each whole number.

3. List each negative number.

4. List each integer.

Consider the number 7,326,549.

5. Which digit is in the thousands column?

6. Which digit is in the hundred thousands column?

7. Round to the nearest hundred.

8. Round to the nearest ten thousand.

9. BIDS A school district received the bids shown in Illustration 1. Which company should be awarded the contract?

Citrus Unified School District Bid 02-9899 CABLING AND CONDUIT INSTALLATION	
Datatel	$2,189,413
Walton Electric	$2,201,999
Advanced Telecorp	$2,175,081
CRF Cable	$2,174,999
Clark & Sons	$2,175,801

ILLUSTRATION 1

10. NUCLEAR POWER The table at the top of the page gives the number of operable nuclear power plants in the United States for the years 1978–1998, in five-year increments. Construct a bar graph using the data.

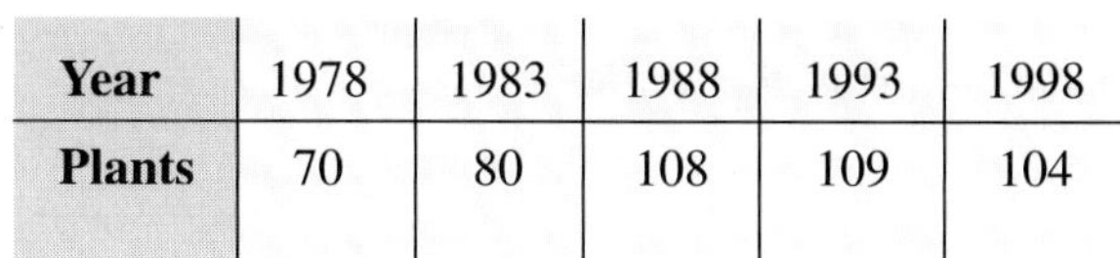

Year	1978	1983	1988	1993	1998
Plants	70	80	108	109	104

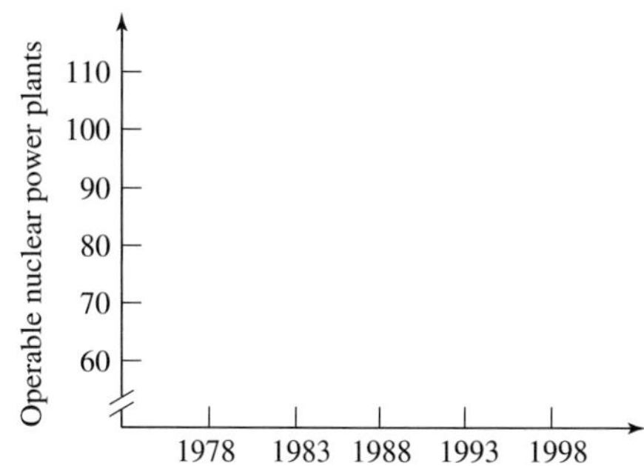

In Exercises 11–14, perform each operation.

11. $237 + 549$

12. $6{,}375 - 2{,}569$

13. $\begin{array}{r} 5{,}369 \\ -\ \ 685 \\ \hline \end{array}$

14. $\begin{array}{r} 7{,}899 \\ +5{,}237 \\ \hline \end{array}$

15. Find the perimeter and the area of the rectangular garden in Illustration 2.

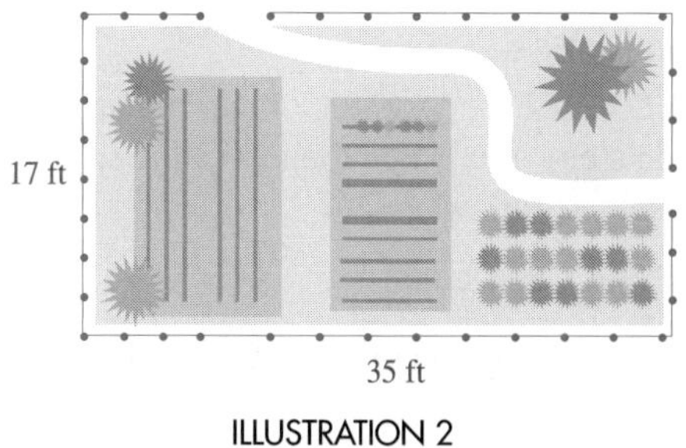

ILLUSTRATION 2

16. In a shipment of 147 pieces of furniture, 27 pieces were sofas, 55 were leather chairs, and the rest were wooden chairs. Find the number of wooden chairs.

In Exercises 17–20, perform each operation.

17. $435 \cdot 27$

18. $1{,}261 \div 97$

19. $\begin{array}{r} 4{,}587 \\ \times\ \ \ 67 \\ \hline \end{array}$

20. $38\overline{)17{,}746}$

21. SHIPPING There are 12 tennis balls in one dozen, and 12 dozen in one gross. How many tennis balls are there in a shipment of 12 gross?

22. Find all of the factors of 18.

In Exercises 23–26, identify each number as a prime, a composite, an even, or an odd.

23. 17

24. 18

25. 0

26. 1

27. Find the prime factorization of 504.

28. Write the expression $11 \cdot 11 \cdot 11 \cdot 11$ using an exponent.

Evaluate each expression.

29. $5^2 \cdot 7$

30. $16 + 2[14 - 3(5 - 4)^2]$

31. $25 + 5 \cdot 5$

32. $\dfrac{16 - 2 \cdot 3}{2 + (9 - 6)}$

33. SPEED CHECK A traffic officer monitored several cars on a city street. She found that the speeds of the cars were as follows:

38, 42, 36, 38, 48, 44

On average, were the drivers obeying the 40-mph speed limit?

34. Tell whether 6 is a solution of the equation $3x - 2 = 16$. Explain why or why not.

Solve each equation and check the result.

35. $50 = x + 37$

36. $a - 12 = 41$

37. $5p = 135$

38. $\dfrac{y}{8} = 3$

In Exercises 39–40, graph each set on a number line.

39. $\{-2, -1, 0, 2\}$

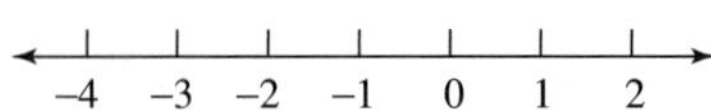

40. The integers greater than -4 but less than 2

−4 −3 −2 −1 0 1 2

41. True or false: $-17 < -16$.

42. Evaluate 3^2 and -3^2.

Evaluate each expression.

43. $-2 + (-3)$

44. $-15 + 10 + (-9)$

45. $-3 - 5$

46. $-15^2 - 2|-3|$

47. $(-8)(-3)$

48. $5(-7)^3$

49. $\dfrac{-14}{-7}$

50. $\dfrac{450}{-9}$

51. $5 + (-3)(-7)$

52. $-20 + 2[12 - 5(-2)(-1)]$

53. $\dfrac{10 - (-5)}{1 - 2 \cdot 3}$

54. $\dfrac{3(-6) - 10}{3^2 - 4^2}$

Solve each equation. Check the result.

55. $-5t + 1 = -14$

56. $\dfrac{x}{-3} - 2 = -2(-2)$

Use the five-step problem-solving strategy.

57. BUYING A BUSINESS When 12 investors decided to buy a bankrupt company, they agreed to assume equal shares of the company's debt of $1,512,444. How much was each person's share?

58. THE MOON The difference in the maximum and the minimum temperatures on the moon's surface is 540°F. The maximum temperature, which occurs at lunar noon, is 261°F. Find the minimum temperature, which occurs just before lunar dawn.

The Language of Algebra

3

ALGEBRA IS THE LANGUAGE OF MATHEMATICS. IN THIS CHAPTER, YOU WILL LEARN MORE ABOUT THINKING AND WRITING IN THIS LANGUAGE, USING ITS MOST IMPORTANT COMPONENT—A VARIABLE.

3.1 Variables and Algebraic Expressions

In this section, you will learn about

- Algebraic expressions • Translating from English to mathematical symbols
- Writing algebraic expressions to represent unknown quantities
- Looking for hidden operations
- Expressions involving more than one operation

INTRODUCTION. In Chapter 1, we introduced the following strategy for problem solving:

1. Analyze the problem.
2. Form an equation.
3. Solve the equation.
4. State the conclusion.
5. Check the result.

In order to form equations, you must represent unknown quantities with variables. Success at doing this depends on your ability to translate English words and phrases into mathematical symbols.

Algebraic expressions

In the equation $1{,}700 = 425 + x$, the notation $425 + x$ is called an **algebraic expression** or, more simply, an **expression.**

Algebraic expressions

> Variables and/or numbers can be combined with the operations of addition, subtraction, multiplication, and division to create **algebraic expressions.**

Here are some examples of algebraic expressions.

$5(2a)$	This algebraic expression is a combination of the numbers 5 and 2, the variable a, and the operation of multiplication.
$x + 2x + 3x$	This algebraic expression involves the variable x, the numbers 2 and 3, and the operations of addition and multiplication.
$-2t^2 \cdot 3t^4$	This algebraic expression contains powers of the variable t, the numbers -2, 2, 3, and 4, and the operation of multiplication.
$5(r - 6)$	This algebraic expression is a combination of the numbers 5 and 6, the variable r, and the operations of subtraction and multiplication.

Algebraic expressions can contain more than one variable. For example, the expressions $3x + 4y$ and $-6m^2n(mn)$ each contain two variables.

Translating from English to mathematical symbols

In order to solve application problems, which are almost always given in words, we must translate those words into mathematical symbols. The following tables list some *key words* and *key phrases* that are used to represent the operations of addition, subtraction, multiplication and division.

Addition

The phrase	translates to the algebraic expression
the *sum* of p and 15	$p + 15$
10 *plus* c	$10 + c$
5 *added to* a	$a + 5$
4 *more than* r	$r + 4$
8 *greater than* A	$A + 8$
S *increased by* 100	$S + 100$
exceeds L *by* 20	$L + 20$

Subtraction

The phrase	translates to the algebraic expression
the difference of 30 and k	$30 - k$
1,000 *minus* R	$1{,}000 - R$
15 *less than* w	$w - 15$
r *decreased by* 5	$r - 5$
T *reduced by* 80	$T - 80$
7 *subtracted from* s	$s - 7$
2,000 *less* c	$2{,}000 - c$

Multiplication

The phrase	translates to the algebraic expression
the *product* of 60 and h	$60h$
10 *times* A	$10A$
twice w	$2w$
$\frac{1}{2}$ *of* t	$\frac{1}{2}t$

Division

The phrase	translates to the algebraic expression
the *quotient* of B and 5	$\frac{B}{5}$
T *divided by* 50	$\frac{T}{50}$
the *ratio* of h to 3	$\frac{h}{3}$
n *split into* 8 equal parts	$\frac{n}{8}$

COMMENT The phrase *greater than* is used to indicate addition. The phrase *is greater than* refers to the symbol $>$. A similar comment applies to the phrases *less than* and *is less than.*

EXAMPLE 1 ***Translating to mathematical symbols.*** Write each phrase as an algebraic expression:

a. The sum of n and 12, **b.** The product of z and 60, and **c.** six less than z.

Solution

a. Key word: *sum* **Translation:** add

The phrase translates to

$$n + 12$$

b. Key word: *product* **Translation:** multiply

The variable z is to be multiplied by 60: $z \cdot 60$. This can be written as

$$60z$$

c. Key phrase: *less than* **Translation:** subtract

Since z is to be made less, we will subtract 6 from it.

$$z - 6$$

Self Check

Write each phrase as an algebraic expression.

a. $\frac{3}{4}$ of k

b. A divided by 8

c. Eighty subtracted from n

Answers: **a.** $\frac{3}{4}k$, **b.** $\frac{A}{8}$, **c.** $n - 80$ ■

COMMENT Be careful when translating a subtraction. For example, we have seen that *six less than* z translates to $z - 6$. It would be incorrect to translate it as $6 - z$, because $6 - z$ and $z - 6$ are not the same. The phrase z *less than six* translates as $6 - z$.

Writing algebraic expressions to represent unknown quantities

In the next two examples, we use the translation skills just discussed to describe unknown numerical quantities.

EXAMPLE 2 ***Writing an algebraic expression.*** Javier deposited d dollars in his checking account. He deposited \$500 more than that in his savings account. How much did he deposit in the savings account?

Self Check

A van weighs p pounds. A car is 1,000 pounds lighter than the van.

Solution

d = the number of dollars deposited in the checking account. To write an expression that describes the amount deposited in his savings account, we look for a key word or phrase in the problem.

Key phrase: *more than* **Translation:** add

The number of dollars deposited in the savings account = $d + 500$.

How many pounds does the car weigh?

Answer: $p - 1,000$ ■

EXAMPLE 3 ***Writing an algebraic expression.*** The pipe in Figure 3-1 is k feet in length. It is to be cut into 5 equally long pieces. How long will each piece be?

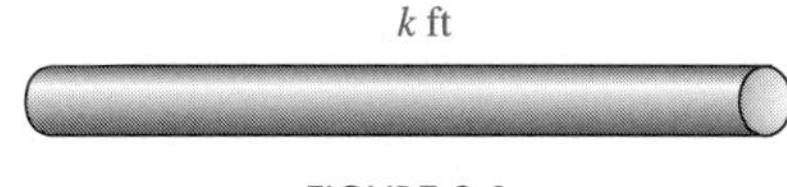

FIGURE 3-1

Solution

k = the length of the original pipe. To describe the length of each piece of the pipe using an algebraic expression, we look for a key word or phrase in the problem.

Key phrase: *equally long pieces* **Translation:** divide

The length of each piece will be = $\frac{k}{5}$ ft.

Self Check

A winning lottery ticket is worth x dollars. The payoff is to be split equally among three friends. Write an algebraic expression that represents each person's share of the prize (in dollars).

Answer: $\frac{x}{3}$ ■

When solving problems in real life, you are rarely told which variable to use. You must decide what the unknown quantities are and how to represent them using variables.

EXAMPLE 4 ***Naming two unknown quantities.*** In Figure 3-2, the baseball card's value is 4 times that of the football card. Choose a variable to represent the value of one card. Then write an expression for the value of the other card.

FIGURE 3-2

Solution

There are two unknown quantities. Since the baseball card's value is related to that of the football card, we will let v = the value of the football card.

Key phrase: 4 *times* **Translation:** multiply by 4

Therefore, $4v$ = the value of the baseball card

Self Check

The sale price of a sweater is \$20 less than the regular price. Choose a variable to represent one price. Then write an expression for the other price.

Answers: r = regular price, $r - 20$ = sale price ■

COMMENT A variable is used to represent an unknown number. Therefore, in the previous example, it would be incorrect to write, "Let v = football card," because the football card is not a number. We need to write, "Let v = the *value* of the football card."

EXAMPLE 5 ***Naming two unknown quantities.*** A 72-inch-long sub sandwich is cut into two pieces. Choose a variable to represent the length of one piece. Then write an expression for the length of the other piece.

Solution

A drawing is helpful in explaining this problem.

If the entire sandwich is 72 inches long, and we let l represent the length of this piece,... the length of this piece is $72 - l$.

72 inches

Self Check

Part of a $500 donation to a college was designated to go to the scholarship fund and the remainder to the building fund. Choose a variable to represent the amount donated to one of the funds. Then write an expression for the amount donated to the other fund.

Answers: s = amount donated to the scholarship fund; $500 - s$ = amount donated to the building fund ■

Looking for hidden operations

Sometimes we must read the phrasing of a problem carefully in order to detect hidden operations.

EXAMPLE 6 ***Hidden operations.*** The Golden Gate Bridge was completed 28 years before the Houston Astrodome was opened. The World Trade Center in New York was built 8 years after the Astrodome. Use algebraic expressions to express the ages (in years) of each of these engineering wonders.

Solution

The ages of the Golden Gate Bridge and the World Trade Center are both related to the age of the Astrodome. Therefore, we will let

x = the age of the Astrodome

Reading the problem carefully, we find that the Golden Gate Bridge was built 28 years before the dome, so its age is more than that of the Astrodome.

Key phrase: *more than* **Translation:** add

In years, the age of the Golden Gate Bridge is $x + 28$.

The Trade Center was built 8 years after the dome, so its age is less than that of the Astrodome.

Key phrase: *less than* **Translation:** subtract

In years, the age of the Trade Center is $x - 8$.

The results are summarized in the table at the left. ■

Engineering feat	Age
Astrodome	x
Golden Gate Bridge	$x + 28$
World Trade Center	$x - 8$

EXAMPLE 7 ***Looking for a pattern.*** How many eggs are there in d dozen?

Solution

Since there are no key words, we must carefully analyze the problem to write an expression that gives the number of eggs in d dozen. It is often helpful to consider some specific cases. For example, let's calculate the number of eggs in 1 dozen, 2 dozen, and 3 dozen. When we write the results in a table (as on the next page), a pattern becomes apparent.

Self Check

Complete the table on the next page. Then use that information to determine how many yards are in f feet.

Number of dozen	Number of eggs
1	$12 \cdot 1 = 12$
2	$12 \cdot 2 = 24$
3	$12 \cdot 3 = 36$
d	$12 \cdot d = 12d$

We multiply the number of dozen by 12 to find the number of eggs.

Therefore, if $d =$ the number of dozen eggs, the number of eggs is $12d$.

Number of feet	Number of yards
3	
6	
9	
f	

Answers: 1, 2, 3; $\frac{f}{3}$ ■

Expressions involving more than one operation

In the previous examples, the algebraic expressions contained one operation. We now examine expressions involving two operations.

EXAMPLE 8 ***An expression involving two operations.*** As Figure 3-3 shows, Alaska is much larger than Vermont. To be exact, the area of Alaska is 380 square miles more than 50 times that of Vermont. Choose a variable to represent one area. Then write an expression for the other area.

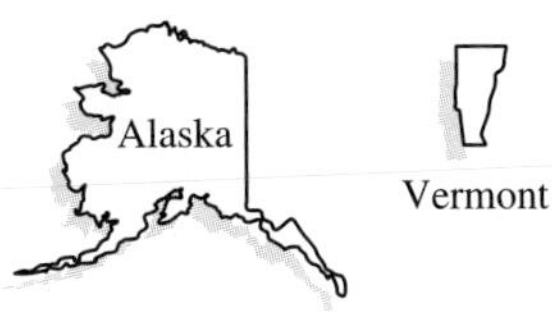

FIGURE 3-3

Solution

Since the area of Alaska is expressed in terms of the area of Vermont, we let $a =$ the area of Vermont.

Key phrase: *more than* **Translation:** add

Key word: *times* **Translation:** multiply

The area of Alaska $= 50a + 380$ square miles.

Self Check

On the second day of her trip, Tamiko drove 100 miles less than twice as far as the first day. Choose a variable to represent the number of miles driven on the first day. Then write an expression for the number of miles driven on the second day.

Answers: $m =$ miles driven on first day; miles driven on second day $= 2m - 100$ ■

STUDY SET Section 3.1

VOCABULARY *Fill in the blanks to make the statements true.*

1. An algebraic ______________ is a combination of variables, numbers, and the operation symbols for addition, subtraction, multiplication, and division.

2. The answer to an addition problem is called the ______________. The answer to a subtraction problem is called the ______________.

3. A ______________ is a letter that is used to stand for a number.

4. The answer to a multiplication problem is called the ______________. The answer to a division problem is called the ______________.

CONCEPTS

5. Write two different algebraic expressions that contain the numbers 10 and 3 and the variable x.

6. Write an equation with one side containing an algebraic expression and the other side the number 20.

7. Illustration 1 shows the commute to work (in miles) for two men who work in the same office. Who lives farther from the office? How much farther?

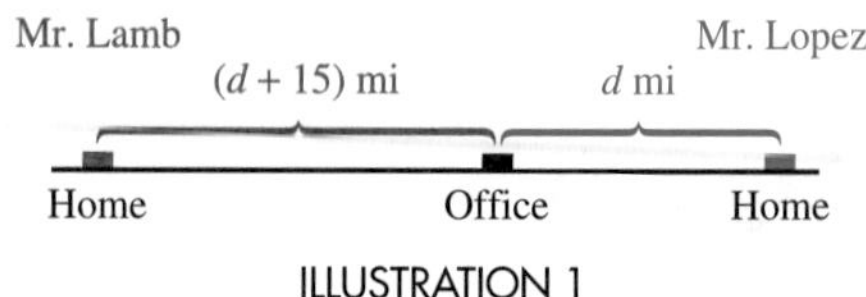

ILLUSTRATION 1

8. See Illustration 2.

a. If we let b represent the height of the birch tree, write an algebraic expression for the height of the elm tree.

b. If we let e stand for the height of the elm tree, write an algebraic expression for the height of the birch tree.

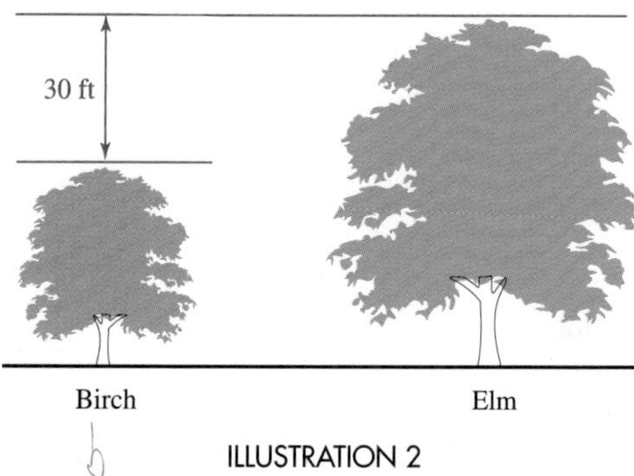

ILLUSTRATION 2

9. In 1999, the business profits of a video rental store were double those of the previous year. In 2000, the profits were triple those of 1998. If the profits in 1998 were p dollars, write algebraic expressions to complete the bar graph in Illustration 3.

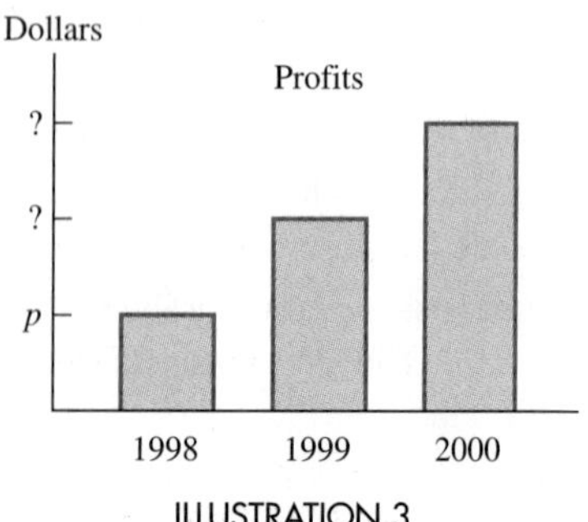

ILLUSTRATION 3

10. Illustration 4 shows the ages of three family members.

	Age
Matthew	x
Sarah	$x - 8$
Joshua	$x + 2$

ILLUSTRATION 4

a. Who is the youngest person shown in the table?

b. Who is the oldest person listed in the table?

c. On whose age are the ages in the table based?

11. On a flight from Dallas to Miami, a jet airliner, which flies at 500 mph in still air, experiences a tail wind of x mph. The tail wind increases the speed of the jet. On the immediate return flight to Dallas, the airliner flies into a head wind of the same strength. The head wind decreases the speed of the jet. Use this information to complete Illustration 5.

Wind conditions	Speed of jet (mph)
In still air	
With the tail wind	
Against the head wind	

ILLUSTRATION 5

12. The weights of two mixtures, measured in ounces, are compared on a balance, as shown in Illustration 6.

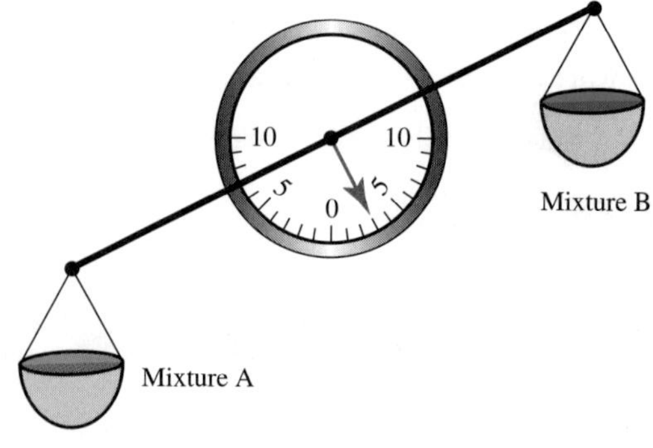

ILLUSTRATION 6

a. Which mixture is heavier? By how much?

b. Choose a variable to represent the weight of one mixture. Then write an expression for the weight of the other mixture.

13. A student figures that she has h hours to study for a government final. She wants to spread the studying evenly over a four-day period. Write an expression for how many hours she should study each day.

14. After all c children complete a Little League tryout, the league officials decide that they have enough players for 8 teams of equal size. Express the number of players on each team.

15. If x inches of tape have been used off the roll shown in Illustration 7, how many inches of tape are left on the roll?

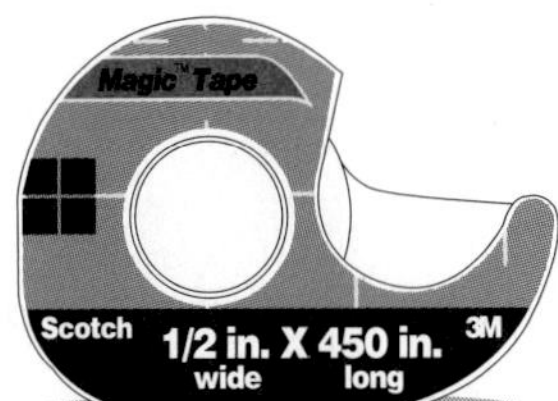

ILLUSTRATION 7

16. Complete each table.

a.

Number of decades	Number of years
1	
2	
3	
d	

b.

Number of inches	Number of feet
12	
24	
36	
i	

NOTATION *Write each expression in another algebraic form.*

17. $x \cdot 8$

18. $5(t)$

19. $10 \div g$

20. $h \div 16$

PRACTICE *Translate each phrase to an algebraic expression.*

21. The difference of x and 9

22. The sum of 12 and p

23. Two-thirds of the population p

24. The product of x and 34

25. r added to six

26. The ratio of i to 100

27. 15 less than d

28. Forty increased by w

29. s subtracted from 1

30. Sixteen minus a

31. Twice the price p

32. T reduced by 50

33. Exceeds the standard s by 14

34. The cost c split five equal ways

35. 35 divided by b

36. The total of 5 and 12 and q

37. x decreased by 2

38. 7 more than the average a

In Exercises 39–42, write a word description of each algebraic expression. (Answers may vary.)

39. $c + 7$

40. $7 - c$

41. $c - 7$

42. $7c$

43. **a.** How many seconds are there in m minutes?
b. In h hours?

44. A man sleeps x hours per day.
a. How many hours does he sleep in a week?
b. In a year?

45. A secretary earns an annual salary of s dollars.
a. Express her salary per month.
b. Express her salary per week.

46. A store manager earns d dollars an hour.
a. How much money will he earn in an 8-hour day?
b. In a 40-hour week?

47. A rope is f feet long.
a. Express its length in inches.
b. Express its length in yards.

48. A chain is y yards long.
a. Express its length in feet.
b. Express its length in inches.

Write an expression for the unknown quantity.

49. The highest decibel reading during a rock concert registered just 5 decibels shy of that of a jet engine. If a jet engine is normally j decibels, what was the decibel reading for the concert?

50. A couple needed to purchase 21 presents for friends and relatives on their holiday gift list. If the husband purchased g presents, how many presents did the wife need to buy?

51. A restaurant owner purchased s six-packs of cola. How many individual cans would this be?

52. The height of a hedge was f feet before a gardener cut 2 feet off the top. What was the height of the trimmed hedge?

53. A pad of yellow legal paper contains p pages. If a lawyer uses 15 pages every day, how many days will one pad last?

54. The projected cost c (in dollars) of a freeway was too low by a factor of 10! What was the actual cost of the freeway?

55. In a recycling drive, a campus ecology club collected t tons of newspaper. A Boy Scout troop then contributed an additional 2 tons. How many tons of newspaper were collected?

56. A graduating class of x people took buses that held 40 students each to an all-night graduation party. How many buses were needed to transport the class?

Choose a variable to represent one unknown. Then write an expression for the other unknown.

57. The rectangle in Illustration 8 is 6 units longer than it is wide. Express the length and width of this rectangle.

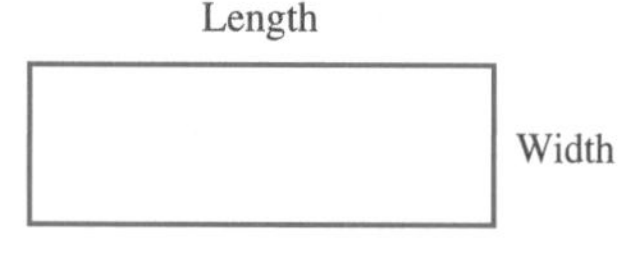

ILLUSTRATION 8

58. The smaller pipe in Illustration 9 takes three times longer to fill the tank than does the larger pipe. Express how long it takes each pipe to fill the tank.

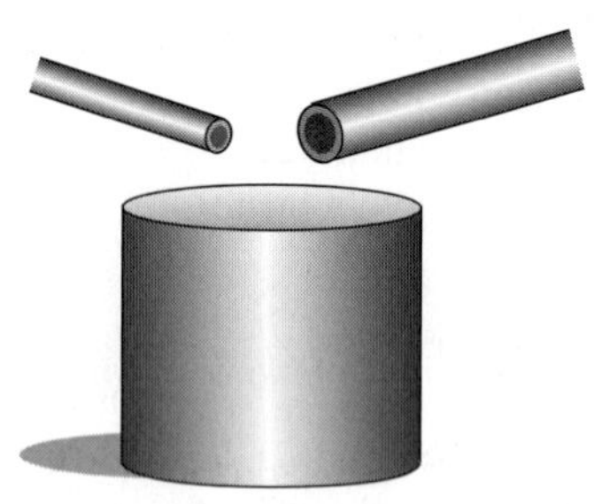

ILLUSTRATION 9

59. The car radiator in Illustration 10 originally contained 6 gallons of coolant before some was drained out. Express the amount that was drained out and the amount that remains in the radiator.

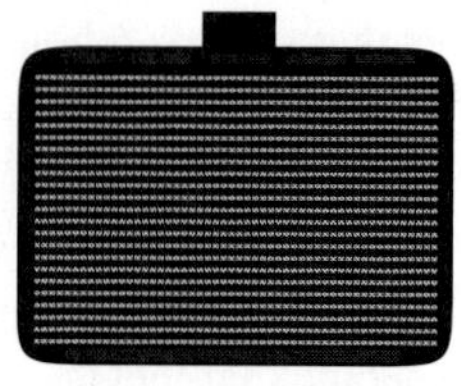

ILLUSTRATION 10

60. During a sale, the regular price of a CD was reduced by \$2. Express the regular price and the new sale price.

Translate each phrase to an algebraic expression.

61. Five more than triple x

62. The quotient of 5 and t is reduced by four.

63. The product of a and 10 is increased by 12.

64. Thirty more than the difference of 78 and d

APPLICATIONS *Choose a variable to represent one unknown. Then write an expression for the other unknown. (Answers may vary.)*

65. PRESIDENTIAL ELECTIONS In 1960, John F. Kennedy was elected President of the United States with a popular vote of only 118,550 votes more than that of Richard M. Nixon. Express how many votes each candidate received.

66. EARTHQUAKES An earthquake with a reading of 7.0 on the Richter scale releases ten times as much energy as an earthquake that registers 6.0 on the scale. Express the amount of energy released by an earthquake of each magnitude.

67. THE BEATLES According to music historians, sales of the Beatles' second most popular single, *Hey Jude,* trail the sales of their most popular single, *I Want to Hold Your Hand,* by 2,000,000 copies. Express the sales of each single.

68. YOUTH SPORTS Illustration 11 shows how participation in organized soccer and volleyball changes as girls and boys enter their teen years.

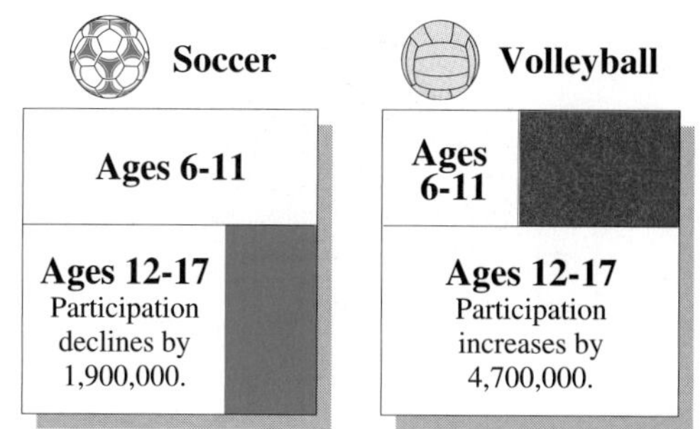

Based on data from Soccer Industry Council of America and American Sports Data

ILLUSTRATION 11

a. Express the number of soccer participants for both age groups.

b. Express the number of volleyball participants for both age groups.

WRITING

69. Explain how variables are used in this section.

70. Explain the difference between the phrases *greater than* and *is greater than.*

71. Suppose you were asked to find the height of a building. What is wrong with the following start?

Let $x =$ building

72. What is an algebraic expression?

REVIEW

73. Find the sum: $-5 + (-6) + 1$.

74. Evaluate: $2 + 3(-3)$.

75. Solve $-x = 4$.

76. Write the related multiplication statement for $\frac{-18}{2} = -9$.

77. List the set of integers.

78. Represent a deficit of $1,200 using a signed number.

79. Find $-3 + (-2) + 7$.

80. Evaluate: $(-5)^3$.

3.2 *Evaluating Algebraic Expressions and Formulas*

In this section, you will learn about

- Evaluating algebraic expressions • Formulas • Formulas from business
- Formulas from science • Formulas from mathematics

Introduction. An algebraic expression is a combination of variables and numbers with the operation symbols of addition, subtraction, multiplication, and division. In this section, we will be replacing the variables in these expressions with specific numbers. Then, using the rules for the order of operations, we will evaluate each expression. We will also study formulas. Like algebraic expressions, formulas involve variables.

Evaluating algebraic expressions

The manufacturer's instructions for installing a kitchen garbage disposal include the diagram in Figure 3-4 (on the next page). Word phrases are used to describe the lengths of the pieces of pipe needed to connect the disposal to the drain line.

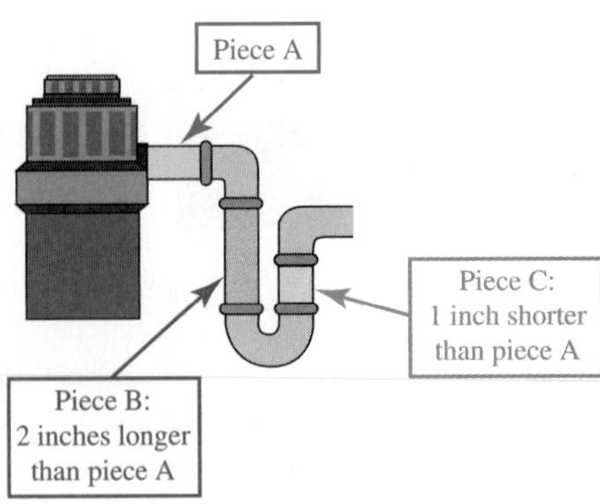

FIGURE 3-4

The instructions tell us that the lengths of pieces B and C are related to the length of piece A. If we let $x =$ the length of piece A, then the lengths of the two other pieces of pipe can be expressed using algebraic expressions, as shown in Figure 3-5.

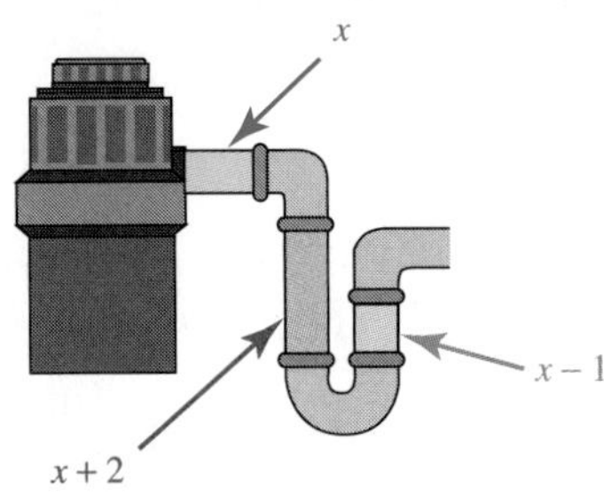

FIGURE 3-5

Since piece B is 2 inches longer than piece A,

$x + 2 =$ the length of piece B

Since piece C is 1 inch shorter than piece A,

$x - 1 =$ the length of piece C

Model	Length of piece A
#101	2 inches
#201	3 inches
#301	4 inches

TABLE 3-1

See Table 3-1, which shows part of the manufacturer's instruction sheet. Suppose that model #201 is being installed. The table tells us that piece A should be 3 inches long. We then find the lengths of the other two pieces of pipe by replacing x with 3 in each of the algebraic expressions.

To find the length of piece B:

Replace x with 3.

$$x + 2 = 3 + 2$$
$$= 5$$

Piece B should be 5 inches long.

To find the length of piece C:

Replace x with 3.

$$x - 1 = 3 - 1$$
$$= 2$$

Piece C should be 2 inches long.

When we replace the variable in an algebraic expression with a specific number and then apply the rules for the order of operations, we are **evaluating the algebraic expression.** In the previous example, we say that we *substituted* 3 for x to find the lengths of the other two pieces of pipe.

COMMENT In this section, we will replace variables with positive and negative numbers. It is a good idea to write parentheses around a number when it is substituted for a variable in an algebraic expression.

EXAMPLE 1 ***Evaluating algebraic expressions.*** Evaluate each expression for $x = 3$: **a.** $2x - 1$ and **b.** $\frac{-x-15}{6}$.

Solution

a. $2x - 1 = 2(3) - 1$ Substitute 3 for x. Use parentheses to show the multiplication.

$= 6 - 1$ Do the multiplication first: $2(3) = 6$.

$= 5$ Do the subtraction.

b. $\frac{-x - 15}{6} = \frac{-(3) - 15}{6}$ Substitute 3 for x. Use parentheses.

$= \frac{-3 - 15}{6}$ $-(3) = -3$.

$= \frac{-3 + (-15)}{6}$ Add the opposite of 15.

$= \frac{-18}{6}$ Do the addition: $-3 + (-15) = -18$.

$= -3$ Do the division.

Self Check

Evaluate each expression for $y = 5$.

a. $5y - 4$

b. $\frac{-y - 15}{5}$

Answers: **a.** 21, **b.** -4 ■

EXAMPLE 2 ***Evaluating algebraic expressions.*** Evaluate each expression for $a = -2$: **a.** $-3a + 4a^2$ and **b.** $-a + 3(1 + a)$.

Solution

a. $-3a + 4a^2 = -3(-2) + 4(-2)^2$ Substitute -2 for a. Use parentheses.

$= -3(-2) + 4(4)$ Evaluate the power: $(-2)^2 = 4$.

$= 6 + 16$ Do the multiplications: $-3(-2) = 6$ and $4(4) = 16$.

$= 22$ Do the addition.

b. $-a + 3(1 + a) = -(-2) + 3[1 + (-2)]$ Substitute -2 for a. Use parentheses.

$= -(-2) + 3(-1)$ Do the addition within the brackets.

$= 2 + (-3)$ Simplify: $-(-2) = 2$. Do the multiplication: $3(-1) = -3$.

$= -1$ Do the addition.

Self Check

Evaluate each expression for $t = -3$.

a. $-2t + 4t^2$

b. $-t + 2(t + 1)$

Answers: **a.** 42, **b.** -1 ■

To evaluate algebraic expressions containing two or more variables, we need to know the value of each variable.

EXAMPLE 3 ***Expressions involving two variables.*** Evaluate $(8hg + 6g)^2$ for $h = -1$ and $g = 5$.

Solution

$(8hg + 6g)^2 = [8(-1)(5) + 6(5)]^2$ Substitute -1 for h and 5 for g.

$= (-40 + 30)^2$ Do the multiplications within the brackets: $8(-1)(5) = -40$ and $6(5) = 30$.

$= (-10)^2$ Do the addition within the parentheses.

$= 100$ Evaluate the power.

Self Check

Evaluate $(5rs + 4s)^2$ for $r = -1$ and $s = 5$.

Answer: 25 ■

Formulas

A **formula** is an equation that is used to state a known relationship between two or more variables. Formulas are used in many fields: economics, physical education, anthropology, biology, automotive repair, and nursing, just to name a few. In this section, we will consider seven formulas from business, science, and mathematics.

Formulas from business

A Formula to Find the Sale Price

If a car that normally sells for $12,000 is discounted $1,500, you can find the sale price using the formula

Sale price	=	original price	−	discount

Using the variables s to represent the sale price, p the original price, and d the discount, this formula can be written as

$$\boxed{s = p - d}$$

To find the sale price of the car, we substitute 12,000 for p, substitute 1,500 for d, and simplify the right-hand side of the equation.

$$\begin{aligned} s &= p - d \\ &= \mathbf{12{,}000} - \mathbf{1{,}500} && \text{Substitute 12,000 for } p \text{ and 1,500 for } d. \\ &= 10{,}500 && \text{Do the subtraction.} \end{aligned}$$

The sale price of the car is $10,500.

A Formula to Find the Retail Price

To make a profit, a merchant must sell a product for more than he paid for it. The price at which he sells the product, called the *retail price,* is the *sum* of what the item cost him and the markup.

Retail price	=	cost	+	markup

Using the variables r to represent the retail price, c the cost, and m the markup, we can write this formula as

$$\boxed{r = c + m}$$

As an example, suppose that a store buys a lamp for $35 and then marks up the cost $20 before selling it. We can find the retail price of the lamp using this formula.

$$\begin{aligned} r &= c + m \\ &= \mathbf{35} + \mathbf{20} && \text{Substitute 35 for } c \text{ and 20 for } m. \\ &= 55 && \text{Do the addition.} \end{aligned}$$

The retail price of the lamp is $55.

A Formula to Find Profit

The profit a business makes is the *difference* of the revenue (the money it takes in) and the costs.

Profit	=	revenue	−	costs

Using the variables p to represent the profit, r the revenue, and c the costs, we have the formula

$$\boxed{p = r - c}$$

As an example, suppose that a charity telethon took in \$14 million in donations but had expenses totaling \$2 million. We can find the profit made by the charity by subtracting the expenses (costs) from the donations (revenue).

$$p = r - c$$
$$= \mathbf{14} - \mathbf{2} \quad \text{Substitute 14 for } r \text{ and 2 for } c.$$
$$= 12 \quad \text{Do the subtraction.}$$

The charity made a profit of \$12 million from the telethon.

Formulas from science

A Formula to Find the Distance Traveled

If we know the rate (speed) at which we are traveling and the time we will be moving at that rate, we can find the distance traveled using the formula

Distance = rate · time

Using the variables d to represent the distance, r the rate, and t the time, we have the formula

$$\boxed{d = rt}$$

EXAMPLE 4 ***Highway speed limits.*** Several state speed limits for trucks are shown below. At each of these speeds, how far would a truck travel in 3 hours?

Solution

To find the distance traveled by a truck in Ohio, we write

$$d = rt$$
$$d = \mathbf{55(3)} \quad \text{55 mph is the rate } r \text{, and 3 hours is the time } t.$$
$$d = 165 \quad \text{Do the multiplication. The units of the answer are miles.}$$

At 55 mph, a truck would travel 165 miles in 3 hours. We can use a table to display the calculations for each state.

	r ·	t =	d
Ohio	55	3	165
Indiana	60	3	180
Kentucky	65	3	195

This column gives the distance traveled, in miles.

Self Check

Nevada's highway speed limit for trucks is 75 mph. How far would a truck travel in 3 hours?

d=rt
d=75(3)
d= 225 mi

Answer: 225 mi

COMMENT When using $d = rt$ to find distance, make sure that the units are similar. For example, if the rate is given in miles per hour, the time must be expressed in hours.

A Formula for Converting Degrees Fahrenheit to Degrees Celsius

Electronic message boards in front of some banks flash two temperature readings. This is because temperature can be measured using the Fahrenheit or the Celsius scale. The Fahrenheit scale is used in the American system of measurement and the Celsius scale in the metric system. The two scales are shown on the thermometers in Figure 3-6. This should help you to see how the two scales are related. There is a formula to convert a Fahrenheit reading F to a Celsius reading C:

$$C = \frac{5(F - 32)}{9}$$

Later we will see that there is a formula to convert a Celsius reading to a Fahrenheit reading.

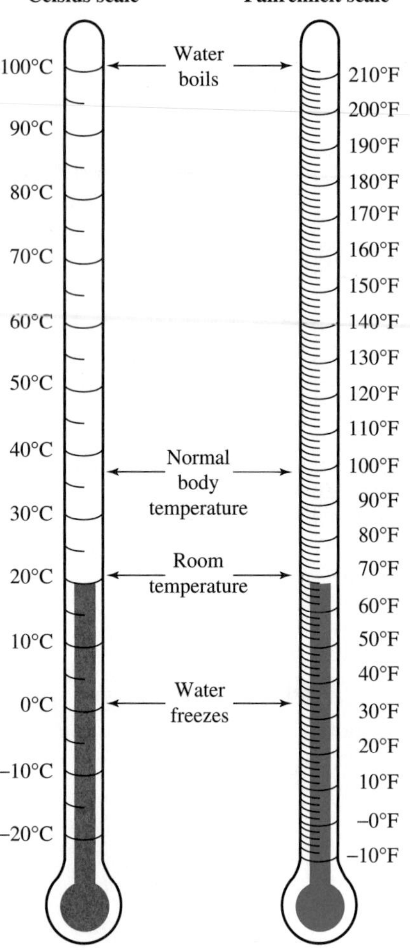

FIGURE 3-6

EXAMPLE 5 ***Converting from degrees Fahrenheit to degrees Celsius.*** The thermostat in an office building was set at 77° F. Convert this setting to degrees Celsius.

Solution

$$C = \frac{5(F - 32)}{9}$$

$= \frac{5(77 - 32)}{9}$ Substitute the Fahrenheit temperature, 77, for F.

$= \frac{5(45)}{9}$ Do the operation within parentheses first: $77 - 32 = 45$.

$= \frac{225}{9}$ Do the multiplication: $5(45) = 225$.

$= 25$ Do the division.

The thermostat is set at 25° C.

Self Check

The record high temperature for New Mexico is 122° F, on June 27, 1994. Convert this temperature to degrees Celsius.

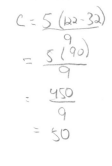

Answer: 50° C

A Formula to Find the Distance an Object Falls

The distance an object falls (in feet) when it is dropped from a height is related to the time (in seconds) that it has been falling by the formula

$$\text{Distance fallen} = 16 \cdot (\text{time})^2$$

Using the variables d to represent the distance and t the time, we have

$$\boxed{d = 16t^2}$$

EXAMPLE 6 ***Finding the distance an object falls.*** Find the distance a camera fell in 6 seconds if it was dropped by a vacationer taking a hot-air balloon ride.

Solution

We can use the formula $d = 16t^2$ to find the distance the camera fell.

$d = 16t^2$

$d = 16(6)^2$ The camera fell for 6 seconds. Substitute 6 for t.

$= 16(36)$ Evaluate the exponential expression: $6^2 = 36$.

$= 576$ Do the multiplication.

The camera fell 576 feet.

Self Check

Find the distance a rock fell in 3 seconds if it was dropped over the edge of the Grand Canyon.

$d = 16t^2$
$d = 16(3)^2$
$d = 16(9)$
$d = 144$

Answer: 144 ft ■

Formulas from mathematics

A Formula to Find the Arithmetic Mean (Average)

The arithmetic mean, or average, of a set of numbers is a value around which the numbers are grouped. To find the arithmetic mean (average), we divide the *sum* of all the values by the *number* of values. Writing this as a formula, we get

$$\text{Mean} = \frac{\text{sum of the values}}{\text{number of values}}$$

Using the variables A to represent the mean (average), S the sum, and n the number of values, we have

$$\boxed{A = \frac{S}{n}}$$

EXAMPLE 7 ***Response time to 911 calls.*** To measure its effectiveness, a police department recorded the length of time between incoming 911 calls and the arrival of a police unit at the scene. The response times for an entire week are listed in Table 3-2. Find the average response time.

Response times	Occurrences
2 min	3
3 min	16
4 min	52
5 min	22

TABLE 3-2

Solution The average response time can be found using the formula $A = \frac{S}{n}$. To find S, we need to find the sum of all the response times. Since the response time of 2 minutes occurred 3 times, we need to add $2 + 2 + 2$. Since the response time of 3 minutes occurred 16 times, we need to add sixteen 3's, and so on. More simply, we can find S by multiplying each response time by the number of occurrences and then adding the results:

$S = 2(3) + 3(16) + 4(52) + 5(22)$ 2 min occurred 3 times, 3 min occurred 16 times, 4 min occurred 52 times, and 5 min occurred 22 times.

$= 6 + 48 + 208 + 110$ Do the multiplications.

$= 372$ Do the additions.

To find n, the number of response times, we add the number of occurrences:

$$n = 3 + 16 + 52 + 22 = 93$$

Replacing S with 372 and n with 93, we have

$$A = \frac{S}{n}$$

$$A = \frac{372}{93}$$

$A = 4$ Do the division.

The average response time was 4 minutes. ■

STUDY SET Section 3.2

VOCABULARY *Fill in the blanks to make the statements true.*

1. A ______________ is an equation that states a known relationship between two or more variables.

2. An algebraic ______________ is a combination of variables, numbers, and the operation symbols for addition, subtraction, multiplication, and division.

3. To evaluate an algebraic expression, we ______________ specific numbers for the variables in the expression and apply the rules for the order of operations.

4. The arithmetic mean or ______________ of a group of numbers is a value around which the numbers are grouped.

CONCEPTS

5. Show the misunderstanding that occurs if we don't write parentheses around -8 when evaluating the expression $2x + 10$ for $x = -8$.

6. a. Which of the formulas studied in this section involve a *difference* of two quantities?

b. Which of the formulas studied in this section involve a *product*?

7. The plans for building a children's swing set are shown in Illustration 1.

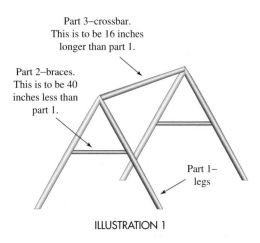

Part 3–crossbar. This is to be 16 inches longer than part 1.

Part 2–braces. This is to be 40 inches less than part 1.

Part 1–legs

ILLUSTRATION 1

a. Choose a variable to represent the length of one part of the swing set. Then write expressions for the lengths of the other two parts.

b. If the builder chooses to have part 1 be 60 inches long, how long should parts 2 and 3 be?

8. A television studio art department plans to construct a series of set decorations out of plywood, using the plan shown in Illustration 2.

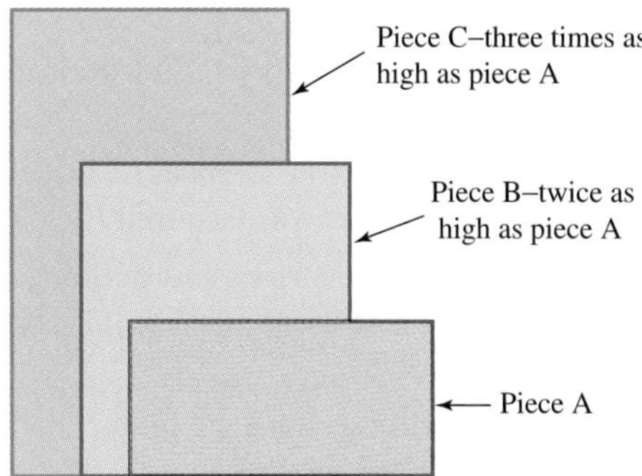

ILLUSTRATION 2

a. Choose a variable to represent the height of one piece of plywood. Then write expressions for the heights of the other two pieces.

b. For the foreground, designers will make piece A 15 inches high. How high should pieces B and C be?

c. For the background, piece A will be 30 inches high. How high should pieces B and C be?

9. A ticket outlet adds a service charge of $2 to the price of every ticket it sells.

a. Complete the pricing chart in Illustration 3.

Ticket price	Service charge	Total cost
20	2	
25	2	
p	2	

ILLUSTRATION 3

b. Write a formula for the total cost T if the price of a ticket is p dollars.

10. Explain why the following instruction is incomplete. Evaluate the algebraic expression $3a^2 - 4$.

11. Complete the chart in Illustration 4 by finding the distance traveled in each instance.

	Rate (mph) ·	Time = (hr)	Distance (mi)
Bike	12	4	
Walking	3	t	
Car	x	3	

ILLUSTRATION 4

12. Illustration 5 shows part of a dashboard. Explain what each instrument measures. How are these measurements mathematically related?

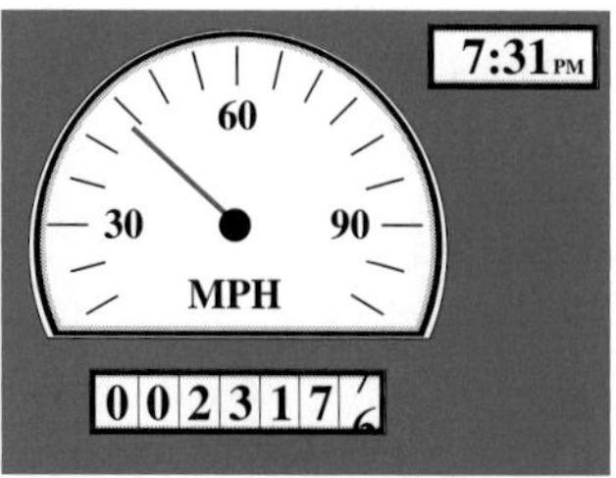

ILLUSTRATION 5

13. What occupation might use a formula that finds

a. target heart rate after a workout

b. gas mileage of a car

c. age of a fossil

d. equity in a home

e. dosage to administer

f. cost-of-living index

14. A car travels at a rate of 65 mph for 15 minutes. What is wrong with the following thinking?

$$d = rt$$
$$d = 65(15)$$
$$d = 975$$

The car travels 975 miles in 15 minutes.

NOTATION

15. Use variables to write the formula that relates each of the quantities listed below.

a. Rate, distance, time

b. Centigrade temperature, Fahrenheit temperature

c. Time, the distance an object falls when dropped

16. Use variables to write the formula that relates each of the quantities listed below.

a. Original price, sale price, discount

b. Number of values, average, sum of values

c. Cost, profit, revenue

d. Markup, retail price, cost

PRACTICE *Evaluate each expression for the given value of the variable.*

17. $3x + 5$ for $x = 4$
18. $1 + 7a$ for $a = 2$
19. $-p$ for $p = -4$
20. $-j$ for $j = -9$
21. $-4t$ for $t = -10$
22. $-12m$ for $m = -6$
23. $\frac{x - 8}{2}$ for $x = -4$
24. $\frac{-10 + y}{-4}$ for $y = -6$
25. $2(p + 9)$ for $p = -12$
26. $3(r - 20)$ for $r = 15$
27. $x^2 - x - 7$ for $x = -5$
28. $a^2 + 3a - 9$ for $a = -3$
29. $8s - s^3$ for $s = -2$
30. $5r + r^3$ for $r = 1$
31. $4x^2$ for $x = 5$
32. $3f^2$ for $f = 3$
33. $3b - b^2$ for $b = -4$
34. $5a - a^2$ for $a = -3$
35. $\frac{24 + k}{3k}$ for $k = 3$
36. $\frac{4 - h}{h - 4}$ for $h = -1$
37. $|6 - x|$ for $x = 50$
38. $|3c - 1|$ for $c = -1$
39. $-2|x| - 7$ for $x = -7$
40. $|x^2 - 7^2|$ for $x = 7$

Evaluate each algebraic expression for the given values of the variables.

41. $\frac{x}{y}$ for $x = 30$ and $y = -10$
42. $\frac{e}{3f}$ for $e = 24$ and $f = -8$
43. $-x - y$ for $x = -1$ and $y = 8$
44. $-a - 5b$ for $a = -9$ and $b = 6$
45. $x(5h - 1)$ for $x = -2$ and $h = 2$
46. $c(2k - 7)$ for $c = -3$ and $k = 4$
47. $b^2 - 4ac$ for $b = -3$, $a = 4$, and $c = -1$
48. $3r^2h$ for $r = 4$ and $h = 2$
49. $x^2 - y^2$ for $x = 5$ and $y = -2$
50. $x^3 - y^3$ for $x = -1$ and $y = 2$
51. $\frac{50 - 6s}{-t}$ for $s = 5$ and $t = 4$
52. $\frac{7v - 5r}{-r}$ for $v = 8$ and $r = 4$
53. $-5abc + 1$ for $a = -2$, $b = -1$, and $c = 3$
54. $-rst + 2t$ for $r = -3$, $s = -1$, and $t = -2$
55. $5s^2t$ for $s = -3$ and $t = -1$
56. $-3k^2t$ for $k = -2$ and $t = -3$
57. $|a^2 - b^2|$ for $a = -2$ and $b = -5$
58. $-|2x - 3y + 10|$ for $x = 0$ and $y = -4$

Use the appropriate formula to answer each question.

59. It costs a snack bar owner 20 cents to make a snow cone. If the markup is 50 cents, what is the price of a snow cone?
60. Find the distance covered by a jet if it travels for 3 hours at 550 mph.
61. A school carnival brought in revenues of \$13,500 and had costs of \$5,300. What was the profit?
62. For the month of June, a florist's cost of doing business was \$3,795. If June revenues totaled \$5,115, what was her profit for the month?
63. A jewelry store buys bracelets for \$18 and marks them up \$5. What is the retail price of a bracelet?
64. A shopkeeper marks up the cost of every item she carries by the amount she paid for the item. If a fan costs her \$27, what does she charge for the fan?
65. Find the distance covered by a car traveling 60 miles per hour for 5 hours.
66. Find the sale price of a pair of skis that normally sells for \$200 but is discounted \$35.
67. Find the Celsius temperature reading if the Fahrenheit reading is 14°.
68. Find the Celsius temperature reading if the Fahrenheit reading is 113°.
69. Find the average for a bowler who rolled scores of 254, 225, and 238.
70. On its first night of business, a pizza parlor brought in \$445. The owner estimated his costs that night to be \$295. What was the profit?
71. Find the distance a ball has fallen 2 seconds after being dropped from a tall building.
72. A store owner buys a pair of pants for \$25 and marks them up \$15 for sale. What is the retail price of the pants?

APPLICATIONS

73. FINANCIAL STATEMENT Use the data in Illustration 6 to complete the financial statement for Avon Products, Inc. (on the next page).

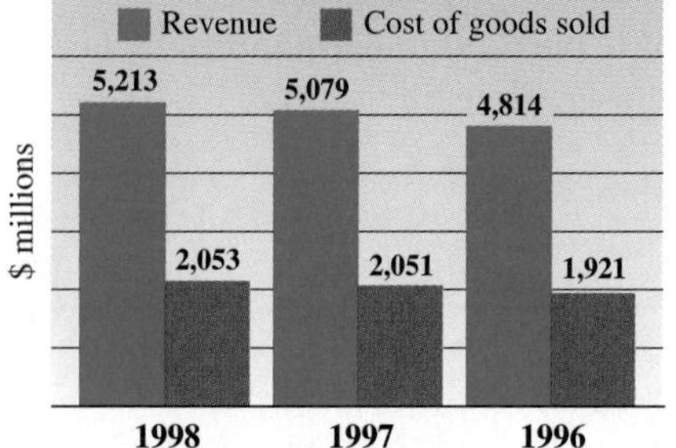

Based on data from *Hoover's Online*

ILLUSTRATION 6

Annual Financials: Income Statement All dollar amounts in millions	Dec. '98	Dec. '97	Dec. '96
Revenue			
Cost of goods sold			
Gross profit			

74. SPREADSHEET A store manager wants to use a spreadsheet to post the prices of sale items for the checkers at the cash registers. (See Illustration 7.) If column B lists the regular price and column C lists the discount, write a formula using column names to have the computer find the sale price. Then fill in column D.

	A	B	C	D
1	Bath towel set	\$25	\$5	
2	Pillows	\$15	\$3	
3	Comforter	\$53	\$11	

ILLUSTRATION 7

75. THERMOMETER SCALE A thermometer manufacturer wishes to scale a thermometer in both degrees Celsius and degrees Fahrenheit. Find the missing Celsius degree measures in Illustration 8.

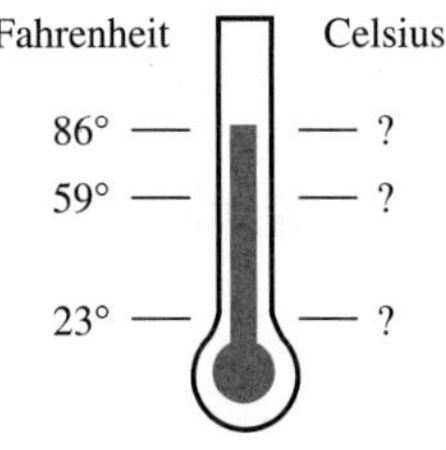

ILLUSTRATION 8

76. DEALER MARKUP A car dealer marks up the cars he sells \$500 above factory invoice (that is, \$500 over what it costs him to purchase the car from the factory).

a. Complete the pricing chart in Illustration 9.

Model	Factory invoice (\$)	Markup (\$)	Price (\$)
Minivan	15,600		
Pickup	13,200		
Convertible	x		

ILLUSTRATION 9

b. Write a formula for the price p of a car if the factory invoice is f dollars.

77. FALLING OBJECT See Illustration 10. First, find the distance in feet traveled by a falling object in 1, 2, 3, and 4 seconds. Enter the results in the middle column. Then find the distance the object traveled over each time interval and enter it in the right-hand column.

Time falling	Distance traveled (ft)	Time intervals
1 sec		Distance traveled from 0 sec to 1 sec
2 sec		Distance traveled from 1 sec to 2 sec
3 sec		Distance traveled from 2 sec to 3 sec
4 sec		Distance traveled from 3 sec to 4 sec

ILLUSTRATION 10

78. DISTANCE TRAVELED

a. When in orbit, the space shuttle travels at a rate of approximately 17,250 miles per hour. How far does it travel in one day?

b. The speed of light is approximately 186,000 miles per second. How far will light travel in one minute?

c. The speed of a sound wave in air is about 1,100 feet per second at normal temperatures. How far does it travel in half a minute?

79. CUSTOMER SATISFACTION SURVEY As customers were leaving a restaurant, they were asked to rate the service they had received. Good service was rated with a 5, fair service with a 3, and poor service with a 1. The tally sheet compiled by the questioner is shown in Illustration 11. What was the restaurant's average score on this survey?

Type of service	Point value	Number
Good	5	𝍸 𝍸 𝍸 𝍸 𝍸 𝍸 𝍸 𝍸 𝍸 𝍸 III
Fair	3	𝍸 𝍸 𝍸 𝍸 𝍸 I
Poor	1	𝍸 IIII

ILLUSTRATION 11

80. SHIPPING Illustration 12 shows the number of ships using the docking facilities of a port one week in July. Find the average number of ships that docked there per day for that week.

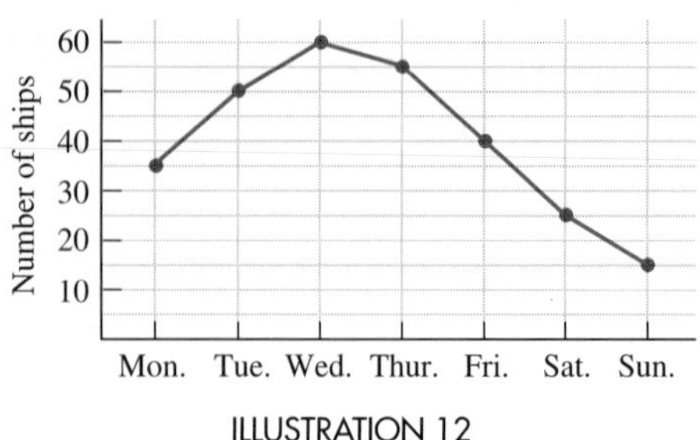

ILLUSTRATION 12

WRITING

81. Explain the error in the student's work shown below.

Evaluate $-a + 3a$ for $a = -6$.

$-a + 3a = -6 + 3(-6)$

$= -6 + (-18)$

$= -24$

82. Explain how we can use a stopwatch to find the distance traveled by a falling object.

83. Write a definition for each of these business words: *revenue, markup,* and *profit.*

84. What is a formula?

85. In this section we *substituted* a number for a variable. List some other uses of the word *substitute* that you encounter in everyday life.

86. Temperature can be measured using the Fahrenheit or the Celsius scale. How do the scales differ?

REVIEW

87. Which of these are prime numbers? 9, 15, 17, 33, 37, 41

88. How can this repeated multiplication be rewritten in simpler form? $2 \cdot 2 \cdot 2 \cdot 2 \cdot 2$

89. Evaluate: $|-2 + (-5)|$.

90. Multiply: $-3(-2)(4)$.

91. In the equation $\frac{x}{3} = -4$, what operation is performed on the variable?

92. Is $t = -6$ a solution of $2t - 3 = 15$? Why or why not?

93. Subtract: $-3 - (-6)$.

94. Which is undefined: division of zero or division by zero?

3.3 *Simplifying Algebraic Expressions and the Distributive Property*

In this section, you will learn about

- Simplifying algebraic expressions involving multiplication
- The distributive property • Distributing a factor of -1
- Extending the distributive property

INTRODUCTION. In mathematics, it is frequently useful to replace something complicated with something that is equivalent and simpler in form. In this section, we will simplify algebraic expressions. The results will be equivalent but simpler expressions.

Simplifying algebraic expressions involving multiplication

To **simplify an algebraic expression,** we use properties of algebra to write the given expression in a simpler form. Two of the properties used to simplify algebraic expressions are the associative and the commutative properties of multiplication. Recall that the associative property allows us to change the grouping of the factors involved in a multiplication. The commutative property allows us to change the order of the factors.

To write $6(5x)$ in a simpler form, we can begin by rewriting it as $6 \cdot (5 \cdot x)$.

$6(5x) = 6 \cdot (5 \cdot x)$	$5x = 5 \cdot x$.
$= (6 \cdot 5)x$	Apply the associative property of multiplication. Instead of grouping 5 with x, group it with 6.
$= 30x$	Do the operation within the parentheses first: $6 \cdot 5 = 30$.

We say that $6(5x)$ simplifies to $30x$. That is, $6(5x) = 30x$.

To verify that $6(5x)$ and $30x$ are **equivalent expressions** (they represent the same number), we can evaluate each expression for several choices of x. For each value of x, the results should be the same.

If $x = 10$

$$\begin{aligned} 6(5x) &= 6[5(\mathbf{10})] & 30x &= 30(\mathbf{10}) \\ &= 6[50] & &= 300 \\ &= 300 \end{aligned}$$

If $x = -3$

$$\begin{aligned} 6(5x) &= 6[5(\mathbf{-3})] & 30x &= 30(\mathbf{-3}) \\ &= 6[-15] & &= -90 \\ &= -90 \end{aligned}$$

EXAMPLE 1 ***Simplifying algebraic expressions involving multiplication.*** Simplify: **a.** $-2(7x)$ and **b.** $-12t(-6)$.

Solution

We use the associative property to regroup the factors of the expression so that the numbers are separated from the variable. After multiplying the numbers, the result is then multiplied by the variable.

a. $-2(7x) = (-2 \cdot 7)x$	Apply the associative property of multiplication to regroup the factors.
$= -14x$	Do the operation within the parentheses first: $-2 \cdot 7 = -14$.
b. $-12t(-6) = -12(-6)t$	Apply the commutative property of multiplication. Change the order of the factors.
$= [-12(-6)]t$	Apply the associative property of multiplication to group the numbers together. Use brackets.
$= 72t$	Do the operation within the brackets first: $-12(-6) = 72$.

Self Check

Simplify each expression.

a. $4 \cdot 8r$

b. $-3y(-5)$

Answers: **a.** $32r$, **b.** $15y$ ■

EXAMPLE 2 ***Simplifying algebraic expressions involving two variables.*** Simplify: **a.** $-4m(-5n)$ and **b.** $2(6y)(-4z)$.

Solution

a. $-4m(-5n) = [-4(-5)](m \cdot n)$	Group the numbers and variables separately, using the commutative and associative properties of multiplication.
$= 20mn$	Do the multiplication within the brackets: $-4(-5) = 20$. Write $m \cdot n$ as mn.
b. $2(6y)(-4z) = [2(6)(-4)](y \cdot z)$	Use the commutative and associative properties to change the order and regroup the factors.
$= -48yz$	Do the multiplication within the brackets: $2(6)(-4) = -48$. Write $y \cdot z$ as yz.

Self Check

Simplify each expression.

a. $-7k(-5t)$

b. $2(4a)(-3d)$

Answers: **a.** $35kt$, **b.** $-24ad$ ■

COMMENT Be careful when using the terms *simplify* and *solve*. In mathematics, we *simplify expressions* and we *solve equations*. We do not simplify equations, nor do we solve expressions.

The distributive property

Another property of algebra that is used to simplify algebraic expressions is the **distributive property.** To introduce this property, we will examine the expression $2(5 + 3)$, which can be evaluated in two ways.

Method 1: Rules for the Order of Operations

Because of the grouping symbols contained in $2(5 + 3)$, the rules for the order of operations require that we compute the *sum* within the parentheses first.

$2(5 + 3) = 2(8)$ Do the addition within the parentheses first: $5 + 3 = 8$.

$= 16$ Do the multiplication.

Method 2: The Distributive Property

The distributive property allows us to evaluate the expression $2(5 + 3)$ in another way. We can distribute the factor of 2 across to the 5 and across to the 3. We then find each of those products separately and add the results.

Distribute the 2.

$2(5 + 3) = 2(5 + 3)$ Each number within the parentheses is multiplied by the factor outside the parentheses.

First product ↓ Second product ↓

$= 2(5) + 2(3)$

$= 10 + 6$ Do the multiplications first: $2(5) = 10$ and $2(3) = 6$.

$= 16$ Do the addition.

Notice that the result using each method is 16.

We now state the distributive property in symbols.

The distributive property

> If a, b, and c represent numbers,
>
> $a(b + c) = ab + ac$

Since subtraction is the same as adding the opposite, the distributive property also holds for subtraction.

The distributive property

> If a, b, and c represent numbers,
>
> $a(b - c) = ab - ac$

EXAMPLE 3 ***The distributive property.*** Apply the distributive property: **a.** $3(s + 7)$ and **b.** $6(x - 1)$.

Solution

a. $3(s + 7) = 3s + 3(7)$ Distribute the multiplication by 3.

$= 3s + 21$ Do the multiplication.

After applying the distributive property to $3(s + 7)$ to obtain $3s + 21$, we say that we have *removed parentheses.*

b. $6(x - 1) = 6x - 6(1)$ Distribute the factor 6.

$= 6x - 6$ Do the multiplication.

Self Check

Apply the distributive property.

a. $5(h + 4)$

b. $9(a - 3)$

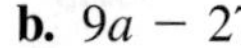

Answers: **a.** $5h + 20$, **b.** $9a - 27$ ■

COMMENT The fact that an expression contains parentheses does not automatically mean that the distributive property can be applied to simplify it. For example, the distributive property does not apply to expressions such as $5(4x)$ or $5(4 \cdot x)$, where a product is multiplied by 5. The distributive property does apply to expressions such as $5(4 + x)$ or $5(4 - x)$, where a sum or difference is multiplied by 5.

EXAMPLE 4 ***Applying the distributive property.*** Remove parentheses: **a.** $-3(4x + 2)$ and **b.** $-9(3 - 2t)$.

Solution

a. $-3(4x + 2) = -3(4x) + (-3)(2)$ Distribute the multiplication by -3.

$= -12x + (-6)$ Do the multiplications.

$= -12x - 6$ Write the answer in simpler form. Adding -6 is the same as subtracting 6.

b. $-9(3 - 2t) = -9(3) - (-9)(2t)$ Apply the distributive property.

$= -27 - (-18t)$ Do the multiplications.

$= -27 + 18t$ Write the answer in simpler form. Add the opposite of $-18t$.

Self Check

Remove parentheses.

a. $-4(6y + 8)$

b. $-7(2 - 8m)$

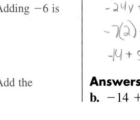

Answers: **a.** $-24y - 32$, **b.** $-14 + 56m$ ■

COMMENT It is common practice to write answers in simplified form. For instance, the answer to Example 4, part a, is expressed as $-12x - 6$ because it involves fewer symbols than $-12x + (-6)$. For the same reason, the answer to Example 4, part b, is given as $-27 + 18t$ instead of $-27 - (-18t)$.

Since multiplication is commutative, we can write the distributive property in either of the following forms:

$$(b + c)a = ba + ca$$
$$(b - c)a = ba - ca$$

EXAMPLE 5 ***The distributive property.*** Multiply **a.** $(5 + 3r)7$ and **b.** $(4 - x)2$.

Solution

a. $(5 + 3r)7 = (5)7 + (3r)7$ Distribute the multiplication by 7.

$= 35 + 21r$ Do the multiplications.

b. $(4 - x)2 = (4)2 - (x)2$ Distribute the multiplication by 2.

$= 8 - 2x$ Do the multiplications.

Self Check

Multiply:

a. $(8 + 7x)5$

b. $(5 - c)3$

Answers: **a.** $40 + 35x$, **b.** $15 - 3c$ ■

Distributing a factor of −1

At first glance, $-(x + 8)$ doesn't appear to be in the proper form to apply the distributive property; the number in front of the parentheses appears to be missing. But the negative sign in front of the parentheses actually represents the number -1.

The negative sign represents -1.

$-(x + 8) = -1(x + 8)$

$= -1(x) + (-1)(8)$ Apply the distributive property. Distribute -1.

$= -x + (-8)$ Do the multiplications.

$= -x - 8$ Adding -8 is the same as subtracting 8.

EXAMPLE 6 ***Distributing a factor of −1.*** Simplify $-(-6 - 2e)$.

Solution

$-(-6 - 2e) = -1(-6 - 2e)$ Rewrite the negative sign in front of the parentheses as -1.

$= -1(-6) - (-1)(2e)$ Distribute the multiplication by -1.

$= 6 - (-2e)$ Do the multiplications.

$= 6 + 2e$ Add the opposite of $-2e$.

Self Check

Simplify $-(-2t + 4)$.

Answer: $2t - 4$ ■

After working several problems like Example 6, you will notice that it is not necessary to show each of the steps. The result can be obtained very quickly by changing the sign of each quantity within the parentheses and dropping the parentheses.

Extending the distributive property

The distributive property can be extended to situations where there are more than two terms within parentheses.

The extended distributive property

> If a, b, c, and d represent numbers,
>
> $a(b + c + d) = ab + ac + ad$

EXAMPLE 7 ***The extended distributive property.*** Remove parentheses: $-6(-3x - 6y + 8)$.

Solution We distribute the multiplication by -6.

$$-6(-3x - 6y + 8) = -6(-3x) - (-6)(6y) + (-6)(8)$$

$= 18x - (-36y) + (-48)$ Do the multiplications.

$= 18x + 36y + (-48)$ Write the subtraction as addition of the opposite of $-36y$, which is $36y$.

$= 18x + 36y - 48$ Adding -48 is the same as subtracting 48.

■

STUDY SET Section 3.3

VOCABULARY *Fill in the blanks to make the statements true.*

1. The ______________ property tells us how to multiply $5(x + 7)$. After doing the multiplication to obtain $5x + 35$, we say that the parentheses have been ______________.

2. To ______________ an algebraic expression means to use algebraic properties to write it in simpler form.

3. When an algebraic expression is simplified, the result is an ______________ expression.

4. We ______________ expressions and we ______________ equations.

CONCEPTS

5. State the distributive property using the variables x, y, and z.

6. Use the variables r, s, and t to state the distributive property of subtraction.

7. The following expressions are examples of the *right* and *left* distributive properties:

$5(w + 7)$ and $(w + 7)5$

Which of the two do you think would be termed the right distributive property?

8. For each of the following expressions, tell whether the distributive property applies.

a. $2(5t)$ **b.** $-2(5 - t)$

c. $5(-2 \cdot t)$ **d.** $(-2t)5$

e. $(2)(-t)5$ **f.** $(5 - t)(-2)$

9. The distributive property can be demonstrated using Illustration 1. Fill in the blanks below. Two groups of 6 plus three groups of 6 is ___ groups of 6. Therefore, ___ $\cdot 2 +$ ___ $\cdot 3 = 6($ ___ $+$ ___ $)$.

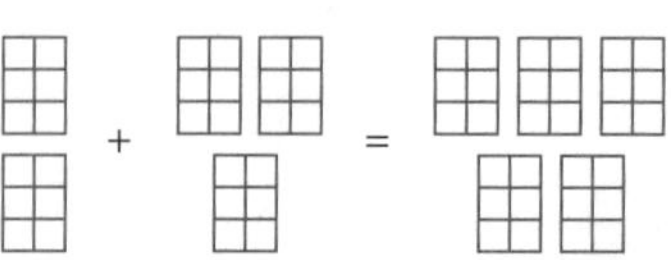

ILLUSTRATION 1

10. a. Simplify $2(5x)$.

b. Remove parentheses: $2(5 + x)$.

11. Write an equivalent expression for $-(y + 9)$ without using parentheses.

12. Explain what the arrows are illustrating.

$-9(y - 7)$

NOTATION *In Exercises 13–16, complete each solution.*

13. Multiply $-5(7n)$.

$-5(7n) = (\square \cdot 7)n$

$= -35n$

14. Multiply $6y(-9)$.

$6y(-9) = 6(\square)y$

$= [6(\square)]y$

$= -54y$

15. Multiply $-9(-4 - 5y)$.

$-9(-4 - 5y) = (\square)(-4) - (\square)(5y)$

$= 36 - (\square)$

$= 36 + 45y$

16. Multiply $4(2a + b - 1)$.

$4(2a + b - 1) = 4(\square) + 4(\square) - \square(1)$

$= 8a + 4b - 4$

17. Write each expression in simpler form, using fewer mathematical symbols.

a. $-(-x)$

b. $x - (-5)$

c. $5x - 10y + (-15)$

d. $5 \cdot x$

18. Tell what number is to be distributed.

a. $-6(x - 2)$ **b.** $(t + 1)(-5)$

c. $(a + 24)8$ **d.** $-(z - 16)$

PRACTICE *Simplify each algebraic expression.*

19. $2(6x)$ **20.** $4(7b)$

21. $-5(6y)$ **22.** $-12(6t)$

23. $-10(-10t)$ **24.** $-8(-6k)$

25. $(4s)3$ **26.** $(9j)7$

27. $2c \cdot 7$ **28.** $11f \cdot 9$

29. $-5 \cdot 8h$ **30.** $-8 \cdot 4d$

31. $-7x(6y)$ **32.** $13a(-2b)$

33. $4r \cdot 4s$ **34.** $7x \cdot 7y$

35. $2x(5y)(3)$ **36.** $4(3z)(4)$

37. $5r(2)(-3b)$ **38.** $4d(5)(-3e)$

39. $5 \cdot 8c \cdot 2$ **40.** $3 \cdot 6j \cdot 2$

41. $(-1)(-2e)(-4)$ **42.** $(-1)(-5t)(-1)$

Use the distributive property to remove parentheses.

43. $4(x + 1)$ **44.** $5(y + 3)$

45. $4(4 - x)$ **46.** $5(7 + k)$

47. $-2(3e + 3)$ **48.** $-5(7t + 2)$

49. $-8(2q - 6)$ **50.** $-5(3p - 8)$

51. $-4(-3 - 5s)$ **52.** $-6(-1 - 3d)$

53. $(7 + 4d)6$ **54.** $(8r + 2)7$

55. $(5r - 6)(-5)$ **56.** $(3z - 7)(-8)$

57. $(-4 - 3d)6$ **58.** $(-4 - 2j)5$

59. $3(3x - 7y + 2)$ **60.** $5(4 - 5r + 8s)$

61. $-3(-3z - 3x - 5y)$ **62.** $-10(5e + 4a + 6t)$

Write each expression without using parentheses.

63. $-(x + 3)$ **64.** $-(5 + y)$

65. $-(4t + 5)$ **66.** $-(8x + 4)$

67. $-(-3w - 4)$ **68.** $-(-6 - 4y)$

69. $-(5x - 4y + 1)$ **70.** $-(6r - 5f + 1)$

Each expression is the result of an application of the distributive property. What was the original algebraic expression?

71. $2(4x) + 2(5)$ **72.** $3(3y) + 3(7)$

73. $-4(5) - 3x(5)$ **74.** $-8(7) - (4s)(7)$

75. $-3(4y) - (-3)(2)$ **76.** $-5(11s) - (-5)(11t)$

77. $3(4) - 3(7t) - 3(5s)$ **78.** $2(7y) + 2(8x) - 2(4)$

WRITING

79. Explain what it means to simplify an algebraic expression. Give an example.

80. Explain the commutative and associative properties of multiplication.

81. Explain how to apply the distributive property.

82. Explain why the distributive property applies to $2(3 + x)$ but does not apply to $2(3x)$.

REVIEW

83. Find $|-6 + 1|$. **84.** Find $-1 - (-4)$.

85. Identify the operation associated with each word: product, quotient, difference, sum.

86. What are the steps used to find the mean (average) of a set of scores?

87. Insert the proper inequality symbol: $-6 \quad -7$.

88. Fill in the blank to make a true statement: To factor a number means to express it as the ______________ of other whole numbers.

89. Which of the following involve area: carpeting a room, fencing a yard, walking around a lake, and painting a wall?

90. Write seven squared and seven cubed.

3.4 *Combining Like Terms*

In this section, you will learn about

- Terms of an algebraic expression • Coefficients of a term • Terms and factors
- Like terms • Combining like terms • Perimeter formulas

INTRODUCTION. In this section, we will see how the distributive property can be used to simplify algebraic expressions that involve addition and subtraction. We will also review the concept of perimeter and write the formulas for the perimeter of a rectangle and a square using variables.

Terms of an algebraic expression

Addition signs break algebraic expressions into smaller parts called **terms.** The expression $3x + 8$ contains two terms, $3x$ and 8.

$$3x + 8$$

Term Term

The addition sign breaks the expression into two terms.

Term

A **term** is a number or a product of a number and one or more variables.

Some examples of terms are

$$6b^2, \quad -15x, \quad -4ac, \quad x, \quad -y, \quad 12$$

If an algebraic expression involves subtraction, the subtraction can be expressed as addition of the opposite. For example, $5x - 6$ can be written in the equivalent form $5x + (-6)$. We then see that $5x - 6$ contains two terms, $5x$ and -6.

EXAMPLE 1 ***Identifying terms.*** List the terms of each expression:
a. $-3x^2 + 5x + 8$, **b.** $-24rs$, and **c.** $x - 5 - 3x + 10$.

Solution

a. $-3x^2 + 5x + 8$ contains three terms: $-3x^2$, $5x$, and 8.

b. $-24rs$ is one term.

c. $x - 5 - 3x + 10$ can be written as $x + (-5) + (-3x) + 10$. It contains four terms: x, -5, $-3x$, and 10.

Self Check

List the terms of each expression.

a. $-12y^2 + y + 10$

b. $-4ab$

c. $9 - m - 6m + 12$

Answers: **a.** $-12y^2$, y, 10, **b.** $-4ab$, **c.** 9, $-m$, $-6m$, 12

Coefficients of a term

A term of an algebraic expression can consist of a single number (called a **constant**), a single variable, or a product of numbers and variables.

Numerical coefficient

In a term that is the product of a number and one or more variables, the numerical factor is called the **numerical coefficient** (or simply the **coefficient**) of the term. The coefficient of a constant term is the constant itself.

In the term $3x$, 3 is the *coefficient,* and x is the *variable* part. Some more examples are shown in Table 3-3.

Term	Coefficient	Variable part
$6b^2$	6	b^2
$-15x$	-15	x
$-4ac$	-4	ac
x	1	x
$-y$	-1	y
25 (constant)	25	none

TABLE 3-3

Notice that when there is no number in front of a variable, the coefficient is understood to be 1. For example, the coefficient of the term x is 1. If there is only a negative (or opposite) sign in front of the variable, the coefficient is understood to be -1. Therefore, $-y$ can be thought of as $-1y$.

EXAMPLE 2 ***Coefficients of terms.*** Identify the coefficient of each term in the expression $-5x^2 + x - 15$.

Solution

First, we write the subtraction as addition of the opposite: $-5x^2 + x - 15 = -5x^2 + x + (-15)$.

Term	Coefficient
$-5x^2$	-5
x	1
-15	-15

Self Check

Identify the coefficient of each term in the expression $6y^3 - y + 7$.

Answers: 6, -1, 7 ■

Terms and factors

It is important to be able to distinguish between the *terms* of an algebraic expression and the *factors* of a term. Terms are separated by an addition sign + and factors are numbers or variables that are multiplied together.

Consider the expression $-2x + 3xy$, which contains two terms.

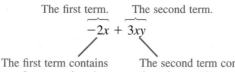

The first term. The second term.

$$-2x + 3xy$$

The first term contains two factors: -2 and x.

The second term contains three factors: 3, x, and y.

EXAMPLE 3 ***Terms and factors.*** Determine whether y is a factor or a term of each expression: **a.** $8 + y$ and **b.** $8y$.

Solution

a. Since y is added to 8, y is a term of $8 + y$.

b. Since y is multiplied by 8, y is a factor of $8y$.

Self Check

Determine whether x is a factor or a term of each expression: **a.** $-5x$ and **b.** $x + 5$.

Answers: **a.** factor, **b.** term

Like terms

The expression $5t - 6t + 10$ contains three terms. The variable part of two of the terms, $5t$ and $-6t$, are identical. We say those two terms are **like** or **similar terms.**

Like terms (similar terms)

Like terms, or **similar terms,** are terms with exactly the same variables and exponents. Any constants in an expression are considered like terms.

Like terms	Unlike terms
$2x$ and $3x$ — Identical variables	$2x$ and $2t$ — Different variables
$-3a^2$, $16a^2$ and $125a^2$ — Identical variable to identical power	$8y$ and $-14y^2$ — Different exponent on the variable y

COMMENT When looking for like terms, don't look at the coefficients of the terms. Consider only their variable parts.

EXAMPLE 4 ***Identifying like terms.*** List the like terms in each of the following expressions.

a. $5a + 6 + 3a$

b. $x^2 - 3x^5 + 2$

c. $-5x^2 + 3 - 1 + x^2$

Solution

a. In $5a + 6 + 3a$, the terms $5a$ and $3a$ have the same variable and the same exponent. They are like terms.

b. $x^2 - 3x^5 + 2$ does not contain any like terms.

c. $-5x^2 + 3 - 1 + x^2$ contains two pairs of like terms. $-5x^2$ and x^2 are like terms, because the variable and the exponent are the same. The numbers 3 and -1 are like terms.

Self Check

List the like terms in each of the following expressions.

a. $10c + 8 + 7$

b. $1 + 3x + x^2 + x^3$

c. $4r + 8t - 3r + t$

Answers: **a.** 8 and 7, **b.** none, **c.** $4r$ and $-3r$, $8t$ and t

Combining like terms

If we are to add or subtract objects, they must be similar. For example, fractions that are to be added must have a common denominator. When adding decimals, we align columns to be sure that we add tenths to tenths, hundredths to hundredths, and so on. The same is true when working with terms of an algebraic expression. They can be added or subtracted only if they are like terms.

The following expression cannot be simplified, because its terms are unlike.

$$3x + 4y$$

Unlike terms.
The variable parts are not identical.

The following expression can be simplified, because it contains like terms.

$$3x + 4x$$

Like terms.
The variable parts are identical.

To simplify an expression containing like terms, we use the distributive property. For example, we can simplify $3x + 4x$ as follows:

$3x + 4x = (3 + 4)x$ The factor x was distributed to 3 and 4.

$= 7x$ Do the addition within the parentheses: $3 + 4 = 7$.

We say that we have *simplified* the expression $3x + 4x$. The result is the equivalent expression $7x$. Simplifying the sum (or difference) of like terms is called **combining like terms.**

EXAMPLE 5 ***Combining like terms.*** Simplify each expression by combining like terms: **a.** $-3x + 7x$ and **b.** $6y - 4y$.

Solution

a. $-3x + 7x = (-3 + 7)x$ Use the distributive property. The factor x was distributed to -3 and 7.

$= 4x$ Do the addition within the parentheses: $-3 + 7 = 4$.

b. $6y - 4y = (6 - 4)y$ Use the distributive property. The factor y was distributed.

$= 2y$ Do the subtraction within the parentheses: $6 - 4 = 2$.

Self Check

Simplify each expression by combining like terms.

a. $-5b + 10b$

b. $12c - 9c$

Answers: **a.** $5b$, **b.** $3c$ ■

The results of Example 5 suggest the following rule.

Combining like terms

> To add or subtract like terms, combine their coefficients and keep the same variables with the same exponents.

EXAMPLE 6 ***Adding like terms.*** Simplify by combining like terms: **a.** $-7x + (-9x)$ and **b.** $6m + 3m$.

Solution

a. $-7x + (-9x) = -16x$ Add the coefficients of the like terms: $-7 + (-9) = -16$. Keep the variable x.

b. $6m + 3m = 9m$ Add the coefficients: $6 + 3 = 9$. Keep the variable m.

Self Check

Simplify by combining like terms.

a. $-5n + (-2n)$

b. $15r + 4r$

Answers: **a.** $-7n$, **b.** $19r$ ■

EXAMPLE 7 ***Subtracting like terms.*** Simplify by combining like terms: **a.** $5n - 3n$, **b.** $16h - 24h$, and **c.** $-4d - (-5d)$.

Solution

a. $5n - 3n = 2n$ Subtract: $5 - 3 = 2$. Keep the variable n.

b. $16h - 24h = -8h$ Subtract: $16 - 24 = 16 + (-24) = -8$. Keep the variable h.

c. $-4d - (-5d) = -4d + 5d$ Add the opposite of $-5d$.

$= 1d$ Add: $-4 + 5 = 1$. Keep the variable d.

$= d$ $1d = d$.

Self Check

Simplify by combining like terms.

a. $25c - 5c$

b. $9w - 15w$

c. $-7p - (-6p)$

Answers: **a.** $20c$, **b.** $-6w$, **c.** $-p$ ■

COMMENT Expressions that involve subtraction from 0 are often incorrectly simplified. For example, $0 - 6x \neq 6x$. To simplify $0 - 6x$, we can use the fact that subtraction is the same as addition of the opposite.

$$\begin{aligned} 0 - 6x &= 0 + (-6x) && \text{Add the opposite of } 6x \text{, which is } -6x. \\ &= -6x && \text{When we add 0 to any number, the number remains the same.} \end{aligned}$$

EXAMPLE 8 ***Combining like terms.*** Simplify $8s - 8S - 5s + S$.

Solution The lowercase s and the capital S are different variables. We rearrange the terms so that like terms are next to each other.

$$\begin{aligned} 8s - 8S - 5s + S &= 8s - 5s - 8S + S && \text{Use the commutative property of addition to get the like terms together.} \\ &= 3s - 7S && \text{Combine like terms: } 8 - 5 = 3 \text{ and keep } s, -8 + 1 = -7 \text{ and keep } S. \end{aligned}$$

■

EXAMPLE 9 ***Simplifying algebraic expressions.*** Simplify $4(x + 3) - 2(x - 1)$.

Solution

$$\begin{aligned} 4(x + 3) - 2(x - 1) &= 4x + 12 - 2x + 2 && \text{Use the distributive property twice.} \\ &= 4x - 2x + 12 + 2 && \text{Use the commutative property of addition to get like terms together.} \\ &= 2x + 14 && \text{Combine like terms: } 4x - 2x = 2x \text{ and } 12 + 2 = 14. \end{aligned}$$

Self Check

Simplify $6(4 + y) - 5(y - 6)$.

Answer: $y + 54$ ■

The expressions in Examples 8 and 9 contained two sets of like terms. In each solution, we rearranged terms so that like terms were next to each other. However, with practice you will be able to combine like terms without having to write them next to each other.

EXAMPLE 10 ***Combining without rearranging.*** Simplify $-5(x - 2) + 8x - 6$.

Solution

$$\begin{aligned} -5(x - 2) + 8x - 6 &= -5x + 10 + 8x - 6 && \text{Distribute } -5. \\ &= 3x + 4 && \text{Combine like terms: } -5x + 8x = 3x \text{ and } 10 - 6 = 4. \end{aligned}$$

Self Check

Simplify $-6(y - 2) + 4y - 1$.

Answer: $-2y + 11$ ■

Perimeter formulas

To develop the formula for the perimeter of a rectangle, we let l = the length of the rectangle and w = width of the rectangle. (See Figure 3-7.) Then

$$\begin{aligned} P &= l + w + l + w && \text{The perimeter is the distance around the rectangle.} \\ &= 2l + 2w && \text{Combine like terms: } l + l = 2l \text{ and } w + w = 2w. \end{aligned}$$

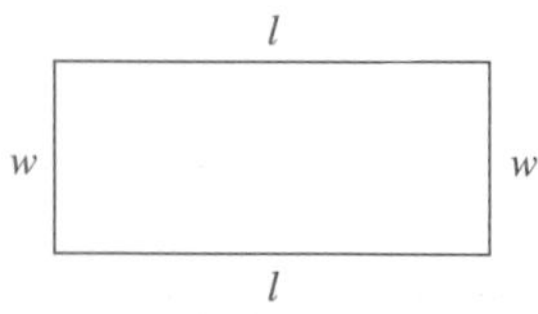

FIGURE 3-7

The perimeter of a rectangle

> The perimeter P of a rectangle with length l and width w is given by
> $$P = 2l + 2w$$

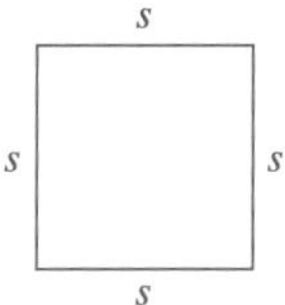

FIGURE 3-8

To develop the formula for the perimeter of a square, we let $s =$ the length of a side of the square. (See Figure 3-8.) Then

$P = s + s + s + s$ Add the lengths of the four sides.

$= 4s$ Combine like terms. Recall that $s = 1s$.

The perimeter of a square

> The perimeter of a square with sides of length s is given by
> $$P = 4s$$

EXAMPLE 11 ***Energy conservation.*** See Figure 3-9. Find the cost to weatherstrip the front door and window of the house if the material costs 20¢ a foot.

Analyze the problem To find the cost of the weatherstripping, we must find the perimeter of the door and the window. The door is in the shape of a rectangle, and the window is in the shape of a square.

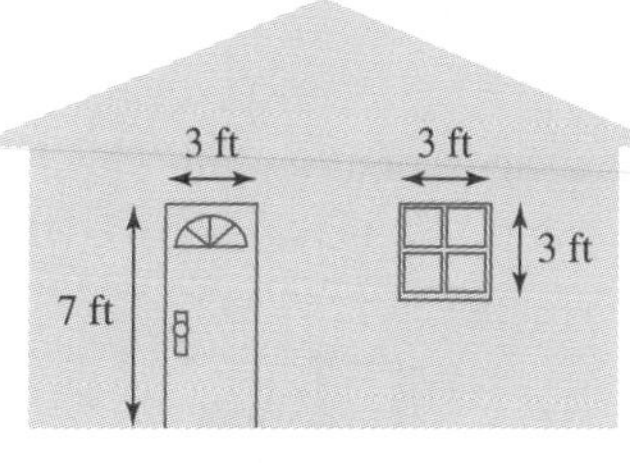

FIGURE 3-9

Form an equation Let P represent the total perimeter and translate the words of the problem into an equation.

The total perimeter	is	the perimeter of the door	plus	the perimeter of the window.	
P	$=$	$2l + 2w$	$+$	$4s$	Write the formulas for the perimeter of a rectangle and a square.

Solve the equation

$P = 2l + 2w + 4s$

$P = 2(7) + 2(3) + 4(3)$ Substitute 7 for l, 3 for w, and 3 for s.

$P = 14 + 6 + 12$ Do the multiplications.

$P = 32$ Do the additions.

State the conclusion The total perimeter is 32 feet. At 20¢ a foot, the total cost will be $(32 \cdot 20)$¢. This is 640¢, or \$6.40.

Check the result We can check the results by estimation. The perimeter is approximately 30 feet, and $30 \cdot 20 = 600$¢, which is \$6. The answer, \$6.40, seems reasonable. ■

STUDY SET Section 3.4

VOCABULARY *Fill in the blanks to make the statements true.*

1. A ______ is a number or a product of a number and one or more variables.
2. In the term $5t$, 5 is called the ______ and t is called the ______ part.
3. The ______ of a geometric figure is the distance around it.
4. A ______ is a general rule that mathematically describes a relationship between two or more variables.
5. $2(x + 3) = 2x + 2(3)$ is an example of the ______ property.
6. Terms with exactly the same variables and exponents are called ______ terms.
7. Simplifying the sum (or difference) of like terms is called ______ like terms.
8. The numbers multiplied together to form a product are called ______.

CONCEPTS

9. Tell whether x is used as a factor or as a term.
   a. $12 + x$
   b. $7x$
   c. $12y + 12x - 6$
   d. $-36xy$
10. Tell whether $6y$ is used as a factor or as a term.
   a. $6yz$ **b.** $10 + 6y$
   c. $9xy + 6y$ **d.** $6y - 18$
11. What is the coefficient of each term?
   a. $11x$ **b.** $8t$
   c. $-4x^2$ **d.** a
   e. $-x$ **f.** $102xy$
12. What is the coefficient of the second term of each expression?
   a. $5x^2 + 6x + 7$
   b. $xy - x + y + 10$
   c. $9y^2 + y + 8$
   d. $5x^3 - 4x^2 + 3x + 1$
13. Complete the table.

Term	Coefficient	Variable part
$6m$		
$-75t$		
w		
$4bh$		

14. Simplify each pair of expressions, if possible.
   a. $5(2x)$ and $5 + 2x$
   b. $6(-7x)$ and $6 - 7x$
   c. $2(3x)(3)$ and $2 + 3x + 3$
   d. $x \cdot x$ and $x + x$
15. When simplifying an algebraic expression, some students use underlining.

   $\underline{\underline{3y}} + \underline{4} + \underline{\underline{5y}} + \underline{8}$

   What purpose does the underlining serve?
16. Tell whether each statement is true or false.
   a. $x = 1x$ **b.** $2x + 0 = 2x$
   c. $-y = -1y$ **d.** $0 - 4c = 4c$
17. Illustration 1 shows the distance (in miles) that two men live from the office. Find the total distance the men travel from home to office.

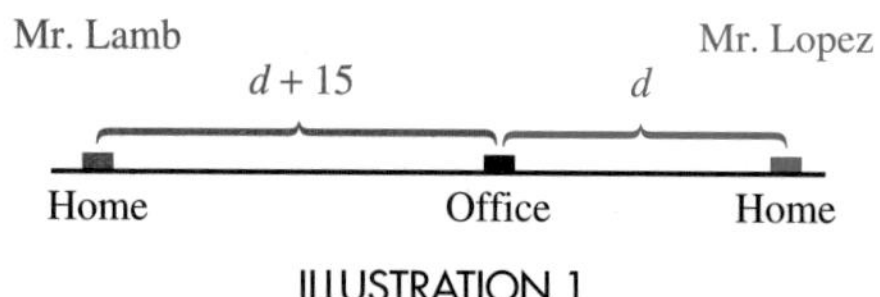

ILLUSTRATION 1

18. The heights of two trees are shown in Illustration 2. Find the sum of their heights.

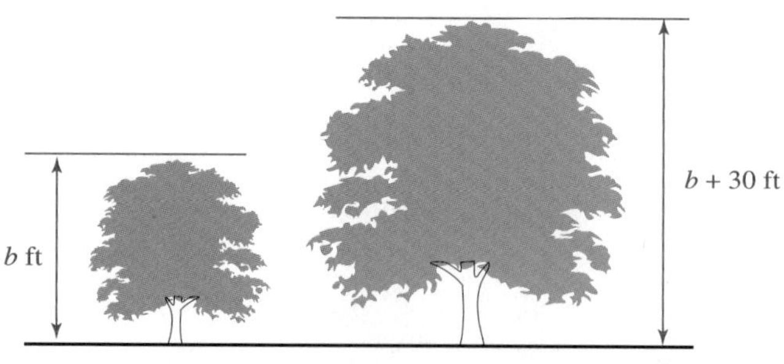

ILLUSTRATION 2

19. What does this diagram illustrate?

   $\underbrace{9x} + \underbrace{5x} = \underbrace{14x}$

20. What does this diagram illustrate?

   $\underbrace{12k} - \underbrace{4k} = \underbrace{8k}$

NOTATION *In Exercises 21–24, complete each solution.*

21. Simplify $5x + 7x$.

   $5x + 7x = (5 + \quad)x$

   $= 12x$

22. Simplify $12w - 16w$.

$12w - 16w = (\square - 16)w$
$= -4w$

23. Simplify $2(x - 1) + 3x$.

$2(x - 1) + 3x = 2x - \square + 3x$
$= 5x - 2$

24. Simplify $-3(1 - b) - b$.

$-3(1 - b) - b = \square + \square - b$
$= -3 + 2b$

25. In the formula $P = 2l + 2w$,
a. what does P represent?
b. what does $2l$ mean?
c. what does $2w$ mean?

26. In the formula $P = 4s$,
a. what does P represent?
b. what does $4s$ mean?

PRACTICE *Identify the terms of each expression.*

27. $3x^2 - 5x + 4$
28. $6 + 12y - y^2$
29. $5 + 5t - 8t + 4$
30. $3x - y - 5x + y$

What exponent must appear in each box to make the terms like terms?

31. $3x^{\square}, -6x^2$
32. $7a^3, 21a^{\square}$
33. $-8h^5, -5h^{\square}$
34. $25n^4, -15n^{\square}$

Simplify by combining like terms, if possible.

35. $6t + 9t$
36. $7r + 5r$
37. $5s - s$
38. $8y - y$
39. $-5x + 6x$
40. $-8m + 6m$
41. $-5d + 9d$
42. $-4a + 12a$
43. $3e - 7e$
44. $2s - 4s$
45. $h - 7$
46. $j - 8$

47. $4z - 10z$
48. $3w - 18w$
49. $-3x - 4x$
50. $-7y - 9y$
51. $2t - 2t$
52. $7r - 7r$
53. $-6s + 6s$
54. $19c + (-19c)$
55. $x + x + x + x$
56. $s - s - s$
57. $2x + 2y$
58. $5a - 5b$

59. $0 - 2y$
60. $0 - 7x$
61. $3a - 0$
62. $10t + 0$
63. $6t + 9 + 5t + 3$
64. $5x + 3 + 5x + 4$

65. $3w - 4 + w - 1$
66. $6y + 6 - y - 1$

67. $-4r + 8R + 2R - 3r + R$
68. $12a - A - a - 8A - a$
69. $-45d - 12a - 5d + 12a$
70. $-m - n - 8m + n$
71. $4x - 3y - 7 + 4x - 2 - y$
72. $2a + 8 - b - 5 + 5a - 9b$

Simplify each expression.

73. $4(x + 1) + 5(6 + x)$
74. $7(1 + y) + 8(2y + 3)$
75. $5(3 - 2s) + 4(2 - 3s)$
76. $6(t - 3) + 9(2 - t)$
77. $-4(6 - 4e) + 3(e + 1)$
78. $-5(7 - 4t) + 3(2 + 5t)$
79. $3t - (t - 8)$
80. $6n - (4n + 1)$
81. $-2(2 - 3x) - 3(x - 4)$
82. $-3(1 - y) - 5(2y - 6)$
83. $-4(-4y + 5) - 6(y + 2)$
84. $-3(-6y - 8) - 4(5 - y)$

APPLICATIONS

85. MOBILE HOME DESIGN The design of a mobile home calls for a six-inch-wide strip of stained pine around the outside of each exterior wall, as shown in brown in Illustration 3. If the pine strip costs 80¢ a running foot, how much will be spent on the pine used for the trim?

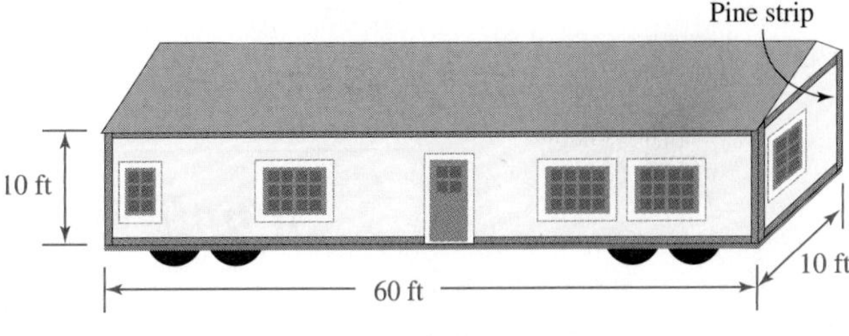

ILLUSTRATION 3

86. LANDSCAPING DESIGN A landscape architect has designed a planter surrounding two birch trees, as shown in Illustration 4 on the next page. The planter is to be outlined with redwood edging in the shape of a rectangle and two squares. If the material costs 17¢ a running foot, how much will the redwood cost for this project?

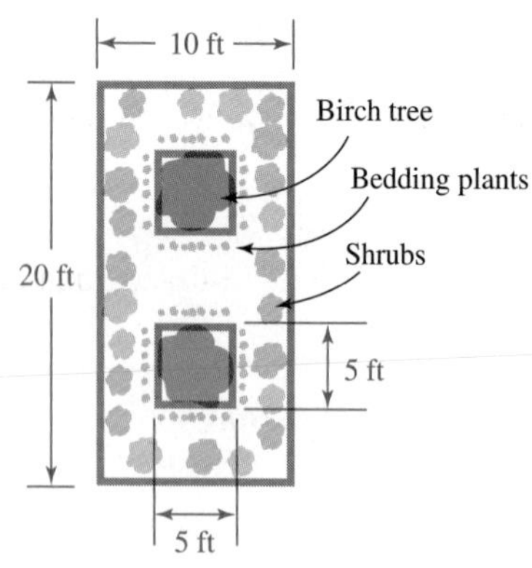

ILLUSTRATION 4

87. PARTY PREPARATIONS The appropriate size of a dance floor for a given number of dancers can be determined from the table shown in Illustration 5. Find the perimeter of each of the dance floors listed.

Slow dancers	Fast dancers	Size of floor (in feet)
8	5	9×9
14	9	12×12
22	15	15×15
32	20	18×18
50	30	21×21

ILLUSTRATION 5

88. COASTAL DRILLING The map in Illustration 6 shows an area of the California coast where oil drilling is planned. Use the scale to estimate the lengths of the sides of the area highlighted on the map. Then find its perimeter.

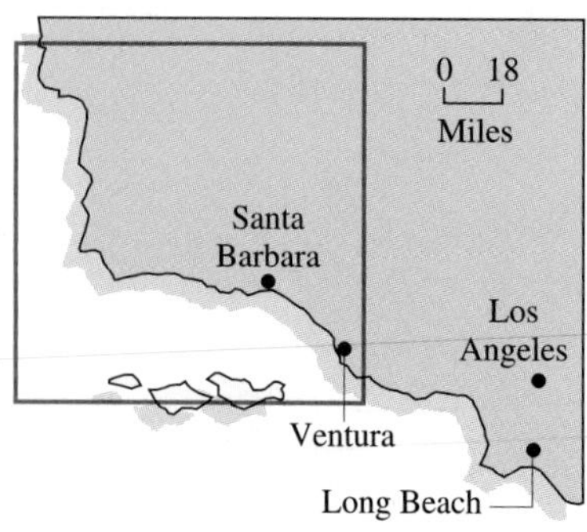

ILLUSTRATION 6

WRITING

89. Explain what it means for two terms to be like terms.

90. Explain what it means to say that the coefficient of x is an understood 1.

91. Explain the difference between a term and a factor. Give some examples.

92. The formula for the perimeter of a rectangle is $P = 2l + 2w$. Explain why we can write this formula in the form $P = 2(l + w)$.

REVIEW

93. Solve $-4t - 3 = -11$.

94. Evaluate: $(-1)(-1)(-1)$.

95. Find the prime factorization of 100.

96. Write $x \cdot x \cdot x \cdot x \cdot x$ as an exponential expression.

97. Fill in the blank to make a true statement. The ______________ of a number is the distance between it and 0 on the number line.

98. State the division property of equality in words.

3.5 *Simplifying Expressions to Solve Equations*

In this section, you will learn about

- Checking solutions • Combining like terms to solve equations
- Variables on both sides of an equation • Removing parentheses
- A strategy for solving equations

Introduction. We must often simplify algebraic expressions to solve equations. Sometimes it will be necessary to combine like terms in order to isolate the variable on one side of the equation. At other times, it will be necessary to apply the distributive property to write an equation in a form that can be solved. In this section, we will discuss both situations.

Checking solutions

Recall that a solution of an equation is any number that satisfies the equation.

EXAMPLE 1 ***Checking a solution.*** Determine whether $x = -3$ is a solution of $5x - 1 = 6x + 8$.

Solution

If $x = -3$ is a solution, we will obtain a true statement when -3 is substituted for x.

$$5x - 1 = 6x + 8$$

$$5(-3) - 1 \stackrel{?}{=} 6(-3) + 8 \quad \text{Substitute } -3 \text{ for } x.$$

$$-15 - 1 \stackrel{?}{=} -18 + 8 \quad \text{Do the multiplications.}$$

$$-16 \neq -10 \quad \text{Simplify each side.}$$

Since $-16 \neq -10$, we conclude that $x = -3$ is not a solution.

Self Check

Is $w = -4$ a solution of $-4w - 6 = -2 - 3w$?

Answer: yes ■

Combining like terms to solve equations

When solving equations, we must often combine like terms before applying the properties of equality.

EXAMPLE 2 ***Combining like terms.*** Solve $7x - 4x = 15$. Check the result.

Solution

We begin by simplifying the left-hand side of the equation.

$$7x - 4x = 15$$

$$3x = 15 \quad \text{Combine like terms: } 7x - 4x = 3x.$$

$$\frac{3x}{3} = \frac{15}{3} \quad \text{To undo the multiplication by 3, divide both sides by 3.}$$

$$x = 5 \quad \text{Do the divisions.}$$

Check:

$$7x - 4x = 15$$

$$7(5) - 4(5) \stackrel{?}{=} 15 \quad \text{Substitute 5 for } x.$$

$$35 - 20 \stackrel{?}{=} 15 \quad \text{Do the multiplications.}$$

$$15 = 15 \quad \text{Do the subtraction.}$$

Since the statement $15 = 15$ is true, 5 is a solution of the equation.

Self Check

Solve $8r - 6r = 16$ and check the result.

Answer: 8 ■

EXAMPLE 3 ***Combining like terms.*** Solve $100 = t - 20 + t$.

Solution

$$100 = t - 20 + t$$

$$100 = 2t - 20 \quad \text{Combine like terms: } t + t = 2t.$$

$$100 + \mathbf{20} = 2t - 20 + \mathbf{20} \quad \text{To undo the subtraction of 20, add 20 to both sides.}$$

$$120 = 2t \quad \text{Do the additions.}$$

$$\frac{120}{\mathbf{2}} = \frac{2t}{\mathbf{2}} \quad \text{To undo the multiplication by 2, divide both sides by 2.}$$

$$60 = t \quad \text{Do the divisions.}$$

$$t = 60 \quad \text{Interchange the sides of the equation.}$$

Verify that 60 satisfies the equation.

Self Check

Solve $155 = d - 1 + d$.

Answer: 78 ■

Variables on both sides of an equation

When solving an equation, we want to isolate the variable on one side of the equation. If variables appear on both sides, we can use the addition (or subtraction) property of equality to get all variable terms on one side and all constant terms on the other.

EXAMPLE 4 ***Eliminating a variable term.*** Solve $8y = 2y + 12$.

Solution

There are variable terms (highlighted in blue) on both sides of the equation. To isolate y on the left-hand side of the equation, we use the subtraction property of equality to eliminate $2y$ on the right-hand side.

$8y = 2y + 12$	
$8y - 2y = 2y + 12 - 2y$	To eliminate $2y$ from the right-hand side, subtract $2y$ from both sides.
$6y = 12$	Combine like terms: $8y - 2y = 6y$ and $2y - 2y = 0$.
$\frac{6y}{6} = \frac{12}{6}$	To undo the multiplication by 6, divide both sides by 6.
$y = 2$	Do the divisions.

Verify that 2 satisfies the equation.

Self Check

Solve $9B = 3B - 18$.

Answer: -3 ■

COMMENT In solving equations, it doesn't matter whether the variable is isolated on the right or the left side of the equation. In Example 4, we could have isolated y on the right-hand side, but this approach would have involved more steps.

EXAMPLE 5 ***Eliminating a variable term.*** Solve $-9 - 6t = 5t + 2$.

Solution

There are variable terms (highlighted in blue) on both sides of the equation. We can either subtract $5t$ from both sides to isolate t on the left side, or we can add $6t$ to both sides to isolate t on the right side. It appears that the computations will be easier if we add $6t$ to both sides.

$-9 - 6t = 5t + 2$	
$-9 - 6t + 6t = 5t + 2 + 6t$	To eliminate $-6t$ from the left-hand side, add $6t$ to both sides.
$-9 = 11t + 2$	Combine like terms: $-6t + 6t = 0$ and $5t + 6t = 11$.
$-9 - 2 = 11t + 2 - 2$	To undo the addition of 2, subtract 2 from both sides.
$-11 = 11t$	Do the subtractions: $-9 - 2 = -11$ and $2 - 2 = 0$.
$\frac{-11}{11} = \frac{11t}{11}$	To undo the multiplication by 11, divide both sides by 11.
$-1 = t$	Do the divisions.
$t = -1$	Interchange the sides of the equation.

Verify that -1 satisfies the equation.

Self Check

Solve $72 - 13d = 12d - 3$.

Answer: 3 ■

Removing parentheses

At times, we must use the distributive property to solve an equation.

EXAMPLE 6 ***Removing parentheses to solve an equation.***
Solve $3(x + 15) = 45$.

Solution

$3(x + 15) = 45$	
$3x + \mathbf{3}(15) = 45$	Distribute the multiplication by 3.
$3x + 45 = 45$	Do the multiplication.
$3x + 45 - \mathbf{45} = 45 - \mathbf{45}$	To undo the addition of 45, subtract 45 from both sides.
$3x = 0$	Do the subtractions: $45 - 45 = 0$
$\frac{3x}{\mathbf{3}} = \frac{0}{\mathbf{3}}$	To undo the multiplication by 3, divide both sides by 3.
$x = 0$	Do the divisions.

Verify that 0 satisfies the equation.

Self Check
Solve $7(t + 5) = -70$.

Answer: -15 ■

A strategy for solving equations

To summarize, when solving an equation we must isolate the variable on one side of the = sign. At times, this requires that we remove parentheses and/or combine like terms. The following steps should be applied, in order, when solving an equation.

Strategy for solving equations

1. Use the distributive property to remove any parentheses.
2. Combine like terms on either side of the equation.
3. Apply the addition or subtraction properties of equality to get the variables on one side of the = sign and the constants on the other.
4. Continue to combine like terms when possible.
5. Undo the operations of multiplication and division to isolate the variable.
6. Check the results.

You won't always have to use all six steps to solve a given equation. If a step doesn't apply, skip it and go to the next step.

EXAMPLE 7 ***Distributing a factor of −1.*** Solve $2x = 2 - (4x + 14)$.

Solution

$2x = 2 - (4x + 14)$	
$2x = 2 - \mathbf{1}(4x + 14)$	Rewrite: $2 - (4x + 14) = 2 - 1(4x + 14)$.
$2x = 2 - 4x - 14$	Use the distributive property to remove parentheses: $-1(4x + 14) = -4x - 14$.
$2x = -12 - 4x$	Combine like terms: $2 - 14 = -12$.
$2x + \mathbf{4x} = -12 - 4x + \mathbf{4x}$	To eliminate $-4x$ from the right-hand side, add $4x$ to both sides.
$6x = -12$	Combine like terms: $2x + 4x = 6x$ and $-4x + 4x = 0$.
$\frac{6x}{\mathbf{6}} = \frac{-12}{\mathbf{6}}$	To undo the multiplication by 6, divide both sides by 6.
$x = -2$	Do the divisions.

Verify that -2 satisfies the equation.

Self Check
Solve $4x = -13 - (3x + 8)$.

Answer: $x = -3$ ■

STUDY SET Section 3.5

VOCABULARY *Fill in the blanks to make the statements true.*

1. To ________ an equation means to find all values of the variable that make the equation a true statement.

2. To ________ a solution means to substitute that value into the original equation to see whether a true statement results.

3. In $2(x + 4)$, to remove parentheses means to apply the ________ property.

4. Algebraic expressions are simplified, and ________ are solved.

5. The phrase "________ like terms" refers to the operations of addition and subtraction.

6. A ________ is a letter that represents a number. A ________ is a number that is fixed and does not change in value.

CONCEPTS

7. Explain why $x = -5$ is not a solution of $5x - 3x = -9$.

8. a. For each equation, circle the terms involving variables:

$$5x + 3x = 8 \quad 5t = 3t + 8 \quad 7 = 5h + 3h - 1$$

b. Which equation has variables on both sides?

9. To solve $6k = 5k - 18$, we need to eliminate $5k$ from the right-hand side. To do this, what should we subtract from both sides?

10. Tell the first step in solving each equation. You do not have to solve it.

a. $2x + 4x = 36$
b. $6x = x + 10$
c. $5(x + 1) = 15$
d. $50 = x + 4 + x$

11. Consider the equation $2x - 8 = 4x - 14$.

a. To solve this equation by isolating x on the left side, what should we subtract from both sides?
b. To solve this equation by isolating x on the right side, what should we subtract from both sides?

12. Fill in the blanks.

$$6 - (d - 4) = 8$$
$$6 - \_\_ (d - 4) = 8$$
$$6 - \_\_ + \_\_ = 8$$

13. a. Simplify $3t - t - 8$.
b. Solve $3t - t = -8$.
c. Evaluate $3t - t - 8$ for $t = -4$.

14. a. Evaluate $2(x + 1)$ for $x = -4$.
b. Simplify $2(x + 1) - 4$.
c. Solve $2(x + 1) = -4$.

NOTATION *Complete each solution to solve the equation.*

15.
$$4x - 2x = -20$$
$$\_\_ = -20$$
$$\frac{2x}{\_\_} = \frac{-20}{\_\_}$$
$$x = -10$$

16.
$$8y - 6 = -2 + 10y$$
$$8y - 6 - \_\_ = -2 + 10y - \_\_$$
$$-6 = -2 + \_\_$$
$$-6 + \_\_ = -2 + 2y + \_\_$$
$$-4 = \_\_$$
$$\frac{-4}{\_\_} = \frac{2y}{\_\_}$$
$$-2 = y$$

17.
$$5(x - 9) = 5$$
$$5x - 5(\_\_) = 5$$
$$5x - \_\_ = 5$$
$$5x - 45 + \_\_ = 5 + \_\_$$
$$\_\_ = 50$$
$$\frac{5x}{\_\_} = \frac{50}{\_\_}$$
$$x = 10$$

18.
$$-2(-1 - x) = 16$$
$$2 + \_\_ = 16$$
$$2 + 2x - \_\_ = 16 - \_\_$$
$$2x = \_\_$$
$$\frac{2x}{2} = \frac{\_\_}{2}$$
$$x = 7$$

PRACTICE *For each equation, determine whether the given number is a solution.*

19. $5f + 8 = 4f + 11; 3$
20. $3r + 8 = 5r - 2; 5$
21. $2(x - 1) = 33; 12$
22. $-6(x + 4) = -40; 8$

Solve each equation by combining like terms.

23. $3x + 6x = 54$
24. $4c + 4c = 16$
25. $6x - 3x = 9$
26. $12b - 10b = 6$
27. $60 = 3v - 5v$
28. $28 = x - 3x$

29. $-28 = -m + 2m$

30. $-120 = -p + 4p$

31. $x + x + 6 = 90$

32. $c + c + 1 = 51$

33. $T + T - 17 = 57$

34. $r + r - 15 = 95$

35. $600 = m - 12 + m$

36. $403 = x - 3 + x$

37. $1{,}500 = b + 30 + b$

38. $8{,}000 = h + 100 + h$

Solve each equation by eliminating a variable term on one side of the equation.

39. $7x = 3x + 8$

40. $4x = 2x + 14$

41. $x - 14 = 2x$

42. $2x - 7 = 3x$

43. $9t - 40 = 14t$

44. $5r - 24 = 8r$

45. $25 + 4j = 9j$

46. $36 + 5j = 9j$

47. $-48 + 12t = 16t$

48. $-28 + 7t = 21t$

49. $-5g - 40 = -15g$

50. $-20s - 20 = -40s$

51. $3s + 1 = 4s - 7$

52. $6v + 2 = 7v - 3$

53. $50a - 1 = 60a - 101$

54. $25y - 2 = 75y - 202$

55. $-7 + 5r = 83 - 10r$

56. $-20 + t = 44 - 7t$

57. $100 - y = 100 + y$

58. $-60 + z = -60 - z$

Solve each equation by removing parentheses.

59. $2(x + 6) = 4$

60. $9(y - 1) = 27$

61. $-16 = 2(t + 2)$

62. $-10 = 5(y - 7)$

63. $-3(2w - 3) = 9$

64. $-4(5t + 2) = -8$

65. $-(c - 4) = 3$

66. $-(6 - 2x) = -8$

67. $4(p - 2) = 0$

68. $10(4s - 4) = 0$

69. $2(4y + 8) = 3(2y - 2)$

70. $3(7 - y) = 3(2y + 1)$

71. $16 - (x + 3) = -13$

72. $10 - (w + 4) = -12$

73. $5 - (7 - y) = -5$

74. $10 - (x - 5) = 40$

75. $2x + 3(x - 4) = 23$

76. $5j + 6(j + 1) = 226$

77. $10q + 3(q - 7) = 18$

78. $2q + 6(q - 4) = 24$

WRITING

79. Explain the error in the work shown below.

Solve $2x = 4x - x$.

$2x = 4x - x$

$2x = 4$

$\frac{2x}{2} = \frac{4}{2}$

$x = 2$

80. Consider $3x = 2x + 9$. Why is it necessary to eliminate one of the variable terms in order to solve for x?

81. What does it mean to *combine like terms*?

82. Explain what it means to *remove parentheses.* When and how is this done?

REVIEW

83. Subtract: $-7 - 9$.

84. Which numbers are *not* factors of 28: 4, 6, 7, 8?

85. Evaluate: $\frac{-8 + 2}{-2 + 4}$.

86. Translate to mathematical symbols: 4 less than x.

87. Find $-(-5)$.

88. Using x and y, illustrate the commutative property of addition.

89. What is the sign of the product of two negative integers?

90. Evaluate: $3 + 4[-4 - 3(-2)]$.

3.6 Problem Solving

In this section, you will learn about

- Problems involving one unknown quantity
- Problems involving two unknown quantities
- Number and value

Introduction. The skills that we have studied in this chapter can now be used to solve more advanced application problems. Once again, we will use the five-step problem-solving strategy as an outline for each solution.

Problems involving one unknown quantity

EXAMPLE 1 ***Volunteer service hours.*** To receive a degree in child development, students at one college must complete 135 hours of volunteer service by working 3-hour shifts at a local preschool. If a student has already volunteered 87 hours, how many more 3-hour shifts must she work to meet the service requirement for her degree?

Analyze the problem

- Students must complete 135 hours of volunteer service.
- Students work 3-hour shifts.
- A student has already completed 87 hours of service.
- Find how many more 3-hour shifts she needs to work.

Form an equation Let $x =$ the number of shifts needed to complete the service requirement. Since each shift is 3 hours long, multiplying 3 by the number of shifts will give the number of additional hours the student needs to volunteer.

The number of hours she has completed	plus	3 times	the number of shifts yet to be completed	is	the number of hours required.
87	$+$	$3 \cdot$	x	$=$	135

Solve the equation

$$87 + 3x = 135$$

$$87 + 3x \mathbf{- 87} = 135 \mathbf{- 87}$$ To undo the addition of 87, subtract 87 from both sides.

$$3x = 48$$ Combine like terms.

$$\frac{3x}{\mathbf{3}} = \frac{48}{\mathbf{3}}$$ To undo the multiplication by 3, divide both sides by 3.

$$x = 16$$ Do the divisions.

State the conclusion The student needs to complete 16 more 3-hour shifts of volunteer service.

Check the result The student has already completed 87 hours. If she works 16 more shifts, each 3 hours long, she will have $16 \cdot 3 = 48$ more hours. Adding the two sets of hours, we get $87 + 48 = 135$. The answer checks. ■

EXAMPLE 2 ***Attorney's fees.*** In return for her services, an attorney and her client split the jury's cash award evenly. After paying her assistant \$1,000, the lawyer ended up making \$10,000 from the case. What was the amount of the award?

Analyze the problem

- The attorney and client split the award evenly.
- The lawyer's assistant was paid \$1,000.
- The lawyer made \$10,000.
- Find the amount of the award.

Form an equation Let $x =$ the amount of the award. In reading the problem, two key phrases help us form an equation.

Key phrase: *split the award evenly* **Translation:** divide by 2

Key phrase: *paying her assistant* \$1,000 **Translation:** subtract \$1,000

We can express the amount of money made by the attorney in two ways.

The award split in half	minus	the amount paid to the assistant	is	$10,000.
$\frac{x}{2}$	$-$	1,000	$=$	10,000

Solve the equation

$$\frac{x}{2} - 1{,}000 = 10{,}000$$

$$\frac{x}{2} - 1{,}000 + \mathbf{1{,}000} = 10{,}000 + \mathbf{1{,}000}$$ To undo the subtraction of 1,000, add 1,000 to both sides.

$$\frac{x}{2} = 11{,}000$$ Combine like terms.

$$\mathbf{2} \cdot \frac{x}{2} = \mathbf{2} \cdot 11{,}000$$ To undo the division by 2, multiply both sides by 2.

$$x = 22{,}000$$ Do the multiplications.

State the conclusion The amount of the award was $22,000.

Check the result If the award of $22,000 is split in half, the attorney's share is $11,000. If $1,000 is paid to her assistant, $10,000 is left for the attorney. The answer checks. ■

Problems involving two unknown quantities

When solving application problems, we normally let the variable stand for the quantity we are asked to find. In the next two examples, each problem contains a second unknown quantity. We will look for a key word or phrase in the problem to help us describe it using an algebraic expression.

EXAMPLE 3 ***Civil service exam.*** A candidate for a position with the FBI scored 12 points higher on the written portion of the civil service exam than she did on her interview. If her combined score was 92, what was her score on the interview?

Analyze the problem

- She scored 12 points higher on the written portion than on the interview.
- Her combined score was 92.
- Find her score on the interview.

Form an equation Since we are told that her score on the written portion was related to her score on the interview, we let $x =$ her score on the interview.

There is a second unknown quantity—her score on the written portion of the exam. We look for a key phrase to help us decide how to represent that score using an algebraic expression.

Key phrase: 12 points *higher* on the written portion than on the interview

Translation: add 12 points to the interview score

So $x + 12 =$ her score on the written part of the test. We can express her overall score in two ways.

The score on the interview	plus	the score on the written part	is	the overall score.
x	$+$	$x + 12$	$=$	92

Solve the equation

$$x + x + 12 = 92$$

$$2x + 12 = 92 \quad \text{Combine like terms: } x + x = 2x.$$

$$2x + 12 - \mathbf{12} = 92 - \mathbf{12} \quad \text{To undo the addition of 12, subtract 12 from both sides.}$$

$$2x = 80 \quad \text{Combine like terms.}$$

$$\frac{2x}{\mathbf{2}} = \frac{80}{\mathbf{2}} \quad \text{To undo the multiplication by 2, divide both sides by 2.}$$

$$x = 40 \quad \text{Do the divisions.}$$

State the conclusion Her score on the interview was 40.

Check the result Her score on the written exam was 12 points higher than her score on the interview. If we add 12 points to 40 points, we find that she received a score of 52 on the written exam. Adding the two scores, we get $40 + 52 = 92$. The answer checks. ■

EXAMPLE 4 ***Playground design.*** After receiving a donation of 400 feet of chain link fencing, the staff of a preschool decided to use it to enclose a playground. Find the width of the playground if it is to be rectangular in shape, with the length three times the width.

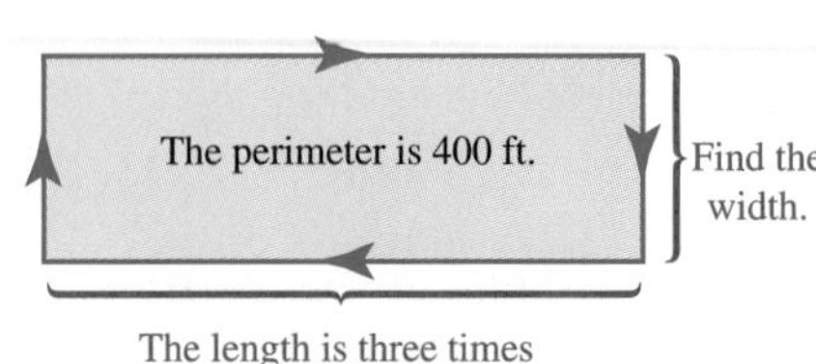
The perimeter is 400 ft.
Find the width.
The length is three times as long as the width.

FIGURE 3-10

Analyze the problem Since 400 feet of fencing will be used, the perimeter of the playground will be 400 feet. Figure 3-10 shows the given information and what we are to find.

Form an equation We will let w = the width of the playground. There is a second unknown quantity: the length of the playground. We look for a key phrase to help us decide how to represent it using an algebraic expression.

Key phrase: length *three times* the width **Translation:** multiply width by 3

So $3w$ = the length of the playground.

The formula for the perimeter of a rectangle is $P = 2l + 2w$. In words, we can write

2 ·	the length of the playground	plus	2 ·	the width of the playground	is	the perimeter.
2 ·	$3w$	+	2 ·	w	=	400

Solve the equation

$$2 \cdot 3w + 2w = 400$$

$$6w + 2w = 400 \quad \text{Do the multiplication: } 2 \cdot 3w = 6w.$$

$$8w = 400 \quad \text{Combine like terms: } 6w + 2w = 8w.$$

$$\frac{8w}{\mathbf{8}} = \frac{400}{\mathbf{8}} \quad \text{Divide both sides by 8.}$$

$$w = 50 \quad \text{Do the divisions.}$$

State the conclusion The width of the playground is 50 feet.

Check the result The length of the playground must be 3(**50**) or 150 feet, since it is three times the width. If we add two lengths and two widths, we get $2(150) + 2(50) = 300 + 100 = 400$. The answer checks. ■

Number and value

Some problems deal with quantities that have a value. In these problems, we must distinguish between the *number of* and the *value of* the unknown quantity. For example, to find the value of 3 quarters, we multiply the number of quarters by the value (in cents) of one quarter. Therefore, the value of 3 quarters is $3 \cdot 25$ cents = 75 cents.

The same distinction must be made if the number is unknown. For example, the value of d dimes is not d cents. The value of d dimes is $d \cdot 10$ cents = $10d$ cents. For problems of this type, we will use the relationship

Number · value = total value

EXAMPLE 5 ***Number and value.*** Suppose Delicious apples sell for 89 cents a pound. Find the cost of **a.** 5 pounds of apples, **b.** p pounds of apples, and **c.** $(p - 2)$ pounds of apples.

Solution

In each case, we will multiply the number of pounds of apples by their value (89 cents a pound).

a. The cost of 5 pounds of apples is $5 \cdot 89$ cents = 445 cents, or \$4.45.

b. The cost of p pounds is $p \cdot 89$ cents = $89p$ cents.

c. The cost of $(p - 2)$ pounds is $(p - 2) \cdot 89$ cents = $89(p - 2)$ cents.

Self Check

A T-bill (Treasury bill) is worth \$10,000. Find the value of

a. two T-bills

b. x T-bills

c. $(x + 3)$ T-bills

Answers: **a.** \$20,000, **b.** $10{,}000x$ dollars, **c.** $10{,}000(x + 3)$ dollars ■

EXAMPLE 6 ***Number and value.*** Ninety-five people attended a movie matinee. Ticket prices were \$6 for adults and \$4 for children. Find the income received from the sale of children's tickets and from the sale of adults' tickets. Use a table to present your results.

Solution

If 95 people attended and if we let c = the number of children's tickets sold, then $95 - c$ = the number of adult tickets sold.

The value of a child's ticket is 4 dollars. Therefore, the income from the sale of c children's tickets is the number times the value: $c \cdot 4$ dollars = $4c$ dollars.

The value of an adult ticket is 6 dollars. The income from the sale of $(95 - c)$ adult tickets is $(95 - c) \cdot 6$ dollars = $6(95 - c)$ dollars.

Type of ticket	Number	Value (\$)	Total value (\$)
Child	c	4	$4c$
Adult	$95 - c$	6	$6(95 - c)$

This column is number · value.

Self Check

A fitness club has 150 members. Monthly membership fees are \$25 for nonseniors and \$15 for senior citizens. Find the income the club receives from nonseniors and from seniors each month. Use the table below to present your results. Let m represent the number of nonseniors.

Member's age	Number	Fee	Total income (\$)

Answers: $25m$ dollars from nonseniors, $15(150 - m)$ dollars from seniors ■

EXAMPLE 7 ***Basketball.*** On a night when they scored 110 points, a basketball team made only 5 free throws (worth 1 point each). The remainder of their points came from two- and three-point baskets. If the number of baskets from the field totaled 45, how many two-point and how many three-point baskets did they make?

Analyze the problem

- The team scored 110 points.
- They made 5 free throws (1 point each).
- They made 45 baskets from the field.
- Find the number of two-point and three-point baskets made.

Form an equation The number of two- and three-point baskets totaled 45. If we let x = the number of three-point baskets made, then $45 - x$ = the number of two-point baskets made. We can now organize the data in a table, as on the next page. We must remember to multiply the *number* of each type of basket by its point *value.*

Type of basket	Number	Value	Total value
Three-point	x	3	$3x$
Two-point	$45 - x$	2	$2(45 - x)$
Free throw	5	1	5

We can express the total number of points scored in two ways.

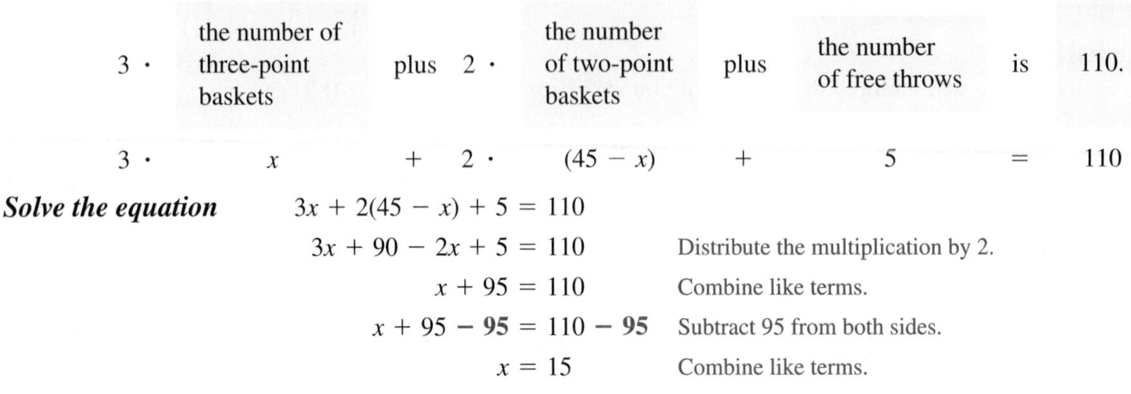

3 · the number of three-point baskets plus 2 · the number of two-point baskets plus the number of free throws is 110.
$3 \cdot x + 2 \cdot (45 - x) + 5 = 110$

Solve the equation

$$3x + 2(45 - x) + 5 = 110$$

$$3x + 90 - 2x + 5 = 110 \quad \text{Distribute the multiplication by 2.}$$

$$x + 95 = 110 \quad \text{Combine like terms.}$$

$$x + 95 - \mathbf{95} = 110 - \mathbf{95} \quad \text{Subtract 95 from both sides.}$$

$$x = 15 \quad \text{Combine like terms.}$$

We can substitute 15 for x in $45 - x$ to find the number of two-point baskets made:

$$45 - x = 45 - \mathbf{15} = 30$$

State the conclusion The basketball team made 15 three-point baskets and 30 two-point baskets.

Check the result If we multiply the number of three-point baskets by their value, we get $15 \cdot 3 = 45$ points. If we multiply the number of two-point baskets by their value, we get $30 \cdot 2 = 60$ points. If we add the number of made free throws to these two subtotals, we get $45 + 60 + 5 = 110$ points. The answers check. ■

STUDY SET Section 3.6

VOCABULARY *Fill in the blanks to make the statements true.*

1. The words *increased by, longer, taller, higher, total,* and *more than* indicate that the operation of ______________ should be used.

2. The words *shorter, less than, fewer than, difference,* and *decreased by* indicate that the operation of ______________ should be used.

CONCEPTS

3. BUSINESS ACCOUNTS Every month, a salesman adds five new accounts. How many new accounts will he add in x months?

4. ANTIQUE COLLECTING Every year, a woman purchases four antique spoons to add to her collection. How many spoons will she purchase in x years?

5. SERVICE STATION See Illustration 1. How many gallons does the smaller tank hold?

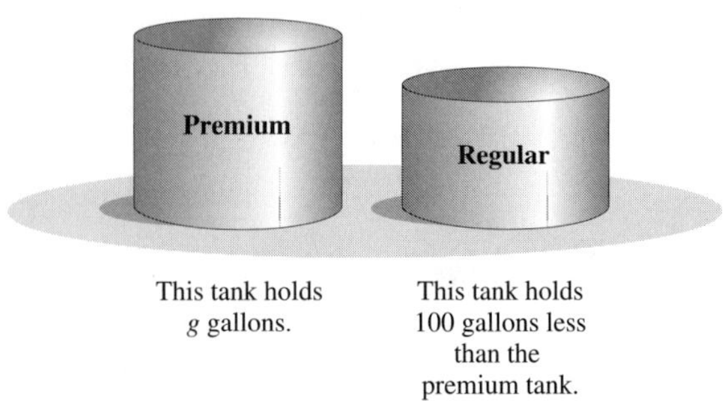

This tank holds g gallons.

ILLUSTRATION 1

6. COLLEGE SCHOLARSHIPS See Illustration 2. How many scholarships were awarded this year?

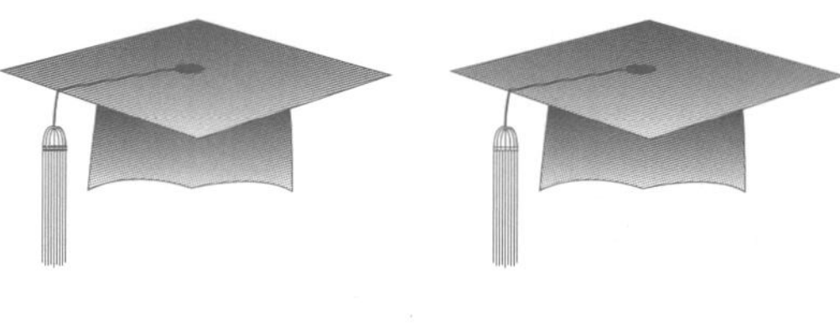

Last year, s scholarships were awarded.

Six more scholarships were awarded this year than last year.

ILLUSTRATION 2

7. OCEAN TRAVEL See Illustration 3. How many miles did the passenger ship travel?

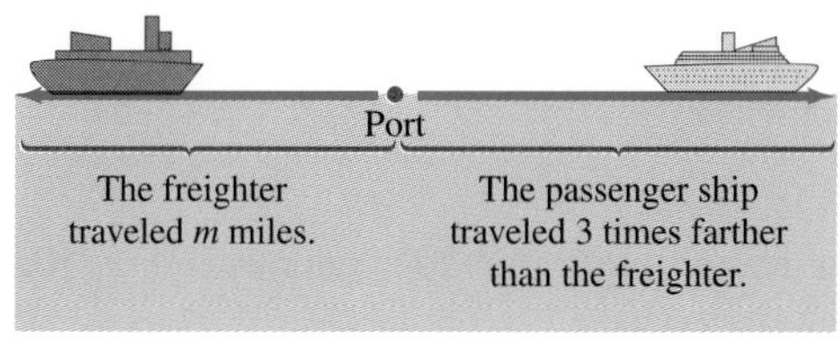

ILLUSTRATION 3

8. TAX REFUND See Illustration 4. How much of the tax refund did the husband get?

A husband and wife received a tax refund of \$$d$.

The couple split the refund equally.

ILLUSTRATION 4

9. The length of a rectangle is twice its width. Complete the labeling of the picture in Illustration 5.

ILLUSTRATION 5

10. Complete this statement about the perimeter of the rectangle shown in Illustration 6.

$2 \cdot \square + 2 \cdot \square = 240$

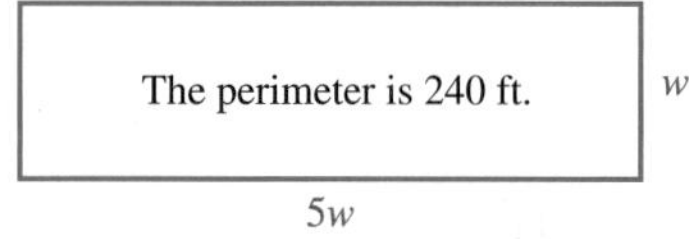

ILLUSTRATION 6

11. COMMISSION A shoe salesman receives a commission for every pair of shoes he sells. Complete the table.

Type of shoe	Number sold	Commission per shoe (\$)	Total commission (\$)
Dress	10	3	
Athletic	12	2	
Child's	x	5	
Sandal	$9 - x$	4	

12. Complete the table.

Type of coin	Number	Value (¢)	Total value (¢)
Nickel	12		
Dime	d		
Quarter	$q + 2$		

13. Illustration 7 shows a rack that contains dress shoes and athletic shoes.

a. How many pairs of shoes are stored in the rack?

b. If there are d pairs of dress shoes in the rack, how many pairs of athletic shoes are there in the rack?

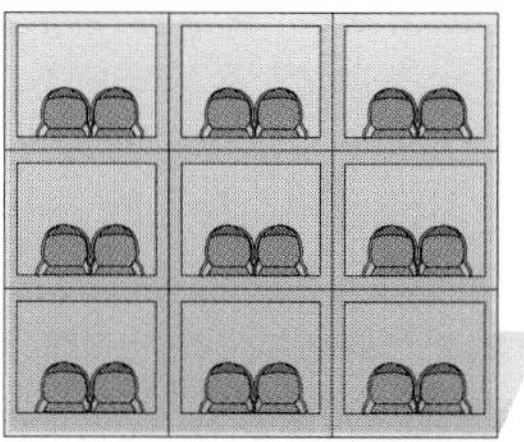

ILLUSTRATION 7

14. A prealgebra quiz is shown in Illustration 8.

a. How many questions were on the quiz?

b. If the student answered c questions correctly, how many did she answer incorrectly?

PREALGEBRA QUIZ
CHAPTER 3

1. 44	6. 250 ft
2. 376	7. 165 mi
3. equal	8. no
4. 9 − x	9. yes
5. 4x	10. simplify

ILLUSTRATION 8

APPLICATIONS *Complete each solution.*

15. AIRLINE SEATING An 88-seat passenger plane has ten times as many economy seats as first-class seats. Find the number of first-class seats.

Analyze the problem

- There are seats on the plane.
- There are times as many economy as first-class seats.
- Find ______________________.

Form an equation Since the number of economy seats is related to the number of first-class seats, we let $x =$ ______________________.

To represent the number of economy seats, look for a key phrase in the problem.

Key phrase: ten times as many

Translation: ______________

So = the number of economy seats.

The number of first-class seats	plus	the number of economy seats	is	88.
x	$+$		$=$	88

Solve the equation

$x + 10x =$

$= 88$

$x = 8$

State the conclusion There are first-class seats.

Check the result If there are 8 first-class seats, there are $\cdot\ 8 = 80$ economy seats. Adding and , we get 88. The answer checks.

16. SUPERMARKET COUPONS A shopper used some 20-cents-off and some 40-cents-off coupons at the supermarket to get a reduction of \$2.60 from her grocery bill. If she used a total of eight coupons, how many of each type did she redeem at the checkout stand?

Analyze the problem

- ¢ and ¢ coupons were redeemed.
- The coupons saved her \$2.60, which is ¢.
- coupons were used.
- Find ______________________________________.

Form an equation The total number of coupons used was 8. If we let $x =$ the number of 20¢ coupons used, then $8 - x =$ ______________________.

20	·	the number of 20¢ coupons used	plus	40	·	the number of 40¢ coupons used	is	260.
		$20x$	$+$				$=$	260

Solve the equation

$20x + 40(8 - x) =$

$20x + \quad - 40x = 260$

$-20x + \quad = 260$

$-20x =$

$x = 3$

If 20¢ coupons were used, then $8 - 3 =$ 40¢ coupons were used.

State the conclusion 20¢ coupons and 40¢ coupons were redeemed.

Check the result The value of 20¢ coupons is $3 \cdot 20 =$ ¢. The value of 40¢ coupons is $5 \cdot 40 =$ ¢. Adding these two subtotals, we get 260¢, which is \$2.60. The answers check.

Form an equation and then solve it to answer each question.

17. BUSINESS ACCOUNTS After beginning a new position with 15 established accounts, a salesman made it his objective to add 5 new accounts every month. His goal was to reach 100 accounts. At this rate, how many months would it take to reach his goal?

18. STUDENT LOAN A student plans to pay back a \$600 loan with monthly payments of \$30. How many payments has she made if the debt has been reduced to \$420?

19. ANTIQUE COLLECTING A woman purchases 4 antique spoons each year. She now owns 56 spoons. In how many years will she have 100 spoons in her collection?

20. CONSTRUCTION To get a heavy-equipment operator's certificate, 48 hours of on-the-job training are required. If a woman has completed 24 hours, and the training sessions last for 6 hours, how many more sessions must she take to get the certificate?

21. APARTMENT RENTAL In renting an apartment with two other friends, Jonathan agreed to pay the security deposit of \$100 himself. The three of them agreed to contribute equally toward the monthly rent. Jonathan's first check to the apartment owner was for \$425. What was the monthly rent for the apartment?

22. TAX REFUND After receiving their tax refund, a husband and wife split the refunded money equally. The husband then gave $50 of his money to charity, leaving him with $70. What was the amount of the tax refund check?

23. BOTTLED WATER DELIVERY A truck driver left the plant carrying 300 bottles of drinking water. His delivery route consisted of office buildings, each of which was to receive 3 bottles of water. The driver returned to the plant at the end of the day with 117 bottles of water on the truck. To how many office buildings did he deliver?

24. CORPORATE DOWNSIZING In an effort to cut costs, a corporation has decided to lay off 5 employees every month until the number of employees totals 465. If 510 people are now employed, how many months will it take to reach the employment goal?

25. SERVICE STATION At a service station, the underground tank storing regular gas holds 100 gallons less than the tank storing premium gas. If the total storage capacity of the tanks is 700 gallons, how much does the premium gas tank hold?

26. COLLEGE SCHOLARSHIPS Because of increased alumni giving, a college scholarship program awarded six more scholarships this year than last year. If a total of 20 scholarships were awarded over the last two years, how many were awarded last year?

27. OCEAN TRAVEL At noon, a passenger ship and a freighter left a port traveling in opposite directions. By midnight, the passenger ship was 3 times farther from port than the freighter was. How far was the freighter from port if the distance between the ships was 84 miles?

28. RADIO STATIONS The daily listening audience of an AM radio station is four times as large as that of its FM sister station. If 100,000 people listen to these two radio stations, how many listeners does the FM station have?

29. INTERIOR DECORATING As part of redecorating, crown molding was installed around the base of the ceiling of a room. (See Illustration 9.) Sixty feet of molding was needed for the project. Find the width of the room if its length is twice the width.

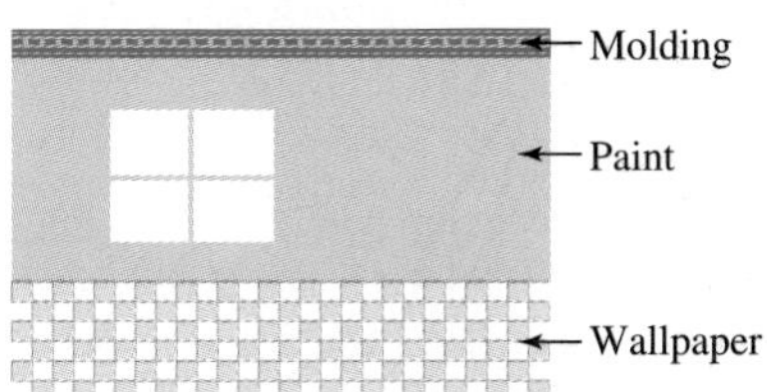

ILLUSTRATION 9

30. SPRINKLER SYSTEM A landscaper buried a water line around a rectangular-shaped lawn to serve as a supply line for a sprinkler system. (See Illustration 10.) The length of the lawn is 5 times its width. If 240 feet of pipe was used to do the job, what is the width of the lawn?

ILLUSTRATION 10

31. COMMERCIALS During a 30-minute television show, a viewer found that the actual program aired a total of 18 minutes more than the time devoted to commercials. How many minutes of commercials were there?

32. CLASS TIME In a biology course, students spend a total of 250 minutes in lab and lecture each week. The lab time is 50 minutes shorter than the lecture time. How many minutes do the students spend in lecture per week?

Form an equation and then solve it to answer each question. Make a table to organize the data.

33. COMMISSION A salesman receives a commission of $3 for every pair of dress shoes he sells. He is paid $2 for every pair of athletic shoes he sells. After selling 9 pairs of shoes in a day, his commission was $24. How many pairs of each kind of shoe did he sell that day?

34. GRADING SCALE For every problem answered correctly on an exam, 3 points are awarded. For every incorrect answer, 4 points are deducted. In a 10-question test, a student scored 16 points. How many correct and incorrect answers did he have on the exam?

35. MOVER'S PAY SCALE A part-time mover's regular pay rate is $6 an hour. If the work involves going up and down stairs, his rate increases to $9 an hour. In one week, he earned $138 and worked 20 hours. How many hours did he work at each rate?

36. PRESCHOOL ENROLLMENT A preschool charges $8 for a child to attend its morning session or $10 to attend the afternoon session. No child can attend both. Thirty children are enrolled in the preschool. If the daily receipts are $264, how many children attend each session?

WRITING

37. Explain what should be accomplished in each of the five steps of the problem-solving strategy studied in this section.

38. Use an example to explain the difference between the number of quarters a person has and the value of those quarters.

39. Write a problem that could be represented by the following equation.

Age of father	plus	age of son	is	50.
x	$+$	$x - 20$	$=$	50

40. Write a problem that could be represented by the following equation.

2 ·	length of a field	plus 2 ·	width of a field	is	600 ft.
$2 \cdot$	$4x$	$+\ 2 \cdot$	x	$=$	600

REVIEW

41. What property is illustrated?

$(2 + 9) + 1 = 2 + (9 + 1)$

42. Solve $4 - x = -8$.

43. Evaluate: -10^2.

44. List the factors of 18.

45. Fill in the blank to make a true statement. Subtraction of a number is the same as ____________ of the opposite of that number.

46. Round 123,808 to the nearest ten thousand.

47. Write this prime factorization using exponents: $2 \cdot 2 \cdot 2 \cdot 5 \cdot 5$.

48. The value of a stock dropped \$3 a day for 6 consecutive days. What was the change in the value of the stock over this period?

KEY CONCEPT

Order of Operations

One of the major objectives of this course is that you thoroughly understand the rules for the order of operations.

The following priority list gives the order in which operations must be performed. Fill in the blanks.

1. Do all calculations within _____________ and other grouping symbols, following the order listed in Steps 2–4 below and working from the _____________ pair to the _____________ pair.

2. Evaluate all _____________ expressions.

3. Do all _____________ and _____________ as they occur from left to right.

4. Do all _____________ and _____________ as they occur from left to right.

When all the grouping symbols have been removed, repeat Steps 2–4 to complete the calculation.

If a fraction bar is present, evaluate the numerator and the denominator _____________. Then do the division indicated by the fraction bar, if possible.

In algebra, we have to apply the rules for the order of operations in many different settings.

Evaluating numerical expressions

5. Evaluate: $-10 + 4 - 3^2$.

6. Evaluate: $\frac{-30}{6} - (-4)3$.

7. Evaluate: $-2(-3) - 12 \div 6 \cdot 3$.

8. Evaluate: $2(-3)^3(4) + (-6)$.

9. Evaluate: $2(4 + 3 \cdot 2)^2 - (-6)$.

10. Evaluate: $-1^2 + 3[6 - (1 - 5)]$.

Evaluating algebraic expressions

11. Evaluate: $1 - 2|8w - w^3|$ for $w = -2$.

12. Evaluate: $\frac{50 - 6s}{-t}$ for $s = 5$ and $t = -4$.

Working with formulas

13. Use the formula $P = 2l + 2w$ to find the perimeter of a rectangle that has a length of 30 feet and a width of 16 feet.

14. WINDOW WASHING Use the formula $d = 16t^2$ to determine the distance a squeegee fell if it took 6 seconds for it to hit the ground after a window washer dropped it.

Simplifying algebraic expressions

15. Simplify $(3x)4 - 2(5x) + x$.

16. Simplify $2(y + 3) - 3(y - 4)$.

Checking a solution of an equation

17. Is $x = -3$ a solution of the equation $15 - 3x = 23$?

18. Is $c = -2$ a solution of the equation $-4(2c + 5) - 4(2c - 5) = 32$?

Solving equations

When solving equations, we use the rules for the order of operations in reverse.

19. Consider $2x - 3 = 13$. In what order should we undo the operations so that we can isolate x on the left side?

20. Consider $-10 = 5 + \frac{m}{-6}$. In what order should we undo the operations so that we can isolate m on the right side?

ACCENT ON TEAMWORK

Section 3.1

TRANSLATING Copy the table below onto a piece of paper. In column 1, write in all the words or phrases you can think of that indicate addition. In column 2, write in all the words or phrases you can think of that indicate subtraction. Continue in this way for each of the remaining columns.

Addition	Subtraction	Multiplication	Division	Equals

Section 3.2

AREA OF A SQUARE Examine Illustration 1. What patterns do you see as the square increases in size? Draw the next four squares of this sequence, labeling them in a similar way.

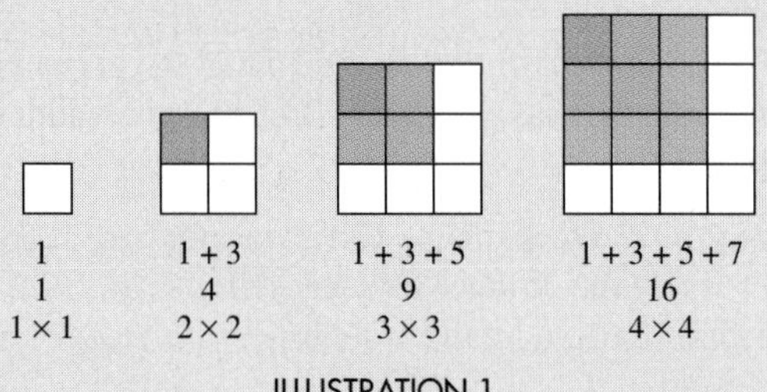

ILLUSTRATION 1

Section 3.3

THE DISTRIBUTIVE PROPERTY Illustration 2 shows three rectangles that are divided into squares. Since the area of the rectangle on the left-hand side of the equals sign can be found by multiplying its width by its length, its area is $4(5 + 3)$ square units. Evaluating this expression, we see that the area shaded in blue is $4(8) = 32$ square units.

The area on the right-hand side of the equals sign is the sum of the areas of the two rectangles: $4(5) + 4(3)$. Evaluating this expression, we see that the area shaded in red is also 32 square units: $4(5) + 4(3) = 20 + 12 = 32$. Therefore,

$$4(5 + 3) = 4(5) + 5(3)$$

Create a similar demonstration of the distributive property using rectangles with dimensions different from those in Illustration 2.

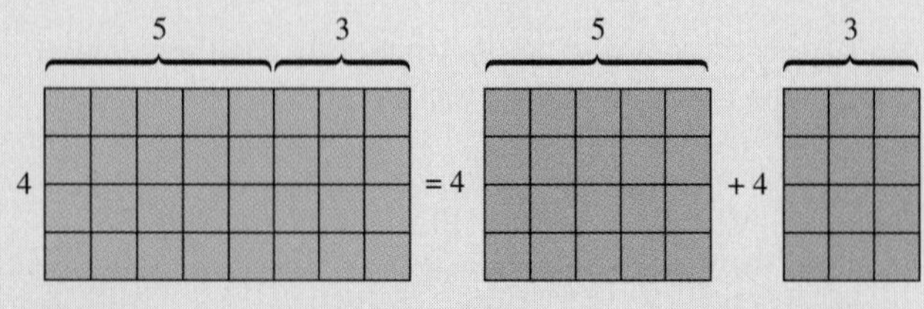

ILLUSTRATION 2

Section 3.4

LIKE TERMS If we are to add or subtract objects, they must be similar. Simplify each of the following expressions by combining like terms. You will have to change some of the units so that you are working with like terms. To do so, use the following conversion facts.

- There are 12 inches in one foot.
- There are 36 inches in one yard.
- There are 3 feet in one yard.
- There are 5,280 feet in one mile.

a. 1 foot + 6 inches **b.** 1 yard + 11 inches

c. 1 mile − 1 foot **d.** 12 feet − 1 yard

e. 1 yard + 1 foot + 5 inches

f. 2 yards + 2 feet − 2 inches

g. 6 inches + 3 feet − 4 inches + 2 feet

Section 3.5

SOLVING EQUATIONS In Sections 1.5 and 1.6, we saw how scales can be used to illustrate the steps used to solve an equation.

a. What equation is being solved in Illustration 3? What is the solution?

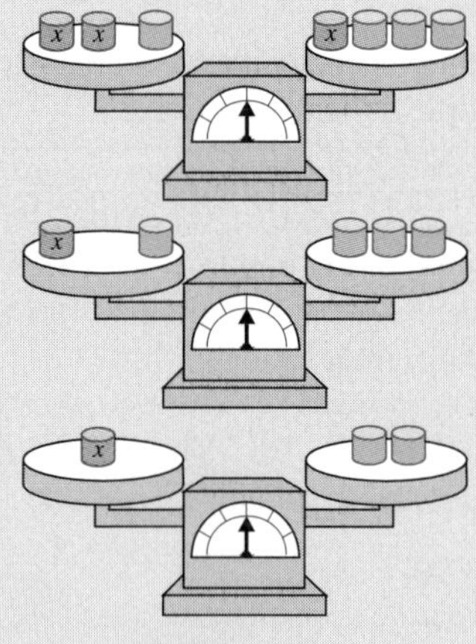

ILLUSTRATION 3

b. Draw a similar series of pictures showing the solution of each of the following equations.

1. $3x = 2x + 4$ **2.** $3x + 1 = 2x + 4$

3. $2x + 6 = 4x$ **4.** $2x + 6 = 4x + 2$

Section 3.6

PROBLEM SOLVING Write a number–value problem that involves two quantities that have different values. Then solve it. See Exercises 33–36 in Study Set 3.6 for some examples.

CHAPTER REVIEW

SECTION 3.1 Variables and Algebraic Expressions

CONCEPTS

A *variable* is used to represent an unknown quantity.

REVIEW EXERCISES

1. a. Illustration 1 shows the distances from two towns to an airport. Which town is closer to the airport? How much closer is it?

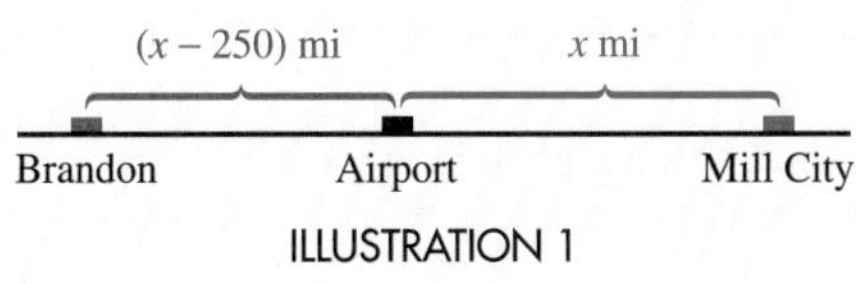

ILLUSTRATION 1

b. See Illustration 2. Let h represent the height of the ladder, and write an algebraic expression for the height of the ceiling in feet.

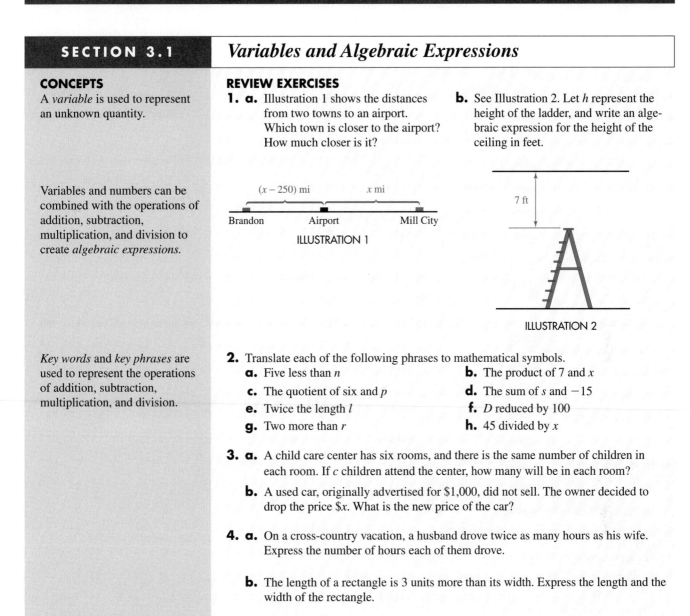

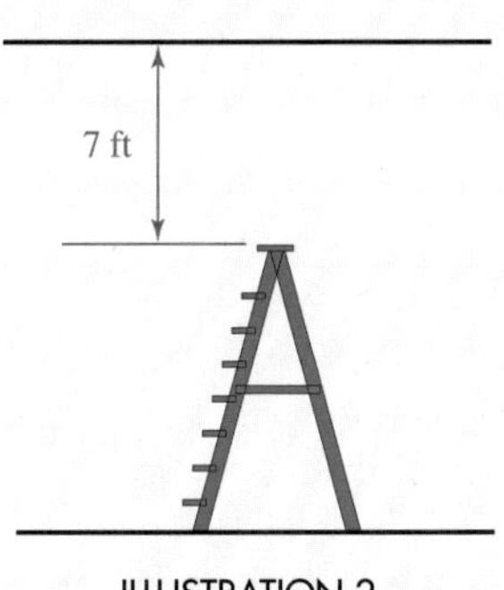

ILLUSTRATION 2

Variables and numbers can be combined with the operations of addition, subtraction, multiplication, and division to create *algebraic expressions.*

Key words and *key phrases* are used to represent the operations of addition, subtraction, multiplication, and division.

2. Translate each of the following phrases to mathematical symbols.

a. Five less than n

b. The product of 7 and x

c. The quotient of six and p

d. The sum of s and -15

e. Twice the length l

f. D reduced by 100

g. Two more than r

h. 45 divided by x

3. a. A child care center has six rooms, and there is the same number of children in each room. If c children attend the center, how many will be in each room?

b. A used car, originally advertised for $1,000, did not sell. The owner decided to drop the price $$x$. What is the new price of the car?

4. a. On a cross-country vacation, a husband drove twice as many hours as his wife. Express the number of hours each of them drove.

b. The length of a rectangle is 3 units more than its width. Express the length and the width of the rectangle.

Sometimes you must rely on common sense and insight to find *hidden operations.*

5. a. How many eggs are in x dozen?

b. d days is how many weeks?

SECTION 3.2 Evaluating Algebraic Expressions and Formulas

When we replace the variable, or variables, in an algebraic expression with a specific number and then apply the rules for the order of operations, we are *evaluating the algebraic expression.*

6. RETAINING WALL Illustration 3 shows the design for a retaining wall. The relationships between the lengths of its important parts are given in words.

a. Use a variable and algebraic expressions to describe the height and the lengths of the upper and lower bases.

b. Suppose engineers determine that a 10-foot-high wall is needed. Find the lengths of the upper and lower bases.

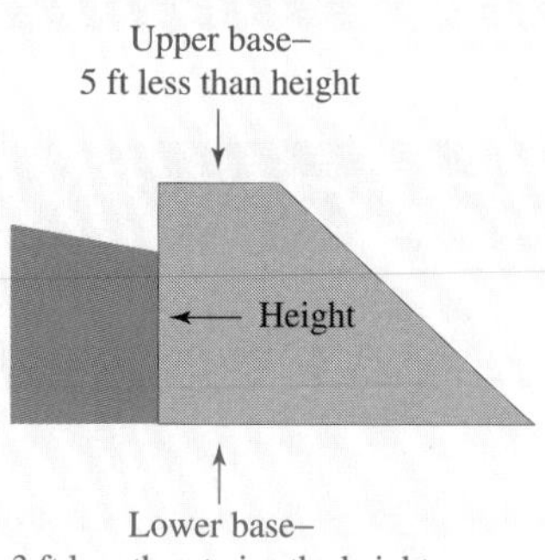

ILLUSTRATION 3

7. Evaluate each algebraic expression.

a. $-2x + 6$ for $x = -3$

b. $\dfrac{6 - a}{1 + a}$ for $a = -2$

c. $b^2 - 4ac$ for $a = 4, b = 6,$ and $c = -4$

d. $\dfrac{-2k^3}{1 - 2 - 3}$ for $k = -2$

A *formula* is a general rule that describes a known relationship between two or more variables.

Distance = rate · time

8. DISTANCE TRAVELED Complete Illustration 4 by finding the distance traveled for a given time at a given rate.

	Rate (mph)	Time (hr)	Distance traveled (mi)
Monorail	65	2	
Subway	38	3	
Train	x	6	
Bus	55	t	

ILLUSTRATION 4

Formulas from business:

Sale price = original price − discount

Retail price = cost + markup

Profit = revenue − costs

9. SALE PRICE Find the sale price of a trampoline that normally sells for \$315 if a \$37 discount is being offered.

10. RETAIL PRICE Find the retail price of a car if the dealer pays \$14,505 and the markup is \$725.

11. ANNUAL PROFIT The bar graph in Illustration 5 shows the revenue and costs for a company for the years 1998 to 2000, in millions of dollars.

a. In which year was there the most revenue?

b. Which year had the largest profit?

c. What can you say about costs over this three-year span?

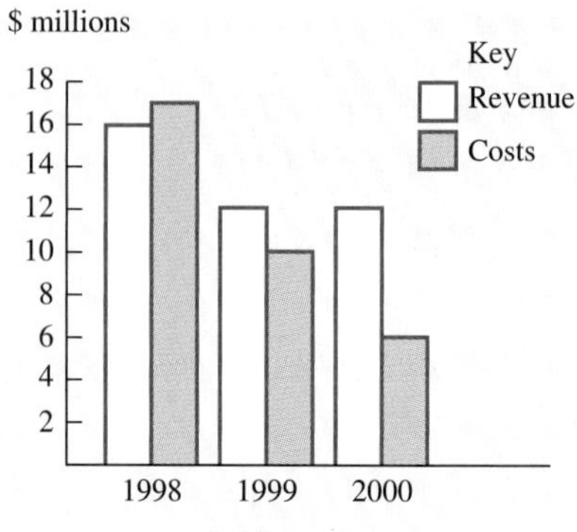

ILLUSTRATION 5

Formulas from science:

$$C = \frac{5(F - 32)}{9}$$

$$\text{Distance fallen} = 16 \cdot (\text{time})^2$$

12. TEMPERATURE CONVERSION At a summer resort, visitors can relax by taking a dip in a swimming pool or a lake. The pool water is kept at a constant temperature of 77° F. The water in the lake is 23° C. Which water is warmer, and by how many degrees Celsius?

13. DISTANCE FALLEN A steelworker accidentally dropped his hammer while working atop a new high-rise building. How far will the hammer fall in 3 seconds?

Formulas from mathematics:

$$\text{Mean} = \frac{\text{sum of values}}{\text{number of values}}$$

14. AVERAGE YEARS OF EXPERIENCE Three generations of Smiths now operate a family-owned real estate office. The grandparents, who started the business, have been realtors for 40 years. Their son and daughter-in-law joined the company as realtors 18 years ago. Their grandson has worked as a realtor for 4 years. What is the average number of years a member of the Smith family has worked at Smith Realty?

SECTION 3.3 Simplifying Algebraic Expressions and the Distributive Property

To *simplify an algebraic expression,* we use properties of algebra to write the expression in simpler form.

15. Simplify each expression.

a. $-2(5x)$ **b.** $-7x(-6y)$ **c.** $4d \cdot 3e \cdot 5$ **d.** $(4s)8$

e. $-1(-e)(2)$ **f.** $7x \cdot 7y$ **g.** $4 \cdot 3k \cdot 7$ **h.** $(-10t)(-10)$

The *distributive property*:

If a, b, and c are numbers, then

$a(b + c) = ab + ac$

$(b + c)a = ba + ca$

$a(b - c) = ab - ac$

$(b - c)a = ba - ca$

$a(b + c + d) = ab + ac + ad$

16. Multiply to remove parentheses.

a. $4(y + 5)$ **b.** $-5(6t + 9)$

c. $(-3 - 3x)7$ **d.** $-3(4e - 8x - 1)$

17. Write an equivalent expression without parentheses.

a. $-(6t - 4)$ **b.** $-(5 + x)$ **c.** $-(6t - 3s + 1)$ **d.** $-(-5a - 3)$

SECTION 3.4 Combining Like Terms

A *term* is a number or a product of a number and one or more variables.

18. Identify the second term and the coefficient of the third term.

a. $5x^2 - 4x + 8$ **b.** $7y - 3y + x - y$

In a term that is a product of a number and one or more variables, the factor that is a number is called the *coefficient* of the term.

19. Tell whether x is used as a factor or a term.

a. $5x - 6y^2$ **b.** $x + 6$ **c.** $6xy$ **d.** $-36 - x + b$

Like terms, or *similar terms,* are terms with exactly the same variables and exponents.

20. Tell whether the following are like terms.

a. $4x, -5x$ **b.** $4x, 4x^2$

c. $3xy, xy$ **d.** $-5b^2c, -5bc^2$

Simplifying the sum (or difference) of like terms is called *combining like terms.*

To combine like terms, combine their coefficients and keep the variables and exponents.

21. Simply by combining like terms.

a. $3x + 4x$ **b.** $-3t - 6t$
c. $2z + (-5z)$ **d.** $6x - x$
e. $-6y - 7y - (-y)$ **f.** $5w - 8 - 4w + 3$
g. $-45d - 2a + 4a - d$ **h.** $5y + 8h - 3 + 7h + 5y + 2$

22. Simplify each expression.

a. $7(y + 6) + 3(2y + 2)$ **b.** $-4(t - 7) - (t + 6)$
c. $5x - 2(x - 6)$ **d.** $6f + 7(12 - 8f)$
e. $0 - 14m$ **f.** $0 + 14m$

The perimeter of a rectangle is given by

$P = 2l + 2w$

The perimeter of a square is given by

$P = 4s$

23. HOLIDAY LIGHTS To decorate a house, lights will be hung around the entire home, as shown in Illustration 6. They will also be placed around the two 5-foot-by-5-foot windows in the front. How many feet of lights will be needed?

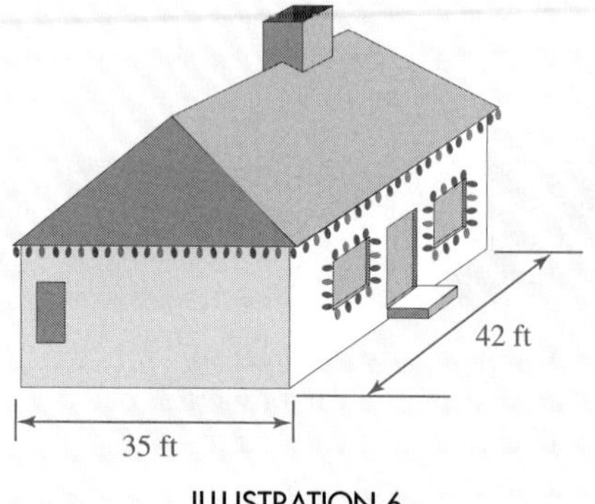

ILLUSTRATION 6

SECTION 3.5 Simplifying Expressions to Solve Equations

To *solve an equation* means to find all values of the variable that, when substituted into the original equation, make the equation a true statement.

24. Is $x = -3$ a solution of $-4x + 6 = 2(x + 12)$? Explain why or why not.

25. Solve each equation. Check all answers.

a. $5a - 3a = -36$ **b.** $3x - 4x = -8$
c. $7x = 3x - 12$ **d.** $5(y - 15) = 0$
e. $3a - (2a - 1) = -2$ **f.** $15 = 5b + 1 + 2b$
g. $-6(2x + 3) = -(5x - 3)$ **h.** $-3(2x + 4) - 4 = -40$

SECTION 3.6 Problem Solving

Be careful to distinguish between the *number* and the *value* of a set of objects.

Total value = number · value

Use the five-step *problem-solving strategy* to solve problems.

1. Analyze the problem.
2. Form an equation.
3. Solve the equation.
4. State the conclusion.
5. Check the result.

26. Complete the chart in Illustration 7.

Type of coin	Number	Value (¢)	Total value (¢)
Dime	6		
Quarter	7		
Penny	x		
Nickel	n		

ILLUSTRATION 7

27. REFRESHMENTS How many cups of coffee are left in the coffee maker shown in Illustration 8, if c cups have already been poured from it?

ILLUSTRATION 8

28. COLD STORAGE A meat locker lowers the temperature of a product 7° Fahrenheit every hour. If placed in the locker, how long would it take freshly ground hamburger to go from a room temperature of 71° F to 29° F?

29. FITNESS The midweek workout for a fitness instructor consists of jogging and bicycling. If she jogs 8 fewer miles than she bikes, and her workout covers a total of 18 miles, how many miles does she bicycle?

30. CAR SHOW Attendance during the first day of a two-day car show was low. On the second day, attendance doubled. If 6,600 people attended the show, what was the attendance on the first day?

31. HEALTH FOOD A fruit juice bar sells two types of drinks: one priced at $3 and the other at $4. One day at lunchtime, business was very brisk. If 50 drinks were sold and the receipts were $185, how many $3 drinks and how many $4 drinks were purchased?

Chapter 3 Test

1. Translate each phrase to mathematical symbols.
   a. 2 less than r
   b. The product of 3, x, and y

2. Together, a couple earns $51,000 a year. If the wife earns e dollars a year, write an expression for the husband's yearly earnings.

3. Evaluate each expression.
   a. $\frac{x - 16}{x}$ for $x = 4$
   b. $2t^2 - 3(t - s)$ for $t = -2$ and $s = 4$

4. DISTANCE TRAVELED Find the distance traveled by a motorist who departed from home at 9:00 A.M. and arrived at his destination at noon, traveling at a rate of 55 miles per hour.

5. PROFIT A craft show promoter had revenues and costs as shown in Illustration 1. Find the profit.

Revenues	Costs
Ticket sales: $40,000 Booth rental: $15,000	Supplies: $13,000 Facility rental fee: $5,000

ILLUSTRATION 1

6. FALLING OBJECT If a tennis ball was dropped from the top of a 200-foot-tall building, would it hit the ground after falling for 3 seconds? If not, how far short of the ground would it be?

7. METER READING Every hour between 8 A.M. and 5 P.M., a technician noted the value registered by a meter in a power plant and recorded that number on a line graph. (See Illustration 2.) Find the average meter value reading for this period.

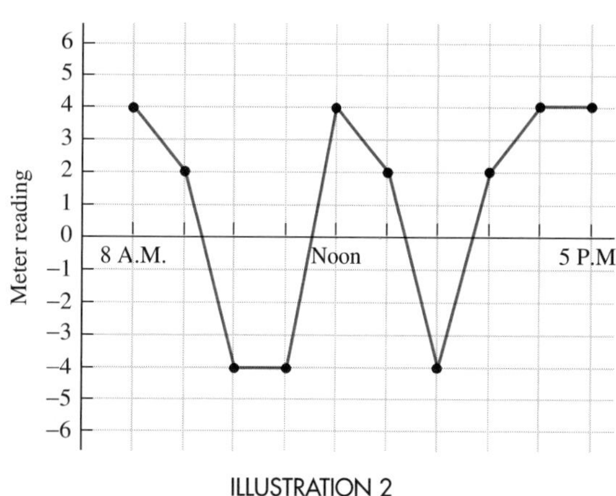

ILLUSTRATION 2

8. COLLEGE LANDMARK Overlund College is going to construct a gigantic block letter O on a foothill slope near campus. The outline of the letter is to be done using redwood edging. (See Illustration 3.) How many feet of edging will be needed?

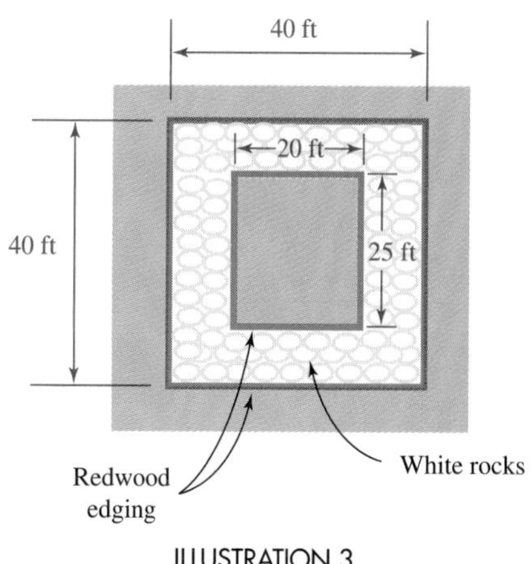

ILLUSTRATION 3

9. AIR CONDITIONING After the air conditioner in a classroom was accidentally left on all night, the room's temperature in the morning was a cool 59° F. What was the temperature in degrees Celsius?

10. Simplify by removing the parentheses.

a. $5(5x + 1)$ **b.** $-6(7 - x)$

c. $-(6y + 4)$ **d.** $3(2a + 3b - 7)$

11. Tell whether x is used as a factor or as a term.

a. $5xy$ **b.** $8y + x + 6$

12. Simplify by combining like terms.

a. $-20y + 6 - 8y + 4$

b. $-t - t - t$

13. a. Identify each term in this algebraic expression: $8x^2 - 4x - 6$.

b. What is the coefficient of the first term?

14. Simplify each expression.

a. $7x + 4x$ **b.** $3c \cdot 4e \cdot 2$

c. $6x - x$ **d.** $-5y(-6)$

e. $0 - 7x$ **f.** $0 + 9y$

15. Simplify the expression $4(y + 3) - 5(2y + 3)$.

16. Solve $5x - 3x = -18$.

17. Solve: $6r = 2r - 12$.

18. Solve: $-45 = 3(1 - 4t)$.

19. Solve $6 - (y - 3) = 19$.

20. a. What is the value of k dimes?

b. What is the value of $p + 2$ twenty-dollar bills?

In Problems 21–22, form an equation and then solve it to answer each question.

21. DRIVING SCHOOL A driver's training program requires students to attend regularly scheduled classroom sessions each afternoon on Monday through Thursday. On Friday, the students take a 2-hour final exam. If the entire program requires 14 hours of a student's time, how long is each classroom session?

22. CABLE TELEVISION In order to receive its broadcasting license, a cable television station was required to broadcast locally produced shows in addition to its nationally syndicated programming. During a typical 24-hour period, the national shows aired for 8 hours more than the local shows. How many hours of local shows were broadcast each day?

23. What are like terms?

24. Let a variable represent the length of one of the fish shown in Illustration 4. Then write an expression for the length of the other fish. Give two possible sets of answers.

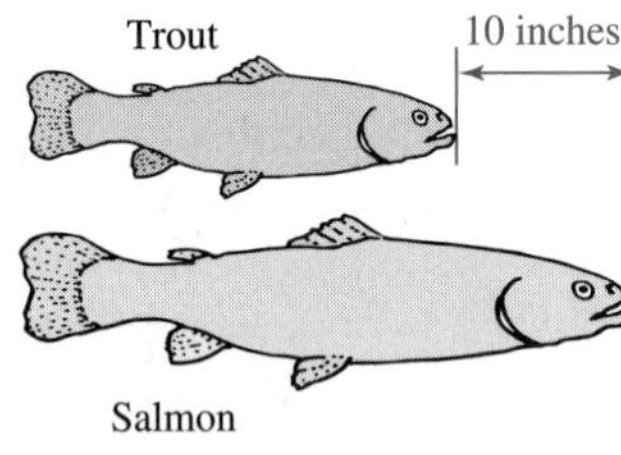

ILLUSTRATION 4

25. Do the instructions *simplify* and *solve* mean the same thing? Explain why or why not.

26. Give an example of an application of the distributive property.

Chapters 1–3 Cumulative Review Exercises

1. GASOLINE In 1999, gasoline consumption in the United States was three hundred fifty-eight million, six hundred thousand gallons a day. Write this number in standard notation.

2. Round 49,999 to the nearest thousand.

In Exercises 3–6, do each operation.

3. $\begin{array}{r} 38{,}908 \\ +15{,}696 \\ \hline \end{array}$

4. $\begin{array}{r} 9{,}700 \\ -5{,}491 \\ \hline \end{array}$

5. $\begin{array}{r} 345 \\ \times\ 67 \\ \hline \end{array}$

6. $23\overline{)2{,}001}$

7. Explain how to check the following result using addition.

$\begin{array}{r} 1{,}142 \\ -\ 459 \\ \hline 683 \end{array}$

8. VIETNAMESE CALENDAR An animal represents each Vietnamese lunar year. Recent Years of the Cat are listed below. If the cycle continues, what year will be the next Year of the Cat?

1915 1927 1939 1951 1963 1975 1987 1999

9. Consider the multiplication statement $4 \cdot 5 = 20$. Show that multiplication is repeated addition.

10. ROOM DIVIDER Four pieces of plywood, each 22 inches wide and 62 inches high, are to be covered with fabric, front and back, to make the room divider shown in Illustration 1. How many square inches of fabric will be used?

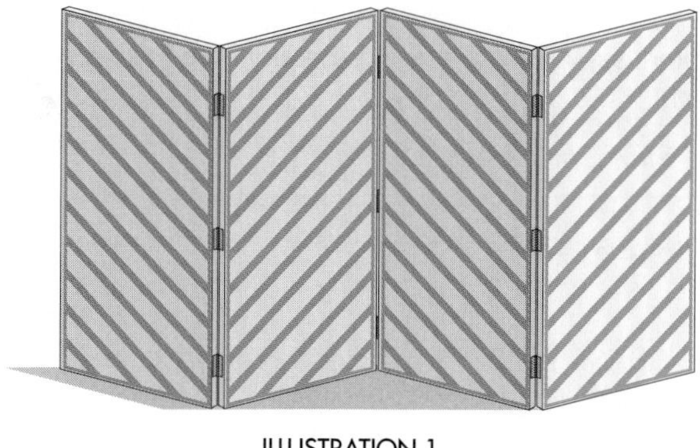

ILLUSTRATION 1

11. **a.** Find the factors of 18.
b. Find the prime factorization of 18.

12. List the first ten prime numbers.

13. Why isn't 27 a prime number?

14. Evaluate: $(9 - 2)^2 - 3^3$.

15. Solve $250 = \frac{y}{2}$.

16. Simplify $-(-6)$.

17. Graph the integers greater than -3 but less than 4.

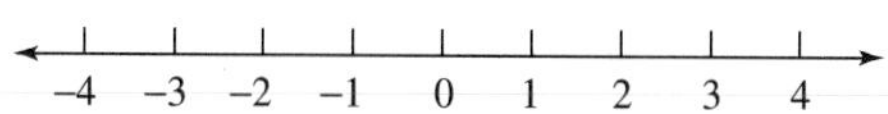

18. Find the absolute value: $|-5|$.

19. Is the statement $-12 > -10$ true or false?

20. ANNUAL NET INCOME Use the following data for the Polaroid Corporation to construct a line graph.

Year	'95	'96	'97	'98	'99
Total net income ($ millions)	−139	15	−127	−51	9

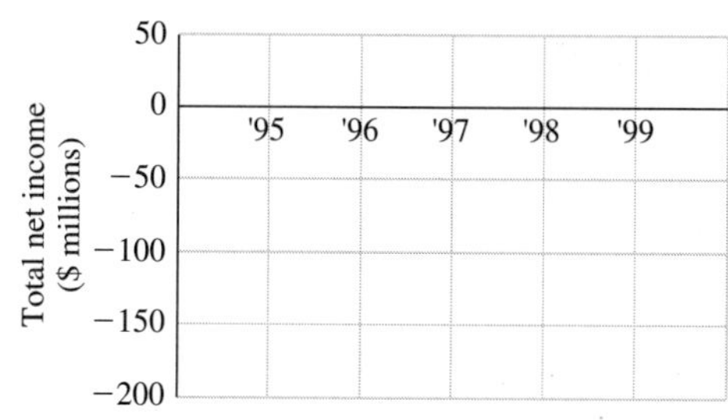

In Exercises 21–24, do each operation.

21. $-25 + 5$

22. $25 - (-5)$

23. $-25(5)(-1)$

24. $\dfrac{-25}{-5}$

25. Evaluate: $\dfrac{(-6)^2 - 1^5}{-4 - 3}$.

26. Evaluate: $-3 + 3[-4 - 4 \cdot 2]^2$.

27. Evaluate -3^2 and $(-3)^2$.

28. PLANETS Mercury orbits closer to the sun than any other planet. Temperatures on Mercury can get as high as 810° F and as low as −290° F. What is the temperature range?

In Exercises 29–30, solve each equation.

29. $-4x + 4 = -24$

30. $-y = 10$.

31. Write an expression illustrating division by 0 and an expression illustrating division of 0. Which is undefined?

32. What property allows us to rewrite $x \cdot 5$ as $5x$?

33. Translate to mathematical symbols: "h increased by 12."

34. TENNIS Write an algebraic expression that represents the length of the handle of the tennis racket shown in Illustration 2.

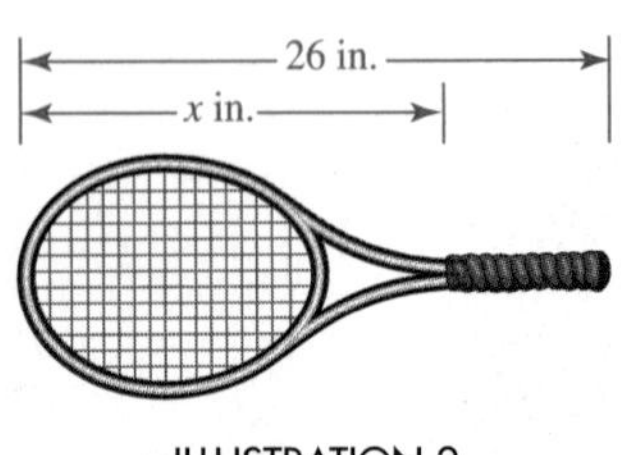

ILLUSTRATION 2

35. Explain the difference between x^2 and $2x$.

36. Evaluate $x^2 - 2x + 1$ for $x = -5$.

37. Complete the table.

	Rate (mph)	Time (hr)	Distance traveled (mi)
Truck	55	4	

38. Remove the parentheses: $5(2x - 7)$.

39. Simplify $-6(-4t)$.

40. Complete the table.

Term	Coefficient
$4a$	
$-2y^2$	
x	
$-m$	

41. Write an expression in which x is used as a term. Then write an expression in which x is used as a factor.

42. What is the value (in cents) of q quarters?

43. Simplify $5b + 8 - 6b - 7$.

In Exercises 44–45, solve each equation and check the result.

44. $8p + 2p - 1 = -11$

45. $7 + 2x = 2 - (4x + 7)$.

46. CLASS TIME In a chemistry course, students spend a total of 300 minutes in lab and lecture each week. The time spent in lab is 50 minutes less than the time spent in lecture. How many minutes do the students spend in lecture each week?

Fractions and Mixed Numbers

4

WHOLE NUMBERS ARE USED TO COUNT OBJECTS. WHEN WE NEED TO REPRESENT PARTS OF A WHOLE, FRACTIONS CAN BE USED.

4.1 The Fundamental Property of Fractions

In this section, you will learn about

- Basic facts about fractions • Equivalent fractions • Simplifying a fraction
- Expressing a fraction in higher terms

INTRODUCTION. There is no better place to start a study of fractions than with *the fundamental property of fractions.* This property is the foundation for two fundamental procedures that are used when working with fractions. But first, we review some basic facts about fractions.

Basic facts about fractions

1. A Fraction Can Be Used to Indicate Equal Parts of a Whole.

In our everyday lives, we often deal with parts of a whole. For example, we talk about parts of an hour, parts of an inch, and parts of a pound.

2. A Fraction is Composed of a Numerator, a Denominator, and a Fraction Bar.

Fraction bar ⟶ $\frac{3}{4}$ ⟵ Numerator / ⟵ Denominator

The denominator (in this case, 4) tells us that a whole was divided into four equal parts. The numerator tells us that we are considering three of those equal parts.

3. Fractions Can Be Proper or Improper.

If the numerator of a fraction is less than its denominator, the fraction is called a **proper fraction.** A proper fraction is less than 1. Fractions whose numerators are greater than or equal to their denominators are called **improper fractions.** An improper fraction is greater than or equal to 1.

Proper fractions	**Improper fractions**
$\frac{1}{4}, \frac{2}{3},$ and $\frac{98}{99}$	$\frac{7}{2}, \frac{98}{97}, \frac{16}{16},$ and $\frac{5}{1}$

EXAMPLE 1 ***Fractional parts of a whole.*** **a.** In Figure 4-1, what fractional part of the barrel is full? **b.** What fractional part is empty?

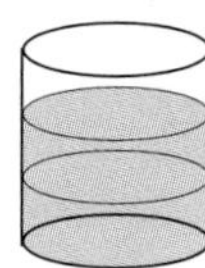

FIGURE 4-1

Solution

The barrel has been divided into three equal parts.

a. Two of the three parts are full. Therefore, the barrel is $\frac{2}{3}$ full.

b. One of the three equal parts is not filled. The barrel is $\frac{1}{3}$ empty.

The fractions $\frac{2}{3}$ and $\frac{1}{3}$ are both proper fractions.

Self Check

a. According to the calendar below, what fractional part of the month has passed? **b.** What fractional part remains?

DECEMBER

X	X	X	X	X	X	X
X	X	X	X	12	13	14
15	16	17	18	19	20	21
22	23	24	25	26	27	28
29	30	31				

Answers: **a.** $\dfrac{11}{31}$, **b.** $\dfrac{20}{31}$ ■

4. The Denominator of a Fraction Cannot Be 0.

$\frac{7}{0}$, $\frac{23}{0}$, and $\frac{0}{0}$ are meaningless expressions. (Recall that $\frac{7}{0}$, $\frac{23}{0}$, and $\frac{0}{0}$ represent *division* by 0, and a number cannot be divided by 0.) However, $\frac{0}{7} = 0$ and $\frac{0}{23} = 0$.

5. The Numerator and the Denominator of a Fraction Can Contain Variables.

Since a variable is a letter that is used to stand for a number, a variable or a combination of variables can appear in the numerator or the denominator of a fraction. Here are several examples of such **algebraic fractions.**

$$\frac{x}{4}, \quad \frac{12}{b}, \quad \frac{x}{y}, \quad \frac{m^2}{2mn}, \quad \frac{2c + d}{3c^3d}$$

6. Fractions Can Be Negative.

There are times when a negative fraction is needed to describe a quantity. For example, if an earthquake causes a road to sink one-half inch, the amount of movement can be represented by $-\frac{1}{2}$ inch.

Negative fractions can be written in three ways. The negative sign can appear in the numerator, in the denominator, or in front of the fraction.

$$\frac{-1}{2} = \frac{1}{-2} = -\frac{1}{2}$$

Negative fractions

If a and b represent positive numbers,

$$\frac{-a}{b} = \frac{a}{-b} = -\frac{a}{b} \quad (b \neq 0)$$

Fractions are often referred to as **rational numbers.** All integers are rational numbers, because every integer can be written as a fraction with a denominator of 1. For example,

$$2 = \frac{2}{1}, \quad -5 = \frac{-5}{1}, \quad \text{and} \quad 0 = \frac{0}{1}$$

Since every integer is also a rational number, the integers are a subset of the rational numbers.

COMMENT Note that not all rational numbers are integers. For example, the rational number $\frac{7}{8}$ is not an integer.

Equivalent fractions

Fractions can look different but still represent the same number. To show this, let's divide the rectangle in Figure 4-2(a) in two ways. In Figure 4-2(b), we divide it into halves (2 equal-sized parts). In Figure 4-2(c), we divide it into fourths (4 equal-sized parts). Notice that one-half of the figure is the same size as two-fourths of the figure.

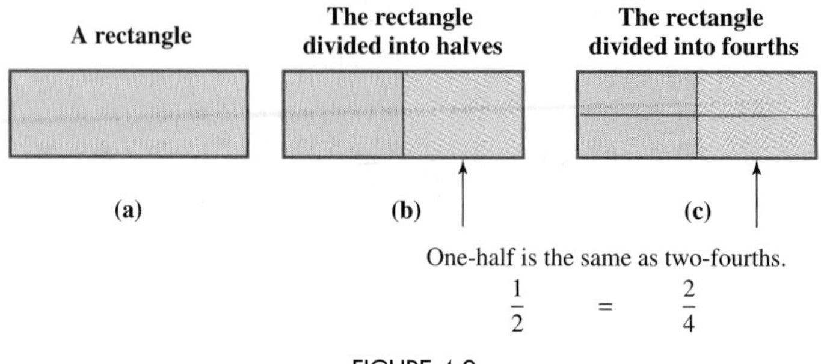

FIGURE 4-2

The fractions $\frac{1}{2}$ and $\frac{2}{4}$ look different, but Figure 4-2 shows that they represent the same amount. We say that they are **equivalent fractions.**

Equivalent fractions

Two fractions are **equivalent** if they represent the same number.

Simplifying a fraction

If we replace a fraction with an equivalent fraction that contains smaller numbers, we are **simplifying** or **reducing the fraction.** To simplify a fraction, we apply the following property.

The fundamental property of fractions

Multiplying or dividing the numerator and the denominator of a fraction by the same nonzero number does not change the value of the fraction. In symbols, for all numbers a, b, and x (provided b and x are not zero),

$$\frac{a}{b} = \frac{a \cdot x}{b \cdot x} \quad \text{and} \quad \frac{a}{b} = \frac{a \div x}{b \div x}$$

As an example, we consider $\frac{24}{28}$. It is apparent that 24 and 28 have a common factor of 4. By the fundamental property of fractions, we can divide the numerator and denominator of this fraction by 4.

$$\frac{24}{28} = \frac{24 \div \mathbf{4}}{28 \div \mathbf{4}}$$

Divide the numerator by 4.

Divide the denominator by 4.

$$= \frac{6}{7} \qquad 24 \div 4 = 6 \text{ and } 28 \div 4 = 7.$$

Thus, $\frac{24}{28} = \frac{6}{7}$. We say that $\frac{24}{28}$ and $\frac{6}{7}$ are equivalent fractions, because they represent the same number.

In practice, we show the previous simplification in a slightly different way.

$\frac{24}{28} = \frac{4 \cdot 6}{4 \cdot 7}$ Once you see a common factor of the numerator and the denominator, factor each of them so that it shows. In this case, 24 and 28 share a common factor of 4.

$= \frac{\overset{1}{\cancel{4}} \cdot 6}{\underset{1}{\cancel{4}} \cdot 7}$ Apply the fundamental property of fractions. Divide the numerator and the denominator by 4 by drawing slashes through the common factors. Use small 1's to represent the result of each division of 4 by 4.

$= \frac{6}{7}$ Multiply in the numerator and in the denominator: $1 \cdot 6 = 6$ and $1 \cdot 7 = 7$.

In the second step of the previous simplification, we say that we *divided out the common factor of 4.*

Simplifying a fraction

We can **simplify** a fraction by factoring its numerator and denominator and then dividing out all common factors in the numerator and denominator.

When a fraction can be simplified no further, we say that it is written in **lowest terms.**

Lowest terms

A fraction is in **lowest terms** if the only factor common to the numerator and denominator is 1.

EXAMPLE 2 ***Simplifying a fraction.*** Simplify to lowest terms: $\frac{25}{75}$.

Solution

The numerator and the denominator have a common factor of 25.

$\frac{25}{75} = \frac{25 \cdot 1}{25 \cdot 3}$ Factor 25 as $25 \cdot 1$ and 75 as $25 \cdot 3$.

$= \frac{\overset{1}{\cancel{25}} \cdot 1}{\underset{1}{\cancel{25}} \cdot 3}$ Divide out the common factor of 25.

$= \frac{1}{3}$ Multiply in the numerator and in the denominator: $1 \cdot 1 = 1$ and $1 \cdot 3 = 3$.

Self Check

Simplify to lowest terms: $\frac{60}{80}$.

Answer: $\frac{3}{4}$ ■

EXAMPLE 3 ***Using prime factorization.*** Simplify $\frac{90}{126}$.

Solution

To find the common factors that will divide out, we prime factor 90 and 126.

$\frac{90}{126} = \frac{5 \cdot 3 \cdot 3 \cdot 2}{7 \cdot 3 \cdot 3 \cdot 2}$ Use the tree method or the division method to prime factor 90 and 126: $90 = 5 \cdot 3 \cdot 3 \cdot 2$ and $126 = 7 \cdot 3 \cdot 3 \cdot 2$.

$= \frac{5 \cdot \overset{1}{\cancel{3}} \cdot \overset{1}{\cancel{3}} \cdot \overset{1}{\cancel{2}}}{7 \cdot \underset{1}{\cancel{3}} \cdot \underset{1}{\cancel{3}} \cdot \underset{1}{\cancel{2}}}$ Divide out the common factors of 3, 3, and 2.

$= \frac{5}{7}$ Multiply in the numerator and in the denominator.

Self Check

Simplify $\frac{42}{150}$.

Answer: $\frac{7}{25}$ ■

COMMENT Negative fractions are simplified in the same way as positive fractions. Just remember to write a negative sign − in each step of the solution.

$$-\frac{45}{72} = -\frac{\overset{1}{\cancel{9}} \cdot 5}{\underset{1}{\cancel{9}} \cdot 8} = -\frac{5}{8}$$

EXAMPLE 4 ***Simplifying an algebraic fraction.*** Simplify $\frac{24xy^2}{16x^2y^2}$.

Solution

To simplify fractions that contain variables, we divide out common numerical factors and common variable factors.

$\frac{24xy^2}{16x^2y^2} = \frac{8 \cdot 3 \cdot xy^2}{8 \cdot 2 \cdot x^2y^2}$ Since 24 and 16 have a common factor of 8, factor 24 as $8 \cdot 3$ and 16 as $8 \cdot 2$.

$= \frac{8 \cdot 3 \cdot x \cdot y \cdot y}{8 \cdot 2 \cdot x \cdot x \cdot y \cdot y}$ Write the variable parts of the numerator and denominator in factored form: $y^2 = y \cdot y$ and $x^2 = x \cdot x$.

$= \frac{\overset{1}{\cancel{8}} \cdot 3 \cdot \overset{1}{\cancel{x}} \cdot \overset{1}{\cancel{y}} \cdot \overset{1}{\cancel{y}}}{\underset{1}{\cancel{8}} \cdot 2 \cdot \underset{1}{\cancel{x}} \cdot x \cdot \underset{1}{\cancel{y}} \cdot \underset{1}{\cancel{y}}}$ Divide out the common factors of 8, x, and y.

$= \frac{3}{2x}$ Multiply in the numerator and in the denominator.

Self Check

Simplify $\frac{45ab^2}{36ab^3}$.

Answer: $\frac{5}{4b}$

■

Expressing a fraction in higher terms

It is sometimes necessary to replace a fraction with an equivalent fraction that involves larger numbers or more complex terms. This is called **expressing the fraction in higher terms** or **building up** the fraction.

For example, to write $\frac{3}{8}$ as an equivalent fraction with a denominator of 40, we can use the fundamental property of fractions and multiply the numerator and denominator by 5.

$\frac{3}{8} = \frac{3 \cdot 5}{8 \cdot 5}$ ← Multiply the numerator by 5. ← Multiply the denominator by 5.

$= \frac{15}{40}$ Do the multiplications in the numerator and in the denominator.

Therefore, $\frac{3}{8} = \frac{15}{40}$.

EXAMPLE 5 ***Expressing a fraction in higher terms.*** Write $\frac{5}{7}$ as an equivalent fraction with a denominator of 28.

Solution

We need to multiply the denominator by 4 to obtain 28. By the fundamental property of fractions, we must multiply the numerator by 4 as well.

$\frac{5}{7} = \frac{5 \cdot 4}{7 \cdot 4}$ Multiply the numerator and denominator by 4.

$= \frac{20}{28}$ Do the multiplication in the numerator and in the denominator.

Self Check

Write $\frac{2}{3}$ as an equivalent fraction with a denominator of 24.

Answer: $\frac{16}{24}$

■

EXAMPLE 6 ***Expressing a whole number as a fraction.*** Write 4 as a fraction with a denominator of 6.

Solution

First, express 4 as a fraction: $4 = \frac{4}{1}$. To obtain a denominator of 6, we need to multiply the numerator and denominator by 6.

$$\frac{4}{1} = \frac{4 \cdot 6}{1 \cdot 6}$$

$$= \frac{24}{6} \quad \text{Do each multiplication: } 4 \cdot 6 = 24 \text{ and } 1 \cdot 6 = 6.$$

Self Check

Write 5 as a fraction with a denominator of 3.

Answer: $\frac{15}{3}$ ■

EXAMPLE 7 ***Writing an algebraic fraction in higher terms.*** Write $\frac{3}{4}$ as an equivalent fraction with a denominator of $28a$.

Solution

We need to multiply the denominator of $\frac{3}{4}$ by $7a$ to obtain $28a$, so we must also multiply the numerator by $7a$.

$$\frac{3}{4} = \frac{3 \cdot 7a}{4 \cdot 7a} \quad \text{Multiply numerator and denominator by } 7a.$$

$$= \frac{21a}{28a} \quad \text{Do each multiplication: } 3 \cdot 7a = 21a \text{ and } 4 \cdot 7a = 28a.$$

Self Check

Write $\frac{2}{5}$ as an equivalent fraction with a denominator of $30x$.

Answer: $\frac{12x}{30x}$ ■

STUDY SET Section 4.1

VOCABULARY *Fill in the blanks to make the statements true.*

1. For the fraction $\frac{7}{8}$, 7 is the ____________ and 8 is the ____________.
2. When we express 15 as $5 \cdot 3$, we say that we have ____________ 15.
3. A ____________ fraction is less than 1. An ____________ fraction is greater than or equal to 1.
4. A fraction is said to be in ____________ terms if the only factor common to the numerator and denominator is 1.
5. Two fractions are ____________ if they have the same value.
6. A ____________ can be used to indicate the number of equal parts of a whole.
7. Multiplying the numerator and denominator of a fraction by a number to obtain an equivalent fraction that involves larger numbers is called expressing the fraction in ____________ terms or ____________ up the fraction.
8. We can ____________ a fraction that is not in lowest terms by applying the fundamental property of fractions. We ____________ out common factors of the numerator and denominator.

CONCEPTS

9. What common factor (other than 1) do the numerator and the denominator have?

   a. $\frac{2}{16}$ **b.** $\frac{6}{9}$ **c.** $\frac{10}{15}$ **d.** $\frac{14}{35}$

10. Given:

$$\frac{15}{35} = \frac{\overset{1}{\cancel{5}} \cdot 3}{7 \cdot \underset{1}{\cancel{5}}}$$

In this work, what do the slashes and small 1's mean?

11. What concept studied in this section is shown by Illustration 1?

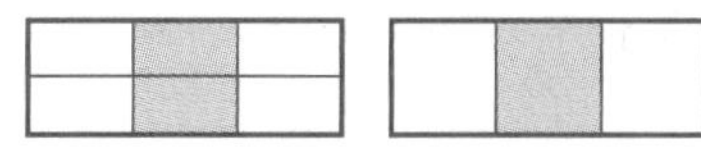

ILLUSTRATION 1

12. Why can't we say that $\frac{2}{5}$ of the figure in Illustration 2 is shaded?

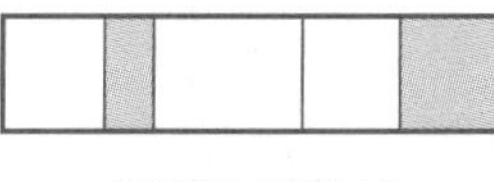

ILLUSTRATION 2

13. a. Explain the difference in the two approaches used to simplify $\frac{20}{28}$.

$$\frac{\overset{1}{\cancel{4}} \cdot 5}{\underset{1}{\cancel{4}} \cdot 7} \quad \text{and} \quad \frac{\overset{1}{\cancel{2}} \cdot \overset{1}{\cancel{2}} \cdot 5}{\underset{1}{\cancel{2}} \cdot \underset{1}{\cancel{2}} \cdot 7}$$

b. Are the results the same?

14. What concept studied in this section does this statement illustrate?

$$\frac{5}{10} = \frac{4}{8} = \frac{3}{6} = \frac{2}{4} = \frac{1}{2}$$

15. Why isn't this a valid application of the fundamental property of fractions?

$$\frac{10}{11} = \frac{2 + 8}{2 + 9} = \frac{\overset{1}{\cancel{2}} + 8}{\underset{1}{\cancel{2}} + 9} = \frac{9}{10}$$

16. Write the fraction $\frac{7}{-8}$ in two other ways.

17. Write as a fraction.

a. 8 **b.** -25

c. x **d.** $7a$

18. Fill in the missing numbers in the following statement:

$$\frac{5 \cdot \square}{9x \cdot \square} = \frac{15}{27x}$$

NOTATION *Complete each solution.*

19. Simplify $\frac{18}{24}$.

$$\frac{18}{24} = \frac{3 \cdot \square \cdot 2}{3 \cdot 2 \cdot \square \cdot 2}$$

$$= \frac{\overset{1}{\cancel{\square}} \cdot 3 \cdot \overset{1}{\cancel{\square}}}{\underset{1}{\cancel{\square}} \cdot 2 \cdot 2 \cdot \underset{1}{\cancel{\square}}}$$

$$= \frac{3}{4}$$

20. Simplify $\frac{60ab^2}{90a^2b}$.

$$\frac{60ab^2}{90a^2b} = \frac{\square \cdot 2 \cdot a \cdot b \cdot b}{\square \cdot 3 \cdot a \cdot a \cdot b}$$

$$= \frac{\overset{1}{\cancel{30}} \cdot \square \cdot \overset{1}{\cancel{a}} \cdot \square \cdot \overset{1}{\cancel{b}}}{\underset{1}{\cancel{30}} \cdot \square \cdot \underset{1}{\cancel{a}} \cdot \square \cdot \underset{1}{\cancel{b}}}$$

$$= \frac{2b}{3a}$$

PRACTICE *Simplify each fraction to lowest terms, if possible.*

21. $\frac{3}{9}$ **22.** $\frac{5}{20}$

23. $\frac{7}{21}$ **24.** $\frac{6}{30}$

25. $\frac{20}{30}$ **26.** $\frac{12}{30}$

27. $\frac{15}{6}$ **28.** $\frac{24}{16}$

29. $-\frac{28}{56}$ **30.** $-\frac{45}{54}$

31. $-\frac{90}{105}$ **32.** $-\frac{26}{78}$

33. $\frac{60}{108}$ **34.** $\frac{75}{125}$

35. $\frac{180}{210}$ **36.** $\frac{76}{28}$

37. $\frac{55}{67}$ **38.** $\frac{41}{51}$

39. $\frac{36}{96}$ **40.** $\frac{48}{120}$

41. $\frac{25x^2}{35x}$ **42.** $\frac{16r^2}{20r}$

43. $\frac{12t}{15t}$ **44.** $\frac{10y}{15y}$

45. $\frac{6a}{7a}$ **46.** $\frac{4c}{5c}$

47. $\frac{7xy}{8xy}$ **48.** $\frac{10ab}{21ab}$

49. $-\frac{10rs}{30}$ **50.** $-\frac{14ab}{28}$

51. $\frac{15st^3}{25xt^3}$ **52.** $\frac{16wx^3}{24x^3y}$

53. $\frac{35r^2t}{28rt^2}$ **54.** $\frac{35m^3n^4}{25m^4n^3}$

55. $\frac{56p^4}{28p^6}$ **56.** $\frac{32k^2}{8k^5}$

Write each fraction as an equivalent fraction with the indicated denominator.

57. $\frac{7}{8}$, denominator 40 **58.** $\frac{3}{4}$, denominator 24

59. $\frac{4}{5}$, denominator 35 **60.** $\frac{5}{7}$, denominator 49

61. $\frac{5}{6}$, denominator 54

62. $\frac{11}{16}$, denominator 32

63. $\frac{1}{2}$, denominator 30

64. $\frac{1}{3}$, denominator 60

65. $\frac{2}{7}$, denominator $14x$

66. $\frac{3}{10}$, denominator $50a$

67. $\frac{9}{10}$, denominator $60t$

68. $\frac{2}{3}$, denominator $27t$

69. $\frac{5}{4s}$, denominator $20s$

70. $\frac{9}{4x}$, denominator $44x$

71. $\frac{2}{15}$, denominator $45y$

72. $\frac{5}{12}$, denominator $36n$

Write each number or algebraic expression as a fraction with the indicated denominator.

73. 3 as fifths

74. 4 as thirds

75. 6 as eighths

76. 3 as sixths

77. $4a$ as ninths

78. $7x$ as fourths

79. $-2t$ as halves

80. $-10c$ as ninths

APPLICATIONS *Use the concept of fraction in answering each question.*

81. COMMUTING How much of the commute from home to work has the motorist in Illustration 3 made?

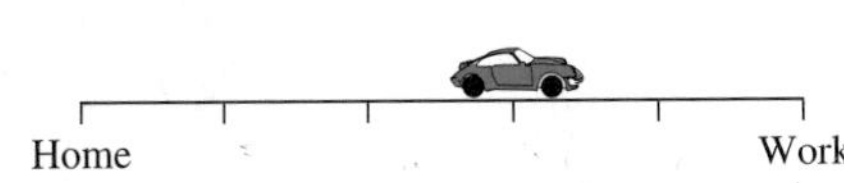

ILLUSTRATION 3

82. TIME CLOCK How much of the hour has passed?

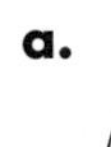

a.

b.

c.

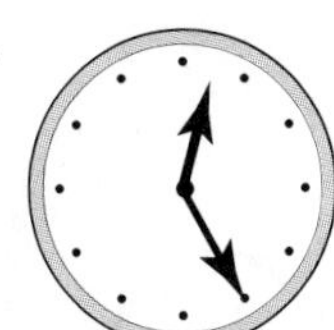

d.

83. SINKHOLE Illustration 4 shows a side view of a depression in the sidewalk near a sinkhole. Describe the movement of the sidewalk using a signed number. (On the tape measure, one inch is divided into 16 equal parts.)

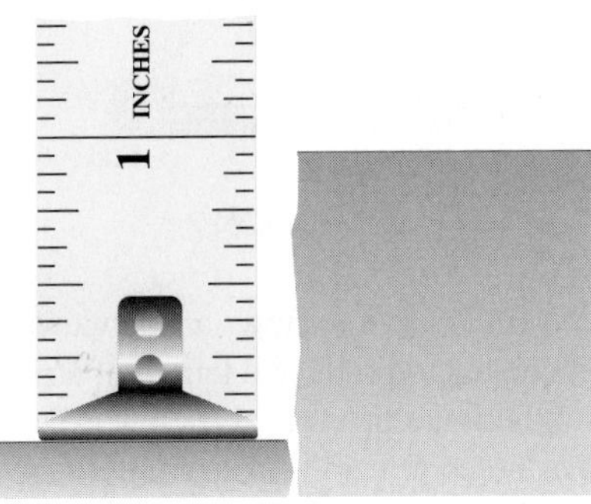

ILLUSTRATION 4

84. POLITICAL PARTIES Illustration 5 shows the political party affiliation of the governors of the 50 states, as of January 1, 2000.

a. What fraction are Democrats?

b. What fraction are Republicans?

c. What fraction are neither Democrat nor Republican?

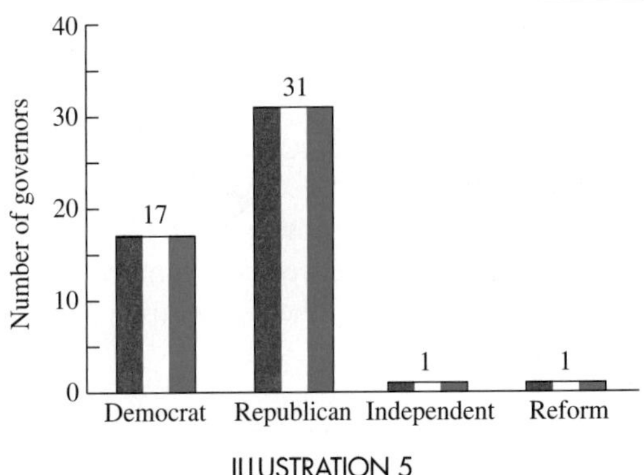

ILLUSTRATION 5

85. PERSONNEL RECORDS Complete the chart in Illustration 6 by finding the amount of the job that will be completed by each person working alone for the given number of hours.

Name	Total time to complete the job alone	Time worked alone	Amount of job completed
Bob	10 hours	7 hours	
Ali	8 hours	1 hour	

ILLUSTRATION 6

86. GAS TANK See Illustration 7. How full does the gauge indicate the gas tank is? How much of the tank has been used?

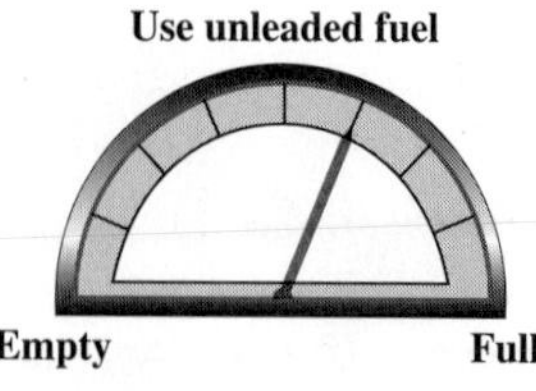

ILLUSTRATION 7

87. MUSIC Illustration 8 shows a side view of the finger position needed to produce a length of string (from the bridge to the fingertip) that gives low C on a violin. To play other notes, fractions of that length are used. Locate these finger positions.

a. $\frac{1}{2}$ of the length gives middle C.

b. $\frac{3}{4}$ of the length gives F above low C.

c. $\frac{2}{3}$ of the length gives G.

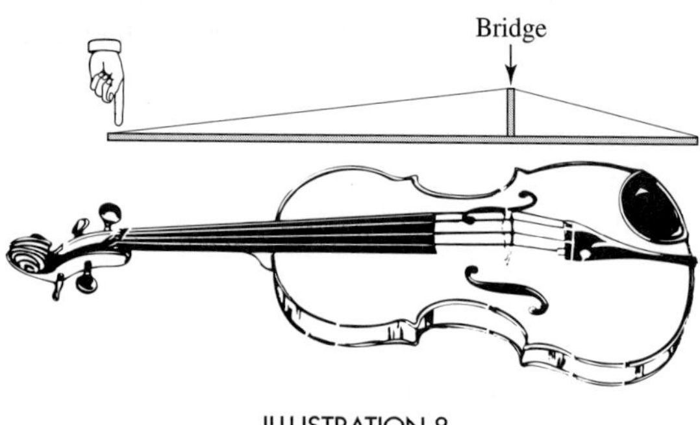

ILLUSTRATION 8

88. RULER Illustration 9 shows a ruler. First, tell how many spaces there are between the numbers 0 and 1. Then tell to what number the arrow is pointing.

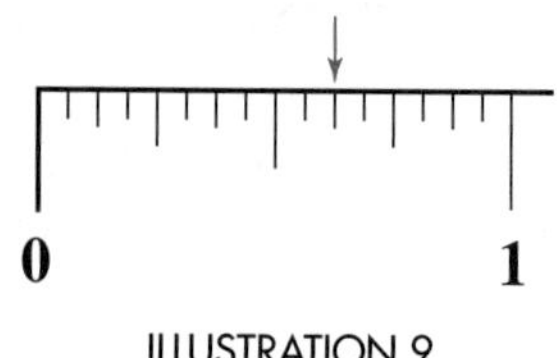

ILLUSTRATION 9

89. MACHINERY The operator of a machine is to turn the dial shown below from setting A to setting B. Express this in two different ways, using fractions of one complete revolution.

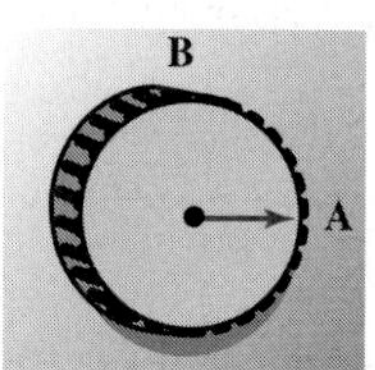

90. EARTH'S ROTATION The Earth rotates about its vertical axis once every 24 hours.

a. What is the significance of $\frac{1}{24}$ of a rotation to us on Earth?

b. What significance does $\frac{24}{24}$ of a revolution have?

91. SUPERMARKET DISPLAY The amount of space to be given each type of snack food in a supermarket display case is expressed as a fraction. Complete the model of the display, showing where the adjustable shelves should be located, and label where each snack food should be stocked.

$\frac{3}{8}$: potato chips

$\frac{2}{8}$: peanuts

$\frac{1}{8}$: pretzels

$\frac{2}{8}$: tortilla chips

92. MEDICAL CENTER Hospital designers have located a nurse's station at the center of a circular building. Show how to divide the surrounding office space so that each medical department has the proper fractional amount allocated to it. (Use the circle graph on the next page.) Label each department.

$\frac{2}{12}$: Radiology

$\frac{5}{12}$: Pediatrics

$\frac{1}{12}$: Laboratory

$\frac{3}{12}$: Orthopedics

$\frac{1}{12}$: Pharmacy

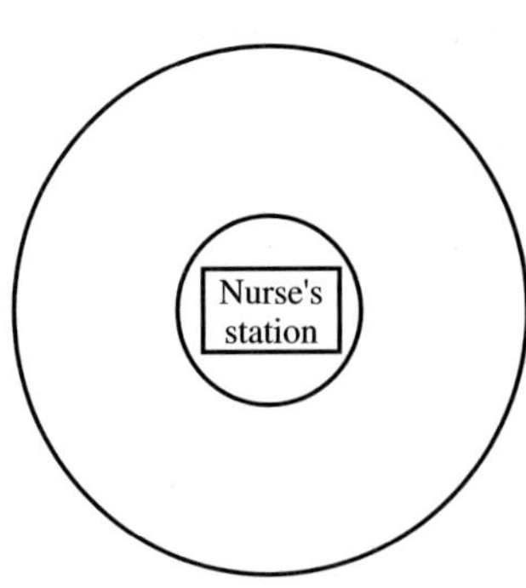

93. CAMERA When the shutter of a camera stays open longer than $\frac{1}{125}$ second, any movement of the camera will probably blur the picture. With this in mind, if a photographer is taking a picture of a fast-moving object, should she select a shutter speed of $\frac{1}{60}$ or $\frac{1}{250}$?

94. GDP The Gross Domestic Product is the official measure of the size of the U.S. economy. It represents the market value of all goods and services that have been bought during a given period of time. The GDP for the second quarter of 1999 is listed below. What is meant by the phrase *second quarter of 1999*?

Second quarter of 1999	\$8,893,300,000,000

WRITING

95. Explain the concept of equivalent fractions.

96. What does it mean for a fraction to be in lowest terms?

97. Explain the difference between three-fourths and three-fifths of a pizza.

98. Explain both parts of the fundamental property of fractions.

REVIEW

99. Solve $-5x + 1 = 16$.

100. Simplify: $4t - 8 + t - 9$.

101. Round 564,112 to the nearest thousand.

102. Give the definition of a prime number.

103. What is the value of d dimes?

104. A father is 24 years older than his son. Use a variable and an algebraic expression to represent their ages.

4.2 *Multiplying Fractions*

In this section, you will learn about

- Multiplying fractions • Simplifying when multiplying fractions
- Multiplying algebraic fractions • Powers of a fraction • Applications

INTRODUCTION. In the next three sections, we will discuss how to add, subtract, multiply, and divide fractions. We begin with the operation of multiplication.

Multiplying fractions

Suppose that a television network is going to take out a full-page ad to publicize its fall lineup of shows. The prime-time shows are to get $\frac{3}{5}$ of the ad space and daytime programming the remainder. Of the space devoted to prime time, $\frac{1}{2}$ is to be used to promote weekend programs. How much of the newspaper page will be used to advertise weekend prime-time programs?

The ad for the weekend prime-time shows will occupy $\frac{1}{2}$ of $\frac{3}{5}$ of the page. This can be expressed as $\frac{1}{2} \cdot \frac{3}{5}$. We can calculate $\frac{1}{2} \cdot \frac{3}{5}$ using a three-step process, illustrated below.

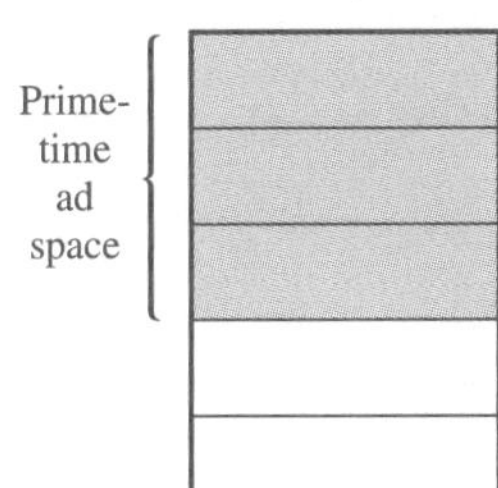

Step 1: We divide the page into fifths and shade three of them. This represents the fraction $\frac{3}{5}$, the amount of the page used to advertise prime-time shows.

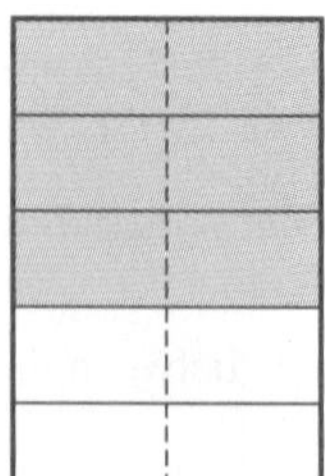

Step 2: Next, we find $\frac{1}{2}$ of the shaded part of the page by dividing the page into halves, using a vertical line.

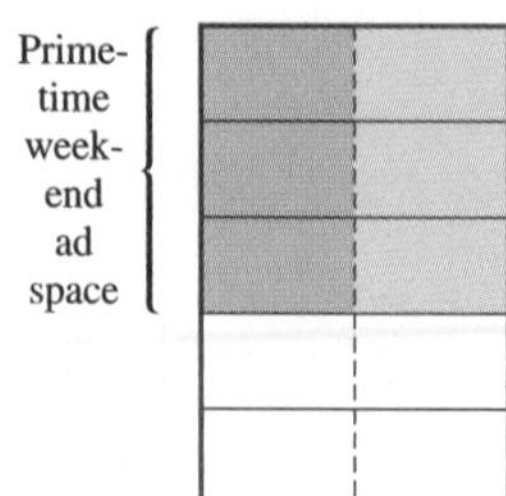

Step 3: Finally, we highlight (in purple) $\frac{1}{2}$ of the shaded parts determined in Step 2. The highlighted parts are 3 out of 10 or $\frac{3}{10}$ of the page. They represent the amount of the page used to advertise the weekend prime-time shows. This leads us to the conclusion that $\frac{1}{2} \cdot \frac{3}{5} = \frac{3}{10}$.

Two observations can be made from this result.

- The numerator of the answer is the product of the numerators of the original fractions.

$$1 \cdot 3 = 3$$

$$\frac{1}{2} \cdot \frac{3}{5} = \frac{3}{10} \quad \text{Answer}$$

$$2 \cdot 5 = 10$$

- The denominator of the answer is the product of the denominators of the original fractions.

These observations suggest the following rule for multiplying two fractions.

Multiplying fractions

> To multiply two fractions, multiply the numerators and multiply the denominators. In symbols, if a, b, c, and d represent numbers,
>
> $$\frac{a}{b} \cdot \frac{c}{d} = \frac{a \cdot c}{b \cdot d} \quad (b \neq 0, d \neq 0)$$

EXAMPLE 1 ***Multiplying fractions.*** Multiply: $\frac{7}{8} \cdot \frac{3}{5}$.

Solution

$\frac{7}{8} \cdot \frac{3}{5} = \frac{7 \cdot 3}{8 \cdot 5}$ Multiply the numerators and multiply the denominators.

$= \frac{21}{40}$ Do the multiplications: $7 \cdot 3 = 21$ and $8 \cdot 5 = 40$.

Self Check

Multiply: $\frac{5}{9} \cdot \frac{2}{3}$.

Answer: $\frac{10}{27}$

The rules for multiplying integers also hold for multiplying fractions. When we multiply two fractions with *like* signs, the product is positive. When we multiply two fractions with *unlike* signs, the product is negative.

EXAMPLE 2 ***Multiplication with a negative fraction.*** Multiply: $-\frac{3}{4}\left(\frac{1}{8}\right)$.

Solution

$$-\frac{3}{4}\left(\frac{1}{8}\right) = -\frac{3 \cdot 1}{4 \cdot 8}$$ Multiply the numerators and multiply the denominators. Since the fractions have unlike signs, the product is negative.

$$= -\frac{3}{32}$$ Do the multiplications: $3 \cdot 1 = 3$ and $4 \cdot 8 = 32$.

Self Check

Multiply: $\frac{5}{6}\left(-\frac{1}{3}\right)$.

Answer: $-\frac{5}{18}$

Simplifying when multiplying fractions

After multiplying two fractions, we should simplify the result, if possible.

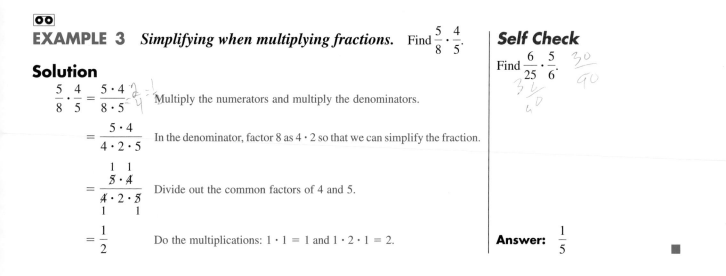

EXAMPLE 3 ***Simplifying when multiplying fractions.*** Find $\frac{5}{8} \cdot \frac{4}{5}$.

Solution

$$\frac{5}{8} \cdot \frac{4}{5} = \frac{5 \cdot 4}{8 \cdot 5}$$ Multiply the numerators and multiply the denominators.

$$= \frac{5 \cdot 4}{4 \cdot 2 \cdot 5}$$ In the denominator, factor 8 as $4 \cdot 2$ so that we can simplify the fraction.

$$= \frac{\overset{1}{\cancel{5}} \cdot \overset{1}{\cancel{4}}}{\underset{1}{\cancel{4}} \cdot 2 \cdot \underset{1}{\cancel{5}}}$$ Divide out the common factors of 4 and 5.

$$= \frac{1}{2}$$ Do the multiplications: $1 \cdot 1 = 1$ and $1 \cdot 2 \cdot 1 = 2$.

Self Check

Find $\frac{6}{25} \cdot \frac{5}{6}$.

Answer: $\frac{1}{5}$

EXAMPLE 4 ***Multiplying three fractions.*** Find $45\left(-\frac{1}{14}\right)\left(-\frac{7}{10}\right)$.

Solution

This expression involves three factors, and one of them is an integer. When multiplying fractions and integers, we express each integer as a fraction with a denominator of 1.

$$45\left(-\frac{1}{14}\right)\left(-\frac{7}{10}\right) = \frac{45}{1}\left(\frac{1}{14}\right)\left(\frac{7}{10}\right)$$ Write 45 as a fraction: $45 = \frac{45}{1}$. The product of two negative numbers is positive.

$$= \frac{45 \cdot 1 \cdot 7}{1 \cdot 14 \cdot 10}$$ Multiply the numerators and multiply the denominators.

$$= \frac{5 \cdot 3 \cdot 3 \cdot 1 \cdot 7}{1 \cdot 7 \cdot 2 \cdot 5 \cdot 2}$$ Factor 45 as $5 \cdot 3 \cdot 3$, factor 14 as $7 \cdot 2$, and factor 10 as $5 \cdot 2$.

$$= \frac{\overset{1}{\cancel{5}} \cdot 3 \cdot 3 \cdot 1 \cdot \overset{1}{\cancel{7}}}{1 \cdot \underset{1}{\cancel{7}} \cdot 2 \cdot \underset{1}{\cancel{5}} \cdot 2}$$ Divide out the common factors of 5 and 7.

$$= \frac{9}{4}$$ Multiply in the numerator and in the denominator.

COMMENT The answer in Example 4 was $\frac{9}{4}$, an improper fraction. In arithmetic, we often write such fractions as mixed numbers. In algebra, it is often more useful to leave $\frac{9}{4}$ in this form. We will discuss improper fractions and mixed numbers in more detail in Section 4.5.

Multiplying algebraic fractions

EXAMPLE 5 ***Multiplication involving a variable.*** Multiply: $\frac{1}{4}(4y)$.

Solution

$$\frac{1}{4}(4y) = \frac{1}{4} \cdot \frac{4y}{1} \quad \text{Write } 4y \text{ as a fraction: } 4y = \frac{4y}{1}.$$

$$= \frac{1 \cdot 4 \cdot y}{4 \cdot 1} \quad \text{Multiply the numerators and multiply the denominators.}$$

$$= \frac{1 \cdot \overset{1}{\cancel{4}} \cdot y}{\underset{1}{\cancel{4}} \cdot 1} \quad \text{Divide out the common factor of 4.}$$

$$= \frac{y}{1} \quad \text{Multiply in the numerator and in the denominator.}$$

$$= y \quad \text{Simplify: } \frac{y}{1} = y.$$

Self Check

Multiply: $\frac{1}{5} \cdot 5m$.

Answer: m ■

To multiply $\frac{1}{2}$ and x, we can express the product as $\frac{1}{2}x$, or we can use the concept of multiplying fractions to write it in a different form.

$$\frac{1}{2} \cdot x = \frac{1}{2} \cdot \frac{x}{1} \quad \text{Write } x \text{ as a fraction: } x = \frac{x}{1}.$$

$$= \frac{1 \cdot x}{2 \cdot 1} \quad \text{Multiply the numerators and multiply the denominators.}$$

$$= \frac{x}{2} \quad \text{Multiply in the numerator and in the denominator.}$$

The product of $\frac{1}{2}$ and x can be expressed as $\frac{1}{2}x$ or $\frac{x}{2}$. Similarly, $\frac{3}{4}t = \frac{3t}{4}$ and $-\frac{5}{16}y = -\frac{5y}{16}$.

EXAMPLE 6 ***Multiplying algebraic fractions.*** Multiply: $\frac{5b}{2} \cdot \frac{4}{7b}$.

Solution

$$\frac{5b}{2} \cdot \frac{4}{7b} = \frac{5b \cdot 4}{2 \cdot 7b} \quad \text{Multiply the numerators and multiply the denominators.}$$

$$= \frac{5 \cdot b \cdot 2 \cdot 2}{2 \cdot 7 \cdot b} \quad \text{Factor 4: } 4 = 2 \cdot 2.$$

$$= \frac{5 \cdot \overset{1}{\cancel{b}} \cdot \overset{1}{\cancel{2}} \cdot 2}{\underset{1}{\cancel{2}} \cdot 7 \cdot \underset{1}{\cancel{b}}} \quad \text{Divide out the common factors of } b \text{ and 2.}$$

$$= \frac{10}{7} \quad \text{Multiply in the numerator and denominator.}$$

Self Check

Multiply: $\frac{5}{12y} \cdot \frac{3y}{8}$.

Answer: $\frac{5}{32}$ ■

EXAMPLE 7 *Multiplying algebraic fractions.* Multiply: $\frac{t^2}{21r} \cdot \frac{14r}{t^4}$.

Solution

$$\frac{t^2}{21r} \cdot \frac{14r}{t^4} = \frac{t^2 \cdot 14r}{21r \cdot t^4}$$ Multiply the numerators and multiply the denominators.

$$= \frac{t \cdot t \cdot 7 \cdot 2 \cdot r}{7 \cdot 3 \cdot r \cdot t \cdot t \cdot t \cdot t}$$ Factor t^2, 14, 21, and t^4.

$$= \frac{\overset{1}{\cancel{t}} \cdot \overset{1}{\cancel{t}} \cdot \overset{1}{\cancel{7}} \cdot 2 \cdot \overset{1}{\cancel{r}}}{\underset{1}{\cancel{7}} \cdot 3 \cdot \underset{1}{\cancel{r}} \cdot \underset{1}{\cancel{t}} \cdot \underset{1}{\cancel{t}} \cdot t \cdot t}$$ Divide out the common factors of t, r, and 7.

$$= \frac{2}{3t^2}$$ Multiply in the numerator and denominator.

Self Check

Multiply: $\frac{2a^3}{5b} \cdot \frac{10ab}{8a^4}$.

Answer: $\frac{1}{2}$

Powers of a fraction

If the base of an exponential expression is a fraction, the exponent tells us how many times to write that fraction as a factor. For example,

$$\left(\frac{2}{3}\right)^2 = \frac{2}{3} \cdot \frac{2}{3} = \frac{2 \cdot 2}{3 \cdot 3} = \frac{4}{9}$$ $\frac{2}{3}$ is used as a factor 2 times.

EXAMPLE 8 *Power of a fraction.* Find $\left(-\frac{4x}{5}\right)^2$.

Solution

Exponents are used to indicate repeated multiplication.

$$\left(-\frac{4x}{5}\right)^2 = \left(-\frac{4x}{5}\right)\left(-\frac{4x}{5}\right)$$ Write $-\frac{4x}{5}$ as a factor 2 times.

$$= \frac{4x \cdot 4x}{5 \cdot 5}$$ The product of two fractions with like signs is positive. Multiply the numerators and multiply the denominators.

$$= \frac{16x^2}{25}$$ Multiply in the numerator and denominator.

Self Check

Find $\left(-\frac{3t}{4}\right)^3$.

$-\frac{27t^3}{64}$

Answer: $-\frac{27t^3}{64}$

Applications

EXAMPLE 9 *House of Representatives.* In the United States House of Representatives, a bill was introduced that would require a $\frac{3}{5}$ vote of the 435 members to authorize any tax increase. Under this requirement, how many representatives would have to vote for a tax increase before it could become law?

Solution

$$\frac{3}{5} \text{ of } 435 = \frac{3}{5} \cdot \frac{435}{1}$$ Here, the word *of* means to multiply. Write 435 as a fraction: $435 = \frac{435}{1}$.

$$= \frac{3 \cdot 435}{5 \cdot 1}$$ Multiply the numerators and multiply the denominators.

$$= \frac{3 \cdot 3 \cdot 5 \cdot 29}{5 \cdot 1}$$ Prime factor 435 as $3 \cdot 5 \cdot 29$.

$$= \frac{3 \cdot 3 \cdot \overset{1}{\cancel{5}} \cdot 29}{\underset{1}{\cancel{5}} \cdot 1}$$ Divide out the common factor of 5.

$$= \frac{261}{1}$$ Multiply in the numerator: $3 \cdot 3 \cdot 1 \cdot 29 = 261$. Multiply in the denominator: $1 \cdot 1 = 1$.

$$= 261$$ Simplify: $\frac{261}{1} = 261$.

It would take 261 representatives voting in favor to pass a tax increase. ■

As Figure 4-3 shows, a triangle has three sides. The length of the base of the triangle can be represented by the letter b and the height by the letter h. The height of a triangle is always perpendicular (makes a square corner) to the base. This is denoted by the symbol ⌉.

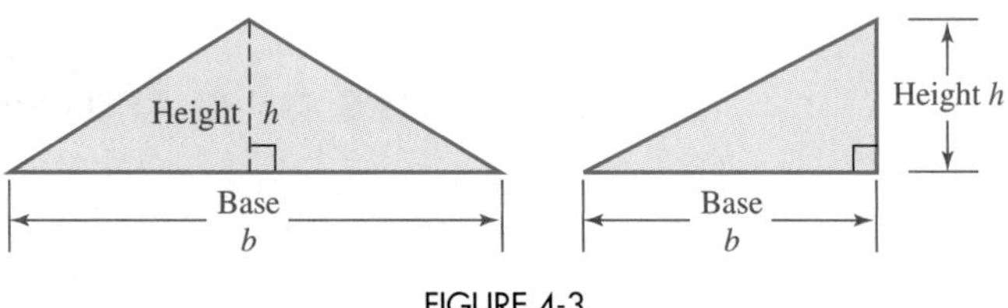

FIGURE 4-3

Recall that the area of a figure is the amount of surface that it encloses. The area of a triangle can be found by using the following formula.

Area of a triangle

The area A of a triangle is one-half the product of its base b and its height h:

$$\text{Area} = \frac{1}{2}(\text{base})(\text{height}) \quad \text{or} \quad A = \frac{1}{2}bh$$

EXAMPLE 10 ***Geography.*** Approximate the area of the state of Virginia using the triangle in Figure 4-4.

Solution We will approximate the area of the state by finding the area of the triangle.

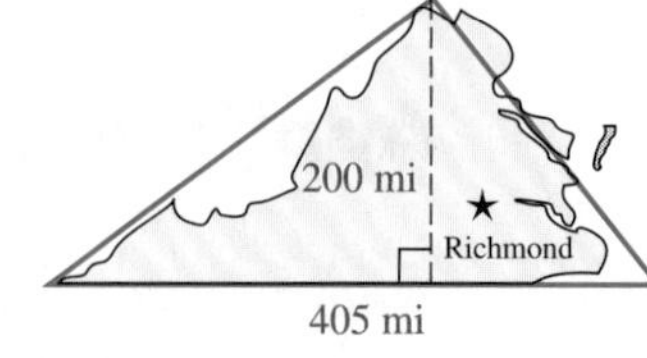

FIGURE 4-4

$$A = \frac{1}{2}bh$$ The formula for the area of a triangle.

$$= \frac{1}{2}(405)(200)$$ Substitute 405 for b and 200 for h.

$$= \frac{1}{2}\left(\frac{405}{1}\right)\left(\frac{200}{1}\right)$$ Write 405 and 200 as fractions.

$$= \frac{1 \cdot 405 \cdot 200}{2}$$ Multiply the numerators. Multiply the denominators.

$$= \frac{1 \cdot 405 \cdot 100 \cdot \overset{1}{\not{2}}}{\underset{1}{\not{2}}}$$ Factor 200 as 100 · 2. Then divide out the common factor of 2.

$$= 40{,}500$$ $405 \cdot 100 = 40{,}500$.

The area of the state of Virginia is approximately 40,500 square miles. ■

STUDY SET Section 4.2

VOCABULARY *Fill in the blanks to make the statements true.*

1. The word *of* in mathematics usually means ______________.

2. The ______________ of a triangle is the amount of surface that it encloses.

3. The result of a multiplication problem is called the ______________.

4. To ______________ a fraction means to divide out common factors of the numerator and denominator.

5. In a triangle, b stands for the length of the ______________ and h stands for the ______________.

6. A ______________ is an equation that mathematically describes a known relationship between two or more variables.

CONCEPTS

7. Find the result when multiplying $\frac{a}{b} \cdot \frac{c}{d}$.

8. Write each of the following as fractions.
a. 4 **b.** −3 **c.** x

9. Use the following rectangle to find $\frac{1}{3} \cdot \frac{1}{4}$.

a. Using vertical lines, divide the given rectangle into four equal parts and lightly shade one of them. What fractional part of the rectangle did you shade?
b. To find $\frac{1}{3}$ of the shaded portion, use two horizontal lines to divide the given rectangle into three equal parts and lightly shade one of them. Into how many equal parts is the rectangle now divided? How many parts have been shaded twice? What is $\frac{1}{3} \cdot \frac{1}{4}$?

10. In the solution at the top of the next column, what initial mistake did the student make that caused him to have to work with such large numbers?

$$\frac{44}{63} \cdot \frac{27}{55} = \frac{44 \cdot 27}{63 \cdot 55}$$
$$= \frac{1{,}188}{3{,}465}$$

11. a. Is the product of two numbers with unlike signs positive or negative?
b. Is the product of two numbers with like signs positive or negative?

12. a. Multiply $\frac{9}{10}$ and 20.
b. When we multiply two numbers, is the product always larger than both those numbers?

13. Tell whether each statement is true or false.
a. $\frac{1}{2}x = \frac{x}{2}$ T **b.** $\frac{2t}{3} = \frac{2}{3}t$
c. $-\frac{3}{8}a = -\frac{3}{8a}$ **d.** $\frac{-4e}{7} = -\frac{4e}{7}$

14. What is the numerator of the result for the multiplication problem shown here?

$$\frac{4}{15} \cdot \frac{3}{4} = \frac{\overset{1}{\not{4}} \cdot \overset{1}{\not{3}}}{5 \cdot \underset{1}{\not{3}} \cdot \underset{1}{\not{4}}}$$

NOTATION *Complete each solution.*

15. Multiply: $\frac{5}{8} \cdot \frac{7}{15}$.

$$\frac{5}{8} \cdot \frac{7}{15} = \frac{5 \cdot 7}{8 \cdot 15}$$
$$= \frac{5 \cdot 7}{8 \cdot 5 \cdot 3}$$
$$= \frac{\overset{1}{\not{5}} \cdot 7}{8 \cdot \underset{1}{\not{5}} \cdot 3}$$
$$= \frac{7}{24}$$

16. Multiply: $\frac{7}{12} \cdot \frac{4}{21}$.

$$\frac{7}{12} \cdot \frac{4}{21} = \frac{7 \cdot 4}{12 \cdot 21}$$

$$= \frac{7 \cdot 4}{4 \cdot 3 \cdot 7 \cdot 3}$$

$$= \frac{\overset{1}{\cancel{7}} \cdot \overset{1}{\cancel{4}}}{\underset{1}{\cancel{4}} \cdot 3 \cdot \underset{1}{\cancel{7}} \cdot 3}$$

$$= \frac{1}{9}$$

49. $-\frac{5}{6} \cdot \frac{6}{5}c$

50. $-\frac{5}{7} \cdot \frac{7}{5}w$

51. $\frac{x}{2} \cdot \frac{4}{9x}$

52. $\frac{c}{5} \cdot \frac{25}{36c}$

53. $4e \cdot \frac{e}{2}$

54. $8x \cdot \frac{2x}{16}$

55. $\frac{5}{8x}\left(\frac{2x^3}{15}\right)$

56. $\frac{3m^3}{16}\left(\frac{4m}{9m^3}\right)$

57. $-\frac{5c}{6cd^2} \cdot \frac{12d^4}{c}$

58. $-\frac{3ef^3}{5b} \cdot \frac{10b}{e^2f}$

59. $-\frac{4h^2}{5}\left(-\frac{15}{16h^3}\right)$

60. $-\frac{2}{21j^2}\left(-\frac{15j}{8}\right)$

PRACTICE *Multiply. Write all answers in lowest terms.*

17. $\frac{1}{4} \cdot \frac{1}{2}$

18. $\frac{1}{3} \cdot \frac{1}{5}$

19. $\frac{3}{8} \cdot \frac{7}{16}$

20. $\frac{5}{9} \cdot \frac{2}{7}$

21. $\frac{2}{3} \cdot \frac{6}{7}$

22. $\frac{5}{12} \cdot \frac{3}{4}$

23. $\frac{14}{15} \cdot \frac{11}{8}$

24. $\frac{5}{16} \cdot \frac{8}{3}$

25. $-\frac{15}{24} \cdot \frac{8}{25}$

26. $-\frac{20}{21} \cdot \frac{7}{16}$

27. $\left(-\frac{11}{21}\right)\left(-\frac{14}{33}\right)$

28. $\left(-\frac{16}{35}\right)\left(-\frac{25}{48}\right)$

29. $\frac{7}{10}\left(\frac{20}{21}\right)$

30. $\left(\frac{7}{6}\right)\frac{9}{49}$

31. $\frac{3}{4} \cdot \frac{4}{3}$

32. $\frac{4}{5} \cdot \frac{5}{4}$

33. $\frac{1}{3} \cdot \frac{15}{16} \cdot \frac{4}{25}$

34. $\frac{3}{15} \cdot \frac{15}{7} \cdot \frac{14}{27}$

35. $\left(\frac{2}{3}\right)\left(-\frac{1}{16}\right)\left(-\frac{4}{5}\right)$

36. $\left(\frac{3}{8}\right)\left(-\frac{2}{3}\right)\left(-\frac{12}{27}\right)$

37. $\frac{5}{6} \cdot 18$

38. $6\left(-\frac{2}{3}\right)$

39. $15\left(-\frac{4}{5}\right)$

40. $-2\left(-\frac{7}{8}\right)$

41. $\frac{5x}{12} \cdot \frac{1}{6x}$

42. $\frac{2t}{3} \cdot \frac{7}{8t}$

43. $\frac{b}{12} \cdot \frac{3}{10b}$

44. $\frac{5c}{8} \cdot \frac{1}{15c}$

45. $\frac{1}{3} \cdot 3d$

46. $\frac{1}{16} \cdot 16x$

47. $\frac{2}{3} \cdot \frac{3s}{2}$

48. $\frac{3}{5} \cdot \frac{5h}{3}$

Multiply and express the product in two ways.

61. $\frac{5}{6} \cdot x$

62. $\frac{2}{3} \cdot y$

63. $-\frac{8}{9} \cdot v$

64. $-\frac{7}{6} \cdot m$

Find each power.

65. $\left(\frac{2}{3}\right)^2$

66. $\left(\frac{3}{5}\right)^2$

67. $\left(-\frac{5}{9}\right)^2$

68. $\left(-\frac{5}{6}\right)^2$

69. $\left(\frac{4m}{3}\right)^2$

70. $\left(\frac{3t}{2}\right)^2$

71. $\left(-\frac{3r}{4}\right)^3$

72. $\left(-\frac{2b}{5}\right)^3$

73. Complete the multiplication table of fractions in Illustration 1.

$\cdot$	$\frac{1}{2}$	$\frac{1}{3}$	$\frac{1}{4}$	$\frac{1}{5}$	$\frac{1}{6}$
$\frac{1}{2}$					
$\frac{1}{3}$					
$\frac{1}{4}$					
$\frac{1}{5}$					
$\frac{1}{6}$					

ILLUSTRATION 1

74. Complete Illustration 2 by finding the original fraction, given its square.

Original fraction squared	Original fraction
$\frac{1}{9}$	
$\frac{1}{100}$	
$\frac{4}{25}$	
$\frac{16}{49}$	
$\frac{81}{36}$	
$\frac{9}{121}$	

ILLUSTRATION 2

Find the area of each triangle.

75.

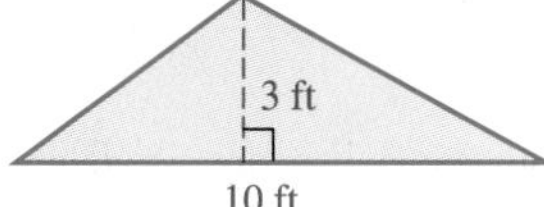

76.

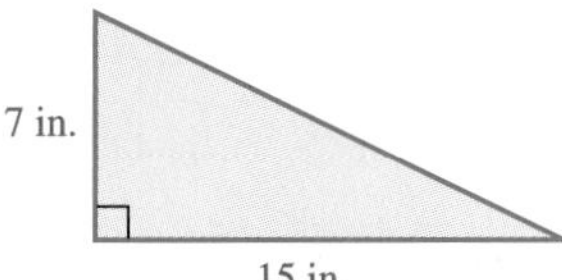

77.

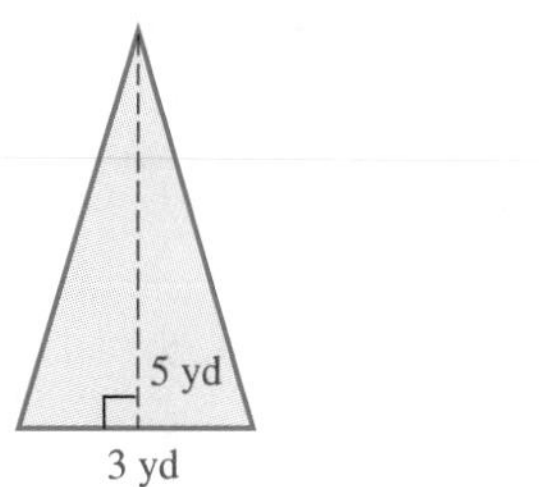

78.

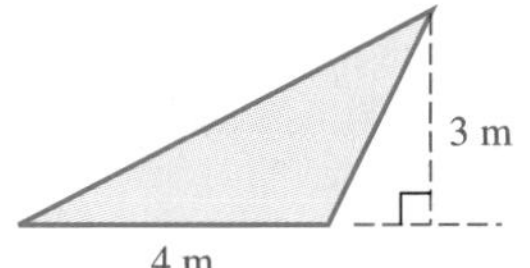

APPLICATIONS

79. THE CONSTITUTION Article V of the United States Constitution requires a two-thirds vote of the House of Representatives to propose a constitutional amendment. The House has 435 members. Find the number of votes needed to meet this requirement.

80. GENETICS Gregor Mendel (1822–1884), an Augustinian monk, is credited with developing a heredity model that became the foundation of modern genetics. In his experiments, he crossed purple-flowered plants with white-flowered plants and found that $\frac{3}{4}$ of the offspring plants had purple flowers and $\frac{1}{4}$ had white flowers. According to this concept, when the group of offspring plants shown in Illustration 3 flower, how many will have purple flowers?

ILLUSTRATION 3

81. TENNIS BALL A tennis ball is dropped from a height of 54 inches. Each time it hits the ground, it rebounds one-third of the previous height it fell. See Illustration 4 and find the three missing rebound heights.

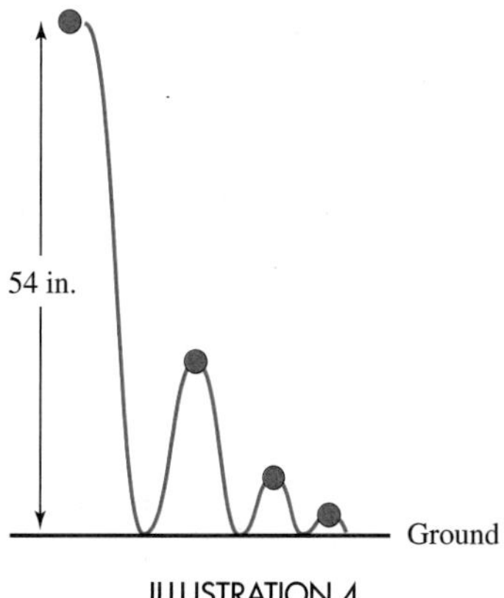

ILLUSTRATION 4

82. ELECTION Illustration 5 shows the final election returns for a city bond measure.

a. How many votes were cast?

b. Find two-thirds of the number of votes cast.

c. Did the bond measure pass?

Measure 1
100% of the precincts reporting

Fire–Police–Paramedics General Obligation Bonds
(Requires two-thirds vote)

Yes 125,599 No 62,801

ILLUSTRATION 5

83. COOKING Use the recipe below, along with the concept of multiplication with fractions, to find how much sugar and molasses are needed to make one dozen cookies.

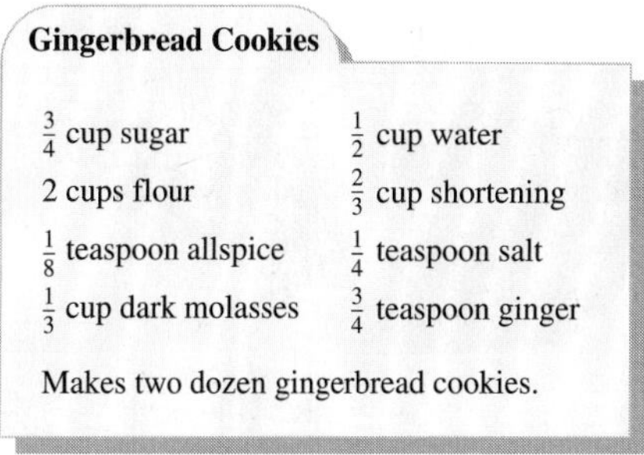

Gingerbread Cookies

$\frac{3}{4}$ cup sugar	$\frac{1}{2}$ cup water
2 cups flour	$\frac{2}{3}$ cup shortening
$\frac{1}{8}$ teaspoon allspice	$\frac{1}{4}$ teaspoon salt
$\frac{1}{3}$ cup dark molasses	$\frac{3}{4}$ teaspoon ginger

Makes two dozen gingerbread cookies.

84. THE EARTH'S SURFACE The surface of the Earth covers an area of approximately 196,800,000 square miles. About $\frac{3}{4}$ of that area is covered by water. Find the number of square miles of the surface covered by water.

85. BOTANY In an experiment, monthly growth rates of three types of plants doubled when nitrogen was added to the soil. Complete Illustration 6 by charting the improved growth rate next to each normal growth rate.

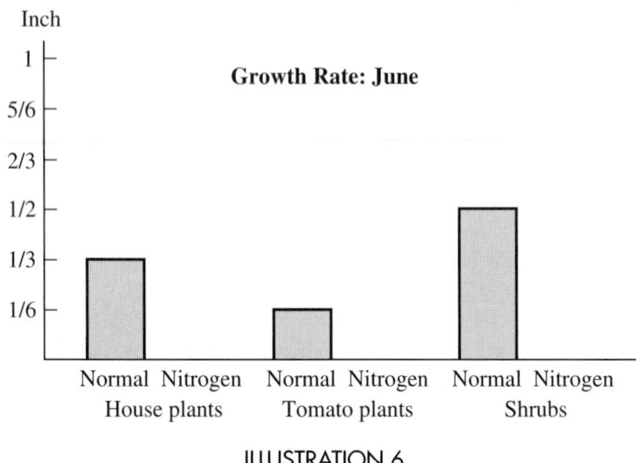

ILLUSTRATION 6

86. STAMPS The best designs in a contest to create a wildlife stamp are shown in Illustration 7. To save on paper costs, the postal service has decided to choose the stamp that has the smaller area. Which one is that?

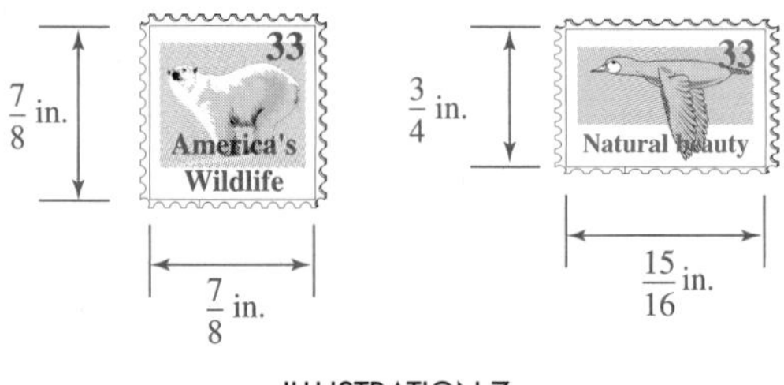

ILLUSTRATION 7

87. THE STARS AND STRIPES Illustration 8 shows a folded U.S. flag. When it is placed on a table as part of an exhibit, how much area will it occupy?

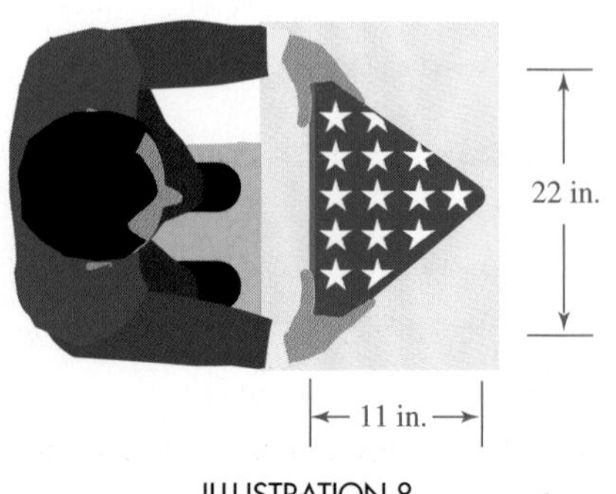

ILLUSTRATION 8

88. WINDSURFING Estimate the area of the sail on the windsurfing board in Illustration 9.

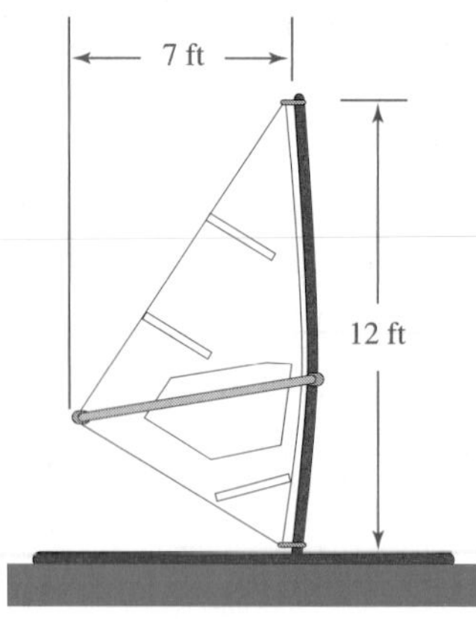

ILLUSTRATION 9

89. TILE DESIGN A design for bathroom tile is shown in Illustration 10. Find the amount of area on a tile that is blue.

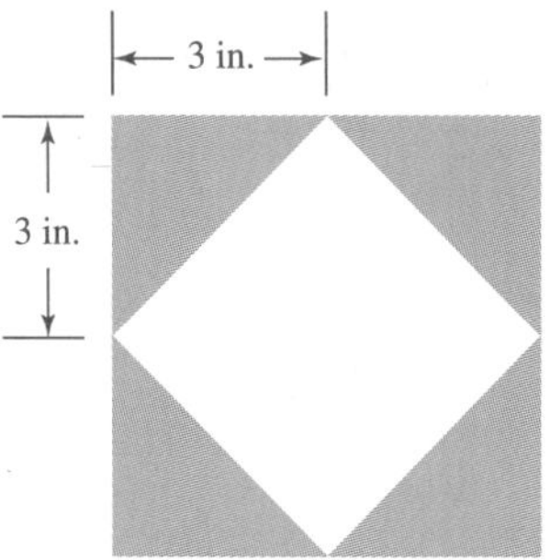

ILLUSTRATION 10

90. GEOGRAPHY Estimate the area of the state of New Hampshire, using the triangle in Illustration 11.

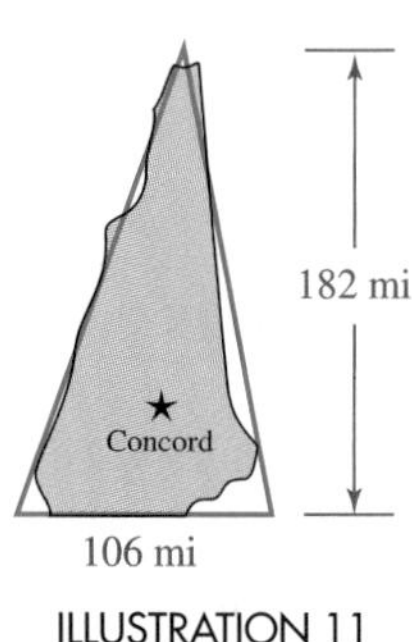

ILLUSTRATION 11

WRITING

91. In mathematics, the word *of* usually means multiply. Give three real-life examples of this usage.

92. Explain how you could multiply the number 5 and another number and obtain an answer that is less than 5.

93. A MAJORITY The definition of the word *majority* is as follows: "a number greater than one-half of the total." Explain what it means when a teacher says, "A majority of the class voted to postpone the test until Monday." Give an example.

94. What does area measure?

REVIEW

95. Apply the distributive property and simplify: $2(x + 7)$.

96. Find the distance covered by a motorist traveling at 60 miles per hour for 3 hours.

97. Is $x = -6$ a solution of $2x + 6 = 6$?

98. Combine like terms: $3x - 8 - 5x + 9$.

99. Find the prime factorization of 125.

100. In each case, tell whether x is used as a factor or as a term.

a. $6x$ **b.** $6y + x$
c. $x + 1$ **d.** $6xy$

4.3 Dividing Fractions

In this section, you will learn about

- Division with fractions • Reciprocals • A rule for dividing fractions
- Dividing algebraic fractions

Introduction. In this section, we will discuss how to divide fractions. We will examine problems involving positive and negative fractions as well as algebraic fractions. The skills you learned in Section 4.2 will be useful in this section.

Division with fractions

Suppose that the manager of a candy store buys large bars of chocolate and divides each one into four equal parts to sell. How many fourths can be obtained from 5 bars?

We are asking, "How many $\frac{1}{4}$'s are there in 5?" To answer the question, we need to use the operation of division. We can represent this division as $5 \div \frac{1}{4}$. See Figure 4-5.

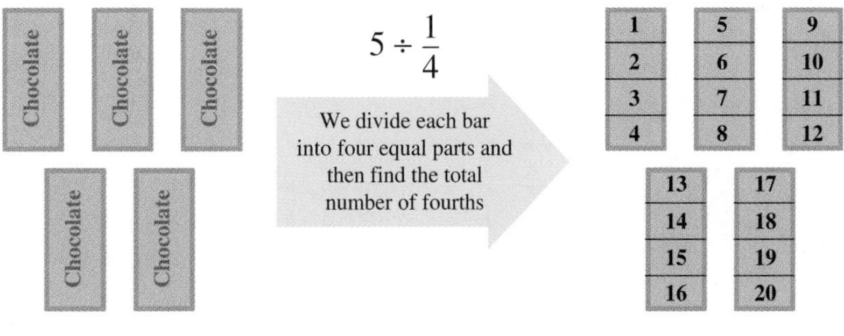

FIGURE 4-5

There are 20 fourths in the 5 bars of chocolate. This result leads to the following observations.

- This division problem involves a fraction: $5 \div \frac{1}{4}$.
- Although we were asked to find $5 \div \frac{1}{4}$, we solved the problem using *multiplication* instead of *division*: $5 \cdot 4 = 20$.

Later in this section, we will see that these observations suggest a rule for dividing fractions. But before we can discuss that rule, we need to introduce a new term.

Reciprocals

Division with fractions involves working with **reciprocals.** To present the concept of reciprocal, we consider the problem $\frac{7}{8} \cdot \frac{8}{7}$.

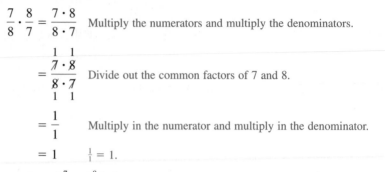

$$\frac{7}{8} \cdot \frac{8}{7} = \frac{7 \cdot 8}{8 \cdot 7}$$ Multiply the numerators and multiply the denominators.

$$= \frac{\overset{1}{\cancel{7}} \cdot \overset{1}{\cancel{8}}}{\underset{1}{\cancel{8}} \cdot \underset{1}{\cancel{7}}}$$ Divide out the common factors of 7 and 8.

$$= \frac{1}{1}$$ Multiply in the numerator and multiply in the denominator.

$$= 1$$ $\frac{1}{1} = 1$.

The product of $\frac{7}{8}$ and $\frac{8}{7}$ is 1.

Whenever the product of two numbers is 1, we say that those numbers are *reciprocals.* Therefore, $\frac{7}{8}$ and $\frac{8}{7}$ are reciprocals. To find the reciprocal of a fraction, *we invert the numerator and the denominator.*

Reciprocals

Two numbers are called **reciprocals** if their product is 1.

COMMENT Zero does not have a reciprocal, because the product of 0 and a number can never be 1.

EXAMPLE 1 ***Reciprocals.*** For each number, find its reciprocal and show that their product is 1: **a.** $\frac{2}{3}$, **b.** $-\frac{3}{4}$, and **c.** 5.

Solution

a. The reciprocal of $\frac{2}{3}$ is $\frac{3}{2}$.

$$\frac{2}{3} \cdot \frac{3}{2} = \frac{\overset{1}{\cancel{2}} \cdot \overset{1}{\cancel{3}}}{\underset{1}{\cancel{3}} \cdot \underset{1}{\cancel{2}}} = 1$$

b. The reciprocal of $-\frac{3}{4}$ is $-\frac{4}{3}$.

$$-\frac{3}{4}\left(-\frac{4}{3}\right) = \frac{\overset{1}{\cancel{3}} \cdot \overset{1}{\cancel{4}}}{\underset{1}{\cancel{4}} \cdot \underset{1}{\cancel{3}}} = 1$$ The product of two fractions with like signs is positive.

c. $5 = \frac{5}{1}$, so the reciprocal of 5 is $\frac{1}{5}$.

$$5 \cdot \frac{1}{5} = \frac{5}{1} \cdot \frac{1}{5} = \frac{\overset{1}{\cancel{5}} \cdot 1}{1 \cdot \underset{1}{\cancel{5}}} = 1$$

Self Check

For each number, find its reciprocal and show that their product is 1.

a. $\frac{3}{5}$

b. $-\frac{5}{6}$

c. 8

Answers: **a.** $\frac{5}{3}$, **b.** $-\frac{6}{5}$, **c.** $\frac{1}{8}$

A rule for dividing fractions

In the candy store example, we saw that we can find $5 \div \frac{1}{4}$ by computing $5 \cdot 4$. That is, division by $\frac{1}{4}$ (a fraction) is the same as multiplication by 4 (its reciprocal).

$5 \div \frac{1}{4} = 5 \cdot 4$

This observation suggests a general rule for dividing fractions.

Dividing fractions

> To divide fractions, multiply the first fraction by the reciprocal of the second fraction. In symbols, if a, b, c, and d represent numbers, then
>
> $$\frac{a}{b} \div \frac{c}{d} = \frac{a}{b} \cdot \frac{d}{c} \quad (b \neq 0, c \neq 0, d \neq 0)$$

For example, to find $\frac{5}{7} \div \frac{3}{4}$, we multiply the first fraction by the reciprocal of the second.

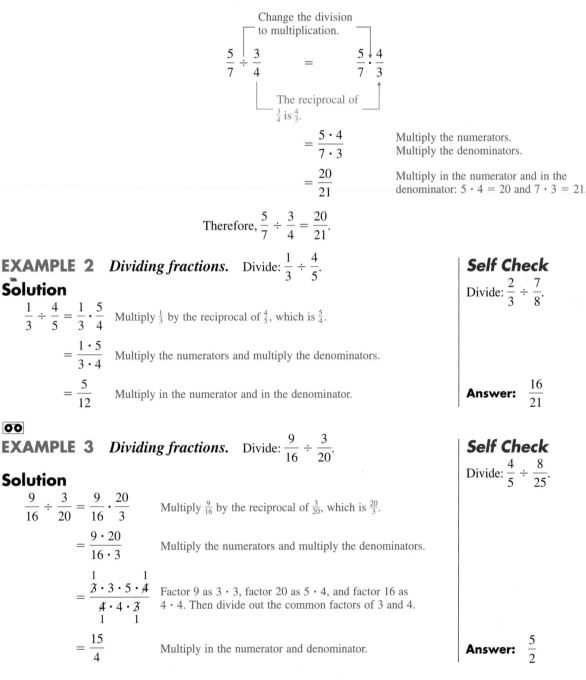

Change the division to multiplication.

$$\frac{5}{7} \div \frac{3}{4} = \frac{5}{7} \cdot \frac{4}{3}$$

The reciprocal of $\frac{3}{4}$ is $\frac{4}{3}$.

$= \dfrac{5 \cdot 4}{7 \cdot 3}$ Multiply the numerators. Multiply the denominators.

$= \dfrac{20}{21}$ Multiply in the numerator and in the denominator: $5 \cdot 4 = 20$ and $7 \cdot 3 = 21$.

Therefore, $\frac{5}{7} \div \frac{3}{4} = \frac{20}{21}$.

EXAMPLE 2 ***Dividing fractions.*** Divide: $\frac{1}{3} \div \frac{4}{5}$.

Solution

$\dfrac{1}{3} \div \dfrac{4}{5} = \dfrac{1}{3} \cdot \dfrac{5}{4}$ Multiply $\frac{1}{3}$ by the reciprocal of $\frac{4}{5}$, which is $\frac{5}{4}$.

$= \dfrac{1 \cdot 5}{3 \cdot 4}$ Multiply the numerators and multiply the denominators.

$= \dfrac{5}{12}$ Multiply in the numerator and in the denominator.

Self Check

Divide: $\frac{2}{3} \div \frac{7}{8}$.

Answer: $\frac{16}{21}$ ■

EXAMPLE 3 ***Dividing fractions.*** Divide: $\frac{9}{16} \div \frac{3}{20}$.

Solution

$\dfrac{9}{16} \div \dfrac{3}{20} = \dfrac{9}{16} \cdot \dfrac{20}{3}$ Multiply $\frac{9}{16}$ by the reciprocal of $\frac{3}{20}$, which is $\frac{20}{3}$.

$= \dfrac{9 \cdot 20}{16 \cdot 3}$ Multiply the numerators and multiply the denominators.

$= \dfrac{\overset{1}{\cancel{3}} \cdot 3 \cdot 5 \cdot \overset{1}{\cancel{4}}}{\underset{1}{\cancel{4}} \cdot 4 \cdot \underset{1}{\cancel{3}}}$ Factor 9 as $3 \cdot 3$, factor 20 as $5 \cdot 4$, and factor 16 as $4 \cdot 4$. Then divide out the common factors of 3 and 4.

$= \dfrac{15}{4}$ Multiply in the numerator and denominator.

Self Check

Divide: $\frac{4}{5} \div \frac{8}{25}$.

Answer: $\frac{5}{2}$ ■

EXAMPLE 4 ***Surfboard design.*** Most surfboards are made of polyurethane foam plastic covered with several layers of fiberglass to keep them water-tight. How many layers are needed to build up a finish three-eighths of an inch thick if each layer of fiberglass has a thickness of one-sixteenth of an inch?

Solution We need to know how many one-sixteenths there are in three-eighths. To answer this question, we will use division and find $\frac{3}{8} \div \frac{1}{16}$.

$\frac{3}{8} \div \frac{1}{16} = \frac{3}{8} \cdot \frac{16}{1}$ Multiply $\frac{3}{8}$ by the reciprocal of $\frac{1}{16}$, which is $\frac{16}{1}$.

$= \frac{3 \cdot 16}{8 \cdot 1}$ Multiply the numerators and multiply the denominators.

$= \frac{3 \cdot \overset{1}{\cancel{8}} \cdot 2}{\underset{1}{\cancel{8}} \cdot 1}$ Factor 16 as $8 \cdot 2$. Then divide out the common factor of 8.

$= \frac{6}{1}$ Multiply in the numerator and denominator.

$= 6$ Simplify: $\frac{6}{1} = 6$.

The number of layers of fiberglass to be applied is 6. ■

EXAMPLE 5 ***Division with a negative fraction.*** Divide: $\frac{1}{6} \div \left(-\frac{1}{18}\right)$.

Solution

When working with divisions involving negative fractions, we use the same rules as for multiplying numbers with like or unlike signs.

$\frac{1}{6} \div \left(-\frac{1}{18}\right) = \frac{1}{6}\left(-\frac{18}{1}\right)$ Multiply $\frac{1}{6}$ by the reciprocal of $-\frac{1}{18}$, which is $-\frac{18}{1}$.

$= -\frac{1 \cdot 18}{6 \cdot 1}$ The product of two fractions with unlike signs is negative. Multiply the numerators and multiply the denominators.

$= -\frac{1 \cdot \overset{1}{\cancel{6}} \cdot 3}{\underset{1}{\cancel{6}} \cdot 1}$ Factor 18 as $6 \cdot 3$. Then divide out the common factor of 6.

$= -\frac{3}{1}$ Multiply in the numerator and denominator.

$= -3$ Simplify: $\frac{3}{1} = 3$.

Self Check

Divide: $\frac{2}{3} \div \left(-\frac{7}{6}\right)$.

Answer: $-\frac{4}{7}$ ■

EXAMPLE 6 ***Division with a fraction and an integer.*** Divide: $-\frac{21}{36} \div (-3)$.

Solution

$-\frac{21}{36} \div (-3) = -\frac{21}{36}\left(-\frac{1}{3}\right)$ Multiply $-\frac{21}{36}$ by the reciprocal of -3, which is $-\frac{1}{3}$.

$= \frac{21 \cdot 1}{36 \cdot 3}$ The product of two fractions with like signs is positive. Multiply the numerators and multiply the denominators.

$= \frac{7 \cdot 3 \cdot 1}{36 \cdot 3}$ Factor 21 as $7 \cdot 3$.

Self Check

Divide: $-\frac{24}{25} \div (-8)$.

$= \dfrac{7 \cdot \overset{1}{\cancel{3}} \cdot 1}{36 \cdot \underset{1}{\cancel{3}}}$ Divide out the common factor of 3.

$= \dfrac{7}{36}$ Multiply in the numerator and in the denominator.

Answer: $\dfrac{3}{25}$ ■

Dividing algebraic fractions

To work problems involving division of algebraic fractions, we must find the reciprocal of an algebraic fraction. We learned earlier that the reciprocal of a numerical fraction is found by inverting its numerator and denominator. The same is true for algebraic fractions. For example, the reciprocal of $\frac{c}{a}$ is $\frac{a}{c}$. Since $x = \frac{x}{1}$, the reciprocal of x is $\frac{1}{x}$.

EXAMPLE 7 ***Dividing algebraic fractions.*** Divide: $\dfrac{2}{3} \div \dfrac{a}{5}$.

Solution

$\dfrac{2}{3} \div \dfrac{a}{5} = \dfrac{2}{3} \cdot \dfrac{5}{a}$ Multiply $\frac{2}{3}$ by the reciprocal of $\frac{a}{5}$, which is $\frac{5}{a}$.

$= \dfrac{2 \cdot 5}{3 \cdot a}$ Multiply the numerators and multiply the denominators.

$= \dfrac{10}{3a}$ Multiply in the numerator and denominator.

Self Check

Divide: $\dfrac{7}{4} \div \dfrac{3}{b}$.

Answer: $\dfrac{7b}{12}$ ■

EXAMPLE 8 ***Dividing algebraic fractions.*** Divide: $\dfrac{15x^2}{8y} \div \dfrac{10x}{y^3}$.

Solution

$\dfrac{15x^2}{8y} \div \dfrac{10x}{y^3} = \dfrac{15x^2}{8y} \cdot \dfrac{y^3}{10x}$ Multiply $\dfrac{15x^2}{8y}$ by the reciprocal of $\dfrac{10x}{y^3}$.

$= \dfrac{15x^2 \cdot y^3}{8y \cdot 10x}$ Multiply the numerators and multiply the denominators.

$= \dfrac{5 \cdot 3 \cdot x \cdot x \cdot y \cdot y \cdot y}{8 \cdot y \cdot 5 \cdot 2 \cdot x}$ Factor 15, x^2, y^3, and 10.

$= \dfrac{\overset{1}{\cancel{5}} \cdot 3 \cdot \overset{1}{\cancel{x}} \cdot x \cdot \overset{1}{\cancel{y}} \cdot y \cdot y}{8 \cdot \underset{1}{\cancel{y}} \cdot \underset{1}{\cancel{5}} \cdot 2 \cdot \underset{1}{\cancel{x}}}$ Divide out the common factors of 5, x, and y.

$= \dfrac{3xy^2}{16}$ Multiply in the numerator and denominator.

Self Check

Divide: $\dfrac{9y^2}{10x} \div \dfrac{18y}{5x}$.

Answer: $\dfrac{y}{4}$ ■

STUDY SET Section 4.3

VOCABULARY *Fill in the blanks to make the statements true.*

1. Two numbers are called ____________ if their product is 1.

2. The result of a division problem is called the ____________.

CONCEPTS

3. Complete this statement:

$\frac{1}{2} \div \frac{2}{3} = \square \cdot \square$

4. Find the reciprocal of each number.

a. $\frac{2}{5}$ **b.** -3

c. x **d.** $\frac{s}{w}$

5. Using horizontal lines, divide each rectangle in Illustration 1 into thirds. What division problem does this illustrate? What is the quotient of that problem?

ILLUSTRATION 1

6. Using horizontal lines, divide each rectangle in Illustration 2 into fifths. What division problem does this illustrate? What is the quotient of that problem?

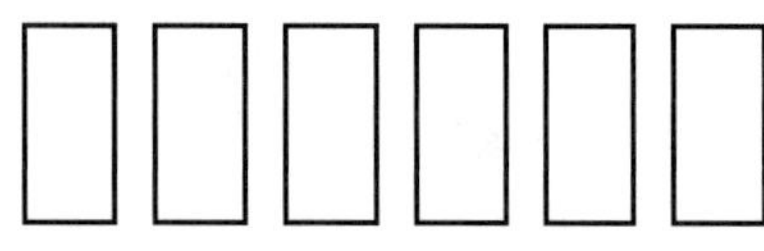

ILLUSTRATION 2

7. Multiply $\frac{4}{5}$ and its reciprocal. What is the result?

8. Multiply $\frac{a}{b}$ and its reciprocal. What is the result?

9. a. Find $15 \div 3$.

b. Rewrite $15 \div 3$ as multiplication by the reciprocal and find the result.

c. Complete this statement: Division by 3 is the same as multiplication by $\square$.

10. a. Find $10 \div \frac{1}{5}$.

b. Complete this statement: Division by $\frac{1}{5}$ is the same as multiplication by $\square$.

NOTATION *Complete each solution.*

11. Divide: $\frac{25}{36} \div \frac{10}{9}$.

$$\frac{25}{36} \div \frac{10}{9} = \frac{25}{36} \cdot \frac{9}{10}$$
$$= \frac{25 \cdot 9}{36 \cdot 10}$$
$$= \frac{5 \cdot 5 \cdot 9}{4 \cdot 9 \cdot 2 \cdot 5}$$
$$= \frac{\overset{1}{\cancel{5}} \cdot 5 \cdot \overset{1}{\cancel{9}}}{4 \cdot \underset{1}{\cancel{9}} \cdot 2 \cdot \underset{1}{\cancel{5}}}$$
$$= \frac{5}{8}$$

12. Divide: $\frac{4a}{9b} \div \frac{8a}{27b}$.

$$\frac{4a}{9b} \div \frac{8a}{27b} = \frac{4a}{9b} \cdot \frac{27b}{8a}$$
$$= \frac{4a \cdot 27b}{9b \cdot 8a}$$
$$= \frac{4 \cdot a \cdot 3 \cdot 9 \cdot b}{9 \cdot b \cdot 4 \cdot 2 \cdot a}$$
$$= \frac{\overset{1}{\cancel{4}} \cdot \overset{1}{\cancel{a}} \cdot 3 \cdot \overset{1}{\cancel{9}} \cdot \overset{1}{\cancel{b}}}{\underset{1}{\cancel{9}} \cdot \underset{1}{\cancel{b}} \cdot \underset{1}{\cancel{4}} \cdot 2 \cdot \underset{1}{\cancel{a}}}$$
$$= \frac{3}{2}$$

PRACTICE *Find each quotient.*

13. $\frac{1}{2} \div \frac{3}{5}$

14. $\frac{5}{7} \div \frac{5}{6}$

15. $\frac{3}{16} \div \frac{1}{9}$

16. $\frac{5}{8} \div \frac{2}{9}$

17. $\frac{4}{5} \div \frac{4}{5}$

18. $\frac{2}{3} \div \frac{2}{3}$

19. $\left(-\frac{7}{4}\right) \div \left(-\frac{21}{8}\right)$

20. $\left(-\frac{15}{16}\right) \div \left(-\frac{5}{8}\right)$

21. $3 \div \frac{1}{12}$

22. $9 \div \frac{3}{4}$

23. $120 \div \frac{12}{5}$

24. $360 \div \frac{36}{5}$

25. $-\frac{4}{5} \div (-6)$

26. $-\frac{7}{8} \div (-14)$

27. $\frac{15}{16} \div 180$

28. $\frac{7}{8} \div 210$

29. $-\frac{9}{10} \div \frac{4}{15}$

30. $-\frac{3}{4} \div \frac{3}{2}$

31. $\frac{9}{10} \div \left(-\frac{3}{25}\right)$

32. $\frac{11}{16} \div \left(-\frac{9}{16}\right)$

33. $-\frac{1}{8} \div 8$

34. $-\frac{1}{15} \div 15$

35. $\frac{15}{32} \div \frac{15}{32}$

36. $-\frac{1}{64} \div \left(-\frac{1}{64}\right)$

37. $\frac{4a}{5} \div \frac{3}{2}$

38. $\frac{2x}{3} \div \frac{3}{2}$

39. $\frac{t}{8} \div \frac{3}{4}$

40. $\frac{x}{9} \div \frac{3}{5}$

41. $\frac{13}{16b} \div \frac{1}{2}$

42. $\frac{7}{8h} \div \frac{6}{7}$

43. $-\frac{15}{32y} \div \frac{3}{4}$

44. $-\frac{7}{10k} \div \frac{4}{5}$

45. $a \div \frac{a}{b}$

46. $\frac{x}{y} \div x$

47. $\frac{x}{y} \div \frac{x}{y}$

48. $\frac{c}{d} \div \frac{c}{d}$

49. $\frac{2s}{3t} \div (-6)$

50. $\frac{x}{4y} \div (-x)$

51. $-\frac{9}{8}x \div \frac{3}{4x^2}$

52. $-\frac{4}{3}t \div \frac{16t^2}{9}$

53. $-8x \div \left(-\frac{4x^3}{9}\right)$

54. $-12y \div \left(-\frac{3y^2}{10}\right)$

55. $-\frac{x^2}{y^3} \div \frac{x}{y}$

56. $-\frac{k}{t^3} \div \frac{k}{t^2}$

57. Find the quotient when $-\frac{26x}{15}$ is divided by $\frac{13}{45x}$.

58. Find the quotient when $\frac{75}{33a}$ is divided by $\frac{25}{11a}$.

APPLICATIONS

59. MARATHON Each lap around a stadium track is $\frac{1}{4}$ mile. How many laps would a runner have to complete to get a 26-mile workout?

60. COOKING A recipe calls for $\frac{3}{4}$ cup of flour, and the only measuring container you have holds $\frac{1}{8}$ cup. How many $\frac{1}{8}$ cups of flour would you need to add to follow the recipe?

61. LASER TECHNOLOGY Using a laser, a technician slices thin pieces of aluminum off the end of a rod that is $\frac{7}{8}$ inch long. How many $\frac{1}{64}$-inch-wide slices can be cut from this rod?

62. FURNITURE A production process applies several layers of a clear acrylic coat to outdoor furniture to help protect it from the weather. If each protective coat is $\frac{3}{32}$ inch thick, how many applications will be needed to build up $\frac{3}{8}$ inch of clear finish?

63. UNDERGROUND CABLE In Illustration 3, which construction proposal will require the fewest days to install underground TV cable from the broadcasting station to the subdivision?

Proposal	*Amount of cable installed per day*	*Comments*
Route 1	$\frac{3}{5}$ of a mile	Longer than Route 2
Route 2	$\frac{2}{5}$ of a mile	Terrain very rocky

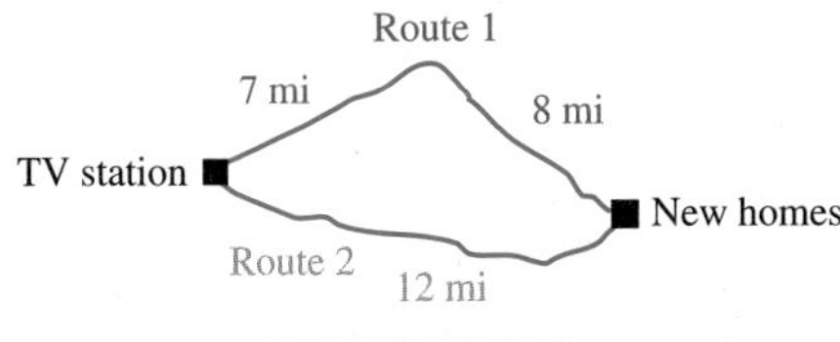

ILLUSTRATION 3

64. PRODUCTION PLANNING The materials used to make a pillow are shown in Illustration 4. Examine the inventory list to decide how many pillows can be manufactured in one production run with the materials in stock.

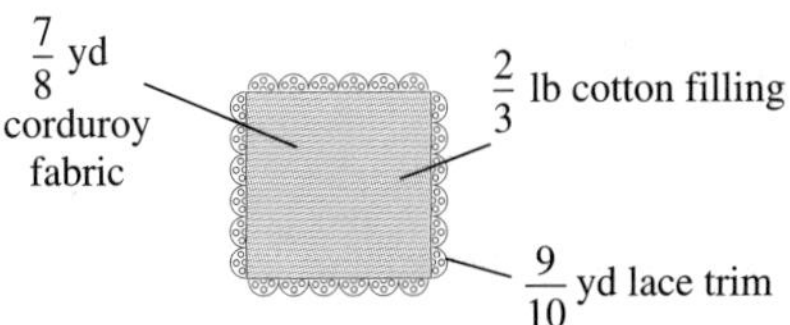

Factory Inventory List

Materials	*Amount in stock*
Lace trim	135 yd
Corduroy fabric	154 yd
Cotton filling	98 lb

ILLUSTRATION 4

65. 3 × 5 CARDS Ninety 3 × 5 cards are shown stacked next to a ruler in Illustration 5.

a. Into how many parts is one inch divided on the ruler?

b. How thick is the stack of cards?

c. How thick is one 3 × 5 card?

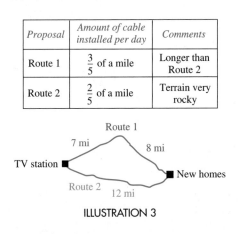

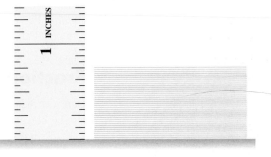

ILLUSTRATION 5

66. COMPUTER PRINTER Illustration 6 shows how the letter E is formed by a dot matrix printer. What is the height of a dot?

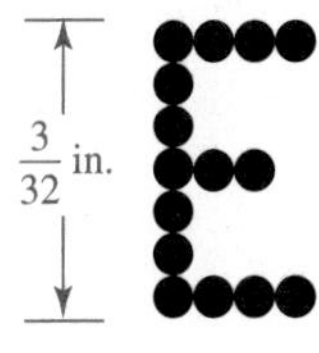

ILLUSTRATION 6

67. FORESTRY A set of forestry maps divides the 6,284 acres of an old-growth forest into $\frac{4}{5}$-acre sections. How many sections do the maps contain?

68. HARDWARE A hardware chain purchases large amounts of nails and packages them in $\frac{9}{16}$-pound bags for sale. How many of these bags of nails can be obtained from 2,871 pounds of nails?

WRITING

69. Explain how to divide two fractions.

70. Explain why 0 does not have a reciprocal.

71. Write an application problem that could be solved by finding $10 \div \frac{1}{5}$.

72. Explain why dividing a fraction by 2 is the same as finding $\frac{1}{2}$ of it.

REVIEW

73. Solve $4x + (-2) = -18$.

74. Distribute and simplify: $-3(x - 4)$.

75. What formula relates costs, profit, and revenue?

76. If a pipe x feet long is divided into 5 equal pieces, how long is each piece?

77. True or false: If equal amounts are subtracted from the numerator and the denominator of a fraction, the result will be an equivalent fraction.

78. Graph each of these numbers on a number line: $-2, 0, |-4|$, and the opposite of 1.

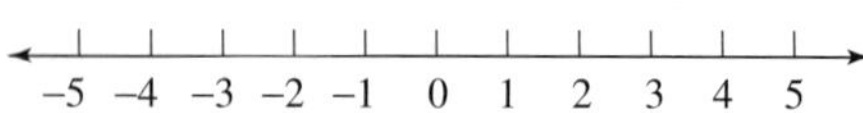

79. Simplify: $-3t + (-5T) + 4T + 8t$.

80. Define the word *variable.*

4.4 Adding and Subtracting Fractions

In this section, you will learn about

- Fractions with the same denominator • Fractions with different denominators
- Finding the LCD • Comparing fractions

INTRODUCTION. In arithmetic and algebra, *we can only add or subtract objects that are similar.* For example, we can add dollars to dollars, but we cannot add dollars to oranges. This concept is important when adding or subtracting fractions.

Fractions with the same denominator

Consider the problem $\frac{3}{5} + \frac{1}{5}$. When we write it in words, it is apparent that we are adding similar objects.

three-**fifths** + one-**fifth**

Similar objects

Because the denominators of $\frac{3}{5}$ and $\frac{1}{5}$ are the same, we say that they have a **common denominator.** Since the fractions have a common denominator, we can add them. Figure 4-6 illustrates the addition process.

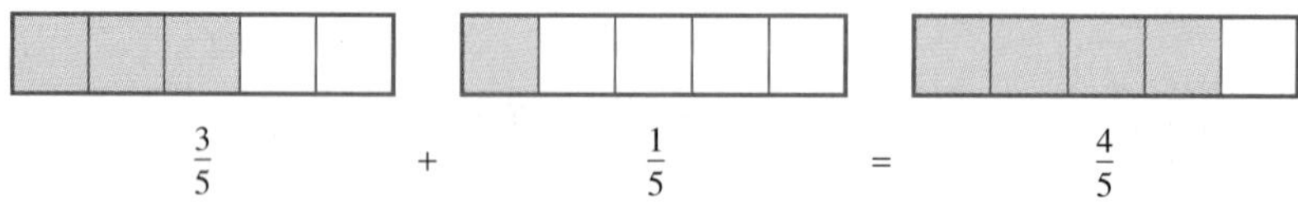

$$\frac{3}{5} + \frac{1}{5} = \frac{4}{5}$$

FIGURE 4-6

We can make some observations about the addition shown in the figure.

The *sum* of the numerators is the numerator of the answer.

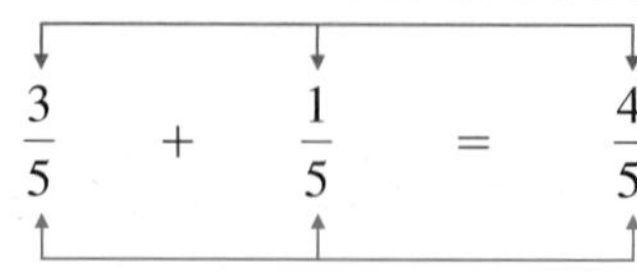

The answer is a fraction that has the *same* denominator as the two fractions that were added.

These observations suggest the following rule.

Adding or subtracting fractions with the same denominators

To add (or subtract) fractions with the same denominators, add (or subtract) their numerators and write that result over the common denominator. Simplify the result, if possible.

In symbols, if a, b, and c represent numbers, then

$$\frac{a}{c} + \frac{b}{c} = \frac{a+b}{c} \quad (c \neq 0) \quad \text{and} \quad \frac{a}{c} - \frac{b}{c} = \frac{a-b}{c} \quad (c \neq 0)$$

EXAMPLE 1 ***Fractions with the same denominator.*** Add: $\frac{1}{8} + \frac{5}{8}$.

Solution

$\frac{1}{8} + \frac{5}{8} = \frac{1+5}{8}$ Add the numerators. Write the sum over the common denominator 8.

$= \frac{6}{8}$ Do the addition: $1 + 5 = 6$. The fraction can be simplified.

$= \frac{3 \cdot \overset{1}{\cancel{2}}}{4 \cdot \underset{1}{\cancel{2}}}$ Factor 6 as $3 \cdot 2$ and 8 as $4 \cdot 2$. Divide out the common factor of 2.

$= \frac{3}{4}$ Multiply in the numerator and in the denominator.

Self Check

Add: $\frac{9}{12} + \frac{1}{12}$.

$\frac{10}{12} = \frac{5}{6}$

Answer: $\frac{5}{6}$ ■

EXAMPLE 2 ***Subtracting fractions.*** Subtract: $-\frac{7}{3} - \left(-\frac{2}{3}\right)$.

Solution

$-\frac{7}{3} - \left(-\frac{2}{3}\right) = -\frac{7}{3} + \frac{2}{3}$ Add the opposite of $-\frac{2}{3}$, which is $\frac{2}{3}$.

$= \frac{-7}{3} + \frac{2}{3}$ Write $-\frac{7}{3}$ as $\frac{-7}{3}$.

$= \frac{-7+2}{3}$ Add the numerators. Write the sum over the common denominator 3.

$= \frac{-5}{3}$ Do the addition: $-7 + 2 = -5$.

$= -\frac{5}{3}$ Rewrite the fraction: $\frac{-5}{3} = -\frac{5}{3}$.

Self Check

Subtract: $-\frac{9}{11} - \left(-\frac{3}{11}\right)$.

$-\frac{9}{11} + \frac{3}{11} = \frac{-6}{11}$

Answer: $-\frac{6}{11}$ ■

EXAMPLE 3 ***Subtracting algebraic fractions.*** Subtract: $\frac{9}{x} - \frac{1}{x}$.

Solution

$\frac{9}{x} - \frac{1}{x} = \frac{9-1}{x}$ Subtract the numerators. Write the difference over the common denominator x.

$= \frac{8}{x}$ Do the subtraction: $9 - 1 = 8$.

Self Check

Subtract: $\frac{11}{b} - \frac{7}{b}$.

$\frac{11}{b} + \frac{-7}{b} = \frac{4}{b}$

Answer: $\frac{4}{b}$ ■

Fractions with different denominators

Now we consider the problem $\frac{3}{5} + \frac{1}{3}$. Since the denominators are different, we cannot add these fractions in their present form.

three-**fifths** + one-**third**
Not similar objects

To add these fractions, we need to find a common denominator. The smallest common denominator (called the **least** or **lowest common denominator**) usually is the easiest common denominator to work with.

Least common denominator

The **least common denominator (LCD)** for a set of fractions is the smallest number each denominator will divide exactly.

In the problem $\frac{3}{5} + \frac{1}{3}$, the denominators are 5 and 3. The numbers 5 and 3 divide many numbers exactly; 30, 45, and 60, to name a few. But the smallest number that 5 and 3 divide exactly is 15. This is the LCD. We will now build each fraction into a fraction with a denominator of 15 by applying the fundamental property of fractions.

$$\frac{3}{5} + \frac{1}{3} = \frac{3 \cdot 3}{5 \cdot 3} + \frac{1 \cdot 5}{3 \cdot 5}$$

We need to multiply this denominator by 3 to obtain 15. We need to multiply this denominator by 5 to obtain 15.

$= \frac{9}{15} + \frac{5}{15}$ Do the multiplications in the numerators and denominators.

$= \frac{9 + 5}{15}$ Add the numerators and write the sum over the common denominator 15.

$= \frac{14}{15}$ Do the addition: $9 + 5 = 14$.

Figure 4-7 shows $\frac{3}{5}$ and $\frac{1}{3}$ expressed as equivalent fractions with a denominator of 15. Once the denominators are the same, the fractions can be added easily.

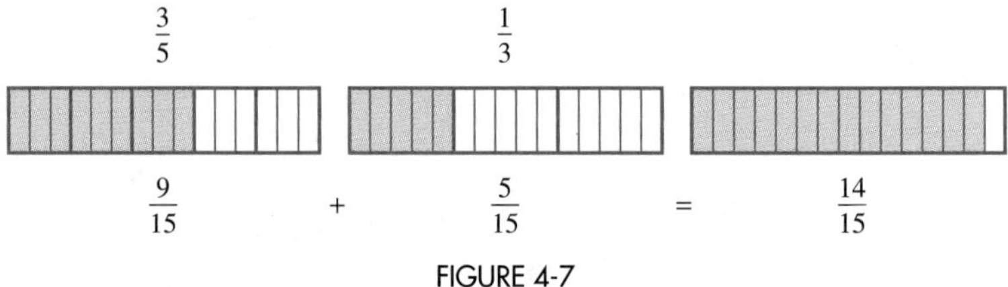

FIGURE 4-7

To add or subtract fractions with different denominators, we follow these steps.

Adding or subtracting fractions with different denominators

1. Find the LCD.
2. Express each fraction as an equivalent fraction with a denominator that is the LCD.
3. Add or subtract the resulting fractions. Simplify the result if possible.

EXAMPLE 4 ***Fractions with different denominators.*** Add: $\frac{1}{8} + \frac{2}{3}$.

Solution

Since the smallest number the denominators 8 and 3 divide exactly is 24, the LCD is 24.

$\frac{1}{8} + \frac{2}{3} = \frac{1 \cdot 3}{8 \cdot 3} + \frac{2 \cdot 8}{3 \cdot 8}$ Apply the fundamental property of fractions to express each fraction in terms of 24ths.

$= \frac{3}{24} + \frac{16}{24}$ Do the multiplications in the numerators and denominators.

Self Check

Add: $\frac{1}{2} + \frac{2}{5}$.

$\frac{1}{2} + \frac{2}{5}$

$\frac{5}{10} + \frac{4}{10} = \frac{9}{10}$

$= \dfrac{3 + 16}{24}$ Add the numerators and write the sum over the common denominator 24.

$= \dfrac{19}{24}$ Do the addition in the numerator: $3 + 16 = 19$.

Answer: $\dfrac{9}{10}$ ■

EXAMPLE 5 ***Adding an integer and a fraction.*** Add: $-5 + \dfrac{1}{4}$.

Solution

$-5 + \dfrac{1}{4} = \dfrac{-5}{1} + \dfrac{1}{4}$ Write -5 as $\frac{-5}{1}$. The smallest number that 1 and 4 divide exactly is 4, so the LCD is 4.

$= \dfrac{-5 \cdot \mathbf{4}}{1 \cdot \mathbf{4}} + \dfrac{1}{4}$ Apply the fundamental property of fractions to express $\frac{-5}{1}$ in terms of 4ths.

$= \dfrac{-20}{4} + \dfrac{1}{4}$ Multiply in the numerator and in the denominator.

$= \dfrac{-20 + 1}{4}$ Write the sum of the numerators over the common denominator 4.

$= \dfrac{-19}{4}$ Do the addition: $-20 + 1 = -19$.

$= -\dfrac{19}{4}$ Rewrite: $\frac{-19}{4} = -\frac{19}{4}$.

Self Check

Add: $-6 + \dfrac{2}{5}$.

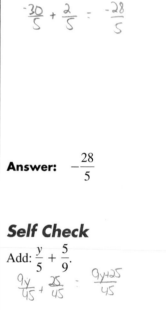

Answer: $-\dfrac{28}{5}$ ■

EXAMPLE 6 ***Adding algebraic fractions.*** Add: $\dfrac{t}{3} + \dfrac{1}{4}$.

Solution

$\dfrac{t}{3} + \dfrac{1}{4} = \dfrac{t \cdot \mathbf{4}}{3 \cdot \mathbf{4}} + \dfrac{1 \cdot \mathbf{3}}{4 \cdot \mathbf{3}}$ The LCD is 12. Express each fraction in terms of 12ths.

$= \dfrac{4t}{12} + \dfrac{3}{12}$ Do the multiplications in the numerators and denominators.

$= \dfrac{4t + 3}{12}$ Write the sum of the numerators over the common denominator 12.

Self Check

Add: $\dfrac{y}{5} + \dfrac{5}{9}$.

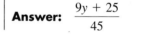

Answer: $\dfrac{9y + 25}{45}$ ■

COMMENT In Example 6, the answer, $\frac{4t + 3}{12}$, does not simplify. Do not make either of the two common mistakes shown below.

$\dfrac{4t + 3}{12} = \dfrac{7t}{12}$ (crossed out) $4t$ and 3 are not like terms. Don't add them.

$\dfrac{4t + \overset{1}{\cancel{3}}}{12} = \dfrac{4t + \overset{1}{\cancel{3}}}{4 \cdot \underset{1}{\cancel{3}}}$ (crossed out) 3 is a term of the numerator, not a factor, and therefore cannot be divided out.

EXAMPLE 7 ***Subtracting algebraic fractions.*** Subtract: $\dfrac{2}{y} - \dfrac{5}{18}$.

Solution

$\dfrac{2}{y} - \dfrac{5}{18} = \dfrac{2 \cdot \mathbf{18}}{y \cdot \mathbf{18}} - \dfrac{5 \cdot \mathbf{y}}{18 \cdot \mathbf{y}}$ The LCD is $18y$. Express each as a fraction with a denominator of $18y$.

$= \dfrac{36}{18y} - \dfrac{5y}{18y}$ Multiply in the numerators and denominators.

$= \dfrac{36 - 5y}{18y}$ Write the difference of the numerators over the common denominator $18y$.

Self Check

Subtract: $\dfrac{6}{t} - \dfrac{4}{9}$.

Answer: $\dfrac{54 - 4t}{9t}$ ■

Finding the LCD

When adding or subtracting fractions with different denominators, the LCD is not always obvious. We will now develop two methods for finding the LCD of a set of fractions. As an example, let's find the LCD of $\frac{3}{8}$ and $\frac{1}{10}$.

Method 1: A **multiple** of a number is the product of that number and a natural number. The multiples of 8 and the multiples of 10 are shown below.

Multiples of 8	Multiples of 10
$8 \cdot 1 = 8$	$10 \cdot 1 = 10$
$8 \cdot 2 = 16$	$10 \cdot 2 = 20$
$8 \cdot 3 = 24$	$10 \cdot 3 = 30$
$8 \cdot 4 = 32$	$10 \cdot 4 = \mathbf{40}$
$8 \cdot 5 = \mathbf{40}$	$10 \cdot 5 = 50$
$8 \cdot 6 = 48$	$10 \cdot 6 = 60$
$8 \cdot 7 = 56$	$10 \cdot 7 = 70$
$8 \cdot 8 = 64$	$10 \cdot 8 = \mathbf{80}$
$8 \cdot 9 = 72$	$10 \cdot 9 = 90$
$8 \cdot 10 = \mathbf{80}$	$10 \cdot 10 = 100$

The multiples that the lists have in common are highlighted in red.

The smallest multiple common to both lists is 40. It is the smallest number that 8 and 10 divide exactly. Therefore, 40 is the LCD of $\frac{3}{8}$ and $\frac{1}{10}$. These observations suggest a method for finding the LCD of a set of fractions.

Finding the LCD by finding multiples

1. List the multiples of each denominator.
2. The smallest multiple common to the lists found in Step 1 is the LCD of the fractions.

Method 2: If the LCD for $\frac{3}{8}$ and $\frac{1}{10}$ is a number that 8 and 10 divide exactly, the prime factorization of the LCD must include the prime factorization of 8 (which is $2 \cdot 2 \cdot 2$) and the prime factorization of 10 (which is $5 \cdot 2$). The smallest number that meets both of these requirements is $2 \cdot 2 \cdot 2 \cdot 5$. Therefore, the LCD is $2 \cdot 2 \cdot 2 \cdot 5 = 40$.

$$\left.\begin{aligned} 8 &= 2 \cdot 2 \cdot 2 \\ 10 &= 5 \cdot 2 \end{aligned}\right\} \text{LCD} = \overbrace{2 \cdot 2 \cdot 2}^{\text{The prime factorization of 8.}} \cdot 5 = 40$$

(The prime factorization of 10: $\underbrace{2 \cdot 5}$.)

In the prime factorization of 8, the factor 2 appears three times. It appears three times in the product $(2 \cdot 2 \cdot 2 \cdot 5)$ that gives the LCD. In the prime factorization of 10, the factor 5 appears once. It appears once in the product that gives the LCD. These observations suggest another method for finding the LCD of a set of fractions.

Finding the LCD using prime factorization

1. Prime factor each denominator.
2. The LCD is a product of prime factors, where each factor is used the greatest number of times it appears in any one factorization found in Step 1.

EXAMPLE 8 ***Finding the LCD using prime factorization.*** Subtract: $\frac{19}{21} - \frac{5}{18}$.

Self Check

Subtract: $\frac{33}{35} - \frac{11}{14}$.

Solution

We use prime factorization to find the LCD.

Step 1: Prime factor each denominator.

$21 = 7 \cdot 3$

$18 = 3 \cdot 3 \cdot 2$

Step 2: The factors 7, 3, and 2 appear in the prime factorizations.

The greatest number of times 7 appears in any one factorization is once.	The greatest number of times 3 appears in any one factorization is twice.	The greatest number of times 2 appears in any one factorization is once.

$$\text{LCD} = 7 \cdot 3 \cdot 3 \cdot 2 = 126$$

$$\frac{19}{21} - \frac{5}{18} = \frac{19 \cdot \mathbf{6}}{21 \cdot \mathbf{6}} - \frac{5 \cdot 7}{18 \cdot 7}$$ Express each fraction in terms of 126ths.

$$= \frac{114}{126} - \frac{35}{126}$$ Do the multiplications in the numerators and in the denominators.

$$= \frac{114 - 35}{126}$$ Write the difference of the numerators over the common denominator 126.

$$= \frac{79}{126}$$ Do the subtraction: $114 - 35 = 79$.

Answer: 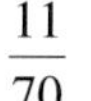 $\frac{11}{70}$ ■

EXAMPLE 9 ***Television viewing habits.*** Students on a college campus were asked to estimate to the nearest hour how much television they watched each day. The results are given in the pie chart in Figure 4-8. For example, the chart tells us that $\frac{1}{4}$ of those responding watched 1 hour per day. Find the fraction of the student body watching from 0 to 2 hours daily.

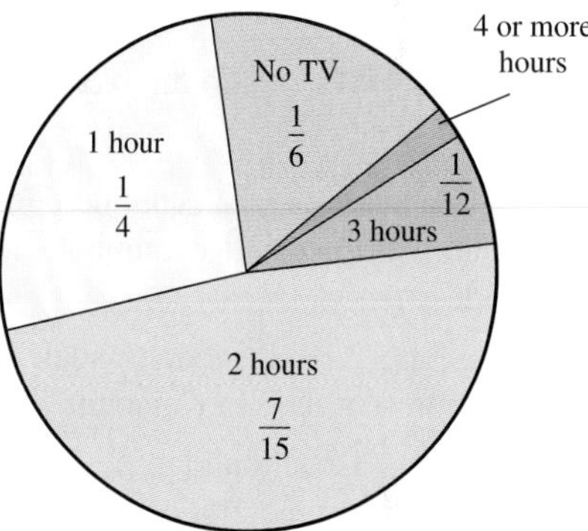

FIGURE 4-8

Solution To answer this question, we need to add $\frac{1}{6}$, $\frac{1}{4}$, and $\frac{7}{15}$. To find the LCD, we prime factor each of the denominators.

$$\left.\begin{aligned} 6 &= 3 \cdot 2 \\ 4 &= 2 \cdot 2 \\ 15 &= 5 \cdot 3 \end{aligned}\right\} \text{LCD} = 5 \cdot 3 \cdot 2 \cdot 2 = 60$$

In any one factorization, the greatest number of times 5 appears is once, the greatest number of times 3 appears is once, and the greatest number of times 2 appears is twice.

$$\frac{1}{6} + \frac{1}{4} + \frac{7}{15} = \frac{1 \cdot \mathbf{10}}{6 \cdot \mathbf{10}} + \frac{1 \cdot \mathbf{15}}{4 \cdot \mathbf{15}} + \frac{7 \cdot \mathbf{4}}{15 \cdot \mathbf{4}}$$ Express each fraction in terms of 60ths.

$$= \frac{10}{60} + \frac{15}{60} + \frac{28}{60}$$ Do the multiplication in the numerators and denominators.

$$= \frac{10 + 15 + 28}{60}$$ Add the numerators. Write the sum over the common denominator 60.

$$= \frac{53}{60}$$ Do the addition: $10 + 15 + 28 = 53$.

The fraction of the student body watching 0 to 2 hours of television daily is $\frac{53}{60}$. ■

Comparing fractions

If fractions have the same denominator, the fraction with the larger numerator is the larger fraction. If their denominators are different, we need to write the fractions with a common denominator before we can make a comparison.

Comparing fractions

> To compare unlike fractions, write the fractions as equivalent fractions with the same denominator—preferably the LCD. Then compare their numerators. The fraction with the larger numerator is the larger fraction.

EXAMPLE 10 ***Comparing fractions.*** Which fraction is larger: $\frac{5}{6}$ or $\frac{7}{8}$?

Solution

To compare these fractions, we express each with the LCD of 24.

$\frac{5}{6} = \frac{5 \cdot 4}{6 \cdot 4}$ $\quad \frac{7}{8} = \frac{7 \cdot 3}{8 \cdot 3}$ Express each fraction in terms of 24ths.

$= \frac{20}{24}$ $\quad = \frac{21}{24}$ Do the multiplications in the numerators and denominators.

Next, we compare the numerators. Since $21 > 20$, we conclude that $\frac{21}{24}$ is greater than $\frac{20}{24}$. Thus, $\frac{7}{8} > \frac{5}{6}$.

Self Check

Which fraction is larger: $\frac{7}{12}$ or $\frac{3}{5}$?

Answer: $\frac{3}{5}$ ■

STUDY SET Section 4.4

VOCABULARY *Fill in the blanks to make the statements true.*

1. The ______ common denominator for a set of fractions is the smallest number each denominator will divide exactly.

2. ______ fractions, such as $\frac{1}{2}$ and $\frac{2}{4}$, are fractions that represent the same amount.

3. To express a fraction in ______ terms, we multiply the numerator and denominator by the same number.

4. ______ up a fraction is the process of multiplying the numerator and the denominator of the fraction by the same number.

CONCEPTS

5. The rule for adding fractions is

$$\frac{a}{c} + \frac{b}{c} = \frac{a+b}{c} \quad (c \neq 0)$$

Fill in the blanks.

This rule tells us how to add fractions having like ______. To find the sum, we add the ______ and then write that result over the ______ denominator.

6. a. Add the indicated fractions.

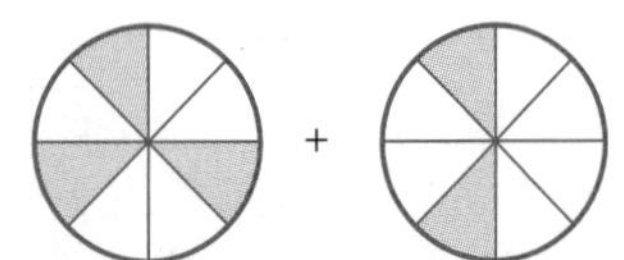

b. Subtract the indicated fractions.

7. Why must we do some preliminary work before doing the following addition?

$$\frac{2}{9} + \frac{2}{5}$$

8. Why must we do some preliminary work before doing the following subtraction?

$$\frac{5}{6} - \frac{5}{18}$$

9. By what are the numerator and the denominator of the following fractions being multiplied?

a. $\frac{5 \cdot 4}{6 \cdot 4}$ **b.** $\frac{1 \cdot c}{3 \cdot c}$

10. Consider $\frac{3}{4}$. By what should we multiply the numerator and denominator of this fraction to express it as an equivalent fraction with the given denominator?

a. 12 **b.** 36

c. $20x$ **d.** $16r$

11. Consider the following prime factorizations:

$$24 = 3 \cdot 2 \cdot 2 \cdot 2$$
$$90 = 5 \cdot 3 \cdot 3 \cdot 2$$

For any one factorization, what is the greatest number of times

a. a 5 appears?

b. a 3 appears?

c. a 2 appears?

12. **a.** List the first ten multiples of 9 and the first ten multiples of 12.

b. What is the LCM of 9 and 12?

13. The denominators of two fractions involved in a subtraction problem have the prime-factored forms $2 \cdot 2 \cdot 5$ and $2 \cdot 3 \cdot 5$. What is the LCD for the fractions?

14. The denominators of three fractions involved in a subtraction problem have the prime-factored forms $2 \cdot 2 \cdot 5$, $2 \cdot 3 \cdot 5$, and $2 \cdot 3 \cdot 3 \cdot 5$. What is the LCD for the fractions?

15. **a.** Divide the figure on the left into fourths and shade one part. Divide the figure on the right into thirds and shade one part. Which shaded part is larger?

b. Express the shaded part of each figure in part a as a fraction. Show that one of those fractions is larger than the other by expressing both in terms of a common denominator and then comparing them.

16. Place a $>$ or $<$ symbol in the blank to make a true statement.

a. $\frac{32}{35}$ ___ $\frac{31}{35}$ **b.** $\frac{7}{8}$ ___ $\frac{31}{32}$

NOTATION *Complete each solution.*

17. Add: $\frac{2}{5} + \frac{1}{3}$.

$$\frac{2}{5} + \frac{1}{3} = \frac{2 \cdot 3}{5 \cdot 3} + \frac{1 \cdot 5}{3 \cdot 5}$$

$$= \frac{6}{15} + \frac{5}{15}$$

$$= \frac{6 + 5}{15}$$

$$= \frac{11}{15}$$

18. Subtract: $\frac{7}{8} - \frac{2}{3}$.

$$\frac{7}{8} - \frac{2}{3} = \frac{7 \cdot 3}{8 \cdot 3} - \frac{2 \cdot 8}{3 \cdot 8}$$

$$= \frac{21}{24} - \frac{16}{24}$$

$$= \frac{21 - 16}{24}$$

$$= \frac{5}{24}$$

PRACTICE *The denominators of two fractions are given. Find the lowest common denominator.*

19. 18, 6 **20.** 15, 3

21. 8, 6 **22.** 10, 4

23. 8, 20 **24.** 14, 21

25. 15, 12 **26.** 25, 30

In Exercises 27–78, do each operation. Simplify when necessary.

27. $\frac{3}{7} + \frac{1}{7}$ **28.** $\frac{16}{25} - \frac{9}{25}$

29. $\frac{37}{103} - \frac{17}{103}$ **30.** $\frac{54}{53} - \frac{52}{53}$

31. $\frac{11}{25} - \frac{1}{25}$ **32.** $\frac{7}{8} - \frac{1}{8}$

33. $\frac{5}{d} + \frac{3}{d}$ **34.** $\frac{17}{x} - \frac{12}{x}$

35. $\frac{1}{4} + \frac{3}{8}$ **36.** $\frac{2}{3} + \frac{1}{6}$

37. $\frac{13}{20} - \frac{1}{5}$ **38.** $\frac{71}{100} - \frac{1}{10}$

39. $\frac{4}{5} + \frac{2}{3}$ **40.** $\frac{1}{4} + \frac{2}{3}$

41. $\frac{1}{8} + \frac{2}{7}$ **42.** $\frac{1}{6} + \frac{5}{9}$

43. $\frac{3}{4} - \frac{2}{3}$ **44.** $\frac{4}{5} - \frac{1}{6}$

45. $\frac{5}{6} - \frac{3}{4}$ **46.** $\frac{7}{8} - \frac{5}{6}$

47. $\frac{16}{25} - \left(-\frac{3}{10}\right)$ **48.** $\frac{3}{8} - \left(-\frac{1}{6}\right)$

49. $-\frac{7}{16} + \frac{1}{4}$ **50.** $-\frac{17}{20} + \frac{4}{5}$

51. $\frac{1}{12} - \frac{3}{4}$ **52.** $\frac{11}{60} - \frac{13}{20}$

53. $-\frac{5}{8} - \frac{1}{3}$ **54.** $-\frac{7}{20} - \frac{1}{5}$

55. $-3 + \frac{2}{5}$ **56.** $-6 + \frac{5}{8}$

57. $-\frac{3}{4} - 5$ **58.** $-2 - \frac{7}{8}$

59. $\frac{7}{8} - \frac{t}{7}$ **60.** $\frac{5}{6} + \frac{c}{7}$

61. $\frac{4}{5} - \frac{2b}{9}$ **62.** $\frac{3}{16} + \frac{4h}{8}$

63. $\frac{4}{7} - \frac{1}{r}$ **64.** $\frac{4}{m} + \frac{2}{7}$

65. $-\frac{5}{9}+\frac{1}{y}$

66. $-\frac{3}{5}+\frac{5}{x}$

67. $\frac{1}{3}+\frac{1}{4}+\frac{1}{5}$

68. $\frac{1}{10}+\frac{1}{8}+\frac{1}{5}$

69. $-\frac{2}{3}+\frac{5}{4}+\frac{1}{6}$

70. $-\frac{3}{4}+\frac{3}{8}+\frac{7}{6}$

71. $\frac{5}{24}+\frac{3}{16}$

72. $\frac{17}{20}-\frac{4}{15}$

73. $-\frac{11}{15}-\frac{2}{9}$

74. $-\frac{19}{18}-\frac{5}{12}$

75. $\frac{7}{25}+\frac{1}{15}$

76. $\frac{11}{20}-\frac{1}{8}$

77. $\frac{4}{27}+\frac{1}{6}$

78. $\frac{8}{9}-\frac{7}{12}$

79. Find the difference of $\frac{11}{60}$ and $\frac{2}{45}$.

80. Find the sum of $\frac{9}{48}$ and $\frac{7}{40}$.

81. Subtract $\frac{5}{12}$ from $\frac{2}{15}$.

82. What is the sum of $\frac{11}{24}$ and $\frac{7}{36}$ increased by $\frac{5}{48}$?

APPLICATIONS

83. BOTANY To assess the effects of smog on tree development, botanists cut down a pine tree and measured the width of the growth rings for the last two years. (See Illustration 1.)

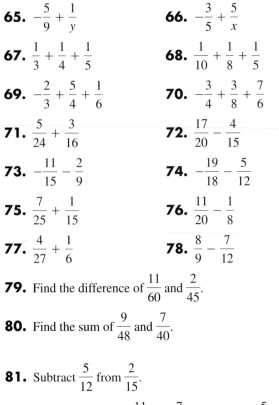

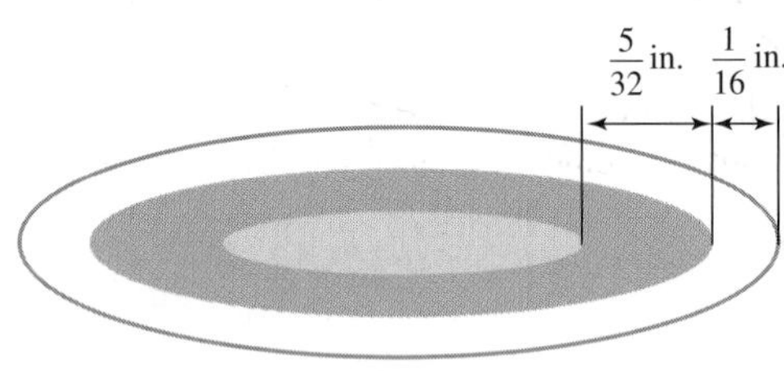

ILLUSTRATION 1

a. What was the growth over this two-year period?

b. What is the difference in the widths of the two rings?

84. MAGAZINE LAYOUT The page design for a magazine cover includes a blank strip at the top, called a header, and a blank strip at the bottom of the page, called a footer. In Illustration 2, how much page length is lost because of the header and footer?

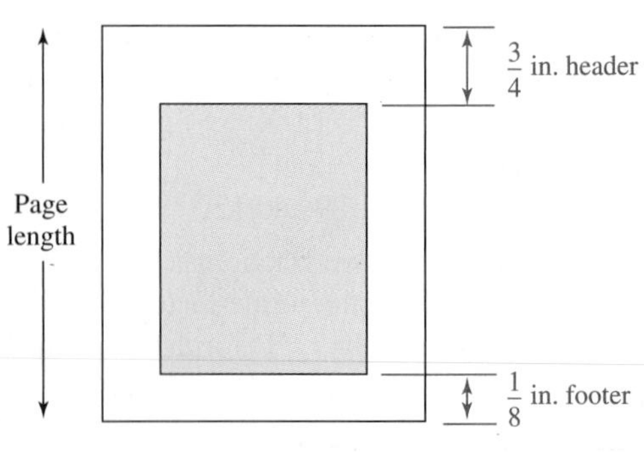

ILLUSTRATION 2

85. FAMILY DINNER A family bought two large pizzas for dinner. Several pieces of each pizza were not eaten, as shown in Illustration 3. How much pizza was left? Could the family have been fed with just one pizza?

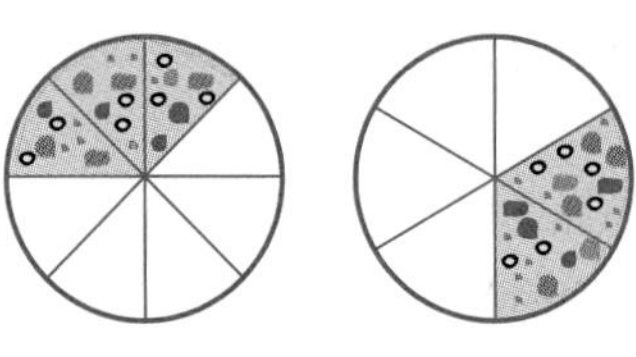

ILLUSTRATION 3

86. GASOLINE BARRELS The contents of two identical-sized barrels are shown in Illustration 4. If they are dumped into an empty third barrel that is the same size, how much of the third barrel will they fill?

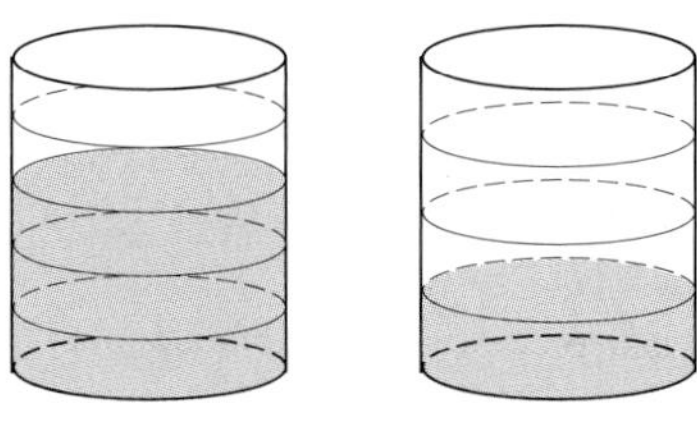

ILLUSTRATION 4

87. WEIGHTS AND MEASURES A consumer protection agency verifies the accuracy of butcher shop scales by placing a known three-quarter-pound weight on the scale and then comparing that to the scale's readout. According to Illustration 5, by how much is this scale off? Does it result in undercharging or overcharging customers on their meat purchases?

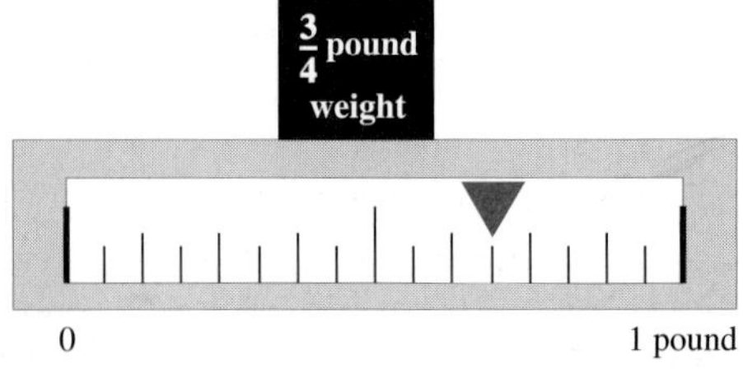

ILLUSTRATION 5

88. WRENCHES A mechanic likes to hang his wrenches above his tool bench in order of narrowest to widest. What is the proper order of the wrenches in Illustration 6?

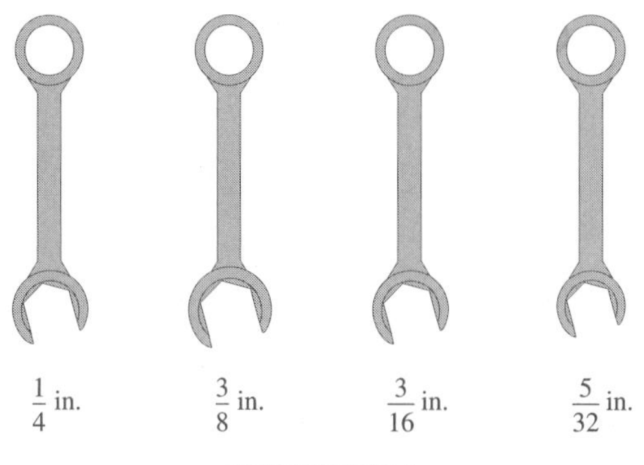

ILLUSTRATION 6

89. HIKING Illustration 7 shows the length of each part of a three-part hike. Rank the lengths from longest to shortest.

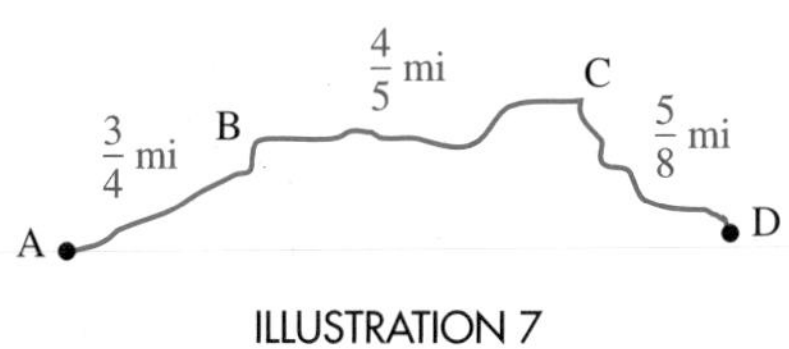

ILLUSTRATION 7

90. FIGURE DRAWING As an aid in drawing the human body, artists divide the body into three parts. Each part is then expressed as a fraction of the total body height. (See Illustration 8.) For example, the torso is $\frac{4}{15}$ of the body height. What fraction of body height is the head?

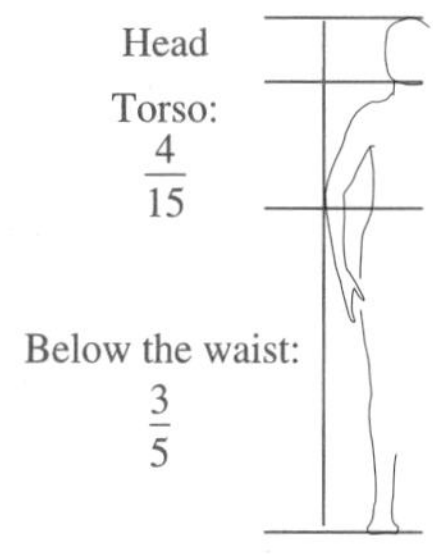

ILLUSTRATION 8

91. STUDY HABITS College students taking a full load were asked to give the average number of hours they studied each day. The results are shown in the pie chart in Illustration 9. What fraction of the students study 2 hours or more daily?

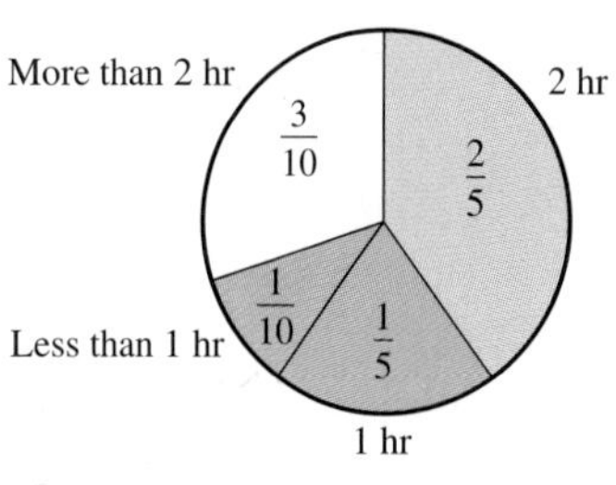

ILLUSTRATION 9

92. MUSICAL NOTES The notes used in music have fractional values. Their names and the symbols used to represent them are shown in Illustration 10(a). In common time, the values of the notes in each measure must add up to 1. Is the measure in Illustration 10(b) complete?

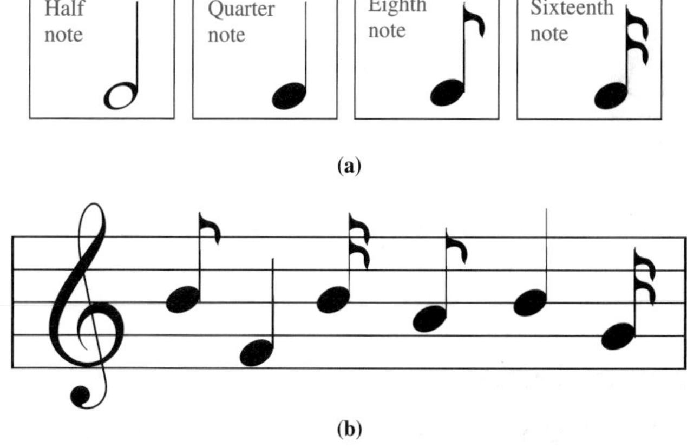

ILLUSTRATION 10

93. GARAGE DOOR OPENER What is the difference in strength between a $\frac{1}{3}$-hp and a $\frac{1}{2}$-hp garage door opener?

94. DELIVERY TRUCK A truck can safely carry a one-ton load. Should it be used to deliver one-half ton of sand, one-third ton of gravel, and one-fifth ton of cement in one trip to a job site?

WRITING

95. How are the procedures for expressing a fraction in higher terms and simplifying a fraction to lowest terms similar, and how are they different?

96. Given two fractions, how do we find their lowest common denominator?

97. How do we compare the relative sizes of two fractions with different denominators?

98. What is the difference between a common denominator and the lowest common denominator?

REVIEW

99. Simplify: $2(2 + x) - 3(x - 1)$.

100. A ball is dropped off the top of a building and falls for 3 seconds. How far does it travel in that time? (*Hint:* $d = 16t^2$.)

101. Translate to mathematical symbols: 5 less than x.

102. What is the formula for finding the area of a rectangle?

103. What is the formula for finding the perimeter of a rectangle?

104. Let $x = -4$. Find $2x^2 - x$.

THE LCM AND THE GCF

As we have seen, the **multiples** of a number can be found by multiplying it successively by 1, 2, 3, 4, 5, and so on. The multiples of 4 and the multiples of 6 are shown below.

$1 \cdot 4 = 4$	$1 \cdot 6 = 6$	
$2 \cdot 4 = 8$	$2 \cdot 6 = \mathbf{12}$	
$3 \cdot 4 = \mathbf{12}$	$3 \cdot 6 = 18$	
$4 \cdot 4 = 16$	$4 \cdot 6 = \mathbf{24}$	Common multiples of 4 and 6 are highlighted in red.
$5 \cdot 4 = 20$	$5 \cdot 6 = 30$	
$6 \cdot 4 = \mathbf{24}$	$6 \cdot 6 = \mathbf{36}$	
$7 \cdot 4 = 28$	$7 \cdot 6 = 42$	
$8 \cdot 4 = 32$	$8 \cdot 6 = 48$	
$9 \cdot 4 = \mathbf{36}$	$9 \cdot 6 = 54$	

Because 12 is the smallest number that is a multiple of both 4 and 6, it is called the **least common multiple (LCM)** of 4 and 6.

Making lists like those shown above can be tedious. A more efficient method to find the least common multiple of several numbers is as follows.

Finding the least common multiple

1. Write each of the numbers in prime-factored form.
2. The least common multiple is a product of prime factors, where each prime factor is used the greatest number of times it appears in any one factorization found in Step 1.

EXAMPLE 1 ***Least common multiple.*** Find the LCM of 24 and 36.

Solution

Step 1: First, we find the prime factorizations of 24 and 36.

$24 = 3 \cdot 2 \cdot 2 \cdot 2$

$36 = 3 \cdot 3 \cdot 2 \cdot 2$

Step 2: The prime factorizations of 24 and 36 contain the prime factors 3 and 2. We use each of these factors the greatest number of times it appears in any one factorization.

The greatest number of times 3 appears in any one factorization is two times.

The greatest number of times 2 appears in any one factorization is three times.

$\text{LCM} = 3 \cdot 3 \cdot 2 \cdot 2 \cdot 2 = 72$

The least common multiple of 24 and 36 is 72.

Self Check

Find the LCM of 18 and 84.

Answer: 252

Because 2 divides 36 exactly and because 2 divides 120 exactly, 2 is called a **common factor** of 36 and 120.

$\frac{36}{2} = 18 \qquad \frac{120}{2} = 60$

The numbers 36 and 120 have other common factors, such as 3 and 6. The **greatest common factor (GCF)** of 36 and 120 is the largest number that is a factor of both. We follow these steps to find the greatest common factor of several numbers.

Finding the greatest common factor

1. Write each of the numbers in prime-factored form.
2. The greatest common factor is the product of the prime factors that are common to the factorizations found in Step 1. If the numbers have no factors in common, the GCF is 1.

EXAMPLE 2 ***Greatest common factor.*** Find the GCF of 36 and 120.

Solution

Step 1: We find the prime factorizations of 36 and 120.

$$36 = 3 \cdot 3 \cdot 2 \cdot 2$$
$$120 = 5 \cdot 3 \cdot 2 \cdot 2 \cdot 2$$

Step 2: One factor of 3 (highlighted in red) and two factors of 2 (highlighted in blue) are common to the factorizations of 36 and 120. To find the GCF, we form their product.

$$\text{GCF} = 3 \cdot 2 \cdot 2 = 12$$

The greatest common factor of 36 and 120 is 12.

Self Check

Find the GCF of 60 and 150.

Answer: 30 ■

STUDY SET The LCM and the GCF

Find the least common multiple of the given numbers.

1. 3, 5
2. 7, 11
3. 8, 14
4. 8, 12
5. 14, 21
6. 16, 20
7. 6, 18
8. 3, 9
9. 44, 60
10. 36, 60
11. 100, 120
12. 120, 180
13. 6, 24, 36
14. 6, 10, 18
15. 18, 54, 63
16. 16, 30, 84

Find the greatest common factor of the given numbers.

17. 6, 9
18. 8, 12
19. 22, 33
20. 15, 20
21. 16, 20
22. 18, 24
23. 25, 100
24. 16, 80
25. 100, 120
26. 120, 180
27. 48, 108
28. 60, 96
29. 18, 24, 36
30. 30, 50, 90
31. 18, 54, 63
32. 28, 42, 84

33. NURSING A nurse, working in an intensive care unit, has to check a patient's vital signs every 45 minutes. Another nurse has to give the same patient his medication every 2 hours. If both nurses are in the patient's room together now, how long will it be until they are once again in the room together?

34. BARBECUES A certain brand of hot dogs comes in packages of 10. A certain brand of hot dog buns comes in packages of 12. For a family reunion barbecue, how many packages of hot dogs and how many packages of hot dog buns should be purchased so that no hot dogs and no buns are wasted?

4.5 *Multiplying and Dividing Mixed Numbers*

In this section, you will learn about

- Mixed numbers • Writing mixed numbers as improper fractions
- Writing improper fractions as mixed numbers
- Graphing fractions and mixed numbers
- Multiplying and dividing mixed numbers

INTRODUCTION. In the next two sections, we will show how to add, subtract, multiply, and divide *mixed numbers.* These numbers are widely used in daily life. Here are a few examples.

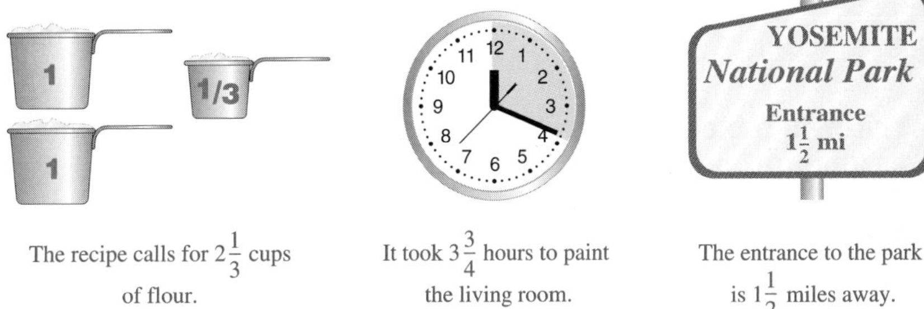

The recipe calls for $2\frac{1}{3}$ cups of flour.

It took $3\frac{3}{4}$ hours to paint the living room.

The entrance to the park is $1\frac{1}{2}$ miles away.

Mixed numbers

A **mixed number** is the *sum* of a whole number and a proper fraction. For example, $2\frac{3}{4}$ is a mixed number.

$$2\frac{3}{4} = 2 + \frac{3}{4}$$

Mixed number ↑ Whole number ↑ Proper fraction ↑

COMMENT Note that $2\frac{3}{4}$ means $2 + \frac{3}{4}$, even though the + sign is not written. Do not confuse $2\frac{3}{4}$ with $2 \cdot \frac{3}{4}$ or $2\left(\frac{3}{4}\right)$, which indicate the multiplication of 2 and $\frac{3}{4}$.

In this section, we will work with negative as well as positive mixed numbers. For example, the negative mixed number $-4\frac{3}{4}$ could be used to represent $4\frac{3}{4}$ feet below sea level. We think of $-4\frac{3}{4}$ as $-4 - \frac{3}{4}$.

Writing mixed numbers as improper fractions

To see that mixed numbers are related to improper fractions, consider $2\frac{3}{4}$. To write $2\frac{3}{4}$ as an improper fraction, we need to find out how many *fourths* it represents. One way is to use the fundamental property of fractions.

$$2\frac{3}{4} = 2 + \frac{3}{4}$$ Write the mixed number $2\frac{3}{4}$ as a sum.

$$= \frac{2}{1} + \frac{3}{4}$$ Write 2 as a fraction: $2 = \frac{2}{1}$.

$= \frac{2 \cdot 4}{1 \cdot 4} + \frac{3}{4}$ Use the fundamental property of fractions to express $\frac{2}{1}$ as a fraction with denominator 4.

$= \frac{8}{4} + \frac{3}{4}$ Do the multiplications in the numerator and denominator.

$= \frac{11}{4}$ Add the numerators: $8 + 3 = 11$. Write the sum over the common denominator, 4.

Thus, $2\frac{3}{4} = \frac{11}{4}$.

We can obtain the same result with far less work. To change $2\frac{3}{4}$ to an improper fraction, we simply multiply 2 by 4 and add 3 to get the numerator, and keep the denominator of 4.

$$2\frac{3}{4} = \frac{2(4) + 3}{4} = \frac{11}{4}$$

This example illustrates the following general rule.

Writing a mixed number as an improper fraction

To write a mixed number as an improper fraction, multiply the whole-number part by the denominator of the fraction and add the result to the numerator. Write this sum over the denominator.

EXAMPLE 1 ***Writing a mixed number as an improper fraction.*** Write the mixed number $5\frac{1}{6}$ as an improper fraction.

Solution

$5\frac{1}{6} = \frac{5(6) + 1}{6}$ Multiply 5 by the denominator 6. Add the numerator 1. Write this sum over the denominator 6.

$= \frac{30 + 1}{6}$ Do the multiplication: $5(6) = 30$.

$= \frac{31}{6}$ Do the addition: $30 + 1 = 31$.

Self Check

Write the mixed number $3\frac{3}{8}$ as an improper fraction.

Answer: $\frac{27}{8}$ ■

To write a negative mixed number in fractional form, ignore the − sign and use the method shown in Example 1 on the positive mixed number. Once that procedure is completed, write a − sign in front of the result. For example, $-3\frac{1}{4} = -\frac{13}{4}$.

Writing improper fractions as mixed numbers

To write an improper fraction as a mixed number, we must find two things: the *whole-number part* and the *fractional part* of the mixed number. To develop a procedure to do this, let's consider the improper fraction $\frac{7}{3}$. To find the number of groups of 3 in 7, we can divide 7 by 3. This will find the whole-number part of the mixed number. The remainder is the numerator of the fractional part of the mixed number.

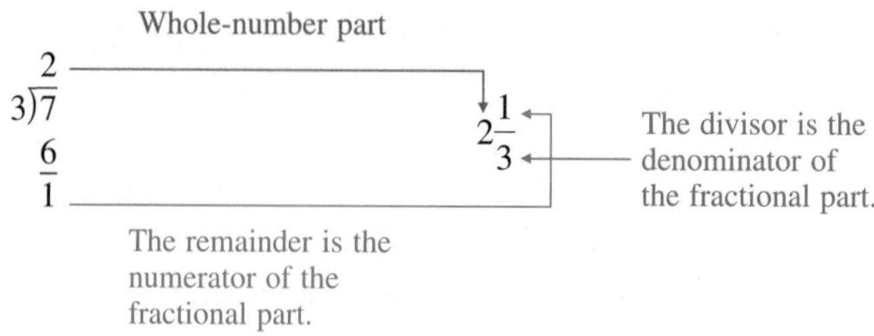

This example suggests the following general rule.

Writing an improper fraction as a mixed number

To write an improper fraction as a mixed number, divide the numerator by the denominator to obtain the whole-number part. The remainder over the divisor is the fractional part.

EXAMPLE 2 ***Writing an improper fraction as a mixed number.*** Write $\frac{29}{6}$ as a mixed number.

Solution

$$\begin{array}{r} 4 \\ 6\overline{)29} \\ \underline{24} \\ 5 \end{array}$$

Divide the numerator by the denominator. The remainder is 5.

Thus, $\frac{29}{6} = 4\frac{5}{6}$.

Self Check

Write $\frac{43}{5}$ as a mixed number.

Answer: $8\frac{3}{5}$

Graphing fractions and mixed numbers

Earlier, we graphed whole numbers and integers on a number line. Fractions and mixed numbers can also be graphed on a number line.

EXAMPLE 3 ***Graphing fractions and mixed numbers.*** Graph $-2\frac{3}{4}, -1\frac{1}{2}, -\frac{1}{8}$, and $\frac{13}{5}$ on a number line.

Solution

To help locate the graph of each number, we make some observations:

- Since $-2\frac{3}{4} < -2$, the graph of $-2\frac{3}{4}$ is to the left of -2.
- The number $-1\frac{1}{2}$ is between -1 and -2.
- The number $-\frac{1}{8}$ is less than 0.
- Expressed as a mixed number, $\frac{13}{5} = 2\frac{3}{5}$.

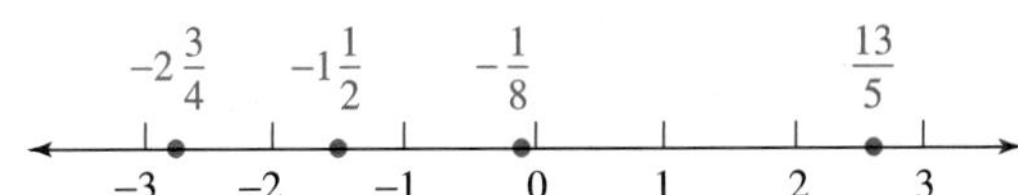

Self Check

Graph $-1\frac{7}{8}, -\frac{2}{3}$, and $\frac{9}{4}$ on a number line.

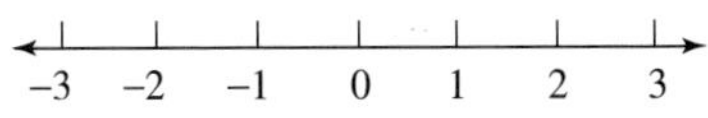

Answer:

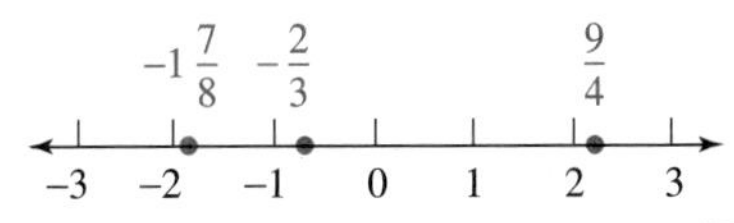

Multiplying and dividing mixed numbers

Multiplying and dividing mixed numbers

To multiply or divide mixed numbers, first change the mixed numbers to improper fractions. Then do the multiplication or division of the fractions.

EXAMPLE 4 ***Multiplying mixed numbers.*** Multiply: $5\frac{1}{5} \cdot 1\frac{2}{13}$.

Solution

$$5\frac{1}{5} \cdot 1\frac{2}{13} = \frac{26}{5} \cdot \frac{15}{13}$$ Write each mixed number as an improper fraction.

$$= \frac{26 \cdot 15}{5 \cdot 13}$$ Multiply the numerators and multiply the denominators.

Self Check

Multiply: $9\frac{3}{5} \cdot 3\frac{3}{4}$.

$= \dfrac{13 \cdot 2 \cdot 5 \cdot 3}{5 \cdot 13}$ Factor 26 as $13 \cdot 2$ and 15 as $5 \cdot 3$.

$= \dfrac{\overset{1}{\cancel{13}} \cdot 2 \cdot \overset{1}{\cancel{5}} \cdot 3}{\underset{1}{\cancel{5}} \cdot \underset{1}{\cancel{13}}}$ Divide out the common factors of 13 and 5.

$= \dfrac{6}{1}$ Multiply in the numerator and denominator.

$= 6$ Simplify: $\frac{6}{1} = 6$.

Answer: 36

EXAMPLE 5 ***Dividing mixed numbers.*** Divide: $-3\frac{3}{8} \div 2\frac{1}{4}$.

Solution

$-3\dfrac{3}{8} \div 2\dfrac{1}{4} = -\dfrac{27}{8} \div \dfrac{9}{4}$ Write each mixed number as an improper fraction.

$= -\dfrac{27}{8} \cdot \dfrac{4}{9}$ Multiply by the reciprocal of $\frac{9}{4}$.

$= -\dfrac{27 \cdot 4}{8 \cdot 9}$ The product of two fractions with unlike signs is negative. Multiply the numerators and multiply the denominators.

$= -\dfrac{\overset{1}{\cancel{9}} \cdot 3 \cdot \overset{1}{\cancel{4}}}{\underset{1}{\cancel{4}} \cdot 2 \cdot \underset{1}{\cancel{9}}}$ Factor 27 as $9 \cdot 3$ and 8 as $4 \cdot 2$. Divide out the common factors of 9 and 4.

$= -\dfrac{3}{2}$ Multiply in the numerator and denominator.

$= -1\dfrac{1}{2}$ Write $-\frac{3}{2}$ as a mixed number.

Self Check

Divide: $3\frac{4}{15} \div \left(-2\frac{1}{10}\right)$.

Answer: $-1\frac{5}{9}$

EXAMPLE 6 ***Government grant.*** If $\$12\frac{1}{2}$ million is to be divided equally among five cities to fund recreation programs, how much will each city receive?

Solution To find the amount received by each city, we divide the grant money by 5.

$12\dfrac{1}{2} \div 5 = \dfrac{25}{2} \div \dfrac{5}{1}$ Write $12\frac{1}{2}$ as an improper fraction, and write 5 as a fraction.

$= \dfrac{25}{2} \cdot \dfrac{1}{5}$ Multiply by the reciprocal of $\frac{5}{1}$.

$= \dfrac{25 \cdot 1}{2 \cdot 5}$ Multiply the numerators and multiply the denominators.

$= \dfrac{\overset{1}{\cancel{5}} \cdot 5 \cdot 1}{2 \cdot \underset{1}{\cancel{5}}}$ Factor 25 as $5 \cdot 5$. Divide out the common factor of 5.

$= \dfrac{5}{2}$ Multiply in the numerator and denominator.

$= 2\dfrac{1}{2}$ Write $\frac{5}{2}$ as a mixed number.

Each city will receive $\$2\frac{1}{2}$ million.

STUDY SET Section 4.5

VOCABULARY *Fill in the blanks to make the statements true.*

1. A ______________ number is the sum of a whole number and a proper fraction.

2. An ______________ fraction is a fraction with a numerator that is greater than or equal to its denominator.

3. To ______________ a number means to locate its position on a number line and highlight it using a heavy dot.

4. Multiplying or dividing the ______________ and ______________ of a fraction by the same nonzero number does not change the value of the fraction.

CONCEPTS

5. What signed number could be used to describe each situation?

a. A temperature of five and one-half degrees below zero

b. One and seven-eighths inches under the finish grade

6. What signed number could be used to describe each situation?

a. A rain total two and three-tenths of an inch lower than the average

b. Three and one-half minutes before liftoff

7. a. In Illustration 1, the divisions on the face of the meter represent fractions. What value is the arrow registering?

b. If the arrow moves two marks to the left, what value will it register?

ILLUSTRATION 1

8. a. In Illustration 2, the divisions on the face of the meter represent fractions. What value is the arrow registering?

b. If the arrow moves up one mark, what value will it register?

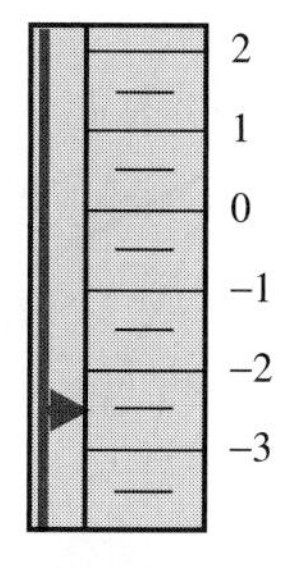

ILLUSTRATION 2

9. What fractions have been graphed on the number line in Illustration 3?

ILLUSTRATION 3

10. What mixed numbers have been graphed on the number line in Illustration 4?

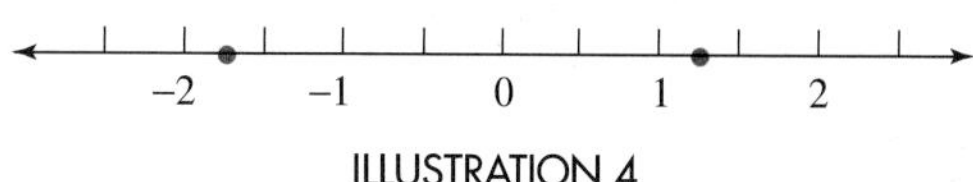

ILLUSTRATION 4

11. DIVING See Illustration 5. Complete the description of the dive by filling in the blank with a mixed number.

Forward ______ somersaults from the pike position

ILLUSTRATION 5

12. PRODUCT LABELING The label in Illustration 6 uses mixed numbers. Write each one as an improper fraction.

ILLUSTRATION 6

13. Draw $\frac{17}{8}$ pizzas.

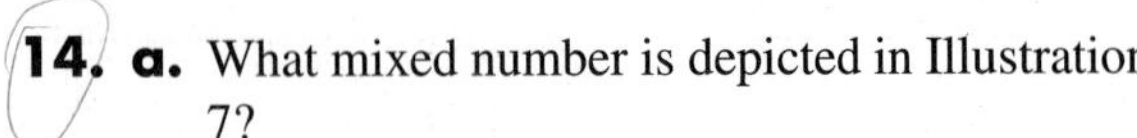

14. **a.** What mixed number is depicted in Illustration 7?

b. What improper fraction is shown in Illustration 7?

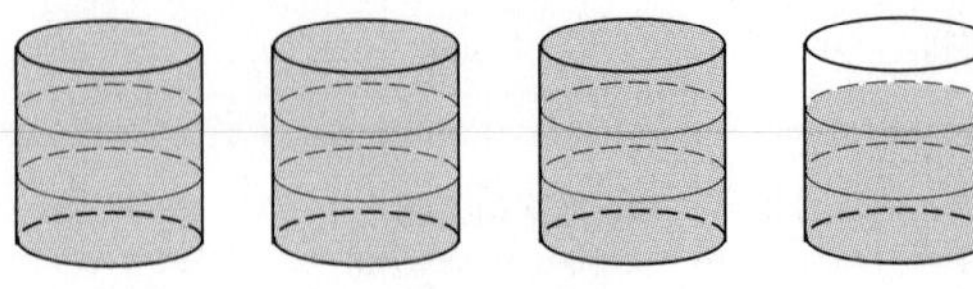

ILLUSTRATION 7

NOTATION *Complete each solution.*

15. Multiply: $-5\frac{1}{4} \cdot 1\frac{1}{7}$.

$$-5\frac{1}{4} \cdot 1\frac{1}{7} = -\frac{21}{4} \cdot \frac{\square}{7}$$
$$= -\frac{21 \cdot \square}{4 \cdot 7}$$
$$= -\frac{\overset{1}{\cancel{7}} \cdot 3 \cdot \overset{1}{\cancel{\square}} \cdot 2}{\underset{1}{\cancel{4}} \cdot \underset{1}{\cancel{7}}}$$
$$= -\frac{\square}{1}$$
$$= -6$$

16. Divide: $-5\frac{5}{6} \div 2\frac{1}{12}$.

$$-5\frac{5}{6} \div 2\frac{1}{12} = -\frac{\square}{6} \div \frac{25}{12}$$
$$= -\frac{35}{6} \cdot \frac{12}{\square}$$
$$= -\frac{35 \cdot 12}{6 \cdot \square}$$
$$= -\frac{\overset{1}{\cancel{5}} \cdot \square \cdot \overset{1}{\cancel{6}} \cdot 2}{\underset{1}{\cancel{6}} \cdot \underset{1}{\cancel{5}} \cdot \square}$$
$$= -\frac{\square}{5}$$
$$= -2\frac{4}{5}$$

PRACTICE *Write each improper fraction as a mixed number. Simplify the result, if possible.*

17. $\frac{15}{4}$ **18.** $\frac{41}{6}$

19. $\frac{29}{5}$ **20.** $\frac{29}{3}$

21. $-\frac{20}{6}$ **22.** $-\frac{28}{8}$

23. $\frac{127}{12}$ **24.** $\frac{197}{16}$

Write each mixed number as an improper fraction.

25. $6\frac{1}{2}$ **26.** $8\frac{2}{3}$

27. $20\frac{4}{5}$ **28.** $15\frac{3}{8}$

29. $-6\frac{2}{9}$ **30.** $-7\frac{1}{12}$

31. $200\frac{2}{3}$ **32.** $90\frac{5}{6}$

Graph each set of numbers on the number line.

33. $\left\{-2\frac{8}{9}, 1\frac{2}{3}, \frac{16}{5}\right\}$

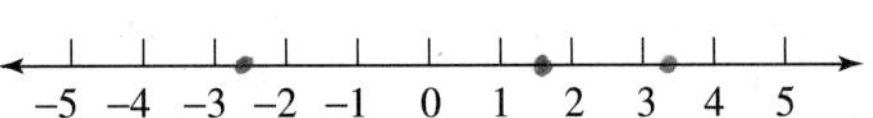

34. $\left\{-\frac{3}{4}, -3\frac{1}{4}, \frac{5}{2}\right\}$

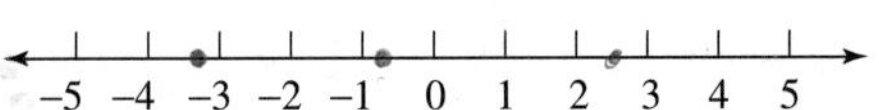

35. $\left\{3\frac{1}{7}, -\frac{98}{99}, -\frac{10}{3}\right\}$

−5 −4 −3 −2 −1 0 1 2 3 4 5

36. $\left\{-2\frac{1}{5}, \frac{4}{5}, -\frac{11}{3}\right\}$

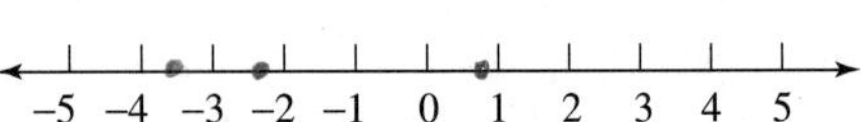

Multiply.

37. $1\frac{2}{3} \cdot 2\frac{1}{7}$ **38.** $2\frac{3}{5} \cdot 1\frac{2}{3}$

39. $-7\frac{1}{2}\left(-1\frac{2}{5}\right)$ **40.** $-4\frac{1}{8}\left(-1\frac{7}{9}\right)$

41. $3\frac{1}{16} \cdot 4\frac{4}{7}$

42. $5\frac{3}{5} \cdot 1\frac{11}{14}$

43. $-6 \cdot 2\frac{7}{24}$

44. $-7 \cdot 1\frac{3}{28}$

45. $2\frac{1}{2}\left(-3\frac{1}{3}\right)$

46. $\left(-3\frac{1}{4}\right)\left(1\frac{1}{5}\right)$

47. $2\frac{5}{8} \cdot \frac{5}{27}$

48. $3\frac{1}{9} \cdot \frac{3}{32}$

49. Find the product of $1\frac{2}{3}$, 6, and $-\frac{1}{8}$.

50. Find the product of $-\frac{5}{6}$, -8, and $-2\frac{1}{10}$.

Evaluate each power.

51. $\left(1\frac{2}{3}\right)^2$

52. $\left(3\frac{1}{2}\right)^2$

53. $\left(-1\frac{1}{3}\right)^3$

54. $\left(-1\frac{1}{5}\right)^3$

Divide.

55. $3\frac{1}{3} \div 1\frac{5}{6}$

56. $3\frac{3}{4} \div 5\frac{1}{3}$

57. $-6\frac{3}{5} \div 7\frac{1}{3}$

58. $-4\frac{1}{4} \div 4\frac{1}{2}$

59. $-20\frac{1}{4} \div \left(-1\frac{11}{16}\right)$

60. $-2\frac{7}{10} \div \left(-1\frac{1}{14}\right)$

61. $6\frac{1}{4} \div 20$

62. $4\frac{2}{5} \div 11$

63. $1\frac{2}{3} \div \left(-2\frac{1}{2}\right)$

64. $2\frac{1}{2} \div \left(-1\frac{5}{8}\right)$

65. $8 \div 3\frac{1}{5}$

66. $15 \div 3\frac{1}{3}$

67. Find the quotient of $-4\frac{1}{2}$ and $2\frac{1}{4}$.

68. Find the quotient of 25 and $-10\frac{5}{7}$.

APPLICATIONS

69. CALORIES A company advertises that its mints contain only $3\frac{1}{5}$ calories apiece. What is the calorie intake if you eat an entire package of 20 mints?

70. CEMENT MIXER A cement mixer can carry $9\frac{1}{2}$ cubic yards of concrete. If it makes 8 trips to a job site, how much concrete will be delivered to the site?

71. SHOPPING In Illustration 8, what is the cost of buying the fruit in the scale?

ILLUSTRATION 8

72. FRAME How much molding is needed to make the square picture frame in Illustration 9?

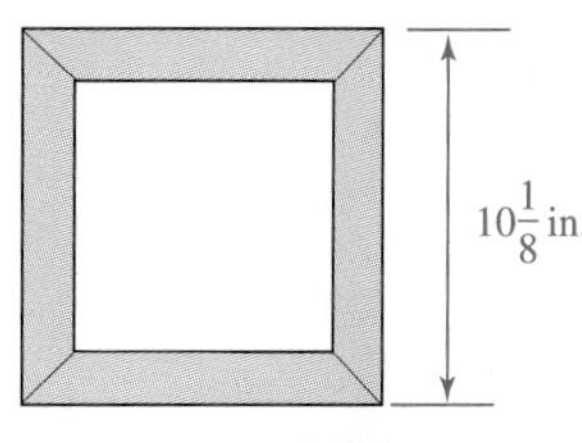

ILLUSTRATION 9

73. SUBDIVISION A developer donated to the county 100 of the 1,000 acres of land she owned. She divided the remaining acreage into $1\frac{1}{3}$-acre lots. How many lots were created?

74. CATERING How many people can be served $\frac{1}{3}$-pound hamburgers if a caterer purchases 200 pounds of ground beef?

75. GRAPH PAPER Mathematicians use specially marked paper, called *graph paper,* when drawing figures. It is made up of $\frac{1}{4}$-inch squares. Find the length and width of the piece of graph paper in Illustration 10.

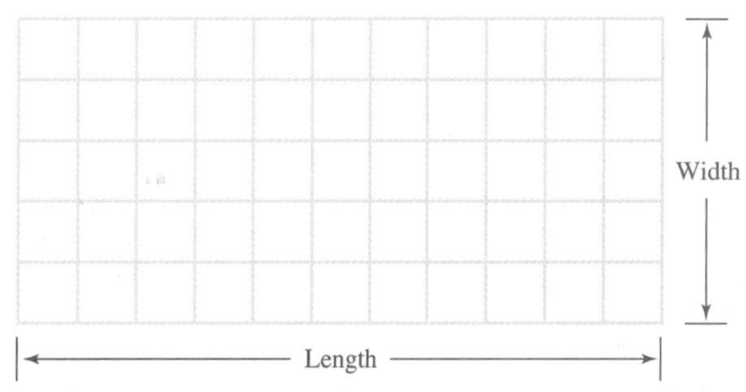

ILLUSTRATION 10

76. LUMBER As shown in Illustration 11 on the next page, 2-by-4's from the lumber yard do not really have dimensions of 2 inches by 4 inches. How wide and how high is the stack of 2-by-4s?

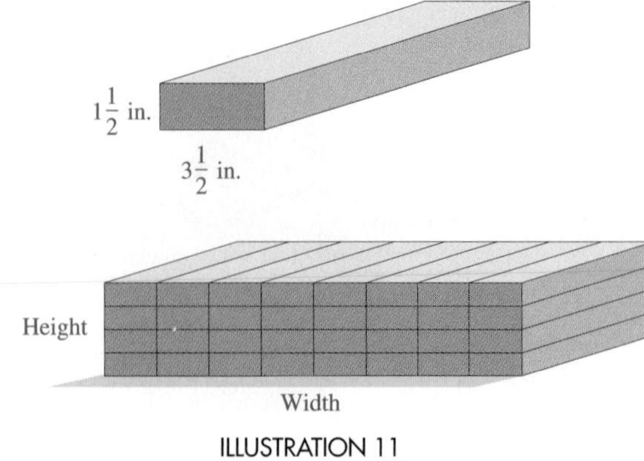

ILLUSTRATION 11

77. EMERGENCY EXIT Illustration 12 shows a sign that marks the emergency exit on a school bus. Find the area of the sign.

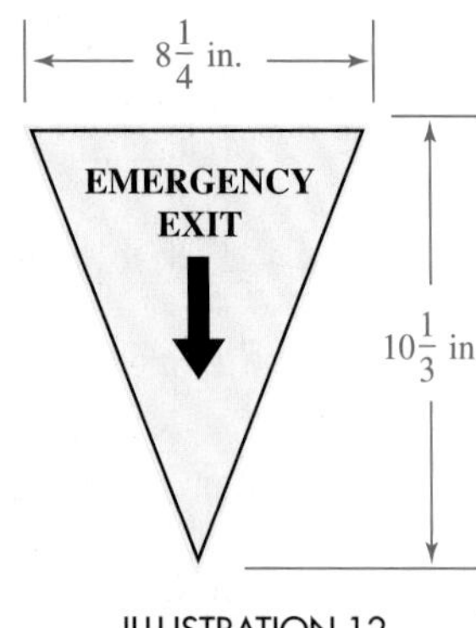

ILLUSTRATION 12

78. HORSE RACING The race tracks on which thoroughbred horses run are marked off in $\frac{1}{8}$-mile-long segments called furlongs. How many furlongs are there in a $1\frac{1}{16}$-mile race?

79. FIRE ESCAPE The fire escape stairway in an office building is shown in Illustration 13. Each riser is $7\frac{1}{2}$ inches high. If each floor is 105 inches high and the building is 43 stories tall, how many steps are there in the stairway?

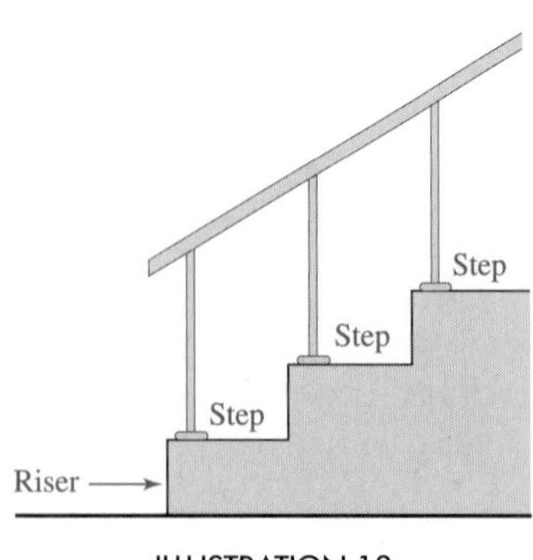

ILLUSTRATION 13

80. LICENSE PLATE Find the area of the license plate in Illustration 14.

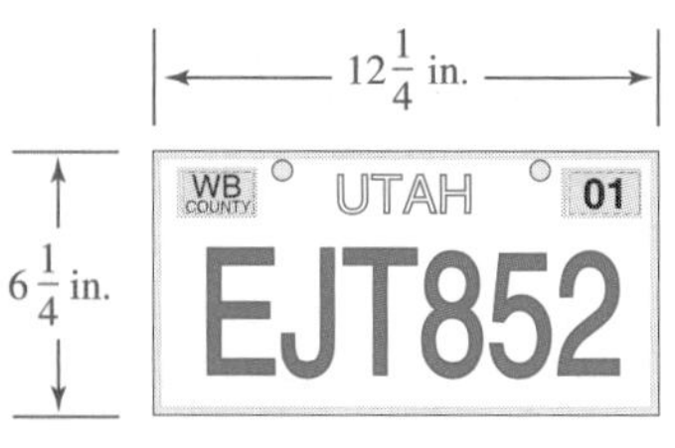

ILLUSTRATION 14

81. SHOPPING ON THE INTERNET A mother is ordering a pair of jeans for her daughter from the screen shown in Illustration 15. If the daughter's height is $60\frac{3}{4}$ in. and her waist is $24\frac{1}{2}$ in., on what size and what cut should the mother point and click?

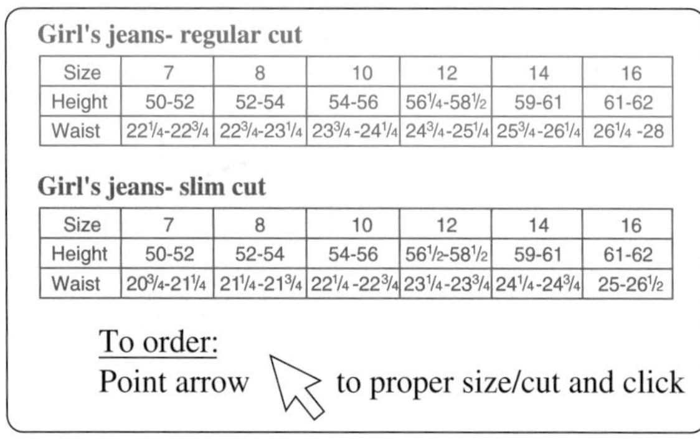

Girl's jeans- regular cut

Size	7	8	10	12	14	16
Height	50-52	52-54	54-56	$56\frac{1}{4}$-$58\frac{1}{2}$	59-61	61-62
Waist	$22\frac{1}{4}$-$22\frac{3}{4}$	$22\frac{3}{4}$-$23\frac{1}{4}$	$23\frac{3}{4}$-$24\frac{1}{4}$	$24\frac{3}{4}$-$25\frac{1}{4}$	$25\frac{3}{4}$-$26\frac{1}{4}$	$26\frac{1}{4}$-28

Girl's jeans- slim cut

Size	7	8	10	12	14	16
Height	50-52	52-54	54-56	$56\frac{1}{2}$-$58\frac{1}{2}$	59-61	61-62
Waist	$20\frac{3}{4}$-$21\frac{1}{4}$	$21\frac{1}{4}$-$21\frac{3}{4}$	$22\frac{1}{4}$-$22\frac{3}{4}$	$23\frac{1}{4}$-$23\frac{3}{4}$	$24\frac{1}{4}$-$24\frac{3}{4}$	25-$26\frac{1}{2}$

To order:
Point arrow to proper size/cut and click

ILLUSTRATION 15

82. SEWING Use the table in Illustration 16 to determine the number of yards of fabric needed:

a. to make a size 16 top if the fabric to be used is 60 inches wide.

b. to make size 18 pants if the fabric to be used is 45 inches wide.

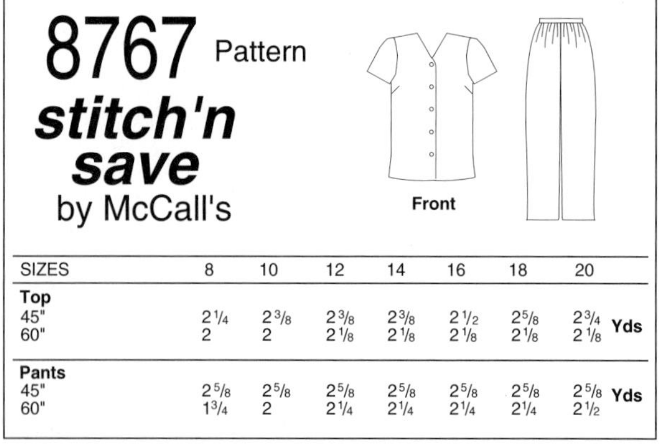

SIZES	8	10	12	14	16	18	20	
Top								
45"	$2\frac{1}{4}$	$2\frac{3}{8}$	$2\frac{3}{8}$	$2\frac{3}{8}$	$2\frac{1}{2}$	$2\frac{5}{8}$	$2\frac{3}{4}$	**Yds**
60"	2	2	$2\frac{1}{8}$	$2\frac{1}{8}$	$2\frac{1}{8}$	$2\frac{1}{8}$	$2\frac{1}{8}$	
Pants								
45"	$2\frac{5}{8}$	$2\frac{5}{8}$	$2\frac{5}{8}$	$2\frac{5}{8}$	$2\frac{5}{8}$	$2\frac{5}{8}$	$2\frac{5}{8}$	**Yds**
60"	$1\frac{3}{4}$	2	$2\frac{1}{4}$	$2\frac{1}{4}$	$2\frac{1}{4}$	$2\frac{1}{4}$	$2\frac{1}{2}$	

ILLUSTRATION 16

WRITING

83. Explain the difference between $2\frac{3}{4}$ and $2\left(\frac{3}{4}\right)$.

84. Give three examples of how you use mixed numbers in daily life.

85. Explain the procedure used to write an improper fraction as a mixed number.

86. Explain the procedure used to multiply two mixed numbers.

REVIEW

87. Evaluate $3^2 \cdot 2^3$.

88. If a represents a number, then $a \cdot 0 =$ ____.

89. Write $8 + 8 + 8 + 8$ as a multiplication.

90. If a square measures 1 inch on each side, what is its area?

91. What operation must be undone to solve for x?

$$\frac{x}{2} = -12$$

92. In the formula $C = \frac{5(F - 32)}{9}$, what do C and F represent?

93. Explain why $3t$ and $3u$ are not like terms.

94. Add: $3x + 3x$.

4.6 Adding and Subtracting Mixed Numbers

In this section, you will learn about

- Adding mixed numbers • Adding mixed numbers in vertical form
- Subtracting mixed numbers

Introduction. In this section, we will discuss three methods for adding and subtracting mixed numbers. The first method works well when the whole-number parts of the mixed numbers are small. The second method works well when the whole-number parts of the mixed numbers are large. The third method uses columns as a way to organize the work.

Adding mixed numbers

We can add mixed numbers by writing them as improper fractions. To do so, we follow these steps.

Adding mixed numbers: method 1

1. Write each mixed number as an improper fraction.
2. Write each improper fraction as an equivalent fraction with a denominator that is the LCD.
3. Add the fractions.
4. Change the result to a mixed number if desired.

EXAMPLE 1 ***Adding mixed numbers.*** Add: $4\frac{1}{6} + 2\frac{3}{4}$.

Solution

$$4\frac{1}{6} + 2\frac{3}{4} = \frac{25}{6} + \frac{11}{4}$$ Write each mixed number as an improper fraction: $4\frac{1}{6} = \frac{25}{6}$ and $2\frac{3}{4} = \frac{11}{4}$.

By inspection, we see that the lowest common denominator is 12.

$$= \frac{25 \cdot 2}{6 \cdot 2} + \frac{11 \cdot 3}{4 \cdot 3}$$ Write each fraction as a fraction with a denominator of 12.

Self Check

Add: $3\frac{2}{3} + 1\frac{1}{5}$.

$= \frac{50}{12} + \frac{33}{12}$ Do the multiplications in the numerators and denominators.

$= \frac{83}{12}$ Add the numerators: $50 + 33 = 83$. Write the sum over the common denominator 12.

$= 6\frac{11}{12}$ Write the improper fraction as a mixed number: $\frac{83}{12} = 6\frac{11}{12}$.

Answer: $4\frac{13}{15}$ ■

We can also add mixed numbers by adding their whole-number parts and their fractional parts. To do so, we follow these steps.

Adding mixed numbers: method 2

1. Write each mixed number as the sum of a whole number and a fraction.
2. Use the commutative property of addition to write the whole numbers together and the fractions together.
3. Add the whole numbers and the fractions separately.
4. Write the result as a mixed number if necessary.

EXAMPLE 2 ***Adding mixed numbers.*** Find the sum: $168\frac{3}{4} + 85\frac{1}{5}$.

Solution

$168\frac{3}{4} + 85\frac{1}{5} = 168 + \frac{3}{4} + 85 + \frac{1}{5}$ Write each mixed number as the sum of a whole number and a fraction.

$= 168 + 85 + \frac{3}{4} + \frac{1}{5}$ Use the commutative property of addition to change the order of the addition.

$= 253 + \frac{3}{4} + \frac{1}{5}$ Add the whole numbers: $168 + 85 = 253$.

$= 253 + \frac{3 \cdot 5}{4 \cdot 5} + \frac{1 \cdot 4}{5 \cdot 4}$ Write each fraction as a fraction with denominator 20.

$= 253 + \frac{15}{20} + \frac{4}{20}$ Multiply in the numerators and denominators.

$= 253 + \frac{19}{20}$ Add the numerators and write the sum over the common denominator 20.

$= 253\frac{19}{20}$ Write the sum as a mixed number.

Self Check

Find the sum: $275\frac{1}{6} + 81\frac{3}{5}$.

Answer: $356\frac{23}{30}$ ■

COMMENT If we use method 1 to add the mixed numbers in Example 2, the numbers we encounter are cumbersome. As expected, the result is the same: $253\frac{19}{20}$.

$168\frac{3}{4} + 85\frac{1}{5} = \frac{675}{4} + \frac{426}{5}$ Write $168\frac{3}{4}$ and $85\frac{1}{5}$ as improper fractions.

$= \frac{675 \cdot 5}{4 \cdot 5} + \frac{426 \cdot 4}{5 \cdot 4}$ The LCD is 20.

$= \frac{3,375}{20} + \frac{1,704}{20}$

$= \frac{5,079}{20}$

$= 253\frac{19}{20}$

Generally speaking, the larger the whole-number parts of the mixed numbers get, the more difficult it becomes to add those mixed numbers using method 1.

Adding mixed numbers in vertical form

By working in columns, we can add mixed numbers quickly. The strategy is the same as in Example 2: Add whole numbers to whole numbers and fractions to fractions.

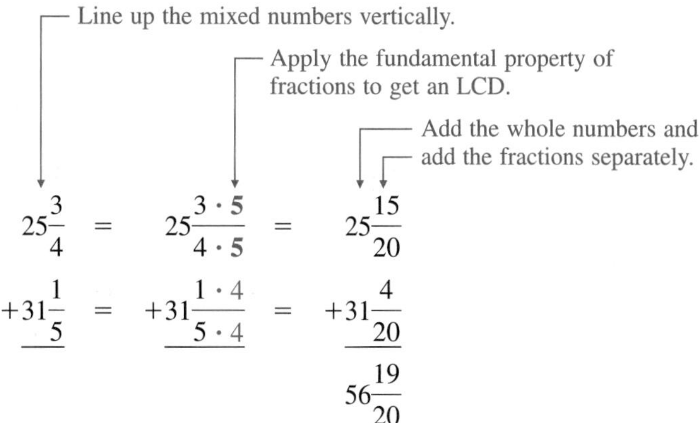

EXAMPLE 3 ***Suspension bridge.*** Find the total length of cable that must be ordered if cables a, d, and e of the suspension bridge in Figure 4-9 are to be replaced. (See the table below.)

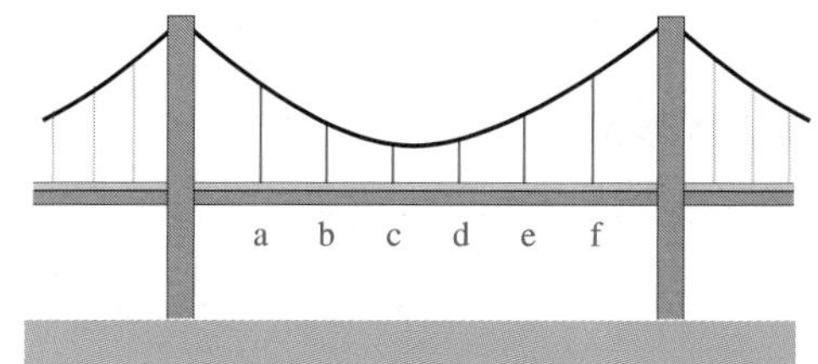

FIGURE 4-9

Bridge Specifications			
Cable	a	b	c
Length (feet)	$75\frac{1}{12}$	$54\frac{1}{6}$	$43\frac{1}{4}$

Solution To find the total length of cable to be ordered, we add the lengths of cables a, d, and e. Because of the symmetric design, cables e and b and cables d and c are the same length.

Length of cable a	plus	length of cable d (or cable c)	plus	length of cable e (or cable b)	equals	the total length needed.
$75\frac{1}{12}$	$+$	$43\frac{1}{4}$	$+$	$54\frac{1}{6}$	$=$	total length

We add the mixed numbers using a vertical format.

$$\begin{aligned} 75\frac{1}{12} &= 75\frac{1}{12} &&= 75\frac{1}{12} \\ 43\frac{1}{4} &= 43\frac{1 \cdot 3}{4 \cdot 3} &&= 43\frac{3}{12} \\ +54\frac{1}{6} &= +54\frac{1 \cdot 2}{6 \cdot 2} &&= +\ 54\frac{2}{12} \\ & && \quad 172\frac{6}{12} = 172\frac{1}{2} \qquad \text{Simplify: } \frac{6}{12} = \frac{1}{2}. \end{aligned}$$

The total length of cable needed for the replacement is $172\frac{1}{2}$ feet. ■

When adding mixed numbers, the sum of the fractions sometimes yields an improper fraction, as in the next example.

EXAMPLE 4 ***Vertical form.*** Add: $45\frac{2}{3} + 96\frac{4}{5}$.

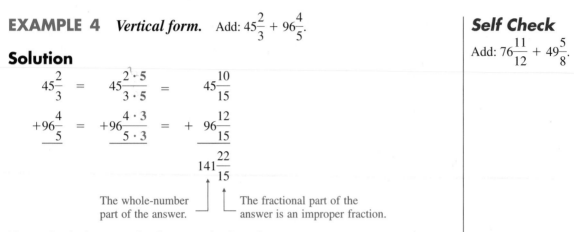

Solution

$$\begin{array}{rcrcr} 45\frac{2}{3} & = & 45\frac{2 \cdot 5}{3 \cdot 5} & = & 45\frac{10}{15} \\ +96\frac{4}{5} & = & +96\frac{4 \cdot 3}{5 \cdot 3} & = & +\ 96\frac{12}{15} \\ \hline & & & & 141\frac{22}{15} \end{array}$$

The whole-number part of the answer. The fractional part of the answer is an improper fraction.

Now write the improper fraction as a mixed number.

$$141\frac{22}{15} = 141 + \frac{22}{15} = 141 + 1\frac{7}{15} = 142\frac{7}{15}$$

Self Check

Add: $76\frac{11}{12} + 49\frac{5}{8}$.

Answer: $126\frac{13}{24}$

Subtracting mixed numbers

Subtracting mixed numbers is similar to adding mixed numbers.

EXAMPLE 5 ***Cooking.*** How much butter is left in a 10-pound tub if $2\frac{2}{3}$ pounds are used for a wedding cake?

Solution The phrase "How much is left?" suggests subtraction.

$$10 - 2\frac{2}{3} = \frac{10}{1} - \frac{8}{3}$$ Write 10 as a fraction: $10 = \frac{10}{1}$. Write $2\frac{2}{3}$ as $\frac{8}{3}$.

By inspection, we see that the LCD is 3.

$$\frac{10}{1} - 2\frac{2}{3} = \frac{10 \cdot 3}{1 \cdot 3} - \frac{8}{3}$$ Write the first fraction with a denominator of 3.

$$= \frac{30}{3} - \frac{8}{3}$$ Do the multiplications in the first fraction.

$$= \frac{30 - 8}{3}$$ Subtract the numerators and write the difference over the common denominator.

$$= \frac{22}{3}$$ Do the subtraction: $30 - 8 = 22$.

$$= 7\frac{1}{3}$$ Write $\frac{22}{3}$ as a mixed number.

There are $7\frac{1}{3}$ pounds of butter left in the tub.

In the next example, the fraction being subtracted *from* is smaller than the fraction being subtracted. Because of this, we will have to borrow.

EXAMPLE 6 ***Borrowing.*** Subtract: $34\frac{1}{5} - 11\frac{2}{3}$.

Solution

We will use the vertical form to subtract. The LCD is 15, so we write each fraction as a fraction with a denominator of 15.

$$\begin{array}{rcrcr} 34\frac{1}{5} & = & 34\frac{1 \cdot 3}{5 \cdot 3} & = & 34\frac{3}{15} \\ -11\frac{2}{3} & = & -11\frac{2 \cdot 5}{3 \cdot 5} & = & -11\frac{10}{15} \end{array}$$

Since $\frac{10}{15}$ is larger than $\frac{3}{15}$, borrow 1 (in the form of $\frac{15}{15}$) from 34 and add it to $\frac{3}{15}$ to obtain $33\frac{3}{15} + \frac{15}{15} = 33\frac{18}{15}$. Then we subtract the fractions and the whole numbers separately.

$$\begin{array}{rcr} 33\frac{3}{15} + \frac{15}{15} & = & 33\frac{18}{15} \\ -11\frac{10}{15} & = & -11\frac{10}{15} \\ \hline & & 22\frac{8}{15} \end{array}$$

Self Check

Subtract: $101\frac{3}{4} - 79\frac{15}{16}$.

Answer: $21\frac{13}{16}$

EXAMPLE 7 ***Borrowing from a whole number.*** Subtract: $419 - 53\frac{11}{16}$.

Solution

We align the numbers vertically and borrow 1 (in the form of $\frac{16}{16}$) from 419. Then we subtract the fractions and subtract the whole numbers separately.

$$\begin{array}{rcr} 419 & = & 418\frac{16}{16} \\ -\ 53\frac{11}{16} & = & -\ 53\frac{11}{16} \\ \hline & & 365\frac{5}{16} \end{array}$$

Self Check

Subtract: $2{,}300 - 129\frac{19}{32}$.

Answer: $2{,}170\frac{13}{32}$

STUDY SET Section 4.6

VOCABULARY *Fill in the blanks to make the statements true.*

1. By the ______________ property of addition, we can add numbers in any order.

2. A ______________ number contains a whole-number part and a fractional part.

3. Consider

$$\begin{array}{rcl} 80\frac{1}{3} & = & 79\frac{1}{3} + \frac{3}{3} \\ -24\frac{2}{3} & = & -24\frac{2}{3} \end{array}$$

To do the subtraction, we ______________ 1 in the form of $\frac{3}{3}$.

4. Fractions that are greater than 1, such as $\frac{11}{8}$, are called ______________ fractions.

CONCEPTS

5. a. For $76\frac{3}{4}$, list the whole-number part and the fractional part.

b. Write $76\frac{3}{4}$ as a sum.

6. Use the commutative property of addition to get the whole numbers together.

$$14 + \frac{5}{6} + 53 + \frac{1}{6}$$

7. What property is being highlighted here?

$$\begin{array}{r} 25\frac{3 \cdot 5}{4 \cdot 5} \\ +31\frac{1 \cdot 4}{5 \cdot 4} \\ \hline \end{array}$$

8. a. The denominators of two fractions, expressed in prime-factored form, are $5 \cdot 2$ and $5 \cdot 3$. Find the LCD for the fractions.

b. The denominators for three fractions, in prime-factored form, are $3 \cdot 5$, $2 \cdot 3$, and $3 \cdot 3$. Find the LCD for the fractions.

9. Simplify.

a. $9\frac{17}{16}$

b. $1{,}288\frac{7}{3}$

c. $16\frac{12}{8}$

d. $45\frac{24}{20}$

10. Consider

$$\begin{array}{r} 108\frac{1}{4} \\ -\ 99\frac{2}{3} \\ \hline \end{array}$$

a. Explain why we will have to borrow if we subtract the mixed numbers in this way.

b. In what form will we borrow a 1 from 108?

NOTATION *Complete each solution.*

11. Add: $70\frac{3}{5} + 39\frac{2}{7}$.

$$\begin{aligned} 70\frac{3}{5} + 39\frac{2}{7} &= 70 + \frac{3}{5} + 39 + \frac{2}{7} \\ &= 70 + 39 + \frac{3}{5} + \frac{2}{7} \\ &= 109 + \frac{3}{5} + \frac{2}{7} \\ &= 109 + \frac{3 \cdot 7}{5 \cdot 7} + \frac{2 \cdot 5}{7 \cdot 5} \\ &= 109 + \frac{21}{35} + \frac{10}{35} \\ &= 109 + \frac{31}{35} \\ &= 109\frac{31}{35} \end{aligned}$$

12. Subtract: $67\frac{3}{8} - 23\frac{2}{3}$.

$$\begin{array}{rcl} 67\frac{3}{8} & = & 67\frac{3 \cdot 4}{8 \cdot 4} \\ -23\frac{2}{3} & = & -23\frac{2 \cdot 8}{3 \cdot 8} \\ \hline \end{array}$$

$$\begin{array}{rcl} 67\frac{9}{24} & = & 66\frac{9}{24} + \frac{33}{24} \\ -23\frac{16}{24} & = & -23\frac{16}{24} \\ \hline \end{array}$$

$$\begin{array}{r} 66\frac{33}{24} \\ -23\ \frac{16}{24} \\ \hline 43\ \frac{17}{24} \end{array}$$

PRACTICE *Find each sum or difference.*

13. $2\frac{1}{5} + 2\frac{1}{5}$

14. $3\frac{1}{3} + 2\frac{1}{3}$

15. $8\frac{2}{7} - 3\frac{1}{7}$

16. $9\frac{5}{11} - 6\frac{2}{11}$

17. $3\frac{1}{4} + 4\frac{1}{4}$

18. $2\frac{1}{8} + 3\frac{3}{8}$

19. $4\frac{1}{6} + 1\frac{1}{5}$

20. $2\frac{2}{5} + 3\frac{1}{4}$

21. $2\frac{1}{2} - 1\frac{1}{4}$

22. $13\frac{5}{6} - 4\frac{2}{3}$

23. $2\frac{5}{6} - 1\frac{3}{8}$

24. $4\frac{5}{9} - 2\frac{1}{6}$

25. $5\frac{1}{2} + 3\frac{4}{5}$

26. $6\frac{1}{2} + 2\frac{2}{3}$

27. $7\frac{1}{2} - 4\frac{1}{7}$

28. $5\frac{3}{4} - 1\frac{3}{7}$

29. $56\frac{2}{5} + 73\frac{1}{3}$

30. $44\frac{3}{8} + 66\frac{1}{5}$

31. $380\frac{1}{6} + 17\frac{1}{4}$

32. $103\frac{1}{2} + 210\frac{2}{5}$

33. $228\frac{5}{9} + 44\frac{2}{3}$

34. $161\frac{7}{8} + 19\frac{1}{3}$

35. $778\frac{5}{7} - 155\frac{1}{3}$

36. $339\frac{1}{2} - 218\frac{3}{16}$

37. $140\frac{5}{6} - 129\frac{4}{5}$

38. $291\frac{1}{4} - 289\frac{1}{12}$

39. $422\frac{13}{16} - 321\frac{3}{8}$

40. $378\frac{3}{4} - 277\frac{5}{8}$

Find each difference.

41. $16\frac{1}{4} - 13\frac{3}{4}$

42. $40\frac{1}{7} - 19\frac{6}{7}$

43. $76\frac{1}{6} - 49\frac{7}{8}$

44. $101\frac{1}{4} - 70\frac{1}{2}$

45. $140\frac{3}{16} - 129\frac{3}{4}$

46. $211\frac{1}{3} - 8\frac{3}{4}$

47. $334\frac{1}{9} - 13\frac{5}{6}$

48. $442\frac{1}{8} - 429\frac{2}{3}$

Find the sum or difference.

49. $7 - \frac{2}{3}$

50. $6 - \frac{1}{8}$

51. $9 - 8\frac{3}{4}$

52. $11 - 10\frac{4}{5}$

53. $4\frac{1}{7} - \frac{4}{5}$

54. $5\frac{1}{10} - \frac{4}{5}$

55. $6\frac{5}{8} - 3$

56. $10\frac{1}{2} - 6$

57. $\frac{7}{3} + 2$

58. $\frac{9}{7} + 3$

59. $2 + 1\frac{7}{8}$

60. $3\frac{3}{4} + 5$

Find each sum.

61. $12\frac{1}{2} + 5\frac{3}{4} + 35\frac{1}{6}$

62. $31\frac{1}{3} + 20\frac{2}{5} + 10\frac{1}{15}$

63. $58\frac{7}{8} + 340 + 61\frac{1}{4}$

64. $191 + 233\frac{1}{16} + 16\frac{5}{8}$

Find each sum or difference.

65. $-3\frac{3}{4} + \left(-1\frac{1}{2}\right)$

66. $-3\frac{2}{3} + \left(-1\frac{4}{5}\right)$

67. $-4\frac{5}{8} - 1\frac{1}{4}$

68. $-2\frac{1}{16} - 3\frac{7}{8}$

APPLICATIONS

69. FREEWAY TRAVEL A freeway exit sign is shown in Illustration 1. How far apart are the Citrus Ave. and Grand Ave. exits?

70. BASKETBALL See Illustration 2. What is the difference in height between the tallest and the shortest of the starting players?

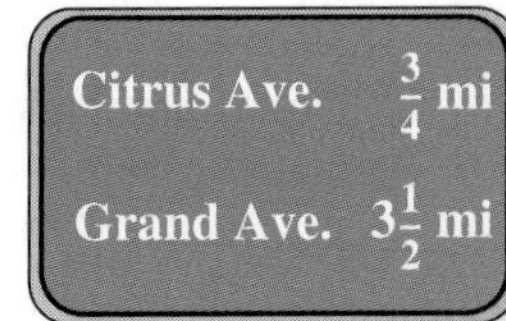

ILLUSTRATION 1

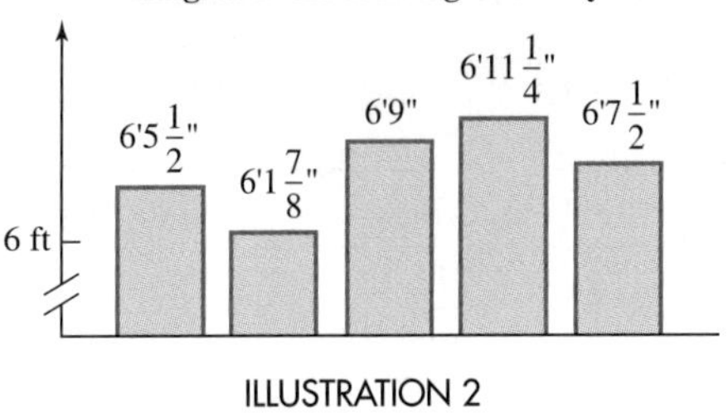

ILLUSTRATION 2

71. TRAIL MIX See Illustration 3. A camper doubles up on the amount of sunflower seeds called for in the recipe. How much trail mix will the adjusted recipe yield?

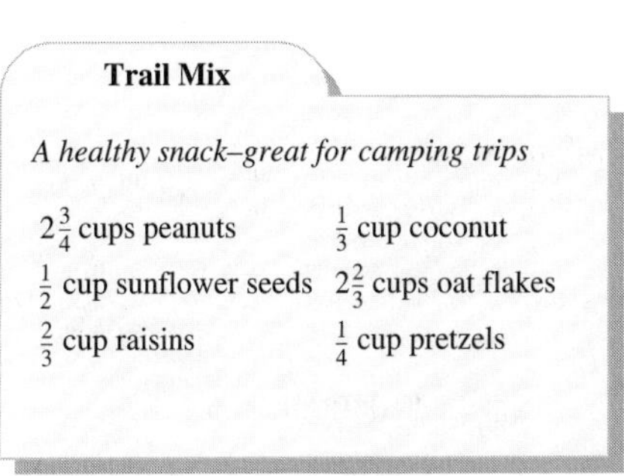

Trail Mix
A healthy snack–great for camping trips
$2\frac{3}{4}$ cups peanuts
$\frac{1}{3}$ cup coconut
$\frac{1}{2}$ cup sunflower seeds
$2\frac{2}{3}$ cups oat flakes
$\frac{2}{3}$ cup raisins
$\frac{1}{4}$ cup pretzels

ILLUSTRATION 3

72. AIR TRAVEL A businesswoman's flight leaves Los Angeles at 8 A.M. and arrives in Seattle at 9:45 A.M.

a. Express the duration of the flight as a mixed number.

b. Upon arrival, she boards a commuter plane at 11:15 A.M., arriving at her final destination at 11:45 A.M. Express the length of this flight as a fraction.

c. Find the total time of these two flights.

73. HOSE REPAIR To repair a bad connector, Ming Lin removes $1\frac{1}{2}$ feet from the end of a 50-foot garden hose. How long is the hose after the repair?

74. SEWING To make some draperies, Liz needs $12\frac{1}{4}$ yards of material for the den and $8\frac{1}{2}$ yards for the living room. If the material comes only in 21-yard bolts, how much will be left over after completing both sets of draperies?

75. SHIPPING A passenger ship and a cargo ship leave San Diego harbor at midnight. During the first hour, the passenger ship travels south at $16\frac{1}{2}$ miles per hour, while the cargo ship is traveling north at a rate of $5\frac{1}{5}$ miles per hour.

a. Complete the chart in Illustration 4 (next page).

b. How far apart are they at 1:00 A.M.?

	Rate (mph)	Time traveling (hr)	Distance traveled (mi)
Passenger ship		1	
Cargo ship		1	

ILLUSTRATION 4

76. HARDWARE See Illustration 5. To secure the bracket to the stock, a bolt and a nut are used. How long should the bolt be?

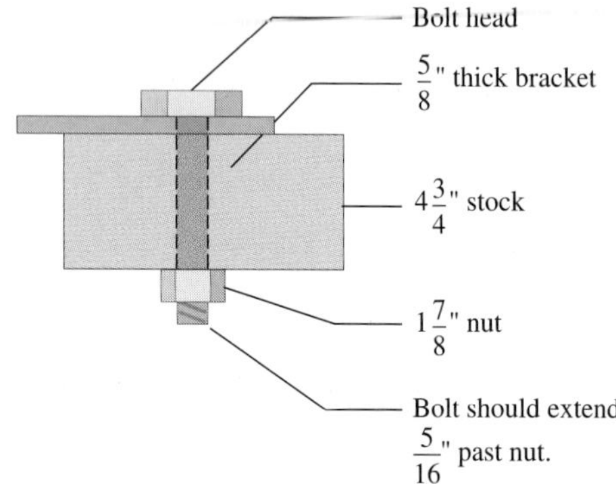

ILLUSTRATION 5

77. SERVICE STATION Use the service station sign in Illustration 6 to answer the following questions.

a. What is the difference in price between the least and most expensive types of gasoline at the self-service pump?

b. How much more is the cost per gallon for full service?

Self Serve	Full Serve
Premium Unleaded	
$169\frac{9}{10}$	$199\frac{9}{10}$
Unleaded	
$159\frac{9}{10}$	$189\frac{9}{10}$
Premium Plus	
$179\frac{9}{10}$	$209\frac{9}{10}$

cents per gallon

ILLUSTRATION 6

78. SEPTUPLETS On November 19, 1997, at Iowa Methodist Medical Center, Bobbie McCaughey gave birth to seven babies. From the information in Illustration 7, find the combined birthweights of the babies.

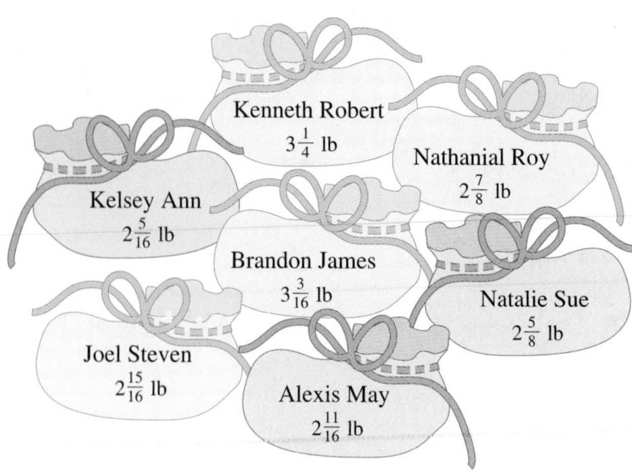

ILLUSTRATION 7

79. WATER SLIDE An amusement park added a new section to a water slide to create a slide $311\frac{5}{12}$ feet long. (See Illustration 8.) How long was the slide before the addition?

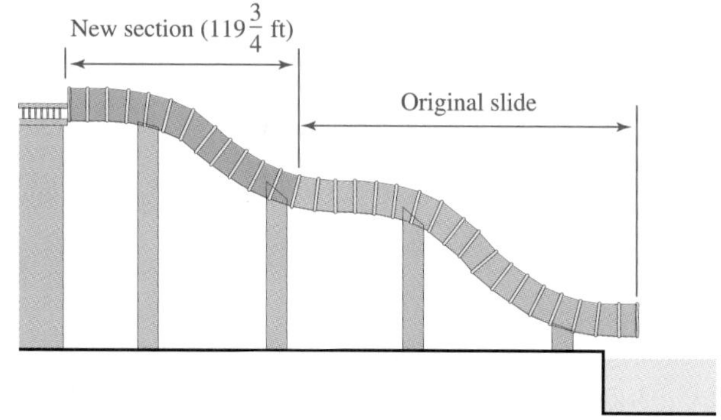

ILLUSTRATION 8

80. JEWELRY A jeweler is to cut a 7-inch-long gold braid into three pieces. He aligns a 6-inch-long ruler directly below the braid and makes the proper cuts. (See Illustration 9.) Find the length of piece 2 of the braid.

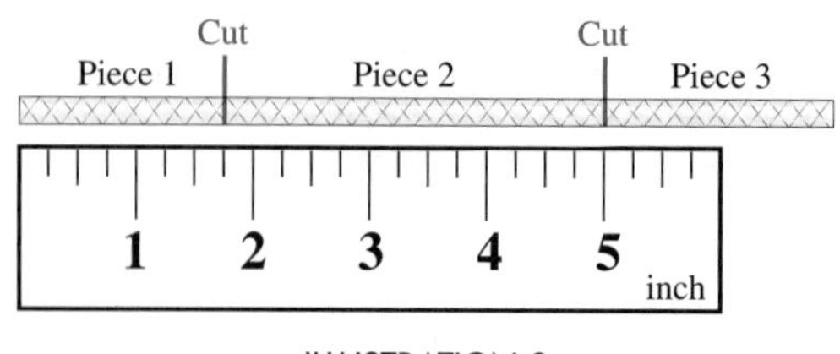

ILLUSTRATION 9

WRITING

81. Of the methods studied to add mixed numbers, which do you like better, and why?

82. When subtracting mixed numbers, when is borrowing necessary? How is it done?

83. Explain how to add $1\frac{3}{8}$ and $2\frac{1}{4}$ if we write each one as an improper fraction.

84. Explain the process of simplifying $12\frac{16}{5}$.

REVIEW

85. Solve $2x - 1 = 3x - 8$.

86. Multiply: $-3(-4)(-5)$.

87. Simplify $8(x + 2) + 3(2 - x)$.

88. In the expression $x + 2$, is x a term or a factor?

89. Find: $-2 - (-8)$.

90. Translate to mathematical symbols: y subtracted from x.

91. What does area measure?

92. Find: $|-12|$.

4.7 *Order of Operations and Complex Fractions*

In this section, you will learn about

- Order of operations • Evaluating algebraic expressions • Complex fractions
- Simplifying complex fractions

INTRODUCTION. In this section, we will evaluate expressions involving fractions and mixed numbers. We will also discuss complex fractions and the methods that are used to simplify them.

Order of operations

The rules for the order of operations are used to evaluate numerical expressions that involve more than one operation.

EXAMPLE 1 *Order of operations.* Evaluate $\frac{3}{4} + \frac{5}{3}\left(-\frac{1}{2}\right)^3$.

Solution

The expression involves the operations of raising to a power, multiplication, and addition. By the rules for the order of operations, we must evaluate the power first, the multiplication second, and the addition last.

$$\frac{3}{4} + \frac{5}{3}\left(-\frac{1}{2}\right)^3 = \frac{3}{4} + \frac{5}{3}\left(-\frac{1}{8}\right)$$ Evaluate the power: $\left(-\frac{1}{2}\right)^3 = -\frac{1}{8}$.

$$= \frac{3}{4} + \left(-\frac{5}{24}\right)$$ Do the multiplication: $\frac{5}{3}\left(-\frac{1}{8}\right) = -\frac{5}{24}$.

$$= \frac{3 \cdot 6}{4 \cdot 6} + \left(-\frac{5}{24}\right)$$ The LCD is 24. Write the first fraction as a fraction with denominator 24.

$$= \frac{18}{24} + \left(-\frac{5}{24}\right)$$ Multiply in the numerator: $3 \cdot 6 = 18$. Multiply in the denominator: $4 \cdot 6 = 24$.

$$= \frac{13}{24}$$ Add the numerators: $18 + (-5) = 13$. Write the sum over the common demominator.

Self Check

Evaluate $\frac{7}{8} + \frac{3}{2}\left(-\frac{1}{4}\right)^2$.

Answer: $\frac{31}{32}$ ■

If an expression contains grouping symbols, we do the operations within the grouping symbols first.

EXAMPLE 2 ***Order of operations.*** Evaluate $\left(\frac{7}{8}-\frac{1}{4}\right) \div \left(-2\frac{3}{16}\right)$.

Solution

$$\left(\frac{7}{8}-\frac{1}{4}\right) \div \left(-2\frac{3}{16}\right) = \left(\frac{7}{8}-\frac{1 \cdot 2}{4 \cdot 2}\right) \div \left(-2\frac{3}{16}\right)$$

Within the first set of parentheses, write $\frac{1}{4}$ as a fraction with denominator 8.

$$= \left(\frac{7}{8}-\frac{2}{8}\right) \div \left(-2\frac{3}{16}\right)$$

Multiply in the numerator: $1 \cdot 2 = 2$. Multiply in the denominator: $4 \cdot 2 = 8$.

$$= \frac{5}{8} \div \left(-2\frac{3}{16}\right)$$

Subtract the numerators and write the difference over the common denominator: $7 - 2 = 5$.

$$= \frac{5}{8} \div \left(-\frac{35}{16}\right)$$

Write the mixed number as an improper fraction.

$$= \frac{5}{8}\left(-\frac{16}{35}\right)$$

Multiply by the reciprocal of $-\frac{35}{16}$.

$$= -\frac{5 \cdot 16}{8 \cdot 35}$$

The product of two fractions with unlike signs is negative. Multiply the numerators and multiply the denominators.

$$= -\frac{\overset{1}{\cancel{5}} \cdot 2 \cdot \overset{1}{\cancel{8}}}{\underset{1}{\cancel{8}} \cdot \underset{1}{\cancel{5}} \cdot 7}$$

Factor 16 as $2 \cdot 8$ and factor 35 as $5 \cdot 7$. Divide out the common factors of 8 and 5.

$$= -\frac{2}{7}$$

Multiply in the numerator: $1 \cdot 2 \cdot 1 = 2$. Multiply in the denominator: $1 \cdot 1 \cdot 7 = 7$. ■

Evaluating algebraic expressions

To evaluate an algebraic expression, we replace its variables with specific numbers and simplify by using the rules for the order of operations.

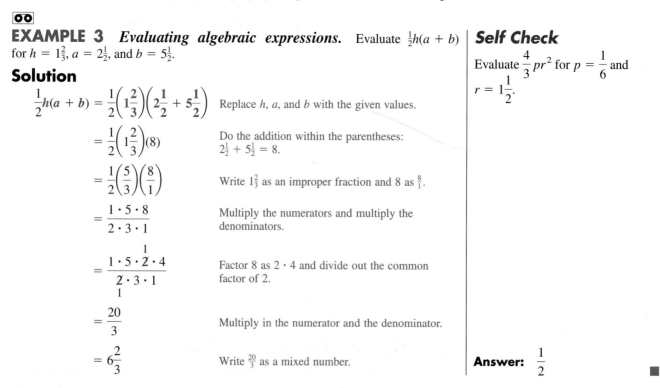

EXAMPLE 3 ***Evaluating algebraic expressions.*** Evaluate $\frac{1}{2}h(a+b)$ for $h = 1\frac{2}{3}$, $a = 2\frac{1}{2}$, and $b = 5\frac{1}{2}$.

Solution

$$\frac{1}{2}h(a+b) = \frac{1}{2}\left(1\frac{2}{3}\right)\left(2\frac{1}{2}+5\frac{1}{2}\right)$$

Replace h, a, and b with the given values.

$$= \frac{1}{2}\left(1\frac{2}{3}\right)(8)$$

Do the addition within the parentheses: $2\frac{1}{2} + 5\frac{1}{2} = 8$.

$$= \frac{1}{2}\left(\frac{5}{3}\right)\left(\frac{8}{1}\right)$$

Write $1\frac{2}{3}$ as an improper fraction and 8 as $\frac{8}{1}$.

$$= \frac{1 \cdot 5 \cdot 8}{2 \cdot 3 \cdot 1}$$

Multiply the numerators and multiply the denominators.

$$= \frac{1 \cdot 5 \cdot \overset{1}{\cancel{2}} \cdot 4}{\underset{1}{\cancel{2}} \cdot 3 \cdot 1}$$

Factor 8 as $2 \cdot 4$ and divide out the common factor of 2.

$$= \frac{20}{3}$$

Multiply in the numerator and the denominator.

$$= 6\frac{2}{3}$$

Write $\frac{20}{3}$ as a mixed number.

Self Check

Evaluate $\frac{4}{3}pr^2$ for $p = \frac{1}{6}$ and $r = 1\frac{1}{2}$.

Answer: $\frac{1}{2}$ ■

EXAMPLE 4 ***Masonry.*** To build a wall, a mason will use blocks that are $5\frac{3}{4}$ inches high, held together with $\frac{3}{8}$-inch-thick layers of mortar. (See Figure 4-10.) If the plans call for 8 layers of blocks, what will be the height of the wall when completed?

Solution To find the height, we must consider 8 layers of blocks and 8 layers of mortar. We will compute the height contributed by one block and one layer of mortar and then multiply that result by 8.

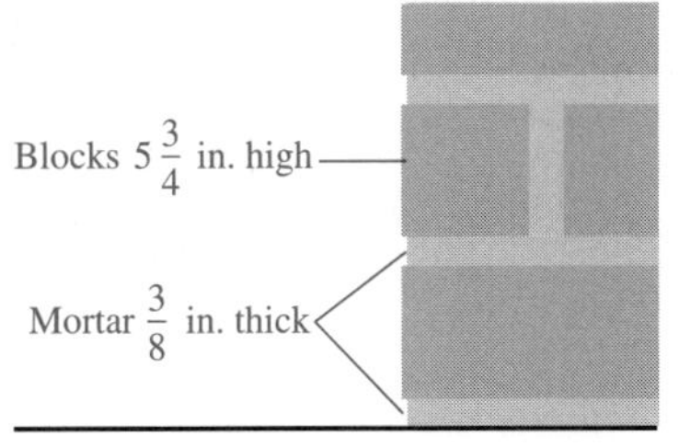

FIGURE 4-10

8 times	(height of 1 block	plus	height of 1 layer of mortar)	equals	height of block wall.
8	($5\frac{3}{4}$	+	$\frac{3}{8}$)	=	height of wall

$$8\left(5\frac{3}{4}+\frac{3}{8}\right) = 8\left(\frac{23}{4}+\frac{3}{8}\right)$$ Write $5\frac{3}{4}$ as the improper fraction $\frac{23}{4}$.

$$= 8\left(\frac{23\cdot \mathbf{2}}{4\cdot \mathbf{2}}+\frac{3}{8}\right)$$ Express $\frac{23}{4}$ in terms of 8ths.

$$= 8\left(\frac{46}{8}+\frac{3}{8}\right)$$ Do the multiplication in the numerator and denominator.

$$= \frac{8}{1}\left(\frac{49}{8}\right)$$ Write 8 as $\frac{8}{1}$. Write the sum of the numerators over the common denominator 8.

$$= \frac{\overset{1}{\cancel{8}}\cdot 49}{\underset{1}{\cancel{8}}}$$ Multiply the numerators and the denominators. Divide out the common factor of 8.

$$= 49$$ Simplify: $\frac{49}{1} = 49$.

The wall will be 49 inches high. ■

Complex fractions

Fractions whose numerators and/or denominators contain fractions are called *complex fractions.* Here is an example.

$$\frac{\frac{3}{4}}{\frac{7}{8}}$$

A fraction in the numerator → $\frac{3}{4}$; ← The main fraction bar; A fraction in the denominator → $\frac{7}{8}$

Complex fraction

A **complex fraction** is a fraction whose numerator or denominator, or both, contain one or more fractions or mixed numbers.

Here are more examples of complex fractions.

$$\frac{\frac{1}{4}-\frac{4}{5}}{2\frac{4}{5}} \qquad \longleftarrow \text{Numerator} \longrightarrow \quad \longleftarrow \text{Main fraction bar} \longrightarrow \quad \longleftarrow \text{Denominator} \longrightarrow \qquad \frac{\frac{1}{3}+\frac{1}{4}}{\frac{1}{3}-\frac{1}{4}}$$

Simplifying complex fractions

To *simplify* complex fractions means to express them as fractions in simplified form.

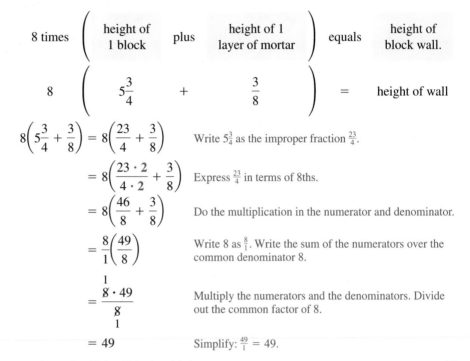

Simplifying a complex fraction: method 1

Write the numerator and the denominator of the complex fraction as single fractions. Then do the indicated division of the two fractions and simplify.

Method 1 is based on the fact that the main fraction bar of the complex fraction indicates division.

$$\frac{\frac{1}{4}}{\frac{2}{5}} \longleftarrow \text{The main fraction bar means "divide the fraction in the numerator by the fraction in the denominator."} \longrightarrow \frac{1}{4} \div \frac{2}{5}$$

EXAMPLE 5 ***Simplifying a complex fraction.*** Simplify: $\dfrac{\frac{1}{4}}{\frac{2}{5}}$.

Solution

Since the numerator and the denominator of this complex fraction are single fractions, we can do the indicated division.

$$\frac{\frac{1}{4}}{\frac{2}{5}} = \frac{1}{4} \div \frac{2}{5} \quad \text{Express the complex fraction as an equivalent division problem.}$$

$$= \frac{1}{4} \cdot \frac{5}{2} \quad \text{Multiply by the reciprocal of } \tfrac{2}{5}.$$

$$= \frac{1 \cdot 5}{4 \cdot 2} \quad \text{Multiply the numerators and multiply the denominators.}$$

$$= \frac{5}{8}$$

Self Check

Simplify $\dfrac{\frac{1}{6}}{\frac{3}{8}}$.

Answer: $\dfrac{4}{9}$ ■

A second method is based on the fundamental property of fractions.

Simplifying a complex fraction: method 2

Multiply the numerator and the denominator of the complex fraction by the LCD of all the fractions that appear in its numerator and denominator. Then simplify.

EXAMPLE 6 ***Simplifying a complex fraction.*** Simplify $\dfrac{-\frac{1}{4} + \frac{2}{5}}{\frac{1}{2} - \frac{4}{5}}$.

Solution Examine the numerator and the denominator of the complex fraction. The fractions involved have denominators of 4, 5, and 2. The LCD of these fractions is 20.

$$\frac{-\frac{1}{4} + \frac{2}{5}}{\frac{1}{2} - \frac{4}{5}} = \frac{20\left(-\frac{1}{4} + \frac{2}{5}\right)}{20\left(\frac{1}{2} - \frac{4}{5}\right)}$$

Apply the fundamental property of fractions. Multiply the numerator and the denominator of the complex fraction by 20. Note how parentheses are used to show this.

$$= \frac{20\left(-\frac{1}{4}\right) + 20\left(\frac{2}{5}\right)}{20\left(\frac{1}{2}\right) - 20\left(\frac{4}{5}\right)}$$ Apply the distributive property in the numerator and in the denominator.

$$= \frac{-5 + 8}{10 - 16}$$ Do the multiplications by 20.

$$= \frac{3}{-6}$$ Do the addition in the numerator and the subtraction in the denominator.

$$= -\frac{1}{2}$$ Simplify.

EXAMPLE 7 ***Simplifying a complex fraction.*** Simplify $\frac{7 - \frac{2}{3}}{4\frac{5}{6}}$.

Solution

Examine the numerator and the denominator of the complex fraction. The fractions have denominators of 3 and 6. The LCD of these fractions is 6.

$$\frac{7 - \frac{2}{3}}{4\frac{5}{6}} = \frac{7 - \frac{2}{3}}{\frac{29}{6}}$$ Express $4\frac{5}{6}$ as an improper fraction.

$$= \frac{6\left(7 - \frac{2}{3}\right)}{6\left(\frac{29}{6}\right)}$$ Apply the fundamental property of fractions. Multiply the numerator and the denominator of the complex fraction by the LCD, 6.

$$= \frac{6(7) - 6\left(\frac{2}{3}\right)}{6\left(\frac{29}{6}\right)}$$ Apply the distributive property in the numerator. Distribute the multiplication by 6.

$$= \frac{42 - 4}{29}$$ Do the multiplications by 6.

$$= \frac{38}{29}$$ Do the subtraction in the numerator.

$$= 1\frac{9}{29}$$ Write $\frac{38}{29}$ as a mixed number.

Self Check

Simplify $\frac{5 - \frac{3}{4}}{1\frac{7}{8}}$.

Answer: $2\frac{4}{15}$

STUDY SET Section 4.7

VOCABULARY *Fill in the blanks to make the statements true.*

1. $\frac{\frac{1}{2}}{\frac{3}{4}}$ is a ______________ fraction.

2. To ______________ an algebraic expression, we substitute specific numbers for the variables in the expression and simplify.

CONCEPTS

3. What division is represented by this complex fraction?

$$\frac{\frac{2}{3}}{\frac{1}{5}}$$

4. Write this division as a complex fraction.

$$-\frac{7}{8} \div \frac{3}{4}$$

5. What is the common denominator of all the fractions in this complex fraction?

$$\frac{\frac{2}{3} - \frac{1}{5}}{\frac{1}{3} + \frac{4}{5}}$$

6. What is the common denominator of all the fractions in this complex fraction?

$$\frac{\frac{1}{8} - \frac{3}{16}}{-5\frac{3}{4}}.$$

7. When this complex fraction is simplified, will the result be positive or negative?

$$\frac{-\frac{2}{3}}{\frac{3}{4}}$$

8. What property is being applied?

$$\frac{1 + \frac{1}{11}}{\frac{1}{2}} = \frac{22\left(1 + \frac{1}{11}\right)}{22\left(\frac{1}{2}\right)}$$

9. What is the LCD of fractions with the denominators 6, 4, and 5?

10. What operations are involved in this numerical expression?

$$5\left(6\frac{1}{3}\right) + \left(-\frac{1}{4}\right)^2$$

NOTATION *Complete each solution.*

11. Simplify $\frac{\frac{1}{8}}{\frac{3}{4}}$.

$$\frac{\frac{1}{8}}{\frac{3}{4}} = \frac{1}{8} \div \frac{3}{4}$$

$$= \frac{1}{8} \cdot \frac{4}{3}$$

$$= \frac{1 \cdot 4}{8 \cdot 3}$$

$$= \frac{1 \cdot \overset{1}{\cancel{4}}}{2 \cdot \underset{1}{\cancel{4}} \cdot 3}$$

$$= \frac{1}{6}$$

12. Simplify $\frac{\frac{1}{6} + \frac{1}{5}}{-\frac{1}{15}}$.

$$\frac{\frac{1}{6} + \frac{1}{5}}{-\frac{1}{15}} = \frac{30\left(\frac{1}{6} + \frac{1}{5}\right)}{30\left(-\frac{1}{15}\right)}$$

$$= \frac{30\left(\frac{1}{6}\right) + 30\left(\frac{1}{5}\right)}{30\left(-\frac{1}{15}\right)}$$

$$= \frac{5 + 6}{-2}$$

$$= \frac{11}{-2}$$

$$= -5\frac{1}{2}$$

PRACTICE *Evaluate each expression.*

13. $\frac{2}{3}\left(-\frac{1}{4}\right) + \frac{1}{2}$

14. $-\frac{7}{8} - \left(\frac{1}{8}\right)\left(\frac{2}{3}\right)$

15. $\frac{4}{5} - \left(-\frac{1}{3}\right)^2$

16. $-\frac{3}{16} - \left(-\frac{1}{2}\right)^3$

17. $-4\left(-\frac{1}{5}\right) - \left(\frac{1}{4}\right)\left(-\frac{1}{2}\right)$

18. $(-3)\left(-\frac{2}{3}\right) - (-4)\left(-\frac{3}{4}\right)$

19. $1\frac{3}{5}\left(\frac{1}{2}\right)^2\left(\frac{3}{4}\right)$

20. $2\frac{3}{5}\left(-\frac{1}{3}\right)^2\left(\frac{1}{2}\right)$

21. $\frac{7}{8} - \left(\frac{4}{5} + 1\frac{3}{4}\right)$

22. $\left(\frac{5}{4}\right)^2 + \left(\frac{2}{3} - 2\frac{1}{6}\right)$

23. $\left(\frac{9}{20} \div 2\frac{2}{5}\right) + \left(\frac{3}{4}\right)^2$

24. $\left(1\frac{2}{3} \cdot 15\right) + \left(\frac{7}{9} \div \frac{7}{81}\right)$

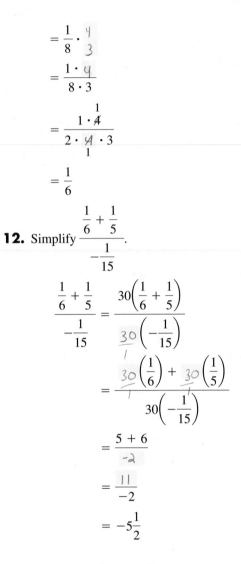

25. $\left(-\frac{3}{4} \cdot \frac{9}{16}\right) + \left(\frac{1}{2} - \frac{1}{8}\right)$

26. $\left(\frac{8}{5} - 1\frac{1}{3}\right) - \left(-\frac{4}{5} \cdot 10\right)$

27. $\left|\frac{2}{3} - \frac{9}{10}\right| \div \left(-\frac{1}{5}\right)$

28. $\left|-\frac{3}{16} \div 2\frac{1}{4}\right| + \left(-2\frac{1}{8}\right)$

29. $\left(2 - \frac{1}{2}\right)^2 + \left(2 + \frac{1}{2}\right)^2$

30. $\left(1 - \frac{3}{4}\right)\left(1 + \frac{3}{4}\right)$

Find $\frac{1}{2}$ of the given number and then square that result. Express your answer as an improper fraction.

31. -7 **32.** -5

33. $\frac{11}{2}$ **34.** $\frac{7}{3}$

Evaluate each algebraic expression for $a = 1\frac{3}{4}$, $b = -\frac{1}{5}$, $r = -1\frac{2}{3}$, and $c = -\frac{2}{3}$.

35. $\frac{1}{3}b^2 + c$ **36.** $\left(-\frac{1}{2}c\right)^3$

37. $-1 - ar$ **38.** $ab - br$

Find the perimeter of each figure.

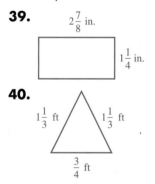

39.

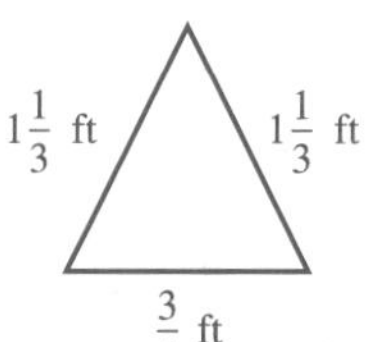

40.

$1\frac{1}{3}$ ft $1\frac{1}{3}$ ft

$\frac{3}{4}$ ft

Simplify each complex fraction.

41. $\dfrac{\frac{2}{3}}{\frac{4}{5}}$ **42.** $\dfrac{\frac{3}{5}}{\frac{9}{25}}$

43. $\dfrac{-\frac{14}{15}}{\frac{7}{10}}$ **44.** $\dfrac{\frac{5}{27}}{-\frac{5}{9}}$

45. $\dfrac{5}{\frac{10}{21}}$ **46.** $\dfrac{6}{\frac{3}{8}}$

47. $\dfrac{-\frac{5}{6}}{-1\frac{7}{8}}$ **48.** $\dfrac{-\frac{4}{3}}{-2\frac{5}{6}}$

49. $\dfrac{\frac{1}{2} + \frac{1}{4}}{\frac{1}{2} - \frac{1}{4}}$ **50.** $\dfrac{\frac{1}{3} + \frac{1}{4}}{\frac{1}{3} - \frac{1}{4}}$

51. $\dfrac{\frac{3}{8} + \frac{1}{4}}{\frac{3}{8} - \frac{1}{4}}$ **52.** $\dfrac{\frac{2}{5} + \frac{1}{4}}{\frac{2}{5} - \frac{1}{4}}$

53. $\dfrac{\frac{1}{5} + 3}{-\frac{4}{25}}$ **54.** $\dfrac{-5 - \frac{1}{3}}{\frac{1}{6} + \frac{2}{3}}$

55. $\dfrac{5\frac{1}{2}}{-\frac{1}{4} + \frac{3}{4}}$ **56.** $\dfrac{4\frac{1}{4}}{\frac{2}{3} + \left(-\frac{1}{6}\right)}$

57. $\dfrac{\frac{1}{5} - \left(-\frac{1}{4}\right)}{\frac{1}{4} + \frac{4}{5}}$ **58.** $\dfrac{\frac{1}{8} - \left(-\frac{1}{2}\right)}{\frac{1}{4} + \frac{3}{8}}$

59. $\dfrac{\frac{1}{3} + \left(-\frac{5}{6}\right)}{1\frac{1}{3}}$ **60.** $\dfrac{\frac{3}{7} + \left(-\frac{1}{2}\right)}{1\frac{3}{4}}$

Evaluate each algebraic expression for $x = -\frac{3}{4}$ and $y = \frac{7}{8}$.

61. $\dfrac{x + y}{2}$ **62.** $\dfrac{x - y}{x + y}$

63. $\left|\dfrac{2x}{y - x}\right|$ **64.** $\left|\dfrac{y^2}{y - 2}\right|$

APPLICATIONS

65. SANDWICH SHOP A sandwich shop sells a $\frac{1}{2}$-pound club sandwich, made up of turkey meat and ham. The owner buys the turkey in $1\frac{3}{4}$-pound packages and the ham in $2\frac{1}{2}$-pound packages. If he mixes a package of each of the meats together, how many sandwiches can he make from the mixture?

66. SKIN CREAM Using a formula of $\frac{1}{2}$ ounce of sun block, $\frac{2}{3}$ ounce of moisturizing cream, and $\frac{3}{4}$ ounce of lanolin, a beautician mixes her own brand of skin cream. She packages it in $\frac{1}{4}$-ounce tubes. How many tubes can be produced using this formula?

67. PHYSICAL FITNESS Two people begin their workouts from the same point on a bike path and travel in opposite directions, as shown in Illustration 1. How far apart are they in $1\frac{1}{2}$ hours? Use the chart to help organize your work.

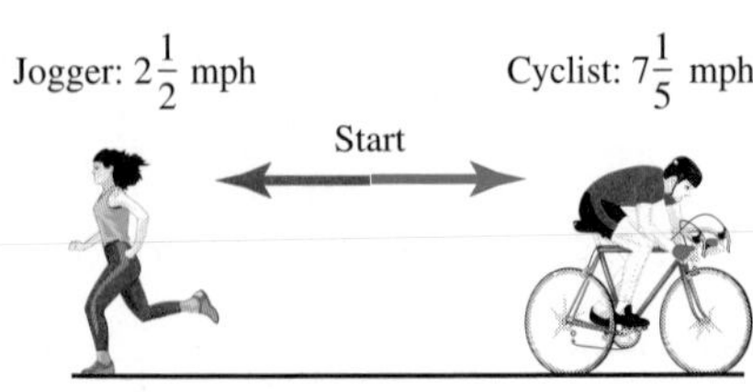

ILLUSTRATION 1

	Rate (mph)	Time (hr)	Distance (mi)
Jogger			
Cyclist			

68. SLEEP Illustration 2 compares the amount of sleep a 1-month-old baby got to the $15\frac{1}{2}$-hour daily requirement recommended by Children's Hospital of Orange County, California. For the week, how far below the baseline was the baby's daily average?

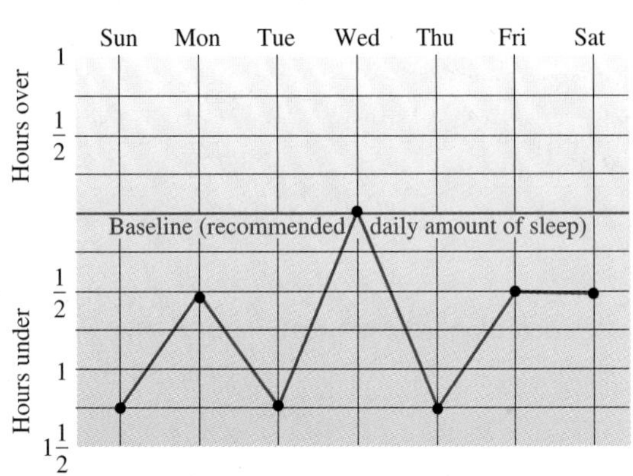

ILLUSTRATION 2

69. POSTAGE RATES Can the advertising package in Illustration 3 be mailed for the one-ounce rate?

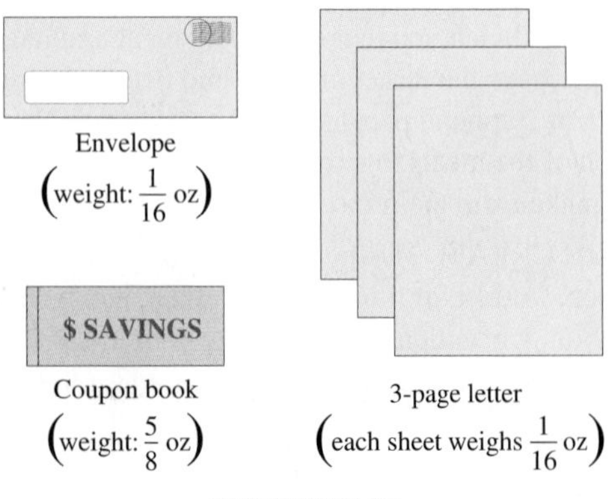

ILLUSTRATION 3

70. PLYWOOD To manufacture a sheet of plywood, several layers of thin laminate are glued together, as shown in Illustration 4. Then an exterior finish is affixed to the top and bottom. How thick is the finished product?

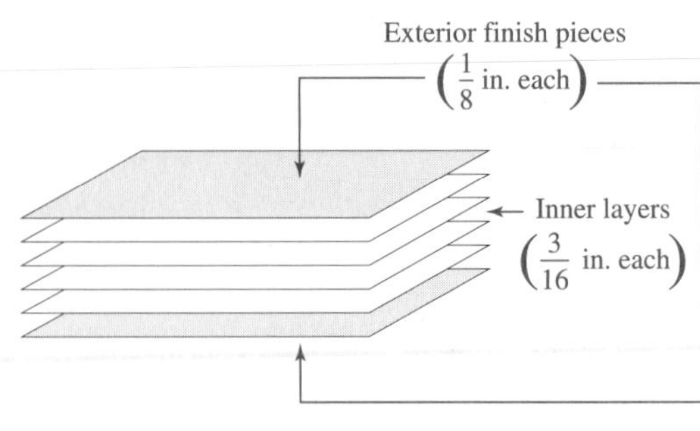

ILLUSTRATION 4

71. PHYSICAL THERAPY After back surgery, a patient undertook a walking program to rehabilitate her back muscles, as specified in Illustration 5. What was the total distance she walked over this three-week period?

Week	Distance per day
#1	$\frac{1}{4}$ mile
#2	$\frac{1}{2}$ mile
#3	$\frac{3}{4}$ mile

ILLUSTRATION 5

72. READING PROGRAM To improve reading skills, elementary school children read silently at the end of the school day for $\frac{1}{4}$ hour on Mondays and for $\frac{1}{2}$ hour on Fridays. For the month of January, how many total hours did the children read silently in class? (See Illustration 6.)

S	M	T	W	T	F	S
	1	2	3	4	5	6
7	8	9	10	11	12	13
14	15	16	17	18	19	20
21	22	23	24	25	26	27
28	29	30	31			

ILLUSTRATION 6

73. AMUSEMENT PARK At the end of a ride at an amusement park, a boat splashes into a pool of water. The time (in seconds) that it takes two pipes to refill the pool is given by

$$\frac{1}{\frac{1}{10}+\frac{1}{15}}$$

Find this time.

74. HIKING A scout troop plans to hike from the campground to Glenn Peak. (See Illustration 7.) Since the terrain is steep, they plan to stop and rest after every $\frac{2}{3}$ mile. With this plan, how many parts will there be to this hike?

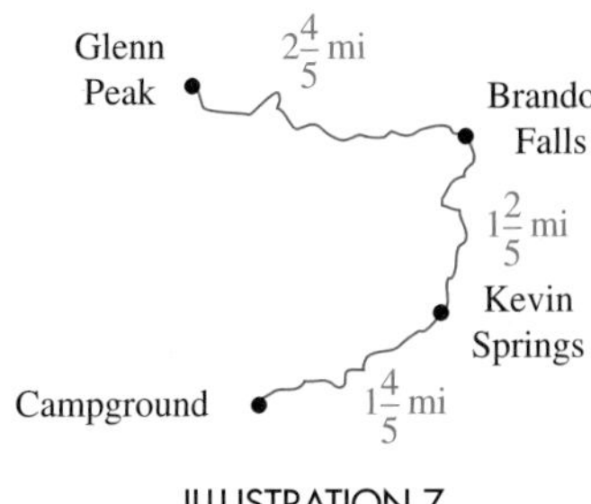

ILLUSTRATION 7

WRITING

75. What is a complex fraction?

76. Explain method 1 for simplifying complex fractions.

77. Write an application problem using a complex fraction, and then solve it.

78. Explain method 2 for simplifying complex fractions.

REVIEW

79. Simplify $-4d - (-7d)$.

80. Solve $8 + 2a = 3 - (4a + 1)$.

81. Translate into mathematical symbols: the product of twice a number and its opposite.

82. List the factors of 24.

83. Evaluate $2 + 3[-3 - (-4 - 1)]$.

84. What is the sign of the quotient of two numbers with unlike signs?

85. Simplify $3 \cdot 3 \cdot 3 \cdot x \cdot x \cdot x \cdot x \cdot x$.

86. In the expression $-4x^2 + 3x - 7$, what is the coefficient of the second term?

4.8 *Solving Equations Containing Fractions*

In this section, you will learn about

- Using reciprocals to solve equations • An alternate method
- The addition and subtraction properties of equality
- Clearing an equation of fractions • The steps to solve equations
- Problem solving with equations

INTRODUCTION. In this section, we will discuss how to solve equations containing fractions, and equations whose solutions are fractions. We will make use of the properties of equality and several concepts from this chapter, including the reciprocal and the LCD.

Using reciprocals to solve equations

In the equation $\frac{3}{4}x = 5$, the variable is multiplied by $\frac{3}{4}$. To undo this multiplication and isolate the variable, we can use the multiplication property of equality and multiply both sides of the equation by the reciprocal of $\frac{3}{4}$.

$\frac{4}{3}\left(\frac{3}{4}x\right) = \frac{4}{3}(5)$ Multiply both sides by the reciprocal of $\frac{3}{4}$, which is $\frac{4}{3}$.

$\left(\frac{4}{3} \cdot \frac{3}{4}\right)x = \frac{4}{3} \cdot \frac{5}{1}$ Use the associative property of multiplication to regroup the factors. Write 5 as $\frac{5}{1}$.

$\frac{\overset{1}{\cancel{4}} \cdot \overset{1}{\cancel{3}}}{\underset{1}{\cancel{3}} \cdot \underset{1}{\cancel{4}}}x = \frac{4 \cdot 5}{3 \cdot 1}$ Multiply the numerators and multiply the denominators. On the left, divide out the common factors of 4 and 3.

$1x = \frac{20}{3}$ Multiply in the numerators and in the denominators.

$x = \frac{20}{3}$ Simplify: $1x = x$.

In algebra, we usually leave a solution to an equation as an improper fraction rather than converting it to a mixed number.

COMMENT We can write expressions such as $\frac{4a}{5}$ and $\frac{-9h}{16}$ in an equivalent form so that the fractional coefficients are more evident.

$$\frac{4a}{5} = \frac{4}{5}a \qquad \frac{-9h}{16} = -\frac{9}{16}h$$

EXAMPLE 1 *Using reciprocals.* Solve $-\frac{7}{8}k = 21$.

Solution

The coefficient of the variable is $-\frac{7}{8}$. To isolate k, we multiply both sides of the equation by the reciprocal of $-\frac{7}{8}$.

$$-\frac{7}{8}k = 21$$

$$-\frac{8}{7}\left(-\frac{7}{8}k\right) = -\frac{8}{7}(21)$$ Multiply both sides by the reciprocal of $-\frac{7}{8}$, which is $-\frac{8}{7}$.

$$1k = -\frac{8}{7}(21)$$ The product of a number and its reciprocal is 1: $-\frac{8}{7}(-\frac{7}{8}) = 1$.

$$k = -\frac{8}{7} \cdot \frac{21}{1}$$ Simplify: $1k = k$. Write 21 as $\frac{21}{1}$.

$$k = -\frac{8 \cdot 3 \cdot \overset{1}{\cancel{7}}}{\underset{1}{\cancel{7}} \cdot 1}$$ The product of two numbers with unlike signs is negative. Multiply the numerators and the denominators. Prime factor 21 and then divide out the common factor of 7.

$$k = -24$$ Multiply in the numerator and the denominator.

Check the result by substituting -24 for k in the original equation.

Self Check

Solve $-\frac{3}{4}t = 15$ and check the result.

Answer: -20 ■

An alternate method

Another method of solving equations such as $\frac{3}{4}x = 5$ uses two steps to isolate the variable. In this method, we consider the variable to be multiplied by 3 and divided by 4. Then, in reverse order, we undo these operations.

$$4\left(\frac{3}{4}x\right) = 4(5)$$ To undo the division by 4, multiply both sides by 4.

$$\left(4 \cdot \frac{3}{4}\right)x = 4(5)$$ Use the associative property to regroup the factors.

$$\left(\frac{\overset{1}{\cancel{4}} \cdot 3}{1 \cdot \underset{1}{\cancel{4}}}\right)x = 4(5)$$ Write 4 as $\frac{4}{1}$, multiply the numerators and the denominators, and divide out the common factor of 4.

$$3x = 20$$ Multiply in the numerator and the denominator.

$$\frac{3x}{3} = \frac{20}{3}$$ To undo the multiplication by 3, divide both sides by 3.

$$x = \frac{20}{3}$$

EXAMPLE 2 ***The two-step method.*** Solve $\frac{3}{5}h = -9$.

Solution

$$\frac{3}{5}h = -9$$

$$5\left(\frac{3}{5}h\right) = 5(-9) \quad \text{To undo the division by 5, multiply both sides by 5.}$$

$$3h = -45 \quad \text{Do the multiplications.}$$

$$\frac{3h}{3} = \frac{-45}{3} \quad \text{To undo the multiplication by 3, divide both sides by 3.}$$

$$h = -15 \quad \text{Do the divisions.}$$

Self Check

Solve $\frac{5}{9}t = -10$ using the two-step method.

$\frac{5}{9}t = -10$

$9\left(\frac{5}{9}t\right) = -10(9)$

$5t = -90$

$\frac{5t}{5} = \frac{-90}{5}$

$t = -18$

Answer: -18

The addition and subtraction properties of equality

The addition property of equality enables us to add the same number to both sides of an equation and obtain an equivalent equation. In the next example, we will use this property to help solve an equation that contains fractions.

EXAMPLE 3 ***The addition property of equality.*** Solve $y - \frac{15}{32} = \frac{1}{32}$.

Solution

To isolate y on the left-hand side, we need to undo the subtraction of $\frac{15}{32}$.

$$y - \frac{15}{32} = \frac{1}{32}$$

$$y - \frac{15}{32} + \mathbf{\frac{15}{32}} = \frac{1}{32} + \mathbf{\frac{15}{32}} \quad \text{Add } \frac{15}{32} \text{ to both sides.}$$

$$y = \frac{16}{32} \quad \text{Simplify: } -\frac{15}{32} + \frac{15}{32} = 0 \text{ and } \frac{1}{32} + \frac{15}{32} = \frac{16}{32}.$$

$$y = \frac{1}{2} \quad \text{Simplify the fraction: } \frac{16}{32} = \frac{\overset{1}{\cancel{16}} \cdot 1}{\underset{1}{\cancel{16}} \cdot 2} = \frac{1}{2}.$$

Self Check

Solve $\frac{11}{16} = a - \frac{1}{16}$.

$\frac{11}{16} + \frac{1}{16} = a - \frac{1}{16} + \frac{1}{16}$

$\frac{12}{16} = a$

$a = \frac{3}{4}$

Answer: $\frac{3}{4}$

EXAMPLE 4 ***The subtraction property of equality.*** Solve $x + \frac{1}{6} = \frac{3}{4}$.

Solution

In this equation, $\frac{1}{6}$ is added to x. We undo this operation by subtracting $\frac{1}{6}$ from both sides.

$$x + \frac{1}{6} = \frac{3}{4}$$

$$x + \frac{1}{6} - \mathbf{\frac{1}{6}} = \frac{3}{4} - \mathbf{\frac{1}{6}} \quad \text{Subtract } \frac{1}{6} \text{ from both sides.}$$

$$x = \frac{3}{4} - \frac{1}{6} \quad \text{Do the subtraction: } \frac{1}{6} - \frac{1}{6} = 0.$$

$$x = \frac{3 \cdot \mathbf{3}}{4 \cdot \mathbf{3}} - \frac{1 \cdot \mathbf{2}}{6 \cdot \mathbf{2}} \quad \text{Use the fundamental property of fractions to write each fraction in terms of the LCD, which is 12.}$$

Self Check

Solve $y + \frac{1}{5} = \frac{2}{3}$.

$y + \frac{1}{5} - \frac{1}{5} = \frac{2}{3} - \frac{1}{5}$

$y = \frac{7}{15}$

$x = \frac{9}{12} - \frac{2}{12}$ Do the multiplications in the numerators and denominators.

$x = \frac{7}{12}$ Subtract the fractions.

Answer: $\frac{7}{15}$ ■

Clearing an equation of fractions

In Example 4, we found an LCD so that we could subtract $\frac{1}{6}$ and $\frac{3}{4}$. We will now discuss a method in which we clear such an equation of fractions.

To clear $x + \frac{1}{6} = \frac{3}{4}$ of fractions, we multiply both sides by the LCD of all fractions that appear in the equation. In this case, the LCD of $\frac{1}{6}$ and $\frac{3}{4}$ is 12.

$$\mathbf{12}\left(x + \frac{1}{6}\right) = \mathbf{12}\left(\frac{3}{4}\right)$$ Multiply both sides by the LCD, 12.

$$12x + 12\left(\frac{1}{6}\right) = 12\left(\frac{3}{4}\right)$$ On the left-hand side, distribute the multiplication by 12.

$$12x + 2 = 9$$ Do the multiplications: $12\left(\frac{1}{6}\right) = 2$ and $12\left(\frac{3}{4}\right) = 9$.

We note that the resulting equation, $12x + 2 = 9$, does not contain fractions. We now complete the solution.

$$12x + 2 - \mathbf{2} = 9 - \mathbf{2}$$ To undo the addition of 2, subtract 2 from both sides.

$$12x = 7$$ Do the subtractions.

$$\frac{12x}{\mathbf{12}} = \frac{7}{\mathbf{12}}$$ To undo the multiplication by 12, divide both sides by 12.

$$x = \frac{7}{12}$$

EXAMPLE 5 ***Clearing an equation of fractions.*** Solve $\frac{3}{4}h - \frac{1}{2} = \frac{5}{8}h$.

Solution

$$\frac{3}{4}h - \frac{1}{2} = \frac{5}{8}h$$

$$\mathbf{8}\left(\frac{3}{4}h - \frac{1}{2}\right) = \mathbf{8}\left(\frac{5}{8}h\right)$$ To clear the equation of fractions, multiply both sides by the LCD of $\frac{3}{4}$, $\frac{1}{2}$, and $\frac{5}{8}$, which is 8.

$$8\left(\frac{3}{4}h\right) - 8\left(\frac{1}{2}\right) = 8\left(\frac{5}{8}h\right)$$ Distribute the multiplication by 8 on the left-hand side.

$$6h - 4 = 5h$$ Do the multiplications: $8\left(\frac{3}{4}\right) = 6$, $8\left(\frac{1}{2}\right) = 4$, and $8\left(\frac{5}{8}\right) = 5$.

$$6h - 4 - \mathbf{5h} = 5h - \mathbf{5h}$$ To eliminate $5h$ from the right-hand side, subtract $5h$ from both sides.

$$h - 4 = 0$$ Combine like terms.

$$h - 4 + \mathbf{4} = 0 + \mathbf{4}$$ To undo the subtraction of 4, add 4 to both sides.

$$h = 4$$ Simplify.

Self Check

Solve $\frac{4}{5}p - \frac{1}{2} = \frac{3}{4}p$ by first clearing it of fractions.

Answer: 10 ■

The steps to solve equations

We can now complete the strategy for solving equations discussed earlier. You won't always have to use all six steps to solve a given equation. If a step doesn't apply, skip it and move to the next step.

Strategy for solving equations

Simplify the equation:

1. Clear the equation of fractions.
2. Use the distributive property to remove any parentheses.
3. Combine like terms on either side of the equation.

Isolate the variable:

4. Use the addition and subtraction properties of equality to get the variables on one side and the constants on the other.
5. Combine like terms when necessary.
6. Undo the operations of multiplication and division to isolate the variable.

Problem solving with equations

EXAMPLE 6 ***Native Americans.*** The United States Constitution requires a population count, called a *census,* to be taken every ten years. In the 1990 census, the population of the Navaho tribe was 225,000. This was about three-fifths of the population of the largest Native American tribe, the Cherokee. What was the population of the Cherokee tribe in 1990?

Analyze the problem

- In 1990, the population of the Navaho tribe was 225,000.
- The population of the Navaho tribe was $\frac{3}{5}$ the population of the Cherokee tribe.
- Find the population of the Cherokee tribe in 1990.

Form an equation Let x = the population of the Cherokee tribe. Next, we look for a key word or phrase in the problem.

Key phrase: *three-fifths of* **Translation:** *multiply by* $\frac{3}{5}$

The population of the Navaho tribe	was	$\frac{3}{5}$	of	the population of the Cherokee tribe.
225,000	=	$\frac{3}{5}$	$\cdot$	x

Solve the equation

$$225{,}000 = \frac{3}{5}x$$

$$\frac{5}{3}(225{,}000) = \frac{5}{3}\left(\frac{3}{5}x\right)$$ To isolate x on the right-hand side, multiply both sides by the reciprocal of $\frac{3}{5}$.

$$375{,}000 = x$$ On the left-hand side, $\frac{5}{3}(225{,}000) = \frac{1{,}125{,}000}{3} = 375{,}000$. On the right-hand side, $\frac{5}{3}\left(\frac{3}{5}x\right) = 1x = x$.

State the conclusion In 1990, the population of the Cherokee tribe was about 375,000.

Check the result Using a fraction to compare the two populations, we have

$$\frac{225{,}000}{375{,}000} = \frac{225}{375} = \frac{75 \cdot 3}{75 \cdot 5} = \frac{3}{5}$$

The answer checks. ■

EXAMPLE 7 ***Filmmaking.*** A movie director has sketched out a "storyboard" for a film that is in the planning stages. On the storyboard, he estimates the amount of time in the film that will be devoted to scenes involving dialogue, action scenes, and scenes used to transition between the two. (See Figure 4-11.) From the information on the storyboard, how long will this film be, in minutes?

Storyboard		Film: "Terminating Force"
Dialogue One-half of film	*Action scenes* One-third of film	*Transition scenes* 20 minutes

FIGURE 4-11

Analyze the problem

- $\frac{1}{2}$ of the film is dialogue.
- $\frac{1}{3}$ of the film is action scenes.
- There are 20 minutes of transition scenes.
- How long is the film?

Form an equation Let $x =$ the length of the film in minutes. To represent the number of minutes for dialogue and action scenes, look for a key word or phrase.

Key phrases: *one-half of, one-third of* **Translation:** multiply

Therefore, $\frac{1}{2}x =$ the number of minutes for dialogue scenes and $\frac{1}{3}x =$ the number of minutes for action scenes.

The time for dialogue scenes	plus	the time for action scenes	plus	the time for transition scenes	is	the total length of the film.
$\frac{1}{2}x$	$+$	$\frac{1}{3}x$	$+$	20	$=$	x

Solve the equation

$$\frac{1}{2}x + \frac{1}{3}x + 20 = x$$

$$\mathbf{6}\left(\frac{1}{2}x + \frac{1}{3}x + 20\right) = \mathbf{6}(x)$$ To clear the equation of fractions, multiply both sides by the LCD, 6.

$$6\left(\frac{1}{2}x\right) + 6\left(\frac{1}{3}x\right) + 6(20) = 6(x)$$ On the left-hand side, distribute the 6.

$$3x + 2x + 120 = 6x$$ Do the multiplications: $6\left(\frac{1}{2}\right) = 3$ and $6\left(\frac{1}{3}\right) = 2$.

$$5x + 120 = 6x$$ Combine like terms.

$$5x + 120 - \mathbf{5x} = 6x - \mathbf{5x}$$ To eliminate $5x$ from the left-hand side, subtract $5x$ from both sides.

$$120 = x$$ Combine like terms.

State the conclusion The length of the film will be 120 minutes.

Check the result If $x = 120$, the time for dialogue scenes is $\frac{1}{2}x = \frac{1}{2} \cdot 120 = 60$ minutes. The time for action scenes is $\frac{1}{3}x = \frac{1}{3} \cdot 120 = 40$ minutes. The time for transition scenes is 20 minutes. Adding the three times, we get $60 + 40 + 20 = 120$ minutes. The answer checks. ■

STUDY SET Section 4.8

VOCABULARY *Fill in the blanks to make the statements true.*

1. To find the ______________ of a fraction, invert the numerator and the denominator.

2. In the term $\frac{5}{12}x$, the number $\frac{5}{12}$ is called the ______________ of the term.

3. The ______________ of a set of fractions is the smallest number each denominator will divide exactly.

4. A ______________ of an equation, when substituted into that equation, makes a true statement.

CONCEPTS

5. Is $x = 40$ a solution of $\frac{5}{8}x = 25$? Explain why or why not.

6. Give the reciprocal of each number.

a. $\frac{7}{9}$ **b.** $-\frac{1}{2}$

7. What is the result when a number is multiplied by its reciprocal?

8. Do each multiplication.

a. $\frac{3}{2}\left(\frac{2}{3}x\right)$ **b.** $-\frac{16}{15}\left(-\frac{15}{16}t\right)$

c. $25\left(\frac{2}{5}\right)$ **d.** $16\left(\frac{3}{8}\right)$

9. Translate to mathematical symbols.

a. Four-fifths of the population p

b. One-quarter of the time t

10. What property is illustrated by the arrows?

$$12\left(y - \frac{1}{4}\right) = 12y - 3$$

11. Explain two ways in which the variable x can be isolated: $\frac{2}{3}x = -4$.

12. What is wrong with this portion of a solution?

$$\frac{x}{6} - \frac{3}{5} = 8$$

$$30\left(\frac{x}{6} - \frac{3}{5}\right) = 8$$

$$30\left(\frac{x}{6}\right) - 30\left(\frac{3}{5}\right) = 8$$

NOTATION *Complete each solution to solve the equation.*

13. $\frac{7}{8}x = 21$

$$8\left(\frac{7}{8}x\right) = 8\,(21)$$

$$x = 24$$

14. $h + \frac{1}{2} = \frac{2}{3}$

$$6\left(h + \frac{1}{2}\right) = 6\left(\frac{2}{3}\right)$$

$$6h + 6\left(\frac{1}{2}\right) = 6\left(\frac{2}{3}\right)$$

$$6h + 3 = 4$$

$$6h + 3 - 3 = 4 - 3$$

$$6h = 1$$

$$\frac{6h}{6} = \frac{1}{6}$$

$$h = \frac{1}{6}$$

15. Tell whether each statement is true or false.

a. $\frac{1}{2}x = \frac{x}{2}$ **b.** $\frac{1}{8}y = 8y$

c. $-\frac{1}{2}x = \frac{-x}{2} = \frac{x}{-2}$ **d.** $\frac{7p}{8} = \frac{7}{8}p$

16. Write the product of $\frac{4}{7}$ and x in two ways.

PRACTICE *Solve each equation.*

17. $\frac{4}{7}x = 16$

18. $\frac{2}{3}y = 30$

19. $\frac{7}{8}t = -28$

20. $\frac{5}{6}c = -25$

21. $-\frac{3}{5}h = 4$

22. $-\frac{5}{6}f = -2$

23. $\frac{2}{3}x = \frac{4}{5}$

24. $\frac{5}{8}y = \frac{10}{11}$

25. $\frac{2}{5}y = 0$

26. $\frac{4}{9}x = 0$

27. $-\frac{5c}{6} = -25$

28. $-\frac{7t}{4} = -35$

29. $\frac{-5f}{7} = -2$

30. $\frac{-3h}{5} = -4$

31. $\frac{5}{8}y = \frac{1}{10}$

32. $\frac{1}{16}x = \frac{5}{24}$

33. $2x + 1 = 0$

34. $3y - 1 = 0$

35. $5x - 1 = 1$

36. $4c + 1 = -2$

37. $6x = 2x - 11$

38. $5t = t - 7$

39. $2(y - 3) = 7$

40. $3(r + 2) = 10$

41. $x - \frac{1}{9} = \frac{7}{9}$

42. $x + \frac{1}{3} = \frac{2}{3}$

43. $x + \frac{1}{9} = \frac{4}{9}$

44. $x - \frac{1}{6} = \frac{1}{6}$

45. $x - \frac{1}{6} = \frac{2}{9}$

46. $y - \frac{1}{3} = \frac{4}{5}$

47. $y + \frac{7}{8} = \frac{1}{4}$

48. $t + \frac{5}{6} = \frac{1}{8}$

49. $\frac{5}{4} + t = \frac{1}{4}$

50. $\frac{2}{3} + y = \frac{4}{3}$

51. $x + \frac{3}{4} = -\frac{1}{2}$

52. $y - \frac{5}{6} = \frac{1}{3}$

53. $\frac{-x}{4} + 1 = 10$

54. $\frac{-y}{6} - 1 = 5$

55. $2x - \frac{1}{2} = \frac{1}{3}$

56. $3y - \frac{2}{5} = \frac{1}{8}$

57. $\frac{1}{2}x - \frac{1}{9} = \frac{1}{3}$

58. $\frac{1}{4}y - \frac{2}{3} = \frac{1}{2}$

59. $5 + \frac{x}{3} = \frac{1}{2}$

60. $4 + \frac{y}{2} = \frac{3}{5}$

61. $\frac{2}{5}x + 1 = \frac{1}{3} + x$

62. $\frac{2}{3}y + 2 = \frac{1}{5} + y$

63. $\frac{x}{3} + \frac{x}{4} = -2$

64. $\frac{y}{6} + \frac{y}{4} = -1$

65. $4 + \frac{s}{3} = 8$

66. $6 + \frac{y}{5} = 1$

67. $\frac{5h}{6} - 8 = 12$

68. $\frac{6a}{7} - 1 = 11$

69. $-4 + 9 + \frac{5t}{12} = 0$

70. $-4 + 10 + \frac{3y}{8} = 0$

71. $-3 - 2 + \frac{4x}{15} = 0$

72. $-1 - 9 + \frac{2y}{15} = 0$

APPLICATIONS *Complete each solution.*

73. TRANSMISSION REPAIR A repair shop found that $\frac{1}{3}$ of its customers with transmission problems needed a new transmission. If the shop installed 32 new transmissions last year, how many customers did the shop have last year?

Analyze the problem

- Only of the customers needed new transmissions.
- The shop installed new transmissions last year.
- Find the number of __________ the shop had last year.

Form an equation

Let x = ________________.

Key phrase: *one-third of*

Translation: __________

$\frac{1}{3}$ of the number of customers last year	was	32.
	=	32

Solve the equation

$$\frac{1}{3}x = 32$$

$$\left(\frac{1}{3}x\right) = (32)$$

$$x =$$

State the conclusion ________________

Check the result If we find $\frac{1}{3}$ of 96, we get . The answer checks.

74. CATTLE RANCHING A rancher is preparing to fence in a rectangular grazing area next to a $\frac{3}{4}$-mile-long lake. See Illustration 1. He has determined that $1\frac{1}{2}$ square miles of land are needed to ensure that overgrazing does not occur. How wide should this grazing area be?

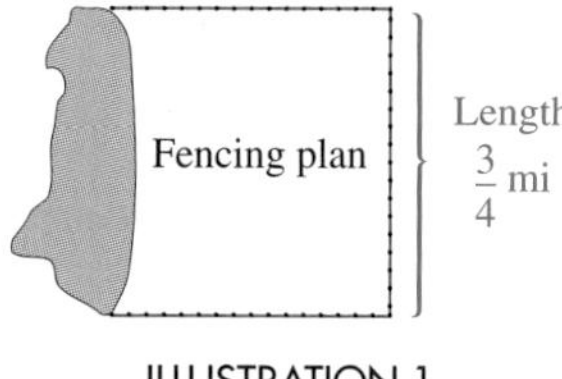

ILLUSTRATION 1

Analyze the problem

- The grazing area is $= \frac{3}{2}$ square miles.
- The length of the rectangle is $\frac{3}{4}$ mile.
- Find the __________ of the grazing area.

Form an equation

Let w = ____________.

Key word: *area* **Translation:** $A =$

The area of the rectangle	is	the length times the width.
	=	

Solve the equation

$$\frac{3}{2} = \frac{3}{4}w$$

$$\left(\frac{3}{2}\right) = \left(\frac{3}{4}w\right)$$

$$= w$$

State the conclusion

____

Check the result If we multiply the length and the width of the rectangular area, we get $\frac{3}{4} \cdot \quad = \frac{3}{2} = 1\frac{1}{2}$ square miles. The answer checks.

Choose a variable to represent the unknown. Then write and solve an equation to answer each question.

75. TOOTH DEVELOPMENT During a checkup, a pediatrician found that only four-fifths of a child's baby teeth had emerged. The mother counted 16 teeth in the child's mouth. How many baby teeth will the child eventually have?

76. GENETICS Bean plants with inflated pods were cross-bred with bean plants with constricted pods. (See Illustration 2.) Of the offspring plants, three-fourths had inflated pods and one-fourth had constricted pods. If 244 offspring plants had constricted pods, how many offspring plants resulted from the cross-breeding experiment?

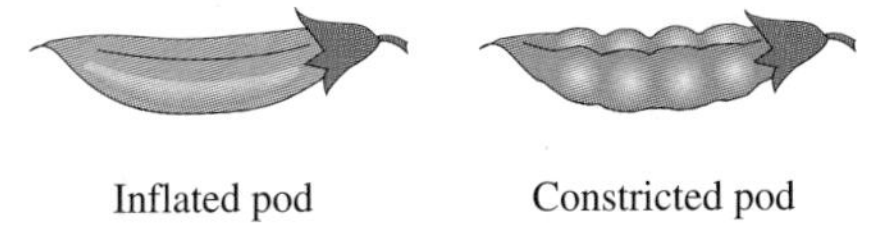
Inflated pod
Constricted pod

ILLUSTRATION 2

77. HOME SALES In less than a month, three-quarters of the homes in a new subdivision were purchased. This left only 9 homes to be sold. How many homes are there in the subdivision?

78. WEDDING GUESTS Of those invited to a wedding, three-tenths were friends of the bride. The friends of the groom numbered 84. How many people were invited to the wedding?

79. TELEPHONE BOOK A telephone book consists of the white pages and the yellow pages. Two-thirds of the book consists of the white pages; the yellow pages number 150. Find the total number of pages in the telephone book.

80. BROADWAY MUSICAL A theater usher at a Broadway musical finds that seven-eighths of the patrons attending a performance are in their seats by show time. The remaining 50 people are seated after the opening number. If the show is always a complete sellout, how many seats does the theater have?

81. SAFETY REQUIREMENT In developing taillights for an automobile, designers must be aware of a safety standard that requires an area of 30 square inches to be visible from behind the vehicle. If the designers want the taillights to be $3\frac{3}{4}$ inches high, how wide must they be to meet safety standards? (See Illustration 3.)

$3\frac{3}{4}$ in.

ILLUSTRATION 3

82. GRAPHIC ARTS A design for a yearbook is shown in Illustration 4. The page is divided into 12 equal parts. The parts that are shaded will contain pictures, and the remainder of the squares will contain copy. If the pictures are to cover an area of 100 square inches, how many square inches are there on the page?

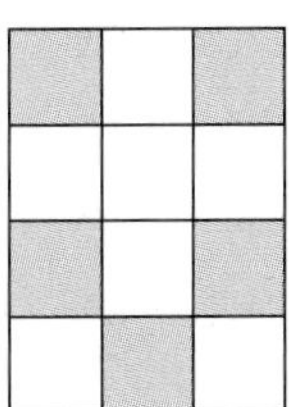

ILLUSTRATION 4

83. CPR CLASS The instructor for a course in CPR (cardiopulmonary resuscitation) has three segments in her lesson plan, as shown in Illustration 5. How many minutes long is the CPR course?

Lecture on subject	Practicing CPR techniques	Legal responsibilities
One-fourth of class	Two-thirds of class	30 min

ILLUSTRATION 5

84. FIREFIGHTING A firefighting crew is composed of three elements, as shown in Illustration 6. How many firefighters are in the crew?

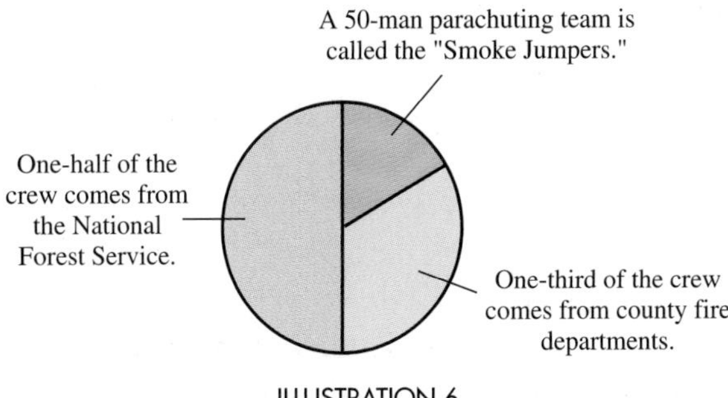
A 50-man parachuting team is called the "Smoke Jumpers."
One-half of the crew comes from the National Forest Service.
One-third of the crew comes from county fire departments.

ILLUSTRATION 6

WRITING

85. What does it mean to isolate the variable when solving an equation?

86. What does it mean to clear an equation of fractions before solving the equation?

87. Which method, the reciprocal method or the two-step method, would you use to solve the equation $\frac{5}{16}t = 15$? Why?

88. Use an example to show why division by a number is the same as multiplying by its reciprocal.

REVIEW

89. Complete this illustration of the distributive property:
$a(b + c) =$ ____________

90. What is the value of q quarters?

91. Convert 41° Fahrenheit to Celsius.

92. Evaluate $\frac{-4 - 8}{3}$.

93. Solve $5x - 3 = 2x + 12$.

94. Solve $10 - (x - 5) = 40$.

95. Round 12,590,767 to the nearest million.

96. In the expression $(-4)^6$, what do we call -4, and what do we call 6?

KEY CONCEPT

The Fundamental Property of Fractions

The **fundamental property of fractions** states that multiplying or dividing the numerator and the denominator of a fraction by the same nonzero number does not change the value of the fraction. This property is used to simplify fractions and to express fractions in higher terms. The following problems review both procedures. Complete each solution.

1. Simplify $\frac{15}{25}$.

Step 1: The numerator and the denominator share a common factor of ___.

Step 2: Apply the fundamental property of fractions. Divide the numerator and the denominator by the common factor ___.

$$\frac{15}{25} = \frac{15 \div 5}{25 \div 5}$$

Step 3: Do the divisions to simplify the fraction.

$$= \frac{3}{5}$$

2. In practice, we often show the simplifying process described in problem 1 in a different form.

Step 1: Factor 15 as 5 · 3 and 25 as 5 · 5.

$$\frac{15}{25} = \frac{5 \cdot 3}{5 \cdot 5}$$

Step 2: The slashes and small 1's indicate that the numerator and the denominator have been divided by 5.

$$= \frac{\overset{1}{\cancel{5}} \cdot 3}{\underset{1}{\cancel{5}} \cdot 5}$$

Step 3: Multiply in the numerator and the denominator.

$$= \frac{3}{5}$$

3. When adding or subtracting fractions and mixed numbers, we often need to express a fraction in higher terms. This is called building up the fraction. Express $\frac{1}{5}$ as a fraction with denominator 35.

Step 1: We must multiply 5 by 7 to obtain 35.

Step 2: Use the fundamental property of fractions. Multiply the numerator and the denominator by 7.

$$\frac{1}{5} = \frac{1 \cdot 7}{5 \cdot 7}$$

Step 3: Multiply in the numerator and the denominator.

$$= \frac{7}{35}$$

4. When adding or subtracting algebraic fractions, we often need to express a fraction in higher terms. Express $\frac{2}{3x}$ as a fraction with a denominator of $18x$.

Step 1: We must multiply $3x$ by 6 to obtain $18x$.

Step 2: Use the fundamental property of fractions. Multiply the numerator and the denominator by 6.

$$\frac{2}{3x} = \frac{2 \cdot 6}{3x \cdot 6}$$

Step 3: Do the multiplication in the numerator and the denominator.

$$= \frac{12x}{18x}$$

ACCENT ON TEAMWORK

Section 4.1

EQUIVALENT FRACTIONS Complete the labeling of each number line using fractions with the same denominator.

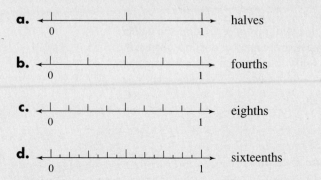

a.  halves

b. 0 — 1 fourths

c. 0 — 1 eighths

d. 0 — 1 sixteenths

FRACTIONS Give everyone in your group a strip of paper that is the same length. (See Illustration 1.) Determine ways to fold the strip of paper into

a. fourths **b.** eighths
c. thirds **d.** sixths

ILLUSTRATION 1

Section 4.2

MULTIPLICATION When we multiply 2 and 4, the answer is greater than 2 and greater than 4. Is this always the case? Is the product of two numbers always greater than either of the two numbers? Explain your answer.

POWERS When we square the number 4, the answer is greater than 4. Is the square of a number always greater than the number? Explain your answer.

Section 4.3

DIVIDING SNACKS Devise a way to divide seven brownies equally among six people.

Section 4.4

ADDING FRACTIONS Without actually doing the addition, explain why $\frac{3}{7} + \frac{1}{4}$ must be less than 1 and why $\frac{4}{7} + \frac{3}{4}$ must be greater than 1.

COMPARING FRACTIONS
a. When 1 is added to the numerator of a fraction, is the result greater than or less than the original fraction? Explain your reasoning.
b. When 1 is added to the denominator of a fraction, is the result greater than or less than the original fraction? Explain your reasoning.

COMPARING FRACTIONS Think of a fraction. Add 1 to its numerator and add 1 to its denominator. Is the resulting fraction greater than, less than, or equal to the original fraction? Explain your reasoning.

Section 4.5

DIVISION WITH MIXED NUMBERS Division can be thought of as repeated subtraction. Use this concept to solve the following problem.

$5\frac{1}{4}$ yards of ribbon needs to be cut into pieces that are $\frac{3}{4}$ of a yard long to form bows. How many bows can be made?

Section 4.6

MIXED NUMBERS Two mixed numbers, A and B, are graphed in Illustration 2. Estimate where on the number line the graph of $A + B$ would lie.

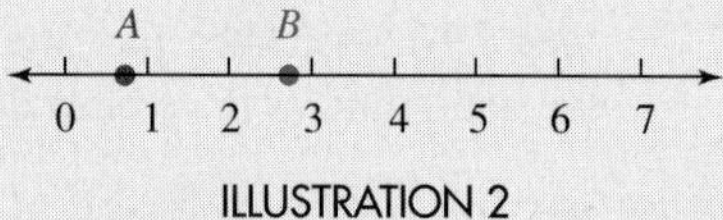

ILLUSTRATION 2

Section 4.7

COMPLEX FRACTION Write a problem that could be solved by simplifying the following.

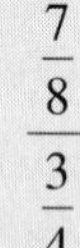

$$\frac{\frac{7}{8}}{\frac{3}{4}}$$

Section 4.8

SOLVING EQUATIONS
a. Solve the equation $\frac{3}{4}x = 15$. Undo the multiplication by $\frac{3}{4}$ by dividing both sides by $\frac{3}{4}$.
b. Do the same for $-\frac{7}{8}x = 21$.

CHAPTER REVIEW

SECTION 4.1 The Fundamental Property of Fractions

CONCEPTS

Fractions are used to indicate equal parts of a whole.

A fraction is composed of a *numerator,* a *denominator,* and a *fraction bar.*

If a and b are positive numbers,

$$\frac{-a}{b} = \frac{a}{-b} = -\frac{a}{b} \quad (b \neq 0)$$

Equivalent fractions represent the same number.

The *fundamental property of fractions:* Dividing the numerator and denominator of a fraction by the same nonzero number does not change the value of the fraction.

To *simplify* a fraction that is not in lowest terms, divide the numerator and denominator by the same number.

A fraction is in *lowest terms* if the only factor common to the numerator and denominator is 1.

The fundamental property of fractions: Multiplying the numerator and denominator of a fraction by a nonzero number does not change its value.

$$\frac{a}{b} = \frac{a \cdot x}{b \cdot x} \quad (b \neq 0, x \neq 0)$$

Expressing a fraction in higher terms results in an equivalent fraction that involves larger numbers or more complex terms.

REVIEW EXERCISES

1. If a woman gets seven hours of sleep each night, what part of a whole day does she spend sleeping?

2. In Illustration 1, why can't we say that $\frac{3}{4}$ of the figure is shaded?

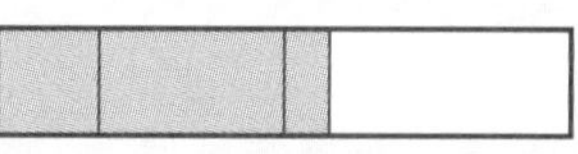

ILLUSTRATION 1

3. Write the fraction $\frac{2}{-3}$ in two other ways.

4. What concept about fractions does Illustration 2 demonstrate?

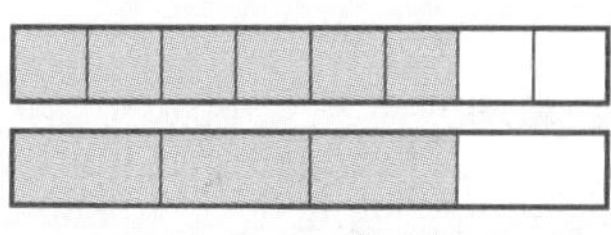

ILLUSTRATION 2

5. Explain the procedure shown here.

$$\frac{4}{6} = \frac{4 \div 2}{6 \div 2} = \frac{2}{3}$$

6. Explain what the slashes and the 1's mean.

$$\frac{4}{6} = \frac{\overset{1}{\cancel{2}} \cdot 2}{\underset{1}{\cancel{2}} \cdot 3} = \frac{2}{3}$$

7. Simplify each fraction to lowest terms.

a. $\frac{15}{45}$ **b.** $\frac{20}{48}$ **c.** $-\frac{63x^2}{84x}$ **d.** $\frac{66m^3n}{108m^4n}$

8. Explain what is being done and why it is valid.

$$\frac{5}{8} = \frac{5 \cdot 2}{8 \cdot 2} = \frac{10}{16}$$

9. Write each fraction or whole number with the indicated denominator (shown in red).

a. $\frac{2}{3}$, 18 **b.** $-\frac{3}{8}$, 16 **c.** $\frac{7}{15}$, $45a$ **d.** 4, 9

SECTION 4.2 Multiplying Fractions

To *multiply two fractions,* multiply their numerators and multiply their denominators.

$$\frac{a}{b} \cdot \frac{c}{d} = \frac{a \cdot c}{b \cdot d}$$
$(b \neq 0, d \neq 0)$

10. Multiply.

a. $\frac{1}{2} \cdot \frac{1}{3}$ **b.** $\frac{2}{5}\left(-\frac{7}{9}\right)$ **c.** $\frac{9}{16} \cdot \frac{20}{27}$ **d.** $\frac{5}{6} \cdot \frac{1}{3} \cdot \frac{18}{25}$

e. $\frac{3}{5} \cdot 7$ **f.** $-4\left(-\frac{9}{16}\right)$ **g.** $3\left(\frac{1}{3}\right)$ **h.** $-\frac{6}{7}\left(-\frac{7}{6}\right)$

11. Tell whether each statement is true or false.

a. $\frac{3}{4}x = \frac{3x}{4}$ **b.** $-\frac{5}{9}e = -\frac{5}{9e}$

12. Multiply.

a. $\frac{3t}{5} \cdot \frac{10}{27t}$ **b.** $-\frac{2}{3}\left(\frac{4}{7}s\right)$ **c.** $\frac{4d^2}{9} \cdot \frac{3}{28d}$ **d.** $9mn\left(-\frac{5}{81n^2}\right)$

An *exponent* indicates repeated multiplication.

13. Evaluate each power.

a. $\left(\frac{3}{4}\right)^2$ **b.** $\left(-\frac{5}{2}\right)^3$ **c.** $\left(\frac{x}{3}\right)^2$ **d.** $\left(-\frac{2c}{5}\right)^3$

In mathematics, the word *of* usually means multiply.

14. GRAVITY ON THE MOON Objects on the moon weigh only one-sixth as much as on Earth. How much will an astronaut weigh on the moon if he weighs 180 pounds on Earth?

The *area of a triangle*:

$$A = \frac{1}{2}bh$$

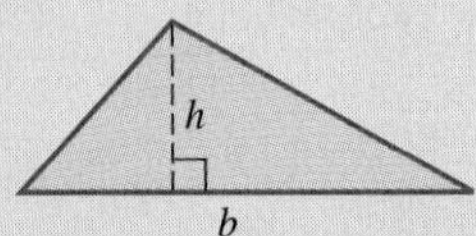

15. Find the area of the triangular sign in Illustration 3.

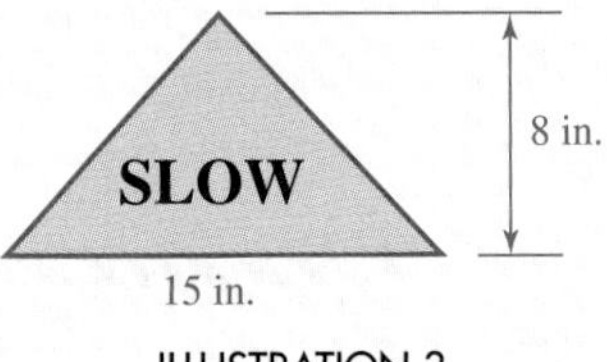

ILLUSTRATION 3

SECTION 4.3 Dividing Fractions

Two numbers are called *reciprocals* if their product is 1.

16. Find the reciprocal of each number.

a. $\frac{1}{8}$ **b.** $-\frac{11}{12}$ **c.** x **d.** $\frac{ab}{c}$

To *divide two fractions,* multiply the first by the reciprocal of the second.

$$\frac{a}{b} \div \frac{c}{d} = \frac{a}{b} \cdot \frac{d}{c}$$
$(b \neq 0, c \neq 0, d \neq 0)$

17. Divide.

a. $\frac{1}{6} \div \frac{11}{25}$ **b.** $-\frac{7}{8} \div \frac{1}{4}$

c. $-\frac{15}{16} \div (-10)$ **d.** $8 \div \frac{16}{5}$

e. $\frac{t}{8} \div \frac{1}{4}$ **f.** $\frac{4a}{5} \div \frac{a}{2}$

g. $-\frac{a}{b} \div \left(-\frac{b}{a}\right)$ **h.** $\frac{2}{3}x \div \left(-\frac{x^2}{9}\right)$

18. GOLD COINS How many $\frac{1}{16}$-ounce coins can be cast from a $\frac{3}{4}$-ounce bar of gold?

SECTION 4.4 Adding and Subtracting Fractions

To add (or subtract) fractions with like denominators, add (or subtract) their numerators and write the result over the common denominator.

$$\frac{a}{c} + \frac{b}{c} = \frac{a+b}{c} \quad (c \neq 0)$$

$$\frac{a}{c} - \frac{b}{c} = \frac{a-b}{c} \quad (c \neq 0)$$

19. Add or subtract.

a. $\frac{2}{7} + \frac{3}{7}$ **b.** $-\frac{3}{5} - \frac{3}{5}$ **c.** $\frac{3}{x} - \frac{1}{x}$ **d.** $\frac{7}{8} + \frac{t}{8}$

20. Explain why we cannot immediately add $\frac{1}{2} + \frac{2}{3}$ without doing some preliminary work.

The *LCD* must include the set of prime factors of each of the denominators.

21. Use prime factorization to find the least common denominator for fractions with denominators of 45 and 30.

To add or subtract fractions with unlike denominators, we must first express them as equivalent fractions with the same denominator, preferably the LCD.

22. Add or subtract.

a. $\frac{1}{6} + \frac{2}{3}$ **b.** $\frac{2}{5} + \left(-\frac{3}{8}\right)$

c. $-\frac{3}{8} - \frac{5}{6}$ **d.** $3 - \frac{1}{7}$

e. $\frac{x}{25} - \frac{3}{10}$ **f.** $\frac{1}{3} + \frac{7}{y}$

g. $\frac{13}{6} - 6$ **h.** $\frac{1}{3} + \frac{1}{4} + \frac{1}{5}$

23. MACHINE SHOP See Illustration 4. How much must be milled off the $\frac{3}{4}$-inch-thick steel rod so that the collar will slip over the end of it?

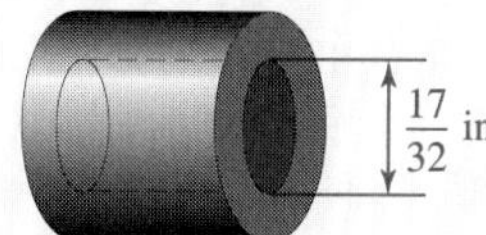

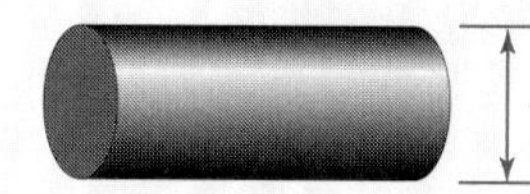

ILLUSTRATION 4

To *compare fractions,* write them as equivalent fractions with the same denominator. Then the fraction with the larger numerator will be the larger fraction.

24. TELEMARKETING In the first hour of work, a telemarketer made 2 sales out of 9 telephone calls. In the second hour, she made 3 sales out of 11 calls. During which hour was the rate of sales to calls better?

SECTION 4.5 Multiplying and Dividing Mixed Numbers

A *mixed number* is the sum of its whole-number part and its fractional part.

25. a. What mixed number is represented in Illustration 5?

b. What improper fraction is represented in Illustration 5?

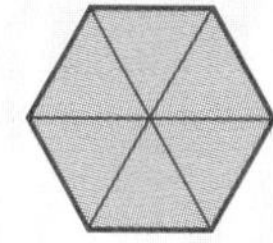
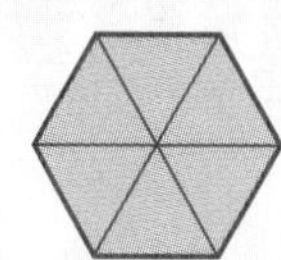
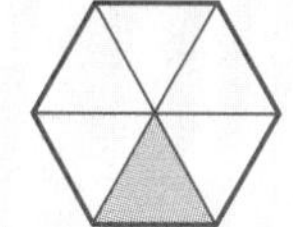

ILLUSTRATION 5

To change an *improper fraction* to a mixed number, divide the numerator by the denominator to obtain the whole-number part. Write the remainder over the denominator for the fractional part.

26. Express each improper fraction as a mixed number or a whole number.

a. $\frac{16}{5}$ **b.** $-\frac{47}{12}$ **c.** $\frac{6}{6}$ **d.** $\frac{14}{6}$

To change a mixed number to an improper fraction, multiply the whole number by the denominator and add the result to the numerator. Write this sum over the denominator.

27. Write each mixed number as an improper fraction.

a. $9\frac{3}{8}$ **b.** $-2\frac{1}{5}$ **c.** $100\frac{1}{2}$ **d.** $1\frac{99}{100}$

28. Graph $-2\frac{2}{3}$, $\frac{8}{9}$, and $\frac{59}{24}$.

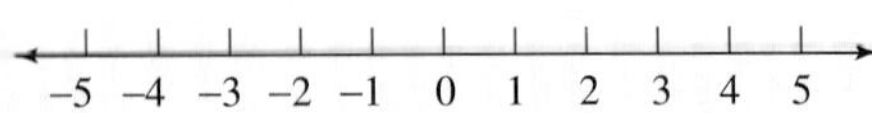

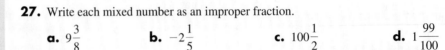

To *multiply* or *divide mixed numbers,* change the mixed numbers to improper fractions and then do the operations as usual.

29. Multiply or divide. Write answers as mixed numbers when appropriate.

a. $-5\frac{1}{4} \cdot \frac{2}{35}$ **b.** $\left(-3\frac{1}{2}\right) \div \left(-3\frac{2}{3}\right)$

c. $\left(-6\frac{2}{3}\right)(-6)$ **d.** $-8 \div 3\frac{1}{5}$

30. CAMERA TRIPOD The three legs of a tripod can be extended to become $5\frac{1}{2}$ times their original length. If each leg is $8\frac{3}{4}$ inches long when collapsed, how long will a leg become when it is completely extended?

SECTION 4.6 *Adding and Subtracting Mixed Numbers*

To add (or subtract) mixed numbers, we can change each to an improper fraction and use the method of Section 4.4.

31. Add or subtract.

a. $1\frac{3}{8} + 2\frac{1}{5}$ **b.** $3\frac{1}{2} + 2\frac{2}{3}$

c. $2\frac{5}{6} - 1\frac{3}{4}$ **d.** $3\frac{7}{16} - 2\frac{1}{8}$

To add mixed numbers, we can add the whole numbers and the fractions separately.

32. PAINTING SUPPLIES In a project to restore a house, painters used $10\frac{3}{4}$ gallons of primer, $21\frac{1}{2}$ gallons of latex paint, and $7\frac{2}{3}$ gallons of enamel. Find the total number of gallons of paint used.

Vertical form can be used to add or subtract mixed numbers.

33. Add or subtract.

a. $133\frac{1}{9} + 49\frac{1}{6}$ **b.** $98\frac{11}{20} + 14\frac{3}{5}$

c. $50\frac{5}{8} - 19\frac{1}{6}$ **d.** $375\frac{3}{4} - 59$

If the fraction being subtracted is larger than the first fraction, we need to *borrow* from the whole number.

34. Subtract.

a. $23\frac{1}{3} - 2\frac{5}{6}$ **b.** $39 - 4\frac{5}{8}$

SECTION 4.7 Order of Operations and Complex Fractions

A *complex fraction* is a fraction whose numerator or denominator, or both, contain one or more fractions or mixed numbers.

To simplify a complex fraction, *Method 1:* The main fraction bar of a complex fraction indicates division.

Method 2: Multiply the numerator and denominator of the complex fraction by the LCD of all the fractions that appear in it.

35. Evaluate each numerical expression.

a. $\frac{3}{4} + \left(-\frac{1}{3}\right)^2\left(\frac{5}{4}\right)$ **b.** $\left(\frac{2}{3} \div \frac{16}{9}\right) - \left(1\frac{2}{3} \cdot \frac{1}{15}\right)$

36. Simplify each complex fraction.

a. $\dfrac{\frac{3}{5}}{-\frac{17}{20}}$ **b.** $\dfrac{\frac{2}{3} - \frac{1}{6}}{-\frac{3}{4} - \frac{1}{2}}$

37. Evaluate each expression for $c = -\frac{3}{4}$, $d = \frac{1}{8}$, and $e = -2\frac{1}{16}$.

a. $d^2 - 2c$ **b.** $-cd + e$ **c.** $e \div (cd)$ **d.** $\dfrac{c - d}{e}$

SECTION 4.8 Solving Equations Containing Fractions

To solve an equation,
Simplify the equation:
1. Clear the equation of fractions.
2. Remove parentheses using the distributive property.
3. Combine like terms.

Isolate the variable:

4. Get the variable on one side and constants on the other using the addition or subtraction properties of equality.
5. Combine like terms.
6. Undo multiplication and division.

38. Solve each equation. Check the result.

a. $\frac{2}{3}x = 16$ **b.** $-\frac{7s}{4} = -49$

c. $\frac{y}{5} = -\frac{1}{15}$ **d.** $2x - 3 = 8$

39. Solve each equation.

a. $\frac{c}{3} - \frac{c}{8} = 2$ **b.** $\frac{5h}{9} - 1 = -3$

c. $4 - \frac{d}{4} = 0$ **d.** $\frac{t}{10} - \frac{2}{3} = \frac{1}{5}$

40. HISTORY TEXTBOOK In writing a history text, the author decided to devote two-thirds of the book to events prior to World War II. The remainder of the book deals with history after the war. If pre–World War II history is covered in 220 pages, how many pages does the textbook have?

Chapter 4 Test

1. See Illustration 1.
a. What fractional part of the plant is above ground?
b. What fractional part of the plant is below ground?

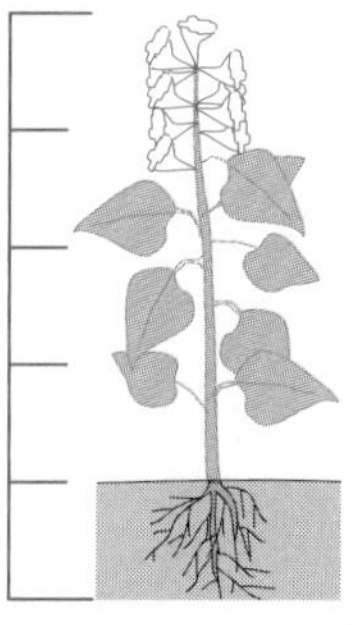
ILLUSTRATION 1

2. Simplify each fraction.
a. $\frac{27}{36}$ **b.** $\frac{72n^2}{180n}$

3. Multiply: $-\frac{3x}{4}\left(\frac{1}{5x^2}\right)$.

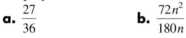

4. COFFEE DRINKERS Of 100 adults surveyed, $\frac{2}{5}$ said they started off their morning with a cup of coffee. Of the 100, how many would this be?

5. Divide: $\frac{4a}{3} \div \frac{a^2}{9}$.

6. Subtract: $\frac{x}{6} - \frac{4}{5}$.

7. Express $\frac{7}{8}$ as an equivalent fraction with denominator $24a$.

8. Graph $2\frac{4}{5}$, $-1\frac{1}{7}$, and $\frac{7}{6}$.

−2 −1 0 1 2 3

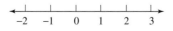

9. SPORTS CONTRACT A basketball player signed a nine-year contract for $\$13\frac{1}{2}$ million. How much is this per year?

10. Evaluate $-2ct^2$ for $c = -2\frac{1}{12}$ and $t = \frac{2}{5}$.

11. Add: $157\frac{5}{9} + 103\frac{3}{4}$.

12. Subtract: $67\frac{1}{4} - 29\frac{5}{6}$.

13. BOXING When Oscar De La Hoya fought Pernell Whitaker, the "Tale of the Tape" shown in Illustration 2 appeared in the sports section of many newspapers. What was the difference in the fighters'
a. weights?
b. chests (expanded)?
c. waists?

Tale of the Tape

De La Hoya		Whitaker
24 yr	Age	33 yr
146 1/2 lb	Weight	146 1/2 lb
5-11	Height	5-6
72 in.	Reach	69 in.
39 in.	Chest (Normal)	37 in.
42 1/4 in.	Chest (Expanded)	39 1/2 in.
31 3/4 in.	Waist	28 in.

ILLUSTRATION 2

14. Add: $-\frac{3}{7} + 2$.

15. SEWING When cutting material for a $10\frac{1}{2}$-inch-wide placemat, a seamstress allows $\frac{5}{8}$ inch at each end for a hem. How wide should the material be cut? See Illustration 3.

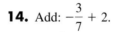

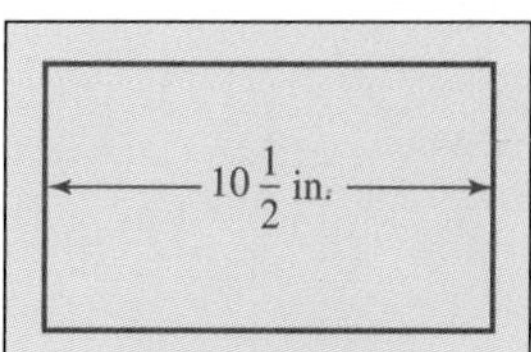

ILLUSTRATION 3

16. In Illustration 4, find the perimeter and the area of the triangle.

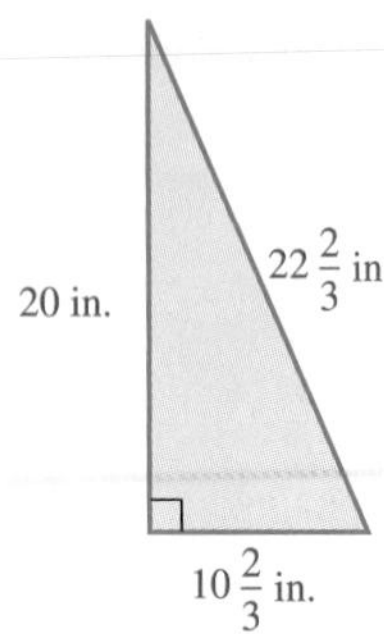

ILLUSTRATION 4

17. Evaluate:

$$\left(\frac{2}{3} \cdot \frac{5}{16}\right) - \left(-1\frac{3}{5} \div 4\frac{4}{5}\right)$$

18. Simplify the complex fraction.

$$\frac{-\frac{5}{6}}{\frac{7}{8}}$$

19. Simplify the complex fraction.

$$\frac{\frac{1}{2} + \frac{1}{3}}{-\frac{1}{6} - \frac{1}{3}}$$

20. Solve each equation.

a. $\frac{x}{3} = 14$ **b.** $-\frac{5}{2}t = 18$

21. Solve $6x - 4 = -3$.

22. Solve $\frac{x}{6} - \frac{2}{3} = \frac{x}{12}$.

23. JOB APPLICANTS Three-fourths of the applicants for a position had previous experience. The number who did not have prior experience was 36. How many people applied?

24. What are the parts of a fraction? What does a fraction represent?

25. Explain what is meant when we say, "The product of any number and its reciprocal is 1."

26. Explain what mathematical concept is being shown.

a. $\frac{6}{8} = \frac{3 \cdot \overset{1}{\cancel{2}}}{4 \cdot \underset{1}{\cancel{2}}} = \frac{3}{4}$

b.

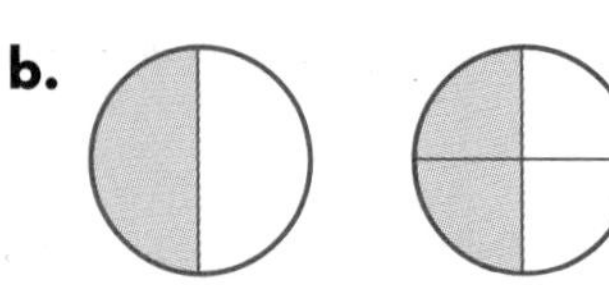

c. $\frac{3}{5} = \frac{3 \cdot 4}{5 \cdot 4} = \frac{12}{20}$

Chapters 1–4 Cumulative Review Exercises

Consider the number 5,434,679.

1. Round to the nearest hundred.

2. Round to the nearest ten thousand.

3. THE STOCK MARKET The graph in Illustration 1 shows the performance of the Dow Jones Industrial Average on the last trading day of 1999. Estimate the highest mark that the market reached. At what time during the day did that occur?

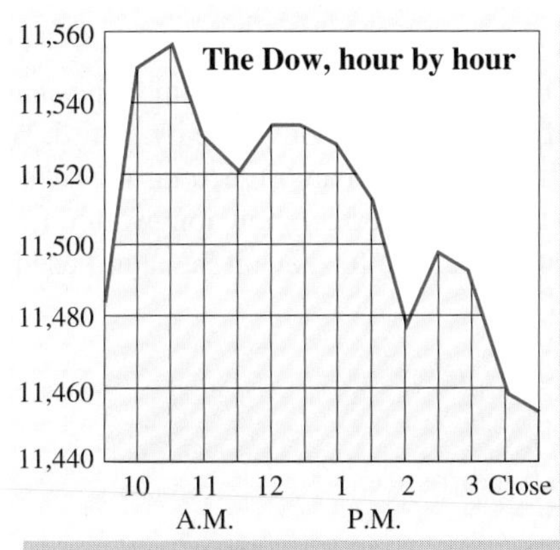

Source: *Los Angeles Times* (December 31, 1999)

ILLUSTRATION 1

4. BANKS As of December 31, 1997, the world's largest bank, with total assets of $691,920,300,000, was the Bank of Tokyo–Mitsubishi Ltd., Japan. In what place value column is the digit 6 located?

Do each operation.

5. $\begin{array}{r} 4{,}679 \\ +3{,}457 \\ \hline \end{array}$

6. $\begin{array}{r} 7{,}897 \\ -4{,}378 \\ \hline \end{array}$

7. $\begin{array}{r} 5{,}345 \\ \times \quad 56 \\ \hline \end{array}$

8. $35\overline{)34{,}685}$

In Exercises 9–10, refer to the rectangular swimming pool shown in Illustration 2.

9. Find the perimeter of the pool.

10. Find the area of the pool's surface.

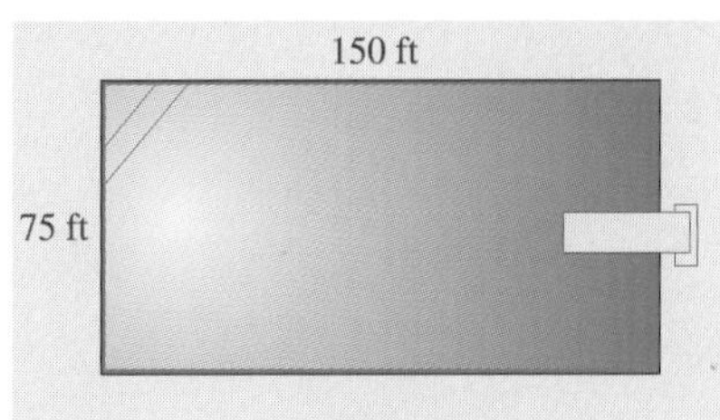

ILLUSTRATION 2

Find the prime factorization of each number.

11. 84

12. 450

13. 360

14. 3,600

Evaluate each expression.

15. $6 + (-2)(-5)$

16. $(-2)^3 - 3^3$

17. $\dfrac{2(-7) + 3(2)}{2(-2)}$

18. $\dfrac{2(3^2 - 4^2)}{-2(3) - 1}$

Translate each expression into a mathematical expression involving the variable x.

19. The sum of a number and 15

20. Eight less than a number

21. The product of a number and 4

22. The quotient obtained when a number is divided by 10

Evaluate each expression when $x = 4$.

23. $2x - 1$

24. $\dfrac{9x}{2} - x$

25. $3x - x^3$

26. $x + 2(x - 7)$

Simplify each expression.

27. $-3(5x)$

28. $-4x(-7x)$

29. $-2(3x - 4)$

30. $-5(3x - 2y + 4)$

Combine like terms.

31. $-3x + 8x$

32. $4a^2 - (-3a^2)$

33. $4x - 3y - 5x + 2y$

34. $-2(3x - 4) + 2x$

In Exercises 35–40, solve each equation. Check the result.

35. $3x + 2 = -13$

36. $-5z - 7 = 18$

37. $\frac{y}{4} - 1 = -5$

38. $\frac{n}{5} + 1 = 0$

39. $6x - 12 = 2x + 4$

40. $3(2y - 8) = -2(y - 4)$

41. OBSERVATION HOURS To get a Master's degree in educational psychology, a student must have 100 hours of observation time at a clinic. If the student has already observed for 37 hours, how many 3-hour shifts must he observe to complete the requirement?

42. GEOMETRY A rectangle is four times as long as it is wide. If its perimeter is 210 feet, find its dimensions.

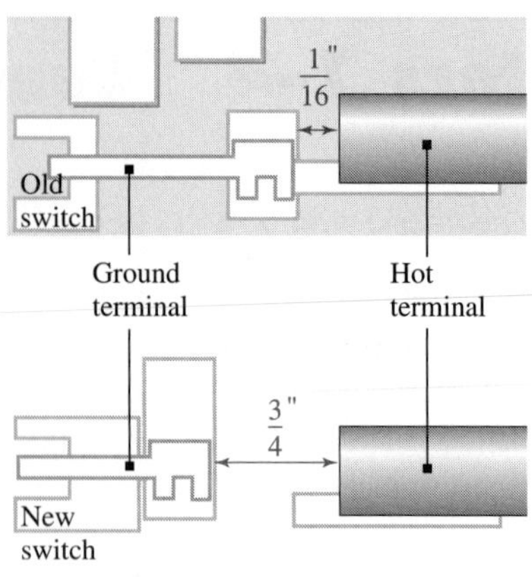

ILLUSTRATION 3

54. SHAVING Advertisements claim that a shaving lotion for men cuts shaving time by a third. When using this lotion, it took a man 60 seconds to shave. If the advertising claim is correct, how long would it normally have taken the man to shave if he hadn't used the special lotion?

Simplify each fraction.

43. $\frac{21}{28}$

44. $\frac{40x^6y^4}{16x^3y^5}$

Do each operation.

45. $\frac{6}{5}\left(-\frac{2}{3}\right)$

46. $\frac{14p^2}{8} \div \frac{7p^3}{2}$

47. $\frac{2}{3} + \frac{3}{4}$

48. $\frac{4}{m} - \frac{3}{5}$

Write each mixed number as an improper fraction.

49. $3\frac{5}{6}$

50. $-6\frac{5}{8}$

In Exercises 51–52, do each operation.

51. $4\frac{2}{3} + 5\frac{1}{4}$

52. $14\frac{2}{5} - 8\frac{2}{3}$

53. FIRE HAZARD Two terminals in an electrical switch were so close that electricity could jump the gap and start a fire. Illustration 3 shows a newly designed switch that will keep this from happening. By how much was the distance between the ground terminal and the hot terminal increased?

Simplify each expression.

55. $\left(\frac{1}{4} - \frac{7}{8}\right) \div \left(-2\frac{3}{16}\right)$

56. $\dfrac{\frac{2}{3} - 7}{4\frac{5}{6}}$

Solve each equation. Check the result.

57. $x + \frac{1}{5} = -\frac{14}{15}$

58. $\frac{3}{4}x = \frac{5}{8}x + \frac{1}{2}$

59. $\frac{2}{3}x = -10$

60. $3y - 8 = 0$

61. Explain the difference between an *expression* and an *equation.*

62. What is a variable?

Decimals

5

DECIMALS PROVIDE ANOTHER WAY TO REPRESENT FRACTIONS AND MIXED NUMBERS. THEY ARE OFTEN USED IN MEASUREMENT, BECAUSE IT IS EASY TO PUT THEM IN ORDER AND TO COMPARE THEM.

5.1 An Introduction to Decimals

In this section, you will learn about

- Decimals • The place value system for decimal numbers
- Reading and writing decimals • Comparing decimals • Rounding

INTRODUCTION. This section introduces the **decimal numeration system**—an extension of the place value system that we used when working with whole numbers. You may not realize it, but you have often worked with the decimal numeration system.

We can use the decimal numeration system to express the car's mileage. The odometer reads 1,537.6 miles.

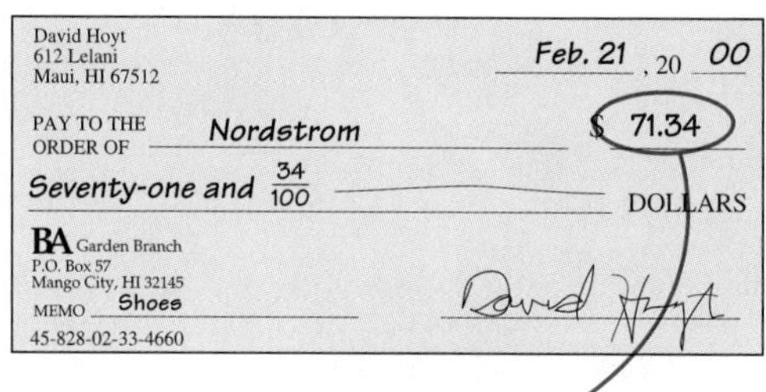

The amount of the check is written using the decimal numeration system.

Decimals

Like fraction notation, decimal notation is used to denote a part of a whole. However, when writing a number in decimal notation, we don't use a fraction bar, nor is a denominator shown. A number written in decimal notation is often called a **decimal.**

In Figure 5-1, a rectangle is divided into 10 equal parts. One-tenth of the figure is shaded. We can use either the fraction $\frac{1}{10}$ or the decimal 0.1 to describe the shaded region. Both are read as "one-tenth."

$$\frac{1}{10} = 0.1$$

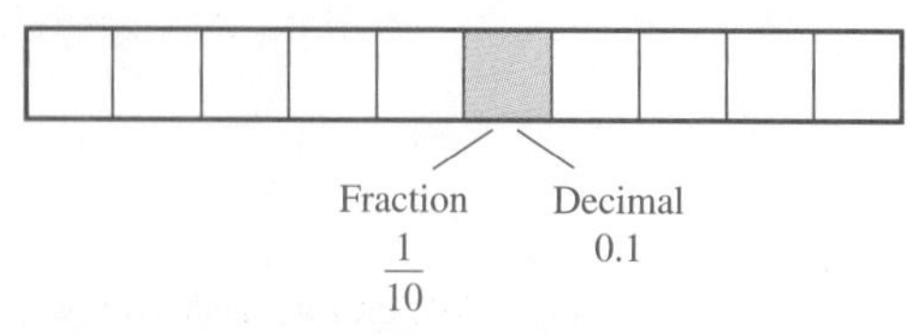

FIGURE 5-1

In Figure 5-2, a square is divided into 100 equal parts. One of the 100 parts is shaded; it can be represented by the fraction $\frac{1}{100}$ or by the decimal 0.01. Both are read as "one one-hundredth."

$$\frac{1}{100} = 0.01$$

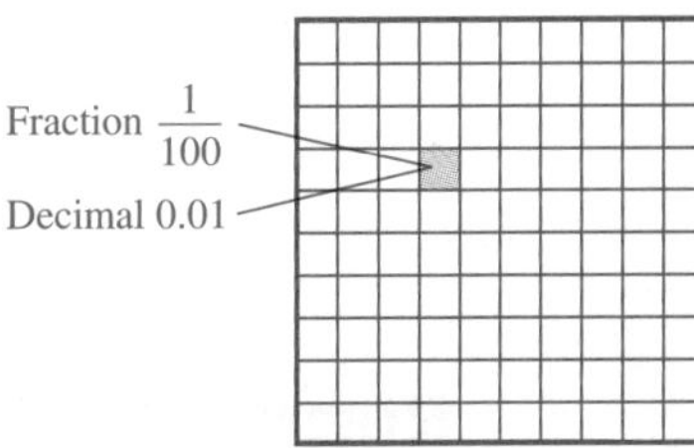

FIGURE 5-2

The place value system for decimal numbers

Decimal numbers are written by placing digits (0, 1, 2, 3, 4, 5, 6, 7, 8, 9) into place value columns that are separated by a **decimal point.** See Figure 5-3. The place value names of all the columns to the right of the decimal point end in "th." The "th" tells us that the value of the column is a fraction whose denominator is a power of ten. Columns to the left of the decimal point have a value greater than or equal to 1; columns to the right of the decimal point have a value less than 1. We can show the value represented by each digit of a decimal by using **expanded notation.**

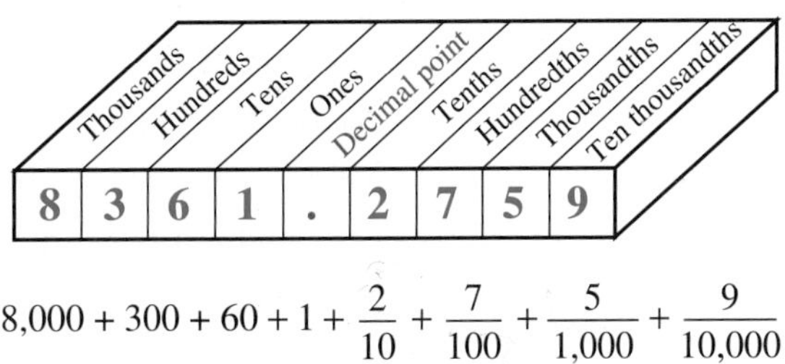

$$8{,}000 + 300 + 60 + 1 + \frac{2}{10} + \frac{7}{100} + \frac{5}{1{,}000} + \frac{9}{10{,}000}$$

Expanded notation

FIGURE 5-3

Decimal points are used to separate the whole-number part of a decimal from its fractional part.

12.37

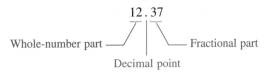

Whole-number part — Decimal point — Fractional part

When there is no whole-number part of a decimal, we can show that by entering a zero to the left of the decimal point.

.85 = 0.85

↑ No whole number part. ↑ Enter a zero here, if desired.

We can write a whole number in decimal notation by placing a decimal point to its right and then adding a zero, or zeros, to the right of the decimal point.

$$99 = 99.0 = 99.00$$

A whole number. Place a decimal point here and enter a zero, or zeros, to the right of it.

Writing additional zeros to the right of the decimal point *following the last digit* does not change the value of the decimal.

$$12.37 = 12.370 = 12.3700$$

These additional zeros do not change the value of the decimal.

Reading and writing decimals

The decimal 12.37 can be read as "twelve point three seven." Another way of reading a decimal states the whole-number part first and then the fractional part.

Reading a decimal

To read a decimal:

1. Look to the left of the decimal point and say the name of the whole number.
2. The decimal point is then read as "and."
3. Say the fractional part of the decimal as a whole number followed by the name of the place value column of the digit that is farthest to the right.

Using this procedure, here is the other way to read 12.37.

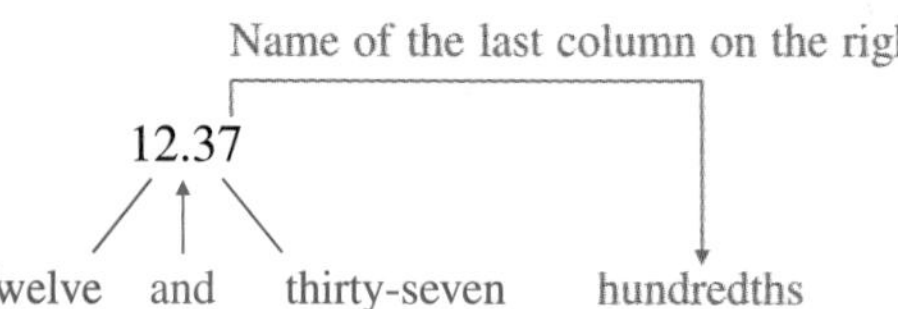

When we read a decimal in this way, it is easy to write it in words and as a mixed number.

Decimal	Words	Mixed number
12.37	Twelve and thirty-seven hundredths	$12\frac{37}{100}$

EXAMPLE 1 ***Writing a decimal in other forms.*** Write each decimal in words and then as a fraction or mixed number. **Do not simplify the fraction.**

a. The world speed record for a human-powered vehicle is 65.484 mph, set in 1986.

b. The smallest fresh-water fish is the dwarf pygmy goby, found in the Philippines. Adult males weigh 0.00014 ounce.

Solution

a. 65.484 is sixty-five and four hundred eighty-four thousandths, or $65\frac{484}{1,000}$.

b. 0.00014 is fourteen hundred-thousandths, or $\frac{14}{100,000}$.

Self Check

Write each decimal in words and then as a mixed number.

a. Sputnik 1, the first artificial satellite, weighed 184.3 pounds.

b. The planet Mercury makes one revolution every 87.9687 days.

Answers: a. One hundred eighty-four and three-tenths, or $184\frac{3}{10}$ **b.** Eighty-seven and nine thousand six hundred eighty-seven ten thousandths, or $87\frac{9,687}{10,000}$ ■

Decimals can be negative. For example, a record low temperature of $-128.6°$ F was recorded in Vostok, Antarctica on July 21, 1983. This is read as "negative one hundred twenty-eight and six tenths." Written as a mixed number, it is $-128\frac{6}{10}$.

Comparing decimals

The relative sizes of a set of decimals can be determined by scanning their place value columns from left to right, column by column, looking for a difference in the digits. For example,

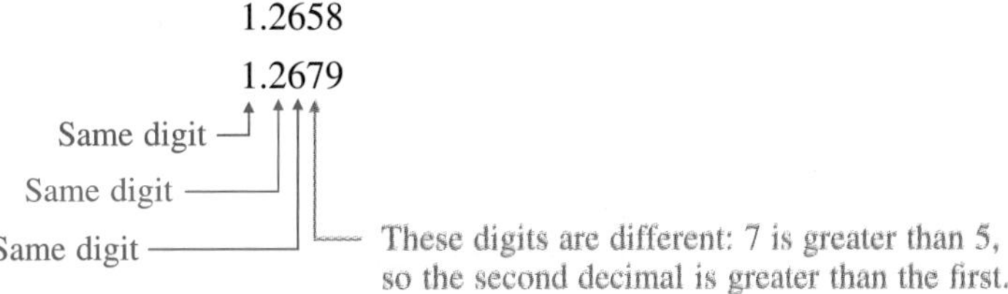

Thus, 1.2679 is greater than 1.2658. We write $1.2679 > 1.2658$.

Comparing positive decimals

> To compare two positive decimals:
> 1. Make sure both numbers have the same number of decimal places to the right of the decimal point. Write any additional zeros necessary to achieve this.
> 2. Compare the digits of each decimal, column by column, working from left to right.
> 3. When two digits differ, the decimal with the greater digit is the greater number.

EXAMPLE 2 ***Comparing positive decimals.*** Which is greater, 54.9 or 54.929?

Solution

54.900 Write two zeros after 9 so that both decimals have the same number of digits to the right of the decimal point.

54.929

↑

Working from left to right, this is the first column in which the digits differ. Since 2 is greater than 0, we can conclude that $54.929 > 54.9$.

Self Check

Which is greater, 113.7 or 113.657?

Answer: 113.7 ■

Comparing negative decimals

> To compare two negative decimals:
> 1. Make sure both numbers have the same number of decimal places to the right of the decimal point. Write any additional zeros necessary to achieve this.
> 2. Compare the digits of each decimal, column by column, working from left to right.
> 3. When two digits differ, the decimal with the smaller digit is the greater number.

EXAMPLE 3 ***Comparing negative decimals.*** Which is greater, -10.45 or -10.419?

Solution

-10.450 Write a 0 after 5 to help in the comparison.

-10.419

↑

Working from left to right, this is the first column in which the digits differ. Since 1 is less than 5, we conclude that $-10.419 > -10.45$.

Self Check

Which is greater, -703.8 or -703.78?

Answer: -703.78 ■

EXAMPLE 4 ***Graphing decimals.*** Graph −1.8, −1.23, −0.3, and 1.89.

Solution

To graph each decimal, we locate its position on the number line and draw a dot. Since −1.8 is to the left of −1.23, we can write $-1.8 < -1.23$.

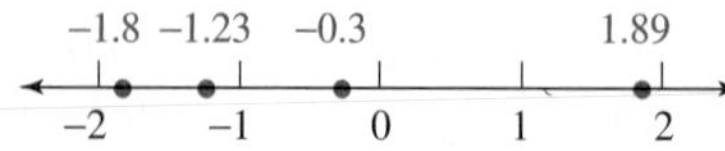

Self Check

Graph −1.1, −0.6, 0.8, and 1.9.

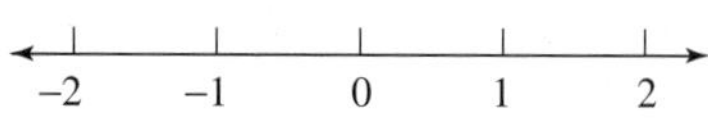

Answer:

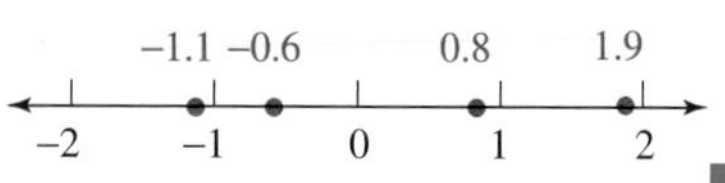

Rounding

When working with decimals, we often round answers to a specific number of decimal places.

Rounding a decimal

1. To round a decimal to a specified decimal place, locate the digit in that place. Call it the *rounding digit.*
2. Look at the *test digit* to the right of the rounding digit.
3. If the test digit is 5 or greater, round up by adding 1 to the rounding digit and dropping all the digits to its right. If the test digit is less than 5, round down by keeping the rounding digit and dropping all the digits to its right.

EXAMPLE 5 ***Chemistry.*** In a chemistry class, a student uses a balance to weigh a compound. The digital readout on the scale shows 1.2387 g. Round this decimal to the nearest thousandth of a gram.

Solution We are asked to round to the nearest thousandth.

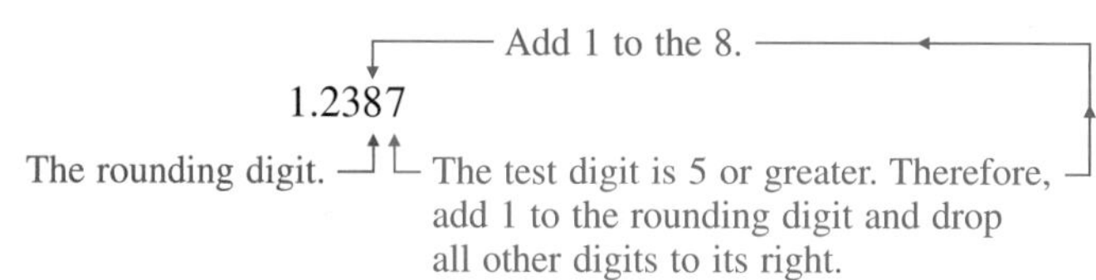

The compound weighs approximately 1.239 g.

EXAMPLE 6 ***Rounding decimals.*** Round each decimal to the indicated place value: **a.** −645.13 to the nearest tenth and **b.** 33.097 to the nearest hundredth.

Solution

a. −645.13

Rounding digit. Since the test digit is less than 5, drop it and all the digits to its right.

The result is −645.1.

b. 33.097

Rounding digit. Since the test digit is greater than 5, we add 1 to 9 and drop all the digits to the right.

33.0~~9~~ (10) Adding a 1 to the 9 requires that we carry a 1 to the tenths column.

When we are asked to round to the nearest hundredth, we must have a digit in the hundredths column, even if it is a zero. Therefore, the result is 33.10.

Self Check

Round each decimal to the indicated place value.

a. −708.522 to the nearest tenth

b. 9.1198 to the nearest thousandth

Answers: **a.** −708.5, **b.** 9.120

STUDY SET Section 5.1

VOCABULARY *Fill in the blanks to make the statements true.*

1. Give the name of each place value column.

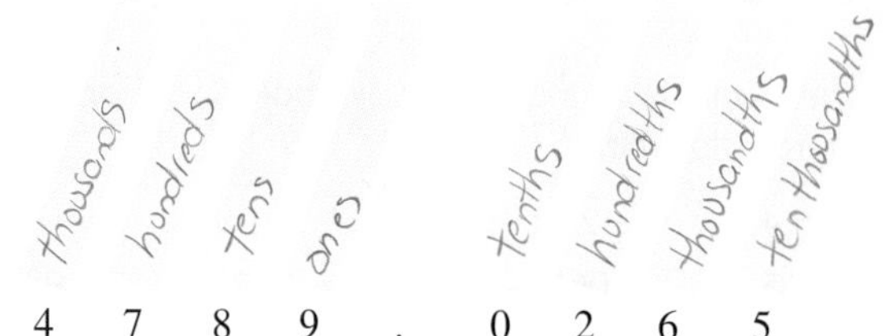

4 7 8 9 . 0 2 6 5

2. We can show the value represented by each digit of the decimal 98.6213 by using expanded notation.

$$98.6213 = 90 + 8 + \frac{6}{10} + \frac{2}{100} + \frac{1}{1{,}000} + \frac{3}{10{,}000}$$

3. We can approximate a decimal number using the process called rounding.

4. When reading 2.37, the decimal point can be read as "point" or "and."

CONCEPTS

5. Consider the decimal 32.415.

a. Write the decimal in words.

b. What is its whole-number part?

c. What is its fractional part?

d. Write the decimal in expanded notation.

6. Write

$400 + 20 + 8 + \frac{9}{10} + \frac{6}{100}$

as a decimal.

7. Graph $\frac{7}{10}$, -0.7, $-3\frac{1}{100}$, and 3.01.

−5 −4 −3 −2 −1 0 1 2 3 4 5

8. Graph −1.21, −3.29, and −4.25.

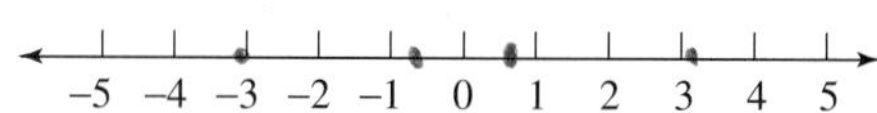

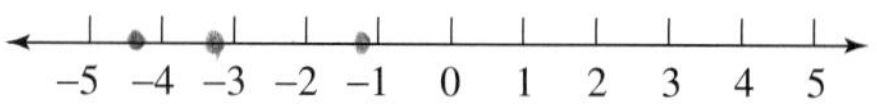

−5 −4 −3 −2 −1 0 1 2 3 4 5

9. True or false?

a. $0.9 = 0.90$

b. $1.260 = 1.206$

c. $-1.2800 = -1.280$

d. $0.001 = .0010$

10. Write each fraction as a decimal.

a. $\frac{9}{10}$ **b.** $\frac{63}{100}$

c. $\frac{111}{1{,}000}$ **d.** $\frac{27}{10{,}000}$

11. Represent the shaded part of the square in Illustration 1 using a fraction and a decimal.

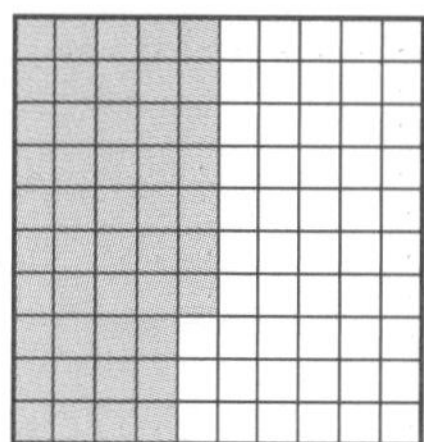

ILLUSTRATION 1

12. Represent the shaded part of the figure in Illustration 2 using a fraction and a decimal.

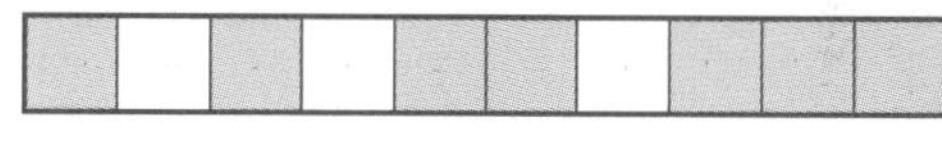

ILLUSTRATION 2

13. The line segment in Illustration 3 is one inch long. Show a length of 0.3 inch on it.

ILLUSTRATION 3

14. Read the meter in Illustration 4. What decimal is indicated by the arrow?

ILLUSTRATION 4

NOTATION

15. Construct a decimal number by writing
0 in the tenths column,
4 in the thousandths column,
1 in the tens column,
9 in the thousands column,
8 in the hundreds column,
2 in the hundredths column,
5 in the ten thousandths column, and
6 in the ones column.

16. Represent each situation using a signed number.
- **a.** A deficit of $15,600.55
- **b.** A river 6.25 feet under flood stage
- **c.** A state budget $6.4 million in the red
- **d.** 3.9 degrees below zero
- **e.** 17.5 seconds prior to liftoff
- **f.** A checking account overdrawn by $33.45

PRACTICE *Write each decimal in words and then as a fraction or mixed number.*

17. 50.1 **18.** 0.73

19. −0.0137 **20.** −76.09

21. 304.0003 **22.** 68.91

23. −72.493 **24.** −31.5013

Write each decimal using numbers.

25. Negative thirty-nine hundredths

26. Negative twenty-seven and forty-four hundredths

27. Six and one hundred eighty-seven thousandths

28. Ten and fifty-six ten-thousandths

Round each decimal to the nearest tenth.

29. 506.098 **30.** 0.441

31. 2.718218 **32.** 3,987.8911

Round each decimal to the nearest hundredth.

33. −0.137 **34.** −808.0897

35. 33.0032 **36.** 64.0059

Round each decimal to the nearest thousandth.

37. 3.14159 **38.** 16.0995

39. 1.414213 **40.** 2,300.9998

Round each decimal to the nearest whole number.

41. 38.901 **42.** 405.64

43. 2,988.399 **44.** 10,453.27

Round each amount to the value indicated.

45. $3,090.28
- **a.** Nearest dollar
- **b.** Nearest ten cents

46. $289.73
- **a.** Nearest dollar
- **b.** Nearest ten cents

Fill in the blanks with the proper symbol ($<$, $>$, or $=$) to make a true statement.

47. −23.45 ___ −23.1 **48.** −301.98 ___ −302.45

49. −.065 ___ −.066 **50.** −3.99 ___ −3.9888

Arrange the decimals in order, from least to greatest.

51. 132.64, 132.6499, 132.6401

52. 0.007, 0.00697, 0.00689

APPLICATIONS

53. WRITING A CHECK Complete the check shown in Illustration 5 by writing in the amount, using a decimal.

Ellen Russell
455 Santa Clara Ave.
Parker, CO 25413 *April 14*, 20 *00*

PAY TO THE ORDER OF *Citicorp* $ ______

One thousand twenty-five and 78/100 DOLLARS

BA Downtown Branch
P.O. Box 2456
Colorado Springs,CO 23712
MEMO *Mortgage* *Ellen Russell*
45-828-02-33-4660

ILLUSTRATION 5

54. MONEY We use a decimal point when working with dollars, but the decimal point is not necessary when working with cents. For each dollar amount in Illustration 6, give the equivalent amount expressed as cents.

Dollars	Cents
$0.50	
$0.05	
$0.55	
$5.00	
$0.01	

ILLUSTRATION 6

55. INJECTIONS A syringe is shown in Illustration 7. Use an arrow to show to what point the syringe should be filled if a 0.38-cc dose of medication is to be administered. (“cc” stands for “cubic centimeters.”)

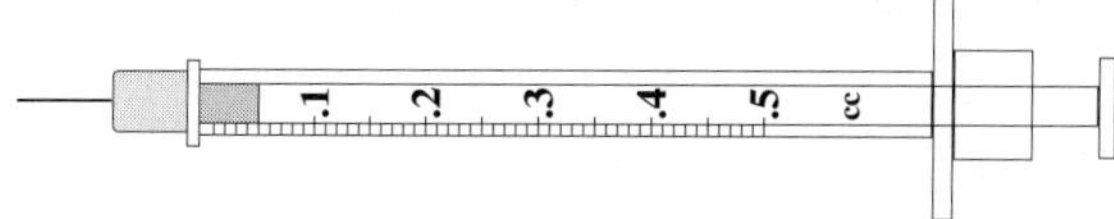

ILLUSTRATION 7

56. LASER The laser used in laser vision correction is so precise that each pulse can remove 39 millionths of an inch of tissue in 12 billionths of a second. Write each of these numbers as decimals.

57. METRIC SYSTEM The metric system is widely used in science to measure length (meters), weight (grams), and capacity (liters). Round each decimal to the nearest hundredth.

a. 1 ft is 0.3048 meter.
b. 1 mi is 1,609.344 meters.
c. 1 lb is 453.59237 grams.
d. 1 gal is 3.785306 liters.

58. WORLD RECORDS As of January, 2000, four American women held world records in swimming. Their times are given below in the form *minutes: seconds.* Round each to the nearest tenth of a second.

100-meter butterfly	Jenny Thompson	0:57.88
200-meter butterfly	Mary T. Meagher	2:05.96
400-meter freestyle	Janet Evans	4:03.85
800-meter freestyle	Janet Evans	8:16.22
1,500-meter freestyle	Janet Evans	15:52.10

59. GEOLOGY Geologists classify types of soil according to the grain size of the particles that make up the soil. The four major classifications are shown below. Complete the chart in Illustration 8 by classifying each sample.

Clay	0.00008 in. and under
Silt	0.00008 in. to 0.002 in.
Sand	0.002 in. to 0.08 in.
Granule	0.08 in. to 0.15 in.

Sample	Location	Size (in.)	Classification
A	riverbank	0.009	
B	pond	0.0007	
C	NE corner	0.095	
D	dry lake	0.00003	

ILLUSTRATION 8

60. MICROSCOPE A microscope used in a lab is capable of viewing structures that range in size from 0.1 to 0.0001 centimeter. Which of the structures listed in Illustration 9 would be visible through this microscope?

Structure	Size (in cm)
bacterium	0.00011
plant cell	0.015
virus	0.000017
animal cell	0.00093
asbestos fiber	0.0002

ILLUSTRATION 9

61. AIR QUALITY Illustration 10 shows the cities with the highest one-hour concentrations of ozone (in parts per million) during the summer of 1999. Rank the cities in order, beginning with the city with the highest reading.

Crestline, California	0.170
Galveston, Texas	0.176
Houston, Texas	0.202
Texas City, Texas	0.206
Westport, Connecticut	0.188
White Plains, New York	0.171

Source: *Los Angeles Times* (August 18, 1999)

ILLUSTRATION 10

62. DEWEY DECIMAL SYSTEM A widely used system for classifying books in a library is the Dewey Decimal System. Books on the same subject are grouped together by number. For example, books about the arts are assigned numbers between 700 and 799. When stacked on the shelves, the books are to be in numerical order, from left to right. How should the titles in Illustration 11 be rearranged to be in the proper order?

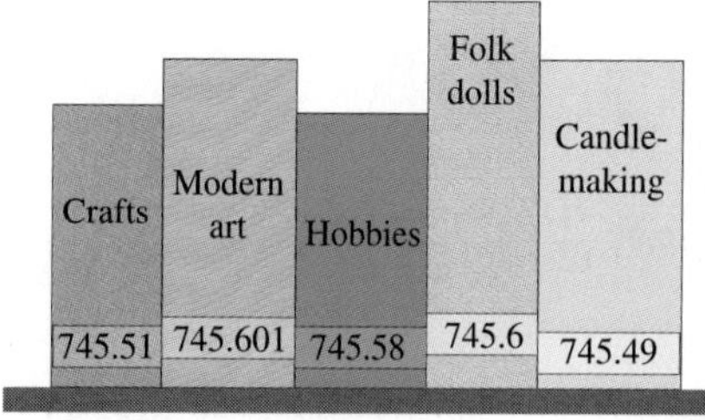

ILLUSTRATION 11

63. OLYMPICS The results of the women's all-around gymnastic competition in the 1988 Los Angeles Olympic Games are shown in Illustration 12 (on the next page). Which gymnasts won the gold, silver, and bronze medals?

Name	Country	Score
Simona Pauca	Romania	78.675
Ma Yanhong	China	77.85
Julianne McNamara	U.S.A.	78.4
Mary Lou Retton	U.S.A.	79.175
Ecaterina Szabo	Romania	79.125
Laura Cutina	Romania	78.3

ILLUSTRATION 12

64. TUNEUP The six spark plugs from the engine of a Nissan Quest were removed, and the spark plug gap was checked. (See Illustration 13.) If vehicle specifications call for the gap to be from 0.031 to 0.035 inch, which of the plugs should be replaced?

Cylinder 1: 0.035 in.
Cylinder 2: 0.029 in.
Cylinder 3: 0.033 in.
Cylinder 4: 0.039 in.
Cylinder 5: 0.031 in.
Cylinder 6: 0.032 in.

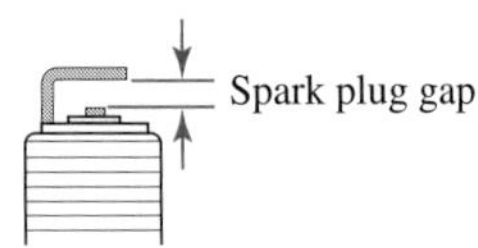

ILLUSTRATION 13

65. E-COMMERCE See Illustration 14. Estimate the loss per share of Amazon.com stock for the third quarter of 1997 and the last quarter of 1998.

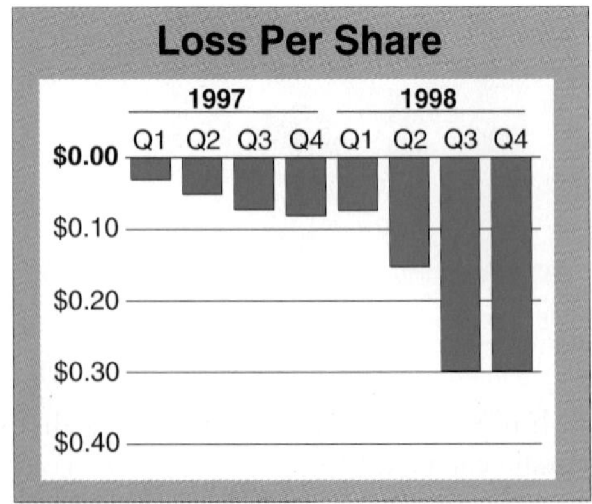

Source: *Los Angeles Times* (January 27, 1999)

ILLUSTRATION 14

66. GASOLINE PRICES Use the data in Illustration 15 to construct a line graph showing the national average retail price per gallon for unleaded regular gasoline (according to *The World Almanac 2000*).

Year	1992	1993	1994	1995	1996	1997	1998
Price (¢)	112.7	110.8	111.2	114.7	123.1	123.4	105.9

ILLUSTRATION 15

WRITING

67. Explain the difference between ten and a tenth.

68. "The more digits a number contains, the larger it is." Is this statement true? Explain your response.

69. How are fractions and decimals related?

70. Explain the benefits of a monetary system that is based on decimals instead of fractions.

71. Illustration 16 shows the unusual notation many service stations use to express the price of a gallon of gasoline. Explain what this notation means.

REGULAR	UNLEADED	UNLEADED +
$1.79\frac{9}{10}$	$1.89\frac{9}{10}$	$1.99\frac{9}{10}$

ILLUSTRATION 16

72. Write a definition for each of these words.

decade *decathlon* *decimal*

REVIEW

73. Add: $75\frac{3}{4} + 88\frac{4}{5}$.

74. Multiply: $\frac{2x}{15}\left(-\frac{5}{4x^2}\right)$

75. Simplify $5R - 3(6 - R)$.

76. Solve $6y + 8 = -y + 9 - y$.

77. Find the area of a triangle with base 16 in. and height 9 in.

78. Express the fraction $\frac{2}{3}$ as an equivalent fraction with a denominator of $3a$.

79. Add: $-2 + (-3) + 4$.

80. Subtract: $-15 - (-6)$.

5.2 *Addition and Subtraction with Decimals*

In this section, you will learn about

- Adding decimals • Subtracting decimals
- Adding and subtracting signed decimals

INTRODUCTION. If we are to add or subtract objects, they must be similar. Federal income tax forms illustrate this concept. (See Figure 5-4.) The boxes on the 1040EZ form ensure that dollars are added to dollars and cents added to cents. In this section, we will show how decimals are added and subtracted using this type of vertical column format.

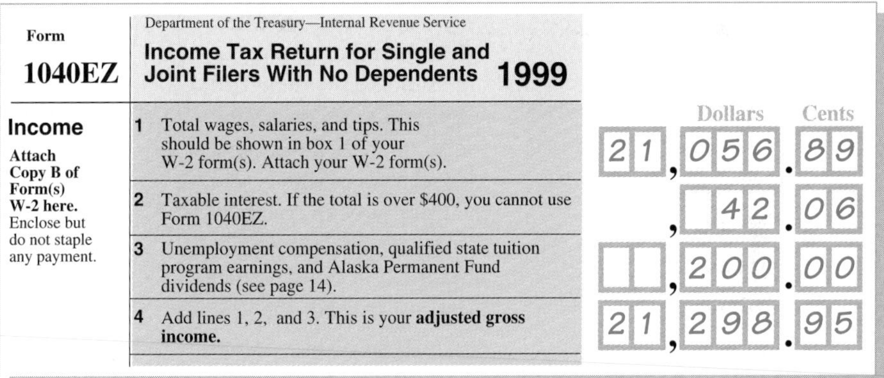

Form 1040EZ — Department of the Treasury—Internal Revenue Service
Income Tax Return for Single and Joint Filers With No Dependents 1999

Income — Attach Copy B of Form(s) W-2 here. Enclose but do not staple any payment.

		Dollars	Cents
1	Total wages, salaries, and tips. This should be shown in box 1 of your W-2 form(s). Attach your W-2 form(s).	21,056	89
2	Taxable interest. If the total is over $400, you cannot use Form 1040EZ.	42	06
3	Unemployment compensation, qualified state tuition program earnings, and Alaska Permanent Fund dividends (see page 14).	200	00
4	Add lines 1, 2, and 3. This is your **adjusted gross income.**	21,298	95

FIGURE 5-4

Adding decimals

When adding decimals, we line up the columns so that ones are added to ones, tenths are added to tenths, hundredths are added to hundredths, and so on. As an example, consider the following problem.

Line up the columns and the decimal points vertically. Then add the numbers.

$$\begin{array}{r} 12.140 \\ 3.026 \\ 4.000 \\ +\ \underline{0.700} \\ 19.866 \end{array}$$

Write the decimal point in the result directly under the decimal points in the problem.

Adding decimals

To add decimal numbers:

1. Line up the decimal points, using the vertical column format.
2. Add the numbers as you would add whole numbers.
3. Write the decimal point in the result directly below the decimal points in the problem.

EXAMPLE 1 ***Adding decimals.*** Add: 1.903 + 0.6 + 8 + 0.78.

Solution

$$\begin{array}{r} \scriptstyle 2 \\ 1.903 \\ 0.600 \\ 8.000 \\ +\ 0.780 \\ \hline 11.283 \end{array}$$

To make the addition by columns easier, write two zeros after 6, a decimal point and three zeros after 8, and one zero after 0.78.

Carry a 2 (shown in blue) to the ones column.

The result is 11.283.

Self Check

Add: 0.07 + 35 + 0.888 + 4.1.

Answer: 40.058

Accent on Technology: Preventing heart attacks

The bar graph in Figure 5-5 shows the number of grams of fiber in a standard serving of each of several foods. It is believed that men can significantly cut their risk of heart attack by eating at least 28 grams of fiber a day. Does this diet meet or exceed the 28-gram requirement?

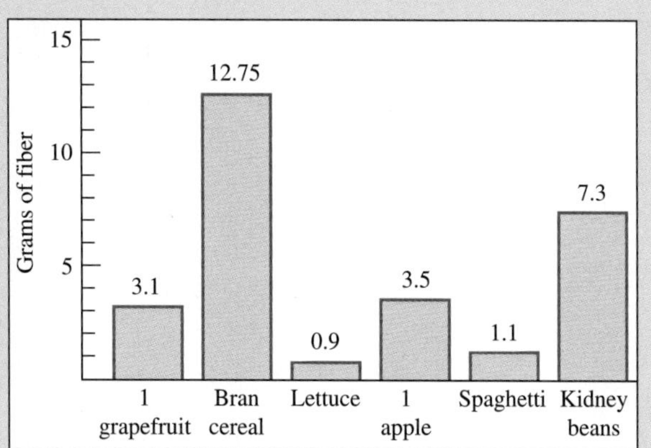

FIGURE 5-5

To find the total fiber intake, we will add the fiber content of each of the foods. We can use a scientific calculator to add the decimals.

Keystrokes 3.1 [+] 12.75 [+] .9 [+] 3.5 [+] 1.1 [+] 7.3 [=] 28.65

Since 28.65 > 28, this diet exceeds the daily fiber requirement of 28 grams.

Subtracting decimals

To subtract decimals, we line up the decimal points and corresponding columns so that we subtract like objects—tenths from tenths, hundredths from hundredths, and so on.

Subtracting decimals

To subtract decimal numbers:

1. Line up the decimal points using the vertical column format.
2. Subtract the numbers as you would subtract whole numbers.
3. Write the decimal point in the result directly below the decimal points in the problem.

EXAMPLE 2 ***Borrowing.*** Subtract: **a.** 279.6 − 138.7 and **b.** 15.4 − 13.059.

Solution

a.

```
   8 16
 279.6
−138.7
 140.9
```

To subtract in the tenths column, borrow 1 one in the form of 10 tenths from the ones column. Add 10 to the 6 in the tenths column, which gives 16 (shown in blue).

b.

```
     9
   3 10 10
 15.4 0 0
−13.0 5 9
  2.3 4 1
```

Add two zeros to the right of 15.4 to make borrowing easier. First, borrow from the tenths column; then borrow from the hundredths column.

Self Check

Subtract:

a. 382.5 − 227.1

b. 30.1 − 27.122

Answers: **a.** 155.4, **b.** 2.978 ■

EXAMPLE 3 ***Conditioning program.*** A 350-pound football player lost 15.7 pounds during the first week of practice. During the second week, he gained 4.9 pounds. What is his weight after the first two weeks of practice?

Solution The word *lost* indicates subtraction. The word *gained* indicates addition.

Beginning weight	minus	first week weight loss	plus	second week weight gain	equals	weight after two weeks of practice.

350 − 15.7 + 4.9 = 334.3 + 4.9 Working from left to right, do the subtraction first: 350 − 15.7 = 334.3.

= 339.2 Do the addition.

The player's weight is 339.2 pounds after two weeks of practice. ■

Accent on Technology: *Weather balloons*

A giant weather balloon is made of neoprene, a flexible rubberized substance, that has an uninflated thickness of 0.011 inch. When the balloon is inflated with helium, the thickness becomes 0.0018 inch.

To find the change in thickness, we need to subtract. We can use a scientific calculator to subtract the decimals.

Keystrokes .011 [−] .0018 [=] 0.0092

After the balloon is inflated, the neoprene loses 0.0092 of an inch in thickness.

Adding and subtracting signed decimals

To add signed decimals, we use the same rules that we used for adding integers.

Adding two decimals

With like signs: Add their absolute values and attach their common sign to the sum.

With unlike signs: Subtract their absolute values (the smaller from the larger) and attach the sign of the number with the larger absolute value to the sum.

EXAMPLE 4 ***Adding signed decimals.*** Add: $-6.1 + (-4.7)$.

Solution

Since the decimals are both negative, we add their absolute values and attach a negative sign to the result.

$-6.1 + (-4.7) = -10.8$ Add the absolute values, 6.1 and 4.7, to get 10.8. Use their common sign.

Self Check

Add: $-5.04 + (-2.32)$.

Answer: -7.36 ■

EXAMPLE 5 ***Adding signed decimals.*** Add: $5.35 + (-12.9)$.

Solution

In this example, the signs are unlike. Since -12.9 has the larger absolute value, we subtract 5.35 from 12.9 to get 7.55, and attach a negative sign to the result.

$5.35 + (-12.9) = -7.55$

Self Check

Add: $-21.4 + 16.75$.

Answer: -4.65 ■

EXAMPLE 6 ***Subtracting signed decimals.*** Subtract: $-4.3 - 5.2$.

Solution

To subtract signed decimals, we can add the opposite of the decimal that is being subtracted.

$-4.3 - 5.2 = -4.3 + (-5.2)$ Add the opposite of 5.2, which is -5.2.

$= -9.5$ Add the absolute values, 4.3 and 5.2, to get 9.5. Attach a negative sign to the result.

Self Check

Subtract: $-1.18 - 2.88$

Answer: -4.06 ■

EXAMPLE 7 ***Subtracting with signed decimals.*** Subtract: $-8.37 - (-6.2)$.

Solution

$-8.37 - (-6.2) = -8.37 + 6.2$ Add the opposite of -6.2, which is 6.2.

$= -2.17$ Subtract the smaller absolute value from the larger, 6.2 from 8.37, to get 2.17. Since -8.37 has the larger absolute value, the result is negative.

Self Check

Subtract: $-2.56 - (-4.4)$.

Answer: 1.84 ■

EXAMPLE 8 ***Grouping symbols.*** Evaluate $-12.2 - (-14.5 + 3.8)$.

Solution

We do the addition within the grouping symbols first.

$-12.2 - (\mathbf{-14.5 + 3.8}) = -12.2 - (\mathbf{-10.7})$ Do the addition: $-14.5 + 3.8 = -10.7$.

$= -12.2 + 10.7$ Add the opposite of -10.7.

$= -1.5$ Do the addition.

Self Check

Evaluate: $-4.9 - (-1.2 + 5.6)$.

Answer: -9.3 ■

STUDY SET Section 5.2

VOCABULARY *Fill in the blanks to make the statements true.*

1. The answer to an addition problem is called the ______________.

2. The answer to a subtraction problem is called the ______________.

3. Every whole number has an unwritten decimal ______________ to its right.

4. To subtract signed decimals, add the ______________ of the decimal that is being subtracted.

CONCEPTS

5. **a.** Add: 0.3 + 0.17.
b. Write 0.3 and 0.17 as fractions. Find a common denominator for the fractions and add them.
c. Express your answer to part b as a decimal.
d. Compare your answers from part a and part c.

6. In the subtraction problem below, we must borrow. How much is borrowed from the 3, and in what form is it borrowed?

```
   2 11
  29.3̸1̸
 −25.1 6
```

PRACTICE *Do each addition.*

7. $\begin{array}{r} 32.5 \\ +\ 7.4 \\ \hline \end{array}$ **8.** $\begin{array}{r} 6.3 \\ +13.5 \\ \hline \end{array}$

9. $\begin{array}{r} 21.6 \\ +33.12 \\ \hline \end{array}$ **10.** $\begin{array}{r} 19.4 \\ +31.95 \\ \hline \end{array}$

11. 12 + 3.9 **12.** 0.01 + 3.6

13. 0.03034 + 0.2003 **14.** 19.9 + 19.9

15. 247.9 + 40 + 0.56 **16.** 0.0053 + 1.78 + 6

17. 45 + 9.9 + 0.12 + 3.02

18. 505.01 + 23 + 0.989 + 12.07

Do each subtraction.

19. $\begin{array}{r} 12.98 \\ -\ 3.45 \\ \hline \end{array}$ **20.** $\begin{array}{r} 1.6 \\ -0.16 \\ \hline \end{array}$

21. $\begin{array}{r} 78.1 \\ -\ 7.81 \\ \hline \end{array}$ **22.** $\begin{array}{r} 202.234 \\ -\ 19.34 \\ \hline \end{array}$

23. 5 − 0.023 **24.** 30 − 11.98

25. 24 − 23.81 **26.** 7.001 − 5.9

Do each addition.

27. −45.6 + 34.7 **28.** −19.04 + 2.4

29. 46.09 + (−7.8) **30.** 34.7 + (−30.1)

31. −7.8 + (−6.5) **32.** −5.78 + (−33.1)

33. −0.0045 + (−0.031) **34.** −90.09 + (−0.087)

Do each subtraction.

35. −9.5 − 7.1 **36.** −7.08 − 14.3

37. 30.03 − (−17.88) **38.** 143.3 − (−64.01)

39. −2.002 − (−4.6) **40.** −0.005 − (−8)

41. −7 − (−18.01) **42.** −63.04 − (−8.911)

Evaluate each expression.

43. 3.4 − 6.6 + 7.3

44. 3.4 − (6.6 + 7.3)

45. (−9.1 − 6.05) − (−51)

46. −9.1 − (−6.05) + 51

47. 16 − (67.2 + 6.27)

48. −43 − (0.032 − 0.045)

49. (−7.2 + 6.3) − (−3.1 − 4)

50. 2.3 + [2.4 − (2.5 − 2.6)]

51. $|-14.1 + 6.9| + 8$

52. $15 - |-2.3 + (-2.4)|$

Add or subtract as indicated.

53. Find the sum of *two and forty-three hundredths* and *five and six-tenths.*

54. Find the difference of *nineteen hundredths* and *six thousandths.*

APPLICATIONS

55. SPORTS PAGE In the sports pages of any newspaper, decimal numbers are used quite often.
a. "German bobsledders set a world record today with a final run of 53.03, finishing ahead of the Italian team by only fourteen thousandths of a second." What was the time for the Italian bobsled team?
b. "The women's figure skating title was decided by only thirty-three hundredths of a point." If the winner's point total was 102.71, what was the second-place finisher's total?

56. NURSING Illustration 1 (on the next page) shows a patient's health chart. A nurse failed to fill in certain portions. (98.6° Fahrenheit is considered normal.) Complete the chart.

Day of week	Patient's temperature	How much above normal
Monday	99.7°	
Tuesday		2.5°
Wednesday	98.6°	
Thursday	100.0°	
Friday		0.9

ILLUSTRATION 1

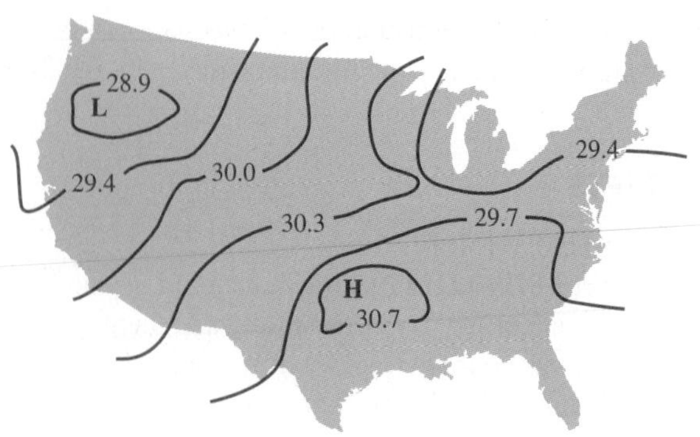

ILLUSTRATION 4

57. VEHICLE SPECIFICATIONS Certain dimensions of a compact car are shown in Illustration 2. What is the wheelbase of the car?

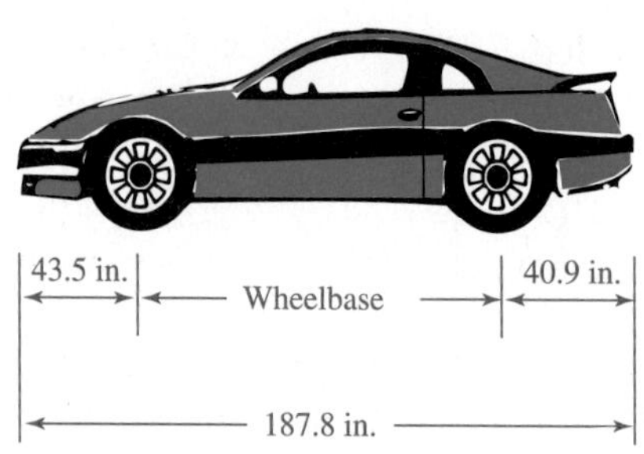

ILLUSTRATION 2

58. pH SCALE The pH scale shown in Illustration 3 is used to measure the strength of acids and bases in chemistry. Find the difference in pH readings between

a. bleach and stomach acid.
b. ammonia and coffee.
c. blood and coffee.

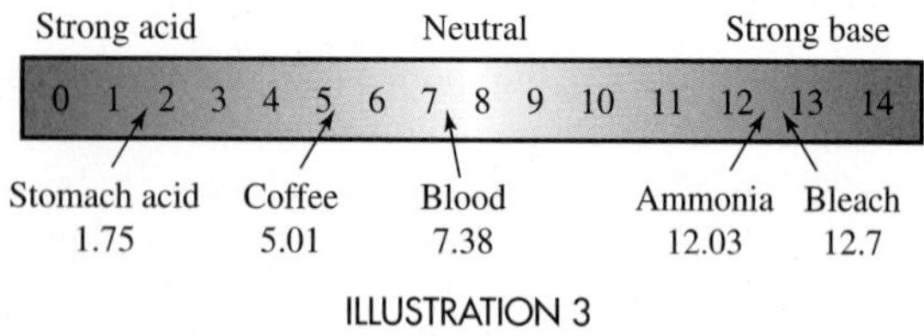

ILLUSTRATION 3

59. BAROMETRIC PRESSURE Barometric pressure readings are recorded on the weather map in Illustration 4. In a low pressure area (L on the map), the weather is often stormy. The weather is usually fair in a high pressure area (H). What is the difference in readings between the areas of highest and lowest pressure? In what part of the country would you expect the weather to be fair?

60. QUALITY CONTROL An electronics company has strict specifications for silicon chips used in a computer. The company will install only chips that are within 0.05 centimeters of the specified thickness. Illustration 5 gives that specification for two types of chip. Fill in the blanks to complete the chart.

Chip type	Thickness specification	Acceptable range	
		Low	High
A	0.78 cm		
B	0.643 cm		

ILLUSTRATION 5

61. OFFSHORE DRILLING A company needs to construct a pipeline from an offshore oil well to a refinery located on the coast. Company engineers have come up with two plans for consideration, as shown in Illustration 6. Use the information in the illustration to complete the table.

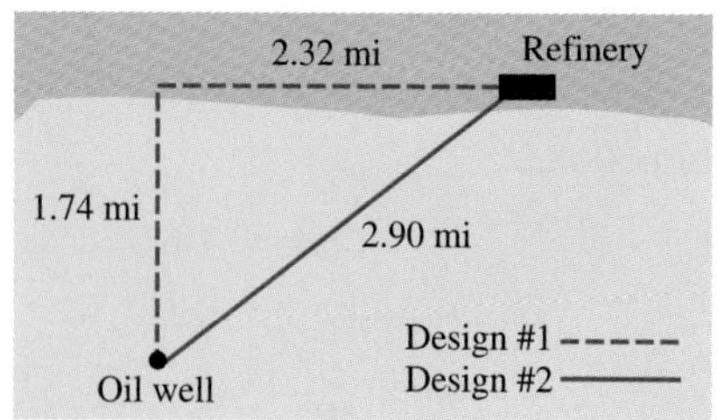

ILLUSTRATION 6

	Pipe underwater (mi)	Pipe underground (mi)	Total pipe (mi)
Design 1			
Design 2			

62. TELEVISION Illustration 7 shows the six most-watched television shows of all time (excluding Super Bowl games).

a. What was the combined total audience of all six shows?

b. How many more people watched the last episode of "MASH" than watched the last episode of "Seinfeld?"

c. How many more people would have had to watch the last "Seinfeld" to move it into a tie for fifth place?

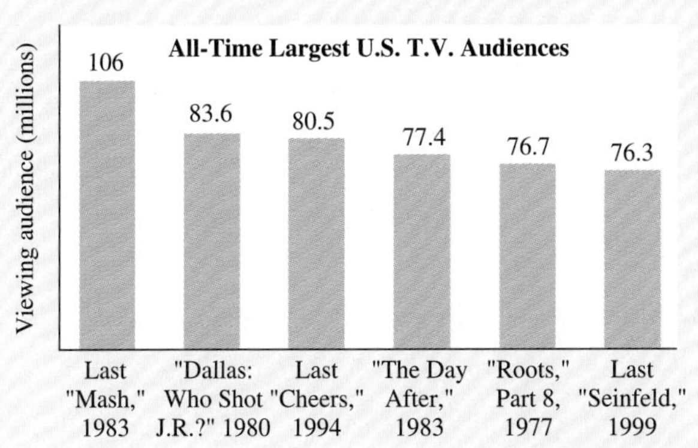

Source: Nielsen Media Research

ILLUSTRATION 7

63. AMERICAN RECORDHOLDERS The late Florence Griffith-Joyner set the United States national and world record in the 100-meter sprint: 10.49 seconds. Jenny Thompson set the national record in the 100-meter freestyle swim: 54.48 seconds. How much faster did Griffith-Joyner run the 100 meters than Thompson swam it?

64. FLIGHT PATH See Illustration 8. Find the added distance a plane must travel to avoid flying through the storm.

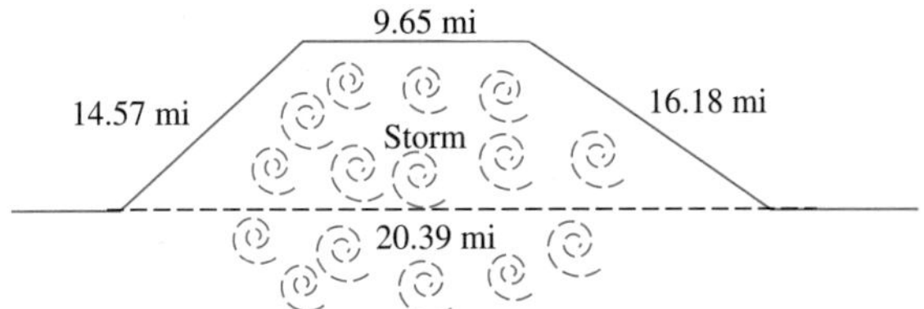

ILLUSTRATION 8

65. DEPOSIT SLIP A deposit slip for a savings account is shown in Illustration 9. Find the subtotal and then the total deposit.

Deposit

Cash	242	50
Checks (properly endorsed)	116	10
	47	93
Total from reverse side	359	16
Subtotal		
Less cash	25	00
Total deposit		

ILLUSTRATION 9

66. MOTION Forces such as water current or wind can increase or decrease the speed of an object in motion. Find the speed of each object.

a. An airplane's speed in still air is 450 mph, and it has a tail wind of 35.5 mph helping it along.

b. A man can paddle a canoe at 5 mph in still water, but he is going upstream. The speed of the current against him is 1.5 mph.

67. THE HOME SHOPPING NETWORK Illustration 10 shows a description of a cookware set that was sold on television.

a. Find the difference between the manufacturer's suggested retail price (MSRP) and the sale price.

b. Including shipping and handling (S & H), how much will the cookware set cost?

Item 229-442	
Continental 9-piece Cookware Set	
Stainless steel	
MSRP	$149.79
HSN Price	$59.85
On Sale	**$47.85**
S & H	$7.95

ILLUSTRATION 10

68. RETAILING Complete Illustration 11 by filling in the retail price of each appliance, given its cost to the dealer and the store markup.

Item	Cost	Markup	Retail price
Refrigerator	$510.80	$105.00	
Washing machine	$189.50	$55.50	
Dryer	$163.99	$$x$	

ILLUSTRATION 11

Evaluate each expression.

69. 2,367.909 + 5,789.0253

70. 0.00786 + 0.3423

71. 9,000.09 − 7,067.445

72. 1 − 0.004999

73. 3,434.768 − (908 − 2.3 + .0098)

74. 12 − (0.723 + 3.05611)

WRITING

75. Explain why we line up the decimal points and corresponding columns when adding decimals.

76. Explain why we can write additional zeros to the right of a decimal such as 7.89 without affecting its value.

77. Explain what is wrong with the work shown below.

Add: 203.56 + 37 + 0.43.

```
203.56
    37
  0.43
204.36
```

78. Consider the addition

$$\begin{array}{r} {}^{2} \\ 23.7 \\ 41.9 \\ +\ 12.8 \\ \hline 78.4 \end{array}$$

Explain the meaning of the small 2 written above the ones column.

REVIEW

79. Add: $44\frac{3}{8} + 66\frac{1}{5}$.

80. Simplify: $\dfrac{-\frac{3}{4}}{\frac{5}{16}}$.

81. Multiply: $\frac{-15}{26} \cdot 1\frac{4}{9}$.

82. Simplify: $2 + 5[-2 - (6 + 1)]$.

5.3 Multiplication with Decimals

In this section, you will learn about

- Multiplying decimals • Multiplying decimals by powers of 10
- Multiplying signed decimals • Order of operations

Introduction. In our study of decimals, we now focus on the operation of multiplication. First, we develop a method used to multiply decimals. Then we use that method to evaluate expressions and to solve problems involving decimals.

Multiplying decimals

To develop a rule for multiplying decimals, we will examine the multiplication $0.3 \cdot 0.17$, finding the product in a roundabout way. First, we will write 0.3 and 0.17 as fractions and multiply them. Then we will express the resulting fraction as a decimal.

$$\begin{aligned} 0.3 \cdot 0.17 &= \frac{3}{10} \cdot \frac{17}{100} && \text{Express 0.3 and 0.17 as fractions.} \\ &= \frac{3 \cdot 17}{10 \cdot 100} && \text{Multiply the numerators and multiply the denominators.} \\ &= \frac{51}{1{,}000} && \text{Multiply in the numerator and denominator.} \\ &= 0.051 && \text{Write } \tfrac{51}{1{,}000} \text{ as a decimal.} \end{aligned}$$

From this example, we can make some observations about multiplying decimals.

- The digits in the answer are found by multiplying 3 and 17.

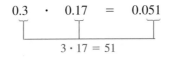

$$0.3 \cdot 0.17 = 0.051$$

$3 \cdot 17 = 51$

- The answer has 3 decimal places. The *sum* of the number of decimal places of the factors 0.3 and 0.17 is also 3.

$$0.3 \cdot 0.17 = 0.051$$

0.3	0.17	0.051
1 decimal place.	2 decimal places.	3 decimal places.

These observations suggest the following rule for multiplying decimals.

Multiplying decimals

To multiply two decimals:

1. Multiply the decimals as if they were whole numbers.
2. Find the total number of decimal places in both factors.
3. Place the decimal point in the result so that the answer has the same number of decimal places as the total found in Step 2.

EXAMPLE 1 ***Multiplying decimals.*** Multiply: $5.9 \cdot 3.4$.

Solution

We ignore the decimal points and multiply the decimals as if they were whole numbers. Initially, we think of this problem as 59 times 34.

```
   59
×  34
  236
 177
 2006
```

To place the decimal point in the product, we find the total number of digits to the right of the decimal points of the factors.

```
  5.9  ← 1 decimal place  }
× 3.4  ← 1 decimal place  } The answer will have 1 + 1 = 2 decimal places.
  236
 177
20.06
```

Locate the decimal point so that the answer has 2 decimal places.

Self Check

Multiply: $2.74 \cdot 4.3$.

Answer: 11.782

When multiplying decimals, it is not necessary to line up the decimal points, as the next example illustrates.

EXAMPLE 2 ***Inserting placeholder zeros.*** Multiply: 1.3(0.005).

Solution

We begin by multiplying 13 by 5.

```
   1.3  ← 1 decimal place   }
×0.005  ← 3 decimal places  } The answer will have 1 + 3 = 4 decimal places.
    65
```

Self Check

Multiply: (0.0002)7.2.

We then place the decimal point in the result.

```
     1.3
× 0.005
  0.0065
```

Add 2 placeholder zeros and position the decimal point so that the product has 4 decimal places.

Answer: 0.00144

Accent on Technology: Heating costs

When billing a household, a gas company converts the amount of natural gas used into units of heat energy called *therms.* The number of therms used by a household in one month and the cost per therm are shown below.

Customer charge . . . . 39 therms @ $0.72264

To find the total charges for the month, we multiply the number of therms by the cost per therm: 39 · 0.72264.

Keystrokes 39 [×] .72264 [=] 28.18296

Rounding to the nearest cent, we see that the total charge is $28.18.

EXAMPLE 3 ***Multiplying a decimal and a whole number.***
Multiply: 234(3.1).

Solution

```
   234   ← No decimal places
×  3.1   ← 1 decimal place
  23 4
 702
 725.4
```

The answer will have 0 + 1 = 1 decimal place.

Locate the decimal point so that the answer has 1 decimal place.

Self Check
Multiply: 178(2.7).

Answer: 480.6

Multiplying decimals by powers of 10

The numbers 10, 100, and 1,000 are called *powers of 10,* because they are the results when we evaluate 10^1, 10^2, and 10^3, respectively. To develop a rule to determine the product when multiplying a decimal and a power of 10, we will multiply 8.675 by three different powers of 10.

Multiply: 8.675 · **10**

```
   8.675
×     10
   0000
  8675
  86.750
```

The answer is 86.75.

Multiply: 8.675 · **100**

```
    8.675
×     100
    0000
   0000
  8675
  867.500
```

The answer is 867.5

Multiply: 8.675 · **1,000**

```
     8.675
×     1000
     0000
    0000
   0000
  8675
  8675.000
```

The answer is 8,675.

We can make some observations about the results.

- In each case, the answer contains the same digits as the factor 8.675.
- When inspecting the answers, the decimal point in the first factor 8.675 appears to be moved to the right by the multiplication process. The number of decimal places it moves depends on the power of 10 by which 8.675 is multiplied.

One zero in 10	Two zeros in 100	Three zeros in 1,000
$8.675 \cdot 10 = 86.75$	$8.675 \cdot 100 = 867.5$	$8.675 \cdot 1{,}000 = 8675$
It moves one place to the right.	It moves two places to the right.	It moves three places to the right.

These observations suggest the following rule.

Multiplying a decimal by a power of 10

> To multiply a decimal by a power of 10, move the decimal point to the right the same number of places as there are zeros in the power of 10.

EXAMPLE 4 ***Multiplying decimals by powers of 10.*** Find the product: **a.** $2.81 \cdot 10$ and **b.** $0.076 \cdot 10{,}000$.

Solution

a. $2.81 \cdot 10 = 28.1$ Since 10 has 1 zero, move the decimal point 1 place to the right.

b. $0.076 \cdot 10{,}000 = 0760.$ Since 10,000 has 4 zeros, move the decimal point 4 places to the right. Write a placeholder zero (shown in blue).

$= 760$

Self Check
Find the product:

a. $0.721 \cdot 100$

b. $6.08\ (1{,}000)$

Answers: **a.** 72.1, **b.** 6,080

EXAMPLE 5 ***Tachometer.*** A tachometer indicates the engine speed of a vehicle, in revolutions per minute (rpm). What engine speed is indicated by the tachometer in Figure 5-6?

Solution

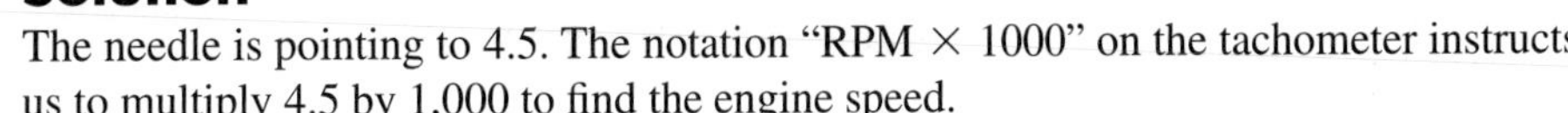

The needle is pointing to 4.5. The notation "RPM × 1000" on the tachometer instructs us to multiply 4.5 by 1,000 to find the engine speed.

FIGURE 5-6

$4.5 \cdot 1{,}000 = 4500$ Since 1,000 has three zeros, move the decimal point three places to the right. Write two placeholder zeros.

$= 4{,}500$

The engine speed is 4,500 rpm.

Multiplying signed decimals

Recall that the product of two numbers with like signs is positive, and the product of two numbers with unlike signs is negative.

EXAMPLE 6 ***Multiplying signed decimals.*** Multiply: **a.** $-1.8(4.5)$ and **b.** $(-1{,}000)(-59.08)$.

Solution

a. Since the decimals have unlike signs, their product is negative.

$-1.8(4.5) = -8.1$ Multiply the absolute values, 1.8 and 4.5, to get 8.1. Make the result negative.

b. Since the decimals have like signs, their product is positive.

$(-1{,}000)(-59.08) = 59{,}080$ Multiply the absolute values, 1,000 and 59.08. Since 1,000 has 3 zeros, move the decimal point 3 places to the right. Write a placeholder zero.

Self Check
Multiply:

a. $6.6(-5.5)$

b. $(-44.968)(-100)$

Answers: **a.** -36.3, **b.** 4,496.8

EXAMPLE 7 ***Evaluating powers of decimals.*** Evaluate: **a.** $(2.4)^2$ and **b.** $(-0.05)^2$.

Solution

a. $(2.4)^2 = 2.4 \cdot 2.4$ Write 2.4 as a factor 2 times.

$= 5.76$ Do the multiplication.

b. $(-0.05)^2 = (-0.05)(-0.05)$ Write -0.05 as a factor 2 times.

$= 0.0025$ Do the multiplication. The product of two decimals with like signs is positive.

Self Check

Evaluate:

a. $(-1.3)^2$

b. $(0.09)^2$

Answers: **a.** 1.69, **b.** 0.0081

Order of operations

In the remaining examples, we apply the rules for the order of operations to evaluate expressions involving decimals.

EXAMPLE 8 ***Order of operations.*** Evaluate $-(0.6)^2 + 5|-3.6 + 1.9|$.

Solution

$-(0.6)^2 + 5|-3.6 + 1.9| = -(0.6)^2 + 5|-1.7|$ Do the addition within the absolute value symbols.

$= -(0.6)^2 + 5(1.7)$ Simplify: $|-1.7| = 1.7$.

$= -0.36 + 5(1.7)$ Find the power: $(0.6)^2 = 0.36$.

$= -0.36 + 8.5$ Do the multiplication: $5(1.7) = 8.5$.

$= 8.14$ Do the addition.

Self Check

Evaluate:

$-2|-4.4 + 5.6| + (-0.8)^2$.

Answer: -1.76

EXAMPLE 9 ***Evaluating algebraic expressions.*** Evaluate $6.28r(h + r)$ for $h = 3.1$ and $r = 6$.

Solution

$6.28r(h + r) = 6.28(\mathbf{6})(\mathbf{3.1} + \mathbf{6})$ Replace r with 6 and h with 3.1.

$= 6.28(6)(9.1)$ Do the addition within the parentheses: $3.1 + 6 = 9.1$.

$= 37.68(9.1)$ Do the multiplication: $6.28(6) = 37.68$.

$= 342.888$ Do the multiplication.

Self Check

Evaluate $1.3pr^3$ for $p = 3.14$ and $r = 3$.

Answer: 110.214

EXAMPLE 10 ***Weekly earnings.*** A cashier's work week is 40 hours. After his daily shift is over, he can work overtime at a rate 1.5 times his regular rate of $7.50 per hour. How much money will he earn in a week if he works 6 hours of overtime?

Solution First, we need to find his overtime rate, which is 1.5 times his regular rate of $7.50 per hour.

$$1.5(7.50) = 11.25$$

His overtime rate is $11.25 per hour.

To find his total weekly earnings, we use the following fact.

The regular rate	times	40 hours	plus	the overtime rate	times	overtime hours worked	equals	his total earnings.

$$7.50(40) + 11.25(6) = 300 + 67.50 \quad \text{Do the multiplications.}$$
$$= 367.50 \quad \text{Do the addition.}$$

The cashier's earnings for the week are $367.50. ■

STUDY SET Section 5.3

VOCABULARY *Fill in each blank to make the statements true.*

1. In the multiplication problem 2.89 · 15.7, the numbers 2.89 and 15.7 are called ______________. The answer, 45.373, is called the ______________.
2. Numbers such as 10, 100, and 1,000 are called ______________ of 10.

CONCEPTS *In Exercises 3–4, fill in each blank to make the statements true.*

3. To multiply decimals, multiply them as if they were ______________ numbers. The number of decimal places in the product is the same as the ______________ of the decimal places of the factors.
4. To multiply a decimal by a power of 10, move the decimal point to the ______________ the same number of decimal places as the number of ______________ in the power of 10.
5. When we move the decimal point to the right, does the decimal number get larger or smaller?
6. Suppose that the result of multiplying two decimals is 2.300. Write this result in simpler form.
7. **a.** Multiply $\frac{3}{10}$ and $\frac{7}{100}$.
   b. Now write both fractions from part a as decimals. Multiply them in that form. Compare your results from parts a and b.
8. **a.** Multiply 0.11 and 0.3.
   b. Now write both decimals in part a as fractions. Multiply them in that form. Compare your results from parts a and b.

PRACTICE *Do each multiplication.*

9. (0.4)(0.2)
10. (0.2)(0.3)
11. (−0.5)(0.3)
12. (0.6)(−0.7)
13. (1.4)(0.7)
14. (2.1)(0.4)
15. (0.08)(0.9)
16. (0.003)(0.9)
17. (−5.6)(−2.2)
18. (−7.1)(−4.1)
19. (−4.9)(0.001)
20. (0.001)(−7.09)
21. (−0.35)(0.24)
22. (−0.85)(0.42)
23. (−2.13)(4.05)
24. (3.06)(−1.82)
25. 16 · 0.6
26. 24 · 0.8
27. −7(8.1)
28. −5(4.7)
29. 0.04(306)
30. 0.02(417)
31. 60.61(−0.3)
32. −70.07 · 0.6
33. −0.2(0.3)(−0.4)
34. −0.1(−2.2)(0.5)
35. 5.5(10)(−0.3)
36. 6.2(100)(−0.8)
37. 4.2 · 10
38. 10 · 7.1
39. 67.164 · 100
40. 708.199 · 100
41. −0.056(10)
42. −100(0.0897)
43. 1,000(8.05)
44. 23.7(1,000)
45. 0.098(10,000)
46. 3.63(10,000)
47. −0.2 · 1,000
48. −1,000 · 1.9

Complete each table.

49.

Decimal	Its square
0.1	0.01
0.2	0.04
0.3	0.09
0.4	0.16
0.5	0.25
0.6	0.36
0.7	0.49
0.8	0.64
0.9	0.81

50.

Decimal	Its cube
0.1	
0.2	
0.3	
0.4	
0.5	
0.6	
0.7	
0.8	
0.9	

Find each power.

51. $(1.2)^2$

52. $(2.3)^2$

53. $(-1.3)^2$

54. $(-2.5)^2$

Evaluate each expression.

55. $-4.6(23.4 - 19.6)$

56. $6.9(9.8 - 8.9)$

57. $(-0.2)^2 + 2(7.1)$

58. $(-6.3)(3) - (1.2)^2$

59. $(-0.7 - 0.5)(2.4 - 3.1)$

60. $(-8.1 - 7.8)(0.3 + 0.7)$

61. $(0.5 + 0.6)^2(-3.2)$

62. $(-5.1)(4.9 - 3.4)^2$

63. $|-2.6| \cdot |-7.2|$

64. $4|-3.1| + 5|-5.5|$

65. $(|-2.6 - 6.7|)^2$

66. $-3|-8.16 + 9.9|$

Evaluate each algebraic expression.

67. $3.14 + 2(d - t)$ for $d = 1.2$ and $t = -6.7$

68. $-8h^2 - rh$ for $r = 2.1$ and $h = -0.02$

69. $t + 0.5rt^2$ for $t = -0.4$ and $r = 100$

70. $1{,}000(x - y)(x + y)$ for $x = 9.8$ and $y = 1.3$

71. $10|a^2 - b^2|$ for $a = -1.1$ and $b = 2.2$

72. $t|r| + t|s|$ for $r = -0.021$, $s = -0.016$, and $t = 100$

APPLICATIONS

73. CONCERT SEATING Two types of tickets were sold for a concert. Floor seating cost $12.50 a ticket, and balcony seats were $15.75.

a. Complete the table in Illustration 1 and find the receipts from each type of ticket.

b. Find the total receipts from the sale of both types of tickets.

Ticket type	Price	Number sold	Receipts
Floor		1,000	
Balcony		100	

ILLUSTRATION 1

74. CITY PLANNING In the city map in Illustration 2, the streets form a grid. They are 0.35 mile apart. Find the distance of each trip.

a. The airport to the Convention Center

b. City Hall to the Convention Center

c. The airport to City Hall

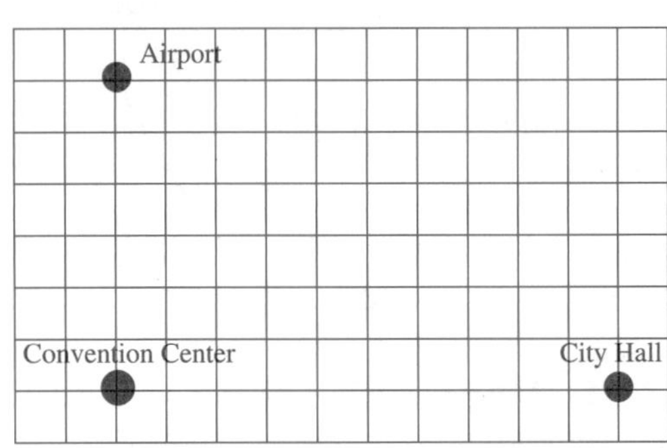

ILLUSTRATION 2

75. STORM DAMAGE After a rainstorm, the saturated ground under a hilltop house began to give way. A survey team noted that the house dropped 0.57 inch initially. In the next two weeks, the house fell 0.09 inch per week. How far did the house fall during this three-week period?

76. WATER USAGE In May, the water level of a reservoir reached its high mark for the year. During the summer months, as water usage increased, the level dropped. In the months of May and June, it fell 4.3 feet each month. In August, because of high temperatures, it fell another 8.7 feet. By September, how far below the year's high mark had the water level fallen?

77. WEIGHTLIFTING The barbell in Illustration 3 is evenly loaded with iron plates. How much plate weight is loaded on the barbell?

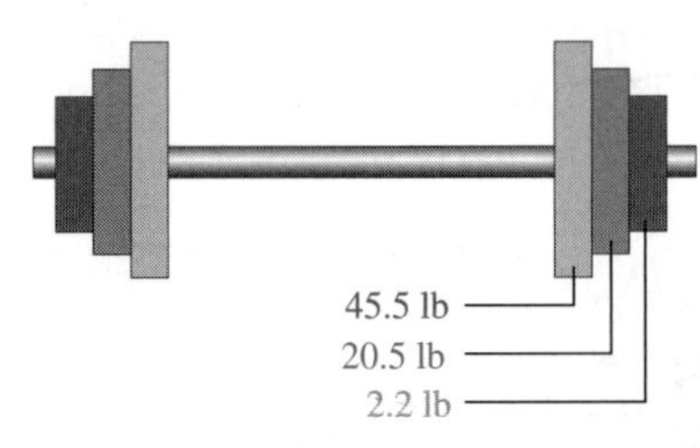

ILLUSTRATION 3

78. PLUMBING BILL In Illustration 4, an invoice for plumbing work is torn. What is the charge for the 4 hours of work? What is the total charge?

Carter Plumbing 100 W. Dalton Ave.		Invoice #210
Standard sevice charge 4 hr @ $40.55/hr		$25.75
Total		

ILLUSTRATION 4

79. BAKERY SUPPLIES A bakery buys various types of nuts as ingredients for cookies. Complete Illustration 5 by filling in the cost of each purchase.

Type of nut	Price per pound	Pounds	Cost
Almonds	$3.25	16	
Walnuts	$2.10	25	
Peanuts	$1.85	x	

ILLUSTRATION 5

80. RETROFIT Illustration 6 shows the width of the three columns of an existing freeway overpass. A computer analysis indicates that each column needs to be increased in width by a factor of 1.4 to ensure stability during an earthquake. According to the analysis, how wide should each of the columns be?

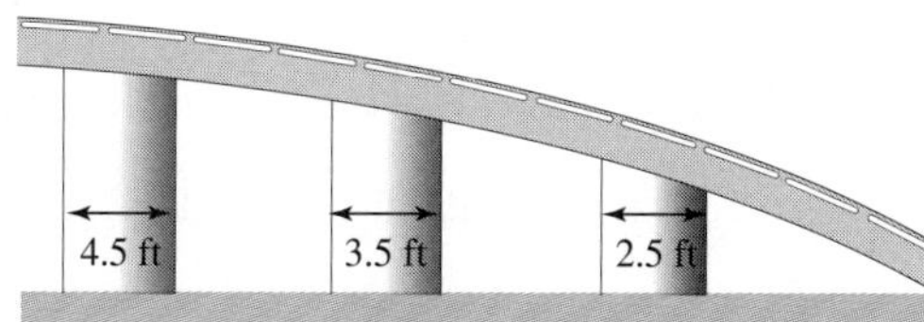

ILLUSTRATION 6

81. SWIMMING POOL CONSTRUCTION Long bricks, called *coping,* can be used to outline the edge of a swimming pool. How many meters of coping will be needed in the construction of the swimming pool shown in Illustration 7?

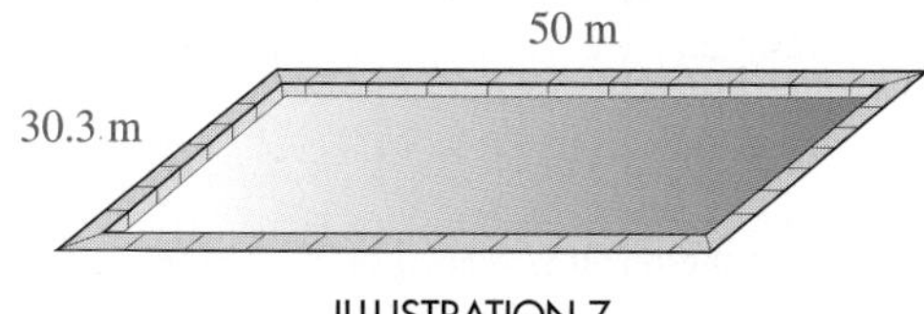

ILLUSTRATION 7

82. SOCCER A soccer goal measures 24 feet wide by 8 feet high. Major League Soccer officials are proposing to increase its width by 1.5 feet and increase its height by 0.75 foot.

a. What is the area of the goal opening now?

b. What would it be if their proposal is adopted?

c. How much area would be added?

83. BIOLOGY DNA is found in cells. It is referred to as the genetic "blueprint." In humans, it determines such traits as eye color, hair color, and height. A model of DNA appears in Illustration 8. If Å = 0.000000004 inch, determine the three dimensions shown in the illustration.

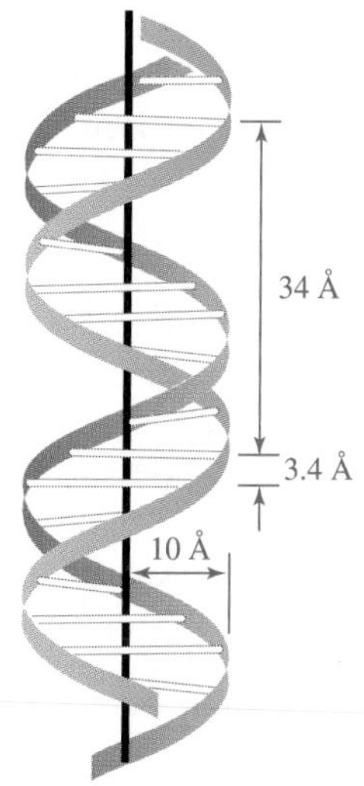

ILLUSTRATION 8

84. TACHOMETER See Illustration 9.

a. To what number is the tachometer needle pointing? Give your estimate in decimal form.

b. What engine speed (in rpm) does the tachometer indicate?

ILLUSTRATION 9

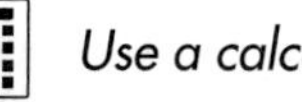 *Use a calculator to answer each problem.*

85. $(-9.0089 + 10.0087)(15.3)$

86. $(-4.32)^3 - 78.969$

87. $(18.18 + 6.61)^2 + (5 - 9.09)^2$

88. $304 - 3.780876(100)$

89. ELECTRIC BILL When billing a household, a utility company charges for the number of kilowatt-hours used. A kilowatt-hour (kwh) is a standard measure of electricity. If the cost of 1 kwh is \$0.14277, what is the electric bill for a household using 719 kwh in a month? Round the answer to the nearest cent.

90. UTILITY TAX Some gas companies are required to tax the number of therms used each month by the customer. What are the taxes collected on a monthly usage of 31 therms if the tax rate is \$0.00566 per therm? Round the answer to the nearest cent.

WRITING

91. Explain how to determine where to place the decimal point in the answer when multiplying two decimals.

92. List the similarities and differences between whole-number multiplication and decimal multiplication.

93. What is a decimal place?

94. What is the purpose of the rules for the order of operations?

REVIEW

95. Solve $\frac{x}{2} - \frac{x}{3} = -2$.

96. Multiply: $3\frac{1}{3}\left(-1\frac{4}{5}\right)$.

97. Write this notation in words: $|-3|$.

98. What is the LCD of fractions with denominators of 4, 5, and 6?

99. Simplify: $-\frac{8}{8}$.

100. Find one-half of 7 and square the result.

5.4 Division with Decimals

In this section, you will learn about

- Dividing a decimal by a whole number • Divisors that are decimals
- Rounding when dividing • Dividing decimals by powers of 10
- Order of operations • Statistics problems

INTRODUCTION. Every division is composed of three parts: the divisor, the dividend, and the quotient.

Long division form

$$\text{Divisor} \longrightarrow 5\overline{)10} \longleftarrow \text{Dividend}, \quad 2 \longleftarrow \text{Quotient}$$

Fraction form

$$\frac{10 \leftarrow \text{Dividend}}{5 \leftarrow \text{Divisor}} = 2 \longleftarrow \text{Quotient}$$

In this section, we examine division problems in which the divisor and/or the dividend are decimals.

Dividing a decimal by a whole number

To use long division to divide 47 by 10, we proceed as follows.

$$\begin{array}{r} 4\frac{7}{10} \\ 10\overline{)47} \\ \underline{40} \\ 7 \end{array}$$

Here the result is written in quotient + $\frac{\text{remainder}}{\text{divisor}}$ form.

To do this same division using decimals, we write 47 as 47.0 and divide as we would divide whole numbers.

$$\begin{array}{r} 4.7 \\ 10\overline{)47.0} \\ \underline{40} \\ 7\,0 \\ \underline{7\,0} \\ 0 \end{array}$$

Note that the decimal point in the result is placed directly above the decimal point of the dividend.

Since $4\frac{7}{10} = 4.7$, either method gives the same result. The second part of this discussion suggests the following method for dividing a decimal by a whole number.

Dividing a decimal by a whole number

1. Write the problem in long division form.
2. Divide as if working with whole numbers.
3. Write the decimal point in the result directly above the decimal point of the dividend. If necessary, additional zeros can be written to the right of the dividend to allow the division to proceed.

EXAMPLE 1 ***Dividing a decimal by a whole number.***
Divide: 71.68 ÷ 28.

Solution

```
      .
28)71.68   Write the decimal point in the answer directly above the decimal point of the dividend.

    2.56
28)71.68   Divide as if working with whole numbers.
   56
   15 6
   14 0
    1 68
    1 68
       0   The remainder is 0.
```

The answer is 2.56. We can check this result by multiplying the divisor and the quotient; their product should equal the dividend. Since $28 \cdot 2.56 = 71.68$, the result is correct.

Self Check
Divide: 101.44 ÷ 32.

Answer: 3.17 ■

EXAMPLE 2 ***Writing extras zeros.*** Divide: 19.2 ÷ 5.

Solution

```
   3.8
5)19.2
  15↓
   4 2
   4 0
     2   All the digits in the dividend have been used, but the remainder is not 0.
```

We can write a zero to the right of 2 in the dividend and continue the division process. Recall that writing additional zeros to the right of the decimal point does not change the value of the decimal.

```
   3.84   Write a zero to the right of the 2 and bring it down.
5)19.20
  15
   4 2
   4 0↓
     20   Continue to divide.
     20
      0   The remainder is 0.
```

The answer is 3.84.

Self Check
Divide: 3.4 ÷ 4.

Answer: 0.85 ■

Divisors that are decimals

When the divisor is a decimal, we change it to a whole number and proceed as in division of whole numbers. To illustrate this procedure, we consider the problem

$0.36\overline{)0.2592}$, where the divisor is a decimal. First, we express the division in another form.

$0.36\overline{)0.2592}$ can be represented by $\dfrac{0.2592}{0.36}$

To write the divisor, 0.36, as a whole number, its decimal point needs to be moved two places to the right. This can be accomplished by multiplying it by 100. However, if the denominator of the fraction is multiplied by 100, the numerator must also be multiplied by 100 so that the fraction maintains the same value.

$$\frac{0.2592}{0.36} = \frac{0.2592 \cdot \mathbf{100}}{0.36 \cdot \mathbf{100}}$$ Multiply numerator and denominator by 100.

$$= \frac{25.92}{36}$$ Multiplying by 100 moves both decimal points two places to the right.

This fraction represents the division problem $36\overline{)25.92}$. From this result, we can make the following observations.

- The division problem $0.36\overline{)0.2592}$ is equivalent to $36\overline{)25.92}$. That is, they have the same answer.
- The decimal points in *both* the divisor and the dividend of the first division problem have been moved two decimal places to the right to create the second division problem.

$0.36\overline{)0.2592}$ becomes $36\overline{)25.92}$

These observations suggest the following rule for division with decimals.

Division with a decimal divisor

To divide with a decimal divisor:

1. Move the decimal point of the divisor so that it becomes a whole number.
2. Move the decimal point of the dividend the same number of places to the right.
3. Divide as if working with whole numbers. Write the decimal point in the answer directly above the decimal point of the dividend.

EXAMPLE 3 ***Dividing decimals.*** Divide: $\dfrac{0.2592}{0.36}$.

Solution

$0.36\overline{)0.25.92}$ Move the decimal point 2 places to the right in the divisor and dividend.

$$\begin{array}{r} 0.72 \\ 36\overline{)25.92} \\ \underline{25\ 2} \\ 72 \\ \underline{72} \\ 0 \end{array}$$

Now divide as with whole numbers. Write the decimal point in the answer directly above the decimal point of the dividend.

The result is 0.72.

Self Check

Divide: $\dfrac{0.6045}{0.65}$.

Answer: 0.93 ■

Rounding when dividing

In Example 3, the division process ended after we obtained a zero from the second subtraction. We say that the division process **terminated.** Sometimes when dividing, the subtractions never give a zero remainder, and the division process continues forever. In such cases, we can round the result.

EXAMPLE 4 ***Rounding when dividing.*** Divide: $\frac{2.35}{0.7}$. Round to the nearest hundredth.

Solution Using long division form, we have $0.7\overline{)2.35}$.

$0.7\overline{)2.3.5}$ To write the divisor as a whole number, move the decimal point one place to the right. Do the same for the dividend. Place the decimal point in the answer directly above the decimal point of the dividend.

$7\overline{)23.500}$ To round to the hundredths column, we must divide to the thousandths column. We write two zeros on the right of the dividend.

$$\begin{array}{r} 3.357 \\ 7\overline{)23.500} \\ \underline{21} \\ 2\,5 \\ \underline{2\,1} \\ 40 \\ \underline{35} \\ 50 \\ \underline{49} \\ 1 \end{array}$$

After dividing to the thousandths column, round to the hundredths column. The rounding digit is 5. The test digit is 7.

To the nearest hundredth, the answer is 3.36. ■

Accent on Technology: The nucleus of a cell

The nucleus of a cell contains vital information about the cell in the form of DNA. The nucleus is very small in size: A typical animal cell has a nucleus that is only 0.00023622 inch across. How many nuclei would have to be laid end-to-end to extend to a length of 1 inch?

To find how many 0.00023622-inch lengths there are in 1 inch, we must use division: 1 ÷ 0.00023622.

Keystrokes 1 [÷] .00023622 [=] [4233.3418]

It would take approximately 4,233 nuclei laid end-to-end to extend to a length of 1 inch.

Dividing decimals by powers of 10

To develop a set of rules for division by a power of 10, we consider the problem 8.13 ÷ 10.

$$\begin{array}{r} 0.813 \\ 10\overline{)8.130} \\ \underline{0} \\ 8\,1 \\ \underline{8\,0} \\ 13 \\ \underline{10} \\ 30 \\ \underline{30} \\ 0 \end{array}$$

Write a zero to the right of the 3.

We note that the quotient, 0.813, and the dividend, 8.13, are the same except for the location of the decimal points. The quotient can be easily obtained by moving the decimal point of the dividend 1 place to the *left*. This observation suggests the following rule for dividing a decimal by a power of 10.

Dividing a decimal by a power of 10

> To divide a decimal by a power of 10, move the decimal point to the left the same number of places as there are zeros in the power of 10.

EXAMPLE 5 ***Dividing decimals by powers of 10.*** Find the quotient: **a.** $16.74 \div 10$ and **b.** $8.6 \div 10{,}000$.

Solution

a. $16.74 \div 10 = 1.674$ Since 10 has 1 zero, move the decimal point 1 place to the left.

b. $8.6 \div 10{,}000 = .00086$ Since 10,000 has 4 zeros, move the decimal point 4 places to the left. Write 3 placeholder zeros.

$= 0.00086$

Self Check

Find the quotient:

a. $721.3 \div 100$

b. $\dfrac{1.07}{1{,}000}$

Answers: **a.** 7.213, **b.** 0.00107

Order of operations

In the next example, we will use the rules for the order of operations to evaluate an expression that involves division by a decimal.

EXAMPLE 6 ***Order of operations.*** Evaluate $\dfrac{2(0.351) + 0.5592}{-0.4}$.

Solution

$$\frac{2(0.351) + 0.5592}{-0.4} = \frac{0.702 + 0.5592}{-0.4}$$ Do the multiplication first: $2(0.351) = 0.702$.

$$= \frac{1.2612}{-0.4}$$ Do the addition: $0.702 + 0.5592 = 1.2612$.

$$= -3.153$$ Do the division. The quotient of two numbers with unlike signs is negative.

Self Check

Evaluate $\dfrac{2.7756 + 3(-0.63)}{-0.8}$.

Answer: -1.107

Statistics problems

Statistics is a branch of mathematics that deals with the analysis of numerical data. It uses three types of averages to describe the "middle" of a collection of numbers: the **mean,** the **median,** and the **mode.**

Mean

The **mean** of several values is the sum of those values divided by the number of values:

$$\text{Mean} = \frac{\text{sum of the values}}{\text{number of values}}$$

Median

The **median** of several values is the middle value. It is found as follows.

1. Arrange the values in increasing order.
2. If there is an odd number of values, choose the middle value.
3. If there is an even number of values, add the middle two values and divide by 2.

Mode

The **mode** of several values is the value that occurs most often.

EXAMPLE 7 ***Machinist's tools.*** The diameters (distances across) of eight stainless steel bearings were found using the vernier calipers shown in Figure 5-7. Find **a.** the mean, **b.** the median, and **c.** the mode of the set of measurements listed below.

3.43 cm, 3.25 cm, 3.48 cm, 3.39 cm, 3.54 cm, 3.48 cm, 3.23 cm, 3.24 cm

FIGURE 5-7

Solution **a.** To find the mean, we add the measurements and divide by the number of values, which is 8.

$$\text{Mean} = \frac{3.43 + 3.25 + 3.48 + 3.39 + 3.54 + 3.48 + 3.23 + 3.24}{8} = 3.38 \text{ cm}$$

b. To find the median, we first arrange the measurements in increasing order:

3.23, 3.24, 3.25, 3.39, 3.43, 3.48, 3.48, 3.54,

Because there is an even number of measurements, the median will be the sum of the middle two values, 3.39 and 3.43, divided by 2. Thus, the median is

$$\text{Median} = \frac{3.39 + 3.43}{2} = \frac{6.82}{2} = 3.41 \text{ cm}$$

c. Since the measurement 3.48 cm occurs most often, it is the mode. ■

STUDY SET Section 5.4

VOCABULARY *Fill in the blanks to make the statements true.*

1. In the division $2.5\overline{)4.075} = 1.63$, the decimal 4.075 is called the ______________, the decimal 2.5 is the ______________, and 1.63 is the ______________.

2. In $\frac{33.6}{0.3}$, the fraction ______________ indicates division.

3. The ______________ of several values is the sum of those values divided by the number of values.

4. The ______________ of several values is the middle value.

5. The ______________ of several values is the value that occurs most often.

6. ______________ is the branch of mathematics that deals with the analysis of numerical data.

CONCEPTS *In Exercises 7–8, fill in the blanks to make the statements true.*

7. To divide by a decimal, move the decimal point of the divisor so that it becomes a ______________ number. The decimal point of the dividend is then moved the same number of places to the ______________. The decimal point in the quotient is written directly ______________ the decimal point of the dividend.

8. To divide a decimal by a power of 10, move the decimal point to the ______________ the same number of decimal places as the number of zeros in the power of 10.

9. Is this statement true or false?

$45 = 45.0 = 45.000$

10. When a decimal is divided by 10, is the answer smaller or larger than the original number?

11. To complete the division $7.8\overline{)14.562}$, the decimal points of the divisor and dividend are moved 1 place to the right. This is equivalent to multiplying the numerator and the denominator of $\frac{14.562}{7.8}$ by what number?

12. **a.** When dividing decimals with like signs, what is the sign of the quotient?
b. When dividing decimals with unlike signs, what is the sign of the quotient?

13. How can we check the result of this division?

$$\frac{1.917}{0.9} = 2.13$$

14. When rounding a decimal to the hundredths column, to what other column must we refer?

15. A student performed the division

$$4.6\overline{)9.522}$$

and obtained the answer 2.07. Without doing the division, check this result. Is it correct?

16. In the division problem below, explain *why* we can write the additional zeros (shown in red) after 5. Doesn't this change the problem?

$$\begin{array}{r} 0.3 \\ 16\overline{)5.50000} \end{array}$$

17. What are the mean, median, and mode of the values 2.3, 2.3, 3.6, 3.8, and 4.5?

18. **a.** Is the mean of a set of values always one of those values?
b. Is the median of a set of values always one of those values?
c. Is the mode of a set of values always one of those values?

NOTATION

19. Explain what the arrows are illustrating.

$$4.67\overline{)32.08.7}$$

20. What is this arrow illustrating?

$$\begin{array}{r} 0.7 \\ 4\overline{)3.100} \\ -2\,8\downarrow \\ 30 \end{array}$$

PRACTICE *Do each division.*

21. $8\overline{)36}$

22. $4\overline{)10}$

23. $-39 \div 4$

24. $-26 \div 8$

25. $49.6 \div 8$

26. $23.5 \div 5$

27. $9\overline{)288.9}$

28. $6\overline{)337.8}$

29. $(-14.76) \div (-6)$

30. $(-13.41) \div (-9)$

31. $\dfrac{-55.02}{7}$

32. $\dfrac{-24.24}{8}$

33. $45\overline{)119.7}$

34. $41\overline{)146.37}$

35. $250.95 \div 35$

36. $241.86 \div 29$

37. $41.6 \div 0.32$

38. $31.8 \div 0.15$

39. $(-199.5) \div (-0.19)$

40. $(-2{,}381.6) \div (-0.26)$

41. $\dfrac{0.0102}{0.017}$

42. $\dfrac{0.0092}{0.023}$

43. $\dfrac{0.0186}{0.031}$

44. $\dfrac{0.416}{0.52}$

Divide and round each result to the nearest tenth.

45. $3\overline{)16}$

46. $7\overline{)20}$

47. $-5.714 \div 2.4$

48. $-21.21 \div 3.8$

Divide and round each result to the nearest hundredth.

49. $12.243 \div 0.9$

50. $13.441 \div 0.6$

51. $0.04\overline{)0.03164}$

52. $0.08\overline{)0.02201}$

Do the division mentally.

53. $7.895 \div 100$

54. $23.05 \div 10$

55. $0.064 \div (-100)$

56. $0.0043 \div (-10)$

57. $1000\overline{)34.8}$

58. $100\overline{)678.9}$

59. $\dfrac{45.04}{10}$

60. $\dfrac{22.32}{100}$

Evaluate each expression. Round each result to the nearest hundredth.

61. $\dfrac{-1.2 - 3.4}{3(1.6)}$

62. $\dfrac{(-1.3)^2 + 6.7}{-0.9}$

63. $\dfrac{40.7(-5.3)}{0.4 - 0.61}$

64. $\dfrac{(0.5)^2 - (0.3)^2}{0.005 + 0.1}$

Evaluate each expression. If an answer is not exact, round it to the nearest hundredth.

65. $\dfrac{5(F - 32)}{9}$ for $F = 48.38$

66. $\dfrac{5(F - 32)}{9}$ for $F = 19.94$

67. $\dfrac{6.7 - x^2 + 1.6}{x^3}$ for $x = 0.3$

68. $\dfrac{a - b}{0.5b - 0.4a}$ for $a = 3.6$ and $b = -1.5$

APPLICATIONS

69. BUTCHER SHOP A meat slicer is designed to trim 0.05-inch-thick pieces from a sausage. If the sausage is 14 inches long, how many slices will result?

70. COMPUTERS A computer can do an arithmetic computation in 0.00003 second. How many of these computations could it do in 60 seconds?

71. HIKING Use the information in Illustration 1 to find the time of arrival for the hiker.

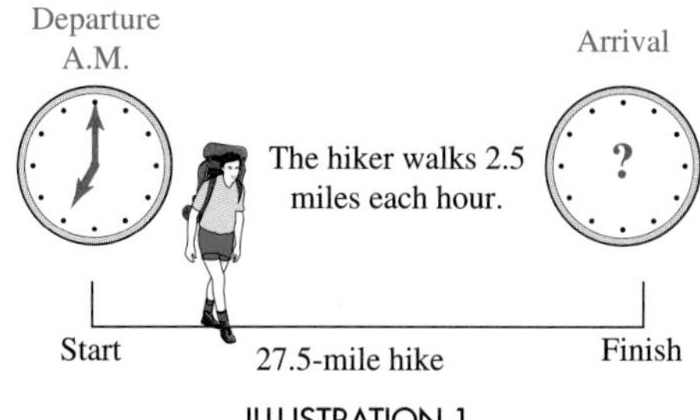

ILLUSTRATION 1

72. VOLUME CONTROL A volume control is shown in Illustration 2. If the distance between the Low and High settings is 21 cm, how far apart are the equally spaced volume settings?

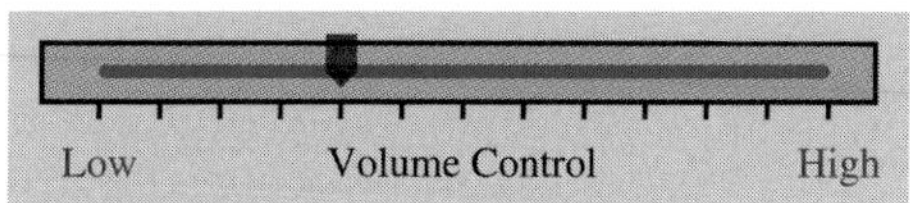

ILLUSTRATION 2

73. SPRAY BOTTLE Production planners have found that each squeeze of the trigger of a spray bottle emits 0.015 ounce of liquid. How many squeezes would there be in an 8.5-ounce bottle?

74. CAR LOAN See the loan statement in Illustration 3. How many more monthly payments must be made to pay off the loan?

American Finance Company	June
Monthly payment:	Paid to date: $547.30
$42.10	Loan balance: $631.50

ILLUSTRATION 3

75. HOURLY PAY Illustration 4 shows the average hours worked and the average weekly earnings of U.S. production workers in 1988 and 1998. What did the average production worker earn per hour in 1988 and in 1998? Round to the nearest cent.

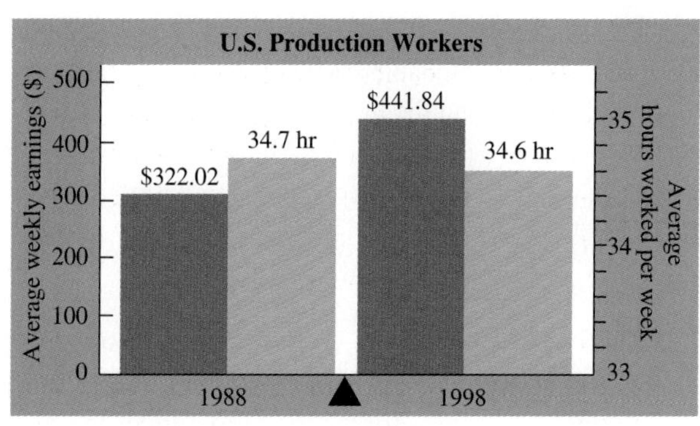

Source: *The World Almanac 2000*

ILLUSTRATION 4

76. PLEASURE TRAVEL Illustration 5 shows the annual number of person-trips of 100 miles or more (one way) for the years 1994–1998, as estimated by the Travel Industry Association of America. Find the mean and the median.

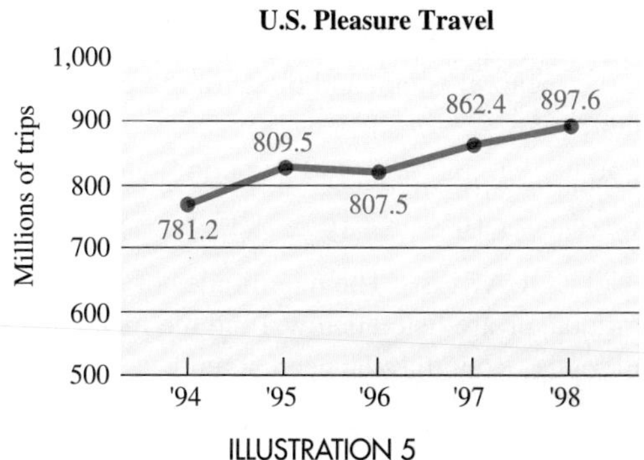

ILLUSTRATION 5

77. OIL WELL Geologists have mapped out the substances through which engineers must drill to reach an oil deposit. (See Illustration 6.) What is the average depth that must be drilled each week if this is to be a four-week project?

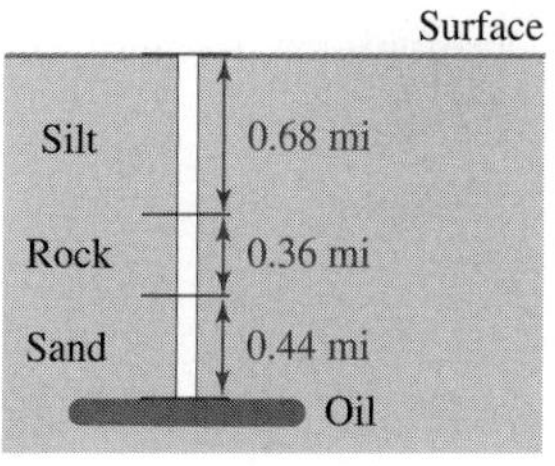

ILLUSTRATION 6

78. INDY 500 Illustration 7 shows the first row of the starting grid for the 1998 Indianapolis 500 automobile race. The drivers' speeds on a qualifying run were used to rank them in this order. What was the mean qualifying speed for the drivers in the first row?

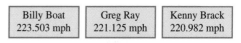

Billy Boat 223.503 mph	Greg Ray 221.125 mph	Kenny Brack 220.982 mph

ILLUSTRATION 7

79. OCTUPLETS In December of 1998, Nkem Chukwu gave birth to eight babies in Texas Children's Hospital. Find the mean and the median of their birth weights.

Ebuka (girl) 24 oz	Odera (girl) 11.2 oz
Chidi (girl) 27 oz	Ikem (boy) 17.5 oz
Echerem (girl) 28 oz	Jioke (boy) 28.5 oz
Chima (girl) 26 oz	Gorom (girl) 18 oz

80. ICE SKATING Listed below are Tara Lipinski's artistic impression scores for the long program of the women's figure skating competition at the 1998 Winter Olympics. Find the mean, median, and mode. Round to the nearest tenth.

Australia	5.8	Germany	5.8	Ukraine	5.9
Hungary	5.8	U.S.	5.8	Poland	5.8
Austria	5.9	Russia	5.9	France	5.9

81. COMPARISON SHOPPING A survey of grocery stores found the price of a 15-ounce box of Cheerios cereal ranging from \$3.89 to \$4.39. (See below.) What are the mean, median, and mode of the prices listed?

\$4.29	\$3.89	\$4.29	\$4.09	\$4.24	\$3.99
\$3.98	\$4.19	\$4.19	\$4.39	\$3.97	\$4.29

82. EARTHQUAKES The magnitudes of 1999's major earthquakes are listed below. Find the mean, median, and mode. Round to the nearest tenth.

1/19/99	New Ireland, Papua New Guinea	7.0
2/6/99	Santa Cruz Islands, S. Pacific Sea	7.3
3/4/99	Celebes Sea, Indonesia	7.1
4/5/99	New Britain, Papua New Guinea	7.4
4/8/99	E. Russia/N.E. China border	7.1
5/10/99	New Britain, Papua New Guinea	7.1
5/16/99	New Britain, Papua New Guinea	7.1
8/17/99	Izmit region, western Turkey	7.4
9/21/99	Taiwan	7.6
9/30/99	Oaxaca, Mexico	7.4
11/12/99	Bolu Province, northwest Turkey	7.2

Round to the nearest hundredth.

83. $\dfrac{8.6 + 7.99 + (4.05)^2}{4.56}$

84. $\dfrac{0.33 + (-0.67)(1.3)^3}{0.0019}$

85. $\left(\dfrac{45.9098}{-234.12}\right)^2 - 4$

86. $\left(\dfrac{6.0007}{3.002}\right) - \left(\dfrac{78.8}{12.45}\right)$

WRITING

87. Explain the process used to divide two numbers when both the divisor and the dividend are decimals.

88. Explain why we must sometimes use rounding when writing the answer to a division problem.

89. The division $0.5\overline{)2.005}$ is equivalent to $5\overline{)20.05}$. Explain what *equivalent* means in this case.

90. 1998 WINTER OLYMPICS Listed below are Michele Kwan's artistic impression scores from the Olympic women's figure skating competition. Explain why it is unnecessary to do any calculations to know her mean score.

Australia	5.9	Germany	5.9	Ukraine	5.9
Hungary	5.9	U.S.	5.9	Poland	5.9
Austria	5.9	Russia	5.9	France	5.9

REVIEW

91. Simplify the complex fraction: $\dfrac{\frac{7}{8}}{\frac{3}{4}}$.

92. Express the fraction $\frac{3}{4}$ as an equivalent fraction with a denominator of 36.

93. List the set of integers.

94. Translate to mathematical symbols: the sum of x and y decreased by 6.

95. Solve: $-\frac{3}{4}A = -9$.

96. Evaluate $\left(\frac{1}{2}\right)^3 - \left(\frac{1}{2}\right)^2$.

97. Simplify $5x - 6(x - 1) - (-x)$.

98. What is the opposite of x?

ESTIMATION

In this section, we will use estimation procedures to approximate the answers to addition, subtraction, multiplication, and division problems involving decimals. You will recall that we use rounding when estimating to help simplify the computations so that they can be performed quickly and easily.

EXAMPLE 1 ***Estimating sums and differences.***

a. Estimate to the nearest ten: $261.76 + 432.94$.

b. Estimate using front-end rounding: $381.77 - 57.01$.

Solution

a. We round each number to the nearest ten.

$$\begin{array}{ccccc} 261.76 & + & 432.94 & & \\ \downarrow & & \downarrow & & \\ 260 & + & 430 & = & 690 \end{array}$$

261.76 rounds to 260, and 432.94 rounds to 430.

The estimate is 690. If we compute 261.76+ 432.94, the sum is 694.7. We can see that our estimate is close; it's just 4.7 less than 694.7. Example 1(a) illustrates the tradeoff when using estimation. The calculations are easier to perform and they take less time, but the answers are not exact.

b. We use front-end rounding.

$$\begin{array}{ccccc} 381.77 & - & 57.01 & & \\ \downarrow & & \downarrow & & \\ 400 & - & 60 & = & 340 \end{array}$$

Each number is rounded to its largest place value: 381.77 to the nearest hundred and 57.01 to the nearest ten.

The estimate is 340.

Self Check

a. Estimate to the nearest ten the sum of 526.93 and 284.03.

b. Estimate using front-end rounding: $512.33 - 36.47$.

Answers: **a.** 810, **b.** 460 ■

EXAMPLE 2 ***Estimating products.***

Estimate each product: **a.** $6.41 \cdot 27$, **b.** $5.2 \cdot 13.91$, and **c.** $0.124 \cdot 98.6$.

Solution

a. We use front-end rounding.

$6.41 \cdot 27 \approx 6 \cdot 30$ The symbol $\approx$ means "is approximately equal to."

The estimate is 180.

b. We use front-end rounding.

$5.2 \cdot 13.91 \approx 5 \cdot 10$

The estimate is 50.

c. Notice that $98.6 \approx 100$.

$0.124 \cdot 98.6 \approx 0.124 \cdot 100$ To multiply a decimal by 100, move the decimal point 2 places to the right.

The estimate is 12.4.

Self Check

a. Estimate the product:

$42 \cdot 17.65$

b. Estimate the product:

$182 \cdot 24.04$

c. Estimate the product:

$979.3 \cdot 2.3485$

Answers: **a.** 800, **b.** 4,000, **c.** 2,348.5 ■

When estimating a quotient, we round the divisor and the dividend so that they will divide evenly. Try to round both numbers up or both numbers down.

EXAMPLE 3 ***Estimating quotients.*** Estimate: $246.03 \div 4.31$.

Solution

4.31 is close to 4. A multiple of 4 close to 246.03 is 240. (Note that both the divisor and dividend were rounded down.)

$246.03 \div 4.31 \approx 240 \div 4$ Do the division in your head.

The estimate is 60.

Self Check

Estimate: $6{,}429.6 \div 7.19$.

Answer: 900

STUDY SET *Use the following information about refrigerators to estimate the answers to each question. Remember that answers may vary, depending on the rounding method used.*

Deluxe model	**Standard model**	**Economy model**
Price: \$978.88	Price: \$739.99	Price: \$599.95
Capacity: 25.2 cubic feet	Capacity: 20.6 cubic feet	Capacity: 18.8 cubic feet
Energy cost: \$6.79 a month	Energy cost: \$5.61 a month	Energy cost: \$4.39 a month

1. How much more expensive is the deluxe model than the standard model?

2. A couple wants to buy two standard models, one for themselves and one for their newly married son and daughter-in-law. What is the total cost?

3. How much less storage capacity does the economy model have than the standard model?

4. The owner of a duplex apartment wants to purchase a standard model for one unit and an economy model for the other. What will be the total cost?

5. A stadium manager has a budget of \$20,000 to furnish the luxury boxes at a football stadium with refrigerators. How many standard models can she purchase for this amount?

6. How many more cubic feet of storage do you get with the deluxe model as compared to the economy model?

7. Three roommates are planning on purchasing the deluxe model and splitting the cost evenly. How much will each have to pay?

8. What is the energy cost per year to run the deluxe model?

9. If you make a \$220 down payment on the standard model, how much of the cost is left to finance?

10. The economy model can be expected to last for 10 years. What would be the total energy cost over that period.

Estimate the answer to each problem. Does the calculator result seem reasonable? That is, does it appear that the problem was entered into the calculator correctly?

11. $25.9 + 345.1 + 0.09$ 347.78

12. $8{,}345.889 - 345.6$ 8000.289

13. $42{,}090.8 + 3{,}303.09$ 45393.89

14. $10.007 - 0.626$ 3.747

15. $9.8(8.8)$ 86.24

16. $\frac{24.56}{2.2}$ 1.116363636

17. $53 \cdot 5.61$ 241.23

18. $89.11 \div 22.707$ 39.24340

5.5 *Fractions and Decimals*

In this section, you will learn about

- Writing fractions as equivalent decimals • Repeating decimals
- Rounding repeating decimals • Graphing fractions and decimals
- Problems involving fractions and decimals

INTRODUCTION. In this section, we will further investigate the relationship between fractions and decimals.

Writing fractions as equivalent decimals

To write $\frac{5}{8}$ as a decimal, we use the fact that $\frac{5}{8}$ indicates the division $5 \div 8$. We can convert $\frac{5}{8}$ to decimal form by doing the division.

$$\begin{array}{r} .625 \\ 8\overline{)5.000} \\ \underline{4\,8} \\ 20 \\ \underline{16} \\ 40 \\ \underline{40} \\ 0 \end{array}$$

Write a decimal point and additional zeros to the right of 5.

0 ← The remainder is zero.

Thus, $\frac{5}{8} = 0.625$.

Writing a fraction as a decimal

To write a fraction as a decimal, divide the numerator of the fraction by its denominator.

EXAMPLE 1 ***Writing a fraction as a decimal.*** Write $\frac{3}{4}$ as a decimal.

Solution

We divide the numerator by the denominator.

$$\begin{array}{r} .75 \\ 4\overline{)3.00} \\ \underline{2\,8} \\ 20 \\ \underline{20} \\ 0 \end{array}$$

Write a decimal point and two zeros to the right of 3.

0 ← The remainder is zero.

Thus, $\frac{3}{4} = 0.75$.

Self Check

Write $\frac{3}{16}$ as a decimal.

Answer: 0.1875 ■

In Example 1, the division process ended because a remainder of 0 was obtained. In this case, we call the quotient, 0.75, a **terminating decimal.**

Repeating decimals

Sometimes, when we are finding a decimal equivalent of a fraction, the division process never gives a remainder of zero. In this case, the result is a **repeating decimal.** Examples of repeating decimals are 0.4444. . . and 1.373737. . . . The three dots tell us that a block of digits repeats in the pattern shown. Repeating decimals can be written using a bar over the repeating block of digits. For example, 0.4444. . . can be written as $0.\overline{4}$, and 1.373737. . . can be written as $1.\overline{37}$.

COMMENT When using an overbar to write a repeating decimal, use the least number of digits necessary to show the repeating block of digits.

$0.333\ldots = 0.\overline{333}$	$6.7454545\ldots = 6.7\overline{454}$
$0.333\ldots = 0.\overline{3}$	$6.7454545\ldots = 6.7\overline{45}$

EXAMPLE 2 *Repeating decimals.* Write $\frac{5}{12}$ as a decimal.

Solution

We use division to find the decimal equivalent.

$$\begin{array}{r}
.4166 \\
12\overline{)5.0000} \\
\underline{4\,8} \\
20 \\
\underline{12} \\
80 \\
\underline{72} \\
80 \\
\underline{72} \\
8
\end{array}$$

Write a decimal point and four zeros to the right of 5.

It is apparent that 8 will continue to reappear as the remainder. Therefore, 6 will continue to reappear in the quotient. Since the repeating pattern is now clear, we may stop the division.

Thus, $\frac{5}{12} = 0.41\overline{6}$.

Self Check

Write $\frac{3}{11}$ as a decimal.

Answer: $0.\overline{27}$ ■

Every fraction can be written as either a terminating decimal or a repeating decimal. For this reason, the set of fractions **(rational numbers)** form a subset of the set of decimals called the set of **real numbers.** The set of real numbers corresponds to *all* points on a number line.

Not all decimals are terminating or repeating decimals. For example,

0.2020020002 . . .

does not terminate, and it has no repeating block of digits. This decimal cannot be written as a fraction with an integer numerator and a nonzero integer denominator. Thus, it is not a rational number. It is an example from the set of **irrational numbers.**

Rounding repeating decimals

When a fraction is written in decimal form, the result is either a terminating or a repeating decimal. Repeating decimals are often rounded to a specified place value.

EXAMPLE 3 *Rounding the decimal equivalent.* Write $\frac{1}{3}$ as a decimal and round to the nearest hundredth.

Solution

First, we divide the numerator by the denominator to find the decimal equivalent of $\frac{1}{3}$.

$$\begin{array}{r}
0.333 \\
3\overline{)1.000} \\
\underline{9} \\
10 \\
\underline{9} \\
10 \\
\underline{9} \\
1
\end{array}$$

Write a decimal point and additional zeros to the right of 1.

We see that the division process never gives a remainder of zero. When we write $\frac{1}{3}$ in decimal form, the result is the repeating decimal 0.333. . . = $0.\overline{3}$.

To find the decimal equivalent of $\frac{1}{3}$ to the nearest hundredth, we proceed as follows.

0.333. . . — Round 0.333 to the nearest hundredth by examining the test digit in the thousandths column.

Since 3 is less than 5, we round down, and $\frac{1}{3} \approx 0.33$. ■

EXAMPLE 4 ***Rounding a decimal equivalent.*** Write $\frac{2}{7}$ as a decimal and round to the nearest thousandth.

Solution

$$\begin{array}{r} .2857 \\ 7\overline{)2.0000} \\ \underline{1\,4} \\ 60 \\ \underline{56} \\ 40 \\ \underline{35} \\ 50 \\ \underline{49} \\ 1 \end{array}$$

Write a decimal point and additional zeros to the right of 2.

To round to the thousandths column, we must divide to the ten thousandths column.

0.2857 — Round 0.2857 to the nearest thousandth by examining the test digit in the ten thousandths column.

Since 7 is greater than 5, we round up, and $\frac{2}{7} \approx 0.286$. Read ≈ as "is approximately equal to."

Self Check

Write $\frac{7}{24}$ as a decimal and round to the nearest thousandth.

Answer: 0.292 ■

Accent on Technology: The fixed-point key

After performing a calculation, a scientific calculator can round the result to a given decimal place. This is done using the *fixed-point key.* As we did in Example 4, let's find the decimal equivalent of $\frac{2}{7}$ and round to the nearest thousandth. This time, we will use a calculator.

Keystrokes First, we set the calculator to round to the third decimal place (thousandths) by pressing [FIX] 3. Then we press 2 [÷] 7 [=] . 0.286

Thus, $\frac{2}{7} \approx 0.286$. To round to the nearest tenth, we would fix 1; to round to the nearest hundredth, we would fix 2, and so on.

If your calculator does not have a fixed-point key, see the owner's manual.

EXAMPLE 5 ***Writing a mixed number as a decimal.*** Write $5\frac{3}{8}$ in decimal form.

Solution

To write a mixed number in decimal form, recall that a mixed number is made up of a whole-number part and a fractional part. Since we can write $5\frac{3}{8}$ as $5 + \frac{3}{8}$, we need only consider how to write $\frac{3}{8}$ as a decimal.

$$\begin{array}{r} .375 \\ 8\overline{)3.000} \\ \underline{2\,4} \\ 60 \\ \underline{56} \\ 40 \\ \underline{40} \\ 0 \end{array}$$

Write a decimal point and three zeros to the right of 3.

Self Check

Write $8\frac{19}{20}$ in decimal form.

Thus, $5\frac{3}{8} = 5 + \frac{3}{8} = 5 + 0.375 = 5.375$. We would obtain the same result if we changed $5\frac{3}{8}$ to the improper fraction $\frac{43}{8}$ and divided 43 by 8.

Answer: 8.95 ■

Graphing fractions and decimals

A number line can be used to show the relationship between fractions and their respective decimal equivalents. Figure 5-8 shows some commonly used fractions that have terminating decimal equivalents. For example, we see from the graph that $\frac{13}{16} = 0.8125$.

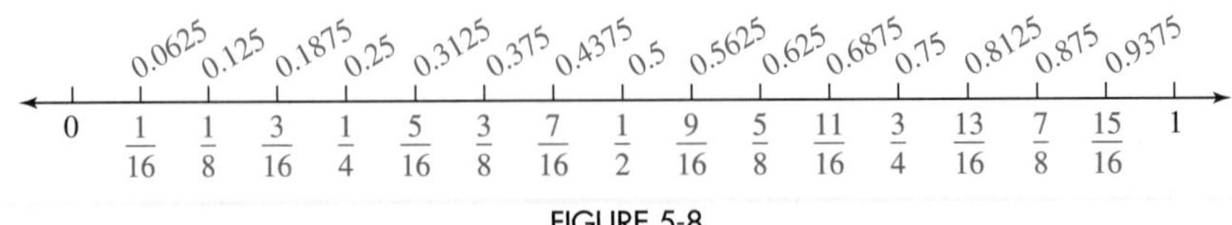

FIGURE 5-8

The number line in Figure 5-9 shows some commonly used fractions that have repeating decimal equivalents.

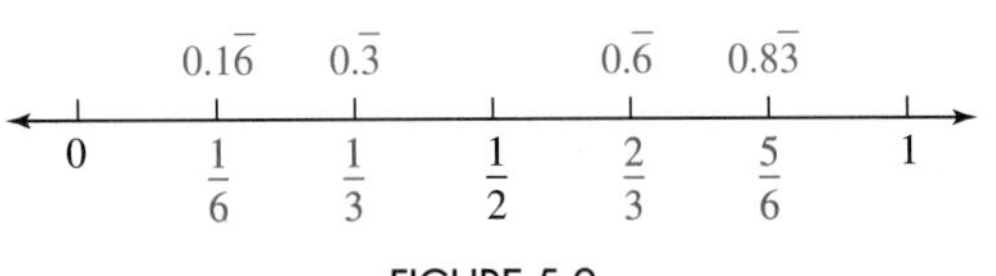

FIGURE 5-9

Problems involving fractions and decimals

Numerical expressions can contain both fractions and decimals. In the following examples, we show how different methods can be used to solve problems of this type.

EXAMPLE 6 ***Expressions containing fractions and decimals.***
Evaluate $\frac{1}{3} + 0.27$ by working in terms of fractions.

Solution

We write 0.27 as a fraction and add it to $\frac{1}{3}$.

$$\frac{1}{3} + 0.27 = \frac{1}{3} + \frac{27}{100} \quad \text{Replace 0.27 with } \tfrac{27}{100}.$$

$$= \frac{1 \cdot \mathbf{100}}{3 \cdot \mathbf{100}} + \frac{27 \cdot \mathbf{3}}{100 \cdot \mathbf{3}} \quad \text{Express each fraction in terms of 300ths.}$$

$$= \frac{100}{300} + \frac{81}{300} \quad \text{Multiply in the numerators and in the denominators.}$$

$$= \frac{181}{300} \quad \text{Add the numerators and write the sum over the common denominator, 300.}$$

Self Check
Evaluate by working in terms of fractions: $0.53 - \frac{1}{6}$.

Answer: $\frac{109}{300}$ ■

EXAMPLE 7 ***Expressions containing fractions and decimals.***
Evaluate $\frac{1}{3} + 0.27$ by working in terms of decimals.

Solution

We have seen that the decimal equivalent of $\frac{1}{3}$ is the repeating decimal 0.333. . . . To add $\frac{1}{3}$ to 0.27, we round 0.333. . . to the nearest hundredth: $\frac{1}{3} \approx 0.33$.

$$\frac{1}{3} + 0.27 \approx 0.33 + 0.27 \quad \text{Approximate } \tfrac{1}{3} \text{ with the decimal 0.33.}$$

$$\approx 0.60 \quad \text{Do the addition.}$$

Self Check
Evaluate by working in terms of decimals: $0.53 - \frac{1}{6}$.

Answer: 0.36 ■

In the previous two examples, we evaluated $\frac{1}{3} + 0.27$ in different ways. In Example 6, we obtained the exact answer, $\frac{181}{300}$. In Example 7, we obtained an approximation, 0.6. It is apparent that the results are in agreement when we write $\frac{181}{300}$ in decimal form: $\frac{181}{300} = 0.60333. \ . \ . \ .$

EXAMPLE 8 ***Expressions containing fractions and decimals.***
Evaluate: $\left(\frac{4}{5}\right)(1.35) + (0.5)^2$.

Solution
It appears simplest to work in terms of decimals. We use division to find the decimal equivalent of $\frac{4}{5}$.

$$\begin{array}{r} .8 \\ 5\overline{)4.0} \\ \underline{4\,0} \\ 0 \end{array}$$

Write a decimal point and one zero to the right of the 4.

Now we use the rules for the order of operations to evaluate the given expression.

$\left(\frac{4}{5}\right)(1.35) + (0.5)^2 = (\mathbf{0.8})(1.35) + (0.5)^2$	Replace $\frac{4}{5}$ with its decimal equivalent, 0.8.
$= (0.8)(1.35) + 0.25$	Find the power: $(0.5)^2 = 0.25$.
$= 1.08 + 0.25$	Do the multiplication: $(0.8)(1.35) = 1.08$.
$= 1.33$	Do the addition.

Self Check
Evaluate: $(-0.6)^2 + (2.3)\left(\frac{1}{8}\right)$.

Answer: 0.6475 ■

EXAMPLE 9 ***Shopping.*** During a trip to the grocery store, a shopper purchased $\frac{3}{4}$ pound of fruit, priced at \$0.88 a pound, and $\frac{1}{3}$ pound of fresh-ground coffee, selling for \$6.60 a pound. Find the total cost of these items.

Solution To find the cost of each item, we multiply the amount purchased by its unit price. Then we add the two individual costs to obtain the total cost.

Cost of fruit	plus	cost of coffee	equals	total cost.
$\left(\frac{3}{4}\right)(0.88)$	$+$	$\left(\frac{1}{3}\right)(6.60)$	$=$	total cost

Because 0.88 is divisible by 4 and 6.60 is divisible by 3, we can work with the decimals and fractions in this form; no conversion is necessary.

$\left(\frac{3}{4}\right)(0.88) + \left(\frac{1}{3}\right)(6.60) = \left(\frac{3}{4}\right)\left(\frac{0.88}{1}\right) + \left(\frac{1}{3}\right)\left(\frac{6.60}{1}\right)$	Express 0.88 as $\frac{0.88}{1}$ and 6.60 as $\frac{6.60}{1}$.
$= \frac{2.64}{4} + \frac{6.60}{3}$	Multiply the numerators and the denominators.
$= 0.66 + 2.20$	Do each division.
$= 2.86$	Do the addition.

The total cost of the items is \$2.86. ■

STUDY SET Section 5.5

VOCABULARY *Fill in the blanks to make the statements true.*

1. The decimal form of the fraction $\frac{1}{3}$ is a ____________ decimal, which is written $0.\overline{3}$ or 0.3333. . . .
2. The decimal form of the fraction $\frac{2}{5}$ is a ____________ decimal, which is written 0.4.
3. The ____________ equivalent of $\frac{1}{16}$ is 0.0625.
4. To write a fraction as a decimal, divide the ____________ of the fraction by its denominator.

CONCEPTS

5. What division is indicated by the fraction $\frac{7}{8}$?
6. Insert the proper symbol < or > in the blank to make the statement true.
   a. $0.\overline{6}$ ___ 0.7 **b.** $0.\overline{6}$ ___ 0.6
7. When rounding 0.272727. . . to the nearest hundredth, is the result larger or smaller than the original number?
8. Write each decimal in fraction form.
   a. 0.7 **b.** 0.77
9. Graph $1\frac{3}{4}$, −0.75, $0.\overline{6}$, and $-3.8\overline{3}$ on the number line.

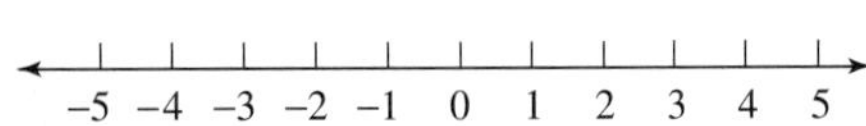

10. Graph $2\frac{7}{8}$, −2.375, $0.\overline{3}$, and $4.1\overline{6}$ on the number line.

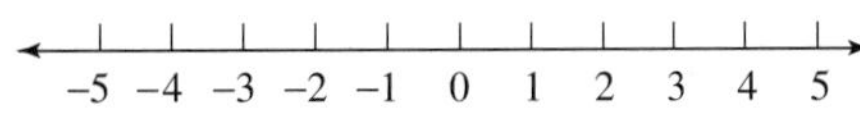

11. Tell whether each statement is true or false.
    a. $\frac{1}{3} = 0.3$ **b.** $\frac{3}{4} = 0.75$
    c. $20\frac{1}{2} = 20.5$ **d.** $\frac{1}{16} = 0.1\overline{6}$
12. When evaluating the expression $0.25 + \left(2.3 + \frac{2}{5}\right)^2$, would it be easier to work in terms of fractions or in terms of decimals?

NOTATION

13. Examine the color portion of the long division in the next column.
    a. Will the remainder ever be zero?
    b. What can be deduced about the decimal equivalent of $\frac{5}{6}$?

```
    .833
 6)5.000
   4 8
     20
     18
      20
```

14. Write each repeating decimal using an overbar.
    a. 0.888. . . **b.** 0.323232. . .
    c. 0.56333. . . **d.** 0.8898989. . .

PRACTICE *Write each fraction in decimal form.*

15. $\frac{1}{2}$
16. $\frac{1}{4}$
17. $-\frac{5}{8}$
18. $-\frac{3}{5}$
19. $\frac{9}{16}$
20. $\frac{3}{32}$
21. $-\frac{17}{32}$
22. $-\frac{15}{16}$
23. $\frac{11}{20}$
24. $\frac{19}{25}$
25. $\frac{31}{40}$
26. $\frac{17}{20}$
27. $-\frac{3}{200}$
28. $-\frac{21}{50}$
29. $\frac{1}{500}$
30. $\frac{1}{250}$

Write each fraction in decimal form. Use an overbar.

31. $\frac{2}{3}$
32. $\frac{7}{9}$
33. $\frac{5}{11}$
34. $\frac{4}{15}$
35. $-\frac{7}{12}$
36. $-\frac{17}{22}$
37. $\frac{1}{30}$
38. $\frac{1}{60}$

Write each fraction in decimal form. Round to the nearest hundredth.

39. $\frac{7}{30}$
40. $\frac{14}{15}$
41. $\frac{17}{45}$
42. $\frac{8}{9}$

Write each fraction in decimal form. Round to the nearest thousandth.

43. $\frac{5}{33}$

44. $\frac{5}{12}$

45. $\frac{10}{27}$

46. $\frac{17}{21}$

Write each fraction in decimal form. Round to the nearest hundredth.

47. $\frac{4}{3}$

48. $\frac{10}{9}$

49. $-\frac{34}{11}$

50. $-\frac{25}{12}$

Write each mixed number in decimal form. Round to the nearest hundredth when the result is a repeating decimal.

51. $3\frac{3}{4}$

52. $5\frac{4}{5}$

53. $-8\frac{2}{3}$

54. $-1\frac{7}{9}$

55. $12\frac{11}{16}$

56. $32\frac{1}{8}$

57. $203\frac{11}{15}$

58. $568\frac{23}{30}$

Fill in the correct symbol (< or >) to make a true statement. (Hint: Express each number as a decimal.)

59. $\frac{7}{8}$ ☐ 0.895

60. 4.56 ☐ $4\frac{2}{5}$

61. $-\frac{11}{20}$ ☐ $-0.\overline{4}$

62. $-9.\overline{09}$ ☐ $-9\frac{1}{11}$

Evaluate each expression. Work in terms of fractions.

63. $\frac{1}{9} + 0.3$

64. $\frac{2}{3} + 0.1$

65. $0.9 - \frac{7}{12}$

66. $0.99 - \frac{5}{6}$

67. $\frac{5}{11}(0.3)$

68. $(0.9)\left(\frac{1}{27}\right)$

69. $\frac{1}{3}\left(-\frac{1}{15}\right)(0.5)$

70. $\left(-0.4\right)\left(\frac{5}{18}\right)\left(-\frac{1}{3}\right)$

Evaluate each expression to the nearest hundredth.

71. $0.24 + \frac{1}{3}$

72. $0.02 + \frac{5}{6}$

73. $5.69 - \frac{5}{12}$

74. $3.19 - \frac{2}{3}$

Evaluate each expression. Work in terms of decimals.

75. $(3.5 + 6.7)\left(-\frac{1}{4}\right)$

76. $\left(-\frac{5}{8}\right)(5.3 - 3.9)$

77. $\left(\frac{1}{5}\right)^2(1.7)$

78. $(2.35)\left(\frac{2}{5}\right)^2$

79. $7.5 - (0.78)\left(\frac{1}{2}\right)$

80. $8.1 - \left(\frac{3}{4}\right)(0.12)$

81. $\frac{3}{8}(-3.2) + (4.5)\left(-\frac{1}{9}\right)$

82. $(-0.8)\left(\frac{1}{4}\right) + \left(\frac{1}{3}\right)(0.39)$

Evaluate each algebraic expression.

83. $\frac{4}{3}pr^3$ for $p = 3.14$ and $r = 3$

84. $\frac{1}{3}pr^2h$ for $p = 3.14$, $r = 6$, and $h = 12$

Write each fraction in decimal form.

85. $\frac{23}{101}$

86. $\frac{1}{99}$

87. $\frac{1,736}{50}$

88. $-\frac{11}{128}$

APPLICATIONS

89. DRAFTING The architect's scale has several measuring edges. The edge marked 16 divides each inch into 16 equal parts. (See Illustration 1.) Find the decimal form for each fractional part of one inch that is highlighted on the scale.

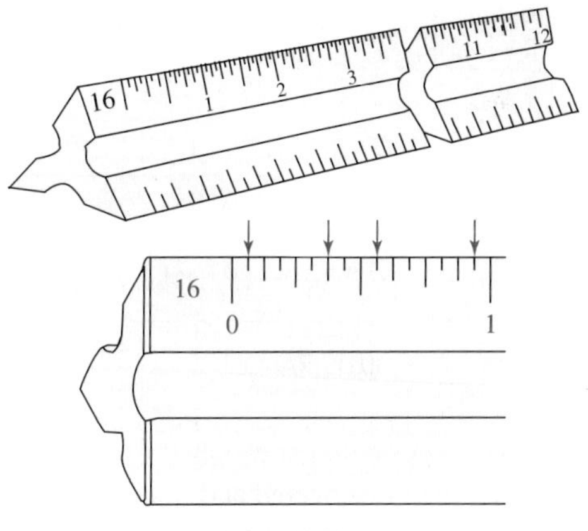

ILLUSTRATION 1

90. FREEWAY SIGNS The freeway sign in Illustration 2 (on the next page) gives the number of miles to the next three exits. Convert the mileages to decimal notation.

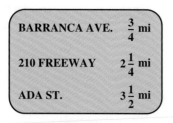

BARRANCA AVE. $\frac{3}{4}$ mi

210 FREEWAY $2\frac{1}{4}$ mi

ADA ST. $3\frac{1}{2}$ mi

ILLUSTRATION 2

91. GARDENING Two brands of replacement line for a lawn trimmer are labeled in different ways. (See Illustration 3.) On one package, the line's thickness is expressed as a decimal; on the other, as a fraction. Which line is thicker?

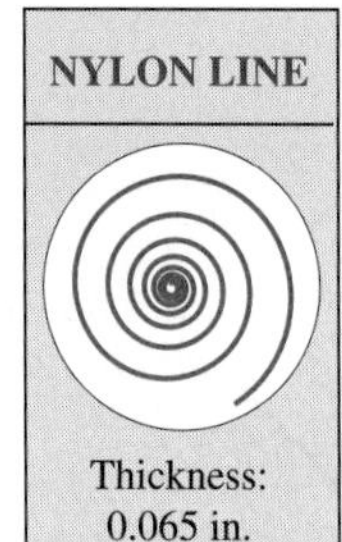

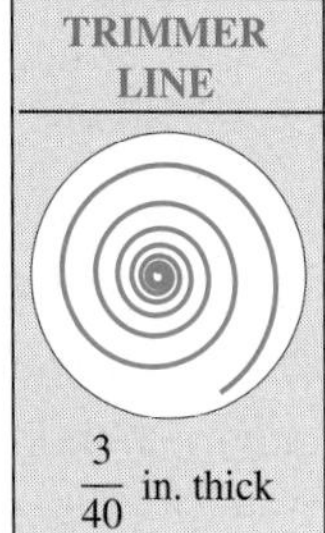

ILLUSTRATION 3

92. AUTO MECHANICS While doing a tuneup, a mechanic checks the gap on one of the spark plugs of a car to be sure it is firing correctly. The owner's manual states that the gap should be $\frac{2}{125}$ inch. The gauge the mechanic uses to check the gap is in decimal notation; it registers 0.025 inch. Is the spark plug gap too large or too small?

93. HORSE RACING In thoroughbred racing, the time a horse takes to run a given distance is measured using fifths of a second. For example, 55^2 (read "fifty-five and two") means $55\frac{2}{5}$ seconds. Illustration 4 lists four split times for a horse. Express the times in decimal form.

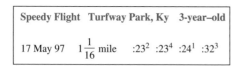

Speedy Flight	Turfway Park, Ky	3-year-old			
17 May 97	$1\frac{1}{16}$ mile	$:23^2$	$:23^4$	$:24^1$	$:32^3$

ILLUSTRATION 4

94. GEOLOGY A geologist weighed a rock sample at the site where it was discovered and found it to weigh $17\frac{7}{8}$ lb. Later, a more accurate digital scale in the laboratory gave the weight as 17.671 lb. What is the difference in the two measurements?

95. WINDOW REPLACEMENT The amount of sunlight that comes into a room depends on the area of the windows in the room. What is the area of the window in Illustration 5?

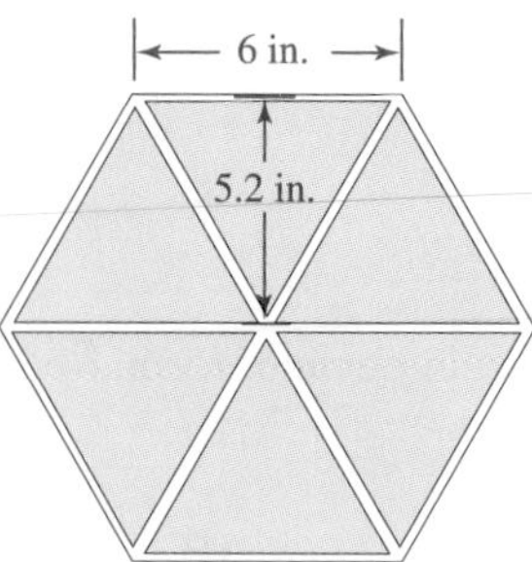

ILLUSTRATION 5

96. FOREST FIRE CONTAINMENT A command post asked each of three fire crews to estimate the length of the fire line they were fighting. Their reports came back in different forms, as indicated in Illustration 6. Find the perimeter of the fire.

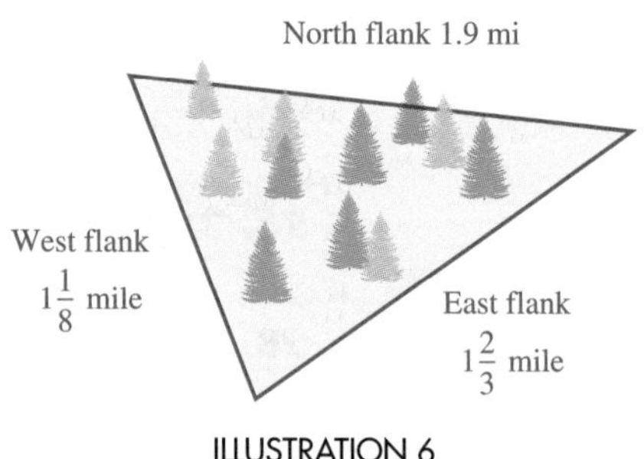

ILLUSTRATION 6

WRITING

97. Explain the procedure used to write a fraction in decimal form.

98. Compare and contrast the two numbers 0.5 and $0.\overline{5}$.

99. A student represented the repeating decimal 0.1333. . . as $0.1\overline{333}$. Is this correct? Explain why or why not.

100. Is 0.10100100010000 . . . a repeating decimal? Explain why or why not.

REVIEW

101. Add: $-2 + (-3) + 10 + (-6)$.

102. Evaluate $-3 + 2[-3 + (2 - 7)]$.

103. Simplify $3T - 4T + 2(-4t)$.

104. In the expression $6x^2 + 3x + 7$, what is the coefficient of the second term?

105. Simplify $4x^2 + 2x^2$.

106. Two pieces of pipe have lengths x and $x + 10$. Which is the longer piece of pipe?

5.6 Solving Equations Containing Decimals

In this section, you will learn about

- Solving equations using the properties of equality
- Simplifying expressions to solve equations • Problem solving with equations

INTRODUCTION. We have studied how to add, subtract, multiply, and divide decimals. We will now use these skills to solve equations containing decimals. We will use the following strategy to solve these equations. However, we will not have to apply Step 1 in any problems in this section.

Strategy for solving equations

To simplify the equation:

1. Clear the equation of any fractions.
2. Use the distributive property to remove parentheses.
3. Combine like terms on either side of the equation.

To isolate the variable:

4. Apply the addition and subtraction properties of equality to get the variables on one side of the equation and the constants on the other.
5. Continue to combine like terms when necessary.
6. Undo the operations of multiplication and division to isolate the variable.

Solving equations using the properties of equality

Recall that the addition and subtraction properties of equality allow us to add the same number to or subtract the same number from both sides of an equation.

EXAMPLE 1 ***Solving equations containing decimals.*** Solve each equation: **a.** $x + 3.5 = 7.8$ and **b.** $y - 1.23 = -4.52$.

Solution

a. To isolate x, we undo the addition of 3.5 by subtracting 3.5 from both sides of the equation.

$$x + 3.5 = 7.8$$

$$x + 3.5 - \mathbf{3.5} = 7.8 - \mathbf{3.5} \quad \text{Subtract 3.5 from both sides.}$$

$$x = 4.3 \quad \text{Simplify.}$$

b. To isolate y, we undo the subtraction of 1.23 by adding 1.23 to both sides of the equation.

$$y - 1.23 = -4.52$$

$$y - 1.23 + \mathbf{1.23} = -4.52 + \mathbf{1.23} \quad \text{Add 1.23 to both sides.}$$

$$y = -3.29 \quad \text{Simplify.}$$

Verify each result.

Self Check

Solve each equation.

a. $4.6 + t = 15.7$

b. $-1.24 = r - 0.04$

Answers: **a.** 11.1, **b.** −1.2

The multiplication property of equality states that we can multiply both sides of an equation by the same nonzero number.

EXAMPLE 2 ***Multiplication property of equality.*** Solve $\frac{m}{2} = -24.8$.

Solution

To isolate m, we undo the division by 2 by multiplying both sides of the equation by 2.

$$\frac{m}{2} = -24.8$$

$$2\left(\frac{m}{2}\right) = 2(-24.8) \quad \text{Multiply both sides by 2.}$$

$$m = -49.6 \quad \text{Do the multiplications.}$$

Check the result.

Self Check

Solve $\frac{y}{3} = -13.11$.

Answer: -39.33 ■

The division property of equality states that we can divide both sides of an equation by the same nonzero number.

EXAMPLE 3 ***Division property of equality.*** Solve $-4.6x = -9.66$.

Solution

To isolate x, we undo the multiplication by -4.6 by dividing by -4.6.

$$-4.6x = -9.66$$

$$\frac{-4.6x}{\mathbf{-4.6}} = \frac{-9.66}{\mathbf{-4.6}} \quad \text{Divide both sides by } -4.6.$$

$$x = 2.1 \quad \text{Do the divisions.}$$

Check the result.

Self Check

Solve $-22.32 = -3.1m$.

Answer: 7.2 ■

Sometimes, more than one property must be used to solve an equation. In the next example, we use the addition property of equality and the division property of equality.

EXAMPLE 4 ***Solving equations containing decimals.***
Solve $8.1y - 6.04 = -13.33$ and check the result.

Solution

The left-hand side involves a multiplication and a subtraction. To solve the equation, we must undo these operations, but in the opposite order. We begin by undoing the subtraction.

$$8.1y - 6.04 = -13.33$$

$$8.1y - 6.04 + \mathbf{6.04} = -13.33 + \mathbf{6.04} \quad \text{To undo the subtraction of 6.04, add 6.04 to both sides.}$$

$$8.1y = -7.29 \quad \text{Simplify: } -13.33 + 6.04 = -7.29.$$

$$\frac{8.1y}{\mathbf{8.1}} = \frac{-7.29}{\mathbf{8.1}} \quad \text{To undo the multiplication of 8.1, divide both sides by 8.1.}$$

$$y = -0.9 \quad \text{Do the divisions.}$$

Check:

$$8.1y - 6.04 = -13.33$$

$$8.1(\mathbf{-0.9}) - 6.04 \stackrel{?}{=} -13.33 \quad \text{Substitute } -0.9 \text{ for } y.$$

$$-7.29 - 6.04 \stackrel{?}{=} -13.33 \quad \text{Do the multiplication: } 8.1(-0.9) = -7.29.$$

$$-13.33 = -13.33 \quad \text{Do the subtraction by adding the opposite: } -7.29 - 6.04 = -7.29 + (-6.04) = -13.33.$$

Since $y = -0.9$ checks, it is a solution.

Self Check

Solve $-4.2h + 3.14 = 1.88$ and check the result.

Answer: 0.3 ■

Simplifying expressions to solve equations

Recall that to *combine like terms* means to simplify the sum (or difference) of like terms.

EXAMPLE 5 ***Combining like terms.*** Simplify $9.9b + 5.4 - 2.6b$.

Solution

This expression involves three terms. We can combine the two that are like terms.

$\mathbf{9.9b} + 5.4 - \mathbf{2.6b} = \mathbf{7.3b} + 5.4$ Subtract: $9.9 - 2.6 = 7.3$. Keep the variable b.

Self Check

Simplify $4.06a - 6.71 - 3.04a$.

Answer: $1.02a - 6.71$ ■

Sometimes we must combine like terms in order to isolate the variable and solve an equation.

EXAMPLE 6 ***Combining like terms to solve an equation.*** Solve $-22.46 + 3.2t + 1.9t = 52$.

Solution

First, we combine the like terms on the left-hand side.

$-22.46 + \mathbf{3.2t} + \mathbf{1.9t} = 52$	
$-22.46 + \mathbf{5.1t} = 52$	Combine like terms: $3.2t + 1.9t = 5.1t$.
$-22.46 + 5.1t + \mathbf{22.46} = 52 + \mathbf{22.46}$	To eliminate -22.46 from the left-hand side, add 22.46 to both sides.
$5.1t = 74.46$	Simplify.
$\frac{5.1t}{\mathbf{5.1}} = \frac{74.46}{\mathbf{5.1}}$	To undo the multiplication by 5.1, divide both sides by 5.1.
$t = 14.6$	Do the divisions.

Self Check

Solve $-1.9 + 2.8x - 1.4x = 12.24$.

Answer: 10.1 ■

EXAMPLE 7 ***Variable terms on both sides of an equation.*** Solve $0.2s - 3 = 0.7s + 1.5$.

Solution

We isolate the variable terms on the right-hand side and isolate the constant terms on the left-hand side of the equation.

$0.2s - 3 = 0.7s + 1.5$	
$0.2s - 3 - \mathbf{0.2s} = 0.7s + 1.5 - \mathbf{0.2s}$	Eliminate $0.2s$ from the left-hand side by subtracting $0.2s$ from both sides.
$-3 = 0.5s + 1.5$	Combine like terms.
$-3 - \mathbf{1.5} = 0.5s + 1.5 - \mathbf{1.5}$	To undo the addition of 1.5, subtract 1.5 from both sides.
$-3 + (-1.5) = 0.5s$	On the left, write the subtraction as addition of the opposite. On the right, simplify.
$-4.5 = 0.5s$	Do the addition: $-3 + (-1.5) = -4.5$.
$\frac{-4.5}{\mathbf{0.5}} = \frac{0.5s}{\mathbf{0.5}}$	To undo the multiplication by 0.5, divide both sides by 0.5.
$-9 = s$	Do the divisions.

Self Check

Solve $5.5 - 6.1b = -5.2b - 5.3$.

Answer: 12 ■

EXAMPLE 8 ***Using the distributive property.*** Solve $5(x + 1.3) = -9.9$.

Solution

First, we remove the parentheses by applying the distributive property.

Self Check

Solve $2(4.1 + c) = -19.4$.

$$5(x + 1.3) = -9.9$$

$$5x + 6.5 = -9.9$$ Distribute the 5: $5 \cdot 1.3 = 6.5$.

$$5x + 6.5 - \mathbf{6.5} = -9.9 - \mathbf{6.5}$$ To undo the addition of 6.5, subtract 6.5 from both sides.

$$5x = -9.9 + (-6.5)$$ On the right-hand side, add the opposite of 6.5.

$$5x = -16.4$$ Do the addition.

$$\frac{5x}{\mathbf{5}} = \frac{-16.4}{\mathbf{5}}$$ To undo the multiplication by 5, divide both sides by 5.

$$x = -3.28$$ Do the divisions.

Answer: -13.8 ■

Problem solving with equations

EXAMPLE 9 ***Business expenses.*** A business decides to rent a copy machine (Figure 5-10) instead of buying one. Under the rental agreement, the company is charged \$65 per month plus 2¢ for every copy made. If the business has budgeted \$125 for copier expenses each month, how many copies can be made before exceeding the budget?

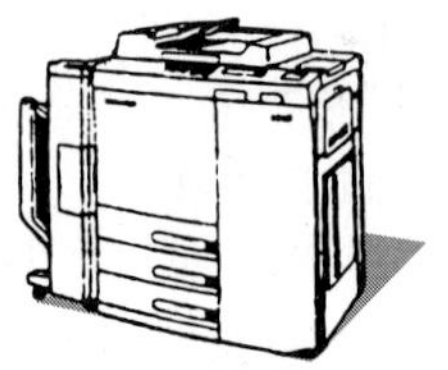

FIGURE 5-10

Analyze the problem

- The basic rental charge is \$65 a month.
- There is a 2¢ charge for each copy made.
- \$125 is budgeted for copier expenses each month.
- We must find the maximum number of copies that can be made each month.

Form an equation Let x = the maximum number of copies that can be made. We can write the amount budgeted for copier expenses in two ways.

The basic fee	plus	the cost of the copies	is	the amount budgeted each month.

We can find the total cost of the copies by multiplying the cost per copy by the maximum number of copies that can be made. Notice that the costs are expressed in terms of dollars and cents. We need to work in terms of one unit, so we write 2¢ as \$0.02 and work in terms of dollars.

65	plus	0.02 ·	the maximum number of copies made	is	125.
65	+	0.02 ·	x	=	125

Solve the equation

$$65 + 0.02x = 125$$

$$65 + 0.02x - \mathbf{65} = 125 - \mathbf{65}$$ To undo the addition of 65, subtract 65 from both sides.

$$0.02x = 60$$ Simplify.

$$\frac{0.02x}{\mathbf{0.02}} = \frac{60}{\mathbf{0.02}}$$ To undo the multiplication by 0.02, divide both sides by 0.02.

$$x = 3{,}000$$ Do the divisions.

State the conclusion The business can make up to 3,000 copies each month without exceeding its budget.

Check the result If we multiply the cost per copy and the maximum number of copies, we get $\$0.02 \cdot 3{,}000 = \$60$. Then we add the \$65 monthly fee: $\$60 + \$65 = \$125$. The answer checks. ■

STUDY SET Section 5.6

VOCABULARY *Fill in the blanks to make the statements true.*

1. To ______________ an equation, we isolate the variable on one side of the equals sign.

2. $4.1(x + 3) = 4.1x + 4.1(3)$ is an example of the use of the ______________ property.

3. In the term $5.65t$, the number 5.65 is called the ______________.

4. A ______________ is a letter that is used to stand for a number.

CONCEPTS

5. Show that $x = 1.7$ is a solution of $2.1x - 6.3 = -2.73$ by checking it.

6. Show that $y = 0.04$ is a solution of $\frac{y}{2} + 0.7 = 0.72$ by checking it.

7. For which problem below does the instruction *simplify* apply?

$7.8x + 9.1 = 12.4$ or $7.8x + 9.1 + 12.4$

8. a. What operations are performed on the variable?

$$\frac{m}{2.1} - 7.4 = 5.6$$

b. In what order should the operations be undone to isolate the variable?

9. Write each amount of money as a dollar amount.

a. 25 cents
b. 1 penny
c. 250 cents
d. 99 cents

10. Why can't the expression $5.6A + 3.4a$ be simplified?

11. What algebraic concept is shown below?

$$3.1(6 - 0.3h)$$

12. Rewrite each subtraction as addition of the opposite.

a. $4.02 - (-1.7)$
b. $y - (-0.6)$

NOTATION *Complete the solution to solve each equation.*

13.

$$0.6s - 2.3 = -1.82$$
$$0.6s - 2.3 + \_\_\_ = -1.82 + \_\_\_$$
$$\_\_\_ = 0.48$$
$$\frac{0.6s}{\_\_\_} = \frac{0.48}{\_\_\_}$$
$$s = 0.8$$

14.

$$\frac{x}{2} = -6.2$$
$$2\left(\frac{x}{2}\right) = 2(\_\_\_)$$
$$x = -12.4$$

PRACTICE *Combine like terms.*

15. $8.7x + 1.4x$

16. $45.1t + 38.6t$

17. $0.05h - 0.03h$

18. $67.89j - 54.73j$

19. $3.1r - 5.5r - 1.3r$

20. $3.8x - 6.5x - 2.4x$

21. $3.2 - 8.78x + 9.1$

22. $25.04 - 5.6w - 12.02$

23. $5.6x - 8.3 - 6.1x + 12.2$

24. $-17.3y - 8.01 + 12.2y - 4.4$

25. $0.05(100 - x) + 0.04x$

26. $0.06(1{,}000 - y) + 0.04y$

Solve each equation.

27. $x + 8.1 = 9.8$

28. $6.75 + y = 8.99$

29. $7.08 = t - 0.03$

30. $14.1 = k - 13.1$

31. $-5.6 + h = -17.1$

32. $-0.05 + x = -1.25$

33. $7.75 = t - (-7.85)$

34. $3.33 = y - (-5.55)$

35. $2x = -8.72$

36. $3y = -12.63$

37. $-3.51 = -2.7x$

38. $-1.65 = -0.5f$

39. $\frac{x}{2.04} = -4$

40. $\frac{y}{2.22} = -6$

41. $\frac{-x}{5.1} = -4.4$

42. $\frac{-t}{8.1} = -3$

43. $\frac{1}{3}x = -7.06$

44. $\frac{1}{5}x = -3.02$

45. $\frac{x}{100} = 0.004$

46. $\frac{y}{1{,}000} = 0.0606$

47. $2x + 7.8 = 3.4$

48. $3x - 1.2 = -4.8$

49. $-0.8 = 5y + 9.2$

50. $-9.9 = 6t + 14.1$

51. $0.3x - 2.1 = 7.2$

52. $0.4a + 3.3 = -5.1$

53. $-1.5b + 2.7 = 1.2$

54. $-2.1x - 3.1 = 5.3$

55. $0.4a - 6 + 0.5a = -5.73$

56. $0.1t - 0.7t + 4 = 3.46$

57. $2(t - 4.3) + 1.2 = -6.2$

58. $3(y - 1.1) + 3.2 = 2.3$

59. $1.2x - 1.3 = 2.4x + 0.02$

60. $-4.4y - 1.3 = -5.1y - 5.08$

61. $53.7t - 10.1 = 46.3t + 4.7$

62. $37.1w + 12.2 = 16.8w + 93.4$

63. $2.1x - 4.6 = 7.3x - 11.36$

64. $4.1y + 5.7 = 6.4y + 0.87$

65. $0.06x + 0.09(100 - x) = 8.85$

66. $0.08(1{,}000 - x) + 0.6x = 72.72$

APPLICATIONS *Complete each solution.*

67. PETITION DRIVE On weekends, a college student works for a political organization, collecting signatures for a petition drive. Her pay is \$15 a day plus 30 cents for each signature she obtains. How many signatures does she have to collect to make \$60 a day?

Analyze the problem

- Her base pay is ___ dollars a day.
- She makes ___ cents for each signature.
- She wants to make ___ dollars a day.
- Find the number of ___ she needs to get.

Form an equation Let x = ______________________

____

We need to work in terms of the same units, so we write 30 cents as ___ .

If we multiply the pay per signature by the number of signatures, we get the money she makes just from collecting signatures. Therefore, ___ = total amount (in dollars) made from collecting signatures.

We can express the money she earns in a day in two ways.

Base pay	+	0.30	·	the number of signatures	is	60.
15	+				=	60

Solve the equation

15 + ___ = ___

___ = 45

$x = 150$

State the conclusion

____

Check the result

If she collects ___ signatures, she will make 0.30 · ___ = ___ dollars from signatures. If we add this to \$15, we get \$60. The answer checks.

68. HIGHWAY CONSTRUCTION A 12.8-mile highway is in its third and final year of construction. In the first year, 2.3 miles of the highway were completed. In the second year, 4.9 miles were finished. How many more miles of the highway need to be completed?

Analyze the problem

- The planned highway is ___ miles long.
- The 1st year, ___ miles were completed.
- The 2nd year, ___ miles were completed.
- Find the number of ___ yet to be completed.

A diagram will help us understand the problem.

12.8-mi highway

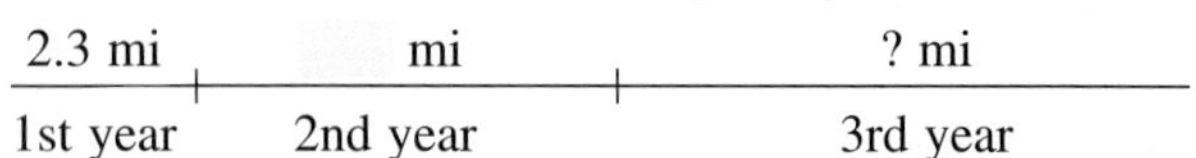

Form an equation

Let x = ______________________

We can express the length of the highway in two ways.

miles 1st year	+	miles 2nd year	+	the number of miles yet to be completed	is	12.8.
	+	4.9	+		=	12.8

Solve the equation

$2.3 + 4.9 + \quad = 12.8$

$\quad + \quad = 12.8$

$x =$

State the conclusion

Check the result

Add: $\quad + \quad + \quad = 12.8$. The answer checks.

Choose a variable to represent the unknown. Then write and solve an equation to answer the question.

69. DISASTER RELIEF After hurricane damage estimated at \$27.9 million, a county looked to three sources for relief. Local agencies contributed \$6.8 million toward the cleanup. A state emergency fund offered another \$12.5 million. When applying for federal government help, how much should the county ask for?

70. TELETHON Midway through a telethon, the donations had reached \$16.7 million. How much more was donated in the second half of the program if the final total pledged was \$30 million?

71. GPA After receiving her grades for the fall semester, a college student noticed that her overall GPA had dropped by 0.18. If her new GPA was 3.09, what was her GPA at the beginning of the fall semester?

72. MONTHLY PAYMENTS A food dehydrator offered on a home shopping channel can be purchased by making 3 equal monthly payments. If the price is \$113.25, how much is each monthly payment?

73. POINTS PER GAME As a senior, a college basketball player's scoring average was double that of her junior season. If she averaged 21.4 points a game as a senior, how many did she average as a junior?

74. NUTRITION One 3-ounce serving of broiled ground beef has 7 grams of saturated fat. This is 14 times the amount of saturated fat in 1 cup of cooked crab meat. How many grams of saturated fat are in 1 cup of cooked crab meat?

75. FUEL EFFICIENCY Each year, the Federal Highway Administration determines the number of vehicle-miles traveled in the country and divides it by the amount of fuel consumed to get an average miles per gallon (mpg). Illustration 1 shows how the figure has changed over the years to reach a high of 16.7 mpg in 1998. What was the average miles per gallon in 1960?

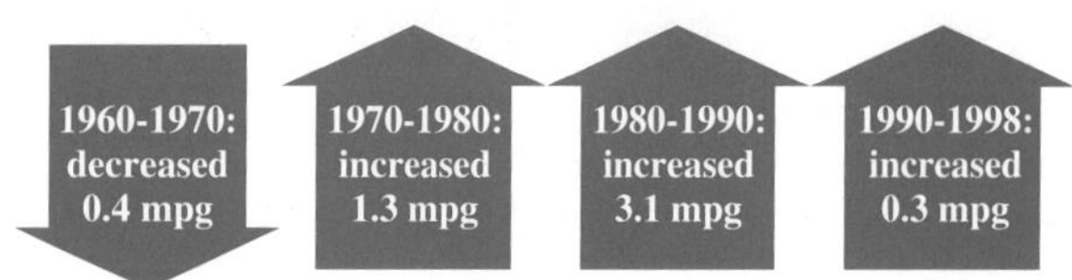

ILLUSTRATION 1

76. RATINGS REPORT Illustration 2 shows the prime-time television ratings for the week of January 3, 2000. If the Fox network ratings had been $\frac{1}{2}$ point higher, there would have been a three-way tie for second place. What prime time rating did Fox have that week?

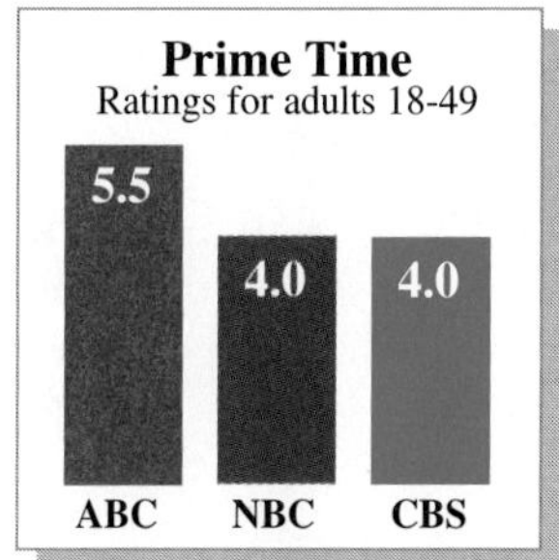

Source: Nielsen Media Research

ILLUSTRATION 2

77. CALLIGRAPHY A city honors its citizen of the year with a framed certificate. A calligrapher charges \$20 for the frame and then 15 cents a word for writing out the proclamation. If the city charter prohibits gifts in excess of \$50, what is the maximum number of words that can be printed on the award?

78. HELIUM BALLOONS The organizer of a jog-a-thon wants an archway of balloons constructed at the finish line of the race. A company charges a \$100 setup fee and then 8 cents for every balloon. How many balloons will be used if \$300 is spent for the decoration?

WRITING

79. Did you encounter any differences in solving equations containing decimals as compared to solving equations containing only integers? Explain your answer.

80. In the following case, why is it rather easy to apply the distributive property?

$$100(0.07x + 5.16)$$

REVIEW

81. Add: $-\frac{2}{3} + \frac{3}{4}$.

82. Add: $-\frac{2}{3} + \frac{1}{x}$.

83. Evaluate $x^3 - y^3$ for $x = -\frac{1}{2}$ and $y = -1$.

84. Multiply: $2\frac{1}{3} \cdot 4\frac{1}{2}$.

85. Evaluate: $\frac{-3 - 3}{-3 + 4}$.

86. Write a complex fraction using $-\frac{4}{5}$ and $\frac{1}{5}$.

5.7 Square Roots

In this section, you will learn about

- Square roots • Evaluating numerical expressions containing radicals
- Square roots of fractions and decimals
- Using a calculator to find square roots • Approximating square roots

INTRODUCTION. There are six basic operations of arithmetic. We have seen the relationships between addition and subtraction and between multiplication and division. In this section, we will explore the relationship between raising a number to a power and finding a root. Decimals will play an important role in this discussion.

Square roots

When we raise a number to the second power, we are squaring it, or finding its **square.**

The square of 6 is 36, because $6^2 = 36$.

The square of -6 is 36, because $(-6)^2 = 36$.

The **square root** of a given number is a number whose square is the given number. For example, the square roots of 36 are 6 and -6, because either number, when squared, yields 36. We can express this concept using symbols.

Square root

A number b is the **square root** of a if $b^2 = a$.

EXAMPLE 1 ***Finding square roots.*** Find the square roots of 49.

Solution

Ask yourself, "What number was squared to obtain 49?"

$7^2 = 49$ and $(-7)^2 = 49$

Thus, 7 and -7 are the square roots of 49.

Self Check

Find the square roots of 64.

Answers: 8 and -8

In Example 1, we saw that 49 has two square roots—one positive and one negative. The symbol $\sqrt{}$ is called a **radical sign** and is used to indicate a positive square root.

When a number, called the **radicand,** is written under a radical sign, we have a **radical expression.** Some examples of radical expressions are

$\sqrt{36}$ $\sqrt{100}$ $\sqrt{144}$ $\sqrt{81}$

To evaluate (or simplify) a radical expression, we need to find the positive square root of the radicand. For example, if we evaluate $\sqrt{36}$ (read as "the square root of 36"), the result is

$$\sqrt{36} = 6$$

because $6^2 = 36$. The negative square root of 36 is denoted $-\sqrt{36}$, and we have

$$-\sqrt{36} = -6$$

EXAMPLE 2 ***Evaluating radical expressions.*** Simplify each expression: **a.** $\sqrt{81}$ and **b.** $-\sqrt{100}$.

Solution

a. $\sqrt{81}$ means the positive square root of 81. Since $9^2 = 81$, we get $\sqrt{81} = 9$.

b. $-\sqrt{100}$ means the negative square root of 100. Since $10^2 = 100$, we get $-\sqrt{100} = -10$.

Self Check

Simplify each radical expression:

a. $\sqrt{144}$ and **b.** $-\sqrt{81}$.

Answers: **a.** 12, **b.** −9 ■

COMMENT Radical expressions such as

$$\sqrt{-36} \qquad \sqrt{-100} \qquad \sqrt{-144} \qquad \sqrt{-81}$$

do not represent real numbers. This is because there are no real numbers that, when squared, yield a negative number.

Be careful to note the difference between pairs of expressions such as $-\sqrt{36}$ and $\sqrt{-36}$. We have seen that $-\sqrt{36}$ does have meaning: $-\sqrt{36} = -6$. On the other hand, $\sqrt{-36}$ does not have meaning as a real number.

Evaluating numerical expressions containing radicals

Numerical expressions can contain radical expressions. When applying the rules for the order of operations, we treat a radical expression as we would a power.

EXAMPLE 3 ***Evaluating expressions containing radicals.*** Evaluate: **a.** $\sqrt{64} + \sqrt{9}$ and **b.** $-\sqrt{25} - \sqrt{4}$.

Solution

a. $\sqrt{64} + \sqrt{9} = 8 + 3$ — Evaluate each radical expression first.
$= 11$ — Do the addition.

b. $-\sqrt{25} - \sqrt{4} = -5 - 2$ — Evaluate each radical expression first.
$= -7$ — Do the subtraction.

Self Check

Evaluate each expression:

a. $\sqrt{121} + \sqrt{1}$ and

b. $-\sqrt{9} - \sqrt{16}$.

Answers: **a.** 12, **b.** −7 ■

EXAMPLE 4 ***Evaluating expressions containing radicals.*** Evaluate **a.** $6\sqrt{100}$ and **b.** $-5\sqrt{16} + 3\sqrt{9}$.

Solution

a. We note that $6\sqrt{100}$ means $6 \cdot \sqrt{100}$.

$6\sqrt{100} = 6(10)$ — Simplify the radical first.
$= 60$ — Do the multiplication.

b. $-5\sqrt{16} + 3\sqrt{9} = -5(4) + 3(3)$ — Simplify each radical first.
$= -20 + 9$ — Do the multiplications.
$= -11$ — Do the addition.

Self Check

Evaluate each expression:

a. $8\sqrt{121}$ and

b. $-6\sqrt{25} + 2\sqrt{36}$.

Answers: **a.** 88, **b.** −18 ■

Square roots of fractions and decimals

So far, we have found square roots of whole numbers. We can also find square roots of fractions and decimals.

EXAMPLE 5 *Square roots of fractions and decimals.*

Simplify **a.** $\sqrt{\frac{25}{64}}$ and **b.** $\sqrt{0.81}$.

Solution

a. $\sqrt{\frac{25}{64}} = \frac{5}{8}$, because $\left(\frac{5}{8}\right)^2 = \frac{25}{64}$.

b. $\sqrt{0.81} = 0.9$, because $(0.9)^2 = 0.81$.

Self Check

Simplify

a. $\sqrt{\frac{16}{49}}$ and **b.** $\sqrt{0.04}$.

Answer: **a.** $\frac{4}{7}$, **b.** 0.2 ■

Using a calculator to find square roots

We can also use a calculator to find square roots.

Accent on Technology: **Finding a square root**

We use the $\boxed{\sqrt{\ }}$ key (square root key) on a scientific calculator to find square roots. For example, to find $\sqrt{729}$, we enter these numbers and press these keys.

Keystrokes 729 $\boxed{\sqrt{\ }}$ | 27 |

We have found that $\sqrt{729} = 27$. To check this result, we need to square 27. This can be done by entering 27 and pressing the $\boxed{x^2}$ key. We obtain 729. Thus, 27 is the square root of 729.

Approximating square roots

Numbers whose square roots are whole numbers are called **perfect squares.** The perfect squares that are less than or equal to 100 are

0, 1, 4, 9, 16, 25, 36, 49, 64, 81, 100

To find the square root of a number that is not a perfect square, we can use a calculator. For example, to find $\sqrt{17}$, we enter these numbers and press the square root key.

17 $\sqrt{\ }$

The display reads 4.123105626. This result is not exact, because $\sqrt{17}$ is a **nonterminating decimal** that never repeats. $\sqrt{17}$ is an **irrational number.** Together, the rational and the irrational numbers form the set of **real numbers.** If we round to the nearest thousandth, we have

$\sqrt{17} \approx 4.123$ Read $\approx$ as "is approximately equal to."

EXAMPLE 6 *Approximating square roots.*

Use a scientific calculator to find each square root. Round to the nearest hundredth.

a. $\sqrt{373}$ **b.** $\sqrt{56.2}$ **c.** $\sqrt{0.0045}$

Solution

a. From the calculator, we get $\sqrt{373} \approx 19.31320792$. Rounding to the nearest hundredth, $\sqrt{373}$ is 19.31.

b. From the calculator, we get $\sqrt{56.2} \approx 7.496665926$. Rounding to the nearest hundredth, $\sqrt{56.2}$ is 7.50.

c. From the calculator, we get $\sqrt{0.0045} \approx 0.067082039$. Rounding to the nearest hundredth, $\sqrt{0.0045}$ is 0.07.

Self Check

Use a scientific calculator to find each square root. Round to the nearest hundredth.

a. $\sqrt{607.8}$

b. $\sqrt{0.076}$

Answers: **a.** 24.65, **b.** 0.28 ■

STUDY SET Section 5.7

VOCABULARY *Fill in the blanks to make the statements true.*

1. When we find what number is squared to obtain a given number, we are finding the square ____________ of the given number.

2. Whole numbers such as 25, 36, and 49 are called ____________ squares because their square roots are whole numbers.

3. The symbol $\sqrt{\ }$ is called a ____________ sign. It indicates that we are to find a ____________ square root.

4. The decimal number that represents $\sqrt{17}$ is a ____________ decimal—it never ends.

5. In $\sqrt{26}$, 26 is called the ____________.

6. The symbol $\approx$ means ____________.

CONCEPTS *In Exercises 7–12, fill in the blanks to make the statements true.*

7. The square of 5 is ____, because $(5)^2 =$ ____.

8. The square of $\frac{1}{4}$ is ____, because $\left(\frac{1}{4}\right)^2 =$ ____.

9. The two square roots of 49 are 7 and -7, because $7^2 = 49$ and -7^2 $= 49$.

10. The two square roots of 4 are 2 and -2, because ____ $= 4$ and ____ $= 4$.

11. Since $\left(\frac{3}{4}\right)^2 = \frac{9}{16}$, we know that $\sqrt{\frac{9}{16}} =$ ____.

12. Since $(0.4)^2 = 0.16$, we know that $\sqrt{0.16} =$ ____.

13. Without evaluating the following square roots, write them in order, from smallest to largest: $\sqrt{23}$, $\sqrt{11}$, $\sqrt{27}$, $\sqrt{6}$.

14. Without evaluating the following square roots, write them in order from smallest to largest: $-\sqrt{13}$, $-\sqrt{5}$, $-\sqrt{17}$, $-\sqrt{37}$.

15. Simplify.

a. $\sqrt{1}$ **b.** $\sqrt{0}$

16. Multiplication can be thought of as the opposite of division. What is the opposite of finding the square root of a number?

In Exercises 17–22, use a calculator.

17. a. Use a calculator to approximate $\sqrt{6}$ to the nearest tenth.

b. Square the result from part a.

c. Find the difference between 6 and the answer to part b.

18. a. Use a calculator to approximate $\sqrt{6}$ to the nearest hundredth.

b. Square the result from part a.

c. Find the difference between the answer to part b and 6.

19. Graph $\sqrt{9}$ and $-\sqrt{5}$ on the number line.

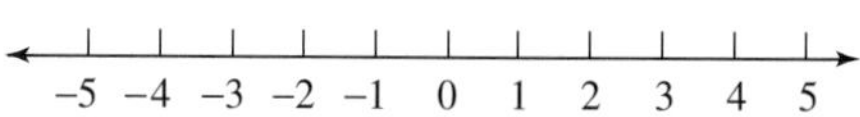

20. Graph $-\sqrt{3}$ and $\sqrt{7}$ on the number line.

−5 −4 −3 −2 −1 0 1 2 3 4 5

21. Between what two whole numbers would each square root be located when graphed on a number line?

a. $\sqrt{19}$ **b.** $\sqrt{87}$

22. Between what two whole numbers would each square root be located when graphed on a number line?

a. $\sqrt{50}$ **b.** $\sqrt{33}$

NOTATION *Complete each solution.*

23. Simplify: $-\sqrt{49} + \sqrt{64}$.

$$-\sqrt{49} + \sqrt{64} = \_\_\_ + \_\_\_$$
$$= 1$$

24. Simplify: $2\sqrt{100} - 5\sqrt{25}$.

$$2\sqrt{100} - 5\sqrt{25} = 2(\_\_\_) - 5(\_\_\_)$$
$$= \_\_\_ - 25$$
$$= -5$$

PRACTICE *Simplify each expression without using a calculator.*

25. $\sqrt{16}$ **26.** $\sqrt{64}$

27. $-\sqrt{121}$ **28.** $-\sqrt{144}$

29. $-\sqrt{0.49}$ **30.** $-\sqrt{0.64}$

31. $\sqrt{0.25}$ **32.** $\sqrt{0.36}$

33. $\sqrt{0.09}$ **34.** $\sqrt{0.01}$

35. $-\sqrt{\frac{1}{81}}$ **36.** $-\sqrt{\frac{1}{4}}$

37. $-\sqrt{\frac{16}{9}}$ **38.** $-\sqrt{\frac{64}{25}}$

39. $\sqrt{\frac{4}{25}}$ **40.** $\sqrt{\frac{36}{121}}$

41. $5\sqrt{36} + 1$ **42.** $2 + 6\sqrt{16}$

43. $-4\sqrt{36} + 2\sqrt{4}$ **44.** $-6\sqrt{81} + 5\sqrt{1}$

45. $\sqrt{\frac{1}{16}} - \sqrt{\frac{9}{25}}$

46. $\sqrt{\frac{25}{9}} - \sqrt{\frac{64}{81}}$

47. $5(\sqrt{49})(-2)$

48. $(-\sqrt{64})(-2)(3)$

49. $\sqrt{0.04} + 2.36$

50. $\sqrt{0.25} + 4.7$

51. $-3\sqrt{1.44}$

52. $-2\sqrt{1.21}$

Use a calculator to complete each square root table. Round to the nearest thousandth when an answer is not exact.

53.

Number	Square root
1	
2	
3	
4	
5	
6	
7	
8	
9	
10	

54.

Number	Square root
10	
20	
30	
40	
50	
60	
70	
80	
90	
100	

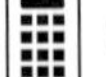

Use a calculator to simplify each of the following.

55. $\sqrt{1,369}$

56. $\sqrt{841}$

57. $\sqrt{3,721}$

58. $\sqrt{5,625}$

Use a calculator to approximate each of the following to the nearest hundredth.

59. $\sqrt{15}$

60. $\sqrt{51}$

61. $\sqrt{66}$

62. $\sqrt{204}$

Use a calculator to approximate each of the following to the nearest thousandth.

63. $\sqrt{24.05}$

64. $\sqrt{70.69}$

65. $-\sqrt{11.1}$

66. $\sqrt{0.145}$

Use a calculator to evaluate each radical expression. If an answer is not exact, round to the nearest ten thousandth.

67. $\sqrt{24,000,201}$

68. $-\sqrt{4.012009}$

69. $-\sqrt{0.00111}$

70. $\sqrt{\frac{27}{44}}$

APPLICATIONS *In Exercises 71–76, square roots have been used to express various lengths. Solve each problem by simplifying any square roots. You may need to use a calculator. If so, round to the nearest tenth.*

71. CARPENTRY Find the length of the slanted side of each roof truss.

a.

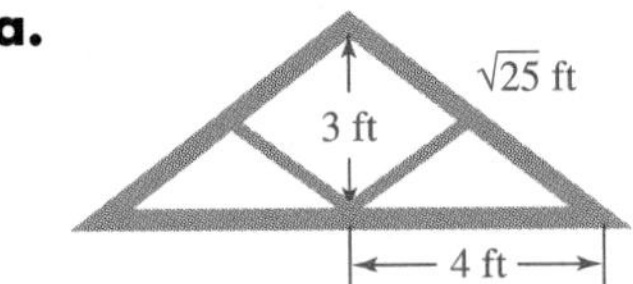

b.

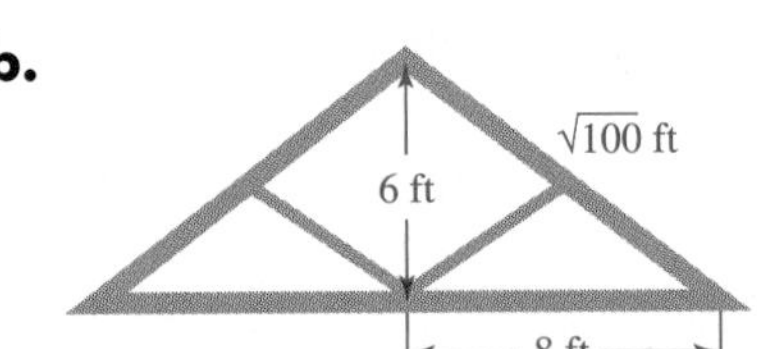

72. RADIO ANTENNA See Illustration 1. How far from the base of the antenna is each guy wire anchored to the ground? (The measurements are in feet.)

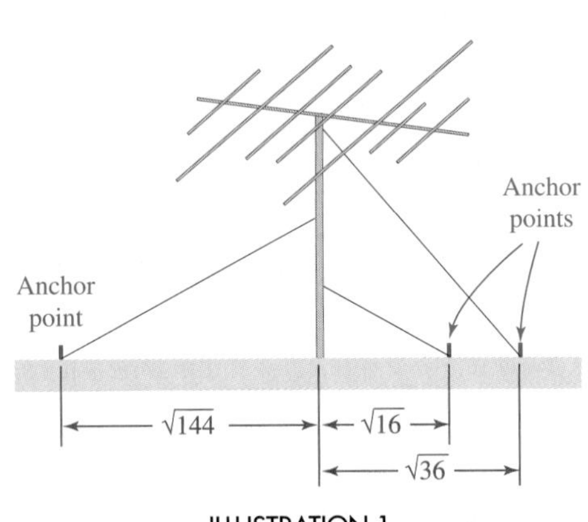

ILLUSTRATION 1

73. BASEBALL DIAMOND Illustration 2 shows some dimensions of a major league baseball field. How far is it from home plate to second base?

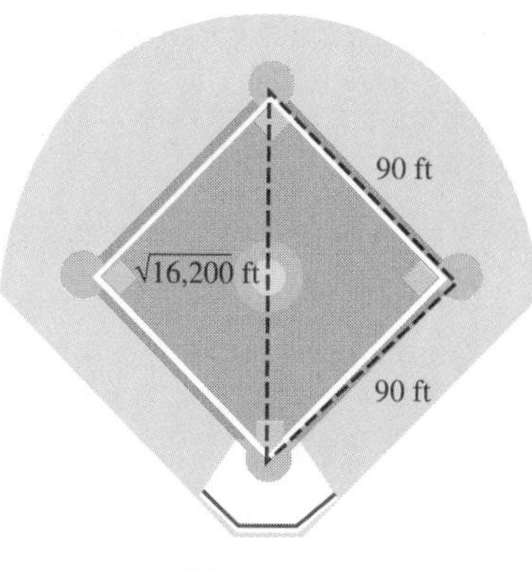

ILLUSTRATION 2

74. SURVEYING Use the imaginary triangles set up by a surveyor to find the length of each lake. (The measurements are in meters.)

a.

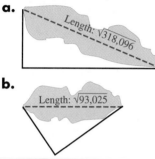

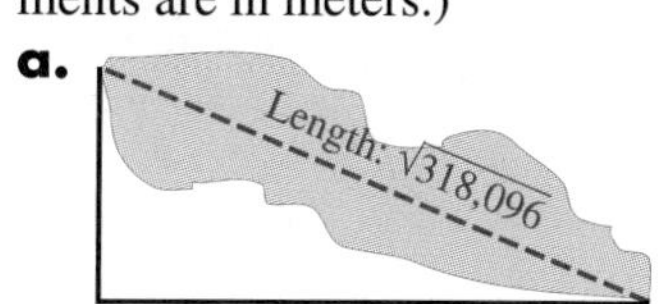

b.

Length: $\sqrt{93,025}$

75. BIG-SCREEN TELEVISION The picture screen on a television set is measured diagonally. What size screen is shown in Illustration 3?

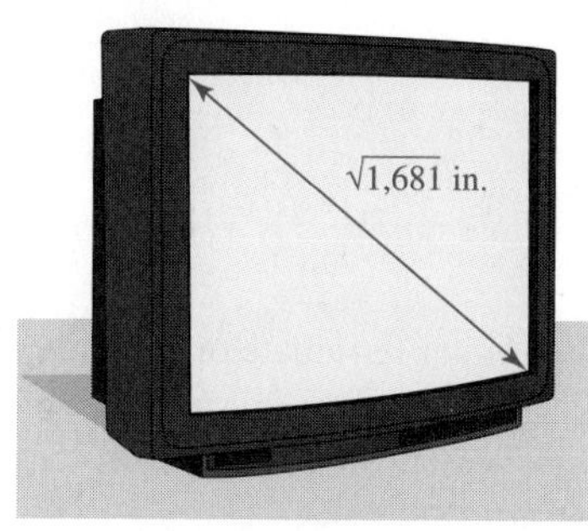

ILLUSTRATION 3

76. LADDER A painter's ladder is shown in Illustration 4. How long are the legs of the ladder?

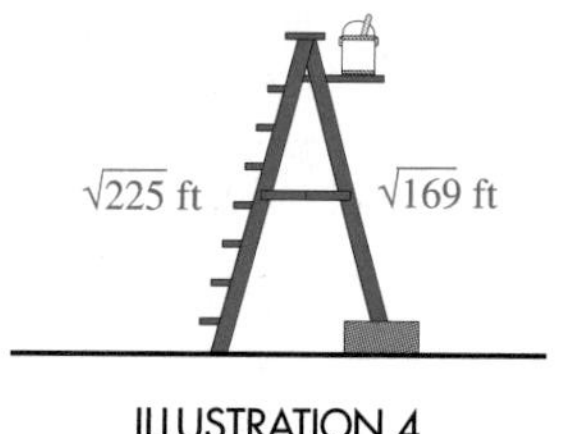

ILLUSTRATION 4

WRITING

77. When asked to find $\sqrt{16}$, one student's answer was 8. Explain his misunderstanding of the concept of square root.

78. Explain the difference between the square and the square root of a number.

79. What is a nonterminating decimal? Use an example in your explanation.

80. What do you think might be meant by the term *cube root*?

81. Explain why $\sqrt{-4}$ does not have meaning as a real number.

82. Is there a difference between $-\sqrt{25}$ and $\sqrt{-25}$? Explain.

REVIEW

83. When solving the equation $2x - 5 = 11$, what operations must be undone in order to isolate the variable?

84. Simplify: $\dfrac{\frac{-2}{3}}{8}$.

85. Evaluate: $5(-2)^2 - \dfrac{16}{4}$.

86. Translate to mathematical symbols: four less than twice x.

87. List the set of whole numbers.

88. Tell whether y is used as a factor or as a term: $5x + y + 3x$.

89. Solve: $8 + \dfrac{a}{5} = 14$.

90. Insert the proper symbol, $<$ or $>$, in the blank to make a true statement: -15 ____ -14.

THE REAL NUMBERS

A **real number** is any number that can be expressed as a decimal. The set of real numbers corresponds to all points on a number line.

Graph each real number on the number line.

1. $\left\{-4, \frac{13}{4}, -0.1, \frac{99}{100}, \sqrt{17}, -2\frac{1}{2}, 1.\overline{3}\right\}$

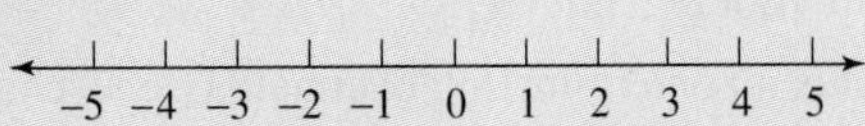

2. 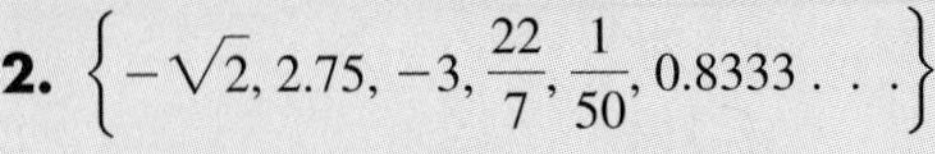
$\left\{-\sqrt{2}, 2.75, -3, \frac{22}{7}, \frac{1}{50}, 0.8333\ldots\right\}$

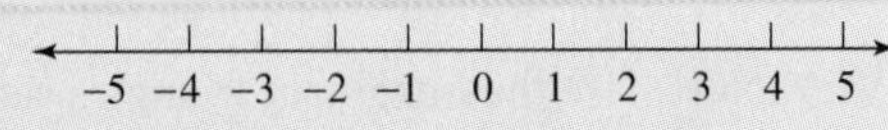

Almost all the types of numbers we have discussed in this book are real numbers. As we have seen, the set of real numbers is made up of several subsets of numbers.

If possible, list the numbers belonging to each set. If it is not possible to list them, define the set in words.

3. Natural numbers

4. Whole numbers

5. Integers

6. Rational numbers

7. Irrational numbers

The diagram below shows how the set of real numbers is made up of two distinct sets: the rational and the irrational numbers. Since every natural number is a whole number, we show the set of natural numbers included in the whole numbers. Because every whole number is an integer, the whole numbers are shown contained in the integers. Since every integer is a rational number, we show the integers included in the rational numbers.

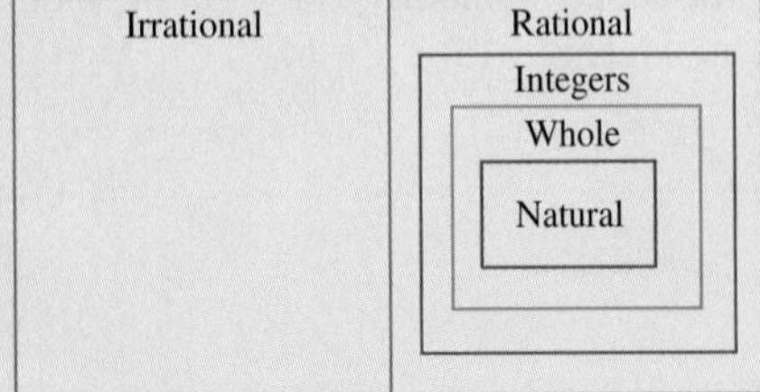

In Exercises 8–17, tell whether each statement is true or false.

8. Every integer is a real number.

9. Every fraction can be written as a terminating decimal.

10. Every real number is a whole number.

11. Some irrational numbers are integers.

12. Some rational numbers are natural numbers.

13. No numbers are both rational and irrational numbers.

14. All real numbers can be graphed on a number line.

15. The set of whole numbers is a subset of the irrational numbers.

16. All decimals either terminate or repeat.

17. Every natural number is an integer.

18. List the numbers in the set $\{-2, -1.2, -\frac{7}{8}, 0, 1\frac{2}{3}, 2.75, \sqrt{23}, 10, 1.161661666\ldots\}$ that are

a. Natural numbers

b. Whole numbers

c. Integers

d. Rational numbers

e. Irrational numbers

f. Real numbers

19. $\sqrt{-25}$ is not a real number. Explain why.

KEY CONCEPT

Simplify and Solve

Two of the instructions most often used in this book are **simplify** and **solve.** In algebra, we *simplify expressions* and *solve equations.*

To simplify an expression, we write it in a more usable form. To do so, we apply the rules of arithmetic as well as algebraic concepts such as the fundamental property of fractions, combining like terms, the distributive property, and the properties of 0 and 1.

To solve an equation means to find the numbers that make the equation true when substituted for its variable. The addition, subtraction, multiplication, and division properties of equality are used to solve equations. We must often simplify expressions on the left and right sides of the equation.

Use the procedures and properties that we have studied to simplify the expression in part a and solve the equation in part b.

Simplify	**Solve**
1. a. $-2x + 3 + 7x - 11$	**b.** $-2x + 3 + 7x - 11 = 7$
2. a. $\frac{x}{2} + \frac{1}{3}$	**b.** $\frac{x}{2} + \frac{1}{3} = \frac{1}{6}$
3. a. $3(0.2y - 1.6) + 0.6y$	**b.** $3(0.2y - 1.6) + 0.6y = -6.6$

ACCENT ON TEAMWORK

Section 5.1

ROUNDING

a. Find all the three-digit numbers that round to 4.7.

b. Find all the four-digit numbers that round to 8.09.

c. Find all the two-digit numbers that round to 0.0.

Section 5.2

VISUAL MODELS Shade a grid like the one in Illustration 1 to compute each addition or subtraction.

a. $0.62 + 0.24$ **b.** $0.45 - 0.41$

c. $0.21 + 0.29$ **d.** $0.98 - 0.18$

e. $0.2 + 0.17$ **f.** $0.57 - 0.3$

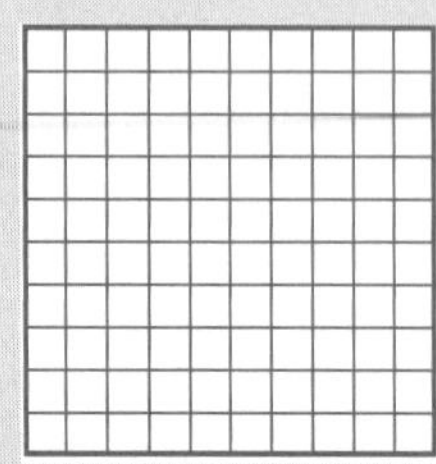

ILLUSTRATION 1

Section 5.3

SEQUENCES Multiplication by 2 is used to form the terms of the sequence 3, 6, 12, 24, 48, 96, ... That is, to form the second term (which is 6), we multiply the first term (which is 3) by 2. To get the third term (which is 12), we multiply the second term by 2. To get the fourth term, we multiply the third term by 2, and so on.

What multiplication is used to form the terms of each of the following sequences?

a. 0.2134, 2.134, 21.34, 213.4, 2,134, 21,340, . . .

b. 0.00005, 0.005, 0.5, 50, 5,000, . . .

c. 3, 0.9, 0.27, 0.081, 0.0243, 0.00729, . . .

d. 0.7, 0.07, 0.007, 0.0007, 0.00007, 0.000007, . . .

Section 5.4

GPA We can use the following steps to calculate a semester grade point average (GPA).

1. Convert each letter grade received to its equivalent point value.

Letter grade	A	B	C	D	F
Grade point value	4.0	3.0	2.0	1.0	0.0

2. For each class, multiply the number of credit hours (units) the class is worth by its point value found in Step 1.

3. Add each of the results from Step 2.

4. Divide the result from Step 3 by the total number of course credit hours (units) taken that semester. Round to the nearest tenth.

Find the GPA for the student whose grade report is shown in Illustration 2.

Course no.	Course title	Units	Grade
101	Intro. Accounting	5.0	C
201	Intro. Psych	3.0	A
102	Spanish II	4.0	B
142	Swimming	1.0	D

ILLUSTRATION 2

Section 5.5

EQUIVALENT DECIMALS A student was asked to write several fractions as decimals. His answers, which are all incorrect, are shown below. What was he doing wrong?

$\frac{2}{5} = 2.5 \qquad \frac{4}{15} = 3.75 \qquad \frac{3}{4} = 1.\overline{3}$

Section 5.6

SOLVING EQUATIONS If an equation contains decimals, we can multiply both sides by a power of 10 to clear the equation of decimals. Multiply both sides of each equation by the appropriate power of 10; then solve it.

a. $0.6x = 3.6$ **b.** $0.8x + 0.4 = 5.2$

c. $0.62x + 1.24 = 5.58$

Section 5.7

A SPIRAL OF ROOTS To do this project, you will need a blank piece of paper, a ruler, a 3×5 index card, and a pencil. Begin by drawing a triangle with two sides 1 inch long, as shown in Illustration 3. Use the corner of the 3×5 card to help draw the "sharp corner" (90-degree angle) of the triangle. Then draw the dashed blue line to complete the first triangle. It is $\sqrt{2}$ inches long.

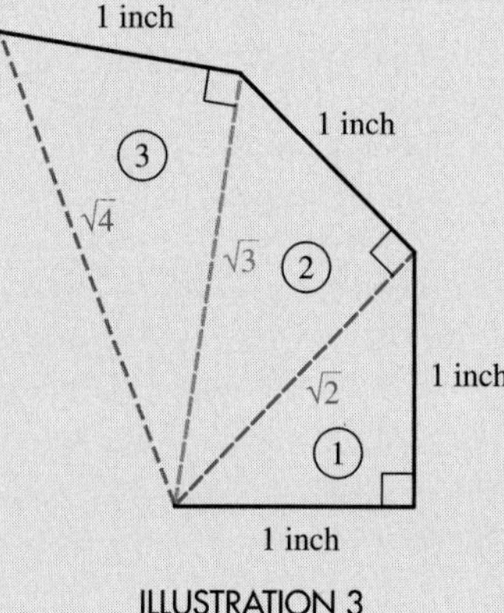

ILLUSTRATION 3

Next, create a second triangle, using one side of the first triangle and drawing another side 1 inch long as shown. Complete the second triangle by drawing the dashed green line. It is $\sqrt{3}$ inches long. Draw a third triangle in a similar fashion. The dashed purple line is $\sqrt{4} = 2$ inches long. Draw a fourth triangle, a fifth triangle, and so on. If the pattern continues, what is the length of the dashed side of each new triangle?

CHAPTER REVIEW

SECTION 5.1 An Introduction to Decimals

CONCEPTS

Decimal notation is used to denote part of a whole.

Expanded notation is used to show the value represented by each digit in the *decimal numeration system.*

$$5.6791 = 5 + \frac{6}{10} + \frac{7}{100} + \frac{9}{1{,}000} + \frac{1}{10{,}000}$$

REVIEW EXERCISES

1. Represent the amount of the square that is shaded in Illustration 1, using a decimal and a fraction.

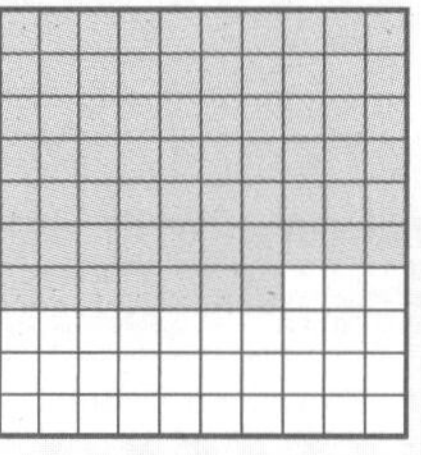

ILLUSTRATION 1

2. In Illustration 2, shade 0.8 of the rectangle.

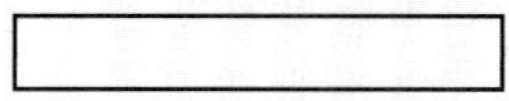

ILLUSTRATION 2

To express a decimal in words, say:

1. the whole number to the left of the decimal point;
2. "and" for the decimal point;
3. the whole number to the right of the decimal point, followed by the name of the last place value column on the right.

3. Write 16.4523 in expanded notation.

4. Write each decimal in words and then as a fraction or mixed number.

a. 2.3 **b.** −15.59

c. 0.0601 **d.** 0.00001

5. Graph 1.55, −0.8, and −2.7 on a number line.

To compare the size of two decimals, compare the digits of each decimal, column by column, working from left to right.

6. VALEDICTORIAN At the end of the school year, the five students listed in Illustration 3 were in the running to be class valedictorian. Rank the students in order by GPA, beginning with the valedictorian.

Name	GPA
Diaz, Cielo	3.9809
Chou, Wendy	3.9808
Washington, Shelly	3.9865
Gerbac, Lance	3.899
Singh, Amani	3.9713

ILLUSTRATION 3

A decimal point and additional zeros may be written to the right of a whole number.

7. True or false: 78 = 78.0.

8. Place the proper symbol (<, >, or =) in the blank to make a true statement.

a. 4.5 4.6 **b.** −2.35 −2.53

c. 10.90 10.9 **d.** 0.027894 0.034

To round a decimal, locate the rounding digit and the test digit.

1. If the test digit is a number less than 5, drop it and all digits to the right of the rounding digit.
2. If it is 5 or larger, add 1 to the rounding digit and drop all digits to its right.

9. Round each decimal to the specified place-value column.

a. 4.578: hundredths
b. 3,706.0895: thousandths
c. −0.0614: tenths
d. 88.12: tenths

SECTION 5.2 *Addition and Subtraction with Decimals*

To add (or subtract) decimals:

1. Line up their decimal points.
2. Add (or subtract) as you would with whole numbers.
3. Write the decimal point in the result directly below the decimal points of the problem.

10. Do each addition or subtraction.

a. $19.5 + 34.4 + 12.8$
b. $3.4 + 6.78 + 35 + 0.008$
c. $68.47 - 53.3$
d. $45.08 - 17.37$

11. Evaluate each expression.

a. $-16.1 + 8.4$
b. $-4.8 - (-7.9)$
c. $-3.55 + (-1.25)$
d. $-15.1 - 13.99$
e. $-8.8 + (-7.3 - 9.5)$
f. $(5 - 0.096) - (-0.035)$

12. SALE PRICE A calculator normally sells for \$52.20. If it is being discounted \$3.99, what is the sale price?

13. MICROWAVE OVEN A microwave oven is shown in Illustration 4. How tall is the window?

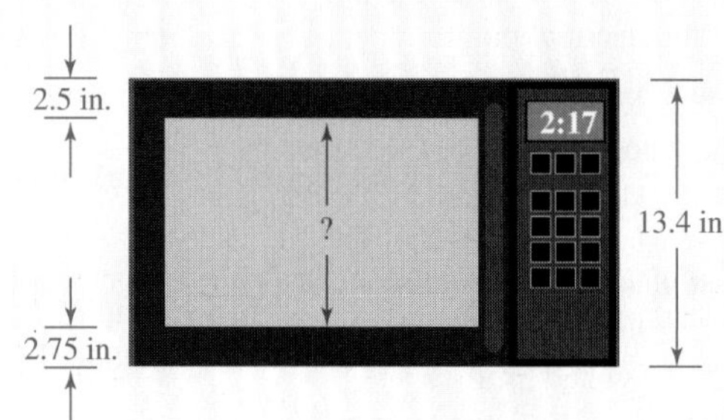

ILLUSTRATION 4

SECTION 5.3 *Multiplication with Decimals*

To multiply decimals:

1. Multiply as if working with whole numbers.
2. Place the decimal point in the result so that the answer has the same number of decimal places as the total number of decimal places of the factors.

14. Do each multiplication.

a. $(-0.6)(0.4)$
b. $2.3 \cdot 0.9$
c. $5.5(-3.1)$
d. $32.45(6.1)$
e. $(-0.003)(-0.02)$
f. $7 \cdot 0.6$

To multiply a decimal by a power of 10, move the decimal point to the right the same number of places as there are zeros in the power of 10.

15. Do each multiplication in your head.

a. $1{,}000(90.1452)$
b. $(-10)(-2.897)(100)$

Exponents are used to represent repeated multiplication.

16. Find each power.

a. $(0.2)^2$ **b.** $(-0.15)^2$ **c.** $(3.3)^2$ **d.** $(0.1)^3$

17. Evaluate each expression.

a. $(0.6 + 0.7)^2 - 12.3$ **b.** $3(7.8) + 2(1.1)^2$

To evaluate an algebraic expression, substitute specific numbers for the variables in the expression and apply the rules for the order of operations.

18. Evaluate the algebraic expression $2pr^2 - h$ for $p = 3.14$, $r = 4$, and $h = 8.1$.

19. WORD PROCESSOR The Page Setup screen for a word processor is shown in Illustration 5. Find the area that can be filled with text on an 8.5-inch-by-11-inch piece of paper if the margins are set as shown.

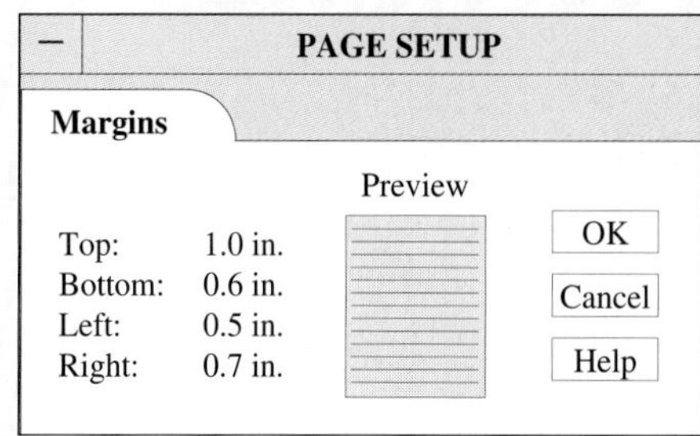

ILLUSTRATION 5

20. AUTO PAINTING A manufacturer uses a three-part process to finish the exterior of the cars it produces.

Step 1: A 0.03-inch-thick rust-prevention undercoat is applied.

Step 2: Three layers of color coat, each 0.015 of an inch thick, are sprayed on.

Step 3: The finish is then buffed down, losing 0.005 of an inch of its thickness.

What is the resulting thickness of the automobile's finish?

SECTION 5.4 *Division with Decimals*

To divide a decimal by a whole number:

1. Divide as if working with whole numbers.
2. Write the decimal point in the result directly above the decimal point in the dividend.

To divide by a decimal:

1. Move the decimal point in the divisor so that it becomes a whole number.
2. Move the decimal point in the dividend the same number of places to the right.
3. Use the process for dividing a decimal by a whole number.

21. Do each division.

a. $12\overline{)15}$ **b.** $-41.8 \div 4$ **c.** $\dfrac{-29.67}{-23}$ **d.** $24.618 \div 6$

22. Do each division.

a. $12.47 \div (-4.3)$ **b.** $\dfrac{0.0742}{1.4}$

c. $\dfrac{15.75}{0.25}$ **d.** $\dfrac{-0.03726}{-0.046}$

23. Divide and round each result to the nearest tenth.

a. $78.98 \div 6.1$ **b.** $\dfrac{-5.338}{0.008}$

24. Evaluate the algebraic expression $\dfrac{5(F - 32)}{9}$ for $F = 68.4$ and round to the nearest hundredth.

25. THANKSGIVING DINNER The cost of purchasing the ingredients for a Thanksgiving turkey dinner for a family of 5 was \$41.70. What was the cost of the dinner per person?

To divide a decimal by a power of 10, move the decimal point to the left the same number of places as there are zeros in the power of 10.

26. Do each division in your head.

a. $89.76 \div 100$ **b.** $\dfrac{0.0112}{-10}$

27. Evaluate the numerical expression $\dfrac{(1.4)^2 + 2(4.6)}{0.5 + 0.3}$.

28. SERVING SIZE Illustration 6 shows the package labeling on a box of children's cereal. Use the information given to find the number of servings.

Nutrition Facts	
Serving size	1.1 ounce
Servings per container	?
Package weight	15.5 ounces

ILLUSTRATION 6

29. TELESCOPE To change the position of a focusing mirror on a telescope, an adjustment knob is used. The mirror moves 0.025 inch with each revolution of the knob. The mirror needs to be moved 0.2375 inch to improve the sharpness of the image. How many revolutions of the adjustment knob does this require?

The three types of *averages*:

1. *Mean* $= \dfrac{\text{sum of values}}{\text{number of values}}$
2. The *median* is the middle value. For an even number of values, add the middle two and divide by 2.
3. The *mode* is the value that occurs most often.

30. BLOOD SAMPLES A medical laboratory technician examined a blood sample under a microscope and measured the sizes (in microns) of the white blood cells. The data are listed below. Find the mean, median, and mode.

7.8 6.9 7.9 6.7 6.8 8.0 7.2 6.9 7.5

31. TOBACCO SETTLEMENT In November of 1998, the country's four largest tobacco companies reached an agreement with 46 states to pay $206.4 billion to cover public health costs related to smoking. The payments to each of the New England states are shown below. Find the median payment.

Connecticut	$3.63 billion	New Hampshire	$1.3 billion
Maine	$1.5 billion	Rhode Island	$1.4 billion
Massachusetts	$8.0 billion	Vermont	$0.81 billion

SECTION 5.5 *Fractions and Decimals*

To write a fraction as a decimal, divide the numerator by the denominator.

32. Write each fraction in decimal form.

a. $\dfrac{7}{8}$ **b.** $-\dfrac{2}{5}$ **c.** $\dfrac{9}{16}$ **d.** $\dfrac{3}{50}$

We obtain either a *terminating* or a *repeating* decimal when using division to write a fraction as a decimal.

33. Write each fraction in decimal form. Use an overbar.

a. $\dfrac{6}{11}$ **b.** $-\dfrac{2}{3}$

An overbar can be used instead of the three dots . . . to represent the repeating pattern in a repeating decimal.

34. Write each fraction in decimal form. Round to the nearest hundredth.

a. $\dfrac{19}{33}$ **b.** $\dfrac{31}{30}$

35. Place the proper symbol ($<$ or $>$) in the blank to make a true statement.

a. $\dfrac{13}{25}$ ___ 0.499 **b.** $-0.\overline{26}$ ___ $-\dfrac{4}{15}$

36. Graph $1\frac{1}{8}$, $-\frac{1}{3}$, $2\frac{3}{4}$, and $-\frac{9}{10}$ on the number line. Label each using its decimal equivalent.

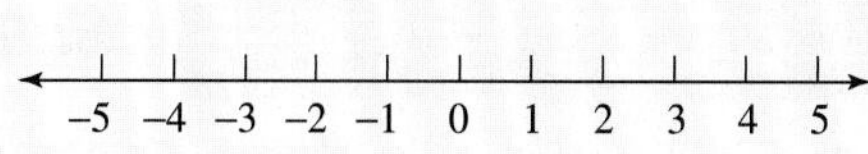

37. Evaluate each numerical expression. Find the exact answer.

a. $\frac{1}{3} + 0.4$

b. $\frac{4}{5}(-7.8)$

c. $\frac{1}{2}(9.7 + 8.9)(10)$

d. $\frac{1}{3}(3.14)(3)^2(4.2)$

38. Evaluate $\frac{4}{3}(3.14)(2)^3$. Round the result to the nearest hundredth.

39. ROADSIDE EMERGENCY In case of trouble, truckers carry reflectors to be placed on the highway shoulder to warn approaching cars of a stalled vehicle. (See Illustration 7.) What is the area of one of these triangular reflectors?

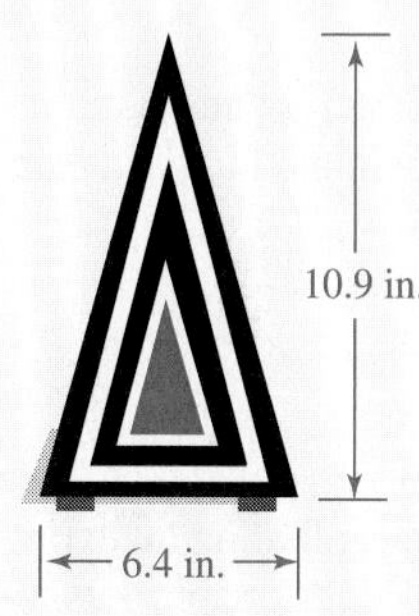

ILLUSTRATION 7

SECTION 5.6 Solving Equations Containing Decimals

The strategy for solving equations containing decimals:

To simplify the equation

1. Use the distributive property to remove any parentheses.
2. Combine any like terms.

To isolate the variable

3. Apply the addition or subtraction properties of equality to get the variable on one side.
4. Continue to combine like terms.
5. Undo the operations of multiplication and division to isolate the variable.

The five-step problem-solving strategy:

1. Analyze the problem.
2. Form an equation.
3. Solve the equation.
4. State the conclusion.
5. Check the result.

40. Solve each equation.

a. $y + 12.4 = -6.01$

b. $0.23 + x = 5$

c. $\frac{x}{1.78} = -3$

d. $-16.1b = -27.37$

41. Is $r = -1.1$ a solution of $-1.3 + 1.2r = 2.4r + 0.02$?

42. Simplify each algebraic expression.

a. $5.7a - 12.4 - 2.9a$

b. $2(0.3t - 0.4) + 3(0.8t - 0.2)$

43. Solve each equation.

a. $1.7y + 1.24 = -1.4y - 0.62$

b. $0.05(1{,}000 - x) + 0.9x = 60.2$

44. BOWLING If it costs $1.45 to rent shoes and 95 cents a game to use a lane, how many games can be bowled for $10?

SECTION 5.7 Square Roots

The number b is a *square root* of a if $b^2 = a$.

45. Fill in the blanks to make the statement true: Two square roots of 64 are 8 and -8, because ____ $= 64$ and ____ $= 64$.

A *radical sign* $\sqrt{}$ is used to indicate a positive square root. The square root of a *perfect square* is a whole number.

46. Simplify each expression without using a calculator.

a. $\sqrt{49}$ **b.** $-\sqrt{16}$ **c.** $\sqrt{100}$ **d.** $\sqrt{0.09}$

e. $\sqrt{\frac{64}{25}}$ **f.** $\sqrt{0.81}$ **g.** $-\sqrt{\frac{1}{36}}$ **h.** $\sqrt{0}$

47. Between what two whole numbers would $\sqrt{83}$ be located when graphed on a number line?

A square root can be approximated using a calculator.

48. Use a calculator to find $\sqrt{11}$ and round to the nearest tenth. Now square the approximation. How close is it to 11?

49. Graph each square root on the number line: $\sqrt{3}$, $-\sqrt{2}$, and $\sqrt{0}$.

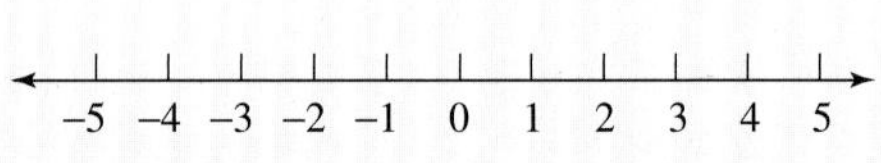

When evaluating an expression containing square roots, treat a radical as you would a power when applying the rules for the order of operations.

50. Evaluate each expression without using a calculator.

a. $-3\sqrt{100}$ **b.** $5\sqrt{0.25}$

c. $-3\sqrt{49} - \sqrt{36}$ **d.** $\sqrt{\frac{9}{100}} + \sqrt{1.44}$

51. Use a calculator to find each square root to the nearest hundredth.

a. $\sqrt{19}$ **b.** $\sqrt{59}$

Chapter 5 Test

1. Express the amount of the square in Illustration 1 that is shaded, using a fraction and a decimal.

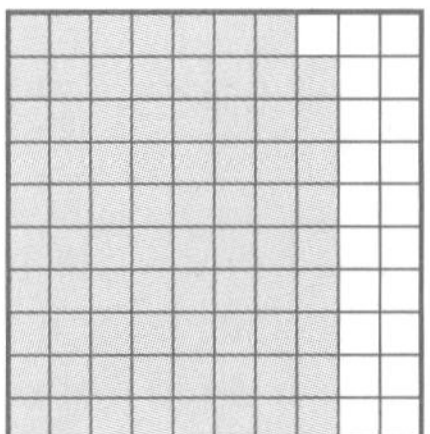

ILLUSTRATION 1

2. WATER PURITY   A county health department sampled the pollution content of tap water in 6 cities, with the results shown in Illustration 2. Rank the cities in order, from dirtiest tap water to cleanest.

City	Pollution parts per million
Monroe	0.0909
Covington	0.0899
Paston	0.0901
Cadia	0.0890
Selway	0.1001

ILLUSTRATION 2

3. Write 0.271 as a fraction.

4. Round to the nearest thousandth: 33.0495.

5. SKATING RECEIPTS   At an ice-skating complex, receipts on Friday were \$30.25 for indoor skating and \$62.25 for outdoor skating. On Saturday, the corresponding amounts were \$40.50 and \$75.75. Find the total receipts for the two days.

6. Do each operation in your head.

a. $567.909 \div 1{,}000$ **b.** $0.00458 \cdot 100$

7. EARTHQUAKE FAULT LINE   After an earthquake, geologists found that the ground on the west side of the fault line had dropped 0.83 inch. The next week, a strong aftershock caused the same area to sink 0.19 inch deeper. How far did the ground on the west side of the fault drop because of the seismic activity?

8. Do each operation.

a. $2 + 4.56 + 0.89 + 3.3$

b. $45.2 - 39.079$

c. $(0.32)^2$

d. $-6.7(-2.1)$

9. NEW YORK CITY   Central Park, which lies in the middle of Manhattan, is the city's best-known park. See Illustration 3. If it is 2.5 miles long and 0.5 mile wide, what is its area?

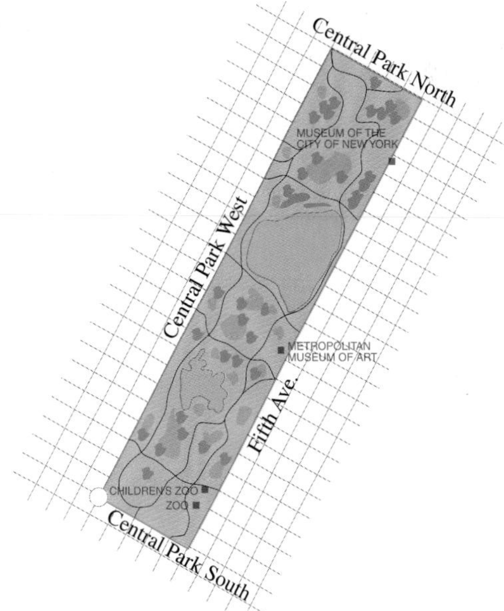

ILLUSTRATION 3

10. TELEPHONE BOOK   To print a telephone book, 565 sheets of paper were used. If the book is 2.3 inches thick, what is the thickness of each sheet of paper? (Round to the nearest thousandth of an inch.)

11. Evaluate $4.1 - (3.2)(0.4)^2$.

12. Write each fraction as a decimal.

a. $\frac{17}{50}$ **b.** $\frac{5}{12}$

13. Do the division and round to the nearest hundredth: $\dfrac{12.146}{-5.3}$.

14. Find $11\overline{)13}$.

15. RATINGS The seven top-rated cable television programs for the week of February 8–14 are given below. What are the mean, median, and mode of the ratings? Round to the nearest tenth.

Show/day/time/network	Rating
1. "WCW Monday," Mon. 9 p.m., TNT	4.5
2. "WCW Monday," Mon. 10 p.m., TNT	4.4
3. "WCW Monday," Mon. 8 p.m., TNT	3.9
4. "WWF Special," Sat. 9 p.m., USA	3.6
5. "WWF Wrestling," Sun. 7 p.m., USA	3.1
6. "Dog Show," Tues. 8 p.m., USA	3.1
7. "WWF Special," Sat. 8 p.m., USA	2.9

16. STATISTICS The graph in Illustration 4 has an asterisk * that refers readers to a note at the bottom. In your own words, complete the explanation of the term *median.*

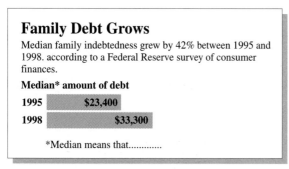

Family Debt Grows

Median family indebtedness grew by 42% between 1995 and 1998. according to a Federal Reserve survey of consumer finances.

Median* amount of debt

1995	**$23,400**
1998	**$33,300**

*Median means that............

Source: *Los Angeles Times* (February 1, 2000)

ILLUSTRATION 4

17. Graph $\frac{3}{8}$ and $-\frac{4}{5}$ on the number line. Label each using its decimal equivalent.

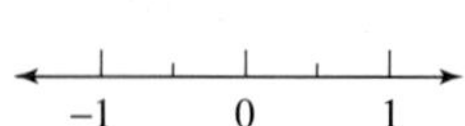

18. Find the exact answer: $\dfrac{2}{3} + 0.7$.

19. Simplify: $6.18s + 8.9 - 1.22s - 6.6$.

20. Simplify: $2.1(x - 3) + 3.1(x - 4)$.

21. Solve each equation.

a. $-2.4t = 16.8$ **b.** $-0.008 + x = 6$

22. Solve: $0.3x - 0.53 = 0.0225 + 1.6x$.

23. CHEMISTRY In a lab experiment, a chemist mixed three compounds together to form a mixture weighing 4.37 g. Later, she discovered that she had forgotten to record the weight of compound C in her notes. Find the weight of compound C used in the experiment.

	Weight
Compound A	1.86 g
Compound B	2.09 g
Compound C	?
Mixture total	4.37 g

24. WEDDING COSTS A printer charges a setup fee of $24 and then 95 cents for each wedding announcement printed (tax included). If a couple has budgeted $100 for printing costs, how many announcements can they have made?

25. Graph $\sqrt{2}$ and $-\sqrt{5}$ on the number line.

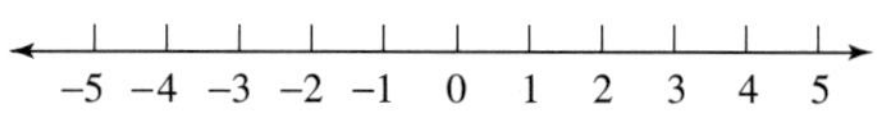

26. Simplify.

a. $-2\sqrt{25} + 3\sqrt{49}$ **b.** $\sqrt{\dfrac{1}{36}} - \sqrt{\dfrac{1}{25}}$

27. Insert the proper symbol (< or >) to make a true statement.

a. $-6.78 \quad -6.79$ **b.** $\dfrac{3}{8} \quad 0.3$

c. $\sqrt{\dfrac{16}{81}} \quad \dfrac{16}{81}$ **d.** $0.\overline{45} \quad 0.45$

28. Simplify each square root.

a. $-\sqrt{0.04}$ **b.** $\sqrt{1.69}$

Chapters 1–5 Cumulative Review Exercises

1. THE EXECUTIVE BRANCH The annual salaries for the President and the Vice President of the United States are \$400,000 and \$181,400, respectively. How much more money does the President make than the Vice President during a four-year term?

2. Use the variables x, y, and z to write the associative property of addition.

3. Find $43\overline{)1{,}203}$.

4. How many thousands are there in one million?

5. Find the prime factorization of 220.

6. List the factors of 20, from least to greatest.

7. List the set of whole numbers.

8. Find $-8 + (-5)$.

9. Fill in the blank to make the statement true: Subtraction is the same as ____________ the opposite.

10. Complete the solution.

$$\begin{aligned}(-6)^2 - 2(5 - 4 \cdot 2) &= (-6)^2 - 2(5 - \quad) \\ &= (-6)^2 - 2(\quad) \\ &= \quad - 2(-3) \\ &= 36 - (\quad) \\ &= 36 + \quad \\ &= 42\end{aligned}$$

11. Consider the division statement $\frac{-15}{-5} = 3$.

What is its equivalent multiplication statement?

12. Find $(-1)^5$.

13. Solve $8 - 2d = -5 - 5$.

14. Solve $0 = 6 + \frac{c}{-5}$.

15. Evaluate $|-7(5)|$.

16. What is the opposite of -102?

17. A chain is x yards long. Express its length in feet.

18. CHECKING ACCOUNT After a deposit of \$995, a student's checking account was still \$105 overdrawn. What was the balance in the account before the deposit?

19. See Illustration 1.

a. Let k represent the length of the key. Write an algebraic expression that represents the length of the match.

b. Let m represent the length of the match. Write an algebraic expression that represents the length of the key.

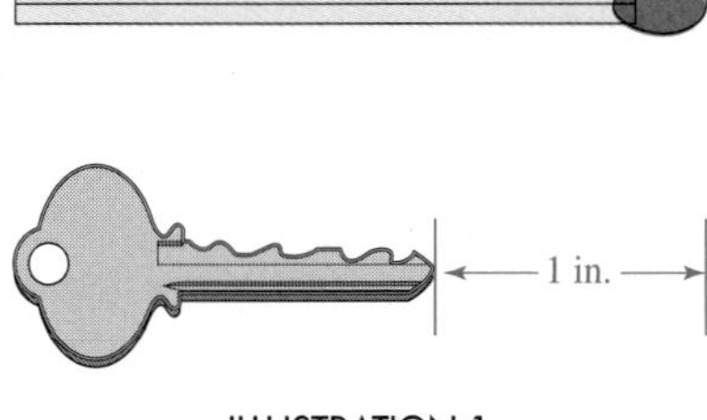

ILLUSTRATION 1

20. The expression $2(4x) + 2(5)$ is the result of an application of the distributive property. What was the original expression?

21. How many terms does the expression $6x^2 - 3x + 18$ have?

22. Simplify $3w - 8w$.

23. Solve $-(5x - 4) + 6(2x - 7) = -3$.

24. See Illustration 2. What fraction of the stripes in the flag are not red?

ILLUSTRATION 2

25. Although the fractions listed below look different, they all represent the same value. What concept does this illustrate?

$$\frac{1}{2} = \frac{2}{4} = \frac{3}{6} = \frac{4}{8} = \frac{5}{10} = \frac{6}{12}$$

26. Simplify $\frac{90x^2}{126x}$.

Perform the indicated operation.

27. $\frac{3}{8} \cdot \frac{7}{16}$

28. $-\frac{15}{8y} \div \frac{10}{y^3}$

29. $\frac{4}{m} + \frac{2}{7}$

30. $-4\frac{1}{4}\left(-4\frac{1}{2}\right)$

31. $76\frac{1}{6} - 49\frac{7}{8}$

32. $\dfrac{\frac{5}{27}}{-\frac{5}{9}}$

33. Solve $\frac{2}{3}y = -30$.

34. Solve $\frac{d}{6} - \frac{2}{3} = \frac{d}{12}$.

35. KITE Find the area of the kite shown in Illustration 3.

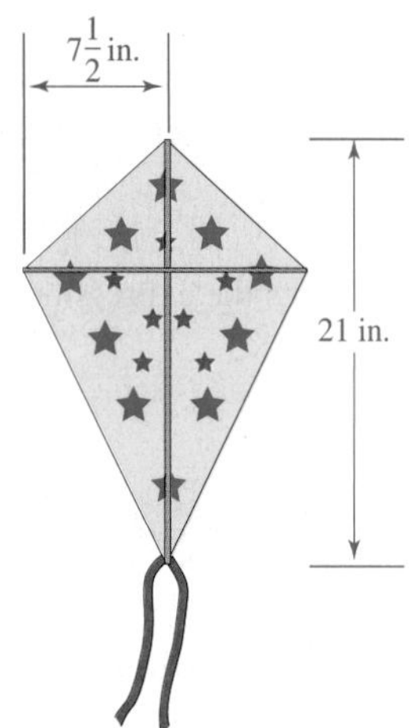

ILLUSTRATION 3

36. Graph each of the numbers in the set $\{-3\frac{1}{4}, 0.75, -1.5, -\frac{9}{8}, 3.8, \sqrt{4}\}$.

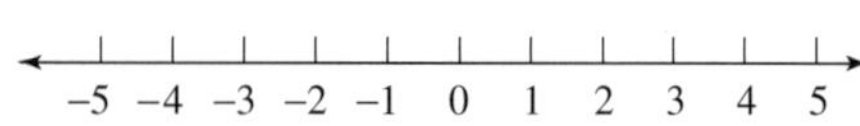

37. GLASS Some electronic and medical equipment uses glass that is only 0.00098 inch thick. Round this number to the nearest thousandth.

38. Place the proper symbol, > or <, in the box to make the statement true.

356.1978 ☐ 356.22

Perform the indicated operation.

39. $-1.8(4.52)$

40. $\frac{-21.28}{-3.8}$

41. $56.012(100)$

42. $\frac{0.897}{10,000}$

43. Evaluate $-9.1 - (-6.05 - 51)$.

44. WEEKLY SCHEDULE Refer to Illustration 4. Determine the number of hours during a week that an adult spends, on average, watching television.

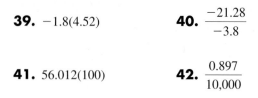

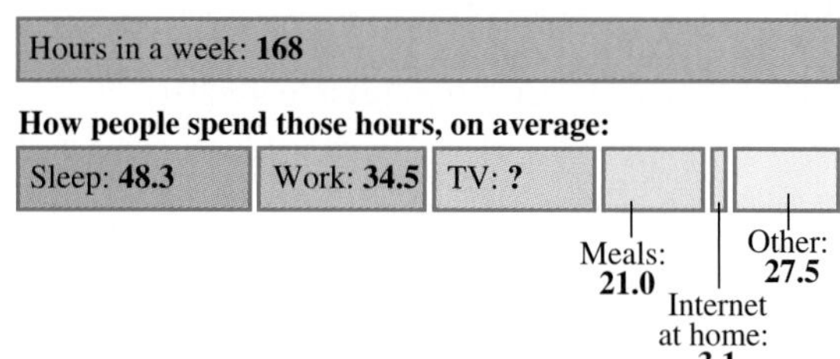

Based on data from the National Sleep Foundation and the United States Bureau of Statistics

ILLUSTRATION 4

45. LITERATURE The novel *Fahrenheit 451,* by Ray Bradbury, is a story about censorship and book burning. Use the formula $C = \frac{5(F - 32)}{9}$ to convert 451° F to degrees Celsius. Round to the nearest tenth of a degree.

46. TEAM GPA The grade point averages of the players on a badminton team are listed below. Find the mean, median, and mode of the team's GPAs.

3.04	4.00	2.75	3.23	3.87	2.20
3.02	2.25	2.99	2.56	3.58	2.75

47. Write $\frac{5}{12}$ as a decimal. Use an overbar.

48. Solve $0.2t - 3 = 0.7t + 1.5$.

49. CONCESSIONAIRE At a ballpark, a vendor is paid \$22 a game plus 35¢ for each bag of peanuts she sells. How many bags of peanuts must she sell to make \$50 a game?

50. Evaluate $-4\sqrt{36} + 2\sqrt{81}$.

Graphing, Exponents, and Polynomials

6

IT IS SAID THAT A PICTURE IS WORTH 1,000 WORDS. IN ALGEBRA, PICTURES OF EQUATIONS ARE GIVEN BY THEIR GRAPHS.

6.1 The Rectangular Coordinate System

In this section, you will learn about

- Equations containing two variables • The rectangular coordinate system
- Graphing ordered pairs of real numbers • Finding coordinates
- An application of the coordinate system

INTRODUCTION. We have seen that business and statistical information is often presented in tables or graphs. In algebra, we also present information that way. For example, Figure 6-1 shows a table and a graph that are related to the equation $d = 4t$, a formula that gives the distance d (in miles) that a woman can walk in a time t (in hours) at a rate of 4 miles per hour.

To find the distance that she can walk in 3 hours, we substitute 3 for t in the formula $d = 4t$ and simplify.

$d = 4t$

$d = 4(3)$ Substitute 3 for t.

$d = 12$ Do the multiplication.

In 3 hours, she can walk 12 miles. This result and others are shown in the table and graph.

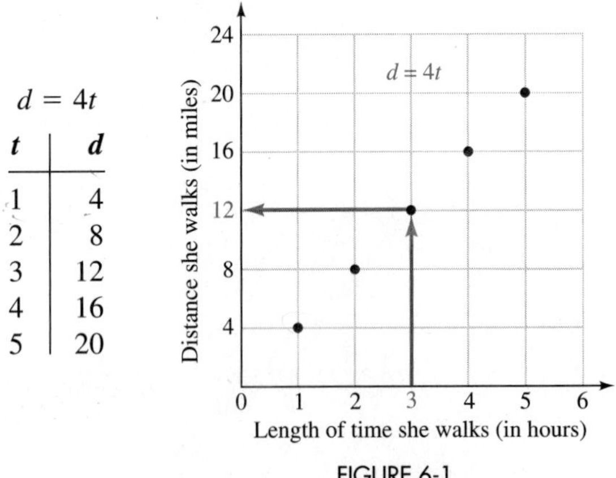

$d = 4t$

t	d
1	4
2	8
3	12
4	16
5	20

FIGURE 6-1

From either the table or the graph, we can see the following:

- In 1 hour, she can walk a distance of 4 miles.
- In 2 hours, she can walk a distance of 8 miles.
- In 3 hours, she can walk a distance of 12 miles.
- In 4 hours, she can walk a distance of 16 miles.
- In 5 hours, she can walk a distance of 20 miles.

In the first two sections of this chapter, we will discuss how to construct tables and graphs like those shown in Figure 6-1.

Equations containing two variables

So far, we have worked with equations containing only one variable. For example, we solved equations such as $2x + 4 = 12$. We will now work with equations that contain two variables, such as $2x + y = 12$.

The equation

$$2x + y = 12$$

sets up a correspondence between x and y. The solutions of this equation are pairs of real numbers. For example, the pair $x = 3$ and $y = 6$ is a solution, because the equation is true when $x = 3$ and $y = 6$.

$$2x + y = 12$$
$$2(3) + \mathbf{6} \stackrel{?}{=} 12 \quad \text{Substitute 3 for } x \text{ and 6 for } y.$$
$$6 + 6 \stackrel{?}{=} 12 \quad \text{Do the multiplication: } 2(3) = 6.$$
$$12 = 12 \quad \text{Do the addition.}$$

Since the pair $x = 3$ and $y = 6$ is a solution, we say that it **satisfies** the equation. The pair $x = 3$ and $y = 6$ can be written in the form (3, 6). When a pair is written in this form, the first number is the x-value, and the second number is the y-value. Since order is important when writing a pair in (x, y) form, we call the pair an **ordered pair.**

COMMENT Don't get confused by this new use of parentheses. The notation (3, 6) represents an ordered pair, whereas 3(6) indicates multiplication.

The ordered pair (2, 8) is also a solution of $2x + y = 12$, because it satisfies the equation.

$$2x + y = 12$$
$$2(2) + \mathbf{8} \stackrel{?}{=} 12 \quad \text{Substitute 2 for } x \text{ and 8 for } y.$$
$$4 + 8 \stackrel{?}{=} 12 \quad \text{Do the multiplication: } 2(2) = 4.$$
$$12 = 12 \quad \text{Do the addition.}$$

EXAMPLE 1 ***Checking a solution.*** Check to see whether $x = -1$ and $y = 12$ is a solution of $2x + y = 12$.

Solution

We substitute -1 for x and 12 for y and see whether the equation is satisfied.

$$2x + y = 12$$
$$2(\mathbf{-1}) + \mathbf{12} \stackrel{?}{=} 12 \quad \text{Substitute } -1 \text{ for } x \text{ and 12 for } y.$$
$$-2 + 12 \stackrel{?}{=} 12 \quad \text{Do the multiplication: } 2(-1) = -2.$$
$$10 = 12 \quad \text{Do the addition.}$$

Since $10 = 12$ is false, the pair does not satisfy the given equation. The ordered pair $(-1, 12)$ is not a solution of $2x + y = 12$.

Self Check

Check to see whether $x = -2$ and $y = 16$ is a solution of $2x + y = 12$.

2(-2) + 16 = 12

-4 + 16 = 12

Answer: yes ■

EXAMPLE 2 *Finding ordered pairs.*

Complete the ordered pairs so that each one satisfies $4x + 2y = 2$: **a.** $(0,\ \ )$ and **b.** $(\ \ , 2)$.

Solution

a. To complete the ordered pair $(0,\ \ )$, we substitute 0 for x and solve for y.

$$\begin{aligned} 4x + 2y &= 2 && \text{The given equation.} \\ 4(0) + 2y &= 2 && \text{Substitute 0 for } x. \\ 0 + 2y &= 2 && \text{Multiply: } 4(0) = 0. \\ 2y &= 2 && \text{Add: } 0 + 2y = 2y. \\ y &= 1 && \text{Divide both sides by 2.} \end{aligned}$$

When $x = 0$, $y = 1$, and the ordered pair $(0, 1)$ satisfies the equation.

b. To complete the ordered pair $(\ \ , 2)$, we substitute 2 for y and solve for x.

$$\begin{aligned} 4x + 2y &= 2 && \text{The given equation.} \\ 4x + 2(2) &= 2 && \text{Substitute 2 for } y. \\ 4x + 4 &= 2 && \text{Multiply: } 2(2) = 4. \\ 4x &= -2 && \text{Subtract 4 from both sides.} \\ x &= -\frac{1}{2} && \text{Divide both sides by 4 and simplify: } \tfrac{-2}{4} = -\tfrac{1}{2}. \end{aligned}$$

When $y = 2$, $x = -\frac{1}{2}$, and the ordered pair $\left(-\frac{1}{2}, 2\right)$ satisfies the equation.

Self Check

Complete the ordered pairs so that each one satisfies $2x - 7y = 14$.

a. $(7,\ \ )$

b. $(\ \ , 1)$

Answers: **a.** $(7, 0)$, **b.** $(10.5, 1)$ ■

The solutions of $4x + 2y = 2$ that we found in Example 2 can be listed in a **table of solutions.**

x	y	(x, y)
0	1	$(0, 1)$
$-\frac{1}{2}$	2	$\left(-\frac{1}{2}, 2\right)$

EXAMPLE 3 *A table of solutions.*

Complete the table of solutions for $3x + 2y = 5$.

x	y	(x, y)
3		$(3,\ \ )$
	4	$(\ \ , 4)$

Solution

If we substitute 3 for x in $3x + 2y = 5$ and solve for y, we get $y = -2$. We then enter -2 in the first row of the table. If we substitute 4 for y in $3x + 2y = 5$ and solve for x, we get $x = -1$. We then enter -1 in the second row of the table. The completed table is as follows:

x	y	(x, y)
3	-2	$(3, -2)$
-1	4	$(-1, 4)$

Self Check

Complete the table of solutions for $2x + 3y = 5$.

x	y	(x, y)
4		$(4,\ \ )$
	-5	$(\ \ , -5)$

Answer:

x	y	(x, y)
4	-1	$(4, -1)$
10	-5	$(10, -5)$

■

We have seen that solutions of an equation containing two variables are ordered pairs and that the ordered pairs can be listed in a table. We will now introduce a way to represent ordered pairs as points on a graph.

The rectangular coordinate system

Ordered pairs of real numbers can be displayed on a grid called a **rectangular coordinate system.** This system is also called the *Cartesian coordinate system* after its developer, René Descartes, a 17th-century French mathematician.

The rectangular coordinate system consists of two number lines, called the ***x*-axis** and the ***y*-axis,** as shown in Figure 6-2(a). The two axes intersect at a point called the **origin,** which is the 0 point on each axis. The positive direction on the x-axis is to the right, and the positive direction on the y-axis is upward.

The two axes divide the coordinate system into four regions, called **quadrants,** which are numbered using Roman numerals, as shown in Figure 6-2(a). The axes are not considered to be in any quadrant.

COMMENT If no scale is given on the x- and y-axes, we assume that the grid lines are one unit apart.

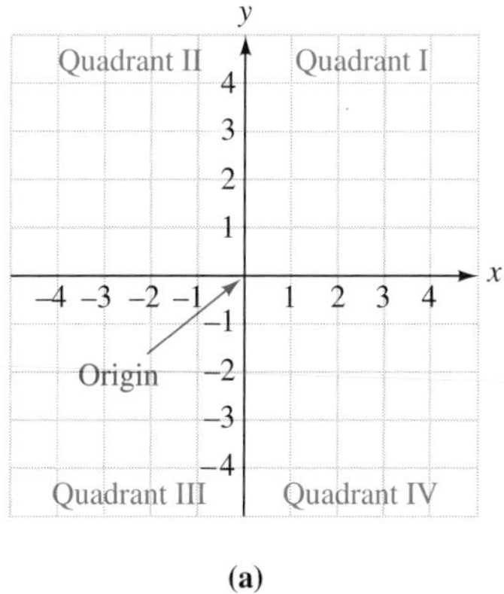

(a)

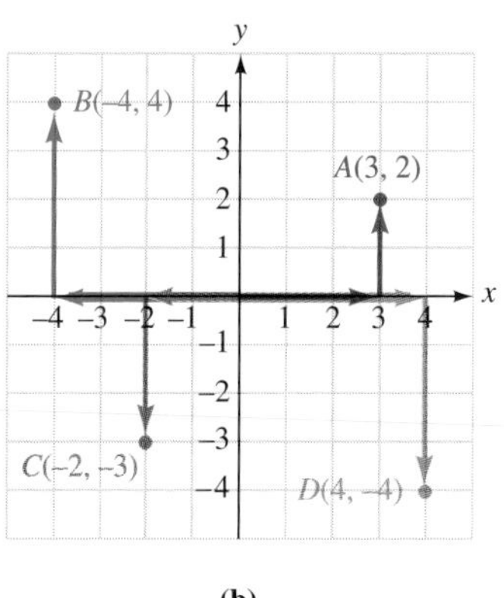

(b)

FIGURE 6-2

Graphing ordered pairs of real numbers

The process of locating an ordered pair on the rectangular coordinate system is called **graphing** or **plotting** the point. In Figure 6-2(b), we plot four ordered pairs.

- To plot the ordered pair (3, 2), we start at the origin and move 3 units to the right along the x-axis and then 2 units up in the y direction. This locates point A, with an ***x*-coordinate** of 3 and a ***y*-coordinate** of 2. The ordered pair (3, 2) gives the **coordinates of point *A*,** and point A is the **graph** of the ordered pair (3, 2). Point A is in quadrant I.
- To plot the point with coordinates $(-4, 4)$, we start at the origin and move 4 units to the left and then 4 units up. This locates point B, which lies in quadrant II.
- To plot the point with coordinates $(-2, -3)$, we start at the origin and move 2 units to the left and then 3 units down. This locates point C, which lies in quadrant III.
- To plot the point with coordinates $(4, -4)$, we start at the origin and move 4 units to the right and then 4 units down. This locates point D, which lies in quadrant IV.

COMMENT The order of the coordinates of a point is important. Point B with coordinates of $(-4, 4)$ is not the same as point D with coordinates of $(4, -4)$.

EXAMPLE 4 ***Graphing ordered pairs of real numbers.*** Graph each ordered pair: (0, 3), (4, 0), $\left(-\frac{5}{2}, 1\right)$, and (2, −1.5).

Solution

Since no scale is indicated on the axes, we assume that they are scaled in units of 1. We note that $-\frac{5}{2} = -2\frac{1}{2}$ and $-1.5 = -1\frac{1}{2}$. The points are graphed in Figure 6-3.

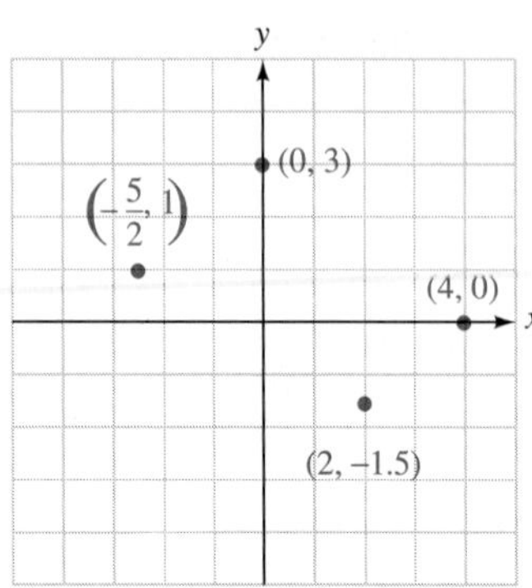

FIGURE 6-3

Self Check

Graph each ordered pair: (−3,0), (0, 4), (2.5, −4), $\left(-\frac{7}{2}, -\frac{5}{2}\right)$.

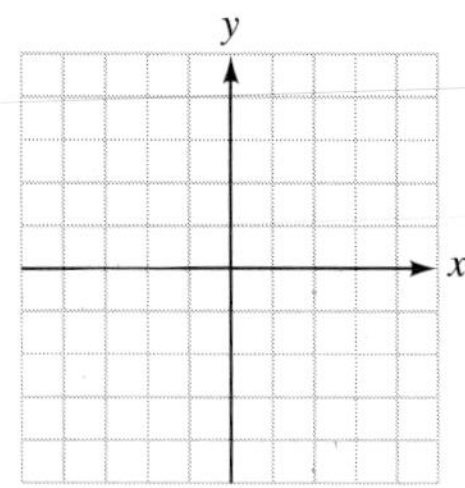

Answers:

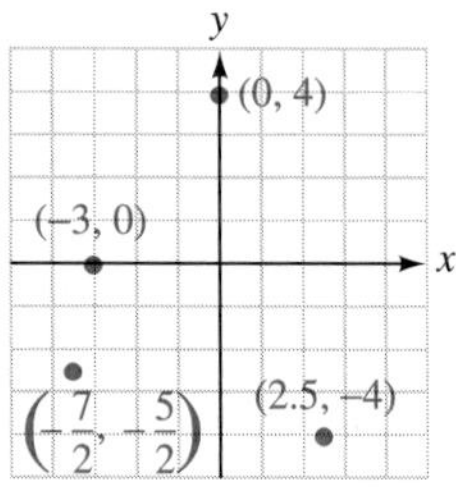

Finding coordinates

To find the coordinates of a point on the rectangular coordinate system, we must determine how far it is to the left or right of the y-axis and how far it is above or below the x-axis.

EXAMPLE 5 ***Determining the coordinates of points.*** Find the coordinates of each point in Figure 6-4.

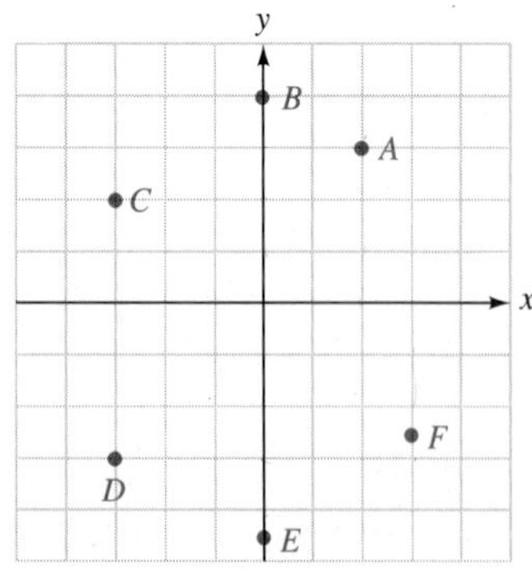

FIGURE 6-4

Solution

Since point A is 2 units to the right of the y-axis and 3 units above the x-axis, its coordinates are (2, 3). The coordinates of the other points are found in the same way. $B(0, 4)$, $C(-3, 2)$, $D(-3, -3)$, $E(0, -4.5)$, $F(3, -2.5)$.

Self Check

Find the coordinates of each point in the graph below.

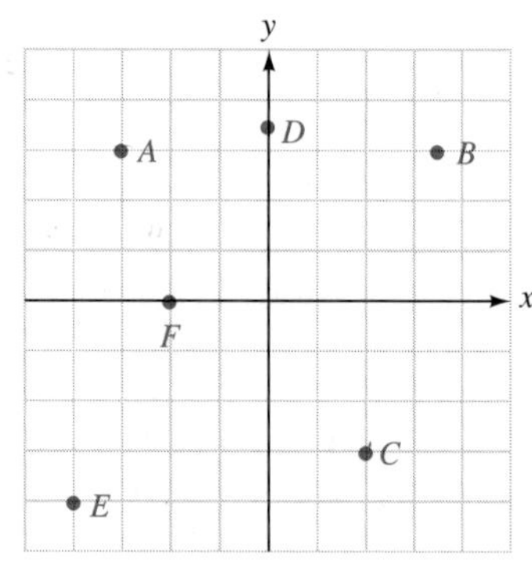

Answers: $A(-3, 3)$, $B(3.5, 3)$, $C(2, -3)$, $D(0, 3.5)$, $E(-4, -4)$, $F(-2, 0)$

An application of the coordinate system

Many cities are laid out in a rectangular grid. For example, on the east side of Rockford, Illinois, all streets run north and south, and all avenues run east and west, as shown in Figure 6-5.

In this rectangular grid system, it is very easy to find an address. For example,

- Don Smith lives at the corner of Fourth Street and Third Avenue.
- Mia Vang lives at the corner of Seventh Street and Fifth Avenue.
- The grocery store is at the corner of Second Street and Sixth Avenue.

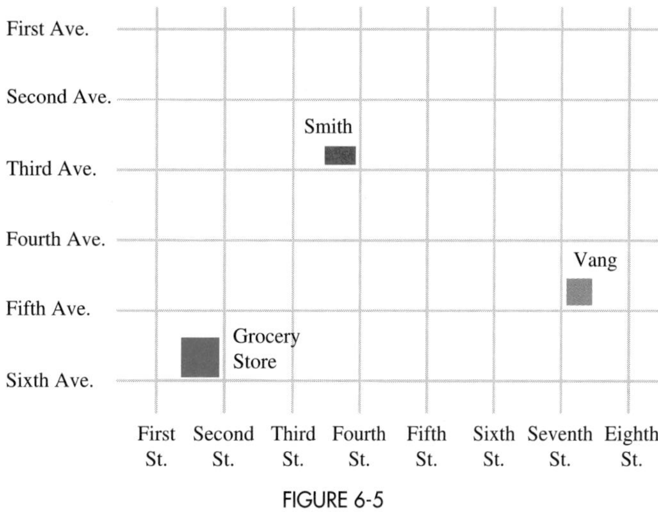

FIGURE 6-5

STUDY SET Section 6.1

VOCABULARY *Fill in the blanks to make the statements true.*

1. The pair of numbers (2, 5) is called an ordered pair.
2. The equation $x + y = 4$ contains two variables. The solution of this equation are pairs of numbers.
3. Since the ordered pair (1, 3) is a solution of $x + y = 4$, we say that (1, 3) Satisfies the equation.
4. The rectangular coordinate System is shown in Illustration 1. It consists of two number lines. Label the x-axis and the y-axis.
5. The rectangular coordinate system is sometimes called the Cartesian coordinate system.

Y

X

ILLUSTRATION 1

6. The x- and y-axes divide the rectangular coordinate system into four regions called quadrants.
7. The point where the x- and y-axes cross is called the origin.
8. In the ordered pair $(-2, 4)$, -2 is the x-coordinate, and 4 is the y-coordinate.

CONCEPTS

9. BURNING CALORIES The table shows the number of calories a 140-pound woman would burn doing light activities such as office work, cleaning house, or playing golf. Create a graph using this data.

Minutes of activity	Calories burned
1	4
2	8
3	12
4	16

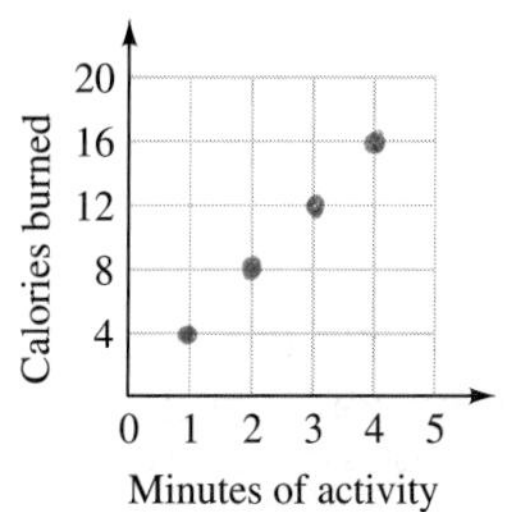

10. MIXING CONCRETE The graph shows the number of shovels of sand that should be used for a given number of shovels of cement when mixing concrete for a walkway. Create a table using this data.

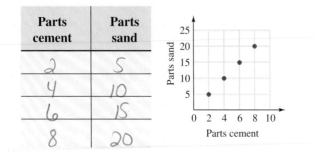

Parts cement	Parts sand
2	5
4	10
6	15
8	20

11. Is (2, 3) a solution of $2x + 3y = 14$? no

12. Is (4, 1) a solution of $3x - 2y = 10$? yes

Fill in the blanks to make the statements true.

13. To plot the point with coordinates $(3, -4)$, we start at the __origin__ and move 3 units to the __right__ and then move 4 units __down__.

14. To plot the point with coordinates $(-2, 3)$, we start at the __origin__ and move 2 units to the __left__ and then move 3 units __up__.

NOTATION *Complete each solution.*

15. For the equation $4x + 3y = 14$, find the value of y when $x = 2$.

$$4x + 3y = 14$$
$$4(2) + 3y = 14$$
$$8 + 3y = 14$$
$$3y = 6$$
$$y = 2$$

16. For the equation $4x - 3y = 12$, find the value of x when $y = 6$.

$$4x - 3y = 12$$
$$4x - 3(6) = 12$$
$$4x - 18 = 12$$
$$4x = 30$$
$$x = 7.5$$

17. Is the point $(3, -\frac{5}{2})$ the same as $(3, -2.5)$? yes

18. List the Roman numerals from 1 to 4. How are they used in this section? I II III IV

PRACTICE *Complete each statement.*

19. $3x + y = 12$
- **a.** If $x = 0$, then $y =$ 12 .
- **b.** If $y = 0$, then $x =$ 4 .
- **c.** If $x = 2$, then $y =$ 6 .

20. $4x + 3y = 24$
- **a.** If $x = 0$, then $y =$ 8 .
- **b.** If $y = 0$, then $x =$ 6 .
- **c.** If $y = 2$, then $x =$ 4.5 .

Complete the ordered pairs so that they are solutions of the equation.

21. $2x + y = 8$: **a.** (0, 8), **b.** (4, 0), and **c.** (3, 2).

22. $x - 3y = 5$: **a.** (2, −1), **b.** (14, 3), and **c.** (26, 7).

Complete each table of solutions.

23. $5x - 4y = 20$

x	y	(x, y)
0	−5	(0, −5)
4	0	(4, 0)
8	5	(8, 5)

24. $7x - y = 21$

x	y	(x, y)
0	−21	(0, −21)
3	0	(3, 0)
2	−7	(2, −7)

Graph each point on the coordinate grid.

25. $A(1, 3)$, $B(-2, 4)$, $C(-3, -2)$, $D(3, -2)$

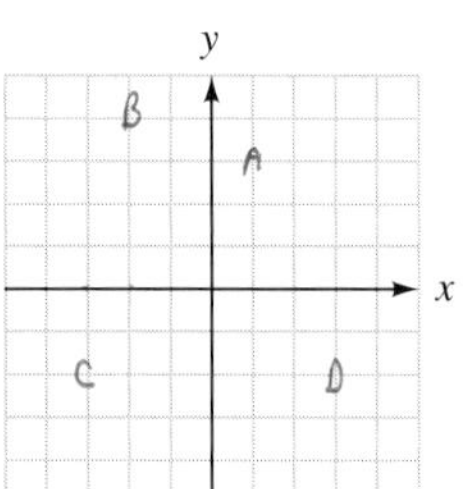

26. $A(-\frac{3}{2}, 2)$, $B(3, -\frac{5}{2})$, $C(0, -3)$, $D(3, 0)$

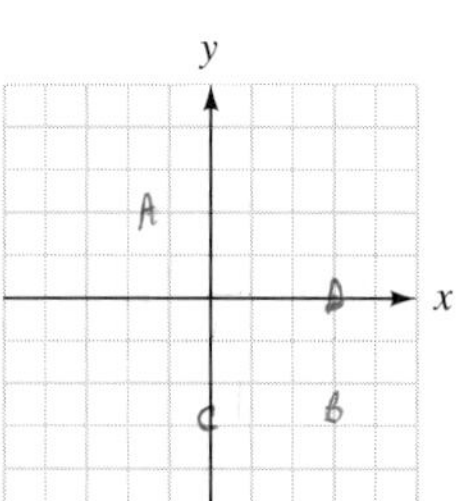

27. $A(-4, -3)$, $B(1.5, 1.5)$, $C(-3.5, 0)$, $D(0, 3.5)$

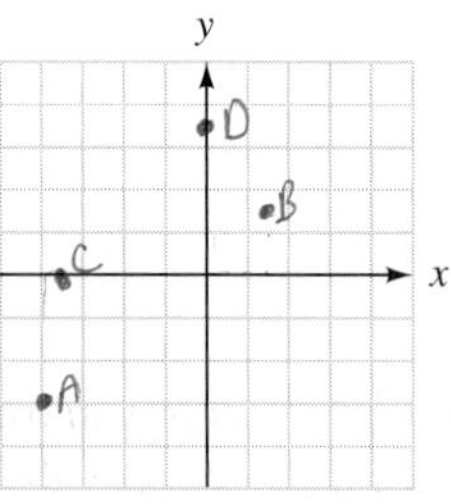

28. $A(0, 0)$, $B\left(-\frac{1}{2}, \frac{5}{2}\right)$, $C(5, -5)$, $D(-5, 5)$

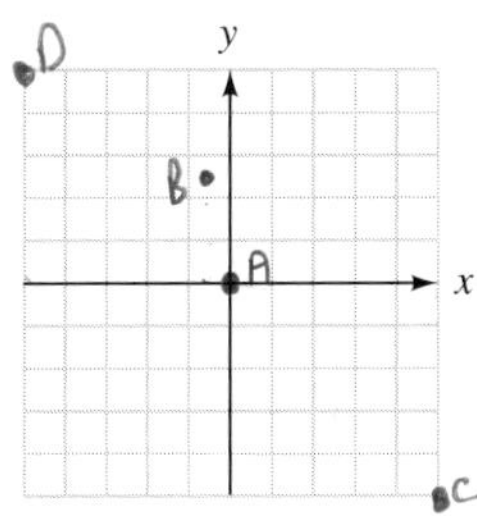

Find the coordinates of each point shown in the graph.

29.

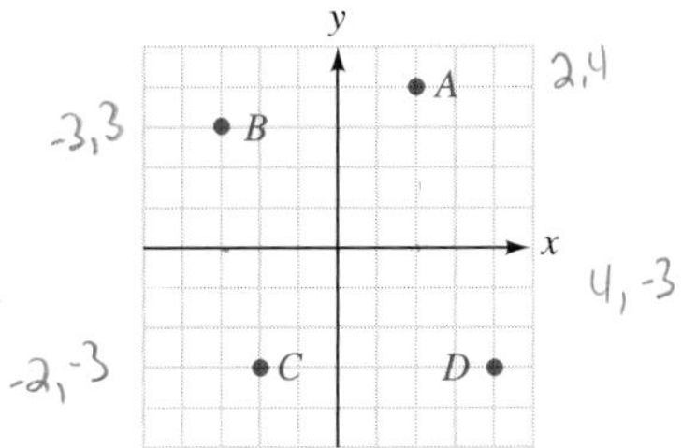

30.

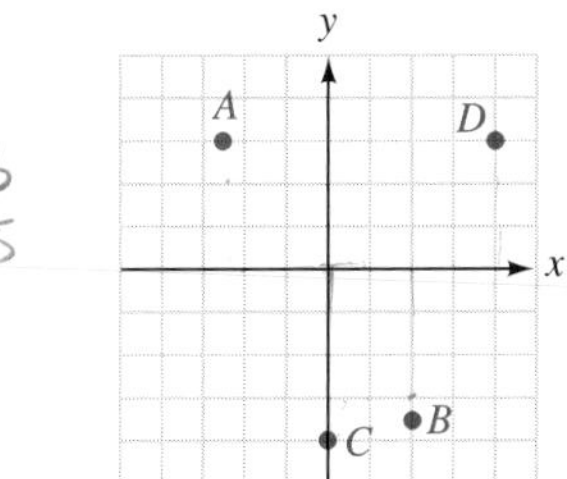

31.

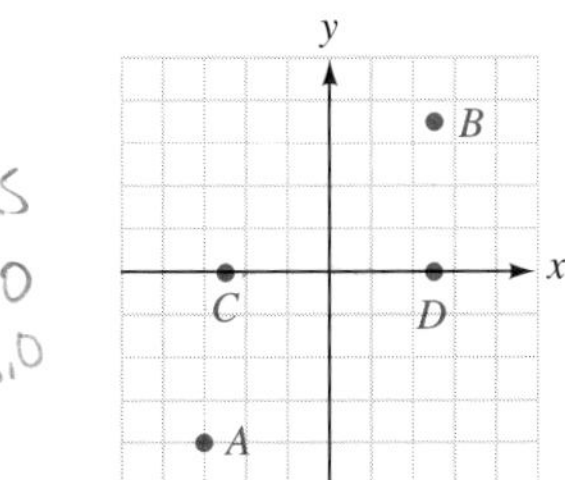

32.

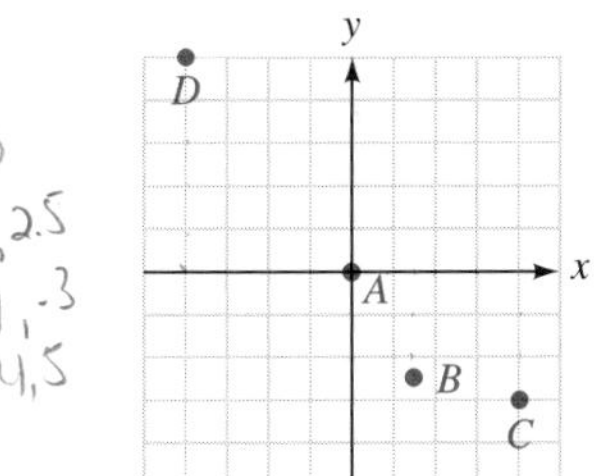

APPLICATIONS

33. ROAD MAPS Road maps usually have a coordinate system to help locate cities. Use the map in Illustration 2 to locate Rockford, Mount Carroll, Harvard, and the intersection of State Highway 251 and U.S. Highway 30. Express each answer in the form (number, letter).

ILLUSTRATION 2

34. BATTLESHIP In the game Battleship, the player uses coordinates to drop depth charges from a battleship to hit a hidden submarine. What coordinates should be used to make three hits on the exposed submarine shown in Illustration 3?

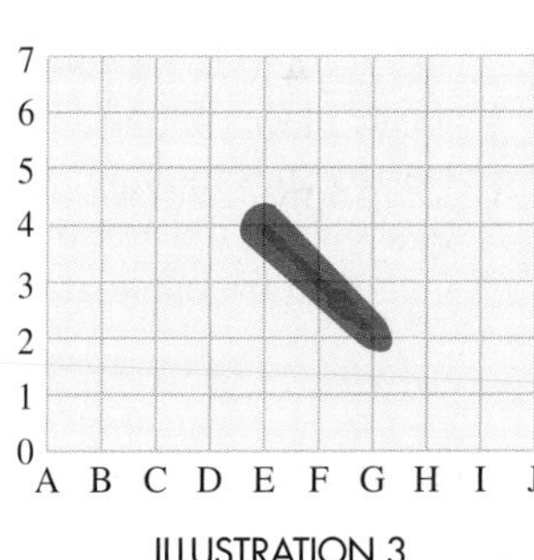

ILLUSTRATION 3

35. EARTHQUAKE DAMAGE The map in Illustration 4 shows the area where damage was caused by an earthquake.

a. Find the coordinates of the epicenter (the source of the quake).

b. Was damage done at the point (4, 5)?

c. Was damage done at the point $(-1, -4)$?

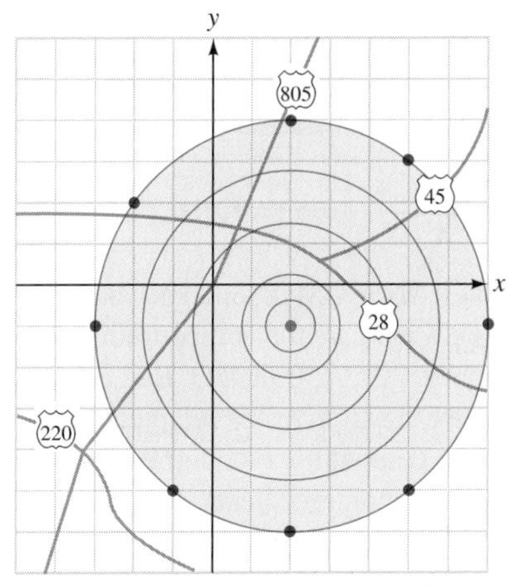

ILLUSTRATION 4

36. AUTOMATION At a factory, a robot can be programmed to make welds on a car chassis. To do this, an imaginary coordinate system is superimposed on the side of the car. (See Illustration 5.) Using the commands Up, Down, Left, and Right, write a set of instructions for the robot arm to move from its initial position to weld the points A, B, C, and D, in that order.

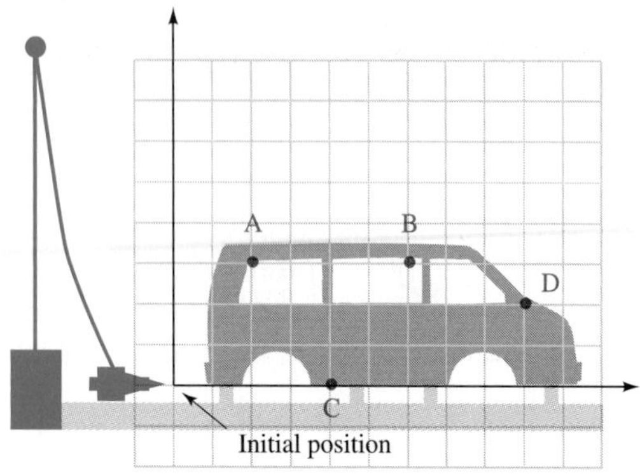

ILLUSTRATION 5

37. THE GLOBE A coordinate system that is used to locate places on the surface of the Earth uses a series of curved lines running north and south and east and west, as shown in Illustration 6. List the cities in order, beginning with the one that is farthest east on this map.

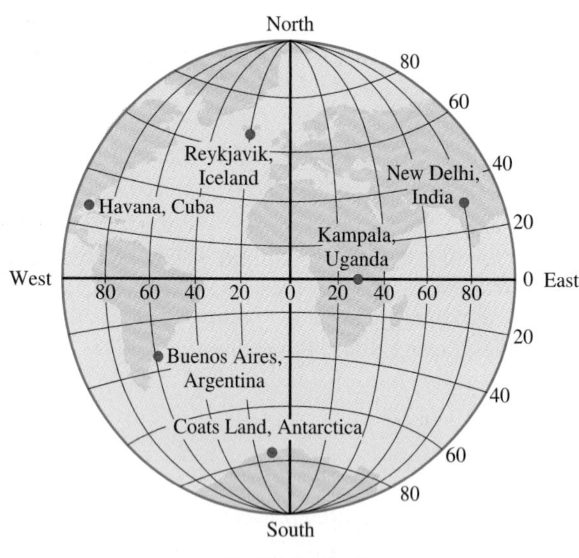

ILLUSTRATION 6

38. BLOOD TRANSFUSIONS The shaded boxes in Illustration 7 indicate the compatibility of the major blood groups, AB, A, B, and O. Red cells of the donor are mixed with serum of the recipient to test for clumping. List the compatible blood groups as ordered pairs of the form (donor, recepient).

	Red cells of donor			
Serum of recipient / group	AB	A	B	O
AB	■	■	■	■
A		■		■
B			■	■
O				■

ILLUSTRATION 7

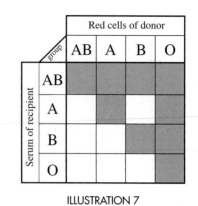

39. COOKING Use the information from the table to complete the graph in Illustration 8 for 2, 4, 6, 8, and 10 servings of instant mashed potatoes.

Number of servings	2	4	6	8	10
Flakes (cups)	$\frac{2}{3}$	$1\frac{1}{3}$	2	$2\frac{2}{3}$	$3\frac{1}{3}$

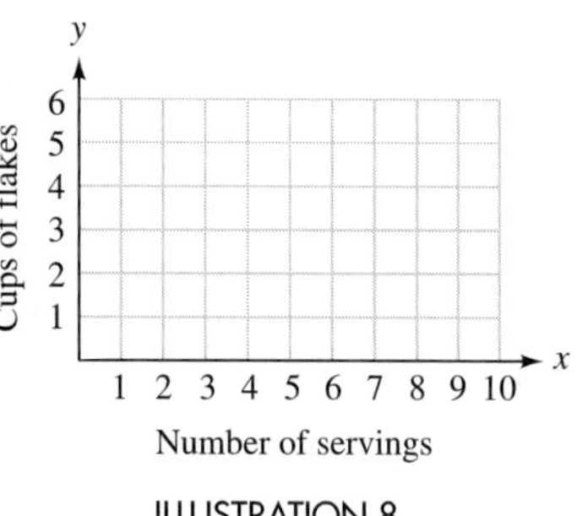

ILLUSTRATION 8

40. ROLLING DICE The red point in Illustration 9(a) represents one of the 36 possible outcomes when two fair dice are rolled a single time. Draw the appropriate number of dots on the top face of each die in Illustration 9(b) to illustrate this outcome.

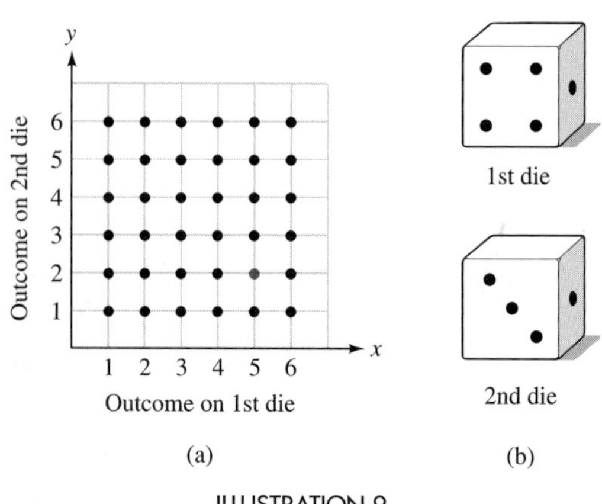

ILLUSTRATION 9

WRITING

41. Explain why the point with coordinates $(-4, 4)$ is not the same as the point with coordinates $(4, -4)$.

42. Explain the difference between $3(6)$, $(3, 6)$, and $(3 + 6)$.

43. Explain how to plot the point with coordinates of $(4, -3)$.

44. Explain why the coordinates of the origin are $(0, 0)$.

REVIEW *Evaluate each expression.*

45. $(-8 - 5) - 3$

46. $-1 - [5 - (-3)]$

47. $(-4)^2 - 3^2$

48. $-5 - \frac{24}{6} - 8(-3)$

Solve each equation and check the solution.

49. $\frac{x}{3} + 3 = 10$

50. $-3x - 4 = 8$

51. $5 - (7 - x) = -5$

52. $2(y + 6) - 4 = 2$

Use a scientific calculator to find each value.

53. $(4^2)^4$

54. $(3^4)^3$

6.2 Graphing Linear Equations

In this section, you will learn about

- Making a table of solutions • Graphing linear equations
- Using intercepts to graph linear equations
- Graphing equations that are solved for y
- Graphing equations of the form $y = b$ and $x = a$

INTRODUCTION. In the previous section, we saw that solutions of equations containing the variables x and y were ordered pairs of real numbers (x, y). We also saw that ordered pairs can be graphed on a rectangular coordinate system. In this section, we will use these skills to see how plotting points can give the graph of an equation.

Making a table of solutions

To find solutions of equations in x and y, we can pick numbers at random, substitute them for x, and find the corresponding values of y. For example, to find an ordered pair that satisfies the equation $2x - 3y = 6$, we can let $x = 0$ and solve for y.

$$2x - 3y = 6$$

$$2(\mathbf{0}) - 3y = 6 \quad \text{Substitute 0 for } x.$$

$$0 - 3y = 6 \quad \text{Do the multiplication: } 2(0) = 0.$$

$$-3y = 6 \quad \text{Do the subtraction: } 0 - 3y = -3y.$$

$$\frac{-3y}{\mathbf{-3}} = \frac{6}{\mathbf{-3}} \quad \text{To undo the multiplication by } -3\text{, divide both sides by } -3.$$

$$y = -2 \quad \text{Do the division: } \tfrac{6}{-3} = -2.$$

x	y	(x, y)
0	−2	(0, −2)

We have found that the ordered pair $(0, -2)$ is a solution of $2x - 3y = 6$. As we find solutions, we will list them in a table of solutions shown at the left.

To find another solution of the equation, we let $x = -3$, and solve for y.

$$2x - 3y = 6$$

$$2(\mathbf{-3}) - 3y = 6 \quad \text{Substitute } -3 \text{ for } x.$$

$$-6 - 3y = 6 \quad \text{Do the multiplication: } 2(-3) = -6.$$

$$-3y = 12 \quad \text{To eliminate the } -6 \text{ from the left-hand side, add 6 to both sides.}$$

x	y	(x, y)
0	−2	(0, −2)
−3	**−4**	**(−3, −4)**

$$\frac{-3y}{\mathbf{-3}} = \frac{12}{\mathbf{-3}} \quad \text{To undo the multiplication by } -3 \text{, divide both sides by } -3.$$

$$y = -4 \quad \text{Do the division: } \tfrac{12}{-3} = -4.$$

A second solution is $(-3, -4)$, and we list it in the table.

We can also find solutions of $2x - 3y = 6$ by picking a number for y and finding the corresponding value of x. For example, we can let $y = 0$ and solve for x.

$$\begin{aligned} 2x - 3y &= 6 \\ 2x - 3(\mathbf{0}) &= 6 && \text{Substitute 0 for } y. \\ 2x - 0 &= 6 && \text{Do the multiplication: } 3(0) = 0. \\ 2x &= 6 && \text{Do the subtraction: } 2x - 0 = 2x. \\ \frac{2x}{\mathbf{2}} &= \frac{6}{\mathbf{2}} && \text{To undo the multiplication by 2, divide both sides by 2.} \\ x &= 3 && \text{Do the division: } \tfrac{6}{2} = 3. \end{aligned}$$

x	y	(x, y)
0	−2	(0, −2)
−3	−4	(−3, −4)
3	**0**	**(3, 0)**

A third solution is $(3, 0)$, which we also add to the table.

If we pick $y = 2$, we have

$$\begin{aligned} 2x - 3y &= 6 \\ 2x - 3(\mathbf{2}) &= 6 && \text{Substitute 2 for } y. \\ 2x - 6 &= 6 && \text{Do the multiplication: } 3(2) = 6. \\ 2x &= 12 && \text{To undo the subtraction of 6, add 6 to both sides.} \\ \frac{2x}{\mathbf{2}} &= \frac{12}{\mathbf{2}} && \text{To undo the multiplication by 2, divide both sides by 2.} \\ x &= 6 && \text{Do the division: } \tfrac{12}{2} = 6. \end{aligned}$$

x	y	(x, y)
0	−2	(0, −2)
−3	−4	(−3, −4)
3	0	(3, 0)
6	**2**	**(6, 2)**

A fourth solution is $(6, 2)$. We enter it in the table.

Since any choice of x will give a corresponding value of y, and any choice of y will give a corresponding value of x, it is apparent that $2x - 3y = 6$ has *infinitely many solutions.* We have found four of them: $(0, -2)$, $(-3, -4)$, $(3, 0)$, and $(6, 2)$.

Graphing linear equations

To graph $2x - 3y = 6$, we plot the ordered pairs that we have just found. See Figure 6-6(a). In Figure 6-6(b), we see that the four points lie on a line.

In Figure 6-6(c), we draw a straight line through the points, because the graph of any solution of $2x - 3y = 6$ will lie on this line. The arrowheads show that the line continues forever in both directions. The line is a picture of all of the solutions of $2x - 3y = 6$, and it is called a **graph** of the equation. Every point on the line is a solution of the equation, and every solution of the equation is on the line. Equations whose graphs are straight lines, such as $2x - 3y = 6$, are called **linear equations.**

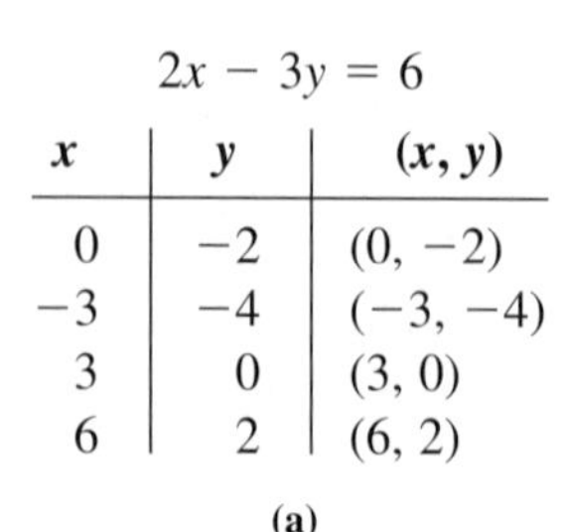

$2x - 3y = 6$

x	y	(x, y)
0	−2	(0, −2)
−3	−4	(−3, −4)
3	0	(3, 0)
6	2	(6, 2)

(a)

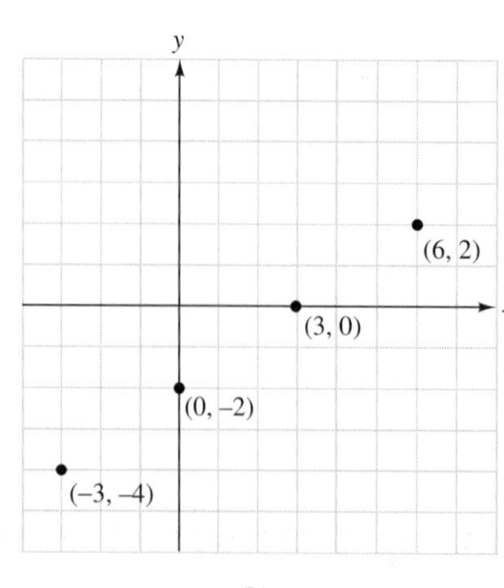

(b)

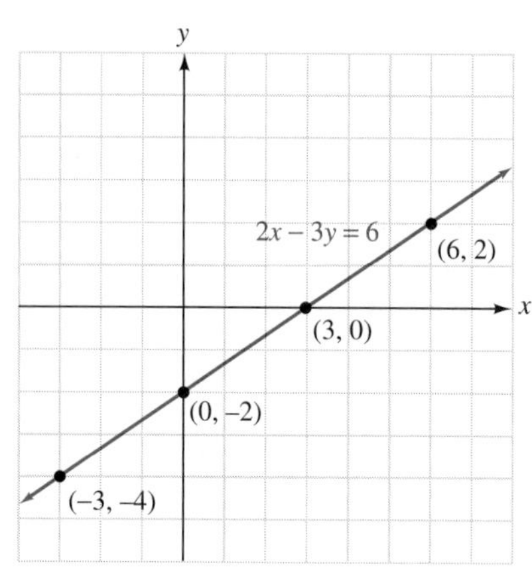

(c)

FIGURE 6-6

When we graphed $2x - 3y = 6$, we did more work than necessary. Since two points determine a line, only two points were necessary to graph the equation. However, it is always a good idea to plot a third point as a check. If the three points do not lie on a straight line, then at least one of them is in error. To graph a linear equation, we will follow these steps.

Strategy for graphing linear equations in x and y

1. Find two pairs (x, y) that satisfy the equation by picking arbitrary numbers for x (or y) and solving for the corresponding values of y (or x). A third point acts as a check.
2. Plot each resulting pair (x, y) on a rectangular coordinate system. If they do not lie on a straight line, check your calculations.
3. Use a straightedge to draw the line passing through the three points. Draw arrowheads to indicate that the line continues forever in both directions.

Using intercepts to graph linear equations

In Figure 6-6(c) on the previous page, the line crosses the y-axis at the point with coordinates $(0, -2)$ and crosses the x-axis at the point with coordinates $(3, 0)$. These points have special names.

y- and x-intercepts

The point where the graph of a linear equation crosses the y-axis is called the **y-intercept.** To find the coordinates of the y-intercept, let $x = 0$ and solve for y.

The point where the graph of a linear equation crosses the x-axis is called the **x-intercept.** To find the coordinates of the x-intercept, let $y = 0$ and solve for x.

EXAMPLE 1 ***Graphing linear equations.*** Graph $3x + 2y = 6$ by plotting the x- and y-intercepts and a third point on the line.

Solution

To find the coordinates of the x-intercept, we substitute 0 for y in $3x + 2y = 6$ and find that $x = 2$. The x-intercept is $(2, 0)$

To find the coordinates of the y-intercept, we substitute 0 for x in $3x + 2y = 6$ and find that $y = 3$. The y-intercept is $(0, 3)$.

To find a third point on the line, we let $x = 1$. Then $y = \frac{3}{2}$.

We list the ordered pairs $(2, 0)$, $(0, 3)$, and $\left(1, \frac{3}{2}\right)$ in a table. To get the graph, we plot each point and join them with a line, as shown in Figure 6-7.

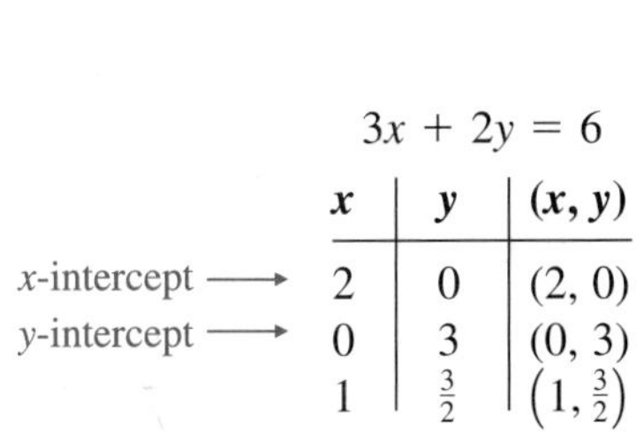

$3x + 2y = 6$

	x	y	(x, y)
x-intercept →	2	0	$(2, 0)$
y-intercept →	0	3	$(0, 3)$
	1	$\frac{3}{2}$	$\left(1, \frac{3}{2}\right)$

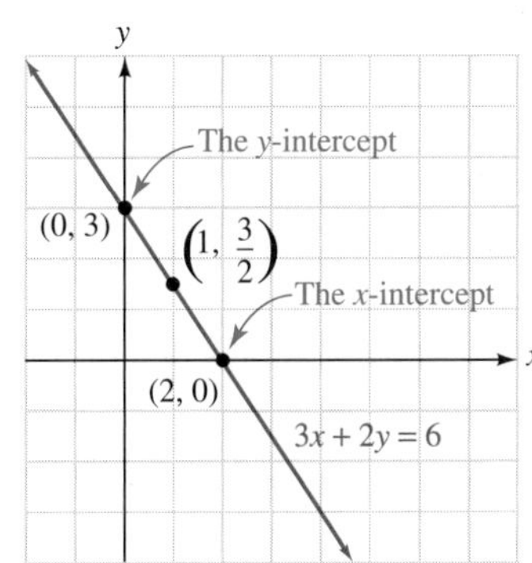

FIGURE 6-7

Self Check

Graph $2x + 3y = 6$.

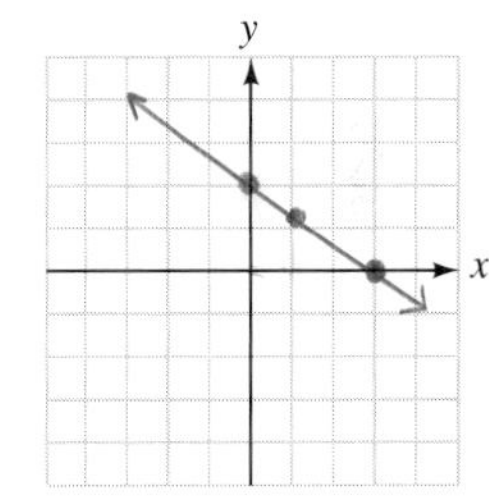

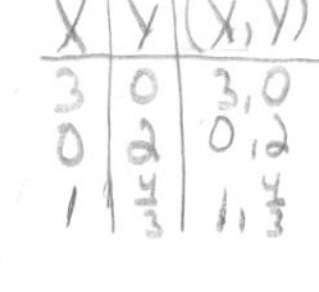

Answer:

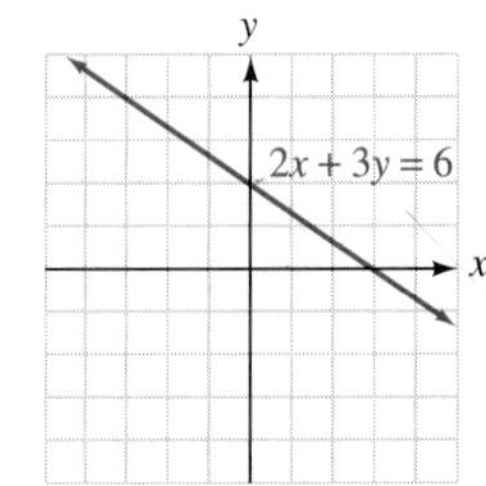

■

Graphing equations that are solved for y

We have graphed the equation $3x + 2y = 6$ and gotten a line. Equations of a line can occur in other forms, such as $y = 2x - 1$. Since this equation is solved for y, the value of y depends on the value chosen for x. For this reason, we call y the **dependent variable** and x the **independent variable.**

To graph $y = 2x - 1$, we will find three ordered pairs (x, y) that satisfy the equation. Since x is the independent variable, it is easiest to pick three values for x and solve for y. If we let $x = 0$, we will find the coordinates of the y-intercept. We may pick any three values for x, but it is a good idea to make $x = 0$ one of those choices, because it makes the computations easy.

x	y	(x, y)
0	−1	(0, −1)

$y = 2x - 1$

$y = 2(\mathbf{0}) - 1$ Substitute 0 for x.

$y = 0 - 1$ Do the multiplication: $2(0) = 0$.

$y = -1$ Do the subtraction.

The y-intercept is $(0, -1)$. If we let $x = 1$, we have

x	y	(x, y)
0	−1	(0, −1)
1	1	(1, 1)

$y = 2x - 1$

$y = 2(\mathbf{1}) - 1$ Substitute 1 for x.

$y = 2 - 1$ Do the multiplication: $2(1) = 2$.

$y = 1$ Do the subtraction.

The ordered pair $(1, 1)$ satisfies the equation. If we let $x = -2$, we have

x	y	(x, y)
0	−1	(0, −1)
1	1	(1, 1)
−2	−5	(−2, −5)

$y = 2x - 1$

$y = 2(\mathbf{-2}) - 1$ Substitute −2 for x.

$y = -4 - 1$ Do the multiplication: $2(-2) = -4$.

$y = -5$ Do the subtraction.

The ordered pair $(-2, -5)$ satisfies the equation. We copy the final table of solutions in Figure 6-8, plot the points, and draw a straight line through them. This line is the graph of the equation.

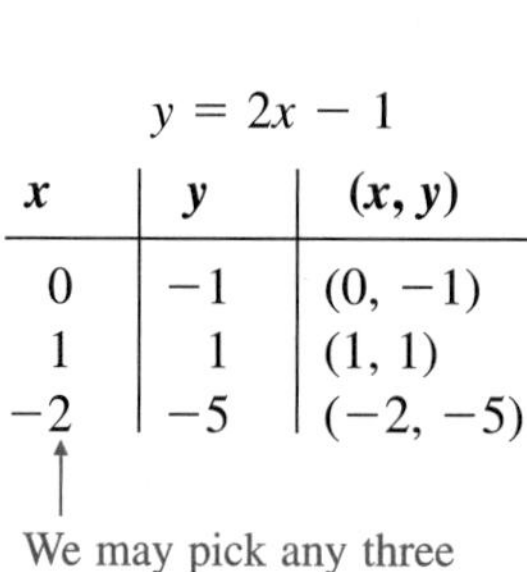

$y = 2x - 1$

x	y	(x, y)
0	−1	(0, −1)
1	1	(1, 1)
−2	−5	(−2, −5)

We may pick any three values for x.

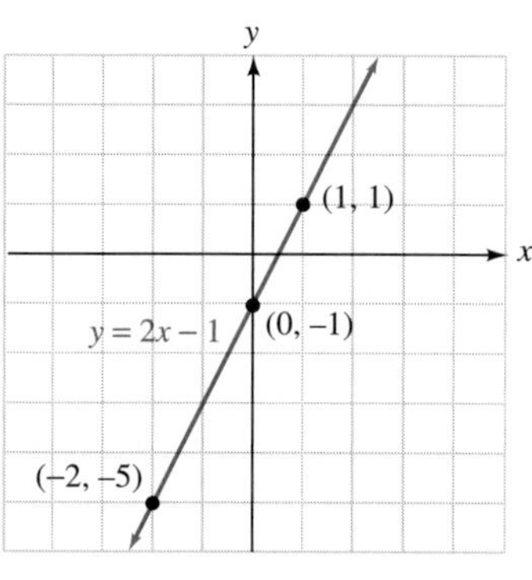

FIGURE 6-8

EXAMPLE 2 *Graphing linear equations.* Graph $y = \frac{1}{2}x + 1$.

Solution

We pick three values for x and then calculate each corresponding value of y. If we let $x = 0$, then

$y = \frac{1}{2}(\mathbf{0}) + 1$ Substitute 0 for x.

$y = 0 + 1$ Do the multiplication.

$y = 1$ The point (0, 1) lies on the graph. It is the y-intercept.

Since each value of x will be multiplied by $\frac{1}{2}$, we choose values for x that can be divided by 2 to make the multiplication easy. If we let $x = 4$, then

Self Check

Graph $y = 3x - 2$.

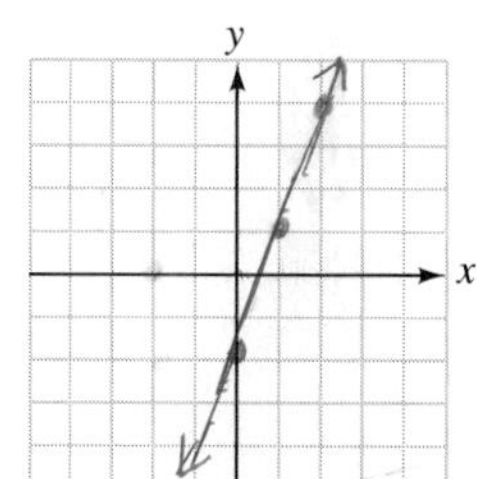

$y = \frac{1}{2}(4) + 1$ Substitute 4 for x.

$y = \frac{4}{2} + 1$ Multiply: $\frac{1}{2}(4) = \frac{1}{2}\left(\frac{4}{1}\right) = \frac{4}{2}$.

$y = 2 + 1$ Divide: $\frac{4}{2} = 2$.

$y = 3$ The point (4, 3) lies on the graph.

If we let $x = -4$, then

$y = \frac{1}{2}(-4) + 1$ Substitute -4 for x.

$y = \frac{-4}{2} + 1$ $\frac{1}{2}(-4) = \frac{1}{2}\left(\frac{-4}{1}\right) = \frac{-4}{2}$.

$y = -2 + 1$ Do the division: $\frac{-4}{2} = -2$.

$y = -1$ The point $(-4, -1)$ lies on the graph.

We list these coordinates in a table of solutions, plot each point, and draw a straight line through them, as shown in Figure 6-9.

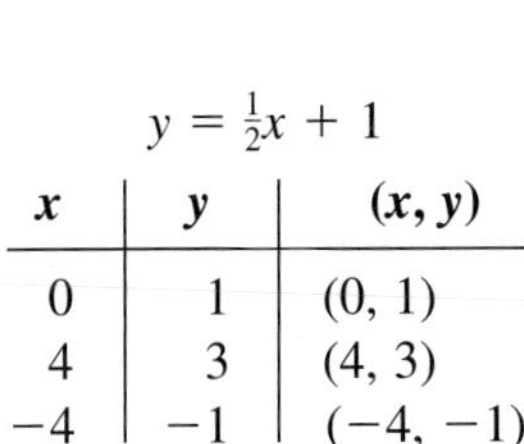

$y = \frac{1}{2}x + 1$

x	y	(x, y)
0	1	(0, 1)
4	3	(4, 3)
−4	−1	(−4, −1)

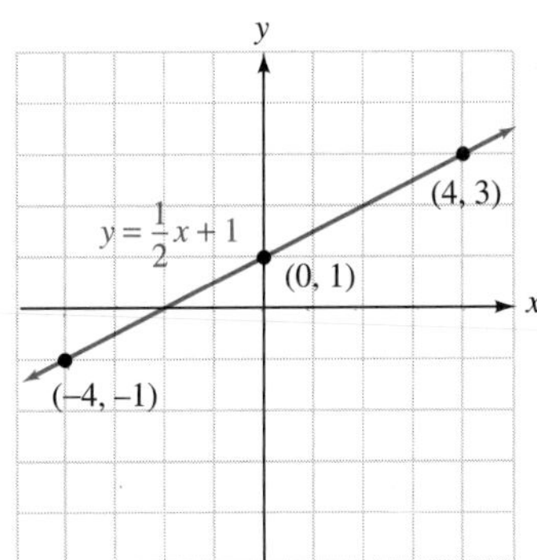

FIGURE 6-9

Answer:

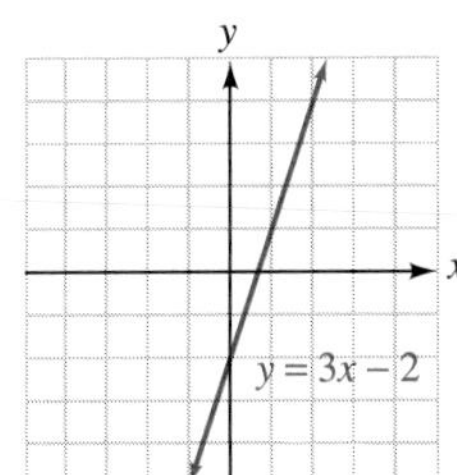

EXAMPLE 3 *Adjusting the scale.* Graph $y = 20x$.

Solution

We begin by arbitrarily (randomly) selecting three values for x: -2, 0, and 2. If $x = -2$, then to calculate the corresponding value of y, we substitute -2 for x in $y = 20x$.

$y = 20x$

$y = 20(-2)$ Substitute -2 for x.

$y = -40$ Do the multiplication.

We see that $x = -2$ and $y = -40$ is a solution of $y = 20x$. In a similar manner, we find the corresponding values for y when x is 0 and 2 and enter them in the table in Figure 6-10 (on the next page).

Because of the sizes of the y-coordinates of the points $(-2, \mathbf{-40})$ and $(2, \mathbf{40})$, we must adjust the scale on the y-axis. (If we used grid lines 1 unit apart, the graph would be enormous.)

One way to accommodate these points is to scale the y-axis in units of 5, 10, or 20. If we choose divisions of 20 units, plot the three solutions from the table, and draw a line through them, we get the graph shown in Figure 6-10 on the next page.

Self Check

Graph $y = -25x + 50$.

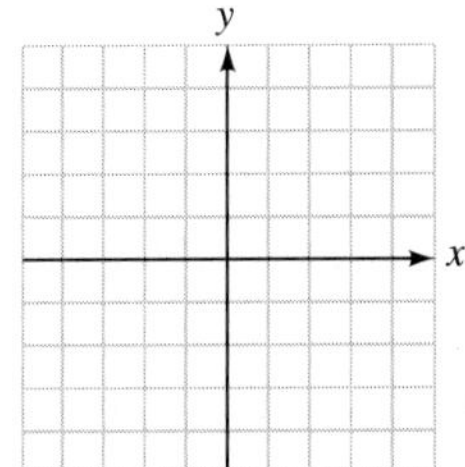

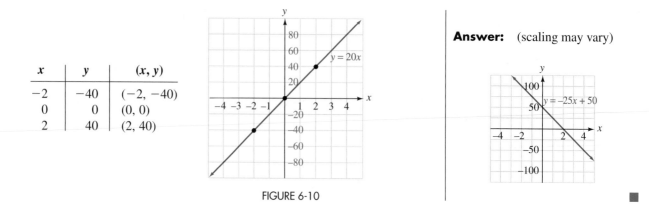

x	y	(x, y)
-2	-40	$(-2, -40)$
0	0	$(0, 0)$
2	40	$(2, 40)$

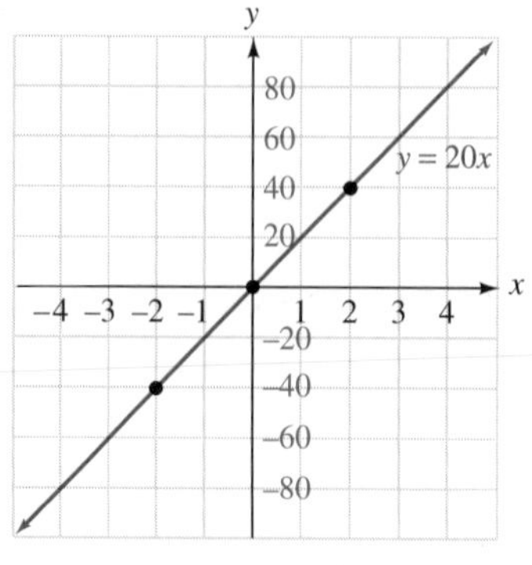

FIGURE 6-10

Answer: (scaling may vary)

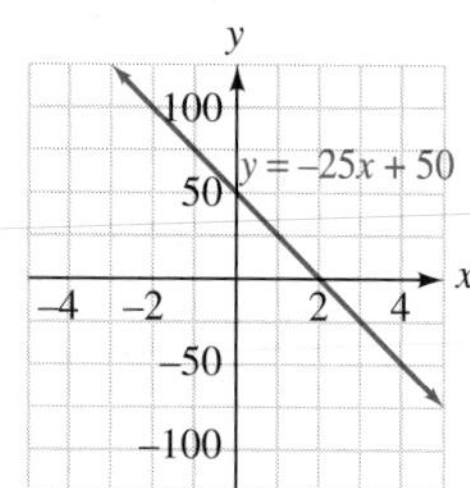

Graphing equations of the form $y = b$ and $x = a$

When a linear equation contains only one variable, such as $y = 4$ or $x = -2$, its graph is either a horizontal or a vertical line.

EXAMPLE 4 ***Graphing linear equations.*** Graph $y = 4$.

Solution

We can write this equation as $0x + y = 4$. Since the coefficient of x is 0, the numbers assigned to x have no effect on y. The value of y is always 4. For example, if $x = 5$, then $y = 4$.

$$0x + y = 4$$

$$0(5) + y = 4 \quad \text{Substitute 5 for } x.$$

$$0 + y = 4 \quad \text{Multiply: } 0(5) = 0.$$

$$y = 4 \quad \text{Add: } 0 + y = y.$$

Similarly, if $x = 2$, then $y = 4$, and if $x = -2$, then $y = 4$. To graph the equation, we plot the points and draw the line, as shown in Figure 6-11.

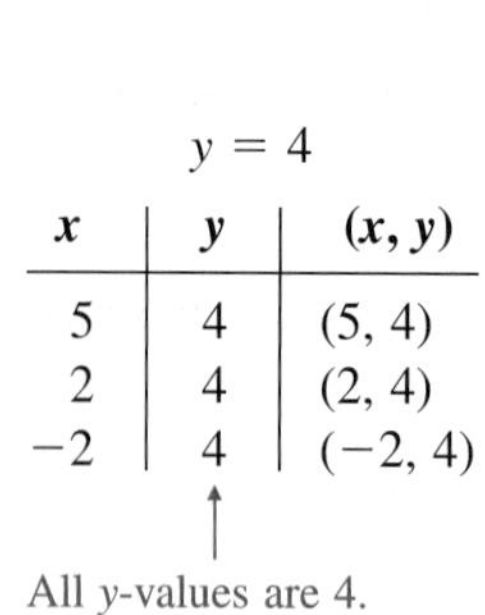

$y = 4$

x	y	(x, y)
5	4	$(5, 4)$
2	4	$(2, 4)$
-2	4	$(-2, 4)$

All y-values are 4.

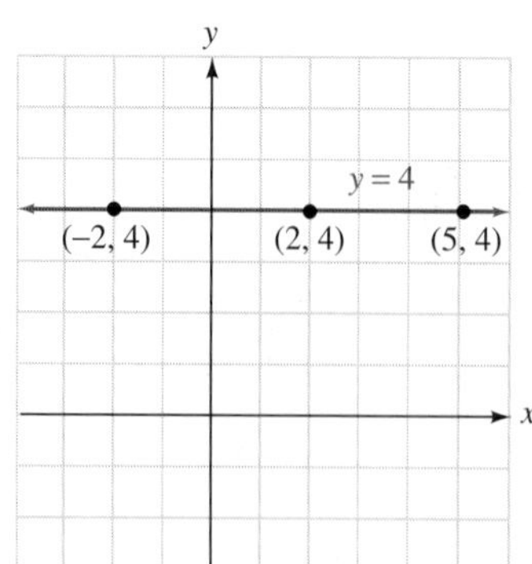

FIGURE 6-11

Self Check

Graph $y = -2$.

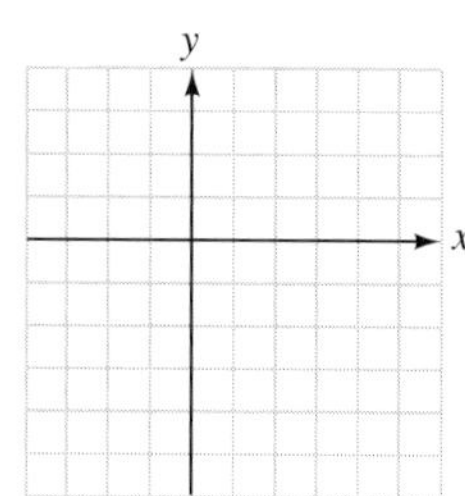

Answer:

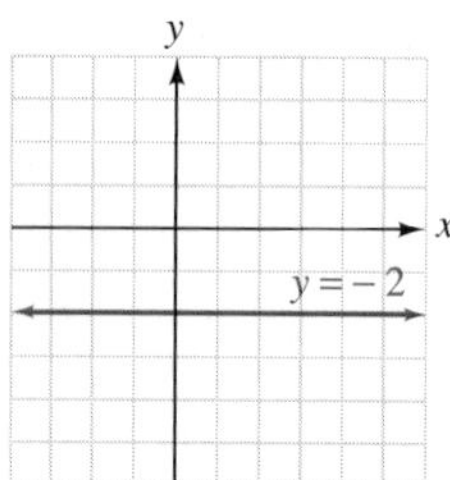

EXAMPLE 5 ***Graphing linear equations.*** Graph $x = -2$.

Solution

We can write this equation as $x + 0y = -2$. Since the coefficient of y is 0, the numbers assigned to y have no effect on x. The value of x is always -2. For example, if $y = 1$, then $x = -2$.

Self Check

Graph $x = 3$.

$$x + 0y = -2$$
$$x + 0(\mathbf{1}) = -2 \quad \text{Substitute 1 for } y.$$
$$x + 0 = -2 \quad \text{Multiply: } 0(1) = 0.$$
$$x = -2 \quad \text{Add: } x + 0 = x.$$

Similarly, if $y = 2$, then $x = -2$, and if $y = -2$, then $x = -2$. To graph the equation, we plot the points and draw the line, as shown in Figure 6-12.

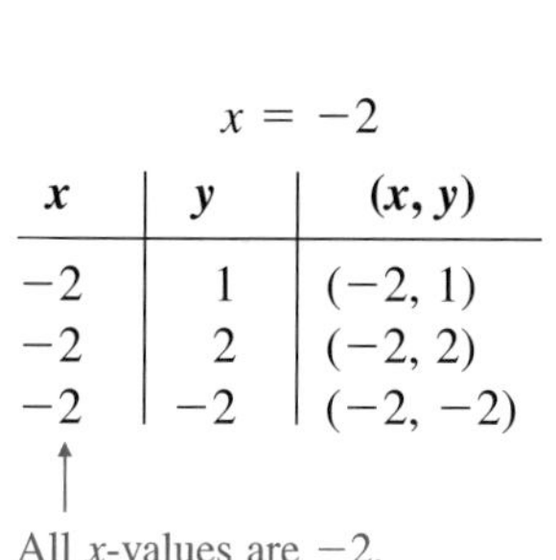

$x = -2$

x	y	(x, y)
-2	1	$(-2, 1)$
-2	2	$(-2, 2)$
-2	-2	$(-2, -2)$

All x-values are -2.

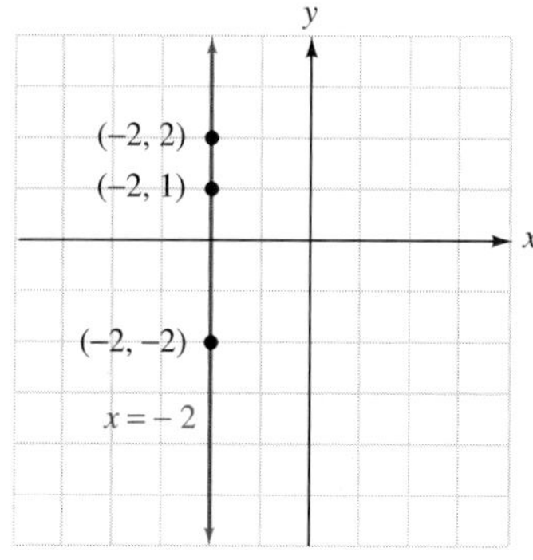

FIGURE 6-12

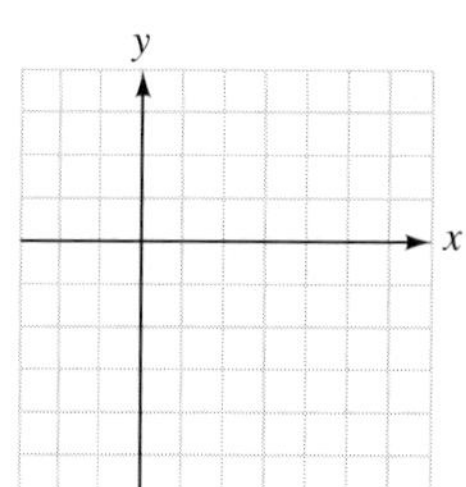

Answer:

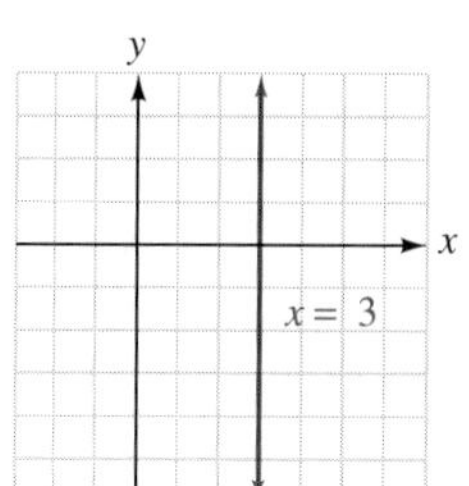

■

The results of Example 4 and the Self Check suggest that the graphs of equations of the form $y = b$ are horizontal lines. The results of Example 5 and the Self Check suggest that the graphs of equations of the form $x = a$ are vertical lines. We now summarize these observations.

Equations of horizontal and vertical lines

> The equation $y = b$ represents a horizontal line that intersects the y-axis at the point $(0, b)$. If $b = 0$, the line is the x-axis.
>
> The equation $x = a$ represents a vertical line that intersects the x-axis at the point $(a, 0)$. If $a = 0$, the line is the y-axis.

STUDY SET Section 6.2

VOCABULARY *Fill in the blanks to make the statements true.*

1. The graph of a linear equation is a ________.

2. If a point lies on a line, its coordinates ________ the equation.

3. The point where the graph of a linear equation crosses the x-axis is called the ________.

4. The y-intercept is the point where the graph of a linear equation crosses the ________.

5. In the equation $y = 7x + 2$, x is called the ________ variable.

6. In the equation $y = 7x + 2$, y is called the ________ variable.

CONCEPTS *In Exercises 7–8, fill in the blanks to make the statements true.*

7. The graph of the equation $y = 3$ is a ________ line.

8. The graph of the equation $x = -2$ is a ________ line.

9. a. What is the y-intercept of the line graphed in Illustration 1?

b. What is its x-intercept?

c. Does the line pass through the point $(4, 3)$?

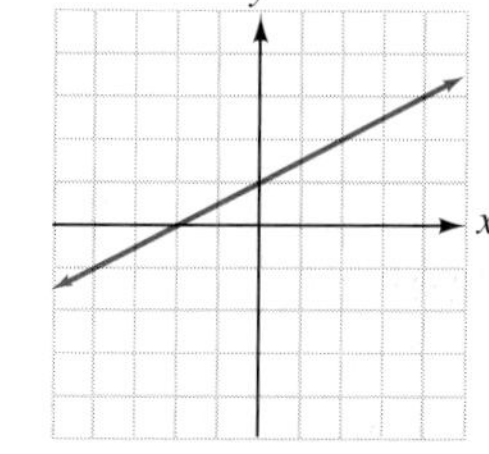

ILLUSTRATION 1

10. a. What is the y-intercept of the line graphed in Illustration 2?

b. What is its x-intercept?

c. Does the line pass through the point $(1, -1)$?

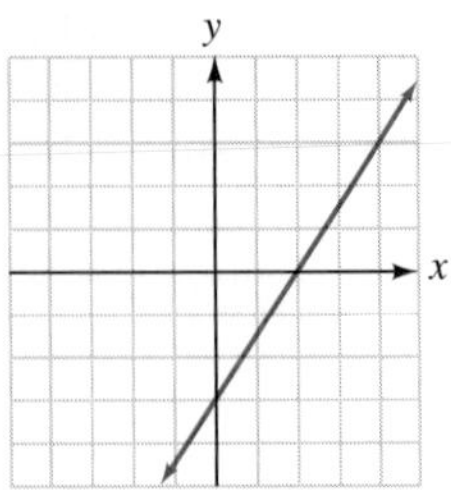

ILLUSTRATION 2

11. What is wrong with the graph of $x - y = 3$ shown in Illustration 3?

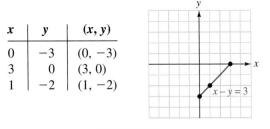

x	y	(x, y)
0	−3	(0, −3)
3	0	(3, 0)
1	−2	(1, −2)

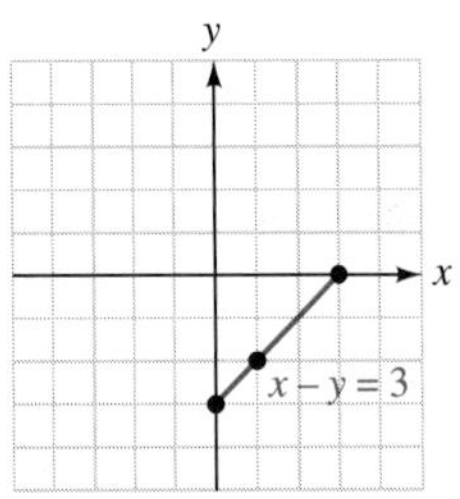

ILLUSTRATION 3

12. To graph $y = -x + 1$, a student constructed a table of solutions and plotted the ordered pairs as shown in Illustration 4. Instead of drawing a crooked line through the points, what should he have done?

$y = -x + 1$

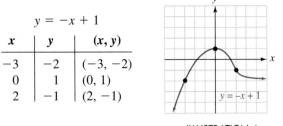

x	y	(x, y)
−3	−2	(−3, −2)
0	1	(0, 1)
2	−1	(2, −1)

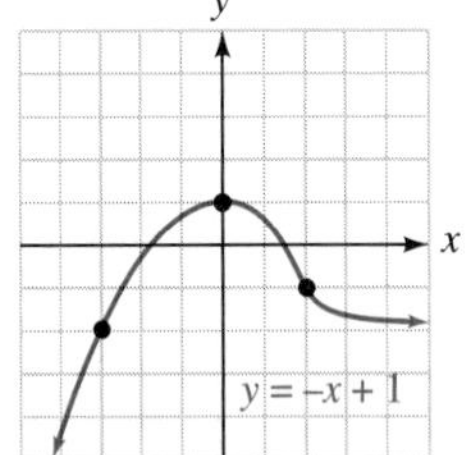

ILLUSTRATION 4

13. The graph of a linear equation is shown in Illustration 5. Determine six solutions of the equation from the graph.

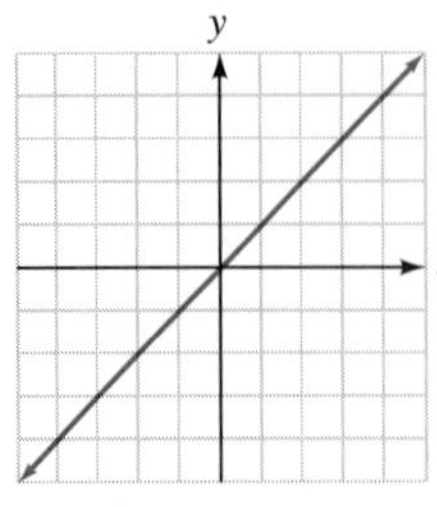

ILLUSTRATION 5

14. The graph of a linear equation is shown in Illustration 6. What three points were apparently plotted to obtain the graph? Show them in the table of solutions.

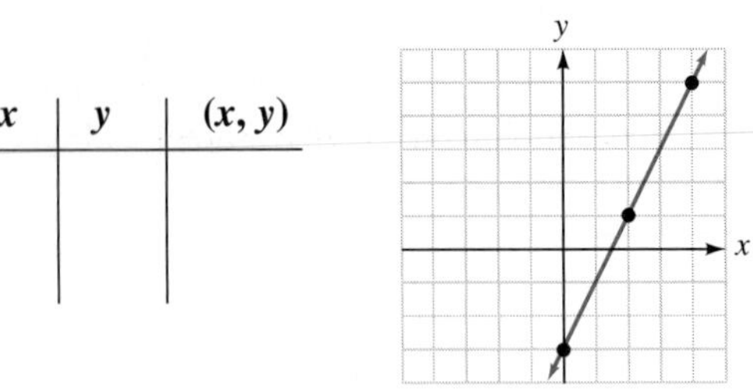

ILLUSTRATION 6

NOTATION *Complete each solution.*

15. Consider the equation $2x - 4y = 8$. Find y if $x = 3$.

$$2x - 4y = 8$$
$$2(\quad) - 4y = 8$$
$$\quad - 4y = 8$$
$$-4y = \quad$$
$$y = -\frac{1}{2}$$

16. Consider the equation $y = \frac{2}{3}x - 1$. Find y when $x = -3$.

$$y = \frac{2}{3}x - 1$$
$$y = \frac{2}{3}(\quad) - 1$$
$$y = \quad - 1$$
$$y = -3$$

PRACTICE *Complete each table of solutions.*

17. $2x - 5y = 10$

x	y	(x, y)
5		
−5		
10		

18. $3x + 4y = 18$

x	y	(x, y)
	0	
	3	
	6	

19. $y = 2x - 3$

x	y	(x, y)
3		
−4		
6		

20. $y = -3x + 4$

x	y	(x, y)
4		
0		
−3		

Find the coordinates of the y- and x-intercepts of the graph of each equation.

21. $x + y = 5$

22. $x - y = 2$

23. $4x + 5y = 20$

24. $3x - 5y = 15$

Complete the table and then graph the equation.

25. $x + y = 5$ (see Exercise 21)

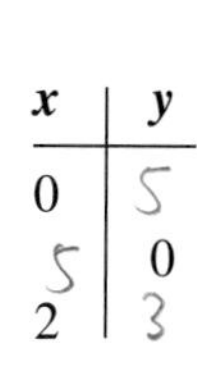

x	y
0	5
5	0
2	3

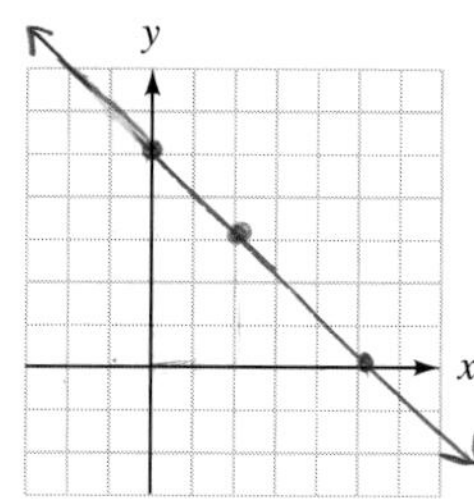

26. $x - y = 2$ (see Exercise 22)

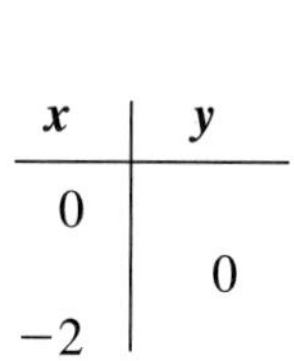

x	y
0	
	0
−2	

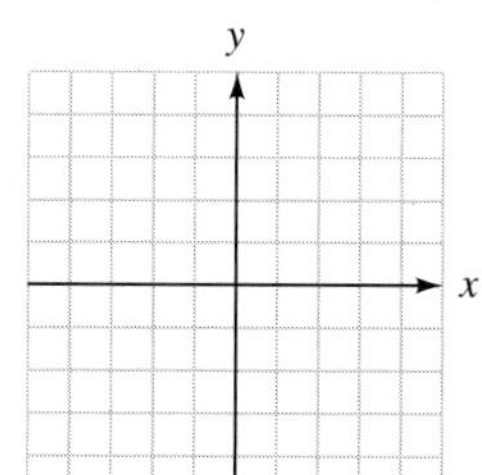

27. $4x + 5y = 20$ (see Exercise 23)

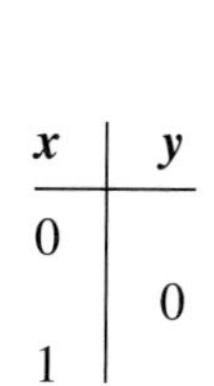

x	y
0	
	0
1	

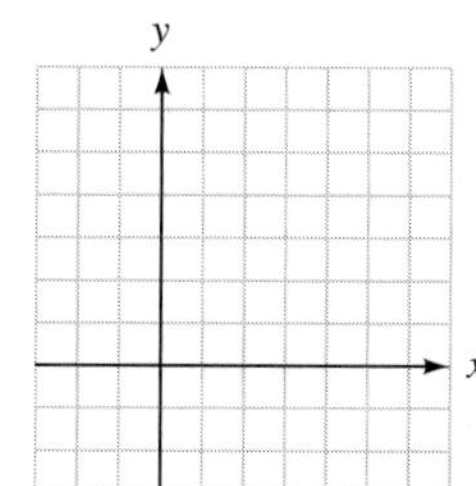

28. $3x - 5y = 15$ (see Exercise 24)

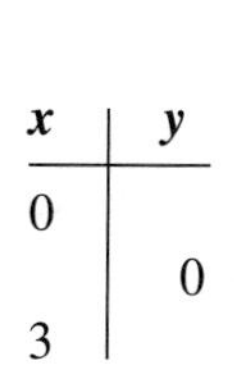

x	y
0	
	0
3	

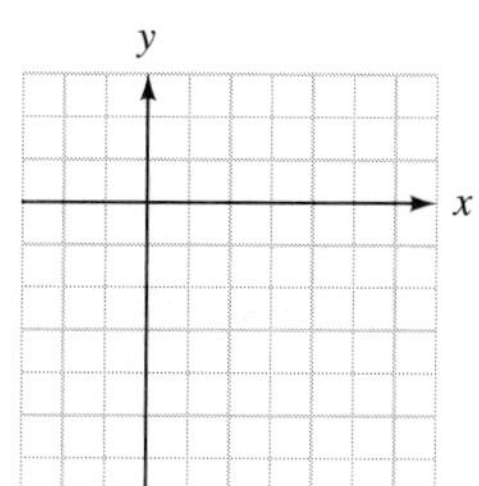

29. $x - 2y = -4$

x	y
0	
	0
4	

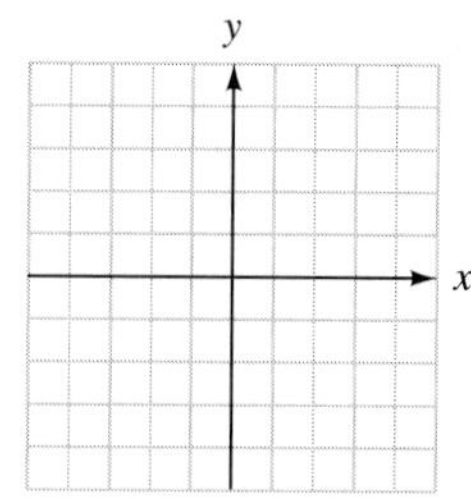

30. $3x + y = -3$

x	y
0	
	0
1	

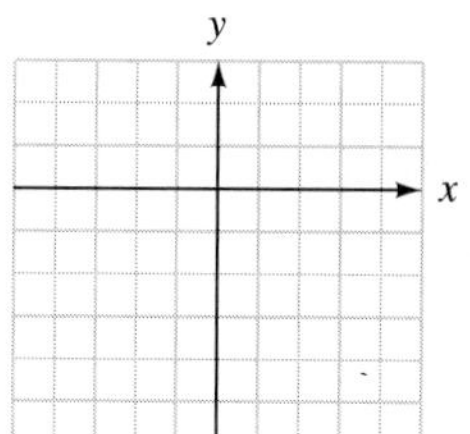

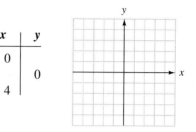

Graph each equation.

31. $y = 2x - 5$

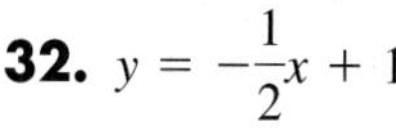

32. $y = -\frac{1}{2}x + 1$

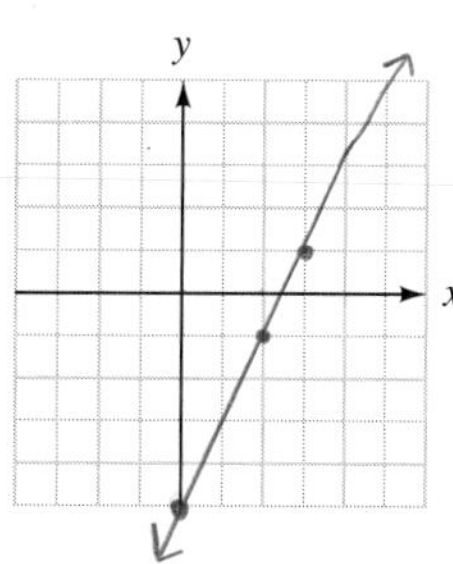

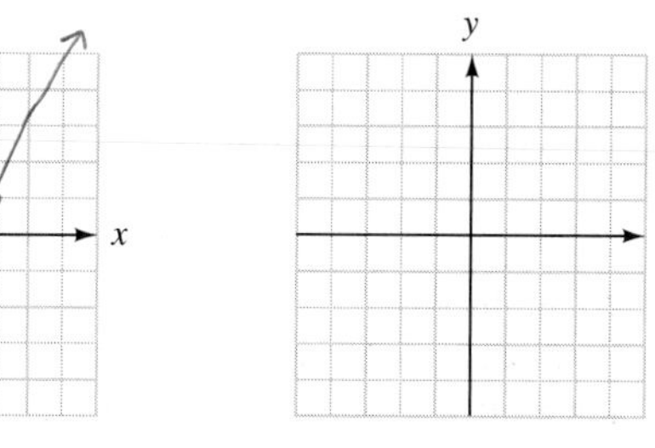

33. $y = -\frac{3}{2}x + 2$

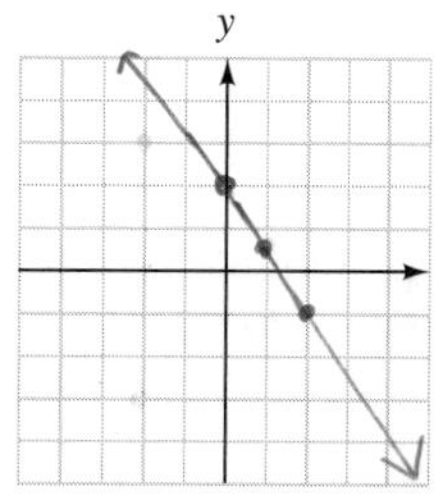

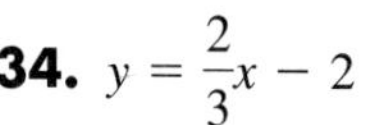

34. $y = \frac{2}{3}x - 2$

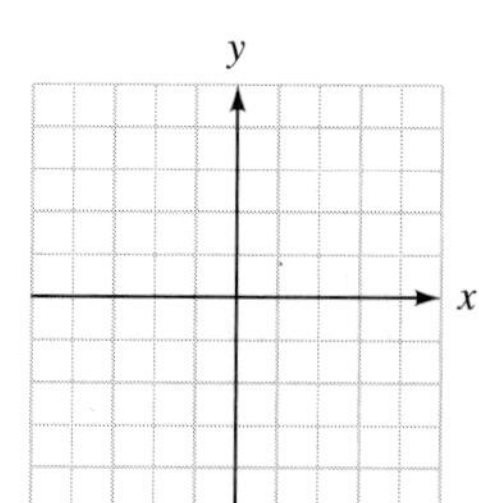

35. $y = 5$

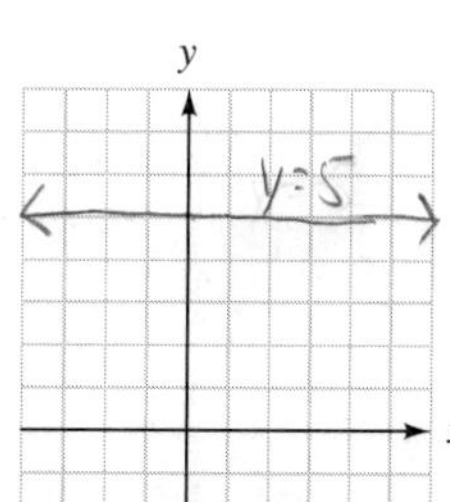

36. $x = -2$

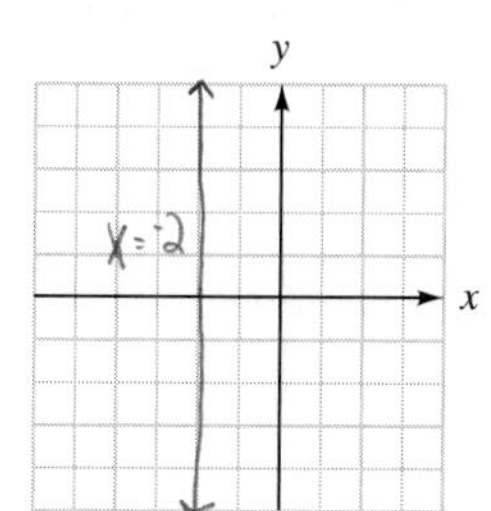

37. $x = 4$

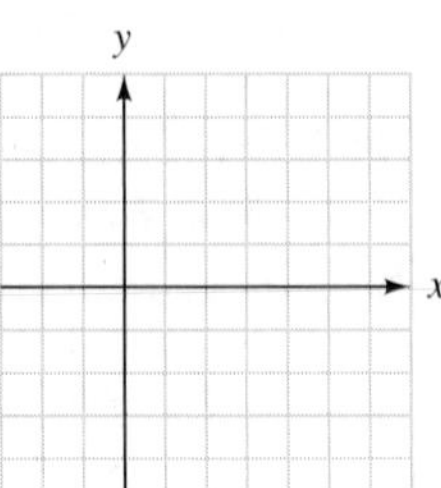

38. $y = -3$

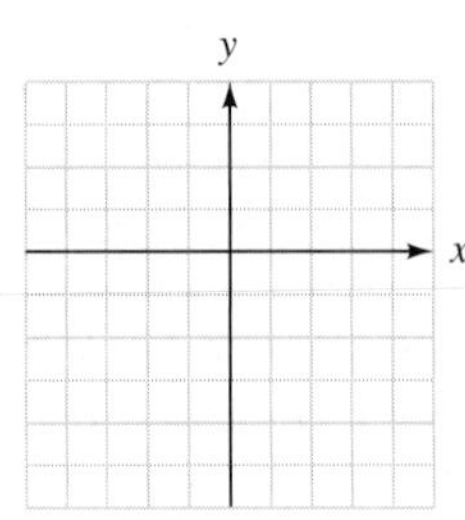

39. $y = 2x$

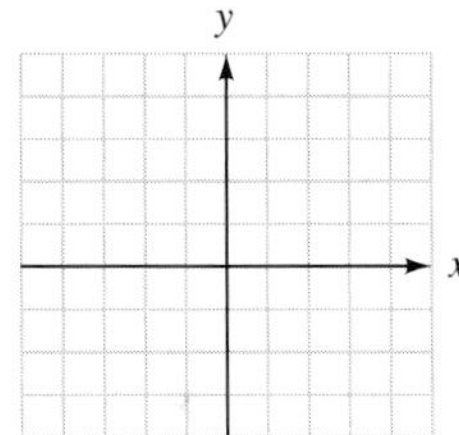

40. $y = -2x$

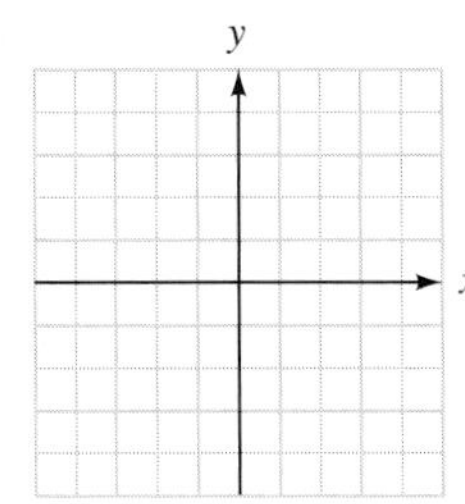

41. $y = \frac{x}{3}$

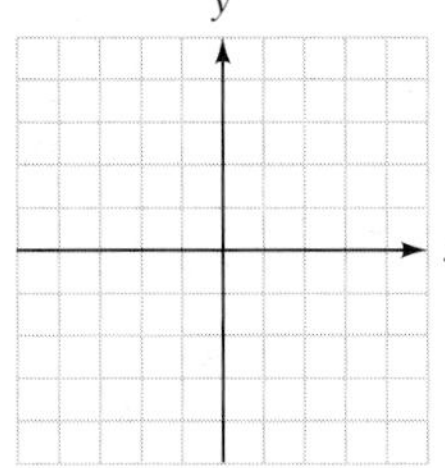

42. $y = x$

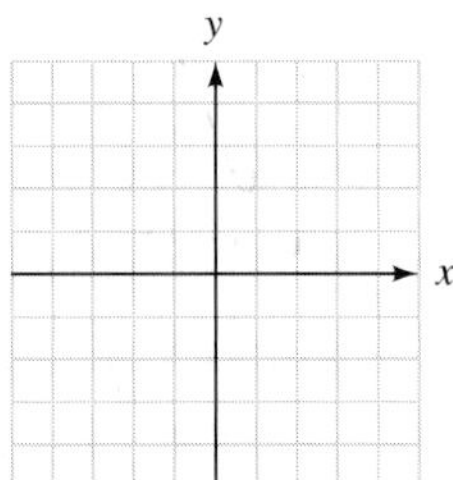

Make a table showing three solutions of each given equation. Determine an appropriate scale for the y-axis and then graph the equation.

43. $y = 100x$

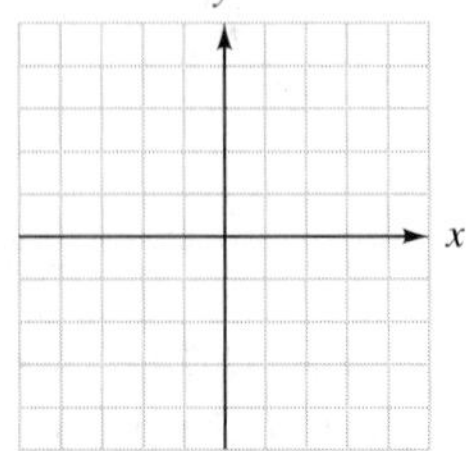

44. $y = -30x$

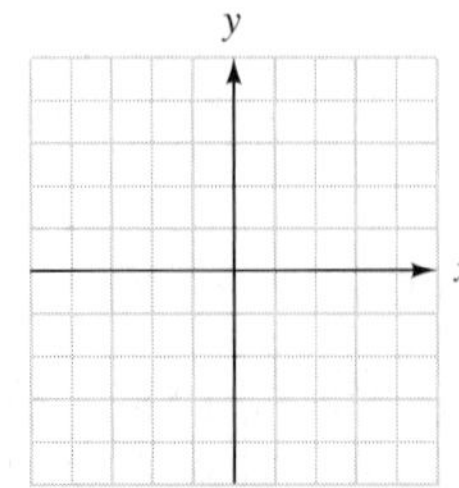

45. $y = -50x - 25$

46. $y = 200x - 400$

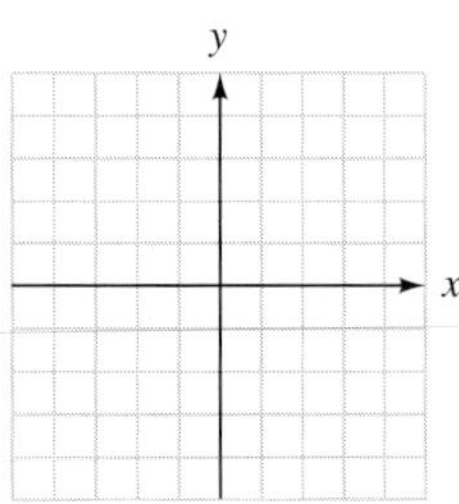

APPLICATIONS

47. HOURLY WAGES The following table gives the amount y (in dollars) that a student can earn by working x hours. Plot the ordered pairs, draw a line connecting the points, and estimate how much the student will earn in 3 hours.

x	y
2	15
4	30
6	45

y
50 45 40 35 30 25 20 15 10 5
Dollars earned
1 2 3 4 5 6 7 8 x
Hours worked

48. VALUE OF A CAR The following table shows the value y (in thousands of dollars) of a car that is x years old. Plot the ordered pairs, draw a line connecting the points, and estimate the value of the car when it is 7 years old.

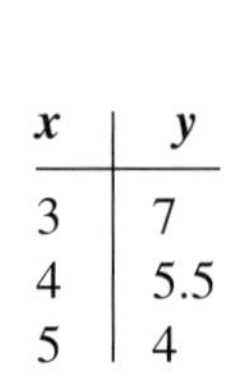

x	y
3	7
4	5.5
5	4

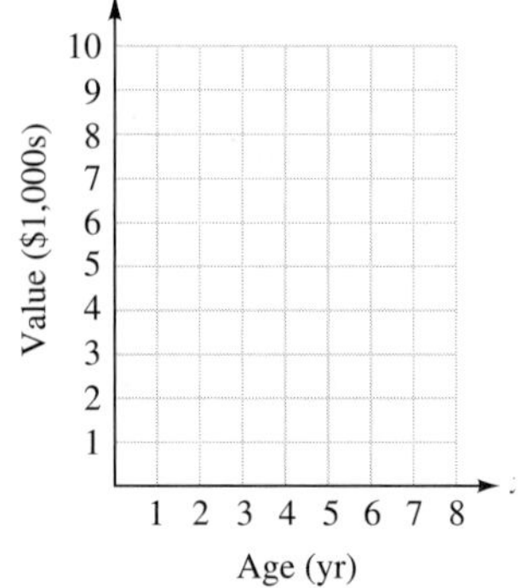

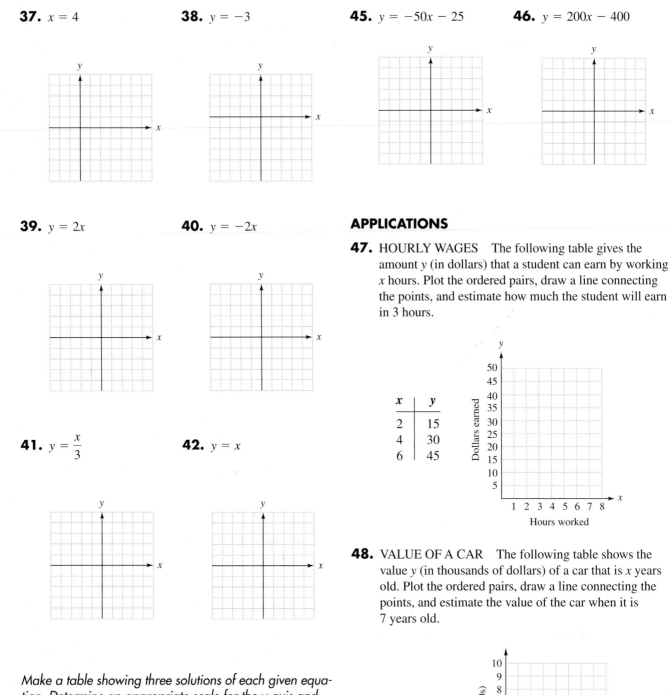

49. DISTANCE, RATE, AND TIME The formula $d = 2t$ gives the distance d (in miles) that a child can walk in a time t (in hours) at the rate of 2 miles per hour. Complete the table of solutions and then graph the equation to get a picture of the relationship between distance and time. (*Hint:* Plot t on the horizontal axis and d on the vertical axis.)

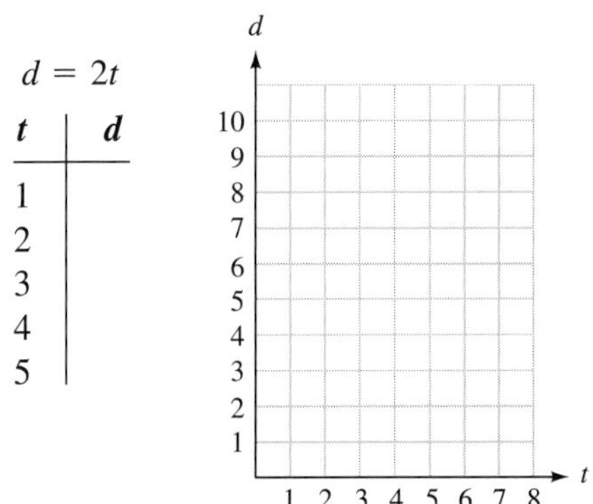

t	d
1	
2	
3	
4	
5	

50. INVESTMENTS If \$100 is invested in a savings account paying 6% per year simple interest, the amount A in the account over a period time t is given by the formula $A = 6t + 100$. Complete the table of solutions and then graph this equation to get a picture of how the account grows over a period of time. (*Hint:* Plot t on the horizontal axis and A on the vertical axis.)

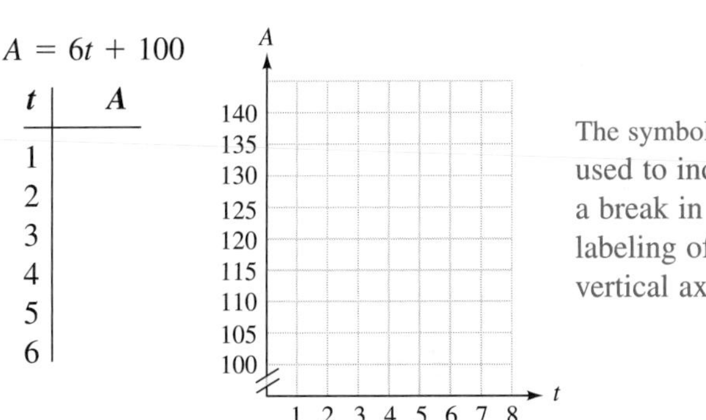

t	A
1	
2	
3	
4	
5	
6	

The symbol ⫽ is used to indicate a break in the labeling of the vertical axis.

51. AIR TRAFFIC CONTROL The equations describing the paths of two airplanes are $y = -\frac{1}{2}x + 3$ and $3y = 2x + 2$. Each equation is graphed on the radar screen in Illustration 7. Is there a possibility of a midair collision? If so, where?

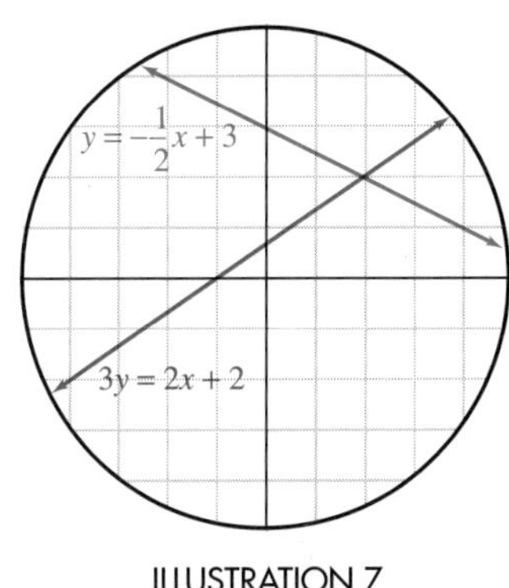

ILLUSTRATION 7

52. TV COVERAGE A television camera is located at $(-2, 0)$, as shown in Illustration 8. (Each unit in the illustration is 1 mile.) The camera is to follow the launch of a space shuttle. As the shuttle rises vertically, the camera can tilt back to a line of sight given by $y = \frac{5}{2}x + 5$. Estimate how many miles the shuttle will be in the camera's view.

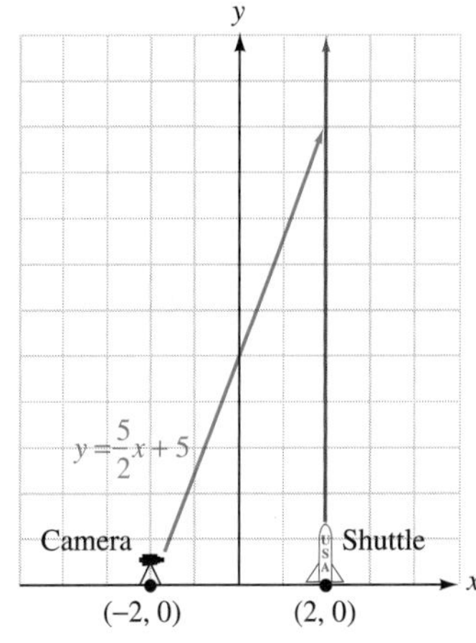

ILLUSTRATION 8

WRITING

53. When we say that $(-2, -1)$ is a solution of $5x - 6y = -4$, what do we mean?

54. What does it mean when we say that the equation $2x + y = 4$ has *infinitely many* solutions?

55. On a quiz, students were asked to graph $y = 3x - 1$. One student made the table of solutions on the left. Another student made the table of solutions on the right. Which table is incorrect? Or could they both be correct? Explain.

x	y	(x, y)
0	−1	(0, −1)
2	5	(2, 5)
3	8	(3, 8)

x	y	(x, y)
−2	−7	(−2, −7)
−1	−4	(−1, −4)
1	2	(1, 2)

56. Explain why x is called the independent variable in the equation $y = -3x - 1$.

57. Why do you think that the point $(0, 4)$ is called the y-intercept of the graph of $3x + 6y = 24$?

58. Why do you think that the point $(6, 0)$ is called the x-intercept of the graph of $3x - 5y = 18$?

REVIEW *Find the prime factorization of each number.*

59. 180

60. 270

Evaluate each expression when $a = -2$ and $b = 3$.

61. $\dfrac{3(b - a)}{5a + 7}$

62. $\dfrac{-2b^2 - b}{b}$

63. LIGHTNING The average flash of lightning lasts 0.25 second. Write the decimal as a fraction in lowest terms.

64. Simplify: $4(a + 1) - 5(6 - a)$.

6.3 *Multiplication Rules for Exponents*

In this section, you will learn about

- The product rule for exponents • The power rule for exponents
- The power rule for products

INTRODUCTION. In previous chapters, we have applied the commutative, associative, and distributive properties to simplify algebraic expressions. We will now discuss three rules that are used to simplify expressions that involve exponents.

The product rule for exponents

We have seen that exponents are used to represent repeated multiplication. For example, x^5 is an exponential expression with base x and an exponent of 5. It is called a *power of x*. Applying the definition of an exponent, we see that

$$x^5 = \underbrace{x \cdot x \cdot x \cdot x \cdot x}_{\text{5 factors of } x}$$

EXAMPLE 1 ***Writing expressions without exponents.*** Write each expression without exponents: **a.** n^3, **b.** $(3h)^2$, and **c.** $(-6b)^4$.

Solution

a. For n^3, the base is n and the exponent is 3. Therefore,

$$n^3 = n \cdot n \cdot n$$

b. For $(3h)^2$, the base is $3h$ and the exponent is 2. Therefore,

$$(3h)^2 = 3h \cdot 3h$$

c. For $(-6b)^4$, the base is $-6b$ and the exponent is 4. Therefore,

$$(-6b)^4 = (-6b)(-6b)(-6b)(-6b)$$

Self Check

Write each expression without exponents:

a. t^4

b. $(12f)^3$

c. $(-2x)^2$

Answers: **a.** $t \cdot t \cdot t \cdot t$, **b.** $12f \cdot 12f \cdot 12f$, **c.** $(-2x)(-2x)$ ■

The expression $x^3 \cdot x^5$ is the *product* of two powers of x. To develop a rule for multiplying them, we will use the fact that an exponent indicates repeated multiplication.

$$x^3 \cdot x^5 = \underbrace{(x \cdot x \cdot x)}_{\text{3 factors of } x}\underbrace{(x \cdot x \cdot x \cdot x \cdot x)}_{\text{5 factors of } x}$$

x^3 means to write x as a factor 3 times. x^5 means to write x as a factor 5 times.

$$= \underbrace{(x \cdot x \cdot x \cdot x \cdot x \cdot x \cdot x \cdot x)}_{\text{8 factors of } x}$$

Do the multiplication to get 8 factors of x.

$$= x^8$$

Since x is used as a factor 8 times, we can write the product as x^8.

Notice that the exponent of the result is the *sum* of the exponents in $x^3 \cdot x^5$.

Sum of the exponents

$$x^3 \cdot x^5 = x^{3+5} = x^8$$

This observation suggests the following rule.

The product rule for exponents

For any number x and any positive integers m and n,

$$x^m \cdot x^n = x^{m+n}$$

To multiply two exponential expressions with the same base, add the exponents and keep the common base.

EXAMPLE 2 ***Simplifying exponential expressions.*** Simplify each product: **a.** $3^4 \cdot 3^7$ **b.** $(y^2)(y^4)$, and **c.** $x^2x^4x^9$.

Solution

a. $3^4 \cdot 3^7 = 3^{4+7}$ Since the bases are the same, add the exponents and keep the common base, which is 3.

$= 3^{11}$ Do the addition: $4 + 7 = 11$.

b. $(y^2)(y^4) = y^{2+4}$ Since the bases are the same, add the exponents and keep the common base, which is y.

$= y^6$ Do the addition: $2 + 4 = 6$.

c. $x^2x^4x^9 = x^{2+4+9}$ Since the bases are the same, add the exponents and keep the common base, which is x.

$= x^{15}$ Do the addition: $2 + 4 + 9 = 15$.

Self Check

Simplify each product:

a. $5^3 \cdot 5^6$

b. $m^2 \cdot m^3$

c. $b^3b^8b^2$

Answers: **a.** 5^9, **b.** m^5, **c.** b^{13}

COMMENT We cannot use the product rule for exponents to simplify an expression such as $x^4 + x^3$, because it is not a product. Nor can we use it to simplify $x^4 \cdot y^3$, because the bases are not the same.

The product rule for exponents can be used to simplify more complicated algebraic expressions involving multiplication.

EXAMPLE 3 ***Using the product rule.*** Simplify each product: **a.** $3a(5a^2)$ and **b.** $-2t^2 \cdot 6t^6$.

Solution

a. $3a(5a^2) = (3 \cdot 5)(a \cdot a^2)$ Apply the commutative and associative properties to change the order and regroup the factors.

$= (3 \cdot 5)(a^1 \cdot a^2)$ Recall that $a = a^1$.

$= 15a^{1+2}$ Do the multiplication: $3 \cdot 5 = 15$. Then add the exponents and keep the common base, which is a.

$= 15a^3$ Do the addition: $1 + 2 = 3$.

b. $-2t^2 \cdot 6t^6 = (-2 \cdot 6)(t^2 \cdot t^6)$ Change the order of the factors and regroup them.

$= -12t^{2+6}$ Do the multiplication: $-2 \cdot 6 = -12$. Then add the exponents and keep the common base.

$= -12t^8$ Do the addition: $2 + 6 = 8$.

Self Check

Simplify each product:

a. $4m \cdot 6m^5$

b. $8r^3(-5r^2)$

Answers: **a.** $24m^6$, **b.** $-40r^5$

Exponential expressions often contain more than one variable.

EXAMPLE 4 ***Working with two variables.*** Simplify the following: **a.** $n^2m \cdot n^8m^3$ and **b.** $4xy^2(-3x^2y^3)$.

Solution

a. $n^2m \cdot n^8m^3 = (n^2 \cdot n^8)(m \cdot m^3)$ Change the order and group the factors with like bases.

$= n^{2+8} \cdot m^{1+3}$ Add the exponents of the like bases. Recall that $m = m^1$.

$= n^{10}m^4$ Do the additions.

b. $4xy^2(-3x^2y^3) = [4(-3)](x \cdot x^2)(y^2 \cdot y^3)$ Group the factors with like bases.

$= -12 \cdot x^{1+2} \cdot y^{2+3}$ Do the multiplication. Add the exponents of the like bases.

$= -12x^3y^5$ Do the additions.

Self Check

Simplify the following:

a. $c^3d^2 \cdot cd^5$

b. $-7a^2b^3(8a^4b^5)$

Answers: **a.** c^4d^7, **b.** $-56a^6b^8$ ■

The power rule for exponents

To develop the power rule for exponents, we consider the expression $(x^2)^5$. Notice that the base, x^2, is raised to a power. Therefore, we are working with a power of a power. We will again rely on the definition of an exponent to find a rule for simplifying this exponential expression.

$(x^2)^5 = x^2 \cdot x^2 \cdot x^2 \cdot x^2 \cdot x^2$ The exponent 5 tells us to write the base x^2 five times.

$= x^{2+2+2+2+2}$ Since the bases are alike, add the exponents and keep the common base.

$= x^{10}$ Do the addition: $2 + 2 + 2 + 2 + 2 = 10$.

Notice that the exponent of the result is the *product* of the exponents in $(x^2)^5$.

Product of the exponents

$$(x^2)^5 = x^{2 \cdot 5} = x^{10}$$

This observation suggests the following rule.

The power rule for exponents

For any number x and any positive integers m and n,

$$(x^m)^n = x^{m \cdot n} \quad \text{or, more simply,} \quad (x^m)^n = x^{mn}$$

To raise an exponential expression to a power, keep the base and multiply the exponents.

EXAMPLE 5 ***The power rule for exponents.*** Simplify each expression: **a.** $(2^3)^7$ and **b.** $(b^5)^3$.

Solution

a. $(2^3)^7 = 2^{3 \cdot 7}$ Apply the power rule for exponents by keeping the base and multiplying the exponents.

$= 2^{21}$ Do the multiplication: $3 \cdot 7 = 21$

b. $(b^5)^3 = b^{5 \cdot 3}$ Keep the base and multiply the exponents.

$= b^{15}$ Do the multiplication: $5 \cdot 3 = 15$.

Self Check

Simplify each expression:

a. $(4^2)^6$

b. $(y^6)^4$

Answers: **a.** 4^{12}, **b.** y^{24} ■

In some cases, when simplifying algebraic expressions involving exponents, two rules of exponents must be applied.

EXAMPLE 6 ***Applying two rules.*** Simplify: **a.** $(n^3)^4(n^2)^5$ and **b.** $(n^2n^3)^5$.

Solution

a. $(n^3)^4(n^2)^5 = n^{3\cdot4} \cdot n^{2\cdot5}$ Keep each base and multiply their exponents.

$= n^{12} \cdot n^{10}$ Do the multiplications: $3 \cdot 4 = 12$ and $2 \cdot 5 = 10$.

$= n^{12+10}$ Since the bases are alike, keep the base and add the exponents.

$= n^{22}$ Do the addition: $12 + 10 = 22$.

b. $(n^2n^3)^5 = (n^{2+3})^5$ Work within the parentheses first. Since the bases are alike, keep the base and add the exponents.

$= (n^5)^5$ Do the addition: $2 + 3 = 5$.

$= n^{5\cdot5}$ Keep the base and multiply the exponents.

$= n^{25}$ Do the multiplication: $5 \cdot 5 = 25$.

Self Check

Simplify:

a. $(x^4)^2(x^3)^3$

b. $(x^4x^2)^3$

Answers: **a.** x^{17}, **b.** x^{18} ■

The power rule for products

The exponential expression $(2x)^4$ has an exponent of 4 and a base of $2x$. The base $2x$ is a product, since $2x = 2 \cdot x$. Therefore, $(2x)^4$ is a power of a product. To find a rule to simplify it, we will again use the definition of exponent.

$(2x)^4 = 2x \cdot 2x \cdot 2x \cdot 2x$ Write the base, $2x$, as a factor 4 times.

$= (2 \cdot 2 \cdot 2 \cdot 2)(x \cdot x \cdot x \cdot x)$ Apply the commutative and associative properties of multiplication to change the order and group like factors.

$= 2^4x^4$ The factor 2 and the factor x are both repeated 4 times. Apply the definition of an exponent.

The result has factors of 2 and x. In the original problem, they were within the parentheses. Each is now raised to the fourth power.

Each factor within the parentheses ends up being raised to the 4th power.

$$(2x)^4 = 2^4x^4$$

This observation suggests the following rule.

The power rule for products

For any numbers x and y, and any positive integer m,

$$(xy)^m = x^my^m$$

To raise a product to a power, raise each factor of the product to that power.

EXAMPLE 7 ***The power rule for products.*** Simplify each expression: **a.** $(8a)^2$ and **b.** $(2bx)^3$.

Solution

a. $(8a)^2 = 8^2a^2$ To raise $8a$ to the 2nd power, raise each factor of the product to the 2nd power.

$= 64a^2$ Find the power: $8^2 = 64$.

b. $(2bx)^3 = 2^3b^3x^3$ To raise $2bx$ to the 3rd power, raise each factor of the product to the 3rd power.

$= 8b^3x^3$ Find the power: $2^3 = 8$.

Self Check

Simplify each expression:

a. $(10c)^2$

b. $(5rs)^3$

Answers: **a.** $100c^2$, **b.** $125r^3s^3$ ■

EXAMPLE 8 ***Applying two rules.*** Simplify each expression:
a. $(10a^2)^3$ and **b.** $(3c^5d^3)^4$.

Solution

a. $(10a^2)^3 = 10^3(a^2)^3$ — To raise $10a^2$ to the 3rd power, raise each factor of the product, 10 and a^2, to the 3rd power.

$= 10^3a^{2\cdot 3}$ — To raise a^2 to a power, keep the base and multiply the exponents.

$= 10^3a^6$ — Do the multiplication: $2 \cdot 3 = 6$.

$= 1{,}000a^6$ — Find the power: $10^3 = 1{,}000$.

b. $(3c^5d^3)^4 = 3^4(c^5)^4(d^3)^4$ — To raise $3c^5d^3$ to the 4th power, raise each factor of the product, 3, c^5, and d^3, to the 4th power.

$= 3^4c^{5\cdot 4}d^{3\cdot 4}$ — To raise c^5 and d^3 to powers, keep the bases and multiply their exponents.

$= 3^4c^{20}d^{12}$ — Do the multiplications: $5 \cdot 4 = 20$ and $3 \cdot 4 = 12$.

$= 81c^{20}d^{12}$ — Find the power: $3^4 = 81$.

Self Check
Simplify each expression:
a. $(3n^2)^3$
b. $(6h^2s^9)^2$

Answers: **a.** $27n^6$, **b.** $36h^4s^{18}$ ■

EXAMPLE 9 ***Applying three rules.*** Simplify $(2a^2)^2(4a^3)^3$.

Solution

$(2a^2)^2(4a^3)^3 = 2^2(a^2)^2 \cdot 4^3(a^3)^3$ — To raise $2a^2$ and $4a^3$ to powers, raise the factors of each product to the appropriate power.

$= 2^2a^{2\cdot 2} \cdot 4^3 \cdot a^{3\cdot 3}$ — To raise a^2 and a^3 to powers, keep the bases and multiply the exponents.

$= 2^2a^4 \cdot 4^3a^9$ — Do the multiplications: $2 \cdot 2 = 4$ and $3 \cdot 3 = 9$.

$= (2^2 \cdot 4^3)(a^4 \cdot a^9)$ — Change the order of the factors and group like bases.

$= (2^2 \cdot 4^3)(a^{4+9})$ — To multiply $a^4 \cdot a^9$, keep the base and add the exponents.

$= (2^2 \cdot 4^3)a^{13}$ — Do the addition: $4 + 9 = 13$.

$= (4 \cdot 64)a^{13}$ — Find the powers: $2^2 = 4$ and $4^3 = 64$.

$= 256a^{13}$ — Do the multiplication: $4 \cdot 64 = 256$.

Self Check
Simplify $(4y^3)^2(3y^4)^3$.

Answer: $432y^{18}$ ■

STUDY SET Section 6.3

VOCABULARY *Fill in the blanks to make the statements true.*

1. In x^n, x is called the ________ and n is called the ________.

2. x^2 is the second ________ of x, or we can read it as "x ________."

3. $x^m \cdot x^n$ is the product of two exponential expressions with ________ bases.

4. $(x^m)^n$ is a power of a ________.

5. $(2x)^n$ is a ________ raised to a power.

6. In x^{m+n}, $m + n$ is the ________ of m and n.

CONCEPTS

7. Represent each repeated multiplication using exponents.
a. $x \cdot x \cdot x \cdot x \cdot x \cdot x \cdot x$
b. $x \cdot x \cdot y \cdot y \cdot y$
c. $3 \cdot 3 \cdot 3 \cdot 3 \cdot a \cdot a \cdot b \cdot b \cdot b$

8. Write each exponential expression as repeated multiplication.
a. a^3b^5
b. $(x^2)^3$
c. $(2a)^3$

9. Write a product of two exponential expressions with like variable bases. Then simplify it using a rule of exponents.

10. Write a power of a product and then simplify it using a rule of exponents.

11. Write a power of a power and then simplify it using a rule of exponents.

12. What algebraic property allows us to change the order of the factors of a multiplication?

13. Simplify each expression using a rule of exponents.
a. $x^m x^n$
b. $(x^m)^n$
c. $(ax)^n$

14. In each case, tell how the expression has been improperly simplified.
a. $2^3 \cdot 2^4 = 2^{12}$
b. $3^3 \cdot 3^4 = 9^7$
c. $(2^3)^4 = 2^7$

15. Write each expression without an exponent.
a. 2^1 **b.** $(-10)^1$ **c.** x^1

16. Find each power.
a. 2^3 **b.** 4^3 **c.** 5^3

17. Simplify each expression, if possible.
a. $x \cdot x$ and $x + x$
b. $x \cdot x^2$ and $x + x^2$
c. $x^2 \cdot x^2$ and $x^2 + x^2$

18. Simplify each expression, if possible.
a. $a \cdot a$ and $a - a$
b. $2ab \cdot ab$ and $2ab - ab$
c. $2ab \cdot 3ab$ and $2ab - 3ab$

19. Simplify each expression, if possible.
a. $4x \cdot x$ and $4x + x$
b. $4x \cdot 3x$ and $4x + 3x$
c. $4x^2 \cdot 3x$ and $4x^2 + 3x$

20. Simplify each expression, if possible.
a. $ab(-2ab)$ and $ab - 2ab$
b. $-2ab(3ab^2)$ and $-2ab + 3ab^2$
c. $-2ab^2(-3ab^2)$ and $-2ab^2 - 3ab^2$

21. Evaluate the exponential expression x^{m+n} for $x = 3$, $m = 2$, and $n = 1$.

22. Evaluate the exponential expression $(x^m)^n$ for $x = 2$, $m = 3$, and $n = 2$.

NOTATION *Complete each solution.*

23. $x^5 \cdot x^7 = x^{\square}$
$= x^{12}$

24. $(x^5)^4 = x^{\square}$
$= x^{20}$

25. $(2x^4)(8x^3) = (2 \cdot 8)(\quad \cdot \quad)$
$= 16x^{\square}$
$= 16x^7$

26. $(2x^2)^3 = 2^3(x^2)^3$
$= 2^3x^{\square}$
$= 2^3x^{\square}$
$= 8x^6$

PRACTICE *Write each expression using one exponent.*

27. $x^2 \cdot x^3$ **28.** $t^4 \cdot t^3$
29. x^3x^7 **30.** y^2y^5
31. $f^5(f^8)$ **32.** $g^6(g^2)$
33. $n^{24} \cdot n^8$ **34.** $m^9 \cdot m^{61}$
35. $l^4 \cdot l^5 \cdot l$ **36.** $w^4 \cdot w \cdot w^3$
37. $x^6(x^3)x^2$ **38.** $y^5(y^2)(y^3)$
39. $2^4 \cdot 2^8$ **40.** $3^4 \cdot 3^2$
41. $5^6(5^2)$ **42.** $(8^3)8^4$

Simplify each product.

43. $2x^2 \cdot 4x$ **44.** $5y \cdot 6y^3$
45. $5t \cdot t^9$ **46.** $f^4 \cdot 3f$
47. $-6x^3(4x^2)$ **48.** $-7y^5(5y^3)$
49. $-x \cdot x^3$ **50.** $8x^6(-x)$
51. $6y(2y^3)3y^4$ **52.** $2d(5d^4)(d^2)$
53. $-2t^3(-4t^2)(-5t^5)$ **54.** $-7k^5(-3k^3)(-2k^9)$

55. $xy^2 \cdot x^2y$ **56.** $s^2t \cdot st$
57. $b^3 \cdot c^2 \cdot b^5 \cdot c^6$ **58.** $h^3 \cdot f^3 \cdot f^2 \cdot h^4$
59. $x^4y(xy)$ **60.** $(ab)(ab^2)$
61. $a^2b \cdot b^3a^2$ **62.** $w^2y \cdot yw^4$
63. $x^5y \cdot y^6$ **64.** $a^7 \cdot b^2a^4$
65. $3x^2y^3 \cdot 6xy$ **66.** $25a^3b \cdot 2ab^5$
67. $xy^2 \cdot 16x^3$ **68.** $mn^4 \cdot 8n^3$
69. $-6f^2t(4f^4t^3)$ **70.** $(-5a^2b^2)(5a^3b^6)$

71. $ab \cdot ba \cdot a^2b$ **72.** $xy \cdot y^2x \cdot x^2y$

73. $-4x^2y(-3x^2y^2)$ **74.** $-2rt^4(-5r^2t^2)$

Simplify each expression.

75. $(x^2)^4$ **76.** $(y^6)^3$
77. $(m^{50})^{10}$ **78.** $(n^{25})^4$
79. $(2a)^3$ **80.** $(3x)^3$

81. $(xy)^4$
82. $(ab)^8$
83. $(3s^2)^3$
84. $(5f^6)^2$
85. $(2s^2t^3)^2$
86. $(4h^5y^6)^2$
87. $(x^2)^3(x^4)^2$
88. $(a^5)^2(a^3)^3$
89. $(c^5)^3 \cdot (c^3)^5$
90. $(y^2)^8 \cdot (y^8)^2$
91. $(2a^4)^2(3a^3)^2$
92. $(5x^3)^2(2x^4)^3$
93. $(3a^3)^3(2a^2)^3$
94. $(6t^5)^2(2t^2)^2$
95. $(x^2x^3)^{12}$
96. $(a^3a^3)^3$
97. $(2b^4b)^5$
98. $(3y^2y^5)^3$

WRITING

99. Explain the difference between x^2 and $2x$.
100. Explain why the rules of exponents do not apply to $x^2 + x^3$.
101. One of the rules of exponents is that the power of a product is the product of the powers. Use a specific example to explain this rule.
102. To find the result when *multiplying* two exponential expressions with like bases, we must *add* the exponents. Explain why this is so.

REVIEW

103. JEWELRY A lot of what we refer to as gold jewelry is actually made of a combination of gold and another metal. For example, 18-karat gold is $\frac{18}{24}$ gold by weight. Simplify this fraction.
104. When evaluated, what is the sign of $(-13)^5$?
105. Divide: $\frac{-25}{-5}$.
106. How much did the temperature change if it went from $-4°$ to $-17°$?
107. Evaluate $2\left(\frac{12}{-3}\right) + 3(5)$.
108. Solve $-4 - 6 = x + 1$.
109. Solve $-x = -12$.
110. Divide: $\frac{0}{10}$.

6.4 Introduction to Polynomials

In this section, you will learn about

- Polynomials • Classifying polynomials • Degree of a polynomial
- Evaluating polynomials • Graphing equations involving polynomials

Introduction. Earlier in this chapter, we graphed the linear equations $y = 2x - 1$ and $y = \frac{1}{2}x + 1$. The expressions $2x - 1$ and $\frac{1}{2}x + 1$ are examples of algebraic expressions called *polynomials.* In this section, we will define polynomials, classify them into groups, and show how to evaluate them. Finally, we will show how to graph equations involving polynomials.

Polynomials

Recall that an **algebraic term,** or simply a **term,** is a number or a product of a number and one or more variables, which may be raised to powers. Some examples of terms are

$$17, \quad 5x, \quad 6t^2, \quad \text{and} \quad -8z^3$$

The coefficients of these terms are 17, 5, 6, and -8, respectively.

Polynomials

A **polynomial** is a term or a sum of terms in which all variables have whole-number exponents.

Some examples of polynomials are

$$0, \quad 8y^2, \quad 2x + 1, \quad 4y^2 - 2y + 3, \quad \text{and} \quad 7a^3 + 2a^2 - a - 1$$

The polynomial $8y^2$ has one term. The polynomial $2x + 1$ has two terms, $2x$ and 1. Since $4y^2 - 2y + 3$ can be written as $4y^2 + (-2y) + 3$, it is the sum of three terms, $4y^2$, $-2y$, and 3.

Classifying polynomials

We classify some polynomials by the number of terms they contain. A polynomial with one term is called a **monomial.** A polynomial with two terms is called a **binomial.** A polynomial with three terms is called a **trinomial.** Some examples of these polynomials are shown in Table 6-1.

Monomials	Binomials	Trinomials
$5x^2$	$2x - 1$	$5t^2 + 4t + 3$
$-6x$	$18a^2 - 4a$	$27x^3 - 6x + 2$
29	$-27z^4 + 7z^2$	$32r^2 + 7r - 12$

TABLE 6-1

EXAMPLE 1 ***Classifying polynomials.*** Classify each polynomial as either a monomial, a binomial, or a trinomial: **a.** $3x + 4$, **b.** $3x^2 + 4x - 12$, and **c.** $25x^3$.

Solution

a. Since $3x + 4$ has two terms, it is a binomial.

b. Since $3x^2 + 4x - 12$ has three terms, it is a trinomial.

c. Since $25x^3$ has one term, it is a monomial.

Self Check

Classify each polynomial as either a monomial, a binomial, or a trinomial:

a. $5x$

b. $8x^2 + 7$

c. $x^2 - 2x - 1$

Answers: **a.** monomial, **b.** binomial, **c.** trinomial ■

Degree of a polynomial

The monomial $7x^3$ is called a **monomial of third degree** or a **monomial of degree 3,** because the variable occurs three times as a factor.

- $5x^2$ is a monomial of degree 2. Because the variable occurs two times as a factor: $x^2 = x \cdot x$.
- $-8x^4$ is a monomial of degree 4. Because the variable occurs four times as a factor: $x^4 = x \cdot x \cdot x \cdot x$.
- $\frac{1}{2}x^5$ is a monomial of degree 5. Because the variable occurs five times as a factor: $x^5 = x \cdot x \cdot x \cdot x \cdot x$.

We define the degree of a polynomial by considering the degrees of each of its terms.

Degree of a polynomial

The **degree of a polynomial** is the same as the degree of its term with largest degree.

For example,

- $x^2 + 5x$ is a binomial of degree 2, because the degree of its term with largest degree (x^2) is 2.
- $4y^3 + 2y - 7$ is a trinomial of degree 3, because the degree of its term with largest degree ($4y^3$) is 3.

- $\frac{1}{2}z + 3z^4 - 2z^2$ is a trinomial of degree 4, because the degree of its term with largest degree ($3z^4$) is 4.

EXAMPLE 2 ***Finding the degree of a polynomial.*** Find the degree of each polynomial: **a.** $-2x + 4$, **b.** $5t^3 + t^4 - 7$, and **c.** $3 - 9z + 6z^2 - z^3$.

Solution

a. Since $-2x$ can be written as $-2x^1$, the degree of the term with largest degree is 1. Thus, the degree of the polynomial is 1.

b. In $5t^3 + t^4 - 7$, the degree of the term with largest degree (t^4) is 4. Thus, the degree of the polynomial is 4.

c. In $3 - 9z + 6z^2 - z^3$, the degree of the term with largest degree ($-z^3$) is 3. Thus, the degree of the polynomial is 3.

Self Check

Find the degree of each polynomial:

a. $3p^3$

b. $17r^4 + 2r^8 - r$

c. $-2g^5 - 7g^6 + 12g^7$

Answers: **a.** 3, **b.** 8, **c.** 7

Evaluating polynomials

When a number is substituted for the variable in a polynomial, the polynomial takes on a numerical value. Finding this value is called **evaluating the polynomial.**

EXAMPLE 3 ***Evaluating a polynomial.*** Evaluate each polynomial when $x = 3$: **a.** $3x - 2$ and **b.** $-2x^2 + x - 3$.

Solution

a.

$3x - 2 = 3(3) - 2$	Substitute 3 for x.
$= 9 - 2$	Multiply: $3(3) = 9$.
$= 7$	Subtract: $9 - 2 = 7$.

b.

$-2x^2 + x - 3 = -2(3)^2 + 3 - 3$	Substitute 3 for x.
$= -2(9) + 3 - 3$	Square 3: $3 \cdot 3 = 9$.
$= -18 + 3 - 3$	Multiply: $-2(9) = -18$.
$= -15 - 3$	Add: $-18 + 3 = -15$.
$= -18$	Subtract: $-15 - 3 = -18$.

Self Check

Evaluate each polynomial when $x = -1$:

a. $-2x^2 - 4$

b. $3x^2 - 4x + 1$

Answers: **a.** -6, **b.** 8

EXAMPLE 4 ***Height of an object.*** The polynomial $-16t^2 + 28t + 8$ gives the height (in feet) of an object t seconds after it has been thrown straight up. Find the height of the object in 1 second.

Solution

To find the height at 1 second, we evaluate the polynomial at $t = 1$.

$-16t^2 + 28t + 8 = -16(1)^2 + 28(1) + 8$	Substitute 1 for t.
$= -16(1) + 28(1) + 8$	Square 1: $1 \cdot 1 = 1$.
$= -16 + 28 + 8$	Multiply: $-16(1) = -16$ and $28(1) = 28$.
$= 12 + 8$	Add: $-16 + 28 = 12$.
$= 20$	Add: $12 + 8 = 20$.

At 1 second, the object is 20 feet above the ground.

Self Check

Find the height of the object in 2 seconds.

Answer: 0 ft

Graphing equations involving polynomials

In the previous section, we graphed the linear equations

$$y = 2x - 1 \quad \text{and} \quad y = \frac{1}{2}x + 1$$

The polynomials $2x - 1$ and $\frac{1}{2}x + 1$ are both of degree 1.

We will now graph some equations involving second-degree polynomials. The method we will use is like that used to graph linear equations earlier in this chapter.

EXAMPLE 5 *Graphing a second-degree polynomial.* Graph $y = 2x^2$.

Solution

To make a table of solutions for $y = 2x^2$, we choose numbers for x and finding the corresponding values of y. If $x = -2$, we get

$y = 2x^2$

$y = 2(-2)^2$ Substitute -2 for x.

$y = 2(4)$ Square -2 first: $(-2)^2 = -2(-2) = 4$.

$y = 8$ Do the multiplication.

Thus, $x = -2$ and $y = 8$ is a solution of $y = 2x^2$. In a similar way, we find the corresponding values for y when x is -1, 0, 1, and 2 and enter them in the table in Figure 6-13. If we plot the ordered pairs and join the points with a smooth curve, we get a U-shaped figure, called a **parabola.**

x	y	(x, y)
-2	8	$(-2, 8)$
-1	2	$(-1, 2)$
0	0	(0, 0)
1	2	(1, 2)
2	8	(2, 8)

↑ We can choose any values for x.

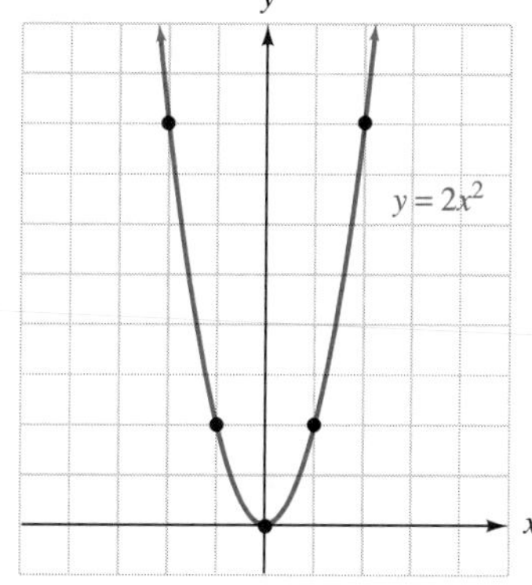

FIGURE 6-13

Self Check

Graph $y = 3x^2$.

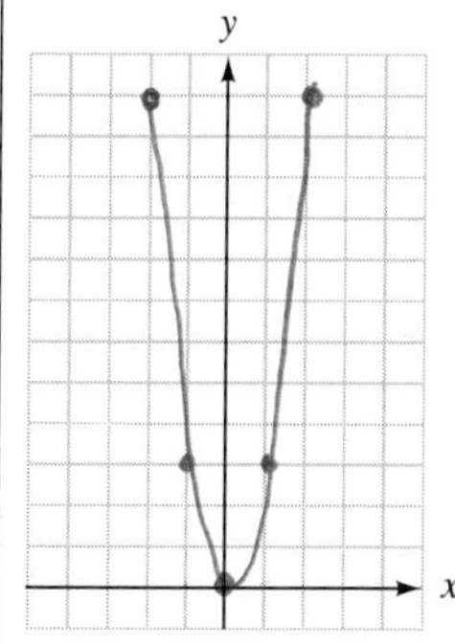

Answer:

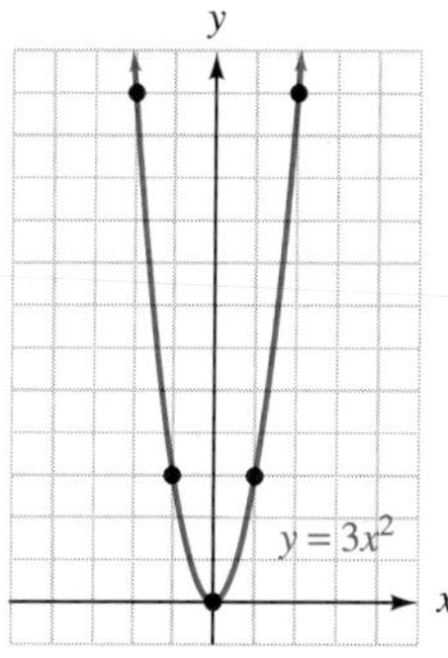

■

COMMENT When an equation is not linear, its graph is not a straight line. To graph nonlinear equations, we must usually plot many points to recognize the shape of the curve.

EXAMPLE 6 *Graphing a second-degree polynomial.* Graph $y = -x^2 + 3$.

Solution

We make a table of values by substituting numbers for x and finding the corresponding values of y. For example, if we substitute -3 for x, we get

$y = -x^2 + 3$

$y = -(-3)^2 + 3$ Don't forget to write the $-$ sign shown in blue.

$y = -9 + 3$ Square -3: $(-3)(-3) = 9$.

$y = -6$ Add: $-9 + 3 = -6$.

The coordinates of this point and others are shown in the table on the next page. To get the graph, we plot each of these points and join them with a smooth curve. (See Figure 6-14.) The graph is a parabola opening downward.

Self Check

Graph $y = x^2 - 1$.

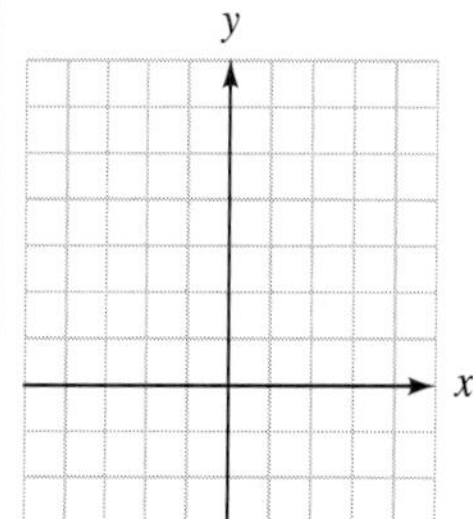

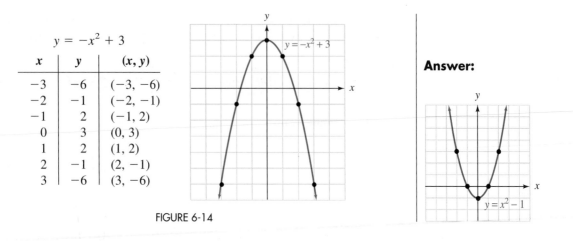

$y = -x^2 + 3$

x	y	(x, y)
−3	−6	(−3, −6)
−2	−1	(−2, −1)
−1	2	(−1, 2)
0	3	(0, 3)
1	2	(1, 2)
2	−1	(2, −1)
3	−6	(3, −6)

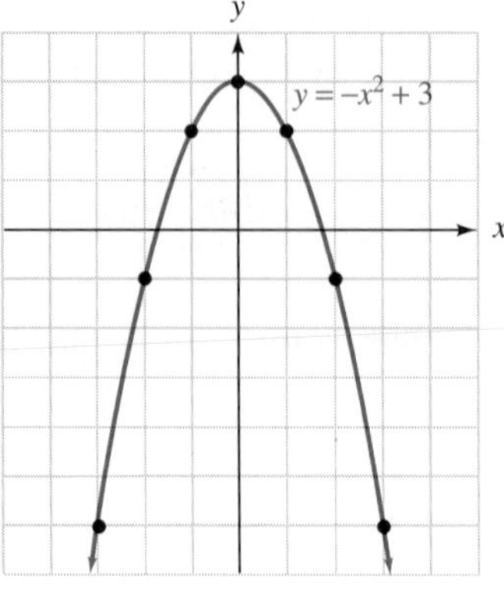

FIGURE 6-14

Answer:

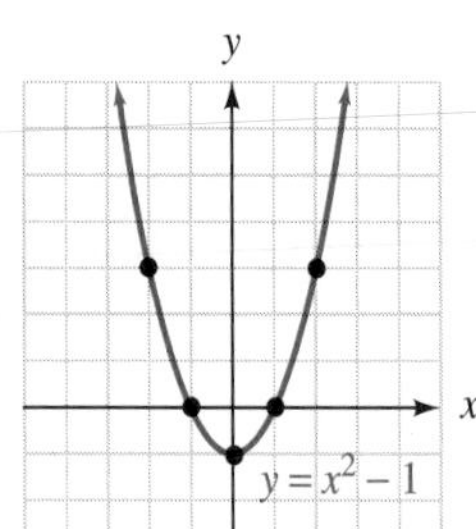

■

STUDY SET Section 6.4

VOCABULARY *Fill in the blanks to make the statements true.*

1. A polynomial with one term is called a monomial.

2. A polynomial with three terms is called a trinomial.

3. A polynomial with two terms is called a binomial.

4. The degree of a polynomial is the Same as the degree of its term with largest degree.

CONCEPTS *Classify each polynomial as a monomial, a binomial, or a trinomial.*

5. $3x^2 - 4$ bi

6. $5t^2 - t + 1$ tri

7. $17e^4$ mono

8. $x^2 + x + 7$ tri

9. $25u^2$ mon

10. $x^2 - 9$ bi

11. $q^5 + q^2 + 1$ tri

12. $4d^3 - 3d^2$ bi

Find the degree of each polynomial.

13. $5x^3$ 3

14. $3t^5 + 3t^2$ 5

15. $2x^2 - 3x + 2$ 2

16. $\frac{1}{2}p^4 - p^2$ 4

17. $2m$ 1

18. $7q - 5$ 1

19. $25w^6 + 5w^7$ 7

20. $p^6 - p^8$ 8

NOTATION *Complete each solution.*

21. Evaluate $3a^2 + 2a - 7$ when $a = 2$.

$$3a^2 + 2a - 7 = 3(2)^2 + 2(2) - 7$$
$$= 3(4) + 4 - 7$$
$$= 12 + 4 - 7$$
$$= 16 - 7$$
$$= 9$$

22. Evaluate $-q^2 - 3q + 2$ when $q = -1$.

$$-q^2 - 3q + 2 = -(1)^2 - 3(-1) + 2$$
$$= -(1) - 3(-1) + 2$$
$$= -1 + 3 + 2$$
$$= 2 + 2$$
$$= 4$$

PRACTICE *Evaluate each polynomial for the given value.*

23. $3x + 4$ when $x = 3$

24. $\frac{1}{2}x - 3$ when $x = -6$

25. $2x^2 + 4$ when $x = -1$

26. $-\frac{1}{2}x^2 - 1$ when $x = 2$

27. $0.5t^3 - 1$ when $t = 4$

28. $0.75a^2 + 2.5a + 2$ when $a = 0$

29. $\frac{2}{3}b^2 - b + 1$ when $b = 3$

30. $3n^2 - n + 2$ when $n = 2$

31. $-2s^2 - 2s + 1$ when $s = -1$

32. $-4r^2 - 3r - 1$ when $r = -2$

Graph each equation by first making a table of solutions.

33. $y = x^2$

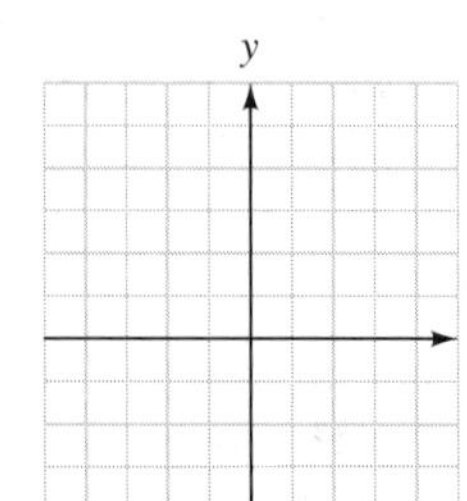

34. $y = -x^2$

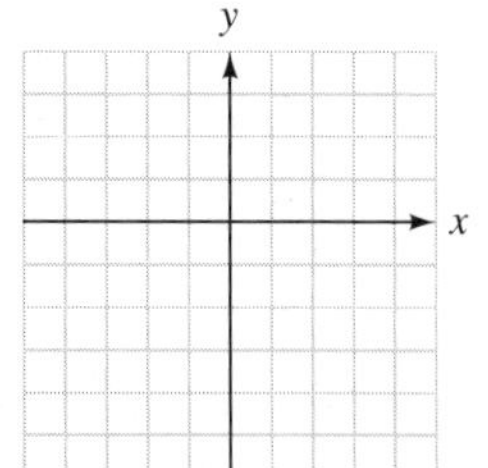

35. $y = \frac{1}{2}x^2$

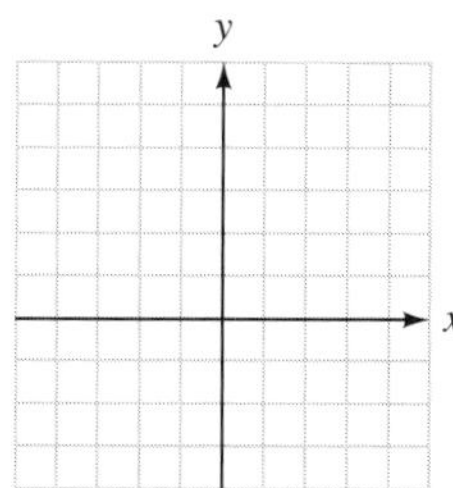

36. $y = \frac{1}{4}x^2$

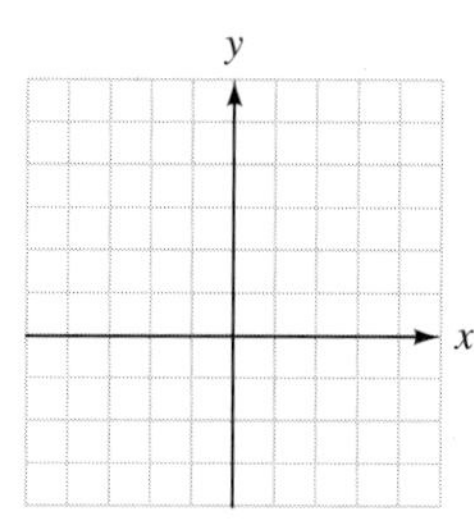

37. $y = -x^2 + 1$

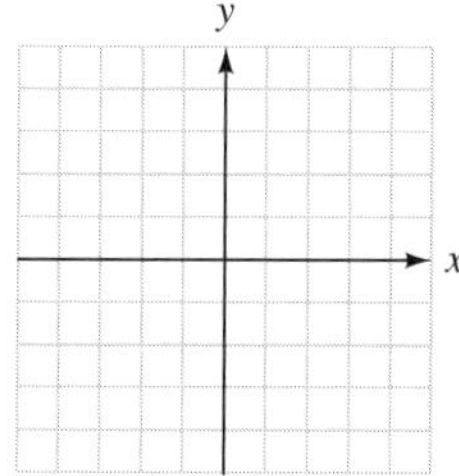

38. $y = x^2 - 4$

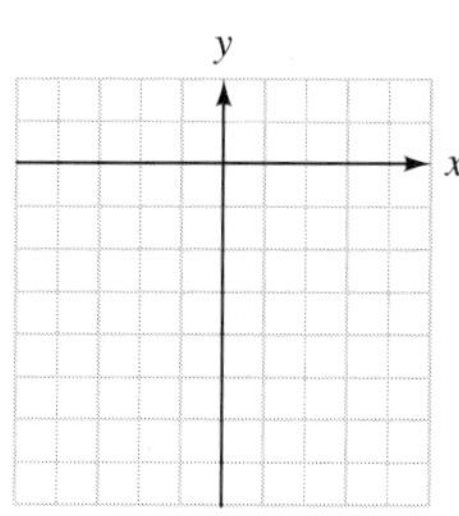

39. $y = 2x^2 - 3$

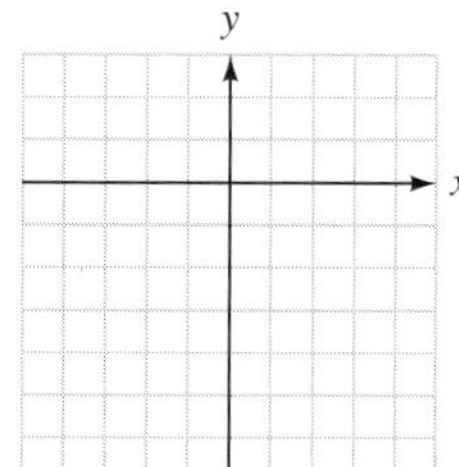

40. $y = -2x^2 + 2$

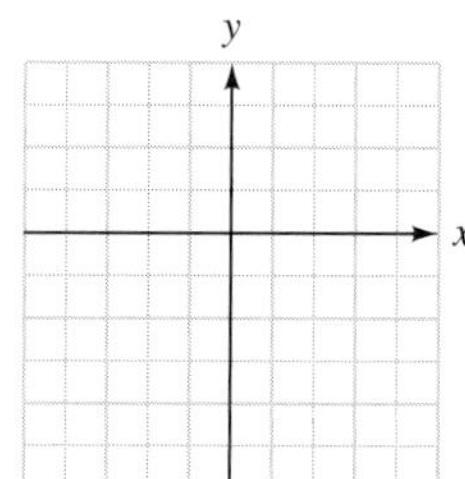

APPLICATIONS *The height h (in feet) of a ball shot straight up with an initial velocity of 64 feet per second is given by the equation $h = -16t^2 + 64t$. Find the height of the ball after the given number of seconds.*

41. 0 second

42. 1 second

43. 2 seconds

44. 4 seconds

The number of feet that a car travels before stopping depends on the driver's reaction time and the braking distance. (See Illustration 1.) For one driver, the stopping distance d is given by the equation $d = 0.04v^2 + 0.9v$, where v is the velocity of the car. Find the stopping distance for each of the following speeds.

45. 30 mph

46. 50 mph

47. 60 mph

48. 70 mph

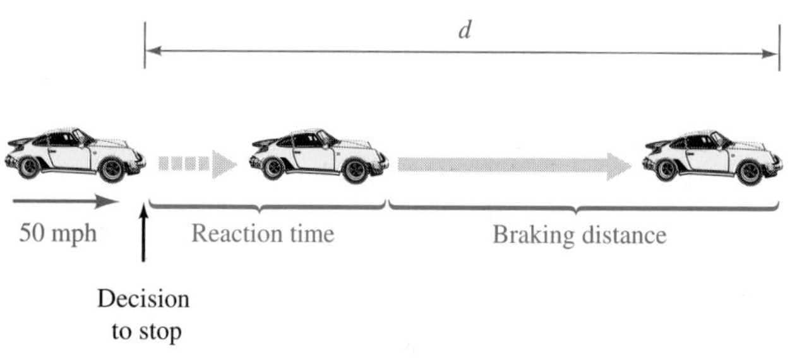

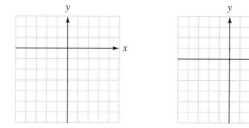

ILLUSTRATION 1

49. SUSPENSION BRIDGE A cable of a suspension bridge hangs in the shape of a parabola. (See Illustration 2.) Use information from the graph to complete the table.

x	0	2	4	−2	−4
y					

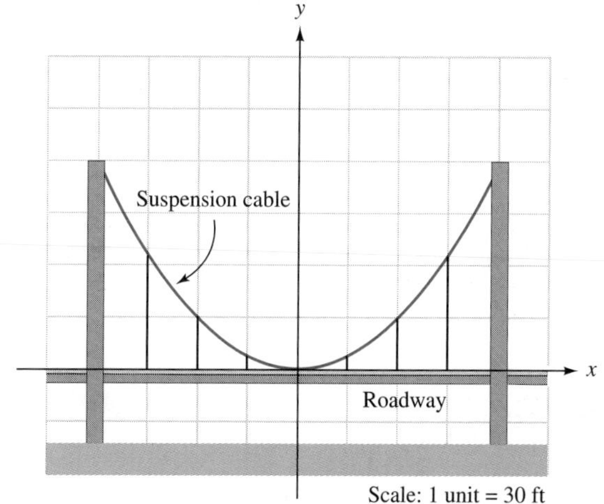

ILLUSTRATION 2

50. FIRE BOAT A stream of water shot from a high-pressure hose on a fire boat travels in the shape of a parabola. (See Illustration 3.) Use information from the graph to complete the table.

x	0	1	2	3	4
y					

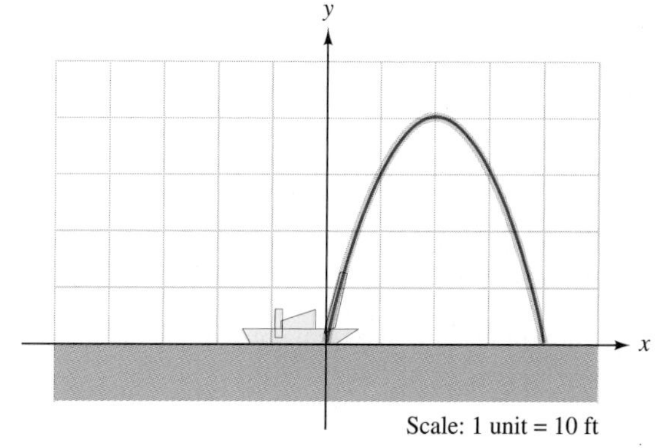

ILLUSTRATION 3

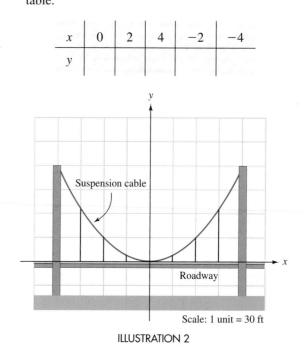

WRITING

51. Explain how to find the degree of the polynomial $2x^3 + 5x^5 - 7x$.

52. Explain how to evaluate the polynomial $-2x^2 - 3$ when $x = 5$.

53. Explain how to graph $y = x^2 - 3$.

54. Graph $y = \frac{1}{2}x^2$, $y = x^2$, and $y = 2x^2$ and explain what happens when the coefficient of x^2 gets larger.

REVIEW *Perform the operations.*

55. $\frac{2}{3} + \frac{4}{3}$

56. $\frac{1}{2} + \frac{2}{3}$

57. $\frac{36}{7} - \frac{23}{7}$

58. $\frac{5}{14} - \frac{4}{21}$

59. $\frac{5}{12} \cdot \frac{18}{5}$

60. $\frac{23}{25} \div \frac{46}{5}$

Solve each equation.

61. $x - 4 = 12$

62. $4z = 108$

63. $2(x - 3) = 6$

64. $3(a - 5) = 4(a + 9)$

6.5 Adding and Subtracting Polynomials

In this section, you will learn about

- Adding polynomials • Subtracting polynomials

INTRODUCTION. Polynomials are the numbers of algebra. They can be added, subtracted, and multiplied just like numbers in arithmetic. In this section, we will show how to find sums and differences of polynomials.

Adding polynomials

Recall that like terms have exactly the same variables and the same exponents. For example, the monomials

$3z^2$ and $-2z^2$ are like terms. Both have the same variable (z) with the same exponent (2).

However, the monomials

$7b^2$ and $8a^2$ are not like terms. They have different variables.

$32p^2$ and $25p^3$ are not like terms. The exponents of p are different.

Also recall that we can combine like terms by adding their coefficients and keeping the same variables and exponents. For example,

$$2y + 5y = (2 + 5)y = 7y \quad \text{and} \quad -3x^2 + 7x^2 = (-3 + 7)x^2 = 4x^2$$

Thus, *to add monomials that are like terms, we add the coefficients and keep the same variables and exponents.*

EXAMPLE 1 ***Adding monomials.*** Add: $5x^3 + 7x^3$.

Solution

Since the monomials are like terms, we add the coefficients and keep the variables and exponents.

$$5x^3 + 7x^3 = 12x^3$$

Self Check

Add: $7y^3 + 12y^3$.

Answer: $19y^3$ ■

EXAMPLE 2 ***Adding monomials.*** Add: $\frac{3}{2}t^2 + \frac{5}{2}t^2 + \frac{7}{2}t^2$.

Solution

Since the monomials are like terms, we add the coefficients and keep the variables and exponents.

$$\frac{3}{2}t^2 + \frac{5}{2}t^2 + \frac{7}{2}t^2 = \left(\frac{3}{2} + \frac{5}{2} + \frac{7}{2}\right)t^2$$

$$= \frac{15}{2}t^2$$ To add the fractions, add the numerators and keep the denominator: $3 + 5 + 7 = 15$.

Self Check

Add: $3.2m^3 + 4.5m^3 + 7.2m^3$.

Answer: $14.9m^3$ ■

To add two polynomials, we write a + sign between them and combine like terms.

EXAMPLE 3 ***Adding binomials.*** Add: $2x + 3$ and $7x - 1$.

Solution

$(2x + 3) + (7x - 1)$ Write a + sign between the binomials.

$= (2x + 7x) + (3 - 1)$ Use the associative and commutative properties to group like terms together.

$= 9x + 2$ Combine like terms.

Self Check

Add: $(5y - 2) + (-3y + 7)$.

Answer: $2y + 5$ ■

The binomials in Example 3 can be added by writing the polynomials so that like terms are in columns.

$$\begin{array}{r} 2x + 3 \\ + \underline{7x - 1} \\ 9x + 2 \end{array}$$ Add the like terms, one column at a time.

EXAMPLE 4 ***Adding a trinomial and a binomial.***
Add: $(5x^2 - 2x + 4) + (3x^2 - 5)$.

Solution

$(5x^2 - 2x + 4) + (3x^2 - 5)$

$= (5x^2 + 3x^2) + (-2x) + (4 - 5)$ Use the associative and commutative properties to group like terms together.

$= 8x^2 - 2x - 1$ Combine like terms.

Self Check

Add: $(2b^2 - 4b) + (b^2 + 3b - 1)$.

Answer: $3b^2 - b - 1$ ■

The polynomials in Example 4 can be added by writing the polynomials so that like terms are in columns.

$$\begin{array}{rrr} 5x^2 & - 2x & + 4 \\ + \; 3x^2 & & - 5 \\ \hline 8x^2 & - 2x & - 1 \end{array}$$ Add the like terms, one column at a time.

EXAMPLE 5 ***Adding trinomials.***
Add: $(3.7x^2 + 4x - 2) + (7.4x^2 - 5x + 3)$.

Solution

$(3.7x^2 + 4x - 2) + (7.4x^2 - 5x + 3)$

$= (3.7x^2 + 7.4x^2) + (4x - 5x) + (-2 + 3)$ Use the associative and commutative properties to group like terms together.

$= 11.1x^2 - x + 1$ Combine like terms.

Self Check

Add:
$(s^2 + 1.2s - 5) + (3s^2 - 2.5s + 4)$.

Answer: $4s^2 - 1.3s - 1$ ■

The trinomials in Example 5 can be added by writing them so that like terms are in columns.

$$\begin{array}{r} 3.7x^2 + 4x - 2 \\ + \underline{\;7.4x^2 - 5x + 3} \\ 11.1x^2 - \;\; x + 1 \end{array}$$ Add the like terms, one column at a time.

Subtracting polynomials

To subtract one monomial from another, we add the opposite of the monomial that is to be subtracted. In symbols, $x - y = x + (-y)$.

EXAMPLE 6 ***Subtracting monomials.*** Subtract: $8x^2 - 3x^2$.

Solution

$8x^2 - 3x^2 = 8x^2 + (-3x^2)$ Add the opposite of $3x^2$.

$= 5x^2$ Add the coefficients and keep the same variable and exponent.

Self Check

Subtract: $6y^3 - 9y^3$.

Answer: $-3y^3$ ■

To subtract polynomials, we also add the opposite. For example, to subtract $3n^2 - 4n + 2$ from $5n^2 + 2n - 3$, we proceed as follows.

$(5n^2 + 2n - 3) - (3n^2 - 4n + 2)$

$= (5n^2 + 2n - 3) + [-(3n^2 - 4n + 2)]$ Add the opposite.

$= (5n^2 + 2n - 3) + (-3n^2 + 4n - 2)$ Use the distributive property to change signs.

$= (5n^2 - 3n^2) + (2n + 4n) + (-3 - 2)$ Use the associative and commutative properties to group like terms together.

$= 2n^2 + 6n - 5$ Combine like terms.

These polynomials can be subtracted by writing them so that like terms are in columns.

$$\begin{array}{r} 5n^2 + 2n - 3 \\ -\underline{(3n^2 - 4n + 2)} \end{array} \longrightarrow \begin{array}{r} 5n^2 + 2n - 3 \\ + \underline{-3n^2 + 4n - 2} \\ 2n^2 + 6n - 5 \end{array}$$ Change signs and add.

EXAMPLE 7 ***Subtracting binomials.*** Subtract: $(3x - 4.2) - (5x + 7.2)$.

Solution

$(3x - 4.2) - (5x + 7.2)$

$= (3x - 4.2) + [-(5x + 7.2)]$ Add the opposite.

$= (3x - 4.2) + (-5x - 7.2)$ Use the distributive property to remove parentheses.

$= (3x - 5x) + (-4.2 - 7.2)$ Use the associative and commutative properties to group like terms together.

$= -2x - 11.4$ Combine like terms.

Self Check

Subtract: $(3.3a - 5) - (7.8a + 2)$.

Answer: $-4.5a - 7$ ■

The binomials in Example 7 can be subtracted by writing them so that like terms are in columns.

$$\begin{array}{r} 3x - 4.2 \\ -\underline{(5x + 7.2)} \end{array} \longrightarrow \begin{array}{r} 3x - \;\; 4.2 \\ + \underline{-5x - \;\; 7.2} \\ -2x - 11.4 \end{array}$$ Change signs and add.

EXAMPLE 8 ***Subtracting trinomials.***
Subtract: $(3x^2 - 4x - 6) - (2x^2 - 6x + 12)$.

Solution

$(3x^2 - 4x - 6) - (2x^2 - 6x + 12)$

$= (3x^2 - 4x - 6) + [-(2x^2 - 6x + 12)]$ Add the opposite.

$= (3x^2 - 4x - 6) + (-2x^2 + 6x - 12)$ Use the distributive property to remove parentheses.

$= (3x^2 - 2x^2) + (-4x + 6x) + (-6 - 12)$ Use the associative and commutative properties to group like terms together.

$= x^2 + 2x - 18$ Combine like terms.

Self Check
Subtract:
$(5y^2 - 4y + 2) - (3y^2 + 2y - 1)$.

Answer: $2y^2 - 6y + 3$ ■

The trinomials in Example 8 can be subtracted by writing them so that like terms are in columns.

$$\begin{array}{r} 3x^2 - 4x - 6 \\ -\underline{(2x^2 - 6x + 12)} \end{array} \longrightarrow \begin{array}{r} 3x^2 - 4x - 6 \\ +\underline{-2x^2 + 6x - 12} \\ x^2 + 2x - 18 \end{array} \quad \text{Change signs and add.}$$

STUDY SET Section 6.5

VOCABULARY *Fill in the blanks to make the statements true.*

1. If two algebraic terms have exactly the same variables and exponents, they are called like terms.

2. $3x^3$ and $3x^2$ are Unlike terms.

CONCEPTS *Fill in the blanks to make a true statement.*

3. To add two monomials, we add the Coefficients and keep the same Variables and exponents.

4. To subtract one monomial from another, we add the opposite of the monomial that is to be subtracted.

Tell whether the monomials are like terms. If they are, combine them.

5. $3y, 4y$ **6.** $3x^2, 5x^2$

7. $3x, 3y$ **8.** $3x^2, 6x$

9. $3x^3, 4x^3, 6x^3$ **10.** $-2y^4, -6y^4, 10y^4$

11. $-5x^2, 13x^2, 7x^2$ **12.** $23, 12x, 25x$

NOTATION *Complete each solution.*

13. Add: $(3x^2 + 2x - 5) + (2x^2 - 7x)$.

$(3x^2 + 2x - 5) + (2x^2 - 7x)$

$= (3x^2 + 2x^2) + (2x - 7x) + (-5)$

$= 5x^2 + (-5x) - 5$

$= 5x^2 - 5x - 5$

14. Subtract: $(3x^2 + 2x - 5) - (2x^2 - 7x)$.

$(3x^2 + 2x - 5) - (2x^2 - 7x)$

$= (3x^2 + 2x - 5) + [-(2x^2 - 7x)]$

$= (3x^2 + 2x - 5) + (-2x^2 + 7x)$

$= (3x^2 - 2x^2) + (2x + 7x) + (-5)$

$= x^2 + 9x - 5$

PRACTICE *Add the polynomials.*

15. $4y + 5y$ **16.** $-2x + 3x$

17. $-8t^2 - 4t^2$ **18.** $15x^2 + 10x^2$

19. $3s^2 + 4s^2 + 7s^2$ **20.** $-2a^3 + 7a^3 - 3a^3$

21. $(3x + 7) + (4x - 3)$ **22.** $(2y - 3) + (4y + 7)$

23. $(2x^2 + 3) + (5x^2 - 10)$

24. $(-4a^2 + 1) + (5a^2 - 1)$

25. $(5x^3 - 4.2x) + (7x^3 - 10.7x)$

26. $(-4.3a^3 + 25a) + (5.8a^3 - 10a)$

27. $(3x^2 + 2x - 4) + (5x^2 - 17)$

28. $(5a^2 - 2a) + (-2a^2 + 3a + 4)$

29. $(7y^2 + 5y) + (y^2 - y - 2)$

30. $(4p^2 - 4p + 5) + (6p - 2)$

31. $(3x^2 - 3x - 2) + (3x^2 + 4x - 3)$

32. $(4c^2 + 3c - 2) + (3c^2 + 4c + 2)$

33. $(3n^2 - 5.8n + 7) + (-n^2 + 5.8n - 2)$

34. $(-3t^2 - t + 3.4) + (3t^2 + 2t - 1.8)$

35. $\begin{array}{r} 3x^2 + 4x + 5 \\ \underline{2x^2 - 3x + 6} \end{array}$

36. $\begin{array}{r} 2x^2 - 3x + 5 \\ \underline{-4x^2 - x - 7} \end{array}$

37. $\begin{array}{r} -3x^2 \qquad - 7 \\ \underline{-4x^2 - 5x + 6} \end{array}$

38. $\begin{array}{r} 4x^2 - 4x + 9 \\ \underline{9x - 3} \end{array}$

39. $\begin{array}{r} -3x^2 + 4x + 25.4 \\ \underline{5x^2 - 3x - 12.5} \end{array}$

40. $\begin{array}{r} -6x^3 - 4.2x^2 + 7 \\ \underline{-7x^3 + 9.7x^2 - 21} \end{array}$

Subtract the polynomials.

41. $32u^3 - 16u^3$

42. $25y^2 - 7y^2$

43. $18x^5 - 11x^5$

44. $17x^6 - 22x^6$

45. $(4.5a + 3.7) - (2.9a - 4.3)$

46. $(5.1b - 7.6) - (3.3b + 5.9)$

47. $(-8x^2 - 4) - (11x^2 + 1)$

48. $(5x^3 - 8) - (2x^3 + 5)$

49. $(3x^2 - 2x - 1) - (-4x^2 + 4)$

50. $(7a^2 + 5a) - (5a^2 - 2a + 3)$

51. $(3.7y^2 - 5) - (2y^2 - 3.1y + 4)$

52. $(t^2 - 4.5t + 5) - (2t^2 - 3.1t - 1)$

53. $(2b^2 + 3b - 5) - (2b^2 - 4b - 9)$

54. $(3a^2 - 2a + 4) - (a^2 - 3a + 7)$

55. $(5p^2 - p + 7.1) - (4p^2 + p + 7.1)$

56. $(m^2 - m - 5) - (m^2 + 5.5m - 7.5)$

57. $\begin{array}{r} 3x^2 + 4x - 5 \\ \underline{-(-2x^2 - 2x + 3)} \end{array}$

58. $\begin{array}{r} 3y^2 - 4y + 7 \\ \underline{-(6y^2 - 6y - 13)} \end{array}$

59. $\begin{array}{r} -2x^2 - 4x + 12 \\ \underline{-(10x^2 + 9x - 24)} \end{array}$

60. $\begin{array}{r} 25x^3 - 45x^2 + 31x \\ \underline{-(12x^3 + 27x^2 - 17x)} \end{array}$

61. $\begin{array}{r} 4x^3 - 3x + 10 \\ \underline{-(5x^3 - 4x - 4)} \end{array}$

62. $\begin{array}{r} 3x^3 + 4x^2 + 12 \\ \underline{-(-4x^3 + 6x^2 - 3)} \end{array}$

APPLICATIONS *Consider the following information: If a house is purchased for \$85,000 and is expected to appreciate \$700 per year, its value y after x years is given by the equation y = 700x + 85,000.*

63. VALUE OF A HOUSE Find the expected value of the house after 10 years.

64. VALUE OF A HOUSE A second house is purchased for \$102,000 and is expected to appreciate \$900 per year. Find an equation that will give the value y of the house after x years.

65. VALUE OF A HOUSE Find the value of the house discussed in Exercise 64 after 12 years.

66. VALUE OF TWO HOUSES Find a single polynomial equation that will give the combined value y of both houses after x years.

67. VALUE OF TWO HOUSES In two ways, find the value of the two houses after 15 years.

a. By substituting into the polynomial equations $y = 700x + 85{,}000$ and $y = 900x + 102{,}000$ and adding.

b. By substituting into the result of Exercise 66.

68. VALUE OF TWO HOUSES In two ways, find the value of the two houses after 25 years.

a. By substituting into the polynomial equations $y = 700x + 85{,}000$ and $y = 900x + 102{,}000$ and adding

b. By substituting into the result of Exercise 66

Consider the following information. A young couple bought two cars, one for \$8,500 and the other for \$10,200. The first car is expected to depreciate \$800 per year and the second car \$1,100 per year.

69. VALUE OF A CAR Write an equation that will give the value y of the first car after x years.

70. VALUE OF A CAR Write an equation that will give the value y of the second car after x years.

71. VALUE OF TWO CARS Find a single equation that will give the value y of both cars after x years.

72. VALUE OF TWO CARS In two ways, find the value of the two cars after 6 years.

WRITING

73. What are *like terms*?

74. Explain how to add two polynomials.

75. Explain how to subtract two polynomials.

76. When two binomials are added, is the result always a binomial? Explain.

REVIEW

77. BASKETBALL SHOES Use the following information to find how much lighter the Kevin Garnett shoe is than the Michael Jordan shoe.

Nike Air Garnett III	Air Jordan XV
Synthetic fade mesh and leather. Sizes: $6\frac{1}{2}$–18.	Full-grain leather upper with woven leather pattern.
Weight: 13.8 oz	Sizes: $6\frac{1}{2}$–18. Weight: 14.6 oz

78. AEROBICS The number of calories burned when doing step aerobics depends on the step height. See Illustration 1. How many more calories are burned during a 10-minute workout using an 8-inch step instead of a 4-inch step?

Step height (in.)	Calories burned per minute
4	4.5
6	5.5
8	6.4
10	7.2

Source: *Reebok Instructor News* (Vol. 4, No. 3, 1991)

ILLUSTRATION 1

79. PANAMA CANAL A ship entering the Panama Canal from the Atlantic Ocean is lifted up 85 feet to Lake Gatun by the Gatun Lock system. See Illustration 2. Then the ship is lowered 31 feet by the Pedro Miguel Lock. By how much must the ship be lowered by the Miraflores Lock system for it to reach the Pacific Ocean water level?

80. CANAL LOCKS See Illustration 2. What is the combined length of the system of locks in the Panama Canal? Express your answer as a mixed number and as a decimal, rounded to the nearest tenth.

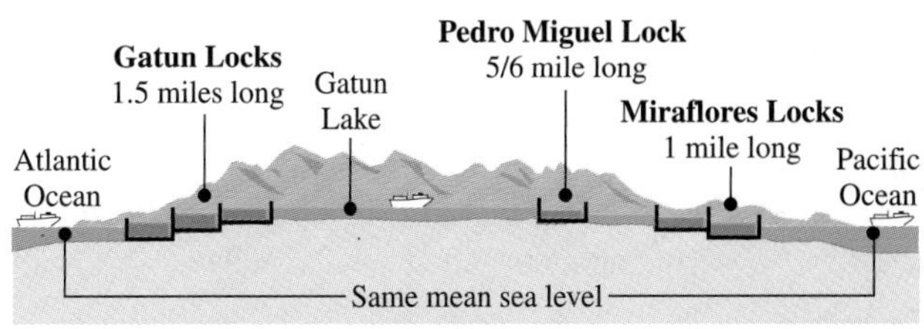

ILLUSTRATION 2

6.6 *Multiplying Polynomials*

In this section, you will learn about

- Multiplying monomials • Multiplying a polynomial by a monomial
- Multiplying a binomial by a binomial • Multiplying a polynomial by a binomial

INTRODUCTION. In this section, we will discuss how to multiply polynomials.

Multiplying monomials

To multiply $4x^2$ by $-2x^3$, we use the commutative and associative properties of multiplication to group the numerical factors and the variable factors and multiply.

$$4x^2(-2x^3) = 4(-2)x^2x^3$$
$$= -8x^5$$

This example suggests the following rule.

Multiplying two monomials

> To multiply two monomials, multiply the numerical factors and then multiply the variable factors.

EXAMPLE 1 ***Multiplying monomials.*** Multiply: **a.** $3y \cdot 6y$ and **b.** $-3x^5(2x^5)$.

Solution

a. $3y \cdot 6y = (3 \cdot 6)(y \cdot y)$ Multiply the numerical factors and multiply the variables.

$= 18y^2$ Multiply: $3 \cdot 6 = 18$ and $y \cdot y = y^2$.

b. $(-3x^5)(2x^5) = (-3 \cdot 2)(x^5 \cdot x^5)$ Multiply the numerical factors and multiply the variables.

$= -6x^{10}$ Multiply: $-3 \cdot 2 = -6$ and $x^5 \cdot x^5 = x^{10}$.

Self Check
Multiply: $-7a^3 \cdot 2a^5$.

$-14a^8$

Answer: $-14a^8$ ■

Multiplying a polynomial by a monomial

To find the product of a polynomial and a monomial, we use the distributive property. To multiply $x + 4$ by $3x$, for example, we proceed as follows:

$3x(x + 4) = 3x(x) + 3x(4)$ Use the distributive property.

$= 3x^2 + 12x$ Multiply the monomials: $3x(x) = 3x^2$ and $3x(4) = 12x$.

The results of this example suggest the following rule.

Multiplying polynomials by monomials

To multiply a polynomial by a monomial, use the distributive property to remove parentheses and simplify.

EXAMPLE 2 ***Multiplying a polynomial by a monomial.*** Multiply: **a.** $2a^2(3a^2 - 4a)$ and **b.** $2x(3x^2 + 2x - 3)$.

Solution

a. $2a^2(3a^2 - 4a)$

$= 2a^2(3a^2) - 2a^2(4a)$ Use the distributive property.

$= 6a^4 - 8a^3$ Multiply: $2a^2(3a^2) = 6a^4$ and $2a^2(4a) = 8a^3$.

b. $2x(3x^2 + 2x - 3)$

$= 2x(3x^2) + 2x(2x) - 2x(3)$ Use the distributive property.

$= 6x^3 + 4x^2 - 6x$ Multiply: $2x(3x^2) = 6x^3$, $2x(2x) = 4x^2$, and $2x(3) = 6x$.

Self Check

Multiply:

a. $3y(5y^3 - 4y)$

b. $5x(3x^2 - 2x + 3)$

$15y^4 - 12y^2$

$15x^3 - 10x^2 + 15x$

Answers: **a.** $15y^4 - 12y^2$, **b.** $15x^3 - 10x^2 + 15x$ ■

Multiplying a binomial by a binomial

To multiply two binomials, we must use the distributive property more than once. For example, to multiply $2x + 3$ by $3x - 5$, we proceed as follows:

$(3x - 5)(2x + 3) = (3x - 5)2x + (3x - 5)3$ Distribute the factor of $3x - 5$ over the two terms within $(2x + 3)$.

$= 2x(3x - 5) + 3(3x - 5)$ Use the commutative property of multiplication.

$= 2x(3x) - 2x(5) + 3(3x) - 3(5)$ Use the distributive property twice.

$= 6x^2 - 10x + 9x - 15$ Do the multiplications.

$= 6x^2 - x - 15$ Combine like terms: $-10x + 9x = -x$.

The results of this example suggest the following rule.

Multiplying a binomial by a binomial

To multiply two binomials, multiply each term of one binomial by each term of the other binomial and combine like terms.

EXAMPLE 3 ***Multiplying a binomial by a binomial.***
Multiply: $(2x - 4)(3x + 5)$.

Self Check

Multiply: $(3x - 2)(2x + 3)$.

Solution

$(2x - 4)(3x + 5)$

$= (2x - 4)3x + (2x - 4)5$	Each term within $(3x + 5)$ is multiplied by $2x - 4$.
$= 3x(2x - 4) + 5(2x - 4)$	Use the commutative property of multiplication.
$= 3x(2x) - 3x(4) + 5(2x) - 5(4)$	Use the distributive property twice.
$= 6x^2 - 12x + 10x - 20$	Do the multiplications.
$= 6x^2 - 2x - 20$	Combine like terms: $-12x + 10x = -2x$.

Answer: $6x^2 + 5x - 6$

EXAMPLE 4 ***Squaring a binomial.*** Find $(5x - 4)^2$.

Solution

In the expression $(5x - 4)^2$, the binomial $5x - 4$ is the base and 2 is the exponent.

$(5x - 4)^2 = (5x - 4)(5x - 4)$	Write $5x - 4$ as a factor two times.
$= (5x - 4)5x - (5x - 4)4$	Distribute the factor of $5x - 4$ over each term within $(5x - 4)$.
$= 5x(5x - 4) - 4(5x - 4)$	Change the order of the factors.
$= 5x(5x) - 5x(4) - 4(5x) - 4(-4)$	Distribute the multiplication by $5x$. Distribute the multiplication by -4.
$= 25x^2 - 20x - 20x + 16$	Do the multiplications.
$= 25x^2 - 40x + 16$	Simplify: $-20x - 20x = -40x$.

Self Check

Find $(5x + 4)^2$.

Answer: $25x^2 + 40x + 16$

COMMENT A common error when squaring a binomial is to square only its first and second terms. For example, it would be incorrect to write

$$(5x - 4)^2 = (5x)^2 - (4)^2 = 25x^2 - 16$$

Multiplying a polynomial by a binomial

We must also use the distributive property more than once to multiply a polynomial by a binomial. For example, to multiply $3x^2 + 3x - 5$ by $2x + 3$, we proceed as follows:

$$\begin{aligned}(2x + 3)(3x^2 + 3x - 5) &= (2x + 3)3x^2 + (2x + 3)3x - (2x + 3)5 \\ &= 3x^2(2x + 3) + 3x(2x + 3) - 5(2x + 3) \\ &= 6x^3 + 9x^2 + 6x^2 + 9x - 10x - 15 \\ &= 6x^3 + 15x^2 - x - 15\end{aligned}$$

EXAMPLE 5 ***Multiplying a trinomial by a binomial.***
Multiply: $(3a + 1)(3a^2 + 2a + 2)$.

Solution

$$\begin{aligned}&(3a + 1)(3a^2 + 2a + 2) \\ &= (3a + 1)3a^2 + (3a + 1)2a + (3a + 1)2 \\ &= 3a^2(3a + 1) + 2a(3a + 1) + 2(3a + 1) \\ &= 3a^2(3a) + 3a^2(1) + 2a(3a) + 2a(1) + 2(3a) + 2(1) \\ &= 9a^3 + 3a^2 + 6a^2 + 2a + 6a + 2 \\ &= 9a^3 + 9a^2 + 8a + 2\end{aligned}$$

Self Check

Multiply: $(x - 2)(3x^2 + 4x + 1)$.

Answer: $3x^3 - 2x^2 - 7x - 2$

We can use a column format to multiply polynomials. To do so, we multiply each term in the top polynomial by each term in the bottom polynomial. To make the addition easy, we will keep like terms in columns. As an example, we can multiply $2x - 4$ by $3x + 2$ as follows.

$$\begin{array}{rrrr} & 2x & - & 4 \\ \times & 3x & + & 2 \\ \hline 3x(2x-4) \longrightarrow \quad 6x^2 & - 12x & & \\ 2(2x-4) \longrightarrow \qquad\quad & + \ 4x & - & 8 \\ \hline 6x^2 & - \ 8x & - & 8 \end{array}$$

EXAMPLE 6 ***Multiplying polynomials.*** Multiply $3a^2 - 4a + 7$ by $2a + 5$.

Solution

$$\begin{array}{rrrr} 3a^2 & - \ 4a & + \ 7 & \\ \times & 2a & + \ 5 & \\ \hline 6a^3 & - \ 8a^2 & + \ 14a & \\ & + \ 15a^2 & - \ 20a & + \ 35 \\ \hline 6a^3 & + \ 7a^2 & - \ 6a & + \ 35 \end{array}$$

Self Check

Multiply $3y^2 - 5y + 4$ by $4y - 3$.

Answer:

$12y^3 - 29y^2 + 31y - 12$ ■

STUDY SET Section 6.6

VOCABULARY *Fill in the blanks to make the statements true.*

1. A polynomial with one term is called a __monomial__.

2. A polynomial with two terms is called a __binomial__.

3. A polynomial with __three__ terms is called a trinomial.

4. $a(b + c) = ab + ac$ illustrates the __distributive__ property.

CONCEPTS *Fill in the blanks to make the statements true.*

5. To multiply two monomials, multiply the __numerical__ factors and then multiply the variable __factors__.

6. To multiply a polynomial by a monomial, use the __distributive__ property to remove parentheses and simplify.

7. To multiply two binomials, multiply each __term__ of one binomial by each term of the other binomial and combine __like__ terms.

8. To multiply a polynomial by a binomial, we must use the distributive __property__ more than once.

NOTATION *Complete each solution.*

9. Multiply: $3x(2x - 5)$.

$$\begin{aligned} 3x(2x - 5) &= 3x(2x) - 3x(5) \\ &= 6x^2 - 15x \end{aligned}$$

10. Multiply: $(3x + 1)(2x^2 - 3x - 2)$.

$$\begin{aligned} &(3x + 1)(2x^2 - 3x - 2) \\ &\quad = (3x + 1)2x^2 - (3x + 1)3x - (3x + 1)2 \\ &\quad = 2x^2(3x + 1) - 3x(3x + 1) - 2(3x + 1) \\ &\quad = 2x^2(3x) + 2x^2(1) - 3x(3x) - 3x(1) \\ &\qquad\quad - 2(3x) - 2(1) \\ &\quad = 6x^3 + 2x^2 - 9x^2 - 3x - 6x - 2 \\ &\quad = 6x^3 - 7x^2 - 9x - 2 \end{aligned}$$

PRACTICE *Find each product.*

11. $(3x^2)(4x^3)$

12. $(-2a^3)(3a^2)$

13. $(3b^2)(-2b)$

14. $(3y)(-y^4)$

15. $(-2x^2)(3x^3)$

16. $(-7x^3)(-3x^3)$

17. $\left(-\frac{2}{3}y^5\right)\left(\frac{3}{4}y^2\right)$

18. $\left(\frac{2}{5}r^4\right)\left(\frac{3}{5}r^2\right)$

19. $3(x + 4)$

20. $-3(a - 2)$

21. $-4(t + 7)$

22. $6(s^2 - 3)$

23. $3x(x - 2)$

24. $4y(y + 5)$

25. $-2x^2(3x^2 - x)$

26. $4b^3(2b^2 - 2b)$

27. $2x(3x^2 + 4x - 7)$

28. $3y(2y^2 - 7y - 8)$

29. $-p(2p^2 - 3p + 2)$

30. $-2t(t^2 - t + 1)$

31. $3q^2(q^2 - 2q + 7)$

32. $4v^3(-2v^2 + 3v - 1)$

33. $(a + 4)(a + 5)$

34. $(y - 3)(y + 5)$

35. $(3x - 2)(x + 4)$

36. $(t + 4)(2t - 3)$

37. $(2a + 4)(3a - 5)$

38. $(2b - 1)(3b + 4)$

Square each binomial.

39. $(2x + 3)^2$

40. $(2y + 5)^2$

41. $(2x - 3)^2$

42. $(2y - 5)^2$

43. $(5t + 1)^2$

44. $(5t - 1)^2$

Multiply the polynomials.

45. $(2x + 1)(3x^2 - 2x + 1)$
46. $(x + 2)(2x^2 + x - 3)$
47. $(x - 1)(x^2 + x + 1)$
48. $(x + 2)(x^2 - 2x + 4)$
49. $(x + 2)(x^2 - 3x + 1)$
50. $(x + 3)(x^2 + 3x + 2)$

Find each product.

51. $\begin{array}{r} 4x + 3 \\ \underline{x + 2} \end{array}$

52. $\begin{array}{r} 5r + 6 \\ \underline{2r - 1} \end{array}$

53. $\begin{array}{r} 4x - 2 \\ \underline{3x + 5} \end{array}$

54. $\begin{array}{r} 6r + 5 \\ \underline{2r - 3} \end{array}$

55. $\begin{array}{r} x^2 - x + 1 \\ \underline{x + 1} \end{array}$

56. $\begin{array}{r} 4x^2 - 2x + 1 \\ \underline{2x + 1} \end{array}$

APPLICATIONS

57. GEOMETRY Express the area of the rectangle in Illustration 1.

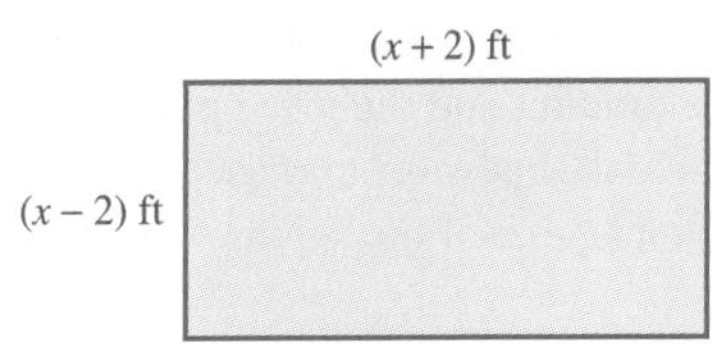

ILLUSTRATION 1

58. SAILING In Illustration 2, the height h of the triangular sail is $4x$ feet, and the base b is $3x - 2$ feet. Express the area of the sail. (*Hint:* The area of a triangle is given by the formula $A = \frac{1}{2}bh$.)

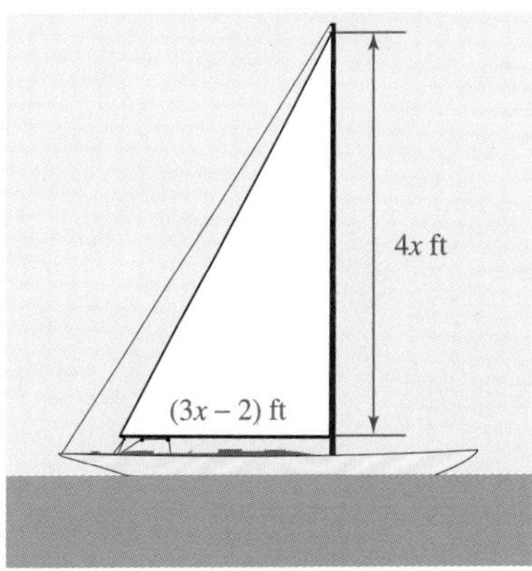

ILLUSTRATION 2

59. ECONOMICS The revenue R received from selling clock radios is the product of their price and the number that are sold. If the price of each radio is given by the formula $-\frac{x}{100} + 30$ and x is the number sold, find a formula that gives the amount of revenue received.

60. ECONOMICS If the pricing formula given in Exercise 59 changes to $-\frac{x}{100} + 40$, find the formula for revenue received.

WRITING

61. Explain how to multiply two binomials.
62. Explain how to find $(2x + 1)^2$.
63. Explain why $(x + 1)^2 \neq x^2 + 1^2$. (Read $\neq$ as "is not equal to.")
64. If two terms are to be added, they have to be like terms. If two terms are to be multiplied, must they be like terms? Explain.

REVIEW

65. THE EARTH It takes 23 hours, 56 minutes, and 4.091 seconds for the Earth to rotate around its axis once. Write 4.091 in words.

66. TAKE-OUT FOOD The sticker in Illustration 3 shows the amount and the price per pound of some spaghetti salad that was purchased at a delicatessen. Find the total price of the salad.

Joan's Spaghetti Salad
303 Foothill Plaza
Plaza Deli
0.78
NET WT. LB.
3.95
PRICE/ LB. $
00.00
TOTAL PRICE $

ILLUSTRATION 3

67. What is $\frac{7}{64}$ in decimal form?

68. Solve $1.7x + 1.24 = -1.4x - 0.62$.

69. Find $56.09 + 78 + 0.567$.

70. Find $-679.4 - (-599.89)$.

71. Evaluate $\sqrt{16} + \sqrt{36}$.

72. Find $103.6 \div 0.56$.

KEY CONCEPT

Graphing

Ordered pairs of real numbers can be graphed on a *rectangular coordinate system.*

Refer to Illustration 1.

1. Label the x- and y-axes.

2. Find the coordinates of point P.

3. Plot the points $(-3, 2)$, $(-3, -2)$, and $(3, -2)$.

4. What are the coordinates of the origin?

5. Label each of the quadrants. In what quadrant do the points have a negative x-coordinate and a positive y-coordinate?

6. Graph the points $(4, 0)$, $(0, 4)$, $(-4, 0)$, and $(0, -4)$.

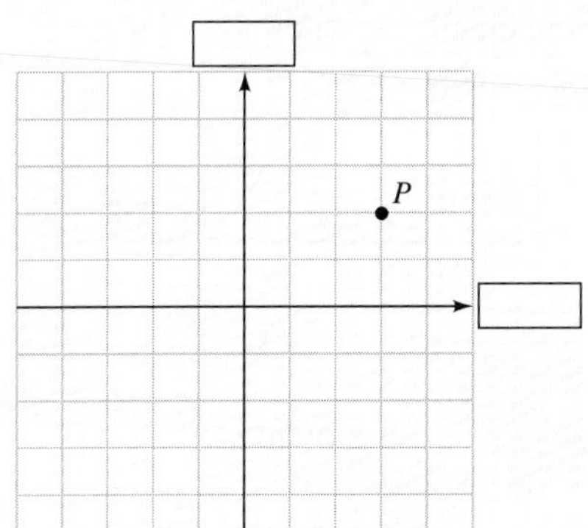

ILLUSTRATION 1

To graph linear equations on a rectangular coordinate system, we must find the coordinates of several points that satisfy the equation.

7. Graph the linear equation $2x - 4y = 8$.

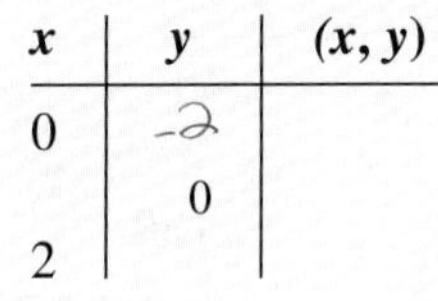

x	y	(x, y)
0	-2	
	0	
2		

Step 1: Complete the table of solutions.

Step 2: Plot the points listed in the table of solutions.

Step 3: Draw a straight line through the points.

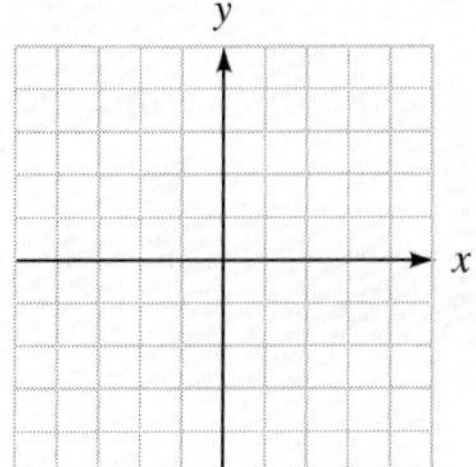

8. Graph the linear equation $y = -2x + 1$.

Step 1: Complete the table of solutions.

x	y	(x, y)
-2		
0		
2		

Step 2: Plot the points listed in the table of solutions.

Step 3: Draw a straight line through the points.

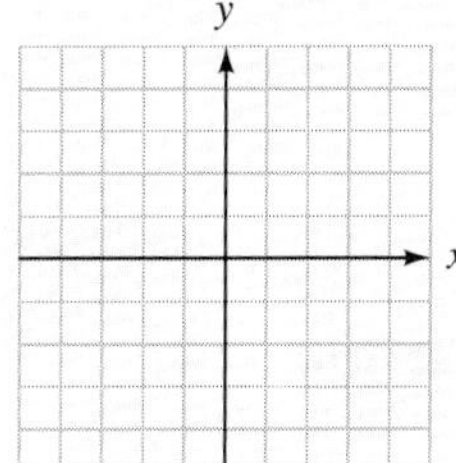

The right-hand side of the equation $y = x^2 + 1$ is a second-degree polynomial. Equations of this type are graphed in the same way as we graph linear equations.

9. Graph $y = x^2 + 1$.

Step 1: Complete the table of solutions.

x	y	(x, y)
-2		
-1		
0		
1		
2		

Step 2: Plot the points listed in the table of solutions.

Step 3: Join the points with a smooth curve.

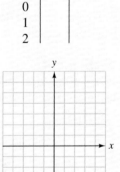

ACCENT ON TEAMWORK

Section 6.1

CAMPUS MAP Get a map of your college campus and use a black marker to draw a rectangular coordinate system on the map. The size of the grid you use will depend on the size of the map. Determine what school landmark should serve as the origin of the coordinate system. List the coordinates of important locations on your campus.

Section 6.2

TRANSLATIONS On the same rectangular coordinate system, sketch the graph of

$$y = x$$
$$y = x + 2$$
$$y = x - 2$$

How are the graphs similar? How are they different?

Section 6.3

RULES FOR EXPONENTS Have one student in your group write the three rules for exponents introduced in Section 6.3 on separate 3×5 cards. Have another student write a word description of the rules on separate cards. Finally, have a third student write an example of the use of the rules on separate cards.

When the three sets of cards are completed, put them together, shuffle them, and then work together as a group to match the symbolic description, the word description, and the example for each of the rules.

Section 6.4

POLYNOMIALS Write a polynomial using the variable x that meets the following conditions:

- It has degree 3.
- It has three terms.
- It does not have an x^2 term.
- The coefficients of the first and second terms are opposites, and the coefficient of the last term is twice the coefficient of the first term.
- When it is evaluated for $x = 0$, the result is 6.

Section 6.5

ADDING AND SUBTRACTING POLYNOMIALS IN VERTICAL FORM Fill in each blank.

a.
$$\begin{array}{r} \square x^2 - \square x + 6 \\ + \; 2x^2 + \square x - 8 \\ \hline 7x^2 + 6x - \square \end{array}$$

b.
$$\begin{array}{r} 12x^2 - 3x - \square \\ - \; 6x^2 + \square x - 7 \\ \hline \square x^2 - 7x - 4 \end{array}$$

Section 6.6

MULTIPLYING POLYNOMIALS To multiply $(2x + 1)(3x - 5)$ using a table, we enter the terms $2x$ and $+1$ of the binomial $2x + 1$ in the leftmost column, as shown in Illustration 1. We enter the terms $3x$ and -5 of the binomial $3x - 5$ in the top row as shown.

Multiply	$3x$	-5
$2x$		
$+1$		

ILLUSTRATION 1

We then multiply each term in the leftmost column by each term in the top row and enter each result in the proper box. To begin, multiply $2x$ and $3x$. The result is shown in red in Illustration 2. Then we multiply $2x$ and -5. The result is shown in blue. Next, we multiply $+1$ and $3x$. The result is shown in green. Finally, we multiply $+1$ and -3. The result is shown in purple.

Multiply	$3x$	-5
$2x$	$6x^2$	$-10x$
$+1$	$+3x$	-5

ILLUSTRATION 2

To complete the process, we combine the like terms along the diagonal and write the final result. See Illustration 3.

Multiply	$3x$	-5	
$2x$	$6x^2$	$-10x$	$+3x - 10x = -7x$
$+1$	$+3x$	-5	

$$(2x + 1)(3x - 5) = 6x^2 - 7x - 5$$

ILLUSTRATION 3

Use a table to find each product.

a. $(4x + 3)(5x - 1)$
b. $(3x - 7)(4x - 3)$
c. $(9x + 2)(8x + 1)$
d. $(6x + 5)(6x - 5)$
e. $(2x^2 + x - 4)(4x - 7)$

CHAPTER REVIEW

SECTION 6.1 The Rectangular Coordinate System

CONCEPTS

A solution of an *equation containing two variables* is an ordered pair of real numbers.

REVIEW EXERCISES

1. a. Check to see whether $x = 2$ and $y = -3$ is a solution of $2x + 5y = -11$.

b. Check to see whether $(-3, 2)$ is a solution of $3x - 5y = 19$.

2. a. Complete the solutions of the equation $3x - 4y = 12$. $(0, \quad)$ and $(-4, \quad)$

b. Complete the table of solutions for $y = -3x - 2$.

x	y	(x, y)
1		(1,)
3		(3,)
−2		(− 2,)

Ordered pairs of real numbers can be graphed on a *rectangular coordinate system.*

The *x-axis* and the *y-axis* divide the coordinate system into four regions, called *quadrants.*

3. a. Graph the points with coordinates of $(2, 3)$, $(-3, 4)$, $(-1.5, -3)$, and $\left(\frac{7}{2}, -1\right)$.

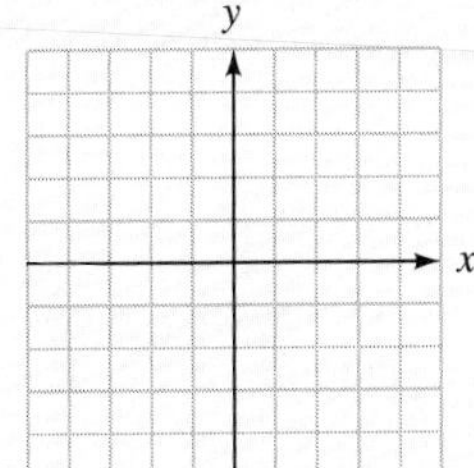

b. Give the coordinates of each point shown in Illustration 1.

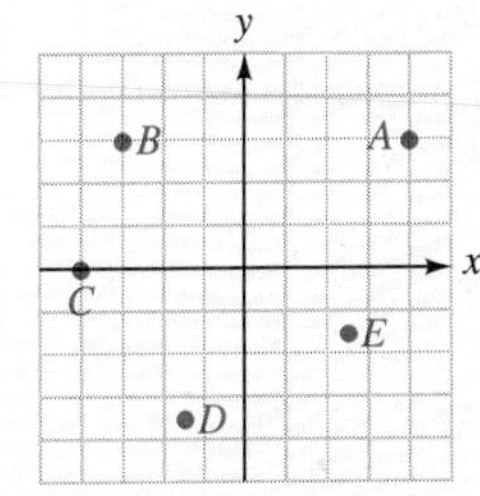

ILLUSTRATION 1

4. In what quadrant does the point $(-3, -4)$ lie?

Coordinate systems have many applications in the real world.

5. Your ticket at the theater is for seat B-10. Locate your seat on the diagram in Illustration 2.

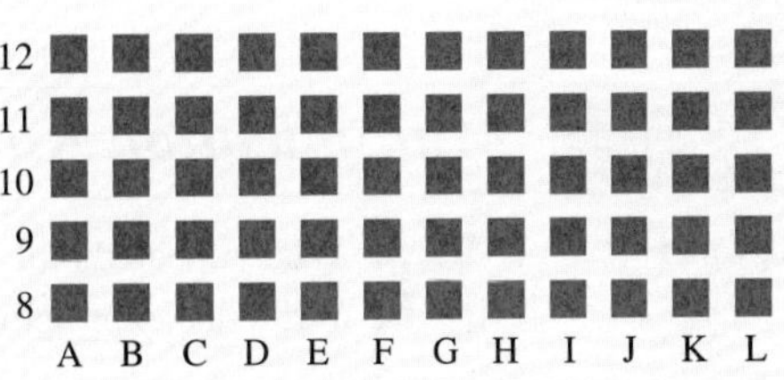

ILLUSTRATION 2

SECTION 6.2 *Graphing Linear Equations*

Equations whose graphs are straight lines are called *linear equations.*

The points where a graph crosses the x- and y-axes, respectively, are called the *x-intercept* and the *y-intercept.*

We can graph a linear equation by constructing a table of solutions, plotting the points, and drawing a straight line through the points.

In the equation $y = \frac{1}{2}x - 1$, x is the *independent variable,* and y is the *dependent variable.*

6. a. Complete the table and then graph $3x - y = 5$. Label the x-intercept and the y-intercept.

x	y
0	
	0
1	

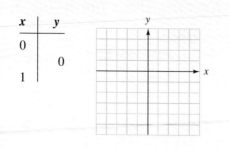

b. Make a table showing three solutions of $y = \frac{1}{2}x - 1$. Then graph it.

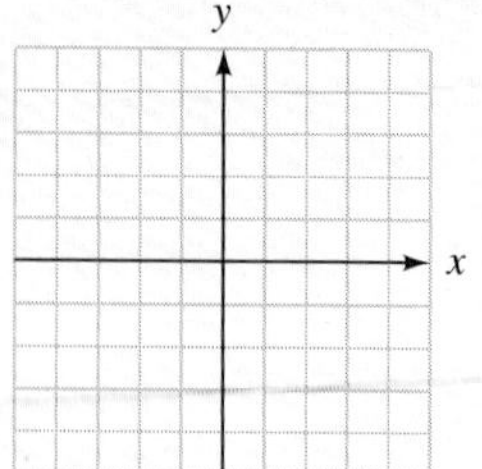

The graph of any equation of the form $y = b$ is a *horizontal line.*

The graph of any equation of the form $x = a$ is a *vertical line.*

7. a. Graph $y = 2$.

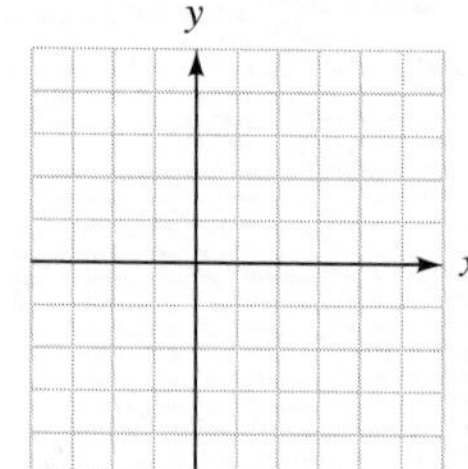

b. Graph $x = 1$.

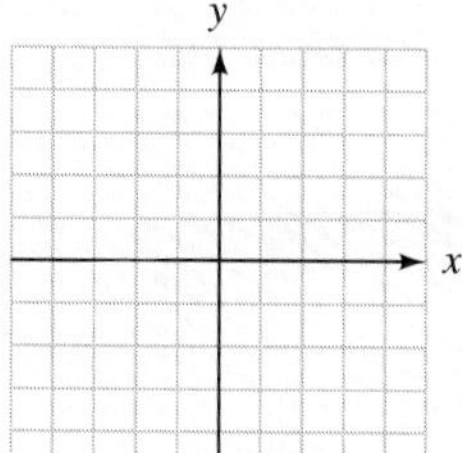

8. The line graphed in Illustration 3 is a picture of all the solutions of $y = x + 1$. Explain.

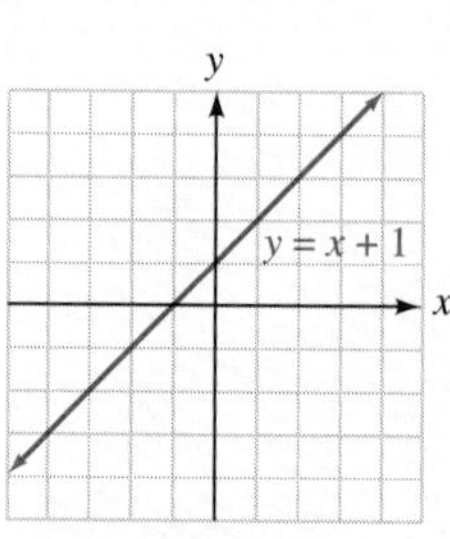

ILLUSTRATION 3

SECTION 6.3 *Multiplication Rules for Exponents*

Exponents represent repeated multiplication.

9. a. What repeated multiplication does $(4h)^3$ represent?

b. Write this expression using exponents: $5 \cdot 5 \cdot d \cdot d \cdot d \cdot m \cdot m \cdot m \cdot m$

The *product rule for exponents*: $x^m x^n = x^{m+n}$

10. Simplify each expression.

a. h^6h^4

b. $t^3(t^5)$

c. $w^2 \cdot w \cdot w^4$

d. $4^7 \cdot 4^5$

The *power rule for exponents*:
$(x^m)^n = x^{m \cdot n}$
The *power rule for products*:
$(xy)^m = x^m y^m$

11. Simplify each product.

a. $2b^2 \cdot 4b^5$ **b.** $-6x^3(4x)$
c. $-2f^2(-4f)(3f^4)$ **d.** $-ab \cdot b \cdot a$
e. $xy^4 \cdot xy^2$ **f.** $(mn)(mn)$
g. $3z^3 \cdot 9m^3z^4$ **h.** $-5cd(4c^2d^5)$

12. Simplify each expression.

a. $(v^3)^4$ **b.** $(3y)^3$
c. $(5t^4)^2$ **d.** $(2a^4b^5)^3$

13. Simplify each expression.

a. $(c^4)^5(c^2)^3$ **b.** $(3s^2)^3(2s^3)^2$
c. $(c^4c^3)^2$ **d.** $(2xx^2)^3$

SECTION 6.4 Introduction to Polynomials

A *monomial* is a polynomial with one term. A *binomial* is a polynomial with two terms. A *trinomial* is a polynomial with three terms.

The *degree* of a polynomial is the same as the degree of its term with largest degree.

A polynomial has a *numerical value* for specific values of its variable.

We can graph *second-degree polynomials* by constructing a table of solutions, plotting the points, and joining them with a smooth curve.

14. Classify each polynomial as a monomial, a binomial, or a trinomial.

a. $3x^2 + 4x - 5$
b. $3t^2$
c. $2x^2 - 1$

15. Give the degree of each polynomial.

a. $3x^2 + 2x^3$
b. $3t^4 - 4t^2 - 3$
c. $3q^2 - 4q^5$

16. a. Evaluate $3x^2 - 2x - 1$ when $x = 2$.
b. Evaluate $2t^2 + t - 2$ when $t = -3$.

17. a. Graph $y = x^2 - 3$. **b.** Graph $y = -\frac{1}{2}x^2 + 3$.

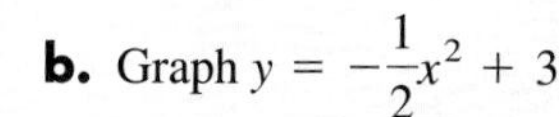

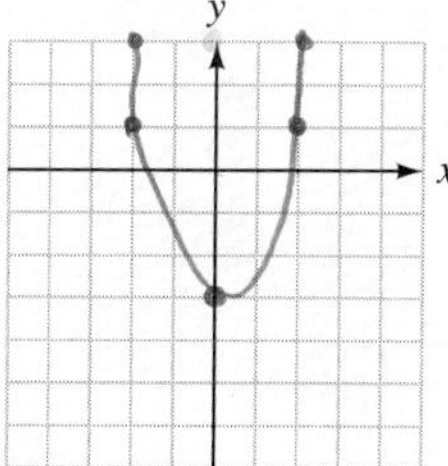

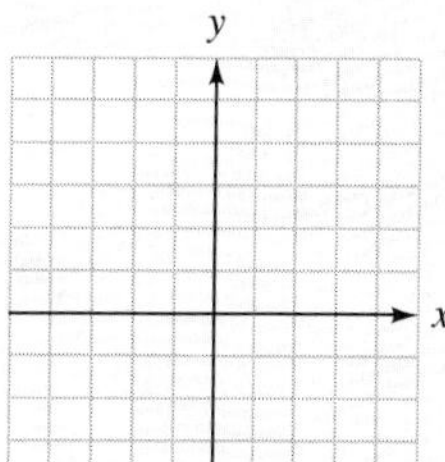

SECTION 6.5 Adding and Subtracting Polynomials

To *add monomials,* add the coefficients and keep the same variables and exponents.

To *add two polynomials,* write + between them and combine like terms.

18. Add the monomials.

a. $3x^3 + 2x^3$ **b.** $\frac{1}{2}p^2 + \frac{5}{2}p^2 + \frac{7}{2}p^2$

19. Add the polynomials.

a. $(3x - 1) + (6x + 5)$ **b.** $(3x^2 - 2x + 4) + (-x^2 - 1)$

20. Add the polynomials.

a. $\begin{array}{r} 5x - 2 \\ \underline{3x + 5} \end{array}$

b. $\begin{array}{r} 3x^2 - 2x + 7 \\ \underline{-5x^2 + 3x - 5} \end{array}$

To *subtract two polynomials,* add the opposite of the polynomial that is to be subtracted.

21. Subtract the monomials.

a. $16p^3 - 9p^3$

b. $4y^2 - 9y^2$

22. Subtract the polynomials.

a. $(2.5x + 4) - (1.4x + 12)$

b. $(3z^2 - z + 4) - (2z^2 + 3z - 2)$

23. Subtract the polynomials.

a. $\begin{array}{r} 5x - 2 \\ \underline{-(3x + 5)} \end{array}$

b. $\begin{array}{r} 3x^2 - 2x + 7 \\ \underline{-(-5x^2 + 3x - 5)} \end{array}$

SECTION 6.6 Multiplying Polynomials

To *multiply two monomials,* multiply the numerical factors and multiply the variable factors.

24. Multiply the monomials.

a. $3x^2 \cdot 5x^3$

b. $(3z^2)(-2z^2)$

To *multiply a polynomial by a monomial,* use the distributive property to remove parentheses and combine like terms.

25. Multiply:

a. $2x^2(3x + 2)$

b. $-5t^3(7t^2 - 6t - 2)$

To *multiply two binomials,* multiply each term of one binomial by each term of the other binomial and combine like terms.

26. Multiply:

a. $(2x - 1)(3x + 2)$

b. $(5t + 4)(7t - 6)$

27. Multiply:

a. $\begin{array}{r} 5x - 2 \\ \underline{\times\ 3x + 5} \end{array}$

b. $\begin{array}{r} 3x + 2 \\ \underline{\times\ 5x - 5} \end{array}$

To *multiply two polynomials,* multiply each term of one polynomial by each term of the other polynomial and combine like terms.

28. Multiply:

a. $(3x + 2)(2x^2 - x + 1)$

b. $(2r - 3)(3r^2 + 2r - 3)$

29. Multiply:

a. $\begin{array}{r} 5x^2 - 2x + 3 \\ \underline{\times \qquad 3x + 5} \end{array}$

b. $\begin{array}{r} 3x^2 - 2x - 1 \\ \underline{\times \qquad 5x - 2} \end{array}$

30. Explain why $(x + 2)^2 \neq x^2 + 4$.

Chapter 6 Test

1. Check to see whether $x = -1$ and $y = 2$ is a solution of $4x + 5y = 6$.

2. Check to see whether $(3, -2)$ is a solution of $3x - 2y = -13$.

3. Complete the ordered pairs so that each one satisfies the equation $x - 2y = 4$.

$(0, \quad), (\quad, 0)$, and $(2, \quad)$

4. Complete the table of solutions for $y = \frac{1}{3}x + 1$.

x	y	(x, y)
0		
3		
-3		

5. PANTS SALE See Illustration 1. List the pant sizes that are not available as ordered pairs of the form (waist, length).

*These pants in your size or they're free. Guaranteed!**

Stonewash jeans, $31.99-$39.99

*Our size guarantee is good only for the following sizes:

	Waist	30	31	32	33	34	36	38
Length	30	X	X	X	X	X	X	
	32		X	X	X	X	X	X
	34			X	X	X	X	X

ILLUSTRATION 1

6. Graph each set of ordered pairs: $(4, 2)$, $(-1, 3)$, $(-2, 0)$, and $(4, -3)$.

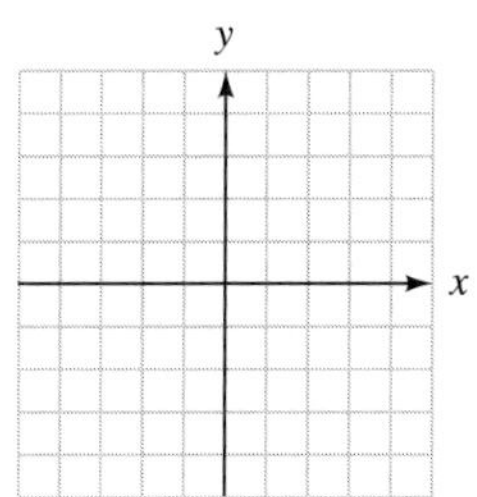

7. Give the coordinates of each point on the graph.

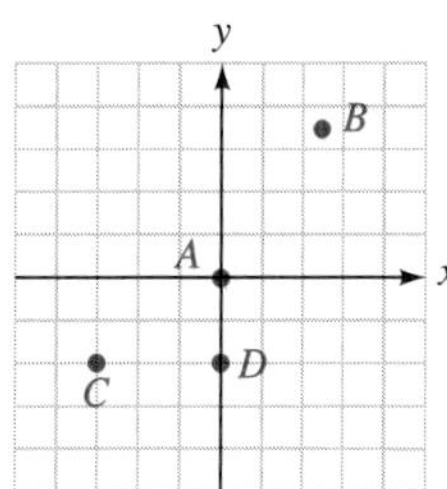

8. Graph the equation $2x - y = 4$.

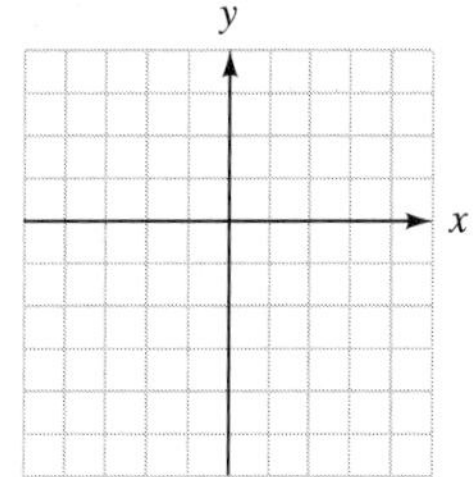

9. **a.** What is the x-intercept of the line graphed in Problem 8?

b. What is the y-intercept of the line graphed in Problem 8?

10. Graph the equation $y = -\frac{3}{2}x - 1$.

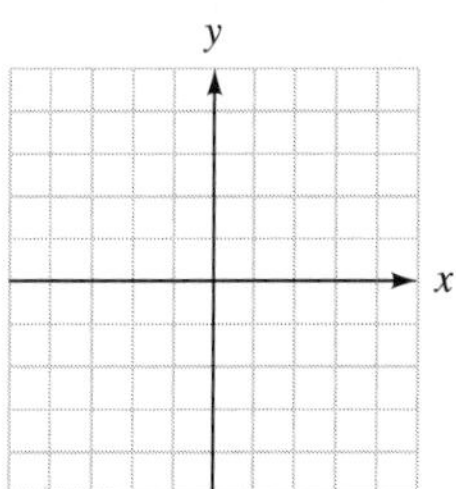

11. Graph the equation $y = -2$.

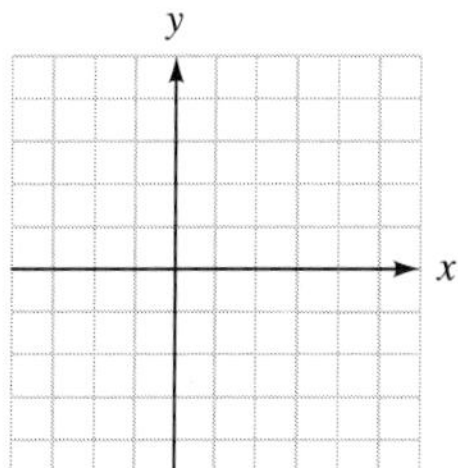

12. Graph the equation $x = 3$.

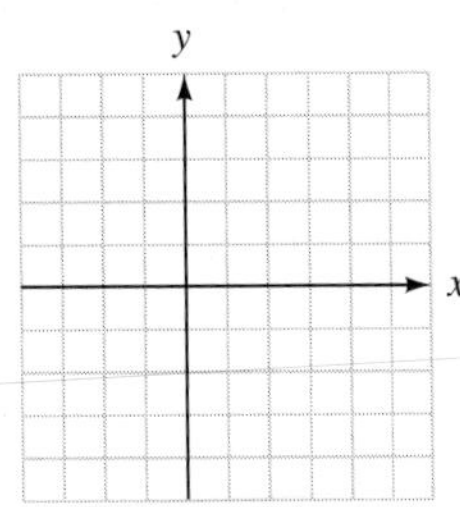

13. Simplify each expression.

a. h^2h^4 **b.** $-7x^3(4x^2)$

c. $b^2 \cdot b \cdot b^5$ **d.** $-3g^2k^3(-8g^3k^{10})$

14. Simplify each expression.

a. $(f^3)^5$ **b.** $(2a^2b)^2$

c. $(x^2)^3(x^3)^3$ **d.** $(x^2x^3)^3$

Classify each polynomial as a monomial, a binomial, or a trinomial.

15. $5x^2 + 4x$

16. $-3x^2 - 2x + 3$

Give the degree of each polynomial.

17. $3t^4 - 2t^3 + 5t^6 - t$

18. $7q^7 + 5q^5 - 8q^2$

Evaluate each polynomial at the given value.

19. $3x^2 - 2x + 4$ when $x = 3$

20. $-2r^2 - r + 3$ when $r = -1$

Graph each equation.

21. $y = 2x^2$

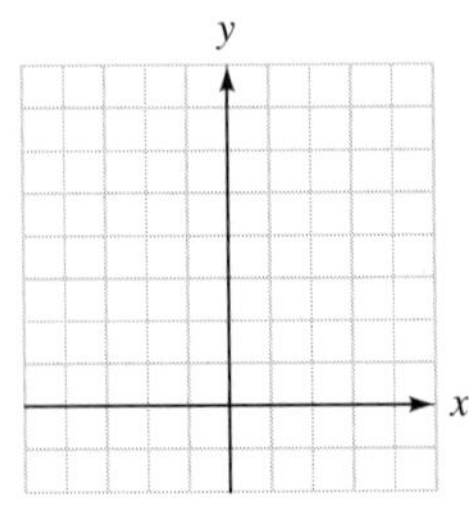

22. $y = -x^2 + 4$

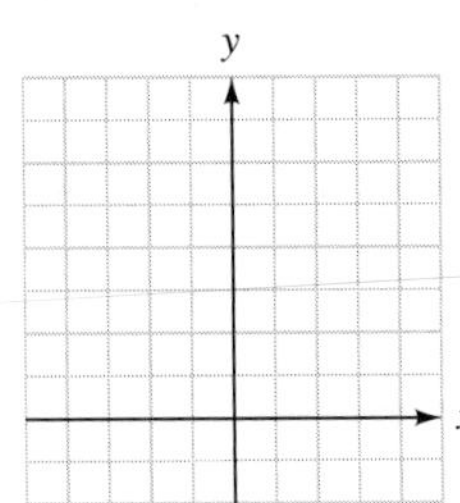

23. Add: $(3x^2 + 2x) + (2x^2 - 5x + 4)$.

24. Add:
$$\begin{array}{r} 4x^2 - 5x + 5 \\ \underline{3x^2 + 7x - 7} \end{array}$$

25. Subtract: $(2.1p^2 - 2p - 2) - (3.3p^2 - 5p - 2)$.

26. Subtract:
$$\begin{array}{r} 3d^2 - 3d + 7.2 \\ -\underline{(-5d^2 + 6d - 5.3)} \end{array}$$

Find each product.

27. $(-2x^3)(4x^2)$

28. $3y^2(y^2 - 2y + 3)$

29. $(2x - 5)(3x + 4)$

30. $(2x - 3)(x^2 - 2x + 4)$

31. Are the points with coordinates $(1, -2)$ and $(-2, 1)$ the same? Explain why or why not.

32. Explain what is meant when we say that the equation $x + y = 8$ has infinitely many solutions.

Chapters 1–6 Cumulative Review Exercises

Consider the number 6,245,867.

1. Round to the nearest thousand.

2. Round to the nearest million.

Find the perimeter of each figure.

3. A rectangle that is 8 meters long and 3 meters wide.

4. A square with sides that are 13 inches long.

5. PARKING The dimensions of various types of rectangular-shaped parking spaces are given in Illustration 1. Complete the table.

Type	Length (ft)	Width (ft)	Area (ft^2)
Standard space	20	9	
Standard space adjacent to a wall	20	10	
Parallel space	25	10	
Compact space	17	8	

ILLUSTRATION 1

6. HEALTH A person's blood pressure is a combination of two measurements, and it is normally written as a fraction of the form $\frac{\text{systolic}}{\text{diastolic}}$. (The fraction is not simplified.) Study the graph in Illustration 2, then complete this sentence:
In healthy persons, blood pressure increases from about ____ in infants, to about ____ at age 30, to about ____ at age 40 and over.

Find the prime factorization of each number.

7. 120

8. 525

9. LAKE TAHOE Because of the growth of algae, a scientific study concluded that the legendary clarity of Lake Tahoe, which straddles the California/Nevada border, could be doomed. Study the data in Illustration 3. How has the visibility changed since the 1960s?

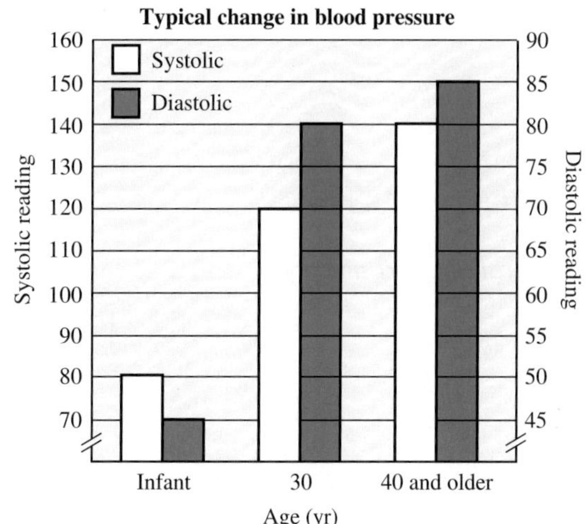

Source: *Microsoft Encarta 98 Encyclopedia*

ILLUSTRATION 2

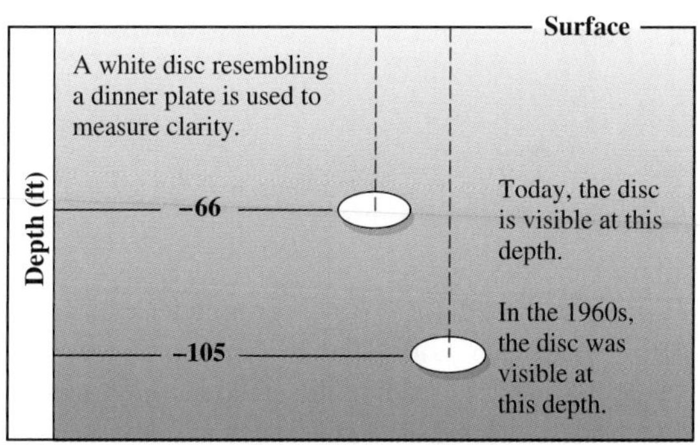

Source: *Los Angeles Times* (February 16, 2000)

ILLUSTRATION 3

10. Evaluate $\frac{6x + x^3}{|x|}$ for $x = -2$.

11. Evaluate $12 - 2[1 - (-8 + 2)]$.

12. Evaluate -3^2.

Combine like terms.

13. $5x - 11x$

14. $-4(x - 3y) + 5x - 2y$

Solve each equation. Check the result.

15. $4x + 3 = 11$

16. $2z + 12 = 6z - 4$

17. $\frac{t}{3} + 2 = -4$

18. $2y + 7 = 2 - (4y + 7)$

Do the operations and simplify.

19. $\frac{5}{10b^3} \cdot 2b^2$

20. $-4\frac{1}{4} \div 4\frac{1}{2}$

21. $34\frac{1}{9} - 13\frac{5}{6}$

22. $\frac{5}{m} - \frac{n}{5}$

Solve each equation. Check the result.

23. $\frac{7}{8}t = -28$

24. $\frac{4}{5}x = \frac{3}{4}x + \frac{1}{2}$

25. PAPER SHREDDER A paper shredder cuts paper into $\frac{1}{4}$-inch-wide strips. If an $8\frac{1}{2} \times 11$ in. piece of notebook paper is fed into the shredder as shown in Illustration 4, into how many strips will it be shredded?

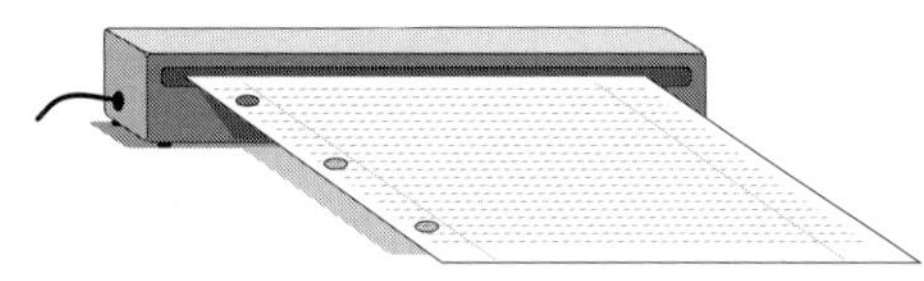

ILLUSTRATION 4

26. Explain why dividing a number by $\frac{1}{4}$ is the same as multiplying it by 4.

27. Round 57.574 to the nearest hundredth.

28. Add: $29.703 + 321.35$.

29. Subtract: $287.23 - 179.97$.

30. Multiply: 7.89×0.27.

31. Divide: $3.8\overline{)17.746}$.

32. Write $\frac{35}{99}$ as a decimal.

Write each number as a decimal. Round to the nearest tenth, if necessary.

33. $5\frac{5}{8}$

34. $-4\frac{7}{9}$

Solve each equation. Check the result.

35. $3.2x = 74.46 - 1.9x$

36. $-5.2x = 108 - 6.1x$

37. $-2(x - 2.1) = -2.4$

38. $\frac{1}{5}x - 2.5 = -17.2$

39. EARTHQUAKES Listed below are the magnitudes of the fifteen largest earthquakes in the United States, according to the U.S. Geological Survey. What are the mean, median, and mode of the listed magnitudes? (Round to the nearest tenth.)

9.2	1964	Alaska	7.9	1812	Missouri
8.8	1957	Alaska	7.9	1857	California
8.7	1965	Alaska	7.9	1868	Hawaii
8.3	1938	Alaska	7.9	1900	Alaska
8.3	1958	Alaska	7.9	1987	Alaska
8.2	1899	Alaska	7.8	1872	California
8.2	1899	Alaska	7.8	1892	California
8.0	1986	Alaska			

40. DRIVING What will be the reading on the odometer in Illustration 5 if a motorist travels at the rate shown on the speedometer for 3.5 hours?

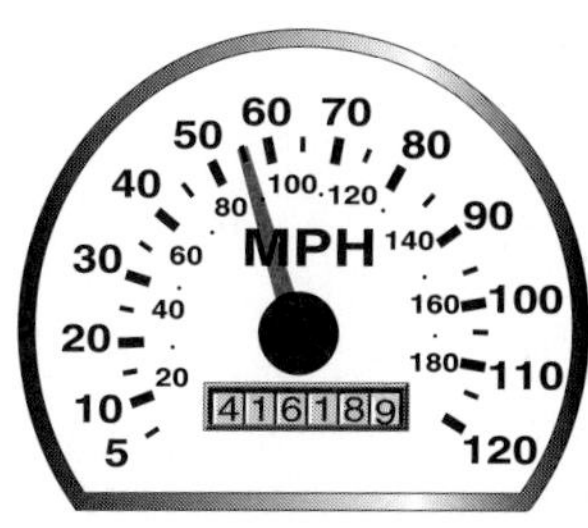

ILLUSTRATION 5

41. PETITION DRIVE A worker for a political organization is to collect signatures for a petition drive. Her pay is \$20 plus 5¢ per signature. How many signatures must she get to earn \$60?

42. CONCERT TICKETS Seven-eighths of the total number of tickets sold for a concert were ordered by mail. The remaining 200 tickets were purchased at the concert hall box office. How many tickets were sold for the concert?

Simplify each expression.

43. $\sqrt{121}$

44. $\sqrt{\frac{81}{4}}$

45. $\sqrt{0.25}$

46. $3\sqrt{144} - \sqrt{49}$

47. Is $(-2, 3)$ a solution of the equation $4x - 5y = -23$?

48. Graph the equation $3x - 4y = 12$.

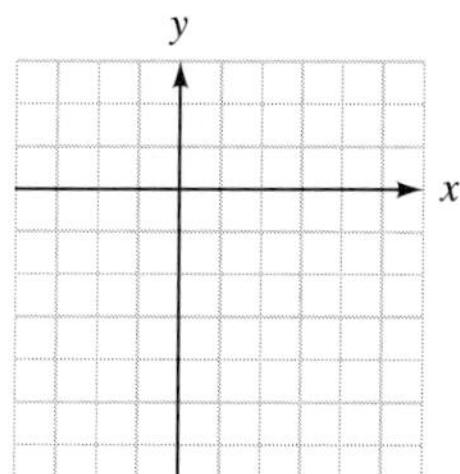

Simplify each product.

49. p^4p^5

50. $(2q^2)(-5q^6)$

51. $(p^3q^2)(p^3q^4)$

52. $(3a^2)^3(-a^3)^3$

Simplify each expression.

53. $(3x^2 - 5x) - (2x^2 + x - 3)$

54. $(2x + 3)(3x - 1)$

7 Percent

PERCENTS ARE BASED ON THE NUMBER 100. THEY OFFER US A STANDARDIZED WAY TO MEASURE AND DESCRIBE MANY SITUATIONS IN OUR DAILY LIVES.

7.1 Percents, Decimals, and Fractions

In this section, you will learn about

- The meaning of percent • Changing a percent to a fraction
- Changing a percent to a decimal • Changing a decimal to a percent
- Changing a fraction to a percent

INTRODUCTION. Percents are a popular way to present numeric information. Stores use them to advertise discounts, manufacturers use them to describe the content of their products, and banks use them to list interest rates for loans and savings accounts. Newspapers are full of statistics presented in percent form. In this section, we introduce percent and show how fractions, decimals, and percents are interrelated.

The meaning of percent

A percent tells us the number of parts per 100. You can think of a percent as the *numerator* of a fraction that has a denominator of 100.

Percent

> **Percent** means parts per one hundred.

In Figure 7-1, 93 out of 100 equal-sized squares are shaded. Thus, $\frac{93}{100}$ or 93 percent of the figure is shaded. The word *percent* can be written using the symbol %, so 93% of Figure 7-1 is shaded.

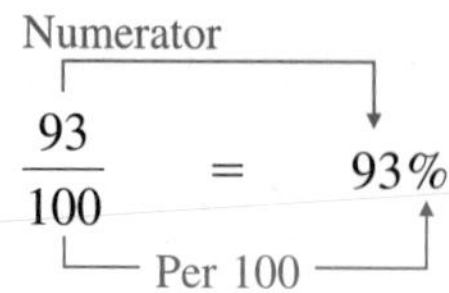

FIGURE 7-1

If the entire grid in Figure 7-1 had been shaded, we would say that 100 out of the 100 squares, or 100%, was shaded. Using this fact, we can determine what percent of the figure is *not* shaded by subtracting the percent of the figure that is shaded from 100%.

$$100\% - 93\% = 7\%$$

7% of Figure 7-1 is not shaded.

Changing a percent to a fraction

To change a percent into an equivalent fraction, we use the definition of percent.

Changing a percent to a fraction

> To change a percent to a fraction, drop the % symbol and write the given number over 100. Then simplify the fraction, if possible.

EXAMPLE 1 ***Changing a percent to a fraction.*** The chemical makeup of the Earth's atmosphere is 78% nitrogen, 21% oxygen, and 1% other gases. Write each percent as a fraction.

Solution

We begin with nitrogen.

$78\% = \frac{78}{100}$ Use the definition of percent. 78% means 78 parts per one hundred. This fraction can be simplified.

$= \frac{39 \cdot \overset{1}{\cancel{2}}}{50 \cdot \underset{1}{\cancel{2}}}$ Factor 78 as $39 \cdot 2$ and 100 as $50 \cdot 2$. Divide out the common factor of 2.

$= \frac{39}{50}$

Nitrogen makes up $\frac{78}{100}$, or $\frac{39}{50}$, of the Earth's atmosphere.

Oxygen makes up 21% or $\frac{21}{100}$ of the earth's atmosphere. Other gases make up 1% or $\frac{1}{100}$ of the atmosphere.

Self Check

An average watermelon is 92% water. Write this percent as a fraction.

Answer: $\frac{23}{25}$ ■

EXAMPLE 2 ***Changing a percent to a fraction.*** In 1998, 46.1% of married women aged 25–54 were employed full-time. Write this percent as a fraction.

Solution

$46.1\% = \frac{46.1}{100}$ Drop the % symbol and write 46.1 over 100.

$= \frac{46.1 \cdot \mathbf{10}}{100 \cdot \mathbf{10}}$ To obtain a whole number in the numerator, multiply by 10. This will move the decimal point 1 place to the right. Multiply the denominator by 10 as well.

$= \frac{461}{1{,}000}$ Do the multiplication in the numerator and in the denominator.

In 1998, 461 out of every 1,000 married women aged 25–54 were employed full-time.

Self Check

In 1978, 26.9% of married women aged 25–54 were employed full-time. Write this percent as a fraction.

Answer: $\frac{269}{1{,}000}$ ■

EXAMPLE 3 ***Changing a percent to a fraction.*** Write $66\frac{2}{3}\%$ as a fraction.

Solution

$66\frac{2}{3}\% = \frac{66\frac{2}{3}}{100}$ Drop the % symbol and write $66\frac{2}{3}$ over 100.

$= 66\frac{2}{3} \div 100$ The fraction bar indicates division.

$= \frac{200}{3} \cdot \frac{1}{100}$ Change $66\frac{2}{3}$ to a mixed number and then multiply by the reciprocal of 100.

$= \frac{2 \cdot 100 \cdot 1}{3 \cdot 100}$ Multiply the numerators and the denominators. Factor 200 as $2 \cdot 100$.

$= \frac{2 \cdot \overset{1}{\cancel{100}} \cdot 1}{3 \cdot \underset{1}{\cancel{100}}}$ Divide out the common factor of 100.

$= \frac{2}{3}$ Multiply in the numerator. Multiply in the denominator.

Self Check

Write $83\frac{1}{3}\%$ as a fraction.

Answer: $\frac{5}{6}$ ■

Changing a percent to a decimal

To write a percent as a decimal, recall that a percent can be written as a fraction with denominator 100, and that a denominator of 100 indicates division by 100.

Consider 14.25%, which means 14.25 parts per 100.

$14.25\% = \frac{14.25}{100}$ Use the definition of percent: write 14.25 over 100.

$= 14.25 \div 100$ The fraction bar indicates division.

$= 0.14.25$ To divide a decimal by 100, move the decimal point 2 places to the left.

$14.25\% = 0.1425$

This example suggests the following procedure.

Changing a percent to a decimal

> To change a percent to a decimal, drop the % symbol and divide by 100 by moving the decimal point 2 places to the left.

EXAMPLE 4 ***The recording industry.*** Figure 7-2 shows that the compact disc has become the format of choice among most consumers. What percent of all music sold is produced on CDs? Write the percent as a decimal.

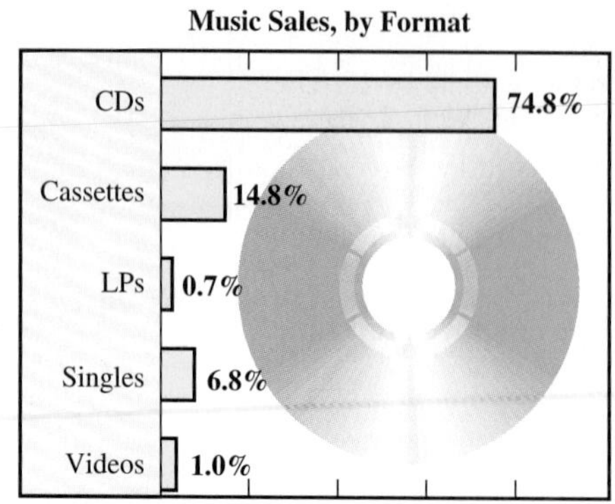

Source: *The New York Times 2000 Almanac*

FIGURE 7-2

Solution

From the graph, we see that 74.8% of all music sold is produced on CDs. To write 74.8% as a decimal, we proceed as follows.

$74.8\% = .74.8$ Drop the percent symbol and divide by 100 by moving the decimal point 2 places to the left.

$= 0.748$ Write a 0 to the left of the decimal point.

Self Check

What percent of all music sold is produced on LPs (long-playing vinyl record albums)? Write the percent as a decimal.

Answer: 0.007

EXAMPLE 5 ***Changing a percent to a decimal.*** Write 310% as a decimal.

Solution

The whole number 310 has an understood decimal point to the right of 0.

$310\% = 310.0\%$ Write a decimal point and a 0 on the right of 310.

$= 3.10.0$ Drop the % symbol and divide by 100 by moving the decimal point 2 places to the left.

$= 3.100$

$= 3.1$ Drop the unnecessary 0's to the right of the 1.

Self Check

Write 600% as a decimal.

Answer: 6

EXAMPLE 6 ***Changing a percent to a decimal.*** The population of the state of Oklahoma is approximately $1\frac{1}{4}\%$ of the population of the United States. Write this percent as a decimal.

Solution

To change a percent to a decimal, we drop the percent symbol and divide by 100 by moving the decimal point 2 places to the left. In this case, however, there is no decimal point in $1\frac{1}{4}\%$ to move. Since $1\frac{1}{4} = 1 + \frac{1}{4}$, and since the decimal equivalent of $\frac{1}{4}$ is 0.25, we can write $1\frac{1}{4}\%$ in an equivalent form as 1.25%.

$1\frac{1}{4}\% = 1.25\%$ Write $1\frac{1}{4}$ as 1.25.

$= 0.01.25$ Drop the % symbol and divide by 100 by moving the decimal point 2 places to the left.

$= 0.0125$

Self Check

Write $15\frac{3}{4}\%$ as a decimal.

Answer: 0.1575

Changing a decimal to a percent

To change a percent to a decimal, we drop the % symbol and move the decimal point 2 places to the left. To write a decimal as a percent, we do the opposite: we move the decimal point two places to the right and insert a % symbol.

Changing a decimal to a percent

> To change a decimal to a percent, multiply the decimal by 100 by moving the decimal point 2 places to the right, and then insert a % symbol.

EXAMPLE 7 ***Changing a decimal to a percent.*** Land areas make up 0.291 of the Earth's surface. Write this decimal as a percent.

Solution

$0.291 = 0.29.1\%$ Multiply the decimal by 100 by moving the decimal point 2 places to the right, and then insert a % symbol.

$= 29.1\%$

Self Check

Write 0.5343 as a percent.

Answer: 53.43%

Changing a fraction to a percent

We will use a two-step process to change a fraction to a percent. First, we write the fraction as a decimal. Then we change that decimal to a percent.

Fraction ⟶ Decimal ⟶ Percent

Changing a fraction to a percent

> To change a fraction to a percent,
> 1. write the fraction as a decimal by dividing its numerator by its denominator;
> 2. multiply the decimal by 100 by moving the decimal point 2 places to the right;
> 3. insert a % symbol.

EXAMPLE 8 ***Changing a fraction to a percent.*** The highest-rated television show of all time was a special episode of M*A*S*H that aired February 28, 1983. Surveys found that three out of every five American households watched this show. Express the rating as a percent.

Solution

3 out of 5 can be expressed as $\frac{3}{5}$. We need to change this fraction to a decimal.

$$\begin{array}{r} 0.6 \\ 5\overline{)3.0} \\ \underline{3\,0} \\ 0 \end{array}$$

Write 3 as 3.0 and then divide the numerator by the denominator.

$\frac{3}{5} = 0.6$ The result is a terminating decimal.

$0.6 = 0.60.\%$ Write a placeholder 0 to the right of the 6. Multiply the decimal by 100 by moving the decimal point 2 places to the right, and then insert a % symbol.

$= 60\%$

60% of American households watched the special episode of M*A*S*H.

Self Check

Write $\frac{7}{8}$ as a percent.

Answer: 87.5%

In Example 8, the result of the division was a terminating decimal. Sometimes when we change a fraction to a decimal, the result of the division is a repeating decimal.

EXAMPLE 9 ***Changing a fraction to a percent.*** Write $\frac{5}{6}$ as a percent.

Solution The first step is to change $\frac{5}{6}$ to a decimal.

```
   0.8333
6)5.0000
  4 8
    20
    18
     20
     18
      20
```

Write 5 as 5.0000. Divide the numerator by the denominator.

$\frac{5}{6} = 0.8333...$ The result is a repeating decimal.

$= 0.83.33...\%$ Change 0.8333. . . to a percent. Multiply the decimal by 100 by moving the decimal point 2 places to the right, and then insert a % symbol.

$= 83.33...\%$ 83.333. . . is a repeating decimal.

We must now decide whether we want an approximation or an exact answer. For an approximation, we can round 83.333. . .% to a specific place value. For an exact answer, we can represent the repeating part of the decimal using an equivalent fraction.

Approximation

$\frac{5}{6} = 83.33...\%$

$\approx 83.3\%$ Round to the nearest tenth.

$\frac{5}{6} \approx 83.3\%$

Exact answer

$\frac{5}{6} = 83.3333...\%$

$= 83\frac{1}{3}\%$ Use the fraction $\frac{1}{3}$ to represent .333. . . .

$\frac{5}{6} = 83\frac{1}{3}\%$ ■

Some percents occur so frequently that it is useful to memorize their fractional and decimal equivalents. Study the information in this table and memorize it for future use.

Percent	Decimal	Fraction	Percent	Decimal	Fraction
1%	0.01	$\frac{1}{100}$	$33\frac{1}{3}\%$	0.3333. . .	$\frac{1}{3}$
10%	0.1	$\frac{1}{10}$	50%	0.5	$\frac{1}{2}$
20%	0.2	$\frac{1}{5}$	$66\frac{2}{3}\%$	0.6666. . .	$\frac{2}{3}$
25%	0.25	$\frac{1}{4}$	75%	0.75	$\frac{3}{4}$

STUDY SET Section 7.1

VOCABULARY *Fill in the blanks to make the statements true.*

1. __Percent__ means parts per one hundred.
2. When changing a fraction to a decimal, the result is either a __approximation__ or a repeating decimal.

CONCEPTS *Fill in the blanks to make the statements true.*

3. To write a percent as a fraction, drop the percent symbol and write the given number over __100__.
4. To change a percent to a decimal, drop the % symbol and divide by 100 by moving the decimal point two places to the __left__.
5. To change a decimal to a percent, multiply the decimal by 100 by moving the decimal point two places to the __right__, and then insert a % symbol.
6. To write a fraction as a percent, first write the fraction as a __decimal__. Then multiply the decimal by 100 by moving the decimal point two places to the __right__, and insert a % symbol.
7. **a.** See Illustration 1. Express the amount of the figure that is shaded as a decimal, a percent, and a fraction.
   b. What percent of the figure is not shaded?

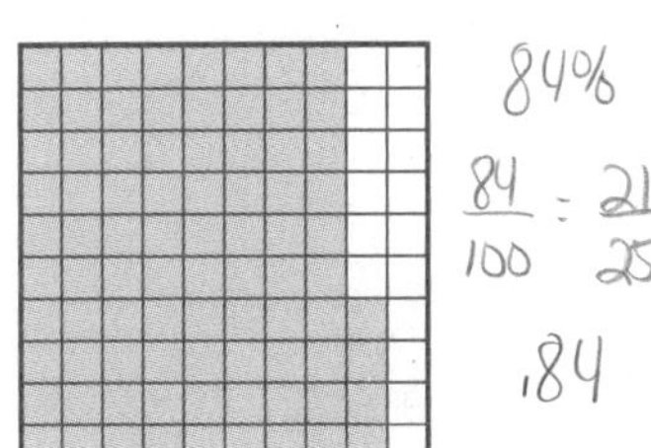

84% $\frac{84}{100} = \frac{21}{25}$ .84

ILLUSTRATION 1

8. In Illustration 2, each set of 100 squares represents 100%. What percent is shaded?

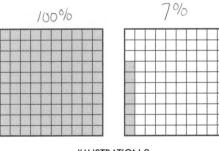

100% 7%

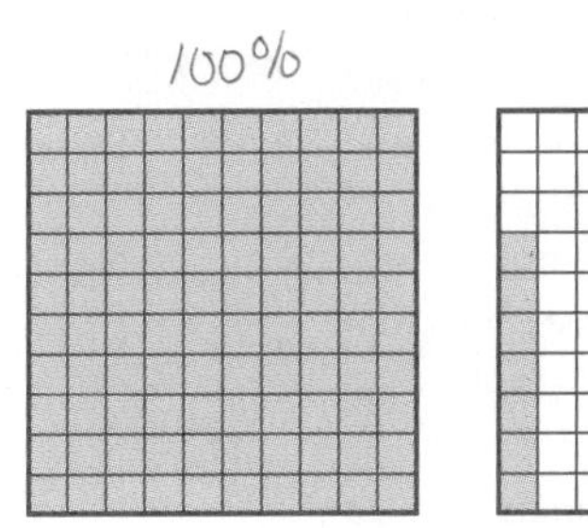

ILLUSTRATION 2

PRACTICE *Change each percent to a fraction. Simplify when necessary.*

9. 17%
10. 31%
11. 5%
12. 4%
13. 60%
14. 40%
15. 125%
16. 210%
17. $\frac{2}{3}\%$
18. $\frac{1}{5}\%$
19. $5\frac{1}{4}\%$
20. $6\frac{3}{4}\%$
21. 0.6%
22. 0.5%
23. 1.9%
24. 2.3%

Change each percent to a decimal.

25. 19%
26. 83%
27. 6%
28. 2%
29. 40.8%
30. 34.2%
31. 250%
32. 600%
33. 0.79%
34. 0.01%
35. $\frac{1}{4}\%$
36. $8\frac{1}{5}\%$

Change each decimal to a percent.

37. 0.93
38. 0.44
39. 0.612
40. 0.727
41. 0.0314
42. 0.0021
43. 8.43
44. 7.03
45. 50
46. 3
47. 9.1
48. 8.7

Change each fraction to a percent.

49. $\frac{17}{100}$
50. $\frac{29}{100}$
51. $\frac{4}{25}$
52. $\frac{47}{50}$
53. $\frac{2}{5}$
54. $\frac{21}{50}$
55. $\frac{21}{20}$
56. $\frac{33}{20}$
57. $\frac{5}{8}$
58. $\frac{3}{8}$
59. $\frac{3}{16}$
60. $\frac{1}{32}$

Find the exact equivalent percent for each fraction.

61. $\frac{2}{3}$

62. $\frac{1}{6}$

63. $\frac{5}{6}$

64. $\frac{4}{3}$

Express each of the given fractions as a percent. Round to the nearest hundredth.

65. $\frac{1}{9}$

66. $\frac{2}{3}$

67. $\frac{5}{9}$

68. $\frac{7}{3}$

APPLICATIONS

69. U.N. SECURITY COUNCIL The United Nations has 188 members. The United States, the Russian Federation, Britain, France, and China, along with 10 other nations, make up the Security Council.
  a. What fraction of the members of the United Nations belong to the Security Council?
  b. Write your answer to part a in percent form. (Round to the nearest one percent.)

70. ECONOMIC FORECAST One economic indicator of the national economy is the number of orders placed by manufacturers. One month, the number of orders rose one-fourth of one percent.
  a. Write this using a % symbol.
  b. Express it as a fraction.
  c. Express it as a decimal.

71. PIANO KEYS Of the 88 keys on a piano, 36 are black.
  a. What fraction of the keys are black?
  b. What percent of the keys are black? (Round to the nearest one percent.)

72. INTEREST RATES Write as a decimal the interest rate associated with each of these accounts.
  a. Home loan: 7.75%
  b. Savings account: 5%
  c. Credit card: 14.25%

73. THE SPINE The human spine consists of a group of bones (vertebrae). (See Illustration 3.)
  a. What fraction of the vertebrae are lumbar?
  b. What percent of the vertebrae are lumbar? (Round to the nearest one percent.)
  c. What percent of vertebrae are cervical? (Round to the nearest one percent.)

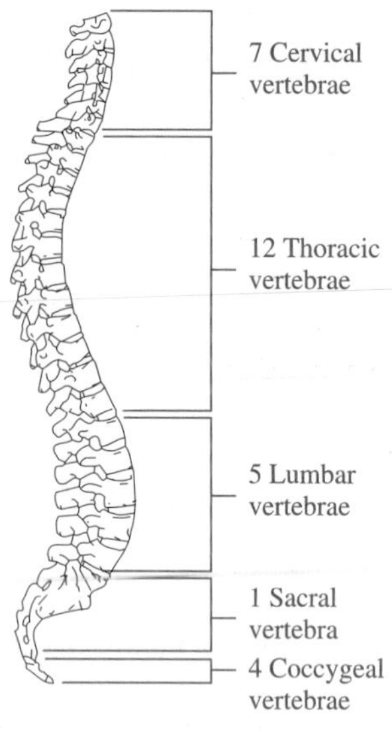

ILLUSTRATION 3

74. REGIONS OF THE COUNTRY The continental United States is divided into seven regions. (See Illustration 4.)
  a. What percent of the 50 states are in the Rocky Mountain region?
  b. What percent of the 50 states are in the Midwestern region?
  c. What percent of the 50 states are not located in any of the seven regions shown here?

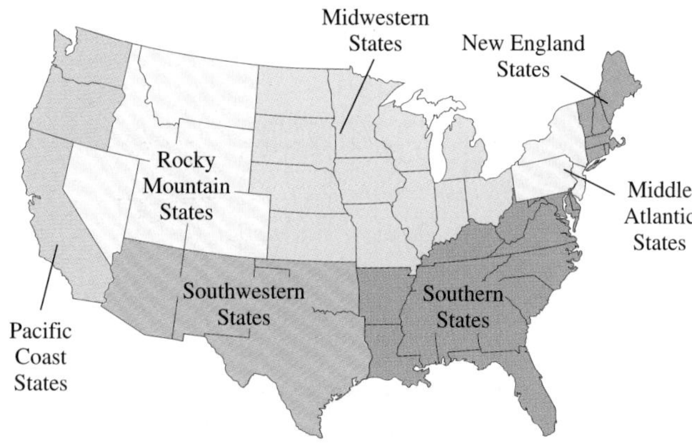

ILLUSTRATION 4

75. STEEP GRADE Sometimes, signs are used to warn truckers when they are approaching a steep grade on the highway. (See Illustration 5.) For a 5% grade, how many feet does the road rise over a 100-foot run?

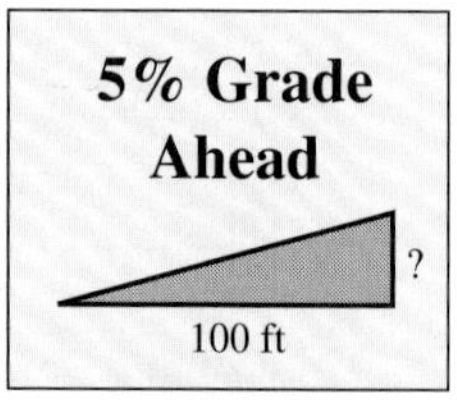

5% Grade Ahead

?

100 ft

ILLUSTRATION 5

76. COMPANY LOGO In Illustration 6, what part of the

company's logo is shaded red? Express your answer as a percent, a fraction, and a decimal. Do not round.

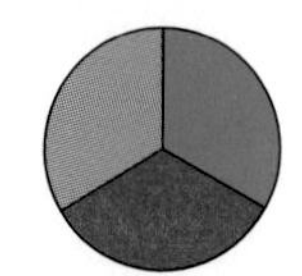

ILLUSTRATION 6

77. IVORY SOAP A popular soap claims to be $99\frac{44}{100}\%$ pure. Write this percent as a decimal.

78. DRUNK DRIVING In most states, it is illegal to drive with a blood alcohol concentration of 0.08% or more. Change this percent to a fraction. Do not simplify. Explain what the numerator and the denominator of the fraction represent.

79. BASKETBALL STANDINGS In the standings, we see that Chicago has won 60 of 67, or $\frac{60}{67}$ of its games. In what form is the team's winning percentage presented in the newspaper? Express it as a percent.

Eastern Conference			
Team	W	L	Pct.
Chicago	60	7	.896

80. WON–LOST RECORD In sports, when a team wins as many as it loses, it is said to be playing "500 ball." Examine the standings and explain the significance of the number 500, using concepts studied in this section.

Eastern Conference			
Team	W	L	Pct.
Orlando	33	33	.500

81. SKIN Illustration 7 shows roughly what percent each section of the body represents of the total skin area. Determine the missing percent, and then complete the bar graph in Illustration 8.

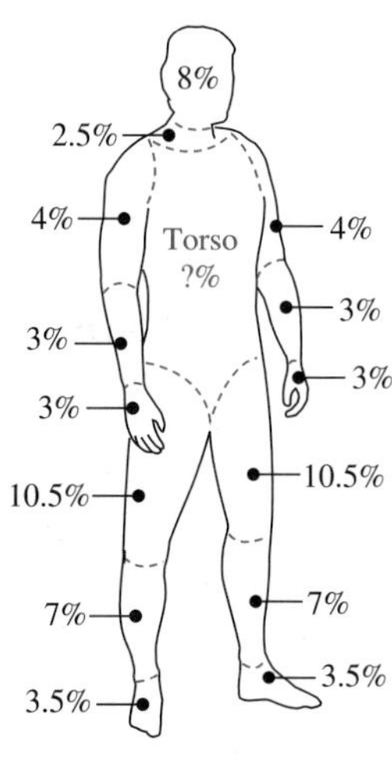

ILLUSTRATION 7

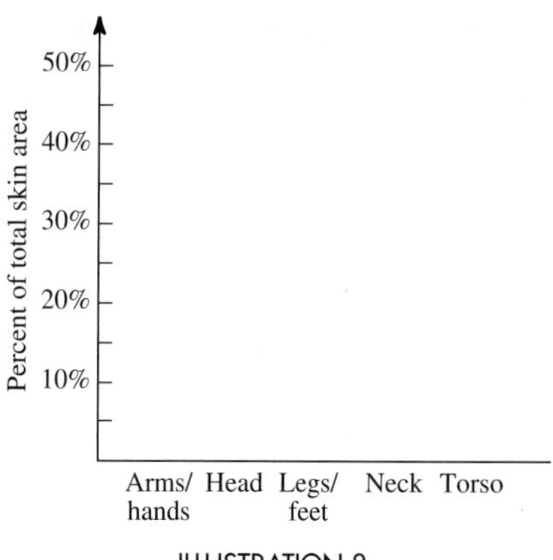

ILLUSTRATION 8

82. RAP MUSIC Illustration 9 shows what percent rap music sales were of total U.S. dollar sales of recorded music for the years 1994–1998. In Illustration 10, construct a line graph using the given data.

1994	1995	1996	1997	1998
7.9%	6.7%	8.9%	10.1%	9.7%

Source: *The New York Times 2000 Almanac*

ILLUSTRATION 9

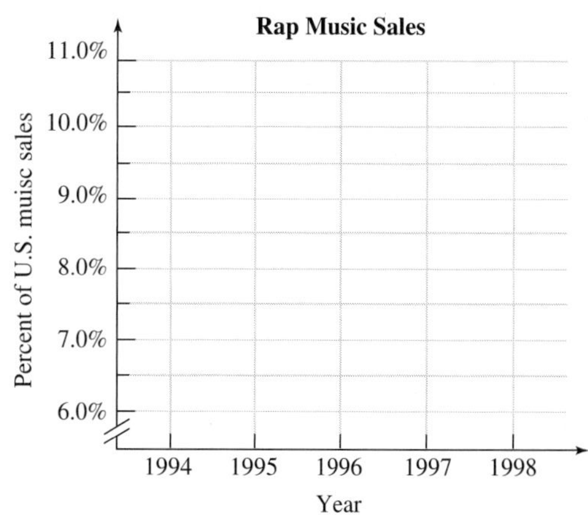

ILLUSTRATION 10

83. CHARITY A 1998 fact sheet released by the American Red Cross stated, "For the past three fiscal years, an average of 92 cents of every dollar spent by the Red Cross went to programs and services to help those in need." What percent of the money spent by the Red Cross went to programs and services?

84. TAXES Santa Anita Thoroughbred Racetrack in Arcadia, California has to pay a one-third of 1% tax on all the money wagered at the track. Write the percent as a fraction.

A calculator may be helpful to solve these problems.

85. BIRTHDAY If the day of your birthday represents $\frac{1}{365}$ of a year, what percent of the year is it? Round to the nearest hundredth of a percent.

86. POPULATION As a fraction, each resident of the United States represents approximately $\frac{1}{270{,}000{,}000}$ of the population. Express this as a percent. Round to one nonzero digit.

WRITING

87. If you were writing advertising, which form do you think would attract more customers: "25% off" or "$\frac{1}{4}$ off"? Explain your reasoning.

88. Many coaches ask their players to give a 110% effort during practices and games. What do you think this means? Is it possible?

89. Explain how to change a fraction to a percent.

90. Explain how an amusement park could have an attendance that is 103% of capacity.

91. HEAVYWEIGHT CHAMPION Muhammad Ali won 92% of his professional boxing matches. Does that mean he had exactly 100 fights, and won 92 of them? Explain your answer.

92. CALCULATORS To change the fraction $\frac{15}{16}$ to a percent, a student used a calculator to divide 15 by 16. The display is shown below.

0.9375

Now what keys should the student press to change this decimal to a percent?

REVIEW

93. Solve $-\frac{2}{3}x = -6$.

94. Add: $\frac{1}{3} + \frac{1}{4} + \frac{1}{2}$.

95. Complete the ordered pairs so that each one satisfies $y = 2x + 3$: (2,), (4,), (0,).

96. Express the area of a square with a side that is x feet long.

97. Multiply: $(x + 1)(x + 2)$.

98. Subtract: $41 - 10.287$.

7.2 Solving Percent Problems

In this section, you will learn about

- Percent problems • Finding the amount • Finding the percent
- Finding the base • Restating the problem
- An alternative approach: the percent formula • Circle graphs

INTRODUCTION. Percent problems occur in three forms. In this section, we will study a single procedure that can be used to solve all three types. It involves the equation-solving skills that we studied earlier.

Percent problems

The articles on the front page of the newspaper in Figure 7-3 suggest three types of percent problems.

- In the labor article, if we want to know how many union members voted to accept the new offer, we would ask:

  What number is 84% of 500?

- In the article on drinking water, if we want to know what percent of the wells are safe, we would ask:

  38 is what percent of 40?

- In the article on new appointees, if we want to know how many examiners are on the State Board, we would ask:

  6 is 75% of what number?

DAILY NEWS

Circulation | Monday, March 23 | 50 cents

Transit Strike Averted!

Labor: 84% of 500-member union votes to accept new offer

Drinking Water
38 of 40 Wells Declared Safe

New Appointees

These six area residents now make up 75% of the State Board of Examiners

FIGURE 7-3

These percent problems have several things in common.

- Each problem contains the word *is.* Here, *is* can be translated to an = sign.
- Each of the problems contains a phrase such as *what number* or *what percent.* In other words, there is an unknown quantity that can be represented by a variable.
- Each problem contains the word *of.* In this context, *of* means multiply.

These observations suggest that each of the percent problems can be translated into an equation. The equation, called a **percent equation,** will contain a variable, and the operation of multiplication will be involved.

Finding the amount

To solve the labor union problem, we translate the words into an equation and then solve it.

What number	is	84%	of	500?	
↓	↓	↓	↓	↓	
x	$=$	84%	$\cdot$	500	Translate to mathematical symbols.

$x = 0.84 \cdot 500$ Change 84% to a decimal: 84% = 0.84.

$x = 420$ Do the multiplication.

We have found that 420 is 84% of 500. That is, 420 union members voted to accept the new offer.

COMMENT When solving percent equations, always write the percent as a decimal or a fraction before performing any calculations. For example, in the previous problem, we wrote 84% as 0.84 before multiplying by 500.

Percent problems involve a comparison of numbers or quantities. In the statement "420 is 84% of 500," the number 420 is called the **amount,** 84% is the **percent,** and 500 is called the **base.** Think of the base as the standard of comparison—it represents the whole of some quantity. The amount is a part of the base, but it can exceed the base when the percent is more than 100%. The percent, of course, has the % symbol.

EXAMPLE 1 ***Finding the amount.*** What number is 160% of 15.8?

Solution

First, we translate the words into an equation.

What number	is	160%	of	15.8?	
↓		↓		↓	
x	$=$	160%	$\cdot$	15.8	x is the amount, 160% is the percent, and 15.8 is the base.

Then we solve the equation.

$x = 1.6 \cdot 15.8$ Change 160% to a decimal: $160\% = 1.6$.

$x = 25.28$ Do the multiplication.

Thus, 25.28 is 160% of 15.8.

Self Check

What number is 240% of 80?

Answer: 192 ■

Finding the percent

In the drinking water problem, we must find the percent. Once again, we translate the words of the problem into an equation and solve it.

38	is	what percent	of	40?	
↓		↓		↓	
38	$=$	x	$\cdot$	40	38 is the amount, x is the percent, and 40 is the base.

$38 = 40x$ Apply the commutative property of multiplication to write $x \cdot 40$ as $40x$.

$\frac{38}{\mathbf{40}} = \frac{40x}{\mathbf{40}}$ To undo the multiplication by 40, divide both sides by 40.

$0.95 = x$ Do the divisions.

$x = 0.95$ Since $0.95 = x$, $x = 0.95$.

$x = 95\%$ To change a decimal to a percent, multiply the decimal by 100 by moving the decimal point 2 places to the right, and then insert a % symbol.

Thus, 38 is 95% of 40. That is, 95% of the wells referred to in the article were declared safe.

EXAMPLE 2 ***Finding the percent.*** 14 is what percent of 32?

Solution

First, we translate the words into an equation.

14	is	what percent	of	32?	
↓		↓		↓	
14	$=$	x	$\cdot$	32	14 is the amount, x is the percent, and 32 is the base.

Self Check

9 is what percent of 16?

Then we solve the equation.

$14 = 32x$ Rewrite the right-hand side: $x \cdot 32 = 32x$.

$\frac{14}{\mathbf{32}} = \frac{32x}{\mathbf{32}}$ To undo the multiplication by 32, divide both sides by 32.

$0.4375 = x$ Do the divisions.

$43.75\% = x$ Change 0.4375 to a percent. Multiply the decimal by 100 by moving the decimal point 2 places to the right, and then insert a % symbol.

Thus, 14 is 43.75% of 32.

Answer: 56.25% ■

Accent on Technology: *Cost of an air bag*

An air bag is estimated to add an additional \$500 to the cost of a car. What percent of the \$16,295 sticker price is the cost of the air bag?

First, we translate the words into an equation.

What percent	of	the \$16,295 sticker price	is	the cost of the air bag?
↓		↓		↓
x	·	16,295	=	500

500 is the amount, x is the percent, and 16,295 is the base.

Then we solve the equation.

$16{,}295x = 500$ $x \cdot 16{,}295 = 16{,}295x$.

$\frac{16{,}295x}{\mathbf{16{,}295}} = \frac{500}{\mathbf{16{,}295}}$ To undo the multiplication by 16,295, divide both sides by 16,295.

$x = \frac{500}{16{,}295}$

To do the division using a calculator, enter these numbers and press these keys.

Keystrokes 500 [÷] 16295 [=] [0.03068425]

This display gives the answer in decimal form. To change it to a percent, we multiply the result by 100 and insert a percent symbol. This moves the decimal point 2 places to the right. If we round to the nearest tenth of a percent, the cost of the air bag is about 3.1% of the sticker price.

Finding the base

In the problem about the State Board of Examiners, we must find the base. As before, we translate the words of the problem into an equation and then solve it.

6	is	75%	of	what number?
↓		↓		↓
6	=	75%	·	x

6 is the amount, 75% is the percent, and x is the base.

$$6 = 0.75x \quad \text{Change 75\% to 0.75.}$$

$$\frac{6}{\mathbf{0.75}} = \frac{0.75x}{\mathbf{0.75}} \quad \text{To undo the multiplication by 0.75, divide both sides by 0.75.}$$

$$8 = x \quad \text{Do the divisions.}$$

Thus, 6 is 75% of 8. That is, there are 8 examiners on the State Board.

EXAMPLE 3 ***Finding the base.*** 31.5 is $33\frac{1}{3}\%$ of what number?

Solution

31.5	is	$33\frac{1}{3}\%$	of	what number?	
↓		↓		↓	
31.5	=	$33\frac{1}{3}\%$	·	x	31.5 is the amount, $33\frac{1}{3}\%$ is the percent, and x is the base.

In this case the computations can be made easier by changing the percent to a fraction instead of to a decimal. We write $33\frac{1}{3}\%$ as a fraction and proceed as follows:

$$31.5 = \frac{1}{3} \cdot x \quad 33\tfrac{1}{3}\% = \tfrac{1}{3}.$$

$$\mathbf{3} \cdot 31.5 = \mathbf{3} \cdot \frac{1}{3}x \quad \text{To isolate } x \text{ on the right-hand side, multiply both sides by 3.}$$

$$94.5 = x \quad \text{Do the multiplications: } 3 \cdot 31.5 = 94.5 \text{ and } 3 \cdot \tfrac{1}{3} = 1.$$

Thus, 31.5 is $33\frac{1}{3}\%$ of 94.5.

Self Check

150 is $66\frac{2}{3}\%$ of what number?

Answer: 225 ■

Restating the problem

Not all percent problems are presented in the form we have been studying. In Example 4, we must examine the given information carefully so that we can restate the problem in the familiar form.

EXAMPLE 4 ***Housing.*** In an apartment complex, 110 of the units are currently being rented. This represents an 88% occupancy rate. How many units are there in the complex?

Solution An occupancy rate of 88% means that 88% of the units are occupied. We restate the problem in the form we have been studying.

110	is	88%	of	what number?	
↓		↓		↓	
110	=	88%	·	x	110 is the amount, 88% is the percent, and x is the base.

Now we solve the equation.

$$110 = 0.88x \quad \text{Change 88\% to a decimal: 88\% = 0.88.}$$

$$\frac{110}{\mathbf{0.88}} = \frac{0.88x}{\mathbf{0.88}} \quad \text{To undo the multiplication by 0.88, divide both sides by 0.88.}$$

$$125 = x \quad \text{Do the divisions.}$$

The complex has 125 units. ■

An alternative approach: the percent formula

In any percent problem, the relationship between the amount, the percent, and the base is as follows: *Amount is percent of base.* This relationship is shown in the **percent formula.**

The percent formula

Amount = percent · base

The percent formula can be used as an alternate way to solve percent problems. With this method, we need to identify the *amount* (the part that is compared to the whole), the *percent* (indicated by the % symbol or the word *percent*), and the *base* (the whole of some quantity, usually following the word *of*).

EXAMPLE 5 ***Finding the amount.*** What number is 160% of 15.8?

Solution

In this example, the percent is 160 and the base is 15.8, the number following the word *of.* We can let A stand for the amount and use the percent formula.

Amount	=	percent	·	base	
↓		↓		↓	
A	=	160%	·	15.8	Substitute 160 for the percent and 15.8 for the base.

The statement $A = 160\% \cdot 15.8$ is an equation, with the amount A being the unknown. We can find the unknown amount by multiplication.

$A = 1.6 \cdot 15.8$ Change 160% to a decimal: $160\% = 1.6$.

$= 25.28$ Do the multiplication.

Thus, 25.28 is 160% of 15.8.

Self Check

What number is 240% of 80?

Answer: 192 ■

EXAMPLE 6 ***Finding the percent.*** 14 is what percent of 32?

Solution

In this example, 14 is the amount and 32 is the base. Once again, we use the percent formula and let p stand for the percent.

Amount	=	percent	·	base	
↓		↓		↓	
14	=	p	·	32	Substitute 14 for the amount and 32 for the base.

The statement $14 = p \cdot 32$ is an equation, with the percent p being the unknown. We can find the unknown percent by division.

Self Check

9 is what percent of 16?

$14 = p \cdot 32$	The equation to solve.
$14 = 32p$	Rewrite the right-hand side: $p \cdot 32 = 32p$.
$\frac{14}{32} = \frac{32p}{32}$	To undo the multiplication by 32, divide both sides by 32.
$0.4375 = p$	$\frac{14}{32} = 0.4375$.
$p = 43.75\%$	To change the decimal to a percent, multiply the decimal by 100 by moving the decimal point 2 places to the right, and then insert a % symbol.

Thus, 14 is 43.75% of 32.

Answer: 56.25%

EXAMPLE 7 ***Finding the base.*** 31.5 is $33\frac{1}{3}\%$ of what number?

Solution

In this example, 31.5 is the amount and $33\frac{1}{3}$ is the percent. To find the base (which we will call b), we form an equation using the percent formula.

Amount	=	percent	·	base	
↓		↓		↓	
31.5	=	$33\frac{1}{3}\%$	·	b	Substitute 31.5 for the amount and $33\frac{1}{3}\%$ for the percent.

The statement $31.5 = 33\frac{1}{3}\% \cdot b$ is an equation, with the base b being the unknown. We can find the unknown base by multiplication.

$31.5 = 33\frac{1}{3}\% \cdot b$	The equation to solve.
$31.5 = \frac{1}{3}b$	$33\frac{1}{3}\% = \frac{33\frac{1}{3}}{100} = \frac{1}{3}$.
$\mathbf{3} \cdot 31.5 = \mathbf{3} \cdot \frac{1}{3}b$	To isolate b on the right-hand side, multiply both sides by 3.
$94.5 = b$	Do the multiplication: $31.5 \cdot 3 = 94.5$.

Thus, 31.5 is $33\frac{1}{3}\%$ of 94.5.

Self Check

150 is $66\frac{2}{3}\%$ of what number?

Answer: 225

Circle graphs

Percents are used with **circle graphs,** or **pie charts,** as a way of presenting data for comparison. In Figure 7-4, the entire circle represents the total amount of electricity generated in the United States in 1998. The pie-shaped pieces of the graph show the relative sizes of the energy sources used to produce the electricity. For example, we see that the greatest amount of electricity (52%) was generated from coal. Note that if we add the percents from all categories (52% + 19% + 15% + 9% + 3% + 2%), the sum is 100%.

The 100 tick marks equally spaced around the circle serve as a visual aid when constructing a circle graph. For example, to represent hydropower as 9%, a line was drawn from the center of the circle to a tick mark. Then we counted off 9 ticks and drew a second line from the center to that tick to complete the pie-shaped wedge.

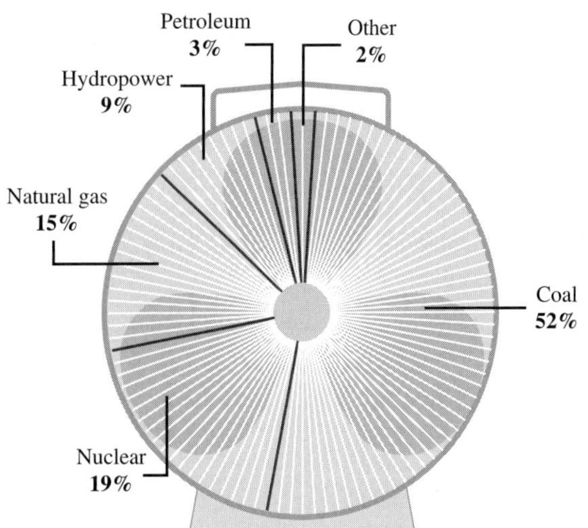

Source: Energy Information Administration

FIGURE 7-4

EXAMPLE 8 ***Presidential election results.*** Results from the 1996 presidential election are shown in Figure 7-5. Use the information to find the number of states won by President Clinton.

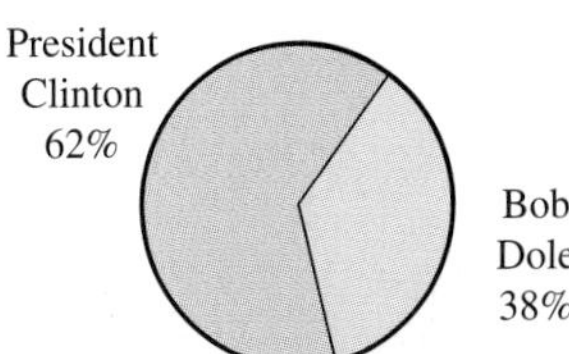

1996 Presidential Election: States won by each candidate

FIGURE 7-5

Solution The circle graph shows that President Clinton was victorious in 62% of the 50 states. We state the problem in the standard form and then translate it into an equation.

What number	is	62%	of	50?	
↓	↓	↓	↓	↓	
x	$=$	62%	$\cdot$	50	Translate the words into an equation.

$x = 0.62 \cdot 50$ Change 62% to a decimal: 62% = 0.62.

$x = 31$ Do the multiplication.

President Clinton won 31 states. ■

STUDY SET Section 7.2

VOCABULARY *Translate each sentence into a percent equation.*

1. What number is 10% of 50?

2. 16 is 55% of what number?

3. 48 is what percent of 47?

4. 12 is what percent of 20?

Fill in the blanks to make the statements true.

5. In a circle graph, pie-shaped wedges are used to show the division of a whole quantity into its component parts.

6. In the statement "45 is 90% of 50," 45 is the amount, 90% is the percent, and 50 is the base.

CONCEPTS

7. When computing with percents, the percent must be changed to a decimal or a fraction. Change each percent to a decimal.

a. 12% **b.** 5.6%

c. 125% **d.** $\frac{1}{4}\%$

8. When computing with percents, the percent must be changed to a decimal or a fraction. Change each percent to a fraction.

a. $33\frac{1}{3}\%$ **b.** $66\frac{2}{3}\%$

c. $16\frac{2}{3}\%$ **d.** $83\frac{1}{3}\%$

9. Without doing the calculation, tell whether 120% of 55 is more than 55 or less than 55.

10. Without doing the calculation, tell whether 12% of 55 is more than 55 or less than 55.

11. Solve each of the following problems in your head.

a. What is 100% of 25?

b. What percent of 132 is 132?

c. What number is 87% of 100?

12. To solve the problem

15 is what percent of 75?

a student wrote a percent equation, solved it, and obtained $x = 0.2$. For her answer, the student wrote

15 is 0.2% of 75.

Explain her error.

13. VIDEO GAMES Illustration 1 shows the market shares of the three main video game systems. What percent of the market does Nintendo 64 have?

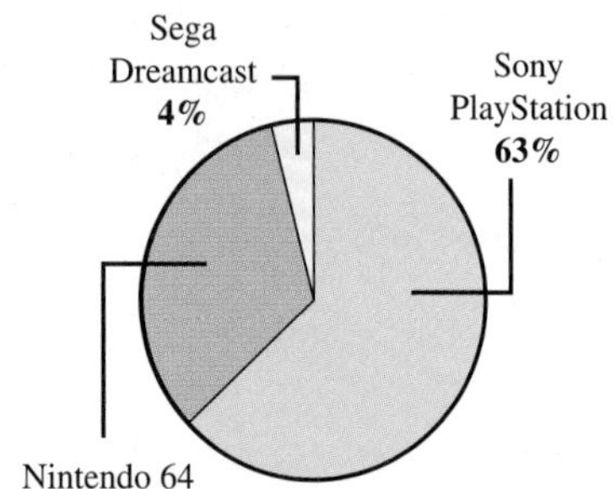

Source: *Los Angeles Times* (December 23, 1999)

ILLUSTRATION 1

14. HOUSING In the last quarter of 1999, approximately 105.3 million housing units in the United States were occupied. Use the data in Illustration 2 to determine what percent were owner-occupied.

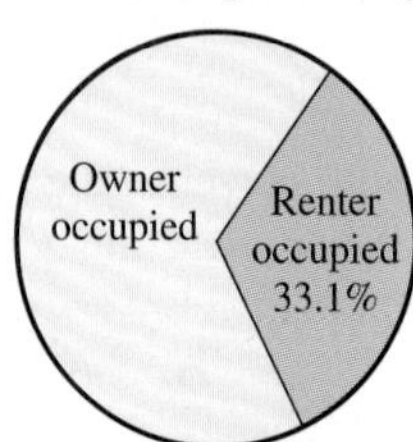

Source: *The New York Times 2000 Almanac*

ILLUSTRATION 2

NOTATION

15. How is each of the following words or phrases translated in this section?

a. of

b. is

c. what number

16. a. Write the repeating decimal shown in the calculator display as a percent. Use an overbar.

0.456666666

b. Round your answer to part a to the nearest hundredth of a percent.

c. Write your answer to part a using a fraction.

PRACTICE *Solve each problem by solving a percent equation.*

17. What number is 36% of 250?

18. What number is 82% of 300?

19. 16 is what percent of 20?

20. 13 is what percent of 25?

21. 7.8 is 12% of what number?

22. 39.6 is 44% of what number?

23. What number is 0.8% of 12?

24. What number is 5.6% of 4,040?

25. 0.5 is what percent of 40,000?

26. 0.3 is what percent of 15?

27. 3.3 is 7.5% of what number?

28. 8.4 is 20% of what number?

29. Find $7\frac{1}{4}\%$ of 600.

30. Find $1\frac{3}{4}\%$ of 800.

31. 102% of 105 is what number?

32. 210% of 66 is what number?

33. $33\frac{1}{3}\%$ of what number is 33?

34. $66\frac{2}{3}\%$ of what number is 28?

35. $9\frac{1}{2}\%$ of what number is 5.7?

36. $\frac{1}{2}\%$ of what number is 5,000?

37. What percent of 8,000 is 2,500?

38. What percent of 3,200 is 1,400?

Use a circle graph to illustrate the given data. A circle divided into 100 sections is provided to aid in the graphing process.

39. Complete Illustration 3 to show what percent of the total U.S. energy consumed was provided by each source in 1997.

Renewable	8%
Nuclear	7%
Coal	23%
Natural gas	24%
Petroleum	38%

Source: *The New York Times 2000 Almanac*

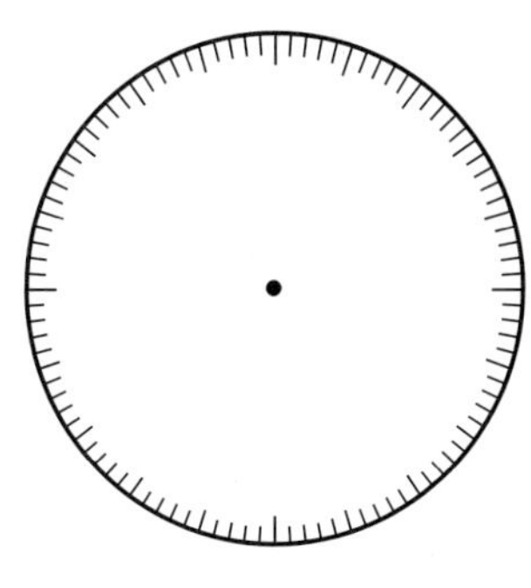

ILLUSTRATION 3

40. GREENHOUSE EFFECT Complete Illustration 4 to show what percent of the total U.S. emissions from human activities came from each greenhouse gas.

Carbon dioxide	82%
Nitrous oxide	6%
Methane	10%
PFCs	2%

Source: *The World Almanac 2000*

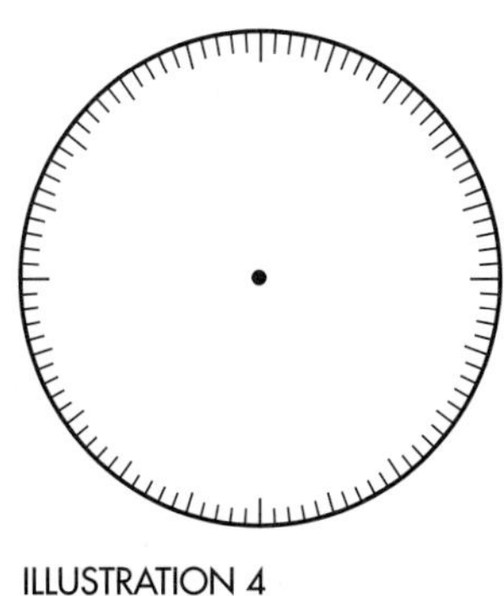

ILLUSTRATION 4

APPLICATIONS

41. CHILD CARE After the first day of registration, 84 children had been enrolled in a new day care center. That represented 70% of the available slots. What was the maximum number of children the center could enroll?

42. RACING PROGRAM One month before a stock car race, the sale of ads for the official race program was slow. Only 12 pages, or just 60% of the available pages, had been sold. What was the total number of pages devoted to advertising in the program?

43. GOVERNMENT SPENDING Illustration 5 shows the breakdown of federal outlays for fiscal year 1998. If the total spending was approximately $1,650 billion, how many dollars were spent on Social Security, Medicare, and other retirement programs?

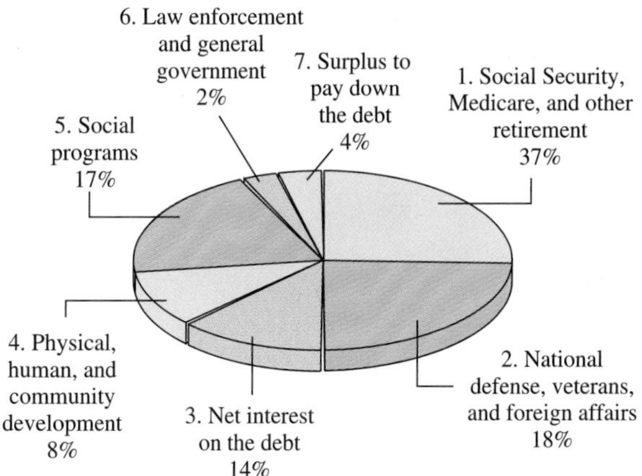

Source: 1999 federal income tax form

ILLUSTRATION 5

44. GOVERNMENT REVENUE Complete the table by finding what percent of total federal government revenue each source provided in 1998. Round to the nearest percent. Then complete the circle graph in Illustration 6.

Total revenue, fiscal year 1998: $1,722 billion		
Source of revenue	**Amount**	**Percent of total**
Social Security, Medicare, unemployment taxes	$572 billion	
Personal income taxes	$829 billion	
Corporate income taxes	$189 billion	
Excise, estate, customs taxes	$132 billion	

Source: 1999 federal income tax form

1998 Federal Revenue

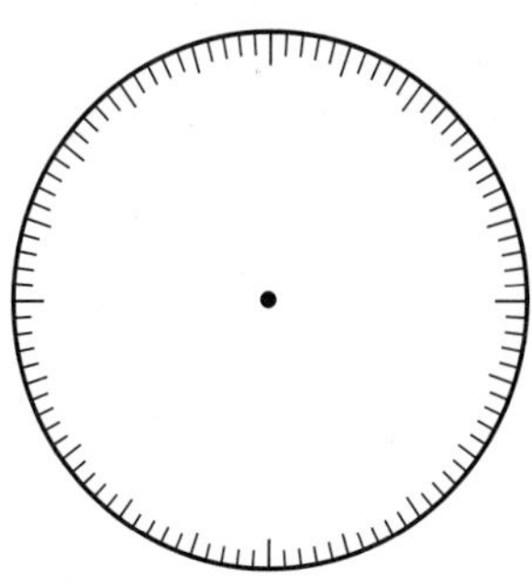

ILLUSTRATION 6

45. THE INTERNET See Illustration 7. The message at the bottom of the screen indicates that 24% of the 50K bytes of information that the user has decided to view have been downloaded to her computer. How many more bytes of information must be downloaded? (50K stands for 50,000.)

ILLUSTRATION 7

46. REBATE A long-distance telephone company offered its customers a rebate of 20% of the cost of all long-distance calls made in the month of July. One customer's calls are listed in Illustration 8. What amount will this customer receive in the form of a rebate?

Date	Time	Place called	Min.	Amount
Jul 4	3:48 P.M.	Denver	47	$3.80
Jul 9	12:00 P.M.	Detroit	68	$7.50
Jul 20	8:59 A.M.	San Diego	70	$9.45

ILLUSTRATION 8

47. PRODUCT PROMOTION To promote sales, a free 6-ounce bottle of shampoo is packaged with every large bottle. (See Illustration 9.) Use the information on the package to find how many ounces of shampoo the large bottle contains.

ILLUSTRATION 9

48. NUTRITION FACTS The nutrition label on a package of corn chips is shown in Illustration 10.

a. How many milligrams of sodium are in one serving of chips?

b. According to the label, what percent of the daily value is this?

c. What daily value of sodium intake is deemed healthy?

Nutrition Facts
Serving Size: 1 oz. (28g/About 29 chips)
Servings Per Container: About 11

Amount Per Serving	
Calories 160	Calories from Fat 90
	% Daily Value
Total fat 10g	**15%**
Saturated fat 1.5 g	**7%**
Cholesterol 0mg	**0%**
Sodium 240mg	**12%**
Total carbohydrate 15g	**5%**
Dietary fiber 1g	**4%**
Sugars less than 1g	
Protein 2g	

ILLUSTRATION 10

49. DRIVER'S LICENSE On the written part of his driving test, a man answered 28 out of 40 questions correctly. If 70% correct is passing, did he pass the test?

50. ALPHABET What percent of the English alphabet do the vowels a, e, i, o, and u make up? (Round to the nearest one percent.)

51. MIXTURES Complete the chart in Illustration 11 to find the number of gallons of sulfuric acid in each of two storage tanks.

Gallons of solution in tank	% sulfuric acid	Gallons of sulfuric acid in tank
60	50%	
40	30%	

ILLUSTRATION 11

52. CUSTOMER GUARANTEE To assure its customers of low prices, the Home Club offers a "10% Plus" guarantee. If the customer finds the same item selling for less somewhere else, he or she receives the difference in price, plus 10% of the difference. A woman bought miniblinds at the Home Club for $120 but later saw the same blinds on sale for $98 at another store. How much can she expect to be reimbursed?

53. MAKING COPIES The zoom key on the control panel of a copier programs it to print a magnified or reduced copy of the original document. If the zoom is set at 180% and the original document contains type that is 1.5 inches tall, what will be the height of the type on the copy?

54. MAKING COPIES The zoom setting for a copier is entered as a decimal: 0.98. Express it as a percent and find the resulting type size on the copy if the original has type 2 inches in height.

55. INSURANCE The cost to repair a car after a collision was \$4,000. The automobile insurance policy paid the entire bill except for a \$200 deductible, which the driver paid. What percent of the cost did he pay?

56. FLOOR SPACE A house has 1,200 square feet on the first floor and 800 square feet on the second floor. What percent of the square footage of the house is on the first floor?

57. A MAJORITY In Los Angeles City Council races, if no candidate receives more than 50% of the vote, a runoff election is held between the first- and second-place finishers. From the election results in Illustration 12, determine whether there must be a runoff election for District 10.

City Council	District 10
Nate Holden	8,501
Madison T. Shockley	3,614
Scott Suh	2,630
Marsha Brown	2,432

ILLUSTRATION 12

58. PORTS In 1997, the busiest port in the United States was the Port of South Louisiana, which handled 183,628,353 tons of goods. Of that amount, 106,846,289 tons were domestic goods, and 76,782,064 tons were foreign. What percent of the total was domestic? Round to the nearest tenth of a percent.

WRITING

59. Explain the relationship in a percent problem between the amount, the percent, and the base.

60. Write a real-life situation that could be described by "9 is what percent of 20?"

61. Explain why 150% of a number is more than the number.

62. Explain why "Find 9% of 100" is an easy problem to solve.

REVIEW

63. Add: $2.78 + 6 + 9.09 + 0.3$.

64. Evaluate: $\sqrt{64} + 3\sqrt{9}$.

65. On a number line, which number is closer to 5, 4.9 or 5.001?

66. Is the x-axis horizontal or vertical?

67. Multiply: $34.5464 \cdot 1{,}000$.

68. Find: $(0.2)^3$.

69. Solve $0.4x + 1.2 = -7.8$.

70. If $d = 4t$, find d when $t = 25$.

7.3 *Applications of Percent*

In this section, you will learn about

- Taxes • Commissions • Percent of increase or decrease
- Discounts

INTRODUCTION. In this section, we discuss four applications of percent. Three of the four (taxes, commissions, and discounts) are directly related to purchasing. A solid understanding of these concepts will make you a better consumer. The fourth application uses percent to describe increases or decreases of such things as unemployment and grocery store sales.

Taxes

The sales receipt in Figure 7-6 (on the next page) gives a detailed account of what items were purchased, how many of each were purchased, and the price of each item.

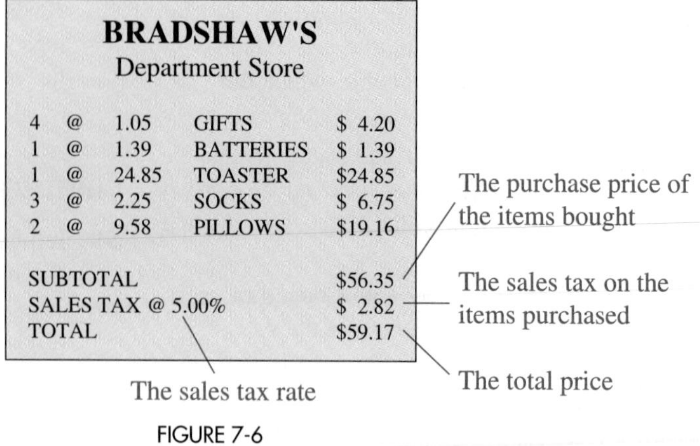

FIGURE 7-6

The receipt shows that the \$56.35 purchase price (labeled *subtotal*) was taxed at a **rate** of 5%. Sales tax of \$2.82 was charged. The sales tax was then added to the subtotal to get the total price of \$59.17.

Finding the total price

Total price = purchase price + sales tax

In Example 1, we verify that the amount of sales tax shown on the receipt in Figure 7-6 is correct.

EXAMPLE 1 ***Finding the sales tax.*** Find the sales tax on a purchase of \$56.35 if the sales tax rate is 5%.

Solution

First we write the problem so that we can translate it into an equation. The rate is 5%. We are to find the amount of the tax.

What number	is	5%	of	56.35?
x	$=$	5%	$\cdot$	56.35

$x = 0.05 \cdot 56.35$ Change 5% to a decimal: 5% = 0.05.

$x = 2.8175$ Do the multiplication.

Rounding to the nearest cent (hundredths), we find that the sales tax would be \$2.82. The sales receipt in Figure 7-6 is correct.

Self Check

What would the sales tax be if the \$56.35 purchase were made in Texas, which has a 6.25% state sales tax?

Answer: \$3.52 ■

In addition to sales tax, we pay many other types of taxes in our daily lives. Income tax, gasoline tax, and Social Security tax are just a few.

EXAMPLE 2 ***Finding the tax rate.*** A waitress found that \$11.04 was deducted from her weekly gross earnings of \$240 for federal income tax. What withholding tax rate was used?

Solution

First, we write the problem in a form that can be translated into an equation. We need to find the tax rate.

Self Check

\$5,250 had to be paid on an inheritance of \$15,000. What is the inheritance tax rate?

11.04	is	what percent	of	240?
11.04	=	x	$\cdot$	240

$11.04 = 240x$ Rewrite the right-hand side: $x \cdot 240 = 240x$.

$\frac{11.04}{\mathbf{240}} = \frac{240x}{\mathbf{240}}$ To undo the multiplication by 240, divide both sides by 240.

$0.046 = x$ Do the divisions.

$4.6\% = x$ Change 0.046 to a percent.

The withholding tax rate was 4.6%.

Answer: 35%

Commissions

Instead of working for a salary or getting paid at an hourly rate, many salespeople are paid on **commission.** They earn an amount based on the goods or services they sell.

EXAMPLE 3 ***Finding a commission.*** The commission rate for a salesperson at an appliance store is 16.5%. What is his commission from the sale of a refrigerator costing \$499.95?

Solution

We write the problem so that it can be translated into an equation. We are to find the amount of the commission.

What number	is	16.5%	of	499.95?
x	=	16.5%	$\cdot$	499.95

$x = 0.165 \cdot 499.95$ Change 16.5% to a decimal: $16.5\% = 0.165$.

$x = 82.49175$ Use a calculator to do the multiplication.

Rounding to the nearest cent (hundredth), we find that the commission is \$82.49.

Self Check

An insurance salesperson receives a 4.1% commission on each \$120 premium paid by a client. What is the amount of the commission on this premium?

Answer: \$4.92

Percent of increase or decrease

Percents can be used to describe how a quantity has changed. For example, consider Figure 7-7, which compares the number of hours of work it took the average U.S. worker to earn enough to buy a dishwasher in 1950 and 1998.

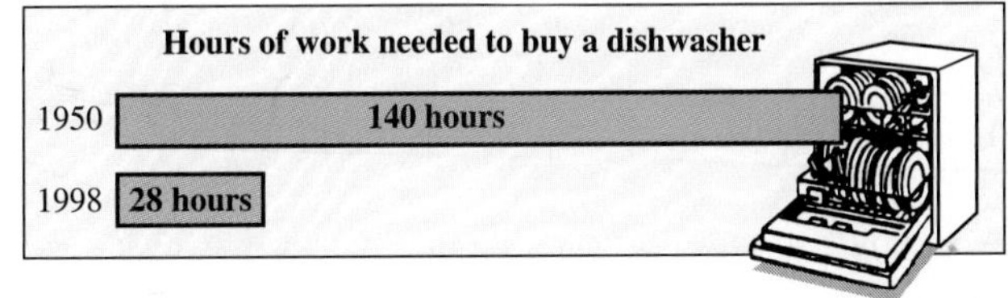

Source: Federal Reserve Bank of Dallas

FIGURE 7-7

From the figure, we see that the number of hours an average American had to work in order to buy a dishwasher has decreased over the years. To describe this decrease using a percent, we first subtract to find the amount of the decrease.

$140 - 28 = 112$ Subtract the hours of work needed in 1998 from the hours of work needed in 1950.

Next, we find what percent of the original number of hours of work needed in 1950 this difference represents.

112	is	what percent	of	140?
112	=	x	·	140

$112 = 140x$ Rewrite the right-hand side: $x \cdot 140 = 140x$.

$\frac{112}{\mathbf{140}} = \frac{140x}{\mathbf{140}}$ To undo the multiplication by 140, divide both sides by 140.

$0.8 = x$ Do the divisions.

$80\% = x$ Change 0.8 to a percent.

From 1950 to 1998, there was an 80% decrease in the number of hours it took the average U.S. worker to earn enough to buy a dishwasher.

Finding the percent of increase or decrease

To find the percent of increase or decrease:

1. Subtract the smaller number from the larger to find the amount of increase or decrease.
2. Find what percent the difference is of the original amount.

EXAMPLE 4 ***Finding the percent of increase.*** A 1996 auction included an oak rocking chair used by President John F. Kennedy in the Oval Office. The chair, originally valued at $5,000, sold for $453,500. Find the percent of increase in the value of the rocking chair.

Solution

First, we find the amount of increase.

$453{,}500 - 5{,}000 = 448{,}500$ Subtract the original value from the price paid at auction.

The rocking chair increased in value by $448,500. Next, we find what percent of the original value the increase represents.

448,500	is	what percent	of	5,000?
448,500	=	x	·	5,000

$448{,}500 = 5{,}000x$ Rewrite the right-hand side: $x \cdot 5{,}000 = 5{,}000x$.

$\frac{448{,}500}{\mathbf{5{,}000}} = \frac{5{,}000x}{\mathbf{5{,}000}}$ To undo the multiplication by 5,000, divide both sides by 5,000.

$89.7 = x$ Do the divisions.

$8{,}970\% = x$ Change 89.7 to a percent.

The Kennedy rocking chair increased in value by an amazing 8,970%.

Self Check

In one school district, the number of home-schooled children increased from 15 to 150 in 4 years. Find the percent of increase.

Answer: 900% ■

EXAMPLE 5 POPULATION DECLINE. Norfolk, Virginia, experienced the greatest percent decrease in population of any major U.S. city over the eight-year period from 1990 to 1998. Use the information in Figure 7-8 to determine the population of Norfolk in 1998.

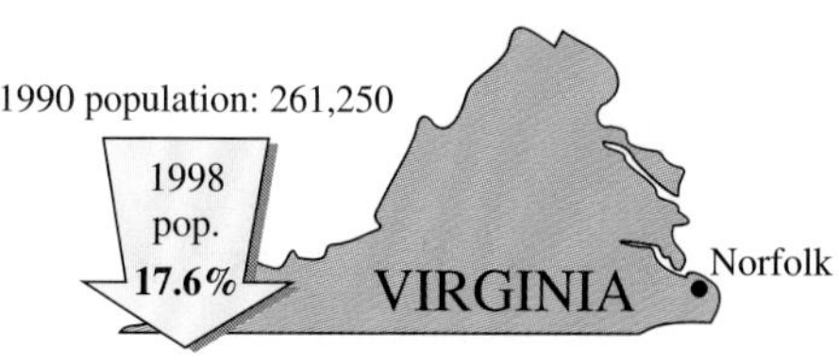

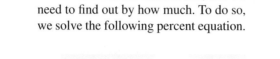

Source: U.S. Bureau of the Census (1999)

FIGURE 7-8

Solution In 1990, the population was 261,250. We are told that the number fell, and we need to find out by how much. To do so, we solve the following percent equation.

What number	is	17.6%	of	261,250?
x	$=$	17.6%	$\cdot$	$261{,}250$

$x = 0.176 \cdot 261{,}250$ Change 17.6% to a decimal: 17.6% = 0.176.

$x = 45{,}980$ Do the multiplication.

From 1990 to 1998, the population decreased by 45,980. To find the city's population in 1998, we subtract the decrease from the population in 1990.

$$261{,}250 - 45{,}980 = 215{,}270$$

In 1998, the population of Norfolk, Virginia was 215,270.

We can solve this problem in another way. If the population of Norfolk decreased by 17.6%, then the population in 1998 was 100% − 17.6% or 82.4% of the population in 1990. Using this approach, we can find the 1998 population directly by solving the following percent equation.

What number	is	82.4%	of	261,250?
x	$=$	82.4%	$\cdot$	$261{,}250$

$x = 0.824 \cdot 261{,}250$ Change 82.4% to a decimal: 82.4% = 0.824.

$x = 215{,}270$ Do the multiplication.

As before, we see that the population of Norfolk in 1998 was 215,270. ■

Discounts

The difference between the original price and the sale price of an item is called the **discount.** If the discount is expressed as a percent of the selling price, it is called the **rate of discount.** We will use the information in the advertisement shown in Figure 7-9 to discuss how to find a discount and how to find a discount rate.

FIGURE 7-9

EXAMPLE 6 ***Finding the discount.*** Find the amount of the discount on the pair of men's basketball shoes shown in Figure 7-9. Then find the sale price.

Solution

To find the discount, we find 25% of the regular price, \$59.80.

What number	is	25%	of	59.80?
x	$=$	25%	$\cdot$	59.80

$x = 0.25 \cdot 59.80$ Change 25% to a decimal: 25% = 0.25.

$x = 14.95$ Do the multiplication.

The discount is \$14.95. To find the sale price, we subtract the amount of the discount from the regular price.

$59.80 - 14.95 = 44.85$

The sale price of the men's basketball shoes is \$44.85.

Self Check

Sunglasses, regularly selling for \$15.40, are discounted 15%. Find the sale price.

Answer: \$13.09 ■

In Example 6, we used the following formula to find the sale price.

Finding the sale price

Sale price = original price − discount

EXAMPLE 7 ***Finding the discount rate.*** What is the rate of discount on the ladies' aerobic shoes advertised in Figure 7-9?

Solution

We can think of this as a percent-of-decrease problem. We first compute the amount of the discount. This decrease in price is found using subtraction.

$39.99 - 21.99 = 18$

The shoes are discounted \$18. Now we find what percent of the original price the discount is.

18	is	what percent	of	39.99?
18	$=$	x	$\cdot$	39.99

$18 = 39.99x$ $x \cdot 39.99 = 39.99x$.

$\frac{18}{\mathbf{39.99}} = \frac{39.99x}{\mathbf{39.99}}$ To undo the multiplication by 39.99, divide both sides by 39.99.

$0.450113 \approx x$ Do the division.

$45.0113\% \approx x$ Change 0.450113 to a percent.

Rounded to the nearest one percent, the discount rate is 45%.

Self Check

An early-bird special at a restaurant offers a \$10.99 prime rib dinner for only \$7.95 if it is ordered before 6 P.M. Find the rate of discount. Round to the nearest one percent.

Answer: 28% ■

STUDY SET Section 7.3

VOCABULARY *Fill in the blanks to make the statements true.*

1. Some salespeople are paid on Commision. It is based on a percent of the total dollar amount of the goods or services they sell.
2. When we use percent to describe how a quantity has increased when compared to its original value, we are finding the percent of increase.
3. The difference between the original price and the sale price of an item is called the discount.
4. The rate of a sales tax is expressed as a percent.

CONCEPTS

5. An organization experiences a 100% increase in membership. Represent the increase in another way.
6. The number of people watching a television show decreased by 50% over a ten-week period. Represent the decrease in another way.

APPLICATIONS *Solve each problem. If a percent answer is not exact, round to the nearest one percent.*

7. STATE SALES TAX The state sales tax rate in Utah is 4.75%. Find the sales tax on a dining room set that sells for $900.
8. STATE SALES TAX Find the sales tax on a pair of jeans costing $40 if they are purchased in Arkansas, which has a sales tax rate of 4.625%.
9. ROOM TAX After checking out of a hotel, a man noticed that the hotel bill included an additional charge labeled *room tax.* If the price of the room was $129 plus a room tax of $10.32, find the room tax rate.
10. EXCISE TAX While examining her monthly telephone bill, a woman noticed an additional charge of $1.24 labeled *federal excise tax.* If the basic service charges for that billing period were $42, what is the federal excise tax rate?
11. SALES RECEIPT Complete the sales receipt in Illustration 1 by finding the subtotal, the sales tax, and the total.

NURSERY CENTER Your one-stop garden supply		
3 @ 2.99	PLANTING MIX	$ 8.97
1 @ 9.87	GROUND COVER	$ 9.87
2 @ 14.25	SHRUBS	$28.50
SUBTOTAL		$ 47.34
SALES TAX @ 6.00%		$ 2.84
TOTAL		$ 50.18

ILLUSTRATION 1

12. SALES RECEIPT Complete the sales receipt in Illustration 2 by finding the prices, the subtotal, the sales tax, and the total.

McCOY'S FURNITURE		
1 @ 450.00	SOFA	$ 450
2 @ 90.00	END TABLES	$ 180
1 @ 350.00	LOVE SEAT	$ 350
SUBTOTAL		$ 980
SALES TAX @ 4.20%		$
TOTAL		$

ILLUSTRATION 2

13. SALES TAX HIKE In order to raise more revenue, some states raise the sales tax rate. How much additional money will be collected on the sale of a $15,000 car if the sales tax rate is raised 1%?
14. FOREIGN TRAVEL Value added tax is a consumer tax imposed on goods and services. Currently, there are VAT systems in place all around the world. (The United States is one of the few industrialized nations not using a value added tax system.) Complete the table by determining the VAT tax a traveler would pay in each country on a dinner costing $20.95.

Country	VAT tax rate	Tax on a $20.95 dinner
Canada	7%	1.47
Germany	16%	3.35
England	17.5%	3.67
Sweden	25%	5.24

15. PAYCHECK Use the information on the paycheck stub in Illustration 3 (on the next page) to find the tax rate for the federal withholding, worker's compensation, and Medicare taxes that were deducted from the gross pay.

6286244	
Issue date: 03-27-00	
GROSS PAY	$360.00
TAXES	
FED. TAX	$ 28.80
WORK. COMP.	$ 4.32
MEDICARE	$ 5.04
NET PAY	$321.84

ILLUSTRATION 3

16. GASOLINE TAX In one state, a gallon of unleaded gasoline sells for $1.89. This price includes federal and state taxes that total approximately $0.54. Therefore, the price of a gallon of gasoline, before taxes, is about $1.35. What is the tax rate on gasoline?

17. OVERTIME Factory management wants to reduce the number of overtime hours by 25%. If the total number of overtime hours is 480 this month, what is the target number of overtime hours for next month?

18. COST-OF-LIVING INCREASE If a woman making $32,000 a year receives a cost-of-living increase of 2.4%, how much is her raise? What is her new salary?

19. REDUCED CALORIES A company advertised its new, improved chips as having 36% fewer calories per serving than the original style. How many calories are in a serving of the new chips if a serving of the original style contained 150 calories?

20. POLICE FORCE A police department plans to increase its 80-person force by 5%. How many additional officers will be hired? What will be the new size of the department?

21. ENDANGERED SPECIES Illustration 4 shows the total number of endangered and threatened plant and animal species for each of the years 1993–1999, as determined by the U.S. Fish and Wildlife Service. When was there a decline in the total? To the nearest percent, find the percent of decrease in the total for that period.

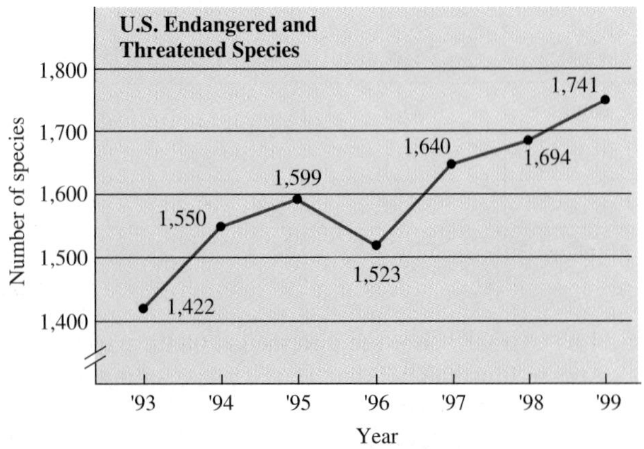

Source: *The New York Times 2000 Almanac*

ILLUSTRATION 4

22. CROP DAMAGE After flooding damaged much of the crop, the cost of a head of lettuce jumped from $0.99 to $2.20. What percent of increase is this?

23. CAR INSURANCE A student paid a car insurance premium of $400 every three months. Then the premium dropped to $360, because she qualified for a good-student discount. What was the percent of decrease in the premium?

24. BUS PASS To increase the number of riders, a bus company reduced the price of a monthly pass from $112 to $98. What was the percent of decrease?

25. LAKE SHORELINE Because of a heavy spring runoff, the shoreline of a lake increased from 5.8 miles to 7.6 miles. What was the percent of increase in the shoreline?

26. BASEBALL Illustration 5 shows the path of a baseball hit 110 mph, with a launch angle of 35 degrees, at sea level and at Coors Field, home of the Colorado Rockies. What is the percent of increase in the distance the ball travels at Coors Field?

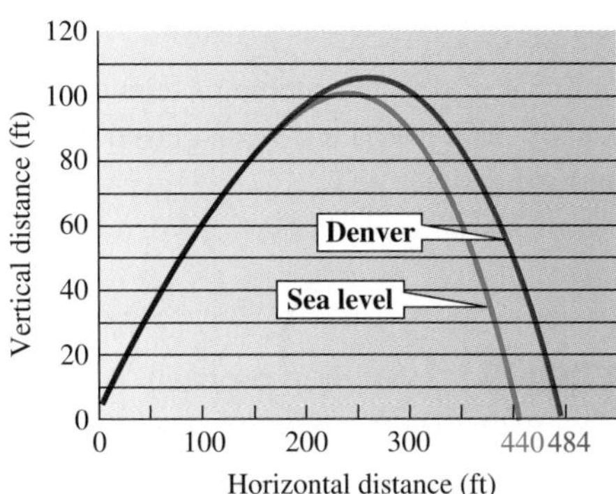

Source: *Los Angeles Times* (September 16, 1996)

ILLUSTRATION 5

27. EARTH MOVING Illustration 6 shows the typical soil volume change during earth moving. (One cubic yard of soil fits in a cube that is 1 yard long, 1 yard wide, and 1 yard high.)

a. Find the percent of increase in the soil volume as it goes through Step 1 of the process.

b. Find the percent of decrease in the soil volume as it goes through Step 2 of the process.

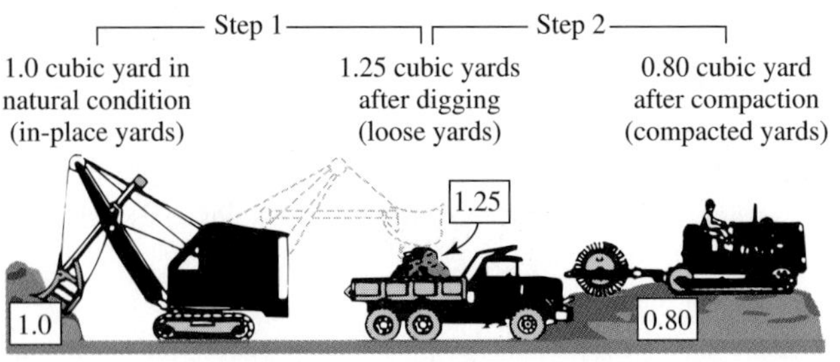

Source: U.S. Department of the Army

ILLUSTRATION 6

28. PARKING The management of a mall has decided to increase the parking area. The plans are shown in Illustration 7. What will be the percent of increase in the parking area once the project is completed?

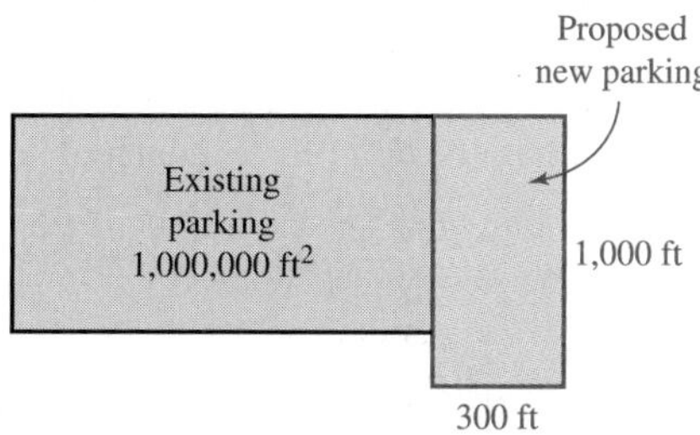

ILLUSTRATION 7

29. REAL ESTATE After selling a house for $98,500, a real estate agent split the 6% commission with another agent. How much did each person receive?

30. MEDICAL SUPPLIES A salesperson for a medical supplies company is paid a commission of 9% for orders under $8,000. For orders exceeding $8,000, she receives an additional 2% in commission on the total amount. What is her commission on a sale of $14,600?

31. SPORTS AGENT A sports agent charges her clients a fee to represent them during contract negotiations. The fee is based on a percent of the contract amount. If the agent earned $37,500 when her client signed a $2,500,000 professional football contract, what rate did she charge for her services?

32. ART GALLERY An art gallery displays paintings for artists and receives a commission from the artist when a painting is sold. What is the commission rate if a gallery received $135.30 when a painting was sold for $820?

33. CONCERT PARKING A concert promoter gets $33\frac{1}{3}\%$ of the revenue the arena receives from its parking concession the night of the performance. How much can the promoter make if 6,000 cars are anticipated and parking is $6 a car?

34. KITCHENWARE PARTY A homemaker invited her neighbors to a kitchenware party to show off cookware and utensils. As party hostess, she received 12% of the total sales. How much was purchased if she received $41.76 for hosting the party?

35. WATCH SALE See Illustration 8. What are the regular price and the rate of discount for the watch that is on sale?

ILLUSTRATION 8

36. STEREO SALE See Illustration 9. What are the regular price and the rate of discount for the stereo system that is on sale?

ILLUSTRATION 9

37. RING SALE What does a ring regularly sell for if it has been discounted 20% and is on sale for $149.99? (*Hint:* The ring is selling for 80% of its regular price.)

38. BLINDS SALE What do vinyl blinds regularly sell for if they have been discounted 55% and are on sale for $49.50? (*Hint:* The blinds are sellng for 45% of their regular price.)

39. VCR SALE What are the sale price and the discount rate for a VCR with remote that regularly sells for $399.97 and is being discounted $50?

40. CAMCORDER SALE What are the sale price and the discount rate for a camcorder that regularly sells for $559.97 and is being discounted $80?

41. REBATE See Illustration 10. Find the discount, the discount rate, and the reduced price for a case of motor oil if a shopper receives the manufacturer's rebate mentioned in the ad.

ILLUSTRATION 10

42. DOUBLE COUPONS See Illustration 11. Find the discount, the discount rate, and the reduced price for a box of cereal that normally sells for $3.29 if a shopper presents the coupon at a store that doubles the value of the coupon.

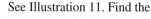

ILLUSTRATION 11

43. TV SHOPPING Determine the Home Shopping Network (HSN) price of the ring described in Illustration 12 if it sells it for 55% off of the retail price. Ignore shipping and handling costs.

Item 169-117
2.75 lb ctw
10K
Blue Topaz Ring
6, 7, 8, 9, 10
Retail value $170
HSN Price
$??.??
S&H $5.95

ILLUSTRATION 12

44. INFOMERCIAL The host of a TV infomercial says that the suggested retail price of a rotisserie grill is $249.95 and that it is now offered "for just 4 easy payments of only $39.95." What is the discount, and what is the discount rate?

WRITING

45. List the pros and cons of working on commission.

46. In Example 6, explain why you get the correct answer for the sale price by finding 75% of the regular price.

47. Explain the difference between a tax and a tax rate.

48. Explain how to find the sale price of an item if you know the regular price and the discount rate.

REVIEW

49. Multiply: $-5(-5)(-2)$.

50. Solve $4(x + 6) = 12$.

51. Evaluate the expression $2a^2(a + b)$ for $a = -2$ and $b = -1$.

52. Let x represent the height of an elm tree. A pine tree is 15 feet taller than the elm. Write an algebraic expression for the height of the pine tree.

53. How many eggs in d dozen?

54. Evaluate $-4 - (-7)$.

55. Evaluate $|-5 - 8|$.

56. A store clerk earns x dollars an hour. How much will she earn in a 40-hour week?

57. Graph each point $A(-3, 4)$, $B(4, 3.5)$, $C\left(-2, -\frac{5}{2}\right)$, $D(0, -4)$, $E\left(\frac{3}{2}, 0\right)$, and $F(3, -4)$.

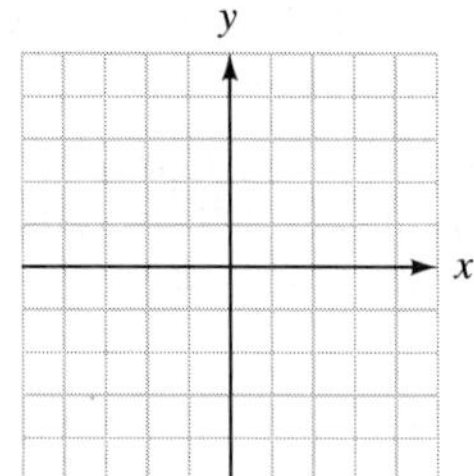

58. Complete the table of solutions for $y = 2x$.

x	y	(x, y)
−2		
0		
3		

ESTIMATION

We will now discuss some estimation methods that can be used when working with percent. To begin, we consider a way to find 10% of a number quickly. Recall that 10% of a number is found by multiplying the number by 10% or 0.1. When multiplying a number by 0.1, we simply move the decimal point one place to the left to find the result.

EXAMPLE 1 ***10% of a number.*** Find 10% of 234.

Solution To find 10% of 234, move the decimal point 1 place to the left.

$$234 = 23.4.0$$

Thus, 10% of 234 is 23.4, or approximately 23. ■

To find 15% of a number, first find 10% of the number. Then find half of that to obtain the other 5%. Finally, add the two results.

EXAMPLE 2 ***Estimating 15% of a number.*** Estimate 15% of 78.

Solution 10% of 78 is 7.8, or about 8. ⟶ 8
Add half of 8 to get the other 5%. ⟶ + 4
12

Thus, 15% of 78 is approximately 12. ■

To find 20% of a number, first find 10% of it and then double that result. A similar procedure can be used when working with any multiple of 10%.

EXAMPLE 3 ***Estimating 20% of a number.*** Estimate 20% of 3,234.15.

Solution 10% of 3,234.15 is 323.415 or about 323. To find 20%, double that.

Thus, 20% of 3,234.15 is approximately 646. ■

EXAMPLE 4 ***1% of a number.*** Find 1% of 0.8.

Solution To find 1% of a number, multiply it by 0.01, because $1\% = 0.01$. When multiplying a number by 0.01, simply move the decimal point two places to the left to find the result.

$$0.8 = .00.8$$

$$= 0.008$$

Thus, 1% of 0.8 is 0.008. ■

EXAMPLE 5 ***50% of a number.*** Find 50% of 2,800,000,000.

Solution To find 50% of a number means to find $\frac{1}{2}$ of that number. To find one-half of a number, simply divide it by 2. Thus, 50% of 2,800,000,000 is $2{,}800{,}000{,}000 \div 2 = 1{,}400{,}000{,}000$. ■

To find 25% of a number, first find 50% of it, then divide that result by 2.

EXAMPLE 6 ***Estimating 25% of a number.*** Estimate 25% of 16,813.

Solution 16,813 is about 16,800. Half of that is 8,400. Thus, 50% of 16,813 is approximately 8,400.

To estimate 25% of 16,813, divide 8,400 by 2. Thus, 25% of 16,813 is approximately 4,200. ■

100% of a number is the number itself. To find 200% of a number, double the number.

EXAMPLE 7 ***Estimating 200% of a number.*** Estimate 200% of 65.198.

Solution 65.198 is about 65. To find 200% of 65, double it. Thus, 200% of 65.198 is approximately $65 \cdot 2$ or 130. ■

STUDY SET *Estimate the answer to each problem.*

1. COLLEGE COURSES 20% of the 815 students attending a small college were enrolled in a science course. How many students is this?

2. SPECIAL OFFER In the grocery store, a 65-ounce bottle of window cleaner was marked "25% free." How many ounces are free?

3. DISCOUNT By how much is the price of a VCR discounted if the regular price of $196.88 is reduced by 30%?

4. TIPPING A restaurant tip is normally 15% of the cost of the meal. Find the tip on a dinner costing $38.64.

5. FIRE DAMAGE An insurance company paid 50% of the $107,809 it cost to rebuild a home that was destroyed by fire. How much did the insurance company pay?

6. SAFETY INSPECTION Of the 2,580 vehicles inspected at a safety checkpoint, 10% had code violations. How many cars had code violations?

7. WEIGHTLIFTING A 158-pound weightlifter can bench press 200% of his body weight. How many pounds can he bench press?

8. TESTING On a 120-question true/false test, 5% of a student's answers were wrong. How many questions did she miss?

9. TRAFFIC STUDY According to an electronic traffic monitor, 20% of the 650 motorists that passed it were speeding. How many of these motorists were speeding?

10. SELLING A HOME A homeowner has been told she will recoup 70% of her $5,000 investment if she paints her home before selling it. What is the potential payback if she paints her home?

Approximate the percent and then estimate the answer to each problem.

11. NO-SHOWS The attendance at a seminar was only 31% of what the organizers had anticipated. If 68 people were expected, how many actually attended the seminar?

12. "A" STUDENTS Of the 900 students in a school, 16% were on the principal's honor roll. How many students were on the honor roll?

13. INTERNET SURVEY Illustration 1 shows an online survey question. How many people voted yes?

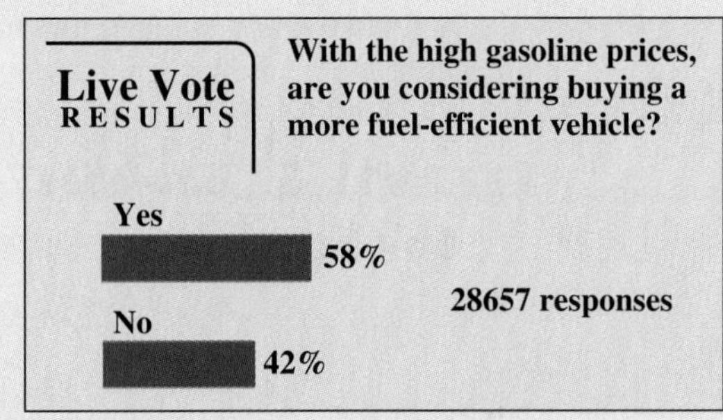

ILLUSTRATION 1

14. MEDICARE The Medicare payroll tax rate is 1.45%. How much Medicare tax will be deducted from a paycheck of $596?

15. VOTING On election day, 48% of the 6,200 workers at the polls were volunteers. How many volunteers helped with the election?

16. BUDGET Each department at a college was asked to cut its budget by 21%. By how much money should the mathematics department budget be reduced if it is currently $4,515?

7.4 Interest

In this section, you will learn about

- Simple interest • Compound interest

Introduction. When money is borrowed, the lender expects to be paid back the amount of the loan plus an additional charge for the use of the money. The additional charge is called **interest.** When money is deposited in a bank, the depositor is paid for the use of the money. The money the deposit earns is also called interest. In general, interest is money that is paid for the use of money.

Simple interest

Interest is calculated in one of two ways: either as **simple interest** or as **compound interest.** We will begin by discussing simple interest. First, we need to introduce some key terms associated with borrowing or lending money.

Principal: the amount of money that is invested, deposited, or borrowed.

Interest rate: a percent that is used to calculate the amount of interest to be paid. It is usually expressed as an annual (yearly) rate.

Time: the length of time (usually in years) that the money is invested, deposited, or borrowed.

The amount of interest to be paid depends on the principal, the rate, and the time. That is why all three are usually mentioned in advertisements for bank accounts, investments, and loans. (See Figure 7-10.)

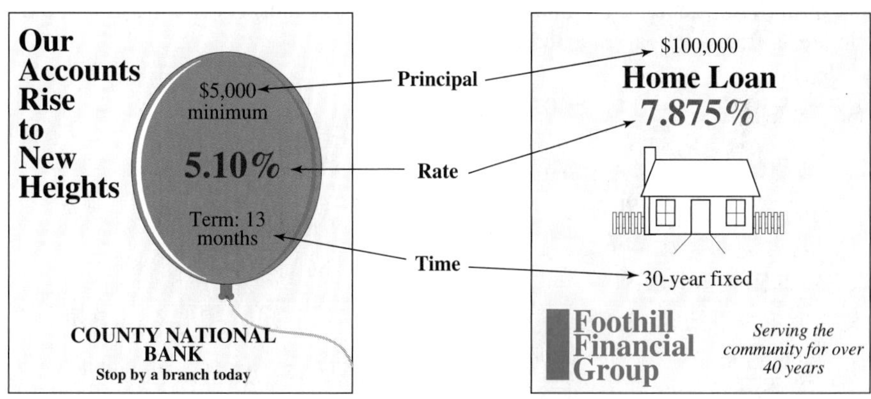

FIGURE 7-10

Simple interest is interest earned on the original principal. It is found by using a formula.

Simple interest formula

Interest = principal · rate · time

or

$I = Prt$

where the rate r is expressed as an annual rate and the time t is expressed in years.

EXAMPLE 1 ***Finding the interest earned.*** $3,000 is invested for 1 year at a rate of 5%. How much interest is earned?

Solution

We will use the formula $I = Prt$ to calculate the interest earned. The principal is $3,000, the interest rate is 5% (or 0.05), and the time is 1 year.

$P = 3{,}000 \qquad r = 5\% = 0.05 \qquad t = 1$ year

$I = Prt$ Write the interest formula.

$I = 3{,}000 \cdot 0.05 \cdot 1$ Substitute the values for P, r, and t.

$I = 150$ Do the multiplication.

The interest earned in 1 year is $150.

The information given in this problem and the result can be presented in a table.

Principal	Rate	Time	Interest earned
$3,000	5%	1 year	$150

Self Check

If $4,200 is invested for 2 years at a rate of 4% annual interest, how much interest is earned?

Answer: $336

When using the formula $I = Prt$, the time must be expressed in years. If the time is given in days or months, we rewrite it as a fractional part of a year. For example, a 30-day investment lasts $\frac{30}{365}$ of a year, since there are 365 days in a year. For a 6-month loan, we express the time as $\frac{6}{12}$ or $\frac{1}{2}$ of a year, since there are 12 months in a year.

EXAMPLE 2 ***Paying off a loan.*** To start a carpet-cleaning business, a couple borrows $5,500 to purchase equipment and supplies. If the loan has a 14% interest rate, how much must they repay at the end of the 90-day period?

Solution

First, we find the amount of interest paid on the loan. We must rewrite the time (90-day period) as a fractional part of a 365-day year.

$P = 5{,}500 \qquad r = 14\% = 0.14 \qquad t = \frac{90}{365}$

$I = Prt$ Write the interest formula.

$I = 5{,}500 \cdot 0.14 \cdot \frac{90}{365}$ Substitute the values for P, r, and t.

$I = \frac{5{,}500}{1} \cdot \frac{0.14}{1} \cdot \frac{90}{365}$ Write 5,500 and 0.14 as fractions.

$I = \frac{69{,}300}{365}$ Use a calculator to multiply the numerators. Multiply the denominators.

$I \approx 189.86$ Use a calculator, do the division. Round to the nearest cent.

The interest on the loan is $189.86. To find how much they must pay back, we add the principal and the interest.

$5{,}500 + 189.86 = 5{,}689.86$

The couple must pay back $5,689.86 at the end of 90 days.

Self Check

How much must be repaid if $3,200 is borrowed at a rate of 15% for 120 days?

Answer: $3,357.81

Compound interest

Most savings accounts pay **compound interest** rather than simple interest. Compound interest is interest paid on accumulated interest. To illustrate this concept, suppose that

\$2,000 is deposited in a savings account at a rate of 5% for 1 year. We can use the formula $I = Prt$ to calculate the interest earned at the end of 1 year.

$I = Prt$

$I = \mathbf{2{,}000 \cdot 0.05 \cdot 1}$ Substitute for P, r, and t.

$I = 100$ Do the multiplication.

Interest of \$100 was earned. At the end of the first year, the account contains the interest (\$100) plus the original principal (\$2,000), for a balance of \$2,100.

Suppose that the money remains in the savings account for another year at the same interest rate. For the second year, interest will be paid on a principal of \$2,100. That is, during the second year, we earn *interest on the interest* as well as on the original \$2,000 principal. Using $I = Prt$, we can find the interest earned in the second year.

$I = Prt$

$I = \mathbf{2{,}100 \cdot 0.05 \cdot 1}$ Substitute for P, r, and t.

$I = 105$ Do the multiplication.

In the second year, \$105 of interest is earned. The account now contains that interest plus the \$2,100 principal, for a total of \$2,205.

As Figure 7-11 shows, we calculated the simple interest two times to find the compound interest.

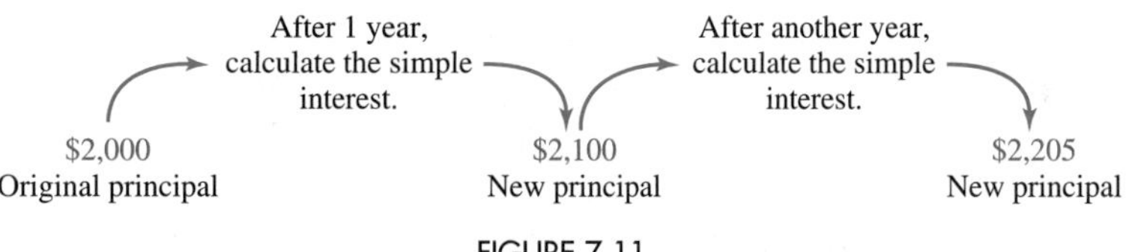

FIGURE 7-11

If we compute the *simple interest* on \$2,000, at 5% for 2 years, the interest earned is $I = 2{,}000 \cdot 0.05 \cdot 2 = 200$. Thus, the account balance would be \$2,200. Comparing the balances, the account earning compound interest will contain \$5 more than the account earning simple interest.

In the previous example, the interest was calculated at the end of each year, or **annually.** When compounding, we can compute the interest in other time increments, such as **semiannually** (twice a year), **quarterly** (four times a year), or even **daily.**

EXAMPLE 3 ***Finding compound interest.*** As a gift for her newborn granddaughter, a grandmother opens a \$1,000 savings account in the baby's name. The interest rate is 4.2%, compounded quarterly. Find the amount of money the child will have in the bank on her first birthday.

Solution If the interest is compounded quarterly, the interest will be computed four times in one year. To find the amount of interest \$1,000 will earn in the first quarter of the year, we use the simple interest formula, where t is $\frac{1}{4}$ of a year.

Interest earned in the first quarter

$P = \mathbf{1{,}000} \qquad r = 4.2\% = \mathbf{0.042} \qquad t = \mathbf{\frac{1}{4}}$

$$I = \mathbf{1{,}000 \cdot 0.042 \cdot \frac{1}{4}}$$

$I = \$10.50$

The interest earned in the first quarter is \$10.50. This now becomes part of the principal for the second quarter:

$\$1{,}000 + \$10.50 = \$1{,}010.50$

To find the amount of interest $1,010.50 will earn in the second quarter of the year, we use the simple interest formula, where t is again $\frac{1}{4}$ of a year.

$$P = \mathbf{1{,}010.50} \qquad r = \mathbf{0.042} \qquad t = \frac{1}{4}$$

$$I = \mathbf{1{,}010.50} \cdot \mathbf{0.042} \cdot \frac{1}{4}$$

$$I \approx \$10.61 \quad \text{(Rounded)}$$

The interest earned in the second quarter is $10.61. This becomes part of the principal for the third quarter:

$$\$1{,}010.50 + \$10.61 = \$1{,}021.11$$

To find the interest $1,021.11 will earn in the third quarter of the year, we proceed as follows.

$$P = \mathbf{1{,}021.11} \qquad r = \mathbf{0.042} \qquad t = \frac{1}{4}$$

$$I = \mathbf{1{,}021.11} \cdot \mathbf{0.042} \cdot \frac{1}{4}$$

$$I \approx \$10.72 \quad \text{(Rounded)}$$

The interest earned in the third quarter is $10.72. This now becomes part of the principal for the fourth quarter:

$$\$1{,}021.11 + \$10.72 = \$1{,}031.83$$

To find the interest $1,031.83 will earn in the fourth quarter, we again use the simple interest formula.

$$P = \mathbf{1{,}031.83} \qquad r = \mathbf{0.042} \qquad t = \frac{1}{4}$$

$$I = \mathbf{1{,}031.83} \cdot \mathbf{0.042} \cdot \frac{1}{4}$$

$$I \approx \$10.83 \quad \text{(Rounded)}$$

The interest earned in the fourth quarter is $10.83. Adding this to the existing principal, we get:

$$\$1{,}031.83 + \$10.83 = \$1{,}042.66$$

The amount that has accumulated in the account after four quarters, or 1 year, is $1,042.66. ■

Computing compound interest by hand is tedious. The **compound interest formula** can be used to find the total amount of money that an account will contain at the end of the term.

Compound interest formula

The total amount A in an account can be found using the formula

$$A = P\left(1 + \frac{r}{n}\right)^{nt}$$

where P is the principal, r is the annual interest rate expressed as a decimal, t is the length of time in years, and n is the number of compoundings in one year.

Accent on Technology: **Compound interest**

A businessman invests \$9,250 at 7.6% interest, to be compounded monthly. To find what the investment will be worth in 3 years, we use the compound interest formula with the following values.

$$P = \$9{,}250, \quad r = 7.6\% = 0.076, \quad t = 3 \text{ years}, \quad n = 12 \text{ times a year (monthly)}$$

We apply the compound interest formula:

$$A = P\left(1 + \frac{r}{n}\right)^{nt}$$ Write the compound interest formula.

$$A = \mathbf{9{,}250}\left(1 + \frac{\mathbf{0.076}}{\mathbf{12}}\right)^{\mathbf{12(3)}}$$ Substitute the values of P, r, t, and n.

$$A = 9{,}250\left(1 + \frac{0.076}{12}\right)^{36}$$ Simplify the exponent: $12(3) = 36$.

To evaluate the expression on the right-hand side of the equation, we enter these numbers and press these keys.

Keystrokes 9250 [×] [(] 1 [+] .076 [÷] 12 [)] [y^x] 36 [=] `11610.43875`

Rounded to the nearest cent, the amount in the account after 3 years will be \$11,610.44.

If your calculator does not have parenthesis keys, calculate the sum inside the parentheses first. Then find the power. Finally, multiply by 9,250.

EXAMPLE 4 ***Compound interest.*** A man deposited \$50,000 in a long-term account at 6.8% interest, compounded daily. How much money will he be able to withdraw in 7 years if the principal is to remain in the bank?

Solution

"Compounded daily" means that compounding will be done 365 times in a year.

$$P = \$50{,}000 \qquad r = 6.8\% = 0.068 \qquad t = 7 \text{ years} \qquad n = 365 \text{ times a year}$$

$$A = P\left(1 + \frac{r}{n}\right)^{nt}$$ Write the compound interest formula.

$$A = \mathbf{50{,}000}\left(1 + \frac{\mathbf{0.068}}{\mathbf{365}}\right)^{365(7)}$$ Substitute the values of P, r, t, and n.

$$A = 50{,}000\left(1 + \frac{0.068}{365}\right)^{2{,}555}$$ $365(7) = 2{,}555$.

$$A \approx 80{,}477.58$$ Use a calculator. Round to the nearest cent.

The account will contain \$80,477.58 at the end of 7 years. To find the amount the man can withdraw, we subtract.

$$80{,}477.58 - 50{,}000 = 30{,}477.58$$

The man can withdraw \$30,477.58 without having to touch the \$50,000 principal.

Self Check

Find the amount of interest \$25,000 will earn in 10 years if it is deposited in an account at 5.99% interest, compounded daily.

Answer: \$20,505.20

STUDY SET Section 7.4

VOCABULARY *Fill in the blanks to make the statements true.*

1. In banking, the original amount of money borrowed or deposited is known as the __principal__.
2. Borrowers pay __interest__ to lenders for the use of their money.
3. The percent that is used to calculate the amount of interest to be paid is called the __interest__ rate.
4. __Compound__ interest is interest paid on accumulated interest.
5. Interest computed only on the original principal is called __Simple__ interest.
6. Percent means parts per __100__.

CONCEPTS

7. When we do calculations with percents, they must be changed to decimals or fractions. Change each percent to a decimal.

   a. 7% **b.** 9.8% **c.** $6\frac{1}{4}\%$

8. Express each of the following as a fraction of a year. Simplify the fraction.

   a. 6 months **b.** 90 days
   c. 120 days **d.** 1 month

9. Complete the table by finding the simple interest earned.

Principal	Rate	Time	Interest earned
$10,000	6%	3 years	

10. Tell how many times a year the interest on a savings account is calculated if the interest is compounded

    a. semiannually **b.** quarterly
    c. daily **d.** monthly

11. **a.** What concept studied in this section is illustrated by the diagram in Illustration 1?
    b. What was the original principal?
    c. How many times was the interest found?
    d. How much interest was earned on the first compounding?
    e. For how long was the money invested?

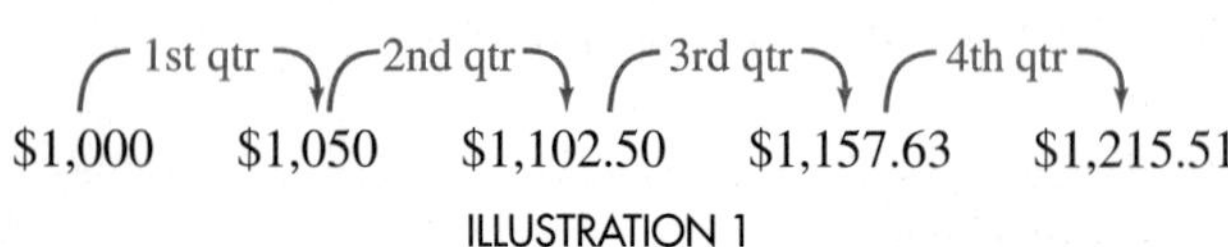

ILLUSTRATION 1

12. $3,000 is deposited in a savings account that earns 10% interest compounded annually. Complete the series of calculations in Illustration 2 to find how much money will be in the account at the end of 2 years.

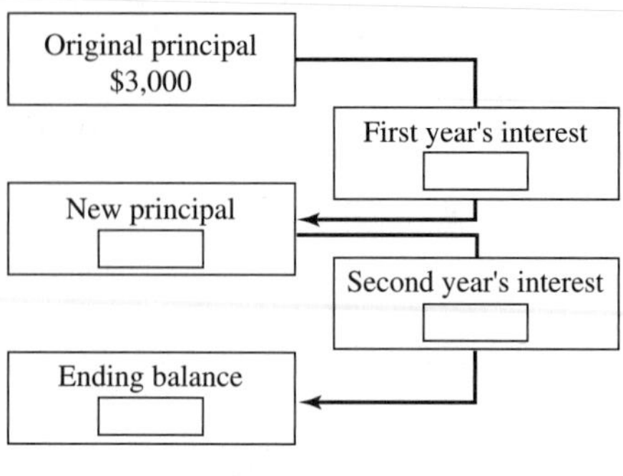

ILLUSTRATION 2

NOTATION

13. In the formula $I = Prt$, what operations are indicated by Prt?

14. In the formula $A = P\left(1 + \frac{r}{n}\right)^{nt}$, how many operations must be performed to find A?

APPLICATIONS *In Exercises 15–26, use simple interest.*

15. RETIREMENT INCOME A retiree invests $5,000 in a savings plan that pays 6% per year. What will the account balance be at the end of the first year?

16. INVESTMENT A developer promised a return of 8% annual interest on an investment of $15,000 in her company. How much could an investor expect to make in the first year?

17. REMODELING A homeowner borrows $8,000 to pay for a kitchen remodeling project. The terms of the loan are 9.2% annual interest and repayment in 2 years. How much interest will be paid on the loan?

18. CREDIT UNION A farmer borrowed $7,000 from a credit union. The money was loaned at 8.8% annual interest for 18 months. How much money did the credit union charge him for the use of the money?

19. MEETING A PAYROLL In order to meet end-of-the-month payroll obligations, a small business had to borrow $4,200 for 30 days. How much did the business have to repay if the interest rate was 18%?

20. CAR LOAN To purchase a car, a man takes out a loan for $2,000. If the interest rate is 9% per year, how much interest will he have to pay at the end of the 120-day loan period?

21. SAVINGS ACCOUNT Find the interest earned on \$10,000 at $7\frac{1}{4}\%$ for 2 years. Use the chart in Illustration 3 to organize your work.

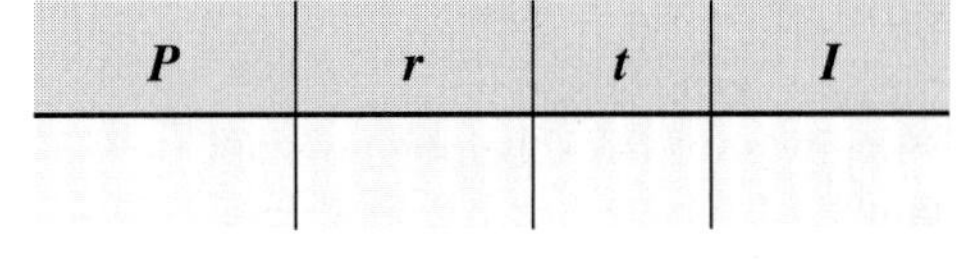

P	*r*	*t*	*I*

ILLUSTRATION 3

22. TUITION A student borrows \$300 from an educational fund to pay for books for spring semester. If the loan is for 45 days at $3\frac{1}{2}\%$ annual interest, what will the student owe at the end of the loan period?

23. LOAN APPLICATION Complete the loan application form in Illustration 4.

Loan Application Worksheet

1. Amount of loan (principal) \$1,200.00
2. Length of loan (time) 2 YEARS
3. Annual percentage rate 8%
4. Interest charged \$192.00
5. Total amount to be repaid \$1392.00
6. Check method of repayment:
   ☐ 1 lump sum ☑ monthly payments

Borrower agrees to pay 24 equal payments of \$58.00 to repay loan.

ILLUSTRATION 4

24. LOAN APPLICATION Complete the loan application form in Illustration 5.

Loan Application Worksheet

1. Amount of loan (principal) \$810.00
2. Length of loan (time) 9 mos.
3. Annual percentage rate 12%
4. Interest charged \$72.90
5. Total amount to be repaid \$882.90
6. Check method of repayment:
   ☐ 1 lump sum ☑ monthly payments

Borrower agrees to pay 9 equal payments of \$98.10 to repay loan.

ILLUSTRATION 5

25. LOW-INTEREST LOAN An underdeveloped country receives a low-interest loan from a bank to finance the construction of a water treatment plant. What must the country pay back at the end of 2 years if the loan is for \$18 million at 2.3%?

26. REDEVELOPMENT A city is awarded a low-interest loan to help renovate the downtown business district. The \$40 million loan, at 1.75%, must be repaid in $2\frac{1}{2}$ years. How much interest will the city have to pay?

A calculator may be helpful in solving these problems.

27. COMPOUNDING ANNUALLY If \$600 is invested in an account that earns 8%, compounded annually, what will the account balance be after 3 years?

28. COMPOUNDING SEMIANNUALLY If \$600 is invested in an account that earns annual interest of 8%, compounded semiannually, what will the account balance be at the end of 3 years?

29. COLLEGE FUND A ninth-grade student opens a savings account that locks her money in for 4 years at an annual rate of 6%, compounded daily. If the initial deposit is \$1,000, how much money will be in the account when she begins college in four years?

30. CERTIFICATE OF DEPOSIT A 3-year certificate of deposit pays an annual rate of 5%, compounded daily. The maximum allowable deposit is \$90,000. What is the most interest a depositor can earn from the CD?

31. TAX REFUND A couple deposits an income tax refund check of \$545 in an account paying an annual rate of 4.6%, compounded daily. What will the size of the account be at the end of 1 year?

32. INHERITANCE After receiving an inheritance of \$11,000, a man deposits the money in an account paying an annual rate of 7.2%, compounded daily. How much money will be in the account at the end of 1 year?

33. LOTTERY Suppose you won \$500,000 in the lottery and deposited the money in a savings account that paid an annual rate of 6% interest, compounded daily. How much interest would you earn each year?

34. CASH GIFT After receiving a \$250,000 cash gift, a university decides to deposit the money in an account paying an annual rate of 5.88%, compounded quarterly. How much money will the account contain in 5 years?

WRITING

35. What is the difference between simple and compound interest?

36. Explain: *Interest is the amount of money paid for the use of money.*

37. On some accounts, banks charge a penalty if the depositor withdraws the money before the end of the term. Why would a bank do this?

38. Explain why it is better for a depositor to open a savings account that pays 5% interest, compounded daily, than one that pays 5% interest, compounded monthly.

REVIEW

39. Simplify $\sqrt{\frac{1}{4}}$.

40. Find: $\left(\frac{1}{4}\right)^2$.

41. Is the point $(2, -3)$ on the line $y = 2x - 10$?

42. Subtract: $32u^3 - 22u^3$.

43. Solve $\frac{2}{3}x = -2$.

44. Divide: $-12\frac{1}{2} \div 5$.

45. How many terms does the polynomial $2x^2 - 3x + 5$ contain?

46. Evaluate $(0.2)^2 - (0.3)^2$.

47. In which quadrant does the point $(-2, -2)$ lie?

48. Multiply: $2x^2(3x + 2)$.

KEY CONCEPT

Equivalent Expressions

Equivalent expressions do not look the same, but they represent the same amount. One of the major objectives of this course is to develop a knack for recognizing situations where an equivalent expression should be written.

Write an equivalent expression for each quantity and describe the concept you applied.

1. $\frac{10}{24}$

2. $3 \cdot x \cdot x \cdot x$

3. $2x + 3x$

4. $-3 - (-8)$

5. $x^3 \cdot x^2$

6. Write 0.125 as a percent.

7. Write $\frac{2}{3}$ as a percent.

8. $\frac{4}{5} - \frac{1}{5}$

9. $4x^2 + 1 - 2x^2$

10. $\frac{\frac{1}{5}}{\frac{3}{4}}$

11. $\frac{6}{6}$

12. Write 5.1% as a decimal.

13. $(-5)(-6)$

14. $\frac{x}{1}$

15. $\frac{2x}{2}$

16. $-x + x$

17. $2 + 3 \cdot 5$

18. $a + 0$

19. $2(x + 5)$

20. $|-4|$

21. $\frac{2}{3} \cdot \frac{3}{2}$

22. $\sqrt{49}$

ACCENT ON TEAMWORK

Section 7.1

M & M'S Give each member of your group a bag of M & M's candies.

a. Determine what percent of the total number of M & M's in your bag are yellow. Do the same for each of the other colors. Enter the results in the table. (Round to the nearest one percent.)

b. Present the data in the table using the circle graph in Illustration 1. Compare your graph to the graphs made by the other members of your group. Do the colors occur in the same percentages in each of the bags?

M & M's color	Percent
Yellow	
Brown	
Green	
Red	
Blue	

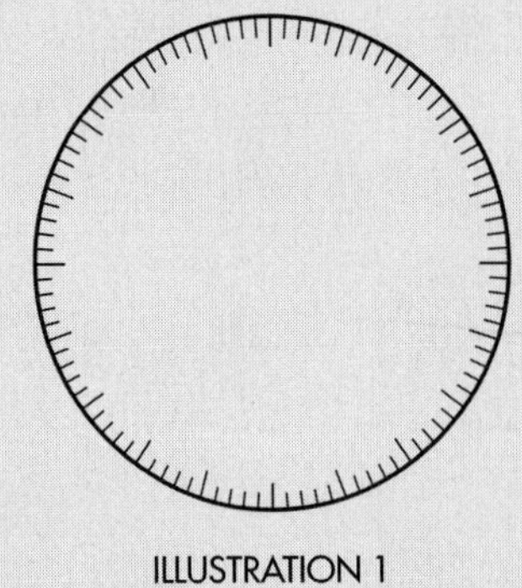

ILLUSTRATION 1

Section 7.2

NUTRITION Have each person in your group bring in a nutrition label like that shown in Illustration 2 and write the name of the food product on the back. Have the members of the group exchange labels. With the label that you receive, determine what percent of the total calories come from fat.

The USDA recommends that no more than 30% of a person's daily calories should come from fat. Which products exceed the recommendation?

Nutrition Facts
Serving Size 1 meal

Amount Per Serving
Calories 560 Calories from Fat 190

	% Daily Value
Total fat 21g	**32%**
Saturated fat 9 g	**43%**
Cholesterol 60mg	**20%**
Sodium 2110mg	**88%**
Total carbohydrate 67g	**22%**
Dietary fiber 7g	**29%**
Sugars less than 25g	
Protein 27g	

ILLUSTRATION 2

Section 7.3

ENROLLMENT From your school's admissions office, get the enrollment figures for the last ten years. Calculate the percent of increase (or decrease) in enrollment for each of the following periods.

- Ten years ago to the present
- Five years ago to the present
- One year ago to the present

NEWSPAPER ADS Have each person in your group find a newspaper advertisement for some item that is on sale. The ad should include only two of the four details listed below.

- The regular price
- The sale price
- The discount
- The discount rate

For example, the ad in Illustration 3 gives the regular price and the sale price, but it doesn't give the discount or the discount rate.

Have the members of your group exchange ads. Determine the two missing details on the ad that you receive. In your group, which item had the highest discount rate?

ILLUSTRATION 3

Section 7.4

INTEREST RATES Recall that interest is money that the borrower pays to the lender for the use of the money. The amount of interest that the borrower must pay depends on the interest rate charged by the lender.

Have members of your group call banks, savings and loans, credit unions, and other financial services to get the lending rates for various types of loans. (See the yellow pages of the phone book.)

Find out what rate is charged by credit cards such as VISA, department stores, and gasoline companies. List the interest rates in order, from greatest to least, and present your findings to the class.

CHAPTER REVIEW

SECTION 7.1 Percents, Decimals, and Fractions

CONCEPTS

Percent means parts per one hundred.

REVIEW EXERCISES

1. Express the amount of each figure that is shaded as a percent, as a decimal, and as a fraction. Each set of squares represents 100%.

a. 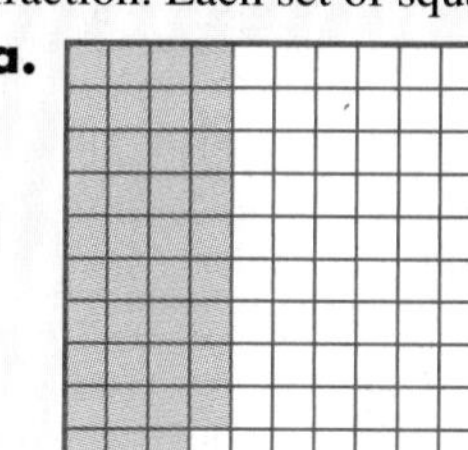 **b.** 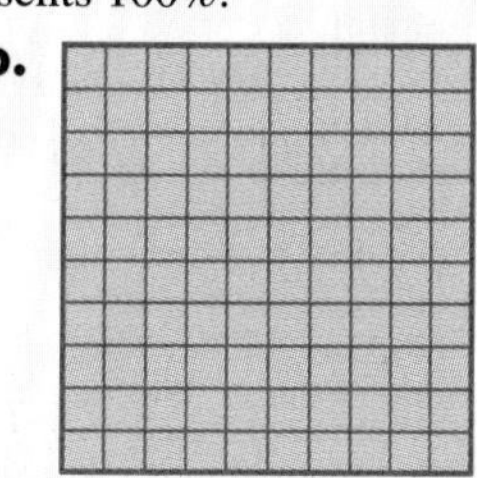 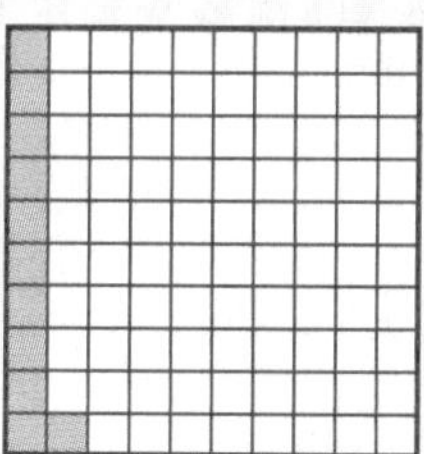

2. In Problem 1, part a, what percent of the figure is not shaded?

To change a percent to a fraction, drop the % symbol and put the given number over 100.

3. Change each percent to a fraction.

a. 15% **b.** 120% **c.** $9\frac{1}{4}\%$ **d.** 0.1%

To change a percent to a decimal, drop the % symbol and divide by 100 by moving the decimal point 2 places to the left.

4. Change each percent to a decimal.

a. 27% **b.** 8% **c.** 155% **d.** $1\frac{4}{5}\%$

To change a decimal to a percent, multiply the decimal by 100 by moving the decimal point 2 places to the right, and then insert a % symbol.

5. Change each decimal to a percent.

a. 0.83 **b.** 0.625 **c.** 0.051 **d.** 6

To change a fraction to a percent, write the fraction as a decimal by dividing its numerator by its denominator. Multiply the decimal by 100 by moving the decimal point 2 places to the right, and then insert a % symbol.

6. Change each fraction to a percent.

a. $\frac{1}{2}$ **b.** $\frac{4}{5}$ **c.** $\frac{7}{8}$ **d.** $\frac{1}{16}$

7. Find the exact percent equivalent for each fraction.

a. $\frac{1}{3}$ **b.** $\frac{5}{6}$

8. Change each fraction to a percent. Round to the nearest hundredth.

a. $\frac{5}{9}$ **b.** $\frac{8}{3}$

9. BILL OF RIGHTS There are 27 amendments to the Constitution of the United States. The first ten are known as the Bill of Rights. What percent of the amendments were adopted after the Bill of Rights? (Round to the nearest one percent.)

10. Explain the difference between one-tenth of one percent and ten percent.

SECTION 7.2 Solving Percent Problems

The percent formula:
Amount = percent · base

11. Identify the amount, the base, and the percent in the statement "15 is $33\frac{1}{3}\%$ of 45."

We can translate a percent problem from words into an equation. A *variable* is used to stand for the unknown number; *is* can be translated to an = sign; and *of* means multiply.

12. Translate the given sentence into a percent equation:

What number is 32% of 96?

13. Solve each percent problem.

a. What number is 40% of 500?

b. 16% of what number is 20?

c. 1.4 is what percent of 80?

d. $66\frac{2}{3}\%$ of 3,150 is what number?

e. Find 220% of 55.

f. What is 0.05% of 60,000?

14. RACING The nitro–methane fuel mixture used to power some experimental cars is 96% nitro and 4% methane. How many gallons of each fuel component are needed to fill a 15-gallon fuel tank?

15. HOME SALES After the first day on the market, 51 homes in a new subdivision had already sold. This was 75% of the total number of homes available. How many homes were originally for sale?

16. HURRICANE DAMAGE 96 of the 110 trailers in a mobile home park were either damaged or destroyed by hurricane winds. What percent is this? (Round to the nearest one percent.)

17. TIPPING The cost of dinner for a family of five at a restaurant was \$36.20. Find the amount of the tip if it should be 15% of the cost of dinner.

A *circle graph* is a way of presenting data for comparison. The sizes of the segments of the circle indicate the percents of the whole represented by each category.

18. AIR POLLUTION Complete Illustration 1 (a circle graph) to show the given data.

Sources of carbon monoxide air pollution	
Transportation vehicles	63%
Fuel combustion in homes, offices, electrical plants	12%
Industrial processes	8%
Solid-waste disposal	3%
Miscellaneous	14%

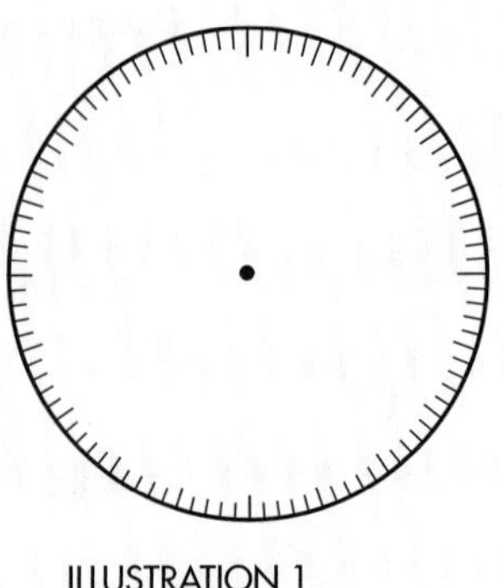

ILLUSTRATION 1

19. EARTH'S SURFACE The surface of the Earth is approximately 196,800,000 square miles. Use the information in Illustration 2 to determine the number of square miles of the Earth's surface that are covered with water.

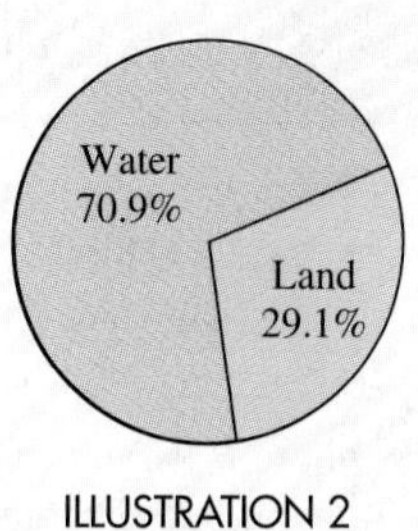

ILLUSTRATION 2

SECTION 7.3 Applications of Percent

To find the total price of an item:
Total price = purchase price + sales tax

20. SALES RECEIPT Complete the sales receipt in Illustration 3.

CAMERA CENTER

35mm Canon Camera	$59.99
SUBTOTAL	$59.99
SALES TAX @ 5.5%	
TOTAL	

ILLUSTRATION 3

21. SALES TAX RATE Find the sales tax rate if the sales tax is $492 on the purchase of an automobile priced at $12,300.

Commission is based on a percent of the total dollar amount of the goods or services sold.

22. COMMISSION If the commission rate is 6%, find the commission earned by an appliance salesperson who sells a washing machine for $369.97 and a dryer for $299.97.

To find *percent of increase or decrease:*

1. Subtract the smaller number from the larger to find the amount of increase or decrease.
2. Find what percent the difference is of the original amount.

23. TROOP SIZE The size of a peace-keeping force was increased from 10,000 to 12,500 troops. What percent of increase is this?

24. GAS MILEAGE Experimenting with a new brand of gasoline in her truck, a woman found that the gas mileage fell from 18.8 to 17.0 miles per gallon. What percent of decrease is this? (Round to the nearest tenth of a percent.)

The difference between the original price and the sale price of an item is called the *discount.*

To find the *sale price:*
Sale price = original price − discount

25. TOOL CHEST See Illustration 4. Use the information in the advertisement to find the discount, the original price, and the discount rate on the tool chest.

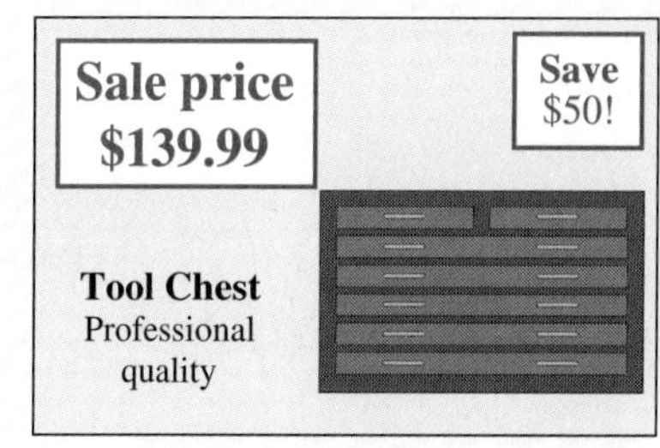

ILLUSTRATION 4

SECTION 7.4 Interest

Simple interest is interest earned on the original principal and is found using the formula

$I = Prt$

where P is the principal, r is the annual interest rate, and t is the length of time in years.

26. Find the interest earned on \$6,000 invested at 8% per year for 2 years. Use the following chart to organize your work.

P	r	t	I

27. CODE VIOLATIONS A business was ordered to correct safety code violations in a production plant. To pay for the needed corrections, the company borrowed \$10,000 at 12.5% for 90 days. Find the total amount that had to be paid after 90 days.

Compound interest is interest earned on interest.

28. MONTHLY PAYMENTS A couple borrows \$1,500 for 1 year at $7\frac{3}{4}\%$ and decides to repay the loan by making 12 equal monthly payments. How much will each monthly payment be?

The compound interest formula:

$$A = P\left(1 + \frac{r}{n}\right)^{nt}$$

where A is the amount in the account, P is the principal, r is the annual interest rate, n is the number of compoundings in one year, and t is the length of time in years.

29. Find the amount of money that will be in a savings account at the end of 1 year if \$2,000 is the initial deposit and the annual interest rate of 7% is compounded semiannually. (*Hint:* Find the simple interest twice.)

30. Find the amount that will be in a savings account at the end of 3 years if a deposit of \$5,000 earns interest at an annual rate of $6\frac{1}{2}\%$, compounded daily.

31. CASH GRANT Each year a cash grant is given to a deserving college student. The grant consists of the interest earned that year on a \$500,000 savings account. What is the cash award for the year if the money is invested at an annual rate of 8.3%, compounded daily?

Chapter 7 Test

1. See Illustration 1. Express the amount of the figure that is shaded as a percent, as a fraction, and as a decimal.

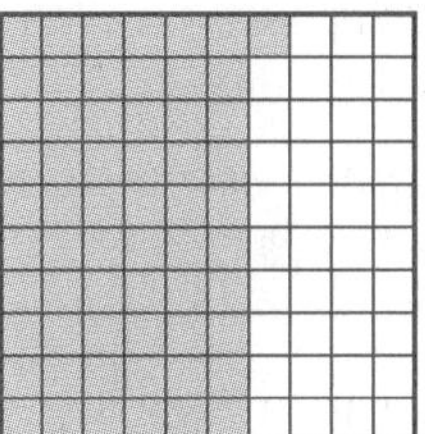

ILLUSTRATION 1

2. In Illustration 2, each set of 100 squares represents 100%. Express as a percent the amount of the figure that is shaded. Then express that percent as a fraction and as a decimal.

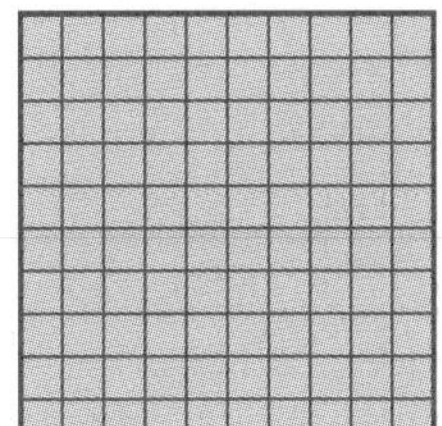 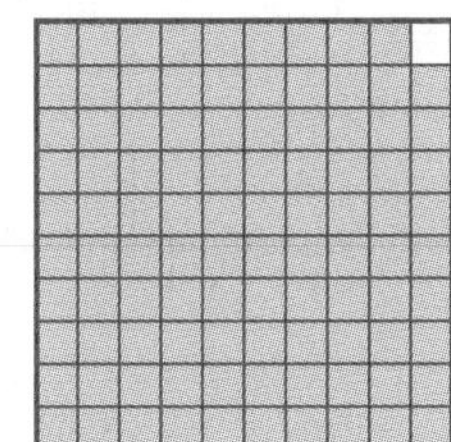

ILLUSTRATION 2

3. Change each percent to a decimal.

a. 67% **b.** 12.3% **c.** $9\frac{3}{4}\%$

4. Change each fraction to a percent.

a. $\frac{1}{4}$ **b.** $\frac{5}{8}$ **c.** $\frac{3}{25}$

5. Change each decimal to a percent.

a. 0.19 **b.** 3.47 **c.** 0.005

6. Change each percent to a fraction.

a. 55% **b.** 0.01% **c.** 125%

7. Change $\frac{7}{30}$ to a percent. Round to the nearest hundredth of a percent.

8. WEATHER REPORT A weatherman states that there is a 40% chance of rain. What are the chances that it will not rain?

9. Find the exact percent equivalent for the fraction $\frac{2}{3}$.

10. Find the exact percent equivalent for the fraction $\frac{1}{4}$.

11. SHRINKAGE See Illustration 3, a label on a new pair of jeans.

a. How much length will be lost due to shrinkage?

b. What will be the resulting length?

WAIST INSEAM
33 34
Expect shrinkage of approximately
3%
in length after the jeans are washed.

ILLUSTRATION 3

12. 65 is what percent of 1,000?

13. TIPPING Find the amount of a 15% tip on a meal costing $25.40.

14. FUGITIVES As of March 15, 2000, 429 of the 458 fugitives who have appeared on the FBI's Ten Most Wanted list have been apprehended or located. What percent is this? Round to the nearest tenth of a percent.

15. SWIMMING WORKOUT A swimmer was able to complete 18 laps before a shoulder injury forced him to stop. This was only 20% of a typical workout. How many laps does he normally complete during a workout?

16. COLLEGE EMPLOYEES The 700 employees at a community college fall into three major categories, as shown in Illustration 4. How many employees are in administration?

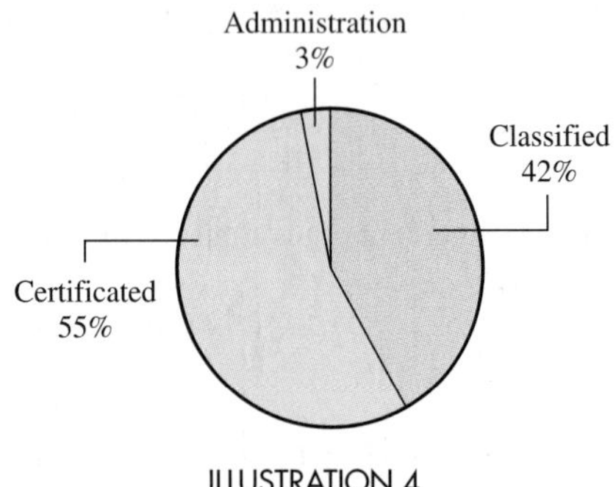

ILLUSTRATION 4

17. What number is 24% of 600?

18. HAIRCUTS Illustration 5 shows the number of minutes it took the average U.S. worker to earn enough to pay for a man's haircut in 1950 and 1998. Find the percent of decrease, to the nearest one percent.

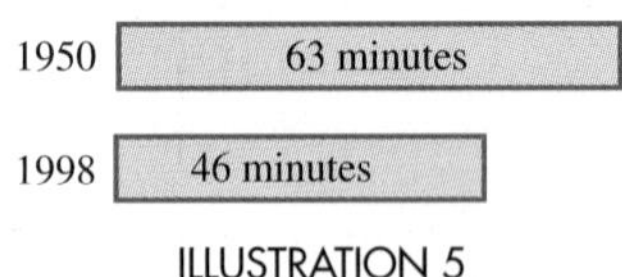

ILLUSTRATION 5

19. HOMEOWNER'S INSURANCE An insurance salesperson receives a 4% commission on the annual premium of any policy she sells. What is her commission on a homeowner's policy if the premium is $898?

20. COST-OF-LIVING INCREASE A teacher earning $40,000 just received a cost-of-living increase of 3.6%. What is the teacher's new salary?

21. CAR WAX SALE A car waxing kit, regularly priced at $14.95, is on sale for $3 off. What are the sale price, the discount, and the rate?

22. POPULATION INCREASE After a new freeway was completed, the population of a city it passed through increased from 12,808 to 15,565 in two years. What percent of increase is this? (Round to the nearest one percent.)

23. Find the simple interest on a loan of $3,000 at 5% per year for 1 year.

24. Find the amount of interest earned on an investment of $24,000 paying an annual rate of 6.4% interest, compounded daily for 3 years.

25. POLITICAL AD Explain what is unclear about the flyer shown in Illustration 6.

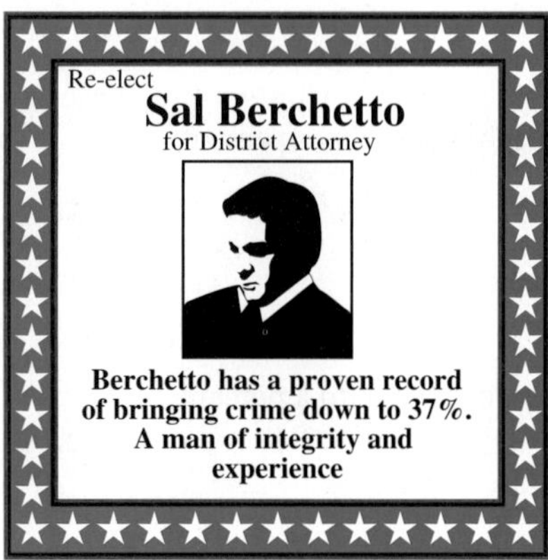

ILLUSTRATION 6

Chapters 1–7 Cumulative Review Exercises

1. SHAQUILLE Use the data in the table to complete Illustration 1 by drawing a line graph to chart the growth of Shaquille O'Neal, the Los Angeles Lakers' center.

Age (yr)	4	6	8	10	12	16	21	28
Weight (lb)	56	82	108	139	192	265	302	315

Based on data from *Los Angeles Times*

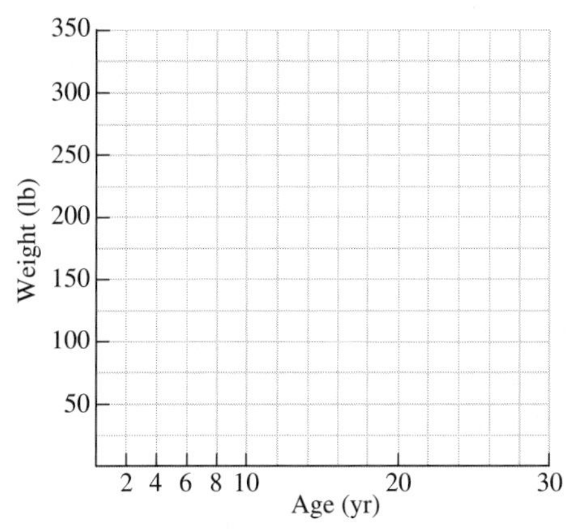

ILLUSTRATION 1

2. State the commutative property of multiplication.

3. **a.** Find the factors of 40.
b. Find the prime factorization of 40.

4. AUTO INSURANCE See the premium comparison in Illustration 2. What is the average (mean) six-month insurance premium?

Allstate	\$2,672	Mercury	\$1,370
Auto Club	\$1,680	State Farm	\$2,737
Farmers	\$2,485	20th Century	\$1,692

Criteria: Six-month premium. Husband, 45, drives a 1995 Explorer, 12,000 annual miles. Wife, 43, drives a 1996 Dodge Caravan, 12,000 annual miles. Son, 17, is an occasional operator. All have clean driving records.

ILLUSTRATION 2

5. PAINTING A square tarp has sides 8 feet long. When it is laid out on a floor, how much area will it cover?

6. Evaluate $-12 - (-5)$.

7. Evaluate $12 - 2[-8 - 2^4(-1)]$.

8. Find $|-55|$.

9. Solve $6 = 2 - 2x$.

10. Translate to mathematical symbols:
16 less than twice the total t

11. FRUIT STORAGE Use the formula $C = \frac{5(F - 32)}{9}$ to complete the label on the box of bananas shown in Illustration 3.

PREMIUM
BANANAS
Keep at 59°F or ?°C
Imported by Pacific Fruit, Inc.

ILLUSTRATION 3

12. Solve $-(3x - 3) = 6(2x - 7)$.

13. SPELLING What fraction of the letters in the word *Mississippi* are vowels?

14. Simplify $\frac{10y}{15y}$.

In Exercises 15–18, do the indicated operation.

15. $-\frac{16a}{35} \cdot \frac{25}{48a^2}$

16. $4\frac{2}{5} \div 11$

17. $\frac{4}{m} + \frac{2}{7}$

18. $34\frac{1}{9} - 13\frac{5}{6}$

19. Solve $\frac{5}{6}y = -25$.

20. Solve $\frac{y}{6} = \frac{y}{12} + \frac{2}{3}$.

In Exercises 21–24, do the indicated operation.

21. $78.1 - 7.81$

22. $2.13(-4.05)$

23. $0.752(1{,}000)$

24. $\frac{241.86}{2.9}$

25. Evaluate $\frac{a - b}{0.5b - 0.4a}$ for $a = 3.6$ and $b = -1.5$. Round to the nearest hundredth.

26. Round 452.0298 to the nearest thousandth.

27. Write $\frac{11}{15}$ as a decimal. Use an overbar.

28. Solve $\frac{y}{2.22} = -5$.

29. Evaluate $3\sqrt{81} - 8\sqrt{49}$.

30. LABOR COST A car repair bill is shown in Illustration 4. The bill is torn, and one line cannot be read. How many hours of labor did it take to repair the car?

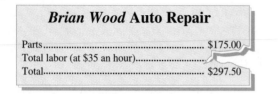

Brian Wood Auto Repair

Parts.. $175.00
Total labor (at $35 an hour)..........................
Total.. $297.50

ILLUSTRATION 4

Do the indicated operation.

31. $(m^2 - m - 5) - (3m^2 + 2m - 8)$

32. $(3x - 2)(x + 4)$

33. $(2y - 5)^2$

Simplify each expression.

34. $y^2 \cdot y^5$

35. $(h^5)^4$

36. $(2a^3b^6)^3$

37. $-7g^5(8g^4)$

38. Graph the points with coordinates $(-1, 3)$, $(0, 1.5)$, $(-4, -4)$, $\left(2, \frac{7}{2}\right)$, and $(4, 0)$.

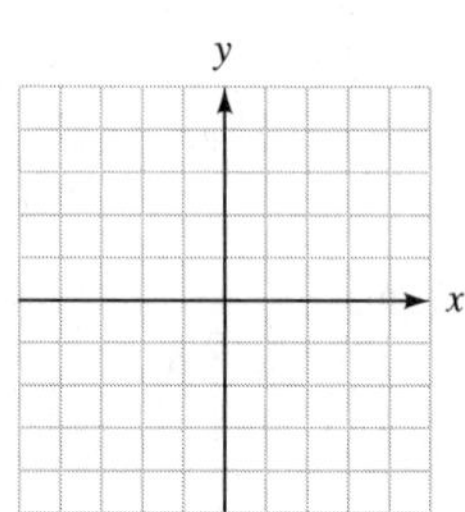

39. Graph $3x - 3y = 9$.

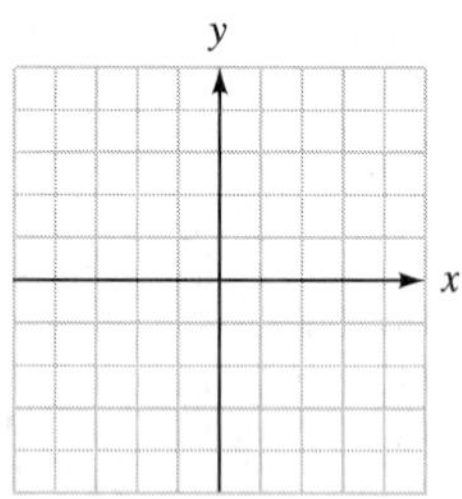

40. Graph $y = -x - 1$.

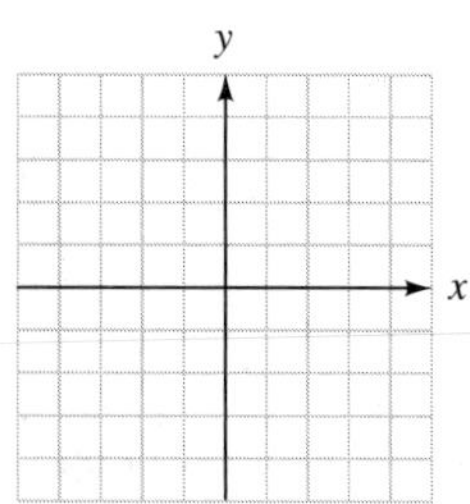

41. Complete the table.

Percent	Decimal	Fraction
	0.29	
47.3%		
		$\frac{7}{8}$

42. 16% of what number is 20?

43. TIPPING Complete the sales draft in Illustration 5 if a 15% tip, rounded up to the nearest dollar, is to be left for the waiter.

STEAK STAMPEDE
Bloomington, MN
Server #12\ AT

VISA 67463777288
NAME DALTON/ LIZ

AMOUNT $75.18
GRATUITY $____________
TOTAL $____________

ILLUSTRATION 5

44. GENEALOGY Through an extensive computer search, a genealogist determined that worldwide, 180 out of every 10 million people had his last name. What percent is this?

45. SAVINGS ACCOUNT Find the simple interest earned on \$10,000 at $7\frac{1}{4}\%$ for 2 years.

Ratio, Proportion, and Measurement

8

RATIOS AND PROPORTIONS ARE USED TO COMPARE QUANTITIES. THEY ARE ALSO USED TO CONVERT FROM ONE UNIT OF MEASUREMENT TO ANOTHER.

8.1 Ratio

In this section, you will learn about

- Ratios • Rates • Unit rates • Unit costs

INTRODUCTION. The concept of *ratio* occurs often in real-life situations.

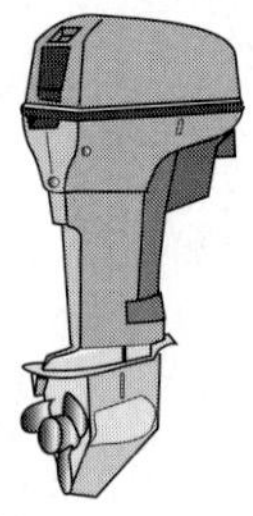

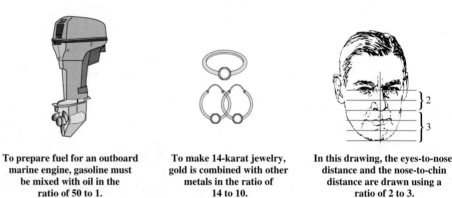

To prepare fuel for an outboard marine engine, gasoline must be mixed with oil in the ratio of 50 to 1.

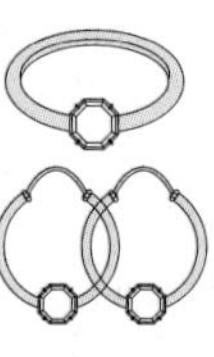

To make 14-karat jewelry, gold is combined with other metals in the ratio of 14 to 10.

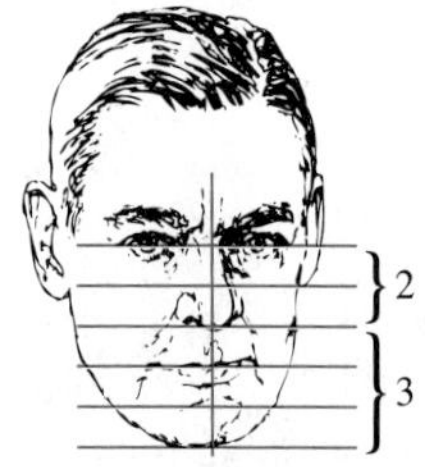

In this drawing, the eyes-to-nose distance and the nose-to-chin distance are drawn using a ratio of 2 to 3.

In this section, we will see how *ratios* and *rates* (a special type of ratio) are used to express relationships between two quantities.

Ratios

Ratios give us a way to compare numerical quantities.

Ratios

A **ratio** is the quotient of two numbers or the quotient of two quantities that have the same units.

There are three ways to write a ratio: as a fraction, as two numbers separated by the word *to,* or as two numbers separated by a colon. For example, the ratios described in the examples above can be written in the following ways:

$\frac{50}{1}$, 14 to 10, and 2:3

- The fraction $\frac{50}{1}$ is read as "the ratio of 50 to 1."
- 14 to 10 is read as "the ratio of 14 to 10."
- 2:3 is read as "the ratio of 2 to 3."

EXAMPLE 1 ***Writing ratios.*** Express each ratio as a fraction in lowest terms: **a.** the ratio of 25 to 10 and **b.** the ratio of 0.3 to 1.2.

Solution

a. To write the phrase "the ratio of 25 to 10" in fractional form, we write 25 as the numerator and 10 as the denominator. Then we simplify the fraction.

$$\frac{25}{10} = \frac{\overset{1}{\cancel{5}} \cdot 5}{\underset{1}{\cancel{5}} \cdot 2}$$ Factor 25 as 5 · 5 and 10 as 5 · 2. Then divide out the common factor of 5.

$$= \frac{5}{2}$$

The ratio 25 to 10 can be written as the fraction $\frac{25}{10}$, which simplifies to $\frac{5}{2}$. Because the fractions $\frac{25}{10}$ and $\frac{5}{2}$ represent equal numbers, they are **equal ratios.**

b. The ratio of 0.3 to 1.2 can be written as the fraction $\frac{0.3}{1.2}$. To write this as a ratio of whole numbers, we need to clear it of decimals. We can do so by multiplying the numerator and denominator by 10.

$$\frac{0.3}{1.2} = \frac{0.3 \cdot \mathbf{10}}{1.2 \cdot \mathbf{10}}$$

$$= \frac{3}{12}$$ Do the multiplications: 0.3 · 10 = 3 and 1.2 · 10 = 12.

$$= \frac{1}{4}$$ Simplify the fraction: $\frac{3}{12} = \frac{\overset{1}{\cancel{3}} \cdot 1}{4 \cdot \underset{1}{\cancel{3}}} = \frac{1}{4}$.

Self Check

Express each ratio as a fraction in lowest terms:

a. the ratio of 9 to 12

b. 3.2:16

Answers: **a.** $\frac{3}{4}$, **b.** $\frac{1}{5}$

EXAMPLE 2 ***Carry-on luggage.*** Airlines allow passengers to carry a piece of luggage onto an airplane only if it will fit in the space shown in Figure 8-1. What is the ratio of the length of the space to its width?

Solution

Since the length of the carry-on luggage space is 24 inches and its width is 10 inches, the ratio of the length to the width is $\frac{24 \text{ inches}}{10 \text{ inches}}$. Common factors and common units should be divided out.

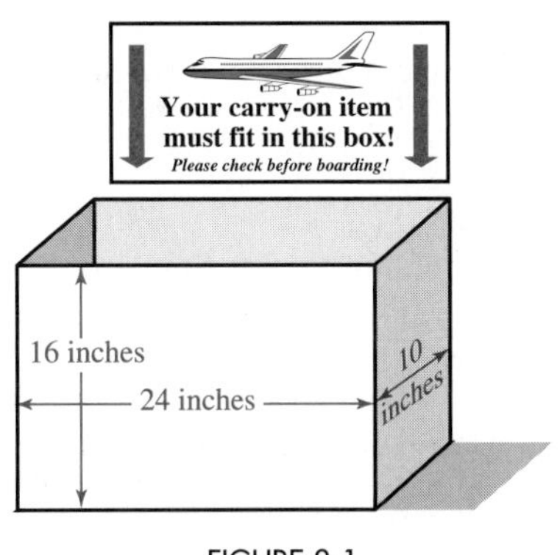

FIGURE 8-1

$$\frac{24 \text{ inches}}{10 \text{ inches}} = \frac{12 \cdot \overset{1}{\cancel{2}} \cancel{\text{inches}}}{5 \cdot \underset{1}{\cancel{2}} \cancel{\text{inches}}}$$ Divide out the common factor of 2 and divide out the common units of inches.

$$= \frac{12}{5}$$

The length-to-width ratio of the carry-on space is $\frac{12}{5}$.

Self Check

What is the ratio of the height to the length of the carry-on space in Figure 8-1?

Answer: $\frac{2}{3}$

EXAMPLE 3 ***Different units.*** Express the phrase *12 ounces to 1 pound* as a ratio in lowest terms.

Solution

Recall that a ratio is a comparison of two quantities with the same units. Since there are 16 ounces in one pound, the phrase *12 ounces to 1 pound* can be expressed as the ratio $\frac{12 \text{ ounces}}{16 \text{ ounces}}$, which simplifies to $\frac{3}{4}$.

Self Check

Express the phrase *2 feet to 1 yard* as a ratio. (*Hint*: 3 feet = 1 yard.)

Answer: $\frac{2}{3}$ ■

Rates

When we compare two different kinds of quantities, we call the comparison a **rate,** and we can write it as a fraction. For example, on the label of the can of paint in Figure 8-2, we see that one quart of paint is needed for every 200 square feet to be painted. Writing this as a rate, we have

$$\frac{1 \text{ quart}}{200 \text{ square feet}}$$ Read as "1 quart per 200 square feet."

When writing a rate, always include the units.

FIGURE 8-2

Rates

A **rate** is a quotient of two quantities with different units.

EXAMPLE 4 ***Writing a rate.*** According to the *Guinness Book of World Records,* a total of 78 inches of snow fell at Mile 47 Camp, Cooper River Division, Arkansas, in a 24-hour period in 1963. What was the rate of snowfall?

Solution

We begin by comparing the amount of snow, 78 inches, to the elapsed time, 24 hours. Then we simplify the fraction.

$$\frac{78 \text{ inches}}{24 \text{ hours}} = \frac{13 \cdot \overset{1}{\cancel{6}} \text{ inches}}{4 \cdot \underset{1}{\cancel{6}} \text{ hours}}$$ Factor 78 and 24. Then divide out the common factor of 6.

The snow fell at a rate of 13 inches per 4 hours: $\frac{13 \text{ inches}}{4 \text{ hours}}$.

Self Check

The fastest-growing flowering plant on record grew 12 feet in 14 days. What was its rate of growth over this period?

Answer: $\frac{6 \text{ feet}}{7 \text{ days}}$ ■

COMMENT Unlike in the fraction $\frac{a}{b}$, b can be zero in a rate. For example, the rate of women to men on the 1999 U.S. Women's World Cup soccer team is expressed as $\frac{20 \text{ women}}{0 \text{ men}}$. Such applications are rare, however.

Unit rates

A **unit rate** is a rate in which the denominator is 1. To illustrate the concept of a unit rate, suppose a driver makes the 354-mile trip from Pittsburgh to Indianapolis in 6 hours. Then the motorist's rate (or more specifically, rate of speed) is given by

$$\frac{354 \text{ miles}}{6 \text{ hours}} = \frac{59 \cdot \overset{1}{\cancel{6}} \text{ miles}}{\underset{1}{\cancel{6}} \cdot 1 \text{ hours}} = \frac{59 \text{ miles}}{1 \text{ hour}}$$

Factor 354 as 59 · 6 and divide out the common factor of 6.

We can also find the unit rate by dividing 354 by 6.

$$\begin{array}{r} 59 \\ 6\overline{)354} \\ \underline{30} \\ 54 \\ \underline{54} \\ 0 \end{array}$$

The unit rate $\frac{59 \text{ miles}}{1 \text{ hour}}$ can be expressed in any of the following forms:

$$59 \frac{\text{miles}}{\text{hour}}, \quad 59 \text{ miles per hour}, \quad 59 \text{ miles/hour}, \quad \text{or} \quad 59 \text{ mph}$$

EXAMPLE 5 ***Finding hourly rates of pay.*** A student earns \$152 for working 16 hours in a bookstore. Find his hourly rate of pay.

Solution

We can write the rate of pay as

$$\text{Rate of pay} = \frac{\$152}{16 \text{ hr}}$$

Compare the amount of money earned to the number of hours worked.

To find the rate of pay for 1 hour of work, we divide 152 by 16.

$$\begin{array}{r} 9.5 \\ 16\overline{)152.0} \\ \underline{144} \\ 8\,0 \\ \underline{8\,0} \\ 0 \end{array}$$

Write a decimal point and a 0 to the right of 2.

The unit rate of pay is $\frac{\$9.50}{1 \text{ hour}}$, which can be written as \$9.50 per hour.

Self Check

Joan earns \$436 per 40-hour week managing a dress shop. Find her hourly rate of pay.

Answer: \$10.90 per hour ■

EXAMPLE 6 ***Energy consumption.*** One household used 795 kilowatt hours (kwh) of electricity during a 30-day period. Find the rate of energy consumption in kilowatt hours per day.

Solution

We can write the rate of energy consumption as

$$\text{Rate of energy consumption} = \frac{795 \text{ kwh}}{30 \text{ days}}$$

To find the unit rate, we divide 795 by 30.

$$\text{Unit rate of energy consumption} = \frac{26.5 \text{ kwh}}{1 \text{ day}}$$

The rate of energy consumption was 26.5 kilowatt hours per day.

Self Check

To heat a house for 30 days, a furnace burned 69 therms of natural gas. Find the rate of gas consumption in therms per day.

Answer: 2.3 therms per day ■

Accent on Technology: Computing gas mileage

A man drove from Houston to St. Louis—a total of 775 miles. Along the way, he stopped for gas three times, pumping 10.5, 11.3, and 8.75 gallons of gas. He started with the tank half full and ended with the tank half full. To find how many miles he got per gallon (mpg), we need to compare the total distance to the total number of gallons of gas consumed.

$$\frac{775 \text{ miles}}{(10.5 + 11.3 + 8.75) \text{ gallons}}$$

We can simplify this rate by entering these numbers and pressing these keys on a scientific calculator.

Keystrokes 775 [÷] [(] 10.5 [+] 11.3 [+] 8.75 [)] [=] 25.368249

To the nearest hundredth, he got 25.37 mpg.

Unit costs

If a store sells 5 pounds of coffee for $18.75, a consumer might want to know what the coffee costs per pound. When we find the cost of one pound of the coffee, we are finding a *unit cost.* To find the unit cost of an item, we begin by comparing its cost to its quantity.

$$\frac{\$18.75}{5 \text{ pounds}}$$

Then we divide the cost by the number of items.

$$5\overline{)18.75} = 3.75$$

The unit cost of the coffee is $3.75 per pound.

EXAMPLE 7 ***Comparison shopping.*** Olives come packaged in a 10-ounce jar, which sells for $2.49, or in a 6-ounce jar, which sells for $1.53. (See Figure 8-3.) Which is the better buy?

Solution

To find the better buy, we must find each unit cost.

$$\frac{\$2.49}{10 \text{ oz}} = \frac{249¢}{10 \text{ oz}}$$ Change $2.49 to 249 cents.

$$= 24.9¢ \text{ per oz}$$ Divide 249 by 10.

$$\frac{\$1.53}{6 \text{ oz}} = \frac{153¢}{6 \text{ oz}}$$ Change $1.53 to 153 cents.

$$= 25.5¢ \text{ per oz}$$ Do the division.

FIGURE 8-3

The unit cost is less when olives are packaged in 10-ounce jars, so that is the better buy.

Self Check

A fast-food restaurant sells a 12-ounce cola for 72¢ and a 16-ounce cola for 99¢. Which is the better buy?

Answer: the 12-oz cola

STUDY SET Section 8.1

VOCABULARY *Fill in the blanks to make the statements true.*

1. A ______________ is a quotient of two numbers or a quotient of two quantities with the same units.

2. A quotient of two quantities with different units is called a ______________.

3. When the price of candy is advertised as \$1.75 per pound, we are told its unit ______________.

4. A ______________ rate is a rate in which the denominator is 1.

CONCEPTS

5. To write the ratio $\frac{15}{24}$ in lowest terms, we divide out any common factors of the numerator and denominator. What common factor do they have?

6. Complete the solution.

Write the ratio $\frac{14}{21}$ in lowest terms.

$$\frac{14}{21} = \frac{7 \cdot 2}{\quad} = \frac{\overset{1}{\cancel{7}} \cdot 2}{\underset{1}{\cancel{7}} \cdot \quad} = \frac{\quad}{\quad}$$

7. Consider the ratio $\frac{0.5}{0.6}$. By what number should we multiply numerator and denominator to make this a ratio of whole numbers?

8. What should be done to write the ratio $\frac{15 \text{ inches}}{22 \text{ inches}}$ in simplest form?

9. Since a ratio is a comparison of quantities with the same units, how should the ratio $\frac{11 \text{ minutes}}{1 \text{ hour}}$ be rewritten?

10. a. Consider the rate $\frac{\$248}{16 \text{ hours}}$. How can we find the unit rate (\$ per hour)?

b. Consider the rate $\frac{\$7.95}{3 \text{ pairs}}$. How can we find the unit cost of a pair of socks?

NOTATION

11. Refer to Illustration 1. Write the ratio of the flag's length to its width using a fraction, using the word *to,* and using a colon.

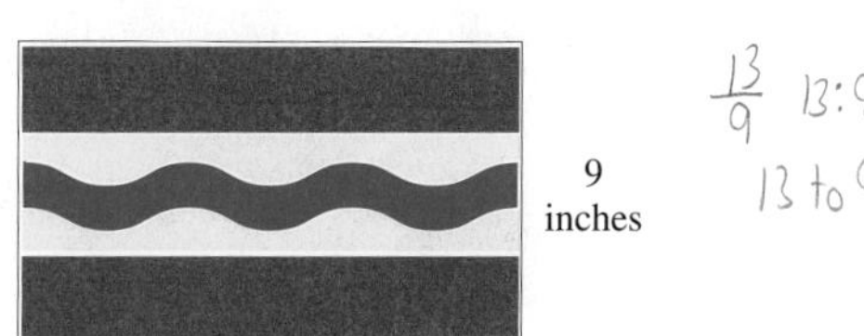

ILLUSTRATION 1

12. The rate $\frac{55 \text{ miles}}{1 \text{ hour}}$ can be expressed as

- 55 ______________ ______________ ______________
- 55 ______________ / ______________
- 55 ______________

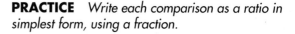

PRACTICE *Write each comparison as a ratio in simplest form, using a fraction.*

13. 5 to 7

14. 3 to 5

15. 17 to 34

16. 19 to 38

17. 22:33

18. 14:21

19. 1.5:2.4

20. 0.9:0.6

21. 7 to 24.5

22. 0.65 to 0.15

23. 4 ounces to 12 ounces

24. 3 inches to 15 inches

25. 12 minutes to 1 hour

26. 8 ounces to 1 pound

27. 3 days to 1 week

28. 4 inches to 2 yards

29. 18 months to 2 years

30. 8 feet to 4 yards

Refer to the monthly budget shown in Illustration 2. Give each ratio in lowest terms.

31. Find the total amount of the budget.

32. Find the ratio of the amount budgeted for rent to the total budget.

33. Find the ratio of the amount budgeted for food to the total budget.

34. Find the ratio of the amount budgeted for the phone to the total budget.

Item	Amount
Rent	\$800
Food	\$600
Gas and electric	\$180
Phone	\$100
Entertainment	\$120

ILLUSTRATION 2

Refer to the list of tax deductions shown in Illustration 3. Give each ratio in lowest terms.

35. Find the total amount of the deductions.

36. Find the ratio of the real estate tax deduction to the total deductions.

37. Find the ratio of the charitable contributions to the total deductions.

38. Find the ratio of the mortgage interest deduction to the union dues deduction.

Item	Amount
Medical expenses	$875
Real estate taxes	$1,250
Charitable contributions	$1,750
Mortgage interest	$4,375
Union dues	$500

ILLUSTRATION 3

Write each rate as a fraction in simplest form.

39. 64 feet in 6 seconds

40. 45 applications for 18 openings

41. 84 made out of 100 attempts

42. 75 days on 20 gallons of water

43. 3,000 students over a 16-year career

44. 16 right compared to 34 wrong

45. 18 beats every 12 measures

46. 1.5 inches as a result of 30 turns

Write each phrase as a unit rate.

47. 60 revolutions in 5 minutes

48. 14 trips every 2 months

49. 12 errors in 8 hours

50. $50,000 paid over 10 years

51. 245 presents for 35 children

52. 108 occurrences in a 12-month period

53. 4,000,000 people living in 12,500 square miles

54. 117.6 pounds of pressure on 8 square inches

Find the unit cost.

55. $3.50 for 50 feet

56. 150 barrels cost $4,950.

57. 65 ounces sell for 78 cents.

58. They charged $48 for 15 minutes.

59. Four of us donated a total of $272.

60. For 7 dozen, you will pay $10.15.

61. $4 billion over a 5-month span

62. 7,020 pesos will buy six stickers.

APPLICATIONS

63. ART HISTORY Leonardo da Vinci drew the human figure shown in Illustration 4 within a square. (All four sides of a square are the same length.) What is the ratio of the length of the man's outstretched arms to his height?

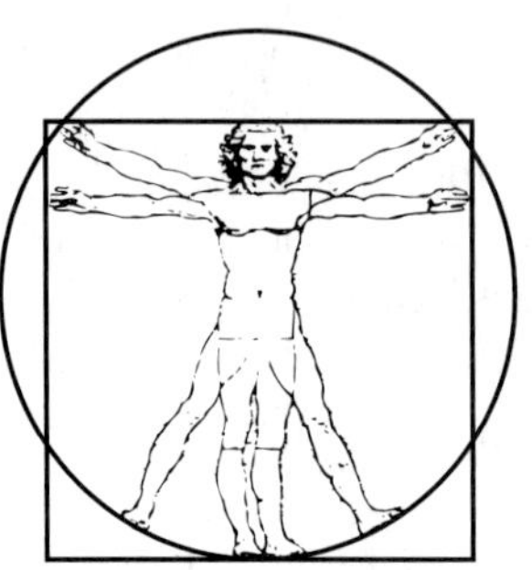

ILLUSTRATION 4

64. CHECKERED FLAG The flag shown in Illustration 5 is composed of squares (All four sides of a square are the same length.) What is the ratio of the width of the flag to its length?

ILLUSTRATION 5

65. GEAR RATIO Refer to Illustration 6. What is the ratio of the number of teeth of the larger gear to the number of teeth of the smaller gear?

ILLUSTRATION 6

66. BANKRUPTCY After declaring bankruptcy, a company could reimburse its creditors only 5¢ on the dollar. Write this as a ratio in lowest terms.

67. COOKING A recipe from Martha Stewart's *Easy Living* magazine is shown in Illustration 7. Write the ratio of sugar to milk as a fraction. You do not have to simplify the ratio.

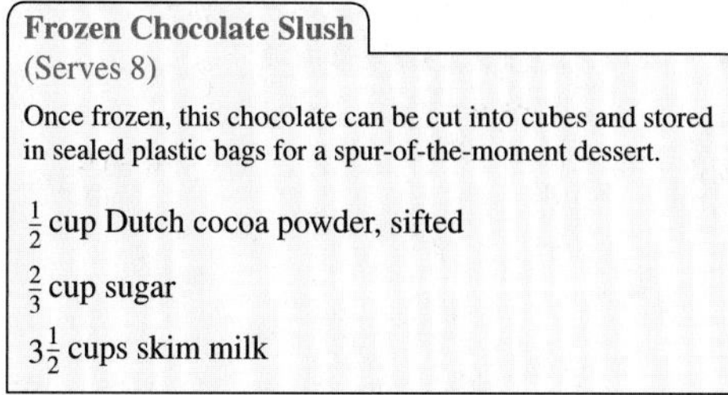

Frozen Chocolate Slush
(Serves 8)
Once frozen, this chocolate can be cut into cubes and stored in sealed plastic bags for a spur-of-the-moment dessert.
$\frac{1}{2}$ cup Dutch cocoa powder, sifted
$\frac{2}{3}$ cup sugar
$3\frac{1}{2}$ cups skim milk

ILLUSTRATION 7

68. HEARING From the graph in Illustration 8, determine the ratio of hearing loss in males as compared to females.

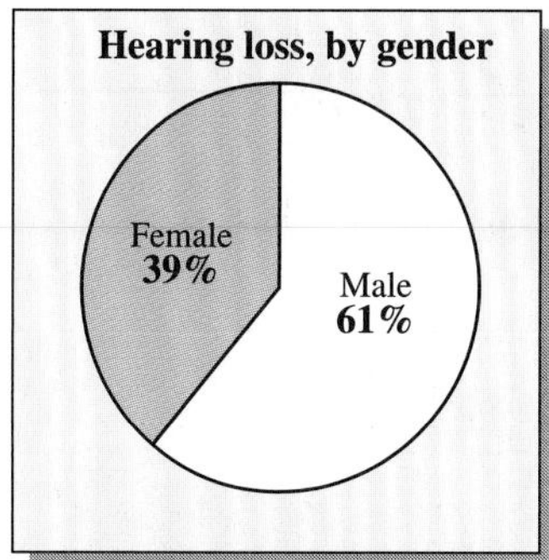

Source: National Academy on an Aging Society

ILLUSTRATION 8

69. SOFTBALL The U.S. women's softball team won the gold medal at the 1996 Olympic games. Hitting statistics for Dot Richardson, the team's shortstop, are shown below. What was her rate of hits (H) to at-bats (AB) during the Olympic competition?

	BA	AB	H	R	2B	3B	HR	RBI
Richardson	.273	33	9	7	2	—	3	7

70. TYPING A secretary typed a document containing 330 words in 5 minutes. How many words per minute did he type?

71. CPR A paramedic performed 125 compressions to 50 breaths on an adult with no pulse. What compressions-to-breaths rate did the paramedic use?

72. INTERNET SALES A web site determined that it had 112,500 hits in one month. Of those visiting the site, 4,500 made purchases. How many did not make a purchase? What was the browser/buyers unit rate for the web site that month?

73. AIRLINE COMPLAINTS An airline had 3.29 complaints for every 1,000 passengers. Write this rate as a fraction of whole numbers.

74. FINGERNAILS On average, fingernails grow 0.02 inch per week. Write this rate using whole numbers.

75. FACULTY–STUDENT RATIO At a college, there are 125 faculty members and 2,000 students. Find the rate of faculty to students. (This is often referred to as the faculty-to-student *ratio,* even though the units are different.)

76. PARKING METER A parking meter requires 25¢ for 20 minutes of parking. What is the unit cost?

77. UNIT COST A driver pumped 17 gallons of gasoline into his tank at a cost of $32.13. Find the unit cost of the gasoline.

78. UNIT COST A 50-pound bag of grass seed costs $222.50. Find the unit cost of grass seed.

79. UNIT COST A 12-ounce can of cranberry juice sells for 84¢. Give the unit cost in cents per ounce.

80. UNIT COST A 24-ounce package of green beans sells for $1.29. Give the unit cost in cents per ounce.

81. COMPARISON SHOPPING A 6-ounce can of orange juice sells for 89¢, and an 8-ounce can sells for $1.19. Which is the better buy?

82. COMPARISON SHOPPING A 30-pound bag of fertilizer costs $12.25, and an 80-pound bag costs $30.25. Which is the better buy?

83. COMPARISON SHOPPING A certain brand of cold and sinus medication is sold in 20-tablet boxes for $4.29 and in 50-tablet boxes for $9.59. Which is the better buy?

84. COMPARISON SHOPPING Which tire shown in Illustration 9 is the better buy?

ILLUSTRATION 9

85. COMPARING SPEEDS A car travels 345 miles in 6 hours, and a truck travels 376 miles in 6.2 hours. Which vehicle is going faster?

86. READING SPEEDS One seventh-grader read a 54-page book in 40 minutes. Another read an 80-page book in 62 minutes. If the books were equally difficult, which student read faster?

87. EMPTYING A TANK An 11,880-gallon tank can be emptied in 27 minutes. Find the rate of flow in gallons per minute.

88. RATE OF PAY Ricardo worked for 27 hours to help insulate a hockey arena. For his work, he received $337.50. Find his hourly rate of pay.

89. AUTO TRAVEL A car's odometer reads 34,746 at the beginning of a trip. Five hours later, it reads 35,071. How far has the car traveled? What is the average rate of speed?

90. RATE OF SPEED An airplane travels from Chicago to San Francisco, a distance of 1,883 miles, in 3.5 hours. Find the rate of speed of the plane.

91. GAS MILEAGE One car went 1,235 miles on 51.3 gallons of gasoline, and another went 1,456 miles on 55.78 gallons. Which car got the better gas mileage?

92. ELECTRICITY RATES In one community, a bill for 575 kilowatt hours of electricity is $38.81. In a second community, a bill for 831 kwh is $58.10. In which community is electricity cheaper?

WRITING

93. Are the ratios 3 to 1 and 1 to 3 the same? Explain why or why not.

94. Give three examples of ratios (or rates) that you have encountered in the past week.

95. How will the topics studied in this section make you a better shopper?

96. What is a unit rate? Give some examples.

REVIEW *Perform each operation.*

97. $3.05 + 17.17 + 25.317$

98. $3.5\overline{)157.85}$

99. $13.2 + 25.07 \cdot 7.16$

100. $\frac{4}{3} - \frac{1}{4}$

101. $5 - 3\frac{1}{4}$

102. Complete the table of solutions for $3x - 5y = 15$. Then graph the equation.

x	y
0	
	0
3	

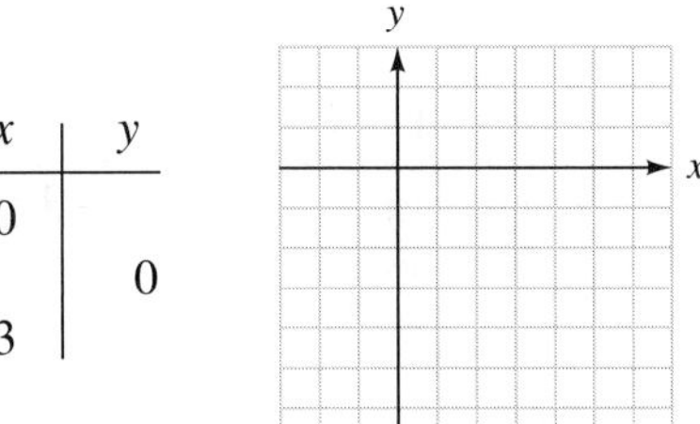

8.2 Proportions

In this section, you will learn about

- Proportions • Means and extremes of a proportion • Solving proportions
- Writing proportions to solve problems

Introduction. Like any tool, a ladder can be dangerous if used improperly. A safety pamphlet states, "When setting up an extension ladder, use the *4-to-1 rule*—For every 4 feet of ladder height, position the legs of the ladder 1 foot away from the base of the wall." The 4-to-1 rule for ladders can be expressed using a ratio.

$$\frac{4 \text{ feet}}{1 \text{ foot}} = \frac{4 \overset{1}{\cancel{\text{feet}}}}{1 \underset{1}{\cancel{\text{foot}}}} = \frac{4}{1}$$

In Figure 8-4, the 4-to-1 rule was used to position the legs of a ladder properly, 3 feet from the base of a 12-foot-high wall. We can write a ratio comparing the ladder's height to its distance from the wall.

$$\frac{12\text{ feet}}{3\text{ feet}} = \frac{12\,\cancel{\text{feet}}^{1}}{3\,\cancel{\text{feet}}_{1}} = \frac{12}{3}$$

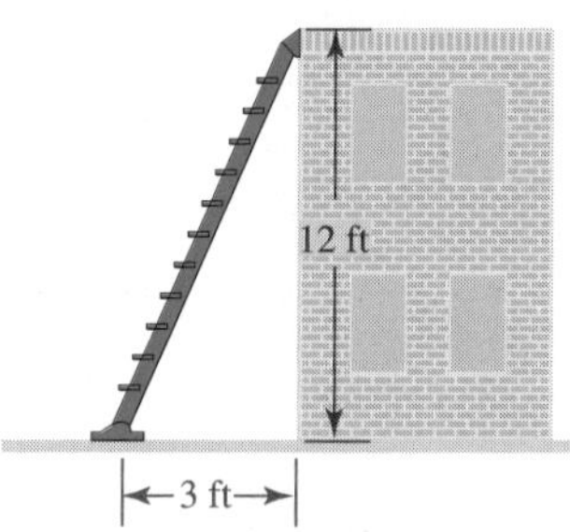

FIGURE 8-4

Since this ratio satisfies the 4-to-1 rule, the two ratios $\frac{4}{1}$ and $\frac{12}{3}$ must be equal. Therefore, we have

$$\frac{4}{1} = \frac{12}{3}$$

Such equations, which show that two ratios are equal, are called *proportions.* In this section, we will introduce the concept of proportion, and we will use proportions to solve many different types of problems.

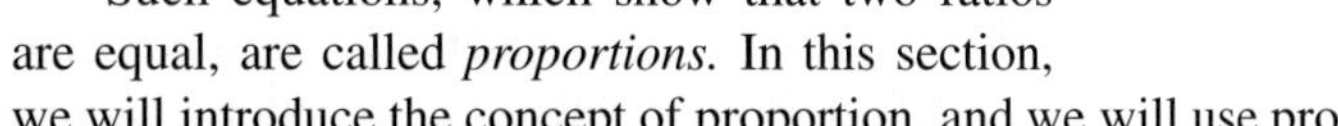

Proportions

Proportion

A **proportion** is a statement that two ratios (or rates) are equal.

Some examples of proportions are

$$\frac{1}{2} = \frac{3}{6}, \quad \frac{3\text{ waiters}}{7\text{ tables}} = \frac{9\text{ waiters}}{21\text{ tables}}, \quad \text{and} \quad \frac{a}{b} = \frac{c}{d}$$

- The proportion $\frac{1}{2} = \frac{3}{6}$ can be read as "1 is to 2 as 3 is to 6."
- The proportion $\frac{3\text{ waiters}}{7\text{ tables}} = \frac{9\text{ waiters}}{21\text{ tables}}$ can be read as "3 waiters are to 7 tables as 9 waiters are to 21 tables."
- The proportion $\frac{a}{b} = \frac{c}{d}$ can be read as "a is to b as c is to d."

The terms of the proportion $\frac{a}{b} = \frac{c}{d}$ are numbered as follows:

First term ⟶ a , c ⟵ Third term
Second term ⟶ b , d ⟵ Fourth term

$$\frac{a}{b} = \frac{c}{d}$$

Means and extremes of a proportion

In any proportion, the first and fourth terms are called the **extremes.** The second and third terms are called the **means.**

In the proportion $\frac{1}{2} = \frac{3}{6}$, 1 and 6 are the **extremes,** and 2 and 3 are the **means.**

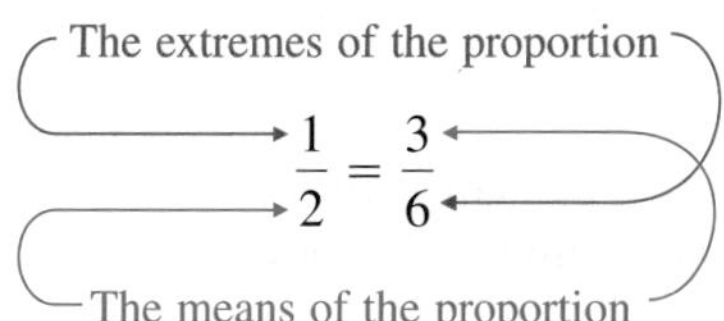

In this proportion, the product of the extremes is equal to the product of the means.

$$1 \cdot 6 = 6 \quad \text{and} \quad 2 \cdot 3 = 6$$

This example illustrates a fundamental property of proportions.

Fundamental property of proportions

In any proportion, the product of the extremes is equal to the product of the means.

In the proportion $\frac{a}{b} = \frac{c}{d}$, a and d are the extremes, and b and c are the means. We can show that the product of the extremes (ad) is equal to the product of the means (bc) by multiplying both sides of the proportion by bd and observing that $ad = bc$.

$$\frac{a}{b} = \frac{c}{d}$$

$$\frac{\mathbf{bd}}{\mathbf{1}} \cdot \frac{a}{b} = \frac{\mathbf{bd}}{\mathbf{1}} \cdot \frac{c}{d}$$ To eliminate the fractions, multiply both sides by $\frac{bd}{1}$.

$$\frac{abd}{b} = \frac{bcd}{d}$$ Multiply the numerators and multiply the denominators.

$$ad = bc$$ Divide out the common factors: $\frac{b}{b} = 1$ and $\frac{d}{d} = 1$.

Since $ad = bc$, the product of the extremes equals the product of the means.

To determine whether an equation is a proportion, we can check to see whether the product of the extremes is equal to the product of the means.

EXAMPLE 1 ***Proportions.*** Determine whether each equation is a proportion: **a.** $\frac{3}{7} = \frac{9}{21}$ and **b.** $\frac{8}{3} = \frac{13}{5}$.

Solution

In each case, we check to see whether the product of the extremes is equal to the product of the means.

a. The product of the extremes is $3 \cdot 21 = 63$. The product of the means is $7 \cdot 9 = 63$. Since the products are equal, the equation is a proportion: $\frac{3}{7} = \frac{9}{21}$.

$3 \cdot 21 = \mathbf{63}$ $\quad 7 \cdot 9 = \mathbf{63}$

$$\frac{3}{7} = \frac{9}{21}$$

The product of the extremes and the product of the means are also known as **cross products.**

b. The product of the extremes is $8 \cdot 5 = 40$. The product of the means is $3 \cdot 13 = 39$. Since the cross products are not equal, the equation is not a proportion: $\frac{8}{3} \neq \frac{13}{5}$.

$8 \cdot 5 = \mathbf{40}$ $\quad 3 \cdot 13 = \mathbf{39}$

$$\frac{8}{3} = \frac{13}{5}$$

Self Check

Determine whether the equation is a proportion:

$$\frac{6}{13} = \frac{18}{39}$$

Answer: yes ■

When two pairs of numbers such as 2, 3 and 8, 12 form a proportion, we say that they are **proportional.** To show that 2, 3 and 8, 12 are proportional, we check to see whether the equation

$$\frac{2}{3} = \frac{8}{12}$$

is a proportion. To do so, we find the product of the extremes and the product of the means:

$2 \cdot 12 = 24$ $\quad 3 \cdot 8 = 24$

Since the cross products are equal, the equation is a proportion, and the numbers are proportional.

EXAMPLE 2 ***Determining whether numbers are proportional.***
Determine whether 3, 7 and 36, 91 are proportional.

Solution

We check to see whether $\frac{3}{7} = \frac{36}{91}$ is a proportion by finding two products:

$3 \cdot 91 = 273$ The product of the extremes.

$7 \cdot 36 = 252$ The product of the means.

Since the cross products are not equal, the numbers are not proportional.

Self Check

Determine whether 6, 11 and 54, 99 are proportional.

Answer: yes ■

Solving proportions

Suppose that we know three terms in the proportion

$$\frac{?}{5} = \frac{24}{20}$$

To find the missing term, we represent it with x, multiply the extremes and multiply the means, set them equal, and solve for x:

$$\frac{x}{5} = \frac{24}{20}$$

$20x = 5 \cdot 24$ In a proportion, the product of the extremes is equal to the product of the means.

$20x = 120$ Do the multiplication: $5 \cdot 24 = 120$.

$\frac{20x}{\mathbf{20}} = \frac{120}{\mathbf{20}}$ To undo the multiplication by 20, divide both sides by 20.

$x = 6$ Do the divisions.

The first term is 6. To check this result, we substitute 6 for x in $\frac{x}{5} = \frac{24}{20}$ and find the cross products.

$\frac{6}{5} \stackrel{?}{=} \frac{24}{20}$ $6 \cdot 20 = \mathbf{120}$
$5 \cdot 24 = \mathbf{120}$

Since the cross products are equal, this is a proportion. The result, 6, is correct.

EXAMPLE 3 ***Solving proportions.*** Solve the proportion $\frac{12}{18} = \frac{3}{x}$ for x.

Solution

$$\frac{12}{18} = \frac{3}{x}$$

$12 \cdot x = 18 \cdot 3$ In a proportion, the product of the extremes equals the product of the means.

$12x = 54$ Multiply: $18 \cdot 3 = 54$.

$\frac{12x}{\mathbf{12}} = \frac{54}{\mathbf{12}}$ To undo the multiplication by 12, divide both sides by 12.

$x = \frac{9}{2}$ Simplify: $\frac{54}{12} = \frac{9 \cdot \overset{1}{\cancel{6}}}{\underset{1}{\cancel{6}} \cdot 2} = \frac{9}{2}$.

Thus, $x = \frac{9}{2}$. Check this result.

Self Check

Solve the proportion $\frac{15}{x} = \frac{20}{32}$ for x. Check the result.

Answer: 24 ■

EXAMPLE 4 ***Solving proportions.*** Find the third term of the proportion $\frac{3.5}{7.2} = \frac{x}{15.84}$.

Solution

$$\frac{3.5}{7.2} = \frac{x}{15.84}$$

$3.5(15.84) = 7.2x$ In a proportion, the product of the extremes equals the product of the means.

$55.44 = 7.2x$ Multiply: $3.5(15.84) = 55.44$.

$\frac{55.44}{\mathbf{7.2}} = \frac{7.2x}{\mathbf{7.2}}$ To undo the multiplication by 7.2, divide both sides by 7.2.

$7.7 = x$ Do the divisions.

The third term is 7.7. Check the result.

Self Check

Find the second term of the proportion $\frac{6.7}{x} = \frac{33.5}{38}$.

Answer: 7.6

Accent on Technology: Solving proportions with a calculator

To solve the proportion in Example 4 with a calculator, we can proceed as follows.

$$\frac{3.5}{7.2} = \frac{x}{15.84}$$

$\frac{3.5(15.84)}{7.2} = x$ Multiply both sides by 15.84 to isolate x.

We can find x by entering these numbers and pressing these keys on a scientific calculator.

Keystrokes 3.5 [×] 15.84 [÷] 7.2 [=] 7.7

Thus, $x = 7.7$.

EXAMPLE 5 ***Solving proportions.*** Solve the proportion $\frac{2a + 1}{4} = \frac{10}{8}$.

Solution

$$\frac{2a + 1}{4} = \frac{10}{8}$$

$8(2a + 1) = 40$ In a proportion, the product of the extremes equals the product of the means.

$16a + 8 = 40$ Distribute the multiplication by 8.

$16a + 8 \mathbf{- 8} = 40 \mathbf{- 8}$ To undo the addition of 8, subtract 8 from both sides.

$16a = 32$ Simplify: $8 - 8 = 0$ and $40 - 8 = 32$.

$\frac{16a}{\mathbf{16}} = \frac{32}{\mathbf{16}}$ To undo the multiplication by 16, divide both sides by 16.

$a = 2$ Do the divisions.

Thus, $x = 2$. Check the result.

Self Check

Solve the proportion $\frac{3m - 1}{2} = \frac{12.5}{5}$.

Answer: 2

Writing proportions to solve problems

We can use proportions to solve many real-world problems. If we are given a ratio (or rate) comparing two quantities, the words of the problem can be translated to a proportion, and we can solve it to find the unknown.

EXAMPLE 6 ***Grocery shopping.*** If 5 apples cost \$1.15, how much will 16 apples cost?

Solution

Let c represent the cost of 16 apples. If we compare the number of apples to their cost, we know that the two rates are equal.

5 apples is to \$1.15 as 16 apples is to \$$c$.

$$\begin{array}{r} \text{5 apples} \longrightarrow \\ \text{Cost of 5 apples} \longrightarrow \end{array} \frac{5}{1.15} = \frac{16}{c} \begin{array}{l} \longleftarrow \text{16 apples} \\ \longleftarrow \text{Cost of 16 apples} \end{array}$$

To find the cost of 16 apples, we solve the proportion for c.

$5 \cdot c = 1.15(16)$ In a proportion, the product of the extremes is equal to the product of the means.

$5c = 18.4$ Do the multiplication: $1.15(16) = 18.4$.

$\frac{5c}{5} = \frac{18.4}{5}$ To undo the multiplication by 5, divide both sides by 5.

$c = 3.68$ Do the divisions.

Sixteen apples will cost \$3.68. To check the result, we substitute 3.68 for c in the proportion and find the cross products.

$$\frac{5}{1.15} \stackrel{?}{=} \frac{16}{3.68} \qquad \begin{aligned} 5 \cdot 3.68 &= \mathbf{18.4} \\ 1.15 \cdot 16 &= \mathbf{18.4} \end{aligned}$$

The cross products are equal. The result of $c = 3.68$ checks.

Self Check

If 9 tickets to a concert cost \$112.50, how much will 15 tickets cost?

Answer: \$187.50

In Example 6, we could have compared the cost of the apples to the number of apples: \$1.15 is to 5 apples as \$$c$ is to 16 apples. This would have led to the proportion

$$\begin{array}{r} \text{Cost of 5 apples} \longrightarrow \\ \text{5 apples} \longrightarrow \end{array} \frac{1.15}{5} = \frac{c}{16} \begin{array}{l} \longleftarrow \text{Cost of 16 apples} \\ \longleftarrow \text{16 apples} \end{array}$$

If we solve this proportion for c, we obtain the same result: $c = 3.68$.

COMMENT When solving problems using proportions, it is a good practice to make sure that the units of the numerators are the same and the units of the denominators are the same. For Example 6, it would be incorrect to write

$$\begin{array}{r} \text{Cost of 5 apples} \longrightarrow \\ \text{5 apples} \longrightarrow \end{array} \frac{1.15}{5} = \frac{16}{c} \begin{array}{l} \longleftarrow \text{16 apples} \\ \longleftarrow \text{Cost of 16 apples} \end{array}$$

EXAMPLE 7 ***Scale drawing.*** A **scale** is a ratio (or rate) that compares the size of a model, drawing, or map to the size of an actual object. On the next page, the airplane in Figure 8-5 is drawn using a scale of 1 inch: 6 feet. This means that 1 inch on the drawing is actually 6 feet on the plane. The distance from wing tip to wing tip (the wingspan) on the drawing is 5 inches. What is the actual wingspan of the plane?

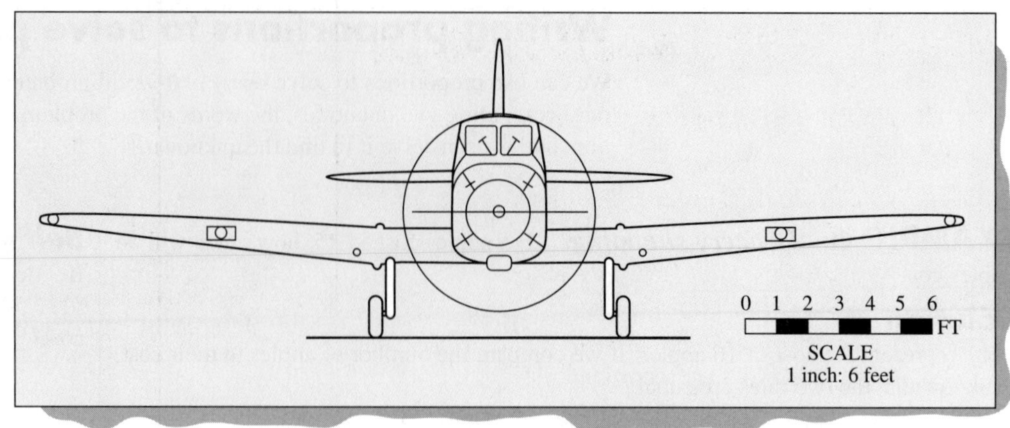

FIGURE 8-5

Solution Let w represent the actual wingspan of the plane. Since 1 inch corresponds to 6 feet as 5 inches corresponds to w feet, we can write the proportion

$$\frac{1}{6} = \frac{5}{w}$$

(Measure on drawing → 1, Measure on plane → 6; 5 ← Measure on drawing, w ← Measure on plane)

$1 \cdot w = 6 \cdot 5$ In a proportion, the product of the extremes is equal to the product of the means.

$w = 30$ Do the multiplications.

The actual wingspan of the plane is 30 feet. Check the result by finding the cross products. ■

EXAMPLE 8 ***Baking.*** A recipe for rhubarb cake calls for $1\frac{1}{4}$ cups of sugar for every $2\frac{1}{2}$ cups of flour. How many cups of flour are needed if the baker intends to use 3 cups of sugar?

Solution

Let f represent the number of cups of flour to be mixed with the sugar. The ratios of the cups of sugar to the cups of flour are equal.

$1\frac{1}{4}$ cups sugar is to $2\frac{1}{2}$ cups flour as 3 cups sugar is to f cups flour. We can write the proportion

$$\frac{1\frac{1}{4}}{2\frac{1}{2}} = \frac{3}{f}$$

(cups sugar → $1\frac{1}{4}$, cups flour → $2\frac{1}{2}$; 3 ← cups sugar, f ← cups flour)

$\frac{1.25}{2.5} = \frac{3}{f}$ Change the fractions to decimals: $1\frac{1}{4} = 1.25$ and $2\frac{1}{2} = 2.5$.

$1.25f = 2.5 \cdot 3$ In a proportion, the product of the extremes is equal to the product of the means.

$1.25f = 7.5$ Do the multiplication: $2.5 \cdot 3 = 7.5$.

$\frac{1.25f}{\mathbf{1.25}} = \frac{7.5}{\mathbf{1.25}}$ To undo the multiplication by 1.25, divide both sides by 1.25.

$f = 6$ Do the divisions.

The baker should use 6 cups of flour.

Self Check

How many cups of sugar will be needed to make several rhubarb cakes that will require a total of 25 cups of flour?

Answer: 12.5 cups ■

STUDY SET Section 8.2

VOCABULARY *Fill in the blanks to make the statements true.*

1. A ______ is a statement that two ratios or rates are equal.
2. In $\frac{1}{2} = \frac{5}{10}$, the terms 1 and 10 are called the ______ of the proportion.
3. In $\frac{5}{6} = \frac{x}{18}$, the terms 6 and x are called the ______ of the proportion.
4. When two pairs of numbers form a proportion, we say that the numbers are ______.

CONCEPTS *Fill in the blanks to make the statements true.*

5. The equation $\frac{a}{b} = \frac{c}{d}$ will be a proportion if the product ______ is equal to the product ______.
6. $\square \cdot 10 = \square$ $\qquad$ $2 \cdot \square = \square$

$$\frac{9}{2} = \frac{45}{10}$$

7. Write each statement as a proportion.
   a. 5 is to 8 as 15 is to 24.
   b. 3 teacher's aides are to 25 children as 12 teacher's aides are to 100 children.
8. Consider the proportion $\frac{3}{4} = \frac{15}{20}$. What are the two cross products?
9. For every 15 feet of chain link fencing, 4 support posts are used. How many support posts will be needed for 300 feet of chain link fence? Which of the following proportions could be used to solve this problem?

   i. $\frac{15}{4} = \frac{300}{x}$ $\qquad$ ii. $\frac{15}{4} = \frac{x}{300}$

   iii. $\frac{4}{15} = \frac{300}{x}$ $\qquad$ iv. $\frac{4}{15} = \frac{x}{300}$
10. Write a problem that could be solved using the following proportion.

ounces of cashews → $\frac{4}{639} = \frac{10}{x}$ ← ounces of cashews
calories → ← calories

NOTATION *Complete each solution.*

11. Solve for x: $\frac{12}{18} = \frac{x}{24}$.

$$12 \cdot 24 = \square$$
$$\square = 18x$$
$$\frac{288}{\square} = \frac{18x}{\square}$$
$$16 = x$$

12. Solve for x: $\frac{14}{x} = \frac{49}{17.5}$.

$$14 \cdot \square = 49x$$
$$\square = 49x$$
$$\frac{245}{\square} = \frac{49x}{\square}$$
$$5 = x$$

PRACTICE *Tell whether each statement is a proportion.*

13. $\frac{9}{7} = \frac{81}{70}$
14. $\frac{5}{2} = \frac{20}{8}$
15. $\frac{7}{3} = \frac{14}{6}$
16. $\frac{13}{19} = \frac{65}{95}$
17. $\frac{9}{19} = \frac{38}{80}$
18. $\frac{40}{29} = \frac{29}{22}$
19. $\frac{10.4}{3.6} = \frac{41.6}{14.4}$
20. $\frac{13.23}{3.45} = \frac{39.96}{11.35}$
21. $\frac{\frac{2}{3}}{\frac{5}{8}} = \frac{\frac{4}{5}}{\frac{9}{16}}$
22. $\frac{\frac{3}{2}}{\frac{8}{9}} = \frac{\frac{1}{4}}{\frac{4}{27}}$
23. $\frac{4\frac{1}{6}}{\frac{12}{7}} = \frac{2\frac{3}{16}}{\frac{9}{10}}$
24. $\frac{2\frac{1}{2}}{\frac{4}{5}} = \frac{3\frac{3}{4}}{\frac{9}{10}}$

Solve for the variable in each proportion. Check each result.

25. $\frac{2}{3} = \frac{x}{6}$
26. $\frac{3}{6} = \frac{x}{8}$
27. $\frac{5}{10} = \frac{3}{c}$
28. $\frac{7}{14} = \frac{2}{x}$
29. $\frac{6}{x} = \frac{8}{4}$
30. $\frac{4}{x} = \frac{2}{8}$
31. $\frac{x}{3} = \frac{9}{3}$
32. $\frac{x}{2} = \frac{18}{6}$
33. $\frac{x+1}{5} = \frac{3}{15}$
34. $\frac{x-1}{7} = \frac{2}{21}$
35. $\frac{x+3}{12} = \frac{-7}{6}$
36. $\frac{x+7}{-4} = \frac{1}{4}$
37. $\frac{4-x}{13} = \frac{11}{26}$
38. $\frac{5-x}{17} = \frac{13}{34}$
39. $\frac{2x+1}{18} = \frac{14}{3}$
40. $\frac{2x-1}{18} = \frac{9}{54}$
41. $\frac{4{,}000}{x} = \frac{3.2}{2.8}$
42. $\frac{0.4}{1.6} = \frac{96.7}{x}$
43. $\frac{\frac{1}{2}}{\frac{1}{5}} = \frac{x}{2\frac{1}{4}}$
44. $\frac{x}{4\frac{1}{10}} = \frac{3\frac{3}{4}}{1\frac{7}{8}}$

APPLICATIONS *Set up and solve a proportion.*

45. SCHOOL LUNCHES A manager of a school cafeteria orders 750 pudding cups. What will the order cost if she purchases them wholesale, 6 cups for $1.75?

46. CLOTHES SHOPPING As part of a spring clearance, a men's store put dress shirts on sale, 2 for $25.98. How much will a businessman pay if he buys five shirts?

47. GARDENING Three packets of garden seeds sell for 98¢. A Girl Scout troop leader needs to purchase three dozen packets. What will they cost?

48. COOKING A recipe for spaghetti sauce requires four 16-ounce bottles of ketchup to make two gallons of sauce. How many bottles of ketchup are needed to make 10 gallons of sauce?

49. BUSINESS PERFORMANCE The bar graph in Illustration 1 shows the yearly costs incurred and the revenue received by a business. How do the ratios of costs to revenue for 1999 and 2000 compare?

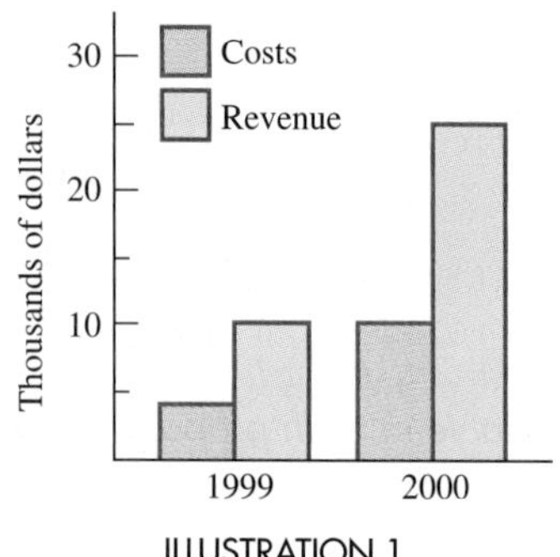

ILLUSTRATION 1

50. RAMP Write a ratio of the rise to the run for each ramp shown in Illustration 2. Set the ratios equal. Is the resulting proportion true? Is one ramp steeper than the other?

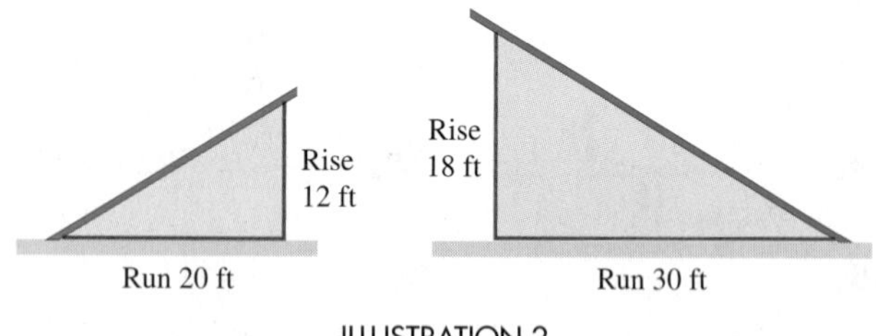

ILLUSTRATION 2

51. MIXING PERFUME A perfume is to be mixed in the ratio of 3 drops of pure essence to 7 drops of alcohol. How many drops of pure essence should be mixed with 56 drops of alcohol?

52. MAKING COLOGNE A cologne can be made by mixing 2 drops of pure essence with 5 drops of distilled water. How much water should be used with 15 drops of pure essence?

53. LAB WORK In a red blood cell count, a drop of the patient's diluted blood is placed on a grid like that shown in Illustration 3. Instead of counting each and every red blood cell in the 25-square grid, a technician just counts the number of cells in the five highlighted squares. Then he or she uses a proportion to estimate the total red blood cell count. If there are 195 red blood cells in the blue squares, about how many red blood cells would there be in the entire grid?

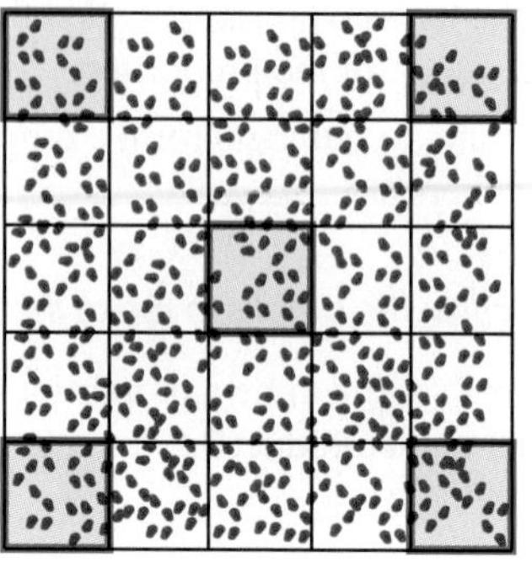

ILLUSTRATION 3

54. DOSAGE The proper dosage of a certain medication for a 30-pound child is shown in Illustration 4. At this rate, what would be the dosage for a 45-pound child?

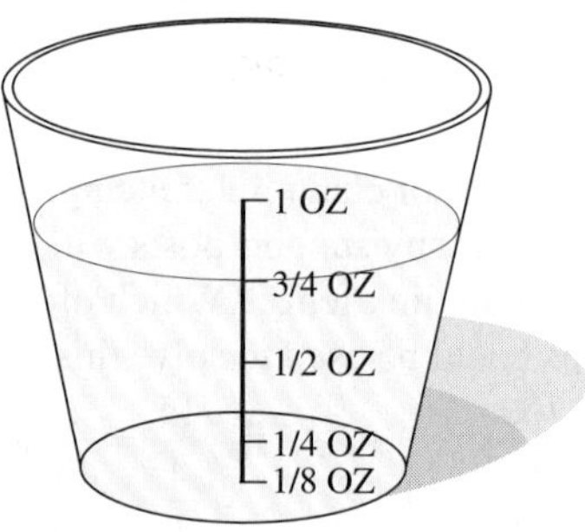

ILLUSTRATION 4

55. MAKING COOKIES A recipe for chocolate chip cookies calls for $1\frac{1}{4}$ cups of flour and 1 cup of sugar. The recipe will make $3\frac{1}{2}$ dozen cookies. How many cups of flour will be needed to make 12 dozen cookies?

56. MAKING BROWNIES A recipe for brownies calls for 4 eggs and $1\frac{1}{2}$ cups of flour. If the recipe makes 15 brownies, how many cups of flour will be needed to make 130 brownies?

57. COMPUTER SPEED Using the *Mathematica 3.0* program, a Dell Dimension XPS R350 (Pentium II) computer can perform a set of 15 calculations in 2.85 seconds. How long will it take the computer to perform 100 such calculations?

58. QUALITY CONTROL Out of a sample of 500 men's shirts, 17 were rejected because of crooked collars.

How many crooked collars would you expect to find in a run of 15,000 shirts?

59. FUEL CONSUMPTION A "high mobility multipurpose wheeled vehicle" is better known as a Hummer. Under normal conditions, a Hummer can travel 325 miles on a full tank (25 gallons) of diesel. How far can it travel on its auxiliary tank, which holds 17 gallons of diesel?

60. ANNIVERSARY GIFT A florist sells a dozen long-stemmed red roses for \$57.99. In honor of their 16th wedding anniversary, a man wants to buy 16 roses for his wife. What will the roses cost?

61. PAYCHECK Billie earns \$412 for a 40-hour week. If she missed 10 hours of work last week, how much did she get paid?

62. STAFFING The school board has determined that there should be 3 teachers for every 50 students. Complete Illustration 5 by filling in the number of teachers needed at each school.

	Glenwood High	Goddard Junior High	Sellers Elementary
Enrollment	2,700	1,900	850
Teachers			

ILLUSTRATION 5

63. BLUEPRINT The scale for the drawing in Illustration 6 tells the reader that a $\frac{1}{4}$-inch length $\left(\frac{1}{4}''\right)$ on the drawing corresponds to an actual size of 1 foot (1′0″). Suppose the length of the kitchen is $2\frac{1}{2}$ inches on the blueprint. How long is the actual kitchen?

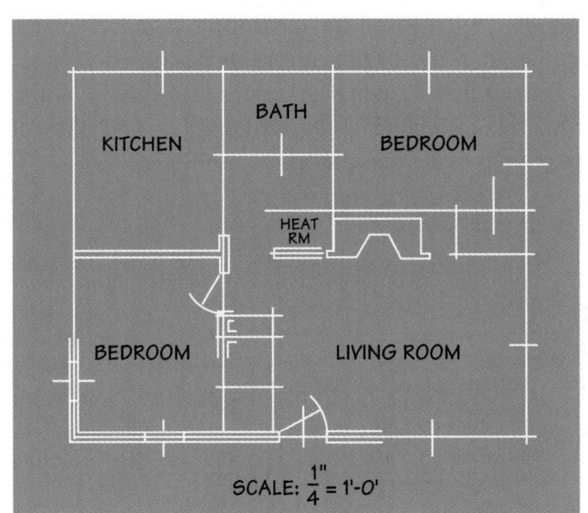

ILLUSTRATION 6

64. DRAFTING In a scale drawing, a 280-foot antenna tower is drawn 7 inches high. The building next to it is drawn 2 inches high. How tall is the actual building?

65. MODEL RAILROADING An HO-scale model railroad engine is 9 inches long. If HO scale is 87 feet to 1 foot, how long is a real engine?

66. MODEL RAILROADING An N-scale model railroad caboose is 4 inches long. If N scale is 169 feet to 1 foot, how long is a real caboose?

67. MINIATURE The ratio in Illustration 7 indicates that 1 inch on the model carousel is equivalent to 160 inches on the actual carousel. How wide should the model be if the actual carousel is 35 feet wide?

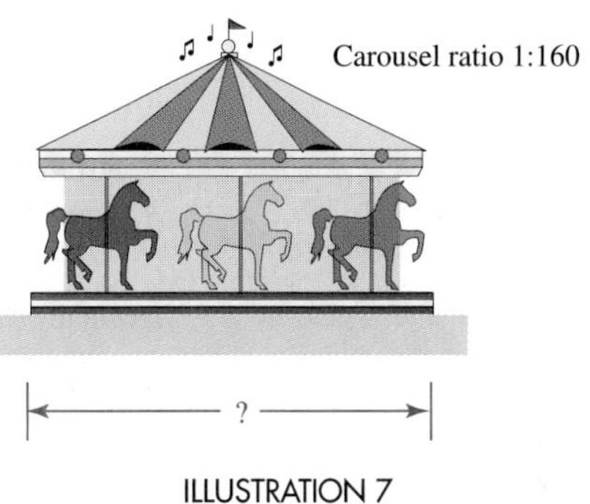

ILLUSTRATION 7

68. MIXING FUEL The instructions on a can of oil intended to be added to lawn mower gasoline read as shown in Illustration 8. Are these instructions correct? (*Hint:* There are 128 ounces in 1 gallon.)

Recommended	Gasoline	Oil
50 to 1	6 gal	16 oz

ILLUSTRATION 8

WRITING

69. Explain the difference between a ratio and a proportion.

70. Explain how to tell whether $\frac{3.2}{3.7} = \frac{5.44}{6.29}$ is a true proportion.

71. SCALE Explain the meaning of the following excerpt from a book about dollhouses.

Today, the internationally recognized scale for dollhouses and miniatures is 1 in. = 1 ft. This is small enough to be defined as a miniature, yet not too small for all details of decoration and furniture to be seen clearly.

72. Write a problem about a situation you encounter in your daily life that could be solved by using a proportion.

REVIEW

73. Change $\frac{9}{10}$ to a percent.

74. Change $\frac{7}{8}$ to a percent.

75. Change $33\frac{1}{3}\%$ to a fraction.

76. Multiply: $(2x - 1)(3x + 2)$.

77. Find $\frac{1}{2}\%$ of 520.

78. SHOPPING Bill purchased a shirt on sale for \$17.50. Find the original cost of the shirt if it was marked down 30%.

8.3 *American Units of Measurement*

In this section, you will learn about

- American units of length • Converting units of length
- American units of weight • American units of capacity • Units of time

INTRODUCTION. Two common systems of measurement are the American (or English) system and the metric system. We will discuss American units in this section and metric units in the next section. Some common American units are *inches, feet, miles, ounces, pounds, tons, cups, pints, quarts,* and *gallons.* These units are used when measuring length, weight, and capacity.

- A newborn baby is 20 inches long.
- The distance from St. Louis to Memphis is 285 miles.
- First-class postage for a letter that weighs less than 1 ounce is 33¢.
- The largest pumpkin ever grown weighed 1,092 pounds.
- Milk is sold in quart and gallon containers.

American units of length

A ruler is one of the most common devices used for measuring distances or lengths. Figure 8-6 shows only a portion of a ruler; most rulers are 12 inches (1 foot) long. Since 12 inches = 1 foot, a ruler is divided into 12 equal distances of 1 inch. Each inch is divided into halves of an inch, quarters of an inch, eighths of an inch, and sixteenths of an inch. Several distances are measured using the ruler shown in Figure 8-6.

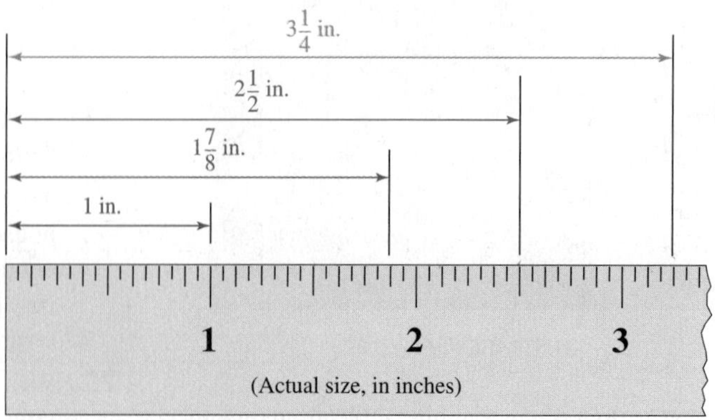

FIGURE 8-6

EXAMPLE 1 ***Measuring the length of a nail.*** To the nearest $\frac{1}{4}$ inch, find the length of the nail in Figure 8-7.

Solution

We place the end of the ruler by one end of the nail and note that the other end of the nail is closer to the $2\frac{1}{2}$-inch mark than to the $2\frac{1}{4}$-inch mark on the ruler. To the nearest quarter-inch, the nail is $2\frac{1}{2}$ inches long.

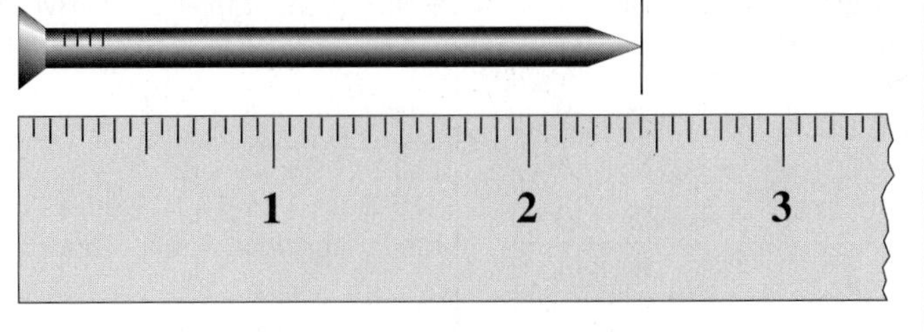

FIGURE 8-7

Self Check

To the nearest $\frac{1}{4}$ inch, find the width of the circle below.

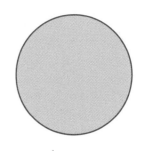

Answer: $1\frac{1}{4}$ in. ■

EXAMPLE 2 ***Measuring the length of a paper clip.*** To the nearest $\frac{1}{8}$ inch, find the length of the paper clip in Figure 8-8.

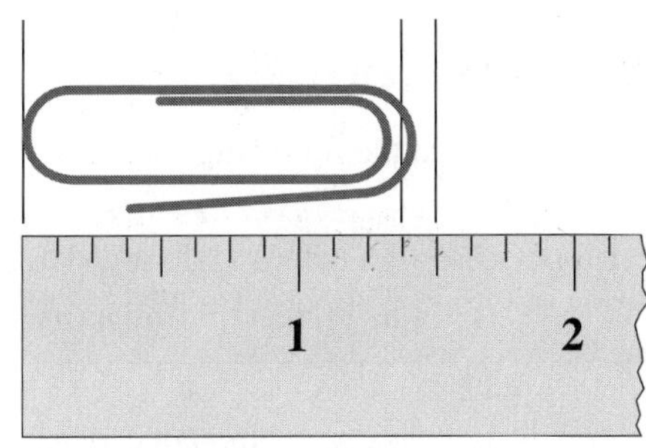

FIGURE 8-8

Solution

We place the end of the ruler by one end of the paper clip and note that the other end is closer to the $1\frac{3}{8}$-inch mark than to the $1\frac{1}{2}$-inch mark on the ruler. To the nearest eighth of an inch, the paper clip is $1\frac{3}{8}$ inches long.

Self Check

To the nearest $\frac{1}{8}$ inch, find the length of the jumbo paper clip below.

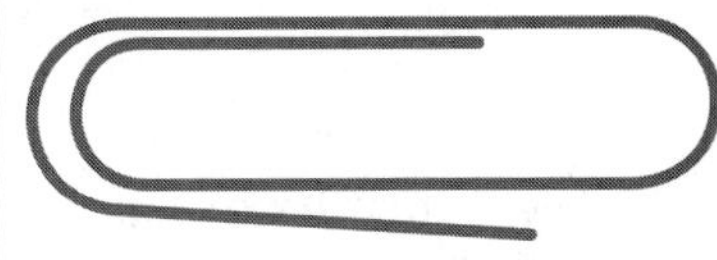

Answer: $1\frac{7}{8}$ in. ■

Each point on a ruler, like each point on a number line, has a number associated with it: the distance between the point and 0. As Example 3 illustrates, the distance between any two points on a ruler (or on a number line) is the difference of the numbers associated with those points.

EXAMPLE 3 ***Measuring the length of a ticket.*** Find the length of the tear-off part of the concert ticket in Figure 8-9.

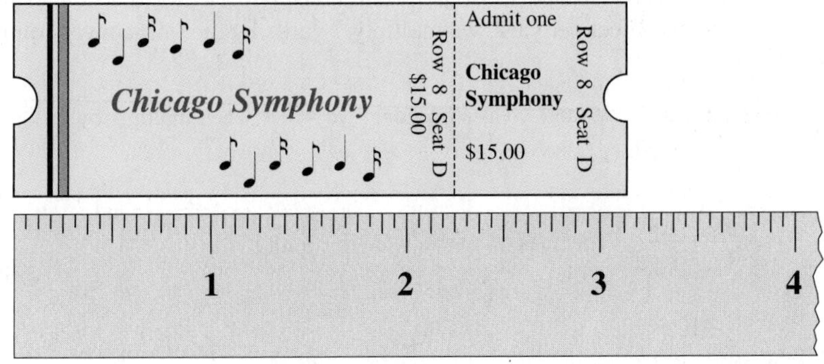

FIGURE 8-9

Solution After placing the end of the ruler by one end of the ticket, we note that the length of the entire ticket is $3\frac{1}{8}$ inches and that the length of the longer part is $2\frac{1}{4}$ inches. Since the length of the tear-off part of the ticket is the *difference* between these two lengths, we must subtract $2\frac{1}{4}$ from $3\frac{1}{8}$.

$3\frac{1}{8} - 2\frac{1}{4} = \frac{25}{8} - \frac{9}{4}$ Change the mixed numbers to improper fractions.

$= \frac{25}{8} - \frac{9 \cdot 2}{4 \cdot 2}$ Write the fractions with a common denominator.

$= \frac{25}{8} - \frac{18}{8}$ Do the multiplication in the numerator and in the denominator.

$= \frac{25 - 18}{8}$ Subtract the fractions.

$= \frac{7}{8}$ Simplify.

The length of the tear-off portion of the ticket is $\frac{7}{8}$ inch. ■

Converting units of length

American units of length are related in the following ways.

American units of length

12 inches (in.) = 1 foot (ft)	36 inches = 1 yard (yd)
3 feet = 1 yard	5,280 feet = 1 mile (mi)

To convert from one unit to another, we use *unit conversion factors.* To find the unit conversion factor between yards and feet, we begin with this fact:

3 ft = 1 yd

If we divide both sides of this equation by 1 yard, we get

$$\frac{3 \text{ ft}}{\mathbf{1 \text{ yd}}} = \frac{1 \text{ yd}}{\mathbf{1 \text{ yd}}}$$

$\frac{3 \text{ ft}}{1 \text{ yd}} = 1$ A number divided by itself is 1: $\frac{1 \text{ yd}}{1 \text{ yd}} = 1$.

The fraction $\frac{3 \text{ ft}}{1 \text{ yd}}$ is called a **unit conversion factor,** because its value is 1. It can be read as "3 feet per yard." Since this fraction is equal to 1, multiplying a length by this fraction does not change its measure; it only changes the *units* of measure.

EXAMPLE 4 ***Converting from yards to feet.*** Convert 7 yards to feet.

Solution

To convert from yards to feet, we must use a unit conversion factor that relates feet to yards. Since there are 3 feet per yard, we multiply 7 yards by the unit conversion factor $\frac{3 \text{ ft}}{1 \text{ yd}}$ to get

$7 \text{ yd} = \frac{7 \text{ yd}}{1} \cdot \frac{\mathbf{3 \text{ ft}}}{\mathbf{1 \text{ yd}}}$ Write 7 yd as a fraction: 7 yd $= \frac{7 \text{ yd}}{1}$. Then multiply by 1: $\frac{3 \text{ ft}}{1 \text{ yd}} = 1$.

$= \frac{7 \cancel{\text{yd}}^{1}}{1} \cdot \frac{3 \text{ ft}}{1 \cancel{\text{yd}}_{1}}$ The units of yards divide out.

$= 7 \cdot 3 \text{ ft}$

$= 21 \text{ ft}$ Multiply: $7 \cdot 3 = 21$.

Seven yards is equal to 21 feet.

Self Check

Convert 9 yards to feet.

Answer: 27 ft ■

Notice that in Example 4, we eliminated the units of yards and introduced the units of feet by multiplying by the appropriate unit conversion factor. In general, a unit conversion factor is a fraction with the following form:

$$\frac{\text{Unit we want to introduce}}{\text{Unit we want to eliminate}} \quad \begin{array}{l}\leftarrow \text{Numerator} \\ \leftarrow \text{Denominator}\end{array}$$

EXAMPLE 5 ***Converting from feet to inches.*** Convert $1\frac{3}{4}$ feet to inches.

Solution

To convert from feet to inches, we must use a unit conversion factor that relates inches to feet. Since there are 12 inches per foot, we multiply $1\frac{3}{4}$ feet by the unit conversion factor $\frac{12 \text{ in.}}{1 \text{ ft}}$ to get

$$1\frac{3}{4} \text{ ft} = \frac{7}{4} \text{ ft} \cdot \frac{\textbf{12 in.}}{\textbf{1 ft}}$$ Write $1\frac{3}{4}$ as an improper fraction: $1\frac{3}{4} = \frac{7}{4}$. Multiply by 1: $\frac{12 \text{ in.}}{1 \text{ ft}} = 1$.

$$= \frac{7}{4} \overset{1}{\cancel{\text{ft}}} \cdot \frac{12 \text{ in.}}{1 \underset{1}{\cancel{\text{ft}}}}$$ The units of feet divide out.

$$= \frac{7 \cdot 12}{4 \cdot 1} \text{ in.}$$ Multiply the fractions.

$$= 21 \text{ in.}$$ Simplify by dividing out the common factors: $\frac{7 \cdot 12}{4 \cdot 1} = \frac{7 \cdot 3 \cdot \overset{1}{\cancel{4}}}{\underset{1}{\cancel{4}} \cdot 1} = 7 \cdot 3 = 21$.

$1\frac{3}{4}$ feet is equal to 21 inches.

Self Check

Convert 1.5 feet to inches.

Answer: 18 in.

Sometimes we must use two unit conversion factors in combination to eliminate the given units while introducing the desired units. The following example illustrates this concept.

Accent on Technology: **Finding the length of a football field in miles**

A football field (including the end zones) is 120 yards long. To find this distance in miles, we set up the problem so that the units of yards divide out and leave us with units of miles. Since there are 3 feet per yard and 5,280 feet per mile, we multiply 120 yards by $\frac{3 \text{ ft}}{1 \text{ yd}}$ and $\frac{1 \text{ mi}}{5{,}280 \text{ ft}}$.

$$120 \text{ yd} = 120 \text{ yd} \cdot \frac{3 \text{ ft}}{1 \text{ yd}} \cdot \frac{1 \text{ mi}}{5{,}280 \text{ ft}}$$ Use two unit conversion factors: $\frac{3 \text{ ft}}{1 \text{ yd}} = 1$ and $\frac{1 \text{ mi}}{5{,}280 \text{ ft}} = 1$.

$$= \frac{120 \overset{1}{\cancel{\text{yd}}}}{1} \cdot \frac{3 \overset{1}{\cancel{\text{ft}}}}{1 \underset{1}{\cancel{\text{yd}}}} \cdot \frac{1 \text{ mi}}{5{,}280 \underset{1}{\cancel{\text{ft}}}}$$ Divide out the units of yards and feet.

$$= \frac{120 \cdot 3}{5{,}280} \text{ mi}$$ Multiply the fractions.

We can do this arithmetic using a scientific calculator by entering these numbers and pressing these keys.

Keystrokes 120 [×] 3 [÷] 5280 [=] 0.0681818

To the nearest hundredth, a football field is 0.07 mile long.

American units of weight

American units of weight are related in the following ways.

American units of weight

16 ounces (oz) = 1 pound (lb)
2,000 pounds = 1 ton

To convert units of weight, we use the following unit conversion factors.

To convert from	Use the unit conversion factor	To convert from	Use the unit conversion factor
pounds to ounces	$\frac{16\text{ oz}}{1\text{ lb}}$	ounces to pounds	$\frac{1\text{ lb}}{16\text{ oz}}$
tons to pounds	$\frac{2{,}000\text{ lb}}{1\text{ ton}}$	pounds to tons	$\frac{1\text{ ton}}{2{,}000\text{ lb}}$

EXAMPLE 6 ***Converting from ounces to pounds.*** Convert 40 ounces to pounds.

Solution

Since there is 1 pound per 16 ounces, we multiply 40 ounces by the unit conversion factor $\frac{1\text{ lb}}{16\text{ oz}}$ to get

$$40\text{ oz} = \frac{40\text{ oz}}{1} \cdot \frac{\mathbf{1\ lb}}{\mathbf{16\ oz}}$$ Write 40 oz as a fraction: 40 oz = $\frac{40\text{ oz}}{1}$. Then multiply by 1: $\frac{1\text{ lb}}{16\text{ oz}} = 1$.

$$= \frac{40\,\overset{1}{\cancel{\text{oz}}}}{1} \cdot \frac{1\text{ lb}}{16\,\underset{1}{\cancel{\text{oz}}}}$$ The units of ounces divide out.

$$= \frac{40}{16}\text{ lb}$$ Multiply the fractions.

There are two ways to complete the solution. First, we can divide out the common factors of the numerator and denominator and then write the result as a mixed number.

$$\frac{40}{16}\text{ lb} = \frac{\overset{1}{\cancel{8}} \cdot 5}{\underset{1}{\cancel{8}} \cdot 2}\text{ lb} = \frac{5}{2}\text{ lb} = 2\frac{1}{2}\text{ lb}$$

A second approach is to divide the numerator by the denominator and express the result as a decimal.

$$\frac{40}{16}\text{ lb} = 2.5\text{ lb}$$ Do the division: $40 \div 16 = 2.5$.

Forty ounces is equal to $2\frac{1}{2}$ lb (or 2.5 lb).

Self Check

Convert 60 ounces to pounds.

Answer: $3\frac{3}{4}$ lb = 3.75 lb ■

EXAMPLE 7 ***Converting from pounds to ounces.*** Convert 25 pounds to ounces.

Solution

Since there are 16 ounces per pound, we multiply 25 pounds by the unit conversion factor $\frac{16\text{ oz}}{1\text{ lb}}$ to get

Self Check

Convert 60 pounds to ounces.

$$25 \text{ lb} = \frac{25 \text{ lb}}{1} \cdot \frac{\mathbf{16 \text{ oz}}}{\mathbf{1 \text{ lb}}}$$ Multiply by 1: $\frac{16 \text{ oz}}{1 \text{ lb}} = 1$.

$$= \frac{25 \cancel{\text{lb}}^{1}}{1} \cdot \frac{16 \text{ oz}}{1 \cancel{\text{lb}}_{1}}$$ The units of pounds divide out.

$$= 25 \cdot 16 \text{ oz}$$

$$= 400 \text{ oz}$$ Multiply: $25 \cdot 16 = 400$.

Twenty-five pounds is equal to 400 ounces.

Answer: 960 oz ■

Accent on Technology: *Finding the weight of a car in pounds*

A BMW 323Ci convertible weighs 1.78 tons. To find its weight in pounds, we set up the problem so that the units of tons divide out and leave us with pounds. Since there are 2,000 pounds per ton, we multiply by $\frac{2{,}000 \text{ lb}}{1 \text{ ton}}$.

$$1.78 \text{ tons} = \frac{1.78 \text{ tons}}{1} \cdot \frac{2{,}000 \text{ lb}}{1 \text{ ton}}$$ Multiply by 1: $\frac{2{,}000 \text{lb}}{1 \text{ ton}} = 1$.

$$= \frac{1.78 \cancel{\text{tons}}^{1}}{1} \cdot \frac{2{,}000 \text{ lb}}{1 \cancel{\text{ton}}_{1}}$$ Divide out the units of tons.

$$= 1.78 \cdot 2{,}000 \text{ lb}$$

We can do this multiplication using a scientific calculator by entering these numbers and pressing these keys.

Keystrokes 1.78 [×] 2000 [=] **3560**

The convertible weighs 3,560 pounds.

American units of capacity

American units of capacity are related as follows.

American units of capacity

1 cup (c) = 8 fluid ounces (fl oz)	1 pint (pt) = 2 cups (c)
1 quart (qt) = 2 pints (pt)	1 gallon (gal) = 4 quarts (qt)

To convert units of capacity, we use the following unit conversion factors.

To convert from	Use the unit conversion factor	To convert from	Use the unit conversion factor
cups to ounces	$\frac{8 \text{ fl oz}}{1 \text{ c}}$	ounces to cups	$\frac{1 \text{ c}}{8 \text{ fl oz}}$
pints to cups	$\frac{2 \text{ c}}{1 \text{ pt}}$	cups to pints	$\frac{1 \text{ pt}}{2 \text{ c}}$
quarts to pints	$\frac{2 \text{ pt}}{1 \text{ qt}}$	pints to quarts	$\frac{1 \text{ qt}}{2 \text{ pt}}$
gallons to quarts	$\frac{4 \text{ qt}}{1 \text{ gal}}$	quarts to gallons	$\frac{1 \text{ gal}}{4 \text{ qt}}$

EXAMPLE 8 ***Using two unit conversion factors.*** If a recipe calls for 3 pints of milk, how many fluid ounces of milk should be used?

Solution

Since there are 2 cups per pint and 8 fluid ounces per cup, we multiply 3 pints by unit conversion factors of $\frac{2\text{ c}}{1\text{ pt}}$ and $\frac{8\text{ fl oz}}{1\text{ c}}$.

$$3\text{ pt} = \frac{3\text{ pt}}{1} \cdot \frac{\mathbf{2\ c}}{\mathbf{1\ pt}} \cdot \frac{\mathbf{8\ fl\ oz}}{\mathbf{1\ c}}$$ Use two unit conversion factors: $\frac{2\text{ c}}{1\text{ pt}} = 1$ and $\frac{8\text{ fl oz}}{1\text{ c}} = 1$.

$$= \frac{3\ \cancel{\text{pt}}}{1} \cdot \frac{2\ \cancel{\text{c}}}{1\ \cancel{\text{pt}}} \cdot \frac{8\text{ fl oz}}{1\ \cancel{\text{c}}}$$ Divide out the units of pints and cups.

$$= 3 \cdot 2 \cdot 8\text{ fl oz}$$

$$= 48\text{ fl oz}$$

Since 3 pints is equal to 48 fluid ounces, 48 fluid ounces of milk should be used.

Self Check

How many pints are in 1 gallon?

Answer: 8 pt

Units of time

Units of time are related in the following ways.

Units of time

1 minute (min) = 60 seconds (sec)	1 hour (hr) = 60 minutes
1 day = 24 hours	

EXAMPLE 9 ***Astronomy.*** A lunar eclipse occurs when the Earth is between the sun and the moon in such a way that the Earth's shadow darkens the moon. See Figure 8-10 (which is not to scale). A total lunar eclipse can last as long as 105 minutes. How many hours is this?

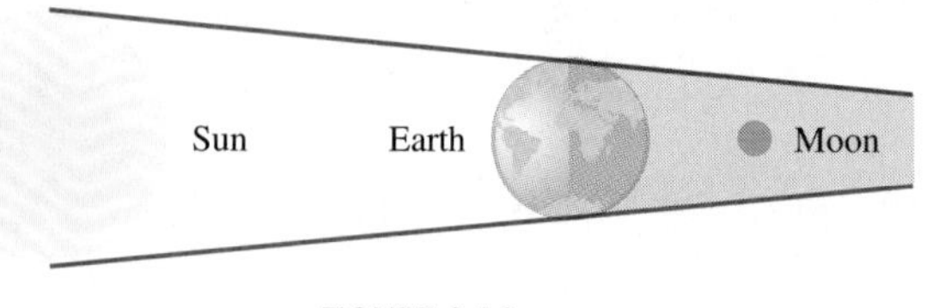

FIGURE 8-10

Solution

Since there is 1 hour for every 60 minutes, we multiply 105 by the unit conversion factor $\frac{1\text{ hr}}{60\text{ min}}$ to get

$$105\text{ min} = \frac{105\text{ min}}{1} \cdot \frac{\mathbf{1\ hr}}{\mathbf{60\ min}}$$ Multiply by 1: $\frac{1\text{ hr}}{60\text{ min}} = 1$.

$$= \frac{105\ \cancel{\text{min}}}{1} \cdot \frac{1\text{ hr}}{60\ \cancel{\text{min}}}$$ The units of minutes divide out.

$$= \frac{105}{60}\text{ hr}$$ Multiply the fractions.

$$= \frac{7 \cdot \overset{1}{\cancel{5}} \cdot \overset{1}{\cancel{3}}}{\underset{1}{\cancel{5}} \cdot \underset{1}{\cancel{3}} \cdot 2 \cdot 2}\text{ hr}$$ Prime factor 105 and 60. Then divide out common factors of the numerator and denominator.

$$= \frac{7}{4}\text{ hr}$$

$$= 1\tfrac{3}{4}\text{ hr}$$ Write $\frac{7}{4}$ as a mixed number.

A total lunar eclipse can last as long as $1\frac{3}{4}$ hours.

Self Check

A solar eclipse (eclipse of the sun) can last as long as 450 seconds. How many minutes is this?

Answer: $7\frac{1}{2}$ min

STUDY SET Section 8.3

VOCABULARY *Fill in the blanks to make the statements true.*

1. Inches, feet, and miles are examples of American units of ______.

2. A ruler is used for measuring ______.

3. The value of any unit conversion factor is ______.

4. Ounces, pounds, and tons are examples of American units of ______.

5. Some examples of American units of ______ are cups, pints, quarts, and gallons.

6. Some units of ______ are seconds, hours, and days.

CONCEPTS *Fill in the blanks to make the statements true.*

7. 12 in. = ___ ft

8. ___ ft = 1 yd

9. 1 mi = ___ ft

10. 1 yd = ___ in.

11. ___ ounces = 1 pound

12. ___ pounds = 1 ton

13. 1 cup = ___ fluid ounces

14. 1 pint = ___ cups

15. 2 pints = ___ quart(s)

16. 4 quarts = ___ gallon

17. 1 day = ___ hours

18. 2 hours = ___ minutes

19. Tell which measurements the arrows point to on the ruler in Illustration 1.

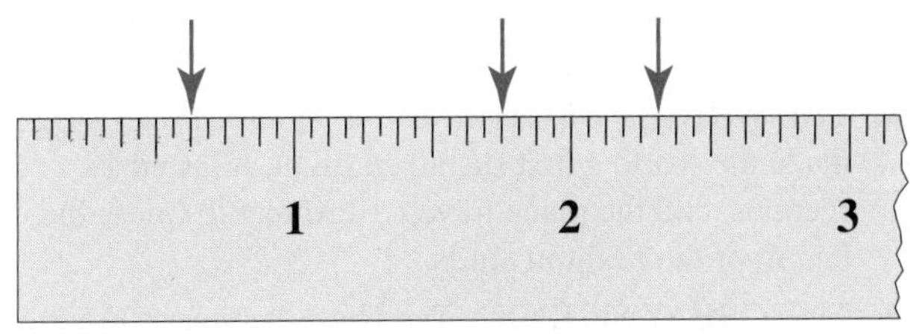

ILLUSTRATION 1

20. Tell which measurements the arrows point to on the ruler in Illustration 2, to the nearest $\frac{1}{8}$ inch.

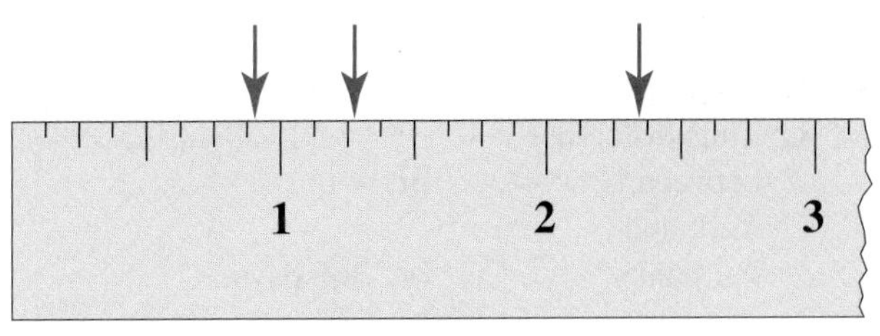

ILLUSTRATION 2

21. Write a unit conversion factor to convert the following.

a. Pounds to tons

b. Quarts to pints

22. Write the two unit conversion factors used to convert the following.

a. Inches to yards

b. Days to minutes

23. Match each item with its proper measurement.

a. Length of the U.S. coastline | **i.** $11\frac{1}{2}$ in.

b. Height of a Barbie doll | **ii.** 4,200 ft

c. Span of the Golden Gate Bridge | **iii.** 53.5 yd

d. Width of a football field | **iv.** 12,383 mi

24. Match each item with its proper measurement.

a. Weight of the men's shot put used in track and field | **i.** $1\frac{1}{2}$ oz

b. Weight of an African elephant | **ii.** 16 lb

c. Amount of gold that is worth $500 | **iii.** 7.2 tons

25. Match each item with its proper measurement.

a. Amount of blood in an adult | **i.** $\frac{1}{2}$ fluid oz

b. Size of the Exxon Valdez oil spill in 1989 | **ii.** 2 cups

c. Amount of nail polish in a bottle | **iii.** 5 qt

d. Amount of flour to make 3 dozen cookies | **iv.** 10,080,000 gal

26. Match each item with its proper measurement.

a. Length of first U.S. manned space flight	**i.** 12 sec
b. A leap year	**ii.** 15 min
c. Time difference between New York and Fairbanks, Alaska	**iii.** 4 hr
d. Length of Wright Brothers' first flight	**iv.** 366 days

NOTATION *Complete each solution.*

27. Convert 12 yards to inches.

$$12 \text{ yd} = 12 \text{ yd} \cdot \frac{36 \text{ in.}}{1 \text{ yd}}$$
$$= 12 \cdot 36 \text{ in.}$$
$$= 432 \text{ in.}$$

28. Convert 1 ton to ounces.

$$1 \text{ ton} = 1 \text{ ton} \cdot \frac{2000 \text{ lb}}{1 \text{ ton}} \cdot \frac{16 \text{ oz}}{1 \text{ lb}}$$
$$= 1 \cdot 2{,}000 \cdot 16 \text{ oz}$$
$$= 32{,}000 \text{ oz}$$

29. Convert 12 pints to gallons.

$$12 \text{ pt} = 12 \text{ pt} \cdot \frac{1 \text{ qt}}{2 \text{ pt}} \cdot \frac{1 \text{ gal}}{4 \text{ qt}}$$
$$= 12 \cdot \frac{1}{2} \cdot \frac{1}{4} \text{ gal}$$
$$= 1.5 \text{ gal}$$

30. Convert 37,440 minutes to days.

$$37{,}440 \text{ min} = 37{,}440 \text{ min} \cdot \frac{1 \text{ hr}}{60 \text{ min}} \cdot \frac{1 \text{ day}}{24 \text{ hr}}$$
$$= \frac{37440}{60 \cdot 24} \text{ days}$$
$$= 26 \text{ days}$$

PRACTICE *Use a ruler with a scale in inches to measure each object to the nearest $\frac{1}{8}$ inch.*

31. The width of a dollar bill

32. The length of a dollar bill

33. The length (top to bottom) of this page

34. The length of the following word: supercalifragilisticexpialidocious

Do each conversion.

35. 4 feet to inches

36. 7 feet to inches

37. $3\frac{1}{2}$ feet to inches

38. $2\frac{2}{3}$ feet to inches

39. 24 inches to feet

40. 54 inches to feet

41. 8 yards to inches

42. 288 inches to yards

43. 90 inches to yards

44. 12 yards to inches

45. 56 inches to feet

46. 44 inches to feet

47. 5 yards to feet

48. 21 feet to yards

49. 7 feet to yards

50. $4\frac{2}{3}$ yards to feet

51. 15,840 feet to miles

52. 2 miles to feet

53. $\frac{1}{2}$ mile to feet

54. 1,320 feet to miles

55. 80 ounces to pounds

56. 8 pounds to ounces

57. 7,000 pounds to tons

58. 2.5 tons to ounces

59. 12.4 tons to pounds

60. 48,000 ounces to tons

61. 3 quarts to pints

62. 20 quarts to gallons

63. 16 pints to gallons

64. 3 gal to fluid ounces

65. 32 fluid ounces to pints

66. 2 quarts to fluid ounces

67. 240 minutes to hours

68. 2,400 seconds to hours

69. 7,200 minutes to days

70. 691,200 seconds to days

APPLICATIONS

71. THE GREAT PYRAMID The Great Pyramid in Egypt is about 450 feet high. Express this distance in yards.

72. THE WRIGHT BROTHERS In 1903, Orville Wright made the world's first sustained flight. It lasted 12 seconds, and the plane traveled 120 feet. Express the length of the flight in yards.

73. THE GREAT SPHINX The Great Sphinx of Egypt is 240 feet long. Express this in inches.

74. HOOVER DAM The Hoover Dam in Nevada is 726 feet high. Express this distance in inches.

75. THE SEARS TOWER The Sears Tower in Chicago has 110 stories and is 1,454 feet tall. To the nearest hundredth, express this height in miles.

76. NFL RECORDS Walter Payton, the Chicago Bears running back, holds the National Football League record for yards rushing in a career: 16,726. How many miles is this? Round to the nearest tenth of a mile.

77. NFL RECORDS When Dan Marino of the Miami Dolphins retired, it was noted that Marino's career passing total was nearly 35 miles! How many yards is this?

78. LEWIS AND CLARK The trail traveled by the Lewis and Clark expedition is shown in Illustration 3. When the expedition reached the Pacific Ocean, Clark estimated that they had traveled 4,162 miles. (It was later determined that his guess was within 40 miles of the actual distance.) Express Clark's estimate of the distance in terms of feet.

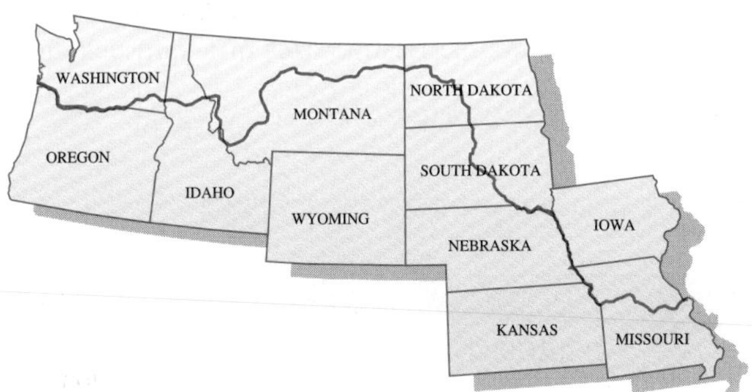

ILLUSTRATION 3

79. WEIGHT OF WATER One gallon of water weighs about 8 pounds. Express this weight in ounces.

80. WEIGHT OF A BABY A newborn baby weighed 136 ounces. Express this weight in pounds.

81. HIPPO An adult hippopotamus can weigh as much as 9,900 pounds. Express this weight in tons.

82. ELEPHANT An adult elephant can consume as much as 495 pounds of grass and leaves in one day. How many ounces is this?

83. BUYING PAINT A painter estimates that he will need 17 gallons of paint for a job. To take advantage of a closeout sale on quart cans, he decides to buy the paint in quarts. How many cans will he need to buy?

84. CATERING How many cups of apple cider can be dispensed from a 10-gallon container of cider?

85. SCHOOL LUNCHES Each student attending Eagle River Elementary School receives one pint of milk for lunch each day. If 575 students attend the school, how many gallons of milk are used each day?

86. RADIATOR The radiator capacity of a piece of earth-moving equipment is 39 quarts. If the radiator is drained and new coolant put in, how many gallons of new coolant will be used?

87. CAMPING How many ounces of camping stove fuel will fit in the container shown in Illustration 4?

ILLUSTRATION 4

88. HIKING A college student walks 11 miles in 155 minutes. To the nearest tenth, how many hours does he walk?

89. SPACE TRAVEL The astronauts of the Apollo 8 mission, which was launched on December 21, 1968, were in space for 147 hours. How many days did the mission take?

90. AMELIA EARHART In 1935, Amelia Earhart became the first woman to fly across the Atlantic Ocean alone, establishing a new record for the crossing: 13 hours and 30 minutes. How many minutes is this?

WRITING

91. Explain how to find the unit conversion factor that will convert feet to inches.

92. Explain how to find the unit conversion factor that will convert pints to gallons.

REVIEW *Round each number as indicated.*

93. 3,673.263; nearest hundred

94. 3,673.263; nearest ten

95. 3,673.263; nearest hundredth

96. 3,673.263; nearest tenth

97. 0.100602; nearest thousandth

98. 0.100602; nearest hundredth

99. 0.09999; nearest tenth

100. 0.09999; nearest one

8.4 Metric Units of Measurement

In this section, you will learn about

- Metric units of length • Converting units of length
- Metric units of mass • Metric units of capacity
- Cubic centimeters

INTRODUCTION. The metric system is the system of measurement used by most countries in the world. All countries, including the United States, use it for scientific purposes. The metric system, like our decimal numeration system, is based on the number 10. For this reason, converting from one metric unit to another is easier than with the American system.

Metric units of length

The basic metric unit of length is the **meter** (m). One meter is approximately 39 inches, slightly more than 1 yard. Figure 8-11 shows the relative sizes of a yardstick and a meterstick.

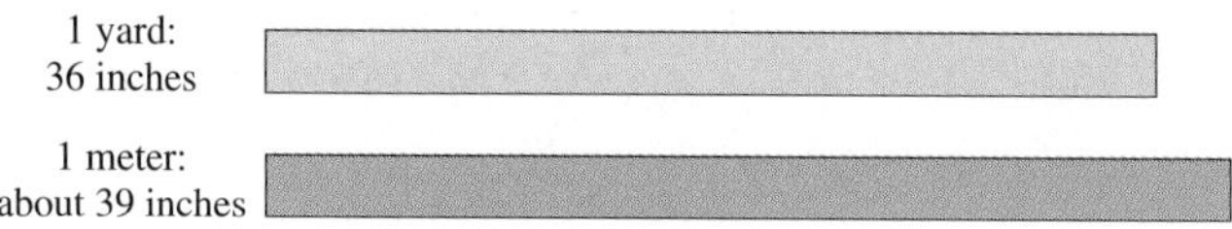

FIGURE 8-11

Larger and smaller units are designated by prefixes in front of this basic unit, *meter*.

deka means tens	*deci* means tenths
hecto means hundreds	*centi* means hundredths
kilo means thousands	*milli* means thousandths

Metric units of length

1 dekameter (dam) = 10 meters. 1 dam is a little less than 11 yards.	**1 decimeter** (dm) = $\frac{1}{10}$ of 1 meter. 1 dm is about the length of your palm.
1 hectometer (hm) = 100 meters. 1 hm is about 1 football field long, plus one end zone.	**1 centimeter** (cm) = $\frac{1}{100}$ of 1 meter. 1 cm is about as wide as the nail of your little finger.
1 kilometer (km) = 1,000 meters. 1 km is about $\frac{3}{5}$ mile.	**1 millimeter** (mm) = $\frac{1}{1,000}$ of 1 meter. 1 mm is about the thickness of a dime.

TABLE 8-1

Figure 8-12 shows a portion of a metric ruler, scaled in centimeters, and a ruler scaled in inches. The rulers are used to measure several lengths.

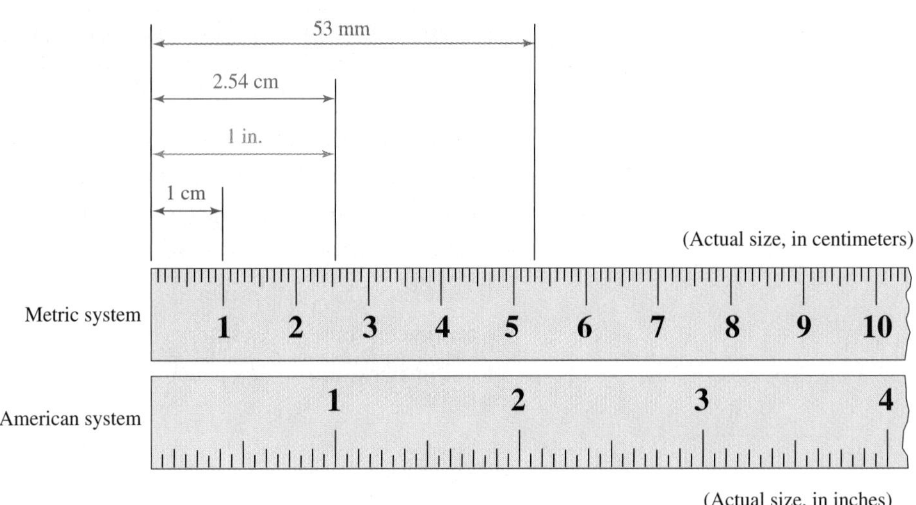

FIGURE 8-12

EXAMPLE 1 ***Measuring the length of a nail.*** To the nearest centimeter, find the length of the nail in Figure 8-13.

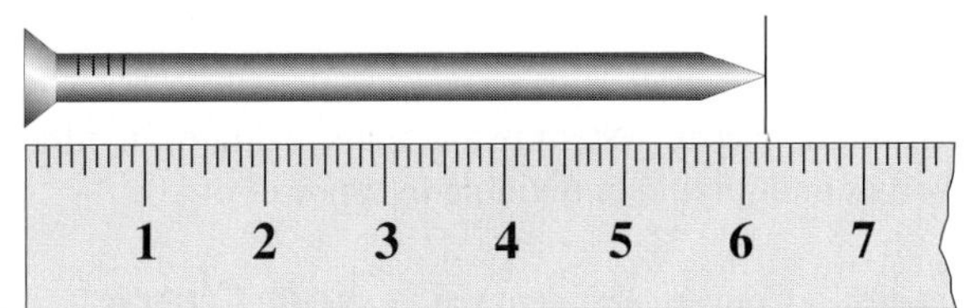

FIGURE 8-13

Solution

We place the end of the ruler by one end of the nail and note that the other end of the nail is closer to the 6-cm mark than to the 7-cm mark on the ruler. To the nearest centimeter, the nail is 6 cm long.

Self Check

To the nearest centimeter, find the width of the circle below.

Answer: 3 cm ■

EXAMPLE 2 ***Measuring the length of a paper clip.*** To the nearest millimeter, find the length of the paper clip in Figure 8-14.

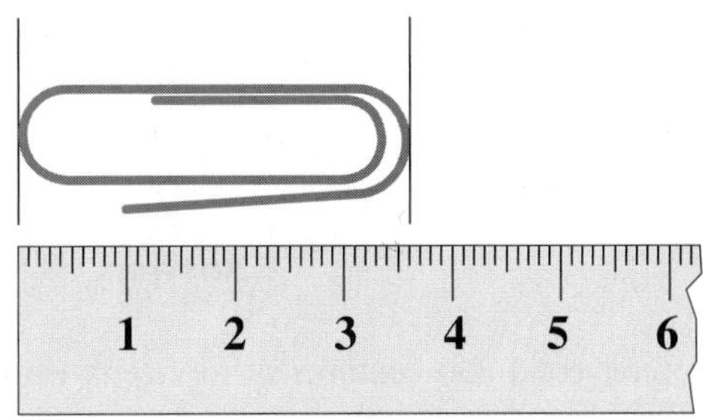

FIGURE 8-14

Solution

On the ruler, each centimeter has been divided into 10 millimeters. We place the end of the ruler by one end of the paper clip and note that the other end is closer to the 36-mm mark than to the 37-mm mark on the ruler. To the nearest millimeter, the paper clip is 36 mm long.

Self Check

To the nearest millimeter, find the length of the jumbo paper clip below.

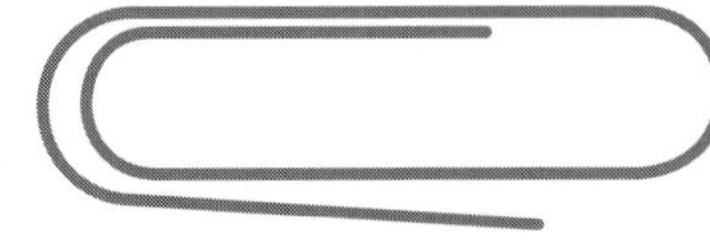

Answer: 47 mm ■

Converting units of length

Metric units of length are related as shown in Table 8-2.

Metric units of length

1 kilometer (km) = 1,000 meters	or	1 meter = $\frac{1}{1,000}$ kilometer
1 hectometer (hm) = 100 meters	or	1 meter = $\frac{1}{100}$ hectometer
1 dekameter (dam) = 10 meters	or	1 meter = $\frac{1}{10}$ dekameter
1 decimeter (dm) = $\frac{1}{10}$ meter	or	1 meter = 10 decimeters
1 centimeter (cm) = $\frac{1}{100}$ meter	or	1 meter = 100 centimeters
1 millimeter (mm) = $\frac{1}{1,000}$ meter	or	1 meter = 1,000 millimeters

TABLE 8-2

We can use the information in the table to write unit conversion factors that can be used to convert metric units of length. For example, in the table we see that

1 meter = 100 centimeters

From this fact, we can write two unit conversion factors:

$$\frac{1 \text{ m}}{100 \text{ cm}} = 1 \quad \text{and} \quad \frac{100 \text{ cm}}{1 \text{ m}} = 1$$

To obtain the first unit conversion factor, divide both sides of the equation 1 m = 100 cm by 100 cm. To obtain the second unit conversion factor, divide both sides by 1 m.

One advantage of the metric system is that multiplying or dividing by a unit conversion factor involves multiplying or dividing by a power of 10.

EXAMPLE 3 ***Changing centimeters to meters.*** Convert 350 centimeters to meters.

Solution

Since there is 1 meter per 100 centimeters, we multiply 350 centimeters by the unit conversion factor $\frac{1 \text{ m}}{100 \text{ cm}}$ to get

$$350 \text{ cm} = \frac{350 \text{ cm}}{1} \cdot \frac{\mathbf{1 \text{ m}}}{\mathbf{100 \text{ cm}}}$$ Multiply by 1: $\frac{1 \text{ m}}{100 \text{ cm}} = 1$.

$$= \frac{350 \, \overset{1}{\cancel{\text{cm}}}}{1} \cdot \frac{1 \text{ m}}{100 \, \underset{1}{\cancel{\text{cm}}}}$$ The units of centimeters divide out.

$$= \frac{350}{100} \text{ m}$$

$$= 3.5 \text{ m}$$ Divide by 100 by moving the decimal point 2 places to the left.

Thus, 350 centimeters = 3.5 meters.

Self Check

Convert 860 centimeters to meters.

Answer: 8.6 m

In Example 3, we converted 350 centimeters to meters using a unit conversion factor. We can also make this conversion by recognizing that all units of length in the metric system are powers of 10 of a meter. Converting from one unit to another is as simple as multiplying by the correct power of 10, or equivalently, by moving a decimal point the correct number of places to the right or left. For example, in the chart below, we see that to convert from centimeters to meters, we move two units to the left.

km hm dam **m** dm **cm** mm

To go from centimeters to meters, we must move 2 places to the left.

If we write 350 centimeters as 350.0 centimeters, we can convert to meters by moving the decimal point two places to the left.

350.0 centimeters = 3.50.0 meters = 3.5 meters

With the unit conversion factor method or the chart method, we get 350 cm = 3.5 m.

COMMENT When using a chart to help make a metric conversion, be sure to list the units from largest to smallest when reading from left to right.

EXAMPLE 4 ***Changing meters to millimeters.*** Convert 2.4 meters to millimeters.

Solution

Since there are 1,000 millimeters per meter, we multiply 2.4 meters by the unit conversion factor $\frac{1{,}000 \text{ mm}}{1 \text{ m}}$ to get

$$2.4 \text{ m} = \frac{2.4 \text{ m}}{1} \cdot \frac{\mathbf{1{,}000 \text{ mm}}}{\mathbf{1 \text{ m}}} \quad \text{Multiply by 1: } \frac{1{,}000 \text{ mm}}{1 \text{ m}} = 1.$$

$$= \frac{2.4 \not{\text{m}}}{1} \cdot \frac{1{,}000 \text{ mm}}{1 \not{\text{m}}} \quad \text{The units of meters divide out.}$$

$$= 2.4 \cdot 1{,}000 \text{ mm}$$

$$= 2{,}400 \text{ mm} \quad \text{Multiply by 1,000 by moving the decimal point 3 places to the right.}$$

Thus, 2.4 meters = 2,400 millimeters.

We can also make this conversion using a chart.

km hm dam **m** dm cm **mm**

From the chart, we see that we should move the decimal point 3 places to the right to convert from meters to millimeters.

2.4 meters = 2.400. millimeters = 2,400 millimeters

Self Check

Convert 5.3 meters to millimeters.

Answer: 5,300 mm ■

EXAMPLE 5 ***Using two unit conversion factors.*** Convert 3.2 kilometers to centimeters.

Solution

To convert to centimeters, we set up the problem so kilometers divide out and leave us with units of centimeters. Since there are 1,000 meters per kilometer and 100 centimeters per meter, we multiply 3.2 kilometers by $\frac{1{,}000 \text{ m}}{1 \text{ km}}$ and $\frac{100 \text{ cm}}{1 \text{ m}}$.

$$3.2 \text{ km} = \frac{3.2 \not{\text{km}}}{1} \cdot \frac{\mathbf{1{,}000} \not{\text{m}}}{\mathbf{1} \not{\text{km}}} \cdot \frac{\mathbf{100 \text{ cm}}}{\mathbf{1} \not{\text{m}}} \quad \text{The units of kilometers and meters divide out.}$$

$$= 3.2 \cdot 1{,}000 \cdot 100 \text{ cm}$$

$$= 320{,}000 \text{ cm} \quad \text{Multiply by 1,000 and 100 by moving the decimal point 5 places to the right.}$$

Thus, 3.2 kilometers = 320,000 centimeters.

Self Check

Convert 5.15 kilometers to centimeters.

Using a chart, we see that the decimal point should be moved 5 places to the right to convert kilometers to centimeters.

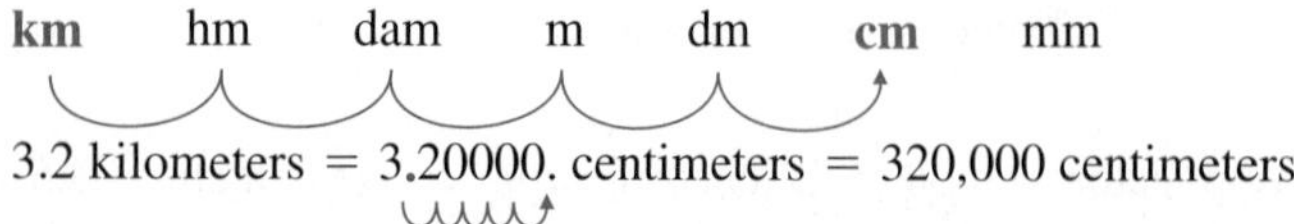

3.2 kilometers = 3.20000. centimeters = 320,000 centimeters

Answer: 515,000 cm

Metric units of mass

The **mass** of an object is a measure of the amount of material in the object. When an object is moved about in space, its mass does not change. One basic unit of mass in the metric system is the **gram (g).** A gram is defined to be the mass of water contained in a cube having sides 1 centimeter long. (See Figure 8-15.)

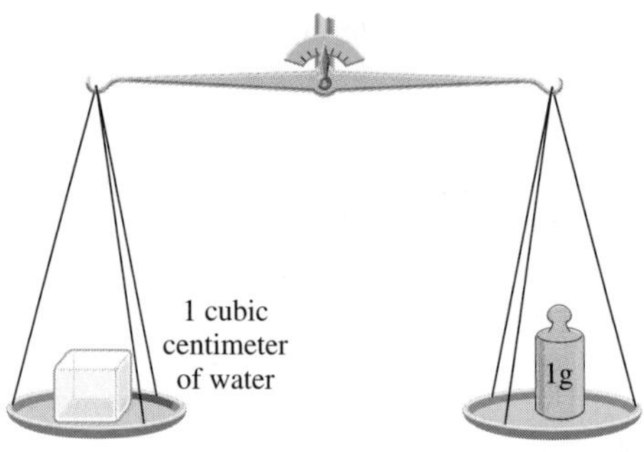

FIGURE 8-15

The **weight** of an object is determined by the Earth's gravitational pull on the object. Since gravitational pull on an object decreases as the object gets farther from Earth, the object weighs less as it gets farther from the Earth's surface. This is why astronauts experience weightlessness in space. However, since most of us remain near the Earth's surface, we will use the words *mass* and *weight* interchangeably. Thus, a mass of 30 grams is said to weigh 30 grams.

Metric units of mass are related as shown in Table 8-3.

Metric units of mass

1 kilogram (kg) = 1,000 grams	or	1 gram = $\frac{1}{1,000}$ kilogram
1 hectogram (hg) = 100 grams	or	1 gram = $\frac{1}{100}$ hectogram
1 dekagram (dag) = 10 grams	or	1 gram = $\frac{1}{10}$ dekagram
1 decigram (dg) = $\frac{1}{10}$ gram	or	1 gram = 10 decigrams
1 centigram (cg) = $\frac{1}{100}$ gram	or	1 gram = 100 centigrams
1 milligram (mg) = $\frac{1}{1,000}$ gram	or	1 gram = 1,000 milligrams

TABLE 8-3

Here are some examples of these units of mass:

- An average bowling ball weighs about 6 kilograms.
- A raisin weighs about 1 gram.
- A certain vitamin tablet contains 450 milligrams of calcium.

We can use the information in Table 8-3 to write unit conversion factors that can be used to convert metric units of mass. For example, in the table we see that

1 kilogram = 1,000 grams

From this fact, we can write two unit conversion factors:

$$\frac{1\text{ kg}}{1{,}000\text{ g}} = 1 \quad \text{and} \quad \frac{1{,}000\text{ g}}{1\text{ kg}} = 1$$

To obtain the first unit conversion factor, divide both sides of the equation 1 kg = 1,000 g by 1,000 g. To obtain the second unit conversion factor, divide both sides by 1 kg.

EXAMPLE 6 ***Changing kilograms to grams.*** Convert 7.2 kilograms to grams.

Solution

To convert to grams, we set up the problem so that the units of kilograms divide out and leave us with the units of grams. Since there are 1,000 grams per 1 kilogram, we multiply 7.2 kilograms by $\frac{1{,}000\text{ g}}{1\text{ kg}}$.

$$7.2\text{ kg} = \frac{7.2\text{ }\cancel{\text{kg}}}{1} \cdot \frac{1{,}000\text{ g}}{1\text{ }\cancel{\text{kg}}}$$ Divide out the units of kilograms.

$$= 7.2 \cdot 1{,}000\text{ g}$$

$$= 7{,}200\text{ g}$$ Do the multiplication by moving the decimal point 3 places to the right.

Thus, 7.2 kilograms = 7,200 grams.

To use a chart to make the conversion, we list the metric units of weight from the largest (kilograms) to the smallest (milligrams).

kg hg dag g dg cg mg

From the chart, we see that we must move the decimal point 3 places to the right to change kilograms to grams.

7.2 kilograms = 7,200. grams = 7,200 grams

Self Check

Convert 5 kilograms to grams.

Answer: 5,000 g

EXAMPLE 7 ***Changing from milligrams to centigrams.*** A bottle of Verapamil, a drug taken for high blood pressure, contains 30 tablets. If each tablet contains 180 mg of active ingredient, how many centigrams of active ingredient are in the bottle?

Solution

Since there are 30 tablets and each one contains 180 mg of active ingredient, there are

$$30 \cdot 180\text{ mg} = 5{,}400\text{ mg}$$

of active ingredient in the bottle.

To convert milligrams to centigrams, we multiply 5,400 milligrams by $\frac{1\text{g}}{1{,}000\text{ mg}}$ and $\frac{100\text{ cg}}{1\text{ g}}$ to get

$$5{,}400\text{ mg} = \frac{5{,}400\text{ }\cancel{\text{mg}}}{1} \cdot \frac{1\text{ }\cancel{\text{g}}}{1{,}000\text{ }\cancel{\text{mg}}} \cdot \frac{100\text{ cg}}{1\text{ }\cancel{\text{g}}}$$ Divide out the units of milligrams and grams.

$$= \frac{5{,}400 \cdot 100}{1{,}000}\text{ cg}$$ Multiply the fractions.

$$= 540\text{ cg}$$ Simplify.

There are 540 centigrams of active ingredient in the bottle.

Using a chart, we see that we must move the decimal point 1 place to the left to convert from milligrams to centigrams.

kg hg dag g dg cg mg

5,400 milligrams = 540.0. centigrams = 540 centigrams

Self Check

One brand name for Verapamil is Isoptin. If a bottle of Isoptin contains 90 tablets, each containing 200 mg of active ingredient, how many centigrams of active ingredient are in the bottle?

Answer: 1,800 cg

Metric units of capacity

In the metric system, one basic unit of capacity is the **liter** (L), which is defined to be the capacity of a cube with sides 10 centimeters long. (See Figure 8-16.) A liter of liquid is slightly more than one quart.

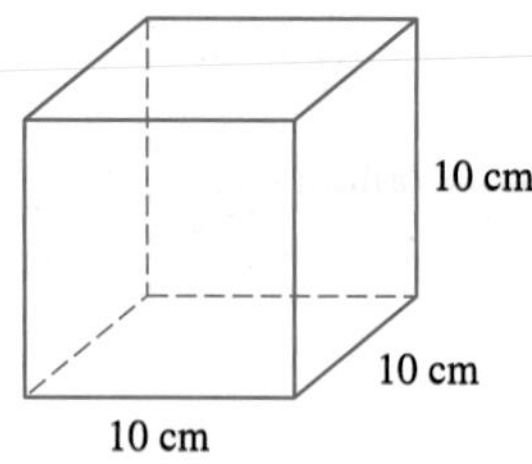

FIGURE 8-16

Metric units of capacity are related as shown in Table 8-4.

Metric units of capacity

1 kiloliter (kL) = 1,000 liters	or	1 liter = $\frac{1}{1,000}$ kiloliter
1 hectoliter (hL) = 100 liters	or	1 liter = $\frac{1}{100}$ hectoliter
1 dekaliter (daL) = 10 liters	or	1 liter = $\frac{1}{10}$ dekaliter
1 deciliter (dL) = $\frac{1}{10}$ liter	or	1 liter = 10 deciliters
1 centiliter (cL) = $\frac{1}{100}$ liter	or	1 liter = 100 centiliters
1 milliliter (mL) = $\frac{1}{1,000}$ liter	or	1 liter = 1,000 milliliters

TABLE 8-4

Here are some examples of these units of capacity:

- Soft drinks are sold in 2-liter plastic bottles.
- The fuel tank of a certain minivan can hold about 75 liters of gasoline.
- Chemists use glass cylinders, scaled in milliliters, to measure liquids.

We can use the information in Table 8-4 to write unit conversion factors that can be used to convert metric units of capacity. For example, in the table we see that

1 liter = 100 centiliters

From this fact, we can write two unit conversion factors:

$$\frac{1\text{ L}}{100\text{ cL}} = 1 \quad \text{and} \quad \frac{100\text{ cL}}{1\text{ L}} = 1$$

EXAMPLE 8 ***Changing from liters to centiliters.*** How many centiliters are there in three 2-liter bottles of cola?

Solution

Three 2-liter bottles of cola contain 6 liters of cola. To convert to centiliters, we set up the problem so that liters divide out and leave us with centiliters. Since there are 100 centiliters per 1 liter, we multiply 6 liters by the unit conversion factor $\frac{100\text{ cL}}{1\text{ L}}$.

$$6\text{ L} = 6\text{ L} \cdot \frac{\mathbf{100\text{ cL}}}{\mathbf{1\text{ L}}} \quad \text{Multiply by 1: } \frac{100\text{ cL}}{1\text{ L}} = 1.$$

$$= \frac{6\,\overset{1}{\cancel{\text{L}}}}{1} \cdot \frac{100\text{ cL}}{1\,\underset{1}{\cancel{\text{L}}}} \quad \text{The units of liters divide out.}$$

$$= 6 \cdot 100\text{ cL}$$

$$= 600\text{ cL}$$

Self Check

How many milliliters are in two 2-liter bottles of cola?

Thus, there are 600 centiliters in three 2-liter bottles of cola.

To make this conversion using a chart, we list the metric units of capacity in order from largest (kiloliter) to smallest (milliliter).

kL hL daL L dL cL mL

From the chart, we see that we should move the decimal point 2 places to the right to convert from liters to centiliters.

6 liters = 6.00. centiliters = 600 centiliters

Answer: 4,000 mL ■

Cubic centimeters

Another metric unit of capacity is the **cubic centimeter,** which is represented by the notation cm^3 or, more simply, cc. One milliliter and one cubic centimeter represent the same capacity.

$$1 \text{ mL} = 1 \text{ cm}^3 = 1 \text{ cc}$$

The units of cubic centimeters are used frequently in medicine. For example, when a nurse administers an injection containing 5 cc of medication, the dosage could also be expressed using milliliters.

5 cc = 5 mL

When a doctor orders that a patient be put on 1,000 cc of dextrose solution, the request could be expressed in several ways.

1,000 cc = 1,000 mL = 1 liter

STUDY SET Section 8.4

VOCABULARY *Fill in the blanks to make the statements true.*

1. *Deka* means 10.

2. *Hecto* means 100.

3. *Kilo* means 1000.

4. *Deci* means $\frac{1}{10}$.

5. *Centi* means $\frac{1}{100}$.

6. *Milli* means $\frac{1}{1000}$.

7. Meters, grams, and liters are units of measurement in the metric system.

8. The weight of an object is determined by the Earth's gravitational pull on the object.

CONCEPTS

9. To the nearest centimeter, tell which measurements the arrows point to on the ruler in Illustration 1.

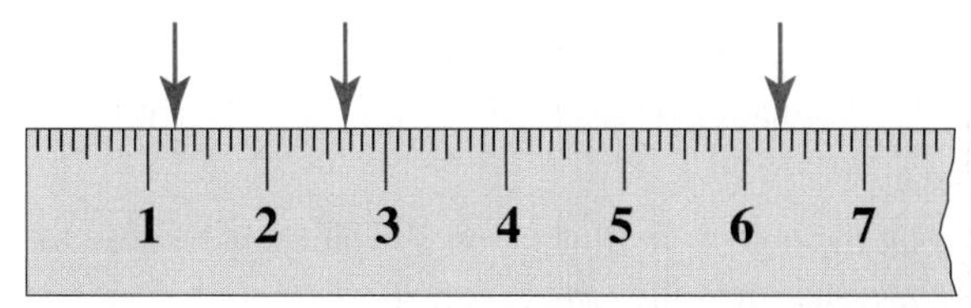

ILLUSTRATION 1

10. To the nearest millimeter, tell which measurements the arrows point to on the ruler in Illustration 2.

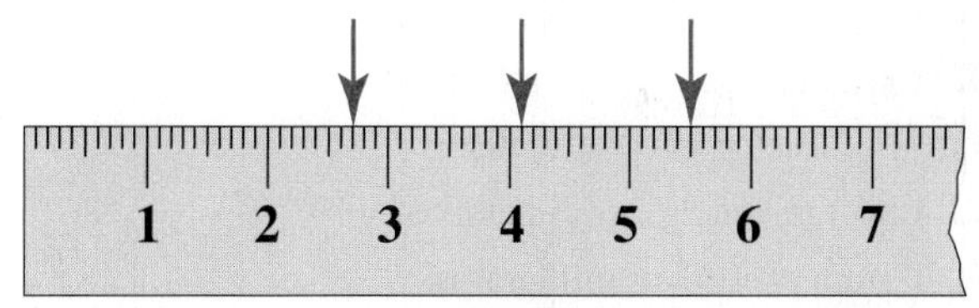

ILLUSTRATION 2

11. Write a unit conversion factor to convert the following.

a. Meters to kilometers
b. Grams to centigrams
c. Liters to milliliters

12. Use the chart to determine how many decimal places and in which direction to move the decimal point when converting the following.

a. Kilometers to centimeters

km hm dam m dm cm mm

b. Milligrams to grams

kg hg dag g dg cg mg

c. Hectoliters to centiliters

kL hL daL L dL cL mL

13. Match each item with its proper measurement.

a. Thickness of a phone book — **i.** 6,275 km
b. Length of the Amazon River — **ii.** 2 m
c. Height of a soccer goal — **iii.** 6 cm

14. Match each item with its proper measurement.

a. Weight of a giraffe — **i.** 800 kg
b. Weight of a paper clip — **ii.** 1 g
c. Active ingredient in an aspirin tablet — **iii.** 325 mg

15. Match each item with its proper measurement.

a. Amount of blood in an adult — **i.** 290,000 kL
b. Cola in an aluminum can — **ii.** 6 L
c. Kuwait's daily production of crude oil — **iii.** 355 mL

16. Of the objects in Illustration 3, which can be used to measure the following?

a. Millimeters
b. Milligrams
c. Milliliters

Balance

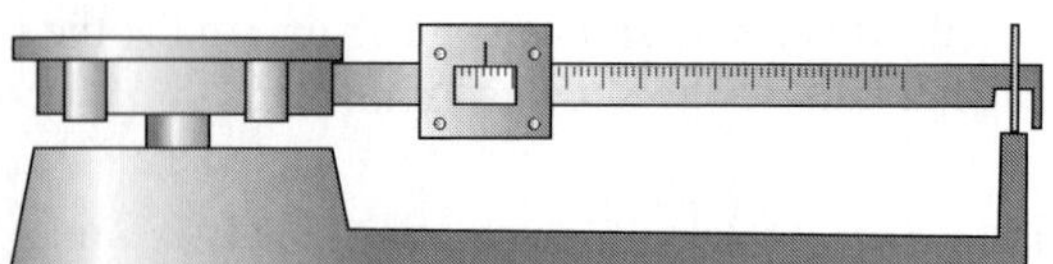

Beaker

Micrometer

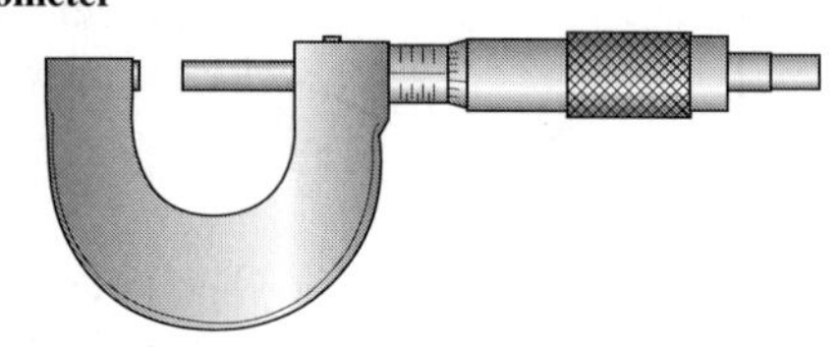

ILLUSTRATION 3

Fill in the blanks to make the statements true.

17. 1 dekameter = ____ meters
18. 1 decimeter = ____ meter
19. 1 centimeter = ____ meter
20. 1 kilometer = ____ meters
21. 1 millimeter = ____ meter
22. 1 hectometer = ____ meters
23. 1 gram = ____ milligrams
24. 100 centigrams = ____ gram
25. 1 kilogram = ____ grams
26. 1 milliliter = ____ cubic centimeter
27. 1 liter = ____ cubic centimeters
28. 1 kiloliter = ____ liters
29. 1 centiliter = ____ liter
30. 1 milliliter = ____ liter
31. 100 liters = ____ hectoliter
32. 10 deciliters = ____ liter

NOTATION *Complete each solution.*

33. Convert 20 centimeters to meters.

$$20 \text{ cm} = 20 \text{ cm} \cdot \frac{\underline{\quad} \text{ m}}{100 \text{ cm}}$$
$$= \frac{20}{\underline{\quad}} \text{ m}$$
$$= 0.2 \text{ m}$$

34. Convert 300 centigrams to grams.

$$300 \text{ cg} = 300 \text{ cg} \cdot \frac{\underline{\quad} \text{ g}}{100 \text{ cg}}$$
$$= \frac{\underline{\quad}}{100} \text{ g}$$
$$= 3 \text{ g}$$

35. Convert 2 kilometers to decimeters.

$$2 \text{ km} = 2 \text{ km} \cdot \frac{\underline{\quad} \text{ m}}{1 \text{ km}} \cdot \frac{10 \text{ dm}}{\underline{\quad} \text{ m}}$$
$$= 2 \cdot \underline{\quad} \cdot 10 \text{ dm}$$
$$= 20{,}000 \text{ dm}$$

36. Convert 3 deciliters to milliliters.

$$3 \text{ dL} = 3 \text{ dL} \cdot \frac{1 \text{ L}}{10 \text{ dL}} \cdot \frac{1000 \text{ mL}}{1 \text{ L}}$$

$$= \frac{3 \cdot 1{,}000}{10} \text{ mL}$$

$$= 300 \text{ mL}$$

PRACTICE *Use a metric ruler to measure each object to the nearest millimeter.*

37. The length of a dollar bill

38. The width of a dollar bill

Use a metric ruler to measure each object to the nearest centimeter.

39. The length (top to bottom) of this page

40. The length of the word antidisestablishmentarianism

Convert each measurement between the given metric units.

41. 3 m = ______ cm

42. 5 m = ______ cm

43. 5.7 m = ______ cm

44. 7.36 km = ______ dam

45. 0.31 dm = ______ cm

46. 73.2 m = ______ dm

47. 76.8 hm = ______ mm

48. 165.7 km = ______ m

49. 4.72 cm = ______ dm

50. 0.593 cm = ______ dam

51. 453.2 cm = ______ m

52. 675.3 cm = ______ m

53. 0.325 dm = ______ m

54. 0.0034 mm = ______ m

55. 3.75 cm = ______ mm

56. 0.074 cm = ______ mm

57. 0.125 m = ______ mm

58. 134 m = ______ hm

59. 675 dam = ______ cm

60. 0.00777 cm = ______ dam

61. 638.3 m = ______ hm

62. 6.77 cm = ______ m

63. 6.3 mm = ______ cm

64. 6.77 mm = ______ cm

65. 695 dm = ______ m

66. 6,789 cm = ______ dm

67. 5,689 m = ______ km

68. 0.0579 km = ______ mm

69. 576.2 mm = ______ dm

70. 65.78 km = ______ dam

71. 6.45 dm = ______ km

72. 6.57 cm = ______ mm

73. 658.23 m = ______ km

74. 0.0068 hm = ______ km

75. 3 g = ______ mg

76. 5 g = ______ cg

77. 2 kg = ______ g

78. 4,000 g = ______ kg

79. 1,000 kg = ______ g

80. 2 kg = ______ cg

81. 500 mg = ______ g

82. 500 mg = ______ cg

83. 3 kL = ______ L

84. 500 mL = ______ L

85. 500 cL = ______ mL

86. 400 L = ______ hL

87. 10 mL = ______ cc

88. 2,000 cc = ______ L

APPLICATIONS

89. SPEED SKATING American Eric Heiden won an unprecedented five gold medals by capturing the men's 500 m, 1,000 m, 1,500 m, 5,000 m and 10,000 m races at the 1980 Winter Olympic Games in Lake Placid, New York. Convert each race length to kilometers.

90. SUEZ CANAL The 163-km-long Suez Canal, shown in Illustration 4, connects the Mediterranean Sea with the Red Sea. It provides a shortcut for ships operating between European and American ports. Convert the length of the Suez Canal to meters.

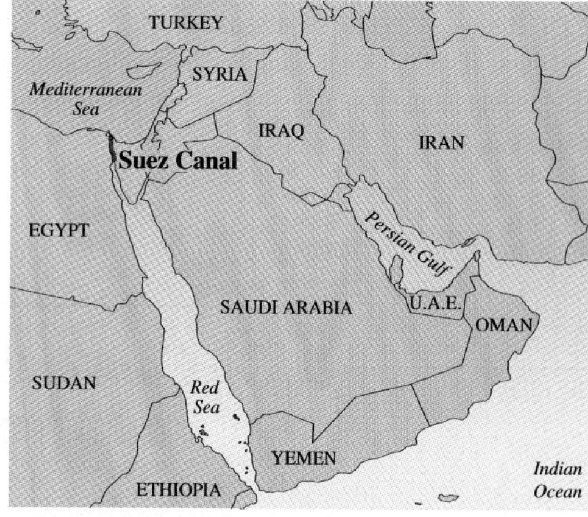

ILLUSTRATION 4

91. HEALTH CARE Blood pressure is measured by a *sphygmomanometer* (Illustration 5, next page). The measurement is read at two points and is expressed, for example, as 120/80. This indicates a *systolic* pressure of 120 millimeters of mercury and a *diastolic* pressure of 80 millimeters of mercury. Convert each measurement to centimeters of mercury.

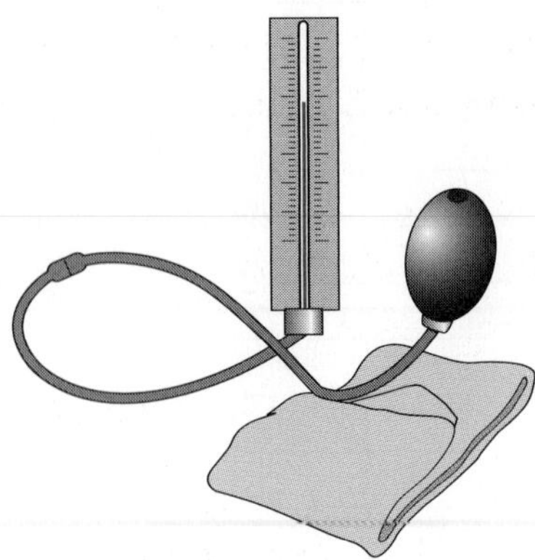

ILLUSTRATION 5

92. THE HANCOCK CENTER The John Hancock Center in Chicago has 100 stories and is 343 meters high. Give this height in hectometers.

93. WEIGHT OF A BABY A baby weighs 4 kilograms. Give this weight in centigrams.

94. JEWELRY A gold chain weighs 1,500 milligrams. Give this weight in grams.

95. CONTAINERS How many deciliters of root beer are there in two 2-liter bottles?

96. BOTTLING How many liters of wine are in a 750-mL bottle?

97. BUYING OLIVES The net weight of a bottle of olives is 284 grams. Find the smallest number of bottles that must be purchased to have at least 1 kilogram of olives.

98. BUYING COFFEE A can of Cafe Vienna has a net weight of 133 grams. Find the smallest number of cans that must be packaged to have at least 1 metric ton of coffee. (*Hint:* 1 metric ton = 1,000 kg.)

99. MEDICINE A bottle of hydrochlorothiazine contains 60 tablets. If each tablet contains 50 milligrams of active ingredient, how many grams of active ingredient are in the bottle?

100. INJECTION Illustration 6 shows a 3cc syringe. Express its capacity using units of milliliters.

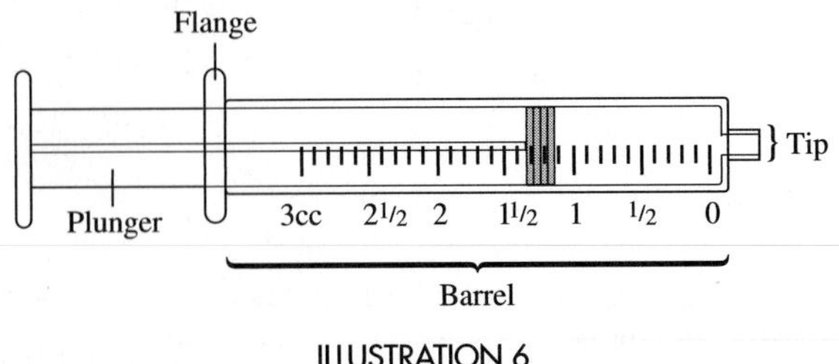

ILLUSTRATION 6

WRITING

101. To change 3.452 kilometers to meters, we can move the decimal point in 3.452 three places to the right to get 3,452 meters. Explain why.

102. To change 7,532 grams to kilograms, we can move the decimal point in 7,532 three places to the left to get 7.532 kilograms. Explain why.

103. A centimeter is one hundredth of a meter. Make a list of other words that begin with the prefix *centi* or *cent* and write a definition for each.

104. List the advantages of the metric system of measurement as compared to the American system. There have been several attempts to bring the metric system into general use in the United States. Why do you think these efforts have been unsuccessful?

REVIEW

105. Find 7% of \$342.72.

106. Add: $(3x - 1) + (6x + 5)$.

107. \$32.16 is 8% of what amount?

108. Divide: $3\frac{1}{7} \div 2\frac{1}{2}$.

109. Simplify $3\frac{1}{7} + 2\frac{1}{2} \cdot 3\frac{1}{3}$.

110. Solve $\frac{x}{5} - 3 = -3$.

8.5 Converting between American and Metric Units

In this section, you will learn about

- Converting between American and metric units
- Comparing American and metric units of temperature

Introduction. It is often necessary to convert between American units and metric units. For example, we must convert units to answer the following questions:

- Which is higher, Pikes Peak (elevation 14,110 feet) or the Matterhorn (elevation 4,478 meters)?
- Does a 2-pound tub of butter weigh more than a 1-kilogram tub?
- Is a quart of soda pop more or less than a liter of soda pop?

In this section, we will discuss how to answer such questions.

Converting between American and metric units

We can convert between American and metric units of length using the table below.

Equivalent Lengths	
American to metric	**Metric to American**
1 in. = 2.54 cm	1 cm = 0.3937 in.
1 ft = 0.3048 m	1 m = 3.2808 ft
1 yd = 0.9144 m	1 m = 1.0936 yd
1 mi = 1.6093 km	1 km = 0.6214 mi

EXAMPLE 1 ***Clothing label.*** Figure 8-17 shows a label sewn into some pants made in Mexico for sale in the United States. Express the waist size to the nearest inch.

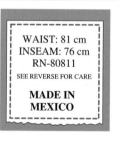

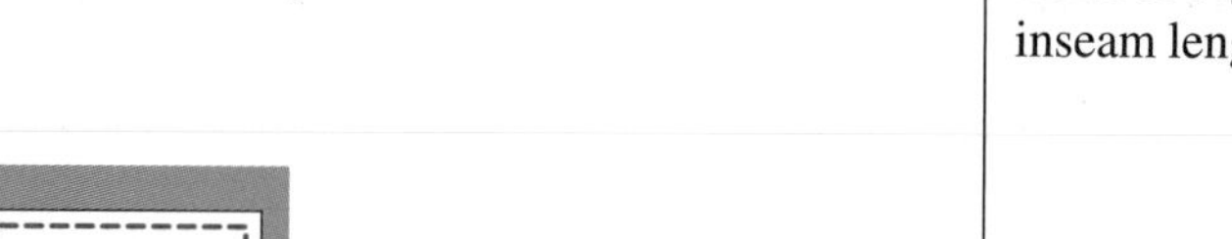

FIGURE 8-17

Solution

We need to convert from metric to American units. From the table, we see that there is 0.3937 inch in 1 centimeter. To make the conversion, we substitute 0.3937 inch for 1 centimeter.

$$\begin{aligned} 81 \text{ centimeters} &= 81\,(\textbf{centimeters}) \\ &= 81(\textbf{0.3937 in.}) \quad \text{Substitute 0.3937 inch for 1 centimeter.} \\ &= 31.8897 \text{ in.} \quad \text{Do the multiplication.} \end{aligned}$$

To the nearest inch, the waist size is 32 inches.

Self Check

Refer to Figure 8-17. What is the inseam length, to the nearest inch?

Answer: 30 in. ■

EXAMPLE 2 ***Mountain elevations.*** Pikes Peak, one of the most famous peaks in the Rocky Mountains, has an elevation of 14,110 feet. The Matterhorn, in the Swiss Alps, rises to an elevation of 4,478 meters. Which mountain is higher?

Solution

To make a comparison, the elevations must be expressed in the same units. We will convert the elevation of Pikes Peak, which is given in feet, to meters.

Self Check

Which is longer, a 500-meter race or a 550-yard race?

14,110 feet = 14,110 (**feet**)
= 14,110 (**0.3048 m**) Substitute 0.3048 meters for 1 foot.
= 4,300.728 m Do the multiplication.

Since the elevation of Pikes Peak is about 4,301 meters, we can conclude that the Matterhorn, with an elevation of 4,478 meters, is higher.

Answer: the 550-yard race ■

We can convert between American units of weight and metric units of mass by using the accompanying table.

Equivalent Weights and Masses

American to metric	Metric to American
1 oz = 28.35 g	1 g = 0.035 oz
1 lb = 0.454 kg	1 kg = 2.2 lb

EXAMPLE 3 ***Changing pounds to grams.*** Change 50 pounds to grams.

Solution

50 lb = 50(**1 lb**)
= 50(**16 oz**) Substitute 16 ounces for 1 pound.
= 50(16)(1 oz)
= 50(16)(28.35 g) Substitute 28.35 grams for 1 ounce.
= 22,680 g Do the multiplication.

Thus, 50 pounds is equal to 22,680 grams.

Self Check
Change 20 kilograms to pounds.

Answer: 44 lb ■

EXAMPLE 4 ***Comparing weights.*** Does a 2-pound tub of butter weigh more than a 1-kilogram tub?

Solution
To decide which contains more butter, we can change 2 pounds to kilograms.

2 lb = 2(**1 lb**)
= 2(**0.454 kg**) Substitute 0.454 kilograms for 1 pound.
= 0.908 kg Do the multiplication.

Since a 2-pound tub weighs only 0.908 kilogram, the 1-kilogram tub weighs more.

Self Check
Who weighs more, a person who weighs 165 pounds or one who weighs 76 kilograms?

Answer: the person who weighs 76 kg ■

We can convert between American and metric units of capacity by using the accompanying table.

Equivalent Capacities

American to metric	Metric to American
1 fl oz = 0.030 L	1 L = 33.8 fl oz
1 pt = 0.473 L	1 L = 2.1 pt
1 qt = 0.946 L	1 L = 1.06 qt
1 gal = 3.785 L	1 L = 0.264 gal

EXAMPLE 5 ***Changing from milliliters to quarts.*** A bottle of 7UP contains 750 milliliters. Convert this measure to quarts.

Solution
We convert milliliters to liters and then liters to quarts.

Self Check
A student bought a 355-mL can of cola. How many ounces of cola does the can contain?

$$
\begin{aligned}
750 \text{ mL} &= 750 \text{ mL} \cdot \frac{1 \text{ L}}{1{,}000 \text{ mL}} && \text{Use a unit conversion factor: } \tfrac{1 \text{ L}}{1{,}000 \text{ mL}} = 1. \\
&= \frac{750}{1{,}000} \text{ L} && \text{The units of mL divide out.} \\
&= \frac{3}{4} \text{ L} && \text{Simplify the fraction: } \tfrac{750}{1{,}000} = \tfrac{3 \cdot \overset{1}{\cancel{250}}}{4 \cdot \underset{1}{\cancel{250}}} = \tfrac{3}{4}. \\
&= \frac{3}{4}(\mathbf{1.06 \text{ qt}}) && \text{Substitute 1.06 quart for 1 liter.} \\
&= 0.795 \text{ qt} && \text{Do the arithmetic.}
\end{aligned}
$$

The bottle contains 0.795 quart.

Answer: 12 oz ■

From the table of equivalent capacities, we see that 1 liter is equal to 1.06 quarts. Thus, one liter of soda pop is more than one quart of soda pop.

EXAMPLE 6 ***Comparison shopping.*** A two-quart bottle of soda pop is priced at \$1.89, and a one-liter bottle is priced at 97¢. Which is the better buy?

Solution

We can convert 2 quarts to liters and find the price per liter of the two-quart bottle.

$$
\begin{aligned}
2 \text{ qt} &= 2(\mathbf{1 \text{ qt}}) \\
&= 2(\mathbf{0.946 \text{ L}}) && \text{Substitute 0.946 liter for 1 quart.} \\
&= 1.892 \text{ L} && \text{Do the multiplication.}
\end{aligned}
$$

Thus, the two-quart bottle contains 1.892 liters. To find the price per liter of the two-quart bottle, we divide $\frac{\$1.89}{1.892}$ to get

$$\frac{\$1.89}{1.892} = \$0.998942917$$

Since the price per liter of the two-quart bottle is a little more than 99¢, the one-liter bottle priced at 97¢ is the better buy.

Self Check

Thirty-four fluid ounces of aged vinegar costs \$3.49. A one-liter bottle of the same vinegar costs \$3.17. Which is the better buy?

Answer: the one-liter bottle ■

Comparing American and metric units of temperature

In the American system, we measure temperature using **degrees Fahrenheit** (°F). In the metric system, we measure temperature using **degrees Celsius** (°C). These two scales are shown on the thermometers in Figure 8-18 (on the next page). From the figure, we can see that

- 212° F = 100° C Water boils.
- 32° F = 0° C Water freezes.
- 5° F = −15° C A cold winter day.
- 95° F = 35° C A hot summer day.

As we have seen, there is a formula that enables us to convert from degrees Fahrenheit to degrees Celsius. There is also a formula to convert from degrees Celsius to degrees Fahrenheit.

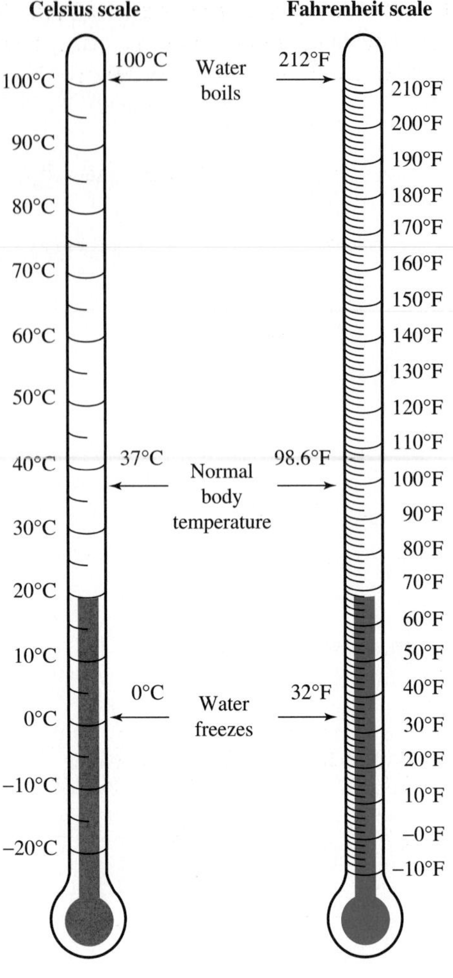

FIGURE 8-18

Conversion formulas for temperature

If F is the temperature in degrees Fahrenheit and C is the corresponding temperature in degrees Celsius, then

$$C = \frac{5(F - 32)}{9} \quad \text{and} \quad F = \frac{9}{5}C + 32$$

An alternate form of the formula $C = \frac{5(F-32)}{9}$ is obtained by distributing the multiplication by 5 in the numerator to get $C = \frac{5F-160}{9}$.

EXAMPLE 7 ***Converting from degrees Fahrenheit to degrees Celsius.*** Warm bath water is 90° F. Find the equivalent temperature in degrees Celsius.

Solution

We substitute 90 for F in the formula $C = \frac{5F - 160}{9}$ and simplify.

$$
\begin{aligned}
C &= \frac{5F - 160}{9} \\
&= \frac{5(90) - 160}{9} && \text{Substitute 90 for } F. \\
&= \frac{450 - 160}{9} && \text{Multiply: } 5(90) = 450. \\
&= 32.222222 && \text{Do the arithmetic.}
\end{aligned}
$$

To the nearest tenth of a degree, the equivalent temperature is 32.2° C.

Self Check

Hot coffee is 110° F. To the nearest tenth of a degree, express this temperature in degrees Celsius.

Answer: 43.3° C

EXAMPLE 8 ***Converting from degrees Celsius to degrees Fahrenheit.*** A dishwasher manufacturer recommends that dishes be rinsed in hot water with a temperature of 60° C. Express this temperature in degrees Fahrenheit.

Solution

We substitute 60 for C in the formula $F = \frac{9}{5}C + 32$ and simplify.

$$F = \frac{9}{5}C + 32$$

$$= \frac{9}{5}(\mathbf{60}) + 32 \quad \text{Substitute 60 for } C.$$

$$= \frac{540}{5} + 32 \quad \text{Multiply: } \tfrac{9}{5}(60) = \tfrac{540}{5}.$$

$$= 108 + 32 \quad \text{Do the division.}$$

$$= 140 \quad \text{Do the addition.}$$

The manufacturer recommends that dishes be rinsed in 140° F water.

Self Check

To see whether a baby has a fever, her mother takes her temperature with a Celsius thermometer. If the reading is 38.8° C, does the baby have a fever? (*Hint:* Normal body temperature is 98.6° F.)

Answer: yes ■

STUDY SET Section 8.5

VOCABULARY *Fill in the blanks to make the statements true.*

1. In the American system, temperatures are measured in degrees ______________.

2. In the metric system, temperatures are measured in degrees ______________.

CONCEPTS

3. Which is longer?
- **a.** A yard or a meter?
- **b.** A foot or a meter?
- **c.** An inch or a centimeter?
- **d.** A mile or a kilometer?

4. Which is heavier?
- **a.** An ounce or a gram?
- **b.** A pound or a kilogram?

5. Which is the greater unit of capacity?
- **a.** A pint or a liter?
- **b.** A quart or a liter?
- **c.** A gallon or a liter?

6. a. What formula is used for changing degrees Celsius to degrees Fahrenheit?

b. What formula is used for changing degrees Fahrenheit to degrees Celsius?

NOTATION *Complete each solution.*

7. Change 4,500 feet to kilometers.

4,500 ft = 4,500 (m)
= m
= 1.3716 km

8. Change 3 kilograms to ounces.

3 kg = 3(lb)
= 3(2.2)(oz)
= 105.6 oz

9. Change 8 liters to gallons.

8 L = 8(gal)
= 2.112 gal

10. Change 70° C to degrees Fahrenheit.

$$F = \frac{9}{5}C + 32$$

$$= \frac{9}{5}(\quad) + 32$$

$$= \quad + 32$$

$$= 158° \text{ F}$$

PRACTICE *Make each conversion. Since most conversions are approximate, answers will vary slightly depending on the method used.*

11. 3 ft = cm

12. 7.5 yd = m

13. 3.75 m = in.

14. 2.4 km = mi

15. 12 km = ______ ft
16. 3,212 cm = ______ ft
17. 5,000 in. = ______ m
18. 25 mi = ______ km
19. 37 oz = ______ kg
20. 10 lb = ______ kg
21. 25 lb = ______ g
22. 7.5 oz = ______ g
23. 0.5 kg = ______ oz
24. 35 g = ______ lb
25. 17 g = ______ oz
26. 100 kg = ______ lb
27. 3 fl oz = ______ L
28. 2.5 pt = ______ L
29. 7.2 L = ______ fl oz
30. 5 L = ______ qt
31. 0.75 qt = ______ mL
32. 3 pt = ______ mL
33. 500 mL = ______ qt
34. 2,000 mL = ______ gal
35. 50° F = ______ C
36. 67.7° F = ______ C
37. 50° C = ______ F
38. 36.2° C = ______ F
39. −10° C = ______ F
40. −22.5° C = ______ F
41. −5° F = ______ C
42. −10° F = ______ C

APPLICATIONS *Since most conversions are approximate, answers will vary slightly depending on the method used.*

43. THE MIDDLE EAST The distance between Jerusalem and Bethlehem is 8 kilometers. To the nearest mile, give this distance in miles.

44. THE DEAD SEA The Dead Sea is 80 kilometers long. To the nearest mile, give this distance in miles.

45. CHEETAH A cheetah can run 112 kilometers per hour. How fast is this in mph?

46. LION A lion can run 50 mph. How fast is this in kilometers per hour?

47. MOUNT WASHINGTON The highest peak of the White Mountains of New Hampshire is Mount Washington, at 6,288 feet. To the nearest tenth, give this height in kilometers.

48. TRACK AND FIELD Track meets are held on an oval track such as that shown in Illustration 1. One lap around the track is usually 400 meters. However, some older tracks in the United States are 440-yard ovals. Are these two types of tracks the same length? If not, which is longer?

ILLUSTRATION 1

49. HAIR GROWTH When hair is short, its rate of growth averages about $\frac{3}{4}$ inch per month. How many centimeters is this a month?

50. KILLER WHALE An adult male killer whale can weigh as much as 12,000 pounds and be as long as 25 feet. Change these measurements to kilograms and meters.

51. WEIGHTLIFTING Illustration 2 shows two International Powerlifting Federation recordholders as of April 13, 2000. Change each metric weight to pounds.

Recordholder	Country	Weight class	Bench press
Vicki Steenrod	USA	82.5 kg	132.5 kg
Jeff Magruder	USA	110.0 kg	270.0 kg

ILLUSTRATION 2

52. WORDS OF WISDOM Refer to the wall hanging in Illustration 3. Convert the first metric weight to ounces and the second to pounds. What famous saying results?

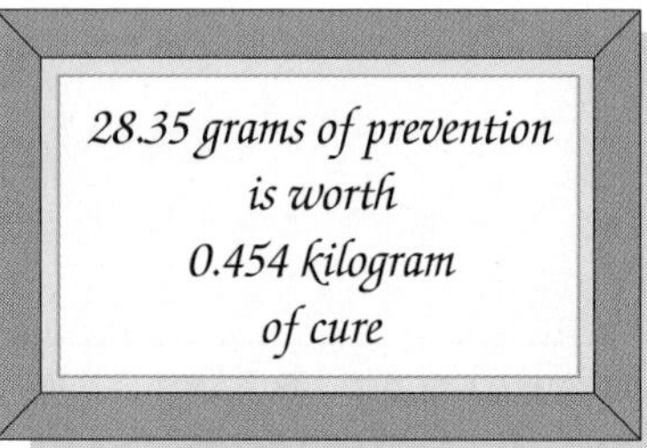

28.35 grams of prevention
is worth
0.454 kilogram
of cure

ILLUSTRATION 3

53. OUNCES AND FLUID OUNCES
a. There are 310 calories in 8 ounces of broiled chicken. Convert 8 ounces to grams.
b. There are 112 calories in a glass of fresh Valencia orange juice that holds 8 fluid ounces. Convert 8 fluid ounces to liters.

54. TRACK AND FIELD A shot-put weighs 7.264 kilograms. Give this weight in pounds.

55. POSTAL REGULATIONS You can mail a package weighing up to 70 pounds via priority mail. Can you mail a package that weighs 32 kilograms by priority mail?

56. HEALTHY EATING Refer to the nutrition label for a packet of oatmeal shown in Illustration 4. Change each circled weight to ounces.

Nutrition Facts
Serving Size: 1 Packet (46g)
Servings Per Container: 10

Amount Per Serving
Calories 170 Calories from Fat 20

	% Daily Value
Total fat 2g	3%
Saturated fat (0.5g)	2%
Polyunsaturated Fat 0.5g	
Monounsaturated Fat 1g	
Cholesterol 0mg	0%
Sodium (250mg)	10%
Total carbohydrate 35g	12%
Dietary fiber 3g	12%
Soluble Fiber 1g	
Sugars 16g	
Protein (4g)	

ILLUSTRATION 4

57. COMPARISON SHOPPING Which is the better buy, 3 quarts of root beer for \$4.50 or 2 liters of root beer for \$3.60?

58. COMPARISON SHOPPING Which is the better buy, 3 gallons of antifreeze for \$10.35 or 12 liters of antifreeze for \$10.50?

59. HOT SPRINGS The thermal springs in Hot Springs National Park in central Arkansas emit water as warm as 143°F. Change this temperature to degrees Celsius.

60. COOKING MEAT Meats must be cooked at high enough temperatures to kill harmful bacteria. According to the USDA and the FDA, the internal temperature for cooked roasts and steaks should be at least 145°F, and whole poultry should be 180°F. Convert these temperatures to degrees Celsius. Round up to the nearest degree.

61. TAKING A SHOWER When you take a shower, which water temperature would you choose: 15° C, 28° C, or 50° C?

62. DRINKING WATER To get a cold drink of water, which temperature would you choose: −2° C, 10° C, or 25° C?

63. SNOWY WEATHER At which temperature might it snow: −5° C, 0° C, or 10° C?

64. RUNNING THE AIR CONDITIONER At which outside temperature would you be likely to run the air conditioner: 15° C, 20° C, or 30° C?

WRITING

65. Explain how to change kilometers to miles.

66. Explain how to change 50° C to degrees Fahrenheit.

67. The United States is the only industrialized country in the world that does not officially use the metric system. Some people claim this is costing American businesses money. Do you think so? Why?

68. What is meant by the phrase *a table of equivalent measures*?

REVIEW *Combine like terms.*

69. $6y + 7 - y - 3$

70. $4x - 3y + 7 + 4x - 2 - 2y$

71. $-3(x - 4) - 2(2x + 6)$

72. $7(4 - t) + 2(3t - 8)$

Simplify each expression.

73. $x \cdot x \cdot x$

74. $a^3 \cdot a^5$

75. $3b(5b)$

76. $(x^2)^5$

KEY CONCEPT

Proportions

A **proportion** is a statement that two ratios or rates are equal.

Fill in the blanks as we set up a proportion to solve a problem.

1. TEACHER'S AIDES For every 15 children on the playground, a child care center is required to have 2 teacher's aides supervising. How many teacher's aides will be needed to supervise 75 children?

Step 1: Let x = the number of

______________________________.

If we compare the number of children to the number of teacher's aides, we know that the two rates must be equal.

15 children are to ____ aides as ____ children are to ____ aides.

Expressing this as a proportion, we have

$$\frac{15}{\quad} = \frac{75}{\quad}$$

Number of children → 15, 75 ← Number of children
Number of aides → ____, ____ ← Number of aides

In the proportion $\frac{15}{2} = \frac{75}{x}$, 15 and x are the *extremes* and 2 and 75 are the *means.* After setting up the proportion, we solve it using the fact that the product of the extremes is equal to the product of the means.

Step 2: Solve for x: $\frac{15}{2} = \frac{75}{x}$.

$\quad \cdot x = 2 \cdot \quad$	The product of the extremes equals the product of the means.
$15x = \quad$	Do the multiplication.
$\frac{15x}{\quad} = \frac{150}{\quad}$	Divide both sides by 15.
$x = \quad$	Simplify.

Ten teacher's aides are needed to supervise 75 children.

Step 3: To check the result, we substitute 10 for x in $\frac{15}{2} = \frac{75}{x}$ and find the cross products.

$15 \cdot 10 = \mathbf{150}$ $\qquad 2 \cdot 75 = \mathbf{150}$

$$\frac{15}{2} \stackrel{?}{=} \frac{75}{10}$$

Since the cross products are equal, 10 is a solution of the proportion.

Set up and solve each problem using a proportion.

2. PARKING A city code requires that companies provide 10 parking spaces for every 12 employees. How many spaces will be needed if a company employs 450 people?

3. MOTION PICTURES Every 2 seconds, 3 feet of motion picture film pass through the projector. How many feet of film are there in a movie that runs for 120 minutes?

ACCENT ON TEAMWORK

Section 8.1

ART HISTORY Illustration 1 shows a drawing made by Leonardo Da Vinci (1452–1519) of a human figure within a square. We see that the man's height and the span of his outstretched arms are the same. That is, their ratio is 1 to 1. Use a tape measure to determine this ratio for each member of your group. Work in terms of inches. Are any ratios exactly 1 to 1?

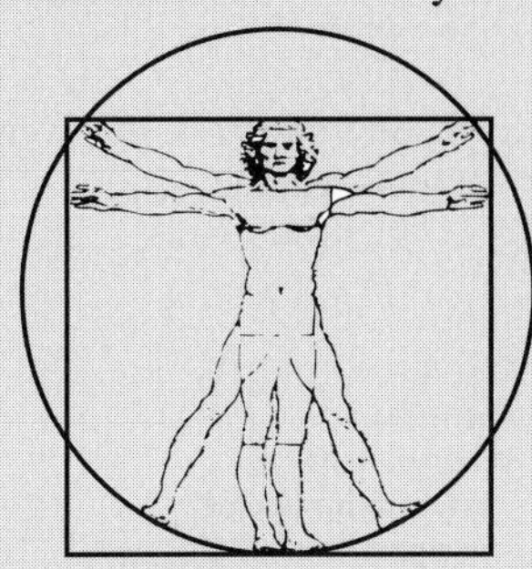

ILLUSTRATION 1

Section 8.2

ENLARGEMENTS Duplicating or enlarging a picture can be done by the *grid-transfer* method. To enlarge the picture of a cook shown in Illustration 2(a), begin by copying the markings in the square in the lower right-hand corner onto its corresponding square in the enlargement grid in Illustration 2(b). One by one, copy the contents of each square of the original picture to its counterpart in the enlargement.

Section 8.3

DISNEY CLASSICS The movie *20,000 Leagues Under the Sea* is a science fiction thriller about Captain Nemo and the crew of the submarine Nautilus. Use the fact that 1 mile $= \frac{1}{3}$ league to express a depth of 20,000 leagues in feet.

Section 8.4

METRIC MISHAP In 1999, NASA lost the \$125 million Mars Climate Orbiter because a Lockheed Martin engineering team used American units of measurement, while NASA's team used the metric system. Use the Internet to research this incident and make a report to your class.

Section 8.5

TRUTH IN LABELING Have each member of your group bring in two items—one whose product label indicates capacity and another that indicates weight. For example, a bottle of shampoo could contain 15 fluid ounces (444 mL) or a can of soup could weigh 1 pound 3 ounces (539 g). Exchange your items with another person in your group. Check the accuracy of each label by converting from American units to metric units.

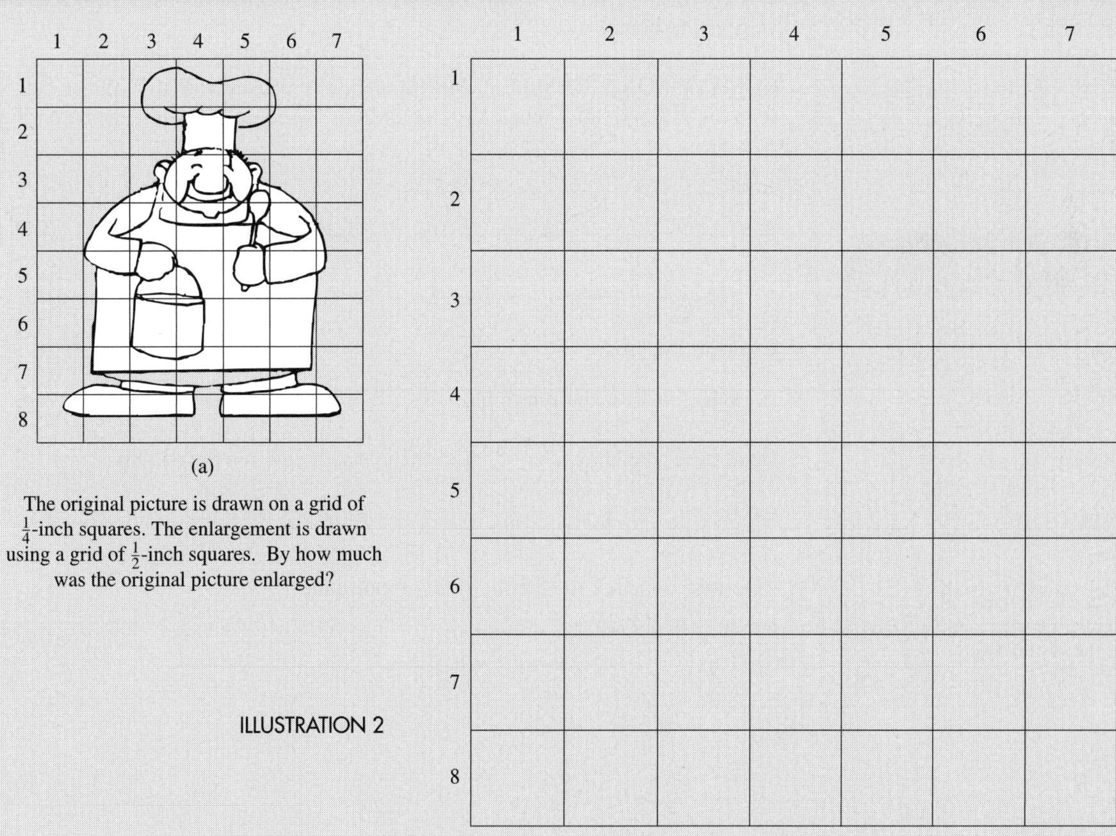

The original picture is drawn on a grid of $\frac{1}{4}$-inch squares. The enlargement is drawn using a grid of $\frac{1}{2}$-inch squares. By how much was the original picture enlarged?

ILLUSTRATION 2

CHAPTER REVIEW

SECTION 8.1 Ratio

CONCEPTS

A *ratio* is a quotient of two numbers or a quotient of two quantities that have the same units.

REVIEW EXERCISES

1. Express each phrase as a ratio in lowest terms.

a. The ratio of 4 inches to 12 inches

b. The ratio of 8 ounces to 2 pounds

c. 21 : 14

d. 24 to 36

A *rate* is a comparison of two quantities with different units.

2. AIRCRAFT Specifications for a Boeing B-52 Stratofortress are given in Illustration 1. What is the ratio of the airplane's wingspan to its length?

Crew: 6
Length: 160 ft
Wingspan: 185 ft
Maximum takeoff weight: 488,000 lb
Maximum speed: 595 mph
Maximum altitude: more than 50,000 ft
Range: 7,500 mi

ILLUSTRATION 1

3. PAY SCALE Find the hourly rate of pay for a student who earned $333.25 for working 43 hours.

A *unit cost* is a comparison of the cost of an item to its quantity.

4. COMPARISON SHOPPING Mixed nuts come packaged in a 12-ounce can, which sells for $4.95, or an 8-ounce can, which sells for $3.25. Which is the better buy?

SECTION 8.2 Proportions

A *proportion* is a statement that two ratios (or rates) are equal.

5. Consider the proportion $\frac{5}{15} = \frac{25}{75}$.

a. Which term is the fourth term?

b. Which term is the second term?

In any proportion, the product of the *extremes* is equal to the product of the *means*.

6. Determine whether each of the following statements is a proportion.

a. $\frac{15}{29} = \frac{105}{204}$

b. $\frac{17}{7} = \frac{204}{84}$

When two pairs of numbers *a*, *b* and *c*, *d* form a proportion, we say that the numbers are *proportional.*

7. Determine whether the numbers are proportional.

a. 5, 9 and 20, 36

b. 7, 13 and 29, 54

If three terms of a proportion are known, we can solve for the missing term.

8. Solve each proportion for the variable.

a. $\frac{12}{18} = \frac{3}{x}$

b. $\frac{4}{b} = \frac{2}{8}$

c. $\frac{c + 1}{5} = \frac{1}{5}$

d. $\frac{3p + 15}{2} = \frac{5}{3}$

9. PICKUP TRUCK A Dodge Ram pickup truck can go 35 miles on 2 gallons of gas. How far can it go on 11 gallons?

10. QUALITY CONTROL In a manufacturing process, 12 parts out of 66 were found to be defective. How many defective parts will be expected in a run of 1,650 parts?

11. SCALE DRAWING Illustration 2 shows an architect's drawing of a kitchen using a scale of $\frac{1}{8}$ inch to 1 foot $\left(\frac{1}{8}'': 1'0''\right)$. On the drawing, the length of the kitchen is $1\frac{1}{2}$ inches. How long is the actual kitchen?

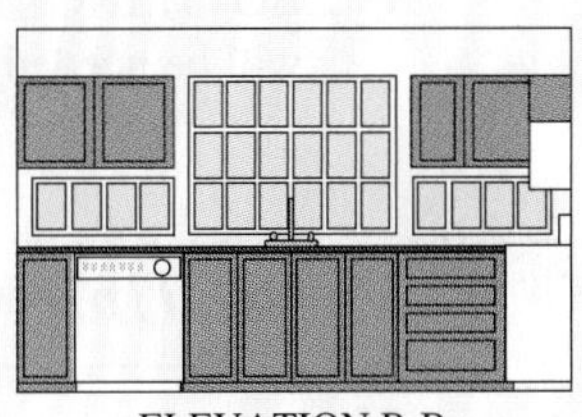

ELEVATION B-B

SCALE: $\frac{1''}{8} = 1'0''$

ILLUSTRATION 2

SECTION 8.3 American Units of Measurement

Common American units of length are *inches, feet, yards,* and *miles.*

12 in. = 1 ft
3 ft = 1 yd
36 in. = 1 yd
5,280 ft = 1 mi

12. Use a ruler to measure the length of the computer mouse in Illustration 3, to the nearest quarter of an inch.

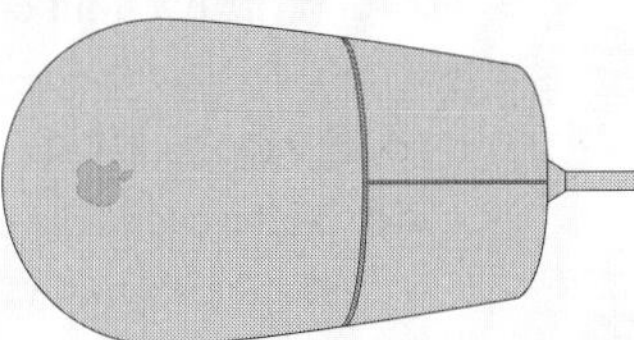

ILLUSTRATION 3

13. Write two unit conversion factors using the fact that 1 mile = 5,280 ft.

14. Make each conversion.

a. 5 yards to feet

b. 6 yards to inches

c. 66 inches to feet

d. 25.5 feet to inches

e. 9,240 feet to miles

f. 1 mile to yards

Common American units of weight are *ounces, pounds,* and *tons.*

16 oz = 1 lb
2,000 lb = 1 ton

15. Make each conversion.

a. 32 ounces to pounds

b. 17.2 pounds to ounces

c. 3 tons to ounces

d. 4,500 pounds to tons

Common American units of capacity are *fluid ounces, cups, pints, quarts,* and *gallons.*

1 c = 8 fl oz
1 pt = 2 c
1 qt = 2 pt
1 gal = 4 qt

16. Make each conversion.

a. 5 pints to fluid ounces
b. 8 cups to gallons
c. 17 quarts to cups
d. 176 fluid ounces to quarts
e. 5 gallons to pints
f. 3.5 gallons to cups

Units of time are *seconds, minutes, hours,* and *days.*

1 min = 60 sec
1 hr = 60 min
1 day = 24 hr

17. Make each conversion.

a. 20 minutes to seconds
b. 900 seconds to minutes
c. 200 hours to days
d. 6 hours to minutes
e. 4.5 days to hours
f. 1 day to seconds

18. SKYSCRAPER The Sears Tower in Chicago is 1,454 feet high. Express this distance in yards.

19. BOTTLING A magnum is a two-quart bottle of wine. How many magnums will be needed to hold 50 gallons of wine?

SECTION 8.4 *Metric Units of Measurement*

Common metric units of length are *millimeter, centimeter, decimeter, meter, dekameter, hectometer,* and *kilometer.*

1 mm = $\frac{1}{1,000}$ m
1 cm = $\frac{1}{100}$ m
1 dm = $\frac{1}{10}$ m
1 dam = 10 m
1 hm = 100 m
1 km = 1,000 m

20. Use a metric ruler to measure the length of the computer mouse in Illustration 4 to the nearest centimeter.

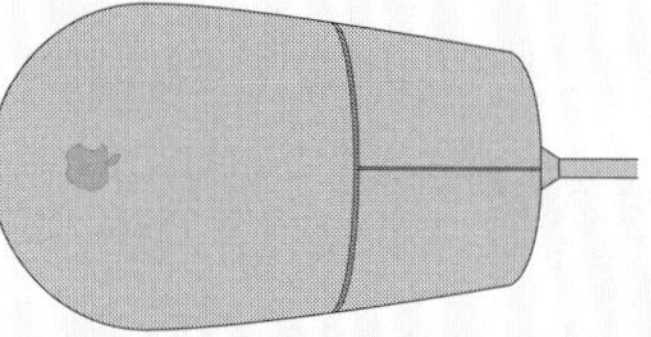

ILLUSTRATION 4

21. Write two unit conversion factors using the fact that 1 km = 1,000 m.

22. Make each conversion.

a. 475 centimeters to meters
b. 8 meters to millimeters
c. 3 dekameters to kilometers
d. 2 hectometers to decimeters
e. 5 kilometers to hectometers
f. 2,500 meters to hectometers

Common metric units of mass are *milligrams, centigrams, grams, kilograms,* and *metric tons.*

1 mg = $\frac{1}{1,000}$ g
1 cg = $\frac{1}{100}$ g
1 g = $\frac{1}{1,000}$ kg

23. Make each conversion.

a. 7 centigrams to milligrams
b. 800 centigrams to grams
c. 5,425 grams to kilograms
d. 5,425 grams to milligrams
e. 7,500 milligrams to grams
f. 5,000 centigrams to kilograms

24. PAIN RELIEVER A bottle of Extra Strength Tylenol contains 100 caplets of 500 milligrams each. How many grams of Tylenol are in the bottle?

Common metric units of capacity are *milliliters, centiliters, deciliters, liters, hectoliters,* and *kiloliters.*

$1 \text{ mL} = \frac{1}{1{,}000} \text{ L}$
$1 \text{ cL} = \frac{1}{100} \text{ L}$
$1 \text{ dL} = \frac{1}{10} \text{ L}$
1 L = 1,000 cc
1 hL = 100 L
1 kL = 1,000 L

25. Make each conversion.

a. 150 centiliters to liters
b. 3,250 liters to kiloliters
c. 1 hectoliter to deciliters
d. 400 milliliters to centiliters
e. 2 kiloliters to hectoliters
f. 4 deciliters to milliliters

26. SURGERY A dextrose solution is being administered to a patient intravenously using the apparatus shown in Illustration 5. How many milliliters of solution does the IV bag hold?

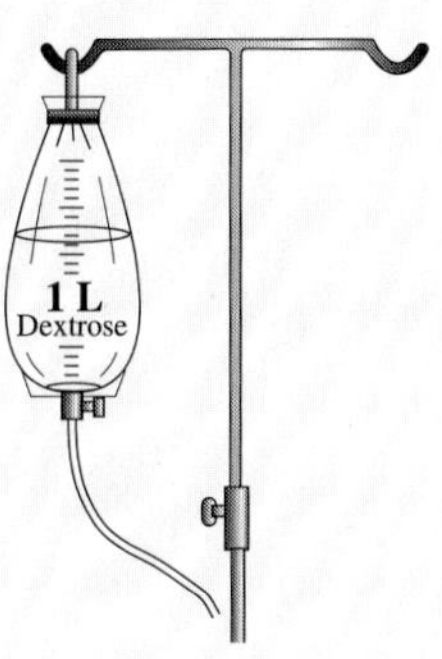

ILLUSTRATION 5

SECTION 8.5 Converting between American and Metric Units

We can convert between American and metric units using the following:

1 in. = 2.54 cm
1 ft = 0.3048 m
1 yd = 0.9144 m
1 mi = 1.6093 km
1 cm = 0.3937 in.
1 m = 3.2808 ft
1 m = 1.0936 yd
1 km = 0.6214 mi

27. SWIMMING Olympic-size swimming pools are 50 meters long. Express this distance in feet.

28. HIGH-RISE BUILDING The World Trade Center is 419 meters high, and the Empire State Building is 1,250 feet high. Which building is taller?

29. WESTERN SETTLERS The Oregon Trail was an overland route pioneers used in the 1840s through the 1870s to reach the Oregon Territory. It stretched 1,930 miles from Independence, Missouri, to Oregon City, Oregon. Find this distance to the nearest kilometer.

30. Michael Jordan is 6 feet, 6 inches tall. Express his height in centimeters.

1 oz = 28.35 g
1 lb = 0.454 kg
1 g = 0.035 oz
1 kg = 2.2 lb

31. Make each conversion.

a. 30 ounces to grams
b. 15 kilograms to pounds
c. 25 pounds to grams (Round to the nearest thousand.)
d. 2,000 pounds to kilograms (Round to the nearest ten.)

1 fl oz = 0.030 L
1 pt = 0.473 L
1 qt = 0.946 L
1 gal = 3.785 L
1 L = 33.8 fl oz
1 L = 2.1 pt
1 L = 1.06 qt
1 L = 0.264 gal

32. POLAR BEAR At birth, polar bear cubs weigh less than human babies—about 910 grams. Convert this to pounds.

33. BOTTLED WATER LaCroix® bottled water can be purchased in bottles containing 17 fluid ounces. Mountain Valley® water can be purchased in half-liter bottles. Which bottle contains more water?

34. COMPARISON SHOPPING One gallon of bleach costs \$1.39. A 5-liter economy bottle costs \$1.80. Which is the better buy?

Two units used to measure temperature are degrees Fahrenheit and degrees Celsius.

$$C = \frac{5F - 160}{9}$$

$$F = \frac{9}{5}C + 32$$

35. Change 77° F to degrees Celsius.

36. Which temperature of water would you like to swim in: 10° C, 30° C, 50° C, or 70° C?

Chapter 8 Test

Write each phrase as a ratio in lowest terms.

1. The ratio of 6 feet to 8 feet

2. The ratio of 8 ounces to 3 pounds

3. COMPARISON SHOPPING Two pounds of coffee can be purchased for \$3.38, and a 5-pound can can be purchased for \$8.50. Which is the better buy?

4. UTILITY COSTS A household used 675 kilowatt hours of electricity during a 30-day month. Find the rate of electric consumption in kilowatt hours per day.

5. CHECKERS What is the ratio of the number of red squares to the number of black squares for the checkerboard shown in Illustration 1? Express your answer in three ways: as a fraction, using a colon, and using the word *to*.

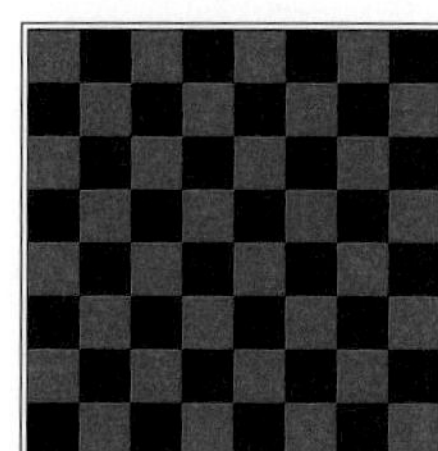

ILLUSTRATION 1

Tell whether each statement is a proportion.

6. $\frac{25}{33} = \frac{350}{460}$

7. $\frac{2.2}{3.5} = \frac{1.76}{2.8}$

8. Are the numbers 7, 15 and 245, 525 proportional?

Solve each proportion.

9. $\frac{x}{3} = \frac{35}{7}$

10. $\frac{15.3}{x} = \frac{3}{12.4}$

11. $\frac{2x + 3}{5} = \frac{5}{1}$

12. $\frac{3}{2z - 1} = \frac{3}{5}$

13. SHOPPING If 13 ounces of tea costs \$2.79, how much would you expect to pay for 16 ounces?

14. COOKING A recipe calls for $\frac{2}{3}$ cup of sugar and 2 cups of flour. How much sugar should be used with 5 cups of flour?

15. Convert 180 inches to feet.

16. TOOLS If a 25-foot tape measure is completely extended, how many yards does it stretch?

17. Convert 10 pounds to ounces.

18. A car weighs 1.6 tons. Find its weight in pounds.

19. How many fluid ounces are in a 1-gallon carton of milk?

20. LITERATURE An excellent work of early science fiction is the book *Around the World in 80 Days* by Jules Verne (1828–1905). Convert 80 days to minutes.

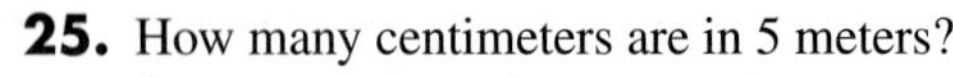

21. A quart and a liter of fruit punch are shown in Illustration 2. Which is the one-liter carton?

ILLUSTRATION 2

22. A yardstick and a meterstick are shown in Illustration 3. Which is the meter stick?

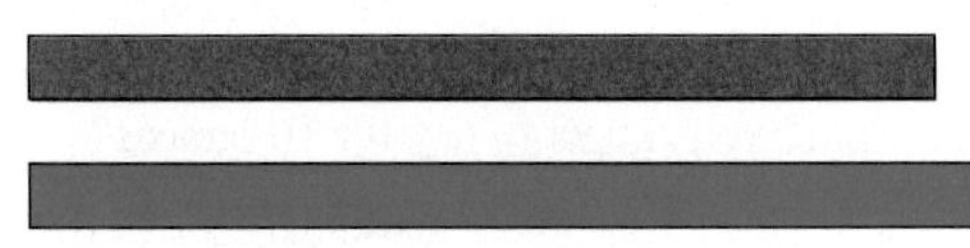

ILLUSTRATION 3

23. An ounce and a gram are placed on the balance shown in Illustration 4. On which side is the gram?

ILLUSTRATION 4

24. SPEED SKATING American Bonnie Blair won gold medals in the women's 500-meter speed skating competitions at the 1988, 1992, and 1994 Winter Olympic Games. Convert the race length to kilometers.

25. How many centimeters are in 5 meters?

26. Convert 8,000 centigrams to kilograms.

27. Convert 70 liters to milliliters.

28. PRESCRIPTION A bottle contains 50 tablets, each containing 150 mg of medicine. How many grams of medicine does the bottle contain?

29. Which is the longer distance: a 100-yard race or an 80-meter race?

30. Which person is heavier: Jim, who weighs 160 pounds, or Ricardo, who weighs 72 kilograms?

31. COMPARISON SHOPPING A two-quart bottle of soda pop costs \$1.73, and a one-liter bottle costs 89¢. Which is the better buy? (*Hint:* 1 quart = 0.946 liter.)

32. COOKING MEAT The USDA recommends that turkey be cooked to a temperature of 83°C. Change this to degrees Fahrenheit. To be safe, round up to the nearest degree. (*Hint:* $F = \frac{9}{5}C + 32$.)

33. What is a scale drawing? Give an example.

34. Explain the benefits of the metric system of measurement as compared to the American system.

Chapters 1–8 Cumulative Review Exercises

1. Write 64,502 in expanded notation.

2. Find $37\overline{)743}$.

3. ENLISTMENT Illustration 1 shows how four of the United States armed services fared in reaching their recruiting goals in 1999. Complete the table. Use a negative number to denote a shortfall of enlistees.

	Goal	Enlistments	Outcome
Air Force	33,800	32,068	
Army	74,500	68,210	
Marines	39,486	39,521	
Navy	52,524	52,595	

Based on data from *Los Angeles Times* (June 23, 2000)

ILLUSTRATION 1

4. Evaluate each expression, if possible.

a. $0 + (-8)$ **b.** $\frac{-8}{0}$

c. $0 - |-8|$ **d.** $\frac{0}{-8}$

e. $0 - (-8)$ **f.** $0(-8)$

5. PROFESSIONAL GOLF Tiger Woods won the 100th U.S. Open in June of 2000 by the largest margin in the history of that tournament. If he shot 12 under par (-12) and the second-place finisher, Miguel Angel Jimenez, shot 3 over par $(+3)$, what was his margin of victory?

6. Evaluate -3^2 and $(-3)^2$.

7. Evaluate $2 + 3[5(-6) - (1 - 10)]$.

8. What is the coefficient of the first term and the second term of $x^2 + 16x - 1$?

9. How many minutes are in h hours?

10. Simplify $3x - 5 - 2x - 2$.

11. Solve $7 + 2x = 2 - (4x + 7)$.

12. Give the formula that relates distance, rate, and time.

13. PHONE BOOKS A driver left a warehouse in the morning with 500 new telephone books loaded on his truck. His delivery route consisted of office buildings, each of which was to receive 5 of the books. The driver returned at the end of the day with 105 books on the truck. To how many office buildings did he deliver?

14. What is the formula for the area of a triangle?

15. Simplify $\frac{16}{20}$.

16. Express $\frac{9}{10}$ as an equivalent fraction with a denominator of $60t$.

17. Divide: $-\frac{7}{8h} \div \frac{7}{8}$.

18. What is $\frac{1}{2}$ of $\frac{1}{2}$?

19. MOTORS What is the difference in horsepower (hp) between the two motors shown in Illustration 2?

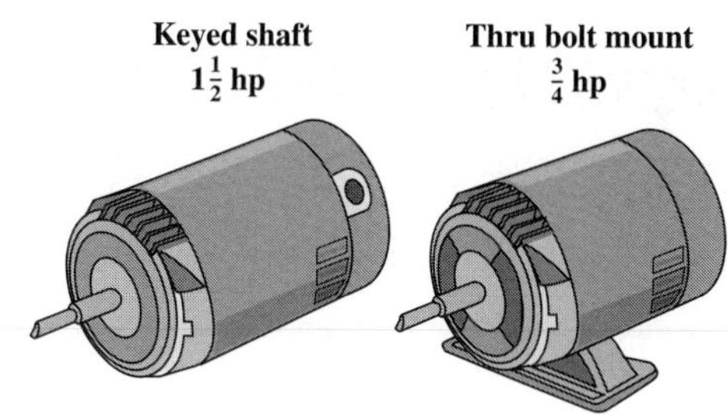

ILLUSTRATION 2

20. Solve $\frac{5}{8}y = \frac{1}{10}$.

21. Solve $\frac{2}{5}y + 1 = \frac{1}{3} + y$.

22. GLOBAL WARMING The average temperature in the United States during March through May of 2000 was a record-setting 55.5°F. That was 0.4 degrees warmer than the previous record, set in 1910. What was the previous spring temperature record?

23. Evaluate the expression $\frac{6.7 - x^2 + 1.6}{-x^3}$ for $x = -0.3$. Round to the nearest hundredth.

24. Write $\frac{1}{12}$ as a decimal.

25. Solve $6(y - 1.1) + 3.2 = -1 + 3y$.

26. Evaluate $3\sqrt{25} + 4\sqrt{4}$.

27. Plot the points $A(-4, -3)$, $B(1.5, 1.5)$, $C(-3, 0)$, $D\left(0, 3\frac{1}{2}\right)$.

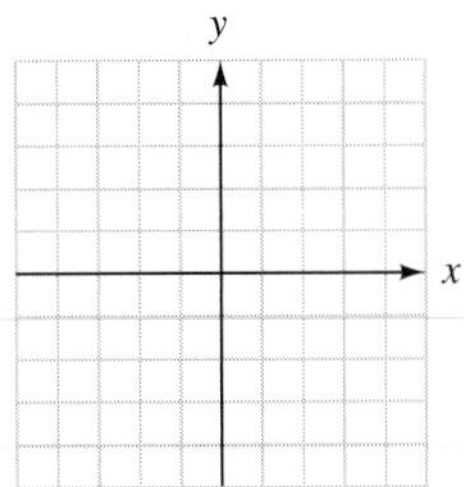

28. Graph each equation.

a. $x - 2y = -4$ **b.** $y = 2x^2$

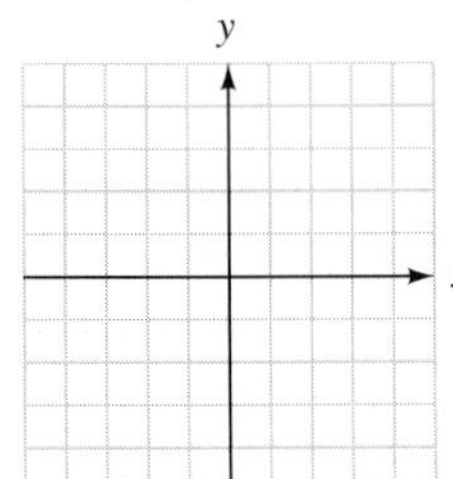

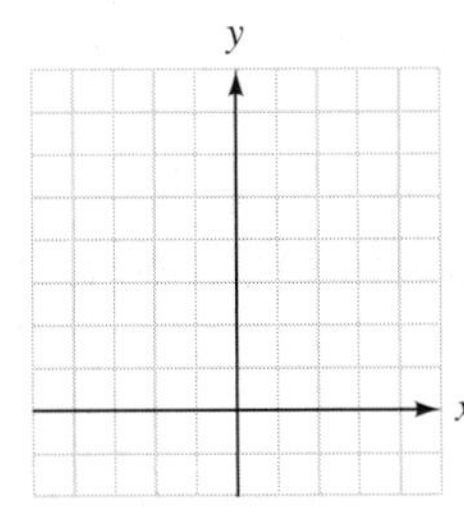

29. Subtract: $(5x^2 - 8x + 1) - (3x^2 - 2x + 3)$.

30. Multiply: $(3x + 2)(2x - 5)$.

31. Simplify each expression.

a. $s^6 \cdot s^7$

b. $(s^6)^7$

c. $(3a^2b^4)^3$

d. $-w^5(8w^3)$

32. 16 is what percent of 24?

33. What is the formula for simple interest?

34. Complete the table.

Percent	Decimal	Fraction
	0.99	
1.3%		
		$\frac{5}{16}$

35. GUITAR SALE What are the regular price and the rate of discount for the guitar shown in Illustration 3?

ILLUSTRATION 3

36. Express the phrase “3 inches to 15 inches” as a ratio in lowest terms.

37. SURVIVAL GUIDE

a. A person can go without food for about 40 days. How many hours is this?

b. A person can go without water for about 3 days. How many minutes is that?

c. A person can go without breathing oxygen for about 8 minutes. How many seconds is that?

38. Convert 40 ounces to pounds.

39. Convert 2.4 meters to millimeters.

40. Convert 320 grams to kilograms.

41. **a.** Which holds more, a 2-liter bottle or a 1-gallon bottle?

b. Which is longer, a meterstick or a yardstick?

42. BUILDING MATERIALS Which is the better buy, a 94-pound bag of cement for $4.48 or a 45-kilogram bag of cement for $4.56?

Introduction to Geometry

9

GEOMETRY COMES FROM THE GREEK WORDS GEO *(MEANING EARTH) AND* METRON *(MEANING MEASURE).*

9.1 Some Basic Definitions

In this section, you will learn about

- Points, lines, and planes • Angles • Adjacent and vertical angles
- Complementary and supplementary angles

INTRODUCTION. In this chapter, we will study two-dimensional geometric figures such as rectangles and circles. In daily life, it is often necessary to find the perimeter or area of one of these figures. For example, to find the amount of fencing that is needed to enclose a circular garden, we must find the perimeter of a circle (called its *circumference*). To find the amount of paint needed to paint a room, we must find the area of its four rectangular walls.

We will also study three-dimensional figures such as cylinders and spheres. To find the amount of space enclosed within these figures, we must find their volumes.

Points, lines, and planes

Geometry is based on three undefined words: **point, line,** and **plane.** Although we will make no attempt to define these words formally, we can think of a point as a geometric figure that has position but no length, width, or depth. Points are always labeled with capital letters. Point A is shown in Figure 9-1(a).

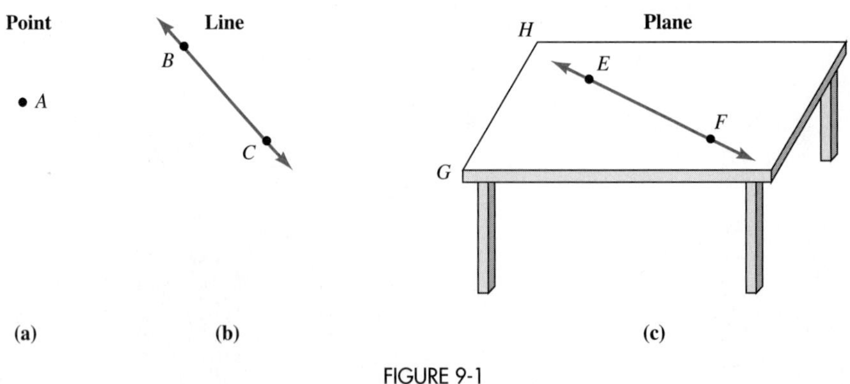

FIGURE 9-1

A line is infinitely long but has no width or depth. Figure 9-1(b) shows line BC, passing through points B and C. A plane is a flat surface, like a table top, that has length and width but no depth. In Figure 9-1(c), line EF lies in the plane GHI.

As Figure 9-1(b) illustrates, points B and C determine exactly one line, the line BC. In Figure 9-1(c), the points E and F determine exactly one line, the line EF. In general, any two points will determine exactly one line.

Other geometric figures can be created by using parts or combinations of points, lines, and planes.

Line segment

> The **line segment** *AB*, denoted as $\overline{AB}$, is the part of a line that consists of points *A* and *B* and all points in between (see Figure 9-2). Points *A* and *B* are the **endpoints** of the segment.

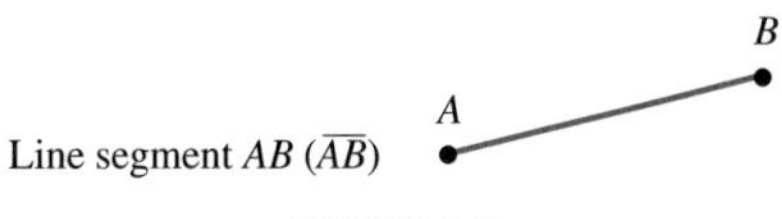

FIGURE 9-2

Every line segment has a **midpoint,** which divides the segment into two parts of equal length. In Figure 9-3, *M* is the midpoint of segment *AB*, because the measure of $\overline{AM}$ (denoted as m($\overline{AM}$)) is equal to the measure of $\overline{MB}$ (denoted as m($\overline{MB}$)).

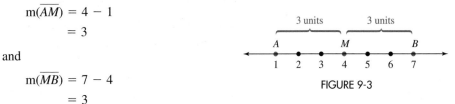

$$m(\overline{AM}) = 4 - 1$$
$$= 3$$

and

$$m(\overline{MB}) = 7 - 4$$
$$= 3$$

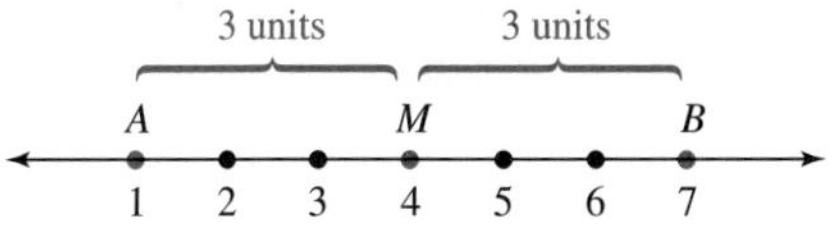

FIGURE 9-3

Since the measure of both segments is 3 units, $m(\overline{AM}) = m(\overline{MB})$.

When two line segments have the same measure, we say that they are **congruent.** Since $m(\overline{AM}) = m(\overline{MB})$, we can write

$\overline{AM} \cong \overline{MB}$ Read $\cong$ as "is congruent to."

Another geometric figure is the *ray,* as shown in Figure 9-4.

Ray

> A **ray** is the part of a line that begins at some point (say, *A*) and continues forever in one direction. Point *A* is the **endpoint** of the ray.

Ray *AB* is denoted as $\overrightarrow{AB}$. The endpoint of the ray is always listed first.

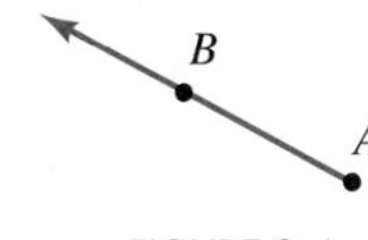

FIGURE 9-4

Angles

Angle

> An **angle** is a figure formed by two rays with a common endpoint. The common endpoint is called the **vertex,** and the rays are called **sides.**

The angle in Figure 9-5 can be denoted as

$\angle BAC$, $\angle CAB$, $\angle A$, or $\angle 1$ The symbol $\angle$ means angle.

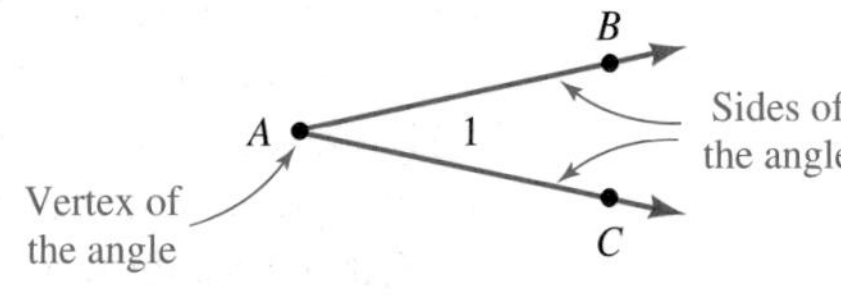

FIGURE 9-5

COMMENT When using three letters to name an angle, be sure the letter name of the vertex is the middle letter.

One unit of measurement of an angle is the **degree.** It is $\frac{1}{360}$ of a full revolution. We can use a **protractor** to measure angles in degrees. See Figure 9-6.

Angle	Measure in degrees
∠*ABC*	30°
∠*ABD*	60°
∠*ABE*	110°
∠*ABF*	150°
∠*ABG*	180°

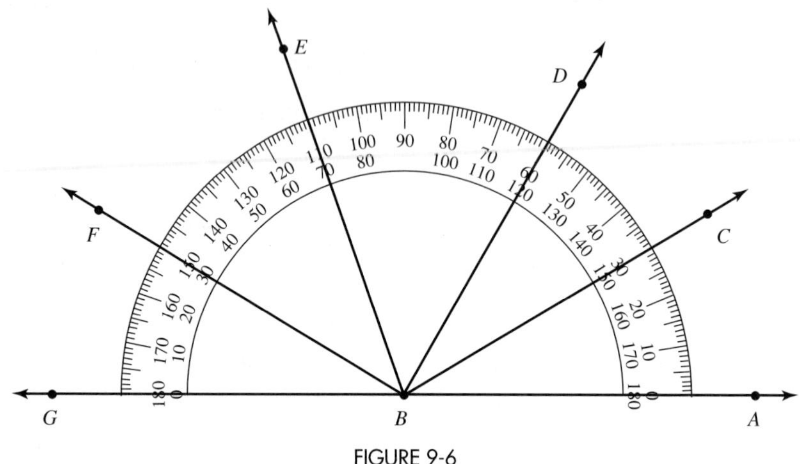

FIGURE 9-6

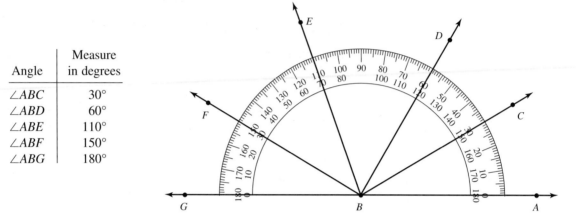

If we read the protractor from left to right, we can see that the measure of ∠*GBF* (denoted as m(∠*GBF*)) is 30°.

When two angles have the same measure, we say that they are congruent. Since m(∠*ABC*) = 30° and m(∠*GBF*) = 30°, we can write

∠*ABC* ≅ ∠*GBF*

We classify angles according to their measure, as in Figure 9-7.

Classification of angles

Acute angles: Angles whose measures are greater than 0° but less than 90°.

Right angles: Angles whose measures are 90°.

Obtuse angles: Angles whose measures are greater than 90° but less than 180°.

Straight angles: Angles whose measures are 180°.

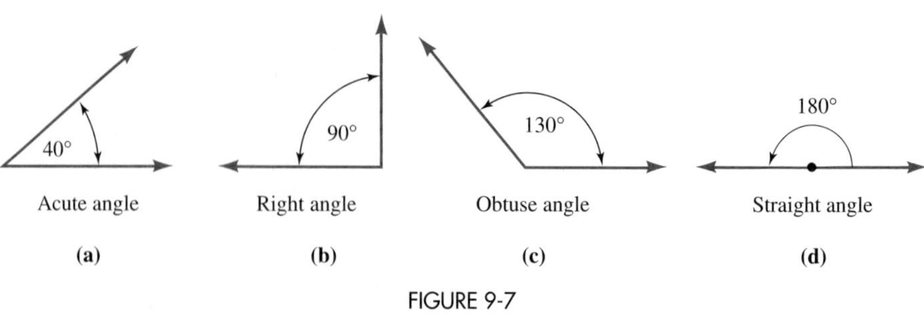

FIGURE 9-7

EXAMPLE 1 ***Classifying angles.*** Classify each angle in Figure 9-8 as an acute angle, a right angle, an obtuse angle, or a straight angle.

Solution Since $m(\angle 1) < 90°$, it is an acute angle.

Since $m(\angle 2) > 90°$ but less than 180°, it is an obtuse angle.

Since $m(\angle BDE) = 90°$, it is a right angle.

Since $m(\angle ABC) = 180°$, it is a straight angle.

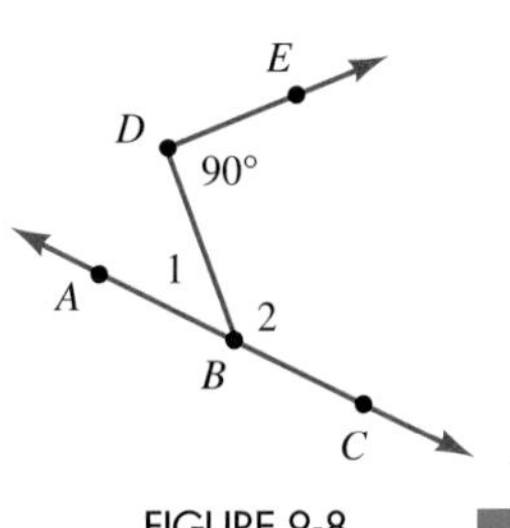

FIGURE 9-8

Adjacent and vertical angles

Two angles that have a common vertex and are side-by-side are called **adjacent angles.**

EXAMPLE 2 ***Evaluating angles.*** Two angles with measures of $x°$ and 35° are adjacent angles. Use the information in Figure 9-9 to find x.

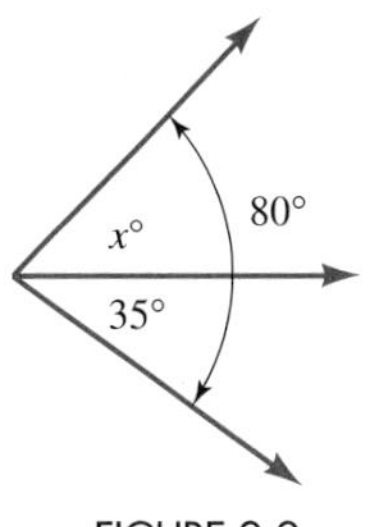

FIGURE 9-9

Solution

We can use algebra to solve this problem. Since the sum of the measures of the angles is 80°, we have

$$x + 35 = 80$$

$$x + 35 - \mathbf{35} = 80 - \mathbf{35}$$ To undo the addition of 35, subtract 35 from both sides.

$$x = 45$$ $35 - 35 = 0$ and $80 - 35 = 45$.

Thus, $x = 45$.

Self Check

In the figure below, find x.

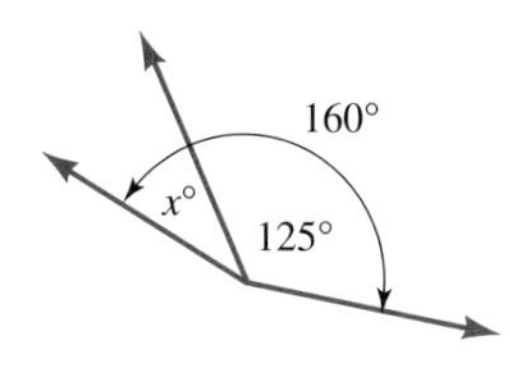

Answer: 35

When two lines intersect, pairs of nonadjacent angles are called **vertical angles.** In Figure 9-10(a), lines l_1 (read as "line l sub 1") and l_2 (read as "line l sub 2") intersect. $\angle 1$ and $\angle 3$ are vertical angles, as are $\angle 2$ and $\angle 4$.

To illustrate that vertical angles always have the same measure, we refer to Figure 9-10(b) with angles having measures of $x°$, $y°$, and 30°. Since the measure of any straight angle is 180°, we have

$$\mathbf{30} + x = 180 \quad \text{and} \quad \mathbf{30} + y = 180$$

$$x = 150 \qquad\qquad y = 150$$ To undo the addition of 30, subtract 30 from both sides.

Since x and y are both 150, $x = y$.

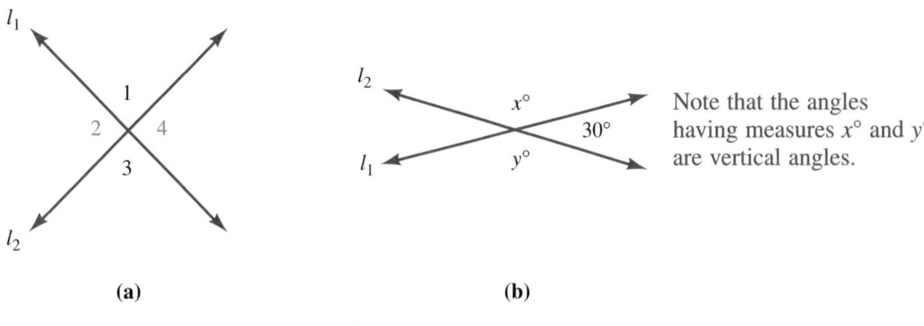

FIGURE 9-10

Property of vertical angles

> Vertical angles are congruent (have the same measure).

EXAMPLE 3 ***Evaluating angles.*** In Figure 9-11, find **a.** m(∠1) and **b.** m(∠3).

Solution

a. The 50° angle and ∠1 are vertical angles. Since vertical angles are congruent, m(∠1) = 50°.

b. Since AD is a line, the sum of the measures of ∠3, the 100° angle, and the 50° angle is 180°. If m(∠3) = x, we have

$$x + 100 + 50 = 180$$

$$x + 150 = 180 \quad \text{Do the addition: } 100 + 50 = 150.$$

$$x = 30 \quad \text{Subtract 150 from both sides.}$$

Thus, m(∠3) = 30°.

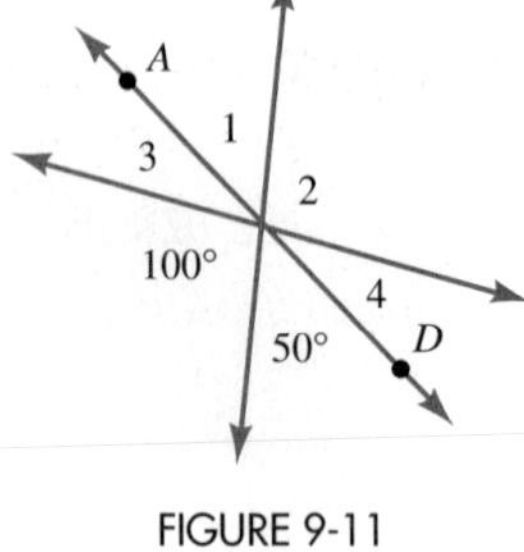

FIGURE 9-11

Self Check

In Figure 9-11, find

a. m(∠2)

b. m(∠4)

Answers: **a.** 100°, **b.** 30°

EXAMPLE 4 ***Evaluating angles.***

In Figure 9-12, find x.

Solution

Since the angles are vertical angles, they have equal measures.

$$4x - 20 = 3x + 15$$

$$x - 20 = 15 \quad \text{To eliminate } 3x \text{ from the right-hand side, subtract } 3x \text{ from both sides.}$$

$$x = 35 \quad \text{To undo the subtraction of 20, add 20 to both sides.}$$

Thus, $x = 35$.

$(3x + 15)°$

$(4x - 20)°$

FIGURE 9-12

Self Check

In the figure below, find y.

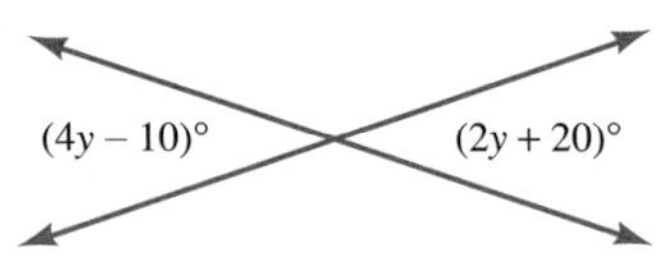

Answer: 15

Complementary and supplementary angles

Complementary and supplementary angles

> Two angles are **complementary angles** when the sum of their measures is 90°.
>
> Two angles are **supplementary angles** when the sum of their measures is 180°.

EXAMPLE 5 ***Complementary and supplementary angles.***

a. Angles of 60° and 30° are complementary angles, because the sum of their measures is 90°. Each angle is the complement of the other.

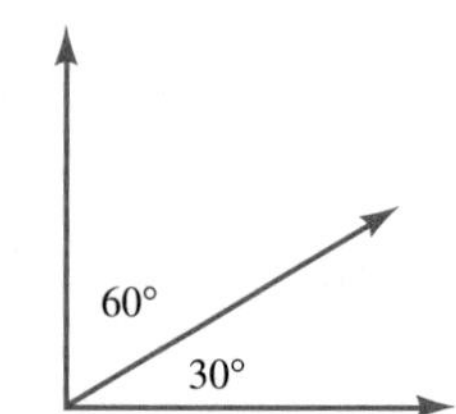

b. Angles of 130° and 50° are supplementary, because the sum of their measures is 180°. Each angle is the supplement of the other.

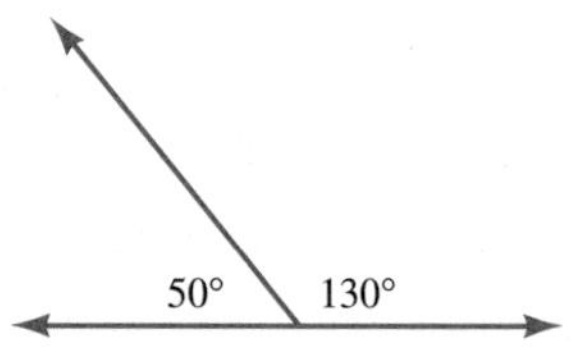

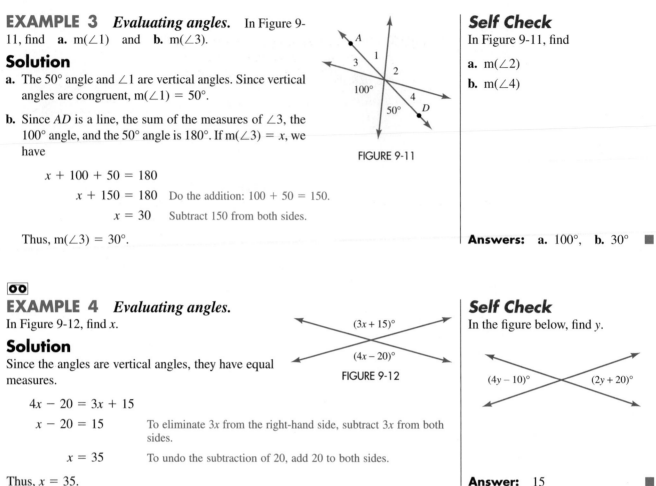

COMMENT The definition of supplementary angles requires that the sum of *two* angles be 180°. Three angles of 40°, 60°, and 80° are not supplementary even though their sum is 180°.

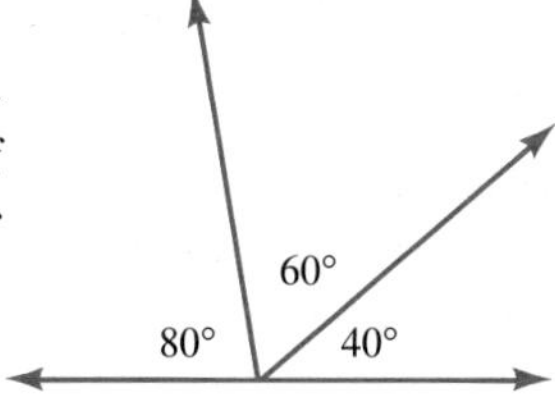

EXAMPLE 6 ***Finding the complement and supplement of an angle.***

a. Find the complement of a 35° angle.

b. Find the supplement of a 105° angle.

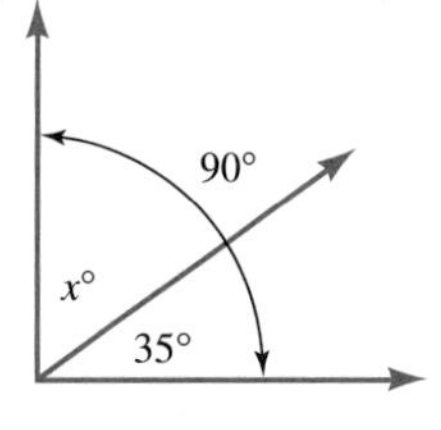

FIGURE 9-13

Solution

a. See Figure 9-13. Let x represent the complement of the 35° angle. Since the angles are complementary, we have

$x + 35 = 90$ The sum of the angles' measures must be 90°.

$x = 55$ To undo the addition of 35, subtract 35 from both sides.

The complement of 35° is 55°.

b. See Figure 9-14. Let y represent the supplement of the 105° angle. Since the angles are supplementary, we have

$y + 105 = 180$ The sum of the angles' measures must be 180°.

$y = 75$ To undo the addition of 105, subtract 105 from both sides.

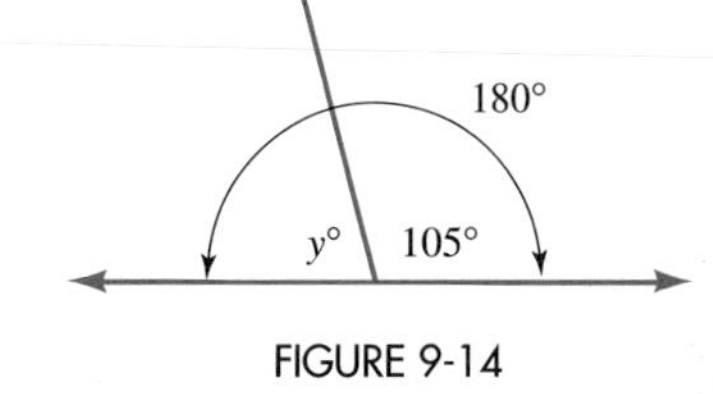

FIGURE 9-14

The supplement of 105° is 75°.

Self Check

a. Find the complement of a 50° angle.

b. Find the supplement of a 50° angle.

Answers: **a.** 40°, **b.** 130° ■

STUDY SET Section 9.1

VOCABULARY *Fill in the blanks to make the statements true.*

1. A line __Segment__ has two endpoints.

2. Two points ________ at most one line.

3. A __midpoint__ divides a line segment into two parts of equal length.

4. An angle is measured in __degrees__.

5. A __protractor__ is used to measure angles.

6. An __acute__ angle is less than 90°.

7. A __right__ angle measures 90°.

8. An __obtuse__ angle is greater than 90° but less than 180°.

9. The measure of a straight angle is __180°__.

10. Adjacent angles have the same vertex and are __Side-by-side__.

11. The sum of two __Supplementary__ angles is 180°.

12. The sum of two complementary angles is __90°__.

CONCEPTS *Refer to Illustration 1 and tell whether each statement is true. If a statement is false, explain why.*

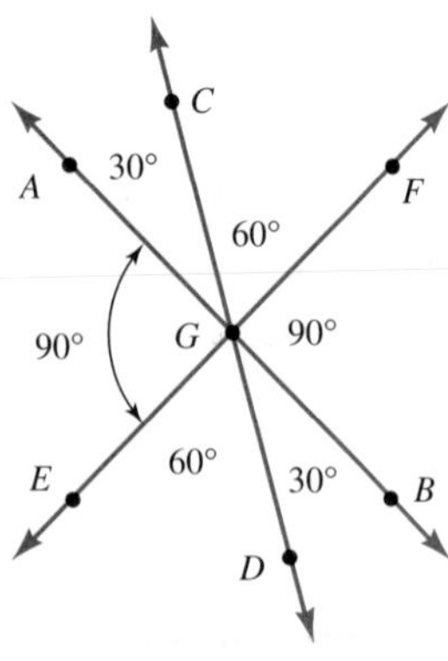

ILLUSTRATION 1

13. $\overrightarrow{GF}$ has point G as its endpoint.

14. $\overline{AG}$ has no endpoints.

15. Line CD has three endpoints.

16. Point D is the vertex of $\angle DGB$.

17. $m(\angle AGC) = m(\angle BGD)$

18. $\angle AGF \cong \angle BGE$

19. $\angle FGB \cong \angle EGA$

20. $\angle AGC$ and $\angle CGF$ are adjacent angles.

Refer to Illustration 1 and tell whether each angle is an acute angle, a right angle, an obtuse angle, or a straight angle.

21. $\angle AGC$

22. $\angle EGA$

23. $\angle FGD$

24. $\angle BGA$

25. $\angle BGE$

26. $\angle AGD$

27. $\angle DGC$

28. $\angle DGB$

Refer to Illustration 2 and tell whether each statement is true. If a statement is false, explain why.

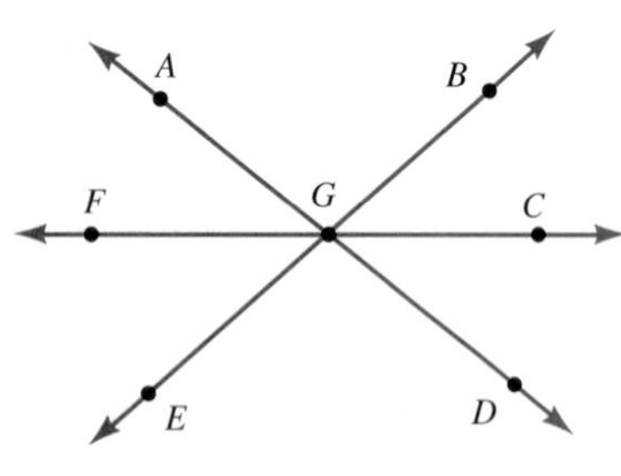

ILLUSTRATION 2

29. $\angle AGF$ and $\angle DGC$ are vertical angles.

30. $\angle FGE$ and $\angle BGA$ are vertical angles.

31. $m(\angle AGB) = m(\angle BGC)$.

32. $\angle AGC \cong \angle DGF$.

Refer to Illustration 3 and tell whether each pair of angles are congruent.

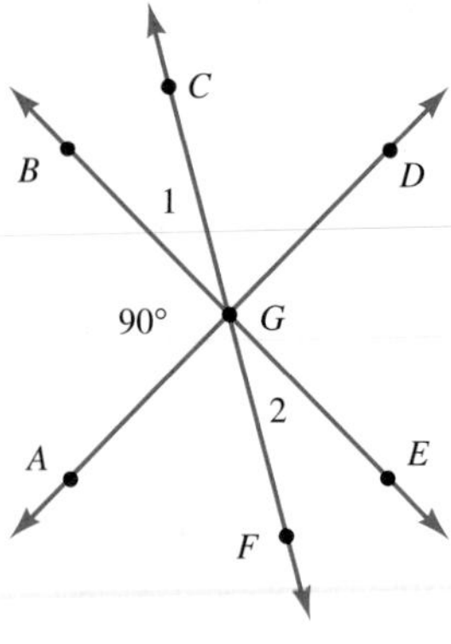

ILLUSTRATION 3

33. $\angle 1$ and $\angle 2$

34. $\angle FGB$ and $\angle CGE$

35. $\angle AGB$ and $\angle DGE$

36. $\angle CGD$ and $\angle CGB$

37. $\angle AGF$ and $\angle FGE$

38. $\angle AGB$ and $\angle BGD$

Refer to Illustration 3 and tell whether each statement is true.

39. $\angle 1$ and $\angle CGD$ are adjacent angles.

40. $\angle 2$ and $\angle 1$ are adjacent angles.

41. $\angle FGA$ and $\angle AGC$ are supplementary.

42. $\angle AGB$ and $\angle BGC$ are complementary.

43. $\angle AGF$ and $\angle 2$ are complementary.

44. $\angle AGB$ and $\angle EGD$ are supplementary.

45. $\angle EGD$ and $\angle DGB$ are supplementary.

46. $\angle DGC$ and $\angle AGF$ are complementary.

NOTATION *Fill in the blanks to make the statements true.*

47. The symbol $\angle$ means ______________.

48. The symbol $\overline{AB}$ is read as "______________ AB."

49. The symbol $\overrightarrow{AB}$ is read as "______________ AB."

50. The symbol ______________ is read as "is congruent to."

PRACTICE *Refer to Illustration 4 and find the length of each segment.*

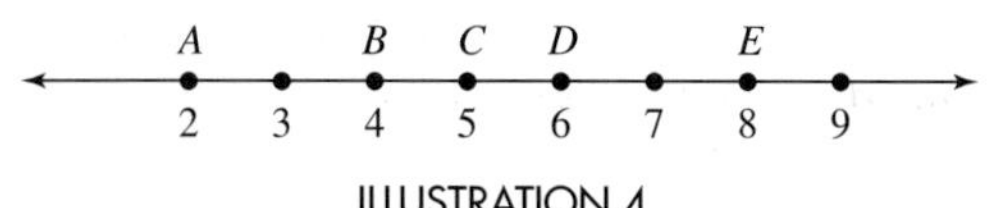

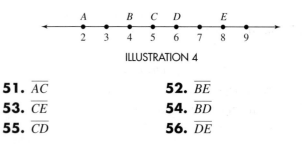

ILLUSTRATION 4

51. $\overline{AC}$

52. $\overline{BE}$

53. $\overline{CE}$

54. $\overline{BD}$

55. $\overline{CD}$

56. $\overline{DE}$

Refer to Illustration 4 and find each midpoint.

57. Find the midpoint of $\overline{AD}$.

58. Find the midpoint of $\overline{BE}$.

Use a protractor to measure each angle.

59. **60.** **61.** **62.**

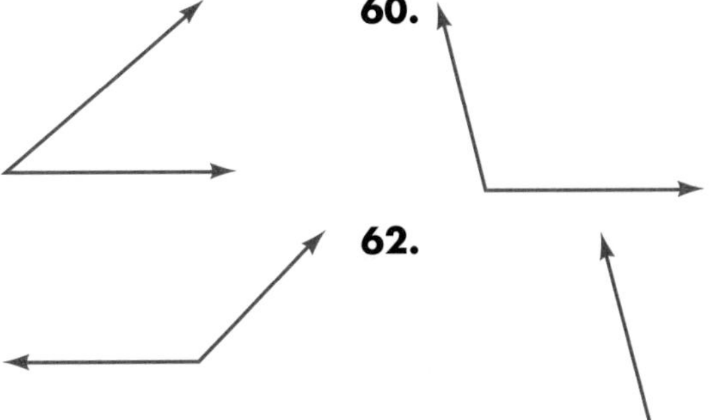

Find x.

63.

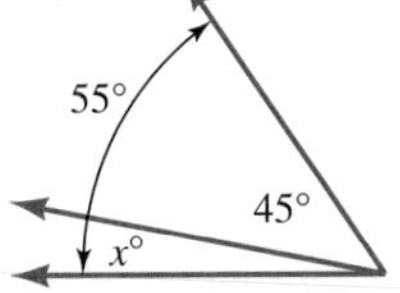

64.

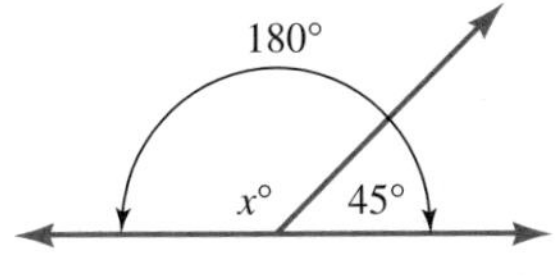

65.

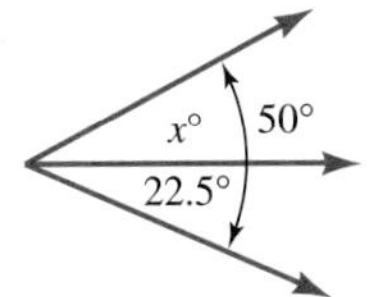

66.

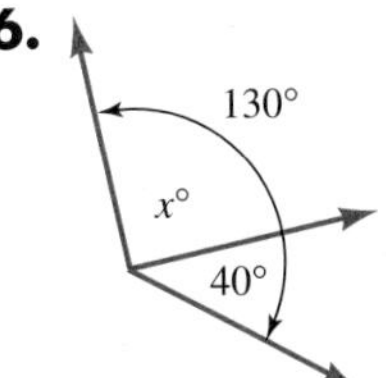

67.

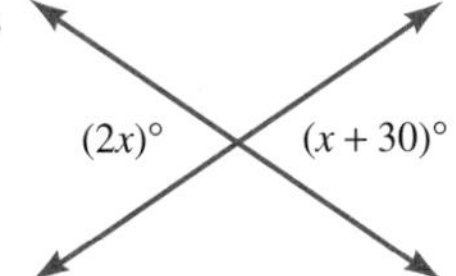

68.

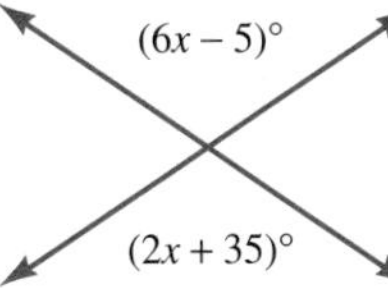

69.

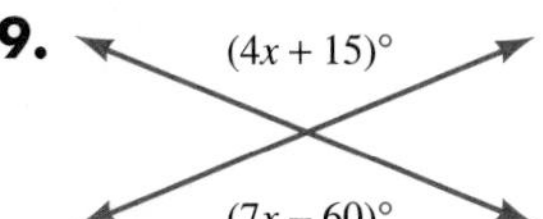

70.

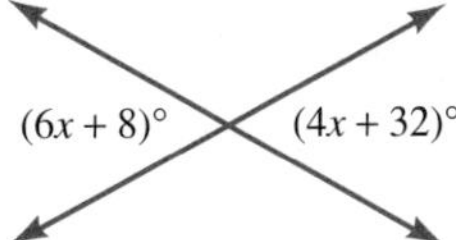

Let x represent the unknown angle measure. Draw a diagram, write an appropriate equation, and solve it for x.

71. Find the complement of a 30° angle.

72. Find the supplement of a 30° angle.

73. Find the supplement of a 105° angle.

74. Find the complement of a 75° angle.

Refer to Illustration 5, in which $m(\angle 1) = 50°$. *Find the measure of each angle or sum of angles.*

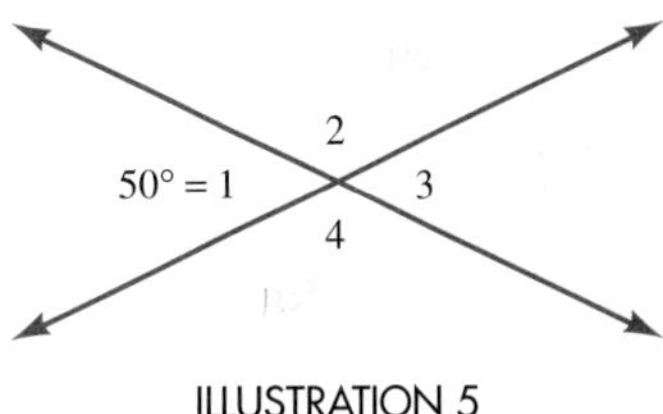

ILLUSTRATION 5

75. $\angle 4$

76. $\angle 3$

77. $m(\angle 1) + m(\angle 2) + m(\angle 3)$

78. $m(\angle 2) + m(\angle 4)$

Refer to Illustration 6, in which $m(\angle 1) + m(\angle 3) + m(\angle 4) = 180°$, $\angle 3 \cong \angle 4$, *and* $\angle 4 \cong \angle 5$. *Find the measure of each angle.*

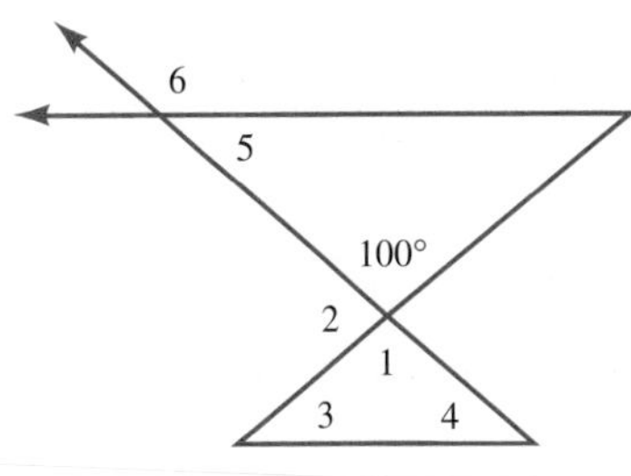

ILLUSTRATION 6

79. $\angle 1$

80. $\angle 2$

81. $\angle 3$

82. $\angle 6$

APPLICATIONS

83. BASEBALL Use the following definition to draw the strike zone for the player shown in Illustration 7.

The strike zone is that area over home plate the upper limit of which is a horizontal line at the midpoint between the top of the shoulders and the top of the uniform pants and the lower level is a line at the hollow beneath the kneecap.

ILLUSTRATION 7

84. PHYSICS Illustration 8 shows a 15-pound block that is suspended with two ropes, one of which is horizontal. Classify each numbered angle in the illustration as either acute, obtuse, or right.

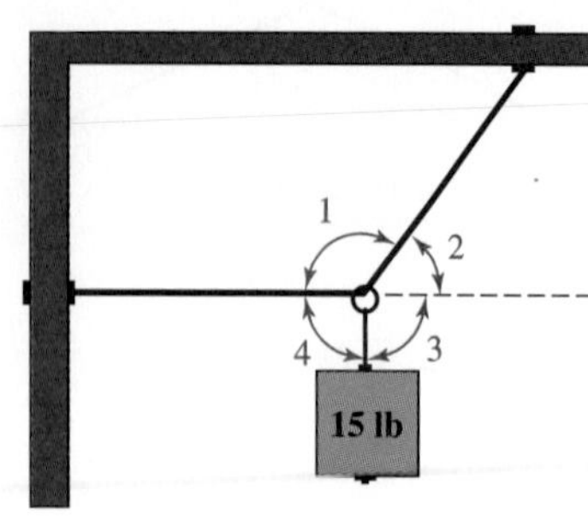

ILLUSTRATION 8

85. SYNTHESIZER Refer to Illustration 9. Find x and y.

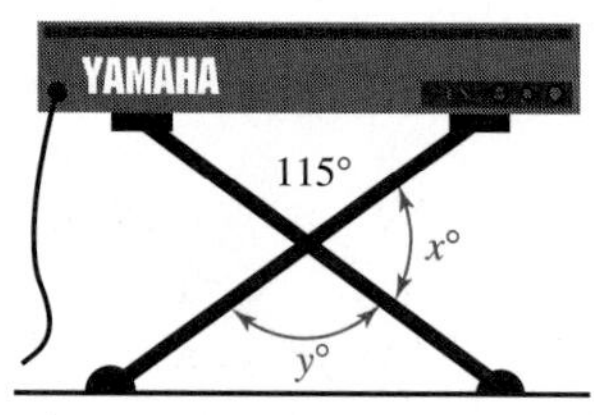

ILLUSTRATION 9

86. AVIATION Refer to Illustration 10. How many degrees from the horizontal position are the wings of the airplane?

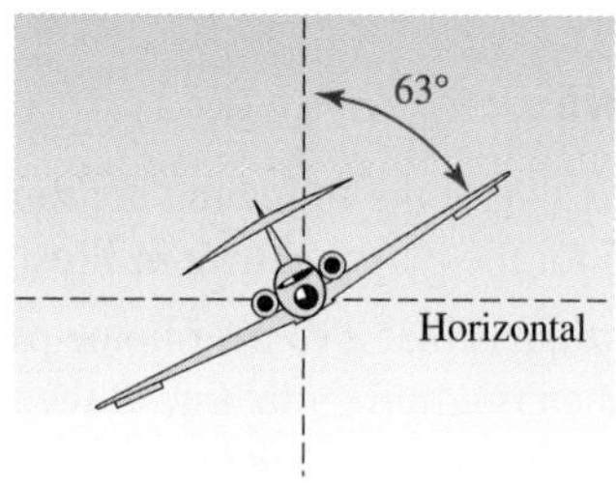

ILLUSTRATION 10

87. GARDENING In Illustration 11, what angle does the handle of the lawn mower make with the ground?

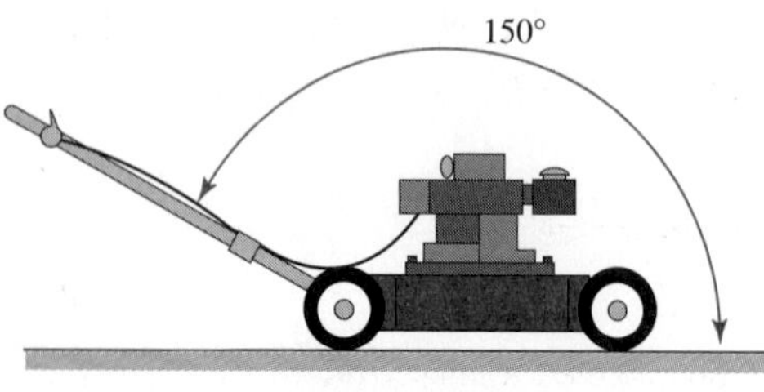

ILLUSTRATION 11

88. MUSICAL INSTRUMENTS Suppose that you are a beginning band teacher describing the correct posture needed to play various instruments. Use the diagrams in Illustration 12 to approximate the angle measure at which each instrument should be held in relation to the student's body: **a.** flute **b.** clarinet **c.** trumpet

ILLUSTRATION 12

WRITING

89. PHRASES Explain what you think each of these phrases means. How is geometry involved?

a. The president did a complete 180-degree flip on the subject of a tax cut.

b. The rollerblader did a "360" as she jumped off the ramp.

90. In the statements below, the ° symbol is used in two different ways. Explain the difference.

85°F and $m(\angle A) = 85°$

91. What is a protractor?

92. Explain the difference between a ray and a line segment.

93. Explain why an angle measuring 105° cannot have a complement.

94. Explain why an angle measuring 210° cannot have a supplement.

REVIEW

95. Find 2^4.

96. Add: $\frac{1}{2} + \frac{2}{3} + \frac{3}{4}$

97. Subtract: $\frac{3}{4} - \frac{1}{8} - \frac{1}{3}$

98. Multiply: $\frac{5}{8} \cdot \frac{2}{15} \cdot \frac{6}{5}$

99. Graph the equation $y = 2x - 5$.

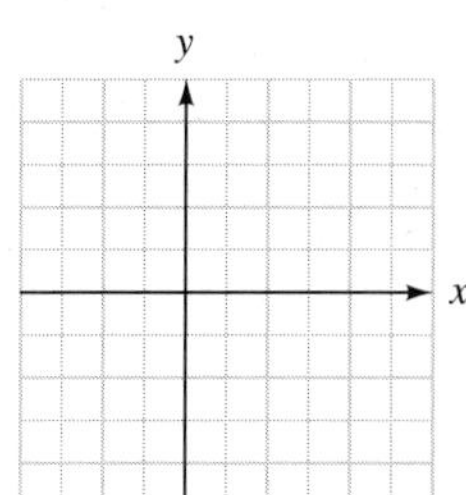

100. What is 7% of 7?

101. Solve the proportion $\frac{x+1}{18} = \frac{12.5}{45}$.

102. Convert 120 yards to feet.

9.2 *Parallel and Perpendicular Lines*

In this section, you will learn about

- Parallel and perpendicular lines • Transversals and angles
- Properties of parallel lines

Introduction. In this section, we will consider *parallel* and *perpendicular* lines. Since parallel lines are always the same distance apart, the railroad tracks shown in Figure 9-15(a) illustrate one application of parallel lines. Figure 9-15(b) shows one of the events of men's gymnastics, the parallel bars. Since perpendicular lines meet and form right angles, the monument and the ground shown in Figure 9-15(c) illustrate one application of perpendicular lines.

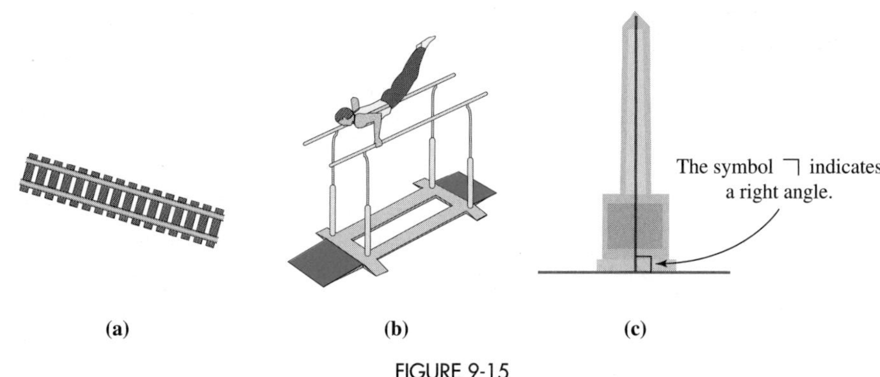

FIGURE 9-15

Parallel and perpendicular lines

If two lines lie in the same plane, they are called **coplanar.** Two coplanar lines that do not intersect are called **parallel lines.** See Figure 9-16(a) on the next page.

Parallel lines

> **Parallel lines** are coplanar lines that do not intersect.

If lines l_1 (l sub 1) and l_2 (l sub 2) are parallel, we can write $l_1 \parallel l_2$, where the symbol $\parallel$ is read as "is parallel to."

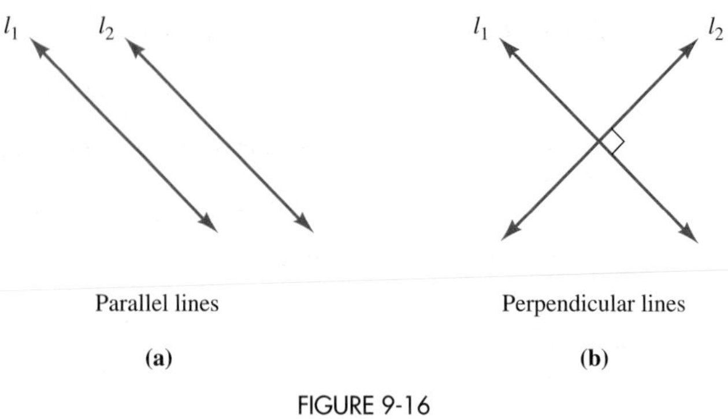

FIGURE 9-16

Perpendicular lines

Perpendicular lines are lines that intersect and form right angles.

In Figure 9-16(b), $l_1 \perp l_2$, where the symbol $\perp$ is read as "is perpendicular to."

Transversals and angles

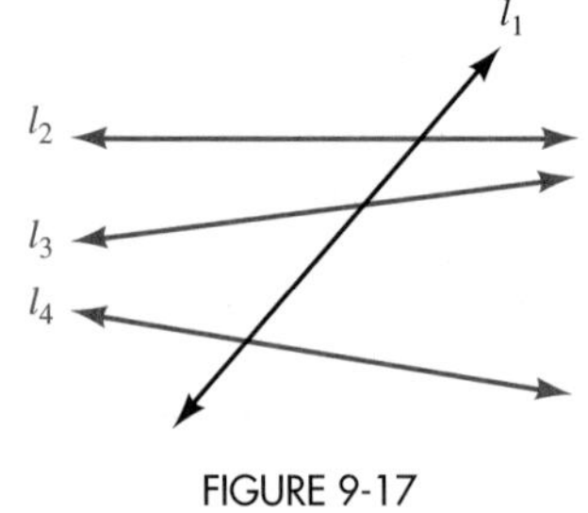

FIGURE 9-17

A line that intersects two or more coplanar lines is called a **transversal.** For example, line l_1 in Figure 9-17 is a transversal intersecting lines l_2, l_3, and l_4.

When two lines are cut by a transversal, the following types of angles are formed.

Alternate interior angles:
$\angle 4$ and $\angle 5$
$\angle 3$ and $\angle 6$

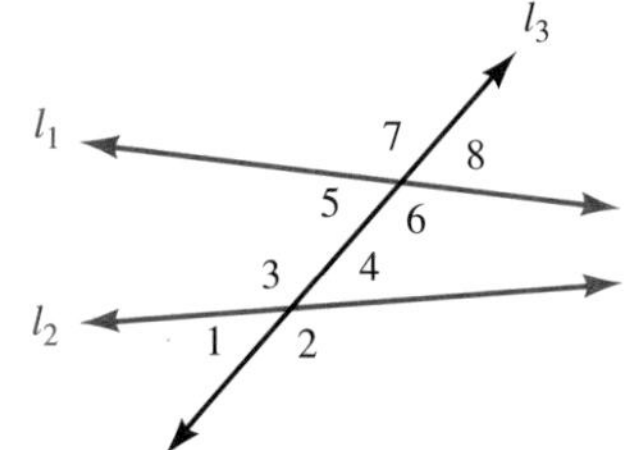

Corresponding angles:
$\angle 1$ and $\angle 5$
$\angle 3$ and $\angle 7$
$\angle 2$ and $\angle 6$
$\angle 4$ and $\angle 8$

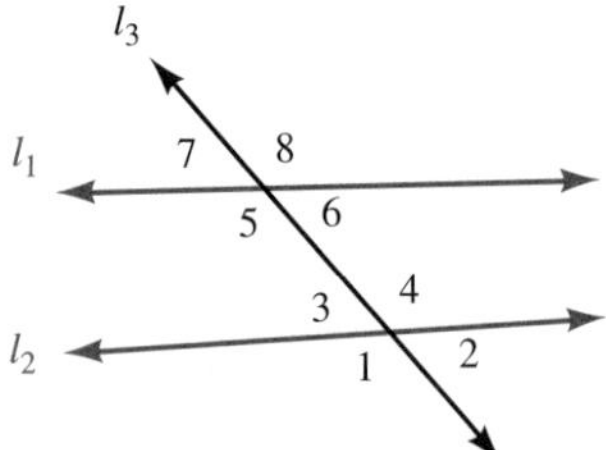

Interior angles:
$\angle 3$, $\angle 4$, $\angle 5$, and $\angle 6$

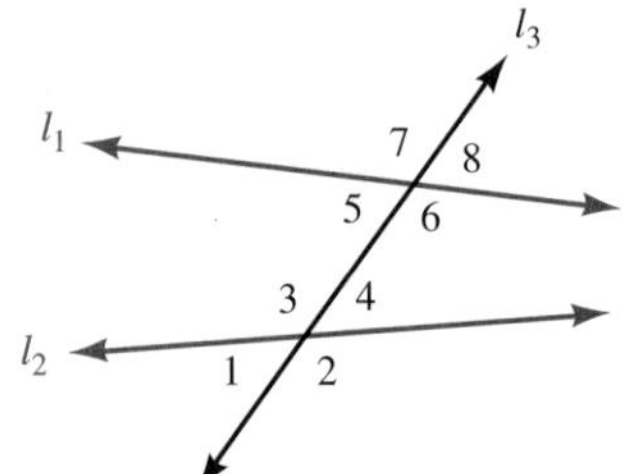

EXAMPLE 1 ***Identifying angles.*** In Figure 9-18, identify **a.** all pairs of alternate interior angles, **b.** all pairs of corresponding angles, and **c.** all interior angles.

Solution

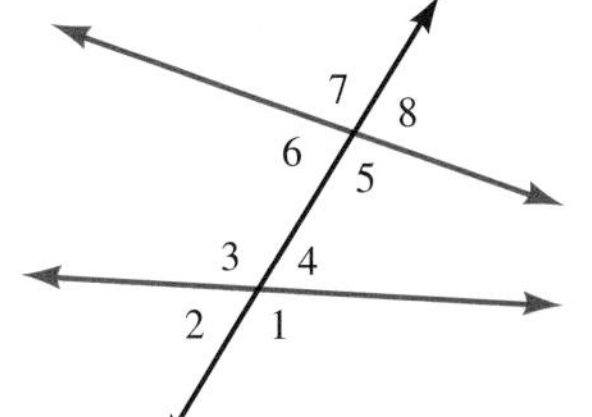

FIGURE 9-18

a. Pairs of alternate interior angles are

$$\angle 3 \text{ and } \angle 5,\ \angle 4 \text{ and } \angle 6$$

b. Pairs of corresponding angles are

$$\angle 1 \text{ and } \angle 5,\ \angle 4 \text{ and } \angle 8,\ \angle 2 \text{ and } \angle 6,\ \angle 3 \text{ and } \angle 7$$

c. Interior angles are

$$\angle 3,\ \angle 4,\ \angle 5, \text{ and } \angle 6$$

■

Properties of parallel lines

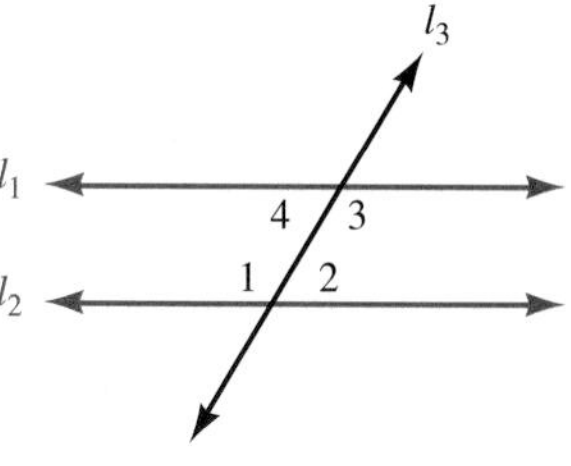

FIGURE 9-19

1. If two parallel lines are cut by a transversal, alternate interior angles are congruent. (See Figure 9-19.) If $l_1 \parallel l_2$, then $\angle 2 \cong \angle 4$ and $\angle 1 \cong \angle 3$.
2. If two parallel lines are cut by a transversal, corresponding angles are congruent. (See Figure 9-20.) If $l_1 \parallel l_2$, then $\angle 1 \cong \angle 5$, $\angle 3 \cong \angle 7$, $\angle 2 \cong \angle 6$, and $\angle 4 \cong \angle 8$.
3. If two parallel lines are cut by a transversal, interior angles on the same side of the transversal are supplementary. (See Figure 9-21.) If $l_1 \parallel l_2$, then $\angle 1$ is supplementary to $\angle 2$ and $\angle 4$ is supplementary to $\angle 3$.

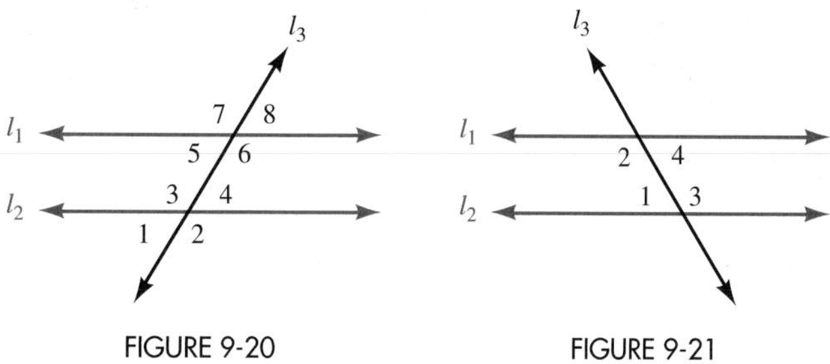

FIGURE 9-20 FIGURE 9-21

4. If a transversal is perpendicular to one of two parallel lines, it is also perpendicular to the other line. (See Figure 9-22.) If $l_1 \parallel l_2$ and $l_3 \perp l_1$, then $l_3 \perp l_2$.
5. If two lines are parallel to a third line, they are parallel to each other. (See Figure 9-23.) If $l_1 \parallel l_2$ and $l_1 \parallel l_3$, then $l_2 \parallel l_3$.

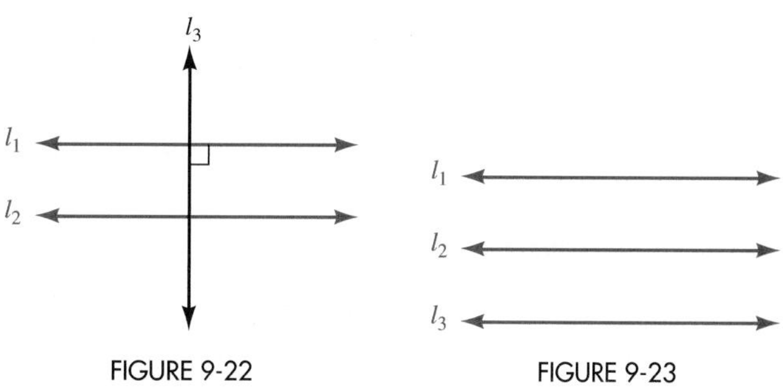

FIGURE 9-22 FIGURE 9-23

EXAMPLE 2 ***Evaluating angles.*** See Figure 9-24 on the next page. If $l_1 \parallel l_2$ and $m(\angle 3) = 120°$, find the measures of the other angles.

Self Check
If $l_1 \parallel l_2$ and $m(\angle 8) = 50°$, find the measures of the other angles. (See Figure 9-24.)

Solution

$m(\angle 1) = 60°$	$\angle 3$ and $\angle 1$ are supplementary.
$m(\angle 2) = 120°$	Vertical angles are congruent: $m(\angle 2) = m(\angle 3)$.
$m(\angle 4) = 60°$	Vertical angles are congruent: $m(\angle 4) = m(\angle 1)$.
$m(\angle 5) = 60°$	If two parallel lines are cut by a transversal, alternate interior angles are congruent: $m(\angle 5) = m(\angle 4)$.
$m(\angle 6) = 120°$	If two parallel lines are cut by a transversal, alternate interior angles are congruent: $m(\angle 6) = m(\angle 3)$.
$m(\angle 7) = 120°$	Vertical angles are congruent: $m(\angle 7) = m(\angle 6)$.
$m(\angle 8) = 60°$	Vertical angles are congruent: $m(\angle 8) = m(\angle 5)$.

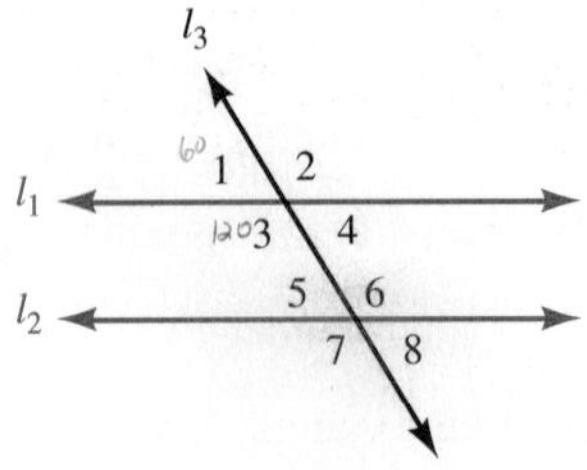

FIGURE 9-24

Answers: $m(\angle 5) = 50°$, $m(\angle 7) = 130°$, $m(\angle 6) = 130°$, $m(\angle 3) = 130°$, $m(\angle 4) = 50°$, $m(\angle 1) = 50°$, $m(\angle 2) = 130°$ ■

EXAMPLE 3 ***Identifying congruent angles.*** See Figure 9-25. If $\overline{AB} \parallel \overline{DE}$, which pairs of angles are congruent?

Solution Since $\overline{AB} \parallel \overline{DE}$, corresponding angles are congruent. So we have

$$\angle A \cong \angle 1 \quad \text{and} \quad \angle B \cong \angle 2$$

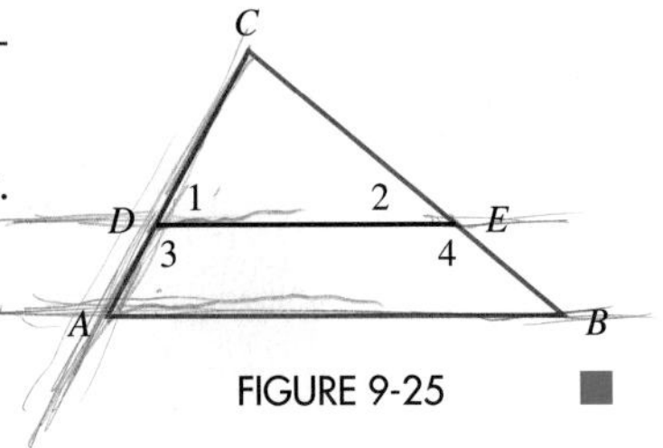

FIGURE 9-25 ■

EXAMPLE 4 ***Using algebra in geometry.*** In Figure 9-26, $l_1 \parallel l_2$. Find x.

Solution

The angles involving x are corresponding angles. Since $l_1 \parallel l_2$, all pairs of corresponding angles are congruent.

$9x - 15 = 6x + 30$	The angle measures are equal.
$3x - 15 = 30$	Subtract $6x$ from both sides.
$3x = 45$	To undo the subtraction of 15, add 15 to both sides.
$x = 15$	To undo the multiplication by 3, divide both sides by 3.

Thus, $x = 15$.

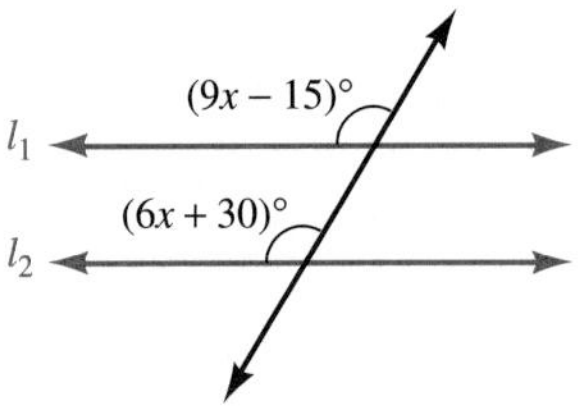

FIGURE 9-26

Self Check

In the figure below, $l_1 \parallel l_2$. Find y.

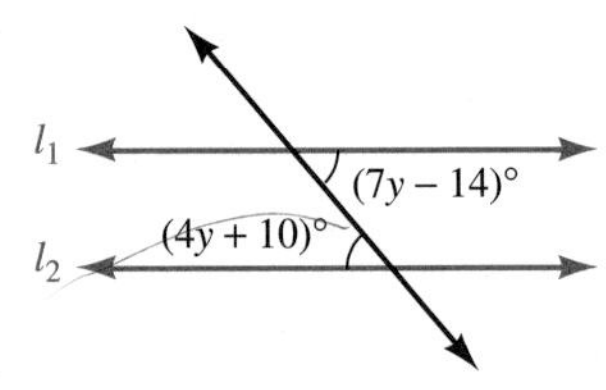

Answer: 8 ■

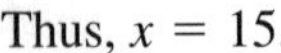

EXAMPLE 5 ***Using algebra in geometry.*** In Figure 9-27, $l_1 \parallel l_2$. Find x.

Solution Since the angles are interior angles on the same side of the transversal, they are supplementary.

$3x - 80 + 3x + 20 = 180$	The sum of the measures of two supplementary angles is 180°.
$6x - 60 = 180$	Combine like terms.
$6x = 240$	To undo the subtraction of 60, add 60 to both sides.
$x = 40$	To undo the multiplication by 6, divide both sides by 6.

Thus, $x = 40$. ■

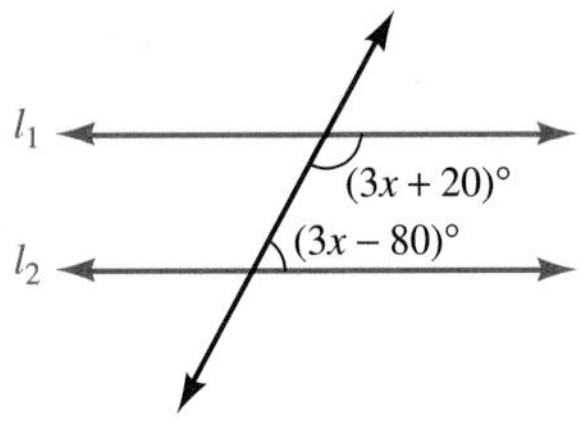

FIGURE 9-27

STUDY SET Section 9.2

VOCABULARY *Fill in the blanks to make the statements true.*

1. Two lines in the same plane are ______________.

2. ______________ lines do not intersect.

3. If two lines intersect and form right angles, they are ______________.

4. A ______________ intersects two or more coplanar lines.

5. In Illustration 1, ∠4 and ∠6 are ______________ interior angles.

6. In Illustration 1, ∠2 and ∠6 are ______________ angles.

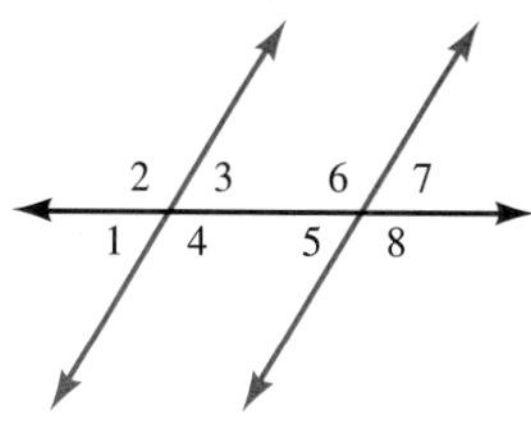

ILLUSTRATION 1

CONCEPTS

7. Which pairs of angles shown in Illustration 1 are alternate interior angles?

8. Which pairs of angles shown in Illustration 1 are corresponding angles?

9. Which angles shown in Illustration 1 are interior angles?

10. In Illustration 2, $l_1 \parallel l_2$. What can you conclude about l_1 and l_3?

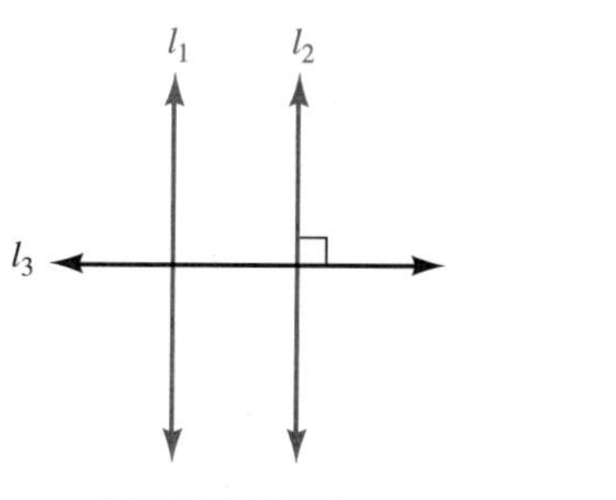

ILLUSTRATION 2

11. In Illustration 3, $l_1 \parallel l_2$ and $l_2 \parallel l_3$. What can you conclude about l_1 and l_3?

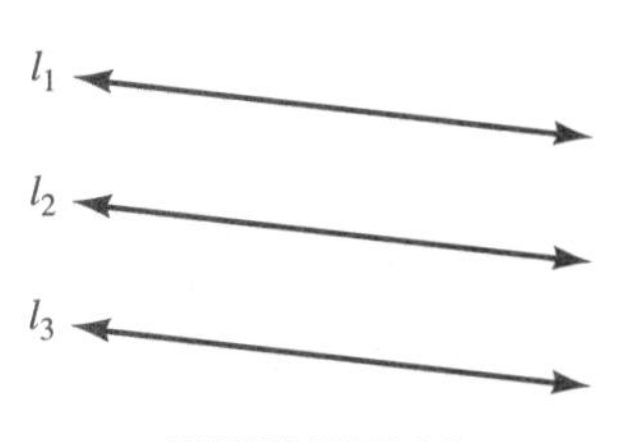

ILLUSTRATION 3

12. In Illustration 4, $\overline{AB} \parallel \overline{DE}$. What pairs of angles are congruent?

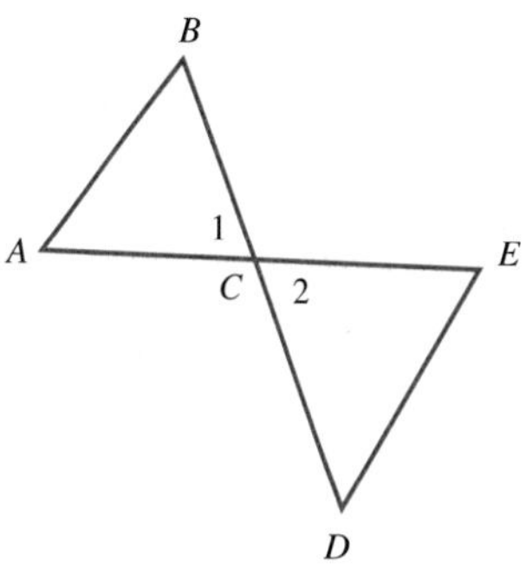

ILLUSTRATION 4

NOTATION *Fill in the blanks to make the statements true.*

13. The symbol ⌐ indicates ______________.

14. The symbol ∥ is read as "______________."

15. The symbol ⊥ is read as "______________."

16. The symbol l_1 is read as "______________."

PRACTICE

17. In Illustration 5, $l_1 \parallel l_2$ and m(∠4) = 130°. Find the measures of the other angles.

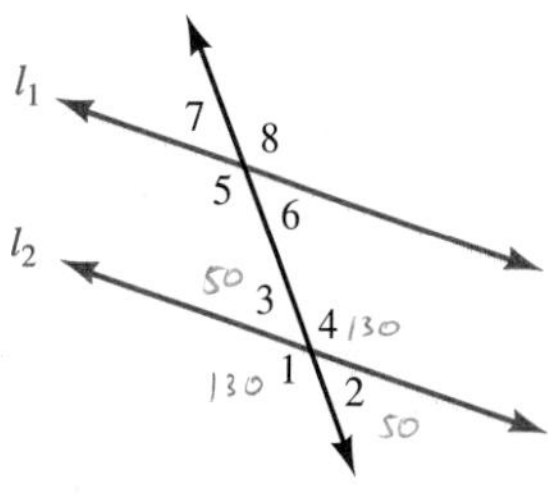

ILLUSTRATION 5

18. In Illustration 6, $l_1 \parallel l_2$ and m(∠2) = 40°. Find the measures of the other angles.

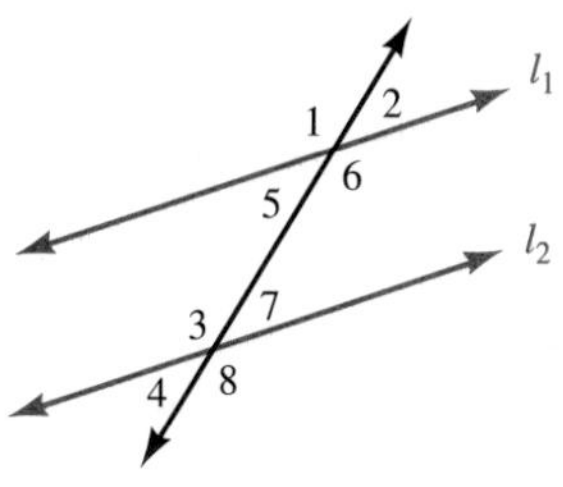

ILLUSTRATION 6

19. In Illustration 7, $l_1 \parallel \overline{AB}$. Find the measure of each angle.

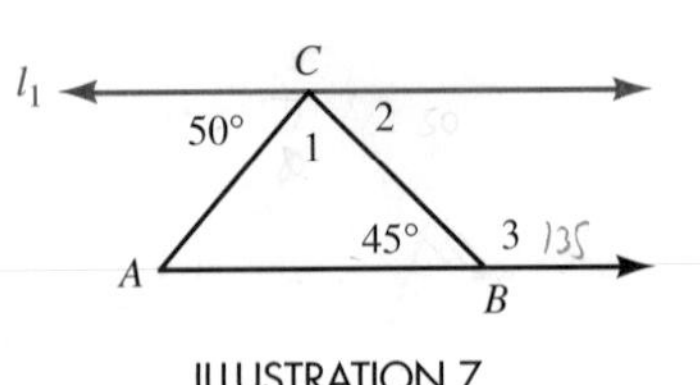

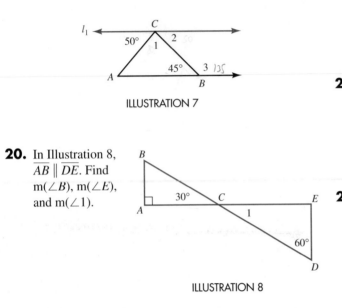

ILLUSTRATION 7

20. In Illustration 8, $\overline{AB} \parallel \overline{DE}$. Find $m(\angle B)$, $m(\angle E)$, and $m(\angle 1)$.

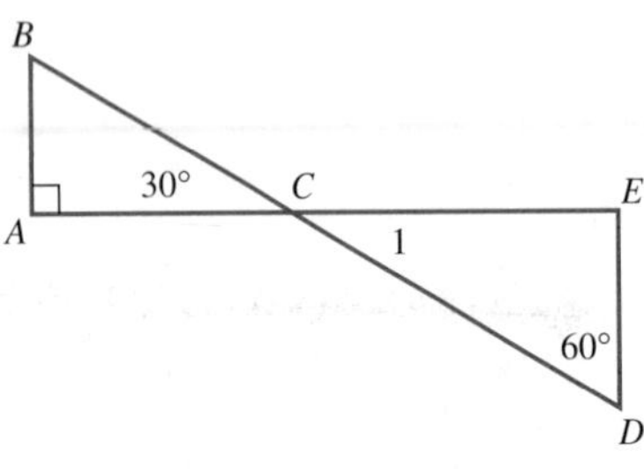

ILLUSTRATION 8

In Exercises 21–24, $l_1 \parallel l_2$. Find x.

21.

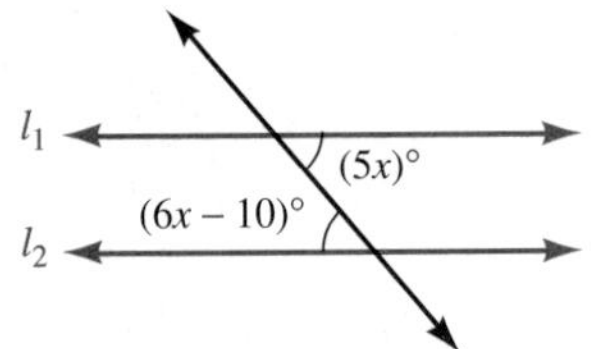

22.

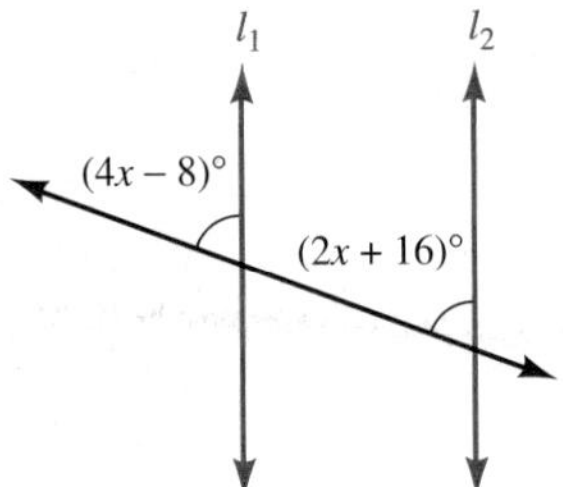

23.

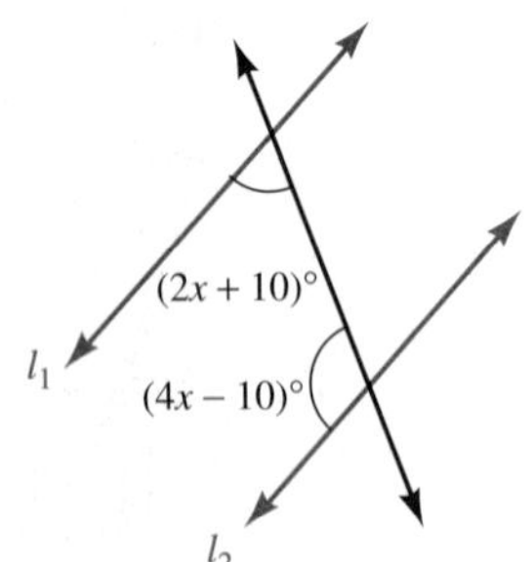

24.

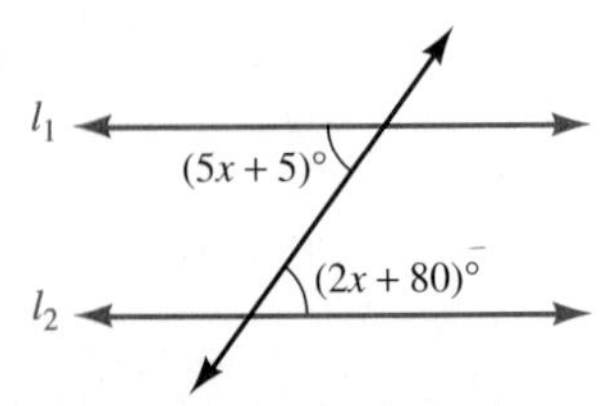

In Exercises 25–28, find x.

25. $l_1 \parallel \overline{CA}$

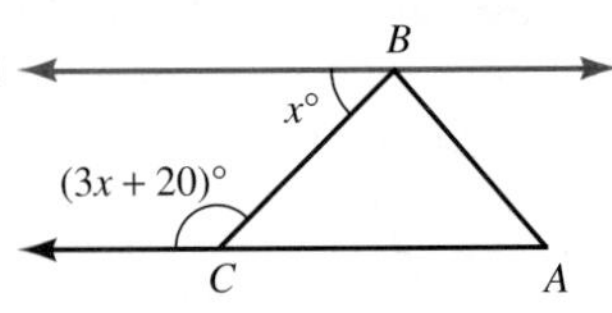

26. $\overline{AB} \parallel \overline{DE}$

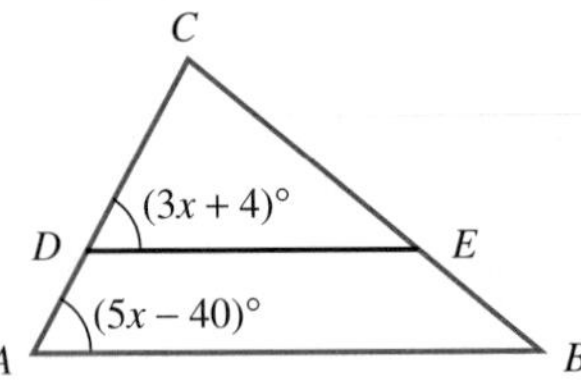

27. $\overline{AB} \parallel \overline{DE}$

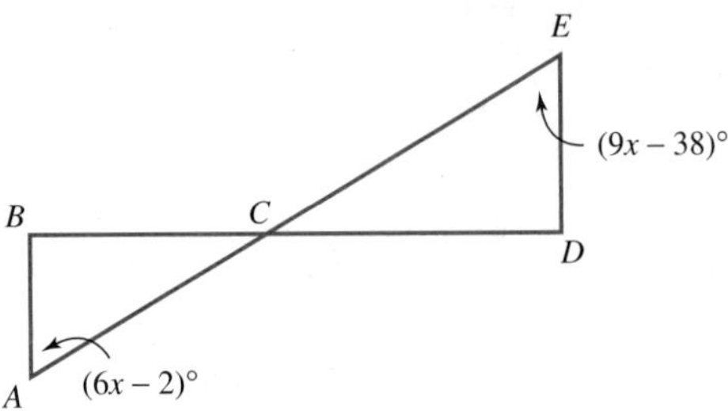

28. $\overline{AC} \parallel \overline{BD}$

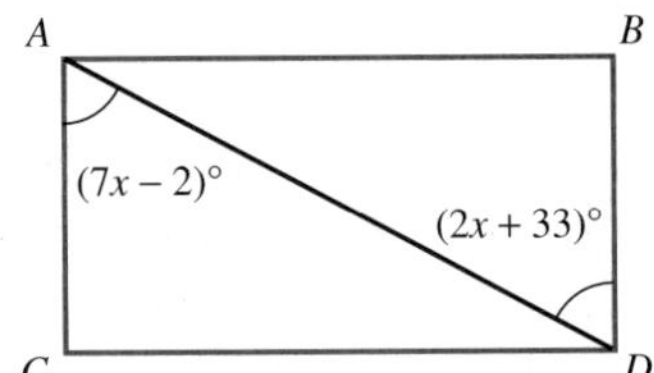

APPLICATIONS

29. CONSTRUCTING PYRAMIDS The Egyptians used a device called a **plummet** to tell whether stones were properly leveled. A plummet, shown in Illustration 9, is made up of an A-frame and a plumb bob suspended from the peak of the frame. How could a builder use a plummet to tell that the stone on the left is not level and that the stones on the right are level?

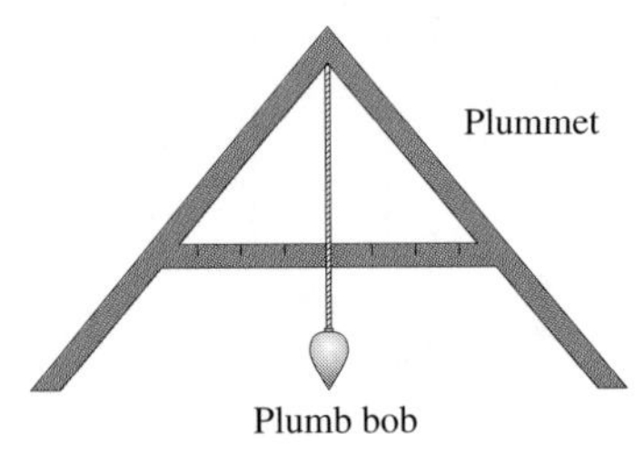

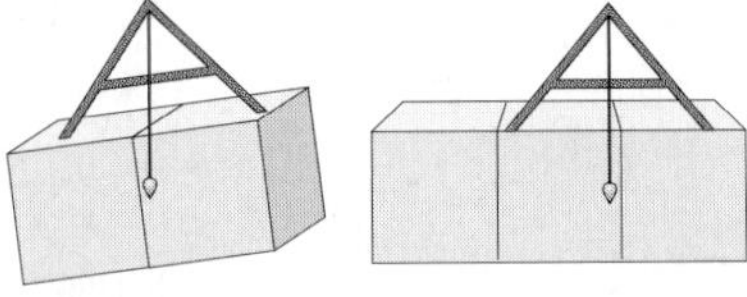
ILLUSTRATION 9

30. DIAGRAMMING SENTENCES English instructors have their students diagram sentences to help teach proper sentence structure. Illustration 10 is a diagram of the sentence *The cave was rather dark and damp.* Point out pairs of parallel and perpendicular lines used in the diagram.

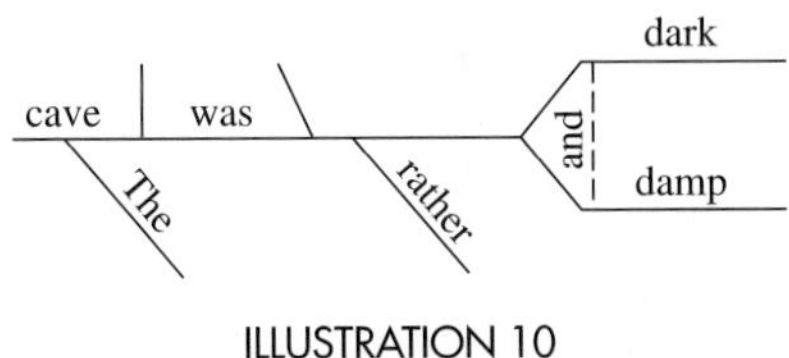

ILLUSTRATION 10

31. LOGO Point out any perpendicular lines that can be found on the BMW company logo shown in Illustration 11.

ILLUSTRATION 11

32. PAINTING SIGNS For many sign painters, the most difficult letter to paint is a capital E, because of all of the right angles involved. See Illustration 12. How many right angles are there?

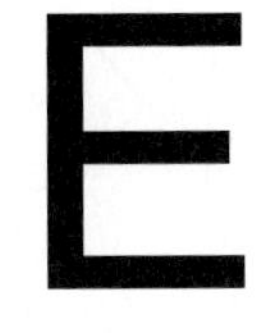

ILLUSTRATION 12

33. HANGING WALLPAPER Explain why the concepts of perpendicular and parallel are both important when hanging wallpaper.

34. TOOLS See Illustration 13. What geometric concepts are seen in the design of the rake?

ILLUSTRATION 13

WRITING

35. PARKING DESIGN Using terms from this chapter, write a paragraph describing the parking layout shown in Illustration 14.

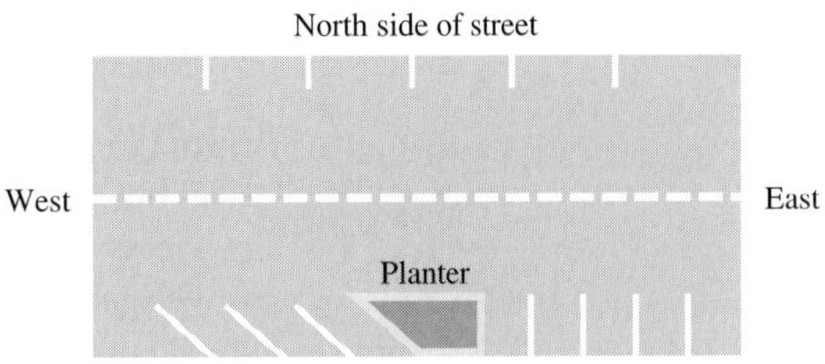

ILLUSTRATION 14

36. In your own words, explain what is meant by each of the following sentences.

a. The hikers were told that the path *parallels* the river.

b. John's quick rise to fame and fortune *paralleled* that of his older brother.

c. The judge stated that the case that was before her court was without *parallel.*

37. Why do you think that $\angle 4$ and $\angle 6$ shown in Illustration 1 are called alternate interior angles?

38. Why do you think that $\angle 4$ and $\angle 8$ shown in Illustration 1 (on page 539) are called corresponding angles?

39. Are pairs of alternate interior angles always congruent? Explain.

40. Are pairs of interior angles always supplementary? Explain.

REVIEW

41. Find 60% of 120.

42. 80% of what number is 400?

43. What percent of 500 is 225?

44. Simplify: $3.45 + 7.37 \cdot 2.98$

45. Is every whole number an integer?

46. Multiply: $2\frac{1}{5} \cdot 4\frac{3}{7}$

47. Express the phrase as a ratio in lowest terms: 4 ounces to 12 ounces

48. Convert 5,400 milligrams to kilograms.

9.3 Polygons

In this section, you will learn about

- Polygons • Triangles • Properties of isosceles triangles
- The sum of the measures of the angles of a triangle • Quadrilaterals
- Properties of rectangles • The sum of the measures of the angles of a polygon

INTRODUCTION. In this section, we will discuss figures called *polygons.* We see these shapes every day. For example, the walls in most buildings are rectangular in shape. We also see rectangular shapes in doors, windows, and sheets of paper.

The gable ends of many houses are triangular in shape, as are the sides of the Great Pyramid in Egypt. Triangular shapes are especially important because triangles are rigid and contribute strength and stability to walls and towers.

The designs used in tile or linoleum floors often use the shapes of a pentagon or a hexagon. Stop signs are in the shape of an octagon.

Polygons

Polygon

A **polygon** is a closed geometric figure with at least three line segments for its sides.

The figures in Figure 9-28 are **polygons.** They are classified according to the number of sides they have. The points where the sides intersect are called **vertices.** If a polygon has sides that are all the same length and angles that are the same, we call it a **regular polygon.**

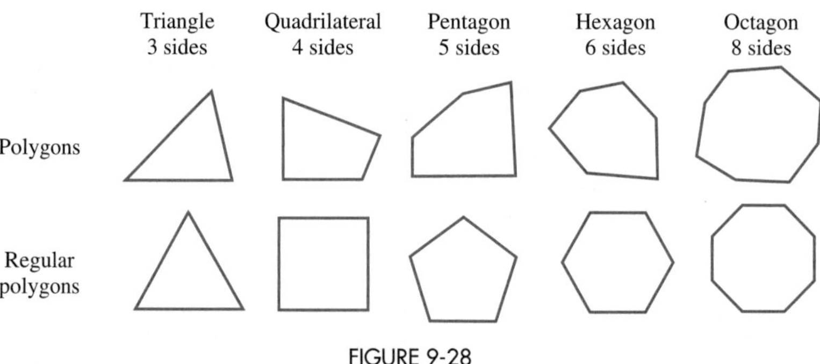

FIGURE 9-28

EXAMPLE 1 ***Vertices of a polygon.*** Give the number of vertices of **a.** a triangle and **b.** a hexagon.

Solution

a. From Figure 9-28, we see that a triangle has three angles and therefore three vertices.

b. From Figure 9-28, we see that a hexagon has six angles and therefore six vertices.

Self Check

Give the number of vertices of

a. a quadrilateral

b. a pentagon

Answers: **a.** 4 **b.** 5 ■

From the results of Example 1, we see that the number of vertices of a polygon is equal to the number of its sides.

Triangles

A **triangle** is a polygon with three sides. Figure 9-29 illustrates some common triangles. The slashes on the sides of a triangle indicate which sides are of equal length.

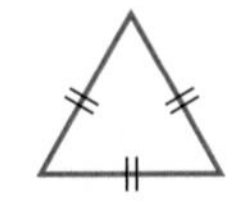

Equilateral triangle (all sides equal length)

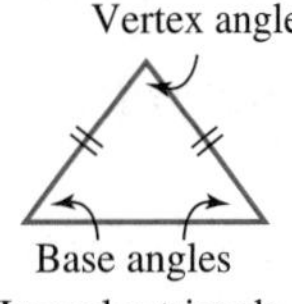

Isosceles triangle (at least two sides of equal length)

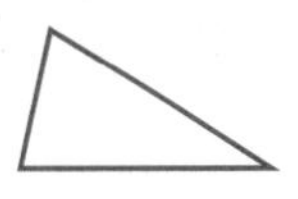

Scalene triangle (no sides equal length)

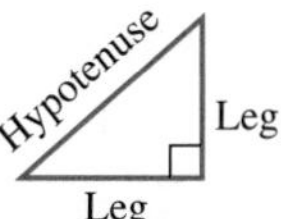

Right triangle (has a right angle)

FIGURE 9-29

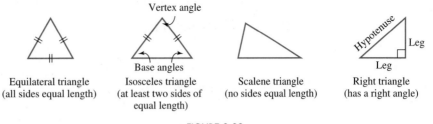

COMMENT Since equilateral triangles have at least two sides of equal length, they are also isosceles. However, isosceles triangles are not necessarily equilateral.

Since every angle of an equilateral triangle has the same measure, an equilateral triangle is also **equiangular.**

In an isosceles triangle, the angles opposite the sides of equal length are called **base angles,** the sides of equal length form the **vertex angle,** and the third side is called the **base.**

The longest side of a right triangle is called the **hypotenuse,** and the other two sides are called **legs.** The hypotenuse of a right triangle is always opposite the 90° angle.

Properties of isosceles triangles

1. Base angles of an isosceles triangle are congruent.
2. If two angles in a triangle are congruent, the sides opposite the angles have the same length, and the triangle is isosceles.

EXAMPLE 2 ***Determining whether a triangle is isosceles.*** Is the triangle in Figure 9-30 an isosceles triangle?

Solution

$\angle A$ and $\angle B$ are angles of the triangle. Since $m(\angle A) = m(\angle B)$, we know that $m(\overline{AC}) = m(\overline{BC})$ and that $\triangle ABC$ (read as "triangle ABC") is isosceles.

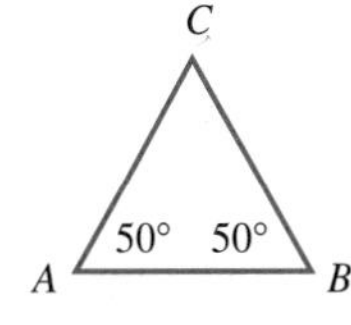

FIGURE 9-30

Self Check

In the figure below, $l_1 \parallel \overline{AB}$. Is the triangle an isosceles triangle?

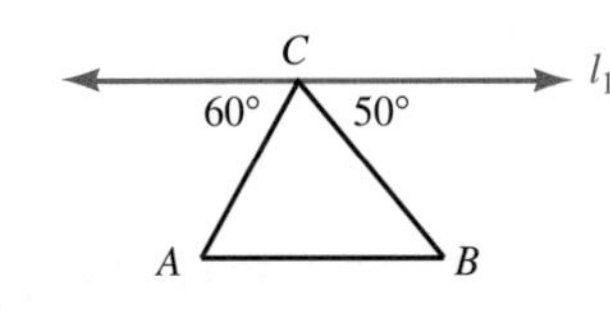

Answer: no

The sum of the measures of the angles of a triangle

If you draw several triangles and carefully measure each angle with a protractor, you will find that the sum of the angle measures in each triangle is 180°.

Angles of a triangle

The sum of the angle measures of any triangle is 180°.

EXAMPLE 3 ***Sum of the angles of a triangle.*** See Figure 9-31. Find x.

FIGURE 9-31

Solution

Since the sum of the angle measures of any triangle is 180°, we have

$$x + 40 + 90 = 180$$

$$x + 130 = 180 \quad \text{Do the addition: } 40 + 90 = 130.$$

$$x = 50 \quad \text{To undo the addition of 130, subtract 130 from both sides.}$$

Thus, $x = 50$.

Self Check

In the figure below, find y.

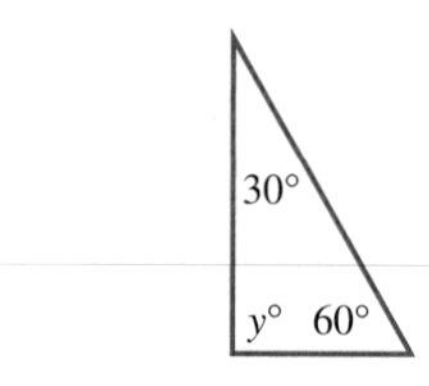

Answer: 90

EXAMPLE 4 ***Vertex angle of an isosceles triangle.*** See Figure 9-32. If one base angle of an isosceles triangle measures 70°, how large is the vertex angle?

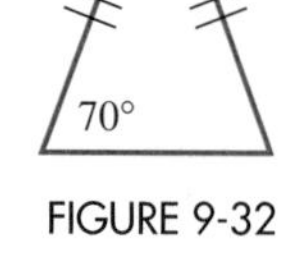

FIGURE 9-32

Solution

Since one of the base angles measures 70°, so does the other. If we let x represent the measure of the vertex angle, we have

$$x + 70 + 70 = 180 \quad \text{The sum of the measures of the angles of a triangle is 180°.}$$

$$x + 140 = 180 \quad 70 + 70 = 140.$$

$$x = 40 \quad \text{To undo the addition of 140, subtract 140 from both sides.}$$

The vertex angle measures 40°.

Quadrilaterals

A **quadrilateral** is a polygon with four sides. Some common quadrilaterals are shown in Figure 9-33.

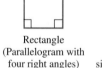

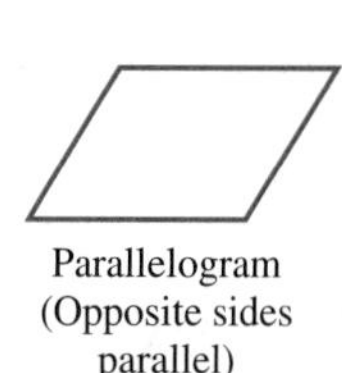

Parallelogram (Opposite sides parallel)

Rectangle (Parallelogram with four right angles)

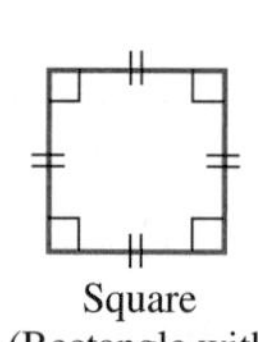

Square (Rectangle with sides of equal length)

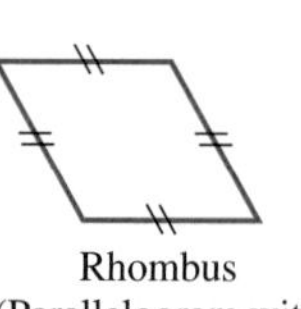

Rhombus (Parallelogram with sides of equal length)

Trapezoid (Exactly two sides parallel)

FIGURE 9-33

Properties of rectangles

1. All angles of a rectangle are right angles.
2. Opposite sides of a rectangle are parallel.
3. Opposite sides of a rectangle are of equal length.
4. The diagonals of a rectangle are of equal length.
5. If the diagonals of a parallelogram are of equal length, the parallelogram is a rectangle.

EXAMPLE 5 ***Squaring a foundation.*** A carpenter intends to build a shed with an 8-by-12-foot base. How can he make sure that the rectangular foundation is "square"?

Solution See Figure 9-34. The carpenter can use a tape measure to find the lengths of diagonals AC and BD. If these diagonals are of equal length, the figure will be a rectangle and have four right angles. Then the foundation will be "square."

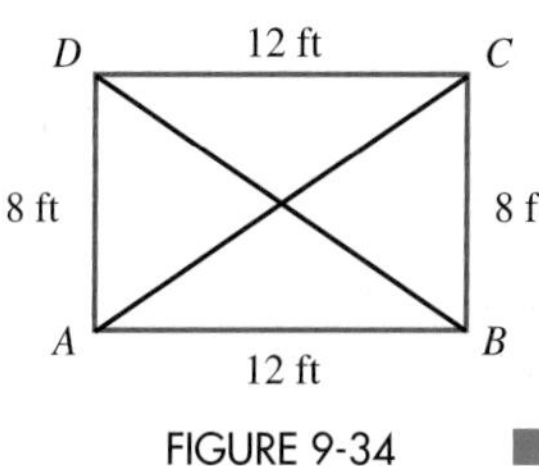

FIGURE 9-34

EXAMPLE 6 ***Properties of rectangles and triangles.*** In rectangle $ABCD$ (Figure 9-35), the length of $\overline{AC}$ is 20 centimeters. Find each measure: **a.** $m(\overline{BD})$, **b.** $m(\angle 1)$, and **c.** $m(\angle 2)$.

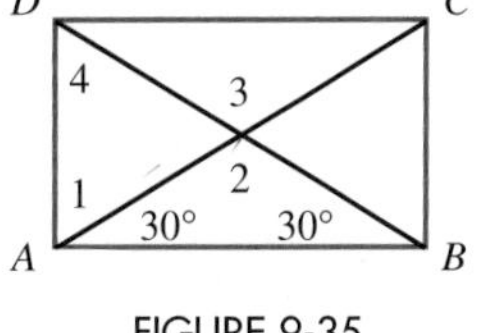

FIGURE 9-35

Solution

a. Since the diagonals of a rectangle are of equal length, $m(\overline{BD})$ is also 20 centimeters.

b. We let $m(\angle 1) = x$. Then, since the angles of a rectangle are right angles, we have

$$x + 30 = 90$$

$$x = 60 \quad \text{To undo the addition of 30, subtract 30 from both sides.}$$

Thus, $m(\angle 1) = 60°$.

c. We let $m(\angle 2) = y$. Then, since the sum of the angle measures of a triangle is 180°, we have

$$30 + 30 + y = 180$$

$$60 + y = 180 \quad \text{Simplify: } 30 + 30 = 60.$$

$$y = 120 \quad \text{To undo the addition of 60, subtract 60 from both sides.}$$

Thus, $m(\angle 2) = 120°$.

Self Check

In rectangle $ABCD$, the length of $\overline{DC}$ is 16 centimeters. Find each measure:

a. $m(\overline{AB})$

b. $m(\angle 3)$

c. $m(\angle 4)$

Answers: **a.** 16 cm **b.** 120° **c.** 60°

The parallel sides of a trapezoid are called **bases,** the nonparallel sides are called **legs,** and the angles on either side of a base are called **base angles.** If the nonparallel sides are the same length, the trapezoid is an **isosceles trapezoid.** In an isosceles trapezoid, the base angles are congruent.

EXAMPLE 7 ***Cross section of a drainage ditch.*** A cross section of a drainage ditch (Figure 9-36) is an isosceles trapezoid with $\overline{AB} \parallel \overline{CD}$. Find x and y.

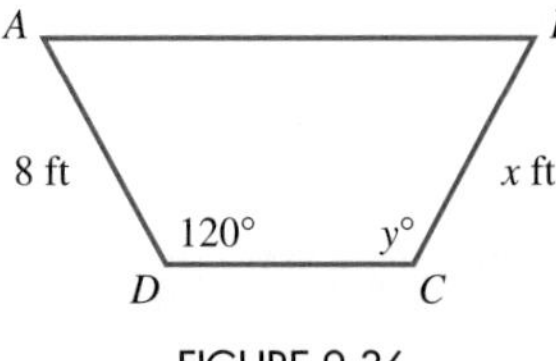

FIGURE 9-36

Solution Since the figure is an isosceles trapezoid, its nonparallel sides have the same length. So $m(\overline{AD})$ and $m(\overline{BC})$ are equal, and $x = 8$.

Since the base angles of an isosceles trapezoid are congruent, $m(\angle D) = m(\angle C)$. Thus, $y = 120$.

The sum of the measures of the angles of a polygon

We have seen that the sum of the angle measures of any triangle is 180°. Since a polygon with n sides can be divided into $n - 2$ triangles, the sum of the angle measures of the polygon is $(n - 2)180°$.

Angles of a polygon

> The sum S, in degrees, of the measures of the angles of a polygon with n sides is given by the formula
>
> $S = (n - 2)180$

EXAMPLE 8 ***Sum of the angles of a pentagon.*** Find the sum of the angle measures of a pentagon.

Solution

Since a pentagon has 5 sides, we substitute 5 for n in the formula and simplify.

$S = (n - 2)180$

$S = (\mathbf{5} - 2)180$ Substitute 5 for n.

$= (3)180$ Do the subtraction within the parentheses.

$= 540$

The sum of the angles of a pentagon is 540°.

Self Check

Find the sum of the angle measures of a quadrilateral.

Answer: 360° ■

EXAMPLE 9 ***Finding the number of sides of a polygon.*** The sum of the measures of the angles of a polygon is 1,080°. Find the number of sides the polygon has.

Solution

To find the number of sides the polygon has, we substitute 1,080 for S in the formula and then solve for n.

$S = (n - 2)180$

$\mathbf{1{,}080} = (n - 2)180$ Substitute 1,080 for S.

$1{,}080 = 180n - 360$ Distribute the multiplication by 180.

$1{,}080 + \mathbf{360} = 180n - 360 + \mathbf{360}$ To undo the subtraction of 360, add 360 to both sides.

$1{,}440 = 180n$ Simplify.

$\frac{1{,}440}{\mathbf{180}} = \frac{180n}{\mathbf{180}}$ To undo the multiplication of 180, divide both sides by 180.

$8 = n$ Do the division: $\frac{1{,}440}{180} = 8$.

The polygon has 8 sides. It is an octagon.

Self Check

The sum of the measures of the angles of a polygon is 720°. Find the number of sides the polygon has.

Answer: 6 ■

STUDY SET Section 9.3

VOCABULARY *Fill in the blanks to make the statements true.*

1. A __regular__ polygon has sides that are all the same length and angles that all have the same measure.
2. A polygon with four sides is called a __quadrilateral__. A __triangle__ is a polygon with three sides.
3. A __hexagon__ is a polygon with six sides.
4. A polygon with five sides is called a __pentagon__.
5. An eight-sided polygon is an __octagon__.
6. The points where the sides of a polygon intersect are called __vertices__.
7. A triangle with three sides of equal length is called an __equilateral__ triangle.
8. An __isoseles__ triangle has two sides of equal length.
9. The longest side of a right triangle is the __hypotenuse__.
10. The __base__ angles of an isosceles triangle have the same measure.

11. A ___parallelogram___ with a right angle is a rectangle.

12. A rectangle with all sides of equal length is a ___Square___.

13. A ___rhombus___ is a parallelogram with four sides of equal length.

14. A ___Trapezoid___ has two sides that are parallel and two sides that are not parallel.

15. The legs of an ___isoseles___ trapezoid have the same length.

16. The ___angles___ of a polygon is the distance around it.

CONCEPTS *Give the number of sides each polygon has and classify it as a triangle, quadrilateral, pentagon, hexagon, or octagon. Then give the number of vertices it has.*

17.

18.

19.
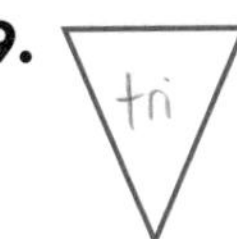

20.
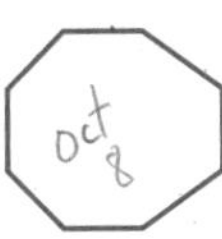

21.
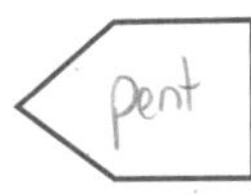

22.

23.

24.
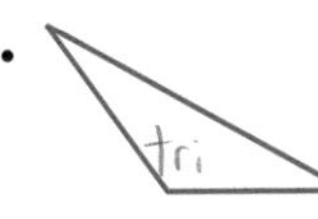

Classify each triangle as an equilateral triangle, an isosceles triangle, a scalene triangle, or a right triangle.

25.

26.
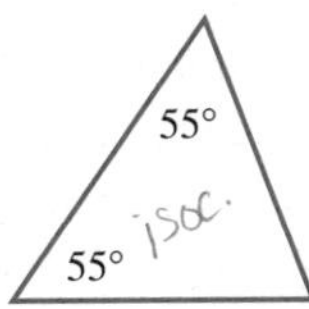

27.
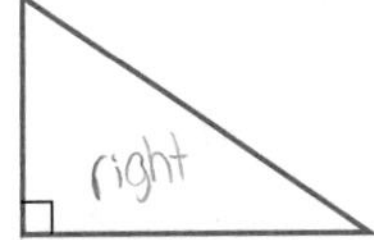

28.
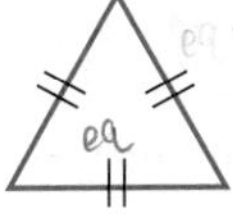

29.
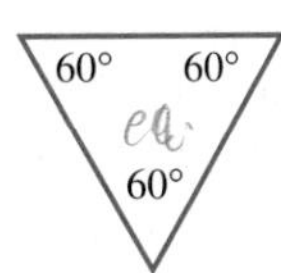

30.
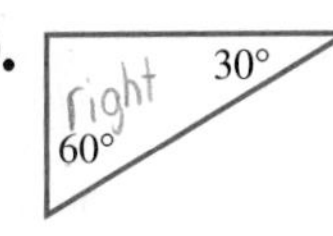

31.
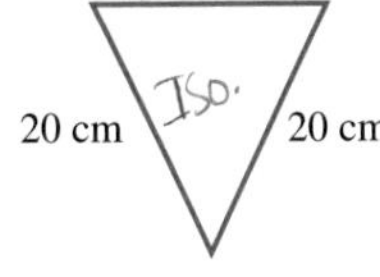

32.
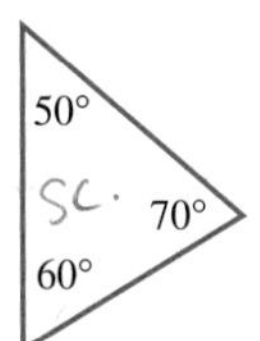

Classify each quadrilateral as a rectangle, a square, a rhombus, or a trapezoid.

33.
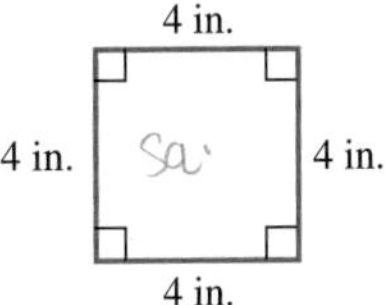

34.
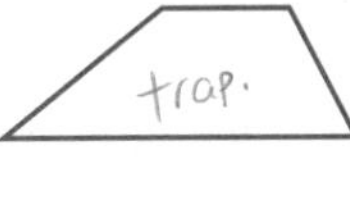

35.
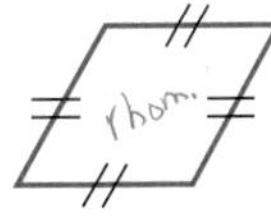

36.
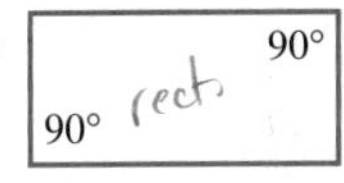

37.

38.
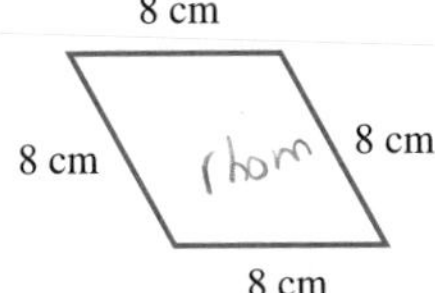

39.

40.
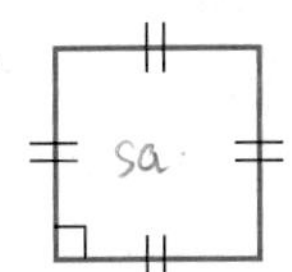

NOTATION *Fill in the blanks to make the statements true.*

41. The symbol $\triangle$ means ___triangle___.

42. The symbol $m(\angle 1)$ means the ___measure___ of angle 1.

PRACTICE *The measures of two angles of $\triangle ABC$ (shown in Illustration 1 on the next page) are given. Find the measure of the third angle.*

43. $m(\angle A) = 30°$ and $m(\angle B) = 60°$
$m(\angle C) =$ ___90°___

44. $m(\angle A) = 45°$ and $m(\angle C) = 105°$
$m(\angle B) =$ ___30°___

45. m(∠*B*) = 100° and m(∠*A*) = 35°
m(∠*C*) = 45°

46. m(∠*B*) = 33° and m(∠*C*) = 77°
m(∠*A*) = 70°

47. m(∠*A*) = 25.5° and m(∠*B*) = 63.8°
m(∠*C*) = 90.7°

48. m(∠*B*) = 67.25° and m(∠*C*) = 72.5°
m(∠*A*) = 40.25°

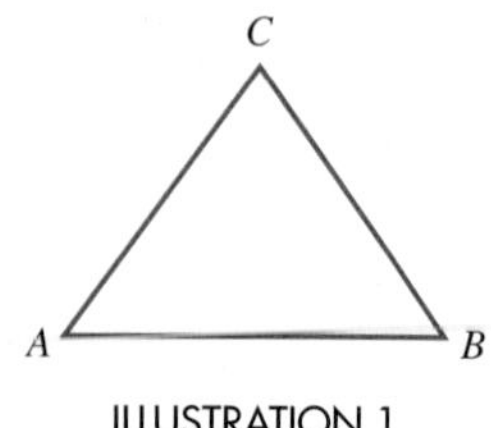

ILLUSTRATION 1

Refer to rectangle ABCD, shown in Illustration 2.

49. m(∠1) = ____________

50. m(∠3) = ____________

51. m(∠2) = ____________

52. If m($\overline{AC}$) is 8 cm, then m($\overline{BD}$) = ____________.

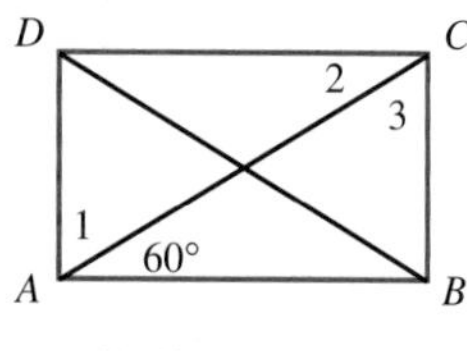

ILLUSTRATION 2

Find the sum of the angle measures of each polygon.

53. A hexagon

54. An octagon

55. A decagon (10 sides)

56. A dodecagon (12 sides)

Find the number of sides a polygon has if the sum of its angle measures is the given number.

57. 900°

58. 1,260°

59. 2,160°

60. 3,600°

APPLICATIONS

61. Give three uses of triangles in everyday life.

62. Give three uses of rectangles in everyday life.

63. Give three uses of squares in everyday life.

64. Give a use of a trapezoid in everyday life.

65. POLYGONS IN NATURE As we see in Illustration 3(a), a starfish has the shape of a pentagon. What polygon shape do you see in each of the other objects in Illustration 3? **b.** Lemon **c.** Chili pepper **d.** Apple

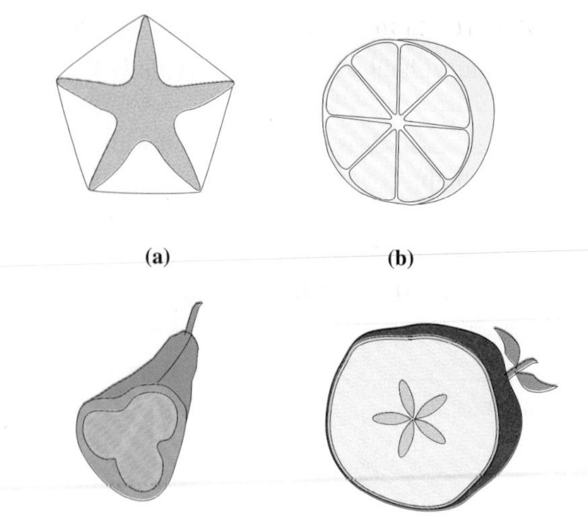

ILLUSTRATION 3

66. FLOWCHART A flowchart shows a sequence of steps to be performed by a computer to solve a given problem. When designing a flowchart, the programmer uses a set of standardized symbols to represent various operations to be performed by the computer. Locate a rectangle, a rhombus, and a parallelogram in the flow chart shown in Illustration 4.

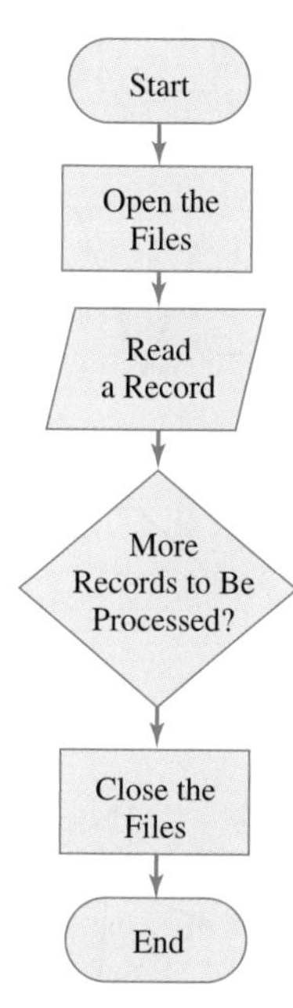

ILLUSTRATION 4

67. CHEMISTRY Polygons are used to represent the chemical structure of compounds graphically. In Illustration 5, what types of polygons are used to represent methylprednisolone, the active ingredient in an antiinflammatory medication?

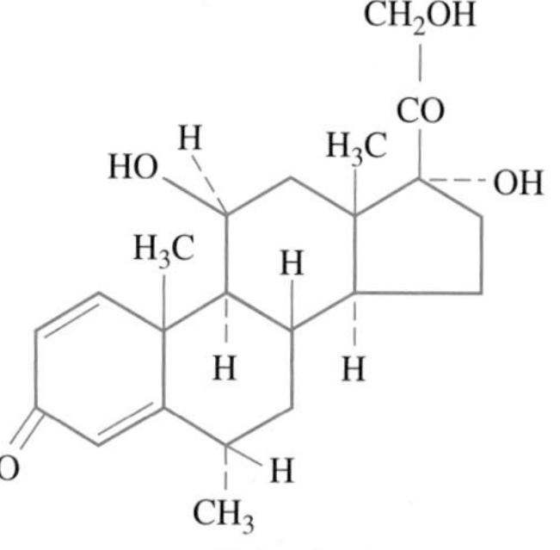

ILLUSTRATION 5

68. PODIUM In Illustration 6, what polygon describes the shape of the upper portion of the podium?

ILLUSTRATION 6

69. EASEL In Illustration 7, show how two of the legs of the easel form the equal sides of an isosceles triangle.

ILLUSTRATION 7

70. AUTOMOBILE JACK Refer to Illustration 8. Show that no matter how high the jack is raised, it always forms two isosceles triangles.

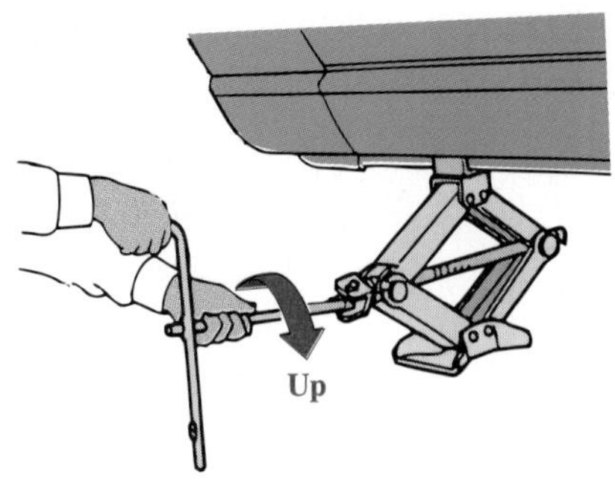

ILLUSTRATION 8

WRITING

71. Explain why a square is a rectangle.

72. Explain why a trapezoid is not a parallelogram.

REVIEW

73. Find 20% of 110.

74. Find 15% of 50.

75. What percent of 200 is 80?

76. 20% of what number is 500?

77. Simplify: $0.85 \div 2(0.25)$.

78. FIRST AID When checking an accident victim's pulse, a paramedic counted 13 beats during a 15-second span. How many beats would be expected in 60 seconds?

9.4 *Properties of Triangles*

In this section, you will learn about

- Congruent triangles • Similar triangles
- The Pythagorean theorem

Introduction. Proportions and triangles are often used to measure distances indirectly. For example, by using a proportion, Eratosthenes (275–195 B.C.) was able to estimate the circumference of the Earth with remarkable accuracy. On a sunny day, we can use properties of similar triangles to calculate the height of a tree while staying safely on the ground. By using a theorem proved by the Greek mathematician Pythagoras (about 500 B.C.), we can calculate the length of the third side of a right triangle whenever we know the lengths of two sides.

Congruent triangles

Triangles that have the same area and the same shape are called **congruent triangles.** In Figure 9-37, triangles ABC and DEF are congruent:

$\triangle ABC \cong \triangle DEF$ Read as "Triangle ABC is congruent to triangle DEF."

Corresponding angles and corresponding sides of congruent triangles are called **corresponding parts.** The notation $\triangle ABC \cong \triangle DEF$ shows which vertices are corresponding parts.

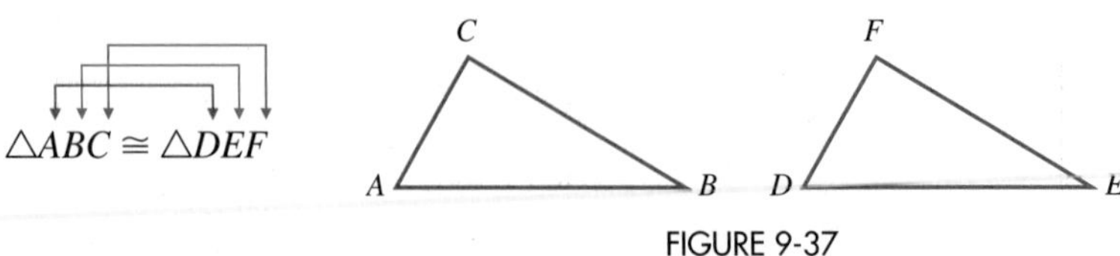

FIGURE 9-37

Corresponding parts of congruent triangles always have the same measure. In the congruent triangles shown in Figure 9-37,

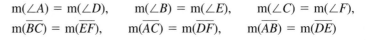

$m(\angle A) = m(\angle D)$, $m(\angle B) = m(\angle E)$, $m(\angle C) = m(\angle F)$,
$m(\overline{BC}) = m(\overline{EF})$, $m(\overline{AC}) = m(\overline{DF})$, $m(\overline{AB}) = m(\overline{DE})$

EXAMPLE 1 ***Corresponding parts of congruent triangles.*** Name the corresponding parts of the congruent triangles in Figure 9-38.

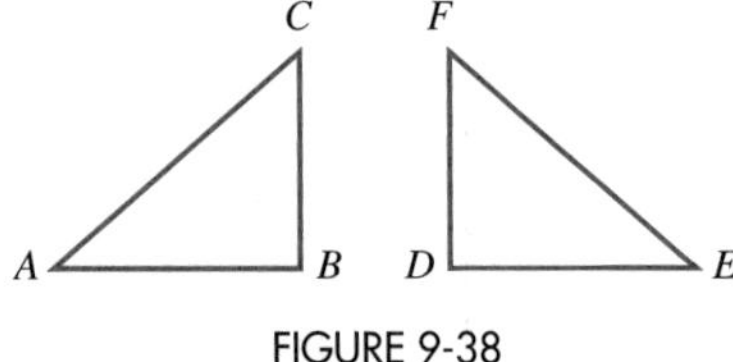

FIGURE 9-38

Solution The corresponding angles are

$\angle A$ and $\angle E$, $\angle B$ and $\angle D$,
$\angle C$ and $\angle F$

Since corresponding sides are always opposite corresponding angles, the corresponding sides are

$\overline{BC}$ and $\overline{DF}$, $\overline{AC}$ and $\overline{EF}$, $\overline{AB}$ and $\overline{ED}$ ■

We will discuss three ways of showing that two triangles are congruent.

SSS property

> If three sides of one triangle are congruent to three sides of a second triangle, the triangles are congruent.

The triangles in Figure 9-39 are congruent because of the SSS property.

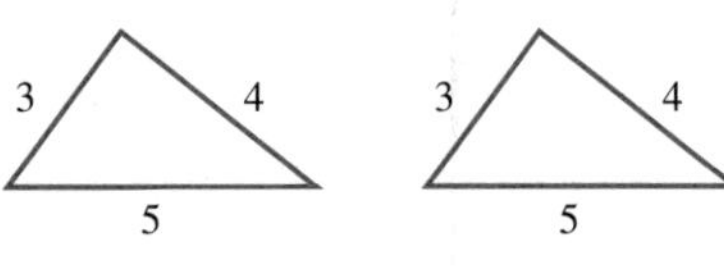

FIGURE 9-39

SAS property

> If two sides and the angle between them in one triangle are congruent, respectively, to two sides and the angle between them in a second triangle, the triangles are congruent.

The triangles in Figure 9-40 are congruent because of the SAS property.

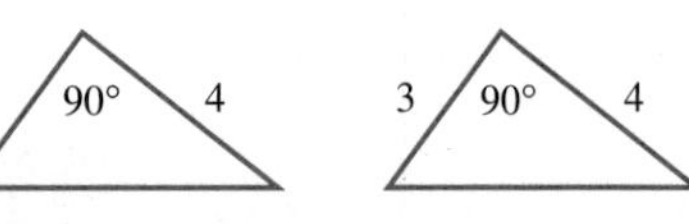

FIGURE 9-40

ASA property

> If two angles and the side between them in one triangle are congruent, respectively, to two angles and the side between them in a second triangle, the triangles are congruent.

The triangles in Figure 9-41 are congruent because of the ASA property.

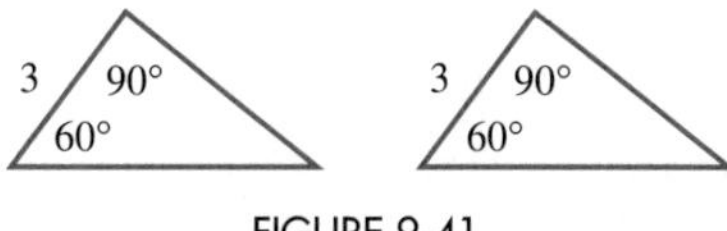

FIGURE 9-41

COMMENT There is no SSA property. To illustrate this, consider the triangles in Figure 9-42. Two sides and an angle of $\triangle ABC$ are congruent to two sides and an angle of $\triangle DEF$. But the congruent angle is *not* between the congruent sides.

We refer to this situation as SSA. Obviously, the triangles are not congruent, because they have different areas.

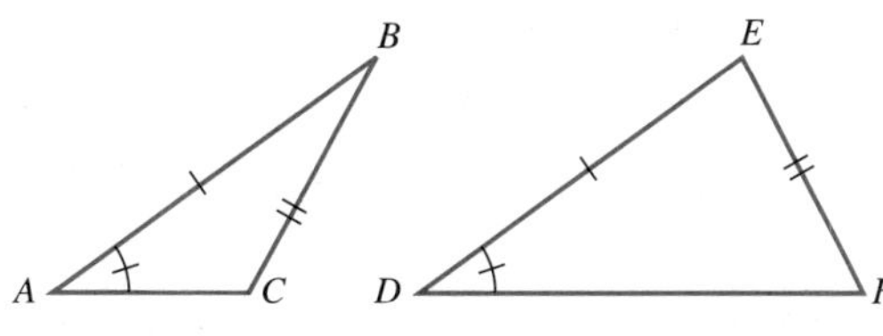

The slash marks indicate congruent parts. That is, the sides with one slash are the same length, the sides with two slashes are the same length, and the angles with one slash have the same measure.

FIGURE 9-42

EXAMPLE 2 ***Determining whether triangles are congruent.*** Explain why the triangles in Figure 9-43 are congruent.

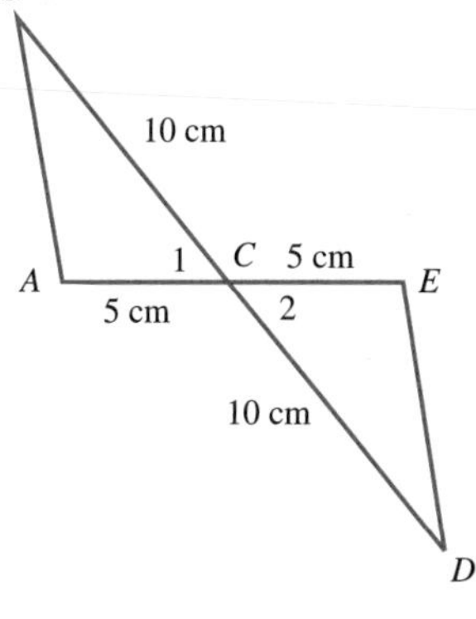

FIGURE 9-43

Solution Since vertical angles are congruent,

$$m(\angle 1) = m(\angle 2)$$

From the figure, we see that

$$m(\overline{AC}) = m(\overline{EC}) \quad \text{and} \quad m(\overline{BC}) = m(\overline{DC})$$

Since two sides and the angle between them in one triangle are congruent, respectively, to two sides and the angle between them in a second triangle, $\triangle ABC \cong \triangle EDC$ by the SAS property. ■

Similar triangles

If two angles of one triangle are congruent to two angles of a second triangle, the triangles will have the same shape. Triangles with the same shape are called **similar triangles.** In Figure 9-44, $\triangle ABC \sim \triangle DEF$ (read the symbol $\sim$ as "is similar to").

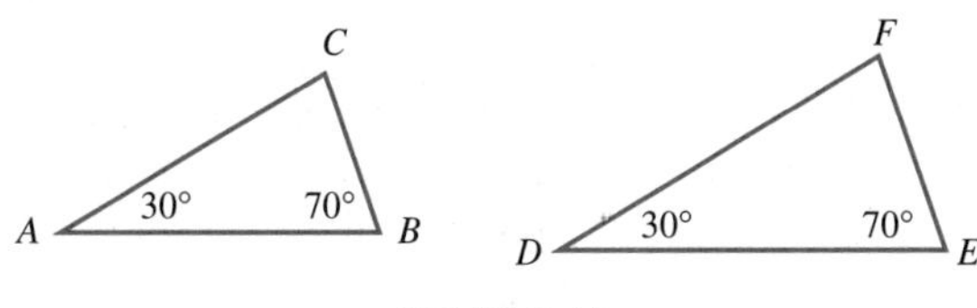

FIGURE 9-44

COMMENT Note that congruent triangles are always similar, but similar triangles are not always congruent.

Property of similar triangles

> If two triangles are similar, all pairs of corresponding sides are in proportion.

In the similar triangles shown in Figure 9-44 on the preceding page, the following proportions are true.

$$\frac{\overline{AB}}{\overline{DE}} = \frac{\overline{BC}}{\overline{EF}}, \quad \frac{\overline{BC}}{\overline{EF}} = \frac{\overline{CA}}{\overline{FD}}, \quad \text{and} \quad \frac{\overline{CA}}{\overline{FD}} = \frac{\overline{AB}}{\overline{DE}}$$

EXAMPLE 3 ***Finding the height of a tree.*** A tree casts a shadow 18 feet long at the same time as a woman 5 feet tall casts a shadow that is 1.5 feet long. (See Figure 9-45.) Find the height of the tree.

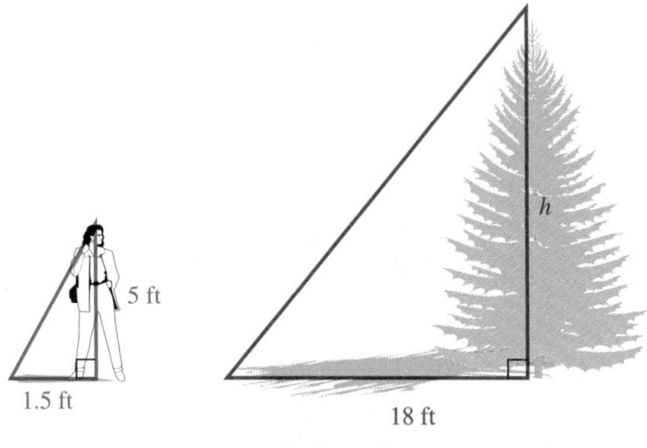

FIGURE 9-45

Solution The figure shows the triangles determined by the tree and its shadow and the woman and her shadow. Since the triangles have the same shape, they are similar, and the lengths of their corresponding sides are in proportion. If we let h represent the height of the tree, we can find h by solving the following proportion.

$\frac{h}{5} = \frac{18}{1.5}$ $\quad \frac{\text{Height of the tree}}{\text{Height of the woman}} = \frac{\text{shadow of the tree}}{\text{shadow of the woman}}$

$1.5h = 5(18)$ In a proportion, the product of the extremes is equal to the product of the means.

$1.5h = 90$ Do the multiplication: $5(18) = 90$.

$h = 60$ To undo the multiplication by 1.5, divide both sides by 1.5.

The tree is 60 feet tall.

The Pythagorean theorem

In the movie *The Wizard of Oz,* the scarecrow was in search of a brain. To prove that he had found one, he recited the Pythagorean theorem.

> *In a right triangle, the square of the hypotenuse is equal to the sum of squares of the other two sides.*

Pythagorean theorem

> If the length of the hypotenuse of a right triangle is c and the lengths of its legs are a and b, then
>
> $$a^2 + b^2 = c^2$$

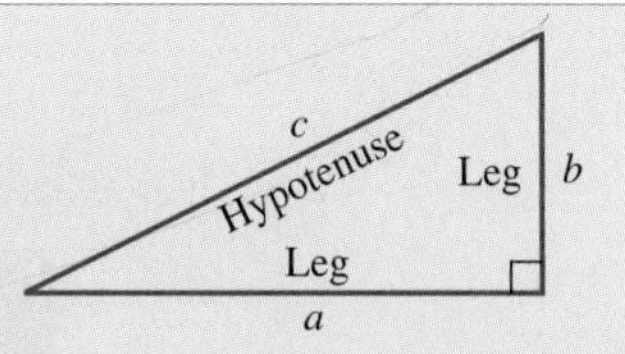

EXAMPLE 4 ***Constructing a high-ropes adventure course.*** A builder of a high-ropes adventure course wants to secure the pole shown in Figure 9-46 by attaching a cable from the anchor stake 8 feet from its base to a point 6 feet up the pole. How long should the cable be?

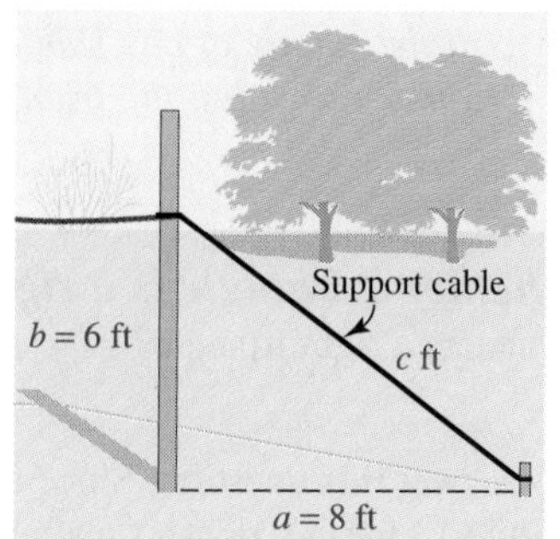

FIGURE 9-46

Solution

The support cable, the pole, and the ground form a right triangle. If we let c represent the length of the cable (the hypotenuse), then we can use the Pythagorean theorem with $a = 8$ and $b = 6$ to find c.

$c^2 = a^2 + b^2$ The Pythagorean theorem.

$c^2 = 8^2 + 6^2$ Substitute 8 for a and 6 for b.

$c^2 = 64 + 36$ Evaluate the exponential expressions.

$c^2 = 100$ Simplify the right-hand side.

To find c, we must find a number that, when squared, is 100. There are two such numbers, one positive and one negative; they are the square roots of 100. Since c represents the length of a support cable, c cannot be negative. For this reason, we need only find the positive square root of 100 to get c.

$c^2 = 100$ The equation to solve.

$c = \sqrt{100}$ The symbol $\sqrt{}$ is used to indicate the positive square root of a number.

$c = 10$ $\sqrt{100} = 10$, because $10^2 = 100$.

The support cable should be 10 feet long.

Self Check

A 26-foot ladder rests against the side of a building. If the base of the ladder is 10 feet from the wall, how far up the side of the building will the ladder reach?

Answer: 24 ft

Accent on Technology: Finding the width of a television screen

The size of a television screen is the diagonal measure of its rectangular screen. (See Figure 9-47.) To find the width of a 27-inch screen that is 17 inches high, we use the Pythagorean theorem with $c = 27$ and $b = 17$.

$$c^2 = a^2 + b^2$$
$$27^2 = a^2 + 17^2$$
$$27^2 - 17^2 = a^2$$

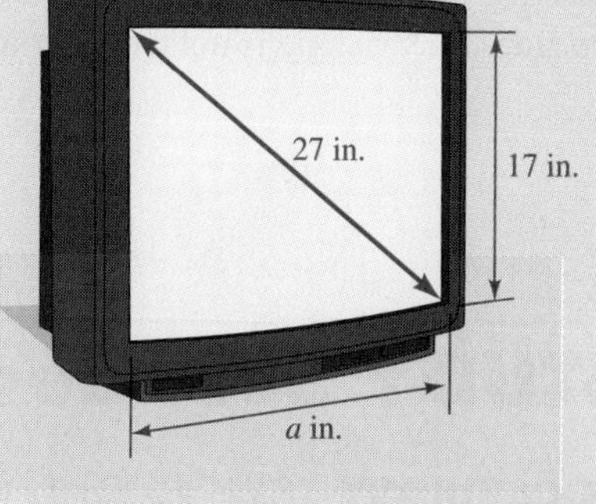

FIGURE 9-47

The variable a represents the width of a television screen, so it must be positive. To find a, we find the positive square root of the result when 17^2 is subtracted from 27^2. Using a radical symbol to indicate this, we have

$$\sqrt{27^2 - 17^2} = a$$

We can evaluate the expression on the left-hand side by entering these numbers and pressing these keys.

Keystrokes 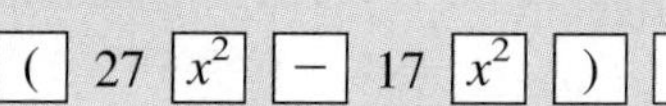 20.97617696

To the nearest inch, the width of the television screen is 21 inches.

It is also true that

If the square of one side of a triangle is equal to the sum of the squares of the other two sides, the triangle is a right triangle.

EXAMPLE 5 ***Determining whether a triangle is a right triangle.***

Is a triangle with sides of 5, 12, and 13 meters a right triangle?

Solution

We can use the Pythagorean theorem to answer this question. Since the longest side of the triangle is 13 meters, we must substitute 13 for c. It doesn't matter which of the two remaining side lengths we substitute for a and which we substitute for b.

$c^2 = a^2 + b^2$ The Pythagorean theorem.

$13^2 \stackrel{?}{=} 5^2 + 12^2$ Substitute 13 for c, 5 for a, and 12 for b.

$169 \stackrel{?}{=} 25 + 144$ Evaluate the exponential expressions.

$169 = 169$ Simplify the right-hand side.

Since the square of the longest side is equal to the sum of the squares of the other two sides, the triangle is a right triangle.

Self Check

Is a triangle with sides of 9, 40, and 41 meters a right triangle?

Answer: yes

EXAMPLE 6 ***Determining whether a triangle is a right triangle.***

Is a triangle with sides of 1, 2, and 3 feet a right triangle?

Solution

We check to see whether the square of the longest side is equal to the sum of the squares of the other two sides.

$c^2 = a^2 + b^2$ The Pythagorean theorem.

$3^2 \stackrel{?}{=} 1^2 + 2^2$ Substitute 3 for c, 1 for a, and 2 for b.

$9 \stackrel{?}{=} 1 + 4$ Evaluate the exponential expressions.

$9 \neq 5$ Simplify the right-hand side.

Since the square of the longest side is not equal to the sum of the squares of the other two sides, the triangle is not a right triangle.

Self Check

Is a triangle with sides of 4, 5, and 6 inches a right triangle?

Answer: no

STUDY SET Section 9.4

VOCABULARY *Fill in the blanks to make the statements true.*

1. Congruent triangles are the same size and the same shape.
2. All Corresponding parts of congruent triangles have the same measure.
3. If two triangles are Similar, they have the same shape.
4. The hypotenuse is the longest side of a right triangle.

CONCEPTS *In Exercises 5–10, tell whether each statement is true. If a statement is false, tell why.*

5. If three sides of one triangle are the same length as three sides of a second triangle, the triangles are congruent.
6. If two sides of one triangle are the same length as two sides of a second triangle, the triangles are congruent.
7. If two sides and an angle of one triangle are congruent, respectively, to two sides and an angle of a second triangle, the triangles are congruent.

8. If two angles and the side between them in one triangle are congruent, respectively, to two angles and the side between them in a second triangle, the triangles are congruent.

9. In a proportion, the product of the means is equal to the product of the extremes.

10. If two angles of one triangle are congruent to two angles of a second triangle, the triangles are similar.

11. Are the triangles shown in Illustration 1 congruent?

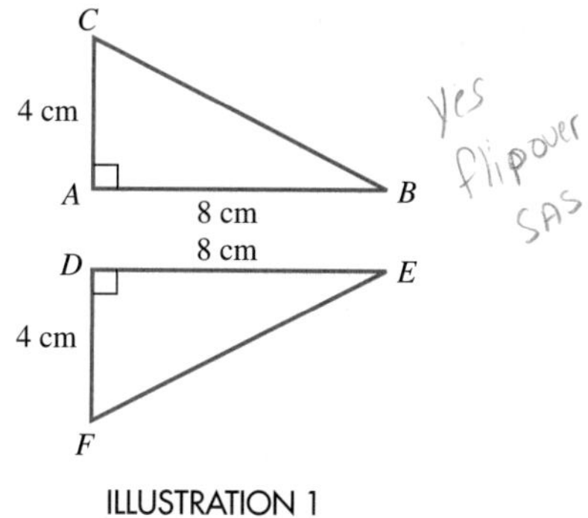

ILLUSTRATION 1

12. Are the triangles shown in Illustration 2 congruent?

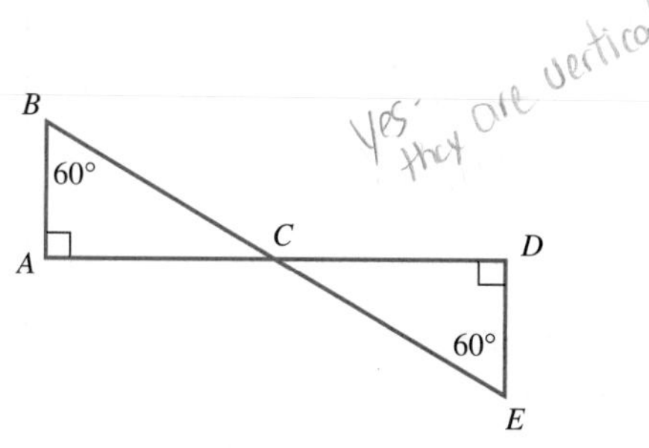

ILLUSTRATION 2

13. Are the triangles shown in Illustration 3 similar?

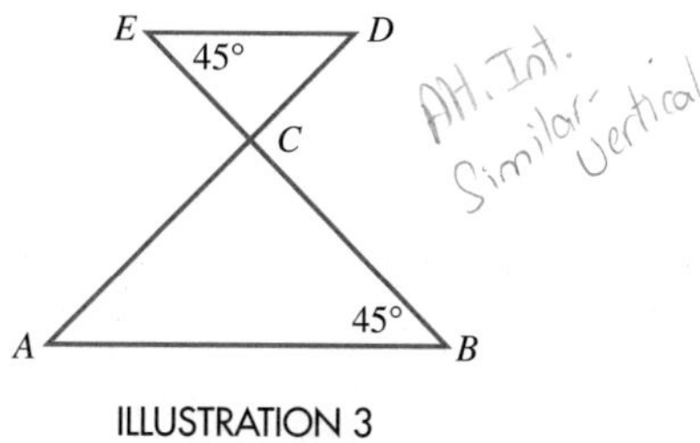

ILLUSTRATION 3

14. Are the triangles shown in Illustration 4 similar?

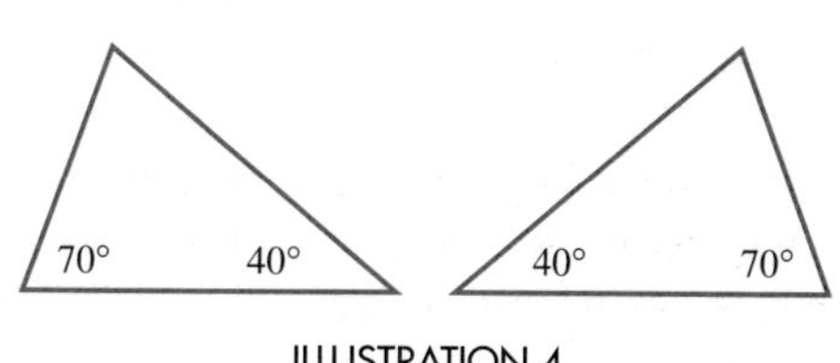

ILLUSTRATION 4

15. The Pythagorean theorem states that for a right triangle, $c^2 = a^2 + b^2$. What do the variables a, b, and c represent?

16. A triangle has sides of length 3, 4, and 5 centimeters. Substitute the lengths into $c^2 = a^2 + b^2$ and show that a true statement results. From the result, what can we conclude about the triangle?

17. Suppose that c represents the length of the hypotenuse of a right triangle and $c^2 = 25$. Fill in the blanks to make the statements true.

To find c, we must find a number that, when squared, is ____ . Since c represents a positive number, we need only find the positive ____________ of 25 to get c.

$c^2 = 25$

$\_\_ = \sqrt{25}$ A radical symbol indicates the positive square root.

$c = \_\_$

18. Solve the equation $\frac{h}{2.6} = \frac{27}{13}$.

NOTATION *Fill in the blanks to make the statements true.*

19. The symbol $\cong$ is read as "____________."

20. The symbol $\sim$ is read as "____________."

PRACTICE *Name the corresponding parts of the congruent triangles.*

21. Refer to Illustration 5.

$\overline{AC}$ corresponds to ____.
$\overline{DE}$ corresponds to ____.
$\overline{BC}$ corresponds to ____.
$\angle A$ corresponds to ____.
$\angle E$ corresponds to ____.
$\angle F$ corresponds to ____.

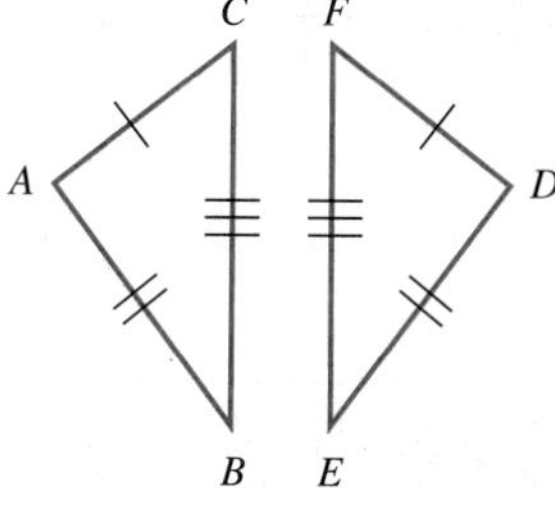

ILLUSTRATION 5

22. Refer to Illustration 6.

$\overline{AB}$ corresponds to ____.
$\overline{EC}$ corresponds to ____.
$\overline{AC}$ corresponds to ____.
$\angle D$ corresponds to ____.
$\angle B$ corresponds to ____.
$\angle 1$ corresponds to ____.

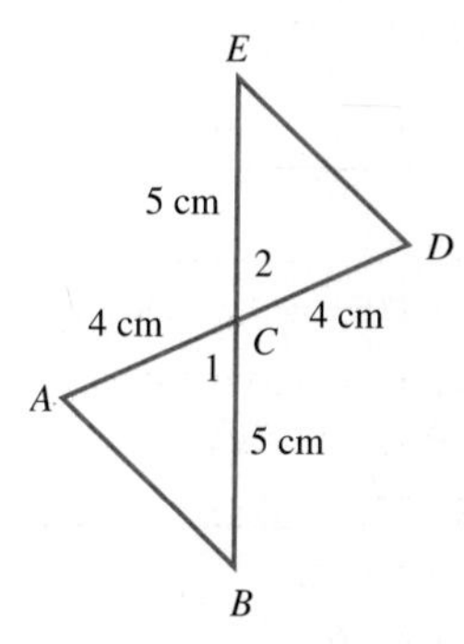

ILLUSTRATION 6

Determine whether each pair of triangles is congruent. If they are, tell why.

23.

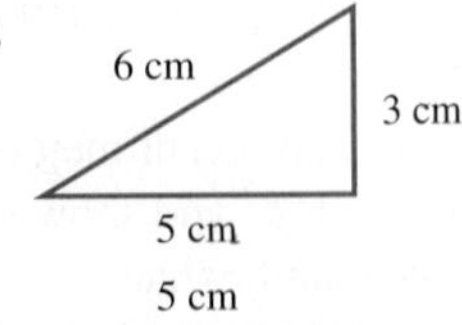

24.

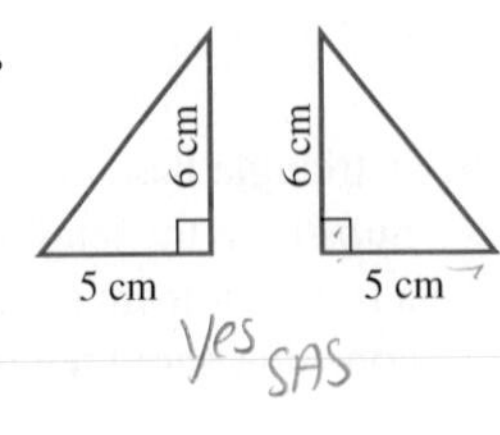

25.

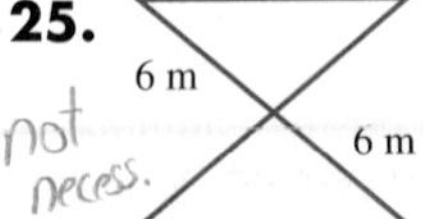

26.

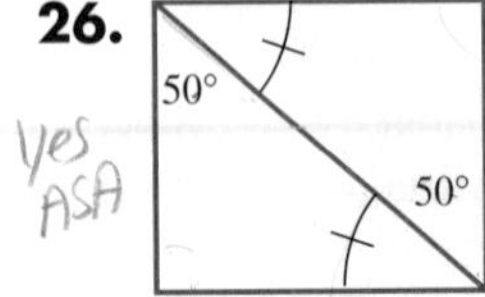

27.

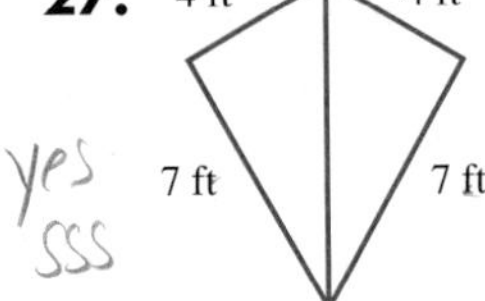

28.

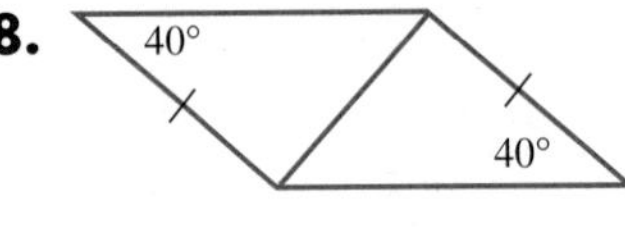
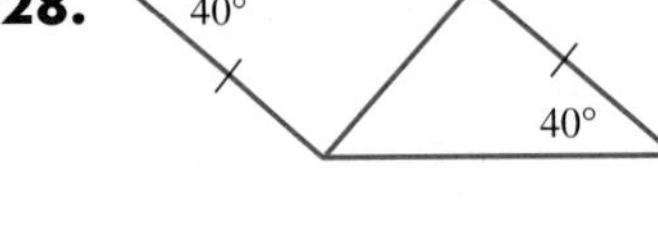

29.

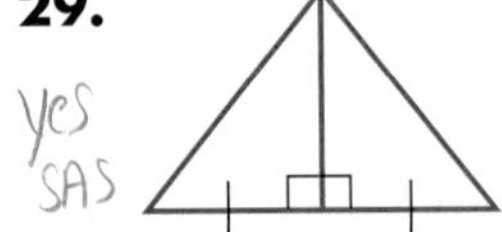

30.

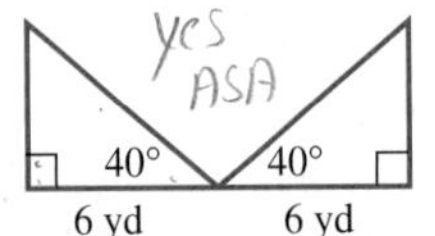

Find x.

31.

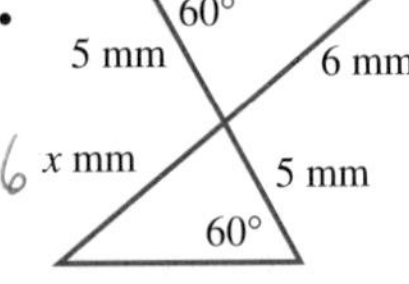

32.

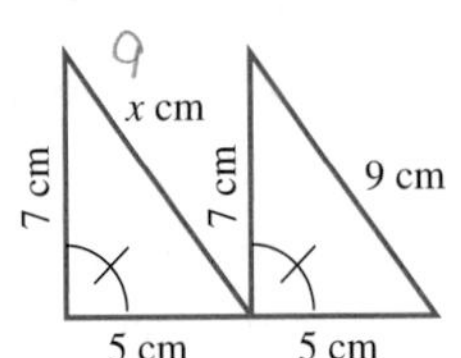

33.

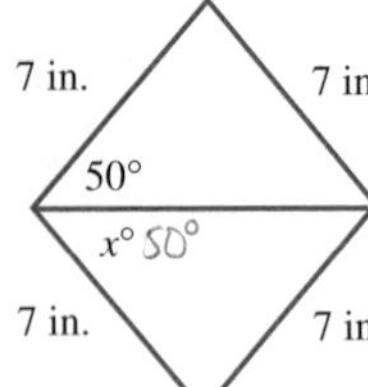

34.

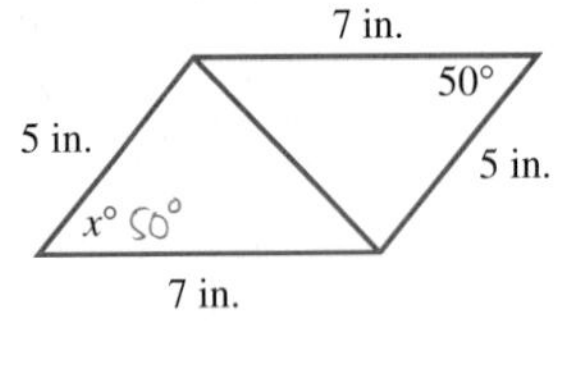

Tell whether the triangles are similar.

35.

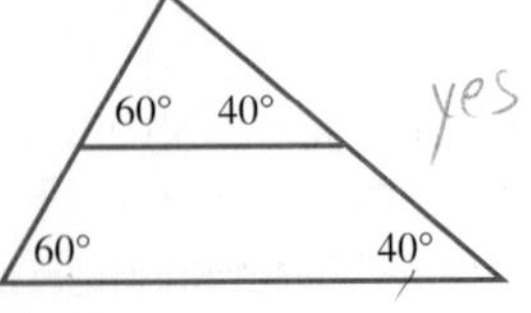

36.

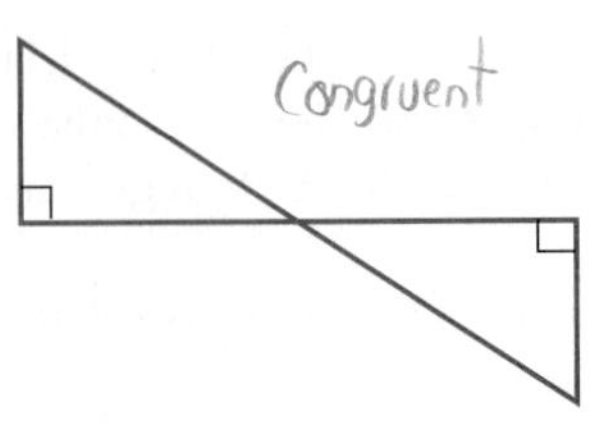

Refer to Illustration 7 and find the length of the unknown side.

37. $a = 3$ and $b = 4$. Find c.

38. $a = 12$ and $b = 5$. Find c.

39. $a = 15$ and $c = 17$. Find b.

40. $b = 45$ and $c = 53$. Find a.

41. $a = 5$ and $c = 9$. Find b.

42. $a = 1$ and $b = 7$. Find c.

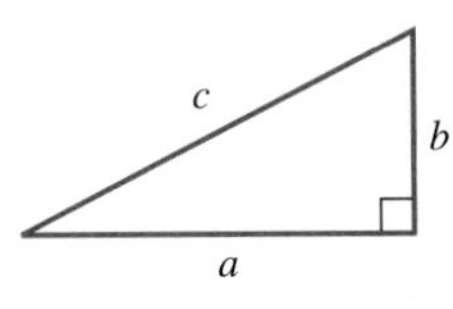

ILLUSTRATION 7

The lengths of the three sides of a triangle are given. Determine whether the triangle is a right triangle.

43. 8, 15, 17

44. 6, 8, 10

45. 7, 24, 26

46. 9, 39, 40

APPLICATIONS *Solve each problem. If an answer is not exact, give the answer to the nearest tenth.*

47. HEIGHT OF A TREE The tree in Illustration 8 casts a shadow 24 feet long when a man 6 feet tall casts a shadow 4 feet long. Find the height of the tree.

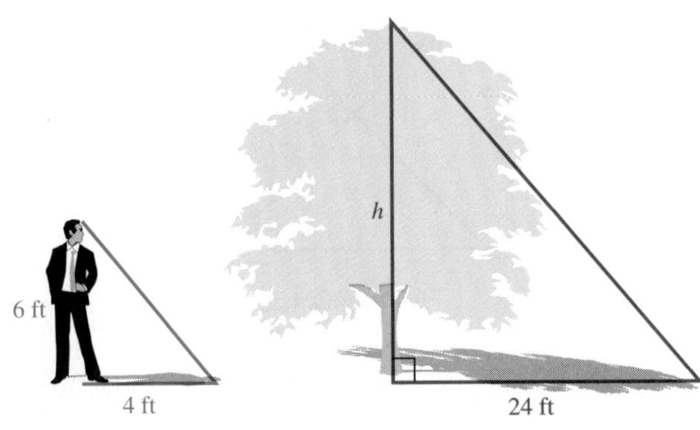

ILLUSTRATION 8

48. HEIGHT OF A BUILDING A man places a mirror on the ground and sees the reflection of the top of a building, as shown in Illustration 9. Find the height of the building.

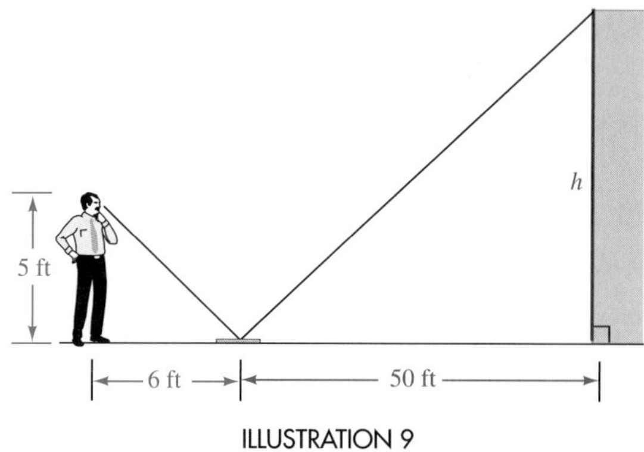

ILLUSTRATION 9

49. WIDTH OF A RIVER Use the dimensions in Illustration 10 to find w, the width of the river.

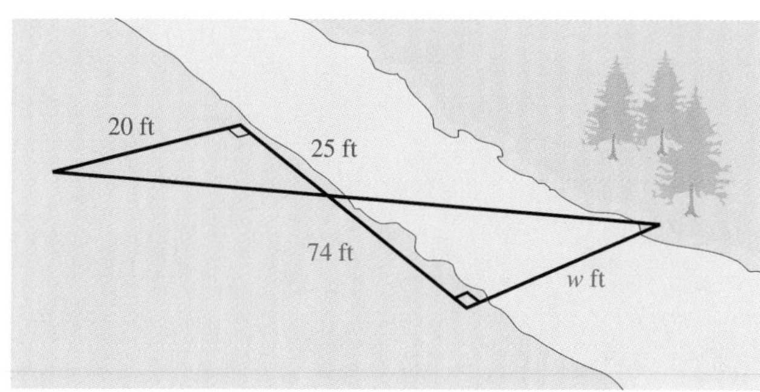

ILLUSTRATION 10

50. FLIGHT PATH The airplane in Illustration 11 ascends 200 feet as it flies a horizontal distance of 1,000 feet. How much altitude is gained as it flies a horizontal distance of 1 mile? (*Hint:* 1 mile = 5,280 feet.)

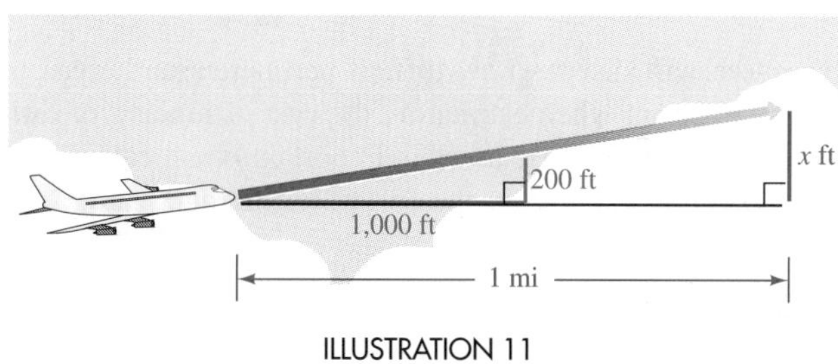

ILLUSTRATION 11

51. FLIGHT PATH An airplane descends 1,200 feet as it files a horizontal distance of 1.5 miles. How much altitude is lost as it flies a horizontal distance of 5 miles?

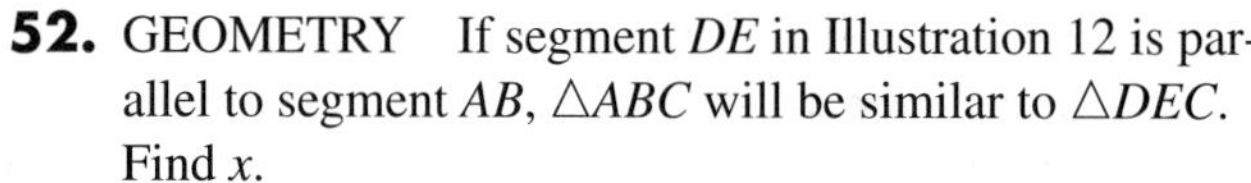

52. GEOMETRY If segment DE in Illustration 12 is parallel to segment AB, $\triangle ABC$ will be similar to $\triangle DEC$. Find x.

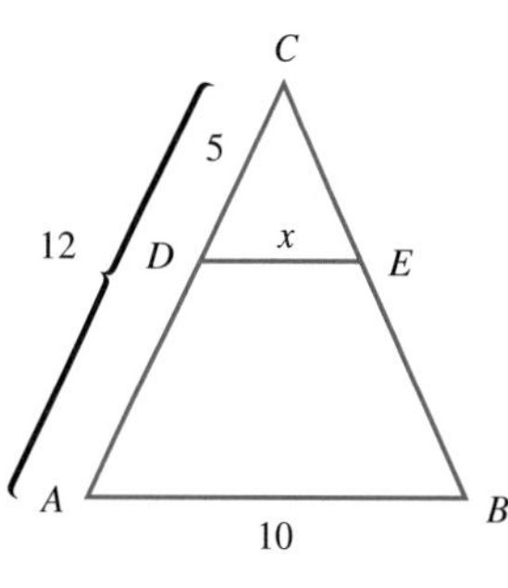

ILLUSTRATION 12

53. ADJUSTING A LADDER A 20-foot ladder reaches a window 16 feet above the ground. How far from the wall is the base of the ladder?

54. LENGTH OF GUY WIRES A 30-foot tower is to be fastened by three guy wires attached to the top of the tower and to the ground at positions 20 feet from its base. How much wire is needed?

55. PICTURE FRAME After gluing and nailing two pieces of picture frame molding together, a frame maker checks her work by making a diagonal measurement. (See Illustration 13.) If the sides of the frame form a right angle, what measurement should the frame maker read on the yardstick?

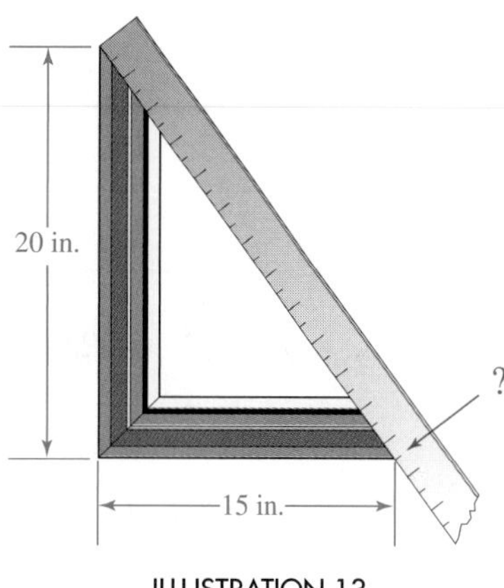

ILLUSTRATION 13

56. CARPENTRY The gable end of the roof shown in Illustration 14 is divided in half by a vertical brace, 8 feet in height. Find the length of the roof line.

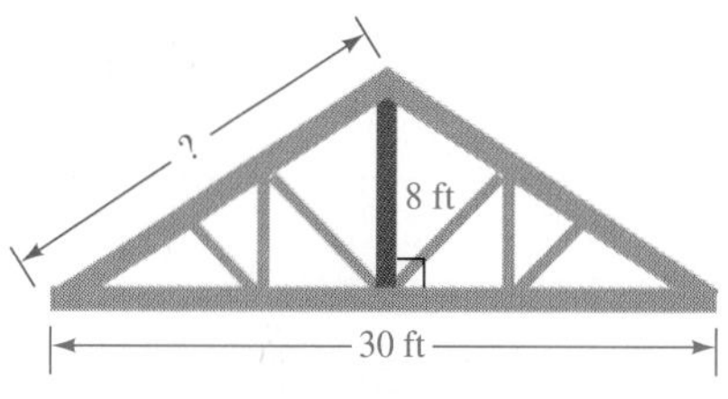

ILLUSTRATION 14

57. BASEBALL A baseball diamond is a square with each side 90 feet long (Illustration 15, next page). How far is it from home plate to second base?

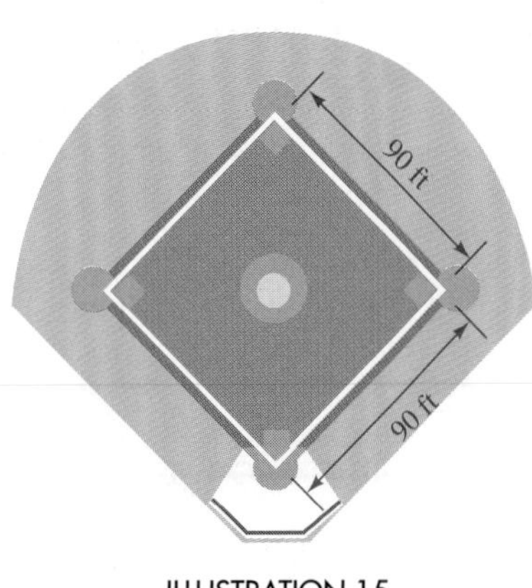

ILLUSTRATION 15

58. TELEVISION What size is the television screen shown in Illustration 16?

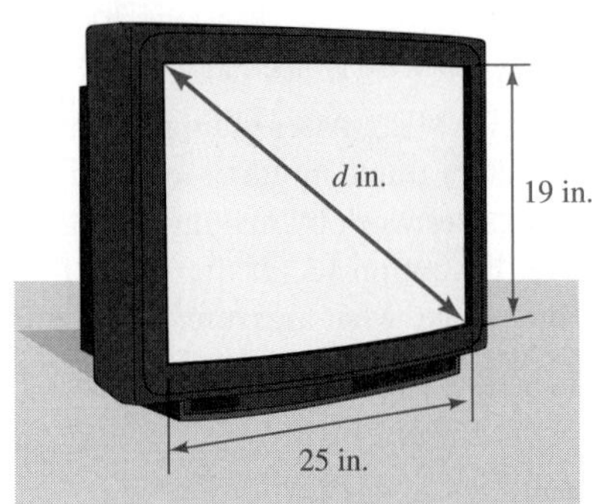

ILLUSTRATION 16

WRITING

59. Explain the Pythagorean theorem.

60. Explain the procedure used to solve the equation $c^2 = 64$. (Assume that c is positive.)

REVIEW *Estimate the answer to each problem.*

61. $\dfrac{0.95 \cdot 3.89}{2.997}$

62. 21% of 42

63. 32% of 60

64. $\dfrac{4.966 + 5.001}{2.994}$

65. 49.5% of 18.1

66. 98.7% of 0.03

9.5 Perimeters and Areas of Polygons

In this section, you will learn how to find

- Perimeters of polygons
- Perimeters of figures that are combinations of polygons
- Areas of polygons
- Areas of figures that are combinations of polygons

Introduction. In this section, we will discuss how to find perimeters and areas of polygons. Finding perimeters is important when estimating the cost of fencing or estimating the cost of woodwork in a house. Finding areas is important when calculating the cost of carpeting, the cost of painting a house, or the cost of fertilizing a yard.

Perimeters of polygons

Recall that the **perimeter** of a polygon is the distance around it. Since a square has four sides of equal length s, its perimeter P is $s + s + s + s$, or $4s$.

Perimeter of a square

If a square has a side of length s, its perimeter P is given by the formula

$$P = 4s$$

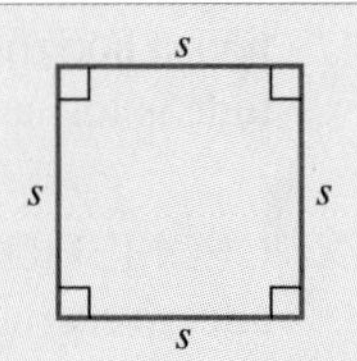

EXAMPLE 1 ***Perimeter of a square.*** Find the perimeter of a square whose sides are 7.5 meters long.

Solution

Since the perimeter of a square is given by the formula $P = 4s$, we substitute 7.5 for s and simplify.

$$P = 4s$$
$$P = 4(\mathbf{7.5})$$
$$P = 30$$

The perimeter is 30 meters.

Self Check

Find the perimeter of a square whose sides are 23.75 centimeters long.

Answer: 95 cm ■

Since a rectangle has two lengths l and two widths w, its perimeter P is $l + l + w + w$, or $2l + 2w$.

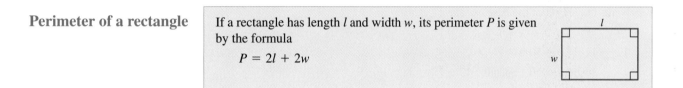

Perimeter of a rectangle

> If a rectangle has length l and width w, its perimeter P is given by the formula
>
> $$P = 2l + 2w$$

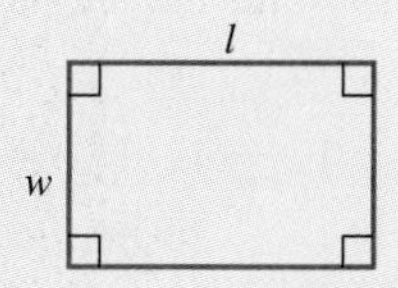

EXAMPLE 2 ***Perimeter of a rectangle.*** Find the perimeter of the rectangle in Figure 9-48.

6 cm
10 cm
FIGURE 9-48

Solution

Since the perimeter is given by the formula $P = 2l + 2w$, we substitute 10 for l and 6 for w and simplify.

$$P = 2l + 2w$$
$$P = 2(\mathbf{10}) + 2(\mathbf{6})$$
$$= 20 + 12$$
$$= 32$$

The perimeter is 32 centimeters.

Self Check

Find the perimeter of the isosceles trapezoid below.

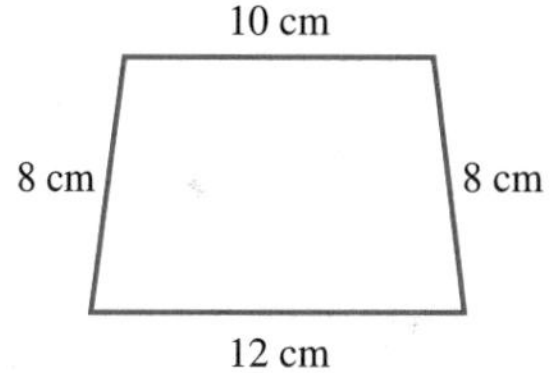

Answer: 38 cm ■

EXAMPLE 3 ***Converting units.*** Find the perimeter of the rectangle in Figure 9-49, in meters.

3 m
80 cm
FIGURE 9-49

Solution

Since 1 meter = 100 centimeters, we can convert 80 centimeters to meters by multiplying 80 centimeters by the unit conversion factor $\frac{1\text{ m}}{100\text{ cm}}$.

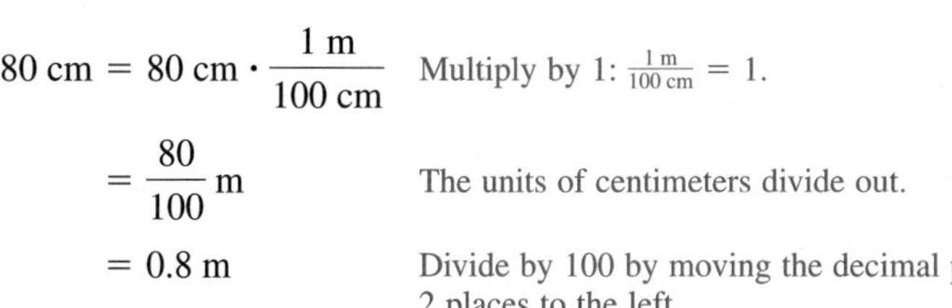

$$80\text{ cm} = 80\text{ cm} \cdot \frac{1\text{ m}}{100\text{ cm}} \quad \text{Multiply by 1: } \frac{1\text{ m}}{100\text{ cm}} = 1.$$

$$= \frac{80}{100}\text{ m} \quad \text{The units of centimeters divide out.}$$

$$= 0.8\text{ m} \quad \text{Divide by 100 by moving the decimal point 2 places to the left.}$$

Self Check

Find the perimeter of the triangle below, in inches.

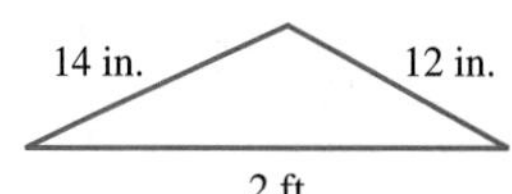

We can now substitute 3 for l and 0.8 for w to get

$$P = 2l + 2w$$
$$P = 2(3) + 2(0.8)$$
$$= 6 + 1.6$$
$$= 7.6$$

The perimeter is 7.6 meters.

Answer: 50 in.

EXAMPLE 4 ***Finding the base of an isosceles triangle.*** The perimeter of the isosceles triangle in Figure 9-50 is 50 meters. Find the length of its base.

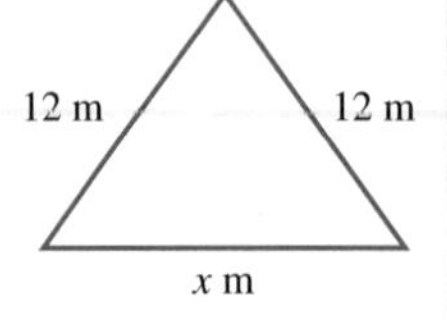

FIGURE 9-50

Solution

Two sides are 12 meters long, and the perimeter is 50 meters. If x represents the length of the base, we have

$$12 + 12 + x = 50$$
$$24 + x = 50 \quad \text{Simplify: } 12 + 12 = 24.$$
$$x = 26 \quad \text{To undo the addition of 24, subtract 24 from both sides.}$$

The length of the base is 26 meters.

Self Check

The perimeter of an isosceles triangle is 60 meters. If one of its sides of equal length is 15 meters long, how long is its base?

Answer: 30 m

Perimeters of figures that are combinations of polygons

Accent on Technology: Perimeter of a figure

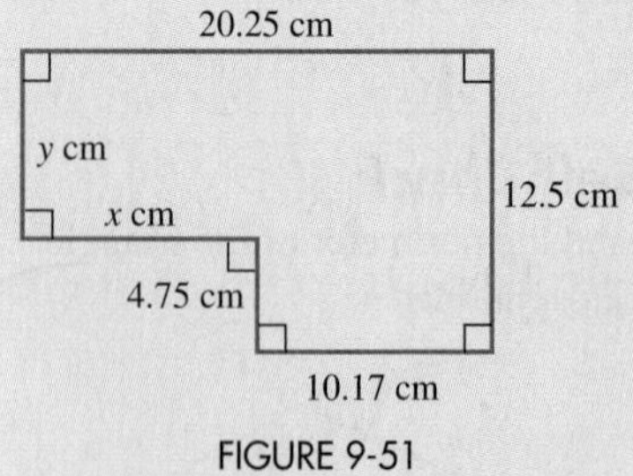

FIGURE 9-51

See Figure 9-51. To find the perimeter, we need to know the values of x and y. Since the figure is a combination of two rectangles, we can use a calculator to see that

$$x = 20.25 - 10.17 \quad \text{and} \quad y = 12.5 - 4.75$$
$$= 10.08 \qquad\qquad = 7.75$$

The perimeter P of the figure is

$$P = 20.25 + 12.5 + 10.17 + 4.75 + x + y$$
$$P = 20.25 + 12.5 + 10.17 + 4.75 + \mathbf{10.08} + \mathbf{7.75}$$

We can use a calculator to evaluate the expression on the right-hand side by entering these numbers and pressing these keys.

Keystrokes 20.25 [+] 12.5 [+] 10.17 [+] 4.75 [+] 10.08 [+] 7.75 [=]

65.5

The perimeter is 65.5 centimeters.

Areas of polygons

Recall that the **area** of a polygon is the measure of the amount of surface it encloses. Area is measured in square units, such as square inches or square centimeters. See Figure 9-52.

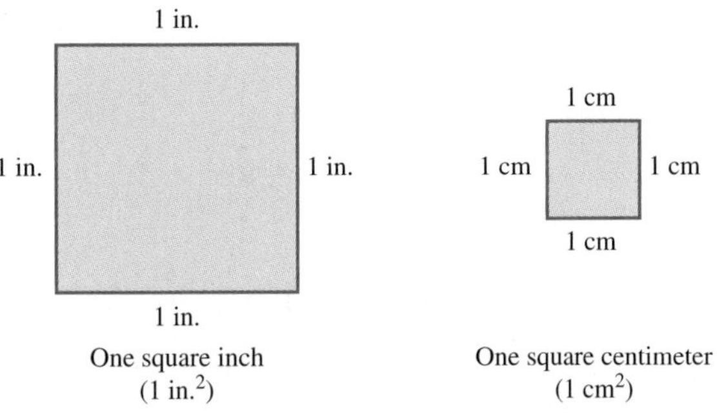

FIGURE 9-52

In everyday life, we often use areas. For example,

- To carpet a room, we buy square yards.
- A can of paint will cover a certain number of square feet.
- To measure vast amounts of land, we often use square miles.
- We buy house roofing by the "square." One square is 100 square feet.

The rectangle shown in Figure 9-53 has a length of 10 centimeters and a width of 3 centimeters. If we divide the rectangle into squares as shown in the figure, each square represents an area of 1 square centimeter—a surface enclosed by a square measuring 1 centimeter on each side. Because there are 3 rows with 10 squares in each row, there are 30 squares. Since the rectangle encloses a surface area of 30 squares, its area is 30 square centimeters, often written as 30 cm^2.

This example illustrates that to find the area of a rectangle, we multiply its length by its width.

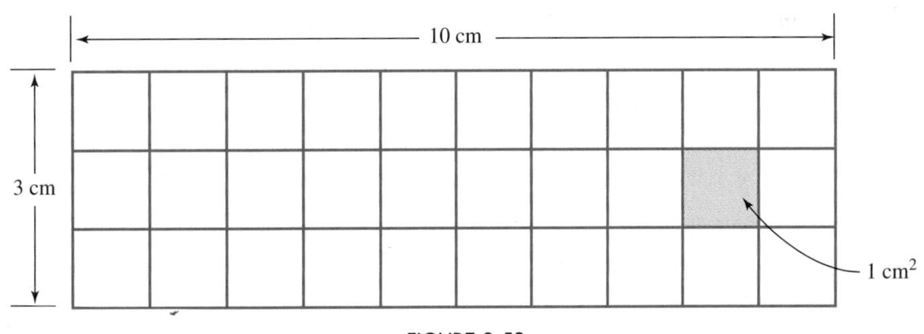

FIGURE 9-53

COMMENT Do not confuse the concepts of perimeter and area. Perimeter is the distance around a polygon. It is measured in linear units, such as centimeters, feet, or miles. Area is a measure of the surface enclosed within a polygon. It is measured in square units, such as square centimeters, square feet, or square miles.

In practice, we do not find areas by counting squares in a figure. Instead, we use formulas for finding areas of geometric figures, as shown in Table 9-1 on the next page.

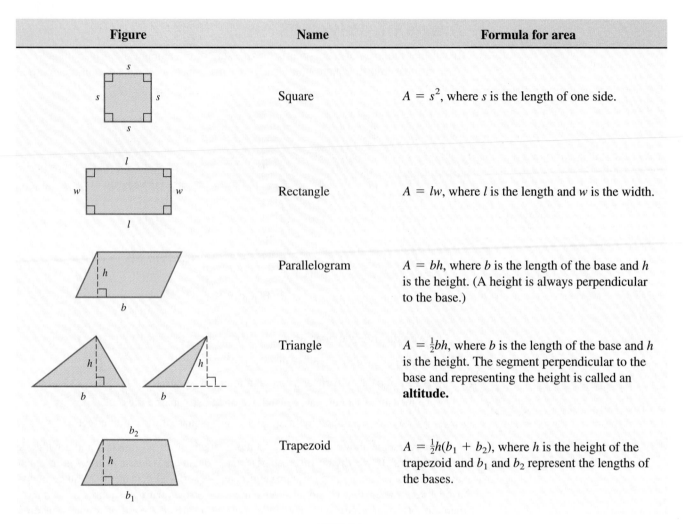

Figure	Name	Formula for area
	Square	$A = s^2$, where s is the length of one side.
	Rectangle	$A = lw$, where l is the length and w is the width.
	Parallelogram	$A = bh$, where b is the length of the base and h is the height. (A height is always perpendicular to the base.)
	Triangle	$A = \frac{1}{2}bh$, where b is the length of the base and h is the height. The segment perpendicular to the base and representing the height is called an **altitude.**
	Trapezoid	$A = \frac{1}{2}h(b_1 + b_2)$, where h is the height of the trapezoid and b_1 and b_2 represent the lengths of the bases.

TABLE 9-1

EXAMPLE 5 ***Area of a square.*** Find the area of the square in Figure 9-54.

Solution

We can see that the length of one side of the square is 15 centimeters. We can find its area by using the formula $A = s^2$ and substituting 15 for s.

$A = s^2$

$A = (15)^2$ Substitute 15 for s.

$A = 225$ Evaluate the exponential expression: $15 \cdot 15 = 225$.

The area of the square is 225 cm^2.

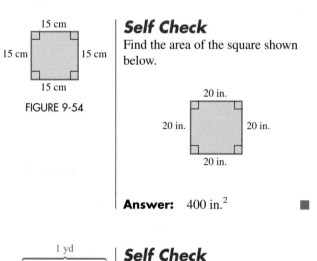

FIGURE 9-54

Self Check

Find the area of the square shown below.

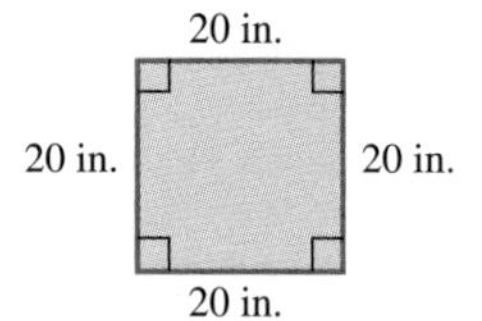

Answer: 400 in.^{2}

EXAMPLE 6 ***Number of square feet in 1 square yard.*** Find the number of square feet in 1 square yard. (See Figure 9-55.)

Solution

Since 3 feet = 1 yard, each side of 1 square yard is 3 feet long.

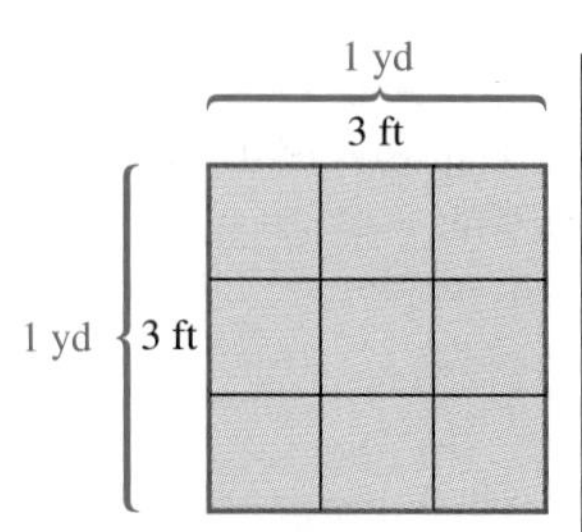

FIGURE 9-55

Self Check

Find the number of square centimeters in 1 square meter.

$$1 \text{ yd}^2 = (\mathbf{1 \text{ yd}})^2$$
$$= (\mathbf{3 \text{ ft}})^2 \quad \text{Substitute 3 feet for 1 yard.}$$
$$= 9 \text{ ft}^2 \quad (3 \text{ ft})^2 = (3 \text{ ft})(3 \text{ ft}) = 9 \text{ ft}^2.$$

There are 9 square feet in 1 square yard.

Answer: $10{,}000 \text{ cm}^2$ ■

EXAMPLE 7 ***Women's sports.*** Field hockey is a team sport in which players use sticks to try to hit a ball into their opponents' goal. Find the area of the rectangular field shown in Figure 9-56. Give the answer in square feet.

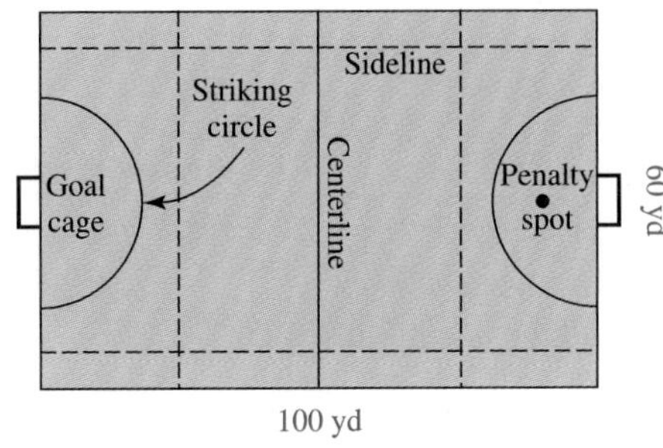

FIGURE 9-56

Self Check

Find the area in square inches of a rectangle with dimensions of 6 inches by 2 feet.

Solution

To find the area in square yards, we substitute 100 for l and 60 for w in the formula for the area of a rectangle, and simplify.

$$A = lw$$
$$A = \mathbf{100}(\mathbf{60})$$
$$= 6{,}000$$

The area is 6,000 square yards. Since there are 9 square feet per square yard, we can convert this number to square feet by multiplying 6,000 square yards by $\frac{9 \text{ ft}^2}{1 \text{ yd}^2}$.

$$6{,}000 \text{ yd}^2 = 6{,}000 \text{ yd}^2 \cdot \frac{9 \text{ ft}^2}{1 \text{ yd}^2} \quad \text{Multiply by the unit conversion factor: } \frac{9 \text{ ft}^2}{1 \text{ yd}^2}.$$
$$= 6{,}000 \cdot 9 \text{ ft}^2 \quad \text{The units of square yards divide out.}$$
$$= 54{,}000 \text{ ft}^2 \quad \text{Multiply: } 6{,}000 \cdot 9 = 54{,}000.$$

The area of the field is $54{,}000 \text{ ft}^2$.

Answer: 144 in.^2 ■

EXAMPLE 8 ***Area of a parallelogram.*** Find the area of the parallelogram in Figure 9-57.

12 ft

5 ft 25 ft

FIGURE 9-57

Solution

The length of the base of the parallelogram is

5 feet + 25 feet = 30 feet

The height is 12 feet. To find the area, we substitute 30 for b and 12 for h in the formula for the area of a parallelogram and simplify.

$$A = bh$$
$$A = \mathbf{30}(\mathbf{12})$$
$$= 360$$

The area of the parallelogram is 360 ft^2.

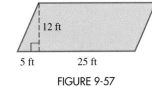

Self Check

Find the area of the parallelogram below.

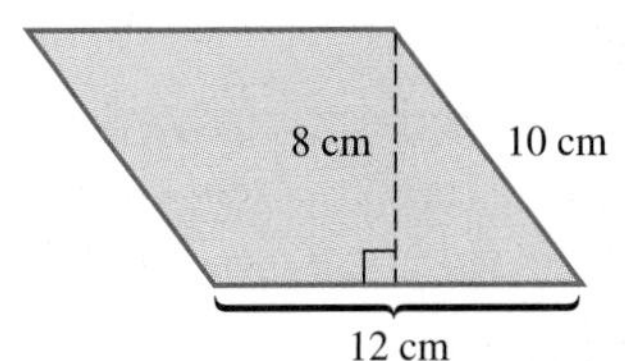

Answer: 96 cm^2 ■

EXAMPLE 9 ***Area of a triangle.*** Find the area of the triangle in Figure 9-58.

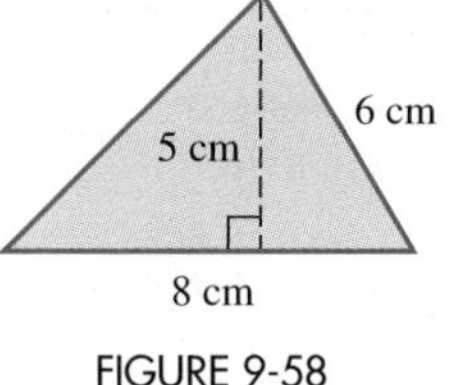

FIGURE 9-58

Solution

We substitute 8 for b and 5 for h in the formula for the area of a triangle, and simplify. (The side having length 6 cm is additional information that is not used to find the area.)

$$A = \frac{1}{2}bh$$

$$A = \frac{1}{2}(8)(5) \quad \text{The length of the base is 8 cm. The height is 5 cm.}$$

$$= 4(5) \quad \text{Do the multiplication: } \tfrac{1}{2}(8) = 4.$$

$$= 20$$

The area of the triangle is 20 cm^2.

Self Check

Find the area of the triangle below.

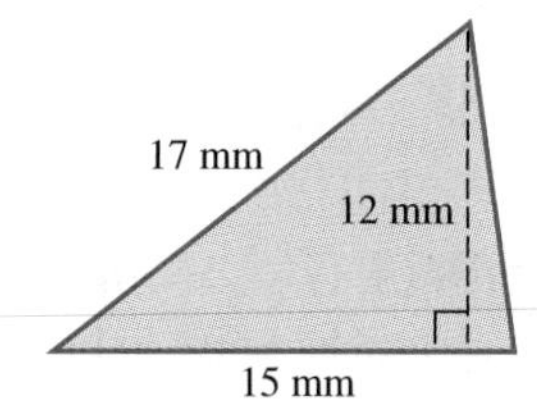

Answer: 90 mm^2

EXAMPLE 10 ***Area of a triangle.*** Find the area of the triangle in Figure 9-59.

Solution

In this case, the altitude falls outside the triangle.

$$A = \frac{1}{2}bh$$

$$A = \frac{1}{2}(9)(13) \quad \text{Substitute 9 for } b \text{ and 13 for } h.$$

$$= \frac{1}{2}\left(\frac{9}{1}\right)\left(\frac{13}{1}\right) \quad \text{Write 9 as } \tfrac{9}{1} \text{ and 13 as } \tfrac{13}{1}.$$

$$= \frac{117}{2} \quad \text{Multiply the fractions.}$$

$$= 58.5 \quad \text{Do the division.}$$

The area of the triangle is 58.5 cm^2.

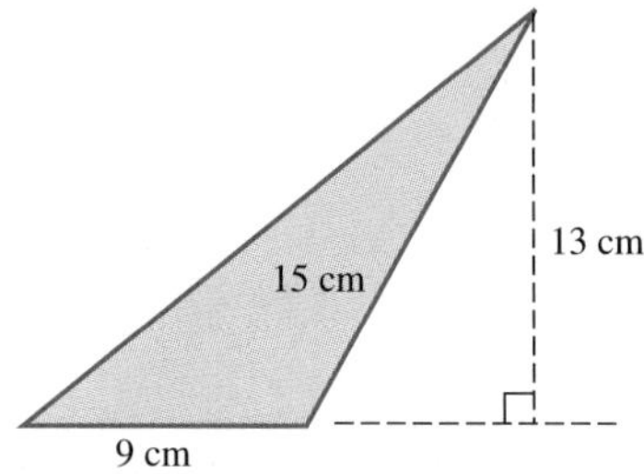

FIGURE 9-59

EXAMPLE 11 ***Area of a trapezoid.*** Find the area of the trapezoid in Figure 9-60.

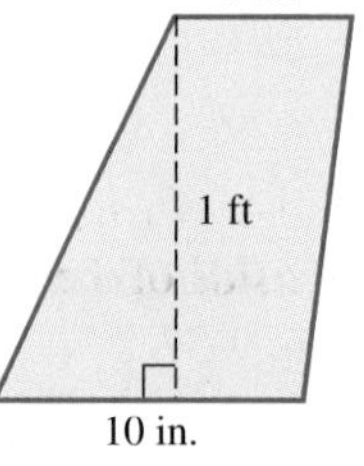

FIGURE 9-60

Solution

In this example, $b_1 = 10$ and $b_2 = 6$. It is incorrect to say that $h = 1$, because the height of 1 foot must be expressed as 12 inches to be consistent with the units of the bases. Thus, we substitute 10 for b_1, 6 for b_2, and 12 for h in the formula for finding the area of a trapezoid and simplify.

$$A = \frac{1}{2}h(b_1 + b_2)$$

$$A = \frac{1}{2}(12)(10 + 6) \quad \text{The length of the lower base is 10 in. The length of the upper base is 6 in. The height is 12 in.}$$

$$= \frac{1}{2}(12)(16) \quad \text{Do the addition within the parentheses.}$$

$$= 6(16) \quad \text{Do the multiplication: } \tfrac{1}{2}(12) = 6.$$

$$= 96$$

The area of the trapezoid is 96 in.^{2}

Self Check

Find the area of the trapezoid below.

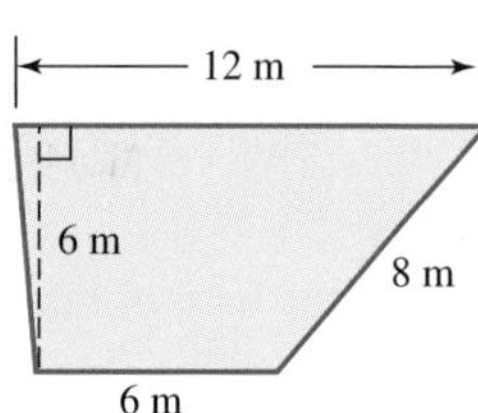

Answer: 54 m^2

Areas of figures that are combinations of polygons

EXAMPLE 12 ***Carpeting a room.*** A living room/dining room area has the floor plan shown in Figure 9-61. If carpet costs \$29 per square yard, including pad and installation, how much will it cost to carpet the room? (Assume no waste.)

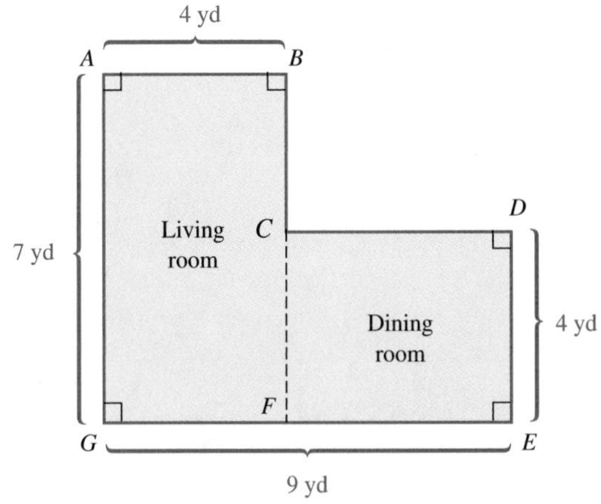

FIGURE 9-61

Solution First we must find the total area of the living room and the dining room:

$A_{\text{total}} = A_{\text{living room}} + A_{\text{dining room}}$

Since $\overline{CF}$ divides the space into two rectangles, the areas of the living room and the dining room are found by multiplying their respective lengths and widths.

$$\begin{aligned}\text{Area of living room} &= lw\\ &= 7(4)\\ &= 28\end{aligned}$$

The area of the living room is 28 yd^2.

To find the area of the dining room, we find its length by subtracting 4 yards from 9 yards to obtain 5 yards. We note that its width is 4 yards.

$$\begin{aligned}\text{Area of dining room} &= lw\\ &= 5(4)\\ &= 20\end{aligned}$$

The area of the dining room is 20 yd^2.

The total area to be carpeted is the sum of these two areas.

$$\begin{aligned}A_{\text{total}} &= A_{\text{living room}} + A_{\text{dining room}}\\ A_{\text{total}} &= 28\text{ yd}^2 + 20\text{ yd}^2\\ &= 48\text{ yd}^2\end{aligned}$$

At \$29 per square yard, the cost to carpet the room will be 48 · \$29, or \$1,392. ■

EXAMPLE 13 ***Area of one side of a tent.*** Find the area of one side of the tent in Figure 9-62.

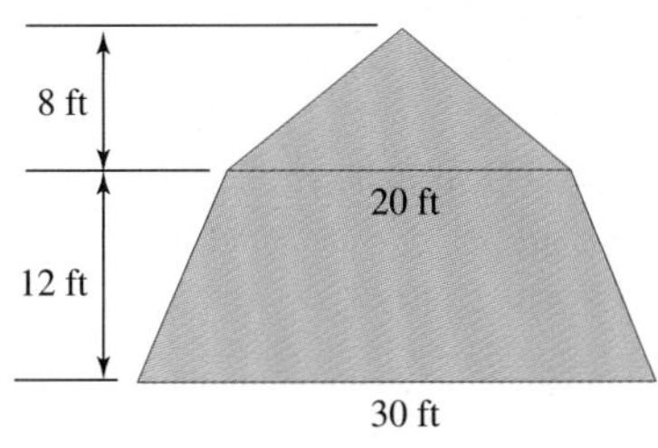

FIGURE 9-62

Solution Each side is a combination of a trapezoid and a triangle. Since the bases of each trapezoid are 30 feet and 20 feet and the height is 12 feet, we substitute 30 for b_1, 20 for b_2, and 12 for h into the formula for the area of a trapezoid.

$$A_{\text{trap.}} = \frac{1}{2}h(b_1 + b_2)$$

$$\begin{aligned}A_{\text{trap.}} &= \frac{1}{2}(12)(30 + 20)\\ &= 6(50)\\ &= 300\end{aligned}$$

The area of the trapezoid is 300 ft^2.

Since the triangle has a base of 20 feet and a height of 8 feet, we substitute 20 for b and 8 for h in the formula for the area of a triangle.

$$A_{\text{triangle}} = \frac{1}{2}bh$$

$$A_{\text{triangle}} = \frac{1}{2}(20)(8)$$

$$= 80$$

The area of the triangle is 80 ft^2.

The total area of one side of the tent is

$$A_{\text{total}} = A_{\text{trap.}} + A_{\text{triangle}}$$

$$A_{\text{total}} = 300 \text{ ft}^2 + 80 \text{ ft}^2$$

$$= 380 \text{ ft}^2$$

The total area is 380 ft^2. ■

STUDY SET Section 9.5

VOCABULARY *Fill in the blanks to make the statements true.*

1. The distance around a polygon is called the perimeter.

2. The perimeter of a polygon is measured in linear units.

3. The measure of the surface enclosed by a polygon is called its surface area.

4. If each side of a square measures 1 foot, the area enclosed by the square is 1 square foot.

5. The area of a polygon is measured in square units.

6. The segment that represents the height of a triangle is called an altitude.

CONCEPTS *Sketch and label each of the figures described.*

7. Two different rectangles, each having a perimeter of 40 in.

8. Two different rectangles, each having an area of 40 in^2.

9. A square with an area of 25 m^2.

10. A square with a perimeter of 20 m.

11. A parallelogram with an area of 15 yd^2.

12. A triangle with an area of 20 ft^2.

13. A figure consisting of a combination of two rectangles whose total area is 80 ft^2.

14. A figure consisting of a combination of a rectangle and a square whose total area is 164 ft^2.

NOTATION *Fill in the blanks to make the statements true.*

15. The formula for the perimeter of a square is P=4s.

16. The formula for the perimeter of a rectangle is P=2l+2w.

17. The symbol 1 in.^{2} means one inch squared.

18. One square meter is expressed as 1m^2.

19. The formula for the area of a square is A=s^2.

20. The formula for the area of a rectangle is A=lw.

21. The formula $A = \frac{1}{2}bh$ gives the area of a triangle.

22. The formula $A = \frac{1}{2}h(b_1 + b_2)$ gives the area of a trapezoid.

PRACTICE *Find the perimeter of each figure.*

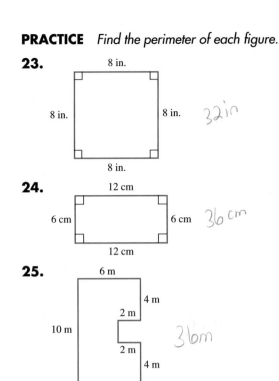

23. 8 in. 8 in. 8 in. 8 in. 32in

24. 12 cm 6 cm 6 cm 12 cm 36 cm

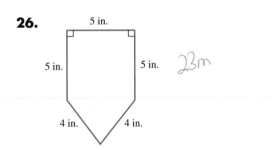

25. 6 m 4 m 2 m 10 m 2 m 4 m 6 m 36m

26. 5 in. 5 in. 5 in. 4 in. 4 in. 23m

27. 6 cm 6 cm 7 cm 8 cm 10 cm 37 cm

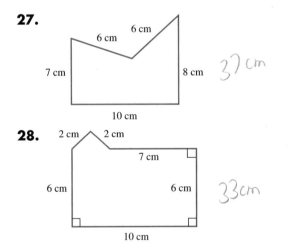

28. 2 cm 2 cm 7 cm 6 cm 6 cm 10 cm 33cm

Solve each problem.

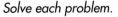

29. Find the perimeter of an isosceles triangle with a base of length 21 centimeters and sides of length 32 centimeters.

30. The perimeter of an isosceles triangle is 80 meters. If the length of one side is 22 meters, how long is the base?

31. The perimeter of an equilateral triangle is 85 feet. Find the length of each side.

32. An isosceles triangle with sides of 49.3 inches has a perimeter of 121.7 inches. Find the length of the base.

Find the area of the shaded part of each figure.

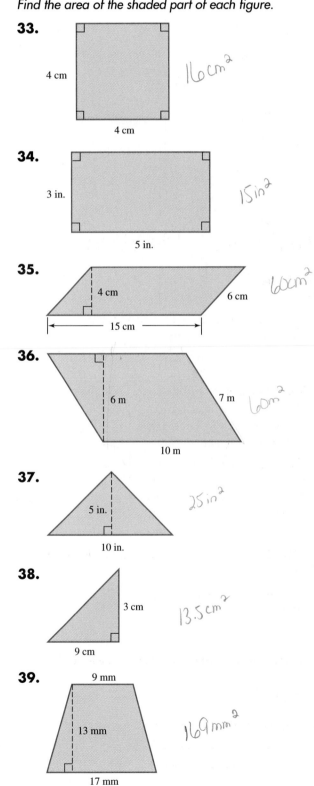

33. 4 cm 4 cm $16cm^2$

34. 3 in. 5 in. $15in^2$

35. 4 cm 6 cm 15 cm $60cm^2$

36. 6 m 7 m 10 m $60m^2$

37. 5 in. 10 in. $25in^2$

38. 3 cm 9 cm $13.5cm^2$

39. 9 mm 13 mm 17 mm $169mm^2$

40.

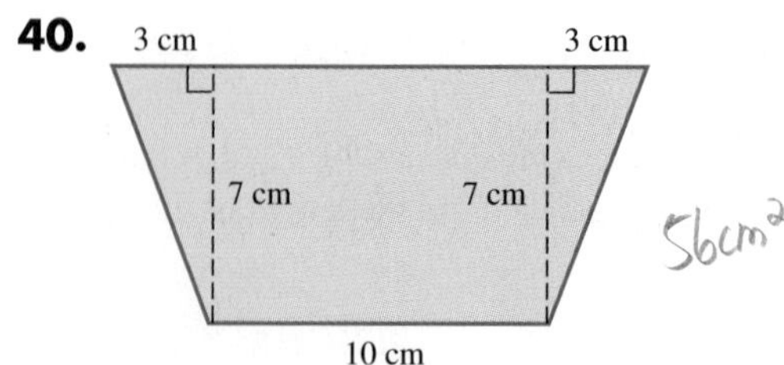

41.

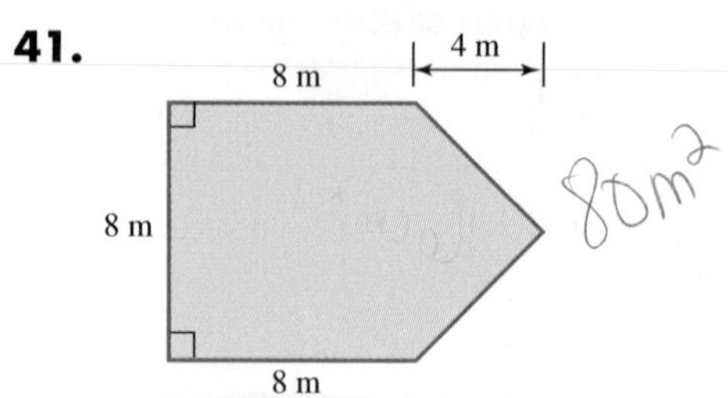

42.

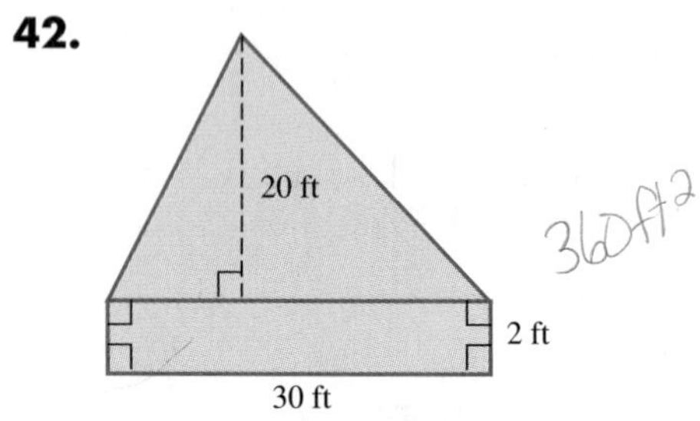

43.

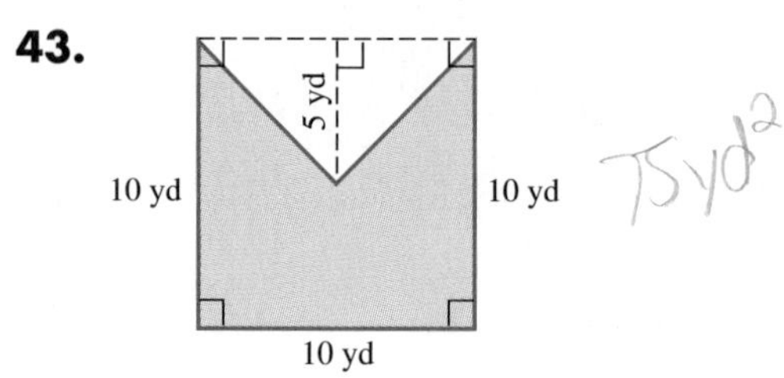

44.

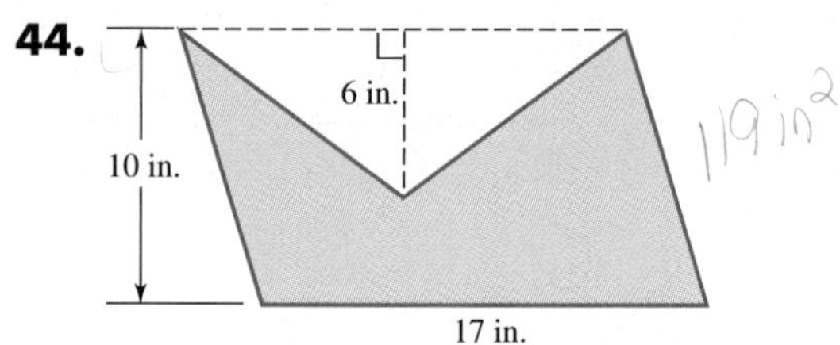

45.

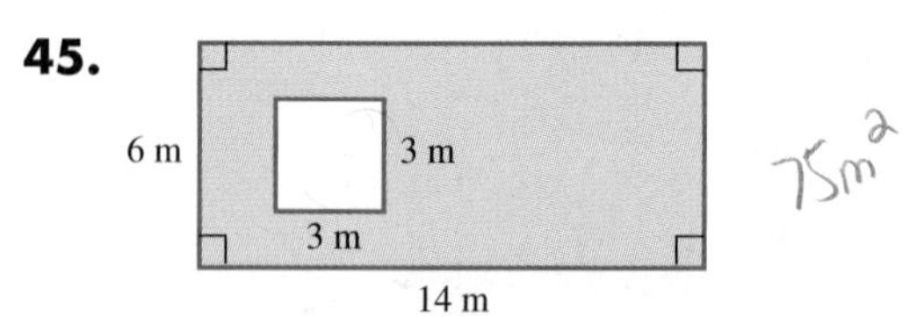

46.

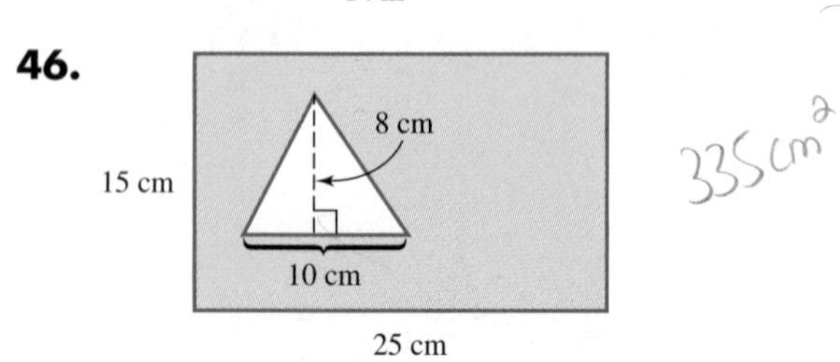

47. How many square inches are in 1 square foot?

48. How many square inches are in 1 square yard?

APPLICATIONS

49. FENCING A YARD A man wants to enclose a rectangular yard with fencing that costs \$12.50 a foot, including installation. Find the cost of enclosing the yard if its dimensions are 110 ft by 85 ft.

50. FRAMING A PICTURE Find the cost of framing a rectangular picture with dimensions of 24 inches by 30 inches if framing material costs \$8.46 per foot, including matting.

51. PLANTING A SCREEN A woman wants to plant a pine-tree screen around three sides of her backyard. (See Illustration 1.) If she plants the trees 3 feet apart, how many trees will she need?

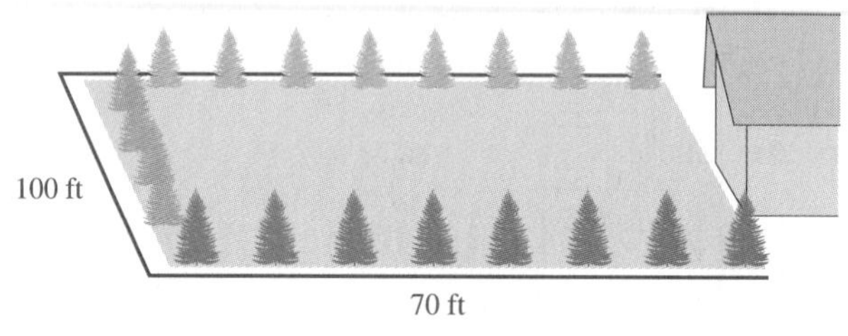

ILLUSTRATION 1

52. PLANTING MARIGOLDS A gardener wants to plant a border of marigolds around the garden shown in Illustration 2, to keep out rabbits. How many plants will she need if she allows 6 inches between plants?

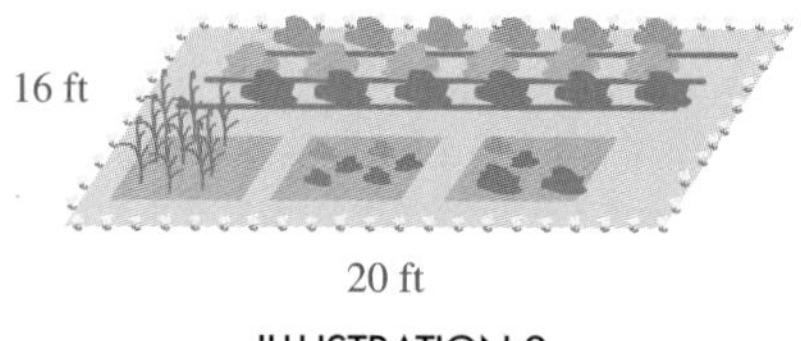

ILLUSTRATION 2

53. BUYING A FLOOR Which is more expensive: A ceramic-tile floor costing \$3.75 per square foot or linoleum costing \$34.95 per square yard?

54. BUYING A FLOOR Which is cheaper: A hardwood floor costing \$5.95 per square foot or a carpeted floor costing \$37.50 per square yard?

55. CARPETING A ROOM A rectangular room is 24 feet long and 15 feet wide. At \$30 per square yard, how much will it cost to carpet the room? (Assume no waste.)

56. CARPETING A ROOM A rectangular living room measures 30 by 18 feet. At \$32 per square yard, how much will it cost to carpet the room? (Assume no waste.)

57. TILING A FLOOR A rectangular basement room measures 14 by 20 feet. Vinyl floor tiles that are 1 ft^2 cost \$1.29 each. How much will the tile cost to cover the floor? (Disregard any waste.)

58. PAINTING A BARN The north wall of a barn is a rectangle 23 feet high and 72 feet long. There are five windows in the wall, each 4 by 6 feet. If a gallon of paint will cover 300 ft^2, how many gallons of paint must the painter buy to paint the wall?

59. MAKING A SAIL If nylon is $12 per square yard, how much would the fabric cost to make a triangular sail with a base of 12 feet and a height of 24 feet?

60. PAINTING A GABLE The gable end of a warehouse is an isosceles triangle with a height of 4 yards and a base of 23 yards. It will require one coat of primer and one coat of finish to paint the triangle. Primer costs $17 per gallon, and the finish paint costs $23 per gallon. If one gallon covers 300 square feet, how much will it cost to paint the gable, excluding labor?

61. GEOGRAPHY See Illustration 3. Use the dimensions of the trapezoid that is superimposed over the state of Nevada to estimate the area of the "Silver State."

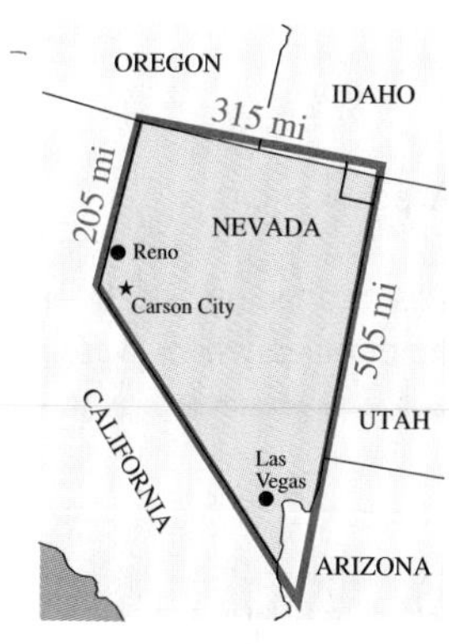

ILLUSTRATION 3

62. COVERING A SWIMMING POOL A swimming pool has the shape shown in Illustration 4. How many square meters of plastic sheeting will be needed to cover the pool? How much will the sheeting cost if it is $2.95 per square meter? (Assume no waste.)

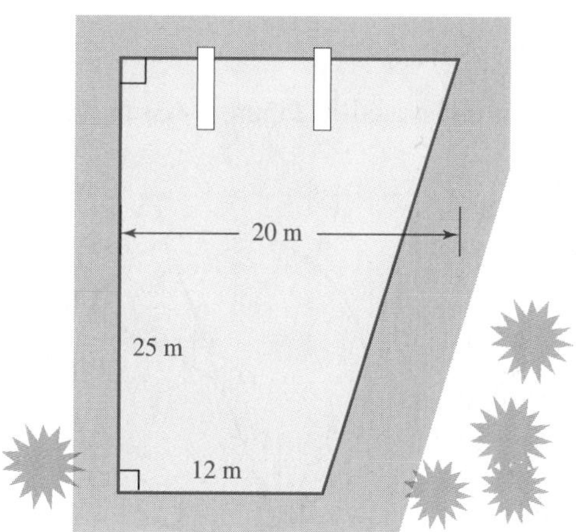

ILLUSTRATION 4

63. CARPENTRY How many sheets of 4-foot-by-8-foot sheetrock are needed to drywall the inside walls on the first floor of the barn shown in Illustration 5? (Assume that the carpenters will cover each wall entirely and then cut out areas for the doors and windows.)

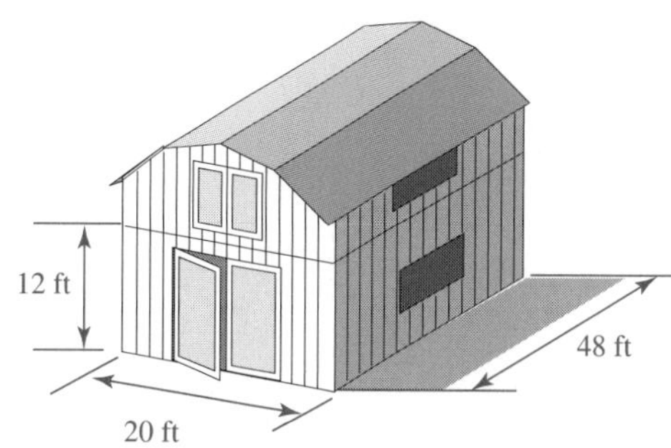

ILLUSTRATION 5

64. CARPENTRY If it costs $90 per square foot to build a one-story home in northern Wisconsin, estimate the cost of building the house with the floor plan shown in Illustration 6.

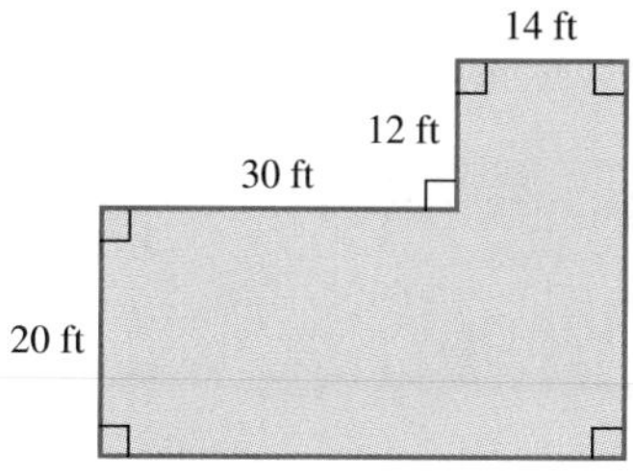

ILLUSTRATION 6

65. DRIVING SAFETY Illustration 7 shows the areas on a highway that a truck driver cannot see in the truck's rear view mirrors. Use the scale to determine the approximate dimensions of each blind spot. Then estimate the area of each of them.

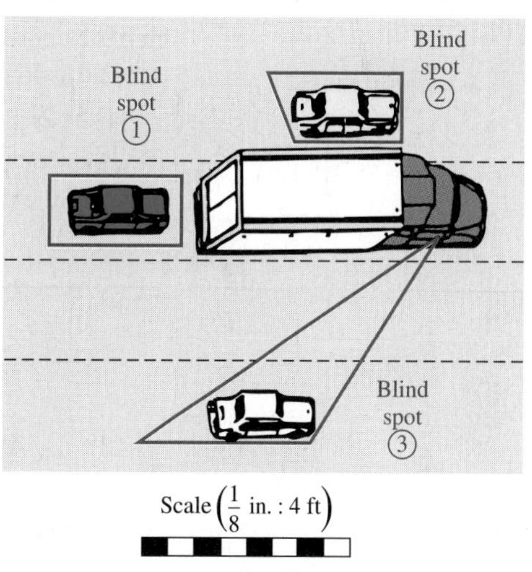

ILLUSTRATION 7

66. ESTIMATING AREA See Illustration 8. Estimate the area of the sole plate of the iron by thinking of it as a combination of a trapezoid and a triangle.

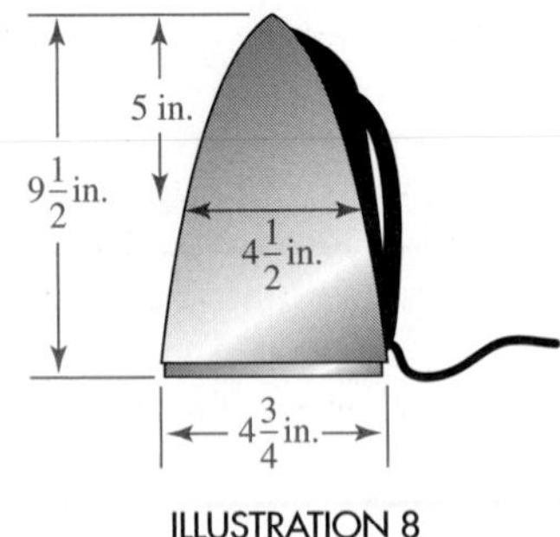

ILLUSTRATION 8

WRITING

67. Explain the difference between perimeter and area.

68. Why is it necessary that area be measured in square units?

REVIEW *Do the calculations. Write all improper fractions as mixed numbers.*

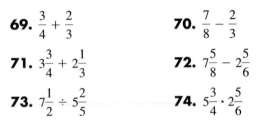

69. $\frac{3}{4} + \frac{2}{3}$

70. $\frac{7}{8} - \frac{2}{3}$

71. $3\frac{3}{4} + 2\frac{1}{3}$

72. $7\frac{5}{8} - 2\frac{5}{6}$

73. $7\frac{1}{2} \div 5\frac{2}{5}$

74. $5\frac{3}{4} \cdot 2\frac{5}{6}$

9.6 Circles

In this section, you will learn about

- Circles • Circumference of a circle • Area of a circle

INTRODUCTION. In this section, we will discuss circles, one of the most useful geometric figures. In fact, the discovery of fire and the circular wheel were two of the most important events in the history of the human race.

Circles

Circle

A **circle** is the set of all points in a plane that lie a fixed distance from a point called its **center.**

A segment drawn from the center of a circle to a point on the circle is called a **radius.** (The plural of *radius* is *radii.*) From the definition, it follows that all radii of the same circle are the same length.

A **chord** of a circle is a line segment connecting two points on the circle. A **diameter** is a chord that passes through the center of the circle. Since a diameter D of a circle is twice as long as a radius r, we have

$$D = 2r$$

Each of the previous definitions is illustrated in Figure 9-63, in which O is the center of the circle.

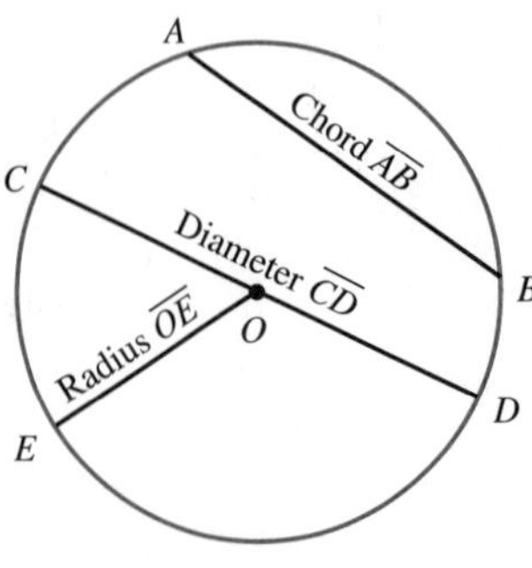

FIGURE 9-63

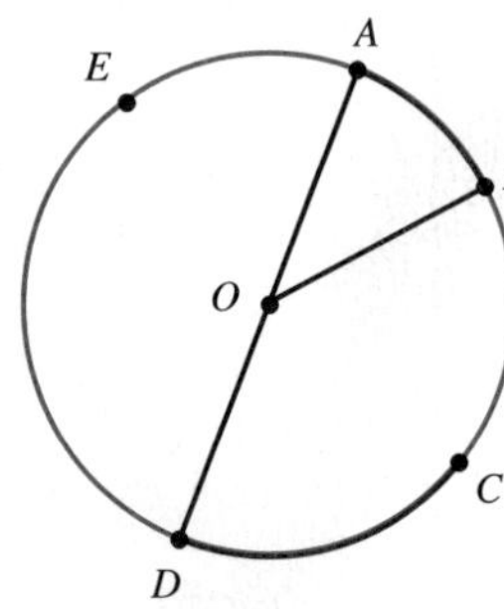

FIGURE 9-64

Any part of a circle is called an **arc.** In Figure 9-64 on the preceding page, the part of the circle from point A to point B is $\overset{\frown}{AB}$, read as "arc AB." $\overset{\frown}{CD}$ is the part of the circle from point C to point D. An arc that is half of a circle is a **semicircle.**

Semicircle

> A **semicircle** is an arc of a circle whose endpoints are the endpoints of a diameter.

If point O is the center of the circle in Figure 9-64, $\overline{AD}$ is a diameter and $\overset{\frown}{AED}$ is a semicircle. The middle letter E is used to distinguish semicircle $\overset{\frown}{AED}$ from semicircle $\overset{\frown}{ABCD}$.

An arc that is shorter than a semicircle is a **minor arc.** An arc that is longer than a semicircle is a **major arc.** In Figure 9-64,

$\overset{\frown}{AB}$ is a minor arc and $\overset{\frown}{ABCDE}$ is a major arc.

Circumference of a circle

Since early history, mathematicians have known that the ratio of the distance around a circle (the circumference) divided by the length of its diameter is approximately 3. First Kings, Chapter 7 of the Bible describes a round bronze tank that was 15 feet from brim to brim and 45 feet in circumference, and $\frac{45}{15} = 3$. Today, we have a better value for this ratio, known as π (pi). If C is the circumference of a circle and D is the length of its diameter, then

$$\pi = \frac{C}{D}, \quad \text{where } \pi = 3.141592653589\ldots$$

$\frac{22}{7}$ and 3.14 are often used as estimates of π.

If we multiply both sides of $\pi = \frac{C}{D}$ by D, we have the following formula.

Circumference of a circle

> The circumference of a circle is given by the formula
>
> $C = \pi D$ where C is the circumference and D is the length of the diameter

Since a diameter of a circle is twice as long as a radius r, we can substitute $2r$ for D in the formula $C = \pi D$ to obtain another formula for the circumference C:

$$C = 2\pi r$$

EXAMPLE 1 ***Circumference of a circle.*** Find the circumference of a circle that has a diameter of 10 centimeters. (See Figure 9-65.)

Solution

We substitute 10 for D in the formula for the circumference of a circle.

$C = \pi D$

$C = \pi(10)$

$C \approx 3.14(10)$ Replace π with an approximation: $\pi \approx 3.14$.

$C \approx 31.4$

The circumference is approximately 31.4 centimeters.

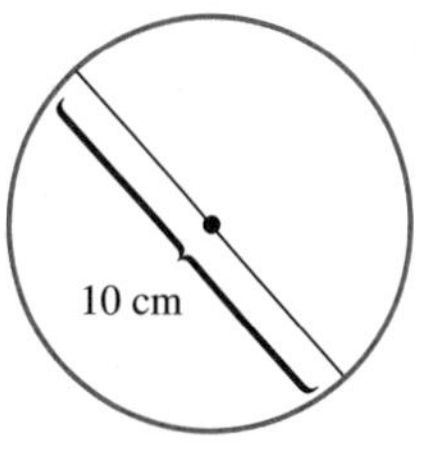

FIGURE 9-65

Self Check

To the nearest tenth, find the circumference of a circle that has a radius of 12 meters.

Answer: 75.4 m

Accent on Technology: **Calculating revolutions of a tire**

When the $\boxed{\pi}$ key on a scientific calculator is pressed (on some models, the $\boxed{\text{2nd}}$ key must be pressed first), an approximation of π is displayed. To illustrate how to use this key, consider the following problem. How many times does a 15-inch tire revolve when a car makes a 25-mile trip?

We first find the circumference of the tire.

$C = \pi D$

$C = \pi(15)$ Substitute 15 for D, the diameter of the tire.

$C = 15\pi$ Normally, we rewrite a product such as $\pi(15)$ so that π is the second factor.

The circumference of the tire is 15π inches.

We then change 25 miles to inches using two unit conversion factors.

$$\frac{25}{1} \text{ miles} \cdot \frac{5{,}280 \text{ feet}}{1 \text{ mile}} \cdot \frac{12 \text{ inches}}{1 \text{ foot}} = 25(5{,}280)(12) \text{ in.}$$

The total distance of the trip is 25(5,280)(12) inches.

Finally, we divide the total distance of the trip by the circumference of the tire to get

$$\text{The number of revolutions of the tire} = \frac{25(5{,}280)(12)}{15\pi}$$

To do this work using a scientific calculator, we enter these numbers and press these keys.

Keystrokes (25 × 5280 × 12) ÷ (15 × π) =

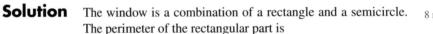

33613.52398

The tire makes about 33,614 revolutions.

EXAMPLE 2 ***Architecture.*** A Norman window is constructed by adding a semicircular window to the top of a rectangular window. Find the perimeter of the Norman window shown in Figure 9-66.

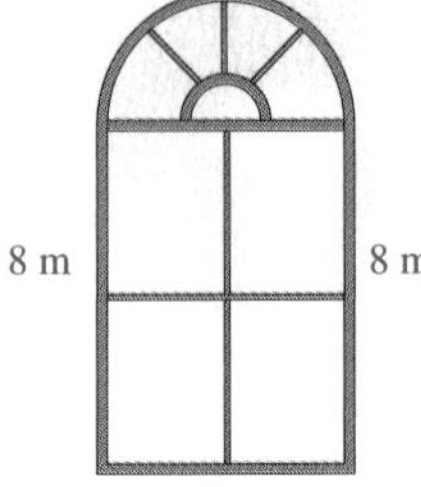

FIGURE 9-66

Solution The window is a combination of a rectangle and a semicircle. The perimeter of the rectangular part is

$P_{\text{rectangular part}} = 8 + 6 + 8 = 22$ Add only 3 sides.

The perimeter of the semicircle is one-half of the circumference of a circle that has a 6-meter diameter.

$$P_{\text{semicircle}} = \frac{1}{2}\pi D$$

$$= \frac{1}{2}\pi(6) \quad \text{Substitute 6 for } D.$$

$$\approx 9.424777961 \quad \text{Use a calculator.}$$

The total perimeter is the sum of the two parts.

$$P_{\text{total}} \approx 22 + 9.424777961$$

$$\approx 31.424777961$$

To the nearest hundredth, the perimeter of the window is 31.42 meters. ■

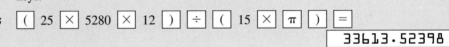

Area of a circle

If we divide the circle shown in Figure 9-67(a) into an even number of pie-shaped pieces and then rearrange them as shown in Figure 9-67(b), we have a figure that looks like a parallelogram. The figure has a base that is one-half the circumference of the circle, and its height is about the same length as a radius of the circle.

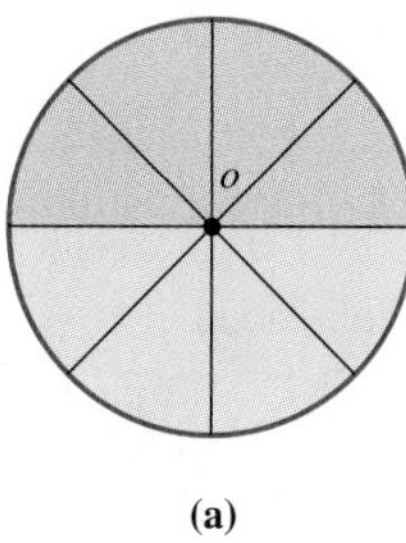

(a)

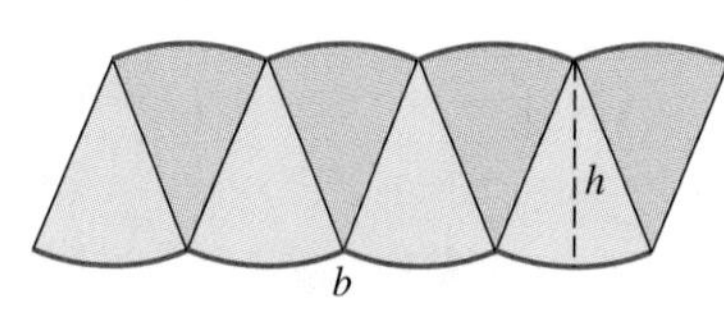

(b)

FIGURE 9-67

If we divide the circle into more and more pie-shaped pieces, the figure will look more and more like a parallelogram, and we can find its area by using the formula for the area of a parallelogram.

$A = \mathbf{bh}$

$= \frac{1}{2}Cr$ Substitute $\frac{1}{2}$ of the circumference for b, and r for the height.

$= \frac{1}{2}(2\pi r)r$ Make a substitution: $C = 2\pi r$.

$= \pi r^2$ Simplify: $\frac{1}{2} \cdot 2 = 1$ and $r \cdot r = r^2$.

Area of a circle

> The **area of a circle** with radius r is given by the formula
>
> $A = \pi r^2$

EXAMPLE 3 ***Area of a circle.*** To the nearest tenth, find the area of the circle in Figure 9-68.

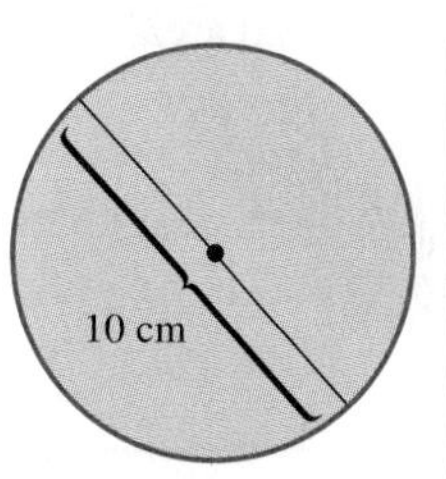

FIGURE 9-68

Solution

Since the length of the diameter is 10 centimeters and the length of a diameter is twice the length of a radius, the length of the radius is 5 centimeters. To find the area of the circle, we substitute 5 for r in the formula for the area of a circle.

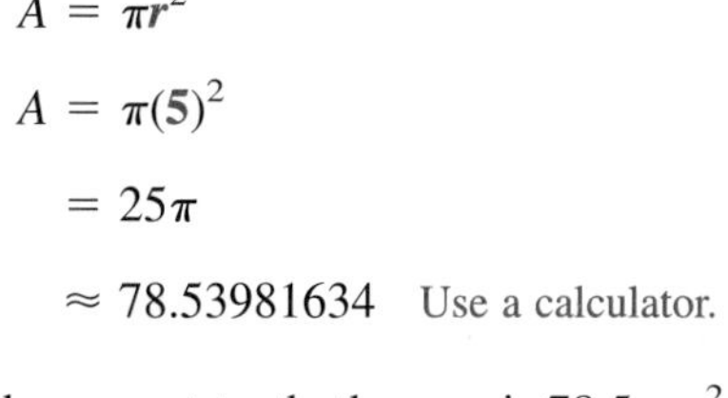

$A = \pi r^2$

$A = \pi(5)^2$

$= 25\pi$

≈ 78.53981634 Use a calculator.

To the nearest tenth, the area is 78.5 cm^2.

Self Check

To the nearest tenth, find the area of a circle with a diameter of 12 feet.

Answer: 113.1 ft^2

Accent on Technology: Painting a helicopter pad

Orange paint is available in gallon containers at \$19 each, and each gallon will cover 375 ft^2. To calculate how much the paint will cost to cover a circular helicopter pad 60 feet in diameter, we first calculate the area of the helicopter pad.

$$A = \pi r^2$$

$$A = \pi(30)^2 \quad \text{Substitute one-half of 60 for } r.$$

$$= 30^2\pi$$

The area of the pad is $30^2\pi$ ft^2. Since each gallon of paint will cover 375 ft^2, we can find the number of gallons of paint needed by dividing $30^2\pi$ by 375.

$$\text{Number of gallons needed} = \frac{30^2\pi}{375}$$

To do this work on a calculator, we enter these numbers and press these keys.

Keystrokes 30 [x^2] [×] [π] [=] [÷] 375 [=]   7.539822369

Because paint comes only in full gallons, the painter will need to purchase 8 gallons. The cost of the paint will be 8(\$19), or \$152.

EXAMPLE 4 ***Finding the area.*** Find the shaded area in Figure 9-69.

Solution The figure is a combination of a triangle and two semicircles. By the Pythagorean theorem, the hypotenuse h of the right triangle is

$$h = \sqrt{6^2 + 8^2} = \sqrt{36 + 64} = \sqrt{100} = 10$$

The area of the triangle is

$$A_{\text{right triangle}} = \frac{1}{2}bh = \frac{1}{2}(6)(8) = \frac{1}{2}(48) = 24$$

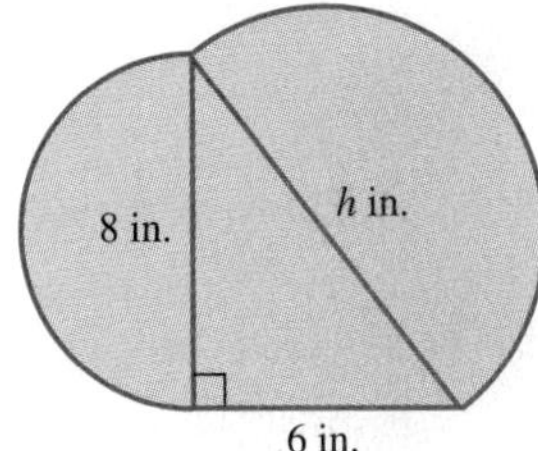

FIGURE 9-69

The area enclosed by the smaller semicircle is

$$A_{\text{smaller semicircle}} = \frac{1}{2}\pi r^2 = \frac{1}{2}\pi(4)^2 = \frac{1}{2}\pi(16) = 8\pi$$

The area enclosed by the larger semicircle is

$$A_{\text{larger semicircle}} = \frac{1}{2}\pi r^2 = \frac{1}{2}\pi(5)^2 = \frac{1}{2}\pi(25) = 12.5\pi$$

The total area is

$$A_{\text{total}} = 24 + 8\pi + 12.5\pi \approx 88.4026494 \quad \text{Use a calculator.}$$

To the nearest hundredth, the area is 88.40 in.^{2} ■

STUDY SET Section 9.6

VOCABULARY *Fill in the blanks to make the statements true.*

1. A segment drawn from the center of a circle to a point on the circle is called a __radius__.

2. A segment joining two points on a circle is called a __chord__.

3. A __diameter__ is a chord that passes through the center of a circle.

4. An arc that is one-half of a complete circle is a __semicircle__.

5. An arc that is shorter than a semicircle is called a __minor__ arc.

6. An arc that is longer than a semicircle is called a ___major___ arc.

7. The distance around a circle is called its ___circumference___.

8. The surface enclosed by a circle is called its ___area___.

CONCEPTS *Refer to Illustration 1.*

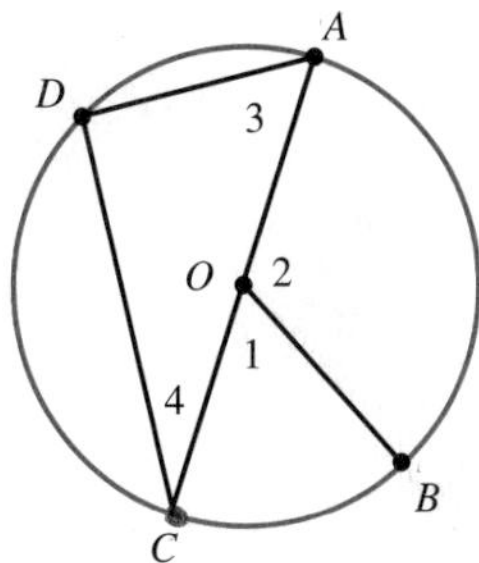

ILLUSTRATION 1

9. Name each radius.

10. Name a diameter.

11. Name each chord.

12. Name each minor arc.

13. Name each semicircle.

14. Name each major arc.

15. If you know the radius of a circle, how can you find its diameter?

16. If you know the diameter of a circle, how can you find its radius?

17. Suppose the two "legs" of the compass shown in Illustration 2 are adjusted so that the distance between the pointed ends is 1 inch. Then a circle is drawn.

a. What will the radius of the circle be?

b. What will the diameter of the circle be?

c. What will the circumference of the circle be?

d. What will the area of the circle be?

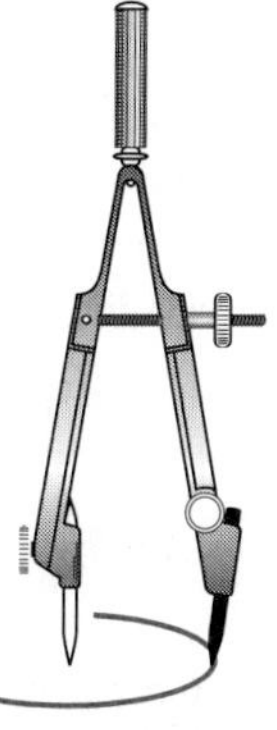
ILLUSTRATION 2

18. Suppose we find the distance around a can and the distance across the can using a measuring tape, as shown in Illustration 3. Then we make a comparison, in the form of a ratio:

$$\frac{\text{The distance around the can}}{\text{The distance across the top of the can}}$$

After we do the indicated division, the result will be close to what number?

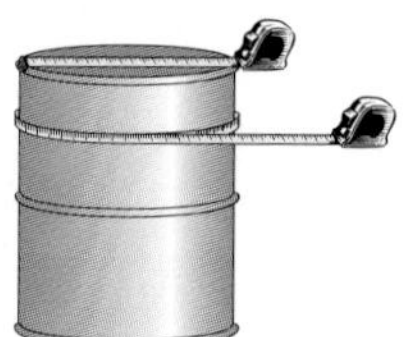
ILLUSTRATION 3

19. When evaluating $\pi(6)^2$, what operation should be performed first?

20. Round $\pi = 3.141592653589...$ to the nearest hundredth.

NOTATION *Fill in the blanks to make the statements true.*

21. The symbol $\overset{\frown}{AB}$ is read as ___arc AB___.

22. To the nearest hundredth, the value of π is ___3.14___.

23. The formula for the circumference of a circle is ________ or ________.

24. The formula $A = \pi r^2$ gives the area of a ________.

25. If C is the circumference of a circle and D is its diameter, then $\frac{C}{D} = \square$.

26. If D is the diameter of a circle and r is its radius, then $D = \square r$.

27. Write $\pi(8)$ in a better form.

28. What does $2\pi r$ mean?

PRACTICE *Solve each problem. Answers may vary slightly depending on which approximation of π is used.*

29. To the nearest hundredth, find the circumference of a circle that has a diameter of 12 inches.

30. To the nearest hundredth, find the circumference of a circle that has a radius of 20 feet.

31. Find the diameter of a circle that has a circumference of 36π meters.

32. Find the radius of a circle that has a circumference of 50π meters.

Find the perimeter of each figure to the nearest hundredth.

33.

8 ft

3 ft

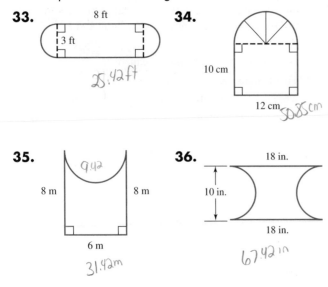

34.

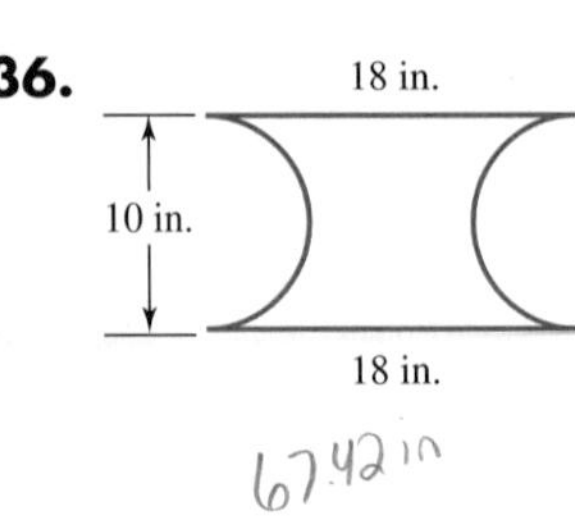

35.

8 m

8 m

6 m

36.

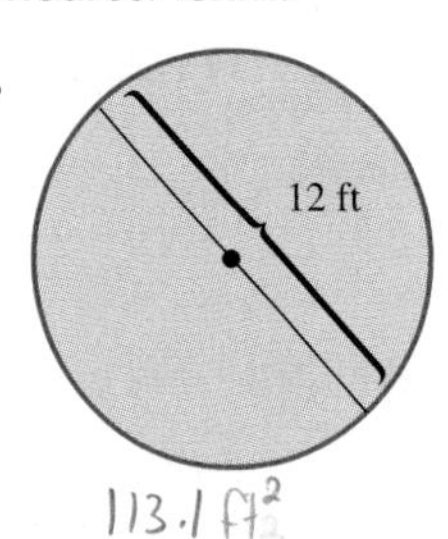

Find the area of each circle to the nearest tenth.

37.

3 in.

38.

12 ft

Find the total area of each figure to the nearest tenth.

39.

6 in.

10 in.

40.

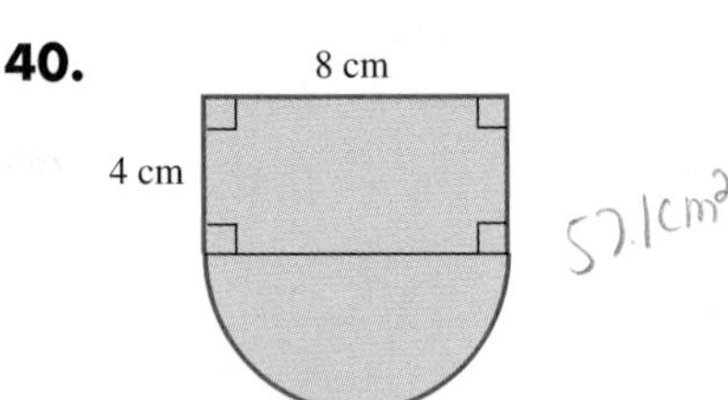

41. **42.**

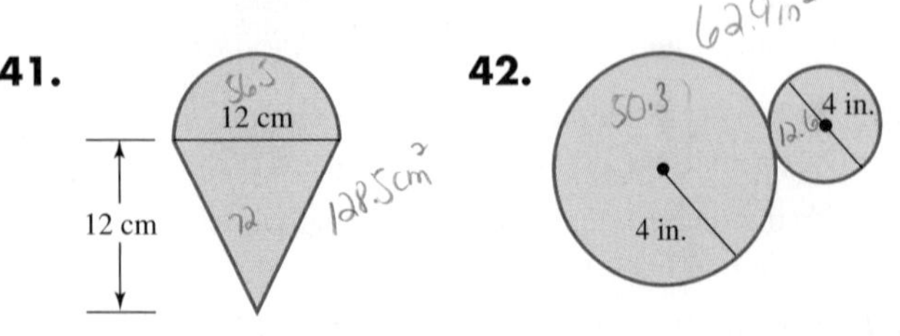

Find the area of each shaded region to the nearest tenth.

43.

4 in.

10 in

44.

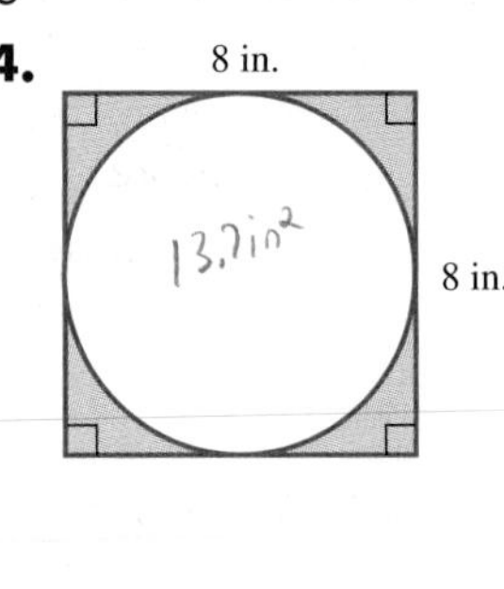

45.

$r = 4$ in.

$h = 9$ in.

13 in.

46.

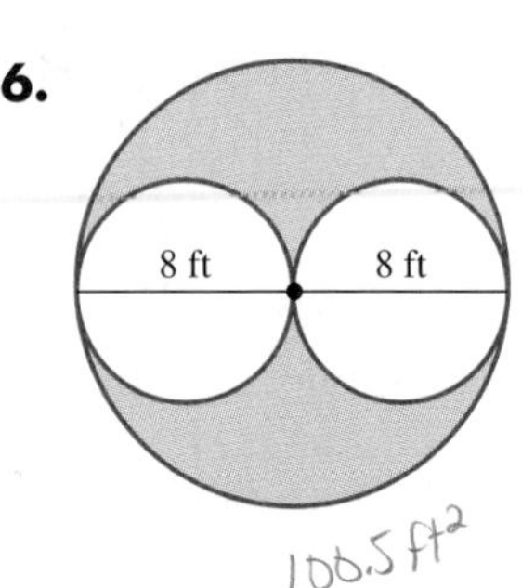

APPLICATIONS *Give each answer to the nearest hundredth. Answers may vary slightly depending on which approximation of π is used.*

47. AREA OF ROUND LAKE Round Lake has a circular shoreline that is 2 miles in diameter. Find the area of the lake.

48. HELICOPTER Refer to Illustration 4. How far does a point on the tip of a rotor blade travel when it makes one complete revolution?

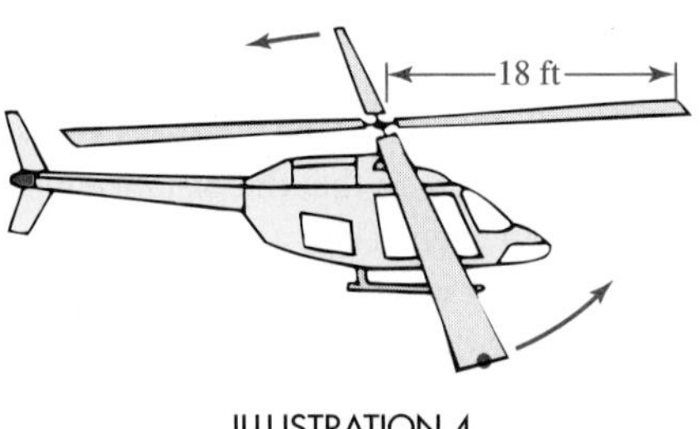

ILLUSTRATION 4

49. GIANT SEQUOIA The largest sequoia tree is the General Sherman Tree in Sequoia National Park in California. In fact, it is considered to be the largest living thing in the world. According to the *Guinness Book of World Records,* it has a circumference of 102.6 feet, measured $4\frac{1}{2}$ feet above the ground. What is the diameter of the tree at that height?

50. TRAMPOLINE See Illustration 5. The distance from the center of the trampoline to the edge of its steel frame is 7 feet. The protective padding covering the springs is 15 inches wide. Find the area of the circular jumping surface of the trampoline, in square feet.

ILLUSTRATION 5

51. JOGGING Joan wants to jog 10 miles on a circular track $\frac{1}{4}$ mile in diameter. How many times must she circle the track?

52. FIXING THE ROTUNDA The rotunda at a state capitol is a circular area 100 feet in diameter. The legislature wishes to appropriate money to have the floor of the rotunda tiled. The lowest bid is $83 per square yard, including installation. How much must the legislature spend?

53. BANDING THE EARTH A steel band is drawn tightly about the Earth's equator. The band is then loosened by increasing its length by 10 feet, and the resulting slack is distributed evenly along the band's entire length. How far above the Earth's surface is the band? (*Hint:* You don't need to know the Earth's circumference.)

54. CONCENTRIC CIRCLES Two circles are called **concentric circles** if they have the same center. Find the area of the band between two concentric circles if their diameters are 10 centimeters and 6 centimeters.

55. ARCHERY See Illustration 6. Find the area of the entire target and the bull's eye. What percent of the area of the target is the bull's eye?

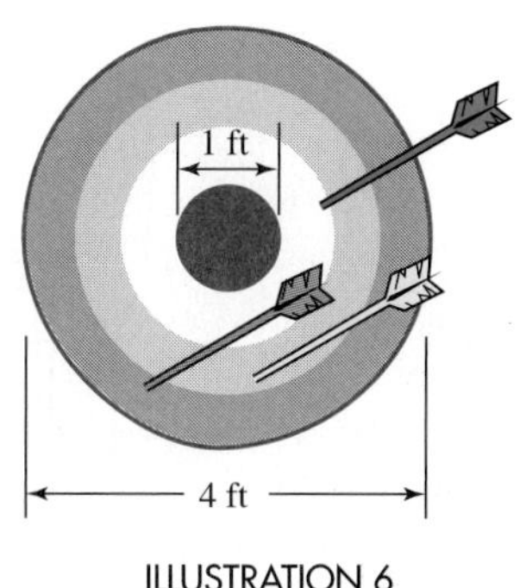

ILLUSTRATION 6

56. LANDSCAPE DESIGN See Illustration 7. How much of the lawn does not get watered by the sprinklers at the center of each circle?

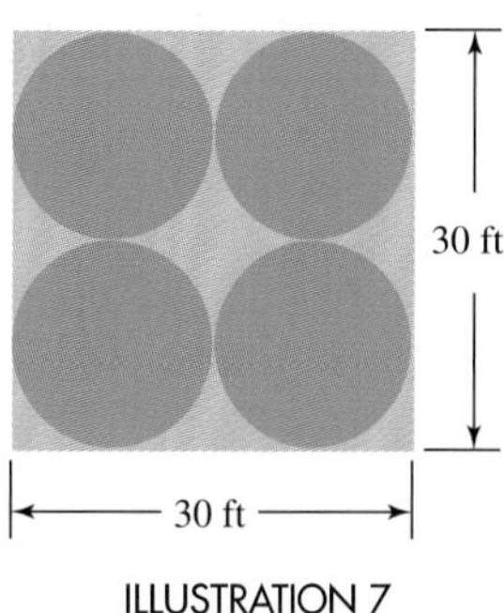

ILLUSTRATION 7

WRITING

57. Explain what is meant by the circumference of a circle.

58. Explain what is meant by the area of a circle.

59. Explain the meaning of π.

60. Distinguish between a major arc and a minor arc.

61. Explain what it means for a car to have a small turning radius.

62. The word *circumference* means the distance around a circle. In your own words, explain what is meant by each of the following sentences.
   a. A boat owner's dream was to *circumnavigate* the globe.
   b. The teenager's parents felt that he was always trying to *circumvent* the rules.
   c. The class was shown a picture of a circle *circumscribed* about an equilateral triangle.

REVIEW

63. Change $\frac{9}{10}$ to a percent.

64. Change $\frac{7}{8}$ to a percent.

65. UNIT COST A 24-ounce package of green beans sells for $1.29. Give the unit cost in cents per ounce.

66. MILEAGE One car went 1,235 miles on 51.3 gallons of gasoline, and another went 1,456 on 55.78 gallons. Which car got the better gas mileage?

67. How many sides does a pentagon have?

68. What is the sum of the measures of the angles of a triangle?

9.7 Surface Area and Volume

In this section, you will learn about

- Volumes of solids • Surface areas of rectangular solids
- Volumes and surface areas of spheres • Volumes of cylinders
- Volumes of cones • Volumes of pyramids

INTRODUCTION. In this section, we will discuss a measure of capacity called **volume.** Volumes are measured in cubic units, such as cubic inches, cubic yards, or cubic centimeters. For example,

- We buy gravel or topsoil by the cubic yard.
- We measure the capacity of a refrigerator in cubic feet.
- We often measure amounts of medicine in cubic centimeters.

We will also discuss surface area. The ability to compute surface area is necessary to solve problems such as calculating the amount of material necessary to make a cardboard box or a plastic beach ball.

Volumes of solids

A **rectangular solid** and a **cube** are two common geometric solids. (See Figure 9-70.)

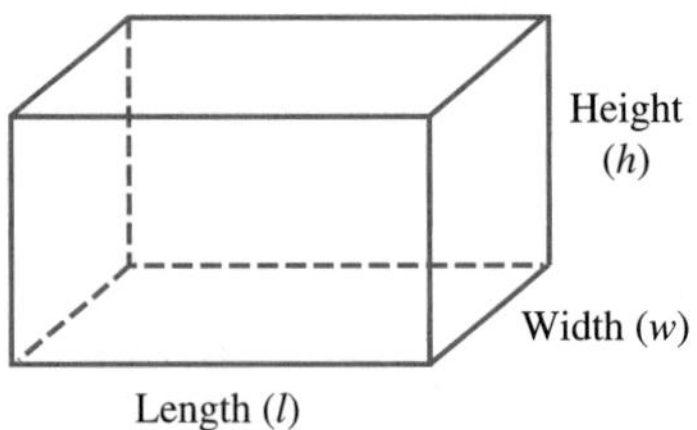

A rectangular solid

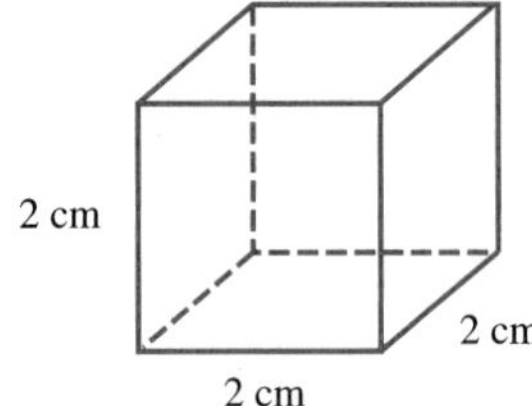

A cube

FIGURE 9-70

The **volume** of a rectangular solid is a measure of the space it encloses. Two common units of volume are cubic inches (in.^{3}) and cubic centimeters (cm^3). (See Figure 9-71.)

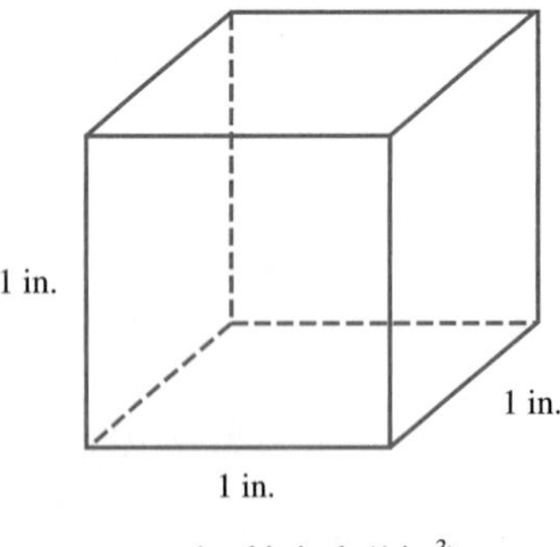

1 cubic inch (1 in.^{3})

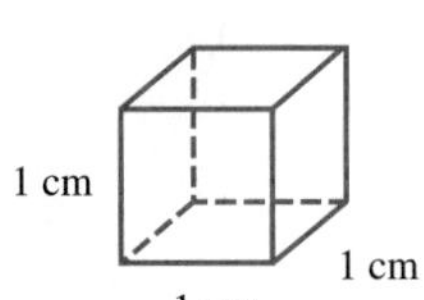

1 cubic centimeter (1 cm^3)

FIGURE 9-71

If we divide the rectangular solid shown in Figure 9-72 into cubes, each cube represents a volume of 1 cm^3. Because there are 2 levels with 12 cubes on each level, the volume of the rectangular solid is 24 cm^3.

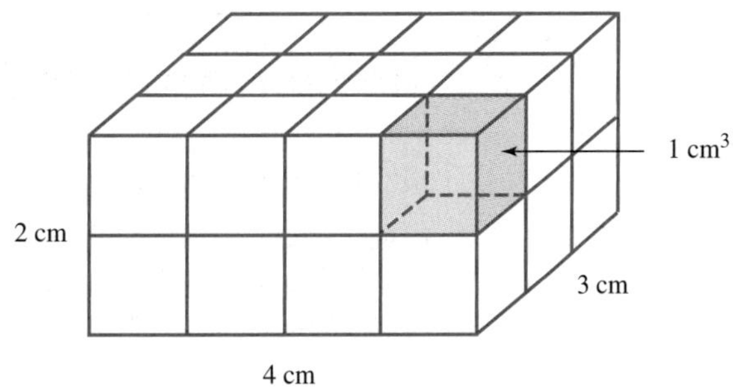

FIGURE 9-72

In practice, we do not find volumes by counting cubes. Instead, we use the formulas shown in Table 9-2.

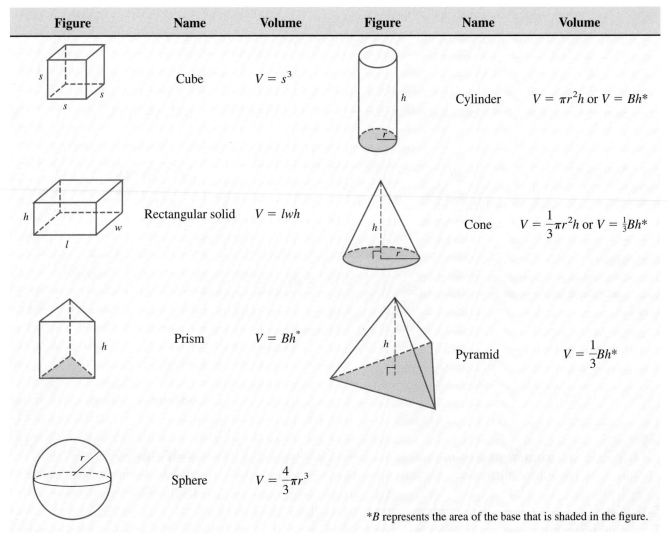

Figure	Name	Volume	Figure	Name	Volume
	Cube	$V = s^3$		Cylinder	$V = \pi r^2 h$ or $V = Bh$*
	Rectangular solid	$V = lwh$		Cone	$V = \frac{1}{3}\pi r^2 h$ or $V = \frac{1}{3}Bh$*
	Prism	$V = Bh$*		Pyramid	$V = \frac{1}{3}Bh$*
	Sphere	$V = \frac{4}{3}\pi r^3$			

**B* represents the area of the base that is shaded in the figure.

TABLE 9-2

COMMENT The height of a geometric solid is always measured along a line perpendicular to its base. In each of the solids in Figure 9-73 (on the next page), *h* is the height.

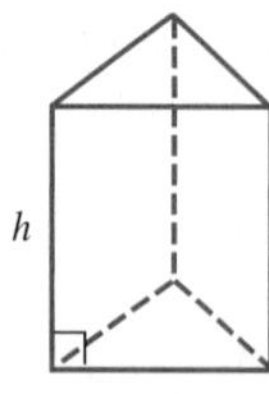

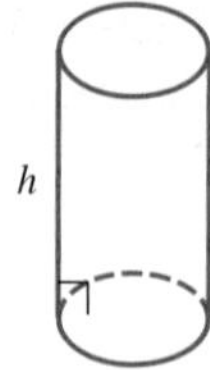

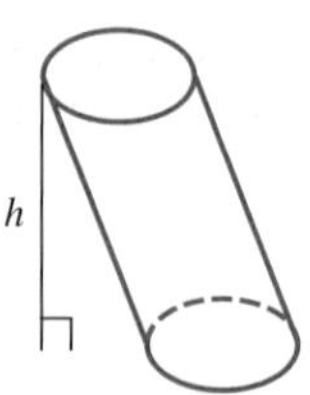

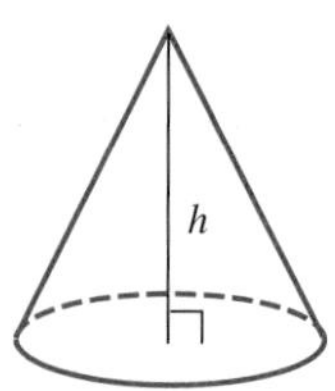

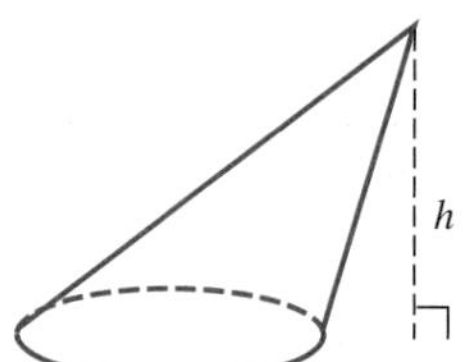

FIGURE 9-73

EXAMPLE 1 ***Number of cubic inches in one cubic foot.*** How many cubic inches are there in 1 cubic foot? (See Figure 9-74.)

Solution

Since a cubic foot is a cube with each side measuring 1 foot, each side also measures 12 inches. Thus, the volume in cubic inches is

$V = s^3$ The formula for the volume of a cube.

$V = (\mathbf{12})^3$ Substitute 12 for s.

$= 1{,}728$

There are 1,728 cubic inches in 1 cubic foot.

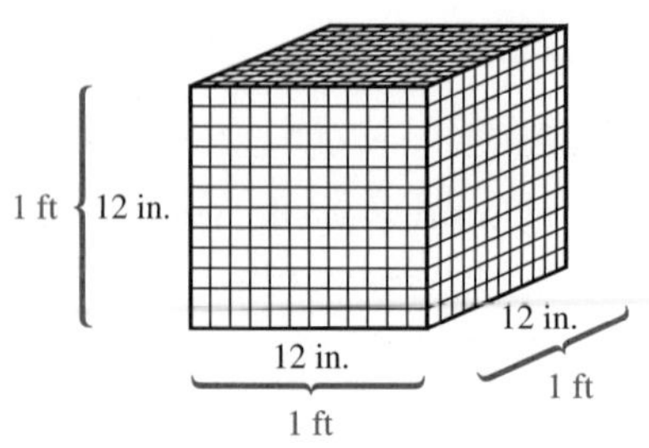

FIGURE 9-74

Self Check

How many cubic centimeters are in 1 cubic meter?

Answer: 1,000,000 cm^3 ■

EXAMPLE 2 ***Volume of an oil storage tank.*** An oil storage tank is in the form of a rectangular solid with dimensions of 17 by 10 by 8 feet. (See Figure 9-75.) Find its volume.

Solution

To find the volume, we substitute 17 for l, 10 for w, and 8 for h in the formula $V = lwh$ and simplify.

$V = lwh$

$V = \mathbf{17}(\mathbf{10})(\mathbf{8})$

$= 1{,}360$

The volume is 1,360 ft^3.

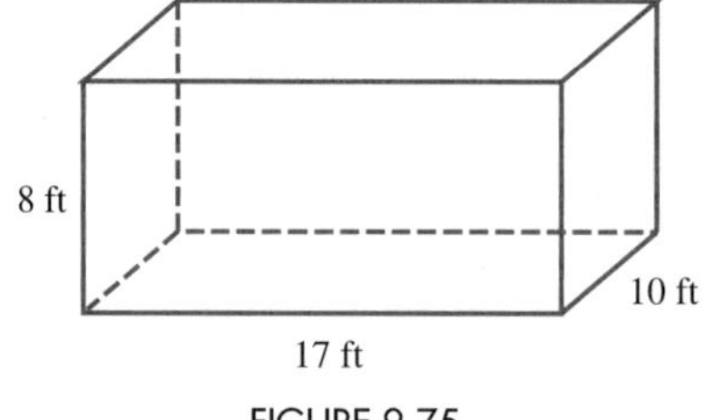

FIGURE 9-75

Self Check

Find the volume of a rectangular solid with dimensions of 8 by 12 by 20 meters.

Answer: 1,920 m^3 ■

EXAMPLE 3 ***Volume of a triangular prism.*** Find the volume of the triangular prism in Figure 9-76.

Solution

The volume of the prism is the area of its base multiplied by its height. Since there are 100 centimeters in 1 meter, the height in centimeters is

$0.5 \text{ m} = 0.5(\mathbf{1\ m})$

$= 0.5(\mathbf{100\ cm})$ Substitute 100 centimeters for 1 meter.

$= 50 \text{ cm}$

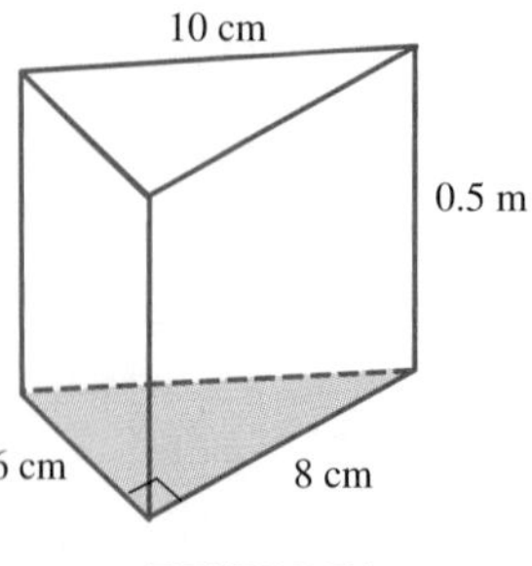

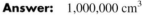

FIGURE 9-76

Self Check

Find the volume of the triangular prism below.

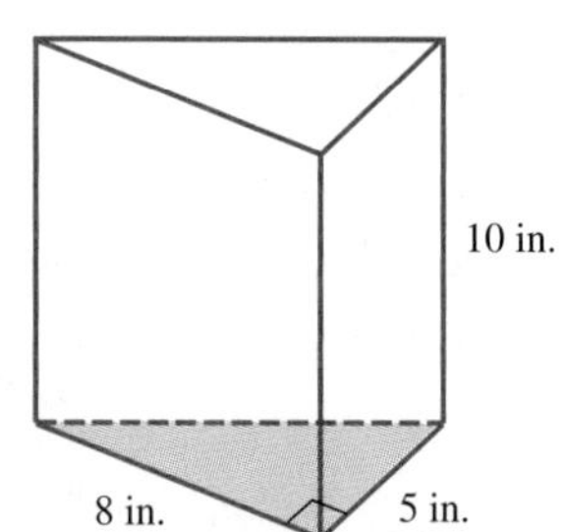

The area of the triangular base is $\frac{1}{2}(6)(8) = 24$ square centimeters. The height of the prism is 50 centimeters. Substituting into the formula for the volume of a prism, we have

$$V = Bh$$
$$V = 24(50)$$
$$= 1{,}200$$

The volume of the prism is 1,200 cm^3.

Answer: 200 in.^{3} ■

Surface areas of rectangular solids

The **surface area** of a rectangular solid is the sum of the areas of its six faces. Figure 9-77 shows how we can unfold the faces of a cardboard box to derive a formula for its surface area (*SA*).

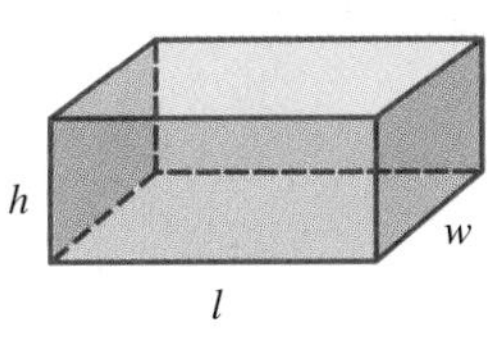

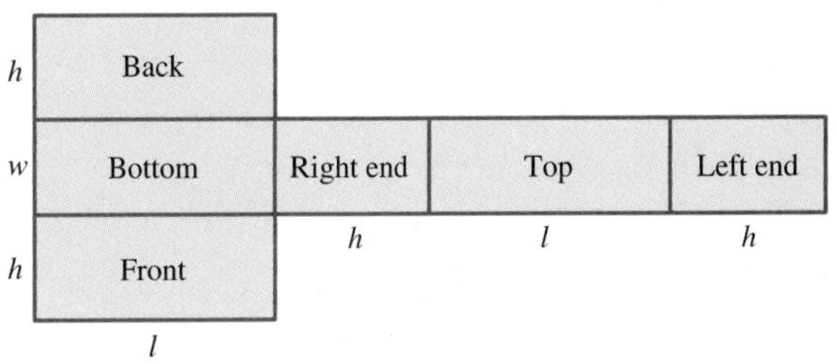

FIGURE 9-77

$$SA = A_{\text{bottom}} + A_{\text{back}} + A_{\text{front}} + A_{\text{right end}} + A_{\text{top}} + A_{\text{left end}}$$
$$SA = lw + lh + lh + hw + lw + hw$$
$$= 2lw + 2lh + 2hw \quad \text{Combine like terms.}$$

Surface area of a rectangular solid

> The surface area of a rectangular solid is given by the formula
>
> $$SA = 2lw + 2lh + 2hw$$
>
> where l is the length, w is the width, and h is the height.

EXAMPLE 4 ***Surface area of an oil tank.*** An oil storage tank is in the form of a rectangular solid with dimensions of 17 by 10 by 8 feet. (See Figure 9-78.) Find the surface area of the tank.

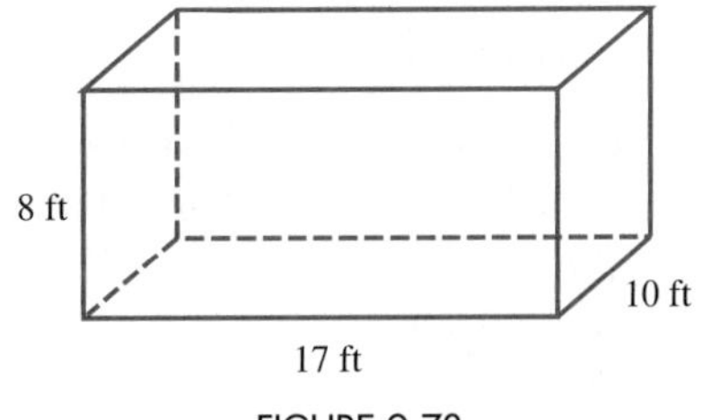

FIGURE 9-78

Solution

To find the surface area, we substitute 17 for l, 10 for w, and 8 for h in the formula for surface area and simplify.

$$SA = 2lw + 2lh + 2hw$$
$$SA = 2(17)(10) + 2(17)(8) + 2(8)(10)$$
$$= 340 + 272 + 160$$
$$= 772$$

The surface area is 772 ft^2.

Self Check

Find the surface area of a rectangular solid with dimensions of 8 by 12 by 20 meters.

Answer: 992 m^2 ■

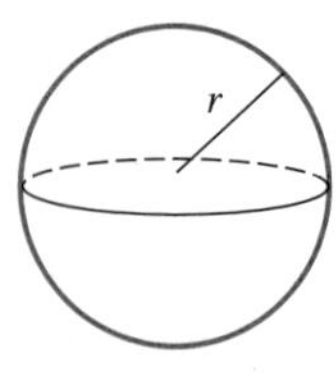

FIGURE 9-79

Volumes and surface areas of spheres

A **sphere** is a hollow, round ball. (See Figure 9-79.) The points on a sphere all lie at a fixed distance r from a point called its *center.* A segment drawn from the center of a sphere to a point on the sphere is called a *radius.*

Accent on Technology: Filling a water tank

See Figure 9-80. To calculate how many cubic feet of water are needed to fill a spherical water tank with a radius of 15 feet, we substitute 15 for r in the formula for the volume of a sphere.

$$V = \frac{4}{3}\pi r^3$$

$$V = \frac{4}{3}\pi(15)^3$$

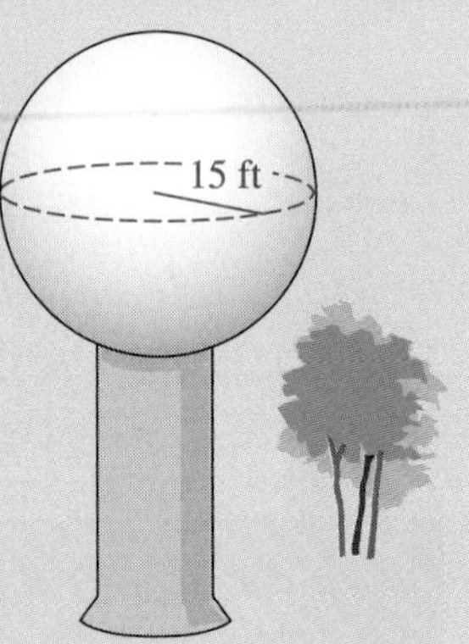

FIGURE 9-80

To do the arithmetic with a scientific calculator, we enter these numbers and press these keys.

Keystrokes 15 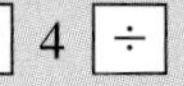 y^x 3 = × 4 ÷ 3 = × π = 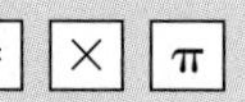 14137.16694

To the nearest tenth, 14,137.2 ft^3 of water will be needed to fill the tank.

There is a formula to find the surface area of a sphere.

Surface area of a sphere

The surface area of a sphere with radius r is given by the formula

$$SA = 4\pi r^2$$

EXAMPLE 5 ***Manufacturing beach balls.*** A beach ball is to have a diameter of 16 inches. (See Figure 9-81.) How many square inches of material will be needed to make the ball? (Disregard any waste.)

Solution Since a radius r of the ball is one-half the diameter, $r = 8$ inches. We can now substitute 8 for r in the formula for the surface area of a sphere.

$$SA = 4\pi r^2$$
$$SA = 4\pi(8)^2$$
$$SA = 4\pi(64)$$
$$SA = 256\pi \qquad \text{Simplify: } 4 \cdot 64 = 256.$$
$$\approx 804.2477193 \qquad \text{Use a calculator.}$$

A little more than 804 in.^{2} of material is needed to make the ball. FIGURE 9-81 ■

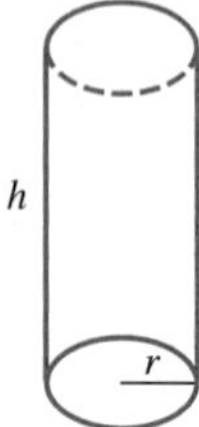

FIGURE 9-82

Volumes of cylinders

A **cylinder** is a hollow figure like a piece of pipe. (See Figure 9-82.)

EXAMPLE 6 Find the volume of the cylinder in Figure 9-83.

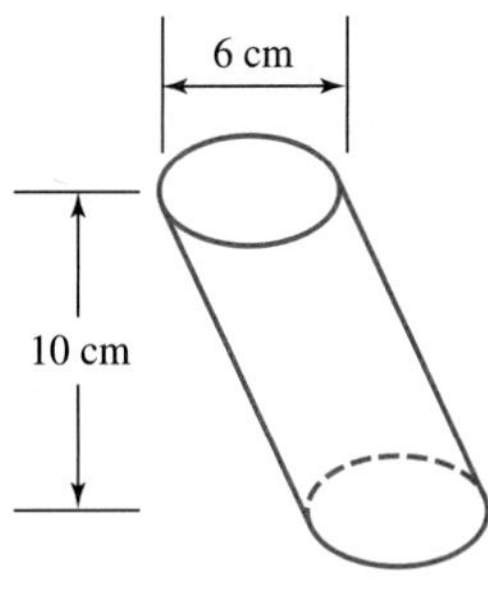

FIGURE 9-83

Solution Since a radius is one-half of the diameter of the circular base, $r = 3$ cm. From the figure, we see that the height of the cylinder is 10 cm. So we can substitute 3 for r and 10 for h in the formula for the volume of a cylinder.

$$V = \pi r^2 h$$
$$V = \pi(3)^2(10)$$
$$= 90\pi \quad \text{Simplify: } (3)^2(10) = 90.$$
$$\approx 282.7433388 \quad \text{Use a calculator.}$$

To the nearest hundredth, the volume of the cylinder is 282.74 cm^3. ■

Accent on Technology: *Volume of a silo*

A silo is a structure used for storing grain. The silo in Figure 9-84 is a cylinder 50 feet tall topped with a **hemisphere** (a half-sphere). To find the volume of the silo, we add the volume of the cylinder to the volume of the dome.

$$\text{Volume}_{\text{cylinder}} + \text{volume}_{\text{dome}} = (\text{Area}_{\text{cylinder's base}})(\text{height}_{\text{cylinder}}) + \frac{1}{2}(\text{volume}_{sphere})$$
$$= \pi r^2 h + \frac{1}{2}\left(\frac{4}{3}\pi r^3\right)$$
$$= \pi r^2 h + \frac{2\pi r^3}{3} \qquad \frac{1}{2}\left(\frac{4}{3}\pi r^3\right) = \frac{1}{2}\cdot\frac{4}{3}\pi r^3 = \frac{4}{6}\pi r^3 = \frac{2\pi r^3}{3}$$
$$= \pi(10)^2(50) + \frac{2\pi(10)^3}{3} \qquad \text{Substitute 10 for } r \text{ and 50 for } h.$$

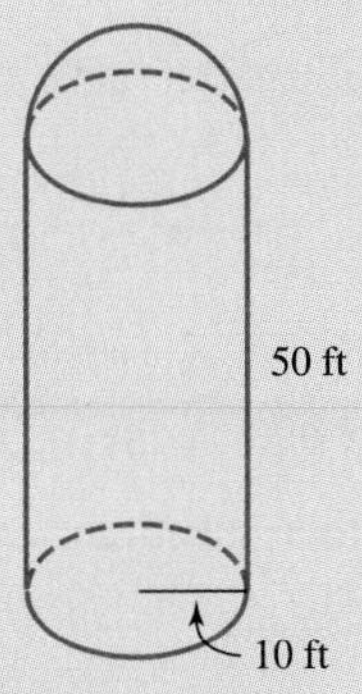

FIGURE 9-84

To do the arithmetic with a scientific calculator, we enter these numbers and press these keys.

Keystrokes [π] [×] 10 [x^2] [×] 50 [=] [+] [(] 2 [×] [π] [×] 10 [y^x] 3 [÷] 3 [)] [=] 17802.35837

The volume of the silo is approximately 17,802 ft^3.

EXAMPLE 7 ***Machining a block of metal.*** See Figure 9-85. Find the volume that is left when the hole is drilled through the metal block.

Solution We must find the volume of the rectangular solid and then subtract the volume of the cylinder. We will think of the rectangular solid and the cylinder as lying on their sides. Thus, the height is 18 cm when we find each volume.

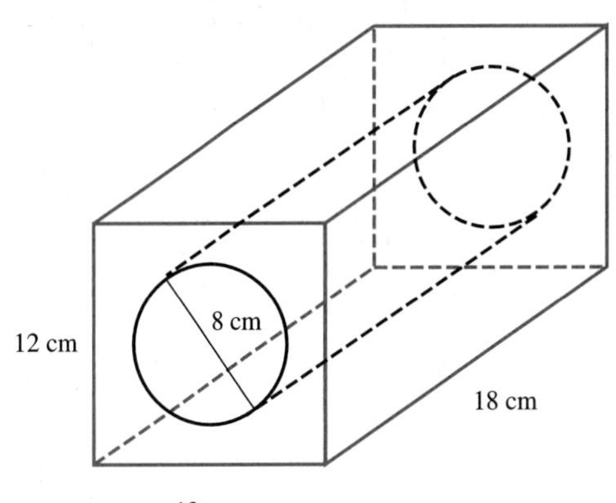

FIGURE 9-85

$$V_{\text{rect. solid}} = lwh$$
$$V_{\text{rect. solid}} = 12(12)(18)$$
$$= 2{,}592$$

$$V_{\text{cylinder}} = \pi r^2 h$$
$$V_{\text{cylinder}} = \pi(4)^2(18)$$
$$= 288\pi$$
$$\approx 904.7786842$$

$$V_{\text{drilled block}} = V_{\text{rect. solid}} - V_{\text{cylinder}}$$
$$\approx 2{,}592 - 904.7786842$$
$$\approx 1{,}687.221316$$

To the nearest hundredth, the volume is 1,687.22 cm^3. ■

Volumes of cones

Two **cones** are shown in Figure 9-86. Each cone has a height h and a radius r, which is the radius of the circular base.

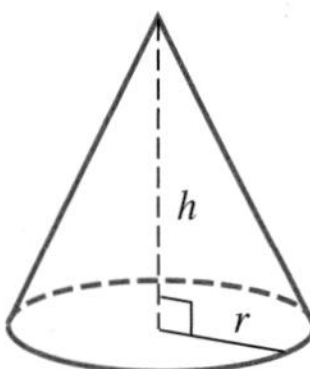

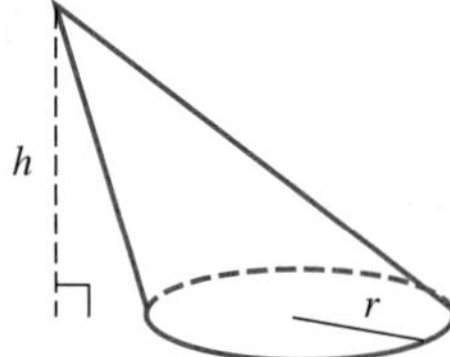

FIGURE 9-86

EXAMPLE 8 ***Volume of a cone.*** To the nearest tenth, find the volume of the cone in Figure 9-87.

Solution Since the radius is one-half of the diameter, $r = 4$ cm. We then substitute 4 for r and 6 for h in the formula for the volume of a cone.

$$V = \frac{1}{3}\pi r^2 h$$

$$V = \frac{1}{3}\pi(4)^2(6)$$

$$V = 32\pi$$
$$\approx 100.5309649$$

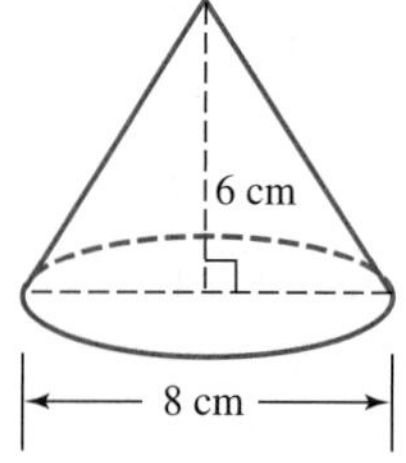

To the nearest tenth, the volume is 100.5 cubic centimeters.

FIGURE 9-87 ■

Volumes of pyramids

Two **pyramids** with a height h are shown in Figure 9-88.

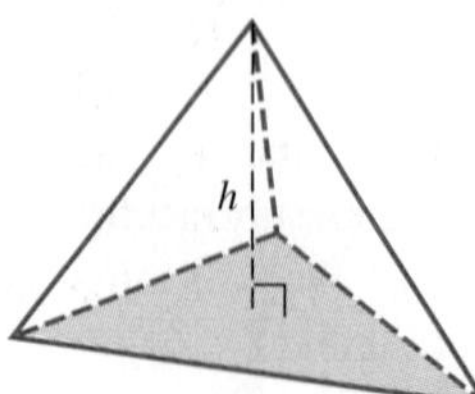

The base is a triangle.

(a)

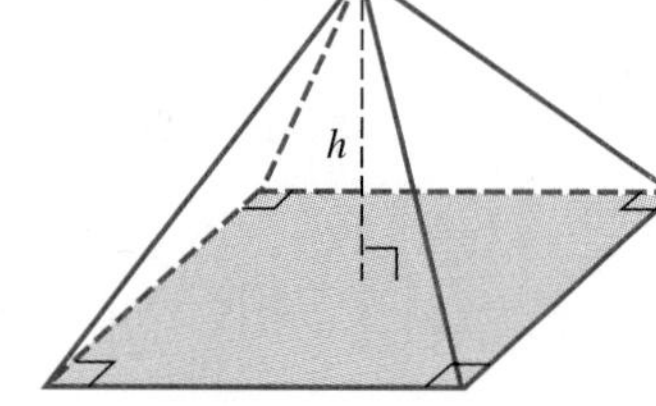

The base is a square.

(b)

FIGURE 9-88

EXAMPLE 9 ***Volume of a pyramid.*** Find the volume of a pyramid that has a square base with each side 6 meters long and a height of 9 meters.

Solution

Since the base is a square with each side 6 meters long, the area of the base is 6^2 m^2, or 36 m^2. We can then substitute 36 for the area of the base and 9 for the height in the formula for the volume of a pyramid.

$$V = \frac{1}{3}Bh$$

$$V = \frac{1}{3}(36)(9)$$

$$= 108$$

The volume of the pyramid is 108 m^3.

Self Check

Find the volume of the pyramid shown below.

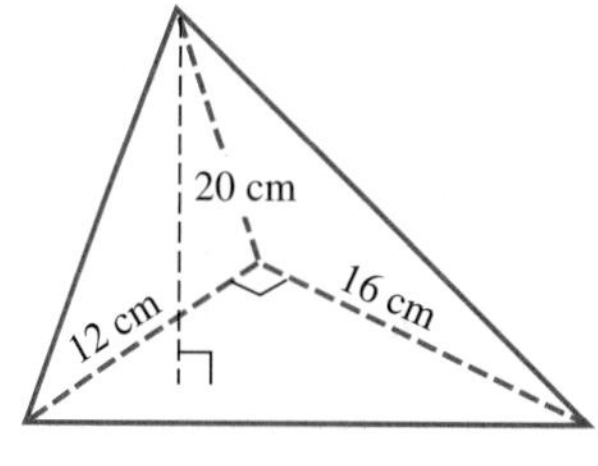

Answer: 640 cm^3 ■

STUDY SET Section 9.7

VOCABULARY *Fill in the blanks to make the statements true.*

1. The space contained within a geometric solid is called its Uolume.

2. A rectangular solid is like a hollow shoe box.

3. A Cube is a rectangular solid with all sides of equal length.

4. The volume of a cube with each side 1 inch long is 1 Cubic inch.

5. The Surface area of a rectangular solid is the sum of the areas of its faces.

6. The point that is equidistant from every point on a sphere is its Center.

7. A Cylinder is a hollow figure like a drinking straw.

8. A radius is one-half of a sphere.

9. A Cone looks like a witch's pointed hat.

10. A figure that has a polygon for its base and that rises to a point is called a pyramid.

CONCEPTS *In Exercises 11–16, write the formula used for finding the volume of each solid.*

11. A rectangular solid $U = lwh$

12. A prism $U = Bh$

13. A sphere $U = \frac{4}{3}\pi r^3$

14. A cylinder $U = \pi r^2 h$ or $U = Bh$

15. A cone $U = \frac{1}{3}\pi r^2 h$ or $U = \frac{1}{3}Bh$

16. A pyramid $U = \frac{1}{3}Bh$

17. Write the formula for finding the surface area of a rectangular solid. $SA = 2lw + 2lh + 2hw$

18. Write the formula for finding the surface area of a sphere. $SA = 4\pi r^2$

19. How many cubic feet are in 1 cubic yard? 27

20. How many cubic inches are in 1 cubic yard? 46,656

21. How many cubic decimeters are in 1 cubic meter? 1000 dm^3

22. How many cubic millimeters are in 1 cubic centimeter?

In Exercises 23–24, which geometric concept (perimeter, circumference, area, volume, or surface area) should be applied to find each of the following?

23. **a.** The size of a room to be air conditioned
b. The amount of land in a national park
c. The amount of space in a refrigerator freezer
d. The amount of cardboard in a shoe box
e. The distance around a checkerboard
f. The amount of material used to make a basketball

24. **a.** The amount of cloth in a car cover
b. The size of a trunk of a car
c. The amount of paper used for a postage stamp
d. The amount of storage in a cedar chest
e. The amount of beach available for sunbathing
f. The distance the tip of a propeller travels

25. In Illustration 1 (next page), the unit of measurement of length that was used to draw the figure was the inch.
a. What is the volume of the figure?
b. What is the area of the front of the figure?
c. What is the area of the base of the figure?

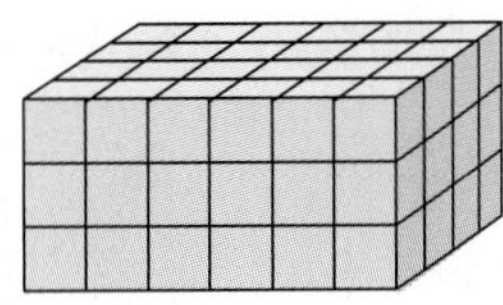

ILLUSTRATION 1

26. The cardboard box shown in Illustration 2 is a cube. Suppose the six faces were unfolded to lie flat on a table. Draw a picture of what this would look like.

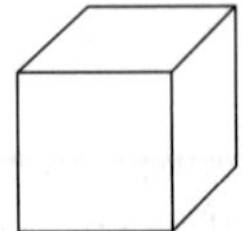

ILLUSTRATION 2

NOTATION *Fill in the blanks to make the statements true.*

27. The notation 1 in.^{3} is read as ~~one cubic inch~~.

28. One cubic centimeter is represented as ~~1 cm^3~~.

PRACTICE *Find the volume of each solid. If an answer is not exact, round to the nearest hundredth. (Answers may vary slightly, depending on which approximation of π is used.)*

29. A rectangular solid with dimensions of 3 by 4 by 5 centimeters.

30. A rectangular solid with dimensions of 5 by 8 by 10 meters.

31. A prism whose base is a right triangle with legs 3 and 4 meters long and whose height is 8 meters.

32. A prism whose base is a right triangle with legs 5 and 12 feet long and whose height is 10 feet.

33. A sphere with a radius of 9 inches.

34. A sphere with a diameter of 10 feet.

35. A cylinder with a height of 12 meters and a circular base with a radius of 6 meters.

36. A cylinder with a height of 4 meters and a circular base with a diameter of 18 meters.

37. A cone with a height of 12 centimeters and a circular base with a diameter of 10 centimeters.

38. A cone with a height of 3 inches and a circular base with a radius of 4 inches.

39. A pyramid with a square base 10 meters on each side and a height of 12 meters.

40. A pyramid with a square base 6 inches on each side and a height of 4 inches.

Find the surface area of each solid. If an answer is not exact, round to the nearest hundredth.

41. A rectangular solid with dimensions of 3 by 4 by 5 centimeters.

42. A cube with a side 5 centimeters long.

43. A sphere with a radius of 10 inches.

44. A sphere with a diameter of 12 meters.

Find the volume of each figure. If an answer is not exact, round to the nearest hundredth. (Answers may vary slightly, depending on which approximation of π is used.)

45.

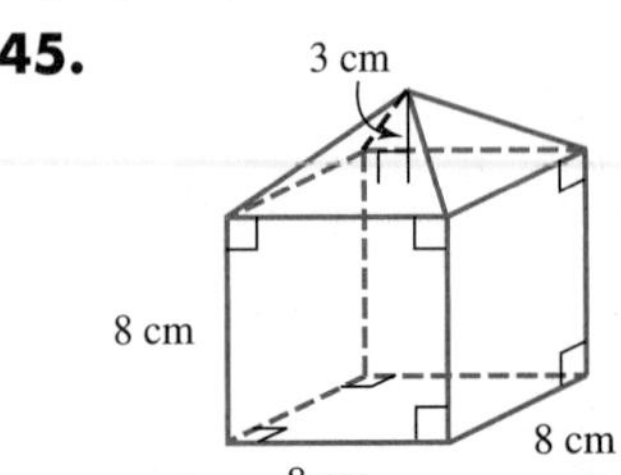

46.

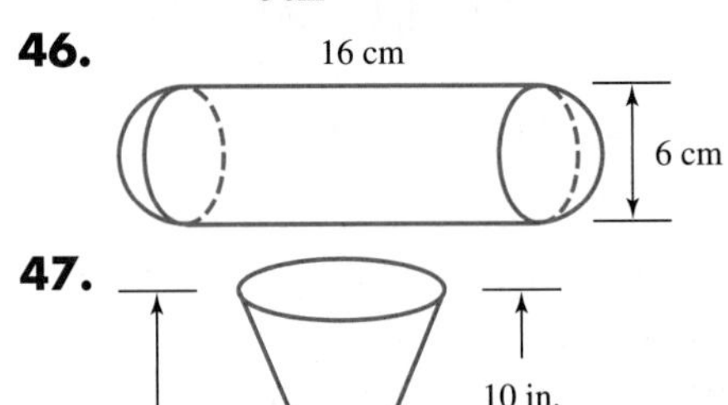

47.

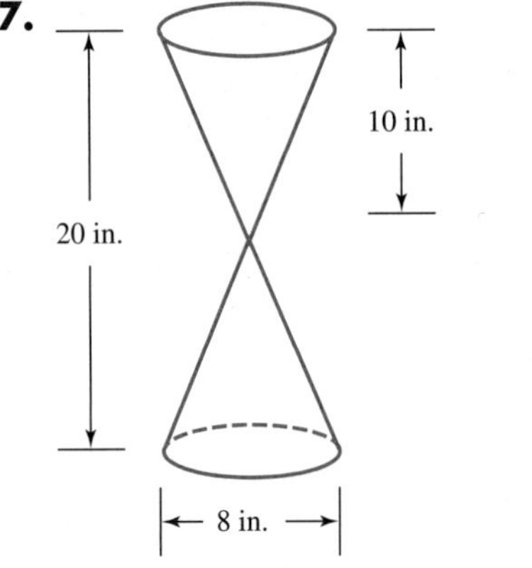

48.

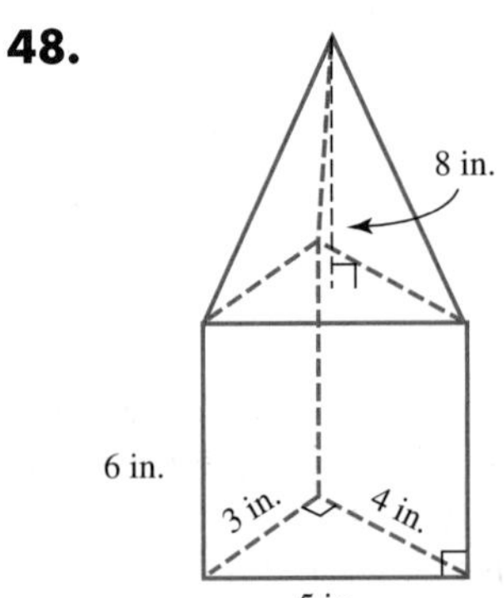

APPLICATIONS *Solve each problem. If an answer is not exact, round to the nearest hundredth.*

49. VOLUME OF A SUGAR CUBE A sugar cube is $\frac{1}{2}$ inch on each edge. How much volume does it occupy?

50. VOLUME OF A CLASSROOM A classroom is 40 feet long, 30 feet wide, and 9 feet high. Find the number of cubic feet of air in the room.

51. WATER HEATER Complete the advertisement for the high-efficiency water heater shown in Illustration 3.

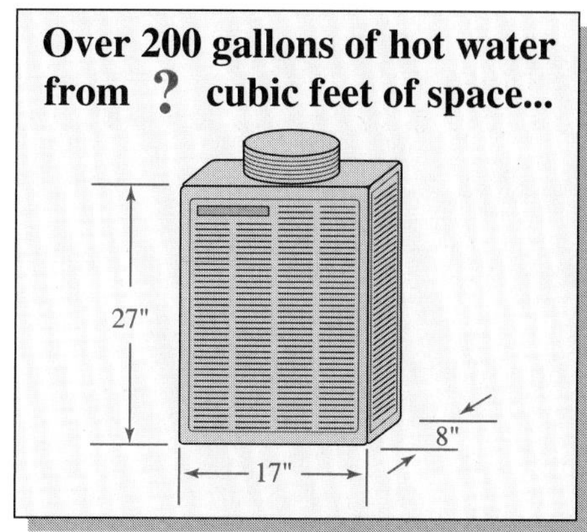

ILLUSTRATION 3

52. REFRIGERATOR CAPACITY The largest refrigerator advertised in a J. C. Penney catalog has a capacity of 25.2 cubic feet. How many cubic inches is this?

53. VOLUME OF AN OIL TANK A cylindrical oil tank has a diameter of 6 feet and a length of 7 feet. Find the volume of the tank.

54. VOLUME OF A DESSERT A restaurant serves pudding in a conical dish that has a diameter of 3 inches. If the dish is 4 inches deep, how many cubic inches of pudding are in each dish?

55. HOT-AIR BALLOON The lifting power of a spherical balloon depends on its volume. How many cubic feet of gas will a balloon hold if it is 40 feet in diameter?

56. VOLUME OF A CEREAL BOX A box of cereal measures 3 by 8 by 10 inches. The manufacturer plans to market a smaller box that measures $2\frac{1}{2}$ by 7 by 8 inches. By how much will the volume be reduced?

57. ENGINE The *compression ratio* of an engine is the volume in one cylinder with the piston at bottom-dead-center (B.D.C.), divided by the volume with the piston at top-dead-center (T.D.C.). From the data given in Illustration 4, what is the compression ratio of the engine? Use a colon to express your answer.

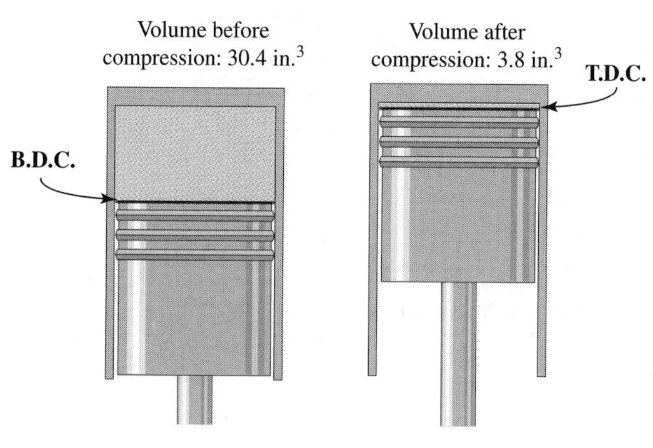

ILLUSTRATION 4

58. LINT REMOVER Illustration 5 shows a handy gadget; it uses a cylinder of sheets of sticky paper that can be rolled over clothing and furniture to pick up lint and pet hair. After the paper is full, that sheet is peeled away to expose another sheet of sticky paper. Find the area of the first sheet by using the formula $LSA = 2\pi rh$, where LSA represents the lateral surface area of the cylinder.

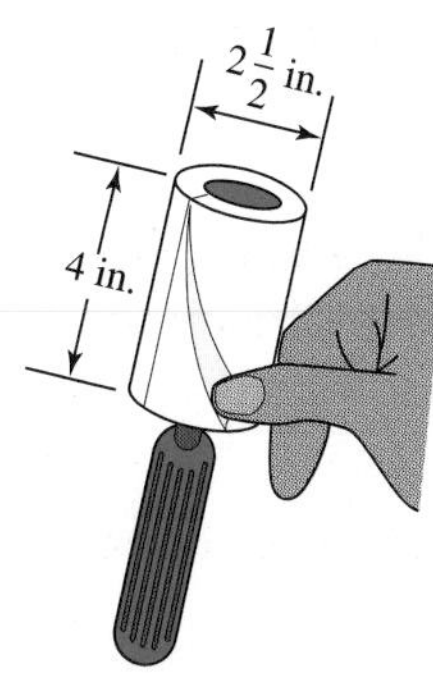

ILLUSTRATION 5

WRITING

59. What is meant by the *volume* of a cube?

60. What is meant by the *surface area* of a cube?

61. Are the units used to measure area different from the units used to measure volume? Explain.

62. The dimensions (length, width, and height) of one rectangular solid are entirely different numbers than the dimensions of another rectangular solid. Would it be possible for the rectangular solids to have the same volume? Explain.

REVIEW

63. Evaluate $-5(5 - 2)^2 + 3$.

64. BUYING PENCILS Carlos bought 6 pencils at \$0.60 each and a notebook for \$1.25. He gave the clerk a \$5 bill. How much change did he receive?

65. Solve $\dfrac{x+7}{-4} = \dfrac{1}{4}$.

66. 38 is what percent of 40?

67. Express the phrase "3 inches to 15 inches" as a ratio in lowest terms.

68. Convert 40 ounces to pounds.

69. Convert 2.4 meters to millimeters.

70. State the Pythagorean theorem.

$4x+28=-4$

$4x+28-28=-4-28$

$4x=-32$

$\frac{4x}{4}=\frac{-32}{4}$

$x=-8$

KEY CONCEPT

Formulas

A **formula** is a mathematical expression that is used to express a relationship between quantities. We have studied formulas used in mathematics, business, and science.

Write a formula describing the mathematical relationship between the given quantities.

1. Distance traveled (d), rate traveled (r), time traveling at that rate (t)
2. Sale price (s), original price (p), discount (d)
3. Perimeter of a rectangle (P), length of the rectangle (l), width of the rectangle (w)
4. Amount of interest earned (I), principal (P), interest rate (r), time the money is invested (t)

Use a formula to solve each problem.

5. Find the area (A) of the triangular lot in Illustration 1.

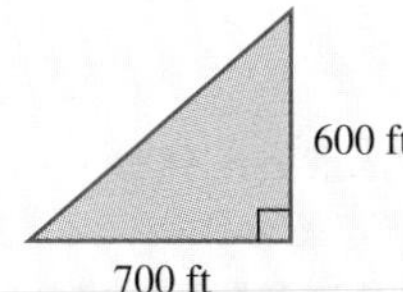

ILLUSTRATION 1

6. Find the volume (V) of the ice chest in Illustration 2.

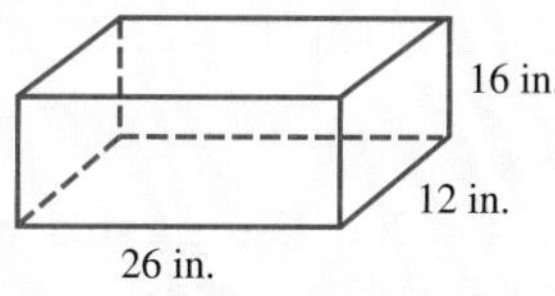

ILLUSTRATION 2

7. Find the retail price (p) of a cookware set that costs the store owner \$45.50 and is marked up \$35.
8. Find the profit (p) made by a school T-shirt sale if revenue was \$14,500 and costs were \$10,200.
9. Find the distance (d) that a rock falls in 3 seconds after being dropped from the edge of a cliff.
10. Find the temperature in degrees Celsius (C) if the temperature in degrees Fahrenheit is 59.

Sometimes we use the same formula to answer several related questions. The results can be displayed in a table.

11. Find the interest earned by each account.

Type of account	Principal	Annual rate earned	Time invested	Interest earned
Savings	\$5,000	5%	3 yr	
Passbook	\$2,250	2%	1 yr	
Trust fund	\$10,000	6.25%	10 yr	

12. Complete the table.

Type of coin	Number	Value (¢)	Total value (¢)
penny	15		
nickel	n		
dime	d		
quarter	q		

ACCENT ON TEAMWORK

Section 9.1

WRITING DIGITS In Illustration 1, the digit 1 is drawn using one angle, and the digit 2 is drawn using two angles. Draw the digit 3 using three angles, the digit 4 using four angles, and so on for all of the digits up to and including 9.

ILLUSTRATION 1

Section 9.2

CONSTRUCTIONS

Step 1: See Illustration 2(a). Using a straightedge, draw $\overline{AB}$. Then place the sharp point of a compass at *A* and draw an arc.

Step 2: With the same compass setting, place the sharp point at *B*. As shown in Illustration 2(b), draw another arc that intersects the arc from Step 1 at two points. Label these points *C* and *D*.

Step 3: Using a straightedge, draw a line through points *C* and *D*. Label the point where line *CD* intersects $\overline{AB}$ as point *E*. Does $m(\overline{AE}) = m(\overline{EB})$?

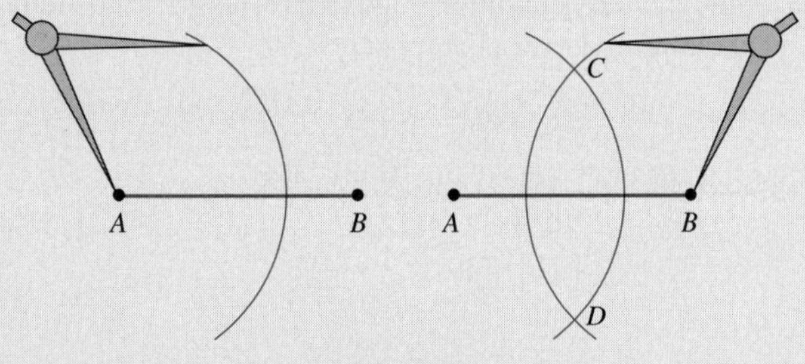

ILLUSTRATION 2

Section 9.3

TANGRAM A tangram is a puzzle in which geometric shapes are arranged to form other shapes. Cut out the pieces in Illustration 3. Assemble them so that they form a square. There should be no gaps, overlaps, or holes.

ILLUSTRATION 3

Section 9.4

CONGRUENT TRIANGLES Draw a triangle on a piece of paper. Then measure the lengths of its sides (with a ruler) and the angle measures (with a protractor). Choose a combination of any three measurements and tell them to your partner. Are the given facts sufficient for your partner to construct a triangle congruent to yours?

Section 9.5

AREA Find the area of the shaded figure on the square grid in Illustration 4.

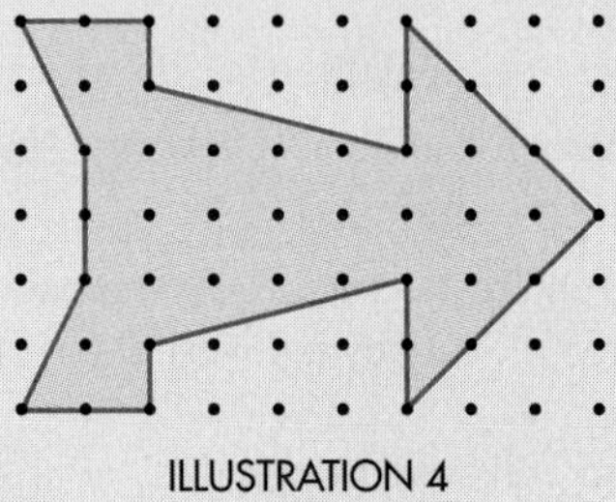

ILLUSTRATION 4

Section 9.6

PI Carefully measure the circumference and the diameter of different-size circles. Record the measurements in a table like the one below. Then use a calculator to find $\frac{C}{D}$. The result should be a number close to π.

Object	Circumference	Diameter	$\frac{C}{D}$
jar	$3\frac{1}{2}$ in.	$1\frac{1}{8}$ in.	3.11

Section 9.7

PYRAMID Cut out, fold, and glue together the pattern shown in Illustration 5. Estimate the volume and surface area of the pyramid.

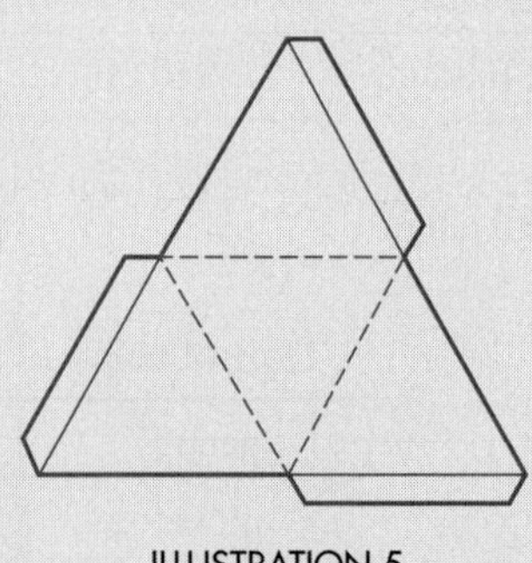

ILLUSTRATION 5

CHAPTER REVIEW

SECTION 9.1 Some Basic Definitions

CONCEPTS

In geometry, we study *points, lines,* and *planes.*

A *line segment* is a part of a line with two endpoints. A *ray* is a part of a line with one endpoint.

An *angle* is a figure formed by two rays with a common endpoint. The common endpoint is called the *vertex* of the angle.

A *protractor* is used to find the measure of an angle.

An *acute angle* is greater than 0° but less than 90°. A *right angle* measures 90°. An *obtuse angle* is greater than 90° but less than 180°. A *straight angle* measures 180°.

Two angles that have the same vertex and are side-by-side are called *adjacent angles.*

REVIEW EXERCISES

1. In Illustration 1, identify a point, a line, and a plane.

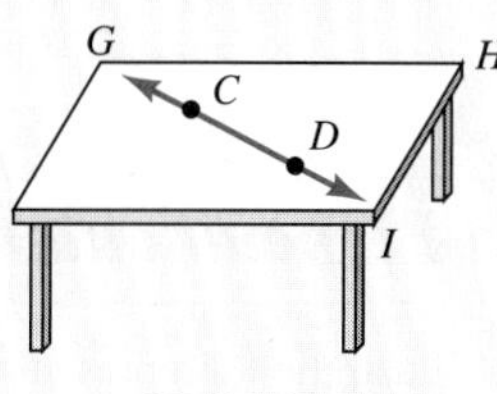

ILLUSTRATION 1

ILLUSTRATION 2

2. In Illustration 2, find $m(\overline{AB})$.

3. In Illustration 3, give four ways to name the angle.

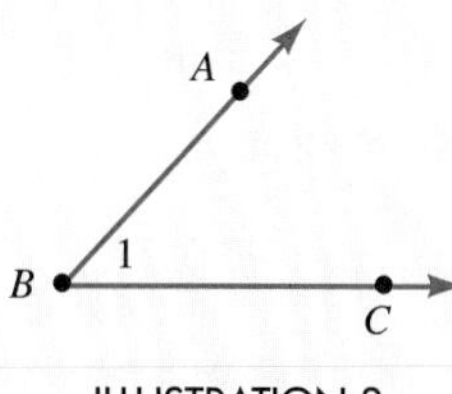

ILLUSTRATION 3

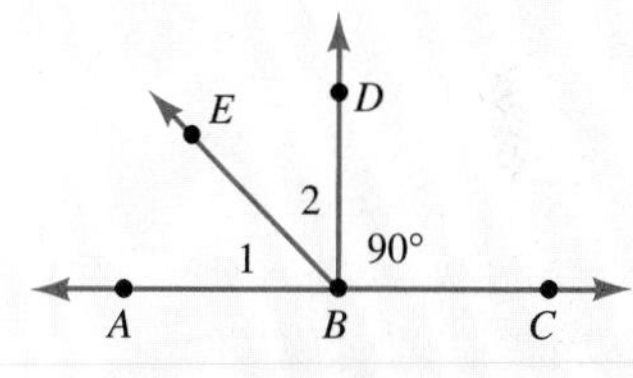

ILLUSTRATION 4

4. In Illustration 3, use a protractor to find the measure of the angle.

5. In Illustration 4, identify each acute angle, right angle, obtuse angle, and straight angle.

6. The measures of several angles are given. Identify each angle as an acute angle, a right angle, an obtuse angle, or a straight angle.

a. $m(\angle A) = 150°$ **b.** $m(\angle B) = 90°$

c. $m(\angle C) = 180°$ **d.** $m(\angle D) = 25°$

7. The two angles shown in Illustration 5 are adjacent angles. Find x.

8. Line AB is shown in Illustration 6. Find y.

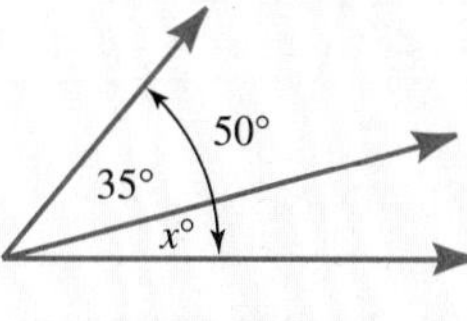

ILLUSTRATION 5

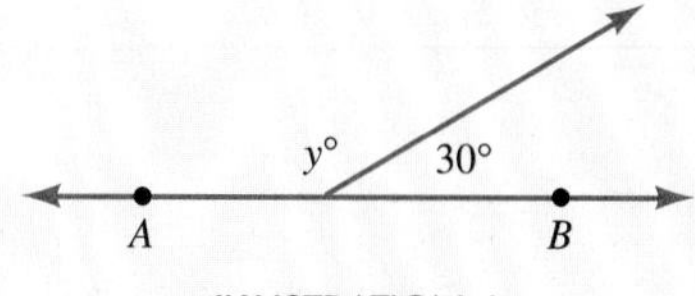

ILLUSTRATION 6

When two lines intersect, pairs of nonadjacent angles are called *vertical angles*.

Vertical angles have the same measure.

9. In Illustration 7, find **a.** m(∠1) and **b.** m(∠2).

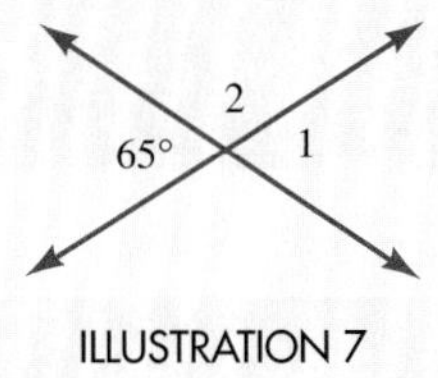

ILLUSTRATION 7

If the sum of two angles is 90°, the angles are *complementary.* If the sum of two angles is 180°, the angles are *supplementary.*

10. Find the complement of an angle that measures 50°.

11. Find the supplement of an angle that measures 140°.

12. Are angles measuring 30°, 60°, and 90° supplementary?

SECTION 9.2 Parallel and Perpendicular Lines

Parallel lines do not intersect. *Perpendicular* lines intersect and make right angles.

A line that intersects two or more *coplanar* lines is called a *transversal.*

13. Which part of Illustration 8 represents parallel lines?

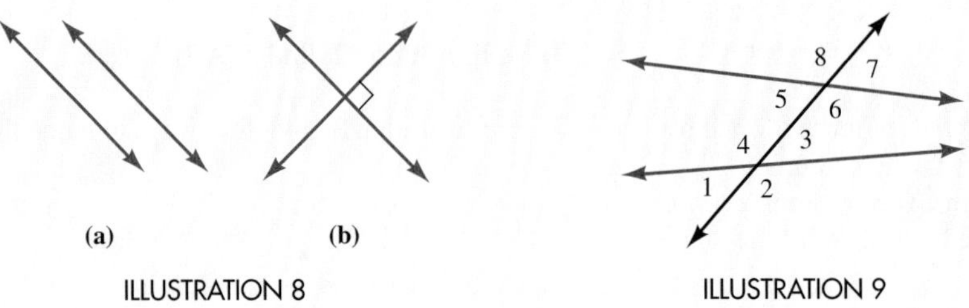

ILLUSTRATION 8 ILLUSTRATION 9

When a transversal intersects two coplanar lines, *alternate interior angles* and *corresponding* angles are formed.

14. Identify all pairs of alternate interior angles shown in Illustration 9.

15. Identify all pairs of corresponding angles shown in Illustration 9.

16. Identify all pairs of vertical angles shown in Illustration 9.

If two parallel lines are cut by a transversal,

1. alternate interior angles are congruent (have equal measures).
2. corresponding angles are congruent.
3. interior angles on the same side of the transversal are supplementary.

17. In Illustration 10, $l_1 \parallel l_2$. Find the measure of each angle.

18. In Illustration 11, $\overline{DC} \parallel \overline{AB}$. Find the measure of each angle.

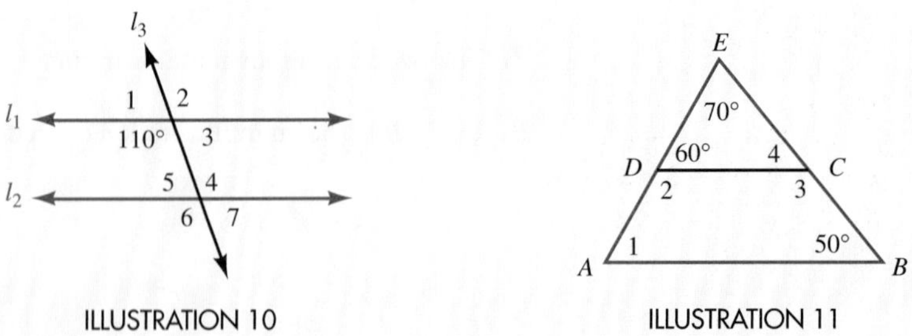

ILLUSTRATION 10 ILLUSTRATION 11

19. In Illustration 12 on the next page, $l_1 \parallel l_2$. Find x.

20. In Illustration 13 on the next page, $l_1 \parallel l_2$. Find x.

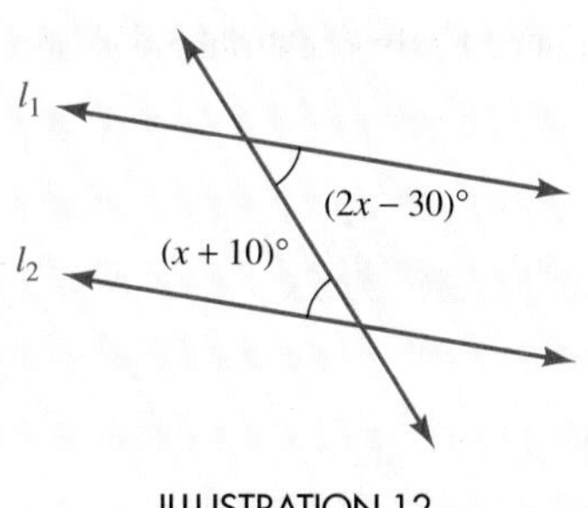

ILLUSTRATION 12

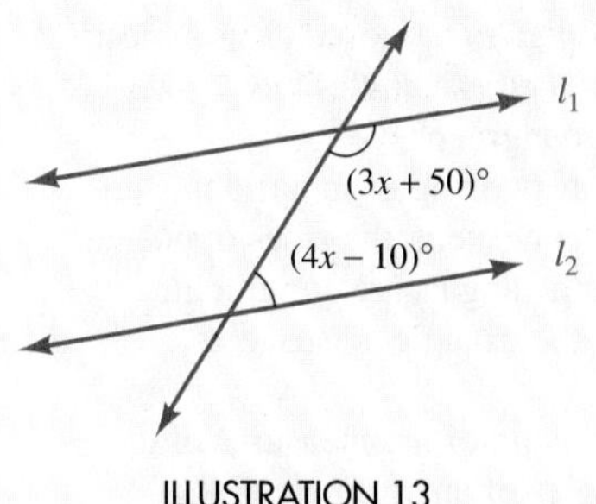

ILLUSTRATION 13

SECTION 9.3 Polygons

A *polygon* is a closed geometric figure. The points at which the sides intersect are called *vertices.* A *regular polygon* has sides that are all the same length and angles that are all the same measure.

Polygons are classified as follows:

Number of sides	Name
3	triangle
4	quadrilateral
5	pentagon
6	hexagon
8	octagon

21. Identify each polygon as a triangle, a quadrilateral, a pentagon, a hexagon, or an octagon.

a.

b.

c.

d.

e.

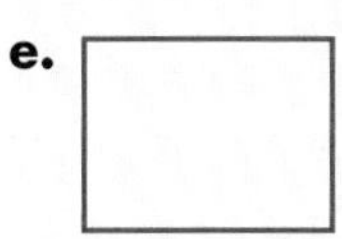

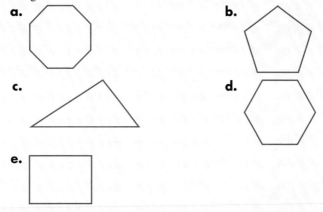

22. Give the number of vertices of each polygon.

a. Triangle
b. Quadrilateral
c. Octagon
d. Hexagon

An *equilateral triangle* has three sides of equal length.
An *isosceles triangle* has at least two sides of equal length.
A *scalene triangle* has no sides of equal length.
A *right triangle* has one right angle.

In an isosceles triangle, the angles opposite the sides of equal length are called *base angles.* The third angle is called the *vertex angle.* The third side is called the *base.*

23. Classify each of the triangles as an equilateral triangle, an isosceles triangle, a scalene triangle, or a right triangle.

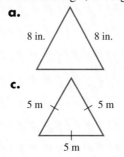

a. 8 in. 8 in.

b.

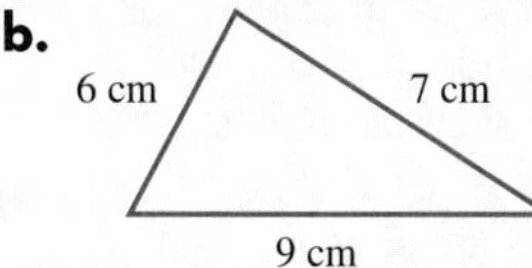

c.

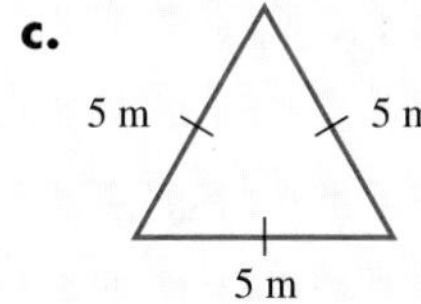

d.

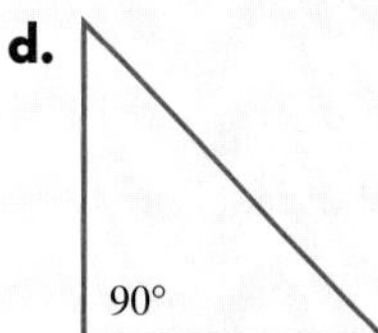

Properties of isosceles triangles:
1. The base angles are congruent.
2. If two angles in a triangle are congruent, the sides opposite the angles are congruent, and the triangle is isosceles.

24. Determine whether each triangle is isosceles.

a.
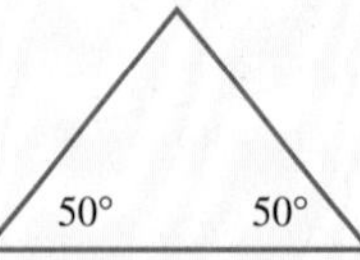

b.
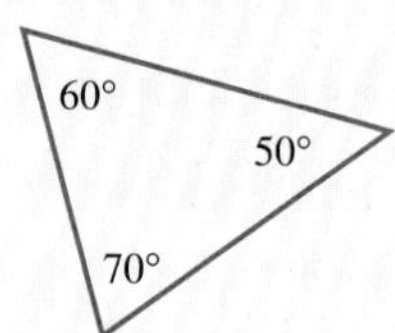

The sum of the measures of the angles of any triangle is 180°.

25. In each triangle, find x.

a.
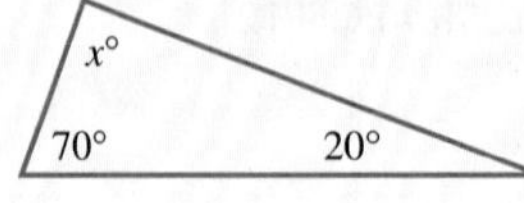

b.
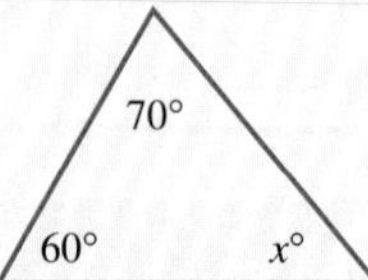

Quadrilaterals are classified as follows:

Property	Name
Opposite sides parallel	parallelogram
Parallelogram with four right angles	rectangle
Rectangle with all sides equal	square
Parallelogram with sides of equal length	rhombus
Exactly two sides parallel	trapezoid

26. If one base angle of an isosceles triangle measures 65°, how large is the vertex angle?

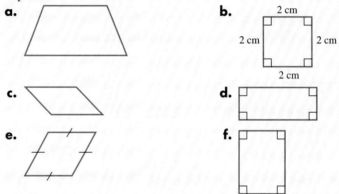

27. If one base angle of an isosceles triangle measures 60°, what can you conclude about the triangle?

28. Classify each quadrilateral as a parallelogram, a rectangle, a square, a rhombus, or a trapezoid.

a.
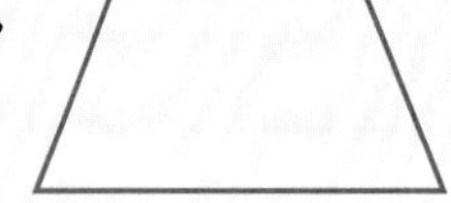

b.
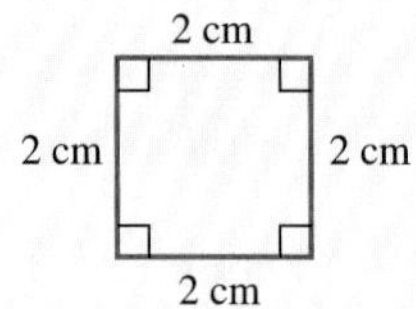

c.

d.

e.
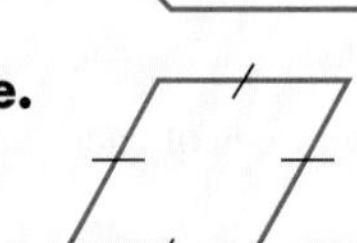

f.
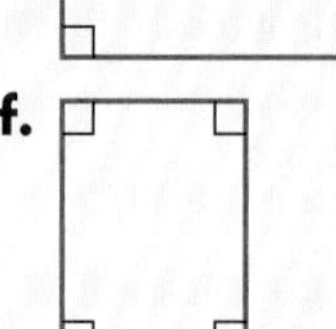

Properties of rectangles:
1. All angles are right angles.
2. Opposite sides are parallel.
3. Opposite sides are of equal length.
4. Diagonals are of equal length.
5. If the diagonals of a parallelogram are of equal length, the parallelogram is a rectangle.

29. In Illustration 14, the length of diagonal $\overline{AC}$ of rectangle $ABCD$ is 15 centimeters. Find each measure.

a. $m(\overline{BD})$ **b.** $m(\angle 1)$ **c.** $m(\angle 2)$

30. In Illustration 14, $ABCD$ is a rectangle. Classify each statement as true or false.

a. $m(\overline{AB}) = m(\overline{DC})$
b. $m(\overline{AD}) = m(\overline{DC})$
c. Triangle ABE is isosceles.
d. $m(\overline{AC}) = m(\overline{BD})$

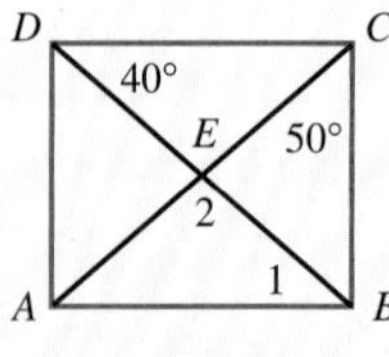

ILLUSTRATION 14

The parallel sides of a trapezoid are called *bases.* The nonparallel sides are called *legs.* If the legs of a trapezoid are of equal length, it is *isosceles.* In an isosceles trapezoid, the angles opposite the sides of equal length are *base angles,* and they are congruent.

31. In Illustration 15, *ABCD* is an isosceles trapezoid. Find each measure.

a. m($\angle B$) **b.** m($\angle C$)

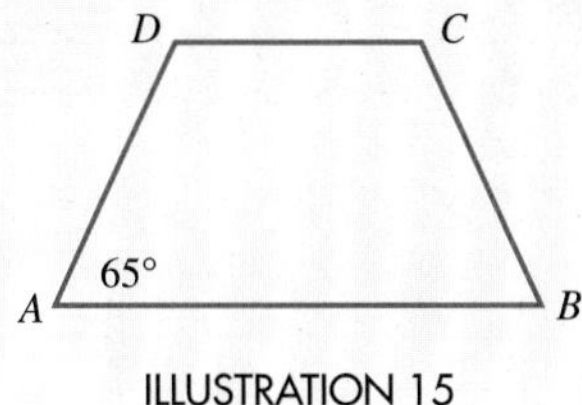

ILLUSTRATION 15

The sum of the measures of the angles of a polygon (in degrees) is given by the formula
$S = (n - 2)180$

32. Find the sum of the angle measures of each polygon.

a. quadrilateral **b.** hexagon

SECTION 9.4 Properties of Triangles

If two triangles have the same size and the same shape, they are *congruent triangles.*

Corresponding parts of congruent triangles have the same measure.

33. See Illustration 16. Complete the list of corresponding parts.

$\angle A$ corresponds to ____________.
$\angle B$ corresponds to ____________.
$\angle C$ corresponds to ____________.
$\overline{AC}$ corresponds to ____________.
$\overline{AB}$ corresponds to ____________.
$\overline{BC}$ corresponds to ____________.

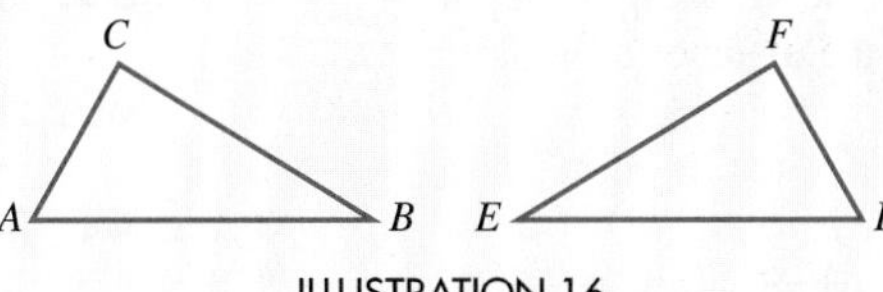

ILLUSTRATION 16

Three ways to show that two triangles are congruent are
1. The SSS property
2. The SAS property
3. The ASA property

34. Tell whether the triangles in each pair are congruent. If they are, tell why.

a.

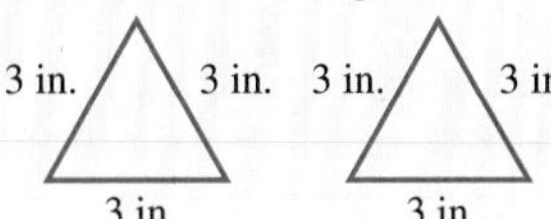

b.

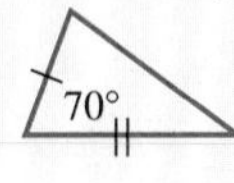

c.

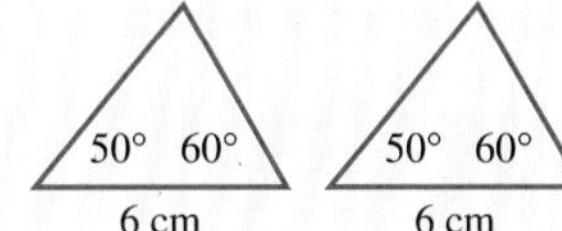

d.

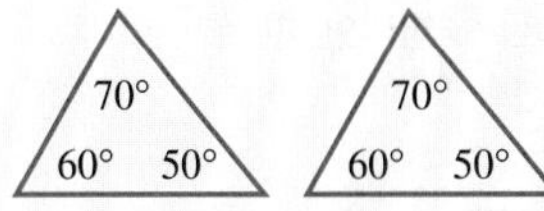

If two triangles have the same shape, they are said to be *similar.* If two angles of one triangle have the same measure as two angles of a second triangle, the triangles are similar.

35. Tell whether the triangles in each pair are similar.

a.

50° 50°
50° 50°

b.

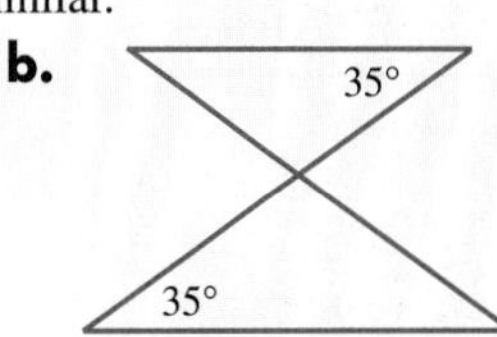

36. If a tree casts a 7-foot shadow at the same time a man 6 feet tall casts a 2-foot shadow, how tall is the tree?

The Pythagorean theorem: If the length of the *hypotenuse* of a right triangle is *c*, and the lengths of its legs are *a* and *b*, then
$a^2 + b^2 = c^2$

37. Refer to Illustration 17 on the next page and find the length of the unknown side.

a. If $a = 5$ and $b = 12$, find c. **b.** If $a = 8$ and $c = 17$, find b.

38. To the nearest tenth, find the height of the television screen shown in Illustration 18.

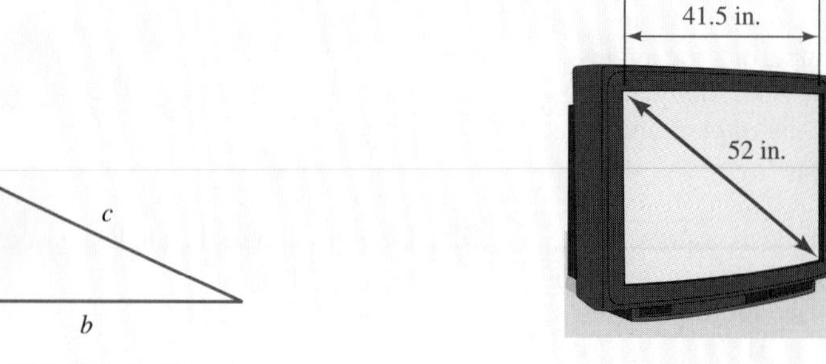

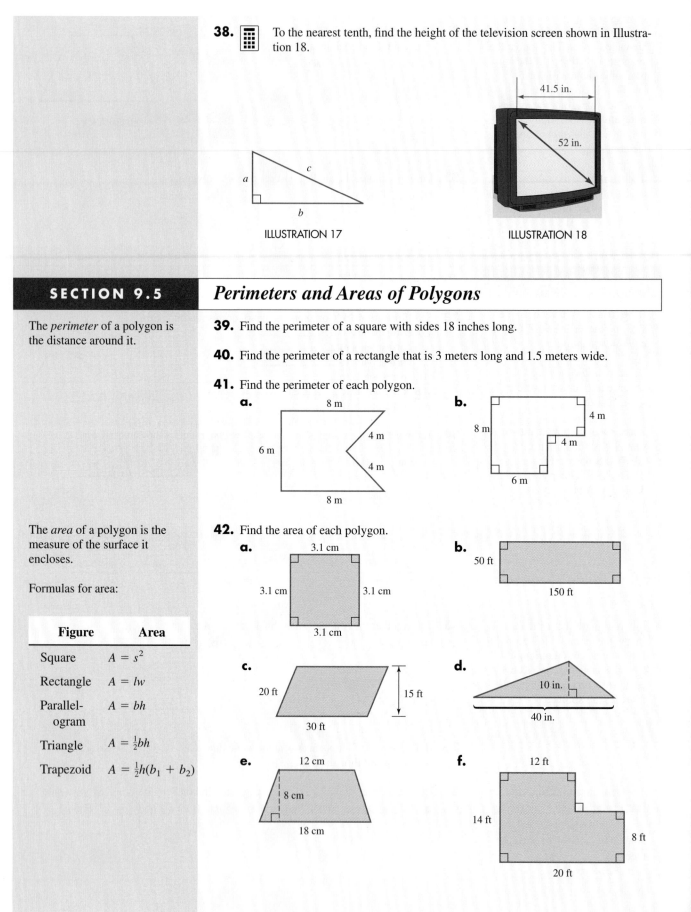

ILLUSTRATION 17

ILLUSTRATION 18

SECTION 9.5 *Perimeters and Areas of Polygons*

The *perimeter* of a polygon is the distance around it.

39. Find the perimeter of a square with sides 18 inches long.

40. Find the perimeter of a rectangle that is 3 meters long and 1.5 meters wide.

41. Find the perimeter of each polygon.

a.

8 m
6 m
4 m
4 m
8 m

b.

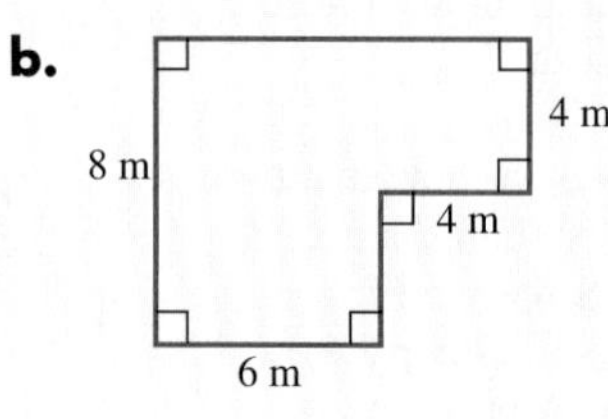

The *area* of a polygon is the measure of the surface it encloses.

Formulas for area:

Figure	Area
Square	$A = s^2$
Rectangle	$A = lw$
Parallelogram	$A = bh$
Triangle	$A = \frac{1}{2}bh$
Trapezoid	$A = \frac{1}{2}h(b_1 + b_2)$

42. Find the area of each polygon.

a.

3.1 cm
3.1 cm
3.1 cm
3.1 cm

b.

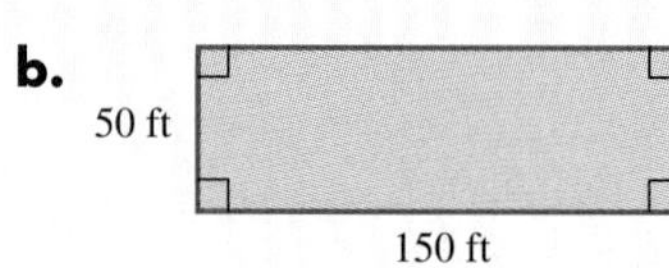

c.

20 ft
15 ft
30 ft

d.

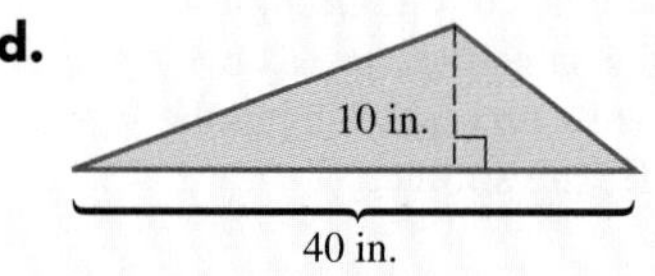

e.

12 cm
8 cm
18 cm

f.

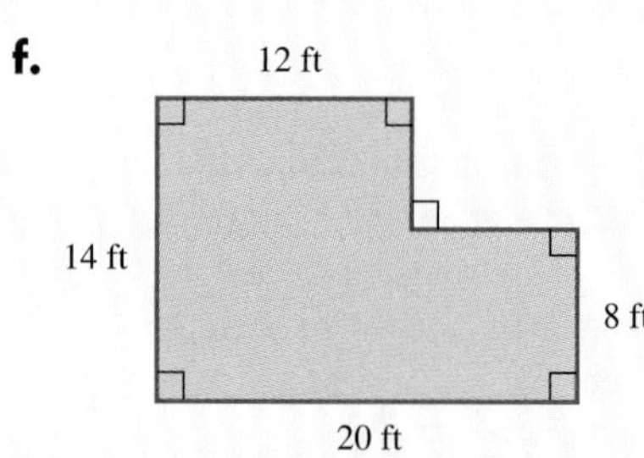

g.

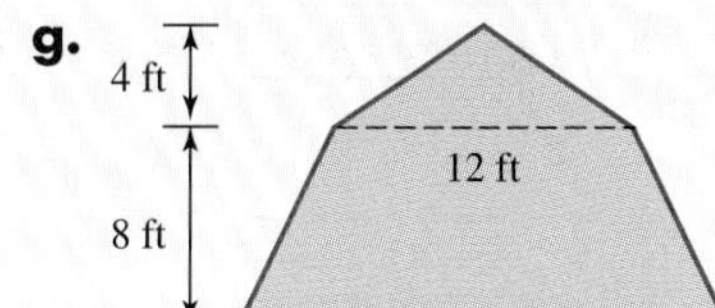

h.

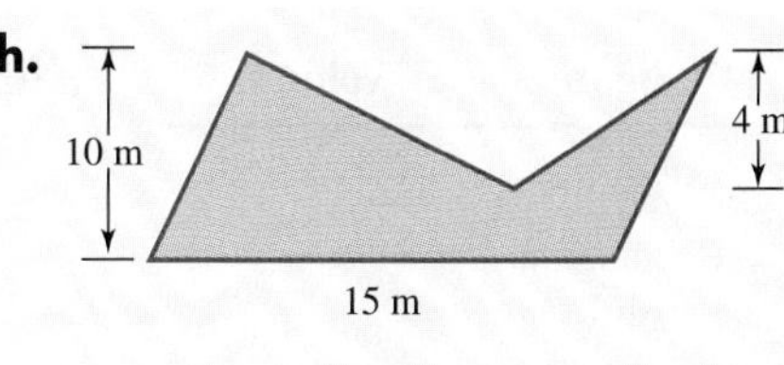

43. How many square feet are there in 1 square yard?

44. How many square inches are in 1 square foot?

SECTION 9.6 Circles

A *circle* is the set of all points in a plane that lie a fixed distance from a point called its *center.* The fixed distance is the circle's *radius.*

A *chord* of a circle is a line segment connecting two points on the circle.
A *diameter* is a chord that passes through the circle's center.

The *circumference* (perimeter) of a circle is given by the formulas
$C = \pi D$ or $C = 2\pi r$

$\pi = 3.14159\ldots$

The *area* of a circle is given by the formula
$A = \pi r^2$

45. Refer to Illustration 19.
a. Name each chord.
b. Name each diameter.
c. Name each radius.
d. Name the center.

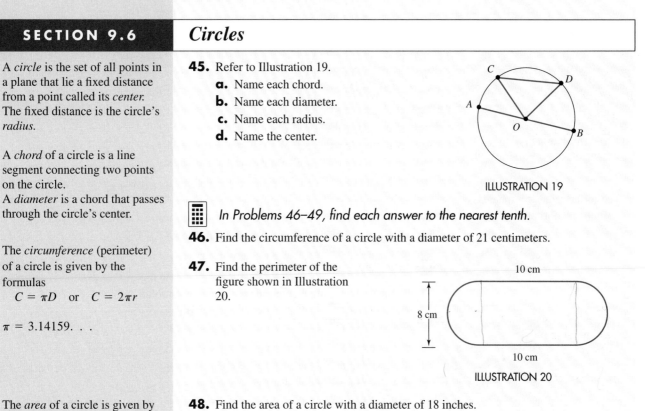

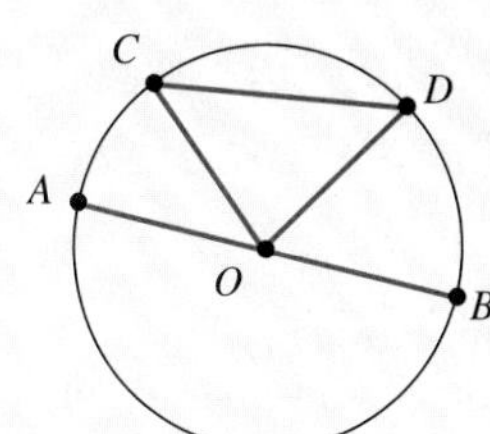

ILLUSTRATION 19

In Problems 46–49, find each answer to the nearest tenth.

46. Find the circumference of a circle with a diameter of 21 centimeters.

47. Find the perimeter of the figure shown in Illustration 20.

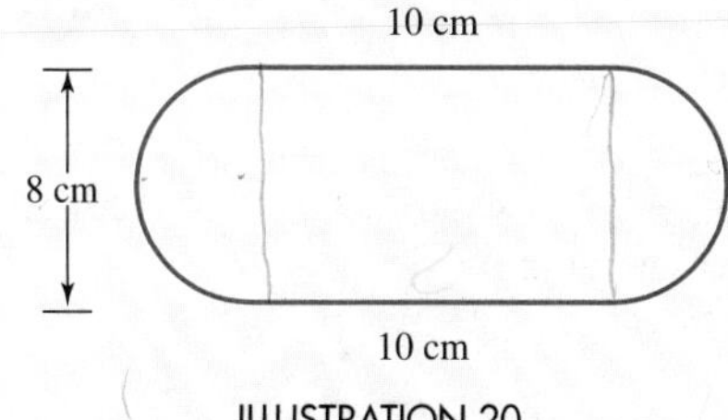

ILLUSTRATION 20

48. Find the area of a circle with a diameter of 18 inches.

49. Find the area of the figure shown in Illustration 20.

SECTION 9.7 Surface Area and Volume

The *volume* of a solid is a measure of the space it occupies.

50. Find the volume of each solid to the nearest unit. (See the table at the top of the next page.)

a.

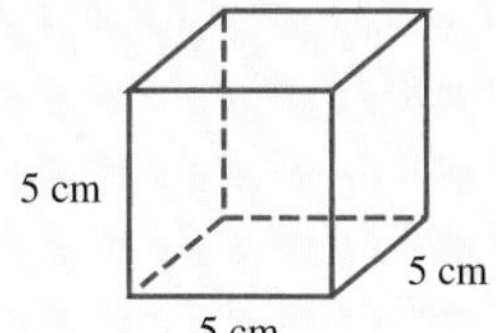

b.

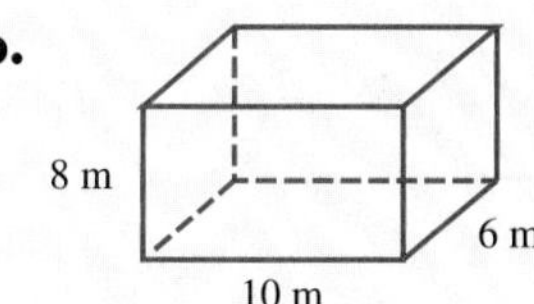

Figure	Volume
Cube	$V = s^3$
Rectangular solid	$V = lwh$
Prism	$V = Bh$*
Sphere	$V = \frac{4}{3}\pi r^3$
Cylinder	$V = \pi r^2 h$
Cone	$V = \frac{1}{3}\pi r^2 h$
Pyramid	$V = \frac{1}{3}Bh$*

*B represents the area of the base.

The *surface area* of a rectangular solid is the sum of the areas of its six faces.

The surface area of a sphere is given by the formula
$SA = 4\pi r^2$

c.

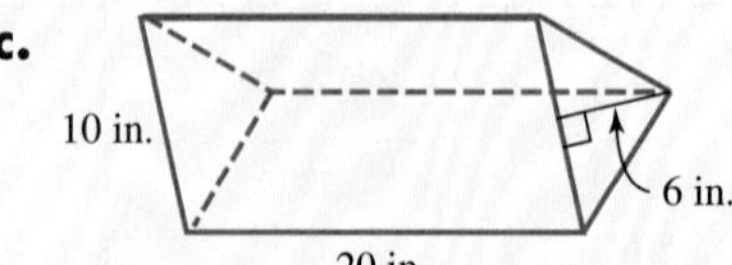

d.

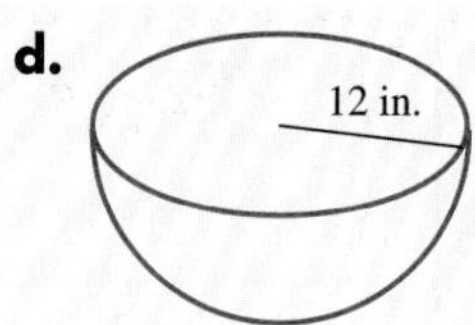

e.

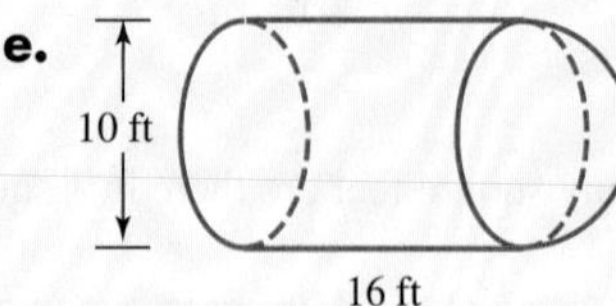

f.

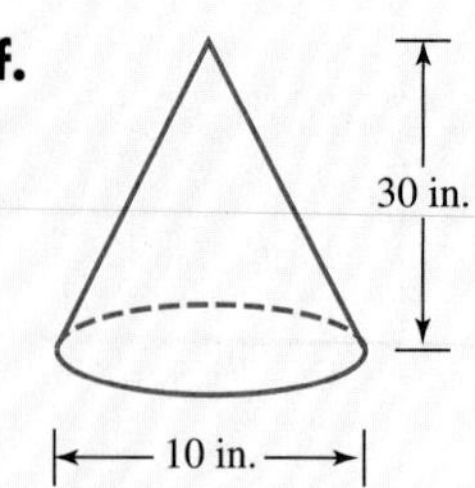

g.

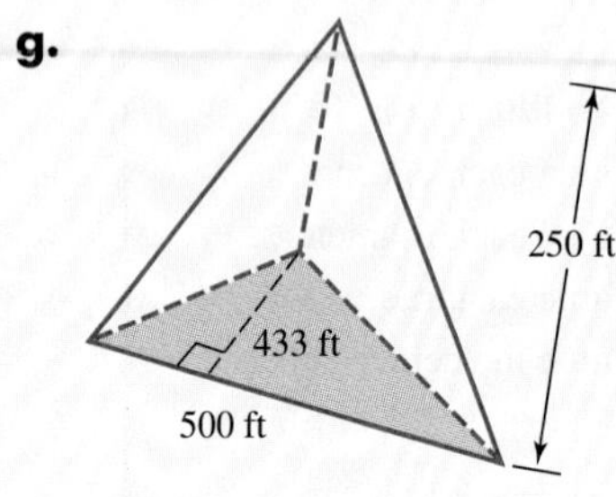

h.

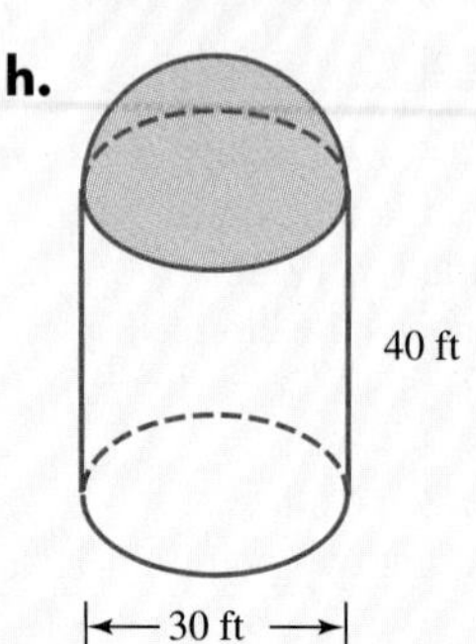

51. How many cubic inches are there in 1 cubic foot?

52. How many cubic feet are there in 2 cubic yards?

53. To the nearest tenth, find the surface area of each solid.

a.

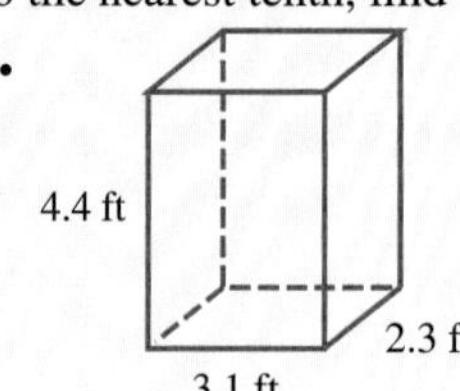

b.

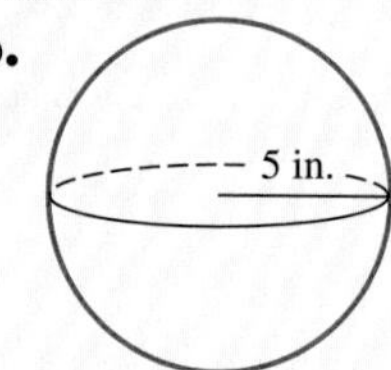

Chapter 9 Test

1. Find m($\overline{AB}$).

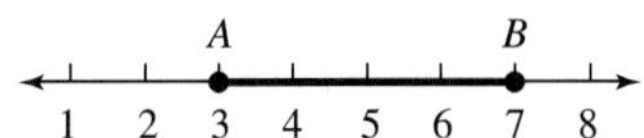

2. Which point is the vertex of $\angle ABC$?

Tell whether each statement is true or false.

3. An angle of 47° is an acute angle.

4. An angle of 90° is a straight angle.

5. An angle of 180° is a right angle.

6. An angle of 132° is an obtuse angle.

7. Find x.

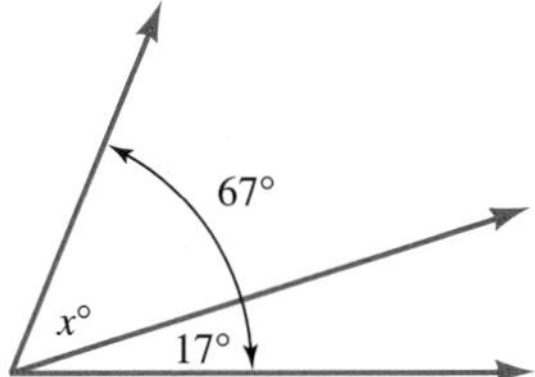

8. Find y.

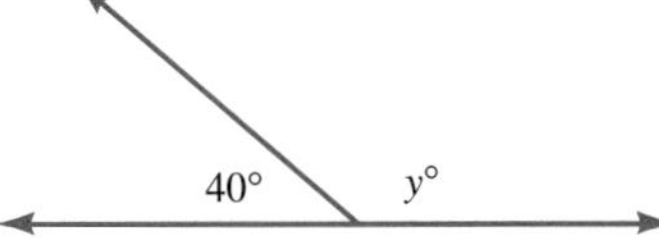

9. Find y.

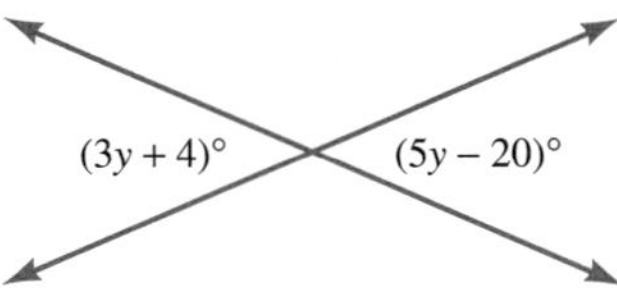

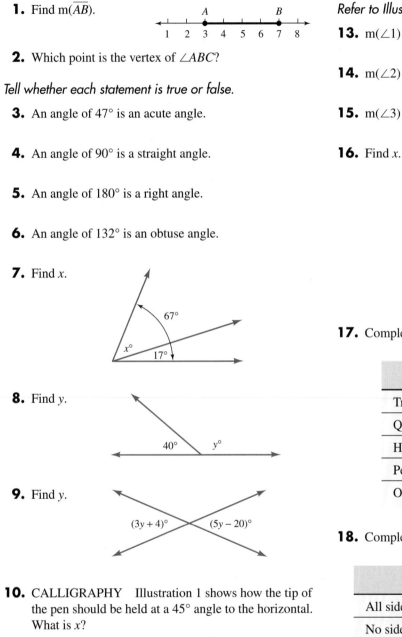

10. CALLIGRAPHY Illustration 1 shows how the tip of the pen should be held at a 45° angle to the horizontal. What is x?

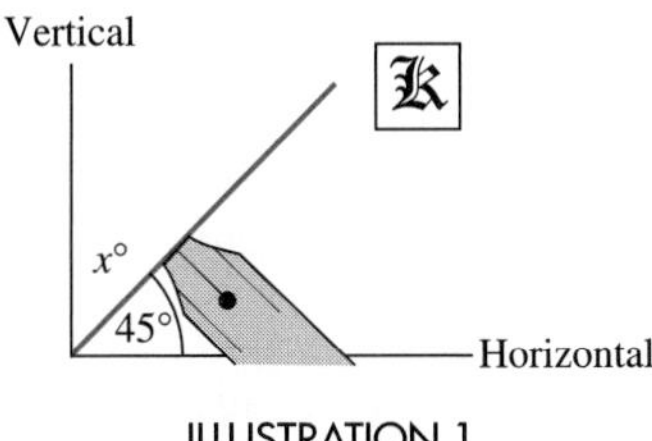

ILLUSTRATION 1

11. Find the complement of an angle measuring 67°.

12. Find the supplement of an angle measuring 117°.

Refer to Illustration 2, in which $l_1 \parallel l_2$.

13. $m(\angle 1) =$ ____________.

14. $m(\angle 2) =$ ____________.

15. $m(\angle 3) =$ ____________.

16. Find x.

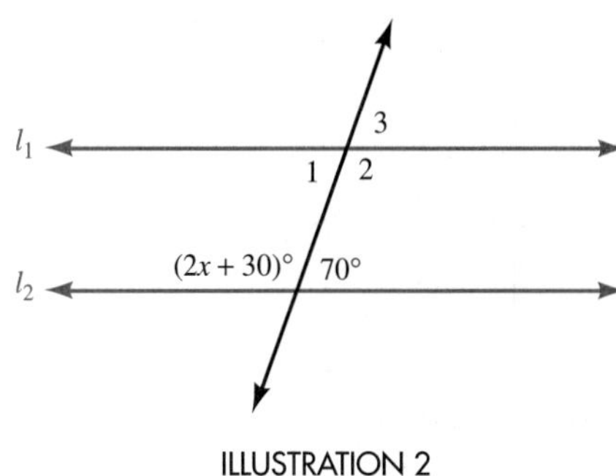

ILLUSTRATION 2

17. Complete the table.

Polygon	Number of sides
Triangle	
Quadrilateral	
Hexagon	
Pentagon	
Octagon	

18. Complete the table about triangles.

Property	Kind of triangle
All sides of equal length	
No sides of equal length	
Two sides of equal length	

In Exercises 19–20, refer to Illustration 3.

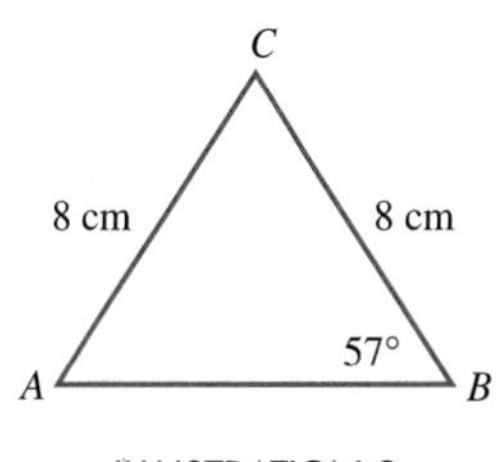

ILLUSTRATION 3

19. Find m($\angle A$).

20. Find m($\angle C$).

21. If the measures of two angles in a triangle are 65° and 85°, find the measure of the third angle.

22. Find the sum of the measures of the angles in a decagon (a ten-sided polygon).

23. In Illustration 4, *ABCD* is a rectangle. Name three pairs of segments with equal lengths.

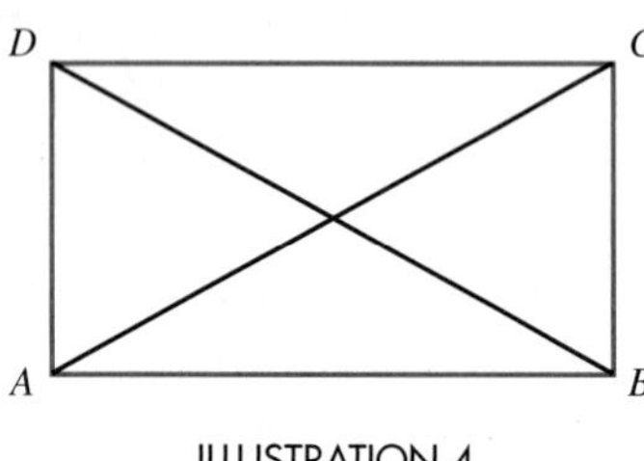

ILLUSTRATION 4

24. In Illustration 5, *ABCD* is an isosceles trapezoid. Find *x*.

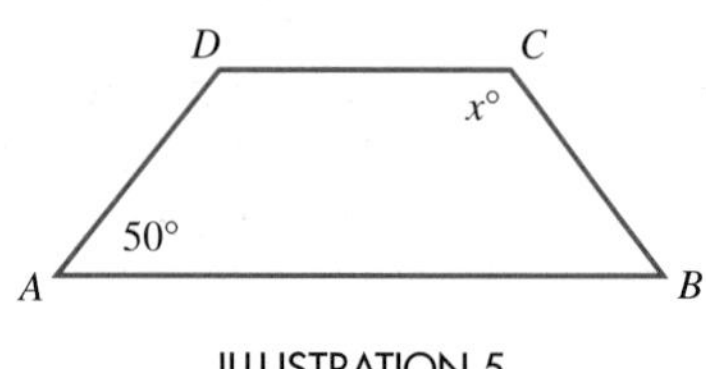

ILLUSTRATION 5

Refer to Illustration 6, in which $\triangle ABC \cong \triangle DEF$.

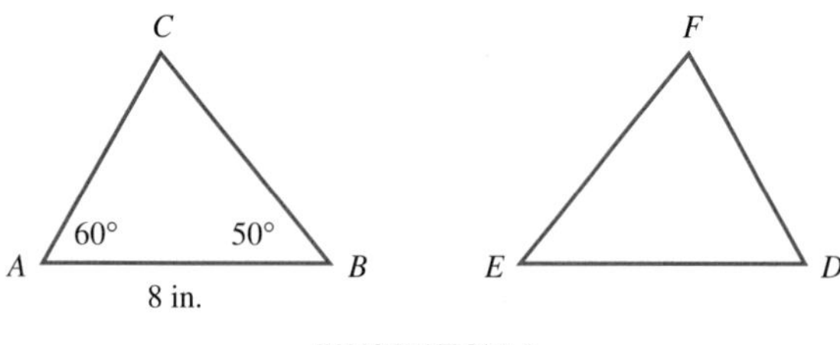

ILLUSTRATION 6

25. Find m($\overline{DE}$).

26. Find m($\angle E$).

Refer to Illustration 7, in which m($\angle A$) = m($\angle D$) *and* m($\angle C$) = m($\angle F$).

27. Find *x*.

28. Find *y*.

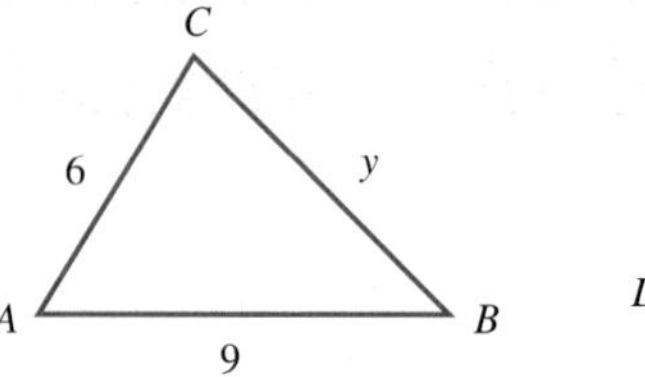

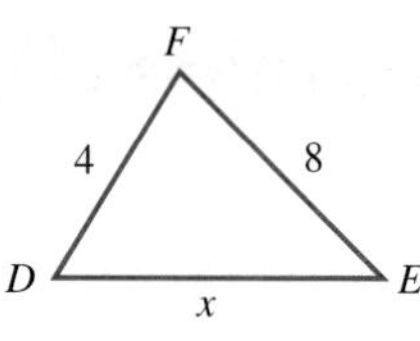

ILLUSTRATION 7

Give each answer to the nearest tenth.

29. A baseball diamond is a square with each side 90 feet long. What is the straight-line distance from third base to first base?

30. Find the area of a triangle with a base 44.5 centimeters long and a height of 17.6 centimeters.

31. Find the area of a trapezoid with a height of 6 feet and bases that are 12.2 feet and 15.7 feet long.

32. THE OLYMPICS Steel rod is to be bent to form the interlocking rings of the Olympic Games symbol, shown in Illustration 8. How many feet of steel rod will be needed to make the symbol if the diameter of each ring is to be 6 feet?

ILLUSTRATION 8

33. Find the area of a circle with a diameter that is 6 feet long.

34. Find the volume of a rectangular solid with dimensions 4.3 by 5.7 by 6.5 meters.

35. Find the volume of a sphere that is 8 meters in diameter.

36. Find the volume of a 10-foot-tall pyramid that has a rectangular base 5 feet long and 4 feet wide.

37. Give a real-life example in which the concept of perimeter is used. Do the same for area and for volume. Be sure to discuss the type of units used in each case.

38. Draw a cube. Explain how to find its surface area.

Chapters 1–9 Cumulative Review Exercises

1. AMUSEMENT PARKS Use the data in the table to construct a bar graph in Illustration 1.

Fatal accidents on amusement park rides							
Year	'93	'94	'95	'96	'97	'98	'99
Number	4	2	3	3	4	5	6

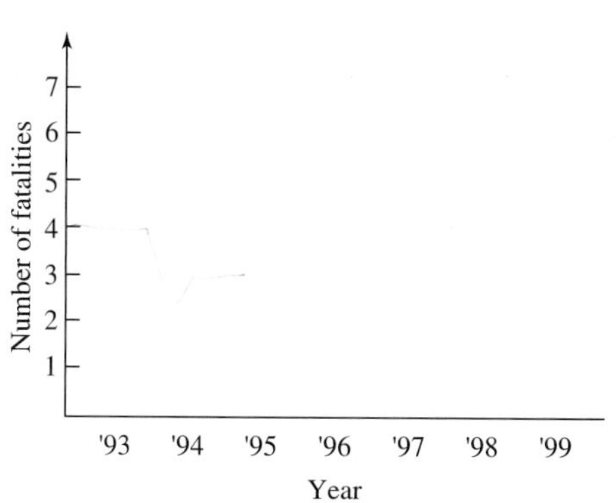

Source: *USA Today* (April 7, 2000)

ILLUSTRATION 1

2. USED CARS Illustration 2 shows an advertisement that appeared in *The Car Trader.* (O.B.O. means "or best offer.")

1969 Ford Mustang. New tires
Must sell!!!! $10,500 O.B.O.

ILLUSTRATION 2

If offers of $8,750, $8,875, $8,900, $8,850, $8,800, $7,995, $8,995, and $8,925 were received, what was the selling price of the car?

3. Subtract: $35{,}021 - 23{,}999$.

4. Divide: $1{,}353 \div 41$.

5. Round 2,109,567 to the nearest thousand.

6. Prime factor 220.

7. Find all the factors of 24.

8. List the set of integers.

9. Evaluate $-10(-2) - 2^3 + 1$.

10. Evaluate $5 - 3[4^2 - (1 + 5 \cdot 2)]$.

11. Evaluate $|-6 - (-3)|$.

12. Evaluate the expression $\dfrac{2x + 3y}{z - y}$ for $x = 2$, $y = -3$, and $z = -4$.

13. What is the difference between an equation and an expression?

14. Simplify $4x - 2(3x - 4) - 5(2x)$.

Solve each equation. Check each result.

15. $3(p + 15) + 4(11 - p) = 0$

16. $5t - 7 = 7t + 13$

17. $-x + 2 = 13$

18. $4 + \dfrac{x}{5} - 6 = -1$

19. SNAILS According to the *Guinness Book of World Records,* in the 1995 World Snail Racing Championships, a snail covered a 13-inch course in 2 minutes. What was the snail's rate in inches per minute?

20. SHOPPING What is the value of x coupons, each of which gives the shopper 50¢ off?

21. Translate to mathematical symbols:
5 less than a number

22. LUMBER To find the number of board feet (b.f.) in a piece of lumber, use the formula

$$\text{b.f.} = \frac{\text{thickness (in.)} \cdot \text{width (in.)} \cdot \text{length (ft)}}{12}$$

Find the number of board feet in the piece of lumber shown in Illustration 3. (*Hint:* the symbol ″ stands for inches and the symbol ′ stands for feet.)

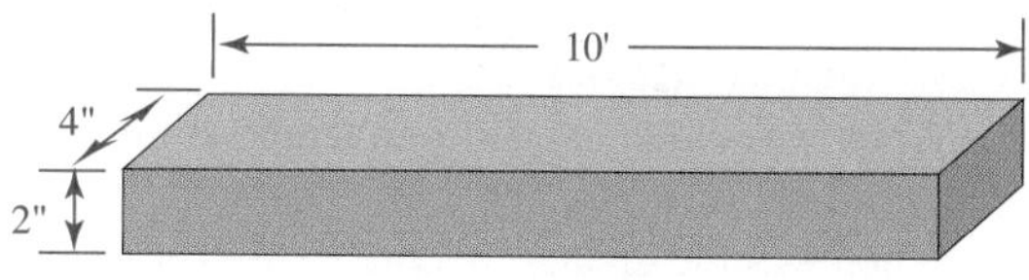

ILLUSTRATION 3

23. Simplify: $\dfrac{35a^2}{28a}$.

24. Add: $45\frac{2}{3} + 96\frac{4}{5}$.

25. Subtract: $\dfrac{x}{4} - \dfrac{3}{5}$.

26. BAKING A 5-pound bag of all-purpose flour contains $17\frac{1}{2}$ cups. A baker uses $3\frac{3}{4}$ cups. How much flour is left?

27. Multiply: $-\dfrac{6}{25}\left(2\dfrac{7}{24}\right)$.

28. Divide: $\dfrac{15}{8q^4} \div \dfrac{45}{8q^3}$.

29. VETERINARY MEDICINE A pet owner was told to use an eye dropper to administer medication to his sick kitten. The cup shown in Illustration 4 contains 8 doses of the medication. Determine the size of a single dose.

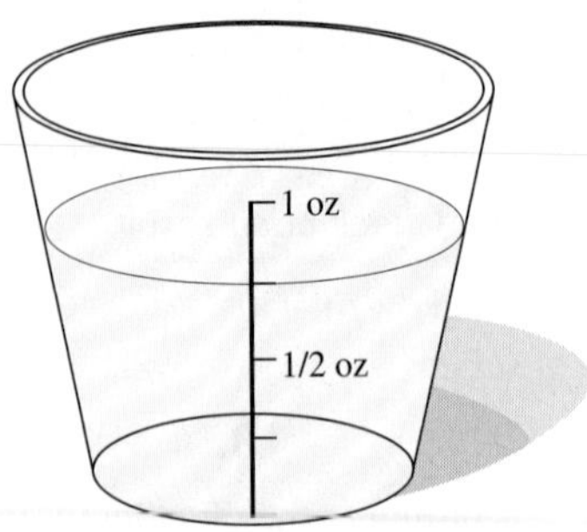

ILLUSTRATION 4

30. Solve $\frac{x}{2} - \frac{1}{9} = \frac{1}{3}$.

31. Solve $\frac{2}{3}q - 1 = -6$.

32. Evaluate $\frac{3}{4} + \left(-\frac{1}{3}\right)^2\left(\frac{5}{4}\right)$.

33. Simplify $\frac{7 - \frac{2}{3}}{4\frac{5}{6}}$.

34. GRAVITY Objects on the moon weigh only one-sixth as much as on Earth. If a rock weighs 3 ounces on the moon, how much does it weigh on Earth?

35. GLOBAL WARMING Illustration 5 is a line graph of the mean global temperature change, as measured by NASA weather balloons.

a. When was the greatest rise in temperature recorded? What was it?

b. When was the greatest decline in temperature recorded? What was it?

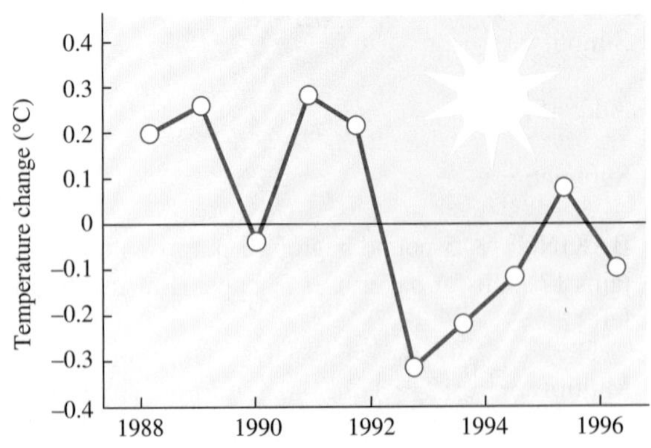

Source: National Aeronautics and Space Administration

ILLUSTRATION 5

36. Graph each member of the set on the number line:

$$\left\{-4\frac{5}{8}, \sqrt{17}, 2.89, \frac{2}{3}, -0.1, -\sqrt{9}, \frac{3}{2}\right\}$$

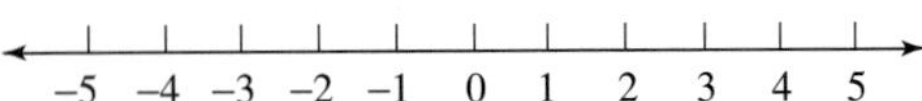

37. Round the number pi to the nearest ten thousandth: $\pi = 3.141592654. \ . \ . \ .$

38. Place the proper symbol (> or <) in the blank: 154.34 ____ 154.33999.

39. Add: $3.4 + 106.78 + 35 + 0.008$.

40. Multiply: $-5.5(-3.1)$.

41. Multiply: $(89.9708)(1{,}000)$.

42. Divide: $\frac{0.0742}{1.4}$.

43. Evaluate $-8.8 + (-7.3 - 9.5)$.

44. Evaluate $\frac{7}{8}(9.7 + 15.8)$.

45. Change $\frac{2}{15}$ to a decimal.

46. Evaluate $\frac{(-1.3)^2 + 6.7}{-0.9}$ and round to the nearest hundredth.

47. DECORATIONS A mother has budgeted $20 for decorations for her daughter's birthday party. She decides to buy a tank of helium for $15.15 and some balloons. If the balloons sell for 5 cents apiece, how many balloons can she buy?

48. Solve $1.7y + 1.24 = -1.4y - 0.62$.

49. Evaluate $2\sqrt{121} - 3\sqrt{64}$.

50. Simplify $\sqrt{\frac{49}{81}}$.

51. TABLE TENNIS The weights (in ounces) of 8 ping-pong balls that are to be used in a tournament are listed below.

0.85	0.87	0.88	0.88
0.85	0.86	0.84	0.85

Find the mean, median, and mode of the weights.

52. Graph each point on the coordinate grid: $A(-4, -3)$, $B(1.5, 1.5)$, $C\left(-\frac{7}{2}, 0\right)$, $D(0, 3.5)$

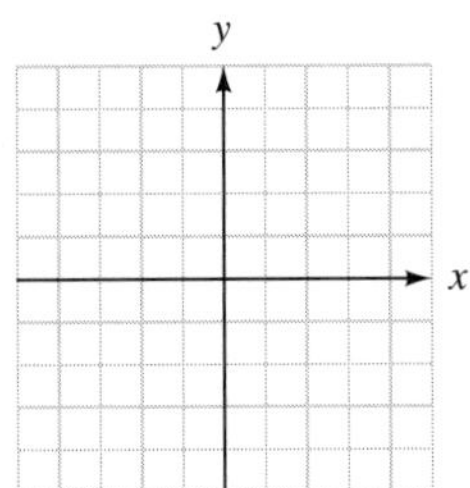

53. What are the coordinates of the origin?

54. Is $(-2, 1)$ a solution of $3x - y = -8$?

Graph each equation.

55. $y = -2x$

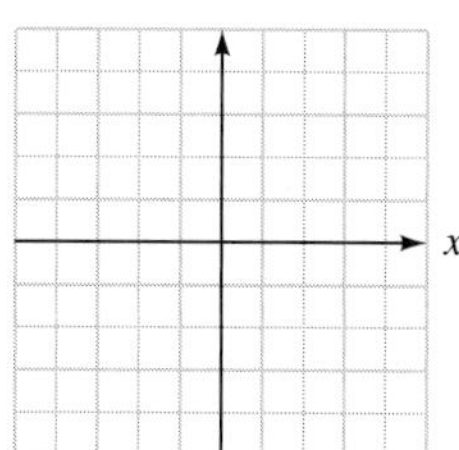

56. $2x - 3y = 14$

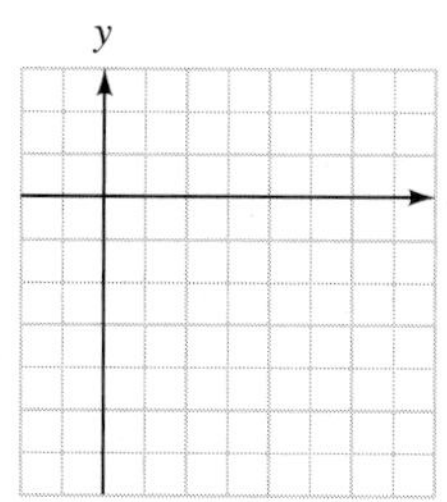

57. $x = 4$

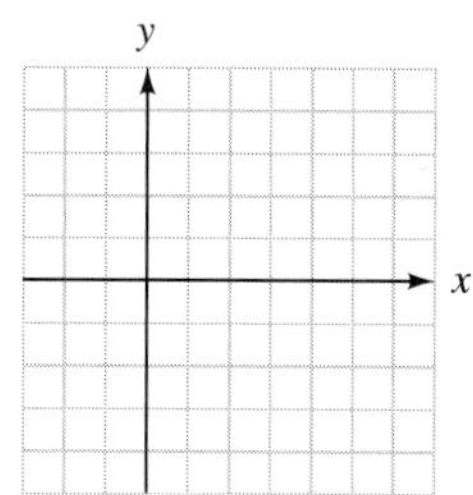

58. $y = x^2$

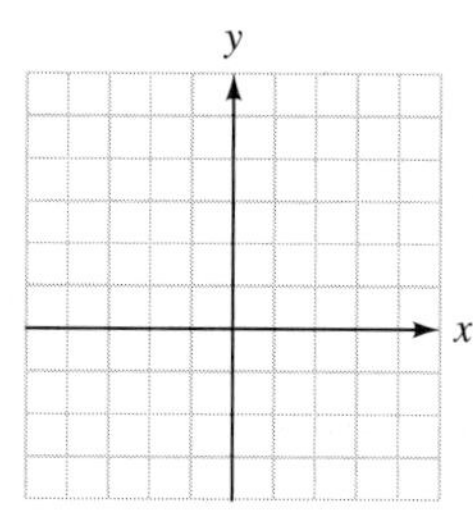

59. Evaluate the polynomial $-3x^2 - 2x$ when $x = -2$.

60. a. Consider $(-3)^2$. What is the base and what is the exponent? Evaluate the expression.

b. Consider -3^2. What is the base and what is the exponent? Evaluate the expression.

Simplify each expression.

61. $s^4 \cdot s^5$

62. $(a^5)^7$

63. $-3h^9(-5h)$

64. $(2b^3c^6)^3$

65. $(y^5)^2(y^4)^3$

66. $x^m \cdot x^n$

67. Classify $3x^2 - 7x + 1$ as a monomial, a binomial, or a trinomial. Then give its degree.

Do the operations.

68. $(5x^2 - 2x + 4) - (3x^2 - 5)$

69. $-3p(2p^2 + 3p - 4)$

70. $(3x + 5)(2x - 1)$

71. $(2y - 7)^2$

72. What percent of the figure in Illustration 6 is shaded? What percent is not shaded?

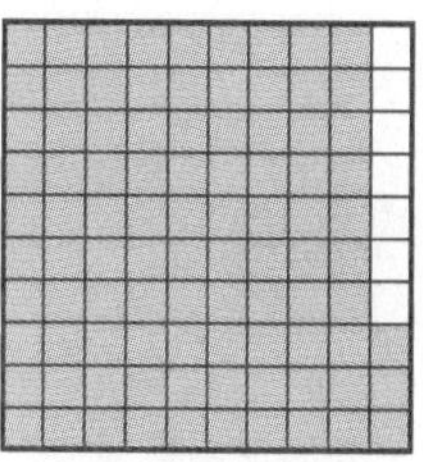

ILLUSTRATION 6

73. What number is 15% of 450?

74. 24.6 is 20.5% of what number?

75. Complete the table.

Percent	Decimal	Fraction
57%		
	0.001	
		$\frac{1}{3}$

76. STUDENT GOVERNMENT In an election for Student Body President, 560 votes were cast. Stan Cisneros received 308 votes, and Amy Huang-Sims received 252 votes. Use a circle graph to show the percent of the vote received by each candidate.

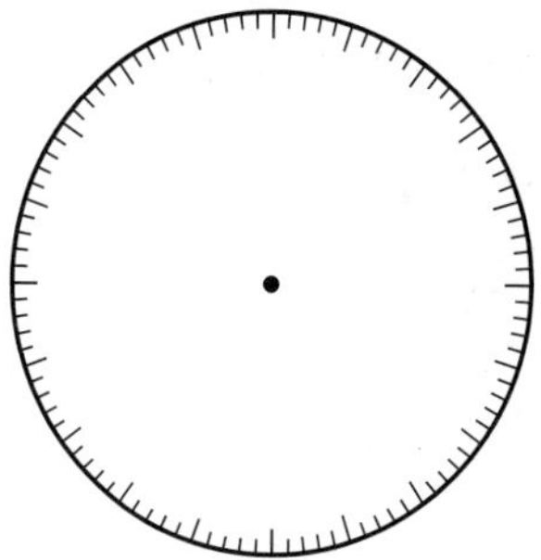

77. SHOPPING See Illustration 7. What is the regular price of the calculator?

ILLUSTRATION 7

78. SALES TAX If the sales tax rate is $6\frac{1}{4}\%$, how much sales tax will be added to the price of a new car selling for $18,550?

79. COLLECTIBLES A German Hummel porcelain figurine, which was originally purchased for $125, was sold by a collector ten years later for $750. What was the percent increase in the value of the figurine?

80. PAYING OFF A LOAN To pay for tuition, a college student borrows $1,500 for two months. If the annual interest rate is 9%, how much will the student have to repay when the loan comes due?

81. SAVING FOR RETIREMENT When he got married, a man invested $5,000 in an account that guaranteed to pay 8% interest, compounded monthly, for 50 years. At the end of 50 years, how much will his account be worth?

Write each phrase as a ratio.

82. 3 centimeters to 7 centimeters

83. 13 weeks to 1 year

84. COMPARISON SHOPPING A dry-erase whiteboard with an area of 400 in.^{2} sells for $24. A larger board, with an area of 600 in.^{2}, sells for $42. Which board is the better buy?

85. Solve the proportion: $\frac{5 - x}{14} = \frac{13}{28}$.

86. INSURANCE CLAIMS In one year, an auto insurance company had 3 complaints per 1,000 policies. If a total of 375 complaints were filed that year, how many policies did the company have?

87. SCALE DRAWING Suppose the house plan shown in Illustration 8 is drawn on a grid of $\frac{1}{4}$-inch squares. How long is the house?

Scale $\frac{1}{4}$ in. : 3 ft

ILLUSTRATION 8

Make each conversion.

88. 168 inches = ____ feet

89. 15 yards = ____ inches

90. 212 ounces = ____ pounds

91. 30 gallons = ____ quarts

92. 25 cups = ____ fluid ounces

93. 738 minutes = ____ hours

94. 654 milligrams = ____ centigrams

95. 500 milliliters = ____ liter

96. 5,890 decimeters = ____ dekameters

97. 75° C = ____ F

98. THE AMAZON The Amazon River enters the Atlantic Ocean through a broad estuary, roughly estimated at 240,000 m in width. Convert the width to kilometers.

99. TENNIS A tennis ball weighs between 57 and 58 g. Express this range in centigrams.

100. OCEAN LINER When it was making the transatlantic cruises from England to America, the Queen Mary got 13 feet to the gallon.

a. How many meters a gallon is this?

b. The fuel capacity of the ship was 3,000,000 gallons. How many liters is this?

101. COOKING What is the weight of a 10-pound ham in kilograms?

102. How many degrees are in a right angle?

103. How many degrees are in an acute angle?

104. Find the supplement of an angle of 105°.

105. Find the complement of an angle of 75°.

Refer to Illustration 9, in which $l_1 \parallel l_2$. Find the measure of each angle.

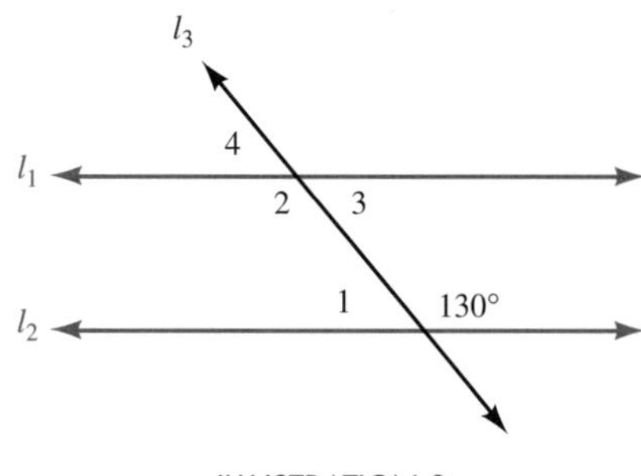

ILLUSTRATION 9

106. m(∠1) **107.** m(∠2)
108. m(∠3) **109.** m(∠4)

Refer to Illustration 10, in which $AB \parallel DE$ and $m(AC) = m(BC)$. Find the measure of each angle.

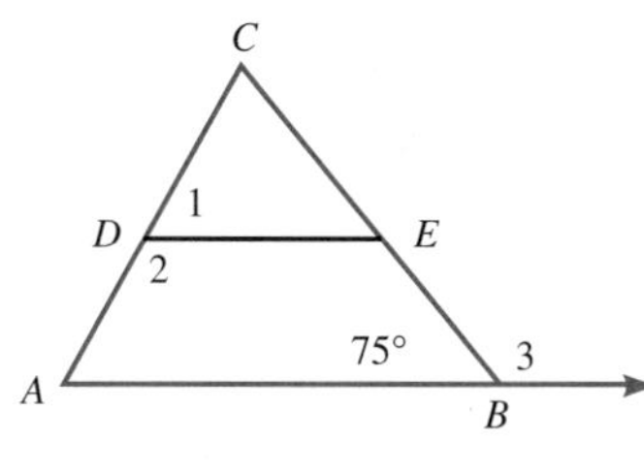

ILLUSTRATION 10

110. m(∠1) **111.** m(∠C)
112. m(∠2) **113.** m(∠3)

114. JAVELIN THROW See Illustration 11. Determine x and y.

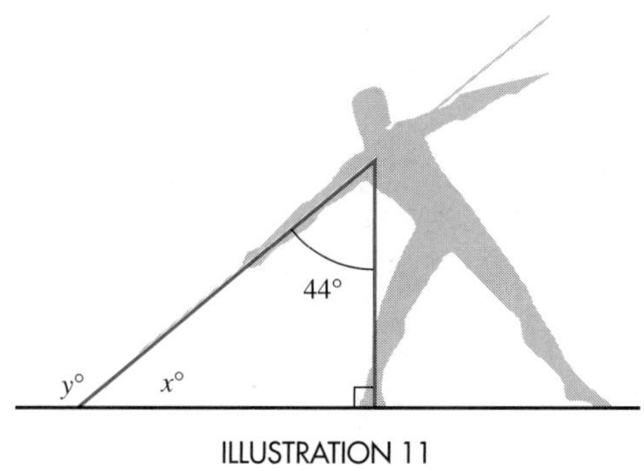

ILLUSTRATION 11

115. Find the sum of the angles of a pentagon.

116. If two sides of a right triangle measure 5 meters and 12 meters, how long is the hypotenuse?

If an answer is not exact, round to the nearest hundredth.

117. Find the perimeter and area of a rectangle with dimensions of 9 meters by 12 meters.

118. Find the area of a triangle with a base that is 14 feet long and an altitude of 18 feet.

119. Find the area of a trapezoid that has bases that are 12 inches and 14 inches long and a height of 7 inches.

120. Find the circumference and area of a circle with a diameter of 14 centimeters.

121. Find the area of the shaded region in Illustration 12, which is created using 2 semicircles.

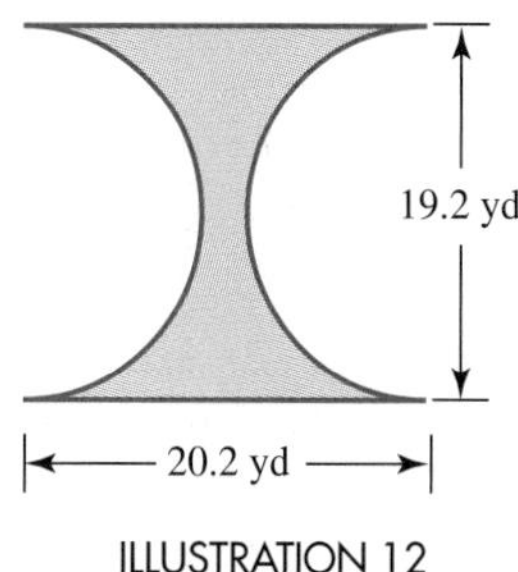

ILLUSTRATION 12

122. Find the volume of a rectangular solid with dimensions of 5 meters by 6 meters by 7 meters.

123. Find the volume of a sphere with a diameter of 10 inches.

124. Find the volume of a cone that has a circular base 8 meters in diameter and a height of 9 meters.

125. Find the volume of a cylindrical pipe that is 20 feet long and 6 inches in diameter.

126. Find the surface area of a block of ice that is in the shape of a rectangular solid with dimensions 15 in. × 24 in. × 18 in.

APPENDIX I INDUCTIVE AND DEDUCTIVE REASONING

In this appendix, you will learn about

- Inductive reasoning • Deductive reasoning

INTRODUCTION. To reason means to think logically. The objective of this appendix is to develop your problem-solving ability by improving your reasoning skills. We will introduce two fundamental types of reasoning that can be applied in a wide variety of settings. They are known as *inductive reasoning* and *deductive reasoning.*

Inductive reasoning

In a laboratory, scientists conduct experiments and observe outcomes. After several repetitions with similar outcomes, the scientist will generalize the results into a statement that appears to be true:

- If I heat water to 212°F, it will boil.
- If I drop a weight, it will fall.
- If I combine an acid with a base, a chemical reaction occurs.

When we draw general conclusions from specific observations, we are using **inductive reasoning.** The next examples show how inductive reasoning can be used in mathematical thinking. Given a list of numbers or symbols, called a *sequence,* we can often find a missing term of the sequence by looking for patterns and applying inductive reasoning.

EXAMPLE 1 ***An increasing pattern.*** Find the next number in the sequence 5, 8, 11, 14, ...

Solution

The terms of the sequence are increasing. To discover the pattern, we find the *difference* between each pair of successive terms.

$8 - 5 = 3$ Subtract the first term from the second term.

$11 - 8 = 3$ Subtract the second term from the third term.

$14 - 11 = 3$ Subtract the third term from the fourth term.

The difference between each pair of numbers is 3. This means that each successive number is 3 greater than the previous one. Thus, the next number in the sequence is $14 + 3$, or 17.

Self Check

Find the next number in the sequence $-3, -1, 1, 3, \ldots$.

Answer: 5 ■

EXAMPLE 2 ***A decreasing pattern.*** Find the next number in the sequence $-2, -4, -6, -8, \ldots$.

Solution

The terms of the sequence are decreasing. Since each successive term is 2 less than the previous one, the next number in the pattern is $-8 - 2$, or -10.

Self Check

Find the next number in the sequence $-0.1, -0.3, -0.5, -0.7, \ldots$.

Answer: -0.9 ■

EXAMPLE 3 ***An alternating pattern.*** Find the next letter in the sequence A, D, B, E, C, F, D, ...

Solution

The letter A is the first letter of the alphabet, D is the fourth letter, B is the second letter, and so on. We can create the following letter–number correspondence:

A → 1
  Add 3.
D → 4
  Subtract 2.
B → 2
  Add 3.
E → 5
  Subtract 2.
C → 3
  Add 3.
F → 6
  Subtract 2.
D → 4

The numbers in the sequence 1, 4, 2, 5, 3, 6, 4, . . . alternate in size. They change from smaller to larger, to smaller, to larger, and so on.

We see that 3 is added to the first number to get the second number. Then 2 is subtracted from the second number to get the third number. To get successive terms in the sequence, we alternately add 3 to one number and then subtract 2 from that result to get the next number.

Applying this pattern, the next number in the numerical sequence would be $4 + 3$, or 7. The next letter in the original sequence would be G, because it is the seventh letter of the alphabet.

Self Check

Find the next entry in the sequence Z, A, Y, B, X, C, ...

Answer: W ■

EXAMPLE 4 ***Two patterns.*** Find the next geometric shape in the sequence below.

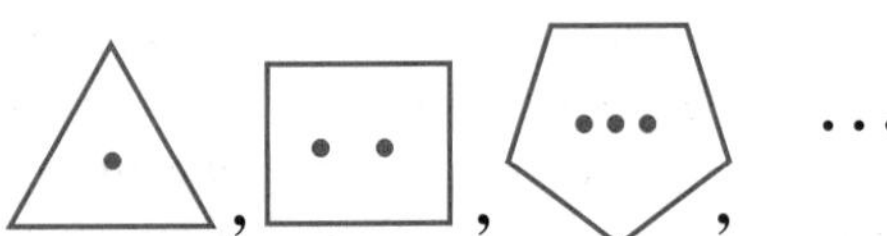

Solution

This sequence has two patterns occurring at the same time. The first figure has three sides and one dot, the second figure has four sides and two dots, and the third figure has five sides and three dots. Thus, we would expect the next figure to have six sides and four dots, as shown in Figure A-1.

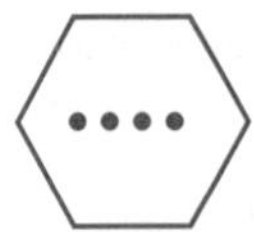

FIGURE A-1

Self Check

Find the next geometric shape in the sequence below.

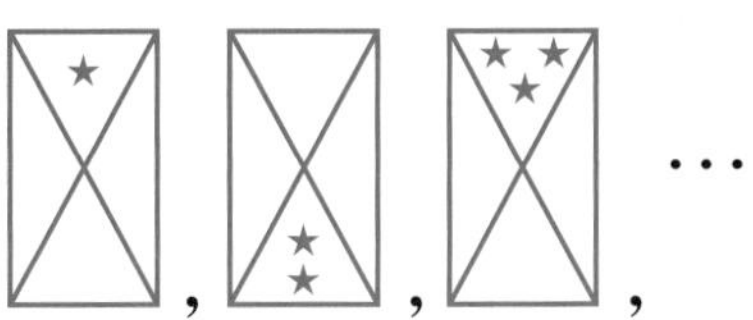

Answer:  ■

EXAMPLE 5 ***A circular pattern.*** Find the next geometric shape in the sequence below.

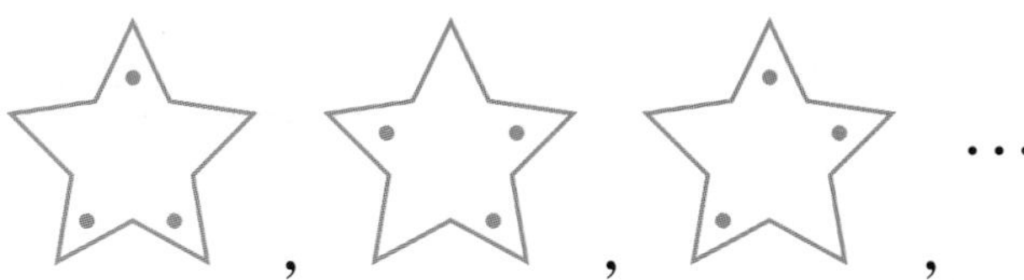

Solution

From figure to figure, we see that each dot moves from one point of the star to the next, in a counterclockwise direction. This is a circular pattern. The next shape in the sequence will be the one shown in Figure A-2.

FIGURE A-2

Self Check

Find the next geometric shape in the sequence below.

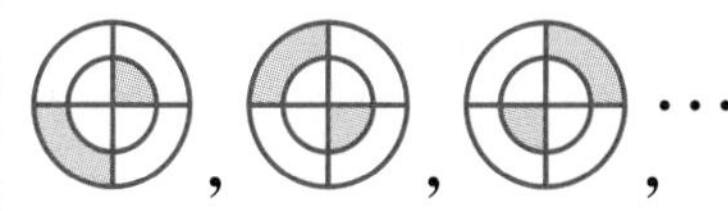

Answer:  ■

Deductive reasoning

As opposed to inductive reasoning, **deductive reasoning** moves from the general case to the specific. For example, if we know that the sum of the angles in any triangle is 180°, we know that the sum of the angles of $\triangle ABC$ is 180°. Whenever we apply a general principle to a particular instance, we are using deductive reasoning.

A deductive reasoning system is built on four elements:

1. **Undefined terms:** terms that we accept without giving them formal meaning
2. **Defined terms:** terms that we define in a formal way
3. **Axioms** or **postulates:** statements that we accept without proof
4. **Theorems:** statements that we can prove with formal reasoning

Many problems can be solved by deductive reasoning. For example, suppose that we plan to enroll in an early-morning algebra class, and that we know that Professors Perry, Miller, and Tveten are scheduled to teach algebra next semester. After some investigating, we find out that Professor Perry teaches only in the afternoon and Professor Tveten teaches only in the evenings. Without knowing anything about Professor Miller, we can conclude that he will be our teacher, since he is the only remaining possibility.

The following examples show how to use deductive reasoning to solve problems.

EXAMPLE 6 ***Scheduling classes.*** Four professors are scheduled to teach mathematics next semester, with the following course preferences:

1. Professors A and B don't want to teach calculus.
2. Professor C wants to teach statistics.
3. Professor B wants to teach algebra.

Who will teach trigonometry?

Solution

The following chart shows each course, with each possible instructor.

Calculus	Algebra	Statistics	Trigonometry
A	A	A	A
B	B	B	B
C	C	C	C
D	D	D	D

Since Professors A and B don't want to teach calculus, we can cross them off the calculus list. Since Professor C wants to teach statistics, we can cross her off every other list. This leaves Professor D as the only person to teach calculus, so we can cross her off every other list. Since Professor B wants to teach algebra, we can cross him off every other list. Thus, the only remaining person left to teach trigonometry is Professor A.

Calculus	Algebra	Statistics	Trigonometry
~~A~~	A	A	A
~~B~~	B	~~B~~	~~B~~
~~C~~	~~C~~	C	~~C~~
D	~~D~~	~~D~~	~~D~~

■

EXAMPLE 7 ***State flags.*** The graph in Figure A-3 gives the number of state flags that feature an eagle, a star, or both. How many state flags have neither an eagle nor a star?

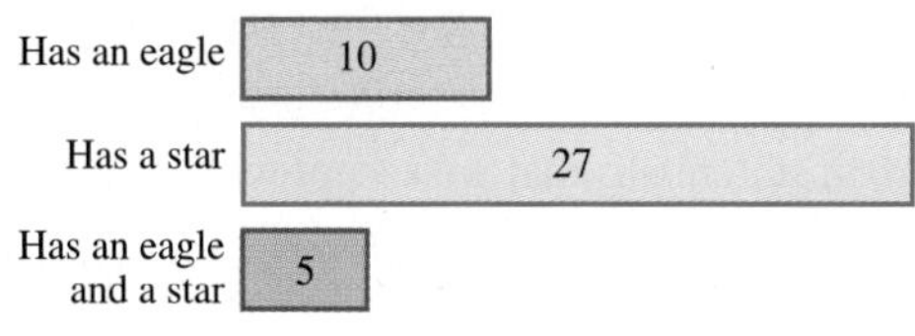

FIGURE A-3

Solution

In Figure A-4(a), the intersection (overlap) of the circles is a way to show that there are 5 state flags that have both an eagle and a star. If an eagle appears on a total of 10 flags, then the left circle must contain 5 more flags outside of the intersection. See Figure A-4(b). If a total of 27 flags have a star, the right circle must contain 22 more flags outside the intersection.

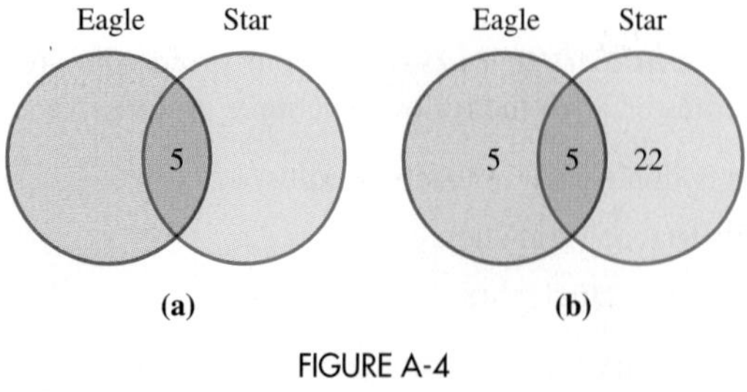

FIGURE A-4

Self Check

Of the 50 cars on a used-car lot, 9 are red, 31 are foreign models, and 6 are red, foreign models. If a customer wants to buy an American model that is not red, how many cars does she have to choose from?

From Figure A-4, we see that 5 + 5 + 22, or 32 flags have an eagle, a star, or both. To find how many flags have neither an eagle nor a star, we subtract this total from the number of state flags, which is 50.

50 − 32 = 18

There are 18 state flags that have neither an eagle nor a star.

Answer: 16

Study Set Appendix I

VOCABULARY *Fill in the blanks to make the statements true.*

1. ______________ reasoning draws general conclusions from specific observations.

2. ______________ reasoning moves from the general case to the specific.

CONCEPTS *Tell whether the pattern shown is increasing, decreasing, alternating, or circular.*

3. 2, 3, 4, 2, 3, 4, 2, 3, 4, . . .

4. 8, 5, 2, −1, . . .

5. −2, −4, 2, 0, 6, . . .

6. 0.1, 0.5, 0.9, 1.3, . . .

7. a, c, b, d, c, e, . . .

8. 

9. ROOM SCHEDULING From the chart, determine what time(s) on a Wednesday morning a practice room in a music building is available. The symbol X indicates that the room has already been reserved.

	M	T	W	Th	F
9 A.M.	X		X	X	
10 A.M.	X	X			X
11 A.M.			X		X

10. COUNSELING QUESTIONNAIRE A group of college students were asked if they were taking a mathematics course and if they were taking an English course. The results are displayed in Illustration 1.

a. How many students were taking a mathematics course and an English course?

b. How many students were taking an English course but not a mathematics course?

c. How many students were taking a mathematics course?

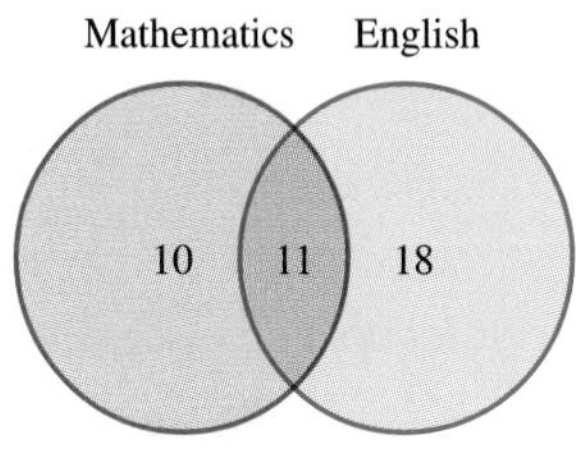

ILLUSTRATION 1

PRACTICE *Find the number that comes next in each sequence.*

11. 1, 5, 9, 13, . . .

12. 15, 12, 9, 6, . . .

13. −3, −5, −8, −12, . . .

14. 5, 9, 14, 20, . . .

15. −7, 9, −6, 8, −5, 7, −4, . . .

16. 2, 5, 3, 6, 4, 7, 5, . . .

17. 9, 5, 7, 3, 5, 1, . . .

18. 1.3, 1.6, 1.4, 1.7, 1.5, 1.8, . . .

19. −2, −3, −5, −6, −8, −9, . . .

20. 8, 11, 9, 12, 10, 13, . . .

21. 6, 8, 9, 7, 9, 10, 8, 10, 11, . . .

22. 10, 8, 7, 11, 9, 8, 12, 10, 9, . . .

Find the figure that comes next in each sequence.

23. 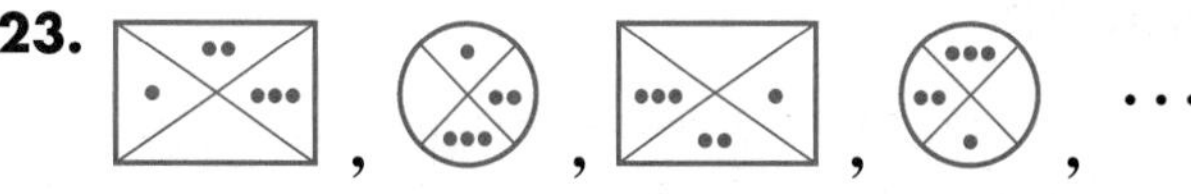

24. 

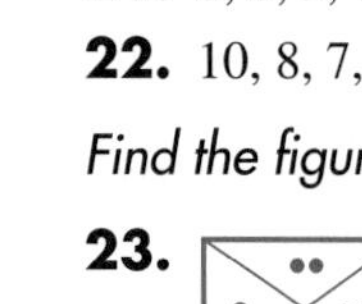

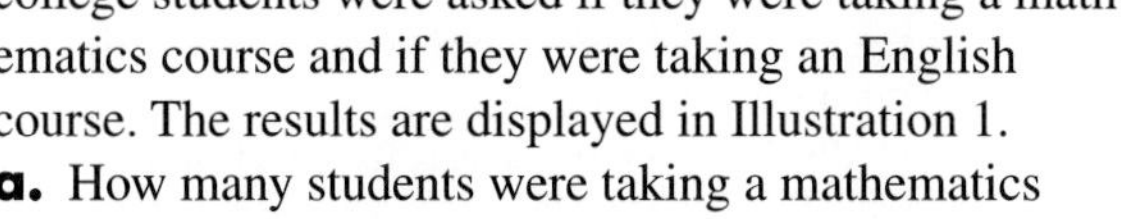

Find the missing figure in each sequence.

25. , , , ?, 

26. , , ?, , 

Find the next letter or letters in the sequence.

27. A, c, E, g, . . .

28. R, SS, TTT, . . .

29. d, h, g, k, j, n, . . .

30. B, N, C, N, D, . . .

What conclusion(s) can be drawn from each set of information?

31. Four people named John, Luis, Maria, and Paula have occupations as teacher, butcher, baker, and candlestick maker.

1. John and Paula are married.
2. The teacher plans to marry the baker in December.
3. Luis is the baker.

Who is the teacher?

32. In a zoo, a zebra, a tiger, a lion, and a monkey are to be placed in four cages numbered from 1 to 4, from left to right. The following decisions have been made:

1. The lion and the tiger should not be side by side.
2. The monkey should be in one of the end cages.
3. The tiger is to be in cage 4.

In which cage is the zebra?

33. A Ford, a Buick, a Dodge, and a Mercedes are parked side by side.

1. The Ford is between the Mercedes and the Dodge.
2. The Mercedes is not next to the Buick.
3. The Buick is parked on the left end.

Which car is parked on the right end?

34. Four divers at the Olympics finished first, second, third, and fourth.

1. Diver A beat diver B
2. Diver C placed between divers B and D.
3. Diver B beat diver D.

In which order did they finish?

35. A green, a blue, a red, and a yellow flag are hanging on a flagpole.

1. The blue flag is between the green and yellow flags.
2. The red flag is next to the yellow flag.
3. The green flag is above the red flag.

What is the order of the flags from top to bottom?

36. Andres, Barry, and Carl each have two occupations: bootlegger, musician, painter, chauffeur, barber, and gardener. From the following facts, find the occupations of each man.

1. The painter bought a quart of spirits from the bootlegger.
2. The chauffeur offended the musician by laughing at his mustache.
3. The chauffeur dated the painter's sister.
4. Both the musician and the gardener used to go hunting with Andres.
5. Carl beat both Barry and the painter at monopoly.
6. Barry owes the gardener $100.

APPLICATIONS

37. JURY DUTY The results of a jury service questionnaire are shown in Illustration 2. Determine how many of the 20,000 respondents have served on neither a criminal court nor a civil court jury.

Jury Service Questionnaire

997	Served on a criminal court jury
103	Served on a civil court jury
35	Served on both

ILLUSTRATION 2

38. ELECTRONIC POLL In Illustration 3, the Internet poll shows that 124 people voted for the first choice, 27 people voted for the second choice, and 19 people voted for both the first and the second choice. How many people clicked the third choice, "Neither"?

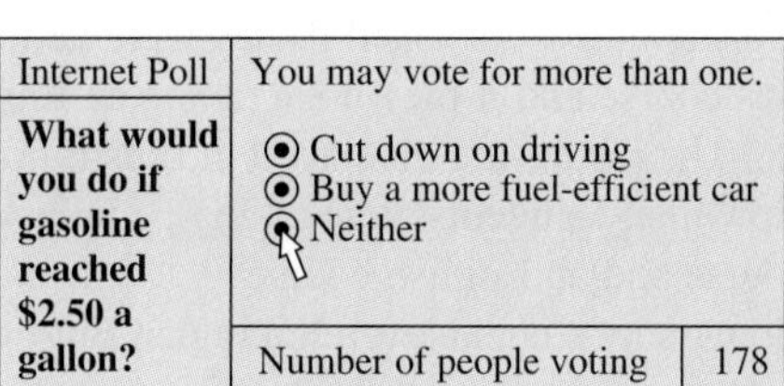

ILLUSTRATION 3

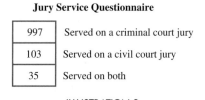

39. THE SOLAR SYSTEM The graph in Illustration 4 shows some important characteristics of the 9 planets in our solar system. How many planets are neither rocky nor have moons?

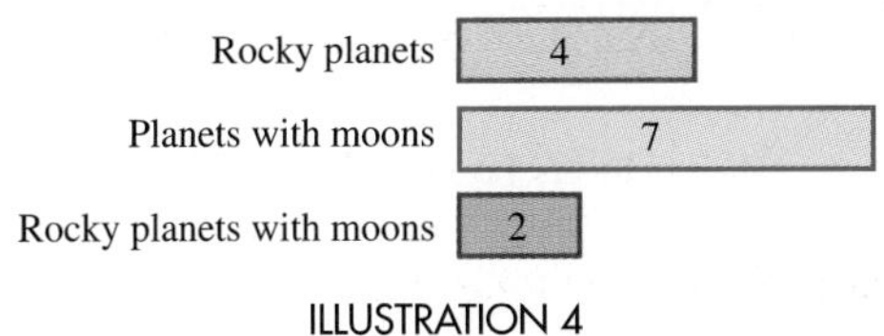

ILLUSTRATION 4

40. Write a problem in such a way that the diagram in Illustration 5 can be used to solve it.

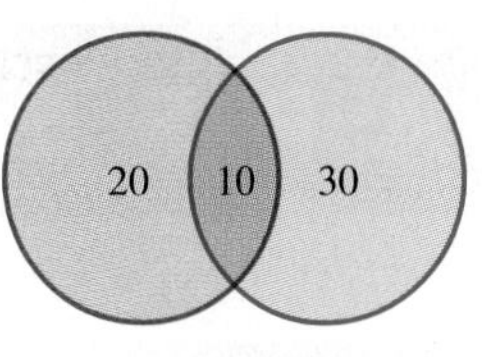

ILLUSTRATION 5

WRITING

41. Describe deductive reasoning.

42. Describe a real-life situation in which you might use deductive reasoning.

43. Describe inductive reasoning.

44. Describe a real-life situation in which you might use inductive reasoning.

APPENDIX REVIEW

APPENDIX I — Inductive and Deductive Reasoning

CONCEPTS

Inductive reasoning draws general conclusions from specific observations.

Deductive reasoning draws specific conclusions by applying general rules.

REVIEW EXERCISES

1. Find the next number in the sequence: 12, 8, 11, 7, 10, . . .

2. Find the next number in the sequence: 5, 9, 17, 33, . . .

3. Find the missing geometric figure in the sequence.

4. Find the missing geometric figure in the sequence.

5. Find the next letter in the sequence:
c, b, a, f, e, d, i, . . .

6. Four animals—a cow, a horse, a pig, and a sheep—are kept in a barn, each in a separate stall.

1. The cow is in the first stall.
2. The pig is between the horse and the sheep.
3. The sheep cannot be next to the cow.

What animal is in the last stall?

7. Jim, Sandra, Juan, and Mary all teach at the same college. One teaches mathematics, one teaches English, one teaches history, and one teaches music.

1. Jim and Sandra eat lunch with the math teacher.
2. Juan and Sandra carpool with the English teacher.
3. Jim is married to the math teacher.
4. Mary works in the same building as the history teacher.

Who is the math teacher?

8. BUYING A CAR A new-car dealership has 103 cars for sale. A computer printout describing the cars in stock states the following:

- 19 cars have a manual transmission (stick shift).
- 53 cars have a CD player.
- 12 cars have a manual transmission and a CD player.

A customer wants to buy a car that has an automatic transmission. She says that she does not want a CD player in the car. How many cars does she have to look at?

APPENDIX II ROOTS AND POWERS

n	n^2	$\sqrt{n}$	n^3	$\sqrt[3]{n}$	n	n^2	$\sqrt{n}$	n^3	$\sqrt[3]{n}$
1	1	1.000	1	1.000	51	2,601	7.141	132,651	3.708
2	4	1.414	8	1.260	52	2,704	7.211	140,608	3.733
3	9	1.732	27	1.442	53	2,809	7.280	148,877	3.756
4	16	2.000	64	1.587	54	2,916	7.348	157,464	3.780
5	25	2.236	125	1.710	55	3,025	7.416	166,375	3.803
6	36	2.449	216	1.817	56	3,136	7.483	175,616	3.826
7	49	2.646	343	1.913	57	3,249	7.550	185,193	3.849
8	64	2.828	512	2.000	58	3,364	7.616	195,112	3.871
9	81	3.000	729	2.080	59	3,481	7.681	205,379	3.893
10	100	3.162	1,000	2.154	60	3,600	7.746	216,000	3.915
11	121	3.317	1,331	2.224	61	3,721	7.810	226,981	3.936
12	144	3.464	1,728	2.289	62	3,844	7.874	238,328	3.958
13	169	3.606	2,197	2.351	63	3,969	7.937	250,047	3.979
14	196	3.742	2,744	2.410	64	4,096	8.000	262,144	4.000
15	225	3.873	3,375	2.466	65	4,225	8.062	274,625	4.021
16	256	4.000	4,096	2.520	66	4,356	8.124	287,496	4.041
17	289	4.123	4,913	2.571	67	4,489	8.185	300,763	4.062
18	324	4.243	5,832	2.621	68	4,624	8.246	314,432	4.082
19	361	4.359	6,859	2.668	69	4,761	8.307	328,509	4.102
20	400	4.472	8,000	2.714	70	4,900	8.367	343,000	4.121
21	441	4.583	9,261	2.759	71	5,041	8.426	357,911	4.141
22	484	4.690	10,648	2.802	72	5,184	8.485	373,248	4.160
23	529	4.796	12,167	2.844	73	5,329	8.544	389,017	4.179
24	576	4.899	13,824	2.884	74	5,476	8.602	405,224	4.198
25	625	5.000	15,625	2.924	75	5,625	8.660	421,875	4.217
26	676	5.099	17,576	2.962	76	5,776	8.718	438,976	4.236
27	729	5.196	19,683	3.000	77	5,929	8.775	456,533	4.254
28	784	5.292	21,952	3.037	78	6,084	8.832	474,552	4.273
29	841	5.385	24,389	3.072	79	6,241	8.888	493,039	4.291
30	900	5.477	27,000	3.107	80	6,400	8.944	512,000	4.309
31	961	5.568	29,791	3.141	81	6,561	9.000	531,441	4.327
32	1,024	5.657	32,768	3.175	82	6,724	9.055	551,368	4.344
33	1,089	5.745	35,937	3.208	83	6,889	9.110	571,787	4.362
34	1,156	5.831	39,304	3.240	84	7,056	9.165	592,704	4.380
35	1,225	5.916	42,875	3.271	85	7,225	9.220	614,125	4.397
36	1,296	6.000	46,656	3.302	86	7,396	9.274	636,056	4.414
37	1,369	6.083	50,653	3.332	87	7,569	9.327	658,503	4.431
38	1,444	6.164	54,872	3.362	88	7,744	9.381	681,472	4.448
39	1,521	6.245	59,319	3.391	89	7,921	9.434	704,969	4.465
40	1,600	6.325	64,000	3.420	90	8,100	9.487	729,000	4.481
41	1,681	6.403	68,921	3.448	91	8,281	9.539	753,571	4.498
42	1,764	6.481	74,088	3.476	92	8,464	9.592	778,688	4.514
43	1,849	6.557	79,507	3.503	93	8,649	9.644	804,357	4.531
44	1,936	6.633	85,184	3.530	94	8,836	9.695	830,584	4.547
45	2,025	6.708	91,125	3.557	95	9,025	9.747	857,375	4.563
46	2,116	6.782	97,336	3.583	96	9,216	9.798	884,736	4.579
47	2,209	6.856	103,823	3.609	97	9,409	9.849	912,673	4.595
48	2,304	6.928	110,592	3.634	98	9,604	9.899	941,192	4.610
49	2,401	7.000	117,649	3.659	99	9,801	9.950	970,299	4.626
50	2,500	7.071	125,000	3.684	100	10,000	10.000	1,000,000	4.642

APPENDIX III
ANSWERS TO SELECTED EXERCISES

Study Set Section 1.1 (page 7)

1. set **3.** expanded **5.** number **7.** 3 **9.** 6
11. whole numbers
13.

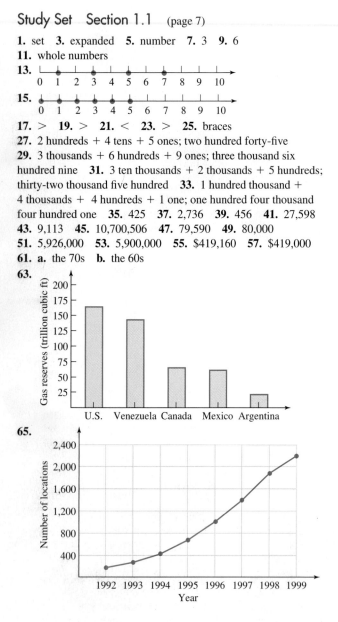

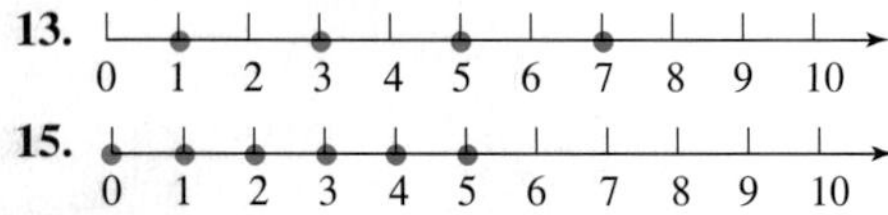

15. 0 1 2 3 4 5 6 7 8 9 10

17. > **19.** > **21.** < **23.** > **25.** braces
27. 2 hundreds + 4 tens + 5 ones; two hundred forty-five
29. 3 thousands + 6 hundreds + 9 ones; three thousand six hundred nine **31.** 3 ten thousands + 2 thousands + 5 hundreds; thirty-two thousand five hundred **33.** 1 hundred thousand + 4 thousands + 4 hundreds + 1 one; one hundred four thousand four hundred one **35.** 425 **37.** 2,736 **39.** 456 **41.** 27,598 **43.** 9,113 **45.** 10,700,506 **47.** 79,590 **49.** 80,000 **51.** 5,926,000 **53.** 5,900,000 **55.** $419,160 **57.** $419,000 **61. a.** the 70s **b.** the 60s
63.

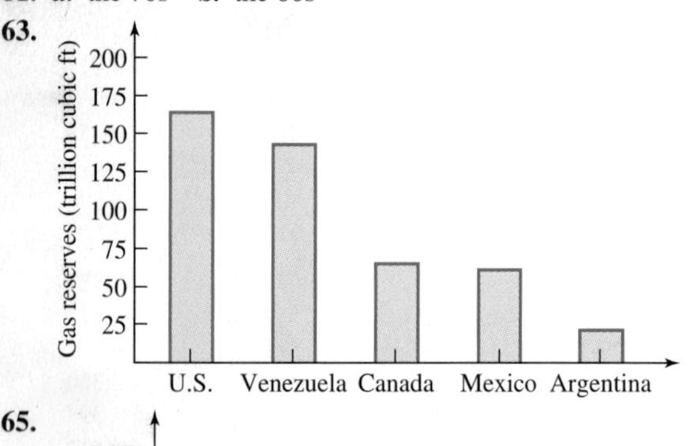

65.

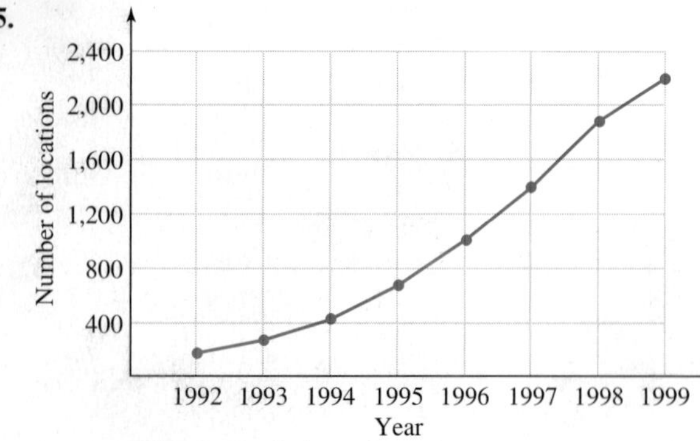

67.

a.

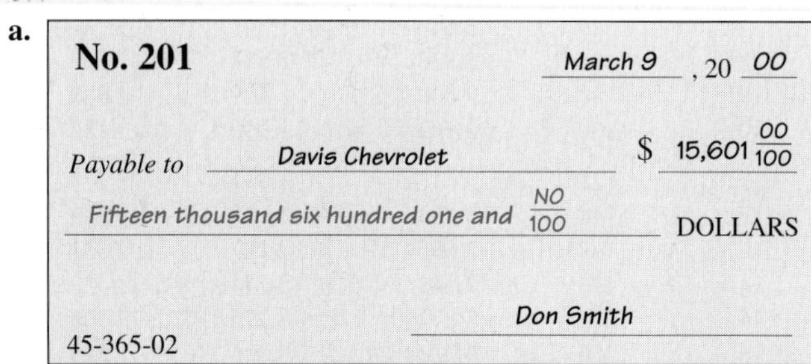

No. 201 March 9, 20 00

Payable to Davis Chevrolet $ 15,601 00/100

Fifteen thousand six hundred one and NO/100 DOLLARS

Don Smith

45-365-02

b.

No. 7890 Aug. 12, 20 00

Payable to Dr. Anderson $ 3,433 46/100

Three thousand four hundred thirty three and 46/100 DOLLARS

Juan Decito

45-828-02

69. 1,865,593; 482,880; 1,503; 269; 43,449
71. a. 299,800,000 m/s **b.** 300,000,000 m/s

Study Set Section 1.2 (page 18)

1. sum, addends **3.** rectangle **5.** difference, subtrahend, minuend **7.** associative **9.** commutative property of addition **11.** associative property of addition **13. a.** $x + y = y + x$ **b.** $(x + y) + z = x + (y + z)$ **15.** 0 **17.** $4 + 3 = 7$ **19.** parentheses **21.** 47 **23.** 38 **25.** 461 **27.** 111 **29.** 150 **31.** 363 **33.** 979 **35.** 1,985 **37.** 10,000 **39.** 15,907 **41.** 1,861 **43.** 5,312 **45.** 88 ft **47.** 68 in. **49.** 3 **51.** 25 **53.** 103 **55.** 65 **57.** 141 **59.** 0 **61.** 24 **63.** 118 **65.** 958 **67.** 1,689 **69.** 10,457 **71.** 303 **73.** 40 **75.** 110 **77.** $18 **79.** 1,750,027 **81.** $213 **83.** 10,057 mi **85. a.** $147,145 **b.** $161,725 **87.** 91 ft **89.** 792 tons **91.** 196 in. **95.** 3 thousands + 1 hundred + 2 tens + 5 ones **97.** 6,354,780 **99.** 6,350,000

Study Set Section 1.3 (page 30)

1. multiplication **3.** commutative **5.** square inch **7.** $4 \cdot 8$
9. Multiply its length by its width. **11. a.** 25 **b.** 62 **c.** 0 **d.** 0

13. $5 \cdot 12$ **15. a.** $\times$, $\cdot$, ( ) **b.** $\overline{)\ \ }$, $\div$, — **17.** square feet **19.** 84 **21.** 324 **23.** 180 **25.** 105 **27.** 7,623 **29.** 1,060 **31.** 2,576 **33.** 20,079 **35.** 2,919,952 **37.** 1,182,116 **39.** 84 in.^{2} **41.** 144 in.^{2} **43.** 8 **45.** 3 **47.** 12 **49.** 13 **51.** 73 **53.** 41 **55.** 205 **57.** 210 **59.** 8 R 25 **61.** 20 R 3 **63.** 30 R 13 **65.** 31 R 28 **67.** $132 **69.** 406 mi **71.** 125,800 **73.** 312 **75.** yes **77.** 72 **79.** 4 **81.** 5 mi **83.** 440 ft **85.** $41 **87.** 9 girls, 24 teams **89.** the square room; the square room **91.** 388 ft^2 **97.** 8 **99.** 872

Study Set Estimation (page 35)

1. no **3.** no **5.** no **7.** approx. 8,900 mi **9.** approx. 30 bags **11.** 1,600,000,000

Study Set Section 1.4 (page 41)

1. factors **3.** factor **5.** composite **7.** prime **9.** base; exponent **11.** $1 \cdot 27$ or $3 \cdot 9$ **13. a.** 44 **b.** 100 **15. a.** 1 and 11 **b.** 1 and 23 **c.** 1 and 37 **d.** They are prime numbers. **17.** yes **19.** 90 **21.** 605 **23.** no **25.** 2 **27.** 2 and 5 **29.** $3 \cdot 5 \cdot 2 \cdot 5$; $5 \cdot 3 \cdot 5 \cdot 2$; they are the same **31.** 13, 8, 7 **33.** 2 **35.** $7 \cdot 7 \cdot 7$ **37.** $3 \cdot 3 \cdot 3 \cdot 3 \cdot 3$ **39.** $5 \cdot 5 \cdot 11$ **41.** 10 **43.** 2^5 **45.** 5^4 **47.** $4^2(5^2)$ **49.** 1, 2, 5, 10 **51.** 1, 2, 4, 5, 8, 10, 20, 40 **53.** 1, 2, 3, 6, 9, 18 **55.** 1, 2, 4, 11, 22, 44 **57.** 1, 7, 11, 77 **59.** 1, 2, 4, 5, 10, 20, 25, 50, 100 **61.** $3 \cdot 13$ **63.** $3^2 \cdot 11$ **65.** $2 \cdot 3^4$ **67.** $2^2 \cdot 5 \cdot 11$ **69.** 2^6 **71.** $3 \cdot 7^2$ **73.** 81 **75.** 32 **77.** 144 **79.** 4,096 **81.** 72 **83.** 3,456 **85.** 12,812,904 **87.** 1,162,213 **91.** 2^2 square units; 3^2 square units; 4^2 square units **97.** 231,000 **99.** 0 **101.** $A = lw$

Study Set Section 1.5 (page 48)

1. parentheses, brackets **3.** evaluate **5.** 3; square, multiply, subtract **7.** multiply, subtract **9.** $2 \cdot 3^2 = 2 \cdot 9$; $(2 \cdot 3)^2 = 6^2$ **11.** 4, 20 **13.** 9, 36 **15.** 27 **17.** 2 **19.** 15 **21.** 25 **23.** 5 **25.** 25 **27.** 18 **29.** 813 **31.** 5,239 **33.** 16 **35.** 5 **37.** 49 **39.** 24 **41.** 13 **43.** 10 **45.** 198 **47.** 18 **49.** 216 **51.** 17 **53.** 191 **55.** 3 **57.** 29 **59.** 14 **61.** 64 **63.** 192 **65.** 74 **67.** 137 **69.** 3 **71.** 21 **73.** 11 **75.** 1 **77.** 10,496 **79.** 2,845 **81.** $2(6) + 4(2) + 2(1)$; $22 **83.** $24 + 6(5) + 10(10) + 12(20) + 2(50) + 100$; $594 **85.** brick: $3(3) + 1 + 1 + 3 + 3(5)$; 29; aphid: $3[1 + 2(3) + 4 + 1 + 2]$; 42 **87.** 79° **89.** 5 **91.** 298 **97.** 7,300 **99.** 9,591

Study Set Section 1.6 (page 56)

1. equal, = **3.** solution, root **5.** equivalent **7.** $y + c$ **9.** addition of 6; subtract 6 from both sides **11.** 8, 8, 16, 24, 16 **13.** yes **15.** no **17.** yes **19.** yes **21.** yes **23.** yes **25.** no **27.** no **29.** no **31.** yes **33.** 10 **35.** 7 **37.** 3 **39.** 4 **41.** 13 **43.** 75 **45.** 740 **47.** 339 **49.** 3 **51.** 5 **53.** 9 **55.** 10 **57.** 1 **59.** 56 **61.** 84 **63.** 105 **65.** 4 **67.** 12 **69.** 8 **71.** 47 **75.** 94,683,948 **77.** 62 **79.** $218,500 **81.** $125 million **83.** 25 units **85.** $190 **93.** 325,780 **95.** 90 **97.** 3

Study Set Section 1.7 (page 64)

1. division **3.** x **5.** $\frac{y}{z}$ **7.** It is being multiplied by 4. Divide by 4. **9. a.** Subtract 5 from both sides. **b.** Add 5 to both sides. **c.** Divide both sides by 5. **d.** Multiply both sides by 5.

11. 3, 3, 4, 12, 4 **13.** 1 **15.** 96 **17.** 3 **19.** 6 **21.** 1 **23.** 2 **25.** 14 **27.** 42 **29.** 75 **31.** 39 **33.** 50 **35.** 49 **37.** 10 **39.** 3 **41.** 2 **43.** 1 **45.** 40 **47.** 1,200 **51.** 390 wpm **53.** 14 **55.** 96 **57.** 32 calls **59.** 55 lb **65.** 48 cm **67.** $2^3 \cdot 3 \cdot 5$ **69.** 72 **71.** 26 mpg

Key Concept (page 67)

1. Let $x =$ the monthly cost to lease the van. **3.** Let $x =$ the width of the field. **5.** Let $x =$ the distance traveled by the motorist. **7.** $a + b = b + a$ **9.** $\frac{b}{1} = b$ **11.** $n - 1 < n$ **13.** $(r + s) + t = r + (s + t)$

Chapter Review (page 69)

1. a. / **b.**

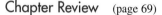

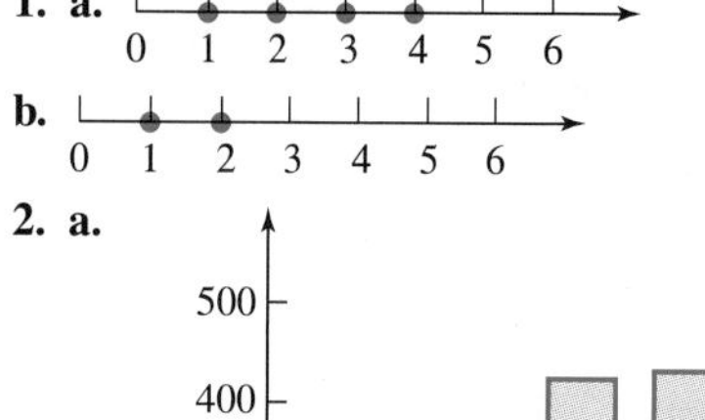

2. a.

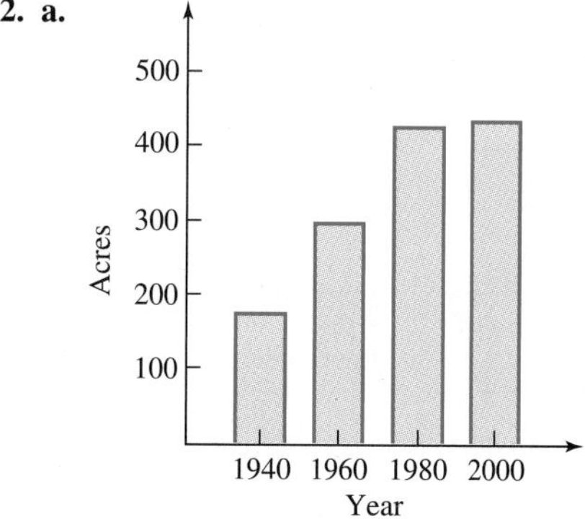

b.

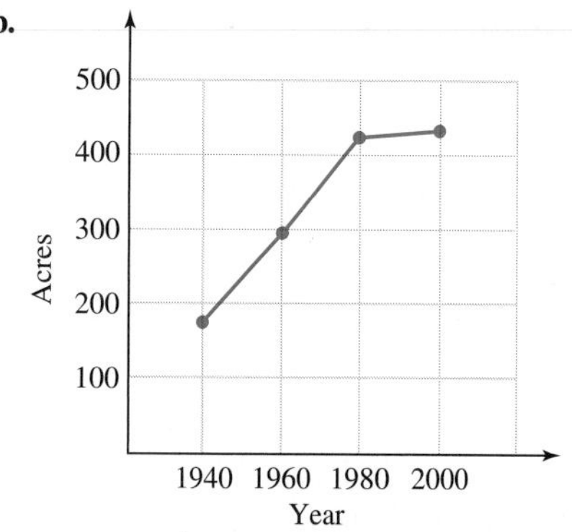

3. a. 6 **b.** 7 **4. a.** 5 hundred thousands + 7 ten thousands + 3 hundreds + 2 ones **b.** 3 ten millions + 7 millions + 3 hundred thousands + 9 thousands + 5 tens + 4 ones **5. a.** 3,207 **b.** 23,253,412 **c.** 16,000,000,000 **6. a.** > **b.** < **7. a.** 2,507,300 **b.** 2,510,000 **c.** 2,507,350 **d.** 2,500,000 **8. a.** 78 **b.** 137 **c.** 55 **d.** 149 **e.** 777 **f.** 2,332 **9. a.** 518 **b.** 6,000 **c.** 1,010 **d.** 24,986 **10. a.** commutative property of addition **b.** associative property of addition **11.** 96 in. **12. a.** 13 **b.** 4 **c.** 11 **d.** 54 **e.** 74 **f.** 2,075 **13. a.** $4 + 2 = 6$ **b.** $5 - 2 = 3$ **14.** $45 **15.** $785 **16.** $23,541 **17. a.** 56 **b.** 56 **c.** 0 **d.** 7 **e.** 560 **f.** 210 **18. a.** 3,297 **b.** 178,704 **c.** 31,684 **d.** 455,544 **19. a.** associative property of multiplication **b.** commutative property of multiplication **20.** $342 **21.** 108 ft, 288 ft^2 **22.** 720 **23. a.** 2 **b.** 15 **c.** undefined **d.** 0 **e.** 21 **f.** 37 **g.** 19 R 6 **h.** 23 R 27 **24.** 16, 25 **25.** 28 **26. a.** 1, 2, 3, 6, 9, 18 **b.** 1, 5, 25 **27. a.** prime **b.** composite **c.** neither **d.** neither

e. composite **f.** prime **28. a.** odd **b.** even **c.** even **d.** odd **29. a.** $2 \cdot 3 \cdot 7$ **b.** $3 \cdot 5^3$ **30. a.** 6^4 **b.** $5^3 \cdot 13^2$ **31. a.** 125 **b.** 121 **c.** 200 **d.** 2,700 **32. a.** 49 **b.** 32 **c.** 75 **d.** 36 **e.** 38 **f.** 24 **g.** 8 **h.** 24 **i.** 53 **j.** 3 **k.** 19 **l.** 7 **33.** $3(6) + 2(5) = 28$ **34.** 201 **35. a.** no **b.** yes **36. a.** y **b.** t **37. a.** 9 **b.** 31 **c.** 340 **d.** 133 **e.** 9 **f.** 14 **g.** 120 **h.** 5 **i.** 7 **j.** 985 **38.** $97,250 **39.** 185 **40. a.** 4 **b.** 3 **c.** 21 **d.** 14 **e.** 21 **f.** 36 **g.** 315 **h.** 425 **i.** 1 **j.** 144 **41.** 24 in. **42.** $128

Chapter 1 Test (page 75)

1. 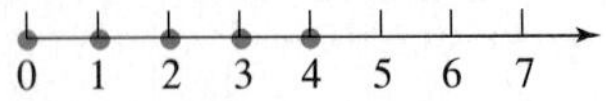

2. 5 thousands + 2 hundreds + 6 tens + 6 ones **3.** 7,507 **4.** 35,000,000 **5.**

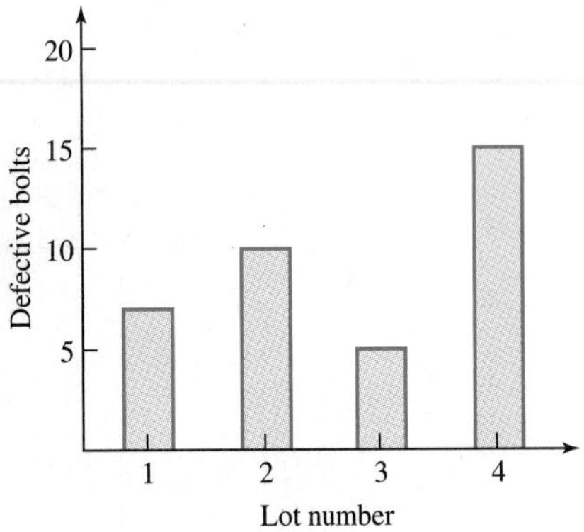

6.

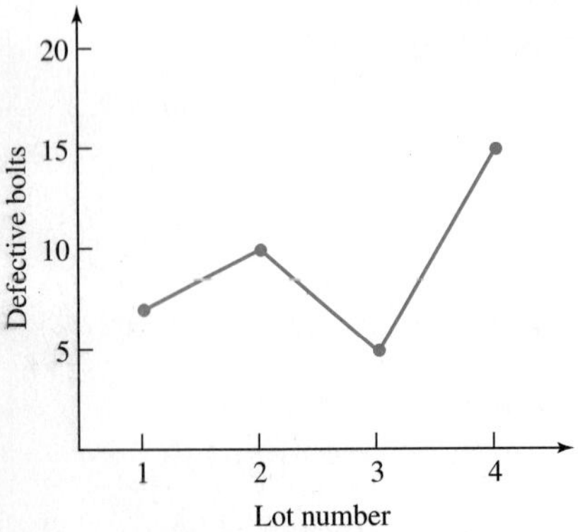

7. > **8.** < **9.** 1,491 **10.** 248 **11.** 58,105 **12.** 942 **13.** $76 **14.** 1, 2, 4, 5, 10, 20 **15.** 424 **16.** 26,791 **17.** 72 **18.** 114 R 57 **19.** 360 ft, 7,875 ft^2 **20.** 47 **21.** 3,456 **22.** $2^2 \cdot 3^2 \cdot 7$ **23.** 29 **24.** 44 **25.** 26 **26.** 39 **27.** yes **28.** 99 **29.** 30 **30.** 11 **31.** 81 **32.** 3,100 **33.** 194 yr **34.** To solve an equation means to find all the values of the variable that, when substituted into the equation, make a true statement.

Study Set Section 2.1 (page 82)

1. negative **3.** number **5.** inequality **7.** opposites **9.** They get smaller. **11.** yes **13.** $15 - 8$ **15.** $15 > 12$ **17. a.** -225 **b.** -10 **c.** -3 **d.** $-12,000$ **19.** -4 **21.** -8 and 2 **23.** -7 **25.** $6 - 4, -6, -(-6)$ (answers may vary) **27. a.** $-(-8)$ **b.** $|-8|$ **c.** $8 - 8$ **d.** $-|-8|$ **29.** 9 **31.** 8 **33.** 14 **35.** -20 **37.** -6 **39.** 203 **41.** 0 **43.** 11 **45.** 4 **47.** 1,201 **49.** 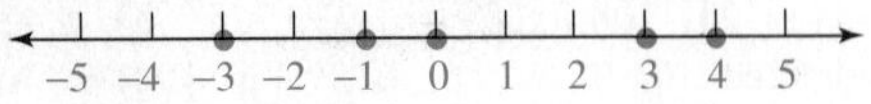

51. 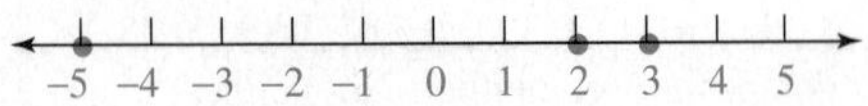

53. <

55. < **57.** > **59.** > **61.** < **63.** < **65.** 2, 3, 2, 0, -3, -7 **67.** peaks: 2, 4, 0; valleys: -3, -5, -2 **69. a.** -1 (1 below par) **b.** -3 (3 below par) **c.** Most of the scores are below par. **71. a.** $-10°$ to $-20°$ **b.** 10° **c.** 10° **73. a.** 200 yr **b.** A.D. **c.** B.C. **d.** the birth of Christ

75.

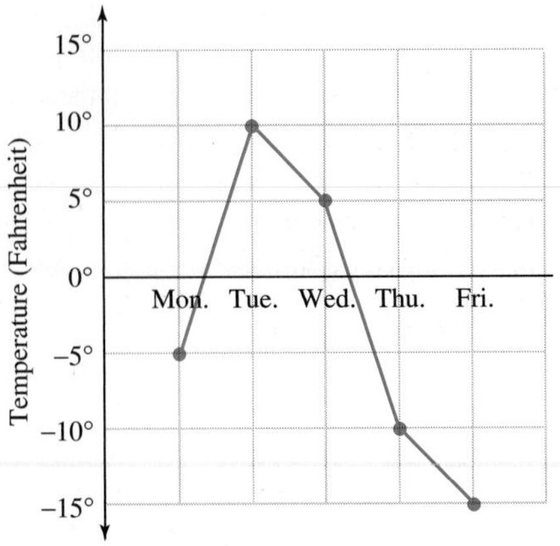

83. 23,500 **85.** 17 **87.** associative property of multiplication

Study Set Section 2.2 (page 92)

1. identity **3.** 3 **5.** -2 **7. a.** yes **b.** yes **9. a.** 7 **b.** 10 **11.** subtract, larger **13.** -18 **15.** 5 **17.** -5 should be within parentheses: $-6 + (-5)$ **19.** 11 **21.** 23 **23.** 0 **25.** -99 **27.** -9 **29.** -10 **31.** 1 **33.** -7 **35.** -20 **37.** 15 **39.** 8 **41.** 2 **43.** -10 **45.** 9 **47.** 8 **49.** -21 **51.** 3 **53.** -10 **55.** -4 **57.** 7 **59.** -21 **61.** -7 **63.** 9 **65.** 0 **67.** 0 **69.** 5 **71.** 0 **73.** -3 **75.** -10 **77.** -1 **79.** -17 **81.** $-8,346$ **83.** $-1,032$ **85.** 3G, -3G **87.** no; $70 shortfall each month **89.** 2% risk **91.** -1, 0 **93.** 7 ft over flood stage **95.** profit: $10 million **101.** 15 ft^2 **103.** 27 **105.** 5^3

Study Set Section 2.3 (page 100)

1. difference **3.** subtraction **5.** 6 **7.** $x + (-y)$ **9.** brackets **11.** $-8 - (-4)$ **13.** 7 **15.** no; $8 - 3 = 5$, $3 - 8 = -5$ **17.** -3, 2 **19.** -2, -10, 6 **21.** 9 **23.** -13 **25.** -10 **27.** -1 **29.** 0 **31.** 8 **33.** 5 **35.** -4 **37.** -4 **39.** -20 **41.** 0 **43.** 0 **45.** -15 **47.** -9 **49.** 3 **51.** 9 **53.** -2 **55.** -10 **57.** -14 **59.** 3 **61.** -8 **63.** -18 **65.** -6 **67.** 10 **69.** -4 **71.** $-2,447$ **73.** 20,503 **75.** $-1,676$ **77.** -120 ft **79.** 16 points **81.** -8 **83.** 1,007 ft **85.** -4 yd **87. a.**

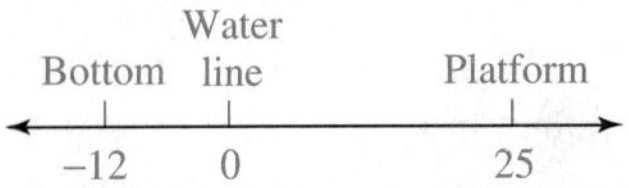

b. 37 ft **89.** No; he will be $244 overdrawn ($-244$). **95.** 3 **97.** 1, 2, 4, 5, 10, 20 **99.** 156 **101.** 4 thousands + 5 hundreds + 2 ones

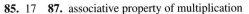

Study Set Section 2.4 (page 108)

1. factors, product **3.** 3, exponent **5.** unlike **7.** commutative **9.** -9, the opposite of that number **11.** pos · pos, pos · neg, neg · pos, neg · neg **13. a.** negative **b.** positive **15. a.** 3 **b.** 12 **c.** 5 **d.** 9 **e.** 10 **f.** 25 **17. a.** 2, 4; 4, 16; 6, 64 **b.** even **19.** 6 **21.** -5 should be in parentheses: $-6(-5)$ **23.** 54 **25.** -15 **27.** -36 **29.** 56 **31.** -20 **33.** -120 **35.** 0 **37.** 6 **39.** 7 **41.** -23 **43.** -48 **45.** 40 **47.** -30 **49.** -60 **51.** -1 **53.** -18 **55.** 0 **57.** 0 **59.** 60 **61.** 16 **63.** -125 **65.** -8 **67.** 81

69. −1 **71.** 1 **73.** 49, −49 **75.** −144, 144 **77.** −59,812 **79.** 43,046,721 **81.** −25,728 **83.** 390,625 **85. a.** plan #1: −30 lb, plan #2: −28 lb **b.** plan #1; the workout time is double that of plan #2 **87. a.** high 2, low −3 **b.** high 4, low −6 **89.** −20° **91.** −20 ft **93.** −$35,718 **99.** 45 **101.** 2,100 **103.** is less than

Study Set Section 2.5 (page 114)

1. quotient, divisor **3.** absolute value **5.** positive **7.** $5(-5) = -25$ **9.** $0(?) = -6$ **11.** $\frac{-20}{5} = -4$ **13. a.** always true **b.** sometimes true **c.** always true **15.** −7 **17.** 2 **19.** 5 **21.** 3 **23.** −20 **25.** −2 **27.** 0 **29.** undefined **31.** −5 **33.** 1 **35.** −1 **37.** 10 **39.** −4 **41.** −3 **43.** 5 **45.** −4 **47.** −5 **49.** −4 **51.** −542 **53.** −16 **55.** −4° **57.** −1,000 ft **59.** −6 (6 games behind) **61.** −$15 **63.** −$1,740 **69.** 104 **71.** $2 \cdot 3 \cdot 5 \cdot 7$ **73.** 56 **75.** 81

Study Set Section 2.6 (page 119)

1. order **3.** grouping **5.** 3; power, multiplication, subtraction **7.** multiplication; subtraction **9.** The base of the first exponential expression is 3; the base of the second is −3. **11.** 4, 20, −20 **13.** 9, −36 **15.** −7 **17.** 1 **19.** −21 **21.** −14 **23.** −7 **25.** −5 **27.** 12 **29.** −14 **31.** 30 **33.** 2 **35.** 15 **37.** −42 **39.** −5 **41.** −3 **43.** 4 **45.** 0 **47.** −14 **49.** 19 **51.** 4 **53.** −3 **55.** 25 **57.** −48 **59.** 44 **61.** 91 **63.** 3 **65.** −5 **67.** 17 **69.** 11 **71.** 8 **73.** 112 **75.** −1,707 **77.** −15 **79.** −200 **81.** −320 **83.** −9,000 **85.** −1,200 **87.** 19 **89.** 11 yd **91.** 60-cent gain **97.** 4 **99.** Add the lengths of all its sides. **101.** no

Study Set Section 2.7 (page 127)

1. solve **3.** x **5. a.** 3 **b.** (−3) **7. a.** multiplication by −2 **b.** addition of −6 **c.** mult. by −4, subtraction of 8 **d.** mult. by −5, addition of −6 **9.** simplify **11.** opposite **13. a.** subtraction of 3 **b.** addition of −6 **15.** −13, 7 **17.** 1, 1, −12, −4, −4, 3 **19.** $-10 \cdot x$ **21.** yes **23.** no **25.** −18 **27.** −14 **29.** 5 **31.** −1 **33.** −9 **35.** −14 **37.** −2 **39.** 0 **41.** −8 **43.** 5 **45.** 2 **47.** −1 **49.** 6 **51.** 0 **53.** 6 **55.** −52 **57.** −7 **59.** −4 **61.** −5 **63.** −2 **65.** 10 **67.** −6 **69.** −3 **71.** 3 **73.** −6 **75.** 54 **77.** 30 **79.** −14 **81.** −3 **83.** −2 **85.** −8 **87.** 15 **91.** 18 ft **93.** −51 yd **95.** 34 **97.** $5 **99.** 29 points **101.** zone −8 **105.** $5 \cdot 5 \cdot 5 \cdot 5 \cdot 5 \cdot 5$ **107.** 12 **109.** $\frac{16}{8}$

Key Concept (page 132)

1. −5 **3.** −30 **5.** +10 or 10 **7.** −205

9. −4 −3 −2 −1 0 1 2 3 4
Negatives Positives
11. $x < y$

13. Like signs: Add their absolute values and attach their common sign to the sum. Unlike signs: Subtract their absolute values, the smaller from the larger, and attach the sign of the number with the larger absolute value to that result. **15.** Divide their absolute values. Like signs: The quotient is positive. Unlike signs: The quotient is negative.

Chapter Review (page 134)

1. a. **b.**

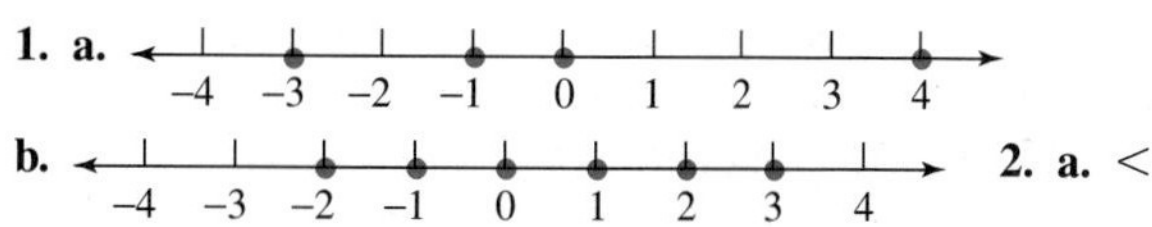

2. a. < **b.** < **c.** > **d.** < **3.** −33 ft **4. a.** −$1,200 **b.** −10 sec **5. a.** 4 **b.** 0 **c.** 43 **d.** −12 **6. a.** negative **b.** the opposite **c.** negative **d.** minus **7. a.** 12 **b.** −8 **c.** 8 **d.** 0 **8. a.** 2 **b.** −4

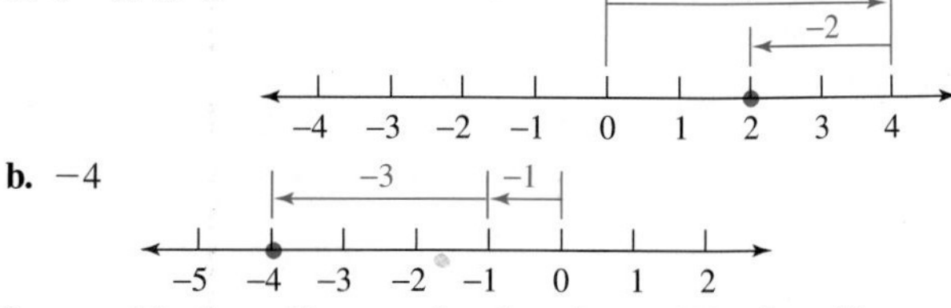

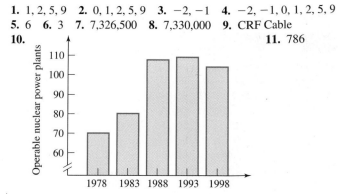

9. a. −10 **b.** −83 **c.** −8 **d.** −1 **e.** 112 **f.** −11 **g.** −3 **h.** −2 **10. a.** −4 **b.** −20 **c.** 0 **d.** 0 **11. a.** 11 **b.** −4 **12.** 65 ft **13. a.** −3 **b.** −21 **c.** 4 **d.** −112 **e.** −6 **f.** 6 **g.** −37 **h.** 30 **14.** adding, opposite **15. a.** −4 **b.** 15 **c.** 6 **d.** −8 **16.** −77 **17.** −225 ft **18.** −1 **19.** Alaska: 180°; Virginia: 140° **20. a.** −45 **b.** 18 **c.** −14 **d.** 376 **e.** −100 **f.** 1 **g.** −25 **h.** −150 **21. a.** −36 **b.** −36 **c.** 0 **d.** 1 **22.** −3, −6, −9 **23. a.** 25 **b.** −32 **c.** 64 **d.** −64 **24.** negative **25.** first expression: base of 2; second: base of −2; −4, 4 **26.** −3 **27. a.** −2 **b.** −5 **c.** −8 **d.** 101 **28. a.** 0 **b.** undefined **c.** 1 **d.** 10 **29.** −2 min **30. a.** −22 **b.** 4 **c.** −43 **d.** 8 **e.** 41 **f.** 0 **g.** −13 **h.** 32 **31. a.** 12 **b.** −16 **c.** −4 **d.** 1 **32. a.** −1 **b.** −4 **33. a.** −70 **b.** 20 **c.** −7,000 **d.** 1,100 **34. a.** yes **b.** no **35. a.** −10 **b.** 12 **c.** −8 **d.** 4 **36. a.** 15 **b.** −4 **37. a.** 3 **b.** −2 **c.** −12 **d.** 0 **38.** −46° **39.** 121 **40.** $8,200

Chapter 2 Test (page 139)

1. a. > **b.** < **c.** < **2.** {. . . , −3, −2, −1, 0, 1, 2, 3, . . .} **3.** Monroe **4.** −5

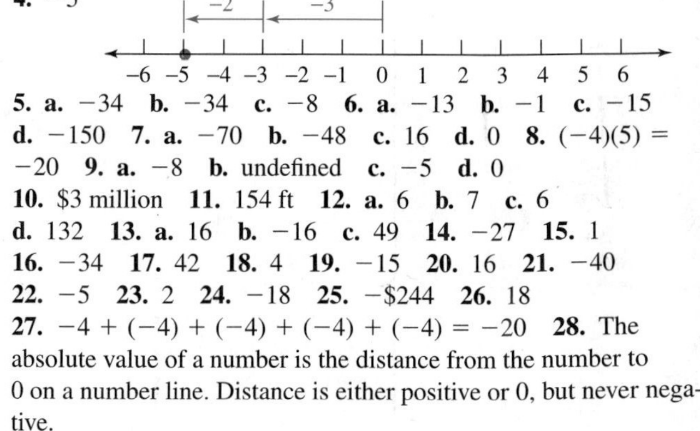

5. a. −34 **b.** −34 **c.** −8 **6. a.** −13 **b.** −1 **c.** −15 **d.** −150 **7. a.** −70 **b.** −48 **c.** 16 **d.** 0 **8.** (−4)(5) = −20 **9. a.** −8 **b.** undefined **c.** −5 **d.** 0 **10.** $3 million **11.** 154 ft **12. a.** 6 **b.** 7 **c.** 6 **d.** 132 **13. a.** 16 **b.** −16 **c.** 49 **14.** −27 **15.** 1 **16.** −34 **17.** 42 **18.** 4 **19.** −15 **20.** 16 **21.** −40 **22.** −5 **23.** 2 **24.** −18 **25.** −$244 **26.** 18 **27.** −4 + (−4) + (−4) + (−4) + (−4) = −20 **28.** The absolute value of a number is the distance from the number to 0 on a number line. Distance is either positive or 0, but never negative.

Cumulative Review Exercises (page 141)

1. 1, 2, 5, 9 **2.** 0, 1, 2, 5, 9 **3.** −2, −1 **4.** −2, −1, 0, 1, 2, 5, 9 **5.** 6 **6.** 3 **7.** 7,326,500 **8.** 7,330,000 **9.** CRF Cable

10.

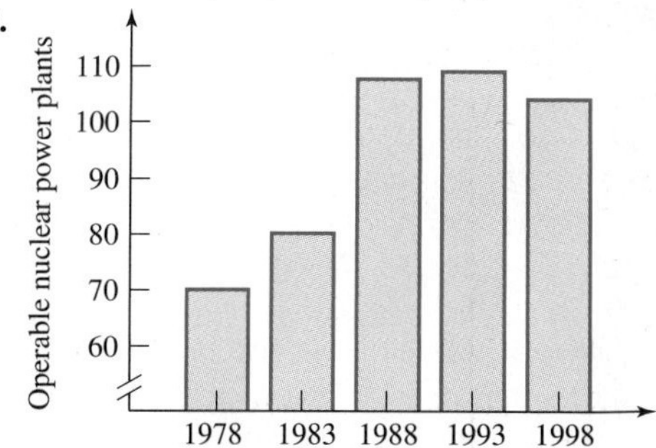

11. 786

12. 3,806 **13.** 4,684 **14.** 13,136 **15.** 104 ft, 595 ft^2 **16.** 65 **17.** 11,745 **18.** 13 **19.** 307,329 **20.** 467 **21.** 1,728 **22.** 1, 2, 3, 6, 9, 18 **23.** prime, odd **24.** composite, even **25.** even **26.** odd **27.** $2^3 \cdot 3^2 \cdot 7$ **28.** 11^4 **29.** 175 **30.** 38 **31.** 50 **32.** 2 **33.** no **34.** yes **35.** 13 **36.** 53 **37.** 27 **38.** 24 **39.** 
40. 
41. true **42.** 9, −9 **43.** −5 **44.** −14 **45.** −8 **46.** −231 **47.** 24 **48.** −1,715 **49.** 2 **50.** −50 **51.** 26 **52.** −16 **53.** −3 **54.** 4 **55.** 3 **56.** −18 **57.** $126,037 **58.** −279° F

Study Set Section 3.1 (page 149)

1. expression **3.** variable **5.** $10 + 3x$, $\frac{10-x}{3}$ (answers may vary) **7.** Mr. Lamb; 15 mi **9.** $2p$, $3p$ **11.** 500, $500 + x$, $500 - x$ **13.** $\frac{h}{4}$ **15.** $450 - x$ **17.** $8x$ **19.** $\frac{10}{g}$ **21.** $x - 9$ **23.** $\frac{2}{3}p$ **25.** $6 + r$ **27.** $d - 15$ **29.** $1 - s$ **31.** $2p$ **33.** $s + 14$ **35.** $\frac{35}{b}$ **37.** $x - 2$ **39.** c increased by 7 **41.** 7 less than c **43. a.** $60m$ **b.** $3{,}600h$ **45. a.** $\frac{s}{12}$ **b.** $\frac{s}{52}$ **47. a.** $12f$ in. **b.** $\frac{f}{3}$ yd **49.** $j - 5$ **51.** $6s$ **53.** $\frac{p}{15}$ days **55.** $t + 2$ **57.** w = width, $w + 6$ = length **59.** g = gal drained out; $6 - g$ = gal remaining **61.** $3x + 5$ **63.** $10a + 12$ **65.** x = votes received by Nixon; $x + 118{,}550$ = votes received by Kennedy **67.** c = number of copies of *I Want to Hold Your Hand* sold; $c - 2{,}000{,}000$ = number of copies of *Hey Jude* sold **73.** −10 **75.** −4 **77.** {. . . , −3, −2, −1, 0, 1, 2, 3, . . .} **79.** 2

Study Set Section 3.2 (page 160)

1. formula **3.** substitute **5.** 2 − 8 + 10; it looks like subtraction **7. a.** x = length part 1; $x - 40$ = length part 2; $x + 16$ = length part 3 (answers may vary) **b.** 20 in. and 76 in. **9. a.** 22, 27 **b.** $T = p + 2$ **11.** 48, $3t$, $3x$ **13. a.** health club instructor **b.** mechanic **c.** paleontologist **d.** realtor **e.** doctor **f.** economist **15. a.** $d = rt$ **b.** $C = \frac{5(F - 32)}{9}$ **c.** $d = 16t^2$ **17.** 17 **19.** 4 **21.** 40 **23.** −6 **25.** −6 **27.** 23 **29.** −8 **31.** 100 **33.** −28 **35.** 3 **37.** 44 **39.** −21 **41.** −3 **43.** −7 **45.** −18 **47.** 25 **49.** 21 **51.** −5 **53.** −29 **55.** −45 **57.** 21 **59.** 70 cents **61.** $8,200 **63.** $23 **65.** 300 mi **67.** −10° C **69.** 239 **71.** 64 ft **73.** 5,213, 5,079, 4,814; 2,053, 2,051, 1,921; 3,160, 3,028, 2,893 **75.** 30°, 15°, −5° **77.** 16, 16 ft; 64, 48 ft; 144, 80 ft; 256, 112 ft **79.** 4 **87.** 17, 37, 41 **89.** 7 **91.** division by 3 **93.** 3

Study Set Section 3.3 (page 168)

1. distributive, removed **3.** equivalent **5.** $x(y + z) = xy + xz$ **7.** $(w + 7)5$ **9.** 5, 6, 6, 2, 3 **11.** $-y - 9$ **13.** −5 **15.** −9, −9, $-45y$ **17. a.** x **b.** $x + 5$ **c.** $5x - 10y - 15$ **d.** $5x$ **19.** $12x$ **21.** $-30y$ **23.** $100t$ **25.** $12s$ **27.** $14c$ **29.** $-40h$ **31.** $-42xy$ **33.** $16rs$ **35.** $30xy$ **37.** $-30br$ **39.** $80c$ **41.** $-8e$ **43.** $4x + 4$ **45.** $16 - 4x$ **47.** $-6e - 6$ **49.** $-16q + 48$ **51.** $12 + 20s$ **53.** $42 + 24d$ **55.** $-25r + 30$ **57.** $-24 - 18d$ **59.** $9x - 21y + 6$ **61.** $9z + 9x + 15y$ **63.** $-x - 3$ **65.** $-4t - 5$ **67.** $3w + 4$ **69.** $-5x + 4y - 1$ **71.** $2(4x + 5)$ **73.** $(-4 - 3x)5$ **75.** $-3(4y - 2)$ **77.** $3(4 - 7t - 5s)$ **83.** 5 **85.** multiplication, division, subtraction, addition **87.** > **89.** carpeting, painting

Study Set Section 3.4 (page 176)

1. term **3.** perimeter **5.** distributive **7.** combining **9. a.** term **b.** factor **c.** factor **d.** factor **11. a.** 11 **b.** 8 **c.** −4 **d.** 1 **e.** −1 **f.** 102 **13.** 6, m; −75, t; 1, w; 4, bh **15.** It helps identify the like terms. **17.** $(2d + 15)$ mi **19.** To add the like terms, add 9 and 5 and keep the variable. **21.** 7 **23.** 2 **25. a.** the perimeter of a rectangle **b.** 2 times the length **c.** 2 times the width **27.** $3x^2$, $-5x$, 4 **29.** 5, $5t$, $-8t$, 4 **31.** 2 **33.** 5 **35.** $15t$ **37.** $4s$ **39.** x **41.** $4d$ **43.** $-4e$ **45.** cannot be simplified **47.** $-6z$ **49.** $-7x$ **51.** 0 **53.** 0 **55.** $4x$ **57.** cannot be simplified **59.** $-2y$ **61.** $3a$ **63.** $11t + 12$ **65.** $2w - 5$ **67.** $-7r + 11R$ **69.** $-50d$ **71.** $8x - 4y - 9$ **73.** $9x + 34$ **75.** $-22s + 23$ **77.** $19e - 21$ **79.** $2t + 8$ **81.** $3x + 8$ **83.** $10y - 32$ **85.** $288 **87.** 36 ft, 48 ft, 60 ft, 72 ft, 84 ft **93.** 2 **95.** $2^2 \cdot 5^2$ **97.** absolute value

Study Set Section 3.5 (page 182)

1. solve **3.** distributive **5.** combine **7.** When we substitute −5 for x, the result is a false statement: −10 = −9. **9.** $5k$ **11. a.** $4x$ **b.** $2x$ **13. a.** $2t - 8$ **b.** −4 **c.** −16 **15.** $2x$, 2, 2 **17.** 9, 45, 45, 45, $5x$, 5, 5 **19.** yes **21.** no **23.** 6 **25.** 3 **27.** −30 **29.** −28 **31.** 42 **33.** 37 **35.** 306 **37.** 735 **39.** 2 **41.** −14 **43.** −8 **45.** 5 **47.** −12 **49.** 4 **51.** 8 **53.** 10 **55.** 6 **57.** 0 **59.** −4 **61.** −10 **63.** 0 **65.** 1 **67.** 2 **69.** −11 **71.** 26 **73.** −3 **75.** 7 **77.** 3 **83.** −16 **85.** −387. 5 **89.** positive

Study Set Section 3.6 (page 188)

1. addition **3.** $5x$ **5.** $g - 100$ **7.** $3m$ **9.** $2w$ **11.** 30, 24, $5x$, $4(9 - x)$ **13. a.** 9 **b.** $9 - d$ **17.** 17 mo **19.** 11 yr **21.** $975 **23.** 61 **25.** 400 gal **27.** 21 mi **29.** 10 ft **31.** 6 min **33.** 6 pairs of dress shoes, 3 pairs of athletic shoes **35.** 14 hr, 6 hr **41.** associative property of addition **43.** −100 **45.** addition **47.** $2^3 \cdot 5^2$

Key Concept (page 193)

1. parentheses, innermost, outermost **3.** multiplications, divisions **5.** −15 **7.** 0 **9.** 206 **11.** −15 **13.** 92 ft **15.** $3x$ **17.** no **19.** Undo the subtraction first. Then undo the multiplication.

Chapter Review (page 195)

1. a. Brandon is closer by 250 mi. **b.** $h + 7$ **2 a.** $n - 5$ **b.** $7x$ **c.** $\frac{6}{p}$ **d.** $s + (-15)$ **e.** $2l$ **f.** $D - 100$ **g.** $r + 2$ **h.** $\frac{45}{x}$ **3. a.** $\frac{c}{6}$ **b.** $1{,}000 - x$ dollars **4. a.** x = hours wife drove; $2x$ = hours husband drove **b.** w = width; length = $w + 3$ **5. a.** $12x$ **b.** $\frac{d}{7}$ **6. a.** h = height of wall, upper base = $h - 5$, lower base = $2h - 3$ **b.** upper base = 5 ft, lower base = 17 ft **7. a.** 12 **b.** −8 **c.** 100 **d.** −4 **8.** 130, 114, $6x$, $55t$ **9.** $278 **10.** $15,230 **11. a.** 1998 **b.** 2000 **c.** They decreased. **12.** The pool is 2° C warmer. **13.** 144 ft **14.** 24 yr **15. a.** $-10x$ **b.** $42xy$ **c.** $60de$ **d.** $32s$ **e.** $2e$ **f.** $49xy$ **g.** $84k$ **h.** $100t$ **16. a.** $4y + 20$ **b.** $-30t - 45$ **c.** $-21 - 21x$ **d.** $-12e + 24x + 3$ **17. a.** $-6t + 4$ **b.** $-5 - x$ **c.** $-6t + 3s - 1$ **d.** $5a + 3$ **18. a.** $-4x$, 8 **b.** $-3y$, 1 **19. a.** factor **b.** term **c.** factor **d.** term **20. a.** yes **b.** no **c.** yes **d.** no **21. a.** $7x$ **b.** $-9t$ **c.** $-3z$ **d.** $5x$ **e.** $-12y$ **f.** $w - 5$ **g.** $-46d + 2a$ **b.** $10y + 15h - 1$ **22. a.** $13y + 48$ **b.** $-5t + 22$ **c.** $3x + 12$ **d.** $-50f + 84$ **e.** $-14m$ **f.** $14m$ **23.** 194 **24.** yes **25. a.** −18 **b.** 8

c. -3 **d.** 15 **e.** -3 **f.** 2 **g.** -3 **h.** 4 **26.** 10, 60; 25, 175; 1, x; 5, $5n$ **27.** $56 - c$ **28.** 6 hr **29.** 13 mi **30.** 2,200 **31.** 15 \$3 drinks, 35 \$4 drinks

Chapter 3 Test (page 201)

1. a. $r - 2$ **b.** $3xy$ **2.** $51{,}000 - e$ = yearly earnings of husband **3. a.** -3 **b.** 26 **4.** 165 mi **5.** \$37,000 **6.** It would be 56 ft short of hitting the ground. **7.** 1 **8.** 250 ft **9.** 15° C **10. a.** $25x + 5$ **b.** $-42 + 6x$ **c.** $-6y - 4$ **d.** $6a + 9b - 21$ **a.** factor **b.** term **12. a.** $-28y + 10$ **b.** $-3t$ **13. a.** $8x^2, -4x, -6$ **b.** 8 **14. a.** $11x$ **b.** $24ce$ **c.** $5x$ **d.** $30y$ **e.** $-7x$ **f.** $9y$ **15.** $-6y - 3$ **16.** -9 **17.** -3 **18.** 4 **19.** -10 **20. a.** $10k$¢ **b.** $20(p + 2)$ dollars **21.** 3 hr **22.** 8 hr **23.** terms with exactly the same variables and exponents **24.** t = length trout, $t + 10$ = length salmon; s = length salmon, $s - 10$ = length trout **25.** No; we simplify expressions, and we solve equations. **26.** $2(x + 3) = 2x + 6$ (answers may vary)

Cumulative Review Exercises (page 203)

1. 358,600,000 gal **2.** 50,000 **3.** 54,604 **4.** 4,209 **5.** 23,115 **6.** 87 **7.** $683 + 459 = 1{,}142$ **8.** 2011 **9.** $4 \cdot 5 = 5 + 5 + 5 + 5 = 20$ **10.** 10,912 in.^{2} **11. a.** 1, 2, 3, 6, 9, 18 **b.** $3^2 \cdot 2$ **12.** 2, 3, 5, 7, 11, 13, 17, 19, 23, 29 **13.** It has factors other than 1 and itself. For example, $27 = 3 \cdot 9$. **14.** 22 **15.** 500 **16.** 6

17.

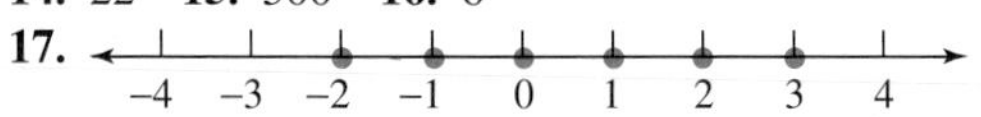

18. 5 **19.** false

20.

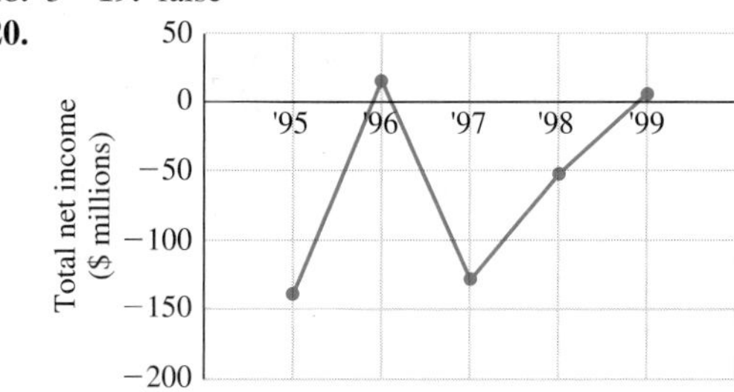

21. -20 **22.** 30 **23.** 125 **24.** 5 **25.** -5 **26.** 429 **27.** $-3^2 = -(3 \cdot 3) = -9; (-3)^2 = (-3)(-3) = 9$ **28.** 1,100°F **29.** 7 **30.** -10 **31.** $\frac{5}{0}; \frac{0}{5}$; division by 0 **32.** the commutative property of multiplication **33.** $h + 12$ **34.** $(26 - x)$ in. **35.** $x^2 = x \cdot x$; $2x = 2 \cdot x$ **36.** 36 **37.** 220 **38.** $10x - 35$ **39.** $24t$ **40.** 4, -2, 1, -1 **41.** $x + 3$; $3x$ (answers may vary) **42.** $25q$¢ **43.** $-b + 1$ **44.** -1 **45.** -2 **46.** 175 min

Study Set Section 4.1 (page 211)

1. numerator, denominator **3.** proper, improper **5.** equivalent **7.** higher, building **9. a.** 2 **b.** 3 **c.** 5 **d.** 7 **11.** equivalent fractions: $\frac{2}{6} = \frac{1}{3}$ **13. a.** In the first case, 20 and 28 were factored. In the second case, they were prime factored. **b.** yes **15.** The 2's in the numerator and denominator aren't common factors. **17. a.** $\frac{8}{1}$ **b.** $-\frac{25}{1}$ **c.** $\frac{x}{1}$ **d.** $\frac{7a}{1}$ **19.** 3, 2, 3, 2, 3, 2 **21.** $\frac{1}{3}$ **23.** $\frac{1}{3}$ **25.** $\frac{2}{3}$ **27.** $\frac{5}{2}$ **29.** $-\frac{1}{2}$ **31.** $-\frac{6}{7}$ **33.** $\frac{5}{9}$ **35.** $\frac{6}{7}$ **37.** in lowest terms **39.** $\frac{3}{8}$ **41.** $\frac{5x}{7}$ **43.** $\frac{4}{5}$ **45.** $\frac{6}{7}$ **47.** $\frac{7}{8}$ **49.** $-\frac{rs}{3}$ **51.** $\frac{3s}{5x}$ **53.** $\frac{5r}{4t}$ **55.** $\frac{2}{p^2}$ **57.** $\frac{35}{40}$ **59.** $\frac{28}{35}$ **61.** $\frac{45}{54}$ **63.** $\frac{15}{30}$ **65.** $\frac{4x}{14x}$ **67.** $\frac{54t}{60t}$ **69.** $\frac{25}{20s}$ **71.** $\frac{6y}{45y}$ **73.** $\frac{15}{5}$ **75.** $\frac{48}{8}$ **77.** $\frac{36a}{9}$ **79.** $-\frac{4t}{2}$ **81.** $\frac{3}{5}$ **83.** $-\frac{15}{16}$ in. **85.** $\frac{7}{10}, \frac{1}{8}$

87.

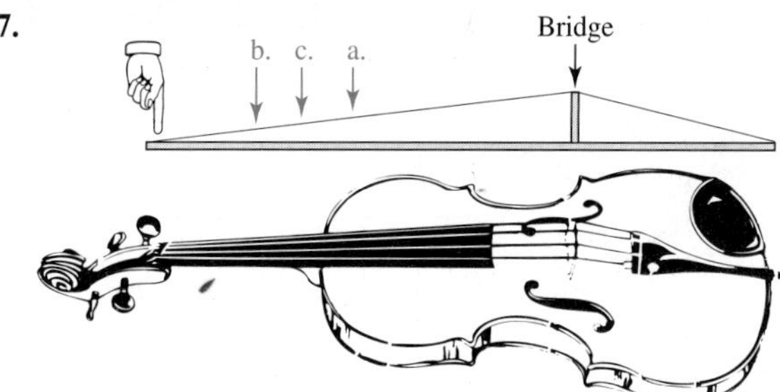

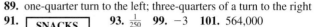

89. one-quarter turn to the left; three-quarters of a turn to the right

91.

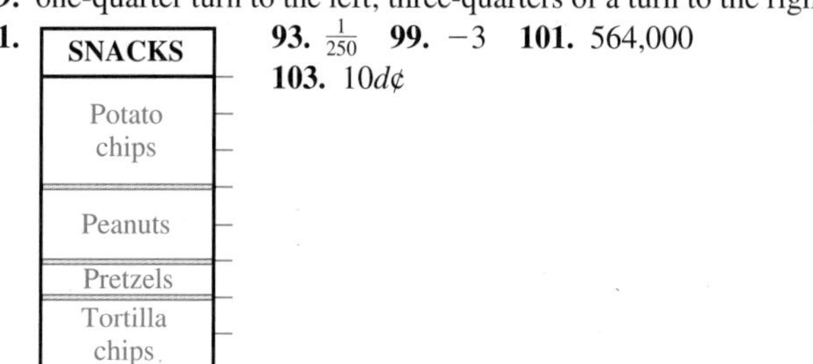

93. $\frac{1}{250}$ **99.** -3 **101.** 564,000 **103.** $10d$¢

Study Set Section 4.2 (page 221)

1. multiply **3.** product **5.** base, height **7.** $\frac{ac}{bd}$

9. **a.** $\frac{1}{4}$ **b.** 12, 1, $\frac{1}{12}$

11. a. negative **b.** positive **13. a.** true **b.** true **c.** false **d.** true **15.** 7, 15, 3, 3 **17.** $\frac{1}{8}$ **19.** $\frac{21}{128}$ **21.** $\frac{4}{7}$ **23.** $\frac{77}{60}$ **25.** $-\frac{1}{5}$ **27.** $\frac{2}{9}$ **29.** $\frac{2}{3}$ **31.** 1 **33.** $\frac{1}{20}$ **35.** $\frac{1}{30}$ **37.** 15 **39.** -12 **41.** $\frac{5}{72}$ **43.** $\frac{1}{40}$ **45.** d **47.** s **49.** $-c$ **51.** $\frac{2}{9}$ **53.** $2e^2$ **55.** $\frac{x^2}{12}$ **57.** $-\frac{10d^2}{c}$ **59.** $\frac{3}{4h}$ **61.** $\frac{5x}{6}, \frac{5}{6}x$ **63.** $-\frac{8v}{9}, -\frac{8}{9}v$ **65.** $\frac{4}{9}$ **67.** $\frac{25}{81}$ **69.** $\frac{16m^2}{9}$ **71.** $-\frac{27r^3}{64}$

73.

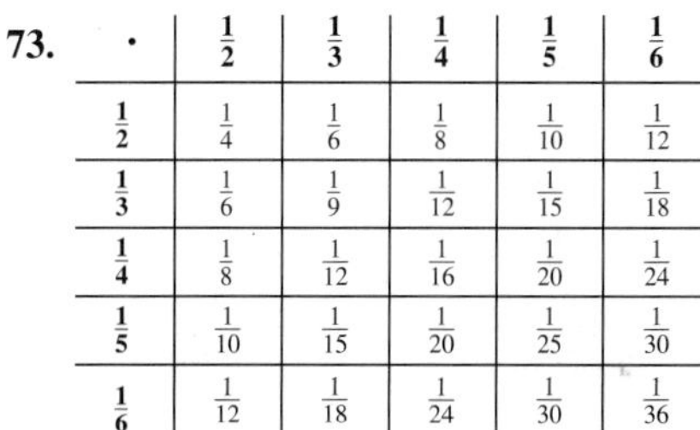

$\cdot$	$\frac{1}{2}$	$\frac{1}{3}$	$\frac{1}{4}$	$\frac{1}{5}$	$\frac{1}{6}$
$\frac{1}{2}$	$\frac{1}{4}$	$\frac{1}{6}$	$\frac{1}{8}$	$\frac{1}{10}$	$\frac{1}{12}$
$\frac{1}{3}$	$\frac{1}{6}$	$\frac{1}{9}$	$\frac{1}{12}$	$\frac{1}{15}$	$\frac{1}{18}$
$\frac{1}{4}$	$\frac{1}{8}$	$\frac{1}{12}$	$\frac{1}{16}$	$\frac{1}{20}$	$\frac{1}{24}$
$\frac{1}{5}$	$\frac{1}{10}$	$\frac{1}{15}$	$\frac{1}{20}$	$\frac{1}{25}$	$\frac{1}{30}$
$\frac{1}{6}$	$\frac{1}{12}$	$\frac{1}{18}$	$\frac{1}{24}$	$\frac{1}{30}$	$\frac{1}{36}$

75. 15 ft^2 **77.** $\frac{15}{2}$ yd^2 **79.** 290 **81.** 18, 6, and 2 in. **83.** $\frac{3}{8}$ cup sugar, $\frac{1}{6}$ cup molasses

85.

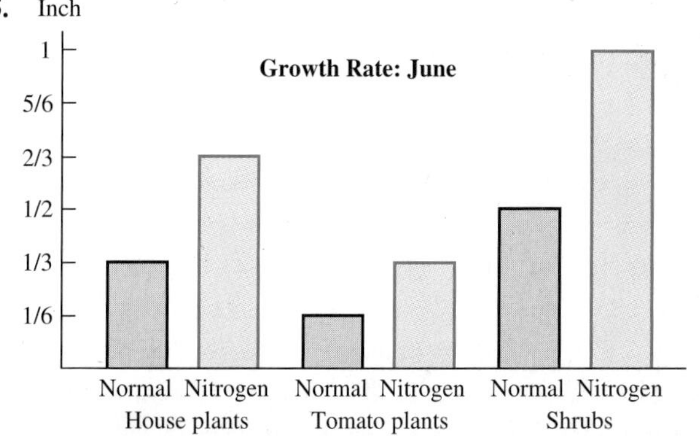

87. 121 in.^{2} **89.** 18 in.^{2} **95.** $2x + 14$ **97.** no **99.** 5^3

Study Set Section 4.3 (page 229)

1. reciprocals **3.** $\frac{1}{2}, \frac{3}{2}$ **5.** $4 \div \frac{1}{3}$, 12 **7.** 1 **9. a.** 5 **b.** 5 **c.** $\frac{1}{3}$ **11.** 9, 10, 9, 10, 5, 5, 9, 9, 5 **13.** $\frac{5}{6}$ **15.** $\frac{27}{16}$ **17.** 1 **19.** $\frac{2}{3}$ **21.** 36 **23.** 50 **25.** $\frac{2}{15}$ **27.** $\frac{1}{192}$ **29.** $-\frac{27}{8}$ **31.** $-\frac{15}{2}$ **33.** $-\frac{1}{64}$ **35.** 1 **37.** $\frac{8a}{15}$ **39.** $\frac{t}{6}$ **41.** $\frac{13}{8b}$ **43.** $-\frac{5}{8y}$ **45.** b **47.** 1 **49.** $-\frac{s}{9t}$ **51.** $-\frac{3x^3}{2}$ **53.** $\frac{18}{x^2}$ **55.** $-\frac{x}{y^2}$ **57.** $-6x^2$ **59.** 104 **61.** 56 **63.** route 1 **65. a.** sixteen parts **b.** $\frac{3}{4}$ in. **c.** $\frac{1}{120}$ in. **67.** 7,855 **73.** -4 **75.** $p = r - c$ **77.** false **79.** $-T + 5t$

Study Set Section 4.4 (page 238)

1. least **3.** higher **5.** denominators, numerators, common **7.** The denominators are unlike. **9. a.** 4 **b.** c **11. a.** once **b.** twice **c.** three times **13.** 60 **15. a.** $\frac{1}{3}$ **b.** $\frac{1}{4} = \frac{3}{12}, \frac{1}{3} = \frac{4}{12}$ **17.** 3, 3, 6, 5, 6, 5 **19.** 18 **21.** 24 **23.** 40 **25.** 60 **27.** $\frac{4}{7}$ **29.** $\frac{20}{103}$ **31.** $\frac{2}{5}$ **33.** $\frac{8}{d}$ **35.** $\frac{5}{8}$ **37.** $\frac{9}{20}$ **39.** $\frac{22}{15}$ **41.** $\frac{23}{56}$ **43.** $\frac{1}{12}$ **45.** $\frac{1}{12}$ **47.** $\frac{47}{50}$ **49.** $-\frac{3}{16}$ **51.** $-\frac{2}{3}$ **53.** $-\frac{23}{24}$ **55.** $-\frac{13}{5}$ **57.** $-\frac{23}{4}$ **59.** $\frac{49 - 8t}{56}$ **61.** $\frac{36 - 10b}{45}$ **63.** $\frac{4r - 7}{7r}$ **65.** $\frac{-5y + 9}{9y}$ **67.** $\frac{47}{60}$ **69.** $\frac{3}{4}$ **71.** $\frac{19}{48}$ **73.** $-\frac{43}{45}$ **75.** $\frac{26}{75}$ **77.** $\frac{17}{54}$ **79.** $\frac{5}{36}$ **81.** $-\frac{17}{60}$ **83. a.** $\frac{7}{32}$ in. **b.** $\frac{3}{32}$ in. **85.** $\frac{17}{24}$; no **87.** $\frac{1}{16}$ lb, undercharge **89.** $\frac{4}{5}, \frac{3}{4}, \frac{5}{8}$ **91.** $\frac{7}{10}$ **93.** $\frac{1}{6}$ hp **99.** $-x + 7$ **101.** $x - 5$ **103.** $P = 2l + 2w$

Study Set The LCM and the GCF (page 244)

1. 15 **3.** 56 **5.** 42 **7.** 18 **9.** 660 **11.** 600 **13.** 72 **15.** 378 **17.** 3 **19.** 11 **21.** 4 **23.** 25 **25.** 20 **27.** 12 **29.** 6 **31.** 9 **33.** 360 min (6 hr)

Study Set Section 4.5 (page 249)

1. mixed **3.** graph **5. a.** $-5\frac{1}{2}°$ **b.** $-1\frac{7}{8}$ in. **7. a.** $-2\frac{2}{3}$ **b.** $-3\frac{1}{3}$ **9.** $-\frac{4}{5}, -\frac{2}{5}, \frac{1}{5}$ **11.** $2\frac{1}{2}$ **13.** **15.** 8, 8, 4, 4, 6 **17.** $3\frac{3}{4}$ **19.** $5\frac{4}{5}$ **21.** $-3\frac{1}{3}$ **23.** $10\frac{7}{12}$ **25.** $\frac{13}{2}$ **27.** $\frac{104}{5}$ **29.** $-\frac{56}{9}$ **31.** $\frac{602}{3}$

33. $-2\frac{8}{9}$, $1\frac{2}{3}$, $\frac{16}{5}$ (−5 −4 −3 −2 −1 0 1 2 3 4 5)

35. $-\frac{10}{3}$, $-\frac{98}{99}$, $3\frac{1}{7}$ (−5 −4 −3 −2 −1 0 1 2 3 4 5)

37. $3\frac{4}{7}$ **39.** $10\frac{1}{2}$ **41.** 14 **43.** $-13\frac{3}{4}$ **45.** $-8\frac{1}{3}$ **47.** $\frac{35}{72}$ **49.** $-1\frac{1}{4}$ **51.** $\frac{25}{9}$ **53.** $-\frac{64}{27}$ **55.** $1\frac{9}{11}$ **57.** $-\frac{9}{10}$ **59.** 12 **61.** $\frac{5}{16}$ **63.** $-\frac{2}{3}$ **65.** $2\frac{1}{2}$ **67.** -2 **69.** 64 calories **71.** \$2.72 **73.** 675 **75.** $2\frac{3}{4}$ in., $1\frac{1}{4}$ in. **77.** $42\frac{5}{8}$ in.^{2} **79.** 602 **81.** size 14, slim cut **87.** 72 **89.** 4(8) **91.** division by 2 **93.** The variables are different.

Study Set Section 4.6 (page 257)

1. commutative **3.** borrow **5. a.** 76, $\frac{3}{4}$ **b.** $76 + \frac{3}{4}$ **7.** the fundamental property of fractions **9. a.** $10\frac{1}{16}$ **b.** $1{,}290\frac{1}{3}$ **c.** $17\frac{1}{2}$ **d.** $46\frac{1}{5}$ **11.** 70, 39, 70, 39, 7, 5, 7, 5, 35, 35, 31 **13.** $4\frac{2}{5}$ **15.** $5\frac{1}{7}$ **17.** $7\frac{1}{2}$ **19.** $5\frac{11}{30}$ **21.** $1\frac{1}{4}$ **23.** $1\frac{11}{24}$ **25.** $9\frac{3}{10}$ **27.** $3\frac{5}{14}$ **29.** $129\frac{11}{15}$ **31.** $397\frac{5}{12}$ **33.** $273\frac{2}{9}$ **35.** $623\frac{8}{21}$ **37.** $11\frac{1}{30}$ **39.** $101\frac{7}{16}$ **41.** $2\frac{1}{2}$ **43.** $26\frac{7}{24}$ **45.** $10\frac{7}{16}$ **47.** $320\frac{5}{18}$ **49.** $6\frac{1}{3}$ **51.** $\frac{1}{4}$ **53.** $3\frac{12}{35}$ **55.** $3\frac{5}{8}$ **57.** $4\frac{1}{3}$ **59.** $3\frac{7}{8}$ **61.** $53\frac{5}{12}$ **63.** $460\frac{1}{8}$ **65.** $-5\frac{1}{4}$ **67.** $-5\frac{7}{8}$ **69.** $2\frac{3}{4}$ mi **71.** $7\frac{2}{3}$ cups **73.** $48\frac{1}{2}$ ft **75. a.** $16\frac{1}{2}, 16\frac{1}{2}$; $5\frac{1}{5}, 5\frac{1}{5}$ **b.** $21\frac{7}{10}$ mi **77. a.** 20¢ **b.** 30¢ **79.** $191\frac{2}{3}$ ft **85.** 7 **87.** $5x + 22$ **89.** 6 **91.** the amount of surface a figure encloses

Study Set Section 4.7 (page 265)

1. complex **3.** $\frac{2}{3} \div \frac{1}{5}$ **5.** 15 **7.** negative **9.** 60 **11.** $\frac{3}{4}, \frac{4}{3}$, 4, 4 **13.** $\frac{1}{3}$ **15.** $\frac{31}{45}$ **17.** $\frac{37}{40}$ **19.** $\frac{3}{10}$ **21.** $-1\frac{27}{40}$ **23.** $\frac{3}{4}$ **25.** $-\frac{3}{64}$ **27.** $-1\frac{1}{6}$ **29.** $8\frac{1}{2}$ **31.** $\frac{49}{4}$ **33.** $\frac{121}{16}$ **35.** $-\frac{49}{75}$ **37.** $1\frac{11}{12}$ **39.** $8\frac{1}{4}$ in. **41.** $\frac{5}{6}$ **43.** $-1\frac{1}{3}$ **45.** $10\frac{1}{2}$ **47.** $\frac{4}{9}$ **49.** 3 **51.** 5 **53.** -20 **55.** 11 **57.** $\frac{3}{7}$ **59.** $-\frac{3}{8}$ **61.** $\frac{1}{16}$ **63.** $\frac{12}{13}$ **65.** $8\frac{1}{2}$ **67.** $2\frac{1}{2}, 1\frac{1}{2}, 3\frac{3}{4}$; $7\frac{1}{5}, 1\frac{1}{2}, 10\frac{4}{5}$; $14\frac{11}{20}$ mi **69.** yes **71.** $10\frac{1}{2}$ mi **73.** 6 sec **79.** $3d$ **81.** $2x(-x)$ **83.** 8 **85.** $27x^5$

Study Set Section 4.8 (page 275)

1. reciprocal **3.** least common denominator **5.** Yes; when 40 is substituted for x, the result is a true statement: $25 = 25$. **7.** 1 **9. a.** $\frac{4}{5}p$ **b.** $\frac{1}{4}t$ **11.** Multiply both sides by $\frac{3}{2}$; or multiply both sides by 3 and then divide both sides by 2. **13.** $\frac{8}{7}, \frac{8}{7}$ **15. a.** true **b.** false **c.** true **d.** true **17.** 28 **19.** -32 **21.** $-\frac{20}{3}$ **23.** $\frac{6}{5}$ **25.** 0 **27.** 30 **29.** $\frac{14}{5}$ **31.** $\frac{4}{25}$ **33.** $-\frac{1}{2}$ **35.** $\frac{2}{5}$ **37.** $-\frac{11}{4}$ **39.** $\frac{13}{2}$ **41.** $\frac{8}{9}$ **43.** $\frac{1}{3}$ **45.** $\frac{7}{18}$ **47.** $-\frac{5}{8}$ **49.** -1 **51.** $-\frac{5}{4}$ **53.** -36 **55.** $\frac{5}{12}$ **57.** $\frac{8}{9}$ **59.** $-\frac{27}{2}$ **61.** $\frac{10}{9}$ **63.** $-\frac{24}{7}$ **65.** 12 **67.** 24 **69.** -12 **71.** $\frac{75}{4}$ **75.** 20 **77.** 36 **79.** 450 **81.** 8 in. **83.** 360 min **89.** $ab + ac$ **91.** 5° C **93.** 5 **95.** 13,000,000

Key Concept (page 279)

1. 5, 5, 5, 5, 5, 5, 5 **3.** 7, 7, 7, 7, 7

Chapter Review (page 281)

1. $\frac{7}{24}$ **2.** The figure is not divided into equal parts. **3.** $-\frac{2}{3}, \frac{-2}{3}$ **4.** equivalent fractions: $\frac{6}{8} = \frac{3}{4}$ **5.** The numerator and denominator of the fraction are being divided by 2. **6.** The numerator and denominator of the fraction are being divided by 2. The answer to each division is 1. **7. a.** $\frac{1}{3}$ **b.** $\frac{5}{12}$ **c.** $-\frac{3x}{4}$ **d.** $\frac{11}{18m}$ **8.** The numerator and denominator of the original fraction are being multiplied by 2 to obtain an equivalent fraction in higher terms. **9. a.** $\frac{12}{18}$ **b.** $-\frac{6}{16}$ **c.** $\frac{21a}{45a}$ **d.** $\frac{36}{9}$ **10. a.** $\frac{1}{6}$ **b.** $-\frac{14}{45}$ **c.** $\frac{5}{12}$ **d.** $\frac{1}{5}$ **e.** $\frac{21}{5}$ **f.** $\frac{9}{4}$ **g.** 1 **h.** 1 **11. a.** true **b.** false **12. a.** $\frac{2}{9}$ **b.** $-\frac{8}{21}s$ **c.** $\frac{d}{21}$ **d.** $-\frac{5m}{9n}$ **13. a.** $\frac{9}{16}$ **b.** $-\frac{125}{8}$ **c.** $\frac{x^2}{9}$ **d.** $-\frac{8c^3}{125}$ **14.** 30 lb **15.** 60 in.^{2} **16. a.** 8 **b.** $-\frac{12}{11}$ **c.** $\frac{1}{x}$ **d.** $\frac{c}{ab}$ **17. a.** $\frac{25}{66}$ **b.** $-\frac{7}{2}$ **c.** $\frac{3}{32}$ **d.** $\frac{5}{2}$ **e.** $\frac{t}{2}$ **f.** $\frac{8}{5}$ **g.** $\frac{a^2}{b^2}$ **h.** $-\frac{6}{x}$ **18.** 12 **19. a.** $\frac{5}{7}$ **b.** $-\frac{6}{5}$ **c.** $\frac{2}{x}$ **d.** $\frac{7 + t}{8}$ **20.** The denominators are not the same. **21.** 90 **22. a.** $\frac{5}{6}$ **b.** $\frac{1}{40}$ **c.** $-\frac{29}{24}$ **d.** $\frac{20}{7}$ **e.** $\frac{2x - 15}{50}$ **f.** $\frac{y + 21}{3y}$ **g.** $-\frac{23}{6}$ **h.** $\frac{47}{60}$ **23.** $\frac{7}{32}$ in. **24.** the second hour **25. a.** $2\frac{1}{6}$ **b.** $\frac{13}{6}$ **26. a.** $3\frac{1}{5}$ **b.** $-3\frac{11}{12}$ **c.** 1 **d.** $2\frac{1}{3}$ **27. a.** $\frac{75}{8}$ **b.** $-\frac{11}{5}$ **c.** $\frac{201}{2}$ **d.** $\frac{199}{100}$

28. $-2\frac{2}{3}$, $\frac{8}{9}$, $\frac{59}{24}$ (number line: −5 −4 −3 −2 −1 0 1 2 3 4 5) **29. a.** $-\frac{3}{10}$ **b.** $\frac{21}{22}$ **c.** 40 **d.** $-2\frac{1}{2}$ **30.** $48\frac{1}{8}$ in. **31. a.** $3\frac{23}{40}$ **b.** $6\frac{1}{6}$ **c.** $1\frac{1}{12}$ **d.** $1\frac{5}{16}$ **32.** $39\frac{11}{12}$ gal **33. a.** $182\frac{5}{18}$ **b.** $113\frac{3}{20}$ **c.** $31\frac{11}{24}$ **d.** $316\frac{3}{4}$ **34. a.** $20\frac{1}{2}$ **b.** $34\frac{3}{8}$ **35. a.** $\frac{8}{9}$ **b.** $\frac{19}{72}$ **36. a.** $-\frac{12}{17}$ **b.** $-\frac{2}{5}$ **37. a.** $\frac{97}{64}$ **b.** $-\frac{63}{32}$ **c.** 22 **d.** $\frac{14}{33}$ **38. a.** 24 **b.** 28 **c.** $-\frac{1}{3}$ **d.** $\frac{11}{2}$ **39. a.** $\frac{48}{5}$ **b.** $-\frac{18}{5}$ **c.** 16 **d.** $\frac{26}{3}$ **40.** 330

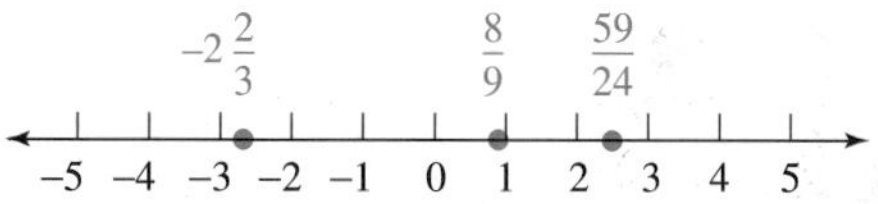

Chapter 4 Test (page 287)

1. a. $\frac{4}{5}$ **b.** $\frac{1}{5}$ **2. a.** $\frac{3}{4}$ **b.** $\frac{2n}{5}$ **3.** $-\frac{3}{20x}$ **4.** 40 **5.** $\frac{12}{a}$ **6.** $\frac{5x - 24}{30}$ **7.** $\frac{21a}{24a}$ **8.** $-1\frac{1}{7}$, $\frac{7}{6}$, $2\frac{4}{5}$ (number line: −2 −1 0 1 2 3) **9.** $\$1\frac{1}{2}$ million **10.** $\frac{2}{3}$ **11.** $261\frac{11}{36}$ **12.** $37\frac{5}{12}$ **13. a.** 0 lb **b.** $2\frac{3}{4}$ in. **c.** $3\frac{3}{4}$ in. **14.** $\frac{11}{7}$ **15.** $11\frac{3}{4}$ in. **16.** perimeter: $53\frac{1}{3}$ in.; area: $106\frac{2}{3}$ in.^{2} **17.** $\frac{13}{24}$ **18.** $-\frac{20}{21}$ **19.** $-\frac{5}{3}$ **20. a.** 42 **b.** $-\frac{36}{5}$ **21.** $\frac{1}{6}$ **22.** 8 **23.** 144 **24.** numerator, fraction bar, denominator; equal parts of a whole, or a division **25.** When we multiply a number, such as $\frac{3}{4}$, and its reciprocal, $\frac{4}{3}$, the result is 1. **26. a.** dividing the numerator and denominator of a fraction by the same number **b.** equivalent fractions: $\frac{1}{2} = \frac{2}{4}$ **c.** multiplying the numerator and denominator of a fraction by the same number

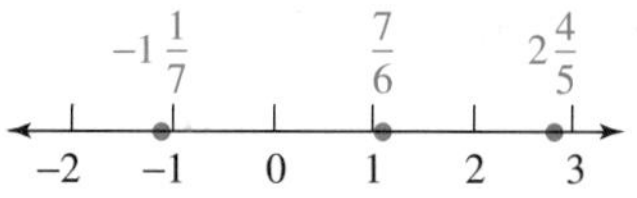

Cumulative Review Exercises (page 289)

1. 5,434,700 **2.** 5,430,000 **3.** 11,555, 10:30 A.M. **4.** hundred billions **5.** 8,136 **6.** 3,519 **7.** 299,320 **8.** 991 **9.** 450 ft **10.** 11,250 ft^2 **11.** $2^2 \cdot 3 \cdot 7$ **12.** $2 \cdot 3^2 \cdot 5^2$ **13.** $2^3 \cdot 3^2 \cdot 5$ **14.** $2^4 \cdot 3^2 \cdot 5^2$ **15.** 16 **16.** −35 **17.** 2 **18.** 2 **19.** $x + 15$ **20.** $x - 8$ **21.** $4x$ **22.** $\frac{x}{10}$ **23.** 7 **24.** 14 **25.** −52 **26.** −2 **27.** $-15x$ **28.** $28x^2$ **29.** $-6x + 8$ **30.** $-15x + 10y - 20$ **31.** $5x$ **32.** $7a^2$ **33.** $-x - y$ **34.** $-4x + 8$ **35.** −5 **36.** −5 **37.** −16 **38.** −5 **39.** 4 **40.** 4 **41.** 21 **42.** 21 ft by 84 ft **43.** $\frac{3}{4}$ **44.** $\frac{5x^3}{2y}$ **45.** $-\frac{4}{5}$ **46.** $\frac{1}{2p}$ **47.** $1\frac{5}{12}$ **48.** $\frac{20 - 3m}{5m}$ **49.** $\frac{23}{6}$ **50.** $-\frac{53}{8}$ **51.** $9\frac{11}{12}$ **52.** $5\frac{11}{15}$ **53.** $\frac{11}{16}$ in. **54.** 90 sec **55.** $\frac{2}{7}$ **56.** $-1\frac{9}{29}$ **57.** $-\frac{17}{15}$ **58.** 4 **59.** −15 **60.** $\frac{8}{3}$ **61.** An expression is a combination of numbers and/or variables with operation symbols. An equation contains an = sign. **62.** a letter that is used to stand for a number

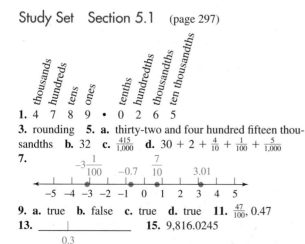

Study Set Section 5.1 (page 297)

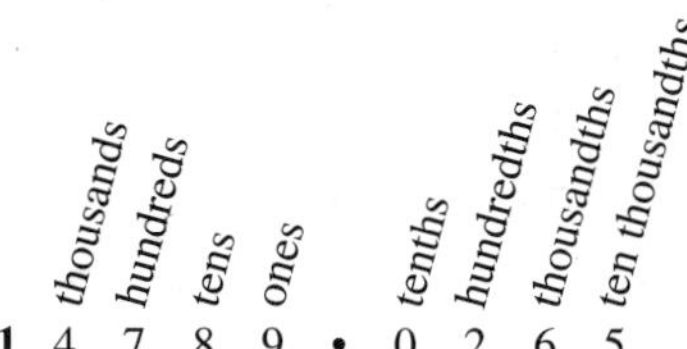

1. 4 7 8 9 • 0 2 6 5 **3.** rounding **5. a.** thirty-two and four hundred fifteen thousandths **b.** 32 **c.** $\frac{415}{1{,}000}$ **d.** $30 + 2 + \frac{4}{10} + \frac{1}{100} + \frac{5}{1{,}000}$ **7.** $-3\frac{1}{100}$, −0.7, $\frac{7}{10}$, 3.01 (number line: −5 −4 −3 −2 −1 0 1 2 3 4 5) **9. a.** true **b.** false **c.** true **d.** true **11.** $\frac{47}{100}$, 0.47 **13.** 0.3 **15.** 9,816.0245 **17.** fifty and one tenth; $50\frac{1}{10}$ **19.** negative one hundred thirty-seven ten thousandths; $-\frac{137}{10{,}000}$ **21.** three hundred four and three ten-thousandths; $304\frac{3}{10{,}000}$ **23.** negative seventy-two and four hundred ninety-three thousandths; $-72\frac{493}{1{,}000}$ **25.** −0.39 **27.** 6.187 **29.** 506.1 **31.** 2.7 **33.** −0.14 **35.** 33.00 **37.** 3.142 **39.** 1.414 **41.** 39 **43.** 2,988 **45. a.** \$3,090 **b.** \$3,090.30 **47.** < **49.** > **51.** 132.64, 132.6401, 132.6499 **53.** \$1,025.78 **55.**

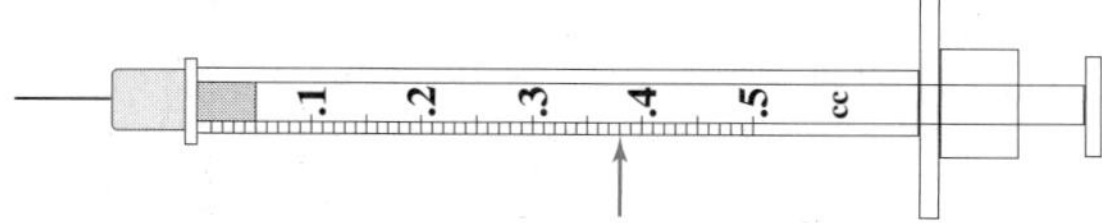

57. a. 0.30 **b.** 1,609.34 **c.** 453.59 **d.** 3.79 **59.** sand, silt, granule, clay **61.** Texas City, Houston, Westport, Galveston, White Plains, Crestline **63.** gold: Retton; silver: Szabo; bronze: Pauca **65.** −\$0.07, −\$0.30 **73.** $164\frac{11}{20}$ **75.** $8R - 18$ **77.** 72 in.^{2} **79.** −1

Study Set Section 5.2 (page 304)

1. sum **3.** point **5. a.** 0.47 **b.** $\frac{3}{10}, \frac{17}{100}, \frac{47}{100}$ **c.** 0.47 **d.** They are the same. **7.** 39.9 **9.** 54.72 **11.** 15.9 **13.** 0.23064 **15.** 288.46 **17.** 58.04 **19.** 9.53 **21.** 70.29 **23.** 4.977 **25.** 0.19 **27.** −10.9 **29.** 38.29 **31.** −14.3 **33.** −0.0355 **35.** −16.6 **37.** 47.91 **39.** 2.598 **41.** 11.01 **43.** 4.1 **45.** 35.85 **47.** −57.47 **49.** 6.2 **51.** 15.2 **53.** 8.03 **55. a.** 53.044 sec **b.** 102.38 **57.** 103.4 in. **59.** 1.8, Texas **61.** 1.74, 2.32, 4.06, 2.90, 0, 2.90 **63.** 43.99 sec **65.** \$765.69, \$740.69 **67. a.** \$101.94 **b.** \$55.80 **69.** 8,156.9343 **71.** 1,932.645 **73.** 2,529.0582 **79.** $110\frac{23}{40}$ **81.** $-\frac{5}{6}$

Study Set Section 5.3 (page 313)

1. factors, product **3.** whole, sum **5.** larger **7. a.** $\frac{21}{1{,}000}$ **b.** $\frac{21}{1{,}000} = 0.021$. They are the same. **9.** 0.08 **11.** −0.15 **13.** 0.98 **15.** 0.072 **17.** 12.32 **19.** −0.0049 **21.** −0.084 **23.** −8.6265 **25.** 9.6 **27.** −56.7 **29.** 12.24 **31.** −18.183 **33.** 0.024 **35.** −16.5 **37.** 42 **39.** 6,716.4 **41.** −0.56 **43.** 8,050 **45.** 980 **47.** −200 **49.** 0.01, 0.04, 0.09, 0.16, 0.25, 0.36, 0.49, 0.64, 0.81 **51.** 1.44 **53.** 1.69 **55.** −17.48 **57.** 14.24 **59.** 0.84 **61.** −3.872 **63.** 18.72 **65.** 86.49 **67.** 18.94 **69.** 7.6 **71.** 36.3 **73. a.** \$12.50, \$12,500, \$15.75, \$1,575 **b.** \$14,075 **75.** 0.75 in. **77.** 136.4 lb **79.** \$52.00, \$52.50, \$1.85x **81.** 160.6 m **83.** 0.000000136 in., 0.0000000136 in., 0.00000004 in. **85.** 15.29694 **87.** 631.2722 **89.** \$102.65 **95.** −12 **97.** the absolute value of negative three **99.** −1

Study Set Section 5.4 (page 321)

1. dividend, divisor, quotient **3.** mean **5.** mode **7.** whole, right, above **9.** true **11.** 10 **13.** Use multiplication to see whether $0.9 \cdot 2.13 = 1.917$. **15.** yes **17.** 3.3, 3.6, 2.3 **19.** moving the decimal points in the divisor and dividend two places to the right **21.** 4.5 **23.** −9.75 **25.** 6.2 **27.** 32.1 **29.** 2.46 **31.** −7.86 **33.** 2.66 **35.** 7.17 **37.** 130 **39.** 1,050 **41.** 0.6 **43.** 0.6 **45.** 5.3 **47.** −2.4 **49.** 13.60 **51.** 0.79 **53.** 0.07895 **55.** −0.00064 **57.** 0.0348 **59.** 4.504

61. −0.96 **63.** 1,027.19 **65.** 9.1 **67.** 304.07 **69.** 280 **71.** 11 hr later: 6 P.M. **73.** 567 **75.** 1988: $9.28; 1998: $12.77 **77.** 0.37 mi **79.** 22.525 oz, 25 oz **81.** $4.15, $4.19, $4.29 **83.** 7.24 **85.** −3.96 **91.** $\frac{7}{6}$ **93.** {. . . , −3, −2, −1, 0, 1, 2, 3, . . .} **95.** 12 **97.** 6

Study Set Estimation (page 326)

1. approx. $240 **3.** approx. 2 cubic feet less **5.** approx. 30 **7.** approx. $330 **9.** approx. $520 **11.** not reasonable **13.** reasonable **15.** reasonable **17.** not reasonable

Study Set Section 5.5 (page 332)

1. repeating **3.** decimal **5.** 7 ÷ 8 **7.** smaller
9.
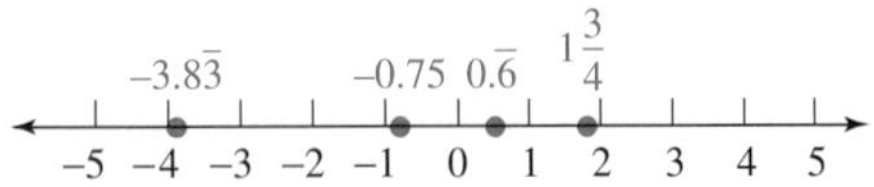

11. a. false **b.** true **c.** true **d.** false **13. a.** no **b.** It is a repeating decimal. **15.** 0.5 **17.** −0.625 **19.** 0.5625 **21.** −0.53125 **23.** $0.\overline{55}$ **25.** $0.\overline{775}$ **27.** −0.015 **29.** 0.002 **31.** $0.\overline{6}$ **33.** $0.\overline{45}$ **35.** $-0.58\overline{3}$ **37.** $0.0\overline{3}$ **39.** 0.23 **41.** 0.38 **43.** 0.152 **45.** 0.370 **47.** 1.33 **49.** −3.09 **51.** 3.75 **53.** −8.67 **55.** 12.6875 **57.** 203.73 **59.** < **61.** < **63.** $\frac{37}{90}$ **65.** $\frac{19}{60}$ **67.** $\frac{3}{22}$ **69.** $-\frac{1}{90}$ **71.** 0.57 **73.** 5.27 **75.** −2.55 **77.** 0.068 **79.** 7.11 **81.** −1.7 **83.** 113.04 **85.** $0.\overline{2277}$ **87.** 34.72 **89.** 0.0625, 0.375, 0.5625, 0.9375 **91.** $\frac{3}{40}$ in. **93.** 23.4 sec, 23.8 sec, 24.2 sec, 32.6 sec **95.** 93.6 in.^{2} **101.** −1 **103.** $-T - 8t$ **105.** $6x^2$

Study Set Section 5.6 (page 339)

1. solve **3.** coefficient **5.** 2.1(1.7) − 6.3 = −2.73 **7.** $7.8x + 9.1 + 12.4$ **9. a.** $0.25 **b.** $0.01 **c.** $2.50 **d.** $0.99 **11.** the distributive property **13.** 2.3, 2.3, 0.6s, 0.6, 0.6 **15.** $10.1x$ **17.** $0.02h$ **19.** $-3.7r$ **21.** $-8.78x + 12.3$ **23.** $-0.5x + 3.9$ **25.** $-0.01x + 5$ **27.** 1.7 **29.** 7.11 **31.** −11.5 **33.** −0.1 **35.** −4.36 **37.** 1.3 **39.** −8.16 **41.** 22.44 **43.** −21.18 **45.** 0.4 **47.** −2.2 **49.** −2 **51.** 31 **53.** 1 **55.** 0.3 **57.** 0.6 **59.** −1.1 **61.** 2 **63.** 1.3 **65.** 5 **69.** $8.6 million **71.** 3.27 **73.** 10.7 **75.** 12.4 mpg **77.** 200 **81.** $\frac{1}{12}$ **83.** $\frac{7}{8}$ **85.** −6

Study Set Section 5.7 (page 345)

1. root **3.** radical, positive **5.** radicand **7.** 25, 25 **9.** $(-7)^2$ **11.** $\frac{3}{4}$ **13.** $\sqrt{6}, \sqrt{11}, \sqrt{23}, \sqrt{27}$ **15. a.** 1 **b.** 0 **17. a.** 2.4 **b.** 5.76 **c.** 0.24
19.
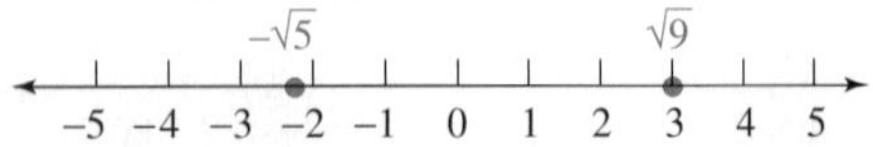

21. a. 4, 5 **b.** 9, 10 **23.** −7, 8 **25.** 4 **27.** −11 **29.** −0.7 **31.** 0.5 **33.** 0.3 **35.** $-\frac{1}{9}$ **37.** $-\frac{4}{3}$ **39.** $\frac{2}{5}$ **41.** 31 **43.** −20 **45.** $-\frac{7}{20}$ **47.** −70 **49.** 2.56 **51.** −3.6 **53.** 1, 1.414, 1.732, 2, 2.236, 2.449, 2.646, 2.828, 3, 3.162 **55.** 37 **57.** 61 **59.** 3.87 **61.** 8.12 **63.** 4.904 **65.** −3.332 **67.** 4,899 **69.** −0.0333 **71. a.** 5 ft **b.** 10 ft **73.** 127.3 ft **75.** 41-inch **83.** subtraction and multiplication **85.** 16 **87.** {0, 1, 2, 3, 4, 5, 6, . . .} **89.** 30

The Real Numbers (page 348)

1.
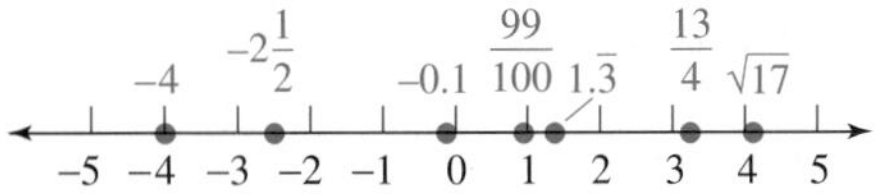

3. {1, 2, 3, 4, 5, . . .} **5.** {. . . , −3, −2, −1, 0, 1, 2, 3, . . .} **7.** nonterminating, nonrepeating decimals; a number that can't be written as a fraction **9.** false **11.** false **13.** true **15.** false **17.** true **19.** There is no real number that, when squared, yields a negative number.

Key Concept (page 349)

1. a. $5x - 8$ **b.** $x = 3$ **3. a.** $1.2y - 4.8$ **b.** $y = -1.5$

Chapter Review (page 351)

1. 0.67, $\frac{67}{100}$ **2.**
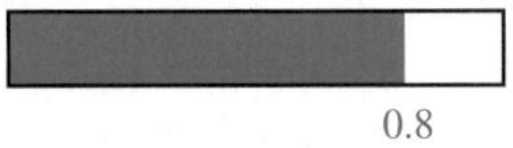

3. $10 + 6 + \frac{4}{10} + \frac{5}{100} + \frac{2}{1,000} + \frac{3}{10,000}$ **4. a.** two and three tenths, $2\frac{3}{10}$ **b.** negative fifteen and fifty-nine hundredths, $-15\frac{59}{100}$ **c.** six hundred one ten-thousandths, $\frac{601}{10,000}$ **d.** one one hundred thousandth, $\frac{1}{100,000}$
5.
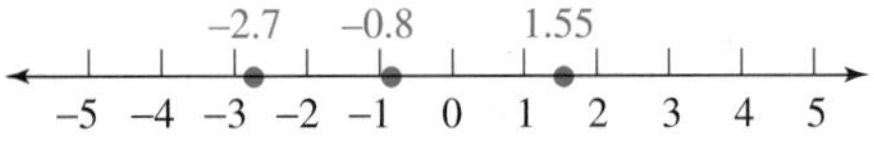

6. Washington, Diaz, Chou, Singh, Gerbac **7.** true **8. a.** < **b.** > **c.** = **d.** < **9. a.** 4.58 **b.** 3,706.090 **c.** −0.1 **d.** 88.1 **10. a.** 66.7 **b.** 45.188 **c.** 15.17 **d.** 27.71 **11. a.** −7.7 **b.** 3.1 **c.** −4.8 **d.** −29.09 **e.** −25.6 **f.** 4.939 **12.** $48.21 **13.** 8.15 in. **14. a.** −0.24 **b.** 2.07 **c.** −17.05 **d.** 197.945 **e.** 0.00006 **f.** 4.2 **15. a.** 90,145.2 **b.** 2,897 **16. a.** 0.04 **b.** 0.0225 **c.** 10.89 **d.** 0.001 **17. a.** −10.61 **b.** 25.82 **18.** 92.38 **19.** 68.62 in.^{2} **20.** 0.07 in. **21. a.** 1.25 **b.** −10.45 **c.** 1.29 **d.** 4.103 **22. a.** −2.9 **b.** 0.053 **c.** 63 **d.** 0.81 **23. a.** 12.9 **b.** −667.3 **24.** 20.22 **25.** $8.34 **26. a.** 0.8976 **b.** −0.00112 **27.** 13.95 **28.** 14 **29.** 9.5 **30.** 7.3 microns, 7.2 microns, 6.9 microns **31.** $1.45 billion **32. a.** 0.875 **b.** −0.4 **c.** 0.5625 **d.** 0.06 **33. a.** $0.\overline{54}$ **b.** $-0.\overline{6}$ **34. a.** 0.58 **b.** 1.03 **35. a.** > **b.** >
36.
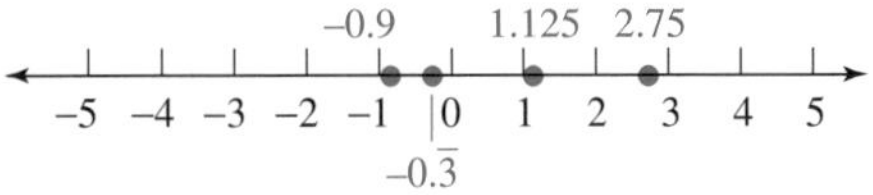

37. a. $\frac{11}{15}$ **b.** −6.24 **c.** 93 **d.** 39.564 **38.** 33.49 **39.** 34.88 in.^{2} **40. a.** −18.41 **b.** 4.77 **c.** −5.34 **d.** 1.7 **41.** yes **42. a.** $2.8a - 12.4$ **b.** $3t - 1.4$ **43. a.** −0.6 **b.** 12 **44.** 9 **45.** $8^2, (-8)^2$ **46. a.** 7 **b.** −4 **c.** 10 **d.** 0.3 **e.** $\frac{8}{5}$ **f.** 0.9 **g.** $-\frac{1}{6}$ **h.** 0 **47.** 9 and 10 **48.** It differs by 0.11.
49.
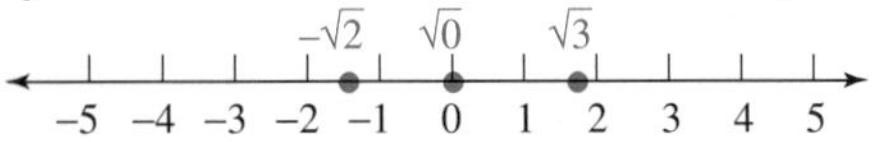

50. a. −30 **b.** 2.5 **c.** −27 **d.** 1.5 **51. a.** 4.36 **b.** 7.68

Chapter 5 Test (page 357)

1. $\frac{79}{100}$, 0.79 **2.** Selway, Monroe, Paston, Covington, Cadia **3.** $\frac{271}{1,000}$ **4.** 33.050 **5.** $208.75 **6. a.** 0.567909 **b.** 0.458 **7.** 1.02 in. **8. a.** 10.75 **b.** 6.121 **c.** 0.1024 **d.** 14.07

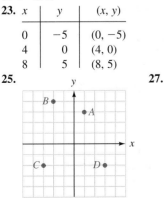

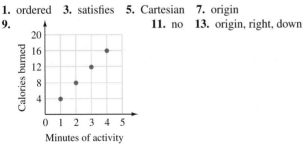

9. 1.25 mi^2 **10.** 0.004 in. **11.** 3.588 **12. a.** 0.34 **b.** $0.41\overline{6}$ **13.** −2.29 **14.** $1.\overline{18}$ **15.** 3.6, 3.6, 3.1 **16.** Half the families had more debt and half had less debt.
17.

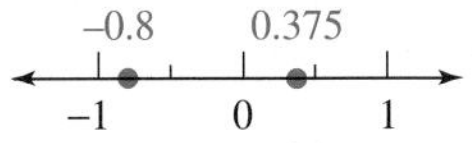

18. $\frac{41}{30}$ **19.** $4.96s + 2.3$
20. $5.2x - 18.7$ **21. a.** −7 **b.** 6.008 **22.** −0.425 **23.** 0.42 g **24.** 80
25.

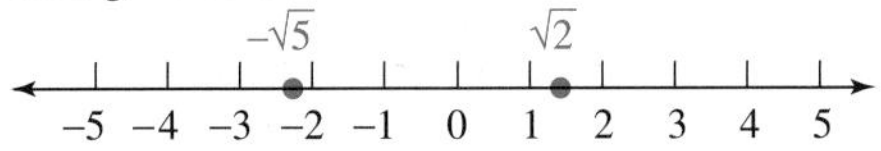

26. a. 11 **b.** $-\frac{1}{30}$ **27. a.** > **b.** > **c.** > **d.** >
28. a. −0.2 **b.** 1.3

Cumulative Review Exercises (page 359)

1. \$874,400 **2.** $(x + y) + z = x + (y + z)$ **3.** 27 R 42 **4.** 1,000 **5.** $11 \cdot 5 \cdot 2^2$ **6.** 1, 2, 4, 5, 10, 20 **7.** {0, 1, 2, 3, 4, 5, . . .} **8.** −13 **9.** adding **10.** 8, −3, 36, −6, 6 **11.** $-15 = -5 \cdot 3$ **12.** −1 **13.** 9 **14.** 30 **15.** 35 **16.** 102 **17.** $3x$ ft **18.** −\$1,100 **19. a.** $(k + 1)$ in. **b.** $(m - 1)$ in. **20.** $2(4x + 5)$ **21.** 3 **22.** $-5w$ **23.** 5 **24.** $\frac{6}{13}$ **25.** equivalent fractions **26.** $\frac{5x}{7}$ **27.** $\frac{21}{128}$ **28.** $-\frac{3y^2}{16}$ **29.** $\frac{28 + 2m}{7m}$ **30.** $19\frac{1}{8}$ **31.** $26\frac{7}{24}$ **32.** $-\frac{1}{3}$ **33.** −45 **34.** 8 **35.** 157.5 in.^{2}
36.

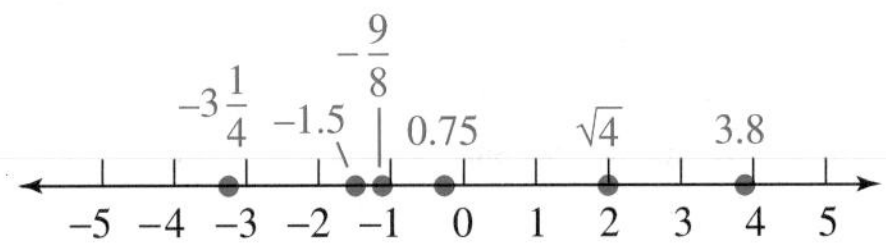

37. 0.001 in. **38.** < **39.** −8.136 **40.** 5.6 **41.** 5,601.2 **42.** 0.0000897 **43.** 47.95 **44.** 33.6 hr **45.** 232.8° C **46.** 3.02, 3.005, 2.75 **47.** $0.41\overline{6}$ **48.** −9 **49.** 80 **50.** −6

Study Set Section 6.1 (page 367)

1. ordered **3.** satisfies **5.** Cartesian **7.** origin
9.

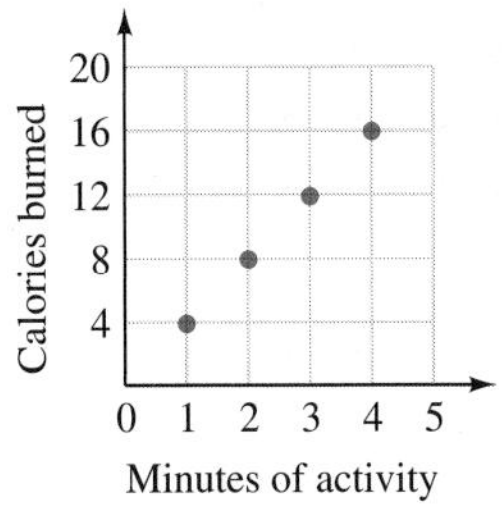

11. no **13.** origin, right, down
15. 2, 8, 6 **17.** yes **19. a.** 12 **b.** 4 **c.** 6 **21. a.** 8 **b.** 4 **c.** 3
23.

x	y	(x, y)
0	−5	(0, −5)
4	0	(4, 0)
8	5	(8, 5)

25.

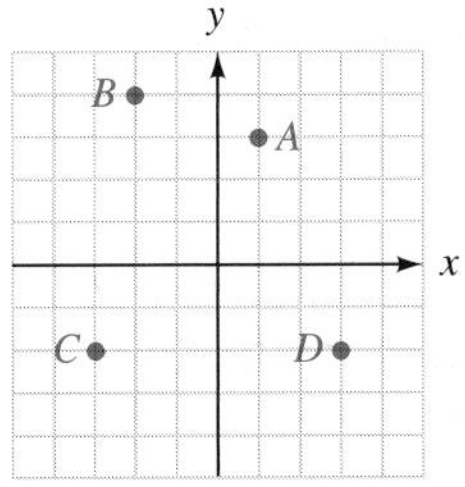

27.

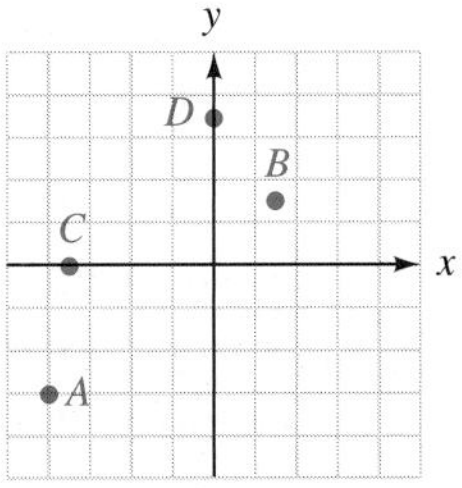

29. $A(2, 4)$, $B(-3, 3)$, $C(-2, -3)$, $D(4, -3)$ **31.** $A(-3, -4)$, $B(2.5, 3.5)$, $C(-2.5, 0)$, $D(2.5, 0)$ **33.** Rockford (5, B), Mount Carroll (1, C), Harvard (7, A), intersection (5, E) **35. a.** (2, −1) **b.** no **c.** yes **37.** New Delhi, Kampala, Coats Land, Reykjavik, Buenos Aires, Havana
39.

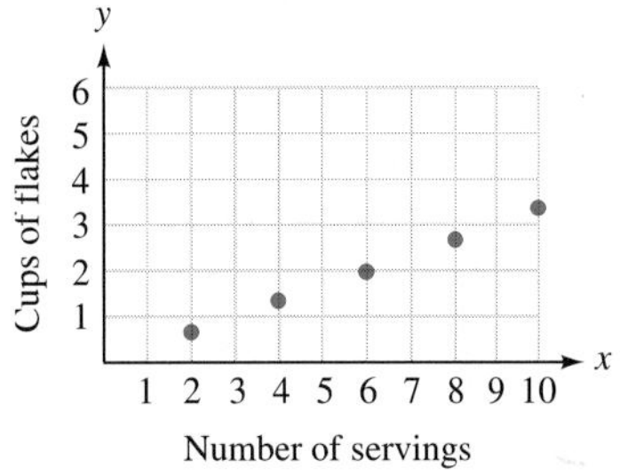

45. −16 **47.** 7 **49.** 21 **51.** −3 **53.** 65,536

Study Set Section 6.2 (page 377)

1. line **3.** x-intercept **5.** independent **7.** horizontal **9. a.** (0, 1) **b.** (−2, 0) **c.** yes **11.** Arrowheads were not drawn on both ends of the line. **13.** (−2, −2), (−1, −1), (0, 0), (1, 1), (2, 2), (3, 3) (answers may vary) **15.** 3, 6, 2 **17.** 0, (5, 0); −4, (−5, −4); 2, (10, 2) **19.** 3, (3, 3); −11, (−4, −11); 9, (6, 9) **21.** y-intercept is (0, 5); x-intercept is (5, 0) **23.** y-intercept is (0, 4); x-intercept is (5, 0) **25.** 5, 5, 3

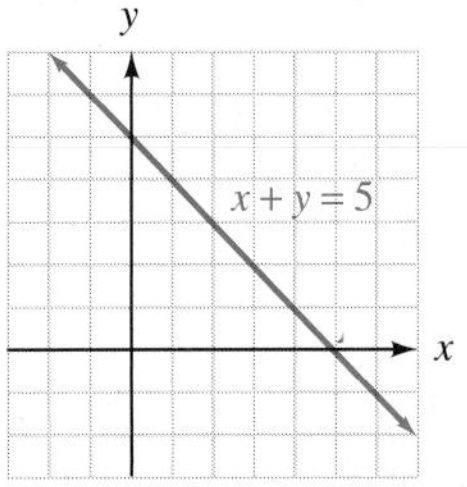

27. 4, 5, $\frac{16}{5}$

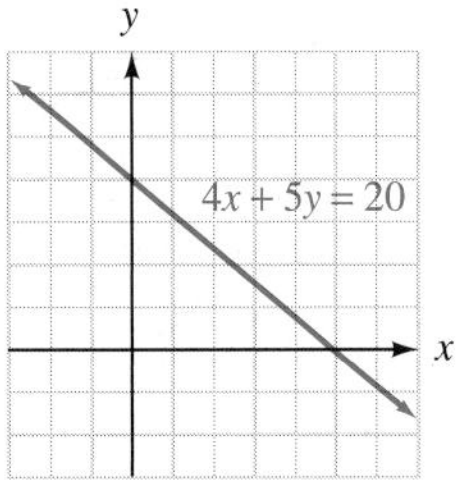

29. 2, −4, 4

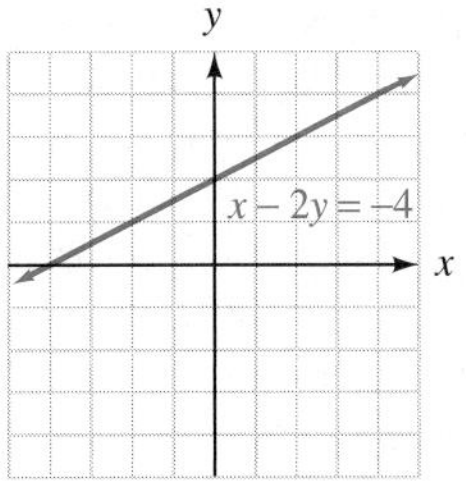

31.

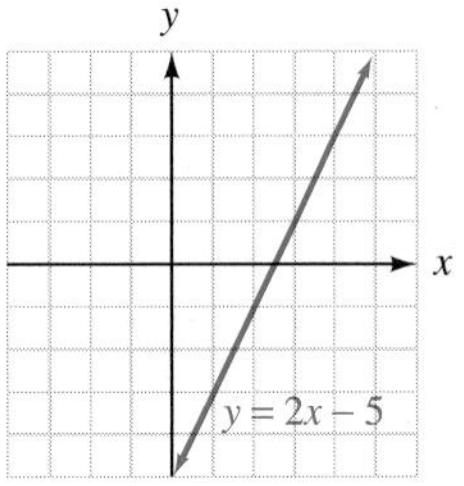

33.

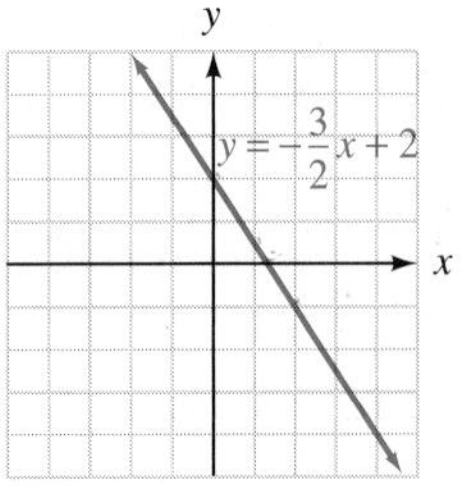

35.

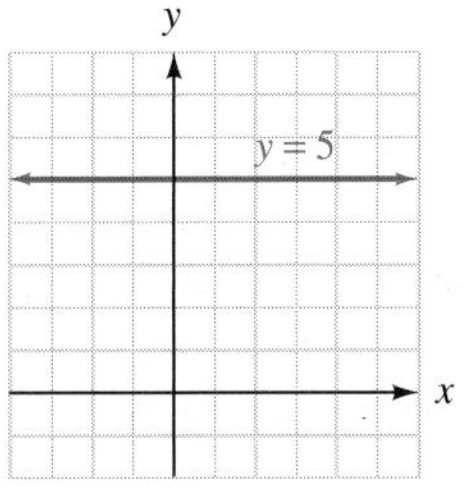

37.

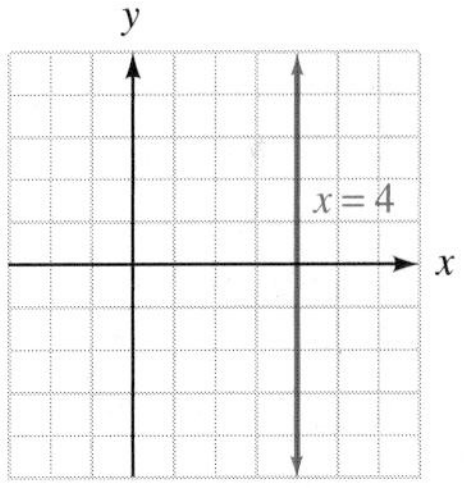

39.

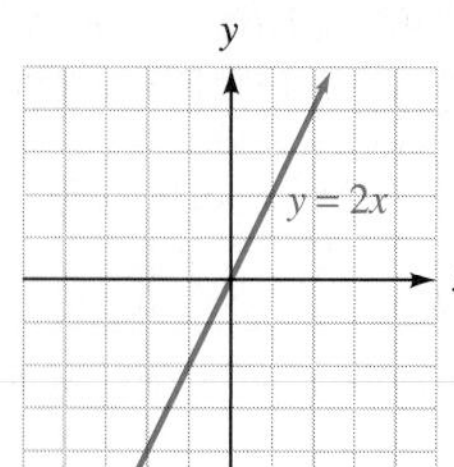

41.

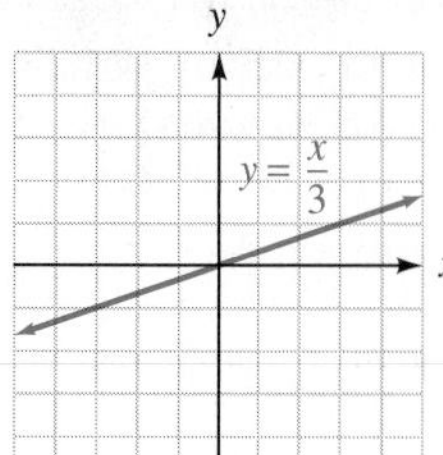

43.

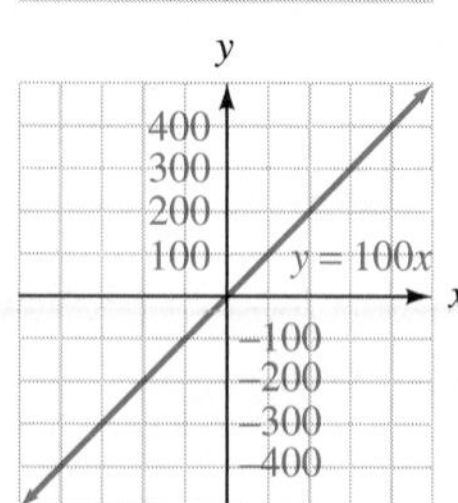

45.

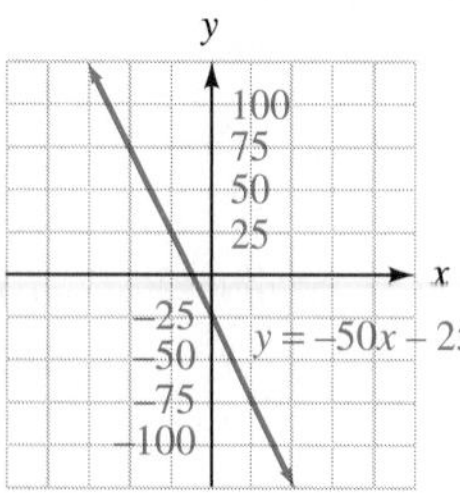

47. \$22.50

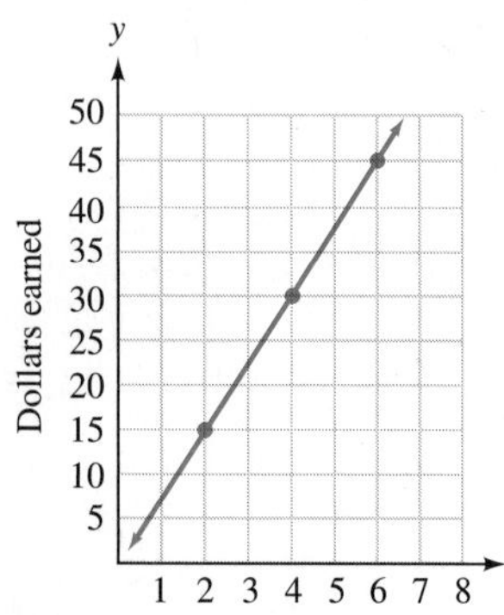

49. 2, 4, 6, 8, 10

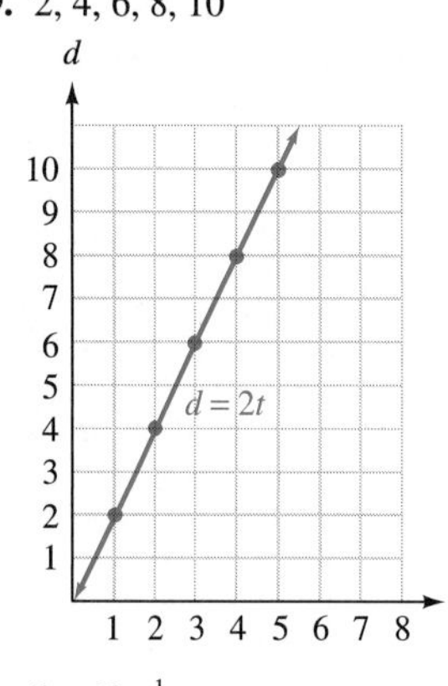

51. yes (2, 2) **59.** $2^2 \cdot 3^2 \cdot 5$ **61.** −5 **63.** $\frac{1}{4}$ sec

Study Set Section 6.3 (page 386)

1. base, exponent **3.** like **5.** product **7. a.** x^7 **b.** x^2y^3 **c.** $3^4a^2b^3$ **9.** $x^2 \cdot x^6 = x^8$ (answers may vary) **11.** $(c^5)^2 = c^{10}$ (answers may vary) **13. a.** x^{m+n} **b.** x^{mn} **c.** a^nx^n **15. a.** 2 **b.** −10 **c.** x **17. a.** x^2; $2x$ **b.** x^3; $x + x^2$ **c.** x^4; $2x^2$ **19. a.** $4x^2$; $5x$ **b.** $12x^2$; $7x$ **c.** $12x^3$; $4x^2 + 3x$ **21.** 27 **23.** 5, 7 **25.** $x^4, x^3, 4, 3$ **27.** x^5 **29.** x^{10} **31.** f^{13} **33.** n^{32} **35.** l^{10} **37.** x^{11} **39.** 2^{12} **41.** 5^8 **43.** $8x^3$ **45.** $5t^{10}$ **47.** $-24x^5$ **49.** $-x^4$ **51.** $36y^8$ **53.** $-40t^{10}$ **55.** x^3y^3 **57.** b^8c^8 **59.** x^5y^2 **61.** a^4b^4 **63.** x^5y^7 **65.** $18x^3y^4$ **67.** $16x^4y^2$ **69.** $-24f^6t^4$ **71.** a^4b^3 **73.** $12x^4y^3$ **75.** x^8 **77.** m^{500} **79.** $8a^3$ **81.** x^4y^4 **83.** $27s^6$ **85.** $4s^4t^6$ **87.** x^{14} **89.** c^{30} **91.** $36a^{14}$ **93.** $216a^{15}$ **95.** x^{60} **97.** $32b^{25}$ **103.** $\frac{3}{4}$ **105.** 5 **107.** 7 **109.** 12

Study Set Section 6.4 (page 392)

1. monomial **3.** binomial **5.** binomial **7.** monomial **9.** monomial **11.** trinomial **13.** 3 **15.** 2 **17.** 1 **19.** 7 **21.** 2, 2, 4, 4, 16 **23.** 13 **25.** 6 **27.** 31 **29.** 4 **31.** 1

33.

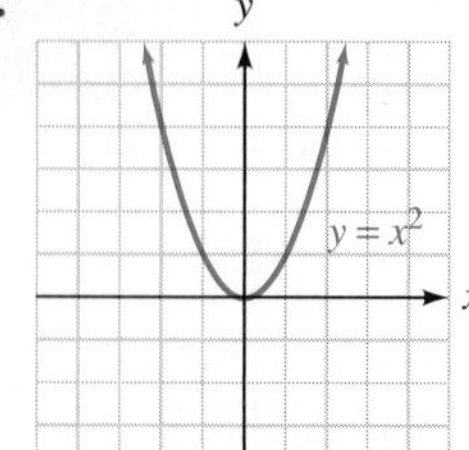

35. $y = \frac{1}{2}x^2$

37.

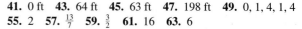

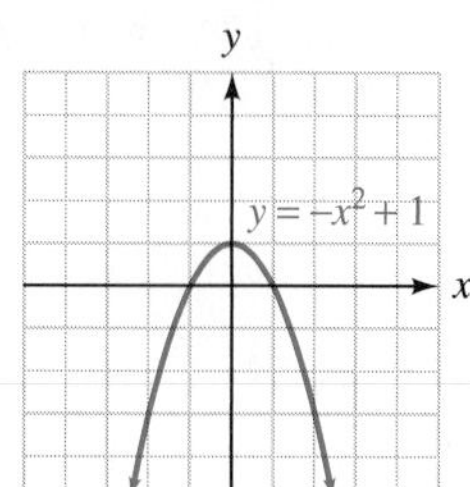

39.

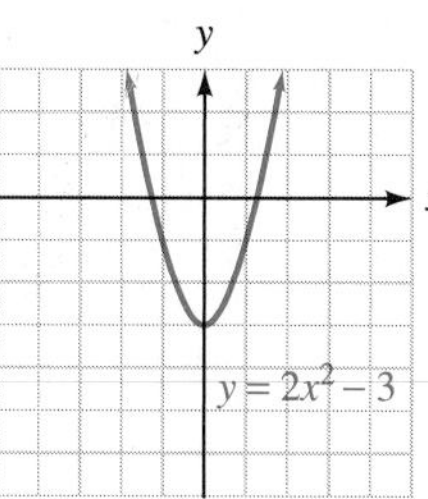

41. 0 ft **43.** 64 ft **45.** 63 ft **47.** 198 ft **49.** 0, 1, 4, 1, 4 **55.** 2 **57.** $\frac{13}{7}$ **59.** $\frac{3}{2}$ **61.** 16 **63.** 6

Study Set Section 6.5 (page 397)

1. like **3.** coefficients, variables **5.** yes, $7y$ **7.** no **9.** yes, $13x^3$ **11.** yes, $15x^2$ **13.** $2x^2, 7x, 5x^2$ **15.** $9y$ **17.** $-12t^2$ **19.** $14s^2$ **21.** $7x + 4$ **23.** $7x^2 - 7$ **25.** $12x^3 - 14.9x$ **27.** $8x^2 + 2x - 21$**29.** $8y^2 + 4y - 2$ **31.** $6x^2 + x - 5$ **33.** $2n^2 + 5$ **35.** $5x^2 + x + 11$ **37.** $-7x^2 - 5x - 1$ **39.** $2x^2 + x + 12.9$ **41.** $16u^3$ **43.** $7x^5$ **45.** $1.6a + 8$ **47.** $-19x^2 - 5$ **49.** $7x^2 - 2x - 5$ **51.** $1.7y^2 + 3.1y - 9$ **53.** $7b + 4$ **55.** $p^2 - 2p$ **57.** $5x^2 + 6x - 8$ **59.** $-12x^2 - 13x + 36$ **61.** $-x^3 + x + 14$ **63.** \$92,000 **65.** \$112,800 **67.** \$211,000 **69.** $y = -800x + 8{,}500$ **71.** $y = -1{,}900x + 18{,}700$ **77.** 0.8 oz **79.** 54 ft

Study Set Section 6.6 (page 402)

1. monomial **3.** 3 **5.** numerical, factors **7.** term, like **9.** $2x$, 5 **11.** $12x^5$ **13.** $-6b^3$ **15.** $-6x^5$ **17.** $-\frac{1}{2}y^7$ **19.** $3x + 12$ **21.** $-4t - 28$ **23.** $3x^2 - 6x$ **25.** $-6x^4 + 2x^3$ **27.** $6x^3 + 8x^2 - 14x$ **29.** $-2p^3 + 3p^2 - 2p$ **31.** $3q^4 - 6q^3 + 21q^2$ **33.** $a^2 + 9a + 20$ **35.** $3x^2 + 10x - 8$ **37.** $6a^2 + 2a - 20$ **39.** $4x^2 + 12x + 9$ **41.** $4x^2 - 12x + 9$ **43.** $25t^2 + 10t + 1$ **45.** $6x^3 - x^2 + 1$ **47.** $x^3 - 1$ **49.** $x^3 - x^2 - 5x + 2$ **51.** $4x^2 + 11x + 6$ **53.** $12x^2 + 14x - 10$ **55.** $x^3 + 1$ **57.** $(x^2 - 4)$ ft² **59.** $R = -\frac{x^2}{100} + 30x$ **65.** four and ninety-one thousandths **67.** 0.109375 **69.** 134.657 **71.** 10

Key Concept (page 405)

1–6.

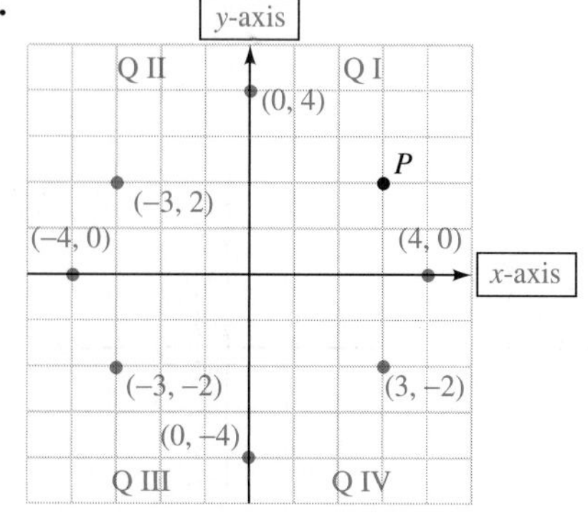

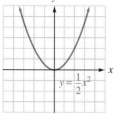

7.

x	y	(x, y)
0	-2	$(0, -2)$
4	0	$(4, 0)$
2	-1	$(2, -1)$

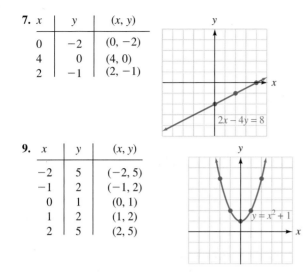

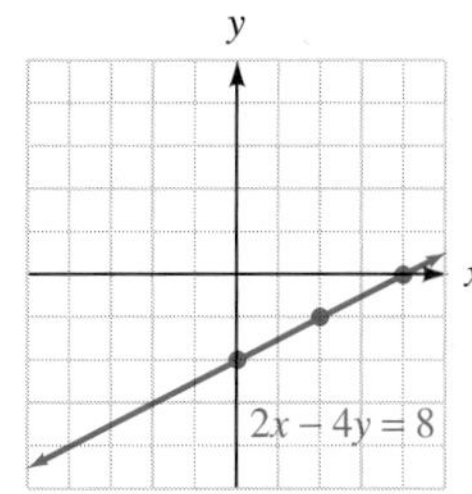

9.

x	y	(x, y)
-2	5	$(-2, 5)$
-1	2	$(-1, 2)$
0	1	$(0, 1)$
1	2	$(1, 2)$
2	5	$(2, 5)$

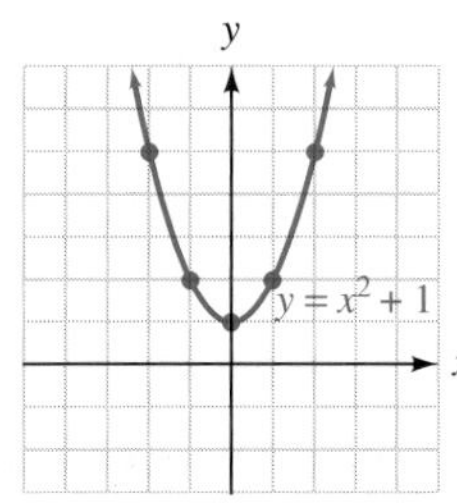

Chapter Review (page 407)

1. a. yes **b.** no **2. a.** $-3, -6$ **b.**

x	y	(x, y)
1	-5	$(1, -5)$
3	-11	$(3, -11)$
-2	4	$(-2, 4)$

3. a.

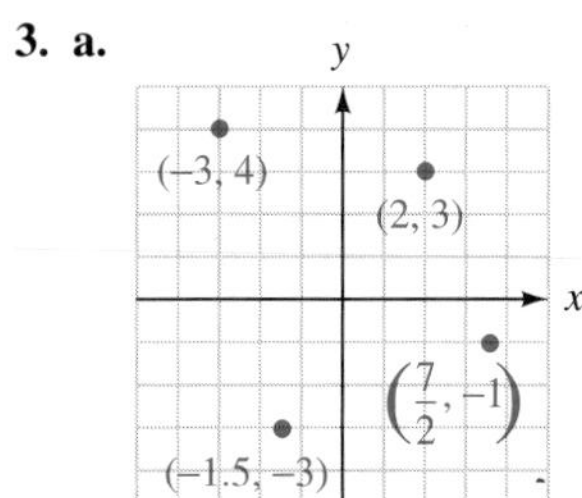

b. $A(4, 3), B(-3, 3), C(-4, 0), D(-1.5, -3.5), E(2.5, -1.5)$

4. III

6. a. $-5, \frac{5}{3}, -2$

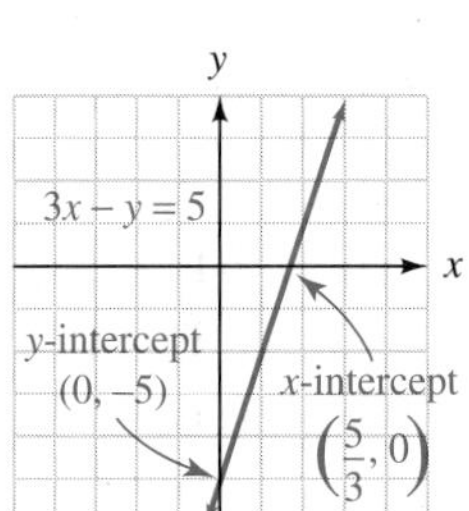

b.

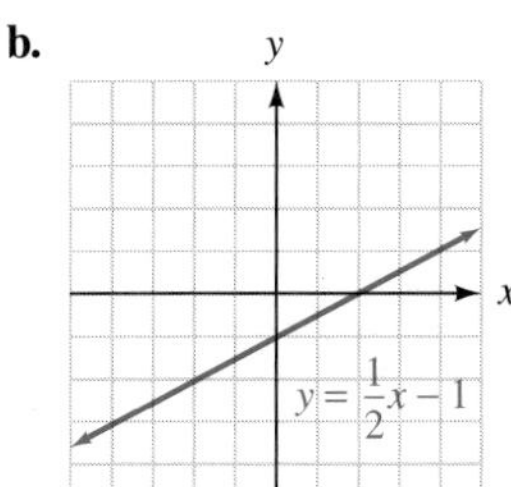

7. a.

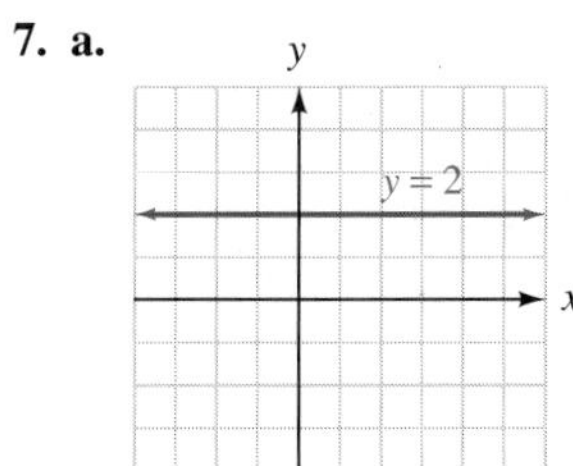

b.

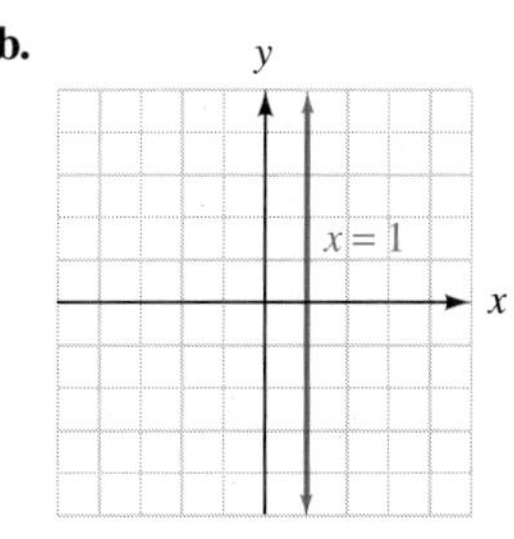

8. The line is made up of infinitely many points. Every point on the line is a solution of the equation, and every solution of the equation is on the line. **9. a.** $4h \cdot 4h \cdot 4h$ **b.** $5^2d^3m^4$ **10. a.** h^{10} **b.** t^8 **c.** w^7 **d.** 4^{12} **11. a.** $8b^7$ **b.** $-24x^4$ **c.** $24f^7$ **d.** $-a^2b^2$ **e.** x^2y^6 **f.** m^2n^2 **g.** $27m^3z^7$ **h.** $-20c^3d^6$ **12. a.** v^{12} **b.** $27y^3$ **c.** $25t^8$ **d.** $8a^{12}b^{15}$ **13. a.** c^{26} **b.** $108s^{12}$ **c.** c^{14} **d.** $8x^9$ **14. a.** trinomial **b.** monomial **c.** binomial **15. a.** 3 **b.** 4 **c.** 5 **16. a.** 7 **b.** 13

17. a. **b.**

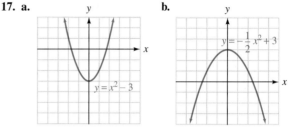

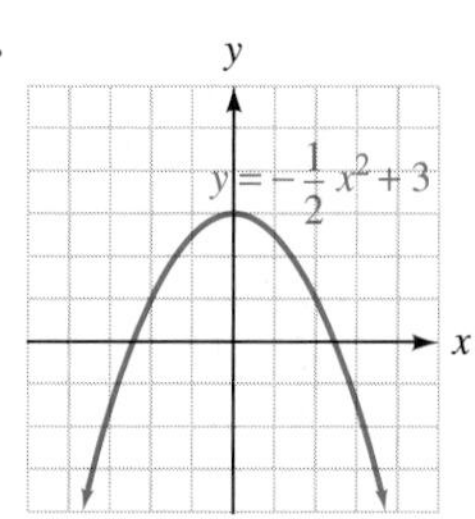

18. a. $5x^3$ **b.** $\frac{13}{2}p^2$ **19. a.** $9x + 4$ **b.** $2x^2 - 2x + 3$ **20. a.** $8x + 3$ **b.** $-2x^2 + x + 2$ **21. a.** $7p^3$ **b.** $-5y^2$ **22. a.** $1.1x - 8$ **b.** $z^2 - 4z + 6$ **23. a.** $2x - 7$ **b.** $8x^2 - 5x + 12$ **24. a.** $15x^5$ **b.** $-6z^4$ **25. a.** $6x^3 + 4x^2$ **b.** $-35t^5 + 30t^4 + 10t^3$ **26. a.** $6x^2 + x - 2$ **b.** $35t^2 - 2t - 24$ **27. a.** $15x^2 + 19x - 10$ **b.** $15x^2 - 5x - 10$ **28. a.** $6x^3 + x^2 + x + 2$ **b.** $6r^3 - 5r^2 - 12r + 9$ **29. a.** $15x^3 + 19x^2 - x + 15$ **b.** $15x^3 - 16x^2 - x + 2$

Chapter 6 Test (page 411)

1. yes **2.** no **3.** $-2, 4, -1$ **4.** $1, (0, 1); 2, (3, 2); 0, (-3, 0)$ **5.** $(30, 32), (30, 34), (31, 34), (38, 30)$

6.

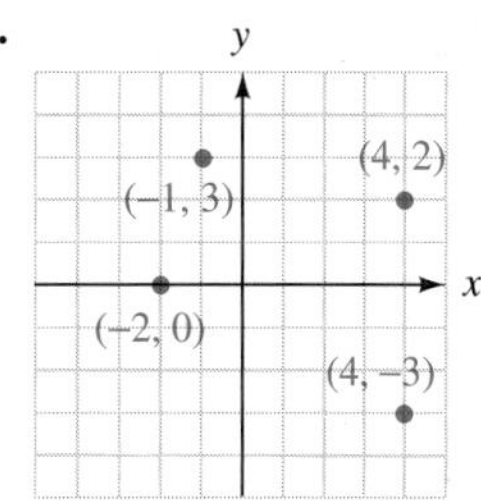

7. $A(0, 0), B(2.5, 3.5), C(-3, -2)$, and $D(0, -2)$

8.

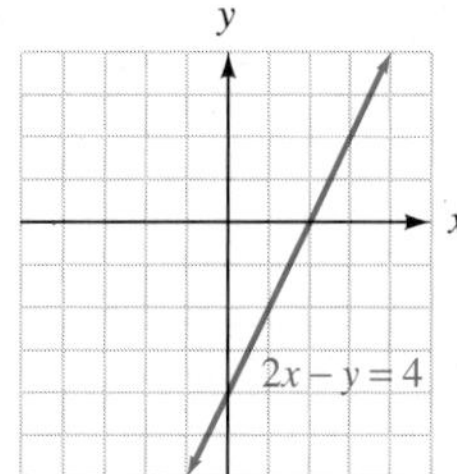

9. a. $(2, 0)$ **b.** $(0, -4)$

10.

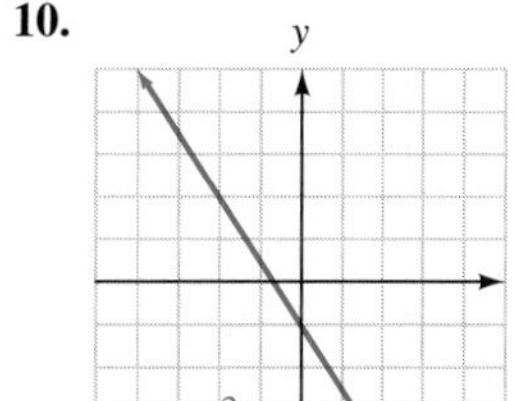

11.

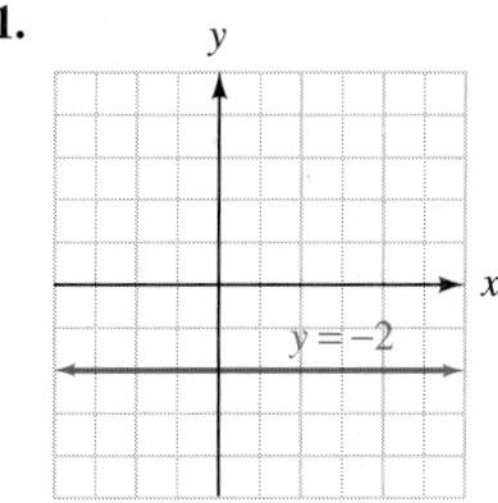

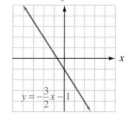

12.

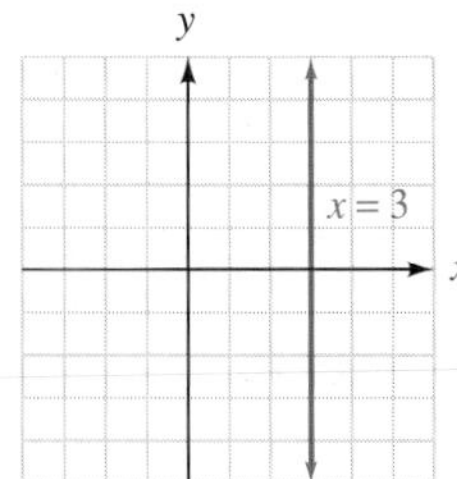

13. a. h^6 **b.** $-28x^5$ **c.** b^8 **d.** $24g^5k^{13}$ **14. a.** f^{15} **b.** $4a^4b^2$ **c.** x^{15} **d.** x^{15} **15.** binomial **16.** trinomial **17.** 6 **18.** 7 **19.** 25 **20.** 2

21.

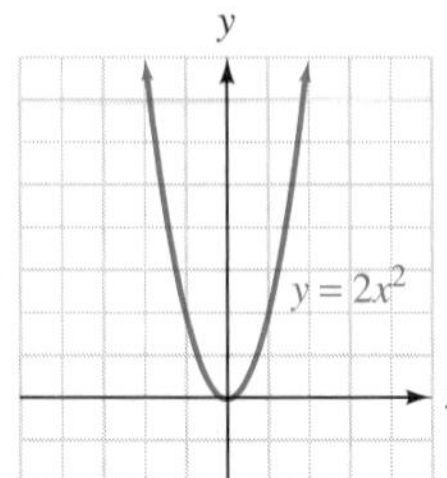

22.

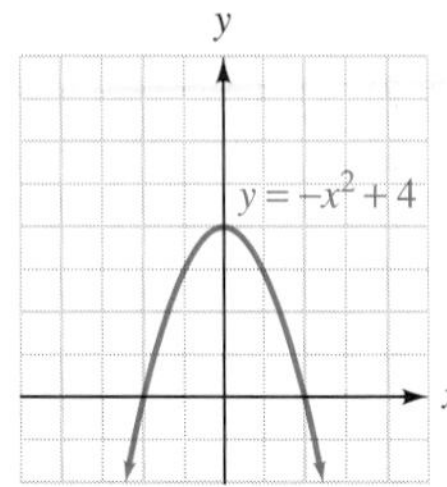

23. $5x^2 - 3x + 4$ **24.** $7x^2 + 2x - 2$ **25.** $-1.2p^2 + 3p$ **26.** $8d^2 - 9d + 12.5$ **27.** $-8x^5$ **28.** $3y^4 - 6y^3 + 9y^2$ **29.** $6x^2 - 7x - 20$ **30.** $2x^3 - 7x^2 + 14x - 12$ **31.** No; $(1, -2)$ lies in quadrant IV, and $(-2, 1)$ lies in quadrant II. **32.** There is an infinite number of ordered pairs that satisfy this equation.

Cumulative Review Exercises (page 413)

1. 6,246,000 **2.** 6,000,000 **3.** 22 m **4.** 52 in. **5.** 180, 200, 250, 136 **6.** $\frac{80}{45}, \frac{120}{80}, \frac{140}{85}$ **7.** $2^3 \cdot 3 \cdot 5$ **8.** $3 \cdot 5^2 \cdot 7$ **9.** Visibility is 39 ft less. **10.** -10 **11.** -2 **12.** -9 **13.** $-6x$ **14.** $x + 10y$**15.** 2 **16.** 4 **17.** -18 **18.** -2 **19.** $\frac{1}{b}$ **20.** $-\frac{17}{18}$ **21.** $20\frac{5}{18}$ **22.** $\frac{25 - mn}{5m}$ **23.** -32 **24.** 10 **25.** 34 **26.** To divide a number by a fraction, we multiply the number by the reciprocal of the fraction. **27.** 57.57 **28.** 351.053 **29.** 107.26 **30.** 2.1303 **31.** 4.67 **32.** $0.\overline{35}$ **33.** 5.6 **34.** -4.8 **35.** 14.6 **36.** 120 **37.** 3.3 **38.** -73.5 **39.** 8.2, 8.0, 7.9 **40.** 41,811.4 **41.** 800 **42.** 1,600 **43.** 11 **44.** $\frac{9}{2}$ **45.** 0.5 **46.** 29 **47.** yes

48.

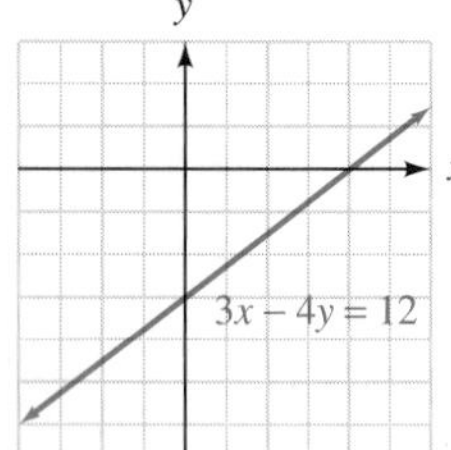

49. p^9 **50.** $-10q^8$ **51.** p^6q^6 **52.** $-27a^{15}$ **53.** $x^2 - 6x + 3$ **54.** $6x^2 + 7x - 3$

Study Set Section 7.1 (page 423)

1. percent **3.** 100 **5.** right **7. a.** 0.84, 84%, $\frac{21}{25}$ **b.** 16% **9.** $\frac{17}{100}$ **11.** $\frac{1}{20}$ **13.** $\frac{3}{5}$ **15.** $\frac{5}{4}$ **17.** $\frac{1}{150}$ **19.** $\frac{21}{400}$ **21.** $\frac{3}{500}$ **23.** $\frac{19}{1,000}$ **25.** 0.19 **27.** 0.06 **29.** 0.408 **31.** 2.5 **33.** 0.0079 **35.** 0.0025 **37.** 93% **39.** 61.2% **41.** 3.14% **43.** 843% **45.** 5,000% **47.** 910% **49.** 17% **51.** 16% **53.** 40% **55.** 105% **57.** 62.5% **59.** 18.75% **61.** $66\frac{2}{3}\%$ **63.** $83\frac{1}{3}\%$ **65.** 11.11% **67.** 55.56% **69. a.** $\frac{15}{188}$ **b.** 8% **71. a.** $\frac{9}{22}$ **b.** 41% **73. a.** $\frac{5}{29}$ **b.** 17% **c.** 24% **75.** 5 ft **77.** 0.9944 **79.** as a decimal; 89.6% **81.** torso: 27.5%

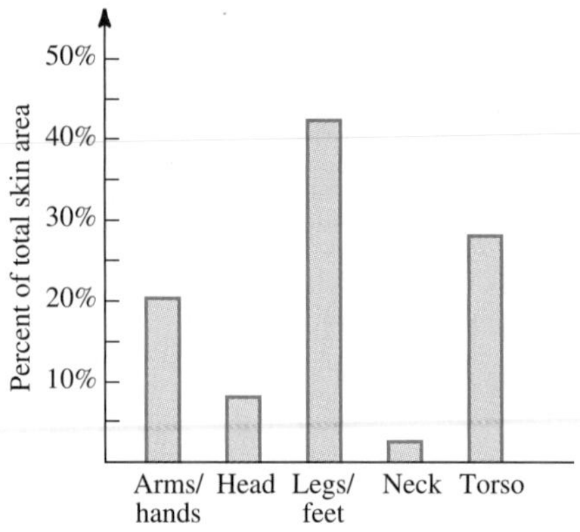

83. 92% **85.** 0.27% **93.** 9 **95.** 7, 11, 3 **97.** $x^2 + 3x + 2$

Study Set Section 7.2 (page 433)

1. $x = 0.10 \cdot 50$ **3.** $48 = x \cdot 47$ **5.** graph **7. a.** 0.12 **b.** 0.056 **c.** 1.25 **d.** 0.0025 **9.** more **11. a.** 25 **b.** 100% **c.** 87 **13.** 33% **15. a.** multiply **b.** equals **c.** x (as a variable) **17.** 90 **19.** 80% **21.** 65 **23.** 0.096 **25.** 0.00125% **27.** 44 **29.** 43.5 **31.** 107.1 **33.** 99 **35.** 60 **37.** 31.25%

39.

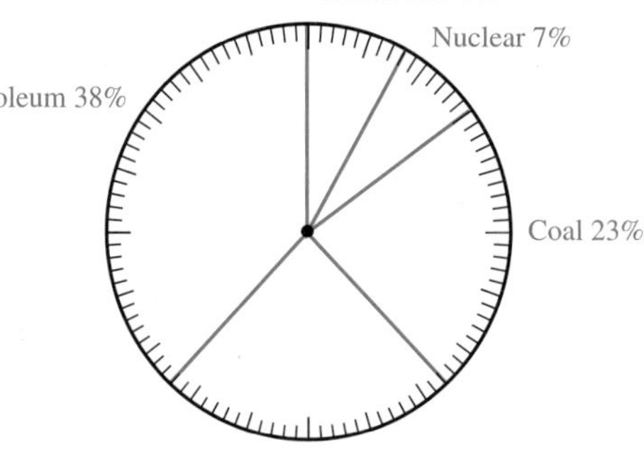

41. 120 **43.** $610.5 billion **45.** 38,000 = 38K **47.** 24 oz **49.** yes **51.** 30, 12 **53.** 2.7 in. **55.** 5% **57.** yes **63.** 18.17 **65.** 5.001 **67.** 34,546.4 **69.** -22.5

Study Set Section 7.3 (page 443)

1. commission **3.** discount **5.** The number of members has doubled. **7.** $42.75 **9.** 8% **11.** $47.34, $2.84, $50.18 **13.** $150 **15.** 8%, 1.2%, 1.4% **17.** 360 hr **19.** 96 calories **21.** 1995–1996; 5% **23.** 10% **25.** 31% **27. a.** 25% **b.** 36% **29.** $2,955 **31.** 1.5% **33.** $12,000 **35.** $39.95, 25% **37.** $187.49 **39.** $349.97, 13% **41.** $3.60, 23%, $11.88 **43.** $76.50 **49.** -50 **51.** -24 **53.** $12d$ **55.** 13

57.

Study Set Estimation (page 448)

1. 164 **3.** \$60 **5.** \$54,000 **7.** 320 lb **9.** 130 **11.** 21 **13.** 18,000 **15.** 3,100

Study Set Section 7.4 (page 454)

1. principal **3.** interest **5.** simple **7. a.** 0.07 **b.** 0.098 **c.** 0.0625 **9.** \$1,800 **11. a.** compound interest **b.** \$1,000 **c.** 4 **d.** \$50 **e.** 1 year **13.** multiplication **15.** \$5,300 **17.** \$1,472 **19.** \$4,262.14 **21.** \$10,000, 0.0725, 2 yr, \$1,450 **23.** \$192, \$1,392, \$58 **25.** \$18.828 million **27.** \$755.83 **29.** \$1,271.22 **31.** \$570.65 **33.** \$30,915.66 **39.** $\frac{1}{2}$ **41.** no **43.** -3 **45.** 3 **47.** III

Key Concept (page 457)

1. $\frac{5}{12}$; simplifying a fraction **3.** $5x$; combining like terms **5.** x^5; when multiplying like bases, add the exponents **7.** $66\frac{2}{3}\%$; divide the numerator by the denominator, move the decimal point two places to the right, and insert a % symbol **9.** $2x^2 + 1$; combining like terms **11.** 1; a number divided by itself is 1 **13.** 30; the product of two negative numbers is positive **15.** x; 2 times x divided by 2 is x **17.** 17; do multiplications before additions **19.** $2x + 10$; distributive property **21.** 1; the product of a number and its reciprocal is 1

Chapter Review (page 459)

1. a. 39%, 0.39, $\frac{39}{100}$ **b.** 111%, 1.11, $1\frac{11}{100}$ **2.** 61% **3. a.** $\frac{3}{20}$ **b.** $\frac{6}{5}$ **c.** $\frac{37}{400}$ **d.** $\frac{1}{1,000}$ **4. a.** 0.27 **b.** 0.08 **c.** 1.55 **d.** 0.018 **5. a.** 83% **b.** 62.5% **c.** 5.1% **d.** 600% **6. a.** 50% **b.** 80% **c.** 87.5% **d.** 6.25% **7. a.** $33\frac{1}{3}\%$ **b.** $83\frac{1}{3}\%$ **8. a.** 55.56% **b.** 266.67% **9.** 63% **10.** $0.1\% = \frac{1}{1,000}$; $10\% = \frac{1}{10}$ **11.** amount: 15, base: 45, percent: $33\frac{1}{3}\%$ **12.** $x = 32\% \cdot 96$ **13. a.** 200 **b.** 125 **c.** 1.75% **d.** 2,100 **e.** 121 **f.** 30 **14.** 14.4 gal nitro, 0.6 gal methane **15.** 68 **16.** 87% **17.** \$5.43 **18.**

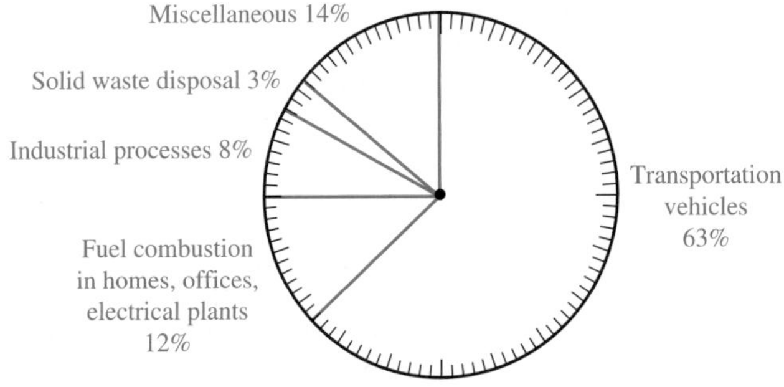

19. 139,531,200 mi^2 **20.** \$3.30, \$63.29 **21.** 4% **22.** \$40.20 **23.** 25% **24.** 9.6% **25.** \$50, \$189.99, 26% **26.** \$6,000, 8%, 2 years, \$960 **27.** \$10,308.22 **28.** \$134.69 **29.** \$2,142.45 **30.** \$6,076.45 **31.** \$43,265.78

Chapter 7 Test (page 463)

1. 61%, $\frac{61}{100}$, 0.61 **2.** 199%, $\frac{199}{100}$, 1.99 **3. a.** 0.67 **b.** 0.123 **c.** 0.0975 **4. a.** 25% **b.** 62.5% **c.** 12% **5. a.** 19% **b.** 347% **c.** 0.5% **6. a.** $\frac{11}{20}$ **b.** $\frac{1}{10,000}$ **c.** $\frac{5}{4}$ **7.** 23.33% **8.** 60% **9.** $66\frac{2}{3}\%$ **10.** 25% **11. a.** 1.02 in. **b.** 32.98 in. **12.** 6.5% **13.** \$3.81 **14.** 93.7% **15.** 90 **16.** 21 **17.** 144 **18.** 27% **19.** \$35.92 **20.** \$41,440 **21.** \$11.95, \$3, 20% **22.** 22% **23.** \$150 **24.** \$5,079.60 **25.** The phrase "bringing crime down to 37%" is unclear. The question that arises is: 37% of what?

Cumulative Review Exercises (page 465)

1.

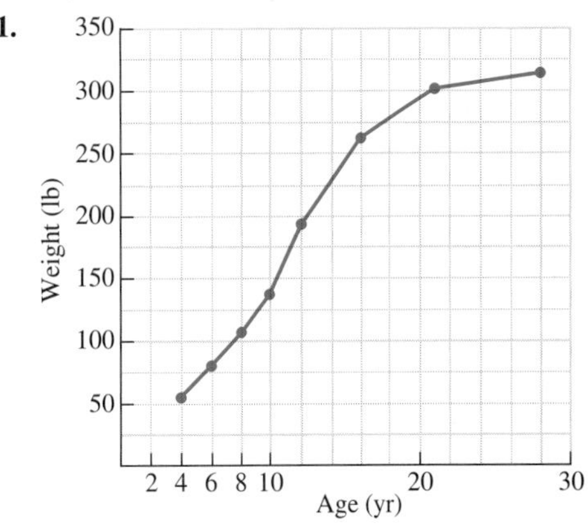

2. If a and b represent numbers, $ab = ba$. **3. a.** 1, 2, 4, 5, 8, 10, 20, 40 **b.** $5 \cdot 2^3$ **4.** \$2,106 **5.** 64 ft^2 **6.** -7 **7.** -4 **8.** 55 **9.** -2 **10.** $2t - 16$ **11.** 15° C **12.** 3 **13.** $\frac{4}{11}$ **14.** $\frac{2}{3}$ **15.** $-\frac{5}{21a}$ **16.** $\frac{2}{5}$ **17.** $\frac{28 + 2m}{7m}$ **18.** $20\frac{5}{18}$ **19.** -30 **20.** 8 **21.** 70.29 **22.** -8.6265 **23.** 752 **24.** 83.4 **25.** -2.33 **26.** 452.030 **27.** $0.7\overline{3}$ **28.** -11.1 **29.** -29 **30.** 3.5 hr **31.** $-2m^2 - 3m + 3$ **32.** $3x^2 + 10x - 8$ **33.** $4y^2 - 20y + 25$ **34.** y^7 **35.** h^{20} **36.** $8a^9b^{18}$ **37.** $-56g^9$ **38.**

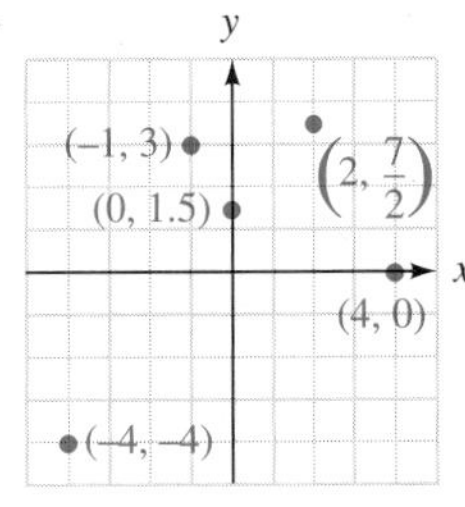

39.

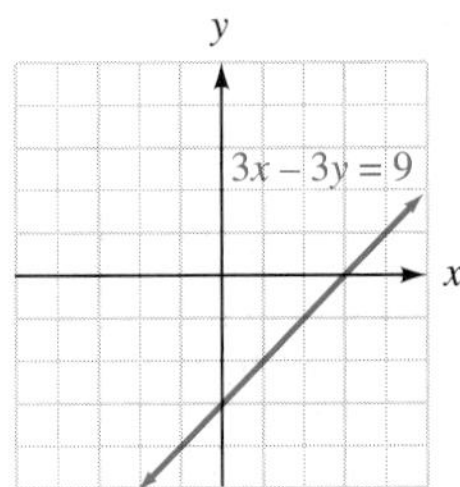

40.

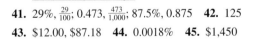

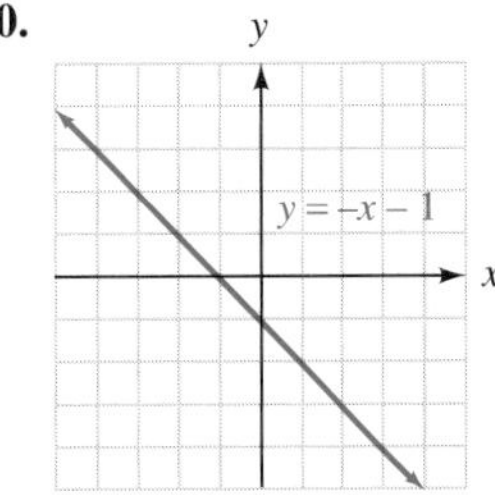

41. 29%, $\frac{29}{100}$; 0.473, $\frac{473}{1,000}$; 87.5%, 0.875 **42.** 125 **43.** \$12.00, \$87.18 **44.** 0.0018% **45.** \$1,450

Study Set Section 8.1 (page 473)

1. ratio **3.** cost **5.** 3 **7.** 10 **9.** $\frac{11 \text{ minutes}}{60 \text{ minutes}} = \frac{11}{60}$ **11.** $\frac{13}{9}$, 13 to 9, 13:9 **13.** $\frac{5}{7}$ **15.** $\frac{1}{2}$ **17.** $\frac{2}{3}$ **19.** $\frac{5}{8}$ **21.** $\frac{2}{7}$ **23.** $\frac{1}{3}$ **25.** $\frac{1}{5}$ **27.** $\frac{3}{7}$ **29.** $\frac{3}{4}$ **31.** \$1,800 **33.** $\frac{1}{3}$ **35.** \$8,750 **37.** $\frac{1}{5}$ **39.** $\frac{32 \text{ ft}}{3 \text{ sec}}$ **41.** $\frac{21 \text{ made}}{25 \text{ attempts}}$ **43.** $\frac{375 \text{ students}}{2 \text{ yr}}$ **45.** $\frac{3 \text{ beats}}{2 \text{ measures}}$ **47.** 12 revolutions per min **49.** 1.5 errors per hr **51.** 7 presents per child **53.** 320 people per square mile **55.** \$0.07 per foot **57.** 1.2 cents per ounce **59.** \$68 per person **61.** \$0.8 billion per month **63.** $\frac{1}{1}$

65. $\frac{3}{2}$ **67.** $\frac{\frac{2}{3}}{3\frac{1}{2}}$ **69.** $\frac{\text{3 hits}}{\text{11 at-bats}}$ **71.** $\frac{\text{5 compressions}}{\text{2 breaths}}$ **73.** $\frac{\text{329 complaints}}{\text{100,000 passengers}}$ **75.** $\frac{\text{1 faculty member}}{\text{16 students}}$ **77.** \$1.89 per gal **79.** 7¢ per oz **81.** the 6-oz can **83.** the 50-tablet boxes **85.** the truck **87.** 440 gal per min **89.** 325 mi, 65 mph **91.** the second car **97.** 45.537 **99.** 192.7012 **101.** $1\frac{3}{4}$

Study Set Section 8.2 (page 483)

1. proportion **3.** means **5.** ad, bc **7. a.** $\frac{5}{8} = \frac{15}{24}$ **b.** $\frac{\text{3 teacher's aides}}{\text{25 children}} = \frac{\text{12 teacher's aides}}{\text{100 children}}$ **9.** i, iv **11.** $18x$, 288, 18, 18 **13.** no **15.** yes **17.** no **19.** yes **21.** no **23.** yes **25.** 4 **27.** 6 **29.** 3 **31.** 9 **33.** 0 **35.** −17 **37.** $-\frac{3}{2}$ **39.** $\frac{83}{2}$ **41.** \$3,500 **43.** 5.625 **45.** \$218.75 **47.** \$11.76 **49.** the same **51.** 24 **53.** 975 **55.** about $4\frac{1}{4}$ **57.** 19 sec **59.** 221 mi **61.** \$309 **63.** 10 ft **65.** 65.25 ft = 65 ft 3 in. **67.** 2.625 in. = $2\frac{5}{8}$ in. **73.** 90% **75.** $\frac{1}{3}$ **77.** 2.6

Study Set Section 8.3 (page 493)

1. length **3.** 1 **5.** capacity **7.** 1 **9.** 5,280 **11.** 16 **13.** 8 **15.** 1 **17.** 24 **19.** $\frac{5}{8}$ in., $1\frac{3}{4}$ in., $2\frac{5}{16}$ in. **21. a.** $\frac{\text{1 ton}}{\text{2,000 lb}}$ **b.** $\frac{\text{2 pt}}{\text{1 qt}}$ **23. a.** iv **b.** i **c.** ii **d.** iii **25. a.** iii **b.** iv **c.** i **d.** ii **27.** 36, 36 **29.** 2, 4, 12 **31.** $2\frac{3}{8}$ in. **33.** $10\frac{3}{4}$ in. **35.** 48 in. **37.** 42 in. **39.** 2 ft **41.** 288 in. **43.** 2.5 yd **45.** $4\frac{2}{3}$ ft **47.** 15 ft **49.** $2\frac{1}{3}$ yd **51.** 3 mi **53.** 2,640 ft **55.** 5 lb **57.** 3.5 tons **59.** 24,800 lb **61.** 6 pt **63.** 2 gal **65.** 2 pt **67.** 4 hr **69.** 5 days **71.** 150 yd **73.** 2,880 in. **75.** 0.28 mi **77.** 61,600 yd **79.** 128 oz **81.** 4.95 tons **83.** 68 **85.** $71\frac{7}{8}$ gal = 71.875 gal **87.** 320 oz **89.** $6\frac{1}{8}$ days = 6.125 days **93.** 3,700 **95.** 3,673.26 **97.** 0.101 **99.** 0.1

Study Set Section 8.4 (page 503)

1. tens **3.** thousands **5.** hundredths **7.** metric **9.** 1 cm, 3 cm, 6 cm **11. a.** $\frac{\text{1 km}}{\text{1,000 m}}$ **b.** $\frac{\text{100 cg}}{\text{1 g}}$ **c.** $\frac{\text{1,000 milliliters}}{\text{1 liter}}$ **13. a.** iii **b.** i **c.** ii **15. a.** ii **b.** iii **c.** i **17.** 10 **19.** $\frac{1}{100}$ **21.** $\frac{1}{1{,}000}$ **23.** 1,000 **25.** 1,000 **27.** 1,000 **29.** $\frac{1}{100}$ **31.** 1 **33.** 1, 100 **35.** 1,000, 1, 1,000 **37.** 156 mm **39.** 28 cm **41.** 300 **43.** 570 **45.** 3.1 **47.** 7,680,000 **49.** 0.472 **51.** 4.532 **53.** 0.0325 **55.** 37.5 **57.** 125 **59.** 675,000 **61.** 6.383 **63.** 0.63 **65.** 69.5 **67.** 5.689 **69.** 5.762 **71.** 0.000645 **73.** 0.65823 **75.** 3,000 **77.** 2,000 **79.** 1,000,000 **81.** 0.5 **83.** 3,000 **85.** 5,000 **87.** 10 **89.** 0.5 km, 1 km, 1.5 km, 5 km, 10 km **91.** 12 cm, 8 cm **93.** 400,000 cg **95.** 40 dL **97.** 4 **99.** 3 g **105.** \$23.99 **107.** \$402 **109.** $11\frac{10}{21}$

Study Set Section 8.5 (page 511)

1. Fahrenheit **3. a.** meter **b.** meter **c.** inch **d.** mile **5. a.** liter **b.** liter **c.** gallon **7.** 0.3048, 1,371.6 **9.** 0.264 **11.** 91.4 **13.** 147.6 **15.** 39,372 **17.** 127 **19.** 1 **21.** 11,350 **23.** 17.6 **25.** 0.6 **27.** 0.1 **29.** 243.4 **31.** 710 **33.** 0.5 **35.** 10° **37.** 122° **39.** 14° **41.** −20.6° **43.** 5 mi **45.** 70 mph **47.** 1.9 km **49.** 1.9 cm **51.** 181.5 lb, 291.5 lb; 242 lb, 594 lb **53. a.** 226.8 g **b.** 0.24 L **55.** no **57.** the 3 quarts **59.** 62° C **61.** 28° C **63.** −5° C and 0° C **69.** $5y + 4$ **71.** $-7x$ **73.** x^3 **75.** $15b^2$

Key Concept (page 514)

3. 10,800 ft

Chapter Review (page 516)

1. a. $\frac{1}{3}$ **b.** $\frac{1}{4}$ **c.** $\frac{3}{2}$ **d.** $\frac{2}{3}$ **2.** $\frac{37}{32}$ **3.** \$7.75 **4.** the 8-oz can **5. a.** 75 **b.** 15 **6. a.** no **b.** yes **7. a.** yes **b.** no **8. a.** 4.5 **b.** 16 **c.** 0 **d.** $-\frac{35}{9}$ **9.** 192.5 mi **10.** 300 **11.** 12 ft **12.** $1\frac{1}{2}$ in. **13.** $\frac{\text{1 mi}}{\text{5,280 ft}} = 1$, $\frac{\text{5,280 ft}}{\text{1 mi}} = 1$ **14. a.** 15 ft **b.** 216 in. **c.** 5.5 ft **d.** 306 in. **e.** 1.75 mi **f.** 1,760 yd **15. a.** 2 lb **b.** 275.2 oz **c.** 96,000 oz **d.** 2.25 tons **16. a.** 80 fl oz **b.** 0.5 gal **c.** 68 c **d.** 5.5 qt **e.** 40 pt **f.** 56 c **17. a.** 1,200 sec **b.** 15 min **c.** $8\frac{1}{3}$ days **d.** 360 min **e.** 108 hr **f.** 86,400 sec **18.** $484\frac{2}{3}$ yd **19.** 100 **20.** 4 cm **21.** $\frac{\text{1 km}}{\text{1,000 m}} = 1$, $\frac{\text{1,000 m}}{\text{1 km}} = 1$ **22. a.** 4.75 m **b.** 8,000 mm **c.** 0.03 km **d.** 2,000 dm **e.** 50 hm **f.** 25 hm **23. a.** 70 mg **b.** 8 g **c.** 5.425 kg **d.** 5,425,000 mg **e.** 7.5 g **f.** 0.05 kg **24.** 50 **25. a.** 1.5 L **b.** 3.25 kL **c.** 1,000 dL **d.** 40 cL **e.** 20 hL **f.** 400 mL **26.** 1,000 mL **27.** 164.04 ft **28.** the World Trade Center **29.** 3,106 km **30.** 198.12 cm **31. a.** 850.5 g **b.** 33 lb **c.** 11,000 g **d.** 910 kg **32.** about 2 lb **33.** LaCroix **34.** the 5-liter bottle **35.** 25° C **36.** 30° C

Chapter 8 Test (page 521)

1. $\frac{3}{4}$ **2.** $\frac{1}{6}$ **3.** the 2-pound can **4.** 22.5 kwh per day **5.** $\frac{1}{1}$, 1 : 1, 1 to 1 **6.** no **7.** yes **8.** yes **9.** 15 **10.** 63.24 **11.** 11 **12.** 3 **13.** \$3.43 **14.** $1\frac{2}{3}$ c **15.** 15 ft **16.** $8\frac{1}{3}$ yd **17.** 160 oz **18.** 3,200 lb **19.** 128 fl oz **20.** 115,200 min **21.** the one on the left **22.** the blue one **23.** the right side **24.** 0.5 km **25.** 500 cm **26.** 0.08 kg **27.** 70,000 mL **28.** 7.5 g **29.** the 100-yd race **30.** Jim **31.** the one-liter bottle **32.** 182° F **33.** A scale is a ratio (or rate) comparing the size of a drawing and the size of an actual object. For example, 1 inch to 6 feet (1 in. : 6 ft). **34.** It is easier to convert from one unit to another in the metric system, because it is based on the number 10.

Cumulative Review Exercises (page 523)

1. 6 ten thousands + 4 thousands + 5 hundreds + 2 ones **2.** 20 R 3 **3.** −1,732, −6,290, 35, 71 **4. a.** −8 **b.** undefined **c.** −8 **d.** 0 **e.** 8 **f.** 0 **5.** 15 shots **6.** −9, 9 **7.** −61 **8.** 1, 16 **9.** $60h$ **10.** $x - 7$ **11.** −2 **12.** $d = rt$ **13.** 79 **14.** $A = \frac{1}{2}bh$ **15.** $\frac{4}{5}$ **16.** $\frac{54t}{60t}$ **17.** $-\frac{1}{h}$ **18.** $\frac{1}{4}$ **19.** $\frac{3}{4}$ hp **20.** $\frac{4}{25}$ **21.** $\frac{10}{9}$ **22.** 55.1° F **23.** 304.07 **24.** $0.08\overline{3}$ **25.** 0.8 **26.** 23 **27.**

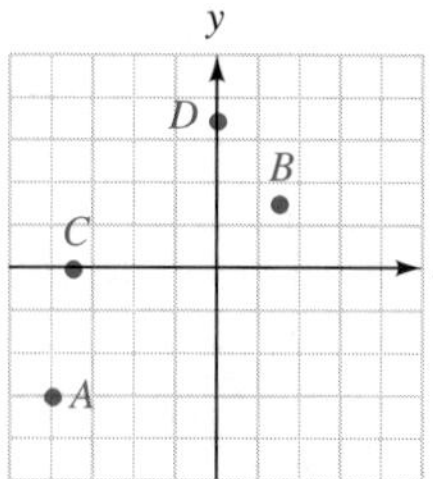

28. a. **b.**

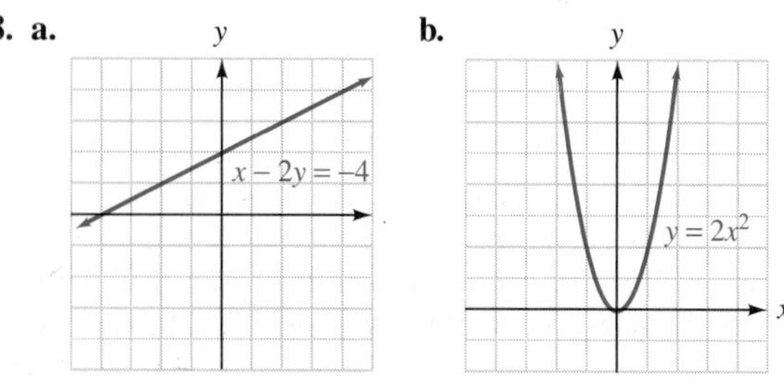

29. $2x^2 - 6x - 2$ **30.** $6x^2 - 11x - 10$ **31. a.** s^{13} **b.** s^{42} **c.** $27a^6b^{12}$ **d** $-8w^8$ **32.** $66\frac{2}{3}\%$ **33.** $I = Prt$ **34.** 99%, $\frac{99}{100}$; 0.013, $\frac{13}{1,000}$; 31.25%, 0.3125 **35.** $427.99; about 30% **36.** $\frac{1}{5}$ **37. a.** 960 hr **b.** 4,320 min **c.** 480 sec **38.** 2.5 lb **39.** 2,400 mm **40.** 0.32 kg **41. a.** 1 gal **b.** a meterstick **42.** the 45-kilogram bag

Study Set Section 9.1 (page 531)

1. segment **3.** midpoint **5.** protractor **7.** right **9.** 180° **11.** supplementary **13.** true **15.** false **17.** true **19.** true **21.** acute **23.** obtuse **25.** right **27.** straight **29.** true **31.** false **33.** yes **35.** yes **37.** no **39.** true **41.** true **43.** true **45.** true **47.** angle **49.** ray **51.** 3 **53.** 3 **55.** 1 **57.** *B* **59.** 40° **61.** 135° **63.** 10 **65.** 27.5 **67.** 30 **69.** 25 **71.** 60° **73.** 75° **75.** 130° **77.** 230° **79.** 100° **81.** 40° **83.**

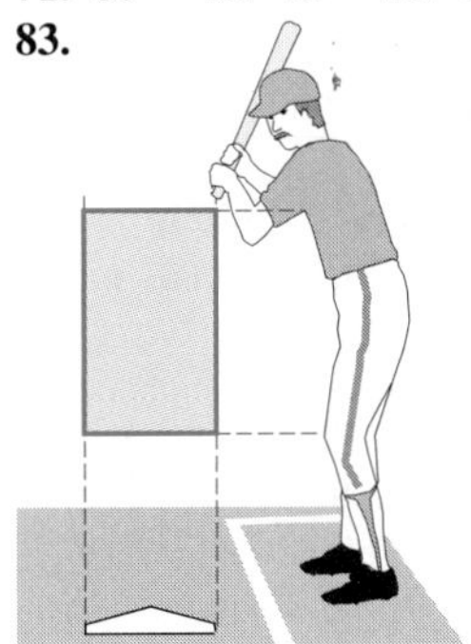

85. 65,115 **87.** 30° **95.** 16 **97.** $\frac{7}{24}$ **99.**

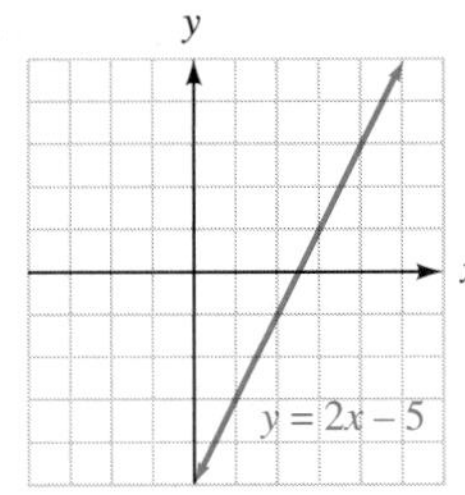

101. 4

Study Set Section 9.2 (page 539)

1. coplanar **3.** perpendicular **5.** alternate **7.** $\angle 4$ and $\angle 6$, $\angle 3$ and $\angle 5$ **9.** $\angle 3$, $\angle 4$, $\angle 5$, $\angle 6$ **11.** They are parallel. **13.** a right angle **15.** is perpendicular to **17.** $m(\angle 1) = 130°$, $m(\angle 2) = 50°$, $m(\angle 3) = 50°$, $m(\angle 5) = 130°$, $m(\angle 6) = 50°$, $m(\angle 7) = 50°$, $m(\angle 8) = 130°$ **19.** $m(\angle A) = 50°$, $m(\angle 1) = 85°$, $m(\angle 2) = 45°$, $m(\angle 3) = 135°$ **21.** 10 **23.** 30 **25.** 40 **27.** 12 **29.** If the stones are level, the plum bob string should pass through the midpoint of the crossbar of the A-frame. **41.** 72 **43.** 45% **45.** yes **47.** $\frac{1}{3}$

Study Set Section 9.3 (page 546)

1. regular **3.** hexagon **5.** octagon **7.** equilateral **9.** hypotenuse **11.** parallelogram **13.** rhombus **15.** isosceles **17.** 4, quadrilateral, 4 **19.** 3, triangle, 3 **21.** 5, pentagon, 5 **23.** 6, hexagon, 6 **25.** scalene triangle **27.** right triangle **29.** equilateral triangle **31.** isosceles triangle **33.** square **35.** rhombus **37.** rectangle **39.** trapezoid **41.** triangle **43.** 90° **45.** 45° **47.** 90.7° **49.** 30° **51.** 60° **53.** 720° **55.** 1,440° **57.** 7 sides **59.** 14 sides **65. b.** octagon **c.** triangle **d.** pentagon **67.** pentagon, hexagon **73.** 22 **75.** 40% **77.** 0.10625

Study Set Section 9.4 (page 554)

1. congruent **3.** similar **5.** true **7.** false **9.** yes **11.** product **13.** yes **15.** *a* and *b* represent the length of the legs; *c* represents the length of the hypotenuse. **17.** 25, square root, *c*, 5 **19.** is congruent to **21.** $\overline{DF}$, $\overline{AB}$, $\overline{EF}$, $\angle D$, $\angle B$, $\angle C$ **23.** yes, SSS **25.** not necessarily **27.** yes, SSS **29.** yes, SAS **31.** 6 mm **33.** 50° **35.** yes **37.** 5 **39.** 8 **41.** $\sqrt{56}$ **43.** yes **45.** no **47.** 36 ft **49.** 59.2 ft **51.** 4,000 ft **53.** 12 ft **55.** 25 in. **57.** 127.3 ft **61.** $1\frac{1}{3}$ **63.** 20 **65.** 9

Study Set Section 9.5 (page 566)

1. perimeter **3.** area **5.** square **7.** length 15 in. and width 5 in.; length 16 in. and width 4 in. (answers may vary) **9.** sides of length 5 m **11.** base 5 yd and height 3 yd (answers may vary) **13.** length 5 ft and width 4 ft; length 20 ft and width 3 ft (answers may vary) **15.** $P = 4s$ **17.** square inch **19.** $A = s^2$ **21.** triangle **23.** 32 in. **25.** 36 m **27.** 37 cm **29.** 85 cm **31.** $28\frac{1}{3}$ ft **33.** 16 cm^2 **35.** 60 cm^2 **37.** 25 in.^{2} **39.** 169 mm^2 **41.** 80 m^2 **43.** 75 yd^2 **45.** 75 m^2 **47.** 144 **49.** $4,875 **51.** 81 **53.** linoleum **55.** $1,200 **57.** $361.20 **59.** $192 **61.** 111,825 mi^2 **63.** 51 **65.** spot 1: $l = 20$ ft, $w = 10$ ft, 200 ft^2; spot 2: $b_1 = 20$ ft, $b_2 = 16$ ft, $h = 10$ ft, 180 ft^2; spot 3: $b = 28$ ft, $h = 28$ ft, 392 ft^2 **69.** $1\frac{5}{12}$ **71.** $6\frac{1}{12}$ **73.** $1\frac{7}{18}$

Study Set Section 9.6 (page 574)

1. radius **3.** diameter **5.** minor **7.** circumference **9.** $\overline{OA}$, $\overline{OC}$, and $\overline{OB}$ **11.** $\overline{DA}$, $\overline{DC}$, and $\overline{AC}$ **13.** $\widehat{ABC}$ and $\widehat{ADC}$ **15.** Double the radius. **17. a.** 1 in. **b.** 2 in. **c.** 2π in. $\approx$ 6.28 in. **d.** π in.$^2 \approx$ 3.14 in.^{2} **19.** Square 6. **21.** arc *AB* **23.** $C = \pi D$, $C = 2\pi r$ **25.** π **27.** 8π **29.** 37.70 in. **31.** 36 m **33.** 25.42 ft **35.** 31.42 m **37.** $A = 28.3$ in.^{2} **39.** 88.3 in.^{2} **41.** 128.5 cm^2 **43.** 27.4 in.^{2} **45.** 66.7 in.^{2} **47.** 3.14 mi^2 **49.** 32.66 ft **51.** 12.73 times **53.** 1.59 ft **55.** 12.57 ft^2; 0.79 ft^2; 6.28% **63.** 90% **65.** 5.375¢ per oz **67.** five

Study Set Section 9.7 (page 585)

1. volume **3.** cube **5.** surface **7.** cylinder **9.** cone **11.** $V = lwh$ **13.** $V = \frac{4}{3}\pi r^3$ **15.** $V = \frac{1}{3}Bh$ or $V = \frac{1}{3}\pi r^2 h$ **17.** $SA = 2lw + 2lh + 2hw$ **19.** 27 ft^3 **21.** 1,000 dm^3 **23. a.** volume **b.** area **c.** volume **d.** surface area **e.** perimeter **f.** surface area **25. a.** 72 in.^{3} **b.** 18 in.^{2} **c.** 24 in.^{2} **27.** 1 cubic inch **29.** 60 cm^3 **31.** 48 m^3 **33.** 3,053.63 in.^{3} **35.** 1,357.17 m^3 **37.** 314.16 cm^3 **39.** 400 m^3 **41.** 94 cm^2 **43.** 1,256.64 in.^{2} **45.** 576 cm^3 **47.** 335.10 in.^{3} **49.** $\frac{1}{8}$ in.^{3} = 0.125 in.^{3} **51.** 2.125 **53.** 197.92 ft^3 **55.** 33,510.32 ft^3 **57.** 8:1 **63.** −42 **65.** −8 **67.** $\frac{1}{5}$ **69.** 2,400 mm

Key Concept (page 589)

1. $d = rt$ **3.** $P = 2l + 2w$ **5.** 210,000 ft^2 **7.** $80.50 **9.** 144 ft **11.** $750, $45, $6,250

Chapter Review (page 591)

1. points *C* and *D*, line *CD*, plane *GHI* **2.** 5 units **3.** $\angle ABC$, $\angle CBA$, $\angle B$, $\angle 1$ **4.** 48° **5.** $\angle 1$ and $\angle 2$ are acute, $\angle ABD$ and $\angle CBD$ are right angles, $\angle CBE$ is obtuse, and $\angle ABC$ is a straight angle. **6. a.** obtuse angle **b.** right angle **c.** straight angle **d.** acute angle **7.** 15 **8.** 150 **9. a.** 65° **b.** 115° **10.** 40° **11.** 40° **12.** no **13.** part a **14.** $\angle 4$ and $\angle 6$, $\angle 3$ and $\angle 5$ **15.** $\angle 1$ and $\angle 5$, $\angle 4$ and $\angle 8$, $\angle 2$ and $\angle 6$, $\angle 3$ and $\angle 7$ **16.** $\angle 1$ and $\angle 3$, $\angle 2$ and $\angle 4$, $\angle 5$ and $\angle 7$, $\angle 6$ and $\angle 8$ **17.** $m(\angle 1) = 70°$, $m(\angle 2) = 110°$, $m(\angle 3) = 70°$, $m(\angle 4) = 110°$, m(∠5) = 70°, m(∠6) = 110°, m(∠7) = 70° **18.** m(∠1) = 60°, m(∠2) = 120°, m(∠3) = 130°, m(∠4) = 50° **19.** 40 **20.** 20 **21. a.** octagon **b.** pentagon **c.** triangle **d.** hexagon **e.** quadrilateral **22. a.** 3 **b.** 4 **c.** 8 **d.** 6 **23. a.** isosceles **b.** scalene **c.** equilateral **d.** right triangle **24. a.** yes **b.** no **25. a.** 90 **b.** 50 **26.** 50° **27.** It is equilateral. **28. a.** trapezoid **b.** square **c.** parallelogram **d.** rectangle **e.** rhombus **f.** rectangle **29. a.** 15 cm **b.** 40° **c.** 100° **30. a.** true **b.** false **c.** true **d.** true **31. a.** 65° **b.** 115° **32. a.** 360° **b.** 720° **33.** ∠D, ∠E, ∠F, $\overline{DF}$, $\overline{DE}$, $\overline{EF}$ **34. a.** congruent, SSS **b.** congruent, SAS **c.** congruent, ASA **d.** not necessarily congruent **35. a.** yes **b.** yes **36.** 21 ft **37. a.** 13 **b.** 15 **38.** 31.3 in. **39.** 72 in. **40.** 9 m **41. a.** 30 m **b.** 36 m **42. a.** 9.61 cm^2 **b.** 7,500 ft^2 **c.** 450 ft^2 **d.** 200 $in.^2$ **e.** 120 cm^2 **f.** 232 ft^2 **g.** 152 ft^2 **h.** 120 m^2 **43.** 9 ft^2 **44.** 144 $in.^2$ **45. a.** $\overline{CD}$, $\overline{AB}$ **b.** $\overline{AB}$ **c.** $\overline{OA}$, $\overline{OC}$, $\overline{OD}$, $\overline{OB}$ **d.** O **46.** 66.0 cm **47.** 45.1 cm **48.** 254.5 $in.^2$ **49.** 130.3 cm^2 **50. a.** 125 cm^3 **b.** 480 m^3 **c.** 600 $in.^3$ **d.** 3,619 $in.^3$ **e.** 1,518 ft^3 **f.** 785 $in.^3$ **g.** 9,020,833 ft^3 **h.** 35,343 ft^3 **51.** 1,728 $in.^3$ **52.** 54 ft^3 **53. a.** 61.8 ft^2 **b.** 314.2 $in.^2$

Chapter 9 Test (page 599)

1. 4 units **2.** B **3.** true **4.** false **5.** false **6.** true **7.** 50 **8.** 140 **9.** 12 **10.** 45 **11.** 23° **12.** 63° **13.** 70° **14.** 110° **15.** 70° **16.** 40 **17.** 3, 4, 6, 5, 8 **18.** equilateral triangle, scalene triangle, isosceles triangle **19.** 57° **20.** 66° **21.** 30° **22.** 1,440° **23.** $m(\overset{\frown}{AB}) = m(\overset{\frown}{DC})$, $m(\overset{\frown}{AD}) = m(\overset{\frown}{BC})$, and $m(\overset{\frown}{AC}) = m(\overset{\frown}{BD})$ **24.** 130° **25.** 8 in. **26.** 50° **27.** 6 **28.** 12 **29.** 127.3 ft **30.** 391.6 cm^2 **31.** 83.7 ft^2 **32.** 94.2 ft **33.** 28.3 ft^2 **34.** 159.3 m^3 **35.** 268.1 m^3 **36.** 66.7 ft^3 **38.** The surface area is 6 times the area of one face of the cube.

Cumulative Review Exercises (page 601)

1.

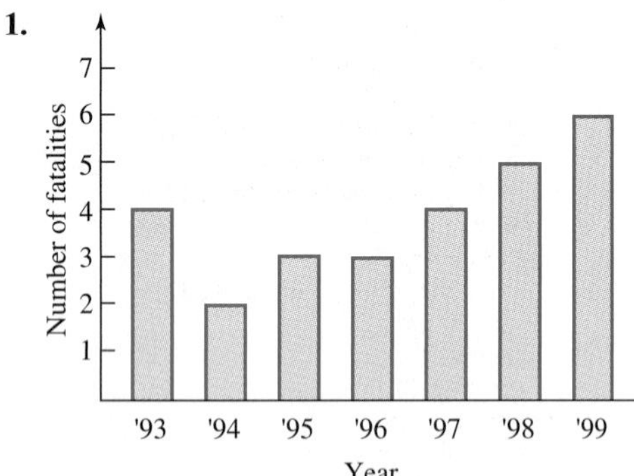

2. $8,995 **3.** 11,022 **4.** 33 **5.** 2,110,000 **6.** $11 \cdot 5 \cdot 2^2$ **7.** 1, 2, 3, 4, 6, 8, 12, 24 **8.** {. . . −3, −2, −1, 0, 1, 2, 3, . . .} **9.** 13 **10.** −10 **11.** 3 **12.** 5 **13.** An equation contains an = sign; an expression does not. **14.** −12x + 8 **15.** 89 **16.** −10 **17.** −11 **18.** 5 **19.** 6.5 in./min **20.** 50x¢ **21.** x − 5 **22.** $6\frac{2}{3}$ b.f. **23.** $\frac{5a}{4}$ **24.** $142\frac{7}{15}$ **25.** $\frac{5x-12}{20}$ **26.** $13\frac{3}{4}$ cups **27.** $-\frac{11}{20}$ **28.** $\frac{1}{3q}$ **29.** $\frac{3}{32}$ fluid oz **30.** $\frac{8}{9}$ **31.** $-\frac{15}{2}$ **32.** $\frac{8}{9}$ **33.** $1\frac{9}{29}$ **34.** 18 oz **35. a.** 1991; about 0.3° C **b.** 1993; about −0.3° C

36.

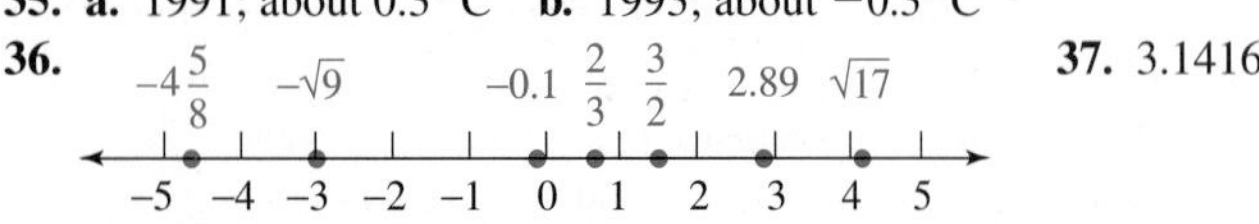

37. 3.1416 **38.** > **39.** 145.188 **40.** 17.05 **41.** 89,970.8 **42.** 0.053 **43.** −25.6 **44.** 22.3125 **45.** $0.1\overline{3}$ **46.** −9.32 **47.** 97 **48.** −0.6 **49.** −2 **50.** $\frac{7}{9}$ **51.** 0.86 oz, 0.855 oz, 0.85 oz

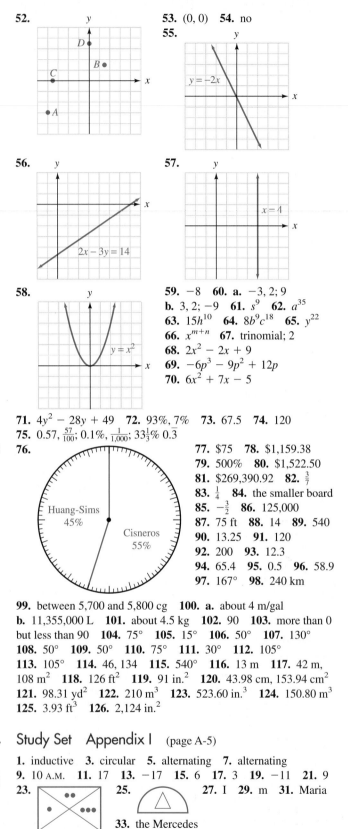

52.

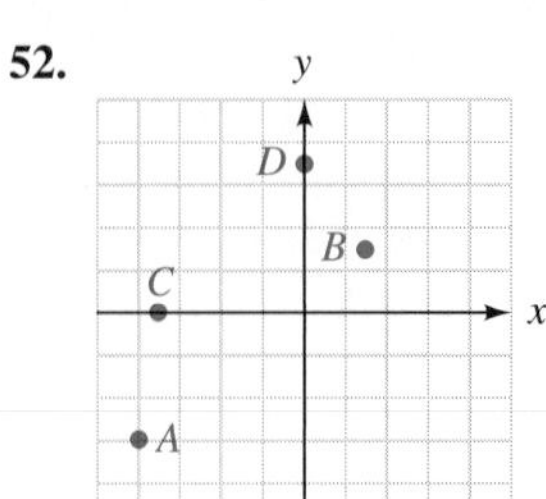

53. (0, 0) **54.** no **55.**

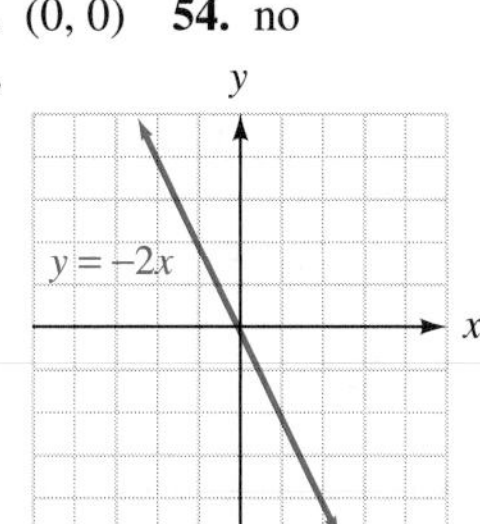

56.

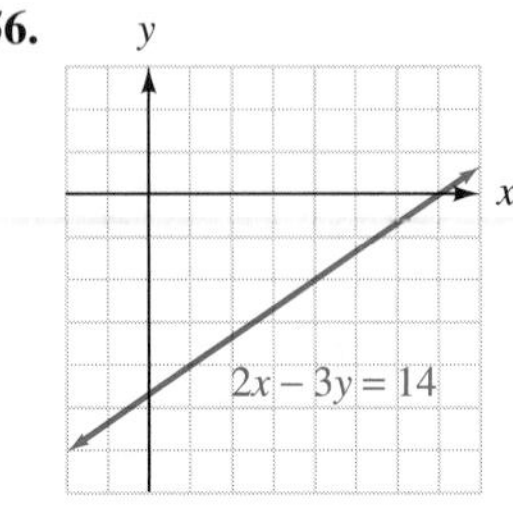

57.

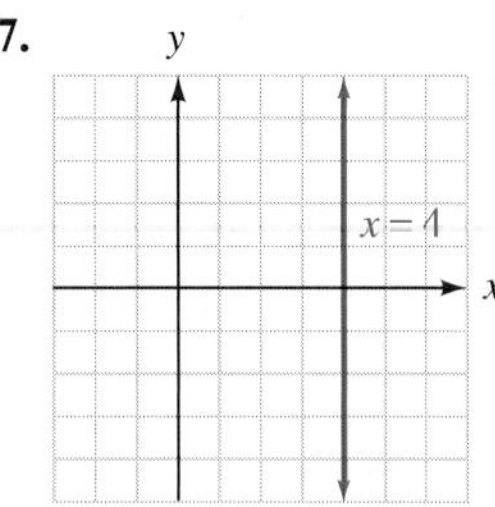

58.

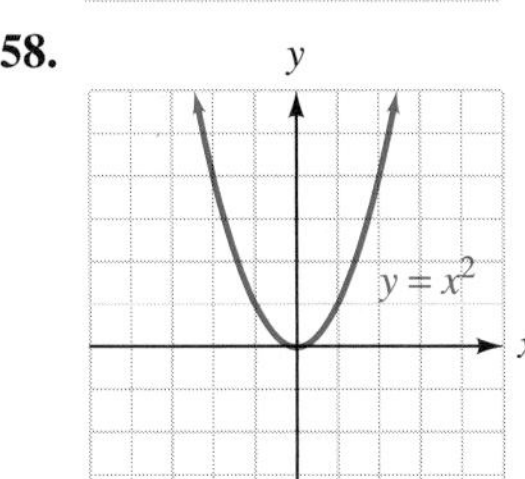

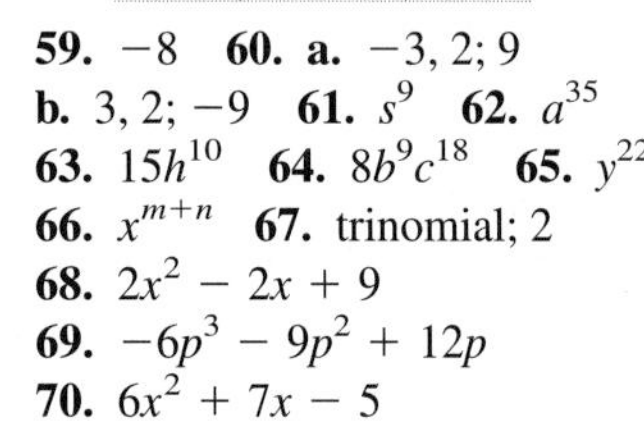

59. −8 **60. a.** −3, 2; 9 **b.** 3, 2; −9 **61.** s^9 **62.** a^{35} **63.** $15h^{10}$ **64.** $8b^9c^{18}$ **65.** y^{22} **66.** x^{m+n} **67.** trinomial; 2 **68.** $2x^2 - 2x + 9$ **69.** $-6p^3 - 9p^2 + 12p$ **70.** $6x^2 + 7x - 5$ **71.** $4y^2 - 28y + 49$ **72.** 93%, 7% **73.** 67.5 **74.** 120 **75.** 0.57, $\frac{57}{100}$; 0.1%, $\frac{1}{1,000}$; $33\frac{1}{3}$% $0.\overline{3}$

76.

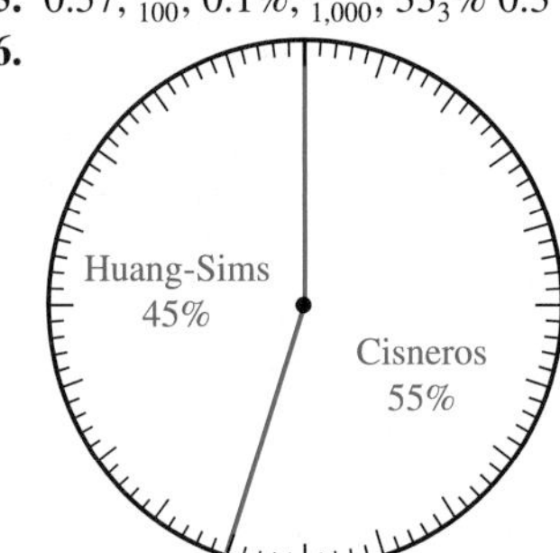

77. $75 **78.** $1,159.38 **79.** 500% **80.** $1,522.50 **81.** $269,390.92 **82.** $\frac{3}{7}$ **83.** $\frac{1}{4}$ **84.** the smaller board **85.** $-\frac{3}{2}$ **86.** 125,000 **87.** 75 ft **88.** 14 **89.** 540 **90.** 13.25 **91.** 120 **92.** 200 **93.** 12.3 **94.** 65.4 **95.** 0.5 **96.** 58.9 **97.** 167° **98.** 240 km **99.** between 5,700 and 5,800 cg **100. a.** about 4 m/gal **b.** 11,355,000 L **101.** about 4.5 kg **102.** 90 **103.** more than 0 but less than 90 **104.** 75° **105.** 15° **106.** 50° **107.** 130° **108.** 50° **109.** 50° **110.** 75° **111.** 30° **112.** 105° **113.** 105° **114.** 46, 134 **115.** 540° **116.** 13 m **117.** 42 m, 108 m^2 **118.** 126 ft^2 **119.** 91 $in.^2$ **120.** 43.98 cm, 153.94 cm^2 **121.** 98.31 yd^2 **122.** 210 m^3 **123.** 523.60 $in.^3$ **124.** 150.80 m^3 **125.** 3.93 ft^3 **126.** 2,124 $in.^2$

Study Set Appendix I (page A-5)

1. inductive **3.** circular **5.** alternating **7.** alternating **9.** 10 A.M. **11.** 17 **13.** −17 **15.** 6 **17.** 3 **19.** −11 **21.** 9 **23.** **25.** **27.** I **29.** m **31.** Maria **33.** the Mercedes **35.** green, blue, yellow, red **37.** 18,935 **39.** 0

Appendix Review (page A-8)

1. 6 **2.** 65 **3.** **4.**

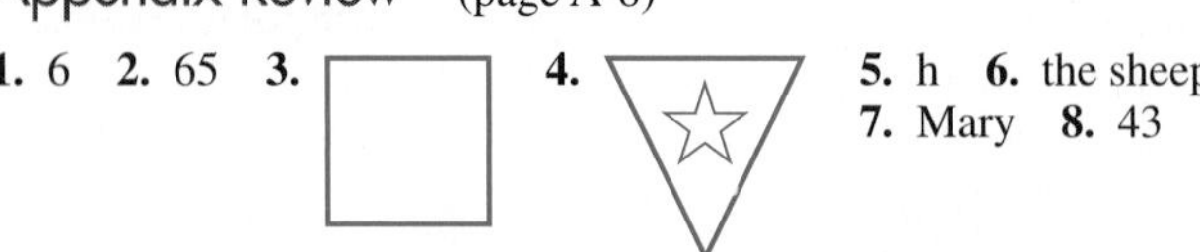

5. h **6.** the sheep **7.** Mary **8.** 43

INDEX